An Invitation to Algebraic Numbers and Algebraic Functions

An Invitation to Algebraic Numbers and Algebraic Functions

Franz Halter-Koch
University of Graz, Austria

CRC Press is an imprint of the
Taylor & Francis Group, an **informa** business
A CHAPMAN & HALL BOOK

First edition published 2020
by CRC Press
6000 Broken Sound Parkway NW, Suite 300, Boca Raton, FL 33487-2742

and by CRC Press
2 Park Square, Milton Park, Abingdon, Oxon, OX14 4RN

***Library of Congress Control Number*: 2020935369**

ISBN: 9781138583610 (hbk)
ISBN: 9780429506550 (ebk)

Typeset in CMR
by Nova Techset Private Limited, Bengaluru & Chennai, India

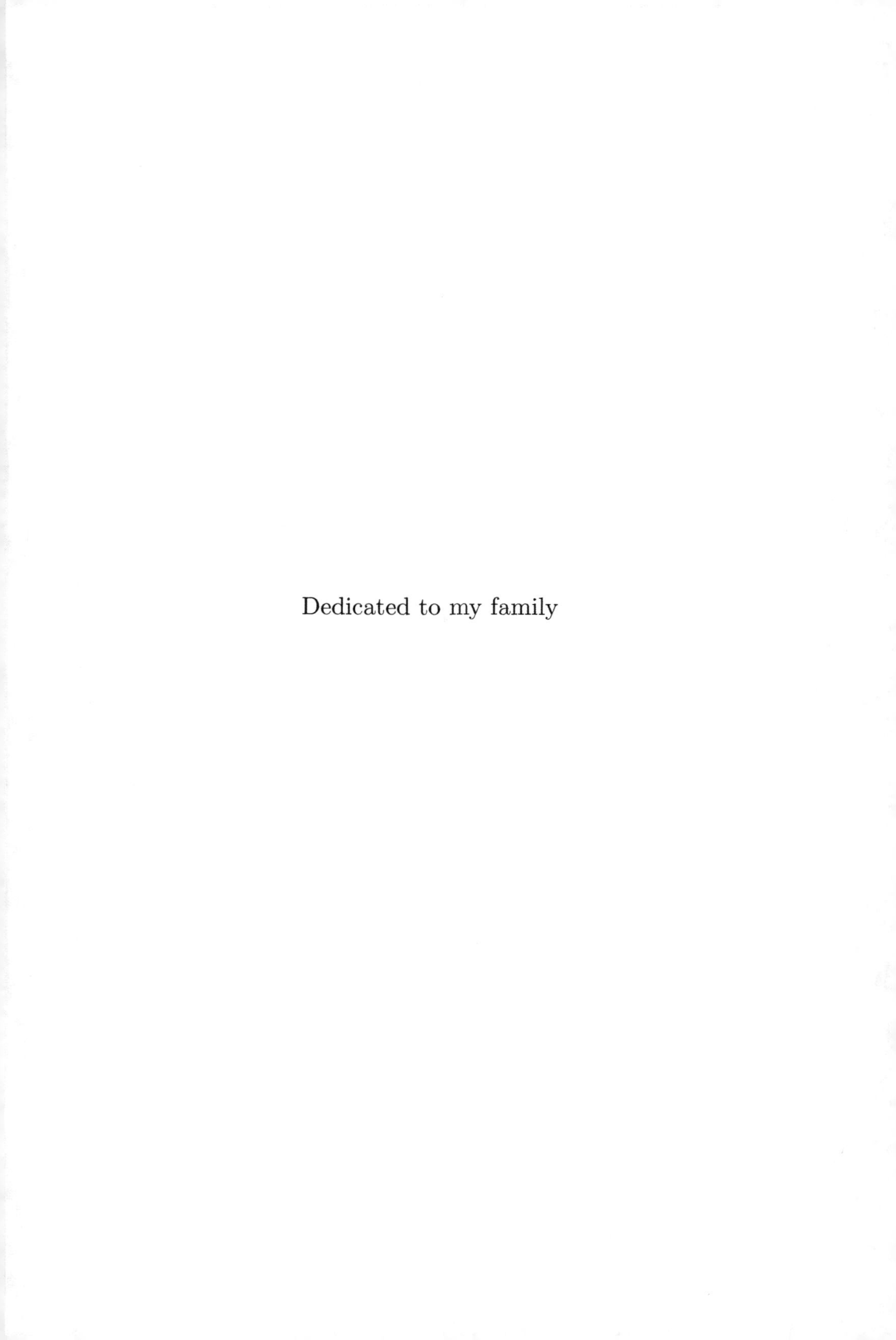

Dedicated to my family

Contents

Preface ix

Notations and Conventions xiii

1 Field Extensions **1**
1.1 Preliminaries on ideals and polynomials 1
1.2 Algebraic field extensions 15
1.3 Normal field extensions 24
1.4 Separable and inseparable field extensions 27
1.5 Galois theory of finite field extensions 36
1.6 Norms, traces, resultants, and discriminants 41
1.7 Finite fields, roots of unity, and cyclic field extensions 53
1.8 Transcendental field extensions 68
1.9 Exercises for Chapter 1 69

2 Dedekind Theory **75**
2.1 Factorial monoids 76
2.2 Factorial domains 84
2.3 Principal ideal domains 88
2.4 Integral elements 1: Ring-theoretic aspects 98
2.5 Integral elements 2: Field-theoretic aspects 103
2.6 Fractional and invertible ideals 108
2.7 Quotient domains and localizations 113
2.8 Dedekind domains 119
2.9 Ray class groups in Dedekind domains 126
2.10 Discrete valuation domains and Dedekind domains 133
2.11 Orders in Dedekind domains 145
2.12 Extensions of Dedekind domains 1: General theory 151
2.13 Extensions of Dedekind domains 2: Galois extensions 164
2.14 Ideal norms and Frobenius automorphisms 175
2.15 Exercises for Chapter 2 184

3 Algebraic Number Fields: Elementary and Geometric Methods **189**
3.1 Complete modules, integral bases and discriminants 190
3.2 Factorization of primes in algebraic number fields 205
3.3 Dirichlet characters and abelian number fields 216

3.4 Quadratic characters and quadratic reciprocity 227
3.5 The finiteness results for algebraic number fields 237
3.6 Class groups of algebraic number fields 253
3.7 The main theorems of classical class field theory 266
3.8 Arithmetic of quadratic orders 270
3.9 Genus theory of quadratic orders 280
3.10 Exercises for Chapter 3 288

4 Elementary Analytic Theory 293
4.1 Euler products and Dirichlet series 293
4.2 Dirichlet L functions 302
4.3 Density of prime ideals 315
4.4 Density results using class field theory 329
4.5 Exercises for Chapter 4 338

5 Valuation Theory 343
5.1 Absolute values 344
5.2 Topology and completion of valued fields 354
5.3 Non-Archimedian valued fields 1 365
5.4 Hensel's lemma, generalizations and applications 377
5.5 Extension of absolute values 391
5.6 Unramified field extensions 407
5.7 Ramified field extensions 417
5.8 Non-Archimedian valued fields 2 422
5.9 Different and discriminant 435
5.10 Higher ramification groups 450
5.11 Exercises for Chapter 5 461

6 Algebraic Function Fields 465
6.1 Field theoretic properties 466
6.2 Divisors 474
6.3 Repartitions and definition of the genus 485
6.4 Weil differentials and the theorem of Riemann-Roch 491
6.5 Algebraic function field extensions 1 498
6.6 Algebraic function field extensions 2 509
6.7 Derivations and differentials 520
6.8 Differentials and Weil differentials 537
6.9 Zeta functions 546
6.10 Exercises for Chapter 6 562

Bibliography 567

Index 571

List of Symbols 579

Preface

The theory of algebraic numbers and algebraic functions is one of the most impressive creations of 19th century mathematics. The theory of algebraic numbers has its origin in C.F. Gauss's famous quadratic reciprocity law and in the subsequent attempts to find and prove higher reciprocity laws. These efforts finally culminated in the development of class field theory in the first half of the 20th century. On the way the arithmetical theory of algebraic numbers was enhanced by various powerful algebraic theories such as E.E. Kummer's theory of ideal numbers culminating in R. Dedekind's ideal theory, L. Kronecker's theory of divisors, and K. Hensel's theory of p-adic numbers, to name but a few. For a thorough presentation of the development of algebraic number theory from the end of the 19th century up to 1950 we refer to the recent book of W. Narkiewicz [51].

The theory of algebraic functions has its origin in B. Riemann's constructions to understand multi-valued complex functions. Among others, it was H. Weber (using the concepts of R. Dedekind and L. Kronecker) who establishsed the algebraic theory lying behind the analytical concepts. In the 20th century this algebraic theory and its close relationship to algebraic number theory were developed further and deepened by the works of E. Artin, H. Hasse, F.K. Schmidt, and A. Weil, to mention only the most prominent ones.

Instead of going further into historical development I turn to the 20th century literature which influenced the actual volume. Above all, it were E. Artin's famous lecture notes [1] in which he presented a unified treatment of the theories of algebraic numbers and of algebraic functions from a valuation-theoretic point of view. In S. Iyanaga's book [31] there is a more involved exposition of this concept, together with deep topological refinements and a thorough presentation of class field theory in a cohomological context. A more elementary account of this approach may be found in [8]. Among the books that aim for a unified treatment of number fields and function fields I appreciate the most elegant and concise but ambitious presentation by A. Weil [62]. Easier accessible but equally highly recommended are the expositions by F. Lorenz [43], H. Koch [36], and H. Hasse [27].

In the present volume we favor a separate treatment of algebraic numbers and algebraic functions using common fundamentals from ideal and valuation theory. This enables us much better to highlight results which are intrinsic for the respective theory. Apart from the books mentioned above, my favored books for the theory of algebraic numbers are [52], [50], and [40]; for the theory of algebraic functions [60], [11], and [48]. Further literature will be given in the respective chapters.

Originally I planned to publish a self-contained volume on class field theory together with a detailed presentation of the necessary requirements. However the material turned out to be too comprehensive to be published in one volume. Thus I decided to split my presentation. The first part contains the classical theory of algebraic numbers and algebraic functions from an ideal-theoretic, an elementary analytic and a valuation-theoretic point of view, together with an exhaustive presentation of the field- and ring-theoretic preliminaries. Based on these foundations, the forthcoming volume on class field theory will contain the main theorems of class field theory for local and global fields together with the necessary additional foundations from topology, homological algebra, and the theory of simple algebras.

Although the contents of the various chapters can be seen from their titles, I give a short sketch of their topics.

Chapter 1 contains (after some preliminaries on polynomials and ideals) the classical theory of (finite) algebraic field extensions, discriminants and resultants. Of course, most of this material can be found in standard textbooks on algebra, but for the convenience of the reader I give a detailed presentation of the most important facts and amend them suitably for the purpose of subsequent applications and references.

Chapter 2 is devoted to the ring-theoretical fundamentals of the theory of algebraic numbers and algebraic functions. The central subject is the theory of Dedekind domains and their extensions, and I give a detailed exposition of the ideal-theoretic fundamentals. Apart from the standard ring-theoretical results I discuss several special topics which are motivated by number-theoretical questions but in fact are of a purely ring-theoretical nature (ray class groups, class groups of non-principal orders, ideal norma, Frobenius automorphisms).

Chapter 3 contains the classical results from the theory of algebraic number fields insofar as they allow a satisfactory approach by only ideal-theoretic or geometric methods. These include special results for quadratic, cubic, cyclotomic, and abelian number fields. I give a brief sketch of the main results of class field theory and discuss in detail the theory of quadratic orders.

In Chapter 4 we discuss analytical results which are indispensable for an advanced understanding of the arithmetic of algebraic number fields. Main topics are various types of density results for prime ideals, the prime ideal theorem, and (using class field theory) R. Dirichlet's theorem on prime ideals in ideal classes. In particular, we highlight the determination of the arithmetic by the class group and the prime ideal theorem for non-principal orders.

Only in Chapter 5 valuation-theoretic methods are introduced, and the theory of valued fields is developed for its own sake in great generality. Then it is applied to the structure theory of $\mathfrak{p}$-adic fields and Laurent series fields and their extensions. In particular, only now do we develop the theory of differents and discriminants both in the local and in the global case, and eventually we investigate higher ramification.

Finally Chapter 6 deals with function fields. Although the presentation here is independent of the number field case, I heavily use the ideal- and valuation-theoretic methods from Chapters 2 and 5. We prove the Riemann-Roch theorem using the concept of Weil differentials, discuss its various applications and investigate the connection between Weil differentials and classical differentials. A final section deals with the theory of function fields having a finite field of constants and their zeta functions, including a proof of the Hasse-Weil theorem.

Each chapter is followed by 20 exercises of different natures. Some are elementary, some are rather tricky (hints are provided) and some contain suggestions for further reading (references are provided). In any case, I avoided the phrase "prove that", since I tacitly presume that all assertions have to be proved.

Preliminaries

Throughout, we assume acquaintance with the basic concepts of algebra: groups, rings, polynomials, and modules including some modern linear algebra (isomorphism theorems, exact sequences, free modules, direct sums etc.). Apart from these basics, I tried to keep the presentation as self-contained as possible. A reader who is acquainted with the first few chapters of S. Lang's book [39], T. W. Hungerford's book [29] or N. Jacobson's book [32] is well prepared for the main part of this volume. Additionally, basic facts from analysis (complex function theory and integration) are used in Chapter 4, and some very elementary basic topological concepts are used in Chapter 5.

Acknowledgements

I thank my colleagues Guenter Lettl, Andreas Reinhart, and Roswitha Rissner who read parts of the manuscript and corrected several misprints and minor errors in an earlier version of the manuscript. I owe special thanks to Florian Kainrath who carefully read most parts of the text and provided several essential improvements. Last not least I give my thanks to Alfred Geroldinger for so many fruitful and encouraging conversations over the years. Although I retired from my university duties several years ago, I appreciate having the facilities of the Mathematical Institute of the University of Graz still at my disposal.

Notations and Conventions

For sets A, B we write $A \subsetneq B$ or $B \supsetneq A$ if A is a proper subset of B. The notions $A \subset B$ and $B \supset A$ include the case $A = B$.

We denote by

- $\mathbb{N}_0$ the set of non-negative integers;
- $\mathbb{N} = \mathbb{N}_0 \setminus \{0\}$ the set of positive integers;
- $\mathbb{P}$ the set of prime numbers;
- $\mathbb{Z}$ the ring of integers, and $\mathbb{Q}$ the field of rational numbers;
- $\mathbb{R}$ the field of real numbers, and $\mathbb{C}$ the field of complex numbers;
- $\mathbb{F}_p = \mathbb{Z}/p\mathbb{Z}$ the field of p elements for a prime p;
- $|A| \in \mathbb{N}_0 \cup \{\infty\}$ the number of elements of a set A;
- $A^\bullet$ the set of non-zero elements of a set A which possibly contains a zero element.

Let A and B be sets. For a map $f\colon A \to B$ and a subset S of A we denote by $f \restriction S\colon S \to B$ the restriction of f to S. For a set A, we denote by $\mathbb{P}(A)$ the set of all subsets of A, by $\mathbb{P}_{\text{fin}}(A)$ the set of all finite subsets of A, and by $\text{card}(A)$ the cardinal number of A. If A is finite, then $\text{card}(A) = |A| \in \mathbb{N}_0$, and if A is infinite, we set $|A| = \infty$. Only in two places (in 1.2.7 and in 2.3.4) we use basic facts of cardinal arithmetic.

We use the self-explaining symbols $\mathbb{R}_{>0}$, $\mathbb{R}_{\ge 0}$, $\mathbb{Q}_{>0}$, $\mathbb{N}_{\ge 2}$ etc. For $a, b \in \mathbb{Z}$, we set $[a, b] = \{x \in \mathbb{Z} \mid a \le x \le b\}$ if there is no danger of confusion with the real interval, and we set $[a, b] = \emptyset$ if $a > b$.

For a real number x, we set

$$\lfloor x \rfloor = \max\{g \in \mathbb{Z} \mid g \le x\}, \quad \lceil x \rceil = \min\{g \in \mathbb{Z} \mid g \ge x\},$$

and we denote by $\text{sgn}(x) \in \{0, \pm 1\}$ its sign. For a complex number z, we denote by $\Re(z)$ its real part, by $\Im(z)$ its imaginary part, and we normalize $\sqrt{z}$ so that $\sqrt{z} \ge 0$ if $z \in \mathbb{R}_{\ge 0}$, and $\Im(\sqrt{z}) > 0$ if $z \in \mathbb{C} \setminus \mathbb{R}_{\ge 0}$. We denote by e Euler's constant and by $\mathsf{i} = \sqrt{-1}$ the imaginary unit.

In asymptotic results, we use simultaneously the O- and o-notation of E. Landau and the $\ll$-notation of A. Vinogradov. To be precise, for functions $f,\, g\colon D \to \mathbb{C}$ we write $f = \mathcal{O}(g)$ or $f \ll g$ if there exists some $M \in \mathbb{R}_{>0}$ such that $|f(x)| \le M|g(x)|$ for all $x \in D$, and if $D \subset \mathbb{R}$, we write $f = o(g)$ if $f(x) = a(x)g(x)$ for some function $a\colon D \to \mathbb{R}$ such that $a(x) \to 0$ (if x goes to ∞ or another specified limit).

More generally, we say that an assertion holds for $x \gg 1$ if there is some $x_0 \ge 1$ such that the assertion holds for all $x \ge x_0$.

We use the symbol $\mathbf{0}$ to denote the set $\{0\}$, and we also write $\mathbf{0} = (0, \ldots, 0)$ for the zero row and $\mathbf{0} = (0, \ldots, 0)^{\mathrm{t}}$ for the zero column provided that the meaning is clear from the context. In a similar way, we use the symbol $\mathbf{1}$. For a multiplicative monoid G we usually denote by $1 = 1_G$ the unit element of G.

We shall frequently use Zorn's Lemma in the following form.

Zorn's lemma. Let $(X, \le)$ be a non-empty partially ordered set such that every chain in X has an upper bound. Then X possesses a maximal element.

1

Field Extensions

In this first chapter we present the classical theory of field extensions and finite Galois theory, and we suitably amend the theory for our requirements. In particular, we put emphasis on the theory of cyclotomic fields and radical extensions, and on the associated elementary number theory. The already mentioned books [39], [29], and [32] serve as standard references.

An eager reader who feels familiar with the subject is invited to skip this chapter and have recourse to it on demand.

1.1 Preliminaries on ideals and polynomials

In this introductory section we gather some frequently used elementary facts concerning rings, modules, ideals, and polynomials, in order to fix notation and for the purpose of a simple reference. The reader should view this section as a picture-book of preliminaries.

By a **semigroup** we mean a non-empty set with an associative law of composition. A **monoid** is a semigroup possessing a neutral element (denoted by 1 in the multiplicative and by 0 in the additive case), and a **submonoid** is always assumed to contain the neutral element. By a **ring** R we always mean an associative ring possessing a unit element $1 = 1_R$, and we assume that $1 \neq 0$. Then $R = (R, +, \cdot, 1)$, we call $(R, +)$ the additive group and $(R, \cdot, 1)$ the multiplicative monoid of R. A **subring** is always assumed to contain the unit element.

Let R be a ring. An element $z \in R$ is called a left (resp. right) zero divisor if there exists some $u \in R^\bullet$ such that $zu = 0$ (resp. $uz = 0$), and it is called a unit (or an invertible element) if there exists some $u \in R$ such that $zu = uz = 1$. We denote by $R^\times$ the group of all invertible elements of R and call it the **unit group** of R. If R is commutative, we denote by $z(R)$ the set of all zero divisors of R. By a **domain** we mean a commutative ring $\mathfrak{o}$ satisfying $z(\mathfrak{o}) = \mathbf{0}$ (then it follows that $0 \neq 1$). A ring R is called a **division ring** if $R^\bullet = R^\times$, and a **field** if it is a commutative divison ring. Every domain $\mathfrak{o}$ possesses (up to isomorphisms) a unique quotient field, denoted by $\mathsf{q}(\mathfrak{o})$. If a domain $\mathfrak{o}$ is contained in a field K, then we tacitly assume that $\mathsf{q}(\mathfrak{o}) \subset K$ (in fact, in this case $\mathsf{q}(\mathfrak{o})$ is the smallest subfield of K containing $\mathfrak{o}$). For a set X and $m, n \in \mathbb{N}$ we denote by $\mathsf{M}_{m,n}(X)$ the set of all $m\times n$-matrices over X, and we set $\mathsf{M}_n(X) = \mathsf{M}_{n,n}(XA)$. We write a matrix $A \in \mathsf{M}_{m,n}(X)$ in the form $A = (a_{j,i})_{j\in[1,m],\, i\in[1,n]}$ and denote by $A^{\mathrm{t}} = (a_{i,j})_{i\in[1,n],\, j\in[1,m]} \in \mathsf{M}_{n,m}(X)$ the transposed matrix.

A **ring homomorphism** $f\colon R \to R'$ is assumed to satisfy $f(1_R) = 1_{R'}$. If R is a division ring, then every ring homomorphism $f\colon R \to R'$ is injective, and if R and R' are both fields, then a ring homomorphism $f\colon R \to R'$ is called a **field homomorphism**.

Let R be a ring. By an **R-module** we mean a unitary left R-module. A $\mathbb{Z}$-module is nothing but an abelian group, and a module over a field is a vector space. If M and M' are R-modules, then $\mathrm{Hom}_R(M, M')$ denotes the abelian group of all R-linear maps (or R-homomorphisms) $f\colon M \to M'$ (with pointwise addition) while $\mathrm{Hom}(M, M') = \mathrm{Hom}_{\mathbb{Z}}(M, M')$ denotes the group of all group homomorphisms. Correspondingly, we set $\mathrm{End}_R(M) = \mathrm{Hom}_R(M, M)$ and $\mathrm{End}(M) = \mathrm{End}_{\mathbb{Z}}(M)$. The ring R itself is an R-module, an R-submodule of R is called a **left ideals**. By an **ideal** per se we always mean a two-sided ideal. If X is a subset of an R-module M, then ${}_R(X)$ denotes the R-submodule generated by X. If $\mathfrak{a}$ is an ideal of R, then

$$\mathfrak{a}M = {}_R(\{ax \mid a \in \mathfrak{a},\ x \in M\}),$$

and if $M = {}_R(X)$, then $\mathfrak{a}M = {}_R(\{ax \mid a \in \mathfrak{a},\ x \in X\})$. If $X = \{x_1, \ldots, x_n\}$, we write ${}_R(X) = {}_R(x_1, \ldots x_n)$, and we call M **finitely generated** if $M = {}_R(x_1, \ldots, x_n)$ for some $x_1, \ldots, x_n \in M$. For an R-module M and a set I we denote by $M^{(I)}$ the R-module of all families $(a_i)_{i\in I}$ where $a_i = 0$ for almost all $i \in I$ (under component-wise operations).

Let M be an R-module. A family $(x_i)_{i\in I}$ in M is called **linearly independent** (over R) if the map

$$\phi\colon R^{(I)} \to M, \quad \text{defined by} \quad \phi((a_i)_{i\in I}) = \sum_{i\in I} a_i x_i, \quad \text{is a monomorphism,}$$

and $(x_i)_{i\in I}$ is called a **basis** of M if ϕ is an isomorphism. A subset B of M is called linearly independent resp. a basis if the family $(b)_{b\in B}$ has this property. An R-module is called **free** if it has a basis.

For a subset X of a group we denote by $\langle X\rangle = {}_{\mathbb{Z}}(X)$ the group generated by X. Every abelian group is a $\mathbb{Z}$-module, and it is called **free** if it is a free $\mathbb{Z}$-module.

We will not be rigid about the notation of direct sums: Usually we will not explicitly distinguish between outer and inner direct sums when the meaning is clear from the context. We demonstrate this procedure by the following (well-known and frequently used) lemmas 1.1.1 and 1.1.2.

Lemma 1.1.1. *Let R be a ring.* An R-module epimorphism $\varphi\colon M \to F$ is said to **split** if there exists an R-homomorphism $j\colon F \to M$ such that $\varphi\circ j = \mathrm{id}_F$.

If this is the case, then j is a monomorphism, and

$$M = \mathrm{Im}(j) \oplus \mathrm{Ker}(\varphi) \cong F \oplus \mathrm{Ker}(\varphi). \tag{$\oplus$}$$

If F is free, then every R-module homomorphism $\varphi\colon M \to F$ splits.
Note that in ($\oplus$) the first direct sum is an inner one and the second direct sum is an outer one.

Proof. Let $\varphi\colon M \to F$ be a split R-module homomorphism and $j\colon F \to M$ an R-homomorphism such that $\varphi\circ j = \mathrm{id}_F$. Then apparently j is a monomorphism. If $x \in F$, then $\varphi(x - j\circ\varphi(x)) = 0$, hence $x = j\circ\varphi(x) + [x - j\circ\varphi(x)]$, and thus $M = \mathrm{Im}(j) + \mathrm{Ker}(\varphi)$. If $x = j(z) \in \mathrm{Im}(j) \cap \mathrm{Ker}(\varphi)$ for some $z \in F$, then it follows that $0 = \varphi(x) = \varphi\circ j(z) = z$ and thus $x = 0$. Hence $\mathrm{Im}(j)\cap\mathrm{Ker}(\varphi) = \mathbf{0}$, the sum is direct, and we obtain $M = \mathrm{Im}(j) \oplus \mathrm{Ker}(\varphi)$ (inner direct sum). The isomorphism $j\colon F \stackrel{\sim}{\to} \mathrm{Im}(j)$ yields an isomorphism

$$j \oplus \mathrm{id}_{\mathrm{Ker}(\varphi)}\colon F \oplus \mathrm{Ker}(\varphi) \stackrel{\sim}{\to} \mathrm{Im}(j) \oplus \mathrm{Ker}(\varphi) = M, \ \ (x, a) \mapsto j(x) + a,$$

between the outer and the inner direct sum.

Let F be free, $(u_i)_{i\in I}$ an R-basis of F, and for $i \in I$ let $x_i \in \varphi^{-1}(u_i)$. Then there exists a (unique) R-homomorphism $j\colon F \to M$ such that $j(u_i) = x_i$ for all $i \in I$. Since $\varphi\circ j(u_i) = u_i$ for all $i \in I$, it follows that $\varphi\circ j = \mathrm{id}_F$. Hence φ splits. □

In contrast to what we have shown above, we shall handle direct products of groups more carefully. For subgroups A, B of an (abelian multiplicative) group G we denote their inner direct product (inside G) by $A{\boldsymbol\cdot}B$ (so that the map $A\times B \to A{\boldsymbol\cdot}B$, $(a, b) \mapsto ab$, is an isomorphism). With this wording, a multiplicative version of 1.1.1 reads as follows.

Lemma 1.1.2. *Let $\varphi\colon M \to F$ be a split epimorphism of (multiplicative) abelian groups, and let $j\colon F \to M$ be a group homomorphism. If $\varphi\circ j = \mathrm{id}_F$, then j is a monomorphism, and $M = \mathrm{Im}(j){\boldsymbol\cdot}\mathrm{Ker}(\varphi)$.*

If F is a free abelian group, then every epimorphism $\varphi\colon M \to F$ splits.

Proof. See 1.1.1. □

Let R be a commutative ring. For a subset X of R we denote by ${}_R(X)$ the ideal generated by X. If $X = \{x_1, \ldots, x_n\}$, then the ideal

$${}_R(X) = {}_R(x_1, \ldots, x_n) = Rx_1 + \ldots + Rx_n$$

is called finitely generated, and if $X = \{x\}$, then ${}_R(x) = Rx = xR$ is called a **principal ideal**. A domain in which every ideal is a principal ideal is called a **principal ideal domain**. Classical examples of principal ideal domains are fields, $\mathbb{Z}$, and polynomial rings $K[X]$ over a field K (see 1.1.8.3(b) below).

A ring R is called **Noetherian** if every ideal is finitely generated. Recall that R is Noetherian if every ascending chain of ideals becomes stationary or, equivalently, if every non-empty set of ideals contains a maximal element (regarding $\subset$). Every finitely generated module over a Noetherian ring is itself

Noetherian (i. e., every submodule is finitely generated, every ascending chain of submodules becomes stationary, and every non-empty subset of submodules contains a maximal element).

An ideal $\mathfrak{a}$ of R is called **proper** if $\mathfrak{a} \ne R$ [that is, $1 \notin \mathfrak{a}$ or $\mathfrak{a} \cap R^\times = \emptyset$]. Finitely many ideals $\mathfrak{a}_1, \ldots, \mathfrak{a}_n$ of R are called **coprime** if $\mathfrak{a}_1 + \ldots + \mathfrak{a}_n = R$. If $\mathfrak{a}$ is an ideal of a commutative ring R, then an element $x \in R$ is called **coprime** to $\mathfrak{a}$ if the ideals xR and $\mathfrak{a}$ are coprime [equivalently, $x + \mathfrak{a} \in (R/\mathfrak{a})^\times$].

Lemma 1.1.3. *Let R be a ring and $m, n \in \mathbb{N}$.*

1. *Let $\mathfrak{a}_1, \ldots, \mathfrak{a}_n, \mathfrak{b}_1, \ldots, \mathfrak{b}_m$ be ideals of R such that $\mathfrak{a}_i + \mathfrak{b}_j = R$ for all $i \in [1, n]$ and $j \in [1, m]$. Then $\mathfrak{a}_1 \cdot \ldots \cdot \mathfrak{a}_n + \mathfrak{b}_1 \cdot \ldots \cdot \mathfrak{b}_m = R$.*

2. *Let $\mathfrak{a}_1, \ldots, \mathfrak{a}_n$ be ideals of R such that $\mathfrak{a}_1 + \ldots + \mathfrak{a}_n = R$ and $k_1, \ldots, k_n \in \mathbb{N}$. Then $\mathfrak{a}_1^{k_1} + \ldots + \mathfrak{a}_n^{k_n} = R$.*

Proof. 1. By induction on $m + n$: For $m = n = 1$ there is nothing to do. Thus let $m + n > 2$, by symmetry we may assume that $n \ge 2$, and let $\mathfrak{a}_1 \cdot \ldots \cdot \mathfrak{a}_{n-1} + \mathfrak{b}_1 \cdot \ldots \cdot \mathfrak{b}_m = R$. If $\mathfrak{a} = \mathfrak{a}_1 \cdot \ldots \cdot \mathfrak{a}_{n-1}$ and $\mathfrak{b} = \mathfrak{b}_1 \cdot \ldots \cdot \mathfrak{b}_m$, then $\mathfrak{a} + \mathfrak{b} = \mathfrak{a}_n + \mathfrak{b} = R$. Hence there exist $a \in \mathfrak{a}$, $a' \in \mathfrak{a}_n$ and $b, b' \in \mathfrak{b}$ such that $a + b = a' + b' = 1$. Then it follows that $1 = (a + b)(a' + b') = aa' + b(a' + b') + ab' \in \mathfrak{a}\mathfrak{a}_n + \mathfrak{b} = \mathfrak{a}_1 \cdot \ldots \cdot \mathfrak{a}_n + \mathfrak{b}_1 \cdot \ldots \cdot \mathfrak{b}_m$.

2. By induction on $k = k_1 + \ldots + k_n$: For $k = n$ there is nothing to do. Thus let $k > n$; by renumbering if necessary we may assume that $k_1 \ge 2$, and we set $\mathfrak{a} = \mathfrak{a}_2^{k_2} + \ldots + \mathfrak{a}_n^{k_n}$. Then $\mathfrak{a}_1^{k_1 - 1} + \mathfrak{a} = \mathfrak{a}_1 + \mathfrak{a} = R$ by the induction hypothesis, hence $R = \mathfrak{a}_1^{k_1} + \mathfrak{a}^2 \subset \mathfrak{a}_1^{k_1} + \mathfrak{a} \subset R$ by 1., and thus eventually $\mathfrak{a}_1^{k_1} + \ldots + \mathfrak{a}_n^{k_n} = R$. □

The following Chinese remainder theorem is an essential tool to represent a commutative ring as a direct product of (simpler) rings.

Theorem 1.1.4 (Chinese remainder theorem). *Let R be a commutative ring and $n \ge 2$. Let $\mathfrak{a}_1, \ldots, \mathfrak{a}_n$ be pairwise coprime ideals of R and $\mathfrak{a} = \mathfrak{a}_1 \cap \ldots \cap \mathfrak{a}_n$. Then $\mathfrak{a} = \mathfrak{a}_1 \cdot \ldots \cdot \mathfrak{a}_n$, there is a ring isomorphism*

$$\Phi \colon R/\mathfrak{a} \stackrel{\sim}{\rightarrow} \prod_{i=1}^n R/\mathfrak{a}_i \quad \text{such that} \quad \Phi(a+\mathfrak{a}) = (a+\mathfrak{a}_1, \ldots, a+\mathfrak{a}_n) \quad \text{for all } a \in R,$$

and Φ induces a group isomorphism

$$\Phi \restriction (R/\mathfrak{a})^\times : (R/\mathfrak{a})^\times \stackrel{\sim}{\rightarrow} \prod_{i=1}^n (R/\mathfrak{a}_i)^\times.$$

In particular:

- *For every n-tuple $(a_1, \ldots, a_n) \in R^n$ there exists a modulo $\mathfrak{a}$ uniquely determined element $a \in R$ satisfying $a \equiv a_i \bmod \mathfrak{a}_i$ for all $i \in [1, n]$.*

- *For* $i \in [1,n]$ *let*

$$\Lambda_i = \{a+\mathfrak{a} \in (R/\mathfrak{a})^\times \mid a \in R,\ a \equiv 1 \bmod \mathfrak{a}_j \ \textit{for all}\ \ j \in [1,n]\setminus\{i\}\}.$$

Then Λ_i *is a subgroup of* $(R/\mathfrak{a})^\times$, *the assignment* $a+\mathfrak{a} \mapsto a+\mathfrak{a}_i$ *defines an isomorphism* $\Phi_i\colon \Lambda_i \xrightarrow{\sim} (R/\mathfrak{a}_i)^\times$, *and* $(R/\mathfrak{a})^\times = \Lambda_1 \cdot \ldots \cdot \Lambda_n$ *(inner direct product). If* $a \in R$ *is coprime to* $\mathfrak{a}$, *and if* $a(i) \in R$ *satisfies* $a(i) \equiv a \bmod \mathfrak{a}_i$ *and* $a(i) \equiv 1 \bmod \mathfrak{a}_j$ *for all* $j \in [1,n] \setminus \{i\}$, *then* $a(i) \in \Lambda_i$,

$$a \equiv a(1) \cdot \ldots \cdot a(n) \bmod \mathfrak{a} \quad \textit{and} \quad \Phi(a+\mathfrak{a}) = (a(1)+\mathfrak{a}_1, \ldots a(n)+\mathfrak{a}_n).$$

Proof. The map

$$\Phi_0\colon R \to R/\mathfrak{a}_1 \times \ldots \times R/\mathfrak{a}_n, \quad \text{defined by} \quad \Phi_0(a) = (a+\mathfrak{a}_1, \ldots, a+\mathfrak{a}_n),$$

is a ring homomorphism, $\mathrm{Ker}(\Phi_0) = \mathfrak{a}$, and as $\mathfrak{a}_1 \cdot \ldots \cdot \mathfrak{a}_n \subset \mathfrak{a}$, it suffices to prove:

> **A.** $\mathfrak{a} \subset \mathfrak{a}_1 \cdot \ldots \cdot \mathfrak{a}_n$, and for every n-tuple $(a_1, \ldots, a_n) \in R^n$ there exists some $a \in R$ satisfying $a \equiv a_i \bmod \mathfrak{a}_i$ for all $i \in [1,n]$ (the uniqueness of a modulo $\mathfrak{a}$ is then obvious).

Indeed, **A** implies that Φ_0 is surjective. Hence it induces a ring isomorphism Φ as asserted, its restriction to unit groups is a group isomorphism, and as

$$(R/\mathfrak{a}_1 \times \ldots \times R/\mathfrak{a}_n)^\times = (R/\mathfrak{a}_1)^\times \times \ldots \times (R/\mathfrak{a}_n)^\times,$$

the restriction $\Phi \restriction (R/\mathfrak{a})^\times$ is a group isomorphism. It induces a representation of $(R/\mathfrak{a})^\times$ as an inner direct product, which is stated explicitly at the end of the theorem and is easily comprehensible.

Proof of **A.** By induction on n.

$n = 2$: Let $e_1 \in \mathfrak{a}_1$ and $e_2 \in \mathfrak{a}_2$ be such that $e_1 + e_2 = 1$. If $a \in \mathfrak{a}_1 \cap \mathfrak{a}_2$, then $a = ae_1 + ae_2 \in \mathfrak{a}_1\mathfrak{a}_2$. If $(a_1, a_2) \in R^2$ and $a = e_2a_1 + e_1a_2$, then

$$a = a_1 + e_1(a_2 - a_1) \equiv a_1 \bmod \mathfrak{a}_1 \quad \text{and} \quad a = a_2 + e_2(a_1 - a_2) \equiv a_2 \bmod \mathfrak{a}_2.$$

$n \geq 3$, $n-1 \to n$: By 1.1.3 we get $\mathfrak{a}_1 \cdot \ldots \cdot \mathfrak{a}_{n-1} + \mathfrak{a}_n = R$. By the induction hypothesis and the case $n = 2$, it follows that

$$\mathfrak{a} = (\mathfrak{a}_1 \cap \ldots \cap \mathfrak{a}_{n-1}) \cap \mathfrak{a}_n = \mathfrak{a}_1 \cdot \ldots \cdot \mathfrak{a}_{n-1} \cap \mathfrak{a}_n = \mathfrak{a}_1 \cdot \ldots \cdot \mathfrak{a}_{n-1} \cdot \mathfrak{a}_n.$$

Thus let $(a_1, \ldots, a_n) \in R^n$. By the induction hypothesis there exists some $b \in R$ satisfying $b \equiv a_i \bmod \mathfrak{a}_i$ for all $i \in [1, n-1]$, and by what we have proved for $n = 2$ there exists some $a \in R$ such that $a \equiv b \bmod \mathfrak{a}_1 \cdot \ldots \cdot \mathfrak{a}_{n-1}$ and $a \equiv a_n \bmod \mathfrak{a}_n$. Overall it follows that $a \equiv a_i \bmod \mathfrak{a}_i$ for all $i \in [1,n]$. $\square$

Let R be a commutative ring. An ideal $\mathfrak{p}$ of R is called a

- **prime ideal** if the residue class ring $R/\mathfrak{p}$ is a domain [equivalently, $\mathfrak{p} \neq R$, and if $a, b \in R$ and $ab \in \mathfrak{p}$, then $a \in \mathfrak{p}$ or $b \in \mathfrak{p}$];
- **maximal ideal** if $\mathfrak{p} \neq R$ and there is no ideal $\mathfrak{c}$ such that $\mathfrak{p} \subsetneq \mathfrak{c} \subsetneq R$ [equivalently, the residue class ring $R/\mathfrak{p}$ is a field]; in this case $\mathsf{k}_\mathfrak{p} = R/\mathfrak{p}$ is called the **residue class field** of $\mathfrak{p}$.

Every maximal ideal of R is a prime ideal, and every prime ideal is proper. The zero ideal $\mathbf{0}$ is a prime ideal of R if and only if R is a domain, and it is a maximal ideal of R if and only if R is a field. We denote by

- $\max(R)$ the set of all maximal ideals of R, and by
- $\mathcal{P}(R)$ the set of all non-zero prime ideals of R.

If $|\max(R)| = 1$, then R is called **local**, and if $\max(R)$ is finite, then R is called **semilocal**.

A subset T of $R^\bullet$ is called **multiplicatively closed** if $1 \in T$ and $TT = T$. An ideal $\mathfrak{p}$ of R is a prime ideal if and only if $R \setminus \mathfrak{p}$ is multiplicatively closed. If $\mathfrak{o}$ is a domain, then $\mathfrak{o}^\bullet$ is a (multiplicative) monoid, and a subset T of $\mathfrak{o}^\bullet$ is multiplicatively closed if and only if T is a submonoid of $\mathfrak{o}^\bullet$.

Theorem 1.1.5. *Let R be a commutative ring.*

1. (Krull's existence theorem) *Let $\mathfrak{a}$ be an ideal of R, T a multiplicatively closed subset of $R^\bullet$ such that $T \cap \mathfrak{a} = \emptyset$, and let Ω be the set of all ideals $\mathfrak{b}$ of R such that $\mathfrak{a} \subset \mathfrak{b}$ and $\mathfrak{b} \cap T = \emptyset$. Then Ω possesses maximal elements (regarding $\subset$), and every maximal element of Ω is a prime ideal.*

2. Every proper ideal of R is contained in a maximal ideal. In particular, if $R \neq \mathbf{0}$, then $\max(R) \neq \emptyset$.

3. R is local if and only if $R \setminus R^\times$ is an ideal of R (and then $R \setminus R^\times$ is the unique maximal ideal of R).

4. Let $\mathfrak{p}$ be a maximal ideal of R and $n \in \mathbb{N}$. Then $R/\mathfrak{p}^n$ is a local ring with maximal ideal $\mathfrak{p}/\mathfrak{p}^n$.

Proof. 1. By Zorn's lemma Ω has a maximal element. Indeed, $(\Omega, \subset)$ is a partially ordered set, $\mathfrak{a} \in \Omega$, and if Σ is a chain in Ω, then its union belongs to Ω and is an upper bound.

Let $\mathfrak{p}$ be a maximal element of Ω and $a, b \in R \setminus \mathfrak{p}$. Then $\mathfrak{p} + aR \notin \Omega$ and $\mathfrak{p} + bR \notin \Omega$, hence $(\mathfrak{p} + aR) \cap T \neq \emptyset$ and $(\mathfrak{p} + aR) \cap T \neq \emptyset$. Let $p, q \in \mathfrak{p}$ and $u, v \in R$ be such that $s = p + au \in T$ and $t = q + bv \in T$. It follows that $st \in T$, hence $st \notin \mathfrak{p}$, and since $st \equiv abuv \mod \mathfrak{p}$ we obtain $ab \notin \mathfrak{p}$. Consequently, $\mathfrak{p}$ is a prime ideal.

2. By 1. with $T = \{1\}$.

3. If $R \setminus R^\times$ is an ideal, then it is the largest proper ideal, and therefore it is the unique maximal ideal of R. Conversely let R be local and $\mathfrak{m}$ the unique maximal ideal of R. Then $\mathfrak{m} \subset R \setminus R^\times$, and we prove that equality holds. If $a \in R \setminus R^\times$, then aR is a proper ideal of R, and by 2. there exists a maximal ideal $\mathfrak{m}'$ of R such that $aR \subset \mathfrak{m}'$. Since $\mathfrak{m} = \mathfrak{m}'$, we get $a \in \mathfrak{m}$.

4. If $a \in R \setminus \mathfrak{p}$, then a is coprime to $\mathfrak{p}$, hence to $\mathfrak{p}^n$ by 1.1.3, and thus we obtain $a + \mathfrak{p}^n \in (R/\mathfrak{p}^n)^\times$. Consequently, $\mathfrak{p}/\mathfrak{p}^n = (R/\mathfrak{p}^n) \setminus (R/\mathfrak{p}^n)^\times$, and thus $R/\mathfrak{p}^n$ is local with maximal ideal $\mathfrak{p}/\mathfrak{p}^n$. □

Theorem 1.1.6 (Prime avoidance). *Let R be a commutative ring, and let S be a subset of R such that $S + S \subset S$ and $SS \subset S$. Let $n \geq 2$, and let $\mathfrak{a}_1, \ldots, \mathfrak{a}_n$ be ideals of R such that all with the exception of at most two of them are prime ideals, and $S \subset \mathfrak{a}_1 \cup \ldots \cup \mathfrak{a}_n$. Then there exists some $i \in [1, n]$ such that $S \subset \mathfrak{a}_i$.*

Proof. By induction on n.

$n = 2$: Assume to the contrary that $S \not\subset \mathfrak{a}_1$ and $S \not\subset \mathfrak{a}_2$. If $x_1 \in S \setminus \mathfrak{a}_1$ and $x_2 \in S \setminus \mathfrak{a}_2$, then $x_1 \in \mathfrak{a}_2$, $x_2 \in \mathfrak{a}_1$, and $x_1 + x_2 \in S \setminus (\mathfrak{a}_1 \cup \mathfrak{a}_2)$, a contradiction.

$n \geq 3$, $n-1 \to n$: For $k \in [1, n]$ we set $\mathfrak{A}_k = \mathfrak{a}_1 \cup \ldots \cup \mathfrak{a}_{k-1} \cup \mathfrak{a}_{k+1} \cup \ldots \cup \mathfrak{a}_n$. If $S \subset \mathfrak{A}_k$ for some $k \in [1, n]$, then the assertion follows from the induction hypothesis. Thus assume that $S \not\subset \mathfrak{A}_k$, and let $x_k \in S \setminus \mathfrak{A}_k \subset \mathfrak{a}_k$ for all $k \in [1, n]$. We may assume that $\mathfrak{a}_1$ is a prime ideal. Then $x_k \notin \mathfrak{a}_1$ for all $k \in [2, n]$, hence $x_2 \cdot \ldots \cdot x_n \notin \mathfrak{a}_1$, and consequently $x_1 + x_2 \cdot \ldots \cdot x_n \in S \setminus \mathfrak{a}_k$ for all $k \in [1, n]$, a contradiction. □

Let R be a commutative ring. A (commutative) R**-algebra** is a ring homomorphism $\varepsilon \colon R \to A$ into a (commutative) ring A such that $\varepsilon(\lambda)a = a\varepsilon(\lambda)$ for all $\lambda \in R$ and $a \in A$. Then A becomes an R-module by means of $\lambda a = \varepsilon(\lambda)a$ for all $\lambda \in R$ and $a \in A$, and $\lambda(ab) = (\lambda a)b = a(\lambda b)$ for all $\lambda \in R$ and $a, b \in A$. On the other hand, if a ring A is also an R-module such that $\lambda(ab) = (\lambda a)b = a(\lambda b)$ for all $\lambda \in R$ and $a, b \in A$, then $\varepsilon_A \colon R \to A$, defined by $\varepsilon_A(\lambda) = \lambda 1_A$ is an R-algebra, and, by abuse of language we call A itself an R-Algebra. If A and B are R-algebras, then an R-algebra homomorphism $\varphi \colon A \to B$ is a ring homomorphism which is also an R-module homomorphism. We denote by $\mathrm{Hom}^R(A, B)$ the set of all R-algebra homomorphisms $\varphi \colon A \to B$, while $\mathrm{Hom}_R(A, B)$ only denotes the set of all R-module homomorphisms.

Every commutative overring A of R is an R-algebra in the natural way. If M is any R-module, then its endomorphism ring $\mathrm{End}_R(M)$ is an R-algebra (where the ring multiplication is given by $(f, g) \mapsto f \circ g$). For $n \in \mathbb{N}$, the matrix ring $\mathsf{M}_n(R)$ is an R-algebra. For every ring A, there is a unique ring homomorphism $\mathbb{Z} \to A$. Consequently, every ring is a $\mathbb{Z}$-algebra.

Let $\varphi \colon R \to R'$ be a ring homomorphism and $\mathfrak{a}'$ an ideal of R'. Then $\varphi^{-1}(\mathfrak{a}')$ is an ideal of R, and φ induces a ring monomorphism

$\varphi'\colon R/\varphi^{-1}(\mathfrak{a}') \to R'/\mathfrak{a}'$. If $\mathfrak{a}'$ is a prime ideal of R', then $\varphi^{-1}(\mathfrak{a}')$ is a prime ideal of R. We usually identify $R/\varphi^{-1}(\mathfrak{a}')$ with its image under φ', and we view $R/\varphi^{-1}(\mathfrak{a}')$ as a subring of $R'/\mathfrak{a}'$.

If R is a subring of R' and $\varphi = (R \hookrightarrow R')$, then $\mathfrak{a} = \varphi^{-1}(\mathfrak{a}') = \mathfrak{a}' \cap R$ for every ideal $\mathfrak{a}'$ of R', and we say that $\mathfrak{a}'$ **lies above** $\mathfrak{a}$. We identify $a+\mathfrak{a} \in R/\mathfrak{a}$ with $a+\mathfrak{a}' \in R'/\mathfrak{a}'$. Then $R/\mathfrak{a}$ is a subring of $R'/\mathfrak{a}'$, and we obtain the following commutative diagram in which the vertical arrows are inclusions.

$$\begin{array}{ccccc} \mathfrak{a}' & \longrightarrow & R' & \longrightarrow & R'/\mathfrak{a}' \\ \uparrow & & \uparrow & & \uparrow \\ \mathfrak{a} = \mathfrak{a}' \cap R & \longrightarrow & R & \longrightarrow & R/\mathfrak{a}. \end{array}$$

If R is a subring of R' and $\mathfrak{a}$ is an ideal of R, we denote by $\mathfrak{a}R' = {}_{R'}(\mathfrak{a})$ the extension ideal of R' generated by $\mathfrak{a}$. Obviously, $\mathfrak{a}R' \cap R \supset \mathfrak{a}$ and $(\mathfrak{a}' \cap R)R' \subset \mathfrak{a}'$ for all ideals $\mathfrak{a}$ of R and $\mathfrak{a}'$ of R'.

Let R be a ring, $\mathfrak{a}$ an ideal of R and $\pi\colon R \to R/\mathfrak{a}$ the residue class homomorphism. Let $\Omega(R, \mathfrak{a})$ be the set of all ideals $\mathfrak{c}$ of R containing $\mathfrak{a}$ and $\Omega(R/\mathfrak{a})$ the set of all ideals of $R/\mathfrak{a}$. Then the map $\Phi\colon \Omega(R, \mathfrak{a}) \to \Omega(R/\mathfrak{a})$, defined by $\Phi(\mathfrak{c}) = \mathfrak{c}/\mathfrak{a}$, is bijective, and $\Phi^{-1}(\mathfrak{c}^*) = \pi^{-1}(\mathfrak{c}^*)$ for all $\mathfrak{c}^* \in \Omega(R/\mathfrak{a})$. If $\mathfrak{c} \in \Omega(R, \mathfrak{a})$, then $R/\mathfrak{c} \cong (R/\mathfrak{a})/\Phi(\mathfrak{c})$ (we identify), and $\mathfrak{c}$ is a prime ideal of R if and only if $\Phi(\mathfrak{c})$ is a prime ideal of $R/\mathfrak{a}$.

We proceed with some remarks on divisibility. For a profound discussion of the subject in the context of (cancellative) monoids we refer to Section 2.1.

Let R be a commutative ring. If $a, b \in R$, then we say that a **divides** b and write $a \mid b$ if $bR \subset aR$. If $Z \subset R$, then an element $d \in R$ is called a **gcd** (greatest common divisor) of Z if dR is the smallest principal ideal containing Z, and an element $e \in R$ is called an **lcm** (lowest common multiple) of Z if eR is the largest principal ideal contained in all principal ideals xR for $x \in Z$. Consequently, if $Z \subset R$ and $d, e \in R$ are such that

$$dR = \sum_{a \in Z} aR \quad \text{and} \quad eR = \bigcap_{a \in Z} aR,$$

then d is a gcd and e is an lcm of Z. In particular, 0 is a gcd and 1 is an lcm of $\emptyset$, and if every ideal of R is a principal ideal, then every subset of R has a gcd and an lcm.

Finitely many elements $a_1, \ldots, a_n \in R$ are called **coprime** if 1 is a gcd of $\{a_1, \ldots, a_n\}$, and in this case we write $(a_1, \ldots, a_n) = 1$. Recall that the principal ideals $a_1R, \ldots, a_nR$ are coprime if there exist $x_1, \ldots, x_n \in R$ such that $a_1x_1 + \ldots + a_nx_n = 1$, and this implies $(a_1, \ldots, a_n) = 1$. The converse is true if ${}_R(a_1, \ldots, a_n)$ is a principal ideal, but false in general (in a polynomial ring $R = K[X, Y]$ over a field K the elements X, Y are coprime but the principal ideals RX, RY are not).

Let $\mathfrak{o}$ be a domain. If $a, b \in \mathfrak{o}$, then $a\mathfrak{o} = b\mathfrak{o}$ if and only if $a\mathfrak{o}^\times = b\mathfrak{o}^\times$. In particular, a gcd and an lcm of a subset of $\mathfrak{o}$ (if they exist) are uniquely determined up to factors from $\mathfrak{o}^\times$. An element $p \in \mathfrak{o}^\bullet$ is called a **prime**

element if $p\mathfrak{o}$ is a prime ideal [equivalently, $p \notin \mathfrak{o}^\times \cup \{0\}$, and if $a,\, b \in \mathfrak{o}$ and $p \,|\, ab$, then $p \,|\, a$ or $p \,|\, b$].

A domain $\mathfrak{o}$ is called **factorial** if every non-zero non-unit $a \in \mathfrak{o}^\bullet \setminus \mathfrak{o}^\times$ is a product of prime elements.

Theorem 1.1.7. *Let $\mathfrak{o}$ be a domain.*

1. *Let $p \in \mathfrak{o}$ be a prime element and $a,\, b \in \mathfrak{o}$ such that $p = ab$. Then $a \in \mathfrak{o}^\times$ or $b \in \mathfrak{o}^\times$.*

2. *Let $m,\, n \in \mathbb{N}_0$, $u \in \mathfrak{o}^\times$, and let $p_1, \ldots, p_n,\, q_1, \ldots, q_m \in \mathfrak{o}$ be prime elements such that $up_1 \cdot \ldots \cdot p_n = q_1 \cdot \ldots \cdot q_m$. Then $m = n$, and after a suitable renumbering there exist $u_1, \ldots, u_n \in \mathfrak{o}^\times$ such that $p_i = u_i q_i$ for all $i \in [1, n]$.*

3. *Let $\mathfrak{o}$ be a principal ideal domain. Then $\mathfrak{o}$ is factorial, and every non-zero prime ideal of $\mathfrak{o}$ is a maximal ideal.*

Proof. 1. As $p \,|\, ab$, it follows that $p \,|\, a$ or $p \,|\, b$, and we may assume that $p \,|\, a$, say $a = pc$ for some $c \in \mathfrak{o}$. Then $p = pcb$, hence $cb = 1$ and $b \in \mathfrak{o}^\times$.

2. We proceed by induction on $m + n$. If $m = 0$ or $n = 0$, then $m = n = 0$ and we are done. Thus suppose that $m,\, n \in \mathbb{N}$. Then $q_1 \,|\, up_1 \cdot \ldots \cdot p_n$, hence $q_1 \,|\, p_i$ for some $i \in [1, n]$, and after renumbering we may assume that $q_1 \,|\, p_1$, say $p_1 = u_1 q_1$ for some $u_1 \in \mathfrak{o}$ and thus $u_1 \in \mathfrak{o}^\times$ by 1. It follows that $uu_1p_2 \cdot \ldots \cdot p_n = q_2 \cdot \ldots \cdot q_m$, and the assertion follows by the induction hypothesis.

3. If $\pi \in \mathfrak{o}$ is a prime element and $u \in \mathfrak{o}^\times$, then πu is a prime element, too. Hence it suffices to prove that every $a \in \mathfrak{o}^\bullet$ has a representation $a = u\pi_1 \cdot \ldots \cdot \pi_n$, where $n \in \mathbb{N}_0$, $u \in \mathfrak{o}^\times$ and $\pi_1, \ldots, \pi_n$ are prime elements. We prove first:

> **A.** Let $a,\, b \in \mathfrak{o}^\bullet$, $n \in \mathbb{N}_0$ and $ab = u\pi_1 \cdot \ldots \cdot \pi_n$ with prime elements $\pi_1, \ldots, \pi_n \in \mathfrak{o}$. Then (after a suitable renumbering) there exists some $k \in [0, n]$ such that $a = u_1\pi_1 \cdot \ldots \cdot u_k$ and $b = u_2\pi_{k+1} \cdot \ldots \cdot \pi_n$, where $u_1,\, u_2 \in \mathfrak{o}^\times$.

Proof of **A.** By induction on n. For $n = 0$ there is nothing to do.

$n \ge 1$, $n - 1 \to n$: If $ab = u\pi_1 \cdot \ldots \cdot \pi_n$, then $\pi_n \,|\, ab$ and thus $\pi_n \,|\, a$ or $\pi_n \,|\, b$, say $\pi_n \,|\, b$. Then $b = \pi_n b'$, $ab' = \pi_n \cdot \ldots \cdot \pi_{n-1}$, and the assertion follows from the induction hypothesis. $\square$[**A.**]

Let T be the set of all elements $a \in \mathfrak{o}^\bullet$ of the form $a = u\pi_1 \cdot \ldots \cdot \pi_n$, where $u \in \mathfrak{o}^\times$, $n \in \mathbb{N}_0$ and $\pi_1, \ldots, \pi_n$ are prime elements of $\mathfrak{o}$. Then T is multiplicatively closed, and it suffices to prove that $T = \mathfrak{o}^\bullet$. If $a,\, b \in \mathfrak{o}^\bullet$ and $ab \in T$, then **A** implies that $a \in T$ and $b \in T$. We assume to the contrary that there exists some $a \in \mathfrak{o}^\bullet \setminus T$. Then $a\mathfrak{o} \cap T = \emptyset$, and by Krull's theorem 1.1.5.1 there exists a prime ideal $\mathfrak{p}$ of $\mathfrak{o}$ such that $\mathfrak{p} \cap T = \emptyset$ and $a\mathfrak{o} \subset \mathfrak{p}$. As $\mathfrak{o}$ is a principal ideal domain, $\mathfrak{p} = p\mathfrak{o}$ for some prime element p of $\mathfrak{o}$, and as $p \in T$, this contradicts $\mathfrak{p} \cap T = \emptyset$.

Now we prove that every non-zero prime ideal is maximal. For this purpose let $\mathfrak{p} = \pi\mathfrak{o} \in \mathcal{P}(\mathfrak{o})$ and $a \in \mathfrak{o}$ such that $\pi\mathfrak{o} \subsetneq a\mathfrak{o}$. Then $\pi = ab$ for some $b \in \mathfrak{o}$ and $\pi \nmid a$. Hence $\pi \mid b$, say $b = \pi b'$ for some $b' \in \mathfrak{o}$, and consequently $ab' = 1$, $a \in \mathfrak{o}^\times$ and $a\mathfrak{o} = \mathfrak{o}$. □

Let R be a commutative ring. We denote by $R[X_1, \ldots, X_n]$ the polynomial ring in the (algebraically independent) indeterminates $X_1, \ldots, X_n$. In a self-explanatory way, $R[X]$, $R[T]$, $R[X,Y]$ or, more generally $R[\boldsymbol{X}]$ (for a set $\boldsymbol{X}$ or a family $\boldsymbol{X} = (X_i)_{i\in I}$ of indeterminates over R) denote polynomial rings. For $\boldsymbol{X} = \emptyset$ we set $R[\emptyset] = R$. Note that

$$R[(X_i)_{i\in I}] = \bigcup_{J\in\mathbb{P}_{\text{fin}}(I)} R[(X_j)_{j\in J}].$$

If $\mathfrak{a}$ is an ideal of R and $\mathfrak{a}[\boldsymbol{X}] = \mathfrak{a}R[\boldsymbol{X}]$ is its extension to $R[\boldsymbol{X}]$, then there is a (natural) ring isomorphism $(R/\mathfrak{a})[\boldsymbol{X}] \xrightarrow{\sim} R[\boldsymbol{X}]/\mathfrak{a}R[\boldsymbol{X}]$, and we identify:

$$(R/\mathfrak{a})[\boldsymbol{X}] = R[\boldsymbol{X}]/\mathfrak{a}R[\boldsymbol{X}].$$

Every homomorphism $\varphi\colon R \to S$ of commutative rings has a (natural) extension to a homomorphism of polynomial rings $\varphi_1\colon R[\boldsymbol{X}] \to S[\boldsymbol{X}]$ satisfying $\varphi_1 \restriction R = \varphi$ and $\varphi_1 \restriction \boldsymbol{X} = \mathrm{id}_{\boldsymbol{X}}$. Throughout, we shall write again φ instead of φ_1, and we tacitly assume that every ring homomorphism is naturally extended to the polynomial rings.

If $\varepsilon\colon R \to A$ is a commutative R-algebra and $\boldsymbol{x} = (x_i)_{i\in I}$ is a family in A, then there exists a unique ring homomorphism $\varepsilon_{\boldsymbol{x}}\colon R[\boldsymbol{X}] \to A$ such that $\varepsilon_{\boldsymbol{x}} \restriction R = \varphi$ and $\varepsilon_{\boldsymbol{x}}(X_i) = x_i$ for all $i \in I$ (Universal property of polynomial rings). If $f \in R[\boldsymbol{X}]$ then we usually set $\varepsilon_{\boldsymbol{x}}(f) = f(\boldsymbol{x})$.

For a commutative ring R we write polynomials $f \in R[X]$ in the form

$$f = \sum_{i\geq 0} a_i X^i, \quad \text{where } a_i \in R \text{ and } a_i = 0 \text{ for all } i \gg 1.$$

If $f \neq 0$ and $n = \max\{i \geq 0 \mid a_i \neq 0\}$, then we call $n = \partial(f)$ the **degree** and a_n the **leading coefficient** of f. If $a_n = 1$, then f is called **monic**. We set $\partial(0) = -\infty$. If f, $g \in R[X]$, then

- $\partial(f+g) \leq \max\{\partial(f), \partial(g)\}$, and equality holds if $\partial(f) \neq \partial(g)$.
- $\partial(fg) \leq \partial(f) + \partial(g)$, and equality holds if the product of the leading coefficients of f and g is not zero, and then this product is the leading coefficient of fg.

Let $\mathfrak{o}$ be a domain and $K = \mathsf{q}(\mathfrak{o})$. Then $\partial(fg) = \partial(f) + \partial(g)$ for all f, $g \in \mathfrak{o}[X]$. Hence $\mathfrak{o}[X]$ is a domain, $\mathfrak{o}[X]^\times = \mathfrak{o}^\times$ and $K(X) = \mathsf{q}(\mathfrak{o}[X])$ is the **rational**

function field in X over K. More generally, if $\boldsymbol{X}$ is any family of indeterminates over $\mathfrak{o}$, then $\mathfrak{o}[\boldsymbol{X}]$ is a domain, $\mathfrak{o}[\boldsymbol{X}]^\times = \mathfrak{o}^\times$, and the field $K(\boldsymbol{X}) = \mathsf{q}(\mathfrak{o}[\boldsymbol{X}]$ is the rational function field in $\boldsymbol{X}$ over K (see Exercise 20).

Let R be a commutative ring and $\boldsymbol{X}$ a family of indeterminates over R. A polynomial $f \in R[\boldsymbol{X}]$ is called **irreducible** (in $R[\boldsymbol{X}]$ or over R) if $f \notin R$ and if there is no factorization $f = gh$ with polynomials $g,\, h \in R[\boldsymbol{X}] \setminus R$. If $f \in R[X] \setminus R$ is not irreducible, then f is called **reducible** (in $R[\boldsymbol{X}]$ or over R).

If $\mathfrak{o}$ is a domain and $f \in \mathfrak{o}[X]$ is monic and reducible, then $f = gh$ with monic polynomials $g,\, h \in \mathfrak{o}[X] \setminus \mathfrak{o}$. Indeed, if $f = gh$ with any $g,\, h \in \mathfrak{o}[X]$ and u is the leading coefficient of g, then $u \in \mathfrak{o}^\times$, and $f = (u^{-1}g)(uh)$ is a factorization of f into two monic polynomials.

Theorem 1.1.8 (Division algorithm for polynomials)**.** *Let R be a commutative ring.*

1. Let $f,\, g \in R[X]^\bullet$, and suppose that the leading coefficient of g is a unit in R. Then there exist uniquely determined polynomials $q,\, r \in R[X]$ such that $f = gq + r$ and $\partial(r) < \partial(g)$.
The equation $f = gq + r$ as above is called the **division with remainder**.

2. Let $f \in R[X]$ and $\alpha \in R$ such that $f(\alpha) = 0$. Then $f = (X - \alpha)g$ for some polynomial $g \in R[X]$, and α is called a **zero** of f. *If R is a domain and $\partial(f) = n \in \mathbb{N}$, then f has at most n zeros.*

3. Let R be a field.

(a) Let $\mathfrak{a} \ne \mathbf{0}$ an ideal of $R[X]$. Then there exists a unique monic polynomial $g \in \mathfrak{a}$ such that $\mathfrak{a} = gR[X]$.

(b) $R[X]$ is a principal ideal domain, and a polynomial $f \in R[X]$ is a prime element of $R[X]$ if and only if it is irreducible.

(c) Let $f,\, g \in R[X]$, $g \ne 0$, and consider the following sequence of length $k \in \mathbb{N}_0$ of divisions with remainder:

$$\begin{aligned}
r_{-1} = f &= gq_0 + r_1\,, && \textit{where}\quad q_0,\, r_1 \in R[X],\ \partial(r_1) < \partial(g),\\
r_0 = g &= r_1q_1 + r_2\,, && \textit{where}\quad q_1,\, r_2 \in R[X],\ \partial(r_2) < \partial(r_1),\\
r_1 &= r_2q_2 + r_3\,, && \textit{where}\quad q_2,\, r_3 \in R[X],\ \partial(r_3) < \partial(r_2),\\
&\ \ \vdots && \qquad\qquad \vdots\\
r_{k-2} &= r_{k-1}q_{k-1} + r_k\,, && \textit{where}\quad q_{k-1},\, r_k \in R[X],\ \partial(r_k) < \partial(r_{k-1}),\\
r_{k-1} &= r_kq_k\,, && \textit{where}\quad q_k \in R[X] \setminus R.
\end{aligned}$$

Then r_k is a gcd *of f and g.*
The above sequence of divisions with remainder is called the **Euclidean algorithm** for f and g.

Proof. 1. *Existence*: If $\partial(g) > \partial(f)$, we set $q = 0$ and $r = f$. Thus assume that $0 \le m = \partial(g) \le \partial(f) = n$, $f = aX^n + f_1$ and $g = uX^m + g_1$, where $a \in R$, $u \in R^\times$, $f_1, g_1 \in R[X]$, $\partial(f_1) < n$ and $\partial(g_1) < m$. We proceed by induction on n. If $n = 0$, then $f = a \in R$, $g = u \in R^\times$, we set $q = u^{-1}a$ and $r = 0$. Thus suppose that $n > 0$ and the assertion holds for all polynomials $f^* \in R[X]$ with $\partial(f^*) < n$. We set $f^* = f - u^{-1}agX^{n-m} = f_1 - u^{-1}ag_1X^{n-m}$. Then $\partial(f^*) < n$, and by the induction hypothesis there exist polynomials $q^*, r^* \in R[X]$ such that $f^* = gq^* + r^*$ and $\partial(r^*) < \partial(g)$. It follows that $f = g(q^* + u^{-1}aX^{n-m}) + r^*$.

Uniqueness: Let $f = gq + r = gq^* + r^*$, where $q, q^*, r, r^* \in R[X]$, $\partial(r) < \partial(g)$ and $\partial(r^*) < \partial(g)$. Then $r - r^* = g(q^* - q)$, and since $\partial(r - r^*) < \partial(g)$, it follows that $q = q^*$ and $r = r^*$.

2. By 1. with $g = X - \alpha$ and induction on $\partial(f)$.

3. Recall that $R[X]^\times = R^\times$.

(a) *Existence*: If $g \in \mathfrak{a}^\bullet$ and a is the leading coefficient of g, then $a^{-1}g$ is a monic polynomial in $\mathfrak{a}$. Let $g \in \mathfrak{a}^\bullet$ be a monic polynomial of smallest degree. If $f \in \mathfrak{a}$, then $f = gq + r$ for some polynomials $q, r \in R[X]$ such that $\partial(r) < \partial(g)$ by 1. Then $r \in \mathfrak{a}$, hence $r = 0$ and therefore $f \in gK[X]$.

Uniqueness: Let $g, g_1 \in R[X]$ be monic polynomials such that $gK[X] = g_1K[X]$. Then $g_1 = cg$ for some $c \in K^\times$, and comparing the leading coefficients we get $c = 1$.

(b) By (a) $R[X]$ is a principal ideal domain. If $f \in R[X]$ is a prime element, then f is irreducible by 1.1.7.1. Thus let f be irreducible. As $R[X]$ is factorial by 1.1.7.3, f is a product of prime elements, and as f is irreducible, it is itself a prime element

(c) By an easy induction we obtain:

- For all $i \in [0, k]$ we have $r_k \,|\, r_{k-i}$ and $r_k \,|\, r_{k-i-1}$. Hence $r_k \,|\, f$ and $r_k \,|\, g$.

- If $d \in R[X[$, $d \,|\, f$ and $d \,|\, g$, then $d \,|\, r_{i-1}$ and $d \,|\, r_i$ for all $i \in [0, k-1]$. Hence $d \,|\, r_k$.

Consequently, r_k is a gcd of f and g. □

Remark 1.1.9. In the principal ideal domains $\mathbb{Z}$ and $K[X]$ (for some field K) the theory of gcd and lcm is special as follows.

1. For every ideal $\mathfrak{a}$ of $\mathbb{Z}$ there exists a unique integer $a \in \mathbb{N}_0$ such that $\mathfrak{a} = a\mathbb{Z}$, and $\mathfrak{a}$ is a prime ideal if and only if a is a prime number. Consequently, if $Z \subset \mathbb{Z}$, then there exist uniquely determined integers $d, e \in \mathbb{N}_0$ such that d is a gcd and e is an lcm of Z, and then we write $d = \gcd(Z)$ and $e = \mathrm{lcm}(Z)$. In particular, if $Z = \{a_1, \ldots, a_n\}$, then we set $d = (a_1, \ldots, a_n)$, and then there exist $x_1, \ldots, x_n \in \mathbb{Z}$ such that $d = a_1x_1 + \ldots + a_nx_n$.

2. Let K be a field. By 1.1.8.3, for every ideal $\mathfrak{a} \ne \mathbf{0}$ of $K[X]$ there exists a unique monic polynomial $g \in K[X]$ such that $\mathfrak{a} = gK[X]$, and $\mathfrak{a}$ is a prime ideal if and only if g is irreducible. Consequently, if $Z \subset K[X]$, then there exist uniquely determined polynomials $d,\ e \in K[X]$ which are either monic or 0 such that d is a gcd and e is an lcm of Z, and then we write $d = \gcd(Z)$ and $e = \operatorname{lcm}(Z)$. In particular, if $Z = \{f_1, \ldots, f_n\}$, then we set $d = (f_1, \ldots, f_n)$, and then there exist $g_1, \ldots, g_n \in \mathbb{Z}$ such that $d = f_1g_1 + \ldots + f_ng_n$.

Corollary 1.1.10.

1. *Let $R \subset R'$ be commutative rings, $f,\ g \in R[X]\backslash R$ and $h \in R'[X]$ such that $f = gh$. If the leading coefficient of g lies in $R^\times$, then $h \in R[X]$.*

2. *Let $K \subset K'$ be fields and $f,\ g \in K[X] \setminus K$.*

(a) *Let $d \in K[X]$ be a* gcd *of f and g in $K[X]$. Then d is a* gcd *of f and g in $K'[X]$, too.*

(b) *Every monic* gcd *of f and g in $K'[X]$ lies in $K[X]$. In particular, f and g are coprime in $K'[X]$ if and only if they are coprime in $K[X]$.*

Proof. 1. By the uniqueness in 1.1.8.1 (with $r = 0$).

2. (a) By 1.1.8.3 the Euclidean algorithm produces a gcd regardless of whether it is calculated in K or in K'.

(b) Let $d \in K'[X]$ be a monic gcd of f and g, and let $d_0 \in K[X]$ be a monic gcd of f and g in $K[X]$. Then d_0 is also a gcd of f and g in $K'[X]$, and therefore $d = d_0$. □

Let $\mathfrak{o}$ be a domain and $\mathfrak{p} \in \mathcal{P}(\mathfrak{o})$ a non-zero prime ideal. A polynomial $f \in \mathfrak{o}[X]$ is called a **$\mathfrak{p}$-Eisenstein polynomial** if

$$f = X^n + c_{n-1}X^{n-1} + \ldots + c_1X + c_0, \text{ where } n \in \mathbb{N},\ c_0, \ldots, c_{n-1} \in \mathfrak{p} \text{ and } c_0 \notin \mathfrak{p}^2.$$

If $\mathfrak{p} = p\mathfrak{o}$ for some prime element $p \in \mathfrak{o}$, then f is called a **p-Eisenstein polynomial**.

Theorem 1.1.11 (Eisenstein's criterion). *Let $\mathfrak{o}$ be a domain and $\mathfrak{p} \in \mathcal{P}(\mathfrak{o})$. Then every $\mathfrak{p}$-Eisenstein polynomial in $\mathfrak{o}[X]$ is irreducible.*

Proof. Let $L = \mathsf{q}(\mathfrak{o}/\mathfrak{p})$, and for a polynomial $q \in \mathfrak{o}[X]$ we denote by

$$\overline{q} = q + \mathfrak{p}\mathfrak{o}[X] \in \mathfrak{o}[X]/\mathfrak{p}\mathfrak{o}[X] = \mathfrak{o}/\mathfrak{p}[X] \subset L[X]$$

the residue class polynomial. Let $f = X^n + c_{n-1}X^{n-1} + \ldots + c_1X + c_0 \in \mathfrak{o}[X]$ be a $\mathfrak{p}$-Eisenstein polynomial, where $n \in \mathbb{N},\ c_0, \ldots, c_{n-1} \in \mathfrak{p}$ and $c_0 \notin \mathfrak{p}^2$. Assume to the contrary that f is reducible. Then f has a factorization $f = gh$ into monic polynomials $g,\, h \in \mathfrak{o}[X] \setminus \mathfrak{o}$, and we set $k = \partial(f) \geq 1$ and $l = \partial(g) \geq 1$. Then $X^n = \overline{f}\overline{g}$, and as $L[X]$ is factorial, we obtain $\overline{f} = X^k$ and $\overline{g} = X^l$. Since $g(0) = \overline{g}(0) = 0$ and $h(0) = \overline{h}(0) = 0$, it follows that $g(0),\, h(0) \in \mathfrak{p}$ and $c_0 = g(0)h(0) \in \mathfrak{p}^2$, a contradiction. □

Theorem 1.1.12 (Gauss's lemma). *Let $f \in \mathbb{Z}[X]$ be a monic polynomial.*

1. *Let $g,\, h \in \mathbb{Q}[X]$ be monic polynomials such that $f = gh$. Then $g,\, h \in \mathbb{Z}[X]$.*

2. *Let f be irreducible in $\mathbb{Z}[X]$. Then f is even irreducible in $\mathbb{Q}[X]$.*

Proof. 1. Let $c,\, d \in \mathbb{N}$ be minimal such that $cg \in \mathbb{Z}[X]$ and $dh \in \mathbb{Z}[X]$. We must prove that $c = d = 1$, and we assume to the contrary that there exists some prime p such that $p \,|\, cd$. For $q \in \mathbb{Z}[X]$ we denote by $\overline{q} = q + p\mathbb{Z}[X] \in \mathbb{Z}[X]/p\mathbb{Z}[X] = \mathbb{F}_p[X]$ the residue class polynomial of q (then $\overline{a} = a + p\mathbb{Z} \in \mathbb{F}_p$ for all $a \in \mathbb{Z}$). Since $\overline{cdf} = (\overline{cg})(\overline{dh}) = 0 \in \mathbb{F}_p[X]$, it follows that $\overline{cg} = 0$ or $\overline{dh} = 0$, and we may assume that $\overline{cg} = 0$. Then $\overline{c} = 0$, hence $p \,|\, c$ and $p^{-1}cg \in \mathbb{Z}[X]$, a contradiction to the minimal choice of c and d.

2. If f is reducible in $\mathbb{Q}[X]$, then we have seen that there exist monic polynomials $g,\, h \in \mathbb{Q}[X]$ such that $f = gh$, and 1. implies $g,\, h \in \mathbb{Z}[X]$, a contradiction. □

Later, in 2.2.4 and 2.2.5 we will see more general forms of Gauss's lemma.

We close by reviewing some notational conventions for ring extensions.

If $R \subset R'$ are commutative rings and $B \subset R'$, then $R[B]$ denotes the smallest subring of R' containing $R \cup B$. If $\varphi,\, \varphi' \colon R' \to \overline{R}$ are ring homomorphisms, then $\varphi \restriction R \cup S = \varphi' \restriction R \cup S$ implies $\varphi \restriction R[S] = \varphi' \restriction R[S]$.

If $R \subset R'$ are fields, then $R(B) = \mathsf{q}(R[B])$ is the smallest subfield of R' containing $R \cup B$. If $\varphi,\, \varphi' \colon R' \to \overline{R}$ are field homomorphisms satisfying $\varphi \restriction R \cup S = \varphi' \restriction R \cup S$, then $\varphi \restriction R(S) = \varphi' \restriction R(S)$.

If $B = B_1 \cup B_2$, then $R[B] = R[B_1][B_2]$ and $R(B) = R(B_1)(B_2)$ (in the case of fields). If $B = \{b_1, \ldots, b_n\}$, we set $R[B] = R[b_1, \ldots, b_n]$ and $R(B) = R(b_1, \ldots, b_n)$ (in the case of fields). The concepts are of finite character, that is,

$$R[B] = \bigcup_{B_0 \in \mathbb{P}_{\mathrm{fin}}(B)} R[B_0] \quad \text{and} \quad R(B) = \bigcup_{B_0 \in \mathbb{P}_{\mathrm{fin}}(B)} R(B_0) \ \text{(in the case of fields).}$$

These definitions are compatible with the usual notation for polynomials. Indeed, if $\boldsymbol{X}$ is a set of indeterminates, then the polynomial ring $R[\boldsymbol{X}]$ is the smallest ring containing $R \cup \boldsymbol{X}$, and if R is a field, then the rational function field $R(\boldsymbol{X}) = \mathsf{q}(R[\boldsymbol{X}])$ is the smallest field containing $R \cup \boldsymbol{X}$.

1.2 Algebraic field extensions

Let R be a ring. Then $R_0 = \{n1_R \mid n \in \mathbb{Z}\}$ is the smallest subring of R and is called the **prime ring** of R. The map $\varphi\colon \mathbb{Z} \to R$, defined by $\varphi(n) = n1_R$, is a ring homomorphism, $\operatorname{Im}(\varphi) = R_0$, $\operatorname{Ker}(\varphi) = n\mathbb{Z}$ for some $n \in \mathbb{N}_0$, and φ induces an isomorphism $\varphi^*\colon \mathbb{Z}/n\mathbb{Z} \xrightarrow{\sim} R_0$. We call $n = \operatorname{char}(R)$ the **characteristic** of R.

If R is a field, then $\mathsf{q}(R_0) = \{a^{-1}b \mid a,\, b \in R_0,\ a \neq 0\} \subset R$ is the smallest subfield of R and is called the **prime field** of R. It is well known that

- either $\operatorname{char}(R) = 0$, $R_0 \cong \mathbb{Z}$ and $\mathsf{q}(R_0) \cong \mathbb{Q}$,
- or $\operatorname{char}(R) = p$ and $R_0 \cong \mathbb{F}_p = \mathbb{Z}/p\mathbb{Z}$ is the field with p elements.

If R is a field, then we usually tacitly assume that either $\mathbb{F}_p \subset R$ or $\mathbb{Q} \subset R$.

If R is a ring and $\operatorname{char}(R) = p$ is a prime, then $x^{p^n} + y^{p^n} = (x+y)^{p^n}$ for all $x,\, y \in R$ and $n \in \mathbb{N}$, and the assignment $x \mapsto x^{p^n}$ is an endomorphism of R.

Theorem 1.2.1. *Let K be a field and G a finite subgroup of $K^\times$. Then G is cyclic, and if $\operatorname{char}(K) = p > 0$, then $p \nmid |G|$.*

Proof. For an element $a \in G$ we denote by $\operatorname{ord}(a) = \operatorname{ord}_G(a) \in \mathbb{N} \cup \{\infty\}$ its order (in G). Let $a \in G$ be an element of maximal order, say $\operatorname{ord}(a) = n$. We assert that $c^n = 1$ for all $c \in G$. Indeed, assume to the contrary that $\operatorname{ord}(c) \nmid n$ for some $c \in G$. Then there exists a prime p such that $\operatorname{ord}(a) = p^\alpha n_0$ and $\operatorname{ord}(c) = p^\beta m_0$, where $\alpha,\, \beta \in \mathbb{N}_0$, $\alpha < \beta$ and $m_0,\, n_0 \in \mathbb{N}$ such that $p \nmid m_0 n_0$. Then $\operatorname{ord}(a^{p^\alpha}) = n_0$, $\operatorname{ord}(c^{m_0}) = p^\beta$ and therefore $\operatorname{ord}(a^{p^\alpha} c^{m_0}) = n_0 p^\beta > n$, a contradiction.

If $G = \langle a \rangle$, we are done. Thus suppose that there exists some $c \in G \setminus \langle a \rangle$. Then $|\{x \in G \mid x^n = 1\}| > |\langle a \rangle| = n$, which is a contradiction, since the polynomial $X^n - 1 \in K[X]$ has at most n zeros (see 1.1.8.2).

If $\operatorname{char}(K) = p > 0$, then the assignment $x \mapsto x^p$ defines a field homomorphism $K \to K$. Hence G has no element of order p, and thus $p \nmid |G|$. $\square$

By a **field extension** L/K we mean a field L together with a subfield K of L (then L is a vector space over K). If L/K is a field extension, then L is called an extension field of K and every field M satisfying $K \subset M \subset L$ is called an intermediate field of L/K. A field extension L/K is called **finitely generated** if $L = K(B)$ for some finite subset B of L.

Let K/K_0 be a field extension, $(K_i)_{i \in I}$ a family of intermediate fields and L the smallest subfield of K which contains all K_i. Then

$$L = K_0\Big(\bigcup_{i \in I} K_i\Big), \quad \text{we set } L = \prod_{i \in I} K_i \subset K,$$

and we call L the **compositum** of the family $(K_i)_{i\in I}$. If $\varphi\colon K \to K'$ is a field homomorphism, then

$$\varphi\Bigl(\prod_{i\in I} K_i\Bigr) = \prod_{i\in K} \varphi(K_i) \subset K'.$$

If I is finite, say $I = [1,n]$, then we denote the compositum of $(K_i)_{i\in[1,n]}$ by $K_1 \cdot \ldots \cdot K_n$.

Let L/K be a field extension. We call $[L\!:\!K] = \dim_K(L) \in \mathbb{N} \cup \{\infty\}$ the **degree** of L/K. The field extension L/K is called **finite** if $[L\!:\!K] < \infty$, and it is called **quadratic** [**cubic**, etc.] if it is of degree 2 [3, etc]. Instead of saying that L/K is finite [resp. quadratic, cubic, etc.] we also say that L is finite [resp. quadratic, cubic, etc.] over K.

Let L/K be a field extension. An element $\alpha \in L$ is called **algebraic** over K if $f(\alpha) = 0$ for some polynomial $f \in K[X]^\bullet$; otherwise α is called **transcendental** over K. Let $\varphi\colon K[X] \to L$ be the (unique) ring homomorphism satisfying $\varphi \restriction K = \mathrm{id}_K$ and $\varphi(X) = \alpha$. Then $\mathrm{Im}(\varphi) = K[\alpha]$, and α is algebraic over K if and only if $\mathrm{Ker}(\varphi) \ne \mathbf{0}$. If α is transcendental over K, then $\varphi\colon K[X] \overset{\sim}{\to} K[\alpha]$ is an isomorphism.

If α is algebraic over K, then there exists a uniquely determined monic polynomial $f \in K[X]$ such that $\mathrm{Ker}(\varphi) = (f)$, and φ induces an isomorphism $K[X]/(f) \overset{\sim}{\to} K[\alpha] \subset L$. Hence $(f) = fK[X]$ is a prime ideal in $K[X]$, and f is irreducible. It is called the **minimal polynomial** and its degree $\partial(f)$ is called the **degree** of α over K. A polynomial $g \in K[X]$ is the minimal polynomial of α over K if and only if g is monic, irreducible, and $g(\alpha) = 0$ [equivalently, g is monic and $g \mid h$ for every polynomial $h \in K[X]$ such that $h(\alpha) = 0$].

If $\alpha,\ \beta \in L$ are algebraic over K, then α and β are called **conjugate** over K if they have the same minimal polynomial over K.

A subset B of L is called **algebraic** over K if every $b \in B$ is algebraic over K. The field extension L/K is called **algebraic** if L is algebraic over K. Otherwise the field extension L/K is called **transcendental**. Instead of saying that a field extension L/K is algebraic resp. transcendental, we also say that L is algebraic resp. transcendental over K.

Theorem 1.2.2. *Let L/K be a field extension. Let $\alpha \in L$ be algebraic over K, $f \in K[X]$ the minimal polynomial of α over K and $\partial(f) = n$. Then there exists an isomorphism $\varphi^*\colon K[X]/(f) \overset{\sim}{\to} K[\alpha]$ satisfying $\varphi^*(g + (f)) = g(\alpha)$ for all $g \in K[X]$ and $\varphi^* \restriction K = \mathrm{id}_K$. Moreover, $K(\alpha) = K[\alpha]$, $[K(\alpha)\!:\!K] = n$, and $(1, \alpha, \ldots, \alpha^{n-1})$ is a K-basis of $K(\alpha)$.*

Proof. We consider the unique ring homomorphism $\varphi\colon K[X] \to L$ satisfying $\varphi \restriction K = \mathrm{id}_K$ and $\varphi(X) = \alpha$. It induces an isomorphism $\varphi^*\colon K[X]/(f) \overset{\sim}{\to} K[\alpha]$ as asserted. As (f) is a maximal ideal (see 1.1.7.3), it follows that $K[X]/(f)$ and thus $K[\alpha]$ is a field, which implies $K(\alpha) = K[\alpha]$.

We eventually prove that $(1, \alpha, \ldots, \alpha^{n-1})$ is a K-basis of $K[\alpha]$. Thus let $x \in K[\alpha]$ and $h \in K[X]$ be such that $x = h(\alpha)$. Then $h = qf + r$ with polynomials $q,\, r \in K[X]$ such that $\partial(r) < n$. If $r = a_0 + a_1X + \ldots a_{n-1}X^{n-1}$, then

$$x = h(\alpha) = r(\alpha) = a_0 + a_1\alpha + \ldots a_{n-1}\alpha^{n-1} \in K + K\alpha + \ldots + K\alpha^{n-1},$$

and we prove that $(1, \alpha, \ldots, \alpha^{n-1})$ is linearly independent over K. If $c_0, \ldots, c_{n-1} \in K$ and $c_0 + c_1\alpha + \ldots c_{n-1}\alpha^{n-1} = 0$, then it follows that $q = a_0 + a_1X + \ldots a_{n-1}X^{n-1} \in K[X]$, $q(\alpha) = 0$, hence $f \mid q$, and since $\partial(q) \le n - 1 < \partial(f)$, we obtain $q = 0$. Consequently, $c_i = 0$ for all $i \in [1, n-1]$. □

Theorem 1.2.3. *Let L/K be a field extension.*

1. Let M be an intermediate field of L/K, $(u_i)_{i\in I}$ a K-basis of M and $(v_j)_{j\in J}$ an M-basis of L. Then $(u_iv_j)_{(i,j)\in I\times J}$ is a K-basis of L, and $[L:K] = [L:M]\,[M:K]$. In particular, L/K is finite if and only if both M/K and L/M are finite.

2. L/K is finite if and only if L/K is algebraic and finitely generated.

3. Let B be a subset of L which is algebraic over K. Then it follows that $K(B) = K[B]$ is algebraic over K.

4. Let N be an extension field of L and M an intermediate field of N/K.

(a) If L/K is (finite) algebraic, then LM/M is also (finite) algebraic.

(b) Let M/K be algebraic, and let $\alpha \in N$ be algebraic over M. Then α is algebraic over K.

(c) Let $M \subset L$. Then L/K is (finite) algebraic if and only if both M/K and L/M are (finite) algebraic.

(d) If L/K and M/K are (finite) algebraic, then LM/K is (finite) algebraic, too.

(e) Let $(L_i)_{i\in I}$ be a family of over K algebraic intermediate fields of N/K and L its compositum. Then L is algebraic over K.

Proof. 1. Since

$$L = \sum_{j\in J} Mv_j = \sum_{j\in J}\Big(\sum_{i\in I} Ku_i\Big)v_j = \sum_{(i,j)\in I\times J} Ku_iv_j,$$

it suffices to prove that $(u_iv_j)_{(i,j)\in I\times J}$ is linearly independent over K. Let $(a_{i,j})_{(i,j)\in I\times J}$ be a family in K such that $a_{i,j}=0$ for almost all $(i,j)\in I\times J$ and

$$0=\sum_{(i,j)\in I\times J} a_{i,j}u_iv_j=\sum_{j\in J}\Big(\sum_{i\in I} a_{i,j}u_i\Big)v_j, \text{ hence } \sum_{i\in I} a_{i,j}u_i=0 \text{ for all } j\in J,$$

and eventually $a_{i,j}=0$ for all $(i,j)\in I\times J$. Hence $[L:K]=[L:M]\,[M:K]$, and L/K is finite if and only if L/M and M/K are both finite.

2. Assume first that $n=[L:K]<\infty$, and let $(u_1,\ldots,u_n)$ be a K-basis of L. Then $L=K(u_1,\ldots,u_n)$ is finitely generated. If $\alpha\in L$, then $(1,\alpha,\ldots,\alpha^n)$ is linearly dependent over K, and thus there is a relation $a_0+a_1\alpha+\ldots+a_n\alpha^n=0$ with $a_0,\ldots,a_n\in K$, not all zero. Then $f=a_0+a_1X+\ldots+a_nX^n\in K[X]^\bullet$, $f(\alpha)=0$, and thus α is algebraic over K. Consequently, L/K is algebraic.

Let now L/K be algebraic and finitely generated, say $L=K(\alpha_1,\ldots,\alpha_n)$ for some $n\in\mathbb{N}_0$. We proceed by induction on n. For $n=0$ there is nothing to do.

$n\ge 1$, $n-1\to n$: If $K'=K(\alpha_1,\ldots,\alpha_{n-1})$, then $[K':K]<\infty$ by the induction hypothesis. As α_n is algebraic over K, it is algebraic over K', and since $L=K'(\alpha_n)$, we obtain $[L:K']<\infty$ by 1.2.2. Hence we obtain $[L:K]=[L:K']\,[K':K]<\infty$ by 1.

3. Assume first that B is finite, say $B=\{x_1,\ldots,x_n\}$ for some $n\in\mathbb{N}_0$. We proceed by induction on n. For $n=0$ there is nothing to do.

$n\ge 1$, $n-1\to n$: If $K'=K(x_1,\ldots,x_{n-1})$, then the induction hypothesis implies $K'=K[x_1,\ldots,x_{n-1}]$ and $[K':K]<\infty$. Since $K(B)=K'(x_n)$ and x_n is algebraic over K', it follows from 1.2.2 that

$$K(B)=K'[x_n]=K[x_1,\ldots,x_n]=K[B] \quad\text{and}\quad [K(B):K']<\infty.$$

Hence $[K(B):K]=[K(B):K'][K':K]<\infty$, and thus $K(B)/K$ is algebraic by 2.

Now let B be arbitrary and $x\in K(B)$. Then $x\in K(B')$ for some finite subset B' of B. Hence x is algebraic over K, and $x\in K(B')=K[B']\subset K[B]$. Hence $K(B)/K$ is algebraic, and $K(B)=K[B]$.

4. (a) As L is algebraic over K, it is algebraic over M, and $LM=M(L)$ is algebraic over M by 3. If $[L:K]<\infty$ and $(u_1,\ldots,u_n)$ is a K-Basis of L, then we obtain $LM=Mu_1+\ldots+Mu_n$, and consequently LM/M is finite, too.

(b) Let $g=b_0+b_1X+\ldots+b_nX^n\in M[X]^\bullet$ be such that $g(\alpha)=0$ and set $M_0=K(b_0,\ldots,b_n)$. Then M_0/K is finite by 1., and α is algebraic over M_0, since $g\in M_0[X]$. As $[M_0(\alpha):K]=[M_0(\alpha):M_0][M_0:K]<\infty$, it follows that α is algebraic over K.

(c) and (d) are now immediate consequences of 1., (a) and (b).

(e) If I if finite, the assertion follows by a simple induction using (d). Thus let I be arbitrary and $\alpha\in L$. Then α is already contained in the compositum of finitely many L_i's and thus is algebraic over K. □

Definition and Theorem 1.2.4. *Let L/K and L'/K be field extensions.*

1. *A field homomorphism $\varphi\colon L \to L'$ is K-linear if and only if $\varphi \restriction K = \mathrm{id}_K$.*
A K-linear field homomorphism $\varphi\colon L \to L'$ is called a **K-homomorphism**, and if φ is bijective, then it is called a **K-isomorphism**. Note that L and L' are K-algebras and that a K-homomorphism is nothing but a K-algebra homomorphism as defined in Section 1.1. Correspondingly we denote by $\mathrm{Hom}^K(L, L')$ the set of all K-homomorphisms $\varphi\colon L \to L'$ (in contrast to $\mathrm{Hom}_K(L, L')$, the set of all K-linear maps $\varphi\colon L \to L'$). We denote by $\mathrm{Gal}(L/K)$ the set of all K-isomorphisms $\varphi\colon L \to L$. It is a subgroup of the group of all automorphisms of L and is called the **Galois group** of L/K. If F is the prime field of K, then $\mathrm{Gal}(K/F)$ is the automorphism group of L.
If $\varphi\colon L \to L'$ is a K-isomorphism, then the map

$$\varphi^*\colon \mathrm{Gal}(L/K) \to \mathrm{Gal}(L'/K), \text{ defined by } \varphi^*(\sigma) = \varphi \circ \sigma \circ \varphi^{-1},$$

is a group isomorphism, and $(\varphi^{-1})^* = (\varphi^*)^{-1}$.

2. *If G is a subgroup of $\mathrm{Gal}(L/K)$, then its* **fixed field**

$$L^G = \{x \in L \mid \sigma(x) = x \text{ for all } \sigma \in G\}$$

is an intermediate field of L/K, and $G \subset \mathrm{Gal}(L/L^G)$. If M is an intermediate field of L/K, then $\mathrm{Gal}(L/M)$ is a subgroup of $\mathrm{Gal}(L/K)$, and $L^{\mathrm{Gal}(L/M)} \supset M$.

3. *Let M be an intermediate field of L/K, $G = \mathrm{Gal}(L/K)$ and $H = \mathrm{Gal}(L/M)$. Then H is a subgroup of G, and if $\sigma \in G$, then $\sigma H \sigma^{-1} = \mathrm{Gal}(L/\sigma M)$.*

4. *Let $\varphi\colon L \to L'$ be a field homomorphism. Then it follows that $\varphi(f(\alpha)) = \varphi(f)(\varphi(\alpha))$ for all $f \in L[X]$ and $\alpha \in L$. If $\varphi \in \mathrm{Hom}^K(L, L')$ and $f \in K[X]$, then $\varphi(f(\alpha)) = f(\varphi(\alpha))$.*

5. *If L/K is algebraic, then $\mathrm{Hom}^K(L, L) = \mathrm{Gal}(L/K)$.*

Proof. 1. If $\varphi \restriction K = \mathrm{id}_K$, then φ is K-linear, since $\varphi(ax) = \varphi(a)\varphi(x) = a\varphi(x)$ for all $a \in K$ and $x \in L$. Conversely, if φ is K-linear, then it follows that $\varphi(a) = a\varphi(1_L) = a\,1_{L'} = a$ for all $a \in K$, hence $\varphi \restriction K = \mathrm{id}_K$.

If $\varphi\colon L \to L'$ is a K-isomorphism, then clearly φ^* is a group isomorphism, and $(\varphi^{-1})^* = (\varphi^*)^{-1}$.

2. Obvious.

3. Clearly H is a subgroup of G. If $\sigma \in G$, then σM is an intermediate field of L/K, and for all $\tau \in G$ we obtain:

$$\tau \in \mathrm{Gal}(L/\sigma M) \iff \tau\sigma(x) = \sigma(x) \text{ for all } x \in M \iff \sigma^{-1}\tau\sigma \in H$$
$$\iff \tau \in \sigma H \sigma^{-1}.$$

4. Recall that $\varphi\colon L \to L'$ extends to an (equally denoted) ring homomorphism $\varphi\colon L[X] \to L'[X]$ satisfying $\varphi(X) = X$. If

$$f = \sum_{\nu=0}^{n} a_\nu X^\nu, \quad \text{then} \quad \varphi(f) = \sum_{\nu=0}^{n} \varphi(a_\nu) X^\nu$$

and

$$\varphi(f(\alpha)) = \varphi\Big(\sum_{\nu=0}^{n} a_\nu \alpha^\nu\Big) = \sum_{\nu=0}^{n} \varphi(a_\nu)\varphi(\alpha)^\nu = \varphi(f)(\varphi(\alpha))\,.$$

If $\varphi \in \mathrm{Hom}^K(L, L')$ and $f \in K[X]$, then $\varphi(f) = f$ and $\varphi(f(\alpha)) = f(\varphi(\alpha))$.

5. Let L/K be algebraic and $\varphi \in \mathrm{Hom}^K(L, L)$. It suffices to prove that $\varphi(L) = L$. If $a \in L$, $f \in K[X]$ is the minimal polynomial of a over K and $E = \{x \in L \mid f(x) = 0\}$, then $\varphi(E) \subset E$ by 4. Hence $\varphi(E) = E$ (since φ is injective and E is finite), and $a \in \varphi(E) \subset \varphi(L)$. □

Theorem 1.2.5 (Kronecker's existence theorem). *Let K be a field, and let $f \in K[X]$ be an irreducible polynomial.*

> *1. Let $\varphi\colon K \stackrel{\sim}{\to} K_1$ be a field isomorphism and $f_1 = \varphi(f) \in K_1[X]$. Let L/K and L_1/K_1 be field extensions, $\alpha \in L$ such that $f(\alpha) = 0$ and $\alpha_1 \in L_1$ such that $f_1(\alpha_1) = 0$. Then there exists a unique field isomorphism $\overline{\varphi}\colon K(\alpha) \to K_1(\alpha_1)$ satisfying $\overline{\varphi} \restriction K = \varphi$ and $\overline{\varphi}(\alpha) = \alpha_1$.*
>
> *2. There exist a field extension L/K and an element $\alpha \in L$ such that $L = K(\alpha)$ and $f(\alpha) = 0$.*
>
> *3. Let L/K be a field extension and $\alpha, \alpha_1 \in L$. Then α and α_1 are conjugate over K if and only if there exists a K-isomorphism $\Phi\colon K(\alpha) \stackrel{\sim}{\to} K(\alpha_1)$ such that $\Phi(\alpha) = \alpha_1$.*

Proof. 1. We may assume that f is monic. Then f_1 is monic, f is the minimal polynomial of α over K, and f_1 is the minimal polynomial of α_1 over K_1. By 1.2.2 there exist isomorphisms

$$\Phi\colon K[X]/(f) \stackrel{\sim}{\to} K(\alpha) \quad \text{and} \quad \Phi_1\colon K_1[X]/(f_1) \stackrel{\sim}{\to} K_1(\alpha_1)$$

satisfying $\Phi \restriction K = \mathrm{id}_K$, $\Phi_1 \restriction K_1 = \mathrm{id}_{K_1}$, $\Phi(X+(f)) = \alpha$ and $\Phi_1(X+(f_1)) = \alpha_1$.

In a natural way φ induces an isomorphism $\varphi^*\colon K[X]/(f) \stackrel{\sim}{\to} K_1[X]/(f_1)$, and then $\overline{\varphi} = \Phi_1 \circ \varphi^* \circ \Phi^{-1}\colon K(\alpha) \stackrel{\sim}{\to} K_1(\alpha_1)$ is an isomorphism satisfying $\overline{\varphi} \restriction K = \varphi$ and $\overline{\varphi}(\alpha) = \alpha_1$. Its uniqueness is obvious.

2. The map $K \to L = K[X]/(f)$, defined by $a \mapsto a + (f)$, is a monomorphism, and we identify K with its image. Then $K \subset L$, and if $\alpha = X+(f) \in L$, then $L = K(\alpha)$ is a field, and $f(\alpha) = f(X) + (f) = 0 \in L$.

3. By 1., applied with $\varphi = \mathrm{id}_K$. □

Let R be a commutative ring and $f \in R[X] \setminus K$. We say that f **splits** in $R[X]$ (or over R) if $f = c(X - \alpha_1) \cdot \ldots \cdot (X - \alpha_n)$ for some $c \in R^\bullet$ and $\alpha_1, \ldots, \alpha_n \in R$.

Definition and Theorem 1.2.6. *Let K be a field.*

1. *The following assertions are equivalent:*

(a) Every (monic) polynomial $f \in K[X] \setminus K$ possesses a zero in K.

(b) Every (monic) polynomial $f \in K[X] \setminus K$ splits in $K[X]$.

(c) There is no algebraic field extension L/K such that $L \neq K$.

If these conditions are fulfilled, K is called **algebraically closed**. An extension field L of K is called an **algebraic closure** of K if L is algebraically closed and L/K is algebraic.

2. *Let L/K be a field extension. The set $\overline{K}_L$ of all $\alpha \in L$ which are algebraic over K is a field.*
$\overline{K}_L$ is called the **(relative) algebraic closure** of K in L. By definition, $\overline{K}_L = L$ if and only if L/K is algebraic, and if $\overline{K}_L = K$, then K is called **(relatively) algebraically closed** in L.

$\overline{K}_L$ is relatively algebraically closed in L, and if L itself is algebraically closed, then $\overline{K}_L$ is an algebraic closure of K.

3. *Let L/K be an algebraic field extension. An extension field $\overline{L}$ of L is an algebraic closure of L if and only if it is an algebraic closure of K.*

Proof. 1. Let $f \in K[X] \setminus K$ be a (monic) polynomial.

(a) $\Rightarrow$ (b) We use induction on $n = \partial(f)$. For $n = 1$ there is nothing to do.

$n \geq 2$, $n - 1 \to n$: If $\alpha \in K$ and $f(\alpha) = 0$, then $f = (X - \alpha)g$ for some (monic) polynomial $g \in K[X]$ such that $\partial(g) = n - 1$. By the induction hypothesis, g splits in $K[X]$, and therefore f splits in $K[X]$, too.

(b) $\Rightarrow$ (c) Let L/K be an algebraic field extension, $\alpha \in L$ and $f \in K[X]$ the minimal polynomial of α over K. As f splits in $K[X]$, we obtain $\alpha \in K$.

(c) $\Rightarrow$ (a) Suppose that $f \in K[X] \setminus K$, and let $h \in K[X]$ be a monic irreducible polynomial such that $h \,|\, f$. By 1.2.5.2 there exists a finite field extension L/K and some $\alpha \in L$ such that $h(\alpha) = 0$. Then $f(\alpha) = 0$, and $L = K$ implies $\alpha \in K$.

2. By 1.2.3.3, $\overline{K}_L = K(\overline{K}_L)$ is algebraic over K. If $\alpha \in L$ is algebraic over $\overline{K}_L$, then α is also algebraic over K by 1.2.3.4(b), hence $\alpha \in \overline{K}_L$, and therefore $\overline{K}_L$ is relatively algebraically closed in L.

Assume now that L is algebraically closed, and let $f \in \overline{K}_L[X] \setminus \overline{K}_L$. Then f has a zero α in L. Since α is algebraic over $\overline{K}_L$, it is algebraic over K, and therefore $\alpha \in \overline{K}_L$. Hence $\overline{K}_L$ is algebraically closed.

3. Obvious by 1.2.3. $\square$

Theorem 1.2.7 (Main theorem on algebraic closures). *Let K be a field.*

1. Let $\varphi\colon K \to K_1$ be a field homomorphism, L/K an algebraic field extension and K_1^* an algebraically closed extension field of K_1. Then there exists a field homomorphism $\overline{\varphi}\colon L \to K_1^*$ such that $\overline{\varphi} \restriction K = \varphi$.

2. Let $\varphi\colon K \xrightarrow{\sim} K_1$ be a field isomorphism, $\overline{K}$ an algebraic closure of K and $\overline{K}_1$ an algebraic closure of K_1. Then there exists a field isomorphism $\overline{\varphi}\colon \overline{K} \xrightarrow{\sim} \overline{K}_1$ such that $\overline{\varphi} \restriction K = \varphi$.

3. Let $\overline{K}$ be an algebraic closure of K. Let L be an intermediate field of $\overline{K}/K$, and let $\varphi \in \mathrm{Hom}^K(L, \overline{K})$. Then there exists some $\sigma \in \mathrm{Gal}(\overline{K}/K)$ such that $\sigma \restriction L = \varphi$.

4. K possesses an up to K-isomorphisms uniquely determined algebraic closure.

Proof. 1. Let Σ be the set of all pairs (M, ψ) consisting of an intermediate field M of L/K and a field homomorphism $\psi\colon M \to K_1^*$ satisfying $\psi \restriction K = \varphi$. For two pairs (M, ψ), $(M', \psi') \in \Sigma$ we define $(M, \psi) \le (M', \psi')$ if $M \subset M'$ and $\psi' \restriction M = \psi$. Then $(K, \varphi) \in \Sigma$, $\le$ is a partial order on Σ, and for every chain in Σ its union is an upper bound in Σ. By Zorn's lemma, Σ possesses a maximal element (M^*, ψ^*), and we shall prove that $M^* = L$. Then $\overline{\varphi} = \psi^*\colon L \to K_1^*$ is a field homomorphism satisfying $\overline{\varphi} \restriction K = \varphi$.

Apparently L is algebraic over M^*. Let $\alpha \in L$ and $f \in M^*[X]$ be the minimal polynomial of α over M^*. Then the polynomial $f_1 = \psi^*(f) \in \psi^*(M^*)[X]$ is irreducible, and there exists some $\alpha_1 \in K_1^*$ such that $f_1(\alpha_1) = 0$. By 1.2.5.1 there exists a unique field homomorphism $\overline{\psi}^*\colon M^*(\alpha) \xrightarrow{\sim} \psi(M^*)(\alpha_1) \hookrightarrow K_1^*$ such that $\overline{\psi}^* \restriction M^* = \psi^*$ and $\overline{\psi}^*(\alpha) = \alpha_1$.

As $(M^*(\alpha), \overline{\psi}^*) \in \Sigma$, $(M^*, \psi^*) \le (M^*(\alpha), \overline{\psi}^*)$ and (M^*, ψ^*) is maximal, it follows that $M^*(\alpha) = M^*$ and $\alpha \in M^*$.

2. By 1. there exists a field homomorphism $\overline{\varphi}\colon \overline{K} \to \overline{K}_1$ such that $\overline{\varphi} \restriction K = \varphi$, and we shall prove that it is surjective. Since $\overline{\varphi}\colon \overline{K} \xrightarrow{\sim} \varphi(\overline{K})$ is a field isomorphism satisfying $\overline{\varphi}(K) = K_1$, it follows that $\overline{\varphi}(\overline{K})$ is an algebraic closure of K_1. If $\alpha \in \overline{K}_1$, then $\overline{\varphi}(\overline{K})(\alpha)/\overline{\varphi}(\overline{K})$ is a finite field extension, and therefore $\overline{\varphi}(\overline{K})(\alpha) = \overline{\varphi}(\overline{K})$ by 1.2.6.1, hence $\alpha \in \overline{\varphi}(\overline{K})$.

3. By 1. there exists some $\sigma \in \mathrm{Hom}^K(\overline{K}, \overline{K})$ such that $\sigma \restriction L = \varphi$, and by 1.2.4.5 it follows that $\sigma \in \mathrm{Gal}(\overline{K}/K)$.

4. Uniqueness up to K-isomorphisms follows from 2., applied with $\varphi = \mathrm{id}_K$. For the existence proof we need some cardinal arithmetic, and we show first:

A. If K^*/K is an algebraic field extension, then it follows that $\operatorname{card}(K^*) \le \operatorname{card}(K[X])$.

Proof of **A.** For $\alpha \in K^*$ let $f_\alpha \in K[X]$ be the minimal polynomial of α over K, and define $\lambda\colon K^* \to K[X]$ by $\lambda(\alpha) = f_\alpha$. Then $\lambda^{-1}(f)$ is finite for all $f \in K[X]$, and as $K[X]$ is infinite, $\operatorname{card}(K^*) = \operatorname{card}(\lambda(K^*)) \le \operatorname{card}(K[X])$. $\square$[**A.**

Let Σ be any set such that $K \subset \Sigma$ and $\operatorname{card}(\Sigma) > \operatorname{card}(K[X])$, and let Ω be the set of all algebraic extension fields L of K whose underlying set belongs to Σ. For $L_1, L_2 \in \Omega$ we define $L_1 \le L_2$ if L_1 is a subfield of L_2. Then $K \in \Omega$, $\le$ is a partial order on Ω, and for every chain in Ω its union is an upper bound in Ω. By Zorn's Lemma Ω possesses a maximal element $\overline{K}$. Then $\overline{K}/K$ is an algebraic field extension, and we prove that $\overline{K}$ is algebraically closed. To do so, let $L/\overline{K}$ be a finite field extension. Then L/K is algebraic, hence $\operatorname{card}(L) \le \operatorname{card}(K[X]) < \operatorname{card}(\Sigma)$, and therefore there exists an injective map $\lambda\colon L \to \Sigma$ such that $\lambda \restriction \overline{K} = \mathrm{id}_{\overline{K}}$. On $\lambda(L)$, there is a unique field structure such that $\lambda\colon L \overset{\sim}{\to} \lambda(L)$ is a field isomorphism and we view $\lambda(L)$ as an extension field of $\overline{K}$ with this field structure. Then $\lambda(L) \in \Omega$, and as $\overline{K}$ is maximal in Ω, it follows that $\lambda(L) = \overline{K}$ and consequently $L = \overline{K}$. Hence $\overline{K}$ is algebraically closed. $\square$

Let K be a field and $F \subset K[X] \setminus K$. An extension field L of K is called a **splitting field** of F (over K) if the following two conditions are satisfied:

- Every polynomial $f \in F$ splits in $L[X]$.
- If S is the set of all zeros of polynomials of F in L, then $L = K(S)$.

By definition, an algebraic closure of K is a splitting field of $K[X] \setminus K$. If $f \in K[X] \setminus K$, then a splitting field of $\{f\}$ is called a splitting field of f.

If $f \in K[X]$, $\partial(f) = n \in \mathbb{N}$ and $c \in K^\times$ is the leading coefficient of f, then L is a splitting field of f if and only if $f = c(X - \alpha_1) \cdot \ldots \cdot (X - \alpha_n)$, where $c \in K^\times$, $\alpha_1, \ldots, \alpha_n \in L$ and $L = K(\alpha_1, \ldots, \alpha_n)$.

If $m \in \mathbb{N}$, $F = \{f_1, \ldots, f_m\} \subset K[X]$ and $f = f_1 \cdot \ldots \cdot f_m$, then L is a splitting field of F if and only if L is a splitting field of f, and then L/K is finite.

Theorem 1.2.8 (Existence and uniqueness of splitting fields). *Let F be a non-empty subset of $K[X] \setminus K$. Then F has an up to K-isomorphisms uniquely determined splitting field over K.*

Proof. Let $\overline{K}$ be an algebraic closure of K and S the set of all zeros of polynomials from F in $\overline{K}$. Then $L = K(S)$ is a splitting field of F. To prove uniqueness, suppose that L' is another splitting field of F, and let $\overline{K}'$ be an algebraic closure of L'. Then $\overline{K}'$ is an algebraic closure of K, by 1.2.7.1

there exists a K-isomorphism $\varphi\colon \overline{K} \to \overline{K}'$, and $\varphi(S)$ is the set of all zeros of polynomials of F in $\overline{K}'$. Hence $K(\varphi(S)) = L'$, and $\varphi \restriction L\colon L \overset{\sim}{\to} L'$ is a K-isomorphism. □

1.3 Normal field extensions

A field extension L/K is called **normal** if it is algebraic and if every (monic) irreducible polynomial $f \in K[X]$ with a zero in L already splits in L. Instead of saying that L/K is normal we also say that L is normal over K.

If $\overline{K}$ is an algebraic closure of K, then $\overline{K}$ is normal over K.

Theorem 1.3.1 (Characterization of normal field extensions). *Let L/K be an algebraic field extension and $\overline{K}$ an algebraically closed extension field of L.*

1. *The following assertions are equivalent:*

   (a) L/K is normal.

   (b) L is a splitting field of a set of (irreducible) polynomials $f \in K[X] \setminus K$.

   (c) If $\varphi \in \mathrm{Hom}^K(L, \overline{K})$, then $\varphi(L) \subset L$ (i.e., $\varphi \in \mathrm{Gal}(L/K)$). In short, $\mathrm{Hom}^K(L, \overline{K}) = \mathrm{Gal}(L/K)$.

   (d) If $\alpha \in L$ and $\beta \in \overline{K}$ are conjugate over K, then $\beta \in L$, and there exists some $\varphi \in \mathrm{Gal}(L/K)$ such that $\varphi(\alpha) = \beta$.

2. *If L/K is finite and normal, then L is a splitting field of some $f \in K[X] \setminus K$.*

3. *Let L/K be normal, M an intermediate field of L/K, and let*

$$\rho\colon \mathrm{Gal}(L/K) \to \mathrm{Hom}^K(M, L) \quad \textit{be defined by} \quad \rho(\varphi) = \varphi \restriction M.$$

Then ρ is surjective.

Proof. (a) $\Rightarrow$ (b) and 2. Let $L = K(B)$, and for $b \in B$ let $f_b \in K[X]$ be the minimal polynomial of b over K. Then L is a splitting field of $\{f_b \mid b \in B\}$. If L/K is finite, then we may assume that B is finite, say $B = \{b_1, \ldots, b_m\}$, and then L is a splitting field of $f_{b_1} \cdot \ldots \cdot f_{b_m}$.

(b) $\Rightarrow$ (c) Let L be a splitting field of some subset F of $K[X] \setminus K$, and suppose that $B = \{a \in L \mid f(a) = 0 \text{ for some } f \in F\}$. Then $L = K(B)$. Let $\varphi \in \mathrm{Hom}^K(L, \overline{K})$. If $b \in B$, $f \in F$ and $f(b) = 0$, then $0 = \varphi(f(b)) = f(\varphi(b))$, and thus $\varphi(b) \in B$. Hence $\varphi(L) = \varphi(K(B)) = K(\varphi(B)) \subset K(B) = L$, and $\varphi \in \mathrm{Gal}(L/K)$ by 1.2.4.5.

(c) $\Rightarrow$ (d) Let $\alpha \in L$ and $\beta \in \overline{K}$ be conjugate over K. By 1.2.5.3 there exists a K-isomorphism $\varphi_0 \colon K(\alpha) \stackrel{\sim}{\to} K(\beta)$ such that $\varphi_0(\alpha) = \beta$, and by 1.2.7.1 there exists a field homomorphism $\varphi \colon L \to \overline{K}$ such that $\varphi \restriction K(\alpha) = \varphi_0$. Then $\varphi(\alpha) = \varphi_0(\alpha) = \beta$ and $\varphi \restriction K = \mathrm{id}_K$. Hence $\varphi \in \mathrm{Hom}^K(L, \overline{K})$ and therefore $\beta = \varphi(\alpha) \in \varphi(L) \subset L$. By 1.2.4.5 it follows that $\varphi \in \mathrm{Gal}(L/K)$.

(d) $\Rightarrow$ (a) Let $f \in K[X]$ be irreducible, $\alpha \in L$, $f(\alpha) = 0$ and $\beta \in \overline{K}$ another zero of f. Then α and β are conjugate over K, hence $\beta \in L$, and therefore f splits in $L[X]$.

3. Let $\overline{K}$ be an algebraic closure of L and $\psi \in \mathrm{Hom}^K(M, L) \subset \mathrm{Hom}^K(M, \overline{K})$. By 1.2.7.1 there exists some $\varphi \in \mathrm{Hom}^K(L, \overline{K})$ such that $\varphi \restriction M = \psi$, by 1. we get $\varphi(L) \subset L$, and thus $\varphi \in \mathrm{Gal}(L/K)$. $\square$

Theorem 1.3.2. *Let L/K be a normal field extension, N an extension field of L and M another intermediate field of N/K.*

1. *LM/M is normal.*
2. *If M/K is normal, then LM/K and $L \cap M/K$ are normal, too.*
3. *Let M be an intermediate field of L/K.*

(a) L/M is normal.

(b) M/K is normal if and only if $\sigma M = M$ for all $\sigma \in \mathrm{Gal}(L/K)$, and then $\mathrm{Gal}(L/M)$ is a normal subgroup of $\mathrm{Gal}(L/K)$.

Proof. Let $\overline{K}$ be an algebraically closed extension field of N. Observe that all pertinent field extensions are algebraic by 1.2.3.

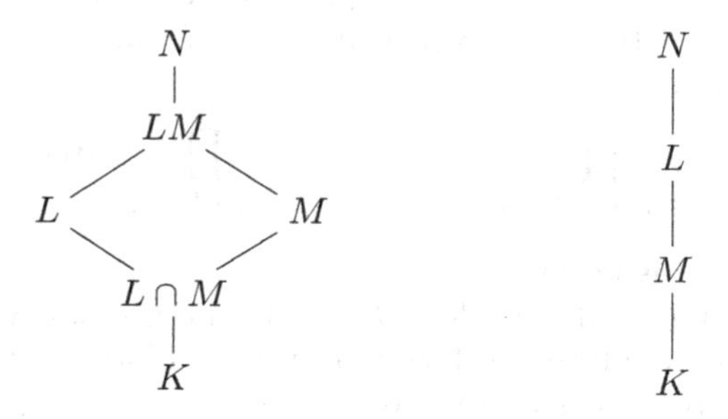

1. If $\varphi \in \mathrm{Hom}^M(LM, \overline{K})$, then $\varphi \restriction L \in \mathrm{Hom}^K(L, \overline{K})$, hence $\varphi(L) \subset L$, and consequently $\varphi(LM) = \varphi(L)\varphi(M) \subset LM$. Hence LM/M is normal.

2. If $\varphi \in \mathrm{Hom}_K(LM, \overline{K})$, then we have $\varphi \restriction L \in \mathrm{Hom}^K(L, \overline{K})$ and thus $\varphi \restriction M \in \mathrm{Hom}^K(M, \overline{K})$. Hence it follows that $\varphi(L) \subset L$, $\varphi(M) \subset M$, thus $\varphi(LM) = \varphi(L)\varphi(M) \subset LM$, and therefore LM/K is normal.

To prove that $L \cap M/K$ is normal, let $f \in K[X]$ be irreducible, and let $\alpha \in L \cap M$ be a zero of f. Then f splits in $L[X]$ and in $M[X]$. Hence f splits in $LM[X]$, and all zeros of f lie in $L \cap M$. Hence f splits in $(L \cap M)[X]$.

3. (a) Since $LM = L$, it follows that L/M is normal by 1.

(b) Let first M/K be normal. If $\sigma \in \mathrm{Gal}(L/K)$, then $\sigma \restriction M \colon M \to L \hookrightarrow \overline{K}$ is a K-homomorphism, hence $\sigma(M) \subset M$ and therefore $\sigma M = M$ by 1.2.4.5.

Assume conversely that $\sigma M = M$ for all $\sigma \in \mathrm{Gal}(L/K)$, and suppose that $\varphi \in \mathrm{Hom}^K(M, \overline{K})$. By 1.2.7.1 there exists some $\overline{\varphi} \in \mathrm{Hom}^K(L, \overline{K})$ such that $\overline{\varphi} \restriction M = \varphi$. As L/K is normal, it follows that $\overline{\varphi} \restriction L \in \mathrm{Gal}(L/K)$ by 1.3.1.1(c) and $\varphi(M) = \overline{\varphi}(M) = M$. Hence M/K is normal.

If M/K is normal, then $\mathrm{Gal}(L/M) = \mathrm{Gal}(L/\sigma M) = \sigma\,\mathrm{Gal}(L/M)\,\sigma^{-1}$ for all $\sigma \in \mathrm{Gal}(L/K)$ by 1.2.4.3, and $\mathrm{Gal}(L/M)$ is a normal subgroup of $\mathrm{Gal}(L/K)$. □

Theorem and Definition 1.3.3. *Let L/K be an algebraic field extension, $\overline{K}$ an algebraically closed extension field of L and*

$$L^* = \prod_{\sigma \in \mathrm{Hom}^K(L, \overline{K})} \sigma(L).$$

1. *L^*/K is algebraic, and L^* is the smallest subfield of $\overline{K}$ which contains L and is normal over K.*
   L^* is called the **normal closure** of L/K in $\overline{K}$.

2. *Suppose that $L = K(B)$, and let $F \subset K[X]$ be the set of all minimal polynomials of elements of B. Then L^* is a splitting field of F. If L/K is finite, then L^*/K is finite, too.*

Proof. 1. As $\sigma(L)/L$ is algebraic for all $\sigma \in \mathrm{Hom}^K(L, \overline{K})$, it follows by 1.2.3.5(e) that L^*/L is algebraic. To prove that L^*/K is normal, let $\varphi \in \mathrm{Hom}^K(L^*, \overline{K})$. If $\sigma \in \mathrm{Hom}^K(L, \overline{K})$, then $\sigma(L) \subset L^*$ and $\varphi\circ\sigma \in \mathrm{Hom}^K(L, \overline{K})$. Therefore we obtain

$$\varphi(L^*) = \prod_{\sigma \in \mathrm{Hom}^K(L, \overline{K})} \varphi\circ\sigma(L) \subset \prod_{\sigma \in \mathrm{Hom}^K(L, \overline{K})} \sigma(L) = L^*,$$

and therefore L^*/K is normal.

Let L' be any intermediate field of $\overline{K}/L$ which is normal over K. In order to prove that $L^* \subset L'$ it suffices to show that $\sigma(L) \subset L'$ for all $\sigma \in \mathrm{Hom}^K(L, \overline{K})$. If $\sigma \in \mathrm{Hom}^K(L, \overline{K})$, then there exists some $\sigma' \in \mathrm{Hom}^K(L', \overline{K})$ such that $\sigma' \restriction L = \sigma$. Since L'/K is normal, we obtain $\sigma'(L') \subset L'$, and therefore $\sigma(L) = \sigma'(L) \subset L'$.

2. Let $L' \subset \overline{K}$ be a splitting field of F. Then $L \subset L'$, L'/K is normal, and $L' \subset L^*$, since every minimal polynomial of an element of B splits in L^*. Hence $L' = L^*$ by the minimality of L^*.

If $[L\colon K] < \infty$, then we may assume that B if finite. Then F is finite and L^* is a splitting field of a single polynomial. Hence L^*/K is finite. □

1.4 Separable and inseparable field extensions

Let R be a commutative ring and $f \in R[X]^\bullet$. An element $a \in R$ is called a **multiple zero** of f if $f = (X-a)^2 g$ for some polynomial $g \in K[X]$, and this holds if and only if $f(a) = f'(a) = 0$.

Definition 1.4.1. Let K be a field.

1. A polynomial $f \in K[X] \setminus K$ is called **separable** if it does not have multiple zeros in some extension field of K. Otherwise it is called **inseparable**. By definition, every non-constant divisor of a separable polynomial is separable.

2. K is called **perfect** if every irreducible polynomial $f \in K[X]$ is separable.

3. Let L/K be a field extension.

(a) An element $\alpha \in L$ is called **separable** over K if it is algebraic over K and its minimal polynomial over K is separable. If α is algebraic but not separable over K, then α is called **inseparable** over K.

(b) L/K is called **separable** if every element of L is separable over K (then L/K is algebraic). If L/K is algebraic and not separable, then L/K is called **inseparable**.

(c) L/K is called **purely inseparable** if every element of $L \setminus K$ is inseparable over K. Note that L/K is both separable and purely inseparable if and only if $L = K$.

(d) If $\operatorname{char}(K) = p > 0$, then an element $\alpha \in L$ is called **purely inseparable** over K if $\alpha^{p^e} \in K$ for some $e \in \mathbb{N}$.

(e) L/K is called **Galois** if it is normal and separable. It is called **abelian** [**cyclic**, **solvable**, a **p-extension** (for some prime p)] if it is Galois and $\operatorname{Gal}(L/K)$ is abelian [a finite cyclic group, a finite solvable group, a finite p-group].

Theorem 1.4.2. *Let K be a field and $f \in K[X] \setminus K$.*

1. The following assertions are equivalent:

(a) $f' = 0$.

(b) $\operatorname{char}(K) = p > 0$, *and* $f = g(X^{p^e})$ *for some* $e \in \mathbb{N}$ *and some polynomial* $g \in K[X]$ *such that* $g' \neq 0$.

2. Let $\overline{K}$ be an algebraically closed extension field of K. Then the following assertions are equivalent:

(a) f is separable.

(b) f and f' are coprime.

(c) $|\{x \in \overline{K} \mid f(x) = 0\}| = \partial(f)$.

3. If f is separable, then $f' \neq 0$. If f is irreducible and $f' \neq 0$, then f is separable.

4. Let $\mathrm{char}(K) = p > 0$. Let f be irreducible and $e \in \mathbb{N}_0$ maximal such that $f = g(X^{p^e})$. Then g is irreducible and separable, and if $e > 0$, then f is inseparable.

Proof. 1. (a) $\Rightarrow$ (b) Assume that

$$f = \sum_{\nu \geq 0} a_\nu X^\nu \quad \text{and} \quad f' = \sum_{\nu \geq 1} \nu a_\nu X^{\nu-1} = 0,$$

where $a_\nu \in K$ and $a_\nu = 0$ for almost all $\nu \geq 0$. It follows that $\nu a_\nu = 0$ for all $\nu \geq 1$, and since $f \notin K$, we obtain $\mathrm{char}(K) = p > 0$ and $a_\nu = 0$ for all $\nu \geq 1$ such that $p \nmid \nu$. Hence

$$f = g(X^p), \quad \text{where} \quad g = \sum_{\nu \geq 0} a_{p\nu} X^\nu.$$

Let $e \in \mathbb{N}$ be maximal such that $f = g(X^{p^e})$ for some polynomial $g \in K[X]$. Then $g \notin K$ and $g' \neq 0$. Indeed, if $g' = 0$, then $g = g_1(X^p)$ for some polynomial $g_1 \in K[X]$, and $f = g_1(X^{p^{e+1}})$ which contradicts the maximal choice of e.

(b) $\Rightarrow$ (a) If $f = g(X^{p^e})$, then $f' = p^e X^{p^e-1} g'(X^{p^e}) = 0$.

2. (a) $\Rightarrow$ (b) Assume to the contrary that there is some $g \in K[X] \setminus K$ such that $g \mid f$ and $g \mid f'$. By 1.2.5 there exist an extension field L of K and an element $\alpha \in L$ such that $g(\alpha) = 0$. But then $f(\alpha) = f'(\alpha) = 0$, α is a multiple zero of f in L and f is inseparable.

(b) $\Rightarrow$ (c) Suppose that $|\{x \in \overline{K} \mid f(x) = 0\}| < \partial(f)$. Then f has a multiple zero $\alpha \in \overline{K}$, and $f(\alpha) = f'(\alpha) = 0$. If $h \in K[X]$ is the minimal polynomial of α over K, then $h \mid f$ and $h \mid f'$. Hence f and f' are not coprime.

(c) $\Rightarrow$ (a) Assume to the contrary that f has a mutiple zero α in some extension field L of K. We may assume that L/K is algebraic, and by 1.2.7.1 there exists a K-homomorphism $\varphi\colon L \to \overline{K}$. Then $\varphi(\alpha)$ is a multiple zero of f in $\overline{K}$, and since f splits in $\overline{K}[X]$, it follows that f has less than n zeros in $\overline{K}$

3. If $f' = 0$, then every zero of f in an extension field of K is a multiple zero. Hence f is inseparable. Assume now that f is irreducible and $f' \neq 0$. If $g \in K[X]$ is a common divisor of f and f', then $\partial(g) \leq \partial(f') < \partial(f)$ and therefore $g \in K^\times$. Hence f and f' are coprime, and f is separable.

4. Clearly, g is irreducible, and if $e > 0$, then $f' = 0$ and f is inseparable by 3. Thus assume that g is inseparable. Then $g' = 0$, hence $g = g_1(X^p)$ for some polynomial $g_1 \in K[X]$ by 1., and then $f = g_1(X^{p^{e+1}})$, a contradiction. □

Theorem 1.4.3. *Let K be a field.*

1. *Assume that* $\text{char}(K) = p > 0$.

(a) *If $a \in K \setminus K^p$ and $n \in \mathbb{N}$, then the polynomial $f = X^{p^n} - a$ is irreducible in $K[X]$ and inseparable.*

(b) *K is perfect if and only if $K = K^p$.*

(c) *Let L/K be a field extension. Let $x \in L$ be purely inseparable over K, and let $n \in \mathbb{N}$ be minimal such that $x^{p^n} \in K$. Then $[K(x){:}K] = p^n$, and x is inseparable over K.*

2. *If K is finite or algebraically closed, or if* $\text{char}(K) = 0$, *then K is perfect.*

Proof. 1. (a) Since $f' = 0$, it suffices to prove that f is irreducible. Let L be an extension field of K containing some x such that $x^{p^n} = a$. Then we get $f = (X - x)^{p^n}$, and we assume to the contrary that f is not irreducible. Then there exists some $k \in [1, p^n - 1]$ such that $g = (X - x)^k \in K[X]$, and it follows that $(-1)^k g(0) = x^k \in K$. Since $(k, p^n) \mid p^{n-1}$, there exist ν, $\mu \in \mathbb{Z}$ such that $p^{n-1} = k\nu + p^n\mu$, and consequently $x^{p^{n-1}} = (x^k)^\nu a^\mu \in K$. Hence $a = (x^{p^{n-1}})^p \in K^p$, a contradiction.

(b) Assume first that $K = K^p$, and let $f \in K[X] \setminus K$ be such that $f' = 0$. By 1.4.2.1 there exists some polynomial $g = a_0 + a_1X + \ldots + a_nX^n \in K[X] \setminus K$ such that $f = g(X^p)$. For every $\nu \in [0, n]$ let $c_\nu \in K$ be such that $a_\nu = c_\nu^p$. Then $f = c_0^p + c_1^pX^p + \ldots + c_n^pX^{np} = (c_0 + c_1X + \ldots + c_nX^n)^p \in K[X]$ is reducible. Hence there is no irreducible inseparable polynomial in $K[X]$, and thus K is perfect.

Conversely, if $K \neq K^p$ and $a \in K \setminus K^p$, then the polynomial $X^p - a$ is irreducible and inseparable by (a). Hence K is not perfect.

(c) Since $a = x^{p^n} \in K \setminus K^p$, the polynomial $f = X^{p^n} - a \in K[X]$ is irreducible and inseparable by (a) and thus it is the minimal polynomial of x. Hence $[K(x){:}K] = p^n$, and x is inseparable over K.

2. If K is finite, then $\text{char}(K) = p > 0$, and the assignment $x \mapsto x^p$ defines an injective and thus bijective map $\varphi\colon K \to K$. Hence $K = K^p$, and K is perfect.

If K is algebraically closed, the every irreducible polynomial in $K[X]$ is linear and thus separable. Hence K is perfect.

If $\text{char}(K) = 0$ and $f \in K[X] \setminus K$, then $f' \neq 0$. Hence K is perfect. □

Theorem and Definition 1.4.4. *Let L/K be an algebraic field extension.*

1. Let $\varphi\colon K \to \overline{K}$ be a field homomorphism into an algebraically closed field $\overline{K}$, and let $\Omega(\varphi, L/K)$ be the set of all field homomorphisms $\overline{\varphi}\colon L \to \overline{K}$ such that $\overline{\varphi}\restriction K = \varphi$. Then the number

$$|\Omega(L/K,\varphi)| \in \mathbb{N} \cup \{\infty\}$$

only depends on the field extension L/K (and not on φ).
$[L\!:\!K]_{\mathsf{s}} = |\Omega(\varphi, L/K)| \in \mathbb{N} \cup \{\infty\}$ is called the **separable degree** of L/K.
In particular, if $\overline{K} \supset K$, then $[L\!:\!K]_{\mathsf{s}} = |\mathrm{Hom}^K(L, \overline{K})|$.

2. Let $L = K(\alpha)$ and $f \in K[X]$ the minimal polynomial of α over K. Then $[L\!:\!K]_{\mathsf{s}}$ is the number of zeros of f in an algebraic closure of K. In particular, $[L\!:\!K]_{\mathsf{s}} \le \partial(f) = [L\!:\!K]$, and equality holds if and only if α is separable over K.

3. Let M be an intermediate field of L/K.

(a) $[L\!:\!K]_{\mathsf{s}} = [L\!:\!M]_{\mathsf{s}}[M\!:\!K]_{\mathsf{s}}$.

(b) If $\alpha \in L$ is separable over K, then α is also separable over M.

(c) L/K is separable if and only if both L/M and M/K are separable.

4. If $[L\!:\!K] < \infty$, then $[L\!:\!K]_{\mathsf{s}} \le [L\!:\!K]$, and equality holds if and only if L/K is separable.

5. Suppose that $L = K(S)$ for some set S of elements which are separable over K. Then L/K is separable.

Proof. 1. Let $\varphi'\colon K \to \overline{K}'$ be another field homomorphism into an algebraically closed field $\overline{K}'$. Let K^* be the relative algebraic closure of $\varphi(K)$ in $\overline{K}$ and $K^{*\prime}$ the relative algebraic closure of $\varphi'(K)$ in $\overline{K}'$. Then K^* is an algebraic closure of $\varphi(K)$, $K^{*\prime}$ is an algebraic closure of $\varphi'(K)$, and the isomorphism $\varphi\circ\varphi'^{-1}\colon \varphi'(K) \overset{\sim}{\to} \varphi(K)$ has an extension to an isomorphism $\Phi\colon K^{*\prime} \overset{\sim}{\to} K^*$. If $\overline{\varphi}\colon L \to \overline{K}$ is a homomorphism satisfying $\overline{\varphi}\restriction K = \varphi$, then $\overline{\varphi}(L)/\varphi(K)$ is algebraic, hence $\overline{\varphi}(L) \subset K^*$, and the homomorphism $\overline{\varphi}' = \Phi^{-1}\circ\overline{\varphi}\colon L \to K^{*\prime} \hookrightarrow \overline{K}'$ satisfies $\overline{\varphi}' \restriction K = \varphi'$. Since conversely $\overline{\varphi}$ is uniquely determined by $\overline{\varphi}'$, the assignment $\overline{\varphi} \mapsto \overline{\varphi}'$ defines an injective map $\Omega(\varphi, L/K) \to \Omega(\varphi', L/K)$, and consequently $|\Omega(\varphi, L/K)| \le |\Omega(\varphi', L/K)|$. Interchanging the roles of φ and φ' yields equality.

2. Let $\overline{K}$ be an algebraic closure of K and $N = \{x \in \overline{K} \mid f(x) = 0\}$. Then $[K(\alpha) : K]_{\mathsf{s}} = |\mathrm{Hom}^K(K(\alpha), \overline{K})|$ by 1. (applied with $\varphi = (K \hookrightarrow \overline{K})$. As the map $\varepsilon\colon \mathrm{Hom}^K(K(\alpha), \overline{K}) \to N$, defined by $\varepsilon(\psi) = \psi(\alpha)$, is bijective, $[K(\alpha)\!:\!K]_{\mathsf{s}} = |N|$ follows.

3. (a) Let $\varphi\colon K \to \overline{K}$ be a field homomorphism into an algebraically closed field $\overline{K}$. Then

$$\Omega(\varphi, L/K) = \biguplus_{\psi\in\Omega(\varphi,M/K)} \Omega(\psi, L/M),$$

and $|\Omega(\psi, L/M)| = [L\colon M]_s$ for all $\psi \in \Omega(M, \varphi)$. Hence it follows that

$$[L\colon K]_s = |\Omega(\varphi, L/K)| = \sum_{\psi \in \Omega(\varphi, M/K)} |\Omega(\psi, L/M)| = [L\colon M]_s |\Omega(\varphi, M/K)| = [L\colon M]_s [M\colon K]_s.$$

(b) Let $\alpha \in L$ be separable over K, $f \in M[X]$ the minimal polynomial of α over M and $h \in K[X]$ the minimal polynomial of α over K. Then h is separable and $f \mid h$. Hence f is separable and α is separable over M.

(c) Let L/K be separable. Then M/K is separable by definition, and L/M is separable by (b). For the converse we first prove 4.

4. Let $L = K(\alpha_1, \ldots, \alpha_m)$, and for $j \in [0, m]$ set $L_j = K(\alpha_1, \ldots, \alpha_j)$. Using 2., we obtain

$$[L\colon K]_s = \prod_{j=1}^{m} [L_j\colon L_{j-1}]_s = \prod_{j=1}^{m} [L_{j-1}(\alpha_j)\colon L_{j-1}]_s \le \prod_{j=1}^{m} [L_{j-1}(\alpha_j)\colon L_{j-1}] = [L\colon K].$$

If L/K is separable, then 3.(b) implies that α_j is separable over L_{j-1} for all $j \in [1, m]$, hence $[L_j\colon L_{j-1}]_s = [L_j\colon L_{j-1}]$ by 2., and therefore $[L\colon K]_s = [L\colon K]$. If L/K is inseparable, we may assume that α_1 is inseparable over K. But then 2. implies that $[K(\alpha_1)\colon K]_s < [K(\alpha_1)\colon K]$ and consequently $[L\colon K]_s < [L\colon K]$.

Now we can complete the proof of 3.(c). Let M/K and L/M be separable, let $\alpha \in L$ and $f \in M[X]$ the minimal polynomial of α over M. Let M_0 be an over K finite intermediate field of M/K such that $f \in M_0[X]$. Then M_0/K is separable, and as f is separable, the field extension $M_0(\alpha)/M_0$ is separable. By means of 1. and 3. we obtain

$$[M_0(\alpha)\colon K]_s = [M_0(\alpha)\colon M_0]_s\, [M_0\colon K]_s = [M_0(\alpha)\colon M_0]\, [M_0\colon K] = [M_0(\alpha)\colon K].$$

Hence $M_0(\alpha)/K$ is separable. In particular, α is separable over K, and therefore L/K is separable.

5. If $\alpha \in L$, then $\alpha \in K(\alpha_1, \ldots, \alpha_m)$ for some $m \in \mathbb{N}$ and $\alpha_1, \ldots, \alpha_m \in S$, and it suffices to prove that $K(\alpha_1, \ldots, \alpha_m)/K$ is separable. We proceed by induction on m. Let $K' = K(\alpha_1, \ldots, \alpha_{m-1})$. Then K'/K is separable by the induction hypothesis and α_m is separable over K'. Hence we obtain $[K'(\alpha_m)\colon K']_s = [K'(\alpha_m)\colon K']$ by 2., $K'(\alpha_m)/K'$ is separable by 4., and eventually $K'(\alpha_m)/K$ is separable by 3.(c). □

Corollary 1.4.5. *Let L/K be an algebraic field extension, N an extension field of L and M another intermediate field of N/K.*

> 1. *Let L/K be (finite) separable [Galois]. Then LM/M is (finite) separable [Galois], and if M is an intermediate field of L/K, then L/M is also separable [Galois].*

2. Let L/K and M/K be (finite) separable [Galois]. Then LM/K and $L \cap M/K$ are (finite) separable [Galois], too.

Proof. By 1.2.3 and 1.3.2 it suffices to prove the assertions for the separability.

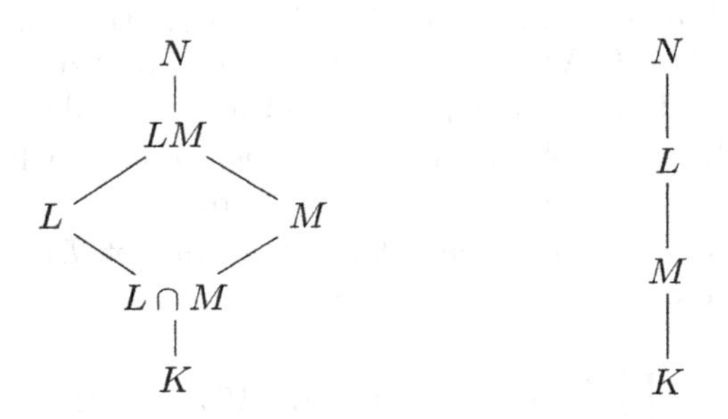

1. L/M is separable by 1.4.4.3(b), and as $LM = M(L)$, it follows from 1.4.4.5 that LM/M is separable.

2. LM/M is separable by 1., and therefore LM/K is separable 1.4.4.3(c). Obviously, $L \cap M/K$ is separable. □

Theorem 1.4.6. *Let K be a field.*

1. K is perfect if and only if every algebraic field extension L/K is separable.

2. Let L/K be an algebraic field extension.

(a) If K is perfect, then L is perfect.

(b) If $[L\!:\!K] < \infty$ and L is perfect, then K is perfect.

Proof. We may assume that $\operatorname{char}(K) = p > 0$.

1. Let K be perfect, L/K an algebraic field extension. Let $x \in L$, and let $f \in K[X]$ be the minimal polynomial of x over K. Then f is separable and thus x is separable over K. Hence L/K is separable.

If K is not perfect, then there exists some $a \in K \setminus K^p$, and if $L = K(x)$, where $x^p = a$, then the inseparable polynomial $X^p - a \in K[X]$ is the minimal polynomial of x over K(see 1.4.3.1(a)). Hence x and thus L/K is inseparable.

2.(a) If L is not perfect, then there exists an inseparable field extension L'/L. But then L'/K is also inseparable by 1.4.4.3(c), and thus K is not perfect.

(b) We use 1.4.3.1.(b). Assume that $[L\!:\!K] < \infty$ and K is not perfect. If $a \in K \setminus K^p$, then $[K(a^{1/p^n})\!:\!K] = p^n$ for all $n \in \mathbb{N}$ by 1.7.9, and therefore ther exists some $n \in \mathbb{N}_0$ such that $a^{1/p^n} \in L$ and $a^{1/p^{n+1}} \notin L$. But then $a^{1/p^n} \in L \setminus L^p$, and L is not perfect. □

Theorem 1.4.7. *Let L/K be an algebraic field extension, and suppose that* $\operatorname{char}(K) = p > 0$.

1. The following assertions are equivalent:

(a) L/K is purely inseparable.

(b) Every element of L is purely inseparable over K.

(c) $L = K(S)$ for some set S of over K purely inseparable elements.

(d) $[L:K]_s = 1$.

2. Let L/K be purely inseparable, and assume that $[L:K] < \infty$. Then $[L:K] = p^n$ for some $n \in \mathbb{N}_0$, and if $[L:K] = p^n$, then $L^{p^n} \subset K$.

Proof. 1. (a) $\Rightarrow$ (b) Let $\alpha \in L$, $f \in K[X]$ the minimal polymomial of α and $e \in \mathbb{N}_0$ maximal such that $f = g(X^{p^e})$ for some $g \in K[X]$. Then g is irreducible and separable by 1.4.2.4, and as $g(\alpha^{p^e}) = 0$ it follows that $\alpha^{p^e} \in K$, and thus α is purely inseparable over K.

(b) $\Rightarrow$ (a) By 1.4.3.1(c) every over K purely inseparable element of L is inseparable over K.

(b) $\Rightarrow$ (c) Obvious.

(c) $\Rightarrow$ (d) Let $\overline{K}$ be an algebraically closed extension field of L, and suppose that $\varphi \in \operatorname{Hom}^K(L, \overline{K})$. If $\alpha \in S$, $e \in \mathbb{N}_0$ and $\alpha^{p^e} \in K$, then $\varphi(\alpha)^{p^e} = \varphi(\alpha^{p^e}) = \alpha^{p^e}$ and thus $\varphi(\alpha) = \alpha$, which implies $\varphi = \mathrm{id}_K$. Hence $[L:K]_s = |\operatorname{Hom}^K(L, \overline{K})| = 1$.

(d) $\Rightarrow$ (b) Assume to the contrary that some $\alpha \in L \setminus K$ is separable over K. Then $1 = [L:K]_s \ge [K(\alpha):K]_s = [K(\alpha):K] > 1$, a contradiction.

2. By 1. we may assume that $L = K(\alpha_1, \ldots, \alpha_m)$ such that $\alpha_j^{p^{n_j}} \in K$ for all $j \in [1, m]$, where $n_j \in \mathbb{N}$. Then 1.4.3.1(c) shows that

$$[L:K] = \prod_{j=1}^{m} [K(\alpha_1, \ldots, \alpha_j) : K(\alpha_1, \ldots, \alpha_{j-1})] \mid p^{n_1 + \ldots + n_m}.$$

If $[L:K] = p^n$ and $x \in L$, then $[K(x):K] \mid p^n$ and therefore $x^{p^n} \in K$, again by 1.4.3.1(c). □

Theorem and Definition 1.4.8. *Let L/K be an algebraic field extension and $\overline{K}$ an algebraically closed extension field of L.*

1. L/K is Galois if and only if L is the splitting field of a set of separable polynomials over K. If L/K is finite, then L/K is Galois if and only if L is a splitting field of a single separable polynomial over K.

2. The compositum of a family of over K normal [separable, Galois] extension fields of K is normal [separable, Galois] over K.

3. Let L/K be separable and L^* its normal closure in $\overline{K}$. Then L^*/K is Galois. We call L^* the **Galois closure** of L/K in $\overline{K}$.

4. Let S be the set of all over K separable elements of L. Then $L_0 = K(S)$ is the largest over K separable intermediate field of L/K, L/L_0 is purely inseparable, and $[L_0\!:\!K] = [L\!:\!K]_{\mathsf{s}}$. If L/K is normal, then L_0/K is Galois, and the restriction $\sigma \mapsto \sigma \restriction L_0$ defines an isomorphism $\mathrm{Gal}(L/K) \stackrel{\sim}{\to} \mathrm{Gal}(L_0/K)$.

L_0 is called the **(relative) separable closure** of K in L, and

$$[L\!:\!K]_{\mathsf{i}} = [L\!:\!L_0]$$

is called the **inseparable degree** of L/K. Apparently we have $[L:K] = [L:K]_{\mathsf{s}}[L:K]_{\mathsf{i}}$. If $\mathrm{char}(K) = 0$, then $[L:K]_{\mathsf{i}} = 1$, and if $\mathrm{char}(K) = p > 0$, then $[L\!:\!K]_{\mathsf{i}}$ is a power of p.

The relative separable closure of K in an algebraic closure of K is called a **separable closure** *of K.*

5. K possesses an (up to K-isomorphisms unique) separable closure K_{sep}. The field extension K_{sep}/K is Galois, and if $\overline{K}$ is an algebraic closure of K containing K_{sep}, then the restriction $\sigma \mapsto \sigma \restriction K_{\mathrm{sep}}$ defines an isomorphism $\mathrm{Gal}(\overline{K}/K) \to \mathrm{Gal}(K_{\mathrm{sep}}/K)$. We identify and call $G_K = \mathrm{Gal}(K_{\mathrm{sep}}/K) = \mathrm{Gal}(\overline{K}/K)$ the **absolute Galois group** of K.

Proof. 1. Obvious by 1.3.1.1.

2. Let L be the compositum of a family $(L_i)_{i\in I}$ of extension fields of K. Assume first that, for all $i \in I$, L_i/K is normal and L_i is the splitting field of a subset F_i of $K[X] \setminus K$. Then L is a splitting field of the union of $(F_i)_{i\in I}$, and thus L/K is normal.

Now let L_i/K be separable for all $i \in I$, and let S be the union of $(L_i)_{i\in I}$. Then $L = K(S)$ and S is separable over K. Hence L/K is separable.

3. Obvious by 1.3.3.2.

4. We may assume that $\mathrm{char}(K) = p > 0$, and we apply 1.4.4. Since $L_0 = K(S)$ is separable over K, it is the largest over K separable intermediate field of L/K. If $\alpha \in L$ is separable over L_0, then $L_0(\alpha)/L_0$ and thus $L_0(\alpha)/K$ is separable. Hence $\alpha \in L_0$, L/L_0 is purely inseparable, $[L:L_0]_{\mathsf{s}} = 1$ by 1.4.7.1, and $[L\!:\!K]_s = [L\!:\!L_0]_{\mathsf{s}}[L_0\!:\!K]_{\mathsf{s}} = [L_0\!:\!K]_{\mathsf{s}} = [L_0\!:\!K]$.

Now let L/K be normal. Let $f \in K[X]$ be irreducible and $\alpha \in L_0$ a zero of f. Then f is separable, f splits in $L[X]$, and all zeros of f lie in L_0. Hence f splits in $L_0[X]$ which shows that L_0/K is normal and thus Galois.

In particular, it follows that $\sigma \restriction L_0 \in \mathrm{Gal}(L_0/K)$ for all $\sigma \in \mathrm{Gal}(L/K)$, and it remains to prove that for every $\sigma_0 \in \mathrm{Gal}(L_0/K)$ there is a unique $\sigma \in \mathrm{Gal}(L/K)$ such that $\sigma \restriction L_0 = \sigma_0$.

Let $\sigma_0 \in \mathrm{Gal}(L_0/K) = \mathrm{Hom}^K(L_0, \overline{K})$ and $\overline{\sigma} \in \mathrm{Gal}(\overline{K}/K)$ be such that $\overline{\sigma} \restriction L_0 = \sigma_0$. Then $\sigma = \overline{\sigma} \restriction L \in \mathrm{Gal}(L/K)$, and $\sigma \restriction L_0 = \sigma_0$. To prove uniqueness, let $\sigma \in \mathrm{Gal}(L/K)$ be such that $\sigma \restriction L_0 = \sigma_0$. If $x \in L$, then $x^{p^e} \in L_0$ for some $e \in \mathbb{N}$, and therefore $\sigma(x)^{p^e} = \sigma(x^{p^e}) = \sigma_0(x^{p^e})$, which implies $\sigma(x) = \sigma_0(x^{p^e})^{1/p^e}$.

5. Obvious by 4. and the uniqueness of an algebraic closure. □

Theorem 1.4.9. *Let $K(\alpha)/K$ be an algebraic field extension, $f \in K[X]$ the minimal polynomial of α over K, $[K(\alpha):K]_{\mathsf{s}} = d$ and $[K(\alpha):K]_{\mathsf{i}} = q$. Let $\overline{K}$ be an algebraically closed extension field of $K(\alpha)$, and suppose that* $\mathrm{Hom}^K(K(\alpha), \overline{K}) = \{\sigma_1, \ldots, \sigma_d\}$ *(where $\sigma_1, \ldots, \sigma_d$ are distinct). Then*

$$f = \prod_{i=1}^{d} (X - \sigma_i(\alpha))^q.$$

Proof. If $\mathrm{char}(K) = 0$, then $q = 1$ and there is nothing to do. Thus suppose that $\mathrm{char}(K) = p > 0$, and let $e \in \mathbb{N}_0$ be maximal such that $f = g(X^{p^e})$ for some (irreducible and separable) polynomial $g \in K[X]$. As $\mathrm{Hom}^K(K(\alpha), \overline{K}) = \{\sigma_1, \ldots, \sigma_d\}$, it follows that $\sigma_1(\alpha), \ldots, \sigma_d(\alpha)$ are the distinct zeros of f, and therefore $\sigma_1(\alpha)^{p^e}, \ldots, \sigma_d(\alpha)^{p^e}$ are the distinct zeros of g. Hence

$$g = \prod_{i=1}^{d} (X - \sigma_i(\alpha)) \quad \text{and} \quad f = \prod_{i=1}^{d} (X^{p^e} - \sigma_i(\alpha)^{p^e}) = \prod_{i=1}^{d} (X - \sigma_i(\alpha))^{p^e}.$$

Now $K(\alpha)/K(\alpha^{p^e}$ is purely inseparable, $K(\alpha^{p^e}/K$ is separable, and

$$[K(\alpha):K] = \partial(f) = p^e \partial(g).$$

Hence $q = p^e = [K(\alpha):K]_{\mathsf{i}}$ and $d = [K(\alpha^{p^e}):K] = [K(\alpha):K]_{\mathsf{s}}$. □

Theorem and Definition 1.4.10 ("Main theorem of algebra"). *$\mathbb{C}$ is algebraically closed.*

We denote by $\overline{\mathbb{Q}}$ the relative algebraic closure of $\mathbb{Q}$ in $\mathbb{C}$ (then $\overline{\mathbb{Q}}$ is an algebraic closure of $\mathbb{Q}$ by 1.2.6.2). The elements of $\overline{\mathbb{Q}}$ are called **algebraic numbers**, and $\overline{\mathbb{Q}}$ itself is called the **field of algebraic numbers**.

There is a wealth of different proofs of this famous theorem of Gauss. A readable overview of the subject is in Wikipedia.

1.5 Galois theory of finite field extensions

Theorem 1.5.1 (Primitive element theorem). *Let L/K be a finite field extension, and suppose that $L = K(\alpha_1, \ldots, \alpha_n)$ for some $n \in \mathbb{N}$ such that $\alpha_2, \ldots, \alpha_n$ are separable over K. Then there exists some $\alpha \in L$ such that $L = K(\alpha)$.*

Proof. If K is finite, then L is finite, too, hence $L^\times = \langle \omega \rangle$ by 1.2.1, and therefore $L = K(\omega)$.

Assume now that K is infinite. We proceed by induction on n. For $n = 1$ there is nothing to do.

$n \geq 2$, $n-1 \to n$: By the induction hypothesis, there exists some $\alpha \in L$ such that $K(\alpha_1, \ldots, \alpha_{n-1}) = K(\alpha)$, and we set $\beta = \alpha_n$. Then $L = K(\alpha, \beta)$, β is separable over K, and we show the existence of some $c \in K$ such that $L = K(\alpha + c\beta)$.

Let $\overline{K}$ be an algebraically closed extension field of L, $f \in K[X]$ the minimal polynomial of α and $g \in K[X]$ the minimal polynomial of β over K. Let

$$f = \prod_{i=1}^{r}(X - \alpha_i) \in \overline{K}[X] \quad \text{and} \quad g = \prod_{j=1}^{s}(X - \beta_j) \in \overline{K}[X],$$

where $\alpha_1 = \alpha$, $\beta_1 = \beta$, and $\beta_1, \ldots, \beta_s$ are distinct. As K is infinite, there exists some $c \in K$ such that $\alpha_i + c\beta_k \neq \alpha + c\beta$ for all $i \in [1, r]$ and $k \in [2, s]$, and we set $\vartheta = \alpha + c\beta$. Then $g(\beta) = 0$, $f(\vartheta - c\beta) = 0$, and β is the unique common zero of the polynomials g and $f(\vartheta - cX)$ in $\overline{K}[X]$. Hence $X - \beta$ is a greatest common divisor of g and $f(\vartheta - cX)$ in $\overline{K}[X]$. Since $g \in K(\vartheta)[X]$ and $f(\vartheta - cX) \in K(\vartheta)[X]$, it follows that $\beta \in K(\vartheta)$ by 1.1.10.2(b), and therefore $K(\alpha, \beta) = K(\vartheta)$. □

Theorem 1.5.2 (Dedekind's independence theorem). *Let K be a field, M a (multiplicative) semigroup, and let $\sigma_1, \ldots, \sigma_n \colon M \to K^\times$ be distinct semigroup homomorphisms. Then $\sigma_1, \ldots, \sigma_n \in K^M$ are linearly independent over K.*

Proof. By induction on n. For $n = 1$ there is nothing to do.

$n \geq 2$, $n - 1 \to n$: Let $\lambda_1, \ldots, \lambda_n \in K$ be such that

$$\lambda_1\sigma_1 + \ldots + \lambda_n\sigma_n = 0 \colon M \to K, \text{ hence } \sum_{i=1}^{n} \lambda_i\sigma_i(x) = 0 \quad \text{for all } x \in M.$$

Let $y \in M$ be such that $\sigma_1(y) \neq \sigma_n(y)$. For all $x \in M$ we obtain

$$0 = \sum_{i=1}^{n} \lambda_i\sigma_i(xy) = \sum_{i=1}^{n} \lambda_i\sigma_i(x)\sigma_i(y), \quad \text{and} \quad 0 = \sum_{i=1}^{n} \lambda_i\sigma_i(x)\sigma_n(y),$$

hence

$$0=\sum_{i=1}^{n-1}\lambda_i\,[\,\sigma_i(y)-\sigma_n(y)\,]\,\sigma_i(x)\quad\text{and therefore}\quad 0=\sum_{i=1}^{n-1}\lambda_i\,[\,\sigma_i(y)-\sigma_n(y)\,]\,\sigma_i.$$

By the induction hypothesis it follows that $\lambda_i\,[\,\sigma_i(y)-\sigma_n(y)\,]=0$ for all $i\in[1,n-1]$, hence $\lambda_1=0$ and consequently $\lambda_2\sigma_2+\ldots+\lambda_n\sigma_n=0$. Again by the induction hypothesis we obtain $\lambda_2=\ldots=\lambda_n=0$. □

Theorem 1.5.3 (Artin). *Let L be a field, G a finite group of automorphisms of L and $K=L^G$. Then L/K is a finite Galois extension, $[L\!:\!K]=|G|$, and* $\mathrm{Gal}(L/K)=G$.

Proof. Let $|G|=n,\ G=\{\sigma_1,\ldots,\sigma_n\},\ \overline{K}$ an algebraically closed extension field of L and $S=\sigma_1+\ldots+\sigma_n\colon L\to L$. Then S is K-linear, $S\neq 0$ by 1.5.2, and $\sigma S(x)=S(x)$ for all $x\in L$ and $\sigma\in G$. It follows that $S(x)\in K$ for all $x\in L$, and consequently $S(L)=K$.

We prove now that any $n+1$ elements of L are linearly dependent over K (then $[L\!:\!K]\le n$). Let $y_1,\ldots,y_{n+1}\in L$. Then the system of linear equations

$$\sum_{\nu=1}^{n+1}\sigma_i^{-1}(y_\nu)\,a_\nu=0\quad\text{for}\quad i\in[1,n]$$

has a non-trivial solution $(a_1,\ldots,a_{n+1})\in L^{n+1}\setminus\{\mathbf{0}\}$, and after renumbering we may assume that $a_1\neq 0$. Since $S(a_1L)=S(L)=K$, there exists some $z\in L$ satisfying $S(a_1z)\neq 0$, and then

$$0=\sum_{i=1}^{n}\sigma_i\Big(\sum_{\nu=1}^{n+1}\sigma_i^{-1}(y_\nu)a_\nu z\Big)=\sum_{\nu=1}^{n+1}\sum_{i=1}^{n}\sigma_i(a_\nu z)y_\nu=\sum_{\nu=1}^{n+1}S(a_\nu z)y_\nu,$$

which shows the linear dependence of $(y_1,\ldots,y_{n+1})$ over K.

By definition, $G\subset\mathrm{Gal}(L/K)$, and therefore

$$|G|\le|\mathrm{Gal}(L/K)|\le|\mathrm{Hom}^K(L,\overline{K})|=[L\!:\!K]_{\mathsf{s}}\le[L\!:\!K]\le n=|G|.$$

Hence $[L:K]=|G|,\ \ G=\mathrm{Gal}(L/K)=\mathrm{Hom}_K(L,\overline{K}),\ \ L/K$ is normal and separable and thus Galois. □

Theorem 1.5.4 (Main theorem of Galois theory for finite field extensions). *Let L/K be a finite field extension and $G=\mathrm{Gal}(L/K)$.*

1. *The following assertions are equivalent:*

(a) *L/K is Galois;* (b) $[L\!:\!K]=|G|$; (c) $K=L^G$.

2. Let L/K be Galois, $\mathcal{F}(L/K)$ the set of all intermediate fields of L/K and $\mathcal{S}(G)$ the set of all subgroups of G. Then the maps

$$\begin{cases} \mathcal{F}(L/K) & \to & \mathcal{S}(G) \\ M & \mapsto & \mathrm{Gal}(L/M) \end{cases} \quad \text{and} \quad \begin{cases} \mathcal{S}(G) & \to & \mathcal{F}(L/K) \\ H & \mapsto & L^H \end{cases}$$

are mutually inverse inclusion-reversing bijections.

3. Let M and M' be intermediate fields of L/K, $H = \mathrm{Gal}(L/M)$ and $H' = \mathrm{Gal}(L/M')$.

- $M \subset M'$ *if and only if* $H \supset H'$.
- $MM' = L^{H \cap H'}$ *and* $H \cap H' = \mathrm{Gal}(L/MM')$.
- $M \cap M' = L^{\langle H \cup H' \rangle}$ *and* $\langle H \cup H' \rangle = \mathrm{Gal}(L/M \cap M')$.

4. Let M be an intermediate field of L/K and $H = \mathrm{Gal}(L/M)$. Then M/K is Galois if and only if H is a normal subgroup of G. In this case the map

$$\rho\colon G \to \mathrm{Gal}(M/K), \quad \textit{defined by} \quad \rho(\sigma) = \sigma \restriction M,$$

is an epimorphism, $\mathrm{Ker}(\rho) = H$, *and ρ induces an isomorphism*

$$\rho^*\colon G/H \xrightarrow{\sim} \mathrm{Gal}(M/K).$$

We identify: $G/H = \mathrm{Gal}(M/K)$ by means of $\sigma H = \sigma \restriction M$ for all $\sigma \in G$.

Proof. Let $\overline{K}$ be an algebraically closed extension field of L.

1. (a) $\Leftrightarrow$ (b) By definition $|G| \le |\mathrm{Hom}^K(L, \overline{K})| = [L\!:\!K]_{\mathrm{s}} \le [L\!:\!K]$. Now $|G| = |\mathrm{Hom}^K(L, \overline{K})|$ if and only if L/K is normal, and $[L\!:\!K]_{\mathrm{s}} = [L\!:\!K]$ if and only if L/K is separable. Hence L/K is Galois if and only if $[L\!:\!K] = |G|$.

(b) $\Leftrightarrow$ (c) As $K \subset L^G \subset L$, we obtain

$$[L\!:\!K] = [L\!:\!L^G]\,[L^G\!:\!K] = |G|\,[L^G\!:\!K] \text{ by 1.5.3.}$$

Hence $K = L^G$ if and only if $[L\!:\!K] = |G|$.

2. If $M \in \mathcal{F}(L/K)$ and $H = \mathrm{Gal}(L/M)$, then L/M is Galois by 1.4.5.1 and $M = L^H$ by 1. If H is a subgroup of G and $M = L^H$, then $\mathrm{Gal}(L/L^H) = H$ by 1.5.3. Therefore the maps in question are mutually inverse bijections, and apparently they are inclusion-reversing.

3. MM' is the smallest intermediate field of L/K which contains both M and M', and $M \cap M'$ is the largest intermediate field of L/K which is contained in both M and M'. On the other hand, $H \cap H'$ is the largest subgroup of G which is contained in both H and H', and $\langle H \cup H' \rangle$ is the smallest subgroup of G which contains both H and H'. Hence the assertions result from 1.

4. By 1.3.2.3(b), M/K is Galois if and only if $\sigma M = M$ for all $\sigma \in G$, and by 1. and 1.2.4.3 this holds if and only if $\mathrm{Gal}(L/\sigma M) = \sigma H \sigma^{-1} = H$ for all $\sigma \in G$. Hence M/K is Galois if and only if H is a normal subgroup of G.

If H is a normal subgroup of G and $\rho\colon G \to \mathrm{Gal}(M/K) = \mathrm{Hom}^K(M, L)$ is defined by $\rho(\sigma) = \sigma \restriction M$, then ρ is an epimorphism by 1.3.1.3, and as $\mathrm{Ker}(\rho) = \mathrm{Gal}(L/M) = H$, it induces an isomorphism $\rho^* : G/H \stackrel{\sim}{\to} \mathrm{Gal}(M/K)$. □

Corollary 1.5.5. *A finite separable field extension has only finitely many intermediate fields.*

Proof. Let L/K be a finite separable field extension and L^* a normal closure of L/K. By 1.4.8.3 and 1.3.3.2, L^*/K is a finite Galois extension. Since its Galois group $\mathrm{Gal}(L^*/K)$ has only finitely many subgroups, 1.5.4.2 implies that L^*/K (and thus also L/K) has only finitely many intermediate fields. □

Theorem 1.5.6 (Shifting theorem of finite Galois theory). *Let L/K be a finite Galois extension, N an extension field of L and M an intermediate field of N/K.*

1. LM/M is a finite Galois extension, the map

$$\rho\colon \mathrm{Gal}(LM/M) \stackrel{\sim}{\to} \mathrm{Gal}(L/L\cap M), \ \textit{defined by}\ \ \rho(\sigma) = \sigma \restriction L,$$

is an isomorphism, and $[LM:M] = [L:L\cap M]$ divides $[L:K]$.

2. Let M/K be another finite Galois extension. Then LM/K is also a finite Galois extension. If moreover $L \cap M = K$, then the map

$$\rho\colon \mathrm{Gal}(LM/K) \to \mathrm{Gal}(L/K)\times\mathrm{Gal}(M/K), \quad \sigma \mapsto (\sigma\restriction L, \sigma\restriction M),$$

is an isomorphism, and $[LM:K] = [L:K]\,[M:K]$.

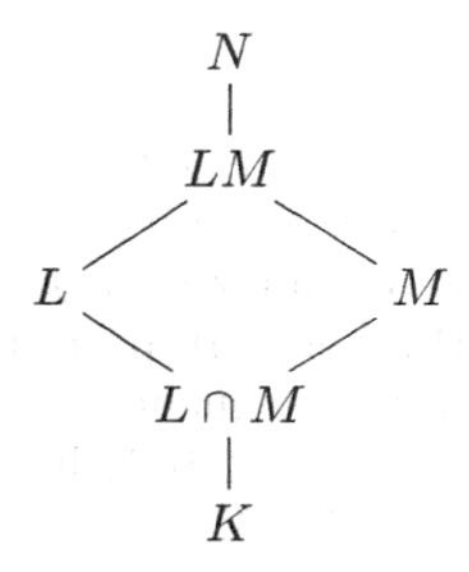

Proof. 1. By 1.4.5.1, LM/M is a finite Galois extension. If $\sigma \in \mathrm{Gal}(LM/M)$, then $\sigma \restriction L \in \mathrm{Hom}^K(L, \overline{K}) = \mathrm{Gal}(L/K)$ and $\sigma \restriction L \cap M = \mathrm{id}_{L\cap M}$. Therefore the assignment $\sigma \mapsto \sigma \restriction L$ yields a homomorphism

$$\rho\colon \mathrm{Gal}(LM/M) \to \mathrm{Gal}(L/L\cap M).$$

If $\sigma \in \ker(\rho)$, then $\sigma \restriction L = \mathrm{id}_L$, and as $\sigma \restriction M = \mathrm{id}_M$, it follows that $\sigma = \mathrm{id}_{LM}$. Hence ρ is injective, and we set $H = \rho(\mathrm{Gal}(LM/M)) \subset \mathrm{Gal}(L/L\cap M)$. Then $L\cap M \subset L^H$, and if $z \in L^H$, then $\sigma(z) = z$ for all $\sigma \in \mathrm{Gal}(LM/M)$ and thus $z \in M$. It follows that $L^H = L\cap M$, $H = \mathrm{Gal}(L/L\cap M)$, and ρ is an isomorphism. In particular,

$$[LM:M] = |\mathrm{Gal}(LM/K)| = |\mathrm{Gal}(L/L\cap M)| = [L:L\cap M] \text{ divides } [L:K].$$

2. By 1.4.5.2, LM/K is a finite Galois field extension. If $L\cap M = K$, then $\rho\colon \mathrm{Gal}(LM/K) \stackrel{\sim}{\to} \mathrm{Gal}(L/K)\times\mathrm{Gal}(M/K)$, defined by $\rho(\sigma) = (\sigma\restriction L, \sigma\restriction M)$, is a monomorphism, and by 1. there exist isomorphisms

$$\begin{cases} \mathrm{Gal}(LM/L) & \stackrel{\sim}{\to} & \mathrm{Gal}(M/K) \\ \tau & \mapsto & \tau\restriction M \end{cases} \quad\text{and}\quad \begin{cases} \mathrm{Gal}(LM/M) & \stackrel{\sim}{\to} & \mathrm{Gal}(L/K) \\ \tau & \mapsto & \tau\restriction L \end{cases}.$$

If $(\tau_1, \tau_2) \in \mathrm{Gal}(L/K)\times\mathrm{Gal}(M/K)$, then there exists some

$$(\sigma_1, \sigma_2) \in \mathrm{Gal}(LM/M)\times\mathrm{Gal}(LM/L) \subset \mathrm{Gal}(LM/K)\times\mathrm{Gal}(LM/K),$$

such that $\sigma_1 \restriction L = \tau_1$ and $\sigma_2 \restriction M = \tau_2$. Hence $\rho(\sigma_1\circ\sigma_2) = (\tau_1, \tau_2)$, ρ is an isomorphism, and

$$[LM:K] = |\mathrm{Gal}(LM/K)| = |\mathrm{Gal}(L/K)|\,|\mathrm{Gal}(M/K)| = [L:K]\,[M:K]. \quad \square$$

In the following Theorem 1.5.8 we trace back to the original definition of Galois groups as permutation groups of the zeros of a polynomial. To this end we must fix our terminology concerning permutations.

Definition and Remarks 1.5.7. Let A be a finite set, $|A| = n \in \mathbb{N}$ and $\mathfrak{S}(A)$ the symmetric group of all **permutations** of A (i. e., the group of all bijective maps $A \to A$). Every permutation $\sigma \in \mathfrak{S}(A)$ has an (up to the order) unique representation in the form

$$\sigma = \prod_{i=1}^{r} (a_{i,1}, \ldots, a_{i,f_i}),$$

where $r,\ f_1, \ldots, f_r \in \mathbb{N}$, $f_1+\ldots+f_r = n$ and $\{a_{i,\nu} \mid \nu \in [1, f_i],\ i \in [1,r]\} = A$. In this representation, the finite sequence $(a_{i,1}, \ldots, a_{i,f})$ denotes the **cycle** of σ, defined by $\sigma^\nu(a_{i,1}) = a_{i,\nu+1}$ for all $\nu \in [1, f_{i-1}]$ and $\sigma^{f_i}(a_{i,1}) = a_{i,1}$. The sets $\{a_{i,1}, \ldots, a_{i,f_i}\}$ are the orbits of σ. In particular, $\{a_{i,1} \mid i \in [1,r],\ f_i = 1\}$ is the set of fixed points of σ. The sequence $(f_1, \ldots, f_r)$ is called the **cycle type** of σ. Two permutations σ_1, $\sigma_2 \in \mathfrak{S}(A)$ have the same cycle type if and only if $\sigma_2 = \tau\sigma_1\tau^{-1}$ for some $\tau \in \mathfrak{S}(A)$.

Definition and Theorem 1.5.8. *Let K be a field, $g \in K[X]$ a separable polynomial, $\partial(g) = n \ge 1$, L a splitting field of g over K, $Z(g) = Z_L(g)$ the set of zeros of g in L and $G = \mathrm{Gal}(L/K)$.*
Then $|Z(g)| = n$, and G operates faithfully on $Z(g)$. Hence the map

$$\iota\colon G \to \mathfrak{S}(Z(g)), \quad \text{defined by} \quad \iota(\sigma) = \sigma \upharpoonright Z(g),$$

is a monomorphism, and the permutation group $\mathrm{Gal}_K(g) = \mathrm{Im}(\iota) \subset \mathfrak{S}(Z(g))$ is called the **Galois group** of g over K. In particular, ι induces a group isomorphism

$$\iota\colon G \xrightarrow{\sim} \mathrm{Gal}_K(g).$$

If we are only interested in the group structure of $\mathrm{Gal}_K(g)$, we may fix a splitting field L of g over K and identify G with $\mathrm{Gal}_K(g)$ (as this is usually done in the literature). We shall use the more precise terminology when we refer to the cycle type of the permutations. Every bijective map $Z(g) \to [1, n]$ induces an isomorphism $\mathrm{Gal}_K(g) \xrightarrow{\sim} \mathfrak{S}_n$ which does not change the cycle type of the permutations. Thus for most investigations we may fix an ordering of $Z(g)$ and view $\mathrm{Gal}_K(g)$ as a subgroup of $\mathfrak{S}_n$.

Assume now that $g = g_1 \cdot \ldots \cdot g_r$, where $r \in \mathbb{N}$, $g_1, \ldots, g_r \in K[X]$ are irreducible and $\partial(g_i) = f_i$ for all $i \in [1, r]$.

1. *$Z(g_1), \ldots, Z(g_r)$ are the orbits of $\mathrm{Gal}_K(g)$. In particular, $\mathrm{Gal}_K(g)$ is transitive if and only if g is irreducible.*

2. *Let L/K be cyclic. If $\mathrm{Gal}_K(g) = \langle\sigma\rangle$, then σ is of type $(f_1, \ldots, f_r)$.*

Proof. 1. Let x, $x' \in Z(g)$. Then x and x' lie in the same orbit of $\mathrm{Gal}_K(g)$ if and only if x and x' are conjugate, and this holds if and only if x and x' are zeros of the same irreducible factor.

2. If $\mathrm{Gal}_K(g) = \langle\sigma\rangle$, then the orbits of $\mathrm{Gal}_K(g)$ are precisely the orbits of σ. If $\sigma = \sigma_1 \cdot \ldots \cdot \sigma_r$ is a product of cycles as in 1.5.7, then $Z(g_1), \ldots, Z(g_r)$ are the orbits of σ, and therefore $(f_1, \ldots, f_r)$ is the cycle type of σ. □

1.6 Norms, traces, resultants, and discriminants

Definition and Remarks 1.6.1. Let K be a field, A a finite-dimensional K-algebra and $n = \dim_K(A) < \infty$. For $a \in A$, we define $\mu_a\colon A \to A$ by $\mu_a(x) = ax$. Then $\mu_a \in \mathrm{End}_K(A)$, and we define the **norm** $\mathsf{N}_{A/K}(a)$ and the **trace** $\mathsf{Tr}_{A/K}(a)$ of a by

$$\mathsf{N}_{A/K}(a) = \det(\mu_a) \quad \text{and} \quad \mathsf{Tr}_{A/K}(a) = \mathrm{trace}(\mu_a)\,.$$

If $\boldsymbol{u} = (u_1, \ldots, u_n)$ is a K-basis of A and $a\boldsymbol{u} = (au_1, \ldots, au_n) = \boldsymbol{u}M$ for some matrix $M \in \mathsf{M}_n(K)$, then $\mathsf{N}_{A/K}(a) = \det(M)$ and $\mathsf{Tr}_{A/K}(a) = \mathrm{trace}(M)$.
If $a, b \in A$ and $\lambda \in K$, then $\mu_{ab} = \mu_a \circ \mu_b$, $\mu_{\lambda a} = \lambda \mu_a$ and $\mu_{a+b} = \mu_a + \mu_b$. Hence

$$\mathsf{N}_{A/K}(ab) = \mathsf{N}_{A/K}(a)\mathsf{N}_{A/K}(b)\,,\ \mathsf{N}_{A/K}(\lambda a) = \lambda^n \mathsf{N}_{A/K}(a)\,,$$

$$\mathsf{Tr}_{A/K} \text{ is } K\text{-linear and } \mathsf{Tr}(\lambda 1_A) = n\lambda.$$

In particular, $\mathsf{N}_{A/K} \restriction A^\times : A^\times \to K^\times$ is a group homomorphism.

Theorem 1.6.2. *Let L/K be a finite field extension, $n = [L\!:\!K]$, $q = [L\!:\!K]_{\mathrm{i}}$, $x \in L$, $d = [K(x)\!:\!K]$, $m = [L\!:\!K(x)]$ and*

$$g = X^d + a_{d-1}X^{d-1} + \ldots + a_1 X + a_0 \in K[X]$$

the minimal polynomial of x over K.

1. $n = md$, $\mathsf{N}_{L/K}(x) = (-1)^n a_0^m$, $\mathsf{N}_{L/K}(-x) = a_0^m$ and $\mathsf{Tr}_{L/K}(x) = -ma_{d-1}$.

2. Let $\overline{K}$ be an algebraically closed extension field of L. Then

$$\mathsf{N}_{L/K}(x) = \prod_{\sigma \in \mathrm{Hom}^K(L,\overline{K})} \sigma(x)^q \quad \text{and} \quad \mathsf{Tr}_{L/K}(x) = q \sum_{\sigma \in \mathrm{Hom}^K(L,\overline{K})} \sigma(x)\,.$$

In particular:

(a) If L/K is separable, then $\mathsf{Tr}_{L/K} \colon L \to K$ is surjective, and if L/K is inseparable, then $\mathsf{Tr}_{L/K} = 0$.

(b) If L/K is Galois and $G = \mathrm{Gal}(L/K)$, then

$$\mathsf{N}_{L/K}(x) = \prod_{\sigma \in G} \sigma(x) \ \text{ and } \ \mathsf{Tr}_{L/K}(x) = \sum_{\sigma \in G} \sigma(x) \quad \text{for all } x \in L.$$

3. Let M be an intermediate field of L/K. Then

$$\mathsf{N}_{L/K} = \mathsf{N}_{M/K} \circ \mathsf{N}_{L/M} \quad \text{and} \quad \mathsf{Tr}_{L/K} = \mathsf{Tr}_{M/K} \circ \mathsf{Tr}_{L/M}\,.$$

Proof. 1. Apparently $n = md$, $\boldsymbol{u} = (1, x, \ldots, x^{d-1})$ is a K-basis of $K(x)$, and $x\boldsymbol{u} = \boldsymbol{u}T$, where

$$T = \begin{pmatrix} 0 & 0 & \ldots & 0 & -a_0 \\ 1 & 0 & \ldots & 0 & -a_1 \\ 0 & 1 & \ldots & 0 & -a_2 \\ . & . & \ldots & . & . \\ 0 & 0 & \ldots & 1 & -a_{d-1} \end{pmatrix}, \ \mathrm{trace}(T) = -a_{d-1} \ \text{ and } \ \det(T) = (-1)^d a_0.$$

If $(v_1,\ldots,v_m) \in L^m$ is a $K(x)$-basis of L, then $(v_1\boldsymbol{u},\ldots,v_m\boldsymbol{u}) \in L^{md}$ is a K-basis of L, and $x\,(v_1\boldsymbol{u},\ldots,v_m\boldsymbol{u}) = (v_1\boldsymbol{u},\ldots,v_m\boldsymbol{u})\,T^{(m)}$, where $T^{(m)} = \operatorname{diag}(T,\ldots,T)$ denotes the diagonal block matrix. Since $\det(T^{(m)}) = \det(T)^m$ and $\operatorname{trace}(T^{(m)}) = m\operatorname{trace}(T)$, we obtain

$$\begin{aligned}\mathsf{N}_{L/K}(x) &= \det(T^{(m)}) = ((-1)^d a_0)^m = (-1)^n a_0^m,\\ \mathsf{Tr}_{L/K}(x) &= \operatorname{trace}(T^{(m)}) = -ma_{d-1}\end{aligned}$$

and $\mathsf{N}_{L/K}(-x) = (-1)^n \mathsf{N}_{L/K}(x) = a_0^m$.

2. Let $q_0 = [K(x)\colon K]_{\mathrm{i}}$ and $H = \operatorname{Hom}^K(K(x),\overline{K})$. Then $|H| = [K(x)\colon K]_{\mathrm{s}}$,

$$q_0|H| = d, \quad \frac{q}{q_0}\,[L\colon K(x)]_{\mathrm{s}} = [L\colon K(x)]_{\mathrm{s}}\,\frac{[L\colon K]\,[K(x)\colon K]_{\mathrm{s}}}{[L\colon K]_{\mathrm{s}}\,[K(x)\colon K]} = [L\colon K(x)] = m,$$

and

$$g = \prod_{\varphi\in H}(X-\varphi(x))^{q_0} \qquad \text{by 1.4.9.}$$

Hence it follows that

$$a_{d-1} = -q_0\sum_{\varphi\in H}\varphi(x) \quad\text{and}\quad a_0 = \prod_{\varphi\in H}(-\varphi(x))^{q_0} = (-1)^d\prod_{\varphi\in H}\varphi(x)^{q_0}\,.$$

Consequently

$$\begin{aligned}\prod_{\sigma\in\operatorname{Hom}^K(L,\overline{K})}\sigma(x)^q &= \prod_{\varphi\in H}\ \prod_{\substack{\sigma\in\operatorname{Hom}^K(L,\overline{K})\\ \sigma\restriction K(x)=\varphi}}\sigma(x)^q = \prod_{\varphi\in H}\varphi(x)^{q\,[L:K(x)]_{\mathrm{s}}}\\ &= [(-1)^d a_0]^{[L:K(x)]_{\mathrm{s}}q/q_0} = (-1)^n a_0^m = \mathsf{N}_{L/K}(x)\end{aligned}$$

and

$$\begin{aligned}q\sum_{\sigma\in\operatorname{Hom}^K(L,\overline{K})}\sigma(x) &= q\sum_{\varphi\in H}\ \sum_{\substack{\sigma\in\operatorname{Hom}^K(L,\overline{K})\\ \sigma\restriction K(x)=\varphi}}\sigma(x) = q\,[L:K(x)]_{\mathrm{s}}\sum_{\varphi\in H}\varphi(x)\\ &= -\frac{q}{q_0}[L\colon K(x)]_{\mathrm{s}}a_{d-1} = -ma_{d-1} = \mathsf{Tr}_{L/K}(x)\,.\end{aligned}$$

(a) If L/K is separable, then $q=1$ and

$$\mathsf{Tr}_{L/K} = \sum_{\sigma\in\operatorname{Hom}^K(L,\overline{K})}\sigma \neq 0$$

by 1.5.2. Since $\mathsf{Tr}_{L/K}\colon L\to K$ is K-linear, it follows that $\mathsf{Tr}_{L/K}(L) = K$. If L/K is inseparable, then $\operatorname{char}(K)\,|\,q$, and therefore $\mathsf{Tr}_{L/K} = 0$.

(b) If L/K is Galois, then $q = 1$ and $\mathrm{Hom}^K(L, \overline{K}) = \mathrm{Gal}(L/K)$.

3. Let $x \in L$, $q_1 = [M:K]_\mathrm{i}$ and $q_2 = [L:M]_\mathrm{i}$. Then $q = q_1q_2$, and

$$\mathsf{N}_{L/K}(x) = \prod_{\sigma \in \mathrm{Hom}^K(L,\overline{K})} \sigma(x)^q = \prod_{\varphi \in \mathrm{Hom}^K(M,\overline{K})} \prod_{\substack{\sigma \in \mathrm{Hom}^K(L,\overline{K}) \\ \sigma \restriction M = \varphi}} \sigma(x)^q .$$

Let $\varphi \in \mathrm{Hom}^K(M, \overline{K})$, and $\widetilde{\varphi} \in \mathrm{Hom}^K(L, \overline{K})$ such that $\widetilde{\varphi} \restriction M = \varphi$. If $\sigma \in \mathrm{Hom}^K(L, \overline{K})$ and $\sigma \restriction M = \varphi$, then

$$\widetilde{\varphi}^{-1} \circ \sigma \colon L \overset{\sigma}{\to} \sigma L \overset{\widetilde{\varphi} \circ \sigma^{-1}}{\to} \widetilde{\varphi}(L) \overset{\widetilde{\varphi}^{-1}}{\to} L \quad \text{satisfies} \quad \widetilde{\varphi}^{-1} \circ \sigma \restriction M = \mathrm{id}_M .$$

Hence the assignment $\sigma \mapsto \widetilde{\varphi}^{-1} \circ \sigma$ defines a bijective map

$$\{\sigma \in \mathrm{Hom}^K(L, \overline{K}) \mid \sigma \restriction M = \varphi\} \to \mathrm{Hom}^M(L, \overline{K}),$$

and we obtain

$$\prod_{\substack{\sigma \in \mathrm{Hom}^K(L,\overline{K}) \\ \sigma \restriction M = \varphi}} \sigma(x)^q = \prod_{\psi \in \mathrm{Hom}^M(L,\overline{K})} \widetilde{\varphi} \circ \psi(x)^{q_2 q_1} = \widetilde{\varphi}\Big(\prod_{\psi \in \mathrm{Hom}^M(L,\overline{K})} \psi(x)^{q_2} \Big)^{q_1}$$

$$= \varphi(\mathsf{N}_{L/M}(x))^{q_1} .$$

Consequently,

$$\mathsf{N}_{L/K}(x) = \prod_{\varphi \in \mathrm{Hom}^K(M,\overline{K})} \varphi(\mathsf{N}_{L/M}(x))^{q_1} = \mathsf{N}_{M/K} \circ \mathsf{N}_{L/M}(x) .$$

The assertion concerning traces follows in the same way. $\square$

Next we develop the theory of resultants and discriminants for arbitrary commutative rings.

Definition 1.6.3. Let R be a commutative ring and $f,\, g \in R[X]^\bullet$ such that $\partial(f) = n$, $\partial(g) = m$ and $m + n > 0$. Suppose that

$$f = a_n X^n + a_{n-1} X^{n-1} + \ldots + a_1 X + a_0$$
$$\text{and} \quad g = b_m X^m + b_{m-1} X^{m-1} + \ldots + b_1 X + b_0,$$

where $a_0, \ldots, a_n,\, b_0, \ldots, b_m \in R$, $a_n \neq 0$ and $b_m \neq 0$. Then we get the matrix equation

$$\begin{pmatrix} X^{m-1}f \\ X^{m-2}f \\ \cdot \\ f \\ X^{n-1}g \\ X^{n-2}g \\ \cdot \\ \cdot \\ g \end{pmatrix} = \begin{pmatrix} a_n & a_{n-1} & \ldots & a_0 & 0 & \cdot & \ldots & 0 \\ 0 & a_n & \ldots & a_1 & a_0 & 0 & \ldots & 0 \\ \cdot & \cdot & \cdot & \cdot & \cdot & \cdot & \cdot & \cdot \\ 0 & 0 & \cdots & a_n & a_{n-1} & \cdot & \ldots & a_0 \\ b_m & b_{m-1} & \ldots & b_1 & b_0 & 0 & \ldots & 0 \\ 0 & b_m & \ldots & \cdot & b_1 & b_0 & \ldots & 0 \\ \cdot & \cdot & \cdot & \cdot & \cdot & \cdot & \cdot & \cdot \\ \cdot & \cdot & \cdot & \cdot & \cdot & \cdot & \cdot & \cdot \\ 0 & 0 & \ldots & \cdot & b_m & b_{m-1} & \ldots & b_0 \end{pmatrix} \begin{pmatrix} X^{n+m-1} \\ X^{n+m-2} \\ \cdot \\ X^n \\ X^{n-1} \\ X^{n-2} \\ \cdot \\ \cdot \\ 1 \end{pmatrix} .$$

The big matrix $\mathcal{A}_{f,g} \in \mathsf{M}_{m+n}(R)$ is called the resultant matrix, and its determinant

$$\mathsf{R}(f,g) = \det(\mathcal{A}_{f,g}) \in R$$

is called that **resultant** of f and g. For $f,\ g \in R^\bullet$ we set $\mathsf{R}(f,g) = 1$, and for every polynomial $h \in R[X]$ we set $\mathsf{R}(h,0) = \mathsf{R}(0,h) = 0$.

For a polynomial $f \in R[X] \setminus R$ of degree $\partial(f) = n$ we define its **discriminant** by

$$\mathsf{D}(f) = (-1)^{\binom{n}{2}}\mathsf{R}(f,f').$$

Immediately from the definition we obtain the following formulas:

- If $f = a_1X + a_0$, then $\mathsf{D}(f) = a_1$.
- If $f = a_2X^2 + a_1X + a_0$, then $\mathsf{D}(f) = a_1^2a_2 - 4a_0a_2^2$.
- If $f = a_3X^3 + a_2X^2 + a_1X + a_0$, then

$$\mathsf{D}(f) = a_1^2a_2^2a_3 - 4a_1^3a_3^2 - 4a_0a_2^3a_3 - 27a_0^2a_3^3 + 18a_0a_1a_2a_3^2.$$

- If $f = X^2 + pX + q$, then $\mathsf{D}(f) = p^2 - 4q$.
- If $f = X^3 + pX + q$, then $\mathsf{D}(f) = -4p^3 - 27q^2$.
- If $m \in \mathbb{N}$, $b \in K^\times$ and $f = X^m - b$, then $\mathsf{D}(f) = (-1)^{(m-1)(m-2)/2}m^mb^{m-1}$.

For a polynomial $f \in R[X]$ we define its (R-)**content ideal** $\mathfrak{c}(f) = \mathfrak{c}_R(f)$ to be the ideal of R generated by the coefficients of f.

Theorem 1.6.4. *Let R be a commutative ring and $f,\ g \in R[X]^\bullet$ polynomials such that $\partial(f) = n$, $\partial(g) = m$ and $m + n > 0$.*

1. *If $h \in R[X]$ and $\partial(h) < m + n$, then there exist polynomials $U,\ V \in \mathfrak{c}(h)R[X]$ such that*

$$\partial(U) < m,\ \partial(V) < n \text{ and } Uf + Vg = \mathsf{R}(f,g)\,h.$$

2. *Assume that $f = pf_0$ and $g = pg_0$ for some monic polynomial $p \in R[X] \setminus R$ and $f_0,\ g_0 \in R[X]$. Then $\mathsf{R}(f,g) = 0$.*

3. *Assume that $f = p^2f_0$ for some monic polynomial $p \in R[X] \setminus R$ and $f_0 \in R[X]$. Then $\mathsf{D}(f) = 0$.*

Proof. 1. Let $\mathcal{A}^{\#}_{f,g}$ be the adjoint matrix of the resultant matrix. Then we obtain

$$\mathcal{A}^{\#}_{f,g}\begin{pmatrix} X^{m-1}f \\ X^{m-2}f \\ \cdot \\ \cdot \\ f \\ X^{n-1}g \\ X^{n-2}g \\ \cdot \\ \cdot \\ g \end{pmatrix} = \mathcal{A}^{\#}_{f,g}\mathcal{A}_{f,g}\begin{pmatrix} X^{n+m-1} \\ X^{n+m-2} \\ \cdot \\ \cdot \\ X^{n} \\ X^{n-1} \\ X^{n-2} \\ \cdot \\ \cdot \\ 1 \end{pmatrix} = \mathsf{R}(f,g)\begin{pmatrix} X^{n+m-1} \\ X^{n+m-2} \\ \cdot \\ \cdot \\ X^{n} \\ X^{n-1} \\ X^{n-2} \\ \cdot \\ \cdot \\ 1 \end{pmatrix}.$$

If $j \in [1, m+n]$ and $(u_{j,m-1}, \ldots, u_{j,0}, v_{j,n-1}, \ldots v_{j,0})$ is the j-th row of $\mathcal{A}^{\#}_{f,g}$, then

$$\Big(\sum_{\mu=0}^{m-1} u_{j,\mu}X^{\mu}\Big)f + \Big(\sum_{\nu=0}^{n-1} v_{j,\nu}X^{\nu}\Big)g = \mathsf{R}(f,g)X^{n+m-j}.$$

If

$$h = \sum_{j=1}^{m+n} c_j X^{n+m-j} \in R[X],$$

then

$$\mathsf{R}(f,g)h = \sum_{j=1}^{m+n} c_j\Big[\Big(\sum_{\mu=0}^{m-1} u_{j,\mu}X^{\mu}\Big)f + \Big(\sum_{\nu=0}^{n-1} v_{j,\nu}X^{\nu}\Big)g\Big] = Uf + Vg$$

where

$$U = \sum_{\mu=0}^{m-1}\Big(\sum_{j=1}^{m+n} c_j u_{j,\mu}\Big)X^{\mu} \in \mathfrak{c}(h)R[X] \quad \text{and}$$

$$V = \sum_{\nu=0}^{n-1}\Big(\sum_{j=1}^{m+n} c_j v_{j,\nu}\Big)X^{\nu} \in \mathfrak{c}(h)R[X].$$

2. By 1. there exist polynomials U, $V \in R[X]$ such that $Uf+Vg = \mathsf{R}(f,g)$. Hence $\mathsf{R}(f,g) \in pR[X] \cap R = \mathbf{0}$.

3. Since $f' = p(2p'f_0 + pf_0')$, we obtain $\mathsf{D}(f) = \pm\mathsf{R}(f,f') = 0$ by 2. □

Theorem 1.6.5. *Let $\varphi\colon R \to \overline{R}$ be a homomorphism of commutative rings. For $p \in R[X]$ we denote by $\overline{p} = \varphi(p) \in \overline{R}[X]$ the image of p. Let f, $g \in R[X]^{\bullet}$, $\partial(f) = n$, $\partial(g) = m$ and $m+n > 0$. Let $a \in R^{\bullet}$ be the leading coefficient of f and $b \in R^{\bullet}$ the leading coefficient of g.*

1. *If* $\overline{a} \neq \overline{0}$ *and* $\overline{b} \neq \overline{0}$, *then* $\overline{\mathsf{R}(f,g)} = \mathsf{R}(\overline{f},\overline{g})$.
2. *If* $n \geq 1$ *and* $\overline{g} = \overline{0}$, *then* $\overline{\mathsf{R}(f,g)} = \overline{0} = \mathsf{R}(\overline{f},\overline{g})$.
3. *If* $\overline{a} = \overline{1}$ *and* $\overline{g} \neq \overline{0}$, *then* $\overline{\mathsf{R}(f,g)} = \mathsf{R}(\overline{f},\overline{g})$.
4. *If* $\overline{a} = \overline{1}$ *and* $\overline{\mathsf{D}(f)} \neq \overline{0}$, *then* $\overline{\mathsf{D}(f)} = \mathsf{D}(\overline{f}\,)$.

Proof. For a matrix $C = (c_{i,j})_{i,j}$ over R we denote by $\overline{C} = (\overline{c}_{i,j})_{i,j}$ the image matrix over $\overline{R}$.

1. If $\overline{a} \neq \overline{0}$ and $\overline{b} \neq \overline{0}$, then $\overline{\mathcal{A}_{f,g}} = \mathcal{A}_{\overline{f},\overline{g}}$, and therefore

$$\overline{\mathsf{R}(f,g)} = \overline{\det(\mathcal{A}_{f,g})} = \det(\,\overline{\mathcal{A}_{f,g}}\,) = \det(\mathcal{A}_{\overline{f},\overline{g}}) = \mathsf{R}(\overline{f},\overline{g}).$$

2. If $n \geq 1$ and $\overline{g} = \overline{0}$, then the matrix $\overline{\mathcal{A}_{f,g}}$ has a zero row, which yields the assertion.

3. Let $\overline{a} = \overline{1}$, and let $e \in [0,m]$ be maximal such that $\overline{b}_e \neq \overline{0}$, so that $e = \partial(\overline{g})$. If $m = e$, the assertion follows from 1. If $e < m$, then $\overline{\mathcal{A}_{f,g}}$ is of the form

$$\overline{\mathcal{A}_{f,g}} = \begin{pmatrix} C_{m-e} & * \\ \mathbf{0}_{n+e,m-e} & \mathcal{A}_{\overline{f},\overline{g}} \end{pmatrix}$$

with an upper triangular matrix $C_{m-e} \in \mathsf{M}_{m-e}(\overline{R})$ with nothing but $\overline{1}$ in the principal diagonal. Hence $\overline{\mathsf{R}(f,g)} = \det(\,\overline{\mathcal{A}_{f,g}}\,) = \det(\mathcal{A}_{\overline{f},\overline{g}}) = \mathsf{R}(\overline{f},\overline{g})$.

4. If $\overline{a} = \overline{1}$ and $\overline{\mathsf{D}(f)} \neq \overline{0}$, then $\mathsf{D}(f) \neq 0$ and therefore $n \geq 1$. By 2. it follows that $\overline{f'} = \overline{f}' \neq \overline{0}$, and by 3. we obtain $\overline{\mathsf{D}(f)} = \mathsf{D}(\overline{f}\,)$. □

Theorem 1.6.6. *Let R be a commutative ring and f, $g \in R[X] \setminus R$, say*

$$f = a_n \prod_{i=1}^{n} (X - \alpha_i) \quad \text{and} \quad g = b_m \prod_{j=1}^{m} (X - \beta_j),$$

where m, $n \in \mathbb{N}$, a_n, $b_m \in R^\bullet$ and $\alpha_1, \ldots, \alpha_n$, $\beta_1, \ldots, \beta_m \in R$. Then

$$\mathsf{R}(f,g) \;=\; a_n^m b_m^n \prod_{j=1}^{m}\prod_{i=1}^{n} (\alpha_i - \beta_j) \;=\; (-1)^{mn} b_m^n \prod_{j=1}^{m} f(\beta_j) \;=\; a_n^m \prod_{i=1}^{n} g(\alpha_i),$$

$$\mathsf{D}(f) \;=\; (-1)^{\binom{n}{2}} a_n^{2n-1} \prod_{\substack{i,\,\nu=1 \\ i\neq\nu}}^{n} (\alpha_i - \alpha_\nu) \;=\; a_n^{2n-1} \prod_{1\leq i<\nu\leq n} (\alpha_\nu - \alpha_i)^2,$$

and if $a_n b_m \neq 0$, then $\mathsf{D}(fg) = \mathsf{D}(f)\mathsf{D}(g)\mathsf{R}(f,g)^2$.

Proof. Let $\widetilde{R} = R[T_1, \ldots, T_n, Y_1, \ldots, Y_m]$ be the polynomial ring in the $m+n$ indeterminants $T_1, \ldots, T_n$, $Y_1, \ldots, Y_m$, set

$$F = a_n \prod_{i=1}^{n} (X - T_i) \in \widetilde{R}[X] \quad \text{and} \quad G = b_m \prod_{j=1}^{m} (X - Y_j) \in \widetilde{R}[X] .$$

In the matrix equation defining $\mathcal{A}_{F,G}$ we replace X successively with $Y_1, \ldots, Y_m$ and $T_1, \ldots, T_n$. Thereby the matrix equation results

$$\begin{pmatrix} Y_1^{m-1}F(Y_1) & \ldots & Y_m^{m-1}F(Y_m) & 0 & 0 & 0 \\ Y_1^{m-2}F(Y_1) & \ldots & Y_m^{m-2}F(Y_m) & 0 & 0 & 0 \\ \ldots & \ldots & \ldots & \ldots & \ldots & \ldots \\ F(Y_1) & \ldots & F(Y_m) & 0 & 0 & 0 \\ 0 & 0 & 0 & T_1^{n-1}G(T_1) & \ldots & T_n^{n-1}G(T_1) \\ 0 & 0 & 0 & T_1^{n-2}G(T_1) & \ldots & T_n^{n-2}G(T_n) \\ \ldots & \ldots & \ldots & \ldots & \ldots & \ldots \\ 0 & 0 & 0 & G(T_1) & \ldots & G(X_n) \end{pmatrix}$$

$$= \mathcal{A}_{F,G} \begin{pmatrix} Y_1^{n+m-1} & \ldots & Y_m^{n+m-1} & T_1^{n+m-1} & \ldots & T_n^{n+m-1} \\ Y_1^{n+m-2} & \ldots & Y_m^{n+m-2} & T_1^{n+m-2} & \ldots & T_n^{n+m-2} \\ \ldots & \ldots & \ldots & \ldots & \ldots & \ldots \\ \ldots & \ldots & \ldots & \ldots & \ldots & \ldots \\ Y_1 & \ldots & Y_m & T_1 & \ldots & T_n \\ 1 & \ldots & 1 & 1 & \ldots & 1 \end{pmatrix} .$$

Calculating determinants and underlying corresponding terms, we obtain

$$\Big(\prod_{j=1}^{m} F(Y_j)\Big)\Big(\underline{(-1)^{\binom{m}{2}} \prod_{1\le j<\mu\le m} (Y_\mu - Y_j)}\Big)\Big(\prod_{i=1}^{n} G(T_i)\Big)\Big(\underline{(-1)^{\binom{n}{2}} \prod_{1\le i<\nu\le n} (T_\nu - T_i)}\Big)$$

$$= \mathsf{R}(F,G)\Big((-1)^{\binom{n+m}{2}} \underline{\prod_{1\le j<\mu\le m} (Y_\mu - Y_j)} \; \underline{\prod_{1\le i<\nu\le n} (T_\nu - T_i)} \prod_{i=1}^{n}\prod_{j=1}^{m} (T_i - Y_j)\Big)$$

and therefore

$$a_n^m \prod_{j=1}^{m}\prod_{i=1}^{n} (Y_j - T_i)\, \underline{b_m^n \prod_{i=1}^{n}\prod_{j=1}^{m} (T_i - Y_j)} = (-1)^K\, \mathsf{R}(F,G) \underline{\prod_{i=1}^{n}\prod_{j=1}^{m} (T_i - Y_j)},$$

where

$$K = \binom{m}{2} + \binom{n}{2} + \binom{m+n}{2} \equiv mn \bmod 2 .$$

It follows that

$$\mathsf{R}(F,G) = a_n^m b_m^n \prod_{i=1}^{n}\prod_{j=1}^{m} (T_i - Y_j) = a_n^m \prod_{i=1}^{n} G(T_i) = (-1)^{mn} b_m^n \prod_{j=1}^{m} F(Y_j) .$$

For $i \in [1, n]$ we have

$$F'(T_i) = a_n \prod_{\substack{j=1 \\ j \neq i}}^{n} (T_i - T_j)$$

and therefore

$$\mathsf{D}(F) \;=\; \mathsf{R}(F, F') \;=\; a_n^{2n-1} \prod_{\substack{j=1 \\ j \neq i}}^{n} (T_i - T_j) \;=\; (-1)^{\binom{n}{2}} a_n^{2n-1} \prod_{1 \le i < \nu \le n} (T_\nu - T_i)^2 .$$

If $a_n b_m \neq 0$, then

$$\begin{aligned}\mathsf{D}(FG) &= (a_n b_m)^{2(m+n-1)} \prod_{1 \le i < \nu \le n} (T_\nu - T_i)^2 \prod_{1 \le j < \mu \le m} (Y_j - Y_\mu)^2 \prod_{i=1}^{n} \prod_{j=1}^{m} (T_i - Y_j)^2 \\ &= \mathsf{D}(F)\mathsf{D}(G)\mathsf{R}(F, G)^2 .\end{aligned}$$

Now the theorem follows by means of the homomorphism $\Phi\colon \widetilde{R} \to R$, defined by $\Phi \restriction R = \mathrm{id}_R$ and $\Phi(T_i) = \alpha_i$ for all $i \in [1, n]$ and $\Phi(Y_j) = \beta_j$ for all $j \in [1, m]$. □

Corollary 1.6.7. *Let K be a field and $f,\, g \in K[X]^\bullet$.*

1. *f and g are coprime if and only if $\mathsf{R}(f, g) \neq 0$.*
2. *g is separable if and only if $\mathsf{D}(g) \neq 0$.*

Proof. Let $\overline{K}$ be an algebraic closure of K,

$$f = a_n \prod_{i=1}^{n} (X - \alpha_i) \quad \text{and} \quad g = b_m \prod_{j=1}^{m} (X - \beta_j),$$

where $m,\, n \in \mathbb{N}$, $a_n,\, b_m \in K^\times$ and $\alpha_1, \ldots, \alpha_n,\, \beta_1, \ldots, \beta_m \in \overline{K}$.

1. By 1.1.10.2(b) f and g are coprime if and only if f and g are coprime in $\overline{K}$, that is, if and only if $\{\alpha_1, \ldots, \alpha_n\} \cap \{\beta_1, \ldots, \beta_m\} = \emptyset$, and by 1.6.6 this holds if and only if $\mathsf{R}(f, g) \neq 0$.

2. By 1.4.2.2 g is separable if and only if g and g' are coprime, and by 1. this is equivalent to $\mathsf{R}(g, g') = \pm \mathsf{D}(g) \neq 0$. □

Definition 1.6.8. Let L/K be a finite field extension and $n = [L\!:\!K]$. For an n-tuple $(u_1, \ldots, u_n) \in L^n$ we define its **discriminant** $d_{L/K}(u_1, \ldots, u_n)$ by

$$d_{L/K}(u_1, \ldots, u_n) = \det(\mathsf{Tr}_{L/K}(u_i u_j))_{i,\, j \in [1,n]} \,.$$

If L/K is inseparable, then $d_{L/K}(u_1, \ldots, u_n) = 0$ for all $(u_1, \ldots, u_n) \in L^n$.

Theorem 1.6.9. *Let L/K be a finite separable field extension, $[L:K] = n$, $\overline{K}$ an algebraically closed extension field of L and $\mathrm{Hom}^K(L, \overline{K}) = \{\sigma_1, \ldots, \sigma_n\}$.*

1. Let $(u_1, \ldots, u_n) \in L^n$ and $\alpha \in L$. Then

$$d_{L/K}(u_1, \ldots, u_n) = \det((\sigma_\nu u_i)_{\nu,\, i\in[1,n]})^2$$

and

$$d_{L/K}(\alpha u_1, \ldots, \alpha u_n) = \mathsf{N}_{L/K}(\alpha)^2 d_{L/K}(u_1, \ldots, u_n).$$

2. Let $L = K(\alpha)$ and $g \in K[X]$ the minimal polynomial of α over K. Then

$$\begin{aligned} d_{L/K}(1, \alpha, \ldots, \alpha^{n-1}) = \mathsf{D}(g) &= \prod_{1\le\nu<\mu\le n} (\sigma_\mu\alpha - \sigma_\nu\alpha)^2 \\ &= (-1)^{\binom{n}{2}}\mathsf{N}_{L/K}(g'(\alpha)) \ne 0\,. \end{aligned}$$

3. Let $\boldsymbol{u} = (u_1, \ldots, u_n)$, $\boldsymbol{v} = (v_1, \ldots, v_n) \in L^n = \mathsf{M}_{1,n}(L)$ and $\boldsymbol{u} = \boldsymbol{v}T$ for some matrix $T \in \mathsf{M}_n(K)$. Then

$$d_{L/K}(u_1, \ldots, u_n) = d_{L/K}(v_1, \ldots, v_n)\,\det(T)^2.$$

4. An n-tuple $(u_1, \ldots, u_n) \in L^n$ is a K-basis of L if and only if $d_{L/K}(u_1, \ldots, u_n) \ne 0$.

Proof. 1. If $U = (\sigma_\nu u_i)_{\nu,\, i\in[1,n]} \in \mathsf{M}_n(\overline{K})$, then

$$\begin{aligned} U^{\mathrm{t}}U &= \Bigl(\sum_{\nu=1}^n \sigma_\nu(u_i)\sigma_\nu(u_j)\Bigr)_{i,\,j\in[1,n]} \\ &= \Bigl(\sum_{\nu=1}^n \sigma_\nu(u_iu_j)\Bigr)_{i,\,j\in[1,n]} = \bigl(\mathsf{Tr}_{L/K}(u_iu_j)\bigr)_{i,\,j\in[1,n]}, \end{aligned}$$

hence $d_{L/K}(u_1, \ldots, u_n) = \det(\mathsf{Tr}_{L/K}(u_iu_j))_{i,\,j\in[1,n]} = \det(U^{\mathrm{t}}U) = \det(U)^2$, and

$$\begin{aligned} d_{L/K}(\alpha u_1, \ldots, \alpha u_n) &= (\det(\sigma_\nu(\alpha u_i))_{\nu,i\in[1,n]})^2 \\ &= \prod_{\nu=1}^n \sigma_\nu(\alpha)^2(\det(\sigma_\nu(u_i))_{\nu,i\in[1,n]})^2 \\ &= \mathsf{N}_{L/K}(\alpha)^2 d_{L/K}(u_1, \ldots, u_n). \end{aligned}$$

2. As $L = K(\alpha)$, it follows that $g = (X - \sigma_1\alpha)\cdot\ldots\cdot(X - \sigma_n\alpha)$, and by 1. we get

$$\begin{aligned} d_{L/K}(1, \alpha, \ldots, \alpha^{n-1}) &= \det\begin{pmatrix} 1 & \sigma_1\alpha & \ldots & \sigma_1\alpha^{n-1} \\ 1 & \sigma_2\alpha & \ldots & \sigma_2\alpha^{n-1} \\ . & . & \ldots & . \\ 1 & \sigma_n\alpha & \ldots & \sigma_n\alpha^{n-1} \end{pmatrix}^2 \\ &= \prod_{1\le\nu<\mu\le n} (\sigma_\mu\alpha - \sigma_\nu\alpha)^2 \\ &= \mathsf{D}(g) \ne 0 \quad \text{(Vandermonde determinant).} \end{aligned}$$

Moreover,

$$g' = \sum_{\nu=1}^{n} \prod_{\substack{i=1 \\ i\neq\nu}}^{n} (X - \sigma_i\alpha) \quad \text{implies} \quad g'(\sigma_\nu\alpha) = \prod_{\substack{i=1 \\ i\neq\nu}}^{n} (\sigma_\nu\alpha - \sigma_i\alpha) \quad \text{for all } \nu \in [1,n]\,,$$

and

$$\begin{aligned} \mathsf{N}_{L/K}(g'(\alpha)) &= \prod_{\nu=1}^{n} \sigma_\nu(g'(\alpha)) = \prod_{\nu=1}^{n} g'(\sigma_\nu\alpha) \\ &= \prod_{\nu=1}^{n} \prod_{\substack{\mu=1 \\ \mu\neq\nu}}^{n} (\sigma_\mu\alpha - \sigma_\nu\alpha) = (-1)^{\binom{n}{2}} \mathsf{D}(g)\,. \end{aligned}$$

3. Since $(\sigma_\nu u_1, \ldots, \sigma_\nu u_n) = (\sigma_\nu v_1, \ldots, \sigma_\nu v_n)\,T$ for all $\nu \in [1,n]$, we get

$$\begin{aligned} d_{L/K}(u_1, \ldots, u_n) &= (\det(\sigma_\nu u_i)_{\nu,\, i\in[1,n]})^2 = (\det(\sigma_\nu v_i)_{\nu,\, i\in[1,n]})^2 \det(T)^2 \\ &= d_{L/K}(v_1, \ldots, v_n) \det(T)^2\,. \end{aligned}$$

4. By 1.5.1, $L = K(\alpha)$ for some $\alpha \in L$. Then $(1, \alpha, \ldots, \alpha^{n-1})$ is a K-basis of L, and $d_{L/K}(1, \alpha, \ldots, \alpha^{n-1}) \neq 0$ by 2. If $(u_1, \ldots, u_n) \in L^n$, then there exists some matrix $T \in \mathsf{M}_n(K)$ such that $(u_1, \ldots, u_n) = (1, \alpha, \ldots, \alpha^{n-1})\,T$, and $(u_1, \ldots, u_n)$ is a K-basis of L if and only if $\det(T) \neq 0$. Now the assertion follows, since $d_{L/K}(u_1, \ldots, u_n) = d_{L/K}(1, \alpha, \ldots, \alpha^{n-1}) \det(T)^2$ by 3. □

Theorem and Definition 1.6.10. *Let L/K be a finite separable field extension.*

1. For every K-basis $(u_1, \ldots, u_n)$ of L there exists a unique K-basis $(u_1^*, \ldots, u_n^*)$ of L such that $\mathsf{Tr}_{L/K}(u_i u_j^*) = \delta_{i,j}$ for all $i,\, j \in [1,n]$. It satisfies

$$d_{L/K}(u_1^*, \ldots, u_n^*) = d_{L/K}(u_1, \ldots, u_n)^{-1}.$$

$(u_1^*, \ldots, u_n^*)$ is called the **dual basis** of $(u_1, \ldots, u_n)$ (with respect to $\mathsf{Tr}_{L/K}$).

2. Let $L = K(\alpha)$, $g = X^n + a_{n-1}X^{n-1} + \ldots + a_1X + a_0 \in K[X]$ the minimal polynomial of α over K, and

$$g = (X - \alpha)(\beta_0 + \beta_1 X + \ldots + \beta_{n-1}X^{n-1}),$$

where $\beta_0, \ldots, \beta_{n-1} \in L$. Then

$$\beta_i = \alpha^{n-i-1} + \sum_{\nu=1}^{n-i-1} a_{i+\nu}\alpha^{\nu-1} \quad \textit{for all} \;\; i \in [0, n-1],$$

and

$$\Big(\frac{\beta_0}{g'(\alpha)},\ldots,\frac{\beta_{n-1}}{g'(\alpha)}\Big) \quad \textit{is the dual basis of} \quad (1,\alpha,\ldots,\alpha^{n-1}).$$

Proof. 1. Let $(u_1,\ldots,u_n)$ be a K-basis of L and $(u_1^*,\ldots,u_n^*)=(u_1,\ldots,u_n)T$, where $T\in \mathsf{GL}_n(K)$. It suffices to prove that $(\mathsf{Tr}_{L/K}(u_iu_j^*))_{i,\,j\in[1,n]}=I$ if and only if $T=(\mathsf{Tr}_{L/K}(u_\nu u_\mu))_{\nu,\,\mu\in[1,n]}^{-1}$. However, if $T=(t_{i,j})_{i,\,j\in[1,n]}\in\mathsf{GL}_n(K)$ and $i,\,j\in[1,n]$, then

$$u_j^*=\sum_{\mu=1}^n u_\mu t_{\mu,j},$$

and consequently

$$\mathsf{Tr}_{L/K}(u_iu_j^*)=\sum_{\mu=1}^n \mathsf{Tr}_{L/K}(u_iu_\mu)t_{\mu,j}=\big[(\mathsf{Tr}_{L/K}(u_\nu u_\mu))_{\nu,\mu\in[1,n]}\,T\,\big]_{i,j}.$$

Hence $(\mathsf{Tr}_{L/K}(u_iu_j^*))_{i,\,j\in[1,n]}=I$ if and only if $T=\big(\mathsf{Tr}_{L/K}(u_\nu u_\mu)\big)_{\nu,\,\mu\in[1,n]}^{-1}$.

2. We set $a_n=1$ and obtain

$$\begin{aligned}g&=(X-\alpha)\sum_{i=0}^{n-1}\beta_iX^i=\sum_{\nu=0}^{n}a_\nu(X^\nu-\alpha^\nu)\\&=(X-\alpha)\sum_{\nu=1}^{n}a_\nu\sum_{i=0}^{\nu-1}\alpha^{\nu-1-i}X^i=(X-\alpha)\sum_{i=0}^{n-1}\Big(\sum_{\nu=i+1}^{n}a_\nu\alpha^{\nu-1-i}\Big)X^i\end{aligned}$$

and therefore

$$\beta_i=\sum_{\nu=i+1}^{n}a_\nu\alpha^{\nu-1-i}=\alpha^{n-i-1}+\sum_{\nu=1}^{n-i-1}a_{i+\nu}\alpha^{\nu-1}\quad\text{for all}\quad i\in[0,n-1].$$

In order to prove that

$$\mathsf{Tr}_{L/K}\Big(\alpha^i\frac{\beta_j}{g'(\alpha)}\Big)=\delta_{i,j}\quad\text{for all}\quad i,\,j\in[0,n-1],$$

we show that

$$\sum_{j=0}^{n-1}\mathsf{Tr}_{L/K}\Big(\alpha^i\frac{\beta_j}{g'(\alpha)}\Big)X^j=X^i\in K[X]\quad\text{for all}\quad i\in[0,n-1].\qquad(*)$$

Let $\overline{K}$ be an algebraic closure of L and $\mathrm{Hom}^K(L,\overline{K})=\{\sigma_1,\ldots,\sigma_n\}$. Then $\sigma_1\alpha,\ldots,\sigma_n\alpha\in\overline{K}$ are distinct and $g=(X-\sigma_1\alpha)\cdot\ldots\cdot(X-\sigma_n\alpha)$. Now it suffices to prove that

$$\sum_{j=0}^{n-1}\mathsf{Tr}_{L/K}\Big(\alpha^i\frac{\beta_j}{g'(\alpha)}\Big)\sigma_l(\alpha)^j=\sigma_l(\alpha)^i\quad\text{for all}\quad l\in[1,n]\quad\text{and}\quad i\in[0,n-1].$$

Indeed, $(*)$ asserts an equality of polynomials of degree less than n, and thus it suffices to check equality for n distinct values. Since

$$\sigma_\nu\Big(\frac{g}{X-\alpha}\Big) = \frac{g}{X-\sigma_\nu(\alpha)} = \sum_{j=0}^{n-1}\sigma_\nu(\beta_j)X^j = \prod_{\substack{k=1\\k\neq\nu}}^{n}(X-\sigma_k(\alpha)) \text{ for all } \nu\in[1,n],$$

it follows that, for all $l\in[1,n]$ and $i\in[0,n-1]$,

$$\begin{aligned}
\sum_{j=0}^{n-1}\mathsf{Tr}_{L/K}\Big(\alpha^i\frac{\beta_j}{g'(\alpha)}\Big)\sigma_l(\alpha)^j &= \sum_{j=0}^{n-1}\sum_{\nu=1}^{n}\sigma_\nu(\alpha^i)\,\frac{\sigma_\nu(\beta_j)}{g'(\sigma_\nu(\alpha))}\,\sigma_l(\alpha^j)\\
&= \sum_{\nu=1}^{n}\frac{\sigma_\nu(\alpha^i)}{g'(\sigma_\nu(\alpha))}\sum_{j=0}^{n-1}\sigma_\nu(\beta_j)\sigma_l(\alpha)^j = \sum_{\nu=1}^{n}\frac{\sigma_\nu(\alpha^i)}{g'(\sigma_\nu(\alpha))}\prod_{\substack{k=1\\k\neq\nu}}^{n}(\sigma_l(\alpha)-\sigma_k(\alpha))\\
&= \frac{\sigma_l(\alpha^i)}{g'(\sigma_l(\alpha))}\,g'(\sigma_l(\alpha)) = \sigma_l(\alpha)^i\,.
\end{aligned}$$

□

1.7 Finite fields, roots of unity, and cyclic field extensions

If E is a finite field, then $\operatorname{char}(E)=p>0$, we assume that $\mathbb{F}_p\subset E$, and if $[E\!:\!\mathbb{F}_p]=k$, then $|E|=p^k$ is a prime power.

Theorem 1.7.1 (Main theorem of finite fields). *Let E be a finite field, $|E|=q$ and* $\operatorname{char}(E)=p$. *Let $\overline{E}$ be an algebraic closure of E, and let $\phi_E\colon\overline{E}\to\overline{E}$ be defined by $\phi_E(x)=x^q$.*

1. *$E^\times$ is cyclic, $x^q=x$ for all $x\in E$, and E is a splitting field of X^q-X (over $\mathbb{F}_p$). In particular, E is (up to isomorphisms) the unique field with q elements* (henceforth denoted by $\mathbb{F}_q$), *and $\phi_E\in\operatorname{Gal}(\overline{E}/E)$.*
   ϕ_E is called the **Frobenius automorphism** over E.

2. *For every $n\in\mathbb{N}$ there exists a unique intermediate field E_n of $\overline{E}/E$ such that $[E_n\!:\!E]=n$.*
   E_n/E is cyclic, $\operatorname{Gal}(E_n/E)=\langle\phi_E\restriction E_n\rangle$, $\phi_{E_n}=\phi_E^n$ and $\mathsf{N}_{E_n/E}E_n=E$.

3. *If $m,n\in\mathbb{N}$, then*

- *$E_n\subset E_m$ if and only if $n\,|\,m$,*
- *$[E_nE_m\!:\!E]=\operatorname{lcm}(n,m)$ and $[E_n\cap E_m\!:\!E]=\gcd(n,m)$.*

Proof. 1. $E^\times$ is cyclic by 1.2.1, and $|E^\times| = q-1$ implies $x^{q-1} = 1$ for all $x \in E^\times$, and consequently $x^q = x$ for all $x \in E$. Hence $\phi_E \restriction E = \mathrm{id}_E$, $\phi_E \in \mathrm{Gal}(\overline{E}/E)$, and E is a splitting field of $X^q - X$ over $\mathbb{F}_p$.

If K is any field with q elements, we may assume that $\mathbb{F}_p \subset K$, and as K is the splitting field of $X^q - X$ over $\mathbb{F}_p$, its uniqueness up to isomorphisms follows from 1.2.8.

2. Let $n \in \mathbb{N}$ and $E_n = \{x \in \overline{E} \mid x^{q^n} = x\}$. Then E_n is an intermediate field of $\overline{E}/E$, and since $X^{q^n} - X \in \mathbb{F}_p[X]$ is separable, we obtain $|E_n| = q^n$ and $[E_n\!:\!E] = n$. If E' is any intermediate field of $\overline{E}/E$ such that $[E'\!:\!E] = n$, then $|E'| = q^n$ and thus $x^{q^n} = x$ for all $x \in E'$ by 1.,and therefore $E' = E_n$.

Since $\phi_E \restriction E = \mathrm{id}_E$ it follows that $\phi_E \restriction E_n \in \mathrm{Gal}(E_n/E)$. If $d \in \mathbb{N}$ and $x \in E_n$, then $\phi_E^d(x) = x^{q^d}$, and therefore

$$\mathrm{ord}(\phi_E \restriction E_n) = n \le |\mathrm{Gal}(E_n/E)| \le [E_n\!:\!E] = n.$$

Hence E_n/E is cyclic, and $\mathrm{Gal}(E_n/E) = \langle \phi_E \restriction E_n \rangle$. If $x \in \overline{E}$, then $\phi_{E_n}(x) = x^{q^n} = \phi_E^n(x)$, and consequently $\phi_{E_n} = \phi_E^n$.

It remains to prove that $\mathsf{N}_{E_n/E}E_n = E$. Let $E_n^\times = \langle \omega \rangle$. Then 1.6.2.2(b) implies

$$\mathsf{N}_{E_n/E}(\omega) = \prod_{\nu=0}^{n-1} \varphi_E^\nu(\omega) = \prod_{\nu=0}^{n-1} \omega^{q^\nu} = \omega^{(q^n-1)/(q-1)}.$$

As $\mathrm{ord}(\omega) = q^n - 1$, we get $\mathrm{ord}(\mathsf{N}_{E_n/E}(\omega)) = q-1$ and therefore $\langle \mathsf{N}_{E_n/E}(\omega) \rangle = E^\times$. Hence $\mathsf{N}_{E_n/E}E_n^\times = E^\times$ and thus $\mathsf{N}_{E_n/E}E_n = E$.

3. If $m \mid n$, say $n = dm$, then $x \in E_m$ implies $\phi_E^n(x) = (\phi_E^m)^d(x) = x$ and thus $x \in E_n$. Conversely, if $E_m \subset E_n$, then

$$n = [E_n\!:\!E] = [E_n\!:\!E_m]\,[E_m\!:\!E] = [E_n\!:\!E_m]\,m$$

and therefore $m \mid n$.

Let $\mathcal{K}$ be the set of all over E finite intermediate fields of $\overline{E}/E$. Then the assignment $K \mapsto [K:E]$ defines a lattice isomorphism $(\mathcal{K}, \subset) \xrightarrow{\sim} (\mathbb{N}, \mid)$. However, if E_1, $E_2 \in \mathcal{K}$, then

$$E_1 \cap E_2 = \inf\{E_1, E_2\} \text{ and } E_1E_2 = \sup\{E_1, E_2\} \text{ in } (\mathcal{K}, \subset),$$

and if m, $n \in \mathbb{N}$, then

$$\gcd(m,n) = \inf\{m,n\} \text{ and } \mathrm{lcm}(m,n) = \sup\{m,n\} \text{ in } (\mathbb{N}, \mid). \qquad \square$$

Before we proceed with the theory of roots of unity and cyclotomic field extensions, we recall from elementary number theory the basic facts concerning the Möbius mu function, the structure of prime residue class groups and the Euler phi function.

Definitions and Theorem 1.7.2. A function $f\colon \mathbb{N} \to C$ into a (multiplicative) abelian monoid C is called **multiplicative** if $f(1) = 1$ and $f(m_1m_2) = f(m_1)f(m_2)$ for all $m_1,\, m_2 \in \mathbb{N}$ such that $(m_1, m_2) = 1$.

The **Möbius mu function** $\mu\colon \mathbb{N} \to \{0, \pm 1\}$ is defined by

$$\mu(n) = \begin{cases} 1 & \text{if } n = 1, \\ (-1)^r & \text{if } r \in \mathbb{N} \text{ and } n \text{ is the product of } r \text{ distinct primes}, \\ 0 & \text{otherwise (that is, if } n \text{ is not squarefree).} \end{cases}$$

By definition, $\mu\colon \mathbb{N} \to \{0, \pm 1\}$ is a multiplicative function.

1. *Let R be a commutative ring , $f\colon \mathbb{N} \to R$ a multiplicative function and $n \in \mathbb{N}$. Then*

$$\sum_{1 \le d \,|\, n} \mu(d)f(d) = \prod_{p \,|\, n}(1 - f(p)), \qquad \sum_{1 \le d \,|\, n} \mu(d) = \begin{cases} 1 & \textit{if} \quad n = 1, \\ 0 & \textit{if} \quad n > 1, \end{cases}$$

and the summatory function $\mathsf{S}f\colon \mathbb{N} \to C$, defined by

$$\mathsf{S}f(n) = \sum_{1 \le d \,|\, n} f(d),$$

is multiplicative, too.

2. *Let C be a (multiplicative) abelian semigroup and $f,\, g\colon \mathbb{N} \to C$. Then*

$$g(n) = \prod_{1 \le d \,|\, n} f(d) \qquad \textit{for all} \quad n \in \mathbb{N}$$

if and only if

$$f(n) = \prod_{1 \le d \,|\, n} g\Big(\frac{n}{d}\Big)^{\mu(d)} \qquad \textit{for all} \quad n \in \mathbb{N}.$$

3. *If $n \in \mathbb{N}$, then*

$$\prod_{1 \le d \,|\, n} \Big(\frac{n}{d}\Big)^{\mu(d)} = \begin{cases} p & \textit{if} \quad n = p^e \quad \textit{for some prime} \quad p \quad \textit{and} \quad e \in \mathbb{N}, \\ 1 & \textit{otherwise} \end{cases}$$

Proof. 1. Let $n = p_1^{e_1} \cdot \ldots \cdot p_r^{e_r}$, where $r \in \mathbb{N}_0$, $p_1, \ldots, p_r$ are distinct primes and $e_1, \ldots, e_r \in \mathbb{N}$. Then

$$\prod_{p \,|\, n}(1 - f(p)) = \prod_{i=1}^{r}(1 - f(p_i)) = \sum_{k=0}^{r}(-1)^k \prod_{1 \le i_1 < \ldots < i_k \le r} f(p_{i_1}) \cdot \ldots \cdot f(p_{i_k})$$

$$= \sum_{k=0}^{r} \prod_{1 \le i_1 < \ldots < i_k \le r} \mu(p_{i_1} \cdot \ldots \cdot p_{i_k}) f(p_{i_1} \cdot \ldots \cdot p_{i_k}) = \sum_{1 \le d \,|\, n} \mu(d)f(d).$$

The second relation follows with $f = 1$ (recall that an empty product has value 1).

To prove that $\mathsf{S}f$ is multiplicative, let $m_1,\ m_2 \in \mathbb{N}$ such that $(m_1, m_2) = 1$. Then

$$\mathsf{S}f(m_1m_2) = \sum_{1\le d\,|\,m_1m_2} f(d) = \sum_{1\le d_1\,|\,m_1}\ \sum_{1\le d_2\,|\,m_2} f(d_1)f(d_2) = \mathsf{S}f(m_1)\mathsf{S}f(m_2).$$

2. Assume first that

$$g(n) = \prod_{1\le d\,|\,n} f(d) \quad \text{for all} \quad n\in\mathbb{N}.$$

Then we obtain

$$\prod_{1\le d\,|\,n} g\Big(\frac{n}{d}\Big)^{\mu(d)} = \prod_{1\le d\,|\,n}\ \prod_{1\le t\,|\,\frac{n}{d}} f(t)^{\mu(d)} = \prod_{1\le t\,|\,n}\ \prod_{1\le d\,|\,\frac{n}{t}} f(t)^{\mu(d)} = f(n),$$

since

$$\sum_{1\le d\,|\,\frac{n}{t}} \mu(d) = \begin{cases} 1 & \text{if} \quad t = n, \\ 0 & \text{otherwise.} \end{cases}$$

Conversely, if

$$f(n) = \prod_{1\le d\,|\,n} g\Big(\frac{n}{d}\Big)^{\mu(d)} \quad \text{for all} \quad n\in\mathbb{N},$$

then (as above)

$$\begin{aligned}\prod_{1\le d\,|\,n} f(d) &= \prod_{1\le d\,|\,n}\ \prod_{1\le t\,|\,d} g\Big(\frac{d}{t}\Big)^{\mu(t)} \underset{[d=td']}{=} \prod_{1\le t\,|\,n}\ \prod_{1\le d'\,|\,\frac{n}{t}} g(d')^{\mu(t)} \\ &= \prod_{1\le d'\,|\,n}\ \prod_{1\le t\,|\,\frac{n}{d'}} g(d')^{\mu(t)} = g(n).\end{aligned}$$

3. Define $f\colon \mathbb{N}\to\mathbb{N}$ by

$$f(n) = \begin{cases} p & \text{if} \quad n = p^e \quad \text{for some} \quad p\in\mathbb{P} \quad \text{and} \quad e\in\mathbb{N}, \\ 1 & \text{otherwise.} \end{cases}$$

Suppose that $n = p_1^{e_1}\cdot\ldots\cdot{}_r^{e_r}$, where $r\in\mathbb{N}_0,\ \ p_1,\ldots,p_r$ are distinct primes and $e_1,\ldots,e_r\in\mathbb{N}$. Then

$$\prod_{1\le d\,|\,n} f(d) = \prod_{i=1}^{r}\prod_{\nu=0}^{e_i} f(p_i^{\nu}) = \prod_{i=1}^{r} p_i^{e_i} = n, \ \text{hence} \ f(n) = \prod_{1\le d\,|\,n}\Big(\frac{n}{d}\Big)^{\mu(d)} \ \text{by 2.} \ \square$$

The **Euler phi function** $\varphi\colon \mathbb{N} \to \mathbb{N}$ is defined by

$$\varphi(m) = |(\mathbb{Z}/m\mathbb{Z})^\times| = |\{a \in [0, m-1] \mid (a, m) = 1\}|.$$

In particular, $\varphi(1) = \varphi(2) = 1$. If $m \in \mathbb{N}$, $a \in \mathbb{Z}$ and $(a, m) = 1$, then we call

$$\operatorname{ord}_m(a) = \operatorname{ord}_{(\mathbb{Z}/m\mathbb{Z})^\times}(a + m\mathbb{Z}) = \min\{f \in \mathbb{N} \mid a^f \equiv 1 \mod m\}$$

the **order of a modulo m**. Obviously, $\operatorname{ord}_m(a) \mid \varphi(m)$.

Theorem 1.7.3 (Structure of residue class groups and the Euler phi function).

1. *The Euler phi function $\varphi\colon \mathbb{N} \to \mathbb{N}$ is multiplicative.*

2. *Let p be an odd prime and $\mathbb{F}_p^\times = (\mathbb{Z}/p\mathbb{Z})^\times = \langle w + p\mathbb{Z}\rangle$ for some $w \in \mathbb{Z}$. Then*
$$(\mathbb{Z}/p^n\mathbb{Z})^\times = \langle w^{p^{n-1}}(p+1)\rangle,$$
$\operatorname{ord}_{p^n}(w^{p^{n-1}}) = p - 1$, $\operatorname{ord}_{p^n}(1+p) = p^{n-1}$, *and $(\mathbb{Z}/p^n\mathbb{Z})^\times$ is a cyclic group of order $\varphi(p^n) = p^{n-1}(p-1)$.*
An integer $w \in \mathbb{Z}$ such that $(\mathbb{Z}/p\mathbb{Z})^\times = \langle w + p\mathbb{Z}\rangle$ is called a **primitive root** modulo p.

3. *If $n \geq 2$, then $(\mathbb{Z}/2^n\mathbb{Z})^\times = \langle -1+2^n\mathbb{Z}, 5+2^n\mathbb{Z}\rangle$, $\varphi(2^n) = 2^{n-1}$,* $\operatorname{ord}_{2^n}(-1) = 2$, $\operatorname{ord}_{2^n}(5) = 2^{n-2}$, *and $(\mathbb{Z}/2^n\mathbb{Z})^\times \cong \mathbb{Z}/2\mathbb{Z} \times \mathbb{Z}/2^{n-2}\mathbb{Z}$.*

4. *If $n \in \mathbb{N}$, then*
$$\varphi(n) = \prod_{p \mid n}\Bigl(1 - \frac{1}{p}\Bigr) \quad \text{and} \quad \sum_{1 \leq d \mid n} \varphi(d) = n.$$

5. *If $m, n \in \mathbb{N}$, $d = \gcd(m, n)$ and $e = \operatorname{lcm}(m, n)$, then $\varphi(m)\varphi(n) = \varphi(d)\varphi(e)$.*

Proof. 1. If $m_1, m_2 \in \mathbb{N}$ and $(m_1, m_2) = 1$, then the Chinese remainder theorem 1.1.4 implies

$$(\mathbb{Z}/m_1m_2\mathbb{Z})^\times \cong (\mathbb{Z}/m_1\mathbb{Z})^\times \times (\mathbb{Z}/m_2\mathbb{Z})^\times,$$

and thus $\varphi(m_1m_2) = \varphi(m_1)\varphi(m_2)$.

2. Apparently $\varphi(p^n) = |\{k \in [0, p^n - 1] \mid p \nmid k\}| = p^n - p^{n-1} = p^{n-1}(p-1)$.

A. $(1+p)^{p^k} \equiv 1 + p^{k+1} \mod p^{k+2}$ for all $k \geq 0$.

Proof of **A.** By induction on k. For $k = 0$ there is nothing to do.
$k \geq 1$, $k - 1 \to k$: The induction hypothesis yields $(1+p)^{p^{k-1}} = 1 + bp^k$ for some $b \in \mathbb{Z}$ such that $b \equiv 1 \mod p$. Then $pk \geq 2k+1 \geq k+2$ and therefore

$$(1+p)^{p^k} = 1+bp^{k+1}+\sum_{i=2}^{p}\binom{p}{i}(bp^k)^i \equiv 1+bp^{k+1} \equiv 1+p^{k+1} \mod p^{k+2}. \quad \square[\mathbf{A}.]$$

By **A** we get $\operatorname{ord}_{p^n}(1+p) = p^{n-1}$, and we assert that $\operatorname{ord}_{p^n}(w^{p^{n-1}}) = p-1$. Indeed, $(w^{p^{n-1}})^{p-1} = w^{\varphi(p^n)} \equiv 1 \bmod p^n$, and if $j \in \mathbb{N}$ and $(w^{p^{n-1}})^j \equiv 1 \bmod p^n$, then $(w^{p^{n-1}})^j \equiv w^j \equiv 1 \bmod p$ (since $w^p \equiv w \bmod p$) and thus $p-1 \mid j$.

Let $\pi\colon (\mathbb{Z}/p^n\mathbb{Z})^\times \to (\mathbb{Z}/p\mathbb{Z})^\times$ be defined by $\pi(a+p^n\mathbb{Z}) = a+p\mathbb{Z}$. Then π is an epimorphism, $\operatorname{Ker}(\pi) \supset \langle (1+p)+p^n\mathbb{Z}\rangle$, and since

$$|\operatorname{Ker}(\pi)| = \frac{|(\mathbb{Z}/p^n\mathbb{Z})^\times|}{|(\mathbb{Z}/p\mathbb{Z})^\times|} = p^{n-1} = \operatorname{ord}((1+p)+p^n\mathbb{Z}),$$

we obtain $\operatorname{Ker}(\pi) = \langle (1+p)+p^n\mathbb{Z}\rangle$.

If $a \in \mathbb{Z}$ and $p \nmid a$, then $a \equiv w^j \bmod p$ for exists some $j \in [0, p-2]$. Let $w_1 \in \mathbb{Z}$ be such that $ww_1 \equiv 1 \bmod p$. Then $aw_1^{p^{n-1}j} \equiv aw_1^j \equiv 1 \bmod p$ and there exists some $k \in [0, p^{n-1}-1]$ such that $aw_1^{p^{n-1}j} \equiv (1+p)^k \bmod p^n$. It follows that $a \equiv (w^{p^{n-1}})^j (1+p)^k \bmod p^n$, hence

$$(\mathbb{Z}/p^n\mathbb{Z})^\times = \langle w^{p^{n-1}} + p^n\mathbb{Z}, (1+p)+p^n\mathbb{Z}\rangle,$$

and $(\operatorname{ord}_{p^n}(w^{p^{n-1}}), \operatorname{ord}_{p^n}(1+p)) = 1$ implies

$$(\mathbb{Z}/p^n\mathbb{Z})^\times = \langle w^{p^{n-1}}(1+p)+p^n\mathbb{Z}\rangle.$$

3. Let $n \ge 2$. Then $\varphi(2^n) = |\{a \in [0, 2^n-1] \mid a \text{ odd}\}| = 2^{n-1}$ and $\operatorname{ord}_{2^n}(-1) = 2$.

B. $5^{2^k} \equiv 1 + 2^{k+2} \bmod 2^{k+3}$ for all $k \ge 0$.

Proof of **B.** By induction on k. For $k=0$ there is nothing to do.

$k \ge 1$, $k-1 \to k$: The induction hypothesis yields $5^{2^{k-1}} = 1 + 2^{k+1}u$ for some odd $u \in \mathbb{Z}$, and thus $5^{2^k} = (1+2^{k+1}u)^2 = 1+2^{k+2}u+2^{2k+2}u^2 \equiv 1+2^{k+2} \bmod 2^{k+3}$, since $2k+2 \ge k+3$. □[**B.**]

By **B** it follows that $\operatorname{ord}_{2^n}(5) = 2^{n-2}$. Let $\pi\colon (\mathbb{Z}/2^n\mathbb{Z})^\times \to (\mathbb{Z}/4\mathbb{Z})^\times$ be the epimorphism defined by $\pi(a+2^n\mathbb{Z}) = a+4\mathbb{Z}$. Then $\operatorname{Ker}(\pi) \supset \langle 5+2^n\mathbb{Z}\rangle$, and since $|\operatorname{Ker}(\pi)| = 2^{n-2} = \operatorname{ord}_{2^n}(5)$, it follows that $\operatorname{Ker}(\pi) = \langle 5+2^n\mathbb{Z}\rangle$.

If $a \in \mathbb{Z} \setminus 2\mathbb{Z}$ and $e \in \{0,1\}$ is such that $(-1)^e a \equiv 1 \bmod 4$, then we obtain $(-1)^e a \equiv 5^k \bmod 2^n$ for some $k \in [0, 2^{n-2}-1]$, hence $a \equiv (-1)^e 5^k \bmod 2^n$, and therefore

$$(\mathbb{Z}/2^n\mathbb{Z})^\times = \langle -1+2^n\mathbb{Z}, 5+2^n\mathbb{Z}\rangle.$$

4. Suppose that $n = p_1^{e_1} \cdot \ldots \cdot p_r^{e_r}$, where $r \in \mathbb{N}_0$, $p_1, \ldots, p_r$ are distinct primes and $e_1, \ldots, e_r \in \mathbb{N}$. Then

$$\varphi(n) = \prod_{i=1}^r \varphi(p_i^{e_i}) = \prod_{i=1}^r p_i^{e_i-1}(p_i-1) = \prod_{i=1}^r p_i^{e_i} \prod_{i=1}^r \Big(1 - \frac{1}{p_i}\Big) = n \prod_{p \mid n} \Big(1 - \frac{1}{p}\Big).$$

The function $\mathsf{S}\varphi\colon \mathbb{N} \to \mathbb{N}$, defined by

$$\mathsf{S}\varphi(n) = \sum_{1 \le d \mid n} \varphi(d),$$

is multiplicative by 1.7.2.1, and thus it suffices to prove that $\mathsf{S}\varphi(p^n) = p^n$ for every prime power p^n. But

$$\mathsf{S}\varphi(p^n) = \sum_{i=0}^{n} \varphi(p^i) = 1 + \sum_{i=1}^{n} p^{i-1}(p-1) = p^n.$$

5. Let $m,\ n \in \mathbb{N}$, $d = \gcd(m,n)$ and $e = \operatorname{lcm}(m,n)$. Then $mn = de$, and

$$\{p \in \mathbb{P} \mid p \,|\, e\} = \{p \in \mathbb{P} \mid p \,|\, m\} \uplus \{p \in \mathbb{P} \mid p \,|\, n,\ p \nmid m\}.$$

Hence

$$\begin{aligned}\frac{\varphi(m)\varphi(n)}{mn} &= \prod_{p\,|\,m}\Big(1-\frac{1}{p}\Big)\prod_{p\,|\,n}\Big(1-\frac{1}{p}\Big) = \prod_{p\,|\,d}\Big(1-\frac{1}{p}\Big)\prod_{p\,|\,m}\Big(1-\frac{1}{p}\Big)\prod_{\substack{p\,|\,n\\ p\nmid m}}\Big(1-\frac{1}{p}\Big)\\ &= \prod_{p\,|\,d}\Big(1-\frac{1}{p}\Big)\prod_{p\,|\,e}\Big(1-\frac{1}{p}\Big) = \frac{\varphi(d)}{d}\,\frac{\varphi(e)}{e} = \frac{\varphi(d)\varphi(e)}{mn}.\end{aligned}$$ □

Let K be a field. An element $x \in K$ is called a **root of unity** if $x^n = 1$ for some $n \in \mathbb{N}$, and then x is called an **n-th root of unity**. The set $\mu_n(K)$ of all n-th roots of unity of K is a cyclic subgroup of $K^\times$ of order dividing n. We denote by $\mu(K)$ the group of all roots of unity of K, and for a prime p, we denote by $\mu_{p^\infty}(K)$ the group of all roots of unity of p-power order in K.

Let $n \in \mathbb{N}$. An n-th root of unity ζ is called **primitive** if $\operatorname{ord}(\zeta) = n$ and K contains a primitive n-th root of unity if and only if $|\mu_n(K)| = n$; then $\mu_n(K) = \langle\zeta\rangle$ for every primitive n-th root of unity $\zeta \in K$. If $\zeta \in K$ is a primitive n-th root of unity, $d \in \mathbb{N}$ and $d \,|\, n$, then $\zeta^{n/d}$ is a primitive d-th root of unity.

If $x \in \mu_n(K)$ and $\kappa = k + n\mathbb{Z} \in \mathbb{Z}/n\mathbb{Z}$, we set $x^\kappa = x^k$, and if $a,\ b \in \mathbb{Z}$, $(a,n) = 1$ and $k \in \mathbb{Z}$ satisfies $ka \equiv b \bmod n$, then we set $x^{b/a} = x^k$. If $\zeta \in \mu_n(K)$ is primitive, then the map $\iota\colon \mathbb{Z}/n\mathbb{Z} \to \mu_n(K)$, defined by $\iota(\kappa) = \zeta^\kappa$, is a group isomorphism, and ζ^κ is primitive if and only if $\kappa \in (\mathbb{Z}/n\mathbb{Z})^\times$.

If $\operatorname{char}(K) = p > 0$, then $\mu_{p^\infty}(K) = \{1\}$, and if $n = p^e m$ for some $e \ge 0$ and $p \nmid m$, then $\mu_n(K) = \mu_m(K)$. If $n \in \mathbb{N}$, $\operatorname{char}(K) \nmid n$ and $\overline{K}$ denotes an algebraic closure of K, then the polynomial $X^n - 1 \in K[X]$ is separable and $\overline{K}$ contains a primitive n-th root of unity.

Note that $\mu_n(\overline{\mathbb{Q}}) = \mu_n(\mathbb{C}) = \{\mathsf{e}^{2\pi\mathsf{i}k/n} \mid k \in [1,n]\} = \langle \mathsf{e}^{2\pi\mathsf{i}/n}\rangle$. The number $\mathsf{e}^{2\pi\mathsf{i}/n}$ is called the **normalized** primitive n-th root of unity.

Theorem 1.7.4. *Let K be a field.*

1. *Let $m \in \mathbb{N}$, $\zeta \in K$ a primitive m-th root of unity and $k \in \mathbb{Z}$. Then ζ^k is a primitive $m/(m,k)$-th root of unity. If $k \not\equiv 0 \bmod m$, then*

$$\sum_{j=0}^{m-1} \zeta^{kj} = 0 \quad \textit{and} \quad \Big|\sum_{j=0}^{N} \zeta^{kj}\Big| \le m \ \textit{ for all } N \in \mathbb{N}.$$

2. Let $m, n \in \mathbb{N}$, $d = \gcd(m, n)$ and $e = \operatorname{lcm}(m, n)$. Then

$$\mu_e(K) = \mu_m(K)\mu_n(K) \quad \text{and} \quad \mu_d(K) = \mu_m(K) \cap \mu_n(K).$$

In particular, if $(m, n) = 1$, then $\mu_{mn}(K) = \mu_m(K) \cdot \mu_n(K)$ (inner direct product).

Proof. 1. Clearly, $(\zeta^k)^{n/(n,k)} = (\zeta^n)^{k/(n,k)} = 1$. If $d \in \mathbb{Z}$ and $(\zeta^k)^d = 1$, then $n \mid kd$, hence

$$\frac{n}{(n,k)} \,\Big|\, \frac{k}{(n,k)}\, d, \text{ and thus } \frac{n}{(n,k)} \,\Big|\, d, \text{ since } \Big(\frac{n}{(n,k)}, \frac{n}{(n,k)}\Big) = 1.$$

Suppose now that $k \not\equiv 0 \bmod m$. Then

$$\sum_{j=0}^{m-1} \zeta^{kj} = \frac{\zeta^{km} - 1}{\zeta^k - 1} = 0.$$

If $N \in \mathbb{N}$, then $N = rm + s$, where $r, s \in \mathbb{N}_0$ and $s < m$, and we obtain

$$\begin{aligned} \Big|\sum_{j=0}^{N} \zeta^{kj}\Big| &= \Big|\sum_{\rho=0}^{r-1}\sum_{j=0}^{m-1} \zeta^{k(\rho m + j)} + \sum_{j=0}^{s} \zeta^{k(rm+j)}\Big| \\ &= \Big|\sum_{\rho=0}^{r-1}\sum_{j=0}^{m-1} \zeta^{kj} + \sum_{j=0}^{s} \zeta^{kj}\Big| \le s + 1 \le m. \end{aligned}$$

2. Clearly, $\mu_d(K) \subset \mu_m(K) \cap \mu_n(K)$ and $\mu_m(K)\mu_n(K) \subset \mu_e(K)$. Suppose that $\mu_e(K) = \langle \zeta \rangle$. Then $\mu_m(K) = \langle \zeta^{e/m} \rangle$ and $\mu_n(K) = \langle \zeta^{e/n} \rangle$. Let $k, l \in \mathbb{Z}$ be such that $d = ln + km$. Then $mn = de$ implies that

$$1 = \frac{ln}{d} + \frac{km}{d} = \frac{le}{m} + \frac{ke}{n},$$

hence $\zeta = (\zeta^{e/m})^l (\zeta^{e/n})^k \in \mu_m(K)\mu_n(K)$ and thus $\mu_e(K) \subset \mu_m(K)\mu_n(K)$. If $\xi \in \mu_m(K) \cap \mu_n(K)$, then $\xi = (\zeta^{e/m})^k$ for some $k \ge 1$ and $\xi^n = 1$. Thus it follows that $\xi^d = \zeta^{dek/m} = \zeta^{nk} = 1$ and therefore $\xi \in \mu_d(K)$. □

Theorem and Definition 1.7.5. *Let K be a field, $\overline{K}$ an algebraic closure of K and F the prime ring of K ($F = \mathbb{Z}$ if $\operatorname{char}(K) = 0$, and $F = \mathbb{F}_p$ if $\operatorname{char}(K) = p > 0$). Let $n \in \mathbb{N}$, $\operatorname{char}(K) \nmid n$ and $\mu_n = \mu_n(\overline{K})$.*

1. If $\zeta \in \mu_n$ is primitive, then $K(\zeta) = K(\mu_n)$ is a splitting field of $X^n - 1$,

$$\Phi_n = \prod_{\substack{\zeta \in \mu_n \\ \zeta \text{ primitive}}} (X - \zeta) \in F[X], \qquad \partial(\Phi_n) = \varphi(n),$$

$$X^n - 1 = \prod_{1 \le d \,|\, n} \Phi_d \qquad \textit{and} \qquad \Phi_n = \prod_{1 \le d \,|\, n} (X^{n/d} - 1)^{\mu(d)}\,.$$

The field $K^{(n)} = K(\mu_n) = K(\zeta)$ is called the **n-th cyclotomic field** over K, and the polynomial $\Phi_n \in F[X]$ is called that **n-th cyclotomic polynomial.**

2. Let $G = \mathrm{Gal}(K^{(n)}/K)$. Then there exists a unique group monomorphism $\theta \colon G \to (\mathbb{Z}/n\mathbb{Z})^\times$ such that $\sigma(\xi) = \xi^{\theta(\sigma)}$ for all $\sigma \in G$ and all $\xi \in \mu_n$.

If Φ_n is irreducible in $K[X]$, then $[K^{(n)} : K] = \varphi(n)$ and θ is an isomorphism. If $\kappa = k + n\mathbb{Z} \in (\mathbb{Z}/n\mathbb{Z})^\times$, then $\sigma_k = \theta^{-1}(\kappa)$ is given by $\sigma_k(\xi) = \xi^k$ for all $\xi \in \mu_n$.

3. The n-th cyclotomic polynomial $\Phi_n \in \mathbb{Q}[X]$ is irreducible.

4. Let K be finite, $|K| = q$, and let $f \in \mathbb{N}$ be minimal such that $q^f \equiv 1 \bmod n$. Then $[K^{(n)} : K] = f$.

Proof. 1. Since

$$\mu_n = \biguplus_{1 \le d \,|\, n} \{\xi \in \mu_n \mid \mathrm{ord}(\xi) = d\},$$

we obtain

$$X^n - 1 = \prod_{\xi \in \mu_n} (X - \xi) = \prod_{1 \le d \,|\, n} \prod_{\substack{\xi \in \mu_n \\ \mathrm{ord}(\xi) = d}} (X - \xi) = \prod_{1 \le d \,|\, n} \Phi_d$$

and therefore

$$\Phi_n = \prod_{1 \le d \,|\, n} (X^{n/d} - 1)^{\mu(d)}$$

by 1.7.2.2(a), applied for the multiplicative monoid $F[X]^\bullet$.

If $\zeta \in \mu_n$ is primitive, then $\mu_n = \langle \zeta \rangle$, and therefore $K(\zeta) = K(\mu_n)$ is a splitting field of $X^n - 1$ over K. Next we prove that $\Phi_n \in F[X]$ by induction on n. Note that $\Phi_1 = X - 1 \in F[X]$. Thus let $n > 1$ and suppose that $\Phi_d \in F[X]$ for all $d < n$. Then

$$X^n - 1 = \Phi_n \prod_{\substack{1 \le d \,|\, n \\ d < n}} \Phi_d \quad \text{and thus} \quad \Phi_n \in F[X] \ \text{ by 1.1.10.1.}$$

2. By 1.4.8.1 it follows that $K^{(n)}/K$ is Galois. If $\zeta \in \mu_n$ is primitive and $\sigma \in G$, then $\sigma(\zeta)$ is again a primitive n-th root of unity, and therefore $\sigma(\zeta) = \zeta^{\theta(\sigma)}$ for some (uniquely determined) residue class $\theta(\sigma) \in (\mathbb{Z}/n\mathbb{Z})^\times$.

If $\xi \in \mu_n$ is arbitrary, then $\xi = \zeta^m$ for some $m \in \mathbb{N}$, and therefore we obtain $\sigma(\xi) = \sigma(\zeta)^m = \zeta^{m\theta(\sigma)} = \xi^{\theta(\sigma)}$. If $\sigma, \tau \in G$ and $\xi \in \mu_n$, then

$$\xi^{\theta(\sigma\tau)} = \sigma\tau(\xi) = \sigma(\xi^{\theta(\tau)}) = \xi^{\theta(\sigma)\theta(\tau)},$$

and therefore $\theta(\sigma\tau) = \theta(\sigma)\theta(\tau)$. Hence $\theta\colon G \to (\mathbb{Z}/n\mathbb{Z})^\times$ is a group homomorphism (its uniqueness is obvious). If $\sigma \in G$ and $\theta(\sigma) = 1 + n\mathbb{Z}$, then $\sigma(\xi) = \xi^{\theta(\sigma)} = \xi$ for all $\xi \in \mu_n$, hence $\sigma = 1$. It follows that θ is a monomorphism. The remaining assertions are obvious.

3. Let $\zeta \in \mu_n(\mathbb{C})$ be primitive and $f \in \mathbb{Q}[X]$ the minimal polynomial of ζ over $\mathbb{Q}$. We assert that $\Phi_n = f$. Thus let $X^n - 1 = fh$ for some monic polynomial $h \in \mathbb{Q}[X]$. Then $f, h \in \mathbb{Z}[X]$ by Gauss's lemma 1.1.12, and it suffices to prove the following two assertions:

A. If p is a prime, $p \nmid n$, $\xi \in \mathbb{C}$ and $f(\xi) = 0$, then $f(\xi^p) = 0$.

B. If $\xi \in \mu_n(\mathbb{C})$ is primitive, then $f(\xi) = 0$.

Indeed, **B** implies $\Phi_n \mid f$ and thus $\Phi_n = f$, since f is irreducible.

Proof of **A.** We assume to the contrary that there exist a prime p such that $p \nmid n$ and some $\xi \in \mathbb{C}$ such that $f(\xi) = 0$ and $f(\xi^p) \neq 0$. Then f is the minimal polynomial of ξ, and since ξ^p is a zero of $X^n - 1$, we obtain $h(\xi^p) = 0$. As ξ is a zero of $h(X^p)$, it follows that $f \mid h(X^p)$ in $\mathbb{Q}[X]$, and if $h(X^p) = fg$, then $g \in \mathbb{Z}[X]$ by 1.1.10.1. For a polynomial $q \in \mathbb{Z}[X]$ we denote by $\overline{q} = q + p\mathbb{Z}[X] \in \mathbb{F}_p[X]$ the residue class polynomial. Then $X^n - \overline{1} = \overline{f}\,\overline{h}$, and since $\overline{a}^p = \overline{a}$ for all $a \in \mathbb{Z}$ it follows that $\overline{h(X^p)} = \overline{h}^p = \overline{f}\overline{g}$. If $\overline{\mathbb{F}}_p$ is an algebraic closure of $\mathbb{F}_p$ and $\alpha \in \overline{\mathbb{F}}_p$ is such that $\overline{f}(\alpha) = 0$, then $\overline{h}(\alpha) = 0$ and α is a multiple zero of $X^n - \overline{1}$, which contradicts the separability of $X^n - \overline{1} \in \mathbb{F}_p[X]$. □[**A.**]

Proof of **B.** We assume to the contrary that there exists some $q \in \mathbb{N}$ such that $(q, n) = 1$ and $f(\zeta^q) \neq 0$, and we assume that q is minimal with this property. As $f(\zeta) = 0$, we have $q > 1$, and we assume that $q = pr$ for some prime p and $r \in \mathbb{N}$. Then $p \nmid n$, and as $r < q$ we obtain $f(\zeta^r) = 0$. Using **A**, it follows that $f((\zeta^r)^p) = f(\zeta^q) = 0$, a contradiction.

4. We apply 1.7.1.2. For $f \in \mathbb{N}$ let K_f be the unique intermediate field of $\overline{K}/K$ such that $[K_f : K] = f$. Then $K^{(n)} = K_f$ for some $f \in \mathbb{N}$. Since $|\mu(K_f)| = q^f - 1$, it follows that $K^{(n)} \subset K_f$ if and only if $n \mid q^f - 1$. Hence $K^{(n)} = K_f$ if and only if $f \in \mathbb{N}$ is minimal such that $n \mid q^f - 1$. □

Corollary 1.7.6. *Let p be a prime and $n \in \mathbb{N}$.*

1. For every prime ring $F \neq \mathbb{F}_p$ we have

$$\Phi_{p^n} = \sum_{j=0}^{p-1} X^{p^{n-1}j} \in F[X].$$

2. If $p \neq 2$, then $\mathbb{Q}^{(p^n)}/\mathbb{Q}$ is a cyclic field extension of degree $p^{n-1}(p-1)$.

3. Let $n \geq 3$ *and* $\zeta = \mathsf{e}^{\pi \mathsf{i}/2^{n-1}} \in \mathbb{Q}^{(2^n)}$ *be the normalized primitive* 2^n*-th root of unity. Then* $\zeta_8 = \zeta^{2^{n-3}} = (1+\mathsf{i})/\sqrt{2}$ *is the normalized primitive 8-th and* $\mathsf{i} = \zeta^{2^{n-2}}$ *is a normalized primitive 4-th root of unity. For* $a \in \mathbb{Z} \setminus 2\mathbb{Z}$ *let* $\sigma_a \in \mathrm{Gal}(\mathbb{Q}^{(2^n)}/\mathbb{Q})$ *be given by* $\sigma_a(\zeta) = \zeta^a$ (see 1.7.5.2). *Then*

$$\mathrm{Gal}(\mathbb{Q}^{(2^n)}/\mathbb{Q}) = \langle \sigma_{-1}, \sigma_5 \rangle \cong \mathbb{Z}/2\mathbb{Z} \times \mathbb{Z}/2^{n-2}\mathbb{Z},$$

and $\mathbb{Q}^{(8)} = \mathbb{Q}(\zeta_8) = \mathbb{Q}(\sqrt{2}, \sqrt{-2})$ *is the fixed field of* $\langle \sigma_{5^2} \rangle$.

(a) $\mathbb{Q}^{(2^n)}$ *has 3 over* $\mathbb{Q}$ *quadratic subfields, namely*

- $\mathbb{Q}(\sqrt{2}) = \mathbb{Q}(\zeta_8 + \zeta_8^{-1})$, *the fixed field of* $\langle \sigma_{-1}, \sigma_{5^2} \rangle$,
- $\mathbb{Q}(\sqrt{-2}) = \mathbb{Q}(\zeta_8 - \zeta_8^{-1})$, *the fixed field of* $\langle \sigma_{-5} \rangle$, *and*
- $\mathbb{Q}(\mathsf{i}) = \mathbb{Q}^{(4)}$, *the fixed field of* $\langle \sigma_5 \rangle$.

(b) $\mathbb{Q}^{(2^n)}$ *has 3 subfields of degree* 2^{n-2} *over* $\mathbb{Q}$, *namely*

- $\mathbb{Q}^{(2^{n-1})} = \mathbb{Q}(\zeta^2)$, *the fixed field of* $\langle \sigma_{1+2^{n-1}} \rangle$,
- $\mathbb{Q}(\zeta + \zeta^{-1})$, *the fixed field of* $\langle \sigma_{-1} \rangle$ *(it is cyclic over* $\mathbb{Q}$, *with Galois group generated by* $\sigma_5 \restriction \mathbb{Q}(\zeta + \zeta^{-1})$*)*.
- $\mathbb{Q}(\zeta - \zeta^{-1})$, *the fixed field of* $\langle \sigma_{-1+2^{n-1}} \rangle$ *(it is cyclic over* $\mathbb{Q}$, *with Galois group generated by* $\sigma_5 \restriction \mathbb{Q}(\zeta - \zeta^{-1})$*)*.

Proof. 1. Let $\overline{K}$ be an algebraically closed field with prime ring $F \neq \mathbb{F}_p$. Then $\mu_{p^n}(\overline{K}) \setminus \mu_{p^{n-1}}(\overline{K})$ is the set of all primitive p^n-th roots of unity in $\overline{K}$, and therefore

$$\Phi_{p^n} = \frac{X^{p^n} - 1}{X^{p^{n-1}} - 1} = \sum_{j=0}^{p-1} X^{p^{n-1}j} \in F[X].$$

2. If $p \neq 2$, then 1.7.5.3 and 1.7.3.3 imply that $\mathrm{Gal}(\mathbb{Q}^{(p^n)}/\mathbb{Q}) \cong (\mathbb{Z}/p^n\mathbb{Z})^\times$ is cyclic of order $p^{n-1}(p-1)$.

3. Again by 1.7.5.3 and 1.7.3.3 it follows that

$$\mathrm{Gal}(\mathbb{Q}^{(2^n)}/\mathbb{Q}) \cong (\mathbb{Z}/2^n\mathbb{Z})^\times = \langle -1 + 2^n\mathbb{Z}, 5 + 2^n\mathbb{Z} \rangle \cong \mathbb{Z}/2\mathbb{Z} \times \mathbb{Z}/2^{n-2}\mathbb{Z}.$$

Hence 1.7.5.2 implies $\mathrm{Gal}(\mathbb{Q}^{(2^n)}/\mathbb{Q}) = \langle \sigma_{-1}, \sigma_5 \rangle$. This group contains 3 elements of order 2, namely σ_{-1}, $\sigma_{-1+2^{n-1}}$ and $\sigma_{1+2^{n-1}}$, and it contains 3 subgroups of order 2^{n-2}, namely $\langle \sigma_{-1}, \sigma_{5^2} \rangle$, $\langle \sigma_{-5} \rangle$ and $\langle \sigma_5 \rangle$. By 1.5.4, $\mathbb{Q}^{(2^n)}$ contains 3 subfields of degree 2^{n-2} and 3 subfields of degree 2 over $\mathbb{Q}$.

It is easily checked that $\zeta_8 = \zeta^{2^{n-3}} = (1+\mathsf{i})/\sqrt{2}$, $\zeta_4 = \zeta_8^2 = \mathsf{i}$, $\zeta_8 + \zeta_8^{-1} = \sqrt{2}$, $\zeta_8 - \zeta_8^{-1} = \mathsf{i}\sqrt{2} = \sqrt{-2}$ and $\mathbb{Q}^{(8)} = \mathbb{Q}(\sqrt{2}, \sqrt{-2})$. For $a \in \mathbb{Z} \setminus 2\mathbb{Z}$ let K_a be the fixed field of $\langle \sigma_a \rangle$; then $[\mathbb{Q}^{(2^n)} : K_a] = \mathrm{ord}_{2^n}(a)$.

Since $\sigma_{5^2}(\zeta_8) = \zeta_8$, it follows that $\mathbb{Q}^{(8)} \subset K_{5^2}$, and equality holds, since $[\mathbb{Q}(2^n) : \mathbb{Q}^{(8)}] = [\mathbb{Q}^{(2^n)} : K_{5^2}] = 2^{n-3}$.

$\sigma_{5^2}(\sqrt{2}) = \sigma_{-1}(\sqrt{2}) = \sqrt{2}$, hence $\mathbb{Q}(\sqrt{2}) \subset K_{5^2} \cap K_{-1}$, which is the fixed field of $\langle \sigma_{5^2}, \sigma_{-1} \rangle$, and equality holds since both fields are quadratic over $\mathbb{Q}$.

$\sigma_{-5}(\sqrt{-2}) = \zeta_8^{-5} - \zeta_8^5 = -\zeta_8^{-1} + \zeta_8 = \sqrt{-2}$, hence $\mathbb{Q}(\sqrt{-2}) \subset K_{-5}$, and equality holds since both fields are quadratic over $\mathbb{Q}$.

$\sigma_5(\mathsf{i}) = \mathsf{i}^5 = \mathsf{i}$, hence $\mathbb{Q}(\mathsf{i}) \subset K_5$, and equality holds since both fields are quadratic over $\mathbb{Q}$.

Since $\sigma_{-1}(\zeta + \zeta^{-1}) = \zeta + \zeta^{-1}$ and $\sigma_{-1+2^{n-1}}(\zeta - \zeta^{-1}) = -\zeta^{-1} + \zeta$, it follows that $\mathbb{Q}(\zeta + \zeta^{-1}) \subset K_{-1}$ and $\mathbb{Q}(\zeta - \zeta^{-1}) \subset K_{-1+2^{n-1}}$. However, as $\zeta^2 - (\zeta \pm \zeta^{-1})\zeta \mp 1 = 0$, we get $[\mathbb{Q}(\zeta) : \mathbb{Q}(\zeta \pm \zeta^{-1})] \le 2$. Hence $\mathbb{Q}(\zeta + \zeta^{-1}) = K_{-1}$ and $\mathbb{Q}(\zeta - \zeta^{-1}) = K_{-1+2^{n-1}}$.

Since $\sigma_{1+2^{n-1}}(\zeta^2) = \zeta^2$, we obtain $\mathbb{Q}^{(2^{n-1})} = \mathbb{Q}(\zeta^2) \subset K_{1+2^{n-1}}$, and since $[\mathbb{Q}^{(2^n)} : \mathbb{Q}^{(2^{n-1})}] = 2$, equality holds.

Obviously, $\mathrm{Gal}(\mathbb{Q}(\zeta + \zeta^{-1})/\mathbb{Q}) = \langle \sigma_5 \restriction \mathbb{Q}(\zeta + \zeta^{-1}) \rangle$.

Since $-1 \equiv (-1 + 2^{n-1})5^{2^{n-3}} \mod 2^n$, we get $\langle \sigma_{-1}, \sigma_5 \rangle = \langle \sigma_{-1+2^{n-1}}, \sigma_5 \rangle$, and therefore $\mathrm{Gal}(\mathbb{Q}(\zeta - \zeta^{-1})/\mathbb{Q}) = \langle \sigma_5 \restriction \mathbb{Q}(\zeta - \zeta^{-1}) \rangle$. □

Theorem 1.7.7. *Let $m, n \in \mathbb{N}$, $d = \gcd(m, n)$ and $e = \operatorname{lcm}(m, n)$. Then*

$$\mathbb{Q}^{(e)} = \mathbb{Q}^{(m)}\mathbb{Q}^{(n)} \quad \textit{and} \quad \mathbb{Q}^{(d)} = \mathbb{Q}^{(m)} \cap \mathbb{Q}^{(n)}.$$

Proof. For $k \in \mathbb{N}$, we set $\mu_k = \mu_k(\mathbb{C})$. By 1.7.4.2 we get $\mu_e = \mu_m\mu_n$ and $\mu_d = \mu_m \cap \mu_n$. Hence $\mathbb{Q}^{(e)} = \mathbb{Q}^{(m)}\mathbb{Q}^{(n)}$ and $\mathbb{Q}^{(d)} \subset \mathbb{Q}^{(m)} \cap \mathbb{Q}^{(n)}$.

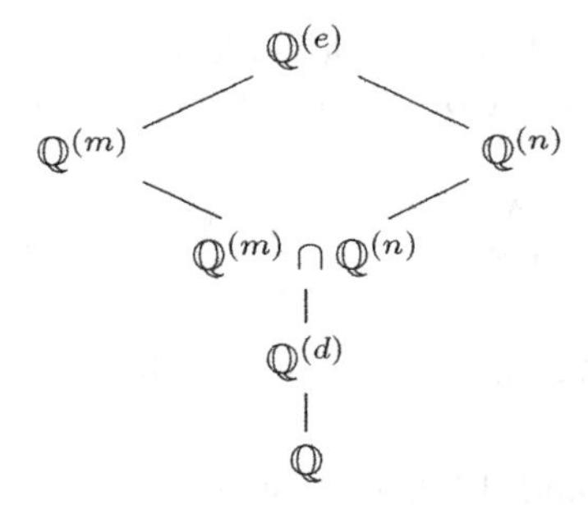

By 1.5.6.2 and 1.7.3.5 we obtain

$$[\mathbb{Q}^{(m)} : \mathbb{Q}^{(m)} \cap \mathbb{Q}^{(n)}] = [\mathbb{Q}^{(e)} : \mathbb{Q}^{(n)}] = \frac{\varphi(e)}{\varphi(n)} = \frac{\varphi(m)}{\varphi(d)} = [\mathbb{Q}^{(m)} : \mathbb{Q}^{(d)}],$$

and therefore $\mathbb{Q}^{(m)} \cap \mathbb{Q}^{(n)} = \mathbb{Q}^{(d)}$. □

Theorem 1.7.8 (Kummer extensions). *Let K be a field, $n \in \mathbb{N}$, and assume that $|\mu_n(K)| = n$.*

1. Let $a \in K^\times$, L a splitting field of $X^n - a$ over K, $\alpha \in L$ such that $\alpha^n = a$ and $G = \mathrm{Gal}(L/K)$. Then $L = K(\alpha)$,

$$X^n - a = \prod_{\zeta \in \mu_n(K)} (X - \zeta\alpha),$$

and the map

$$\chi\colon G \to \mu_n(K), \quad \text{defined by } \chi(\sigma) = \frac{\sigma(\alpha)}{\alpha} \quad \text{for all } \sigma \in G,$$

is a group monomorphism which does not depend on the choice of α. In particular, L/K is cyclic and $[L{:}K] \mid n$.
The symbol $\sqrt[n]{\alpha}$ is ambiguous and denotes any element α such that $\alpha^n = a$. We use it if (as in the above situation) the precise value does not matter. Thus we write (unambiguously) $L = K(\sqrt[n]{a})$, and we call $K(\sqrt[n]{a})/K$ a **Kummer extension**.

2. Let L/K be a cyclic field extension such that $[L : K] \mid n$. Then $L = K(\sqrt[n]{a})$ for some $a \in K$.

Proof. 1. Since $\{\zeta\alpha \mid \zeta \in \mu_n(K)\} = \{x \in L \mid x^n = a\}$, the polynomial $X^n - a$ splits as asserted, and $L = K(\alpha)$. For every $\sigma \in G$ there exists a unique root of unity $\chi(\sigma) \in \mu_n(K)$ such that $\sigma(\alpha) = \chi(\sigma)\alpha$, and $\chi(\sigma)$ does not depend on α. Indeed, if $\alpha_1 \in L$ and $\alpha_1^n = a$, then $\alpha_1 = \xi\alpha$ for some $\xi \in \mu_n(K)$, and $\sigma(\alpha_1) = \xi\sigma(\alpha) = \xi\chi(\sigma)\alpha = \chi(\sigma)\alpha_1$.

If $\sigma, \tau \in G$, then $\chi(\sigma\tau)\alpha = \sigma\tau(\alpha) = \sigma(\chi(\tau)\alpha) = \chi(\tau)\sigma(\alpha) = \chi(\tau)\chi(\sigma)\alpha$. Hence $\chi(\sigma\tau) = \chi(\sigma)\chi(\tau)$, and therefore $\chi\colon G \to \mu_n(K)$ is a group homomorphism. If $\sigma \in \mathrm{Ker}(\chi)$, then $\sigma(\alpha) = \alpha$ and thus $\sigma = \mathrm{id}_L$ since $L = K(\alpha)$. Consequently, χ is a monomorphism.

2. Suppose that $G = \mathrm{Gal}(L/K) = \langle\sigma\rangle$, $[L{:}K] = m \mid n$, and let $\xi \in \mu_m(K)$ be a primitive m-th root of unity. Since

$$\Bigl(\sum_{j=0}^{m-1} \xi^{-j}\sigma^j : L \to L\Bigr) \neq 0 \quad \text{by 1.5.2,}$$

there exists some $\beta \in L$ such that

$$\alpha = \sum_{j=0}^{m-1} \xi^{-j}\sigma^j(\beta) \neq 0.$$

It follows that

$$\sigma(\alpha) = \sum_{j=0}^{m-1} \xi^{-j}\sigma^{j+1}(\beta) = \sum_{j=1}^{m} \xi^{-j+1}\sigma^j(\beta) = \xi\alpha,$$

hence $\sigma(\alpha^m) = \alpha^m$ and therefore $a_0 = \alpha^m \in K$.

As $\alpha \in L^\times$, we get $K(\alpha) \subset L$ and suppose that $\mathrm{Gal}(L/K(\alpha)) = \langle \sigma^d \rangle$, where $d \in [0, m-1]$ and $d \,|\, m$. Then $\alpha = \sigma^d(\alpha) = \xi^d \alpha$, hence $\xi^d = 1$ and therefore $d = 0$. Thus we obtain $\mathrm{Gal}(L/K(\alpha)) = \mathbf{1}$, and consequently $L = K(\alpha) = K(\sqrt[m]{a_0}) = K(\sqrt[n]{a})$ if $a = a_0^{n/m}$. □

Next we consider the irreducibility of the polynomial $X^n - a$ without assumptions concerning roots of unity.

Theorem 1.7.9 (Capelli - Redei). *Let L/K be a field extension, $n \in \mathbb{N}$, $\alpha \in L$ and $\alpha^n = a \in K^\times$. Then $[K(\alpha) : K] = n$ (and thus the polynomial $X^n - a \in K[X]$ is irreducible) if and only if the following conditions are fulfilled:*

- *$a \notin K^p$ for all primes p dividing n.*
- *$a \notin -4K^4$ provided that $4 \,|\, n$.*

Proof. If p is a prime such that $p \,|\, n$ and $a \in K^p$, say $n = mp$ and $a = b^p$, where $m \in \mathbb{N}$ and $b \in K$, then

$$X^n - a = X^{mp} - b^p = (X^m - b) \sum_{j=0}^{p-1} X^{m(p-1-j} b^j,$$

and thus $X^n - a$ is reducible. If $4 \,|\, n$ and $a \in -4K^4$, say $n = 4m$ and $a = -4b^4$, where $m \in \mathbb{N}$ and $b \in K$, then

$$X^n - a = X^{4m} + 4b^4 = (X^{2m} + 2bX^m + 2b^2)(X^{2m} - 2bX^m + 2b^2),$$

and thus $X^n - a$ is reducible.

Assume now that the conditions for $a \in K^\times$ are fulfilled. We use induction on n to prove that $[K(\alpha) : K] = n$. For $n = 1$ there is nothing to do. Thus we assume that $n \geq 2$ and the assertion holds for all $n' < n$. We proceed in 3 steps.

I. $n = p$ is a prime. We assume to the contrary that $[K(\alpha) : K] = q < p$. Then $a^q = \mathsf{N}_{K(\alpha)/K}(a) = \mathsf{N}_{K(\alpha)/K}(\alpha^p) \in K^p$, and since $(q, p) = 1$, there exist $x, y \in \mathbb{Z}$ such that $px + qy = 1$, and consequently $a = (a^x)^p (a^q)^y \in K^p$, a contradiction.

II. $n = p^r m$ for some prime $p \neq 2$, $r \geq 1$ and $m \geq 2$ (not necessarily coprime to p). We set $\alpha_1 = \alpha^{p^r}$. Then $\alpha_1^m = a$, $[K(\alpha_1) : K] = m$ by the induction hypothesis, and $X^m - a$ is the minimal polynomial of α_1 over K. It is now sufficient to prove that $\alpha_1 \notin K(\alpha_1)^p$, since then again the induction hypothesis implies $[K(\alpha) : K(\alpha_1)] = p^r$ and $[K(\alpha) : K] = p^r m = n$.

Assume to the contrary that $\alpha_1 = \beta^p$ for some $\beta \in K(\alpha_1)$. By 1.6.2.1 we obtain $\mathsf{N}_{K(\alpha_1)/K}(\alpha_1) = (-1)^{m+1} a = \mathsf{N}_{K(\alpha_1)/K}(\beta)^p \in K^p$, and thus $a \in K^p$ (since $p \neq 2$), a contradiction.

III. $n = 2^r$ for some $r \geq 2$. We set $\alpha_1 = \alpha^{2^{r-1}}$. Then $\alpha_1^2 = a \notin K^2$, hence $[K(\alpha_1) : K] = 2$ by I., and it suffices to prove that $[K(\alpha) : K(\alpha_1)] = 2^{r-1}$. We assume to the contrary that $[K(\alpha) : K(\alpha_1)] < 2^{r-1}$. By the induction hypothesis it follows that either $\alpha_1 \in K(\alpha_1)^2$, or $r \geq 3$ and $\alpha_1 \in -4K(\alpha_1)^4$. In any case we obtain $\alpha_1 = \varepsilon\beta^2$ for some $\varepsilon \in \{\pm 1\}$ and $\beta \in K(\alpha_1)$. It follows that $-a = -\alpha_1^2 = \mathsf{N}_{K(\alpha_1)/K}(\alpha_1) = b^2$, where $b = \mathsf{N}_{K(\alpha_1)/K}(\beta) \in K^\times$, and if $j = \alpha_1 b^{-1}$, then $j^2 = -1$ and $K(\alpha_1) = K(j) = K + Kj$.

Since $0 = \alpha^{2^r} - a = (\alpha^{2^{r-1}} + jb)(\alpha^{2^{r-1}} - jb)$, we obtain $\alpha^{2^{r-1}} = \pm jb$. Since $[K(\alpha){:}K(\alpha_1)] < 2^{r-1}$, the induction hypothesis implies either $\pm jb \in K(\alpha_1)^2$, or $r \geq 3$ and $\pm jb \in -4K(\alpha_1)^4$. Since $-4K(\alpha_1)^4 \subset K(\alpha_1)^2$, it follows that $\pm jb = \gamma^2$ for some $\gamma \in K(\alpha_1) = K + Kj$, say $\gamma = u + vj$ for some $u, v \in K$. Then $\pm jb = u^2 - v^2 + 2uvj$, hence $u^2 - v^2 = 0$ and $\pm b = 2uv$. Finally this implies $a = -b^2 = -4u^2v^2 = -4u^4 \in -4K^4$, a contradiction. □

Theorem 1.7.10 (Artin - Schreier). *Let K be a field,* $\mathrm{char}(K) = p > 0$, *$\mathbb{F}_p \subset K$ and $\wp(X) = X^p - X \in \mathbb{F}_p[X]$.*

1. *Let $a \in K$, L a splitting field of $\wp(X) - a$ over K, $x \in L$ such that $\wp(x) = a$ and $G = \mathrm{Gal}(L/K)$. Then $L = K(x)$,*

$$\wp(X) - a = \prod_{c \in \mathbb{F}_p} (X - x - c),$$

and the map $\chi\colon G \to \mathbb{F}_p$, defined by $\chi(\sigma) = \sigma x - x$, is a group monomorphism which does not depend on the choice of x. In particular, L/K is cyclic, and if $x \notin K$, then $\wp(X) - a \in K[X]$ is irreducible and $[L{:}K] = p$.

If $L = K(x)$ as above, we set $L = K(\wp^{-1}a)$, and we call $K(\wp^{-1}a)/K$ an **Artin-Schreier extension**.

2. *Let L/K be a cyclic field extension of degree p. Then there exists some $a \in K$ such that $L = K(\wp^{-1}a)$.*

Proof. 1. If $c \in \mathbb{F}_p$, then $c^p = c$ and therefore $\wp(x + c) = \wp(x) = a$. Hence the polynomial $\wp(X) - a$ splits as asserted, and $L = K(x)$. For every $\sigma \in G$ there exists a unique $\chi(\sigma) \in \mathbb{F}_p$ such that $\sigma(x) = x + \chi(\sigma)$, and $\chi(\sigma)$ does not depend on x. Indeed, if $x_1 \in L$ and $\wp(x_1) - a = 0$, then $x_1 = x + b$ for some $b \in \mathbb{F}_p$, and $\sigma(x_1) = \sigma(x) + b = x + \chi(\sigma) + b = x_1 + \chi(\sigma)$.

If $\sigma, \tau \in G$, then $x + \chi(\sigma\tau) = \sigma\tau(x) = \sigma(x + \chi(\tau)) = \sigma(x) + \chi(\tau) = x + \chi(\sigma) + \chi(\tau)$, hence $\chi(\sigma\tau) = \chi(\sigma) + \chi(\tau)$, and $\chi\colon G \to \mathbb{F}_p$ is a homomorphism. Since $\mathbb{F}_p$ is cyclic of order p, it follows that either $\chi = 0$ or χ is an isomorphism. If $x \in K$, then $L = K$, $G = \mathbf{1}$ and $\chi = 0$. If $x \notin K$, then $L \neq K$, χ is an isomorphism, $[L{:}K] = p$, and $\wp(X) - a$ is the minimal polynomial of x over K.

2. Suppose that $G = \mathrm{Gal}(L/K) = \langle\sigma\rangle$. By 1.5.2 we obtain

$$\Big(\sum_{j=0}^{p-1} \sigma^j : L \to L\Big) \neq 0.$$

Let $y \in L$ such that

$$c = \sum_{j=0}^{p-1} \sigma^j(y) \in L^\times \quad \text{and set} \quad x = \frac{1}{c}\sum_{j=0}^{p-1} j\sigma^j(y) \in L.$$

Then

$$x - \sigma x = \frac{1}{c}\Big[\sum_{j=0}^{p-1} j\sigma^j y - \sum_{j=1}^{p} (j-1)\sigma^j y\Big] = \frac{1}{c}\Big[\sigma y + y + \sum_{j=2}^{p-1} \sigma^j y\Big] = 1.$$

Now $\sigma x = x - 1 \neq x$ implies $x \notin K$, and $\sigma\wp(x) = \wp(\sigma x) = \wp(x+1) = \wp(x)$ implies $\wp(x) \in K$, and consequently $L = K(x) = K(\wp^{-1}a)$. □

In Exercise **9** we show how to generalize Theorem 1.7.10 for cyclic field extensions of degee p^n for $n > 1$.

1.8 Transcendental field extensions

In this section we give a brief overview of Steinitz's theory of general field extensions without proofs. For a detailed treatment and proofs we refer to [32, Vol. II, Ch. 8.12] and [3, Ch. V, §14].

Definition 1.8.1. Let L/K be a field extension.

1. A family $\boldsymbol{x} = (x_i)_{i\in I}$ in L is called **algebraically independent** over K if for every equally indexed family of indeterminates $\boldsymbol{X} = (X_i)_{i\in I}$ over K and every polynomial $f \in K[\boldsymbol{X}]$ a relation $f(\boldsymbol{x}) = 0$ implies $f = 0$. In addition, $\boldsymbol{x}$ is called a **transcendence basis** of L/K if $L/K(\boldsymbol{x})$ is algebraic.

2. A subset B of L is called **algebraically independent** over K if the family $(b)_{b\in B}$ is algebraically independent over K, and it is called a **transcendence basis** of L/K if $L/K(B)$ is algebraic.

By definition, a subset B of L is algebraically independent over K if and only if every finite subset is algebraically independent over K. The extension L/K is algebraic if and only if the empty set is a transcendence basis of L/K.

Theorem 1.8.2. *Let L/K be a field extension.*

1. Let B_0 be an over K algebraically independent subset of L. Then there exists a transcendence basis B of L/K such that $B_0 \subset B$.

2. Let B_1 be a subset of L such that $L = K(B_1)$. Then there exists a transcendence basis B of L/K such that $B \subset B_1$.

3. There exists a transcendence basis of L/K, and any two transcendence bases of L/K have the same cardinality.

Definition 1.8.3. Let L/K a field extension and B a transcendence basis of L/K. Then

$$\mathsf{tr}(L/K) = |B| \in \mathbb{N}_0 \cup \{\infty\}$$

is called the **transcendence degree** of L/K.

By definition, a field extension L/K is algebraic if and only if $\mathsf{tr}(L/K) = 0$.

Theorem 1.8.4. *Let L/K be a field extension and M an intermediate field of L/K. Then $\mathsf{tr}(L/K) = \mathsf{tr}(M/K) + \mathsf{tr}(L/M)$.*

1.9 Exercises for Chapter 1

1. Let K be a field, $f \in K[X]$, $n = \partial(f) \ge 1$ and L a splitting field of f over K. Then $[L\colon K] \le n!$.

2. Let L/K be a finite Galois extension, $f \in K[X]$ an irreducible polynomial and $f = f_1 \cdot \ldots \cdot f_n$ with irreducible polynomials $f_1, \ldots, f_n \in L[X]$. Then the polynomials $f_1, \ldots, f_n$ all have the same degree.

3. Let L/K be a quadratic field extension.

a. If $\mathrm{char}(K) \ne 2$, then $L = K(\sqrt{a})$ for some $a \in K^\times$, and at it the coset $aK^\times \in K^\times/K^{\times 2}$ is uniquely determined by L.

b. If $\mathrm{char}(K) = 2$, then $L = K(\wp^{-1}(a))$ for some $a \in K$, and at it the coset $a + \wp K \in K/\wp K$ is uniquely determined by L.

4. This exercise provides an alternative proof of the existence of an algebraic closure. Let K be a field and P the set of all monic irreducible polynomials in $K[X]$.

a. Let L/K be an algebraic field extension such that every $f \in P$ has a root in L. Then L is an algebraic closure of K (the separable case is easy, use 1.5.1).

b. Let $\boldsymbol{X} = (X_f)_{f \in P}$ be a family of indeterminates over K. Then $K[\boldsymbol{X}]$ contains a maximal ideal $\mathfrak{m}$ such that $\{f(X_f) \mid f \in P\} \subset \mathfrak{m}$, and we set $K^* = K[\boldsymbol{X}]/\mathfrak{m}$. The map $i\colon K \mapsto K^*$, $x \mapsto x + \mathfrak{m}$, is a monomorphism, and if we identify K with $i(K)$, then K^* is an algebraic closure of K.

5. This exercise provides an explicit form of the Chinese remainder theorem. Let $m_1, \ldots, m_n \in \mathbb{N}$ be pairwise coprime, $m = m_1 \cdot \ldots \cdot m_n$, and for $i \in [1, n]$ let $m = m_i m_i^*$. Then $(m_1^*, \ldots, m_n^*) = 1$, and we fix $a_1, \ldots, a_n \in \mathbb{Z}$ such that $a_1 m_1^* + \ldots + a_n m_n^* = 1$. Then the maps

$$\Psi\colon \prod_{i=1}^{n} \mathbb{Z}/m_i\mathbb{Z} \to \mathbb{Z}/m\mathbb{Z} \quad \text{and} \quad \Theta\colon \mathbb{Z}/m\mathbb{Z} \to \prod_{i=1}^{n} \mathbb{Z}/m_i\mathbb{Z},$$

defined by

$$\Psi(t_1 + m_1\mathbb{Z}, \ldots, t_n + m_n\mathbb{Z}) = \sum_{i=1}^{n} t_i m_i^* + m\mathbb{Z} \quad \text{for all} \quad (t_1, \ldots, t_n) \in \mathbb{Z}^n$$

and

$$\Theta(x + m\mathbb{Z}) = (a_1 x + m_1\mathbb{Z}, \ldots, a_n x + m_n\mathbb{Z}) \quad \text{for all} \quad x \in \mathbb{Z},$$

are mutually inverse isomorphisms of additive groups, and

$$\Psi\Big(\prod_{i=1}^{n} (\mathbb{Z}/m_i\mathbb{Z})^\times\Big) = (\mathbb{Z}/m\mathbb{Z})^\times.$$

6. Let K be a field, $\operatorname{char}(K) \neq 2$, $f \in K[X] \setminus K$ a separable polynomial, $\partial(f) = n$ and L a splitting field of f over K.

a. $\sqrt{\mathsf{D}(f)} \in L$, and $\sigma\sqrt{\mathsf{D}(f)} = \operatorname{sgn}(\sigma)\sqrt{\mathsf{D}(f)}$ for all $\sigma \in \operatorname{Gal}_K(f)$. In particular, $\mathsf{D}(f) \in K^2$ if and only if $\operatorname{Gal}_K(f) \subset \mathfrak{A}_n$.

b. Let f be irreducible. If L/K is cyclic and $[L:K] \equiv 0 \bmod 2$, then $\mathsf{D}(f) \notin K^2$.

c. Let f be irreducible, $\partial(f) = 3$, $L = K(\alpha)$ and $f(\alpha) = 0$. Then $L(\sqrt{\mathsf{D}(f)})$ is a splitting field of f over K.

7. Let K be a field, $\operatorname{char}(K) \neq 2$ and $d \in K \setminus K^2$. Let $K_1 = K(\sqrt{d})$, $\alpha = a + b\sqrt{d} \in K_1 \setminus K_1^2$ (where $a, b \in K$) and $K_2 = K_1(\sqrt{\alpha})$. Set $\alpha' = a - \sqrt{d}$ and $r = \mathsf{N}_{K_1/K}(\alpha)$.

a. K_2/K is Galois if and only if $r \in K_1^2$.

b. Let $r \in K_1^2$. Then K_2/K is cyclic if and only if $r \notin K^2$.

c. Assume that $-1 \notin K^2$. Then there exists a cyclic field extension L/K of degree 4 such that $K_1 \subset L$ if and only if $K^2 + K^2 \neq K^2$. More precisely if $b \in K^2 + K^2 \setminus K^2$, then $K(\sqrt{b + \sqrt{b}})$ is a cyclic field extension of degree 4 of K.

8. Witt vectors (see [57, Ch. V]). Let p be a prime, and let $(X_n)_{n \geq 0}$, $(Y_n)_{n \geq 0}$ sequences of indeterminates. For $n \in \mathbb{N}_0$ let

$$W_n = W_n(X_0, \ldots, X_n) = \sum_{i=0}^{n} p^i X_i^{p^{n-i}} \in \mathbb{Z}[X_0, \ldots, X_n].$$

In particular, $W_0 = X_0$, $W_1 = X_0^p + pX_1, \ldots$ and

$$W_n(X_0, \ldots, X_n) = W_{n-1}(X_0^p, \ldots, X_{n-1}^p) + p^n X_n \quad \text{for all } n \in \mathbb{N}.$$

Let R be a commutative ring, and let $\mathsf{W}(R)$ be the set of all sequences $x = (x_n)_{n \geq 0}$ in R. For a sequence $x \in \mathsf{W}(R)$ we define the associated **ghost sequence** $x^{(\cdot)} = (x^{(n)})_{n \geq 0}$ by $x^{(n)} = W_n(x_0, \ldots, x_n)$ for all $n \geq 0$.

a. Let $x, y \in \mathsf{W}(R)$, $n \in \mathbb{N}_0$ and $r \in \mathbb{N}$. Then $x_i \equiv y_i \mod p^r R$ for all $i \in [0, n]$ if and only if $x^{(i)} \equiv y^{(i)} \mod p^{r+i} R$ for all $i \in [0, n]$.

b. If $p \in R^\times$, then the map $(\cdot)\colon \mathsf{W}(R) \to \mathsf{W}(R)$, $x \mapsto x^{(\cdot)}$ is bijective.

c. For every $n \in \mathbb{N}_0$ there exist uniquely determined polynomials

$$S_n,\ P_n \in \mathbb{Z}[X_0, \ldots, X_n, Y_0, \ldots, Y_n]$$

such that

$$W_n(S_0, \ldots, S_n) = W_n(X_0, \ldots, X_n) + W_n(Y_0, \ldots, Y_n)$$

and

$$W_n(P_0, \ldots, P_n) = W_n(X_0, \ldots, X_n) W_n(Y_0, \ldots, Y_n)$$

for all $n \in \mathbb{N}_0$. Hint: Use **b** with $R = p^{-1}\mathbb{Z}[X_0, \ldots, X_n, Y_0, \ldots, Y_n]$ and then **a** to get rid of the denominators.
For sequences $x, y \in \mathsf{W}(R)$ we define the $x + y \in \mathsf{W}(R)$ and $x \cdot y \in \mathsf{W}(R)$ by

$$(x+y)_n = S_n(x_0, \ldots, x_n, y_0, \ldots, y_n) \quad \text{and} \quad (x \cdot y)_n = P_n(x_0, \ldots, x_n, y_0, \ldots, y_n),$$

or, equivalently, $(x+y)^{(n)} = x^{(n)} + y^{(n)}$ and $(x \cdot y)^{(n)} = x^{(n)} y^{(n)}$. In particular, $(x+y)_0 = x_0 + y_0$, $(x \cdot y)_0 = x_0 y_0$, $(x+y)_1 = x_1 + y_1 + [x_0^p + y_0^p - (x_0 + y_0)^p]/p$ and $(x \cdot y)_1 = p x_1 y_1 + x_0^p y_1 + x_1 y_0^p$.

Then $\mathsf{W}(R) = (\mathsf{W}(R), +, \cdot)$ is a commutative ring, called the **Witt ring** over R.

The embedding $\sigma\colon R \to \mathsf{W}(R)$, defined by $\sigma(a) = (a, 0, 0, \ldots)$, satisfies $\sigma(ab) = \sigma(a) \cdot \sigma(b)$ for all $a, b \in R$. Every ring homomorphism $S \to R$ induces (in the natural way) a ring homomorphism $\mathsf{W}(S) \to \mathsf{W}(R)$. In particular, if S is a subring of R, then $\mathsf{W}(S)$ is a subring of $\mathsf{W}(R)$.

d. Let R be a ring and $p = \text{char}(R)$ a prime. Define $\mathsf{V}, \mathsf{F}\colon \mathsf{W}(R) \to \mathsf{W}(R)$ by

$$\mathsf{V}(x)_n = \begin{cases} 0 & \text{if } n = 0, \\ x_{n-1} & \text{if } n > 0 \end{cases} \quad \text{and} \quad \mathsf{F}(x)_n = x_n^p \text{ for all } n \geq 0 \text{ and } x \in \mathsf{W}(R).$$

We call V the shift ("Verschiebung") and F the Frobenius. These maps satisfy

$$\mathsf{F}(x+y) = \mathsf{F}(x) + \mathsf{F}(y), \quad \mathsf{F}(x \cdot y) = \mathsf{F}(x) \cdot \mathsf{F}(y), \quad \mathsf{V}(x+y) = \mathsf{V}(x) + \mathsf{V}(y)$$

and

$$\mathsf{VF}(x) = \mathsf{FV}(x) = (0, x_0^p, x_1^p, \ldots) = px \ (= x + \ldots + x).$$

In particular, $\mathrm{char}(\mathsf{W}(R)) \neq p$.
Hint: R is the epimorphic image of a suitable polynomial domain $\mathbb{Z}[\boldsymbol{X}]$, and $\mathsf{W}(\mathbb{Z}[\boldsymbol{X}])$ is a subring of $\mathsf{W}(\mathbb{Q}[\boldsymbol{X}])$. Hence it suffices to prove the relations in $\mathsf{W}(\mathbb{Q}[\boldsymbol{X}])$, and there we have $x = y$ if and only if $x^{(\cdot)} = y^{(\cdot)}$.

9. Witt vectors and Galois theory (Exercise **8** continued, see [57, Ch. VI.4]). Let K be a field, $\mathrm{char}(K) = p > 0$ and $m \in \mathbb{N}$.

a. $J_m = \{x \in \mathsf{W}(K) \mid x_0 = x_1 = \ldots = x_{m-1} = 0\}$ is an ideal of $\mathsf{W}(K)$, and

$$\mathsf{W}_m(K) = \mathsf{W}(K)/J_m = \{(x_0, \ldots, x_{m-1}) + J_m \mid x_0, \ldots, x_{m-1} \in K\}.$$

There is a monomorphism $\mathbb{Z}/p^m\mathbb{Z} \to \mathsf{W}_m(K)$, and $\mathsf{W}_m(\mathbb{F}_p) \cong \mathbb{Z}/p^m\mathbb{Z}$.

b. Let L/K be a finite field extension. Then the inclusion $K \hookrightarrow L$ induces a monomorphism $\mathsf{W}_m(K) \to \mathsf{W}_m(L)$. We identify $\mathsf{W}_m(K) \subset \mathsf{W}_m(L)$. For $y \in \mathsf{W}_m(L)$ and $\sigma \in \mathrm{Gal}(L/K)$ we define $\wp(y) \in \mathsf{W}_m(L)$ by $\wp(y)_n = \wp(y_n)$ and $\sigma(y)_n = \sigma(y_n)$. If $y,\ y' \in \mathsf{W}_m(L)$, then $\wp y = \wp y'$ if and only if $y' \in y + \mathsf{W}_m(\mathbb{F}_p)$.

c. Let $\overline{K}$ be an algebraic closure of K and $x \in \mathsf{W}_m(K)$. Then there exists some $y \in \mathsf{W}_m(\overline{K})$ such that $\wp(y) = x$. Hint: Induction on m using the identity

$$\wp(y_0, \ldots, y_{m-2}, X) = (\wp(y_0), \ldots, \wp(y_{m-2}), X^p - X + \gamma) \quad \text{for some } \ \gamma \in K(y_0, \ldots, y_{m-2}).$$

d. Let $x \in \mathsf{W}_m(K)$, $y = (y_0, \ldots, y_{m-1}) + J_m \in \mathsf{W}_m(\overline{K})$ such that $\wp(y) = x$ and $L = K(y) = K(y_0, \ldots, y_{m-1})$. Then L/K is a cyclic field extension, and the map $\varphi\colon \mathrm{Gal}(L/K) \to \mathsf{W}_m(\mathbb{F}_p)$, defined by $\varphi(\sigma) = \sigma y - y \in \mathsf{W}_m(\mathbb{F}_p)$ is a monomorphism (compare with 1.7.10).

10. Let K be a field, $\mathrm{char}(K) \neq 2,\, 3$ and $a \in K \setminus K^3$. Then $K(\sqrt{-3}, \sqrt[3]{a})$ is the splitting field of $X^3 - a$ over K.

11. Let K be a field, $\mathrm{char}(K) \neq 2$, $r \in \mathbb{N}$ and L/K a Galois field extension such that $\mathrm{Gal}(L/K) \cong (\mathbb{Z}/2\mathbb{Z})^n$. Then

$$L = K(\sqrt{a_1}, \ldots, \sqrt{a_n}) = K(\sqrt{a_1} + \ldots + \sqrt{a_n}) \text{ for some } a_1, \ldots, a_n \in K^\times.$$

At it the subgroup $\langle x_1, \ldots, x_n \rangle K^{\times 2}$ of $K^\times / K^{\times 2}$ is uniquely determined by L.

12. Let L/K be a field extension and $x,\ y \in L^\times$ such that x is separable and y is purely inseparable over K. Then $K(x, y) = K(x + y) = K(xy)$.

13. Let $m \in \mathbb{N}$.

a. If $m \equiv 1 \bmod 2$, then $\Phi_{2m}(X) = \Phi_m(-X)$.

b. If p is a prime, then

$$\Phi_m(X^p) = \begin{cases} \Phi_m(X)\Phi_{mp}(X) & \text{if } \ p \nmid m, \\ \Phi_{mp}(X) & \text{if } \ p \mid m. \end{cases}$$

c. If $n > 1$, then $\Phi_n(X^{-1})X^{\varphi(n)} = \Phi_n(X)$.

14. Let q be a prime power and $f \in \mathbb{F}_q[X]$ an irreducible polynomial. Then $f \mid X^{q^n} - 1$ if and only if $\partial(f) \mid n$.

15. Let p be a prime, $n \in \mathbb{N}$, and for $a \in \mathbb{Z} \setminus p\mathbb{Z}$ let $\mathrm{ord}_p(a)$ be the order of $a + p\mathbb{Z}$ in $\mathbb{F}_p^\times$. Then $\Phi_n(a) \equiv 0 \bmod p$ if and only if $\mathrm{ord}_p(a) = n$.

16. Let $n \in \mathbb{N}$, and let H be a proper subgroup of $(\mathbb{Z}/n\mathbb{Z})^\times$. Then there exist infinitely many primes p such that $p + n\mathbb{Z} \notin H$. Hint: Imitate Euclid's proof for the existence of infinitely many primes.

17. Let L/K be a field extension, let $f,\, g \in K[X]$ be irreducible and $\alpha,\, \beta \in L$ such that $f(\alpha) = g(\beta) = 0$. Then

$$\mathsf{N}_{K(\alpha)/K}(g(\alpha)) = (-1)^{\partial(f)\partial(g)}\mathsf{N}_{K(\beta)/K}(f(\beta)).$$

18. Let K be a field, p a prime, $\mathrm{char}(K) \ne p$, $L = K(\zeta)$, where ζ is a primitive p-th root of unity, $[L\!:\!K] = m$, $\mathrm{Gal}(L/K) = \langle\tau\rangle$ and $\tau(\zeta) = \zeta^g$. Let $a \in L \setminus L^p$ and $N = L(\sqrt[p]{a})$. Then N/K is Galois if and only if $\tau(a) = a^r c^p$ for some $r \in [1, p-1]$ and $c \in L^\times$, and then

$$\mathrm{Gal}(L/K) = \langle \sigma, \tau \mid \sigma^p = \tau^m = 1,\ \sigma\tau = \tau\sigma^k\rangle, \quad \text{where } kg \equiv r \bmod p.$$

In particular, N/K is abelian (and then cyclic) if and only if $g \equiv r \bmod p$.

19. Let K be a field, $\mathrm{char}(K) \ne 2$, $m \in \mathbb{N}$, ζ a primitive 2^m-th root of unity and $a \in K \setminus K^2$. Then $L = K(\zeta,\, \sqrt[2^m]{a})$ is a splitting field of $X^{2^m} - a$ over K. Determine $[L\!:\!K]$ (there are special cases: $a \in -4K^4$ and $\pm 2 \in K$).

20. Let $\boldsymbol{X} = (X_i)_{i \in I}$ be a non-empty family of indeterminates over R.

a. $R[\boldsymbol{X}]$ is not a field.

b. $R[\boldsymbol{X}]$ is a principal ideal domain if and only if R is a field and $|I| = 1$.

c. If $\mathfrak{o}$ is a domain and $K = \mathsf{q}(\mathfrak{o})$, then $\mathfrak{o}[\boldsymbol{X}]$ is a domain, $\mathfrak{o}[\boldsymbol{X}]^\times = \mathfrak{o}^\times$ and $\mathsf{q}(\mathfrak{o}[\boldsymbol{X}]) = K(\boldsymbol{X})$.

2

Dedekind Theory

This chapter contains the ring-theoretic tools for R. Dedekind's foundation of algebraic number theory. The main topics are contained in almost every book on algebraic number theory or multiplicative ideal theory; in particular, I recommend [39], [40], [18], [41], [50], [59], [64]. In all these volumes R. Dedekind's ideal theory is only one topic among others, and an interested reader will find there a wealth of valuable additional material.

The first Section 2.1 highlights the purely multiplicative character of elementary arithmetic (for a thorough treatment of multiplicative ideal theory in this vein we refer to [21]). The core of this chapter are Sections 2.8, 2.10, 2.12 and 2.13, which contain the classical theory of Dedekind domains and their extensions. Sections 2.2 to 2.8 are devoted to ring- and module-theoretic preliminaries (such as factorial domains, modules over principal ideal domains, integral ring extensions, fractional and invertible ideals) in order to keep the presentation self-contained. In particular, I point out the various versions of Gauss's lemma in Section 2.2 and the Galois-theoretic aspects of integral ring extensions in Section 2.5.

Besides the basic concepts we deal with some special topics with number-theoretical relevance. In Section 2.9 we introduce and investigate congruences with signatures and ray class groups, and in Section 2.11 we deal with the ideal theory of (non-principal) orders and connect it with the theory of ray class groups. This connection is the basis for the prime ideal theorem in non-principal orders of algebraic number fields which we shall prove in 4.4.4. We refer to [17] and [20] for preliminary versions of the material of these two sections in the present abstract form. In 2.12.9 and 2.12.10 we present the chararacterization of conductor ideals due to C. Prabpayak and G. Lettl [53] based on a classical paper of P. Furtwängler [16].

Finally in 2.14 we investigate ideal norms and introduce the fundamental concept of Frobenius automorphisms. As a main result we highlight 2.14.9 where we address the connection between the factorization type of a prime ideal and the cycle type of its Frobenius automorphism.

As already mentioned we postpone the theory of differents and discriminants to Chapter 5 where this can be done much better using methods from valuation theory.

2.1 Factorial monoids

Throughout, a **monoid** is a (commutative, associative and, unless otherwise specified, multiplicative) semigroup D possessing a unit element $1 \in D$, such that the cancellation law holds for all $a, b, c \in D$ (this means, that $ab = ac$ implies $b = c$). We adhere to the convention that empty products equal 1. A subset H of D is called **multiplicatively closed** or a **submonoid** if $1 \in H$ and $HH = H$. We denote by $D^\times$ the group of invertible elements of D, called the **unit group** of D, and we call D **reduced** if $D^\times = \{1\}$. A group Q is called a **quotient group** of D if D is a submonoid of Q and $Q = \{a^{-1}b \mid a, b \in D\}$. Every monoid D possesses an (up to natural isomorphisms) uniquely determined quotient group which we denote by $\mathsf{q}(D)$. The construction of a quotient group and the proof of its uniqueness are literally the same as that for the quotient field of a domain. If G is a (multiplicative) group, then G is a monoid such that $G^\times = \mathsf{q}(G) = G$, and for a submonoid D of G we always assume that $\mathsf{q}(D) = \{a^{-1}b \mid a, b \in D\} \subset G$.

If U is a subgroup of a monoid D, then clearly $U \subset D^\times$, and the coset space $D/U = \{aU \mid a \in D\}$ is a monoid by virtue of $(aU)(bU) = abU$ for all $a, b \in D$. It follows that $(D/U)^\times = D^\times/U$ and $\mathsf{q}(D/U) = \mathsf{q}(D)/U$. In particular, the monoid $D_{\text{red}} = D/D^\times$ is reduced; it is called the **associated reduced monoid** of D.

A map $\varphi\colon D \to D'$ of monoids is called a (**monoid**) **homomorphism** if $\varphi(1) = 1$ and $\varphi(ab) = \varphi(a)\varphi(b)$ for all $a, b \in D$. Every monoid homomorphism $\varphi\colon D \to D'$ has a unique extension to a group homomorphism $\mathsf{q}(\varphi)\colon \mathsf{q}(D) \to \mathsf{q}(D')$, called the **quotient homomorphism** of φ.

For many purposes it is advantageous to enlarge a monoid D by a zero element. Then we set $D^\circ = D \cup \{0\}$ and define $a0 = 0a = 0$ for all $a \in D^\circ$.

These are the simplest examples:

$\mathbb{N}$ is a reduced monoid, $\mathsf{q}(\mathbb{N}) = \mathbb{Q}_{>0}$, $\mathbb{N}^\circ = \mathbb{N}_0$ and $\mathsf{q}(\mathbb{N})^\circ = \mathbb{Q}_{\ge 0}$. If $\mathfrak{o}$ is a domain, then $\mathfrak{o}^\bullet$ equipped with the ring multiplication is a monoid, $\mathfrak{o} = \mathfrak{o}^{\bullet\circ}$, $\mathfrak{o}^\times = \mathfrak{o}^{\bullet\times}$, and $\mathsf{q}(\mathfrak{o}^\bullet) = \mathsf{q}(\mathfrak{o})^\times$ is the multiplicative group of the quotient field $\mathsf{q}(\mathfrak{o})$ of $\mathfrak{o}$.

Let D be a monoid. If $a \in D^\circ$, then the subset aD° of D° is called the **principal ideal** of D generated by a. If $a, b \in D^\circ$, then a and b are called **associated** and we write $a \simeq b$ if $aD^\circ = bD^\circ$ [equivalently, $aD^\times = bD^\times$]. Note that $aD^\circ = D^\circ$ if and only if $a \in D^\times$, and $aD^\circ = \mathbf{0}$ if and only if $a = 0$. The set $\mathcal{H}'(D)$ of all non-zero principal ideals of D is a monoid by virtue of $(aD^\circ)(bD^\circ) = abD^\circ$ for all $a, b \in D$, and it is easily checked that the assignment $aD^\times \mapsto aD^\circ$ defines an isomorphism $D_{\text{red}} \overset{\sim}{\to} \mathcal{H}'(D)$. If $\mathfrak{o}$ is a domain, then $\mathcal{H}'(\mathfrak{o}^\bullet)$ is nothing but the monoid of non-zero principal ideals of $\mathfrak{o}$.

Let $a, b \in D^\circ$. We say that a **divides** b (in D) if $bD^\circ \subset aD^\circ$ and then we write $a \mid b$ (or $a \mid_D b$ if necessary). Note that $a \mid b$ if and only if $a = bc$ for some

$c \in D^\circ$, and that a and b are associates if and only if $a \mid b$ and $b \mid a$. Moreover, $a \mid 1$ if and only if $a \in D^\times$, $a \mid 0$ for all $a \in D^\circ$, and $0 \mid a$ if and only if $a = 0$.

Let X be a subset of D°. An element $d \in D^\circ$ is called a **gcd** (a greatest common divisor) of X if dD° is the smallest principal ideal of D containing X [equivalently, dD° is the smallest principal ideal such that $d \mid a$ for all $a \in X$]. In particular, 0 is a gcd of the empty set.

An element $e \in D^\circ$ is called an **lcm** (a least common multiple) of X if eD° is the largest principal ideal of D contained in all principal ideals aD° for $a \in X$ [equivalently, eD° is the largest principal ideal such that $a \mid e$ for all $a \in X$]. In particular, 1 is an lcm of the empty set.

If $d \in D^\circ$ is a gcd [an lcm] of X, then the coset $dD^\times$ is the set of all gcd [all lcm] of X. In particular, if D is reduced, then a gcd resp. an lcm of a subset X of D is unique (provided that it exists), and then we denote it by $\gcd(X)$ resp. $\operatorname{lcm}(X)$. Finitely many elements $a_1, \ldots, a_n \in D^\circ$ are called **coprime** if 1 is a gcd of $\{a_1, \ldots, a_n\}$; we denote this by $(a_1, \ldots, a_n) = 1$.

Definition and Remarks 2.1.1. Let D be a monoid.

1. An element $u \in D$ is called

- an **atom** if $u \notin D^\times$ and, for all $a, b \in D$, $u = ab$ implies $a \in D^\times$ or $b \in D^\times$;
- a **prime element** if $u \notin D^\times$ and, for all $a, b \in D$, $u \mid ab$ implies $u \mid a$ or $u \mid b$.

2. D is called

- **atomic** if every $a \in D \setminus D^\times$ is a product of atoms;
- **factorial** if every $a \in D \setminus D^\times$ is a product of prime elements.

3. Let $a \in D$. We say that a has a **factorization into atoms** provided that $a \simeq u_1 \cdot \ldots u_n$ for some $n \in \mathbb{N}_0$ and atoms $u_1, \ldots, u_n$. Two factorizations $a \simeq u_1 \cdot \ldots \cdot u_n$ and $a \simeq v_1 \cdot \ldots \cdot v_m$ of a into atoms are called **not essentially distinct** if $m = n$ and there exists some permutation $\sigma \in \mathfrak{S}_n$ such that $u_i \simeq v_{\sigma(i)}$ for all $i \in [1, n]$. We say that a **has a unique factorization into atoms** if a has a factorization into atoms and any two such factorizations are not essentially distinct.

All factorization properties of an element $a \in D$ mentioned hitherto do not depend on the individual element a but rather only on the principal ideal $aD^\circ \in \mathcal{H}'(D)$ resp. on the class of associates $aD^\times \in D_{\mathrm{red}}$. For example: $a \mid_D b$ if and only if $aD^\circ \mid_{\mathcal{H}'(D)} bD^\circ$ resp. $aD^\times \mid_{D_{\mathrm{red}}} bD^\times$; a is an atom [a prime element] of D if and only if aD° resp. $aD^\times$ is an atom [a prime element] of $\mathcal{H}'(D)$ resp. D_{red}; D is atomic [factorial] if and only if $\mathcal{H}'(D)$ resp. D_{red} is atomic [factorial]. We phrase this fact by saying that D, D_{red} and $\mathcal{H}'(D)$ **have the same factorization properties**.

If $D = \mathfrak{o}^\bullet$ is the multiplicative monoid of a domain, then a concept of divisibility theory is said to apply for the domain $\mathfrak{o}$ if and only if it applies for the monoid $\mathfrak{o}^\bullet$. In this sense all concepts considered in Section 1.1 coincide with the corresponding concepts in the current section.

In particular, a domain $\mathfrak{o}$ (and thus its multiplictive monoid $\mathfrak{o}^\bullet$) has the same factorization properties as the monoid $\mathcal{H}'(\mathfrak{o})$ of its non-zero principal ideals.

Theorem 2.1.2 (Factorization theorem). *Let D be a monoid.*

1. *Every prime element of D is an atom. In particular, every factorial monoid is atomic.*

2. *Let $n \in \mathbb{N}_0$, $a, b \in D$, and suppose that $ab \simeq \pi_1 \cdot \ldots \cdot \pi_n$ with prime elements $\pi_1, \ldots, \pi_n$. Then there exists (after a suitable renumbering of $\pi_1, \ldots, \pi_n$) some $k \in [0, n]$ such that $a \simeq \pi_1 \cdot \ldots \cdot \pi_k$ and $b \simeq \pi_{k+1} \cdot \ldots \cdot \pi_n$.*

3. *Any two factorizations of an element $a \in D$ into prime elements are not essentially distinct.*

4. *D is factorial if and only if every $a \in D$ has a unique factorization into atoms.*

Proof. 1., 2. and 3. The proofs are literally the same as those given in 1.1.7.

4. If D is factorial and $a \in D$, then a has a factorization into prime elements, hence into atoms, and any two such factorizations are not essentially distinct by 3. Conversely, assume that every $a \in D$ has a unique factorization into atoms. It suffices to prove that every atom is a prime element. Thus let u be an atom of D, and let $a, b \in D$ be such that $u \,|\, ab$, say $ab = uc$ for some $c \in D$. Let $a \simeq u_1 \cdot \ldots \cdot u_n$, $b \simeq v_1 \cdot \ldots \cdot v_m$ and $c \simeq w_1 \cdot \ldots \cdot w_k$ be factorizations of a, b and c into atoms. Then $u_1 \cdot \ldots \cdot u_n v_1 \cdot \ldots \cdot v_m \simeq u w_1 \cdot \ldots \cdot w_k$, and by the uniqueness of factorizations it follows that either $u \simeq u_i$ for some $i \in [1, n]$ (and then $u \,|\, a$) or $u \simeq v_j$ for some $j \in [1, m]$ (and then $u \,|\, b$). □

Theorem 2.1.3. *A monoid satisfying the ascending chain condition for principal ideals is atomic. In particular, every Noetherian domain is atomic.*

Proof. Let D be a monoid which satisfies the ascending chain condition for principal ideals. We prove first that every non-empty set of principal ideals of D possesses a maximal element. Indeed, assume that there exists a non-empty set Ω of principal ideals without a maximal element. Then there exists a function $\phi \colon \Omega \to \Omega$ such that $\phi(J) \supsetneq J$ for all $J \in \Omega$. If $J_0 \in \Omega$ is arbitrary and $J_{n+1} = \phi(J_n)$ for all $n \in \mathbb{N}_0$, then $(J_n)_{n \ge 0}$ is a non-terminating ascending chain of principal ideals.

Assume that D is not atomic. Let Ω be the set of all principal ideals aD° built with elements $a \in D$ having no factorization into atoms, and let aD° be a maximal element of Ω. Then a is not an atom and thus there exist $b, c \in D \setminus D^\times$ such that $a = bc$. But then $aD^\circ \subsetneq bD^\circ$ and $aD^\circ \subsetneq cD^\circ$. Hence both b and c have factorizations into atoms, and thus also a has a factorization into atoms, a contradiction. □

There is a substantial theory concerning the various phenomena which occur when unique factorization fails. We cannot go into details here and refer the interested reader to [17].

We proceed with the theory of factorial monoids.

Definition and Theorem 2.1.4 (π-adic valuations). *Let D be a factorial monoid, $Q = \mathsf{q}(D)$ and $\pi \in D$ a prime element. Then every $z \in Q$ has a representation $z = \pi^k a^{-1} b$, where $k \in \mathbb{Z}$, $a, b \in D$ and $\pi \nmid ab$. In this representation the exponent k is uniquely determined by the principal ideals zD° and πD°.*

We call $k = v_\pi(z)$ the **π-adic value** of z, and we set $v_\pi(0) = \infty$. The map

$$v_\pi \colon Q^\circ \to \mathbb{Z} \cup \{\infty\}$$

is called that **π-adic valuation** of Q°. For the hitherto undefined symbol ∞ we set $a + \infty = \infty + a = \infty$ and $a \le \infty$ for all $a \in \mathbb{Z} \cup \{\infty\}$.

If $x, y \in Q^\circ$, then $v_\pi(xy) = v_\pi(x) + v_\pi(y)$, and $v_\pi \restriction Q \colon Q \to \mathbb{Z}$ is a surjective group homomorphism. If π and π' are associated prime elements of D, then $v_\pi = v_{\pi'}$.

Proof. Let first $z \in D$. Then $z = u\pi_1 \cdot \ldots \cdot \pi_n$, where $u \in D^\times$, $n \in \mathbb{N}_0$, and $\pi_1, \ldots, \pi_n$ are prime elements of D. After a suitable renumbering we may assume that there is some $l \in [0, n]$ such that $\pi_i \simeq \pi$ for all $i \in [1, l]$ and $\pi_i \not\simeq \pi$ for all $i \in [l+1, n]$. It follows that $z = \pi^l b$ for some $b \in D$ such that $\pi \nmid b$.

Now let $z = x^{-1} y \in Q$ for some $x, y \in D$, $x = \pi^l a$ and $y = \pi^{l'} b$, where $l, l' \in \mathbb{N}_0$, $a, b \in D$ and $\pi \nmid ab$. If $k = l' - l$, then $z = \pi^k a^{-1} b$, and we must prove that k only depends on the principal ideals zD and πD.

Thus let $\pi_1 = \pi u$, $z_1 = zv$, where $u, v \in D^\times$, $z = \pi^k a^{-1} b$ and $z_1 = \pi_1^{k_1} a_1^{-1} b_1$, where $k, k_1 \in \mathbb{Z}$ and $a, b, a_1, b_1 \in D$ such that $\pi \nmid ab$ and $\pi_1 \nmid a_1 b_1$. Then $\pi \nmid a_1 b_1$, and we may assume that $k \ge k_1$. Since $z_1 = (\pi u)^{k_1} a_1^{-1} b_1 = \pi^k a^{-1} b v$, we obtain $\pi^{k-k_1} a_1 b v = u^{k_1} a b_1$. If $k > k_1$, then $\pi \mid u^{k_1} a b_1$, a contradiction.

It is now easily checked that $v_\pi(xy) = v_\pi(x) + v_\pi(y)$ for all $x, y \in Q^\circ$. Consequently, $v_\pi \restriction Q \colon Q \to \mathbb{Z}$ is a group homomorphism, and since $v_\pi(\pi) = 1$, it is surjective. $\square$

Definition 2.1.5. Let D be a factorial monoid and $Q = \mathsf{q}(D)$. A set P of prime elements of D is called a set of **representatives for the prime elements** of D if for every prime element π of D there exists a unique $\pi_0 \in P$ such that $\pi \simeq \pi_0$.

Theorem 2.1.6 (Arithmetic in factorial monoids, I). *Let D be a factorial monoid, $Q = \mathsf{q}(D)$ and P a set of representatives for the prime elements of D.*

1. Let $z \in Q$. Then $v_\pi(z) = 0$ for almost all $\pi \in P$, there exists some $u \in D^\times$ such that

$$z = u \prod_{\pi \in P} \pi^{v_\pi(z)},$$

and for every $v \in D^\times$ and every family $(n_\pi)_{\pi \in P} \in \mathbb{Z}^{(P)}$,

$$z = v \prod_{\pi \in P} \pi^{n_\pi} \quad \text{implies} \quad v = u \text{ and } n_\pi = v_\pi(z) \text{ for all } \pi \in P.$$

Moreover, $D^\circ = \{z \in Q^\circ \mid v_\pi(z) \geq 0 \text{ for all } \pi \in P\}$, and the assignment $z \mapsto (u, (v_\pi(z))_{\pi \in P})$ defines a monoid isomorphism $D \xrightarrow{\sim} D^\times \times \mathbb{N}_0^{(P)}$ and a group isomorphism $Q \xrightarrow{\sim} D^\times \times \mathbb{Z}^{(P)}$. In particular, $Q = D^\times \cdot \langle P \rangle$.

2. If $z \in Q^\circ$, then there exist up to associates uniquely determined elements $a \in D$ and $b \in D^\circ$ such that $(a, b) = 1$ and $z = a^{-1}b$.

3. Let $a, b \in D^\circ$. Then

- *$a \,|\, b$ (and thus $bD^\circ \subset aD^\circ$) if and only if $v_\pi(a) \leq v_\pi(b)$ for all $\pi \in P$;*
- *$a \simeq b$ (and thus $bD^\circ = aD^\circ$) if and only if $v_\pi(a) = v_\pi(b)$ for all $\pi \in P$.*

In particular, every $b \in D$ has up to associates only finitely many divisors.

4. Let $a, b, c \in D^\circ$ be such that $a \,|\, bc$ and $(a, b) = 1$. Then $a \,|\, c$.

Proof. 1. First let $z \in D$. Then $z = u_0\pi_1 \cdot \ldots \cdot \pi_n$, where $u_0 \in D^\times$, $n \in \mathbb{N}_0$. and $\pi_1, \ldots, \pi_n$ are prime elements of D. If $\pi \in P$, then it follows that $v_\pi(z) = |\{i \in [1, n] \mid \pi_i \simeq \pi\}|$, and gathering the occurring unit factors we obtain

$$z = u \prod_{\pi \in P} \pi^{v_\pi(z)} \quad \text{for some} \quad u \in D^\times.$$

In particular, $v_\pi(z) \geq 0$ for all $\pi \in P$, and $v_\pi(z) = 0$ for almost all $\pi \in P$.

Now let $z \in Q$ and $z = a^{-1}b$ for some $a, b \in D$, say

$$a = u' \prod_{\pi \in P} \pi^{v_\pi(a)} \quad \text{and} \quad b = u'' \prod_{\pi \in P} \pi^{v_\pi(b)}, \quad \text{where} \quad u', u'' \in D^\times.$$

Then we obtain

$$z = a^{-1}b = u'^{-1}u'' \prod_{\pi \in P} \pi^{-v_\pi(a) + v_\pi(b)} = u \prod_{\pi \in P} \pi^{v_\pi(z)}, \quad \text{where} \quad u = u'^{-1}u'' \in D^\times.$$

To prove uniqueness, we assume that

$$z = u \prod_{\pi \in P} \pi^{v_\pi(z)} = v \prod_{\pi \in P} \pi^{n_\pi} \quad \text{for some} \quad v \in D^\times \quad \text{and} \quad (n_\pi)_{\pi \in P} \in \mathbb{Z}^{(P)}.$$

Then it follows that

$$v \prod_{\substack{\pi \in P \\ n_\pi > v_\pi(z)}} \pi^{n_\pi - v_\pi(z)} = u \prod_{\substack{\pi \in P \\ n_\pi < v_\pi(z)}} \pi^{v_\pi(z) - n_\pi},$$

and this is only possible if $\{\pi \in P \mid n_\pi > v_\pi(z)\} = \{\pi \in P \mid n_\pi < v_\pi(z)\} = \emptyset$ and $v = u$.

We have already seen that $D \subset \{z \in Q \mid v_\pi(z) \ge 0 \text{ for all } \pi \in P\}$. The reverse inclusion is obvious, and $D^\circ = \{z \in Q^\circ \mid v_\pi(z) \ge 0 \text{ for all } \pi \in P\}$, since $v_\pi(0) = \infty$ for all $\pi \in P$.

2. If $z = 0$, then necessarily $b = 0$ and $a \in D^\times$. If $z \ne 0$, then $z = a^{-1}b$ for some $a, b \in D$ such that $(a, b) = 1$ if and only if there exist $u, v \in D^\times$ such that

$$a = u \prod_{\substack{\pi \in P \\ v_\pi(z) < 0}} \pi^{-v_\pi(z)} \quad \text{and} \quad b = v \prod_{\substack{\pi \in P \\ v_\pi(z) > 0}} \pi^{v_\pi(z)}.$$

Hence existence and uniqueness up to associates of a and b follow.

3. If $a \mid b$, then $b = ac$ for some $c \in D^\circ$ and $v_\pi(b) = v_\pi(a) + v_\pi(c) \ge v_\pi(a)$ for all $\pi \in P$. Assume now that $v_\pi(b) \ge v_\pi(a)$ for all $\pi \in P$. If $b = 0$, then $v_\pi(b) = \infty$ for all $p \in P$, and obviously $a \mid b$. If $b \ne 0$, then $v_\pi(a) \le v_\pi(b) < \infty$ for all $\pi \in P$, hence $a \ne 0$,

$$a^{-1}b = u \prod_{\pi \in P} \pi^{v_\pi(b) - v_\pi(a)} \in D \quad \text{for some} \quad u \in D^\times, \quad \text{and} \quad b = a(a^{-1}b).$$

Hence $a \mid b$. The remaining assertions are now obvious.

4. We prove that $v_\pi(a) \le v_\pi(c)$ for all $\pi \in P$. Let $\pi \in P$. If $v_\pi(a) = 0$, we are done. If $v_\pi(a) > 0$, then $v_\pi(b) = 0$ and $v_\pi(a) \le v_\pi(bc) = v_\pi(b) + v_\pi(c) = v_\pi(c)$. □

An element $a \in D$ is called **squarefree** if $q^2 \nmid a$ for all $q \in D \setminus D^\times$. If $a \in Q$ and $a_0 \in D$ is squarefree such that $a = q^2 a_0$ for some $q \in Q$, then we call a_0 a **squarefree kernel** of a.

Theorem 2.1.7 (Arithmetic in factorial monoids, II). *Let D be a factorial monoid, $Q = \mathsf{q}(D)$ and P a set of representatives for the prime elements of D.*

1. *Let X be a non-empty subset of D° and $d, e \in D^\circ$.*

(a) *d is a* gcd *of X if and only if $v_\pi(d) = \min\{v_\pi(a) \mid a \in X\}$ for all $\pi \in P$. In particular, X possesses an up to associates uniquely determined* gcd *in D°.*

(b) e is an lcm *of X if and only if*

- *either $\{aD^\circ \mid a \in X\}$ is infinite and $e = 0$,*
- *or $\{aD^\circ \mid a \in X\}$ is finite and*

$$v_\pi(e) = \max\{v_\pi(a) \mid a \in X\} \text{ for all } \pi \in P.$$

In particular, X possesses an up to associates uniquely determined lcm *in D°.*

2. Let $n \in \mathbb{N}$, and for $i \in [1, n]$ let X_i be a subset of D°, $d_i \in D^\circ$ a gcd *and $e_i \in D^\circ$ an* lcm *of X_i. Then $d_1 \cdot \ldots \cdot d_n$ is a* gcd *and $e_1 \cdot \ldots \cdot e_n$ is an* lcm *of the set $X_1 \cdot \ldots \cdot X_n$.*

3. Let $(X_i)_{i \in I}$ be a family of subsets of D°, and for every $i \in I$ let d_i be a gcd *and e_i an* lcm *of X_i. Let*

$$X = \bigcup_{i \in I} X_i, \quad D = \{d_i \mid i \in I\} \text{ and } E = \{e_i \mid i \in I\}.$$

Then every gcd *of D is a* gcd *of X, and every* lcm *of E is an* lcm *of X.*

4. Let $a \in Q$ and $a_0 \in D$.

(a) a_0 is squarefree if and only if $v_\pi(a_0) \le 1$ for all $\pi \in P$.

(b) Suppose that

$$a = u \prod_{\pi \in P} \pi^{v_\pi(a)}, \quad \text{where } u \in D^\times.$$

Then a_0 is a squarefree kernel of a if and only if

$$a_0 = v^2 u \prod_{\substack{\pi \in P \\ v_\pi(a) \equiv 1 \bmod 2}} \pi \quad \text{for some } v \in D^\times.$$

In particular, a possesses an up to factors from $D^{\times 2}$ uniquely determined squarefree kernel.

Proof. 1. Let $d,\, e \in D^\circ$ be such that

- $v_\pi(d) = \min\{v_\pi(a) \mid a \in X\}$ for all $\pi \in P$,
- $v_\pi(e) = \max\{v_\pi(a) \mid a \in X\}$ for all $\pi \in P$ if the set $\{aD^\circ \mid a \in X\}$ is finite,
- $e = 0$ if the set $\{aD^\circ \mid a \in X\}$ is infinite.

We must prove that dD° is the smallest principal ideal such that $aD^\circ \subset dD^\circ$ for all $a \in X$, and eD° is the largest principal ideal such that $aD^\circ \supset eD^\circ$ for all $a \in X$.

If $a \in X$, then $v_\pi(d) \le v_\pi(a) \le v_\pi(e)$ for all $\pi \in P$, and therefore it follows that $eD^\circ \subset aD^\circ \subset dD^\circ$. Now let $d_1,\, e_1 \in D^\circ$ such that $e_1D^\circ \subset aD^\circ \subset d_1D^\circ$ for all $a \in X$. Then $v_\pi(d_1) \le v_\pi(a)$ for all $a \in X$ and $\pi \in P$, hence $v_\pi(d_1) \le v_\pi(d)$ for all $\pi \in P$ and thus $dD^\circ \subset d_1D^\circ$. If $\{aD^\circ \mid a \in X\}$ is infinite, then $e_1 = 0$ and thus $e_1D = \mathbf{0} \subset eD^\circ$. If $\{aD^\circ \mid a \in X\}$ is finite, then $v_\pi(e) \le v_\pi(e_1)$ for all $\pi \in P$ and therefore $e_1D^\circ \subset eD$.

2., 3. and 4. follow with some notational effort immediately from 1. □

Corollary 2.1.8. *Let D be a factorial monoid, P a set of representatives for the prime elements of D, $n \in \mathbb{N}$ and $a_1, \ldots, a_n \in Q^\circ$, not all equal 0. For $c \in Q$, the following assertions are equivalent:*

> *(a)* $\{c^{-1}a_1, \ldots, c^{-1}a_n\} \subset D^\circ$ *and* $(c^{-1}a_1, \ldots, c^{-1}a_n) = 1$.
>
> *(b)* $v_\pi(c) = \min\{v_\pi(a_1), \ldots, v_\pi(a_n)\}$ *for all* $\pi \in P$.

In particular, there exists an up to associates unique element $c \in D$ such that (a) *holds, and if $\{a_1, \ldots, a_n\} \subset D^\circ$, then c is a* gcd *of $a_1, \ldots, a_n$.*

Proof. Obvious by 2.1.7. □

Definition and Theorem 2.1.9. A monoid D is called **free** if it is factorial and reduced. In this case, the set P of prime elements of D is called a **basis** of D. If D is free, then every subset X of D° possesses an unique gcd, denoted by $\gcd(X)$, and a unique lcm, denoted by $\operatorname{lcm}(X)$.

Let D be a free monoid with basis P and $Q = \mathsf{q}(D)$.

> *1. Q is a free abelian group with basis P. In particular, every $z \in Q$ has a unique representation*
>
> $$z = \prod_{\pi \in P} \pi^{n_\pi}, \quad \text{where } n_\pi = v_\pi(z) \text{ for all } \pi \in P,$$
>
> *and the assignment $z \mapsto (v_\pi(z))_{\pi \in P}$ defines a group isomorphism $Q \xrightarrow{\sim} \mathbb{Z}^{(P)}$ and a monoid isomorphism $D \xrightarrow{\sim} \mathbb{N}_0^{(P)}$.*
> In particular, D is (up to isomorphisms) uniquely determined by P and is denoted by $\mathcal{F}(P)$.
>
> *2. Let M be a monoid, $L = \mathsf{q}(M)$ and $f_0 \colon P \to M$ a map. Then there exists a unique group homomorphism $f \colon Q \to L$ satisfying $f \restriction P = f_0$, and then $f(D) \subset M$.*

Proof. 1. follows by 2.1.6.1, and then 2. is obvious. □

2.2 Factorial domains

Let $\mathfrak{o}$ be a domain and $K = \mathsf{q}(\mathfrak{o})$. Then $\mathfrak{o}^\bullet$ is a multiplicative monoid, and $\mathfrak{o} = \mathfrak{o}^{\bullet\circ}$. We use all arithmetical concepts and results from Section 2.1. In particular, an element $\pi \in \mathfrak{o}^\bullet$ is a prime element of $\mathfrak{o}$ if and only if $\pi\mathfrak{o} \in \mathcal{P}(\mathfrak{o})$, and $\mathfrak{o}$ is a factorial domain if and only if $\mathfrak{o}^\bullet$ is a factorial monoid.

In 1.1.7.3 we have seen that every principal ideal domain is factorial. Hence $\mathbb{Z}$, every field and every polynomial domain $K[X]$ over a field K are factorial.

Let K be a field. Then $\emptyset$ is a set of representatives for the prime elements of K, $K[X]^\times = K^\times$, and the set $\mathsf{P}(K,X)$ of all monic irreducible polynomials is a set of representatives for the prime elements of $K[X]$. Every $f \in K[X]^\bullet$ has an (up to the order of the factors) unique factorization $f = uf_1 \cdot \ldots \cdot f_n$, where $u \in K^\times$, $n \in \mathbb{N}_0$ and $f_1, \ldots, f_n \in \mathsf{P}(K,X)$. In particular, the set $\mathsf{D}(K,X)$ of all monic polynomials in $K[X]$ is a free monoid with basis $\mathsf{P}(K,X)$.

Clearly, $\mathbb{Z}^\times = \{\pm 1\}$, and the set $\mathbb{P}$ of all primes is a set of representatives for the prime elements of $\mathbb{Z}$. If $a \in \mathbb{Q}^\times$, then a has a unique factorization

$$a = \operatorname{sgn}(a) \prod_{p \in \mathbb{P}} p^{v_p(a)},$$

and a has a unique squarefree kernel a_0, given by

$$a_0 = \operatorname{sgn}(a) \prod_{\substack{p \in \mathbb{P} \\ v_p(a) \equiv 1 \mod 2}} p.$$

Moreover, $\mathbb{N} = \mathcal{F}(\mathbb{P})$, $\mathsf{q}(\mathbb{N}) = \mathbb{Q}_{>0} \cong \mathbb{P}^{(\mathbb{Z})}$ and $\mathbb{Q}^\times = \{\pm 1\} \cdot \langle \mathbb{P} \rangle$ (compare 2.1.6.1, 2.1.7.4 and 2.1.9).

Definition and Theorem 2.2.1. *Let K be a field.* A **discrete valuation** of K is a surjective map $v \colon K \to \mathbb{Z} \cup \{\infty\}$ such that the following assertions hold for all $x, y \in K$:

- $v(x) = \infty$ if and only if $x = 0$;
- $v(xy) = v(x) + v(y)$;
- $v(x+y) \geq \min\{v(x), v(y)\}$.

Let $v \colon K \to \mathbb{Z} \cup \{\infty\}$ be a discrete valuation.

1. *$v \restriction K^\times : K^\times \to \mathbb{Z}$ is a group epimorphism.*

2. Let $x, y \in K$. Then $v(-x) = v(x)$, $v(x - y) \geq \min\{v(x), v(y)\}$, and if $v(x) \neq v(y)$, then $v(x + y) = \min\{v(x), v(y)\}$. If $x \in \mu(K)$ is a root of unity, then $v(x) = 0$.

Proof. 1. By definition.

2. If $x \in \mu(K)$ and $n \in \mathbb{N}$ is such that $x^n = 1$, then $0 = v(x^n) = nv(x)$ and therefore $v(x) = 0$. Hence $v(-x) = v(-1) + v(x) = v(x)$, and consequently $v(x - y) \geq \min\{v(x), v(-y)\} = \min\{v(x), v(y)\}$. If $v(x) < v(y)$, then

$$v(x) = v((x + y) - y) \geq \min\{v(x + y), v(y)\} \geq \min\{v(x), v(y)\} = v(x),$$

hence $v(x) = \min\{v(x + y), v(y)\} = v(x + y)$. □

Theorem 2.2.2. *Let $\mathfrak{o}$ be a factorial domain, $K = \mathsf{q}(\mathfrak{o})$ and $\pi \in \mathfrak{o}$ a prime element. Then $v_\pi \colon K \to \mathbb{Z} \cup \{\infty\}$ is a discrete valuation.*

Proof. By 2.1.4 it suffices to prove that $v_\pi(x + y) \geq \min\{v_\pi(x), v_\pi(y)\}$ for all $x, y \in K$. This is obvious if $xy = 0$. Thus let $x, y \in K^\times$, say $v_\pi(x) = m$, $v_\pi(y) = n$ and $m \leq n$. Then $x = \pi^m ac^{-1}$ and $y = \pi^n bd^{-1}$ for some $a, b, c, d \in \mathfrak{o}$ such that $\pi \nmid abcd$, and

$$x + y = \pi^m \Big(\frac{a}{c} + \pi^{n-m}\frac{b}{d}\Big) = \pi^m \,\frac{ad + \pi^{n-m}bc}{cd}.$$

We set $ad + \pi^{n-m}bc = \pi^\delta e$, where $\delta \in \mathbb{N}_0$, $e \in \mathfrak{o}$ and $\pi \nmid e$. Then we get $x + y = \pi^{m+\delta}e(cd)^{-1}$, and as $\pi \nmid ecd$ it follows that

$$v_\pi(x + y) = m + \delta \geq m = \min\{v_\pi(x), v_\pi(y)\}.$$ □

In the remainder of this section we consider polynomials over factorial domains. We prove a series of closely related results which are cited as Gauss's lemma in the literature and generalize 1.1.12.

Definition 2.2.3. Let $\mathfrak{o}$ be a factorial domain, $K = \mathsf{q}(\mathfrak{o})$ and

$$f = a_n X^n + a_{n-1}X^{n-1} + \ldots + a_0 \in K[X]^\bullet.$$

1. f is called **primitive** (with respect to $\mathfrak{o}$) if $f \in \mathfrak{o}[X]$ and $(a_0, \ldots, a_n) = 1$. In particular, if $f \in K^\times$, then f is primitive if and only if $f \in \mathfrak{o}^\times$.

2. An element $c \in K^\times$ is called a **content** of f (with respect to $\mathfrak{o}$) if $c^{-1}f$ is primitive.

If $\mathfrak{o}$ is a principal ideal domain and $f \in \mathfrak{o}[X]^\bullet$, then an element $c \in \mathfrak{o}^\bullet$ is a content of f if and only if the principal ideal $c\mathfrak{o}$ is a content ideal of f as defined in 1.6.3. In general however the concepts differ.

Theorem 2.2.4 (Gauss's lemma, I). *Let $\mathfrak{o}$ be a factorial domain and $K = \mathsf{q}(\mathfrak{o})$. Let $n \in \mathbb{N}_0$, $f = a_nX^n + a_{n-1}X^{n-1} + \ldots + a_0 \in K[X]^\bullet$ and $c \in K^\times$.*

1. *c is a content of f if and only if $v_\pi(c) = \min\{v_\pi(a_0), \ldots, v_\pi(a_n)\}$ for all prime elements π of $\mathfrak{o}$. In particular, f possesses an up to factors from $\mathfrak{o}^\times$ uniquely determined content, and f is primitive if and only if 1 is a content of f.*

2. *Let c be a content of f.*

   (a) $f \in \mathfrak{o}[X]$ if and only if $c \in \mathfrak{o}$.

   (b) If f is monic, then $c^{-1} \in \mathfrak{o}$.

Proof. 1. By 2.1.8.

2. (a) Clearly, $f \in \mathfrak{o}[X]$ if and only if $v_\pi(c) = \min\{v_\pi(a_0), \ldots, v_\pi(a_n)\} \geq 0$ for all prime elements $\pi \in \mathfrak{o}$, and this holds if and only if $c \in \mathfrak{o}$.

(b) If f is monic, then $v_\pi(c) \leq v_\pi(1) = 0$, hence $v_\pi(c^{-1}) = -v_\pi(c) \geq 0$ for all prime elements π of $\mathfrak{o}$ and thus $c^{-1} \in \mathfrak{o}$. □

Theorem 2.2.5 (Gauss's lemma, II). *Let $\mathfrak{o}$ be a factorial domain, and let $K = \mathsf{q}(\mathfrak{o})$. For every prime element π of $\mathfrak{o}$ we denote by $\rho_\pi \colon \mathfrak{o}[X] \to (\mathfrak{o}/\pi\mathfrak{o})[X]$ the residue class epimorphism.*

1. *A polynomial $f \in \mathfrak{o}[X]^\bullet$ is primitive if and only if $\rho_\pi(f) \neq 0$ for all prime elements π of $\mathfrak{o}$.*

2. *Let $f, g \in K[X]^\bullet$. Let $c \in K^\times$ be a content of f and $d \in K^\times$ a content of g. Then cd is a content of fg. In particular, if f and g are primitive, then fg is also primitive.*

3. *If $f \in \mathfrak{o}[X]$ is irreducible (in $\mathfrak{o}[X]$), then f is also irreducible in $K[X]$.*

4. *If $f, g \in \mathfrak{o}[X]$, $f \mid g$ in $K[X]$ and f is primitive, then $f \mid g$ in $\mathfrak{o}[X]$.*

Proof. 1. Let $f = a_nX^n + a_{n-1}X^{n-1} + \ldots + a_0 \in \mathfrak{o}[X]^\bullet$. By 2.2.4.1 f is primitive if and only if $\min\{v_\pi(a_0), \ldots, v_\pi(a_n)\} = 0$ for all prime elements π of $\mathfrak{o}$, and for a prime element π of $\mathfrak{o}$ this holds if and only if $v_\pi(a_i) = 0$ for some $i \in [0, n]$, that is, if and only if $\rho_\pi(f) \neq 0$.

2. The polynomials $f_1 = c^{-1}f \in \mathfrak{o}[X]$ and $g_1 = d^{-1}g \in \mathfrak{o}[X]$ are primitive, and we assume that f_1g_1 is not primitive. Then there exists some prime element π of $\mathfrak{o}$ such that $\rho_\pi(f_1g_1) = \rho_\pi(f_1)\rho_\pi(g_1) = 0$, and as $(\mathfrak{o}/\pi\mathfrak{o})[X]$ is a domain, it follows that $\rho_\pi(f_1) = 0$ or $\rho_\pi(g_1) = 0$, a contradiction. Hence $f_1g_1 = (cd)^{-1}(fg)$ is primitive and cd is a content of fg.

3. Assume to the contrary that a polynomial $f \in \mathfrak{o}[X]$ is reducible in $K[X]$, say $f = gh$, where $g, h \in K[X] \setminus K$. Let $c \in K^\times$ be a content of g and $d \in K^\times$ a content of h. Then $c^{-1}g \in \mathfrak{o}[X]$, $d^{-1}h \in \mathfrak{o}[X]$ and $cd \in \mathfrak{o}$, since cd is a content of f. Now the factorization $f = cd(c^{-1}g)(d^{-1}h)$ shows that f is also reducible in $\mathfrak{o}[X]$.

4. Let $f, g \in \mathfrak{o}[X]$ and $h \in K[X]$ such that $g = fh$. Let $c \in K^\times$ be a content of h. If f is primitive, then 1 is a content of f, and therefore c is a content of g. Hence we obtain $c \in \mathfrak{o}$, $h \in \mathfrak{o}[X]$ and $f \mid g$ in $\mathfrak{o}[X]$. □

Theorem 2.2.6 (Gauss's lemma, III). *Let $\mathfrak{o}$ be a factorial domain, and let $K = \mathsf{q}(\mathfrak{o})$. Then $\mathfrak{o}[X]$ is factorial, and an element $u \in \mathfrak{o}[X]$ is a prime element if and only if it satisfies one of the following two conditions:*

- *u is a prime element of $\mathfrak{o}$.*
- *u is primitive and irreducible in $K[X]$.*

Proof. If u is a prime element of $\mathfrak{o}$, then $\mathfrak{o}[X]/u\mathfrak{o}[X] = (\mathfrak{o}/u\mathfrak{o})[X]$ is a domain, and thus u is also a prime element of $\mathfrak{o}[X]$. Assume now that $u \in \mathfrak{o}[X] \setminus \mathfrak{o}$ is primitive and irreducible in $K[X]$ (hence a prime element of $K[X]$), and let $g, h \in \mathfrak{o}[X]$ be such that $u \mid gh$ in $\mathfrak{o}[X]$. Then $u \mid gh$ in $K[X]$, hence $u \mid g$ or $u \mid h$ in $K[X]$ and by 2.2.5.4 also in $\mathfrak{o}[X]$. Hence u is a prime element of $\mathfrak{o}[X]$.

We show now that every $f \in \mathfrak{o}[X]^\bullet$ has a factorization $f = \varepsilon q_1 \cdot \ldots \cdot q_n$, where $\varepsilon \in \mathfrak{o}[X]^\times = \mathfrak{o}^\times$, $n \in \mathbb{N}_0$, and $q_1, \ldots, q_n$ are prime elements of $\mathfrak{o}[X]$. For $f \in \mathfrak{o}^\bullet$ this is obvious. If $f \in \mathfrak{o}[X] \setminus \mathfrak{o}$, then f has in $K[X]$ a factorization $f = g_1 \cdot \ldots \cdot g_r$ into irreducible polynomials $g_1, \ldots, g_r \in K[X]$. For $i \in [1, r]$ let $c_i \in K^\times$ be a content of g_i. Then $f = c_1 \cdot \ldots \cdot c_r(c_1^{-1}g_1) \cdot \ldots \cdot (c_r^{-1}g_r)$, the polynomials $c_i^{-1}g_i \in \mathfrak{o}[X]$ are primitive and irreducible in $K[X]$, and as $c_1 \cdot \ldots \cdot c_r$ is a content of f, it follows that $c_1 \cdot \ldots \cdot c_r \in \mathfrak{o}^\bullet$. Now we write $c_1 \cdot \ldots \cdot c_r$ as a product of prime elements of $\mathfrak{o}$ and obtain a factorization of f into prime elements of $\mathfrak{o}[X]$. In particular, every prime element of $\mathfrak{o}[X]$ is of the asserted form. □

Theorem 2.2.7 (Gauss's lemma, IV). *Let $\mathfrak{o}$ be a factorial domain, and let $K = \mathsf{q}(\mathfrak{o})$.*

1. *Let $g, h \in K[X]$ be monic polynomials such that $gh \in \mathfrak{o}[X]$. Then it follows that $g \in \mathfrak{o}[X]$ and $h \in \mathfrak{o}[X]$.*

2. *Let $f, g \in \mathfrak{o}[X]$ be polynomials without a common factor in $\mathfrak{o}[X] \setminus \mathfrak{o}$. Then f and g are coprime in $K[X]$, and there exist polynomials $p, q \in \mathfrak{o}[X]$ satisfying $pf + qg \in \mathfrak{o}^\bullet$.*

Proof. 1. Let $c \in K^\times$ be a content of g and $d \in K^\times$ a content of h. Then cd is a content of f, hence $cd \in \mathfrak{o}$, $c^{-1} \in \mathfrak{o}$ and $d^{-1} \in \mathfrak{o}$ by 2.2.5.2 and 2.2.4.2(a). Since $c = (cd)d^{-1} \in \mathfrak{o}$ and $d = (cd)c^{-1} \in \mathfrak{o}$, it follows that $g \in \mathfrak{o}[X]$ and $h \in \mathfrak{o}[X]$.

2. We prove first that f and g are coprime in $K[X]$. Assume to the contrary that $f = qf_0$ and $g = qg_0$ for polynomials $q \in K[X] \setminus K$ and $f_0, g_0 \in K[X]^\bullet$. Let $c \in K^\times$ be a content of q, $c_0 \in K^\times$ a content of f_0 and $d_0 \in K^\times$ a content of g_0. Then $c^{-1}q \in \mathfrak{o}[X] \setminus \mathfrak{o}$, cc_0 is a content of f, cd_0 is a content of g and therefore $cc_0, cd_0 \in \mathfrak{o}$. Since $f = (c^{-1}q)(cf_0)$, $g = (c^{-1}q)(cg_0)$, $cf_0 = (cc_0)(c_0^{-1}f_0) \in \mathfrak{o}[X]$ and $cg_0 = (cd_0)(d_0^{-1}g_0 \in \mathfrak{o}[X]$, it follows that $c^{-1}q \in \mathfrak{o}[X] \setminus \mathfrak{o}$ is a common factor of f and g in $\mathfrak{o}[X]$, a contradiction.

As $K[X]$ is a principal ideal domain, there exist polynomials $p_1, q_1 \in K[X]$ such that $p_1 f + q_1 g = 1$. If $r \in \mathfrak{o}^\bullet$ is such that $p = rp_1$ and $q = rq_1$ lie in $\mathfrak{o}[X]$, then $pf + qg = r$. □

2.3 Principal ideal domains

The main subjects of this section are the theory of partial fractions and the theory of free modules over a principal ideal domain. Note that hitherto $\mathbb{Z}$ and the polynomial domains over fields are the only examples of principal ideal domains. As further examples we discuss Euclidean quadratic orders in Exercise 15.

Theorem 2.3.1 (Partial fractions). *Let $\mathfrak{o}$ be a principal ideal domain, let $K = \mathsf{q}(\mathfrak{o})$, and let P be a set of representatives for the prime elements of $\mathfrak{o}$.*

1. Let $x = a^{-1}b \in K$, where $a, b \in \mathfrak{o}$, $(a, b) = 1$ and $a = p_1^{e_1} \cdot \ldots \cdot p_r^{e_r}$ such that $r, e_1, \ldots, e_r \in \mathbb{N}$ and $p_1, \ldots, p_r \in P$. Then there exist $b_1, \ldots, b_r \in \mathfrak{o}$ such that $p_i \nmid b_i$ for all $i \in [1, r]$ and

$$x = \sum_{i=1}^{r} \frac{b_i}{p_i^{e_i}}.$$

Thereby the residue classes $b_1 + p_1^{e_1}\mathfrak{o}, \ldots, b_r + p_r^{e_r}\mathfrak{o}$ are uniquely determined.

2. For $p \in P$ and $n \in \mathbb{N}$ let $R(p^n) \subset \mathfrak{o}$ be a set of representatives for $\mathfrak{o}/p^n\mathfrak{o}$. Then every $x \in K$ has a unique representation

$$x = \sum_{p \in P} \frac{b_p}{p^{n_p}} + b$$

where $b \in \mathfrak{o}$, $(b_p, n_p) = (0,0)$ for almost all $p \in P$, and for all $p \in P$ either $(b_p, n_p) = (0,0)$ or $n_p \in \mathbb{N}$, $b_p \in R_p$ and $p \nmid b_p$.

3. For $p \in P$ let

$$(K/\mathfrak{o})_p = \{x \in K/\mathfrak{o} \mid p^n x = 0 \ \textit{for some} \ n \in \mathbb{N}\} = \bigcup_{n\geq 1} p^{-n}\,\mathfrak{o}/\mathfrak{o}$$

be the p-component of $K/\mathfrak{o}$. Then

$$K/\mathfrak{o} = \bigoplus_{p \in P} (K/\mathfrak{o})_p.$$

Proof. 1. *Existence.* By induction on r. For $r = 1$, there is nothing to do. $r \geq 2$, $r-1 \to r$: If $p = p_1^{e_1}$ and $q = p_2^{e_2} \cdot \ldots \cdot p_r^{e_r}$, then $(p,q) = 1$ and there exist $u, v \in \mathfrak{o}$ such that $up + vq = 1$. Then $(bu, q) = (bv, p) = 1$ and

$$x = \frac{b}{pq} = \frac{b(up+vq)}{pq} = \frac{bu}{q} + \frac{bv}{p} = \frac{bv}{p_1^{e_1}} + \sum_{i=2}^{r} \frac{b_i}{p_i^{e_i}}$$

by the induction hypothesis, where $b_2, \ldots, b_r \in \mathfrak{o}$ and $p_i \nmid b_i$ for all $i \in [2, r]$.

Uniqueness. Let $b_1, \ldots, b_r, b_1', \ldots, b_r' \in \mathfrak{o}$ be such that $p_i \nmid b_i b_i'$ for all $i \in [1, r]$ and

$$x = \sum_{i=1}^{r} \frac{b_i}{p_i^{e_i}} = \sum_{i=1}^{r} \frac{b_i'}{p_i^{e_i}}, \quad \text{hence} \quad \frac{b_1 - b_1'}{p_1^{e_1}} = \sum_{i=2}^{r} \frac{b_i' - b_i}{p_i^{e_i}}.$$

Then

$$(b_1 - b_1') \prod_{i=2}^{r} p_i^{e_i} = p_1^{e_1} \sum_{i=2}^{r} (b_i - b_i') \prod_{\substack{j=2 \\ j \neq i}}^{r} p_j^{e_j}$$

and therefore $p_1^{e_1} \mid b_1 - b_1'$. In the same way, we get $p_i^{e_i} \mid b_i - b_i'$ for all $i \in [2, r]$.

2. and 3. are useful reformulations of 1. □

Now we turn to the most important case of a polynomial domain over a field. In this case more precise results are possible.

Theorem 2.3.2 (Partial fractions of rational functions). *Let K be a field.*

1. Let $f, g \in K[X]$ and $\partial(g) \geq 1$. Then there exists a unique sequence $(f_n)_{n\geq 1}$ in $K[X]$ such that $\partial(f_n) < \partial(g)$ for all $n \geq 0$, $f_n = 0$ for almost all $n \geq 0$, and

$$f = \sum_{n \geq 0} f_n g^n.$$

2. Let $\mathsf{P}(K,X)$ *be the set of all monic irreducible polynomials in* $K[X]$*. Then every* $z \in K(X)$ *has a unique representation*

$$z = \sum_{p \in \mathsf{P}(K,X)} \sum_{j \ge 1} \frac{b_{p,j}}{p^j} + g.$$

where $g \in K[X]$*, and* $b_{p,j} \in K[X]$ *such that* $\partial(b_{p,j}) < \partial(p)$ *for all* $(p,j) \in \mathsf{P}(K,X) \times \mathbb{N}$*, and* $b_{p,j} = 0$ *for almost all* $(p,j) \in \mathsf{P}(K,X) \times \mathbb{N}$*.*

Proof. 1. *Existence.* We proceed by induction on $\partial(f)$. If $\partial(f) < \partial(g)$, we set $f_0 = f$ and $f_n = 0$ for all $n \ge 1$. Thus let $\partial(f) \ge \partial(g)$ and $f = gf^* + f_0$, where f^*, $f_0 \in K[X]$ and $\partial(f_0) < \partial(g)$. Since $\partial(g) \ge 1$, we get $\partial(f^*) < \partial(f)$, and by the induction hypothesis there exists a sequence $(f_n)_{n \ge 1}$ in $K[X]$ such that $\partial(f_n) < \partial(g)$ for all $n \ge 1$, $f_n = 0$ for almost all $n \ge 1$, and

$$f^* = \sum_{n \ge 1} f_n g^{n-1}, \quad \text{which implies} \quad f = \sum_{n \ge 0} f_n g^n.$$

Uniqueness. Suppose that

$$f = \sum_{n \ge 0} f_n g^n = \sum_{n \ge 0} f_n^* g^n$$

where f_n, $f_n^* \in K[X]$ such that $\partial(f_n) < \partial(g)$ and $\partial(f_n^*) < \partial(g)$ for all $n \ge 0$, $f_n = f_n^* = 0$ for almost all $n \ge 0$, and $m = \inf\{n \ge 0 \mid f_n \ne f_n^*\} < \infty$. Then $f_m \ne f_m^*$, and $g^m(gh + f_m) = g^m(gh^* + f_m^*)$ for some polynomials h, $h^* \in K[X]$, which contradicts the uniqueness of the division with remainder.

2. *Existence.* Let $z = a^{-1}b$, where a, $b \in K[X]$, such that $(a,b) = 1$ and $a = p_1^{e_1} \cdot \ldots \cdot p_r^{e_r}$, where $r \in \mathbb{N}_0$, $p_1, \ldots, p_r \in \mathsf{P}(K,X)$ and $e_1, \ldots, e_r \in \mathbb{N}$. If $r = 0$, then $z = g \in K[X]$ and we are done. If $r \ge 1$, then from 2.3.1 we obtain a representation

$$z = \sum_{i=1}^{r} \frac{b_i}{p_i^{e_i}}, \quad \text{where } b_i \in K[X] \text{ and } p_i \nmid b_i \text{ for all } i \in [1,r]. \qquad (\#)$$

For $i \in [1,r]$, 1. implies

$$b_i = \sum_{j \ge 0} b_{i,j} p_i^j,$$

where $b_{i,j} \in K[X]$, $\partial(b_{i,j}) < \partial(p_i)$ for all $j \ge 0$ and $b_{i,j} = 0$ for all $j > e_i$. Hence

$$\frac{b_i}{p_i^{e_i}} = \sum_{j=0}^{e_i} \frac{b_{i,j} p_i^j}{p_i^{e_i}} = \sum_{j=0}^{e_i - 1} \frac{b_{i,j}}{p_i^{e_i - j}} + b_{i,e_i}.$$

Inserting this expression into (#) yields a representation of z in the desired form.

Uniqueness. Assume that

$$z = \sum_{p \in \mathsf{P}(K,X)} \sum_{j \geq 1} \frac{b_{p,j}}{p^j} + g = \sum_{p \in \mathsf{P}(K,X)} \sum_{j \geq 1} \frac{b^*_{p,j}}{p^j} + g^*,$$

where $g,\ g^*,\ b_{p,j},\ b^*_{p,j} \in K[X]$, $\partial(b_{p,j}) < \partial(p)$, $\partial(b^*_{p,j}) < \partial(p)$ for all and $b_{p,j} = b^*_{p,j} = 0$ for almost all $(p,j) \in \mathsf{P}(K,X) \times \mathbb{N}_0$. If $b_{p,j} = b^*_{p,j}$ for all $(p,j) \in \mathsf{P}(K,X) \times \mathbb{N}_0$, then $g = g^*$ and we are done. Thus suppose that there exists some $p_1 \in \mathsf{P}(K,X)$ such that $b_{p_1,j} \neq b^*_{p_1,j}$ for some $j \geq 0$, and let $m = \max\{j \geq 0 \mid b_{p_1,j} \neq b^*_{p_1,j}\}$. Then there exist some $N \in \mathbb{N}$, $r \geq 2$ and $p_2, \ldots, p_r \in \mathsf{P}(K,X) \setminus \{p_1\}$ such that

$$\sum_{j=1}^{m} \frac{b_{p_1,j} - b^*_{p_1,j}}{p_1^j} = \sum_{i=2}^{r} \sum_{j=1}^{N} \frac{b^*_{p_i,j} - b_{p_i,j}}{p_i^j} + g^* - g.$$

This relation has the form

$$\frac{1}{p_1^m}(b_{p_1,m} - b^*_{p_1,m} + p_1 h) = \frac{1}{(p_2 \cdot \ldots \cdot p_r)^N} h^*$$

for some $h,\ h^* \in K[X]$ and thus $(p_2 \cdot \ldots \cdot p_r)^N (b_{p_1,m} - b^*_{p_1,m} + p_1 h) = p_1^m h^*$, hence $p_1 \mid (b_{p_1,m} - b^*_{p_1,m})$, a contradiction, since $\partial(b_{p_1,m} - b^*_{p_1,m}) < \partial(p_1)$. □

Now we turn towards module theory. We start with some general remarks concerning modules over a commutative ring R. Let M be an R-module. Recall that a family $(u_i)_{i \in I}$ in M is **linearly independent** over R if the map

$$\phi\colon R^{(I)} \to M, \quad \text{defined by } \phi((a_i)_{i \in I}) = \sum_{i \in I} a_i u_i,$$

is an a monomorphism, and it is an (R-)**basis** of M if ϕ is an isomorphism. A subset B of M is linearly independent resp. a basis of M if the family $(b)_{b \in B}$ has this property.

It is well known that two bases of a vector space have the same cardinality. We prove that this also holds for modules over a commutative ring.

Theorem 2.3.3. *Let R be a commutative ring and F a free R-module.*

1. *Let $(u_i)_{i \in I}$ be an R-basis of F and $\mathfrak{a}$ an ideal of R. Then the family $(u_i + \mathfrak{a}F)_{i \in I}$ is an $R/\mathfrak{a}$-basis of $F/\mathfrak{a}F$.*

2. *If B and B' are R-bases of M, then* $\operatorname{card}(B) = \operatorname{card}(B')$.

Proof. 1. Recall that

$$\mathfrak{a}F = \Big\{ \sum_{i \in I} a_i u_i \ \Big|\ (a_i)_{i \in I} \in \mathfrak{a}^{(I)} \Big\} \quad \text{and} \quad F/\mathfrak{a}F = {}_{R/\mathfrak{a}}(\{u_i + \mathfrak{a}F \mid i \in I\}).$$

If $(\lambda_i + \mathfrak{a})_{i\in I} \in (R/\mathfrak{a})^{(I)}$ is such that

$$0 = \sum_{i\in I}(\lambda_i + \mathfrak{a})(u_i + \mathfrak{a}F) = \sum_{i\in I} \lambda_i u_i + \mathfrak{a}F, \quad \text{then} \sum_{i\in I} \lambda_i u_i \in \mathfrak{a}F,$$

hence $\lambda_i \in \mathfrak{a}$ and thus $\lambda_i + \mathfrak{a} = 0 \in R/\mathfrak{a}$ for all $i \in I$.

2. By 1.1.5.2 R has a maximal ideal $\mathfrak{m}$. Let $\pi\colon F \to F/\mathfrak{m}F$ be the residue class epimorphism. By 1. the restrictions $\pi \restriction B$ and $\pi \restriction B'$ are injective. Since $\pi(B)$ and $\pi(B')$ are $R/\mathfrak{m}$-bases of the vector space $M/\mathfrak{m}M$, we get $\text{card}(\pi(B)) = \text{card}(\pi(B'))$, and therefore $\text{card}(B) = \text{card}(B')$. □

Now let $\mathfrak{o}$ be a domain, $K = \mathsf{q}(\mathfrak{o})$ and M an $\mathfrak{o}$-module. For $a \in M$, the ideal $\text{Ann}_{\mathfrak{o}}(a) = \{\lambda \in \mathfrak{o} \mid \lambda a = 0\}$ of $\mathfrak{o}$ is called the **annihilator** of a, and the $\mathfrak{o}$-submodule $M_{\text{tor}} = \{a \in M \mid \text{Ann}_{\mathfrak{o}}(a) \neq \mathbf{0}\}$ is called the **torsion submodule** of M. If $M = M_{\text{tor}}$, then M is called a **torsion module**, and if $M_{\text{tor}} = \mathbf{0}$, then M is called **torsion-free**.

The tensor product $K \otimes_{\mathfrak{o}} M$ is a vector space over K, and M_{tor} is the kernel of the natural homomorphism $M \to K \otimes_{\mathfrak{o}} M$, given by $a \mapsto 1 \otimes a$. We call

$$\text{rk}(M) = \text{rk}_{\mathfrak{o}}(M) = \dim_K(K \otimes_{\mathfrak{o}} M) \in \mathbb{N}_0 \cup \{\infty\}$$

the **rank** of M (over $\mathfrak{o}$). By definition, $\text{rk}(M)$ is the supremum of all lengths n of linearly independent n-tuples $(u_1, \ldots, u_n) \in M^n$. We recall the following basic properties of the rank:

- If $\mathbf{0} \to M' \to M \to M'' \to \mathbf{0}$ is an exact sequence of $\mathfrak{o}$-modules, then

$$\text{rk}(M) = \text{rk}(M') + \text{rk}(M'').$$

- If $(u_1, \ldots, u_n)$ is an ($\mathfrak{o}$-)basis of M, then $\text{rk}(M) = n$.
- M is a torsion module if and only if $\text{rk}(M) = 0$.

An $\mathfrak{o}$-module M is torsion-free if and only if it is contained in a vector space U over K. In this case we denote by $KM = {}_K(M)$ the vector space spanned by M in U, linear (in)dependence over $\mathfrak{o}$ is the same as linear (in)dependence over K, $\text{rk}_{\mathfrak{o}}(M) = \dim_K(KM)$, and there is a unique K-linear isomorphism $\iota\colon K \otimes_{\mathfrak{o}} M \to KM$ satisfying $\iota(\lambda \otimes m) = \lambda m$ for all $\lambda \in K$ and $m \in M$. We identify: $K \otimes_{\mathfrak{o}} M = KM$. .

Theorem 2.3.4 (Free modules over principal ideal domains). *Let $\mathfrak{o}$ be a principal ideal domain, M a free $\mathfrak{o}$-module with basis B and N an $\mathfrak{o}$-submodule of M. Then N is a free $\mathfrak{o}$-module and possesses some basis B' such that* $\text{card}(B') \le \text{card}(B)$.

Proof. Let Ω be the set of all triples (E, E', φ), where $E' \subset E \subset B$, and $\varphi\colon E' \to N \cap {}_{\mathfrak{o}}(E)$ is such that $\varphi(E')$ is an $\mathfrak{o}$-basis of $N \cap {}_{\mathfrak{o}}(E)$. Then $\Omega \neq \emptyset$, since $(\emptyset, \emptyset, \emptyset) \in \Omega$, and for (E, E', φ), $(F, F', \psi) \in \Omega$ we define

$$(E, E', \varphi) \le (F, F', \psi) \quad \text{if} \quad E \subset F\,, \; E' \subset F' \; \text{ and } \; \psi \,|\, E' = \varphi\,.$$

We shall prove:

A. $(\Omega, \le)$ has a maximal element.

B. If (E, E', φ) is a maximal element of Ω, then $E = B$.

Suppose that **A** and **B** hold, and let (B, E', φ) be a maximal element of Ω. Then $B' = \varphi(E')$ is an $\mathfrak{o}$-basis of N, and $\mathrm{card}(B') \le \mathrm{card}(E') \le \mathrm{card}(B)$. □

Proof of **A.** Clearly, $(\Omega, \le)$ is a partially ordered non-empty set. We shall apply Zorn's lemma, and for that we prove that every chain in Ω has an upper bound. Let $\mathcal{S} = \{(E_\lambda, E'_\lambda, \varphi_\lambda \mid \lambda \in \Lambda\}$ be a chain in Ω,

$$E = \bigcup_{\lambda \in \Lambda} E_\lambda, \quad E' = \bigcup_{\lambda \in \Lambda} E'_\lambda \quad \text{and} \quad \varphi = \bigcup_{\lambda \in \Lambda} \varphi_\lambda.$$

We prove that $(E, E', \varphi) \in \Omega$ (then it is an upper bound of $\mathcal{S}$). By definition we obtain $E' \subset E \subset B$,

$$\varphi\colon E' \to \bigcup_{\lambda \in \Lambda} N \cap {}_{\mathfrak{o}}(E_\lambda) \hookrightarrow N \cap {}_{\mathfrak{o}}\Big(\bigcup_{\lambda \in \Lambda} E_\lambda\Big) = N \cap {}_{\mathfrak{o}}(E) \quad \text{is a map,}$$

and we must prove that $\varphi(E')$ is an $\mathfrak{o}$-basis of $N \cap {}_{\mathfrak{o}}(E)$. Since $\{\varphi(E'_\lambda) \mid \lambda \in \Lambda\}$ is a chain of linearly independent sets, its union

$$\varphi(E') = \bigcup_{\lambda \in \Lambda} \varphi(E'_\lambda) \subset N \cap {}_{\mathfrak{o}}(E)$$

is again linearly independent. If $x \in N \cap {}_{\mathfrak{o}}(E)$, then there exists some $\lambda \in \Lambda$ such that $x \in N \cap {}_{\mathfrak{o}}(E_\lambda) = {}_{\mathfrak{o}}(\varphi(E'_\lambda)) \subset {}_{\mathfrak{o}}(\varphi(E'))$, and therefore ${}_{\mathfrak{o}}(\varphi(E')) = N \cap {}_{\mathfrak{o}}(E)$. □[**A.**]

Proof of **B.** Let (E, E', φ) be a maximal element of Ω, and assume to the contrary that $E \subsetneq B$. Let $b \in B \setminus E$, $\overline{b} = b + (N + {}_{\mathfrak{o}}(E)) \in M/(N + {}_{\mathfrak{o}}(E))$ and consider the $\mathfrak{o}$-ideal

$$I = \mathrm{Ann}_{\mathfrak{o}}(\overline{b}) = \{\lambda \in \mathfrak{o} \mid y + \lambda b \in N \;\text{ for some } y \in {}_{\mathfrak{o}}(E)\}.$$

CASE 1: $I = \mathbf{0}$. If $x \in {}_{\mathfrak{o}}(E \cup \{b\}) \cap N = ({}_{\mathfrak{o}}(E) + \mathfrak{o}b) \cap N$, then $x = y + \lambda b$ for some $y \in {}_{\mathfrak{o}}(E)$ and $\lambda \in \mathfrak{o}$, and therefore $\lambda b \in N + {}_{\mathfrak{o}}(E)$. Hence $\lambda = 0$, $x \in {}_{\mathfrak{o}}(E)$, and therefore $N \cap {}_{\mathfrak{o}}(E \cup \{b\}) = N \cap {}_{\mathfrak{o}}(E)$, which implies that $(E \cup \{b\}, E', \varphi) \in \Omega$, a contradiction to the maximality of (E, E', φ).

CASE 2: $I = \alpha\mathfrak{o}$ for some $\alpha \in \mathfrak{o}^\bullet$. Let $y_0 \in {}_{\mathfrak{o}}(E)$ be such that $y_0 + \alpha b \in N$, set $\overline{E} = E \cup \{b\}$, $\overline{E}' = E' \cup \{b\}$, and define $\overline{\varphi}\colon \overline{E}' \to N \cap {}_{\mathfrak{o}}(\overline{E})$ by $\overline{\varphi} \restriction E = \varphi$ and $\overline{\varphi}(b) = y_0 + \alpha b \in {}_{\mathfrak{o}}(E) + \mathfrak{o}b = {}_{\mathfrak{o}}(\overline{E})$. Then $\overline{E}' \subset \overline{E} \subset B$, and we shall prove that $(\overline{E}, \overline{E}', \overline{\varphi}) \in \Omega$, which again contradicts the maximality of (E, E', φ). For this we must prove that $\overline{\varphi}(\overline{E}') = \varphi(E') \cup \{y_0 + \alpha b\}$ is an $\mathfrak{o}$-basis of $N \cap ({}_{\mathfrak{o}}(E) + \mathfrak{o}b)$.

1) $\varphi(E') \cup \{y_0 + \alpha b\}$ is linearly independent: Let $n \in \mathbb{N}$ and $e_1, \ldots, e_n \in E'$ such that $\varphi(e_1), \ldots, \varphi(e_n)$ are distinct and $\lambda_1, \ldots, \lambda_n \in \mathfrak{o}$. If

$$\sum_{i=1}^{n} \lambda_i \varphi(e_i) + \lambda(y_0 + \alpha b) = 0,$$

then $\lambda(y_0 + \alpha b) \in {}_{\mathfrak{o}}(\varphi(E')) \subset {}_{\mathfrak{o}}(E)$ and thus $\lambda\alpha b \in {}_{\mathfrak{o}}(E)$ since $y_0 \in {}_{\mathfrak{o}}(E)$. As $E \cup \{b\}$ is linearly independent, we get $\lambda\alpha = 0$. Hence $\lambda = 0$, and since $\varphi(E')$ is linearly independent, we finally obtain $\lambda_i = 0$ for all $i \in [1, n]$.

2) ${}_{\mathfrak{o}}(\varphi(E') \cup \{y_0 + \alpha b\}) = N \cap ({}_{\mathfrak{o}}(E) + \mathfrak{o}b)$. If $z \in N \cap ({}_{\mathfrak{o}}(E) + \mathfrak{o}b)$, then $z = y + \lambda b$, where $y \in {}_{\mathfrak{o}}(E)$ and $\lambda \in \mathfrak{o}$, and consequently $\lambda \in I$, say $\lambda = \alpha\beta$ for some $\beta \in \mathfrak{o}$. It follows that $z - \beta(y_0 + \alpha b) = y - \beta y_0 \in N \cap {}_{\mathfrak{o}}(E) = {}_{\mathfrak{o}}(\varphi(E'))$ and therefore $z \in {}_{\mathfrak{o}}(\varphi(E') \cup \{y_0 + \alpha b\})$. □[**B.**]

Theorem 2.3.5 (Finitely generated modules over principal ideal domains). *Let $\mathfrak{o}$ be a principal ideal domain.*

1. Let F be a finitely generated free $\mathfrak{o}$-module, $\mathrm{rk}_{\mathfrak{o}} F = n$, and let N be an $\mathfrak{o}$-submodule of F. Then N is free, and $\mathrm{rk}_{\mathfrak{o}} N = m \le n$. Also there exist an $\mathfrak{o}$-basis $(u_1, \ldots, u_n)$ of F and $d_1, \ldots, d_m \in \mathfrak{o}^\bullet$ satisfying $d_1\mathfrak{o} \supset d_2\mathfrak{o} \supset \ldots \supset d_m\mathfrak{o}$ such that $(d_1u_1, \ldots, d_mu_m)$ is an $\mathfrak{o}$-basis of N. The ideals $d_1\mathfrak{o}, \ldots, d_m\mathfrak{o}$ are uniquely determined by N.

2. Let M be a finitely generated $\mathfrak{o}$-module. Then there exist integers $m, n \in \mathbb{N}_0$ such that $m \le n$, $x_1, \ldots, x_n \in M$ and $d_1, \ldots, d_m \in \mathfrak{o}^\bullet$ such that $M = {}_{\mathfrak{o}}(x_1, \ldots, x_n)$, $d_1\mathfrak{o} \subset d_2\mathfrak{o} \subset \ldots \subset d_m\mathfrak{o} \subsetneq \mathfrak{o}$,

$$\mathrm{Ann}_{\mathfrak{o}}(x_i) = \begin{cases} d_i\mathfrak{o} & \text{if } i \in [1, m], \\ 0 & \text{if } i \in [m+1, n], \end{cases}$$

$$\text{and} \quad M \cong_{\mathfrak{o}} \mathfrak{o}/d_1\mathfrak{o} \oplus \ldots \oplus \mathfrak{o}/d_m\mathfrak{o} \oplus \mathfrak{o}^{n-m}.$$

In these assertions n is the minimal number of generators of M, $M_{\mathrm{tor}} = \mathfrak{o}x_1 \oplus \ldots \oplus \mathfrak{o}x_m$, $n - m = \mathrm{rk}_{\mathfrak{o}}(M)$, and the ideals $d_1\mathfrak{o}, \ldots, d_m\mathfrak{o}$ are uniquely determined by M. In particular, if M is torsion-free, then M is free.

3. Let F be an $\mathfrak{o}$-module, $n \in \mathbb{N}$ and $u_1, \ldots, u_n \in F$ such that $F = \mathfrak{o}u_1 + \ldots + \mathfrak{o}u_n$. Then $\mathrm{rk}_{\mathfrak{o}}(F) \le n$, and equality holds if and only if $(u_1, \ldots, u_n)$ is an $\mathfrak{o}$-basis of F.

Proof. 1. *Existence*: By induction on $n = \mathrm{rk}_{\mathfrak{o}}(F)$. If $n = 0$ or $N = \mathbf{0}$, there is nothing to do. Thus suppose that $N \neq \mathbf{0}$.

$n \geq 1$, $n-1 \to n$: Let $F^* = \mathrm{Hom}_{\mathfrak{o}}(F, \mathfrak{o})$ be the dual $\mathfrak{o}$-module. If $f \in F^*$, then $f(N)$ is an ideal of $\mathfrak{o}$, and we choose $f_1 \in F^*$, $d_1 \in \mathfrak{o}$ and $x_1 \in N$ such that $f_1(N) = d_1\mathfrak{o}$ is a maximal element of the set $\{f(N) \mid f \in F^*\}$ and $f_1(x_1) = d_1$. We prove first:

A. $0 \neq x_1 \in d_1F$.

Proof of **A.** Let $(e_1, \ldots, e_n)$ be a basis of F, and let $(e_1^*, \ldots, e_n^*)$ be the dual basis of F^* satisfying $e_i^*(e_j) = \delta_{i,j}$ for all $i, j \in [1, n]$. If $x \in N^\bullet$, then $x = \lambda_1 e_1 + \ldots + \lambda_n e_n$, where $\lambda_1, \ldots, \lambda_n \in \mathfrak{o}$, and there exists some $j \in [1, n]$ such that $\lambda_j = e_j^*(x) \neq 0$, hence $e_j^*(N) \neq \mathbf{0}$, and therefore $d_1 \neq 0$ and $x_1 \neq 0$. Let $x_1 = a_1e_1 + \ldots + a_ne_n$, where $a_1, \ldots, a_n \in \mathfrak{o}$. For $i \in [1, n]$, let q_i be a gcd of a_i and d_1, and set $q_i = a_ib_i + d_1c_i$ for some b_i, $c_i \in \mathfrak{o}$. Then $g_i = b_ie_i^* + c_if_1 \in F^*$ and $g_i(x_1) = b_ia_i + c_id_1 = q_i$. It follows that $f_1(N) = d_1\mathfrak{o} \subset q_i\mathfrak{o} \subset g_i(N)$ and thus $f_1(N) = g_i(N)$ by the maximality of $f_1(N)$. Hence we obtain $a_i \in q_i\mathfrak{o} = d_1\mathfrak{o}$ for all $i \in [1, n]$ and consequently $x_1 \in d_1F$. □[**A.**]

Let $u_1 \in F$ be such that $x_1 = d_1u_1$. Since $f_1(x_1) = d_1 = d_1f_1(u_1)$ we obtain $f_1(u_1) = 1$. We set $F_1 = \mathrm{Ker}(f_1) \subset F$, $N_1 = N \cap F_1 \subset N$, and then we prove:

B. $F = \mathfrak{o}u_1 \oplus F_1$ and $N = \mathfrak{o}d_1u_1 \oplus N_1$.

Proof of **B.** If $x \in F$, then $x - f_1(x)u_1 \in \mathrm{Ker}(f_1) = F_1$, hence $x \in \mathfrak{o}u_1 + F_1$, and therefore $F = \mathfrak{o}u_1 + F_1$. If $x \in N$, then $f_1(x)u_1 \in d_1\mathfrak{o}u_1 = \mathfrak{o}x_1 \subset N$ and consequently $x - f_1(x)u_1 \in F_1 \cap N = N_1$, hence $x \in \mathfrak{o}d_1u_1 + N_1$ and therefore $N = \mathfrak{o}d_1u_1 + N_1$. It remains to prove that $\mathfrak{o}u_1 \cap F_1 = \mathbf{0}$ (then it follows that $\mathfrak{o}d_1u_1 \cap N_1 = \mathbf{0}$ and both sums are direct). If $x \in \mathfrak{o}u_1 \cap F_1$, then $x = \lambda u_1$, where $\lambda \in \mathfrak{o}$, and $0 = f_1(x) = \lambda$, hence $x = 0$. □[**B.**]

By 2.3.4 F_1 is a free $\mathfrak{o}$-module, and $1 \leq k = \mathrm{rk}_{\mathfrak{o}}(F_1) \leq n$. Let $(v_2, \ldots v_{k+1})$ be a basis of F_1. Then $(u_1, v_2, \ldots, v_{k+1})$ is a basis of F and therefore $k = n-1$. The induction hypothesis yields some $m \in [1, n]$, a basis $(u_2, \ldots, u_n)$ of F_1 and elements $d_2, \ldots, d_m \in \mathfrak{o}^\bullet$ such that $d_2\mathfrak{o} \supset \ldots \supset d_m\mathfrak{o}$ and $(d_2u_2, \ldots, d_mu_m)$ is a basis of N_1. Then $(u_1, \ldots, u_n)$ is a basis of F, $(d_1u_1, \ldots, d_mu_m)$ is a basis of N, and it remains to prove that $d_1\mathfrak{o} \supset d_2\mathfrak{o}$ (we will give the uniqueness proof only after the proof of 2.).

Let d be a gcd of d_1 and d_2, and let a_1, $a_2 \in \mathfrak{o}$ be such that $d = a_1d_1 + a_2d_2$. If $(u_1^*, \ldots, u_n^*)$ is the dual basis of $(u_1, \ldots, u_n)$, then $g = a_1u_1^* + a_2u_2^* \in F^*$, and if $u = d_1u_1 + d_2u_2$, then $u \in N$ and $g(u) = d$. Hence $d_1\mathfrak{o} \subset d\mathfrak{o} \subset g(N)$, and the maximality of $d_1\mathfrak{o}$ implies $d_1\mathfrak{o} = d\mathfrak{o} \supset d_2\mathfrak{o}$.

2. Let F be a finitely generated free $\mathfrak{o}$-module of minimal rank admitting a surjective $\mathfrak{o}$-homomorphism $g\colon F \to M$, and set $N = \mathrm{Ker}(g)$. We may assume that $n = \mathrm{rk}(F) \geq 1$. By 1. there exist a basis $(u_1, \ldots, u_n)$ of F, some $m \in [0, n]$ and $d_1, \ldots, d_m \in \mathfrak{o}^\bullet$ such that $d_1\mathfrak{o} \supset d_2\mathfrak{o} \supset \ldots \supset d_m\mathfrak{o}$ and $(d_1u_1, \ldots, d_mu_m)$

is a basis of N. If $x_i = g(u_i)$, then $M = {}_{\mathfrak o}(x_1, \ldots, x_n)$, and by the minimality of F we get $x_i \neq 0$ for all $i \in [1, n]$. In particular, $d_1\mathfrak o \subsetneq \mathfrak o$, and we set $d_i = 0$ for all $i \in [m+1, n]$. Then $\operatorname{Ann}_{\mathfrak o}(x_i) = d_i\mathfrak o$ and $\mathfrak o x_i \cong_{\mathfrak o} \mathfrak o/d_i\mathfrak o$ for all $i \in [1, n]$. Hence it suffices to show that, for all $(a_1, \ldots, a_n) \in \mathfrak o^n$,

$$\sum_{i=1}^{n} a_i x_i = 0 \quad \text{implies} \quad a_i \in d_i\mathfrak o \ \text{ for all } \ i \in [1, n]$$

(then $M = \mathfrak o x_1 \oplus \ldots \oplus \mathfrak o x_n \cong_{\mathfrak o} \mathfrak o/d_1\mathfrak o \oplus \ldots \oplus \mathfrak o/d_m\mathfrak o \oplus \mathfrak o^{n-m}$). If $a_1, \ldots, a_n \in \mathfrak o$ are such that $a_1x_1 + \ldots + a_nx_n = 0$, then $a_1u_1 + \ldots + a_nu_n \in N$, hence $a_i \in d_i\mathfrak o$ and therefore $a_ix_i = 0$ for all $i \in [1, n]$.

It remains to prove that the ideals $d_1\mathfrak o, \ldots, d_m\mathfrak o$ are uniquely determined by M. By definition, n is the minimal number of generators of M; hence $m = n - \operatorname{rk}_{\mathfrak o}(M)$ is uniquely determined by M, and as $M_{\text{tor}} = \mathfrak o x_1 \oplus \ldots \oplus \mathfrak o x_m$, it suffices to prove:

> **I.** Let A be an $\mathfrak o$-module such that $A \cong \mathfrak o/d_1\mathfrak o \oplus \ldots \oplus \mathfrak o/d_m\mathfrak o$, where $m \in \mathbb N_0$ and $d_1, \ldots, d_m \in \mathfrak o^\bullet$ be such that $\mathfrak o \supsetneq d_1\mathfrak o \supset \ldots \supset d_m\mathfrak o$. Then the ideals $d_1\mathfrak o, \ldots, d_m\mathfrak o$ are uniquely determined by A.

Proof of **I.** By induction on m. Let $A_1, \ldots, A_m$ be submodules of A such that $A = A_1 \oplus \ldots \oplus A_m$ and $A_j \cong \mathfrak o/d_j\mathfrak o$ for all $j \in [1, m]$. Then $d_m\mathfrak o = \operatorname{Ann}_{\mathfrak o}(A)$, and since $A/A_m \cong \mathfrak o/d_1\mathfrak o \oplus \ldots \oplus \mathfrak o/d_{m-1}\mathfrak o$, the uniqueness of $d_1\mathfrak o, \ldots, d_{m-1}\mathfrak o$ follows from the induction hypothesis. □[**I.**]

Proof of uniqueness in 1. Let $(u_1, \ldots, u_n)$ be a basis of F, $m \in [0, n]$, $k \in [0, m]$ and $d_1, \ldots, d_m \in \mathfrak o^\bullet$ such that

$$d_1\mathfrak o = \ldots = d_k\mathfrak o \supsetneq d_{k+1}\mathfrak o \supset \ldots \supset d_m\mathfrak o,$$

and $(d_1u_1, \ldots, d_mu_m)$ is an $\mathfrak o$-basis of N. Then it follows that $m = \operatorname{rk}_{\mathfrak o}(N)$ and $F/N \cong \mathfrak o/d_{k+1}\mathfrak o \oplus \ldots \oplus \mathfrak o/d_m\mathfrak o$. Hence k and the ideals $d_{k+1}\mathfrak o, \ldots, d_m\mathfrak o$ are uniquely determined by 2.

3. Let $(e_1, \ldots, e_n)$ be the canonical basis of $\mathfrak o^n$, let $\pi \colon \mathfrak o^n \to F$ be the unique epimorphism satisfying $\pi(e_i) = u_i$ for all $i \in [1, n]$, and $K = \operatorname{Ker}(\pi)$. By 1., K is a free $\mathfrak o$-module, and $\operatorname{rk}_{\mathfrak o}(K) = m \in [0, n]$. Hence $\operatorname{rk}_{\mathfrak o}(F) = n - m \le n$. If $(u_1, \ldots, u_n)$ is an $\mathfrak o$-basis of F, then $\operatorname{rk}_{\mathfrak o}(F) = n$ by the very definition. Thus assume that $\operatorname{rk}_{\mathfrak o}(F) = n$. Then $m = 0$, hence $K = \mathbf 0$, π is an isomorphism, and thus $(u_1, \ldots, u_n)$ is an $\mathfrak o$-basis of F. □

We reformulate and amend Theorem 2.3.5 in the important special case of abelian groups. Recall that an abelian group A is a $\mathbb Z$-module; it is called **free** if it is a free $\mathbb Z$-module, we set $\operatorname{rk}(A) = \operatorname{rk}_{\mathbb Z}(A)$, and by a **basis** of A we mean a $\mathbb Z$-basis.

Theorem 2.3.6 (Structure of finitely generated abelian groups). *Let A be a finitely generated abelian group.*

1. Let A be free, $\mathrm{rk}(A) = n$, and let B be a subgroup of A. Then B is free, $\mathrm{rk}(B) = m \le n$, and there exist a basis $(u_1, \ldots, u_n)$ of A and $e_1, \ldots, e_m \in \mathbb{N}$ such that $e_1 \,|\, e_2 \,|\, \ldots \,|\, e_m$ and $(e_1u_1, \ldots, e_mu_m)$ is a basis of B. In particular,

$$A/B \cong \mathbb{Z}^{n-m} \oplus \mathbb{Z}/e_1\mathbb{Z} \oplus \ldots \oplus \mathbb{Z}/e_m\mathbb{Z},$$

and A/B is finite if and only if $\mathrm{rk}(A) = \mathrm{rk}(B)$.

2. There exist uniquely determined integers $r, t \in \mathbb{N}_0$ and $e_1, \ldots, e_t \in \mathbb{N}$ such that

$$1 < e_1 \,|\, e_2 \,|\, \ldots \,|\, e_t \quad \text{and} \quad A \cong \mathbb{Z}^r \oplus \mathbb{Z}/e_1\mathbb{Z} \oplus \ldots \oplus \mathbb{Z}/e_t\mathbb{Z}.$$

3. Let A be free, B a subgroup of A, $\mathrm{rk}(A) = \mathrm{rk}(B) = n \in \mathbb{N}$, $\boldsymbol{u} = (u_1, \ldots, u_n)$ a basis of A, $\boldsymbol{v} = (v_1, \ldots, v_n)$ a basis of B, and let $T \in \mathsf{M}_n(\mathbb{Z})$ such that $\boldsymbol{v} = \boldsymbol{u}T$. Then $(A:B) = |\det(T)|$. In particular, if $t \in \mathbb{Z}$, then $(A{:}tA) = |t|^n$.

Proof. 1. and 2. hold by 2.3.5.

3. By 1. there exist a basis $\boldsymbol{w} = (w_1, \ldots, w_n)$ of A and $e_1, \ldots, e_n \in \mathbb{N}$ such that $(e_1w_1, \ldots, e_nw_n)$ is a basis of B. Then $A/B \cong \mathbb{Z}/e_1\mathbb{Z} \oplus \ldots \oplus \mathbb{Z}/e_n\mathbb{Z}$, and consequently $(A{:}B) = |A/B| = e_1 \cdot \ldots \cdot e_n$. Since $\boldsymbol{w}$ and $\boldsymbol{u}$ are bases of A, and $\boldsymbol{v}$ and $(e_1w_1, \ldots, e_nw_n) = \boldsymbol{w}\,\mathrm{diag}(e_1, \ldots, e_n)$ are bases of B, there exist matrices $U, V \in \mathsf{GL}_n(\mathbb{Z})$ such that $\boldsymbol{w} = \boldsymbol{u}U$ and $\boldsymbol{w}\,\mathrm{diag}(e_1, \ldots, e_n) = \boldsymbol{v}V$. Consequently $\boldsymbol{u}TV = \boldsymbol{v}V = \boldsymbol{w}\,\mathrm{diag}(e_1, \ldots, e_n) = \boldsymbol{u}U\,\mathrm{diag}(e_1, \ldots, e_n)$, and thus eventually $TV = U\,\mathrm{diag}(e_1, \ldots, e_n)$ and $|\det(T)| = e_1 \cdot \ldots \cdot e_n$, since $|\det(U)| = |\det(V)| = 1$. □

We close the section with a formulation of the (well-known) structure theorem for finite abelian groups which is now an immediate consequence.

Theorem 2.3.7 (Structure of finite abelian groups). *Let G be a (multiplicative) finite abelian group.*

1. There exist uniquely determined integers $k \in \mathbb{N}_0$ and $e_1, \ldots, e_k \in \mathbb{N}$ such that $1 < e_1 \,|\, e_2 \,|\, \ldots \,|\, e_k$, and G is an (inner) direct product $G = G_1 \cdot \ldots \cdot G_k$, where G_i is a cyclic subgroup of G of order e_i for all $i \in [1, k]$.

2. Suppose that $|G| = p_1^{b_1} \cdot \ldots \cdot p_r^{b_r}$, where $r \in \mathbb{N}_0$, $p_1, \ldots, p_r$ are distinct primes and $b_1, \ldots, b_r \in \mathbb{N}$. Then $G = P_1 \cdot \ldots \cdot P_r$ is an (inner) direct product of subgroups P_i of G such that $|P_i| = p_i^{b_i}$ for all $i \in [1, r]$. At it P_i is the p_i-Sylow subgroup consisting of all elements of p_i-power order of G.

Proof. 1. Obvious by 2.3.6.2.

2. Let $G = G_1 \cdot \ldots \cdot G_k$ be as in 1. For $j \in [1,k]$, let

$$e_j = |G_j| = \prod_{i=1}^{r} p_i^{l_{i,j}}, \quad \text{where } \ l_{i,j} \in [0, b_j] \ \text{ for all } \ i \in [1,r].$$

Then it follows by the Chinese remainder theorem that

$$G_j \cong \mathbb{Z}/e_j\mathbb{Z} \cong \prod_{i=1}^{r} \mathbb{Z}/p_i^{l_{i,j}}\mathbb{Z} \quad \text{for all } \ j \in [1,r],$$

and consequently

$$G = P_1 \cdot \ldots \cdot P_r, \quad \text{where } \ P_i \cong \prod_{j=1}^{k} \mathbb{Z}/p_i^{l_{i,j}} \quad \text{for all } \ i \in [1,r]. \qquad \square$$

2.4 Integral elements 1: Ring-theoretic aspects

Let R and S be commutative rings such that $R \subset S$. An element $x \in S$ is called **integral** over R if there exists a monic polynomial $f \in R[X]$ such that $f(x) = 0$. Explicitly, this means that there is a relation of the form

$$x^n + a_{n-1}x^{n-1} + \ldots + a_1x + a_0 = 0, \text{ where } n \in \mathbb{N} \text{ and } \ a_0, \ldots, a_{n-1} \in R.$$

Every such relation is called an **integral equation** for x over R. We denote by $\mathrm{cl}_S(R)$ the set of all elements of S which are integral over R and call it the **integral closure** of R in S. Note that $R \subset \mathrm{cl}_S(R)$ (an element $x \in R$ is the zero of the monic polynomial $X - x \in R[X]$). If $\mathrm{cl}_S(R) = S$ then we say that S is **integral** over R. If $\mathrm{cl}_S(R) = R$ then R is called **(relatively) integrally closed** in S. If R and S are fields, then S is integral over R if and only if S is algebraic over R.

A domain $\mathfrak{o}$ is called **integrally closed** if it is integrally closed in its quotient field. If $\mathfrak{o}$ is an integrally closed domain and $K = \mathsf{q}(\mathfrak{o})$, then all roots of unity of K lie in $\mathfrak{o}$.

Theorem 2.4.1. *Every factorial domain is integrally closed.*

Proof. Let $\mathfrak{o}$ be a factorial domain, $K = \mathsf{q}(\mathfrak{o})$, $\ x \in \mathrm{cl}_K(\mathfrak{o})$, and let

$$x^n + a_{n-1}x^{n-1} + \ldots + a_1x + a_0 = 0 \quad (\text{for some } n \in \mathbb{N} \text{ and } a_0, \ldots, a_{n-1} \in R)$$

be an integral equation for x over $\mathfrak{o}$. By 2.1.6.2 there exist elements $a \in \mathfrak{o}^\bullet$ and $b \in \mathfrak{o}$ such that $x = a^{-1}b$ and $(a,b) = 1$. Clearing denominators in the integral equation, we obtain $0 = b^n + aa_{n-1}b^{n-1} + \ldots + a^{n-1}a_1b + a^na_0 = b^n + ac$ for some $c \in \mathfrak{o}$. If $x \notin \mathfrak{o}$, then $a \notin \mathfrak{o}^\times$, and there exists some prime element $\pi \in \mathfrak{o}$ such that $\pi \,|\, a$. As $b^n = -ac$ it follows that $\pi \,|\, b$, which contradicts $(a,b) = 1$. □

Theorem 2.4.2. *Let R and S be commutative rings such that $R \subset S$, $x \in S$, and let M be an $R[x]$-submodule of S such that $\mathrm{Ann}_{R[x]}(S) = \mathbf{0}$ and M is a finitely generated R-module. Then x is integral over R.*

Proof. Let $M = {}_R(b_1, \ldots, b_m)$, where $m \in \mathbb{N}$ and $b_1, \ldots, b_m \in M$. If $j \in [1,m]$, then $xb_j \in M$ and therefore

$$xb_j = \sum_{\mu=1}^{m} a_{j,\mu}b_\mu, \quad \text{where } a_{j,\mu} \in R.$$

If $A = (x\delta_{j,\mu} - a_{j,\mu})_{j,\mu\in[1,m]} \in \mathsf{M}_m(S)$ and $\boldsymbol{b} = (b_1, \ldots, b_m)^{\mathrm{t}} \in \mathsf{M}_{m,1}(S)$, then the above equation gains the matrix form $A\boldsymbol{b} = \mathbf{0}$ and thus we obtain $\det(A)\boldsymbol{b} = A^\# A\boldsymbol{b} = \mathbf{0}$ (where $A^\#$ denotes the adjoint matrix). If $g \in R[X]$ is the characteristic polynomial of the matrix $(a_{j,\mu})_{j,\mu\in[1,m]}$, then $\det(A) = g(x)$, and as $g(x)b_j = 0$ for all $j \in [1,m]$, it follows that $g(x)M = \mathbf{0}$. Hence $g(x) = 0$, and this is an integral equation for x over R. □

Theorem 2.4.3. *Let R and S be commutative rings such that $R \subset S$.*

1. *Let $x \in S$ be integral over R and*

$$x^n + a_{n-1}x^{n-1} + \ldots + a_1x + a_0 = 0,$$

where $n \in \mathbb{N}$ and $a_0, \ldots, a_{n-1} \in R$. Then $R[x] = {}_R(1, x, \ldots x^{n-1})$.

2. *Let $m \in \mathbb{N}_0$ and $x_1, \ldots, x_m \in S$. Then the following assertions are equivalent:*

(a) $x_1, \ldots, x_m$ are integral over R.

(b) $R[x_1, \ldots, x_m]$ is a finitely generated R-module.

(c) $R[x_1, \ldots, x_m]$ is integral over R.

In particular:

- *An element $x \in S$ is integral over R if and only if $R[x]$ is a finitely generated R-module.*

- *If S is a finitely generated R-module, then S is integral over R.*

Proof. 1. We must prove that $x^\nu \in R + Rx + \ldots + Rx^{n-1}$ for all $\nu \ge 0$, and we use induction on ν. For $\nu \le n-1$ there is nothing to do. Thus suppose that $\nu \ge n$ and $x_j \in R + Rx + \ldots + Rx^{n-1}$ for all $j \in [0, \nu-1]$. Then

$$x^n = -\sum_{j=1}^{n} a_{n-j}x^{n-j} \quad \text{implies } x^\nu = x^{\nu-n}x^n = -\sum_{j=1}^{n} a_{n-j}x^{\nu-j} \in \sum_{i=0}^{n-1} Rx^i.$$

2. It suffices to prove the equivalence of (a), (b) and (c).

(a) $\Rightarrow$ (b) Induction on m. For $m = 0$ there is nothing to do.

$m \ge 1$, $m-1 \to m$: From the induction hypothesis we obtain

$$R' = R[x_1, \ldots, x_{m-1}] = Rb_1 + \ldots + Rb_k \quad \text{for some } k \in \mathbb{N} \text{ and } b_1, \ldots, b_k \in R'.$$

Set $x = x_m$, and let $x^n + a_{n-1}x^{n-1} + \ldots + a_1x + a_0 = 0$ be an integral equation for x over R. Then $R[x] = R + Rx + \ldots + Rx^{n-1}$ by 1., and

$$R[x_1, \ldots, x_m] = R'[x] = R'R[x] = \Big(\sum_{\mu=1}^{k} Rb_\mu\Big)\Big(\sum_{i=0}^{n-1} Rx^i\Big) = \sum_{\mu=1}^{k}\sum_{i=0}^{n-1} Rb_\mu x^i.$$

(b) $\Rightarrow$ (c) If $x \in R[x_1, \ldots, x_m]$, then $xR[x_1, \ldots, x_m] \subset R[x_1, \ldots, x_m]$, and therefore x is integral over R by 2.4.2.

(c) $\Rightarrow$ (a) Obvious. □

Theorem 2.4.4. *Let R, S and T be commutative rings such that $R \subset S \subset T$ and S is integral over R. If $x \in T$ is integral over S, then x is integral over R. In particular, if T is integral over S, then T is integral over R.*

Proof. We use 2.4.3. Let $x \in T$, and let $x^n + a_{n-1}x^{n-1} + \ldots + a_1x + a_0 = 0$ be an integral equation for x over S. Then $R' = R[a_0, \ldots, a_{n-1}]$ is integral over R, and x is integral over R'. Then R' is a finitely generated R-module, and $R'[x]$ is a finitely generated R'-module, say

$$R' = Rb_1 + \ldots + Rb_m \text{ and } R'[x] = R'c_1 + \ldots R'c_k$$

for some $m, k \in \mathbb{N}$, $b_1, \ldots, b_m \in R'$ and $c_1, \ldots, c_k \in R'[x]$. From this we obtain

$$R'[x] = \sum_{\mu=1}^{m}\sum_{i=1}^{k} Rb_\mu c_i.$$

Hence $R'[x]$ is a finitely generated R-module, and thus x is integral over R. □

Theorem 2.4.5. *Let R and S be commutative rings such that $R \subset S$. Then the integral closure $\mathrm{cl}_S(R)$ of R in S is a ring which is integrally closed in S. In particular, if S is a domain, then $\mathrm{cl}_S(R)$ is an integrally closed domain.*

Recall from 1.4.10 that we denoted by $\overline{\mathbb{Q}}$ the field of algebraic numbers. An algebraic number $z \in \overline{\mathbb{Q}}$ is called an **algebraic integer** if it is integral over $\mathbb{Z}$. The integrally closed domain $\overline{\mathbb{Z}} = \mathrm{cl}_{\overline{\mathbb{Q}}}(\mathbb{Z})$ is called the **domain of algebraic integers**.

Proof. If $x,\ y \in \mathrm{cl}_S(R)$, then $R[x,y]$ is a finitely generated R-module, and thus it is integral over R by 2.4.3.2. Since $\{1, x-y, xy\} \subset R[x,y] \subset \mathrm{cl}_S(R)$ for all $x,\ y \in \mathrm{cl}_S(R)$, it follows that $\mathrm{cl}_S(R)$ is a subring of S. If $x \in S$ is integral over $\mathrm{cl}_S(R)$, then x is integral over R by 2.4.4 and thus $x \in \mathrm{cl}_S(R)$. Consequently, $\mathrm{cl}_S(R)$ is integrally closed in S. □

Theorem 2.4.6. *Let R and S be commutative rings such that $R \subset S$.*

1. *Let $\sigma\colon S \to S'$ be a ring homomorphism. If $x \in S$ is integral over R, then $\sigma(x)$ is integral over $\sigma(R)$.*

2. *Let S be integral over R, $\mathfrak{A}$ an ideal of S and $\mathfrak{a} = \mathfrak{A} \cap R$. Then $S/\mathfrak{A}$ is integral over $R/\mathfrak{a}$.*

Proof. 1. Let $x \in S$, and let $x^n + a_{n-1} + \ldots + a_1 x + a_0 = 0$ be an integral equation for x over R. Then

$$\varphi(x)^n + \varphi(a_{n-1})\varphi(x)^{n-1} + \ldots + \varphi(a_1)\varphi(x) + \varphi(a_0) = 0$$

is an integral equation for $\varphi(x)$ over $\varphi(R)$.

2. By 1., applied with the residue class homomorphism $\sigma\colon S \to S/\mathfrak{A}$. □

In the following results we investigate the behavior of ideals and in particular of prime ideals in integral ring extensions.

Theorem 2.4.7 (Krull-Cohen-Seidenberg). *Let R and S be commutative rings such that $R \subset S$ and S is integral over R.*

1. *If $\mathfrak{A}$ is an ideal of S and $\mathfrak{A} \not\subset z(S)$, then $\mathfrak{A} \cap R \ne \mathbf{0}$.*

2. *Let $\mathfrak{P}$ be a prime ideal of S and $\mathfrak{p} = \mathfrak{P} \cap R$.*

   (a) If $\mathfrak{A}$ is an ideal of S, $\mathfrak{P} \subset \mathfrak{A}$ and $\mathfrak{A} \cap R = \mathfrak{p}$, then $\mathfrak{A} = \mathfrak{P}$.

   (b) $\mathfrak{p} \in \max(R)$ if and only if $\mathfrak{P} \in \max(S)$.

3. *Let $\mathfrak{p}$ be a prime ideal of R. Then there exists a prime ideal $\mathfrak{P}$ of S such that $\mathfrak{P} \cap R = \mathfrak{p}$, and if $\mathfrak{P}$, $\mathfrak{P}'$ are prime ideals of S such that $\mathfrak{P} \subset \mathfrak{P}'$ and $\mathfrak{P} \cap R = \mathfrak{P}' \cap R$, then $\mathfrak{P} = \mathfrak{P}'$.*

4. *If $\mathfrak{a}$ is a proper ideal of R, then $\mathfrak{a}S$ is a proper ideal of S.*

5. *$S^\times \cap R = R^\times$.*

Proof. 1. Let $\mathfrak{A}$ be an ideal of S, $x \in \mathfrak{A} \setminus z(S)$ and $n \in \mathbb{N}$ be minimal such that there exists an integral equation $x^n + a_{n-1}x^{n-1} + \ldots + a_1 x + a_0 = 0$ for x over R. Then $a_0 \ne 0$ and $a_0 \in \mathfrak{A} \cap R$.

2. (a) Let $\overline{R} = R/\mathfrak{p}$ and $\overline{S} = S/\mathfrak{P}$. Then $\overline{R} \subset \overline{S}$, $\overline{S}$ is integral over $\overline{R}$ by 2.4.6.2, $\overline{\mathfrak{A}} = \mathfrak{A}/\mathfrak{P}$ is an ideal of $\overline{S}$ and $\overline{\mathfrak{A}} \cap \overline{R} = \mathbf{0}$. Since $\overline{S}$ is a domain, we obtain $\overline{\mathfrak{A}} = \mathbf{0}$ by 1. and therefore $\mathfrak{A} = \mathfrak{P}$.

(b) Let first $\mathfrak{p} \in \max(R)$ and $\mathfrak{A}$ an ideal of S such that $\mathfrak{P} \subset \mathfrak{A} \subsetneq S$. Since $\mathfrak{p} \subset \mathfrak{A} \cap R \subsetneq R$, it follows that $\mathfrak{A} \cap R = \mathfrak{p}$ and thus $\mathfrak{A} = \mathfrak{P}$ by (a), and $\mathfrak{P} \in \max(S)$.

Now let $\mathfrak{P} \in \max(S)$, $\overline{S} = S/\mathfrak{P}$ and $\overline{R} = R/\mathfrak{p}$. Then $\overline{S}$ is integral over $\overline{R}$, and $\overline{S}$ is a field. We must prove that $\overline{R}$ is a field, too. Thus let $a \in \overline{R}^\bullet$. Then $a^{-1} \in \overline{S}^\bullet$, and we consider an integral equation

$$a^{-n} + a_{n-1}a^{-(n-1)} + \ldots + a_1a^{-1} + a_0 = 0$$

for a^{-1} over $\overline{R}$. We multiply it by a^n and obtain

$$1 = -a(a_{n-1} + \ldots + a^{n-2}a_1 + a^{n-1}a_0).$$

Hence $a \in \overline{R}^\times$, and thus $\overline{R}$ is a field.

3. Let Ω be the set of all ideals $\mathfrak{A}$ of S such that $\mathfrak{A} \cap R \subset \mathfrak{p}$ (or, equivalently $\mathfrak{A} \cap (R \setminus \mathfrak{p}) = \emptyset$). Since $R \setminus \mathfrak{p}$ is a multiplicatively closed subset of $S^\bullet$, Krull's Theorem 1.1.5.1 implies that Ω contains maximal elements and that every maximal element of Ω is a prime ideal. Let $\mathfrak{P}$ be a maximal element of Ω, and suppose that $\mathfrak{P} \cap R \subsetneq \mathfrak{p}$. If $u \in \mathfrak{p} \setminus \mathfrak{P}$, then $(\mathfrak{P} + uS) \cap (R \setminus \mathfrak{p}) \neq \emptyset$, and we let $p \in \mathfrak{P}$ and $s \in S$ be such that $x = p + us \in R \setminus \mathfrak{p}$. Let

$$s^n + a_{n-1}s^{n-1} + \ldots + a_1s + a_0 = 0$$

be an integral equation for s over R. Multiplying it by u^n yields $(us)^n + a_{n-1}u(us)^{n-1} + \ldots + a_1u^{n-1}(us) + a_0u^n = 0$, and as $us \equiv x \mod \mathfrak{P}$ we obtain $x^n + a_{n-1}ux^{n-1} + \ldots + a_1u^{n-1}x + a_0u^n \in \mathfrak{P} \cap R \subset \mathfrak{p}$. Now $u \in \mathfrak{p}$ implies $x^n \in \mathfrak{p}$ and finally $x \in \mathfrak{p}$, a contradiction.

If $\mathfrak{P}$, $\mathfrak{P}'$ are prime ideals of S such that $\mathfrak{P} \cap R = \mathfrak{P}' \cap R$ and $\mathfrak{P} \subset \mathfrak{P}'$, then 2.(a) implies $\mathfrak{P} = \mathfrak{P}'$.

4. If $\mathfrak{a}$ is a proper ideal of R, then by Krull's theorem 1.1.5.2 there exists a prime ideal $\mathfrak{p} \supset \mathfrak{a}$, by 3. there exists a prime ideal $\mathfrak{P}$ of S such that $\mathfrak{p} \subset \mathfrak{P}$, and then $\mathfrak{a}S \subset \mathfrak{p}S \subset \mathfrak{P} \subsetneq S$.

5. Obviously, $R^\times \subset R \cap S^\times$. If $x \in R \setminus R^\times$, then $xR \neq R$, hence $xS \neq S$ by 4. and therefore $x \notin S^\times$. □

Let R be a ring. The **(Krull) dimension** $\dim(R) \in \mathbb{N}_0 \cup \{\infty\}$ is the supremum of all $l \in \mathbb{N}_0$ for which there exists a chain of prime ideals

$$\mathfrak{p}_0 \subsetneq \mathfrak{p}_1 \subsetneq \ldots \subsetneq \mathfrak{p}_l \text{ in } R.$$

If $\mathfrak{o}$ is a domain, then

- $\dim(\mathfrak{o}) = 0$ if and only if $\mathfrak{o}$ is a field (then $\max(\mathfrak{o}) = \{\mathbf{0}\}$ and $\mathcal{P}(\mathfrak{o}) = \emptyset$);
- $\dim(\mathfrak{o}) = 1$ if and only if $\mathcal{P}(\mathfrak{o}) = \max(\mathfrak{o})$.

If $\mathfrak{o}$ is a principal ideal domain, then $\dim(\mathfrak{o}) \leq 1$ by 1.1.7.3.

Theorem 2.4.8. *Let R and S be commutative rings such that $R \subset S$ and S is integral over R. Then* $\dim(R) = \dim(S)$.

Proof. Let $l \in \mathbb{N}_0$ and $\mathfrak{P}_0 \subsetneq \mathfrak{P}_1 \subsetneq \ldots \subsetneq \mathfrak{P}_l$ be a chain of prime ideals in S. Then $\mathfrak{P}_0 \cap R \subset \mathfrak{P}_1 \cap R \subset \ldots \subset \mathfrak{P}_l \cap R$ is a chain of prime ideals in R, and $\mathfrak{P}_{i-1} \cap R \neq \mathfrak{P}_i \cap R$ for all $i \in [1, l]$ by 2.4.7.3. Hence $\dim(S) \le \dim(R)$, and it remains to prove:

> **A.** If $l \in \mathbb{N}_0$ and $\mathfrak{p}_0 \subsetneq \mathfrak{p}_1 \subsetneq \ldots \subsetneq \mathfrak{p}_l$ is a chain of prime ideals in R, then there exists a chain of prime ideals $\mathfrak{P}_0 \subsetneq \mathfrak{P}_1 \subsetneq \ldots \subsetneq \mathfrak{P}_l$ in S such that $\mathfrak{P}_i \cap R = \mathfrak{p}_i$ for all $i \in [0, l]$.

Indeed, **A** implies $\dim(R) \le \dim(S)$, and thus equality holds. □

Proof of **A.** By induction on l. For $l = 0$ this is just 2.4.7.3.

$l \ge 1$, $l-1 \to l$: Let $\mathfrak{P}_0 \subsetneq \mathfrak{P}_1 \subsetneq \ldots \subsetneq \mathfrak{P}_{l-1}$ be a chain of prime ideals in S satisfying $\mathfrak{P}_i \cap R = \mathfrak{p}_i$ for all $i \in [0, l-1]$. Then $S/\mathfrak{P}_{l-1}$ is integral over $R/\mathfrak{p}_{l-1}$, and $\mathfrak{p}_l/\mathfrak{p}_{l-1}$ is a prime ideal of $R/\mathfrak{p}_{l-1}$. By 2.4.7.3 there exists a prime ideal $\mathfrak{P}_l^*$ of $S/\mathfrak{P}_{l-1}$ such that $\mathfrak{P}_l^* \cap (R/\mathfrak{p}_{l-1}) = \mathfrak{p}_l/\mathfrak{p}_{l-1}$. However, $\mathfrak{P}_l^* = \mathfrak{P}_l/\mathfrak{P}_{l-1}$ for some prime ideal $\mathfrak{P}_l$ of S such that $\mathfrak{P}_l \supset \mathfrak{P}_{l-1}$, and $\mathfrak{P}_l/\mathfrak{P}_{l-1} \cap (R/\mathfrak{p}_{l-1}) = \mathfrak{p}_l/\mathfrak{p}_{l-1}$ implies $\mathfrak{P}_l \cap R = \mathfrak{p}_l$. □[**A.**]

2.5 Integral elements 2: Field-theoretic aspects

In this section we study the structure of integral ring extensions inside finite field extensions. This situation is ubiquitous in the theory of algebraic number fields and algebraic function fields.

Theorem 2.5.1. *Let $\mathfrak{o}$ be an integrally closed domain, $K = \mathsf{q}(\mathfrak{o})$, L/K a finite field extension and $\mathfrak{O} = \mathrm{cl}_L(\mathfrak{o})$.*

1. *$\mathfrak{O}$ is an integrally closed domain,*

$$L = \mathsf{q}(\mathfrak{O}) = \{q^{-1}x \mid x \in \mathfrak{O},\ q \in \mathfrak{o}^\bullet\},$$

$\mathfrak{O} \cap K = \mathfrak{o}$, and every $\mathfrak{O}$-submodule $\mathfrak{A} \neq \mathbf{0}$ of L contains a K-basis of L.

2. *Let $\mathfrak{A}$ be a finitely generated $\mathfrak{o}$-submodule of L. Then there exists some $q \in \mathfrak{o}^\bullet$ such that $q\mathfrak{A} \subset \mathfrak{O}$.*

3. *Let $a \in \mathfrak{O}$ and $g \in K[X]$ be the minimal polynomial of a over K. Then $g \in \mathfrak{o}[X]$, $\mathsf{N}_{L/K}(a) \in \mathfrak{o}$, $\mathsf{Tr}_{L/K}(a) \in \mathfrak{o}$, and $d_{L/K}(u_1, \ldots, u_n) \in \mathfrak{o}$ for every n-tuple $(u_1, \ldots, u_n) \in \mathfrak{O}^n$.*

4. *Let L/K be separable, $\mathfrak{A}$ an $\mathfrak{o}$-submodule of $\mathfrak{O}$, $(u_1, \ldots, u_n) \in \mathfrak{A}^n$ a K-basis of L and $(u_1^*, \ldots, u_n^*)$ its dual* (see 1.6.10). *Then we have $\mathfrak{A} \subset {}_{\mathfrak{o}}(u_1^*, \ldots, u_n^*)$, and if $\mathfrak{o}$ is Noetherian, then $\mathfrak{A}$ is a finitely generated $\mathfrak{o}$-module.*

5. *Let $\mathfrak{o}$ be Noetherian, L/K separable and $\mathfrak{A}$ an ideal of $\mathfrak{O}$. Then $\mathfrak{A}$ is a finitely generated $\mathfrak{o}$-module. In particular, $\mathfrak{O}$ is a finitely generated $\mathfrak{o}$-module and a Noetherian domain.*

Proof. 1. By 2.4.5, $\mathfrak{O}$ is an integrally closed domain, and as $\mathfrak{o}$ is integrally closed, we get $\mathfrak{O} \cap K = \mathfrak{o}$. Obviously, $\{q^{-1}x \mid x \in \mathfrak{O},\ q \in \mathfrak{o}^\bullet\} \subset \mathsf{q}(\mathfrak{O}) \subset L$, and we assert that for every $z \in L$ there exists some $q \in \mathfrak{o}^\bullet$ such that $qz \in \mathfrak{O}$.

Let $z \in L$, and let $f = X^d + a_{d-1}X^{d-1} + \ldots + a_1X + a_0 \in K[X]$ be the minimal polynomial of z over K. If $q \in \mathfrak{o}^\bullet$ is such that $qa_i \in \mathfrak{o}$ for all $i \in [0, d-1]$, then $(qz)^d + (qa_{d-1})(qz)^{d-1} + \ldots + (q^{d-1}a_1)(qz) + q^d a_0 = 0$ is an integral equation for qz over $\mathfrak{o}$, which implies $qz \in \mathfrak{O}$.

Let $\mathfrak{A} \neq \mathbf{0}$ be an $\mathfrak{O}$-submodule of L, $a \in \mathfrak{A}^\bullet$, $(u_1, \ldots, u_n)$ a K-basis of L and $q \in \mathfrak{o}^\bullet$ such that $(qu_1, \ldots, qu_n) \in \mathfrak{O}^n$. Then $(qu_1a, \ldots, qu_na)$ is a K-basis of L in $\mathfrak{A}$.

2. By 1., there exists a K-basis $(u_1, \ldots, u_n) \in \mathfrak{O}^n$, and we suppose that $\mathfrak{A} = {}_{\mathfrak{o}}(v_1, \ldots, v_k)$. For $i \in [1, k]$, let

$$v_i = \sum_{\nu=1}^{n} c_{i,\nu}u_\nu, \quad \text{where } c_{i,\nu} \in K.$$

If $q \in \mathfrak{o}^\bullet$ is such that $qc_{i,\nu} \in \mathfrak{o}$ for all $i \in [1, k]$ and $\nu \in [1, n]$, then $qv_i \in \mathfrak{O}$ for all $i \in [1, k]$ and therefore $q\mathfrak{A} \subset \mathfrak{O}$.

3. Let $\overline{K}$ be an algebraically closed extension field of L, and suppose that $[L : K(a)] = m$, $[K(a) : K]_{\mathsf{s}} = d$, $[K(a) : K]_{\mathsf{i}} = q$, $[L : K] = n = dqm$ and $\mathrm{Hom}^K(K(a), \overline{K}) = \{\sigma, \ldots, \sigma_d\}$. Then

$$g = \prod_{i=1}^{d}(X - \sigma_i(a))^q = X^{dq} + a_{dq-1}X^{dq-1} + \ldots + a_0 \in K[X]$$

by 1.4.9. By 1.6.2.1 we get $\mathsf{N}_{L/K}(a) = (-1)^n a_0^m$ and $\mathsf{Tr}_{L/K} = -ma_{dq-1}$. By 2.4.6.1, $\sigma_1(a), \ldots, \sigma_d(a)$ are integral over $\mathfrak{o}$. Hence $g \in \mathfrak{o}[X]$, $\mathsf{N}_{L/K}(a) \in \mathfrak{o}$, $\mathsf{Tr}_{L/K}(a) \in \mathfrak{o}$, and $d_{L/K}(u_1, \ldots, u_n) \in \mathfrak{o}$ for every n-tuple $(u_1, \ldots, u_n) \in \mathfrak{O}^n$.

4. If $z \in \mathfrak{A}$ and $z = a_1u_1^* + \ldots + a_nu_n^*$ for some $a_1, \ldots, a_n \in K$, then $u_iz \in \mathfrak{O}$,

$$\mathsf{Tr}_{L/K}(u_iz) = \sum_{\nu=1}^{n} a_\nu \mathsf{Tr}_{L/K}(u_iu_\nu^*) = a_i \in \mathfrak{o} \quad \text{for all } i \in [1, n] \quad \text{by 3.},$$

and $z \in \mathfrak{o}u_1^* + \ldots + \mathfrak{o}u_n^* = {}_{\mathfrak{o}}(u_1^*, \ldots, u_n^*)$. Hence $\mathfrak{A} \subset {}_{\mathfrak{o}}(u_1^*, \ldots, u_n^*)$, and if $\mathfrak{o}$ is Noetherian, then $\mathfrak{A}$ is a finitely generated $\mathfrak{o}$-module (since it is an $\mathfrak{o}$-submodule of a finitely generated $\mathfrak{o}$-module).

5. As $\mathfrak{A}$ contains a K-basis of L by 1., it is a finitely generated $\mathfrak{o}$-module by 4. Hence every non-zero ideal of $\mathfrak{O}$ is a finitely generated $\mathfrak{o}$-module and therefore a finitely generated ideal of $\mathfrak{O}$. Consequently, $\mathfrak{O}$ is a Noetherian domain and a finitely generated $\mathfrak{o}$-module. □

Theorem 2.5.2. *Let $\mathfrak{o}$ be an integrally closed domain, $K = \mathsf{q}(\mathfrak{o})$, and let $f \in \mathfrak{o}[X]$ be monic. If $f = gh$, where $g,\, h \in K[X]$ are monic, then $g,\, h \in \mathfrak{o}[X]$. In particular, if f is irreducible in $\mathfrak{o}[X]$, then f is irreducible in $K[X]$.*

Proof. Let L be a splitting field of f over K, $\mathfrak{O} = \mathrm{cl}_L(\mathfrak{o})$ and $f = gh$ for some monic polynomials $g,\, h \in K[X]$. Suppose that $f = (X - x_1) \cdot \ldots \cdot (X - x_n)$, where $n \in \mathbb{N}$ and $x_1, \ldots, x_n \in L$. After renumbering we may assume that there is some $k \in [0, n]$ such that

$$g = (X - x_1) \cdot \ldots \cdot (X - x_k) \quad \text{and} \quad h = (X - x_{k+1}) \cdot \ldots \cdot (X - x_n).$$

As $f \in \mathfrak{o}[X]$, it follows that $x_1, \ldots, x_n \in \mathfrak{O}$ and therefore $g,\, h \in \mathfrak{O}[X] \cap K[X] = \mathfrak{o}[X]$ (since $\mathfrak{O} \cap K = \mathfrak{o}$). □

Remarks and Definitions 2.5.3. Let $\mathfrak{o}$ be an integrally closed domain and $K = \mathsf{q}(\mathfrak{o})$. Let $\mathfrak{p} \in \max(\mathfrak{o})$, $\mathsf{k}_\mathfrak{p} = \mathfrak{o}/\mathfrak{p}$, L/K a finite field extension, $\mathfrak{O} = \mathrm{cl}_L(\mathfrak{o})$, $\mathfrak{P} \in \max(\mathfrak{O})$ such that $\mathfrak{P} \cap \mathfrak{o} = \mathfrak{p}$ and $\mathsf{k}_\mathfrak{P} = \mathfrak{O}/\mathfrak{P}$. Then $\mathfrak{o} = \mathfrak{O} \cap K$, $\mathfrak{p} = \mathfrak{P} \cap K$, and for $a \in \mathfrak{o}$ we identify (as usual) $a + \mathfrak{p} \in \mathsf{k}_\mathfrak{p}$ with $a + \mathfrak{P} \in \mathsf{k}_\mathfrak{P}$, so that $\mathsf{k}_\mathfrak{p} \subset \mathsf{k}_\mathfrak{P}$. We call $\mathsf{k}_\mathfrak{P}/\mathsf{k}_\mathfrak{p}$ the **residue class extension** and its degree

$$f(\mathfrak{P}/\mathfrak{p}) = [\mathsf{k}_\mathfrak{P} : \mathsf{k}_\mathfrak{p}] = \dim_{\mathsf{k}_\mathfrak{p}} \mathsf{k}_\mathfrak{P} \in \mathbb{N} \cup \{\infty\}$$

the **(residue class) degree** of $\mathfrak{P}$ over $\mathfrak{p}$.

From now on let L/K be Galois and $G = \mathrm{Gal}(L/K)$. If $\sigma \in G$, then $\sigma\mathfrak{O} = \mathfrak{O}$ by 2.4.6.1, $\sigma \restriction \mathfrak{O} \colon \mathfrak{O} \to \mathfrak{O}$ is a ring isomorphism and $\sigma\mathfrak{P} \in \max(\mathfrak{O})$. Moreover, σ induces a $\mathsf{k}_\mathfrak{p}$-isomorphism

$$\sigma_\mathfrak{P} \colon \mathsf{k}_\mathfrak{P} \to \mathsf{k}_{\sigma\mathfrak{P}} \quad \text{by means of} \quad \sigma_\mathfrak{P}(a + \mathfrak{P}) = \sigma(a) + \sigma\mathfrak{P} \quad \text{for all } a \in \mathfrak{O},$$

and consequently $f(\sigma\mathfrak{P}/\mathfrak{p}) = f(\mathfrak{P}/\mathfrak{p})$. We call $\sigma_\mathfrak{P}$ the **residue class isomorphism** induced by σ. The isotropy group $G_\mathfrak{P} = \{\sigma \in G \mid \sigma\mathfrak{P} = \mathfrak{P}\}$ is called the **decomposition group** and its fixed field $D_\mathfrak{P} = L^{G_\mathfrak{P}}$ is called the **decomposition field** of $\mathfrak{P}$ over K. If $\sigma \in G_\mathfrak{P}$, then $\sigma_\mathfrak{P} \in \mathrm{Gal}(\mathsf{k}_\mathfrak{P}/\mathsf{k}_\mathfrak{p})$ is called the **residue class automorphism** induced by σ. Apparently, the assignment $\sigma \mapsto \sigma_\mathfrak{P}$ defines a group homomorphism $\rho_\mathfrak{P} \colon G_\mathfrak{P} \to \mathrm{Gal}(\mathsf{k}_\mathfrak{P}/\mathsf{k}_\mathfrak{p})$.

Theorem 2.5.4. *Let $\mathfrak{o}$ be an integrally closed domain, $K = \mathsf{q}(\mathfrak{o})$, L/K a finite Galois extension, $G = \mathrm{Gal}(L/K)$ and $\mathfrak{O} = \mathrm{cl}_L(\mathfrak{o})$. Let $\mathfrak{p} \in \max(\mathfrak{o})$, and let $\Omega(\mathfrak{p})$ be the set of all $\mathfrak{P}' \in \max(\mathfrak{O})$ lying above $\mathfrak{p}$ and $\mathfrak{P} \in \Omega(\mathfrak{p})$.*

1. *$\Omega(\mathfrak{p}) = \{\sigma\mathfrak{P} \mid \sigma \in G\}$. In particular:*

- *$\Omega(\mathfrak{p})$ is finite, and G operates transitively on $\Omega(\mathfrak{p})$.*
- *$G_{\sigma\mathfrak{P}} = \sigma G_\mathfrak{P} \sigma^{-1}$ for all $\sigma \in G$.*
- *If $(G : G_\mathfrak{P}) = r$ and $G/G_\mathfrak{P} = \{\sigma_1 G_\mathfrak{P}, \ldots, \sigma_r G_\mathfrak{P}\}$, then it follows that $\Omega(\mathfrak{p}) = \{\sigma_1\mathfrak{P}, \ldots, \sigma_r\mathfrak{P}\}$.*

2. The residue class extension $\mathsf{k}_{\mathfrak{P}}/\mathsf{k}_{\mathfrak{p}}$ is normal, and the group homomorphism $\rho_{\mathfrak{P}} \colon G_{\mathfrak{P}} \to \mathrm{Gal}(\mathsf{k}_{\mathfrak{P}}/\mathsf{k}_{\mathfrak{p}})$, defined by $\rho_{\mathfrak{P}}(\sigma) = \sigma_{\mathfrak{P}}$, is surjective.

Proof. 1. Assume to the contrary that there is some $\mathfrak{P}' \in \Omega(\mathfrak{p}) \setminus \{\sigma\mathfrak{P} \mid \sigma \in G\}$. Then 1.1.6 implies

$$\mathfrak{P}' \not\subset \bigcup_{\sigma \in G} \sigma\mathfrak{P},$$

and thus there exists some $a \in \mathfrak{P}'$ such that $\sigma^{-1}a \notin \mathfrak{P}$ for all $\sigma \in G$. Then

$$\mathsf{N}_{L/K}(a) = \prod_{\sigma \in G} \sigma^{-1}a \in \mathfrak{P}' \cap \mathfrak{o} = \mathfrak{p}, \quad \text{but} \quad \mathsf{N}_{L/K}(a) \notin \mathfrak{P}, \quad \text{a contradiction.}$$

The remaining assertions are purely group-theoretic consequences of the fact that G operates transitively on $\Omega(\mathfrak{p})$.

2. For $h \in \mathfrak{O}[X]$ we denote by $\overline{h} = h + \mathfrak{P}\mathfrak{O}[X] \in \mathsf{k}_{\mathfrak{P}}[X]$ the residue class polynomial of h. Let $\psi \in \mathsf{k}_{\mathfrak{p}}[X]$ be a monic irreducible polynomial with a zero $\alpha \in \mathsf{k}_{\mathfrak{P}}$. Let $a \in \mathfrak{O}$ be such that $\overline{a} = \alpha$, and let $g \in \mathfrak{o}[X]$ be the minimal polynomial of a over K (see 2.5.1.3). As L/K is normal, g splits in $L[X]$, hence in $\mathfrak{O}[X]$, and therefore $\overline{g}$ splits in $\mathsf{k}_{\mathfrak{P}}[X]$. Since $\overline{g}(\overline{a}) = \overline{g(a)} = 0$, it follows that $\psi \mid \overline{g}$, and therefore ψ splits in $\mathsf{k}_{\mathfrak{P}}[X]$, too. Hence $\mathsf{k}_{\mathfrak{P}}/\mathsf{k}_{\mathfrak{p}}$ is normal, and it remains to prove that $\rho_{\mathfrak{P}}$ is surjective.

Let $\mathsf{k}'_{\mathfrak{P}}$ be the separable closure of $\mathsf{k}_{\mathfrak{p}}$ in $\mathsf{k}_{\mathfrak{P}}$ and $a \in \mathfrak{O}$ such that $\mathsf{k}'_{\mathfrak{P}} = \mathsf{k}_{\mathfrak{p}}(\overline{a})$. Let $g \in \mathfrak{o}[X]$ be the minimal polynomial of a over K and $\psi \in \mathsf{k}_{\mathfrak{p}}[X]$ the minimal polynomial of $\overline{a}$ over $\mathsf{k}_{\mathfrak{p}}$. Then g splits in $\mathfrak{O}[X]$, $\psi \mid \overline{g}$, and after renumbering we may assume that

$$g = (X-a)(X-a_2)\cdot\ldots\cdot(X-a_m) \quad \text{and} \quad \psi = (X-\overline{a})(X-\overline{a}_2)\cdot\ldots\cdot(X-\overline{a}_k),$$

where $m,\ k \in \mathbb{N}$, $k \le m$ and $a_2, \ldots, a_m \in \mathfrak{O}$.

The map $\mathrm{Gal}(\mathsf{k}_{\mathfrak{P}}/\mathsf{k}_{\mathfrak{p}}) \to \mathrm{Gal}(\mathsf{k}'_{\mathfrak{P}}/\mathsf{k}_{\mathfrak{p}})$, defined by $\tau \mapsto \tau \restriction \mathsf{k}'_{\mathfrak{P}}$, is an isomorphism by 1.4.8.4. Hence an automorphism $\tau \in \mathrm{Gal}(\mathsf{k}_{\mathfrak{P}}/\mathsf{k}_{\mathfrak{p}})$ is uniquely determined by its value $\tau(\overline{a})$. If $\tau \in \mathrm{Gal}(\mathsf{k}_{\mathfrak{P}}/\mathsf{k}_{\mathfrak{p}})$ and $\tau(\overline{a}) = \overline{a}_j$ for some $j \in [1,k]$, then there exists some $\sigma \in G$ such that $\sigma(a) = a_j$, hence $\sigma_{\mathfrak{P}}(\overline{a}) = \overline{a}_j = \tau(\overline{a})$, and consequently $\sigma_{\mathfrak{P}} = \tau$. □

We now apply Theorem 2.5.4 to derive a reduction principle for Galois groups. We use the terminology introduced in 1.5.7 and 1.5.8.

Theorem 2.5.5. *Let $\mathfrak{o}$ be an integrally closed domain, $K = \mathsf{q}(\mathfrak{o})$, $\mathfrak{p} \in \max(\mathfrak{o})$, $\mathsf{k}_{\mathfrak{p}} = \mathfrak{o}/\mathfrak{p}$ and $g \in \mathfrak{o}[X] \setminus \mathfrak{o}$ a monic polynomial such that the residue class polynomial $\overline{g} = g + \mathfrak{p}\mathfrak{o}[X] \in \mathsf{k}_{\mathfrak{p}}[X]$ is separable. Let L be a splitting field of g over K, $\mathfrak{O} = \mathrm{cl}_L(\mathfrak{o})$, $\mathfrak{P} \in \max(\mathfrak{O})$ such that $\mathfrak{P} \cap \mathfrak{o} = \mathfrak{p}$, $\mathsf{k}_{\mathfrak{p}} \subset \mathsf{k}_{\mathfrak{P}} = \mathfrak{O}/\mathfrak{P}$, and for $x \in \mathfrak{O}$ we denote by $\overline{x} = x + \mathfrak{P} \in \mathsf{k}_{\mathfrak{P}}$ its residue class.*

1. Let $Z(g)$ (resp. $Z(\overline{g})$) be the set of zeros of g (resp. $\overline{g}$) in $\mathfrak{O}$ (resp. $\mathsf{k}_{\mathfrak{P}}$). Then the assignment $x \mapsto \overline{x}$ induces a bijective map $Z(g) \to Z(\overline{g})$. The field $\mathsf{k}_{\mathfrak{p}}^* = \mathsf{k}_{\mathfrak{p}}(Z(\overline{g}))$ is the splitting field of $\overline{g}$ and the separable closure of $\mathsf{k}_{\mathfrak{p}}$ in $\mathsf{k}_{\mathfrak{P}}$. Let $\iota\colon G = \mathrm{Gal}(L/K) \overset{\sim}{\to} \mathrm{Gal}_K(g)$ be the isomorphism given by $\iota(\sigma) = \sigma \restriction Z(g)$, and consider the image $\mathrm{Gal}_{K,\mathfrak{P}}(g) = \iota(G_{\mathfrak{P}}) \subset \mathrm{Gal}_K(g)$. Then the map

$$\overline{\rho}_{\mathfrak{P}}\colon \mathrm{Gal}_{K,\mathfrak{P}}(g) \to \mathrm{Gal}_{\mathsf{k}_{\mathfrak{p}}}(\overline{g}), \quad \textit{defined by } \overline{\rho}_{\mathfrak{P}}(\sigma) = \sigma_{\mathfrak{P}},$$

where $\sigma_{\mathfrak{P}}(\overline{x}) = \overline{\sigma(x)}$ for all $x \in Z(f)$, is an isomorphism.

2. Let $\mathsf{k}_{\mathfrak{p}}^*/\mathsf{k}_{\mathfrak{p}}$ be cyclic, and suppose that $\overline{g} = \psi_1 \cdot \ldots \cdot \psi_r$ with monic irreducible polynomials $\psi_1, \ldots, \psi_r \in \mathsf{k}_{\mathfrak{p}}[X]$. Then $\mathrm{Gal}_K(g)$ contains a permutation σ of type $(\partial(\psi_1), \ldots \partial(\psi_r))$.

Proof. 1. Note that $Z(g) \subset \mathfrak{O}$, since g is monic. If $g = (X - x_1) \cdot \ldots \cdot (X - x_n)$, then $\overline{g} = (X - \overline{x}_1) \cdot \ldots \cdot (X - \overline{x}_n)$, and $\overline{x}_1, \ldots, \overline{x}_n$ are distinct, since $\overline{g}$ is separable. Hence the assignment $x \mapsto \overline{x}$ defines a bijective map $Z(g) \to Z(\overline{g})$. By definition, $\mathsf{k}_{\mathfrak{p}}^* = \mathsf{k}_{\mathfrak{p}}(\overline{x}_1, \ldots, \overline{x}_n)$ is the splitting field of $\overline{g}$ in $\mathsf{k}_{\mathfrak{P}}$, and thus it is Galois over $\mathsf{k}_{\mathfrak{p}}$. The homomorphism $\rho_{\mathfrak{P}}\colon G_{\mathfrak{P}} \to \mathrm{Gal}(\mathsf{k}_{\mathfrak{P}}/\mathsf{k}_{\mathfrak{p}})$, defined by $\rho_{\mathfrak{P}}(\sigma) = \sigma_{\mathfrak{P}}$, is an epimorphism by 2.5.4.2, and since $\rho_{\mathfrak{P}}(\sigma)(Z(\overline{g})) \subset Z(\overline{g})$ for all $\sigma \in G_{\mathfrak{P}}$ by the very definition, we obtain $\mathrm{Im}(\rho_{\mathfrak{P}}) \subset \mathrm{Gal}(\mathsf{k}_{\mathfrak{p}}^*/\mathsf{k}_{\mathfrak{p}})$. Hence $\mathrm{Gal}(\mathsf{k}_{\mathfrak{p}}^*/\mathsf{k}_{\mathfrak{p}}) = \mathrm{Gal}(\mathsf{k}_{\mathfrak{P}}/\mathsf{k}_{\mathfrak{p}})$, $\mathsf{k}_{\mathfrak{p}}^*$ is the separable closure of $\mathsf{k}_{\mathfrak{p}}$ in $\mathsf{k}_{\mathfrak{P}}$, and there is a commutative diagram

$$\begin{array}{ccc} G_{\mathfrak{P}} & \xrightarrow{\rho_{\mathfrak{P}}} & \mathrm{Gal}(\mathsf{k}_{\mathfrak{p}}^*/\mathsf{k}_{\mathfrak{p}}) \\ {\scriptstyle\iota}\downarrow & & \downarrow{\scriptstyle\overline{\iota}} \\ \mathrm{Gal}_{K,\mathfrak{P}}(g) & \xrightarrow{\overline{\rho}_{\mathfrak{P}}} & \mathrm{Gal}_{\mathsf{k}_{\mathfrak{p}}}(\overline{g}), \end{array}$$

where $\overline{\rho}_{\mathfrak{P}}$ is as above, and $\overline{\iota}(\tau) = \tau \restriction Z(\overline{g})$ for all $\tau \in \mathrm{Gal}(\mathsf{k}_{\mathfrak{p}}^*/\mathsf{k}_{\mathfrak{p}})$. Since ι and $\overline{\iota}$ are isomorphisms, $\overline{\rho}_{\mathfrak{P}}$ is surjective, and we assert that it is also injective. Considering that, let $\sigma \in \mathrm{Gal}_{K,\mathfrak{P}}(g) \setminus \{\mathrm{id}_{Z(g)}\}$. Then there exist $x, y \in Z(g)$ such that $\sigma x = y \neq x$, hence $\overline{\rho}_{\mathfrak{P}}(\sigma)(\overline{x}) = \sigma_{\mathfrak{P}}(\overline{x}) = \overline{y} \neq \overline{x}$, and consequently $\overline{\rho}_{\mathfrak{P}}(\sigma) \neq \mathrm{id}_{Z(\overline{g})}$.

2. By assumption, $\mathrm{Gal}_{\mathsf{k}_{\mathfrak{p}}}(\overline{g}) \cong \mathrm{Gal}(\mathsf{k}_{\mathfrak{p}}^*/\mathsf{k}_{\mathfrak{p}})$, and this group is cyclic, say $\mathrm{Gal}_{\mathsf{k}_{\mathfrak{p}}}(\overline{g}) = \langle\sigma\rangle$. By 1.5.8.2, σ is of type $(\partial(\psi_1), \ldots \partial(\psi_r))$. □

Example 2.5.6. Let $n \in \mathbb{N}$, $n \geq 4$. We use the method developed in 2.5.5 to construct a monic irreducible polynomial $g \in \mathbb{Z}[X]$ such that $\mathrm{Gal}_{\mathbb{Q}}(g) \cong \mathfrak{S}_n$. For this we need the following facts:

- For every prime p there exists a monic irreducible polynomial in $\mathbb{F}_p[X]$ of degree n. Indeed, $[\mathbb{F}_{p^n} : \mathbb{F}_p] = n$, and if $\mathbb{F}_{p^n}^\times = \langle x\rangle$, then the minimal polynomial of x over $\mathbb{F}_p$ is a monic irreducible polynomial of degree n.

- If $i_1, \ldots, i_{n-1} \in [1, n]$ are distinct and $k \in [1, n-1]$, then it follows that $\mathfrak{S}_n = \langle (i_1, \ldots, i_{n-1}), (i_k, i_n) \rangle$ (this is an easy exercise on permutations).

For a polynomial $g \in \mathbb{Z}[X]$ and a prime p we set $\overline{g}[p] = g + p\mathbb{Z}[X] \in \mathbb{F}_p[X]$. Now let $g_1, g_2, g_3 \in \mathbb{Z}[X]$ be monic polynomials of degree n such that

- $\overline{g}_1[2]$ is irreducible in $\mathbb{F}_2[X]$;
- $\overline{g}_2[3] = \varphi\psi$, where $\varphi, \psi \in \mathbb{F}_3[X]$ are irreducible, $\partial(\varphi) = n-1$ and $\partial(\psi) = 1$;
- $\overline{g}_3[5] = \varphi\psi$ or $\overline{g}_3[5] = \varphi\psi\psi'$, where $\varphi, \psi, \psi' \in \mathbb{F}_5[X]$ are irreducible such that $\partial(\varphi) = 2$ and $\partial(\psi) \equiv \partial(\psi') \equiv 1 \mod 2$.

We consider the polynomial

$$g = -15g_1 + 10g_2 + 6g_3 \in \mathbb{Z}[X].$$

g is monic, $\partial(g) = n$, $\overline{g}[2] = \overline{g}_1[2]$, $\overline{g}[3] = \overline{g}_2[3]$, $\overline{g}[5] = \overline{g}_3[5]$, and we assert that $\mathrm{Gal}_{\mathbb{Q}}(g) \cong \mathfrak{S}_n$. As all finite field extensions of a finite field are cyclic, we may apply 2.5.5.2.

Let $Z(g) = \{x_1, \ldots, x_n\}$ and define $\iota \colon \mathrm{Gal}_{\mathbb{Q}}(g) \to \mathfrak{S}_n$ by $\sigma(x_\nu) = x_{\iota(\sigma)(\nu)}$ for all $\sigma \in \mathrm{Gal}_{\mathbb{Q}}(g)$ and $\nu \in [1, n]$. Then ι is a monomorphism and we identify $\mathrm{Gal}_{\mathbb{Q}}(g)$ with its image in $\mathfrak{S}_n$ (so that $\mathrm{Gal}_{\mathbb{Q}}(g) \subset \mathfrak{S}_n$).

As $\overline{g}[2]$ is irreducible, it is separable, g is irreducible, too, and thus $\mathrm{Gal}_{\mathbb{Q}}(g)$ is transitive by 1.5.8.1. By 2.5.5.2 $\mathrm{Gal}_{\mathbb{Q}}(g)$ contains a cycle of order $n-1$ and a permutation of the form $\tau\rho$, where $\mathrm{supp}(\tau) \cap \mathrm{supp}(\rho) = \emptyset$, $\mathrm{ord}(\tau) = 2$, and ρ is either a cycle of odd order or the product of two disjoint cycles of odd order. In any case $d = \mathrm{ord}(\rho) \equiv 1 \mod 2$ and $\tau = (\tau\rho)^d \in \mathrm{Gal}_{\mathbb{Q}}(g)$, say $\tau = (j_1, j_2)$, where $1 \le j_1 < j_2 \le n$. Being transitive, $\mathrm{Gal}_{\mathbb{Q}}(g)$ contains a permutation σ such that $\sigma(j_2) = i_n$. Then $(k, i_n) = \sigma(j_1, j_2)\sigma^{-1} \in \mathrm{Gal}_{\mathbb{Q}}(g)$, where $k = \sigma(j_1) \in [1, n-1]$, hence $(i_1, \ldots, i_{n-1}) \in \mathrm{Gal}_{\mathbb{Q}}(g)$ and $(k, i_n) \in \mathrm{Gal}_{\mathbb{Q}}(g)$, which implies $\mathrm{Gal}_{\mathbb{Q}}(g) = \mathfrak{S}_n$. □

2.6 Fractional and invertible ideals

In this section we develop the basic tools of classical multiplicative ideal theory which are necessary for a thorough understanding of domains of algebraic numbers and functions, and in particular of Dedekind domains.

Definition and Remarks 2.6.1. Let $\mathfrak{o}$ be a domain and $K = \mathsf{q}(\mathfrak{o})$.

1. Let X be a subset of K and $\mathfrak{a} = {}_{\mathfrak{o}}(X)$ the $\mathfrak{o}$-submodule of K generated by X. We define

$$X^{-1} = (\mathfrak{o}:X) = \{x \in K \mid xX \subset \mathfrak{o}\} = \bigcap_{z \in X^\bullet} z^{-1}\mathfrak{o} = \bigcap_{\substack{c \in K^\times \\ c^{-1} \in X}} c\mathfrak{o}.$$

Then X^{-1} is an $\mathfrak{o}$-submodule of K, and $\mathfrak{a}^{-1} = X^{-1}$. Moreover,

$$(\mathfrak{a}^{-1})^{-1} = (X^{-1})^{-1} = \bigcap_{\substack{c \in K^\times \\ c^{-1} \in X^{-1}}} c\mathfrak{o} = \bigcap_{\substack{c \in K \\ X \subset c\mathfrak{o}}} c\mathfrak{o}.$$

Indeed, if $c \in K^\times$, then $c^{-1} \in X^{-1}$ if and only if $X \subset c\mathfrak{o}$. Note that $X \subset \{0\}$ implies $X^{-1} = K$ and $(X^{-1})^{-1} = K^{-1} = \{0\}$.

2. If $\mathfrak{a}$ and $\mathfrak{b}$ are $\mathfrak{o}$-submodules of K, then we define their product (as for ideals) by

$$\mathfrak{a}\mathfrak{b} = \{a_1b_1 + \ldots + a_nb_n \mid n \in \mathbb{N},\ a_1, \ldots, a_n \in \mathfrak{a},\ b_1, \ldots, b_n \in \mathfrak{b}\}$$

and their quotient by

$$(\mathfrak{b}:\mathfrak{a}) = \{x \in K \mid x\mathfrak{a} \subset \mathfrak{b}\} = \bigcap_{a \in \mathfrak{a}} a^{-1}\mathfrak{b}.$$

Clearly, $\mathfrak{a} + \mathfrak{b}$, $\mathfrak{a}\mathfrak{b}$ and $(\mathfrak{b}:\mathfrak{a})$ are again $\mathfrak{o}$-submodules of K. Addition and multiplication of $\mathfrak{o}$-submodules of K are commutative and associative laws of composition. The zero module $\mathbf{0}$ is the zero of the addition, and $\mathfrak{o}$ (the **unit ideal** $\mathfrak{o} = \mathbf{1}$) is the neutral element of the multiplication. For that reason it is common in multiplicative ideal theory to set $\mathfrak{o} = \mathbf{1}$. For $\mathfrak{o}$-submodules $\mathfrak{a}$, $\mathfrak{b}$, $\mathfrak{c}$ of K it is easily checked that $\mathfrak{a}(\mathfrak{b} + \mathfrak{c}) = \mathfrak{a}\mathfrak{b} + \mathfrak{a}\mathfrak{c}$, $((\mathfrak{a}:\mathfrak{b}):\mathfrak{c}) = (\mathfrak{a}:\mathfrak{b}\mathfrak{c})$ and $(\mathfrak{b}:\mathfrak{a})\mathfrak{a} \subset \mathfrak{b}$ (hence in particular $\mathfrak{a}(\mathfrak{o}:\mathfrak{a}) \subset \mathfrak{o}$). If $x \in K^\times$, then $x\mathfrak{a} = (x\mathfrak{o})\mathfrak{a}$ is again an $\mathfrak{o}$-submodule of K and $x(\mathfrak{b}:\mathfrak{a}) = (x\mathfrak{b}:\mathfrak{a}) = (\mathfrak{b}:x^{-1}\mathfrak{a})$.

3. An $\mathfrak{o}$-submodule $\mathfrak{a}$ of K is called

- a **fractional ideal** of $\mathfrak{o}$ if $\mathfrak{a} \neq \mathbf{0}$ and $(\mathfrak{o}:\mathfrak{a}) \neq \mathbf{0}$;
- $(\mathfrak{o}$-)**invertible** if $\mathfrak{a}(\mathfrak{o}:\mathfrak{a}) = \mathfrak{o}$ (then $\mathfrak{a}$ is a fractional ideal of $\mathfrak{o}$).

If $\mathfrak{a}$ is an invertible fractional ideal of $\mathfrak{o}$, then we call $\mathfrak{a}^{-1} = (\mathfrak{o}:\mathfrak{a})$ the **inverse** of $\mathfrak{a}$. If $x \in K^\times$, then $x\mathfrak{o}$ is an invertible fractional ideal, and $(x\mathfrak{o})^{-1} = x^{-1}\mathfrak{o}$. We call $x\mathfrak{o}$ the **fractional principal ideal** of $\mathfrak{o}$ generated by x.

Every non-zero ideal of $\mathfrak{o}$ is a fractional ideal. Therefore non-zero ideals of $\mathfrak{o}$ are sometimes called **integral ideals**, and by an **invertible ideal** we mean an invertible integral ideal.

4. We denote by

- $\mathcal{J}(\mathfrak{o})$ the set of all ($\mathfrak{o}$-)invertible fractional ideals of $\mathfrak{o}$;

- $\mathcal{J}'(\mathfrak{o}) = \{\mathfrak{a} \in \mathcal{J}(\mathfrak{o}) \mid \mathfrak{a} \subset \mathfrak{o}\}$ the set of all ($\mathfrak{o}$-)invertible (integral) ideals of $\mathfrak{o}$;
- $\mathcal{H}(\mathfrak{o}) = \{a\mathfrak{o} \mid a \in K^\times\}$ the set of all fractional principal ideals of $\mathfrak{o}$;
- $\mathcal{H}'(\mathfrak{o}) = \mathcal{H}(\mathfrak{o}) \cap \mathcal{J}'(\mathfrak{o}) = \{a\mathfrak{o} \mid a \in \mathfrak{o}^\bullet\}$ the set of all non-zero (integral) principal ideals of $\mathfrak{o}$.

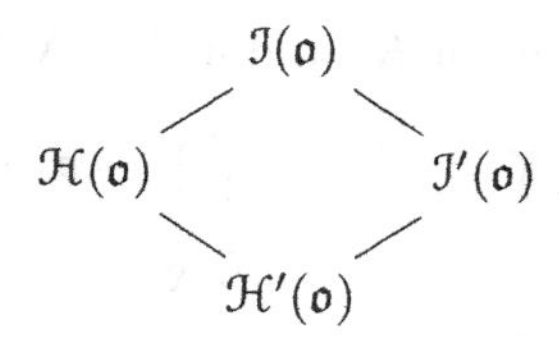

If $\mathfrak{o} = K$ is a field, then K is the only non-zero $\mathfrak{o}$-submodule of K.

In the subsequent two theorems we gather those elementary properties of fractional and invertible ideals which we will use in the sequel without further reference.

Theorem 2.6.2 (Fractional ideals). *Let $\mathfrak{o}$ be a domain and $K = \mathsf{q}(\mathfrak{o})$.*

1. An $\mathfrak{o}$-submodule $\mathfrak{a}$ of K is a fractional ideal if and only if $\mathfrak{a} \neq \mathbf{0}$ and there exists some $a \in \mathfrak{o}^\bullet$ such that $a\mathfrak{a} \subset \mathfrak{o}$.

2. If $\mathfrak{a}$ is a fractional ideal of $\mathfrak{o}$ and $x \in K^\times$, then $x\mathfrak{a} = (x\mathfrak{o})\mathfrak{a}$ is a fractional ideal of $\mathfrak{o}$. Moreover, if $\mathfrak{a}$ is invertible, then $x\mathfrak{a}$ is invertible and $(x\mathfrak{a})^{-1} = x^{-1}\mathfrak{a}^{-1}$.

3. Let $\mathfrak{a}$ and $\mathfrak{b}$ be fractional ideals of $\mathfrak{o}$. Then $\mathfrak{a}+\mathfrak{b}$, $\mathfrak{a}\mathfrak{b}$, $\mathfrak{a}\cap\mathfrak{b}$, $(\mathfrak{b}:\mathfrak{a})$ and in particular $(\mathfrak{o}:\mathfrak{a})$ are fractional ideals of $\mathfrak{o}$.

4. Let $\mathfrak{a} \neq \mathbf{0}$ be an $\mathfrak{o}$-submodule of K.

(a) If $\mathfrak{a}$ is finitely generated, then $\mathfrak{a}$ is a fractional ideal of $\mathfrak{o}$.

(b) If $\mathfrak{o}$ is Noetherian and $\mathfrak{a}$ is a fractional ideal of $\mathfrak{o}$, then $\mathfrak{a}$ is a finitely generated $\mathfrak{o}$-module.

Proof. 1. Let $\mathfrak{a}$ be an $\mathfrak{o}$-submodule of K. If $\mathfrak{a}$ is a fractional ideal of $\mathfrak{o}$, then $\mathfrak{a} \neq \mathbf{0}$, and if $x \in (\mathfrak{o}:\mathfrak{a})^\bullet$, then $x = ab^{-1}$ for some $a,\, b \in \mathfrak{o}^\bullet$, and then it follows that $a\mathfrak{a} = bx\mathfrak{a} \subset b\mathfrak{o} \subset \mathfrak{o}$. The converse is obvious.

2. Obvious.

3. Let $a,\, b,\, c,\, d \in \mathfrak{o}^\bullet$ such that $a \in \mathfrak{a}$, $b \in \mathfrak{b}$, $c\mathfrak{a} \subset \mathfrak{o}$ and $d\mathfrak{b} \subset \mathfrak{o}$. Then $a \in \mathfrak{a}+\mathfrak{b}$ and $bd(\mathfrak{a}+\mathfrak{b}) \subset \mathfrak{o}$, $ab \in \mathfrak{a}\mathfrak{b}$ and $bd\mathfrak{a}\mathfrak{b} \subset \mathfrak{o}$, $cadb \in \mathfrak{a}\cap\mathfrak{b}$ and $c(\mathfrak{a}\cap\mathfrak{b}) \subset \mathfrak{o}$, $b \in (\mathfrak{b}:\mathfrak{a})$ and $bc(\mathfrak{b}:\mathfrak{a}) \subset \mathfrak{o}$. Hence $\mathfrak{a}+\mathfrak{b}$, $\mathfrak{a}\mathfrak{b}$, $\mathfrak{a}\cap\mathfrak{b}$ and $(\mathfrak{b}:\mathfrak{a})$ are fractional ideals of $\mathfrak{o}$ by 1.

4. (a) If $\mathfrak{a} = \mathfrak{o}a_1 + \ldots + \mathfrak{o}a_n$ and $a \in \mathfrak{o}^\bullet$ is such that $aa_i \in \mathfrak{o}$ for all $i \in [1,n]$, then $a\mathfrak{a} \subset \mathfrak{o}$. Hence $\mathfrak{a}$ is a fractional ideal of $\mathfrak{o}$ by 1.

(b) Let $a \in \mathfrak{o}^\bullet$ such that $a\mathfrak{a} \subset \mathfrak{o}$. Then $a\mathfrak{a}$ is an ideal of $\mathfrak{o}$, hence finitely generated, say $a\mathfrak{a} = {}_\mathfrak{o}(b_1, \ldots, b_m)$ for some $m \in \mathbb{N}$ and $b_1, \ldots, b_m \in \mathfrak{o}$, and then we obtain $\mathfrak{a} = {}_\mathfrak{o}(a^{-1}b_1, \ldots, a^{-1}b_m)$. □

Theorem 2.6.3 (Invertible ideals). *Let $\mathfrak{o}$ be a domain, $K = \mathsf{q}(\mathfrak{o})$, and let $\mathfrak{a}$ and $\mathfrak{b}$ be $\mathfrak{o}$-submodules of K.*

1. If $\mathfrak{a}$ is invertible, then $\mathfrak{a}$ is a finitely generated $\mathfrak{o}$-module, and $(\mathfrak{a}\colon\mathfrak{a}) = \mathfrak{o}$.

2. Let $\mathfrak{a}$ be invertible. Then $\mathfrak{b} \subset \mathfrak{a}$ if and only if $\mathfrak{b} = \mathfrak{a}\mathfrak{c}$ for some ideal $\mathfrak{c}$ of $\mathfrak{o}$, and then $\mathfrak{c} = \mathfrak{a}^{-1}\mathfrak{b}$ is the unique ideal of $\mathfrak{o}$ satisfying $\mathfrak{b} = \mathfrak{a}\mathfrak{c}$.

3. If $\mathfrak{a}\mathfrak{b} = \mathfrak{o}$, then $\mathfrak{a}$ is invertible, and $\mathfrak{b} = \mathfrak{a}^{-1}$. In particular, if $\mathfrak{a}$ is invertible, then $\mathfrak{a}^{-1}$ is invertible, and $(\mathfrak{a}^{-1})^{-1} = \mathfrak{a}$.

4. $\mathfrak{a}\mathfrak{b}$ is invertible if and only if both $\mathfrak{a}$ and $\mathfrak{b}$ are invertible, and then it follows that $(\mathfrak{a}\mathfrak{b})^{-1} = \mathfrak{a}^{-1}\mathfrak{b}^{-1}$.

Proof. 1. Let $\mathfrak{a}$ be invertible. Since $1 \in \mathfrak{a}\mathfrak{a}^{-1}$, there exist $a_1, \ldots, a_n \in \mathfrak{a}$ and $c_1, \ldots, c_n \in \mathfrak{a}^{-1}$ such that $1 = a_1c_1 + \ldots + a_nc_n$. If $x \in \mathfrak{a}$, then $c_ix \in \mathfrak{o}$ for all $i \in [1, n]$ and $x = c_1xa_1 + \ldots + c_nxa_n \in {}_\mathfrak{o}(a_1, \ldots, a_n)$, hence $\mathfrak{a} \subset {}_\mathfrak{o}(a_1, \ldots, a_n)$, and consequently $\mathfrak{a} = {}_\mathfrak{o}(a_1, \ldots, a_n)$.

Obviously, $\mathfrak{o} \subset (\mathfrak{a}:\mathfrak{a})$, and if $x \in (\mathfrak{a}:\mathfrak{a})$, then it follows that $x\mathfrak{a} \subset \mathfrak{a}$ and $x \in x\mathfrak{o} = x\mathfrak{a}\mathfrak{a}^{-1} \subset \mathfrak{a}\mathfrak{a}^{-1} = \mathfrak{o}$. Hence $(\mathfrak{a}\colon\mathfrak{a}) = \mathfrak{o}$.

2. If $\mathfrak{b} \subset \mathfrak{a}$, then $\mathfrak{c} = \mathfrak{a}^{-1}\mathfrak{b} \subset \mathfrak{a}^{-1}\mathfrak{a} = \mathfrak{o}$ and $\mathfrak{a}\mathfrak{c} = \mathfrak{b}$. On the other hand, if $\mathfrak{c}$ is any ideal of $\mathfrak{o}$ such that $\mathfrak{b} = \mathfrak{a}\mathfrak{c}$, then $\mathfrak{b} \subset \mathfrak{a}$ and $\mathfrak{c} = \mathfrak{a}^{-1}\mathfrak{b}$.

3. If $\mathfrak{a}\mathfrak{b} = \mathfrak{o}$, then $\mathfrak{b} \subset (\mathfrak{o}:\mathfrak{a}) = (\mathfrak{o}:\mathfrak{a})\mathfrak{a}\mathfrak{b} \subset \mathfrak{o}\mathfrak{b} = \mathfrak{b}$ and therefore $\mathfrak{b} = (\mathfrak{o}:\mathfrak{a})$. Hence $\mathfrak{a}$ is invertible and $\mathfrak{b} = \mathfrak{a}^{-1}$. If $\mathfrak{a}$ is invertible, then $\mathfrak{a}\mathfrak{a}^{-1} = \mathfrak{o}$, and thus $\mathfrak{a}^{-1}$ is invertible and $(\mathfrak{a}^{-1})^{-1} = \mathfrak{a}$.

4. If $\mathfrak{a}\mathfrak{b}$ is invertible, then $\mathfrak{a}\mathfrak{b}(\mathfrak{a}\mathfrak{b})^{-1} = \mathfrak{a}[\mathfrak{b}(\mathfrak{a}\mathfrak{b})^{-1}] = \mathfrak{b}[\mathfrak{a}(\mathfrak{a}\mathfrak{b})^{-1}] = \mathfrak{o}$, and thus both $\mathfrak{a}$ and $\mathfrak{b}$ are invertible. If both $\mathfrak{a}$ and $\mathfrak{b}$ are invertible, then $(\mathfrak{a}\mathfrak{b})\mathfrak{a}^{-1}\mathfrak{b}^{-1} = \mathfrak{a}\mathfrak{a}^{-1}\mathfrak{b}\mathfrak{b}^{-1} = \mathfrak{o}$. Hence $\mathfrak{a}\mathfrak{b}$ is invertible, and $(\mathfrak{a}\mathfrak{b})^{-1} = \mathfrak{a}^{-1}\mathfrak{b}^{-1}$. □

Corollary 2.6.4. *Let $\mathfrak{o}$ be a domain. Then $\mathcal{J}'(\mathfrak{o})$ and $\mathcal{H}'(\mathfrak{o})$ are reduced monoids, $\mathcal{J}(\mathfrak{o}) = \mathsf{q}(\mathcal{J}'(\mathfrak{o}))$ and $\mathcal{H}(\mathfrak{o}) = \mathsf{q}(\mathcal{H}'(\mathfrak{o}))$ are their quotient groups, $\mathcal{H}(\mathfrak{o})$ is a subgroup of $\mathcal{J}(\mathfrak{o})$, and the map*

$$\partial\colon K^\times \to \mathcal{J}(\mathfrak{o}), \quad \textit{defined by} \quad \partial(a) = a\mathfrak{o},$$

is a group homomorphism such that $\mathrm{Ker}(\partial) = \mathfrak{o}^\times$ *and* $\mathrm{Im}(\partial) = \mathcal{H}(\mathfrak{o})$.

Proof. By 2.6.3, $\mathcal{J}(\mathfrak{o})$ is a (multiplicative) group with neutral element $\mathbf{1} = \mathfrak{o}$, and for every $\mathfrak{a} \in \mathcal{J}(\mathfrak{o})$ the invertible ideal $\mathfrak{a}^{-1}$ is its group-theoretical inverse. If $\mathfrak{a}, \mathfrak{b} \in \mathcal{J}'(\mathfrak{o})$, then $\mathfrak{ab} \in \mathcal{J}'(\mathfrak{o})$, and $\mathfrak{ab} = \mathfrak{o}$ implies $\mathfrak{a} = \mathfrak{b} = \mathfrak{o}$. Hence $\mathcal{J}'(\mathfrak{o})$ is a reduced submonoid of $\mathcal{J}(\mathfrak{o})$. If $\mathfrak{a} \in \mathcal{J}(\mathfrak{o})$, then there exists some $a \in \mathfrak{o}^\bullet$ such that $a\mathfrak{a} \subset \mathfrak{o}$. Hence $\mathfrak{a} = (a\mathfrak{o})^{-1}(a\mathfrak{a})$, and since $a\mathfrak{o}, a\mathfrak{a} \in \mathcal{J}'(\mathfrak{o})$ it follows that $\mathcal{J}(\mathfrak{o}) = \mathsf{q}(\mathcal{J}'(\mathfrak{o}))$. The remaining assertions are obvious. □

If $\mathfrak{o}$ is a domain and $\mathfrak{a} \in \mathcal{J}(\mathfrak{o})$, then (in view of 2.6.4) the powers $\mathfrak{a}^n$ are defined for all $n \in \mathbb{Z}$, and the usual calculation rules hold. In particular, $\mathfrak{a}^0 = \mathbf{1} = \mathfrak{o}$, and $(\mathfrak{a}^{-1})^k = (\mathfrak{a}^k)^{-1}$ for all $k \in \mathbb{Z}$.

In the following definition we introduce the ideal class group of a domain. In the case of algebraic number fields and algebraic function fields it will prove to be an important invariant of the arithmetic.

Definition 2.6.5. Let $\mathfrak{o}$ be a domain and $K = \mathsf{q}(\mathfrak{o})$. The factor group

$$\mathcal{C}(\mathfrak{o}) = \mathcal{J}(\mathfrak{o})/\mathcal{H}(\mathfrak{o})$$

is called the **Picard group** or **ideal class group** of $\mathfrak{o}$. If $\mathfrak{a} \in \mathcal{J}(\mathfrak{o})$, then we denote by $[\mathfrak{a}] = \mathfrak{a}\mathcal{H}(\mathfrak{o}) \in \mathcal{C}(\mathfrak{o})$ the **ideal class** of $\mathfrak{a}$. If $\mathfrak{a}, \mathfrak{b} \in \mathcal{J}(\mathfrak{o})$, then $[\mathfrak{a}] = [\mathfrak{b}]$ if and only if $\mathfrak{b} = \lambda\mathfrak{a}$ for some $\lambda \in K^\times$. If $\mathfrak{a} \in \mathcal{J}(\mathfrak{o})$ and $a \in \mathfrak{a}^{-1}$, then $a\mathfrak{a} \in \mathcal{J}'(\mathfrak{o})$ and $[\mathfrak{a}] = [a\mathfrak{a}] \in \mathcal{C}(\mathfrak{o})$. Hence $\mathcal{C}(\mathfrak{o}) = \{[\mathfrak{a}] \mid \mathfrak{a} \in \mathcal{J}'(\mathfrak{o})\}$.

The definition of the ideal class group gives rise to the exact sequence

$$\mathbf{1} \to \mathfrak{o}^\times \hookrightarrow K^\times \xrightarrow{\partial} \mathcal{J}(\mathfrak{o}) \xrightarrow{\pi} \mathcal{C}(\mathfrak{o}) \to \mathbf{1},$$

where $\partial(a) = a\mathfrak{o}$ for all $a \in K^\times$, and $\pi(\mathfrak{a}) = [\mathfrak{a}]$ for all $\mathfrak{a} \in \mathcal{J}(\mathfrak{o})$.

Theorem 2.6.6 (Invertible ideals in semilocal domains). *Le $\mathfrak{o}$ be a semilocal domain. Then every invertible fractional ideal of $\mathfrak{o}$ is principal. Consequently, we obtain $\mathcal{J}(\mathfrak{o}) = \mathcal{H}(\mathfrak{o})$ and $\mathcal{C}(\mathfrak{o}) = \mathbf{1}$.*

Proof. Let $\mathfrak{a} \in \mathcal{J}(\mathfrak{o})$ and $\max(\mathfrak{o}) = \{\mathfrak{p}_1, \ldots, \mathfrak{p}_r\}$ (where $r \in \mathbb{N}$ and $\mathfrak{p}_1, \ldots, \mathfrak{p}_r$ are distinct). For $j \in [1, r]$ we set

$$\mathfrak{p}_j^* = \bigcap_{\substack{i=1 \\ i \neq j}}^{r} \mathfrak{p}_i.$$

If $j \in [1, r]$, then $\mathfrak{p}_j^* \not\subset \mathfrak{p}_j$, hence $\mathfrak{a}\mathfrak{p}_j^* \not\subset \mathfrak{a}\mathfrak{p}_j$. We choose $a_j \in \mathfrak{a}\mathfrak{p}_j^* \setminus \mathfrak{a}\mathfrak{p}_j$, and set $a = a_1 + \ldots + a_r$. Then $a \in \mathfrak{a}$, and if $j \in [1, r]$, then

$$a - a_j = \sum_{\substack{i=1 \\ i \neq j}}^{r} a_i \in \mathfrak{a}\mathfrak{p}_j, \text{ and } a_j \notin \mathfrak{a}\mathfrak{p}_j \quad \text{implies} \quad a \notin \mathfrak{a}\mathfrak{p}_j \text{ and } a\mathfrak{a}^{-1} \not\subset \mathfrak{p}_j.$$

Since $a\mathfrak{a}^{-1} \subset \mathfrak{a}\mathfrak{a}^{-1} = \mathfrak{o}$ and $a\mathfrak{a}^{-1}$ is not contained in any maximal ideal of $\mathfrak{o}$, Krull's theorem 1.1.5.2 implies $a\mathfrak{a}^{-1} = \mathfrak{o}$ and $\mathfrak{a} = a\mathfrak{a}^{-1}\mathfrak{a} = a\mathfrak{o}$. □

Theorem 2.6.7 (Fractional ideals in ring extensions). *Let $\mathfrak{o}$ and $\mathfrak{O}$ be domains such that $\mathfrak{o} \subset \mathfrak{O}$, and $K = \mathsf{q}(\mathfrak{o}) \subset \mathsf{q}(\mathfrak{O}) \subset L$.*

1. *If $\mathfrak{a}$ and $\mathfrak{b}$ are $\mathfrak{o}$-submodules of K, then*

$$(\mathfrak{a}\mathfrak{b})\mathfrak{O} = \mathfrak{a}(\mathfrak{b}\mathfrak{O}) = (\mathfrak{a}\mathfrak{O})(\mathfrak{b}\mathfrak{O}).$$

2. *Let $\mathfrak{a}$ be a fractional ideal of $\mathfrak{o}$. Then $\mathfrak{a}\mathfrak{O}$ is a fractional ideal of $\mathfrak{O}$. If moreover $\mathfrak{a}$ is $\mathfrak{o}$-invertible, then $\mathfrak{a}\mathfrak{O}$ is $\mathfrak{O}$-invertible, and $(\mathfrak{a}\mathfrak{O})^{-1} = \mathfrak{a}^{-1}\mathfrak{O}$. In particular, the ideal extension*

$$j_{\mathfrak{O}/\mathfrak{o}} \colon \mathcal{J}(\mathfrak{o}) \to \mathcal{J}(\mathfrak{O}), \quad \text{defined by } j_{\mathfrak{O}/\mathfrak{o}}(\mathfrak{a}) = \mathfrak{a}\mathfrak{O},$$

is a group homomorphism and induces a group homomorphism

$$j^*_{\mathfrak{O}/\mathfrak{o}} \colon \mathcal{C}(\mathfrak{o}) \to \mathcal{C}(\mathfrak{O}), \quad \text{given by } j^*_{\mathfrak{O}/\mathfrak{o}}([\mathfrak{a}]) = [\mathfrak{a}\mathfrak{O}] \text{ for all } \mathfrak{a} \in \mathcal{J}(\mathfrak{o}).$$

Proof. 1. Let $\mathfrak{a}$ and $\mathfrak{b}$ be $\mathfrak{o}$-submodules of K. Then $\mathfrak{a}\mathfrak{b} \subset (\mathfrak{a}\mathfrak{O})(\mathfrak{b}\mathfrak{O})$ and therefore $(\mathfrak{a}\mathfrak{b})\mathfrak{O} \subset (\mathfrak{a}\mathfrak{O})(\mathfrak{b}\mathfrak{O})$. On the other hand, if $x \in (\mathfrak{a}\mathfrak{O})(\mathfrak{b}\mathfrak{O})$, then x is a sum of elements of the form

$$\Big(\sum_{i=1}^{n} a_i x_i\Big)\Big(\sum_{j=1}^{m} b_j y_j\Big) = \sum_{i=1}^{n}\sum_{j=1}^{m} a_i b_j x_i y_j,$$

where $m, n \in \mathbb{N}$, $a_i \in \mathfrak{a}$, $b_j \in \mathfrak{b}$ and $x_i, y_j \in \mathfrak{O}$. As every such element lies in $\mathfrak{a}\mathfrak{b}\mathfrak{O}$, it follows that $x \in \mathfrak{a}\mathfrak{b}\mathfrak{O}$.

2. If $a \in \mathfrak{o}^\bullet$ is such that $a\mathfrak{a} \subset \mathfrak{o}$, then $a\mathfrak{a}\mathfrak{O} \subset \mathfrak{O}$. Hence $\mathfrak{a}\mathfrak{O}$ is a fractional ideal of $\mathfrak{O}$. If $\mathfrak{a}$ is $\mathfrak{o}$-invertible, then $\mathfrak{O} = \mathfrak{a}\mathfrak{a}^{-1}\mathfrak{O} = (\mathfrak{a}\mathfrak{O})(\mathfrak{a}^{-1}\mathfrak{O})$ by 1. Hence $\mathfrak{a}\mathfrak{O}$ is $\mathfrak{O}$-invertible, and $(\mathfrak{a}\mathfrak{O})^{-1} = \mathfrak{a}^{-1}\mathfrak{O}$. Consequently, $j_{\mathfrak{O}/\mathfrak{o}}$ and $j^*_{\mathfrak{O}/\mathfrak{o}}$ are group homomorphisms. Note that $j^*_{\mathfrak{O}/\mathfrak{o}}$ need not be injective, even if $j_{\mathfrak{O}/\mathfrak{o}}$ is. □

2.7 Quotient domains and localizations

Let $\mathfrak{o}$ be a domain, $K = \mathsf{q}(\mathfrak{o})$, U a vector space over K and T a submonoid of $\mathfrak{o}^\bullet$. For a subset X of U we set

$$T^{-1}X = \{t^{-1}x \mid t \in T,\ x \in X\} \subset U.$$

Obviously, $X \subset T^{-1}X$, and if $n \in \mathbb{N}$, $x_1, \ldots, x_n \in T^{-1}X$ and $TX \subset X$, then there exists some $t \in T$, such that $tx_i \in X$ for all $i \in [1, n]$ (existence of a common denominator).

If $U = K$, then $T^{-1}\mathfrak{o}$ is a domain such that $\mathfrak{o} \subset T^{-1}\mathfrak{o} \subset K$, and the usual calculation rules for fractions hold: If b, $c \in \mathfrak{o}$ and a, $d \in T$, then

$$\frac{b}{a} + \frac{c}{d} = \frac{bd + ac}{ad} \quad \text{and} \quad \frac{b}{a}\frac{c}{d} = \frac{bc}{ad}.$$

In particular, $T \subset T^{-1}T \subset (T^{-1}\mathfrak{o})^\times$ and $(\mathfrak{o}^\bullet)^{-1}\mathfrak{o} = K$.

If M is an $\mathfrak{o}$-submodule of U, then $T^{-1}M$ is a $T^{-1}\mathfrak{o}$-submodule of U, again with the usual calculation rules: If a, $b \in M$, $x \in \mathfrak{o}$ and s, $t \in T$, then

$$\frac{a}{s} + \frac{b}{t} = \frac{ta + sb}{st} \quad \text{and} \quad \frac{x}{s}\frac{a}{t} = \frac{xa}{st}.$$

If $T \subset \mathfrak{o}^\times$, then $T^{-1}M = M$.

A submonoid T of $\mathfrak{o}$ is called **saturated** if $st \in T$ implies $s \in T$ and $t \in T$ for all s, $t \in \mathfrak{o}$.

In the following theorem we gather the elementary properties of quotients which will be used in the sequel without further reference.

Theorem 2.7.1. *Let $\mathfrak{o}$ be a domain, $K = \mathsf{q}(\mathfrak{o})$ and T a submonoid of $\mathfrak{o}^\bullet$.*

1. Let U be a vector space over K, and let M and N be $\mathfrak{o}$-submodules of U.

(a) $T^{-1}M \cap T^{-1}N = T^{-1}(M \cap N)$.

(b) $T^{-1}M = (T^{-1}\mathfrak{o})M$, and if $(M_i)_{i\in I}$ is a family of $\mathfrak{o}$-submodules of M such that M is the (direct) sum of the family $(M_i)_{i\in I}$, then $T^{-1}M$ is the (direct) sum of the family $(T^{-1}M_i)_{i\in I}$.

(c) If $(x_i)_{i\in I}$ is a family in M such that $M = {}_{\mathfrak{o}}(\{x_i \mid i \in I\})$ [$(x_i)_{i\in I}$ is an $\mathfrak{o}$-basis of M], then $T^{-1}M = {}_{T^{-1}\mathfrak{o}}(\{x_i \mid i \in I\})$ [$(x_i)_{i\in I}$ is a $T^{-1}\mathfrak{o}$-basis of $T^{-1}M$]. In particular, if M is a finitely generated [resp. free] $\mathfrak{o}$-module, then $T^{-1}M$ is a finitely generated [resp. free] $T^{-1}\mathfrak{o}$-module.

(d) If M is a finitely generated $\mathfrak{o}$-module, then $M \subset T^{-1}N$ if and only if $tM \subset N$ for some $t \in T$.

2. Let $\mathfrak{a}$ and $\mathfrak{b}$ be $\mathfrak{o}$-submodules of K. Then it follows that $T^{-1}(\mathfrak{a}\mathfrak{b}) = (T^{-1}\mathfrak{a})(T^{-1}\mathfrak{b})$, and

$$(T^{-1}\mathfrak{a} : T^{-1}\mathfrak{b}) = (T^{-1}\mathfrak{a} : \mathfrak{b}) \supset T^{-1}(\mathfrak{a} : \mathfrak{b}),$$

with equality if $\mathfrak{b}$ is finitely generated.

3. If $\mathfrak{a}$ is a (finitely generated) (fractional) ideal of $\mathfrak{o}$, then $T^{-1}\mathfrak{a}$ is a (finitely generated) (fractional) ideal of $T^{-1}\mathfrak{o}$.

4. If T is saturated, then $(T^{-1}\mathfrak{o})^\times = T^{-1}T = \mathsf{q}(T)$.

Proof. 1. Straightforward.

2. $T^{-1}(\mathfrak{a}\mathfrak{b}) = (\mathfrak{a}\mathfrak{b})T^{-1}\mathfrak{o} = (\mathfrak{a}T^{-1}\mathfrak{o})(\mathfrak{b}T^{-1}\mathfrak{o}) = (T^{-1}\mathfrak{a})(T^{-1}\mathfrak{b}$ by 2.6.7.1. Obviously, $(T^{-1}\mathfrak{a}:T^{-1}\mathfrak{b}) = (T^{-1}\mathfrak{a}:\mathfrak{b})$, and since $(\mathfrak{a}:\mathfrak{b})\mathfrak{b} \subset \mathfrak{a}$, it follows that $T^{-1}(\mathfrak{a}:\mathfrak{b})T^{-1}\mathfrak{b} \subset T^{-1}\mathfrak{a}$ and therefore $T^{-1}(\mathfrak{a}:\mathfrak{b}) \subset (T^{-1}\mathfrak{a}:T^{-1}\mathfrak{b})$.

Now let $\mathfrak{b} = {}_{\mathfrak{o}}(b_1,\ldots,b_m)$ and $x \in (T^{-1}\mathfrak{a}:T^{-1}\mathfrak{b})$. Since $xb_j \in T^{-1}\mathfrak{a}$ for all $j \in [1,m]$, there exists some $t \in T$ such that $(tx)b_j \in \mathfrak{a}$ for all $j \in [1,m]$, hence $tx\mathfrak{b} \subset \mathfrak{a}$ and $x \in T^{-1}(\mathfrak{b}:\mathfrak{a})$.

3. Obvious.

4. Clearly, $T^{-1}T = \mathsf{q}(T) \subset (T^{-1}\mathfrak{o})^{\times}$. Thus let T be saturated, and suppose that $x = t^{-1}a \in (T^{-1}\mathfrak{o})^{\times}$, where $a \in \mathfrak{o}$ and $t \in T$. Then there exist elements $b \in \mathfrak{o}$ and $s \in T$ such that $(t^{-1}a)(s^{-1}b) = 1$. Hence $ab = st \in T$, and therefore $a \in T$ and $x \in T^{-1}T$. □

Theorem 2.7.2. *Let $\mathfrak{o}$ and $\mathfrak{O}$ be domains such that $\mathfrak{o} \subset \mathfrak{O} \subset \mathsf{q}(\mathfrak{o})$ and $T = \mathfrak{O}^{\times} \cap \mathfrak{o}$. Suppose that for all $a,\, b \in \mathfrak{o}$ there exists some $n \in \mathbb{N}$ such that the ideal ${}_{\mathfrak{o}}(a,b)^n$ is a principal ideal. Then $\mathfrak{O} = T^{-1}\mathfrak{o}$.*

Proof. Evidently, $T^{-1}\mathfrak{o} \subset \mathfrak{O}$. Thus let $z = b^{-1}a \in \mathfrak{O}$, where $b \in \mathfrak{o}^{\bullet}$ and $a \in \mathfrak{o}$. By assumption, there exists some $n \in \mathbb{N}$ such that

$${}_{\mathfrak{o}}(a,b)^n = {}_{\mathfrak{o}}(a^n, a^{n-1}b, \ldots, ab^{n-1}, b^n) = c\mathfrak{o} \quad \text{for some} \quad c \in \mathfrak{o}.$$

We set $b^n = uc$, $ab^{n-1} = vc$ and $c = d_0a^n + d_1a^{n-1}b + \ldots + d_{n-1}ab^{n-1} + d_nb^n$, where u, v, $d_0,\ldots,d_n \in \mathfrak{o}$. Then

$$u^{-1} = b^{-n}c = d_0z^n + d_1z^{n-1} + \ldots + d_{n-1}z + d_n \in \mathfrak{O},$$

hence $u \in \mathfrak{O}^{\times} \cap \mathfrak{o} = T$ and $z = u^{-1}v \in T^{-1}\mathfrak{o}$. □

Theorem 2.7.3. *Let $\mathfrak{o}$ be a domain and T a submonoid of $o^{\bullet}$.*

1. *If $\mathfrak{a}$ is an ideal of $\mathfrak{o}$, then $T^{-1}\mathfrak{a}$ is an ideal of $T^{-1}\mathfrak{o}$, and we have $T^{-1}\mathfrak{a} = T^{-1}\mathfrak{o}$ if and only if $\mathfrak{a} \cap T = \emptyset$.*

2. *If $\mathfrak{A}$ is an ideal of $T^{-1}\mathfrak{o}$, then $\mathfrak{A} \cap \mathfrak{o}$ is an ideal of $\mathfrak{o}$, and $T^{-1}(\mathfrak{A} \cap \mathfrak{o}) = \mathfrak{A}$.*

3. *If $\mathfrak{o}$ is Noetherian, then $T^{-1}\mathfrak{o}$ is Noetherian, too.*

4. *Let $\mathfrak{a}$ be an $\mathfrak{o}$-submodule of $\mathsf{q}(\mathfrak{o})$, $\mathfrak{c}$ an ideal of $\mathfrak{o}$ such that $T \cap \mathfrak{p} = \emptyset$ for all maximal ideals $\mathfrak{p}$ of $\mathfrak{o}$ containing $\mathfrak{c}$, and $\mathfrak{b} = \mathfrak{c}\mathfrak{a}$. Then the inclusion $\mathfrak{a} \hookrightarrow T^{-1}\mathfrak{a}$ induces an $\mathfrak{o}$-isomorphism*

$$j\colon \mathfrak{a}/\mathfrak{b} \xrightarrow{\sim} T^{-1}\mathfrak{a}/T^{-1}\mathfrak{b}, \quad \textit{given by } j(a+\mathfrak{b}) = a + T^{-1}\mathfrak{b} \textit{ for all } a \in \mathfrak{a}.$$

Proof. 1. Evidently, $T^{-1}\mathfrak{a} = \mathfrak{a}T^{-1}\mathfrak{o}$ is an ideal of $T^{-1}\mathfrak{o}$ and $\mathfrak{a} \subset T^{-1}\mathfrak{a} \cap \mathfrak{o}$. If $T^{-1}\mathfrak{a} = T^{-1}\mathfrak{o}$, then $1 \in T^{-1}\mathfrak{a}$, hence $1 = t^{-1}a$ for some $t \in T$ and $a \in \mathfrak{o}$, and therefore $t = a \in \mathfrak{a} \cap T$. Conversely, if $t \in \mathfrak{a} \cap T$, then $1 = t^{-1}t \in T^{-1}\mathfrak{a}$ and $T^{-1}\mathfrak{a} = T^{-1}\mathfrak{o}$.

2. Apparently $\mathfrak{A}\cap\mathfrak{o}$ is an ideal of $\mathfrak{o}$ and $T^{-1}(\mathfrak{A}\cap\mathfrak{o})\subset\mathfrak{A}$. To prove equality, let $a = t^{-1}c \in \mathfrak{A}$, where $t \in T$ and $c \in \mathfrak{A}$. Then $c = ta \in \mathfrak{A}\cap\mathfrak{o}$ and $a = t^{-1}c \in T^{-1}(\mathfrak{A}\cap\mathfrak{o})$.

3. Let $\mathfrak{o}$ be Noetherian and $\mathfrak{A}$ an ideal of $T^{-1}\mathfrak{o}$. Then $\mathfrak{A} = T^{-1}(\mathfrak{A}\cap\mathfrak{o})$ by 2., $\mathfrak{A}\cap\mathfrak{o}$ is a finitely generated ideal of $\mathfrak{o}$ and thus $\mathfrak{A}$ is finitely generated by 2.7.1.3.

4. If $t \in T$, then $\mathfrak{c} + t\mathfrak{o}$ lies in no maximal ideal of $\mathfrak{o}$, and thus $\mathfrak{c} + t\mathfrak{o} = \mathfrak{o}$. It suffices to prove that $\mathfrak{b} = T^{-1}\mathfrak{b}\cap\mathfrak{a}$ and $T^{-1}\mathfrak{a}\subset T^{-1}\mathfrak{b}+\mathfrak{a}$.

Apparently, $\mathfrak{b}\subset\mathfrak{a}\cap T^{-1}\mathfrak{b}$. Thus let $a = t^{-1}b \in \mathfrak{a}\cap T^{-1}\mathfrak{b}$, where $t \in T$ and $b \in \mathfrak{b}$. If $c \in \mathfrak{c}$ and $u \in \mathfrak{o}$ are such that $c + tu = 1$, then it follows that $a = ca + tua = ca + ub \in \mathfrak{b}$.

Now let $x = t^{-1}a \in T^{-1}\mathfrak{a}$, where $t \in T$ and $a \in \mathfrak{a}$. If again $c \in \mathfrak{c}$ and $u \in \mathfrak{o}$ are such that $c + tu = 1$, then $t^{-1}a = t^{-1}ac + au \in T^{-1}\mathfrak{b}+\mathfrak{a}$. □

Definition 2.7.4. Let $\mathfrak{o}$ be a domain, $K = \mathsf{q}(\mathfrak{o})$, U a vector space over K and X a subset of U. If $\mathfrak{p}$ is a prime ideal of $\mathfrak{o}$, then $T = \mathfrak{o}\setminus\mathfrak{p}$ is a saturated submonoid of $\mathfrak{o}$, and we call

$$X_\mathfrak{p} = (\mathfrak{o}\setminus\mathfrak{p})^{-1}X$$

the **localization** of X at $\mathfrak{p}$.

If M is an $\mathfrak{o}$-submodule of U, then $M_0 = \mathfrak{o}^{\bullet -1}M = KM$ is the vector space spanned by M over K, and $\mathfrak{o}_0 = \mathfrak{o}^{\bullet -1}\mathfrak{o} = K$. If $\mathfrak{o} = \mathbb{Z}$ and $\mathfrak{p} = p\mathbb{Z}$ for some prime p, we set $M_{(p)} = M_{p\mathbb{Z}}$.

In particular, $\mathbb{Z}_{(p)} = \{s^{-1}a \mid s \in \mathbb{Z}\setminus p\mathbb{Z},\ a \in \mathbb{Z}\} \subset \mathbb{Q}$ is the domain of ***p*-integral** rational numbers.

Theorem 2.7.5. *Let $\mathfrak{o}$ be a domain, $K = \mathsf{q}(\mathfrak{o})$ and U a vector space over K. If M is an $\mathfrak{o}$-submodule of U, then*

$$M = \bigcap_{\mathfrak{p}\in\max(\mathfrak{o})} M_\mathfrak{p}.$$

In particular, if M and N are $\mathfrak{o}$-submodules of U such that $M_\mathfrak{p}\subset N_\mathfrak{p}$ [resp. $M_\mathfrak{p} = N_\mathfrak{p}$] for all $\mathfrak{p}\in\max(\mathfrak{o})$, then $M\subset N$ [resp. $M = N$].

Proof. By definition, $M \subset M_\mathfrak{p}$ for all $\mathfrak{p}\in\max(\mathfrak{o})$. Thus let $c \in U$ and $c \in M_\mathfrak{p}$ for all $\mathfrak{p}\in\max(\mathfrak{o})$. For every $\mathfrak{p}\in\max(\mathfrak{o})$ there exists some $s \in \mathfrak{o}\setminus\mathfrak{p}$ such that $sc \in M$. Hence it follows that $\mathfrak{a} = \{a \in \mathfrak{o} \mid ac \in M\} \not\subset \mathfrak{p}$ but as $\mathfrak{a}$ is an ideal of $\mathfrak{o}$, Krull's theorem 1.1.5.2 implies $\mathfrak{a} = \mathfrak{o}$ and $c \in M$. □

Theorem 2.7.6 (Prime ideals in quotient domains). *Let $\mathfrak{o}$ be a domain.*

1. *Let T be a submonoid of $\mathfrak{o}^\bullet$.*

(a) $\mathcal{P}(T^{-1}\mathfrak{o}) = \{T^{-1}\mathfrak{p} \mid \mathfrak{p} \in \mathcal{P}(\mathfrak{o}),\ \mathfrak{p} \cap T = \emptyset\}$.

(b) *Let* $\mathfrak{p} \in \mathcal{P}(\mathfrak{o})$ *and* $\mathfrak{p} \cap T = \emptyset$. *Then* $T^{-1}\mathfrak{p} \cap \mathfrak{o} = \mathfrak{p}$ *and* $(T^{-1}\mathfrak{o})_{T^{-1}\mathfrak{p}} = \mathfrak{o}_{\mathfrak{p}}$.

2. *If* $\mathfrak{p} \in \mathcal{P}(\mathfrak{o})$, *then* $\mathfrak{o}_\mathfrak{p}$ *is a local domain, and* $\mathfrak{p}_\mathfrak{p} = \mathfrak{p}\mathfrak{o}_\mathfrak{p}$ *is its maximal ideal.*

3. *Let* $\mathfrak{p} \in \max(\mathfrak{o})$ *and* T *a submonoid of* $\mathfrak{o}^\bullet$ *such that* $T \cap \mathfrak{p} = \emptyset$.

(a) $T^{-1}\mathfrak{p} \in \max(T^{-1}\mathfrak{o})$.

(b) *Let* $n \in \mathbb{Z}$, $k \in \mathbb{N}$, *and suppose that* $n \geq 0$ *if* $\mathfrak{p}$ *is not invertible. Then* $T^{-1}\mathfrak{p}^{n+k} \cap \mathfrak{p}^n = \mathfrak{p}^{n+k}$, *and the inclusion* $\mathfrak{p}^n \hookrightarrow T^{-1}\mathfrak{p}^n$ *induces an isomorphism of additive abelian groups*

$$\psi_{n,k} \colon \mathfrak{p}^n/\mathfrak{p}^{n+k} \xrightarrow{\sim} T^{-1}\mathfrak{p}^n/T^{-1}\mathfrak{p}^{n+k},$$

given by $\psi_{n,k}(a + \mathfrak{p}^{n+k}) = a + T^{-1}\mathfrak{p}^{n+k}$ *for all* $a \in \mathfrak{p}^n$. *In particular,* $\psi_{0,k} \colon \mathfrak{o}/\mathfrak{p}^k \xrightarrow{\sim} T^{-1}\mathfrak{o}/T^{-1}\mathfrak{p}^k$ *is a ring isomorphism.*

We identify: $\mathfrak{p}^n/\mathfrak{p}^{n+k} = T^{-1}\mathfrak{p}^n/T^{-1}\mathfrak{p}^{n+k}$. In particular: $\mathfrak{o}/\mathfrak{p}^k = T^{-1}\mathfrak{o}/T^{-1}\mathfrak{p}^k$, $\mathsf{k}_{T^{-1}\mathfrak{p}} = \mathsf{k}_\mathfrak{p}$ and $\mathfrak{p}^n/\mathfrak{p}^{n+1} = T^{-1}\mathfrak{p}^n/T^{-1}\mathfrak{p}^{n+1}$.

Proof. 1. (a) Let $\mathfrak{p} \in \mathcal{P}(\mathfrak{o})$ and $\mathfrak{p} \cap T = \emptyset$. Then $T^{-1}\mathfrak{p}$ is a proper ideal of $T^{-1}\mathfrak{o}$, and we assume that $x, y \in T^{-1}\mathfrak{o}$ and $xy \in T^{-1}\mathfrak{p}$. Let $x = s^{-1}a$, $y = t^{-1}b$ and $xy = w^{-1}c$, where $s, t, w \in T$, $a, b \in \mathfrak{o}$ and $c \in \mathfrak{p}$. Then it follows that $wab = stc \in \mathfrak{p}$, and since $w \notin \mathfrak{p}$ we get $a \in \mathfrak{p}$ or $b \in \mathfrak{p}$ and thus $x \in T^{-1}\mathfrak{p}$ or $y \in T^{-1}\mathfrak{p}$. Hence $T^{-1}\mathfrak{p} \in \mathcal{P}(T^{-1}\mathfrak{o})$. Conversely, if $\mathfrak{P} \in \mathcal{P}(T^{-1}\mathfrak{o})$, then $\mathfrak{P} \cap T = \emptyset$, $\mathfrak{P} \cap \mathfrak{o} \in \mathcal{P}(\mathfrak{o})$ and $\mathfrak{P} = T^{-1}(\mathfrak{P} \cap \mathfrak{o})$ by 2.7.3.2.

(b) Clearly $\mathfrak{p} \subset T^{-1}\mathfrak{p} \cap \mathfrak{o}$. If $a \in T^{-1}\mathfrak{p} \cap \mathfrak{o}$, then $a = t^{-1}c$ for some $t \in T$ and $c \in \mathfrak{p}$. Hence $ta = c \in \mathfrak{p}$ and thus $a \in \mathfrak{p}$.

Clearly, $\mathfrak{o} \setminus \mathfrak{p} \subset T^{-1}\mathfrak{o} \setminus T^{-1}\mathfrak{p}$ implies $\mathfrak{o}_\mathfrak{p} \subset (T^{-1}\mathfrak{o})_{T^{-1}\mathfrak{p}}$. Thus assume that $x = u^{-1}v \in (T^{-1}\mathfrak{o})_{T^{-1}\mathfrak{p}}$, where $u = t^{-1}a \in T^{-1}\mathfrak{o}$ and $v = s^{-1}b$ such that $s, t \in T$, $a, b \in \mathfrak{o}$ and $b \notin \mathfrak{p}$, then $tb \notin \mathfrak{p}$ and $x = (tb)^{-1}(sa) \in \mathfrak{o}_\mathfrak{p}$.

2. $\mathcal{P}(\mathfrak{o}_\mathfrak{p}) = \{\mathfrak{p}\mathfrak{o}_\mathfrak{p}\}$ by 1.

3. (a) Let $\mathfrak{P}$ be an ideal of $T^{-1}\mathfrak{o}$ such that $T^{-1}\mathfrak{p} \subset \mathfrak{P} \subsetneq T^{-1}\mathfrak{o}$. Then it follows that $\mathfrak{p} = T^{-1}\mathfrak{p} \cap \mathfrak{o} \subset \mathfrak{P} \cap \mathfrak{o} \subsetneq \mathfrak{o}$, hence $\mathfrak{p} = \mathfrak{P} \cap \mathfrak{o}$, and consequently $T^{-1}\mathfrak{p} = T^{-1}(\mathfrak{P} \cap \mathfrak{o}) = \mathfrak{P}$.

(b) By 2.7.3.4, applied with $\mathfrak{a} = \mathfrak{p}^n$ and $\mathfrak{c} = \mathfrak{p}^k$ (note that $\mathfrak{p}$ is the only maximal ideal containing $\mathfrak{p}^k$). Apparently, $\psi_{0,k}$ is a ring isomorphism. □

Theorem 2.7.7 (Local-global principle for invertibility). *Let* $\mathfrak{o}$ *be a domain and* $\mathfrak{a}$ *a fractional ideal of* $\mathfrak{o}$. *Then* $\mathfrak{a}$ *is* ($\mathfrak{o}$-)*invertible if and only if* $\mathfrak{a}$ *is finitely generated, and* $\mathfrak{a}_\mathfrak{p}$ *is principal for all* $\mathfrak{p} \in \mathcal{P}(\mathfrak{o})$.

Proof. If $\mathfrak{a}$ is invertible, then $\mathfrak{a}$ is finitely generated by 2.6.3.1, $\mathfrak{a}_\mathfrak{p}$ is $\mathfrak{o}_\mathfrak{p}$-invertible by 2.6.7.2, and $\mathfrak{a}_\mathfrak{p}$ is principal by 2.6.6. Thus suppose that $\mathfrak{a}$ is finitely generated, $\mathfrak{a}_\mathfrak{p}$ is $\mathfrak{o}_\mathfrak{p}$-invertible for all $\mathfrak{p} \in \mathcal{P}(\mathfrak{o})$, and yet $\mathfrak{a}$ is not invertible. Then $\mathfrak{a}(\mathfrak{o}:\mathfrak{a}) \subsetneq \mathfrak{o}$, and by 1.1.5.2 there exists some $\mathfrak{p} \in \mathcal{P}(\mathfrak{o})$ such that $\mathfrak{a}(\mathfrak{o}:\mathfrak{a}) \subset \mathfrak{p}$. It follows that $\mathfrak{o}_\mathfrak{p} = \mathfrak{a}_\mathfrak{p}\mathfrak{a}_\mathfrak{p}^{-1} = \mathfrak{a}_\mathfrak{p}(\mathfrak{o}_\mathfrak{p}:\mathfrak{a}_\mathfrak{p}) = (\mathfrak{a}(\mathfrak{o}:\mathfrak{a}))_\mathfrak{p} \subset \mathfrak{p}\mathfrak{o}_\mathfrak{p}$, since $\mathfrak{a}$ by 2.7.1.2, a contradiction. □

Theorem 2.7.8 (Integrality in quotient domains). *Let $\mathfrak{o}$ and $\mathfrak{O}$ be domains such that $\mathfrak{o} \subset \mathfrak{O}$, and let T be a submonoid of $\mathfrak{o}^\bullet$.*

1. *$\mathrm{cl}_{T^{-1}\mathfrak{O}}(T^{-1}\mathfrak{o}) = T^{-1}\mathrm{cl}_\mathfrak{O}(\mathfrak{o})$.*
2. *If $\mathfrak{o}$ is integrally closed, then $T^{-1}\mathfrak{o}$ is integrally closed, too.*
3. *Let $\mathfrak{p} \in \mathcal{P}(\mathfrak{o})$ such that $\mathfrak{p} \cap T = \emptyset$. Then*

$$\{\mathfrak{P}' \in \mathcal{P}(T^{-1}\mathfrak{O}) \mid \mathfrak{P}' \cap T^{-1}\mathfrak{o} = T^{-1}\mathfrak{p}\} = \{T^{-1}\mathfrak{P} \mid \mathfrak{P} \in \mathcal{P}(\mathfrak{O}),\ \mathfrak{P} \cap \mathfrak{o} = \mathfrak{p}\}.$$

Proof. 1. Let first $z = t^{-1}x \in \mathrm{cl}_{T^{-1}\mathfrak{O}}(T^{-1}\mathfrak{o}) \subset T^{-1}\mathfrak{O}$, where $t \in T$ and $z \in \mathfrak{O}$, and let $z^n + u_{n-1}z^{n-1} + \ldots + u_1 z + u_0 = 0$ be an integral equation for z over $T^{-1}\mathfrak{o}$. Let $a_0, \ldots, a_{n-1} \in \mathfrak{o}$ and $s \in T$ such that $u_i = s^{-1}a_i$ for all $i \in [0, n-1]$. Then the integral equation takes the form

$$\left(\frac{x}{t}\right)^n + \frac{a_{n-1}}{s}\left(\frac{x}{t}\right)^{n-1} + \ldots + \frac{a_1}{s}\frac{x}{t} + \frac{a_0}{s} = 0,$$

and after multiplication by $t^n s^n$ we get

$$(xs)^n + \sum_{i=1}^{n} a_{n-i}t^i s^{i-1}(sx)^{n-i} = 0,$$

an integral equation of sx over $\mathfrak{o}$, and therefore $z = (st)^{-1}sx \in T^{-1}\mathrm{cl}_\mathfrak{O}(\mathfrak{o})$.

Conversely, if $z = t^{-1}x \in T^{-1}\mathrm{cl}_\mathfrak{O}(\mathfrak{o})$ for some $t \in T$ and $x \in \mathrm{cl}_\mathfrak{O}(\mathfrak{o})$, and if $x^n + a_{n-1}x + \ldots + a_1 x + a_0 = 0$ is an integral equation for x over $\mathfrak{o}$, then

$$\left(\frac{x}{t}\right)^n + \frac{a_{n-1}}{t}\left(\frac{x}{t}\right)^{n-1} + \ldots + \frac{a_1}{t^{n-1}}\left(\frac{x}{t}\right) + \frac{a_0}{t^n} = 0$$

is an integral equation for $z = t^{-1}x$ over $T^{-1}\mathfrak{o}$.

2. If $K = \mathsf{q}(\mathfrak{o})$, then $T^{-1}K = K$, and if $\mathfrak{o} = \mathrm{cl}_K(\mathfrak{o})$, then $T^{-1}\mathfrak{o} = \mathrm{cl}_K(T^{-1}\mathfrak{o})$ by 1.

3. If $\mathfrak{P}' \in \mathcal{P}(T^{-1}\mathfrak{O})$ and $\mathfrak{P}' \cap T^{-1}\mathfrak{o} = T^{-1}\mathfrak{p}$, then $\mathfrak{P} = \mathfrak{P}' \cap \mathfrak{O} \in \mathcal{P}(\mathfrak{O})$ and $\mathfrak{P} \cap \mathfrak{o} = \mathfrak{P}' \cap \mathfrak{O} \cap \mathfrak{o} = \mathfrak{P}' \cap T^{-1}\mathfrak{o} \cap \mathfrak{o} = T^{-1}\mathfrak{p} \cap \mathfrak{o} = \mathfrak{p}$.

Conversely, if $\mathfrak{P} \in \mathcal{P}(\mathfrak{O})$ and $\mathfrak{P} \cap \mathfrak{o} = \mathfrak{p}$, then $\mathfrak{P}' = T^{-1}\mathfrak{P} \in \mathcal{P}(T^{-1}\mathfrak{O})$ and $T^{-1}\mathfrak{p} = \mathfrak{P}' \cap T^{-1}\mathfrak{o}$. □

Corollary 2.7.9 (Integrality in localizations). *Let $\mathfrak{o}$ and $\mathfrak{O}$ be domains such that $\mathfrak{o} \subset \mathfrak{O}$ and $L = \mathsf{q}(\mathfrak{O})$.*

1. *Let $\mathfrak{p} \in \mathcal{P}(\mathfrak{o})$ and let $\mathfrak{A}$ be an $\mathfrak{O}$-submodule of $\mathsf{q}(\mathfrak{O})$. Then $\mathrm{cl}_{\mathfrak{O}_\mathfrak{p}}(\mathfrak{o}_\mathfrak{p}) = \mathrm{cl}_{\mathfrak{O}}(\mathfrak{o})_\mathfrak{p}$,*

$$\{\mathfrak{P}' \in \mathcal{P}(\mathfrak{O}_\mathfrak{p}) \mid \mathfrak{P}' \cap \mathfrak{o}_\mathfrak{p} = \mathfrak{p}\mathfrak{o}_\mathfrak{p}\} = \{\mathfrak{P}\mathfrak{O}_\mathfrak{p} \mid \mathfrak{P} \in \mathcal{P}(\mathfrak{O}),\ \mathfrak{P} \cap \mathfrak{o} = \mathfrak{p}\},$$

and if $\mathfrak{A}$ is an $\mathfrak{O}$-submodule of L, then

$$\mathfrak{A}_\mathfrak{p} = \mathfrak{A}\mathfrak{O}_\mathfrak{p} = \bigcap_{\substack{\mathfrak{P} \in \mathcal{P}(\mathfrak{O}) \\ \mathfrak{P} \cap \mathfrak{o} = \mathfrak{p}}} \mathfrak{A}_\mathfrak{P}.$$

2. *$\mathfrak{o}$ is integrally closed if and only if $\mathfrak{o}_\mathfrak{p}$ is integrally closed for all $\mathfrak{p} \in \max(\mathfrak{o})$.*

Proof. 1. The first two assertions follow by 2.7.8.3. If $\mathfrak{A}$ is an $\mathfrak{O}$-submodule of L, then $\mathfrak{A}_\mathfrak{p} = \mathfrak{A}\mathfrak{O}_\mathfrak{p}$, and

$$\mathfrak{A}_\mathfrak{p} = \bigcap_{\mathfrak{P}' \in \mathcal{P}(\mathfrak{O}_\mathfrak{p})} (\mathfrak{A}_\mathfrak{p})_{\mathfrak{P}'} = \bigcap_{\substack{\mathfrak{P} \in \mathfrak{P}(\mathfrak{O}) \\ \mathfrak{P} \cap \mathfrak{o} = \mathfrak{p}}} (\mathfrak{A}_\mathfrak{p})_{\mathfrak{P}\mathfrak{O}_\mathfrak{p}} = \bigcap_{\substack{\mathfrak{P} \in \mathfrak{P}(\mathfrak{O}) \\ \mathfrak{P} \cap \mathfrak{o} = \mathfrak{p}}} \mathfrak{A}_\mathfrak{P}$$

by 2.7.5 and 2.7.6.1(b).

2. If $\mathfrak{o}$ is integrally closed and $\mathfrak{p} \in \max(\mathfrak{o})$, then $\mathfrak{o}_\mathfrak{p}$ is integrally closed by 2.7.8.2. Assume now that $\mathfrak{o}_\mathfrak{p}$ is integrally closed for all $\mathfrak{p} \in \max(\mathfrak{o})$. If $\mathfrak{p} \in \max(\mathfrak{o})$, then $K_\mathfrak{p} = K$, and by 1. and 2.7.5 we obtain

$$\mathrm{cl}_K(\mathfrak{o}) = \bigcap_{\mathfrak{p} \in \max(\mathfrak{o})} \mathrm{cl}_K(\mathfrak{o})_\mathfrak{p} = \bigcap_{\mathfrak{p} \in \max(\mathfrak{o})} \mathrm{cl}_K(\mathfrak{o}_\mathfrak{p}) = \bigcap_{\mathfrak{p} \in \max(\mathfrak{o})} \mathfrak{o}_\mathfrak{p} = \mathfrak{o}. \qquad \square$$

2.8 Dedekind domains

Definition and Theorem 2.8.1 (Dedekind domains). *Let $\mathfrak{o}$ be a domain which is not field. Then the following assertions are equivalent:*

(a) $\mathfrak{o}$ is Noetherian, integrally closed, and $\dim(\mathfrak{o}) = 1$.

(b) Every non-zero ideal of $\mathfrak{o}$ is invertible.

(c) Every fractional ideal of $\mathfrak{o}$ is invertible.

If these conditions are fulfilled, then $\mathfrak{o}$ is called a **Dedekind domain**. Recall that for $\mathfrak{p} \in \mathcal{P}(\mathfrak{o})$ we denote by $\mathsf{k}_\mathfrak{p} = \mathfrak{o}/\mathfrak{p}$ the residue class field.

Proof. (a) $\Rightarrow$ (b) We assume that (a) holds and prove first:

> **A.** Every non-zero ideal of $\mathfrak{o}$ contains a product of non-zero prime ideals.
>
> **B.** If $\mathfrak{p} \in \mathcal{P}(\mathfrak{o})$, then $(\mathfrak{o}:\mathfrak{p}) \supsetneq \mathfrak{o}$.

Proof of **A.** Assume to the contrary that there exists a non-zero ideal $\mathfrak{a}$ of $\mathfrak{o}$ which does not contain a product of non-zero prime ideals. As $\mathfrak{o}$ is Noetherian, we may assume that $\mathfrak{a}$ is maximal with this property. Then $\mathfrak{a} \notin \mathcal{P}(\mathfrak{o})$, and as $\mathfrak{a} \neq \mathfrak{o}$, there exist $b, c \in \mathfrak{o} \setminus \mathfrak{a}$ such that $bc \in \mathfrak{a}$. Since $\mathfrak{a} \subsetneq \mathfrak{a} + b\mathfrak{o}$ and $\mathfrak{a} \subsetneq \mathfrak{a} + c\mathfrak{o}$, there exist $r, s \in \mathbb{N}$ and $\mathfrak{p}_1, \ldots, \mathfrak{p}_r, \mathfrak{q}_1, \ldots, \mathfrak{q}_s \in \mathcal{P}(\mathfrak{o})$ such that $\mathfrak{p}_1 \cdot \ldots \cdot \mathfrak{p}_r \subset \mathfrak{a} + bR$, $\mathfrak{q}_1 \cdot \ldots \cdot \mathfrak{q}_s \subset \mathfrak{a} + cR$, and consequently $\mathfrak{p}_1 \cdot \ldots \cdot \mathfrak{p}_r \mathfrak{q}_1 \cdot \ldots \cdot \mathfrak{q}_s \subset (\mathfrak{a} + b\mathfrak{o})(\mathfrak{a} + c\mathfrak{o}) \subset \mathfrak{a}$, a contradiction. $\square$[**A.**]

Proof of **B.** Let $\mathfrak{p} \in \mathcal{P}(\mathfrak{o})$ and $a \in \mathfrak{p}^\bullet$. By **A** there exist some $r \in \mathbb{N}$ and $\mathfrak{p}_1, \ldots, \mathfrak{p}_r \in \mathcal{P}(\mathfrak{o})$ such that $\mathfrak{p}_1 \cdot \ldots \cdot \mathfrak{p}_r \subset a\mathfrak{o} \subset \mathfrak{p}$, and we let r be minimal with this property. After renumbering we may assume that $\mathfrak{p}_1 \subset \mathfrak{p}$ and thus $\mathfrak{p}_1 = \mathfrak{p}$, since $\mathfrak{p}_1$ is a maximal ideal. The minimal choice of r implies that $\mathfrak{p}_2 \cdot \ldots \cdot \mathfrak{p}_r \not\subset a\mathfrak{o}$. If $b \in \mathfrak{p}_2 \cdot \ldots \cdot \mathfrak{p}_r \setminus a\mathfrak{o}$, then $a^{-1}b \notin \mathfrak{o}$ and $b\mathfrak{p} \subset \mathfrak{p}_1 \cdot \ldots \cdot \mathfrak{p}_r \subset a\mathfrak{o}$, hence $a^{-1}b\mathfrak{p} \subset \mathfrak{o}$ and therefore $a^{-1}b \in (\mathfrak{o}:\mathfrak{p}) \setminus \mathfrak{o}$. Since obviously $\mathfrak{o} \subset (\mathfrak{o}:\mathfrak{p})$, we get $\mathfrak{o} \subsetneq (\mathfrak{o}:\mathfrak{p})$. $\square$[**B.**]

Now we prove (b) and assume that there exists a non-zero ideal $\mathfrak{a}$ of $\mathfrak{o}$ which is not invertible. As $\mathfrak{o}$ is Noetherian, we may assume that $\mathfrak{a}$ is maximal with this property. By Krull's theorem 1.1.5.2 there exists some $\mathfrak{p} \in \mathcal{P}(\mathfrak{o})$ such that $\mathfrak{p} \supset \mathfrak{a}$. Then $\mathfrak{a} \subset \mathfrak{a}(\mathfrak{o}:\mathfrak{p}) \subset \mathfrak{p}(\mathfrak{o}:\mathfrak{p}) \subset \mathfrak{o}$, and we assert that $\mathfrak{a} \subsetneq \mathfrak{a}(\mathfrak{o}:\mathfrak{p})$. Indeed, if $\mathfrak{a} = \mathfrak{a}(\mathfrak{o}:\mathfrak{p})$ and $x \in (\mathfrak{o}:\mathfrak{p})$ then $x\mathfrak{a} \subset \mathfrak{a}$, hence x is integral over $\mathfrak{o}$ by 2.4.2 and thus $x \in \mathfrak{o}$. Hence $(\mathfrak{o}:\mathfrak{p}) = \mathfrak{o}$, which contradicts **B**. Now $\mathfrak{a} \subsetneq \mathfrak{a}(\mathfrak{o}:\mathfrak{p}) \subset \mathfrak{o}$ implies that $\mathfrak{a}(\mathfrak{o}:\mathfrak{p})$ is invertible, and thus $\mathfrak{a}$ is invertible by 2.6.3.4, a contradiction.

(b) $\Rightarrow$ (c) Let $\mathfrak{a}$ be a fractional ideal of $\mathfrak{o}$ and $a \in \mathfrak{o}^\bullet$ such that $a\mathfrak{a} \subset \mathfrak{o}$. Then $a\mathfrak{a} \neq$ is a non-zero ideal of $\mathfrak{o}$, hence invertible, and thus $\mathfrak{a}$ is invertible, too.

(c) $\Rightarrow$ (a) By 2.6.3.1, every invertible ideal is finitely generated. Hence $\mathfrak{o}$ is Noetherian.

If $x \in \mathsf{q}(\mathfrak{o})$ is integral over $\mathfrak{o}$, then $\mathfrak{o}[x]$ is a finitely generated $\mathfrak{o}$-module by 2.4.3.1. Hence $\mathfrak{o}[x]$ is a fractional ideal of $\mathfrak{o}$ by 2.6.2.4(a) and thus invertible. Therefore $\mathfrak{o}[x] = \mathfrak{o}[x]\mathfrak{o}[x] = \mathfrak{o}[x]\mathfrak{o}$ implies $\mathfrak{o}[x] = \mathfrak{o}$, hence $x \in \mathfrak{o}$, and thus $\mathfrak{o}$ is integrally closed.

It remains to prove that every non-zero prime ideal is maximal. Assume to the contrary that there exist $\mathfrak{p}, \mathfrak{q} \in \mathcal{P}(\mathfrak{o})$ such that $\mathfrak{p} \subsetneq \mathfrak{q}$. Then we obtain $\mathfrak{p} = \mathfrak{q}(\mathfrak{q}^{-1}\mathfrak{p})$, $\mathfrak{q}^{-1}\mathfrak{p} \subset \mathfrak{o}$ and $\mathfrak{q} \not\subset \mathfrak{p}$, hence $\mathfrak{q}^{-1}\mathfrak{p} \subset \mathfrak{p}$ and $\mathfrak{q}^{-1} \subset \mathfrak{p}\mathfrak{p}^{-1} = \mathfrak{o}$, a contradiction. $\square$

Theorem 2.8.2. *Let* $\mathfrak{o}$ *be a Dedekind domain,* $K = \mathsf{q}(\mathfrak{o})$, $\mathfrak{a} \in \mathcal{J}(\mathfrak{o})$ *a fractional ideal of* $\mathfrak{o}$ *and* $X \subset K$. *Then* $\mathfrak{a} = {}_{\mathfrak{o}}(X)$ *if and only if* $\mathfrak{a} = (X^{-1})^{-1}$.

Proof. If $\mathfrak{a} = {}_{\mathfrak{o}}(X)$, then $\mathfrak{a} = (\mathfrak{a}^{-1})^{-1} = (X^{-1})^{-1}$. Thus suppose now that $\mathfrak{a} = (X^{-1})^{-1}$ and let $\mathfrak{b} = {}_{\mathfrak{o}}(X)$. Then $(\mathfrak{b}^{-1})^{-1} = (X^{-1})^{-1} = \mathfrak{a}$, hence $\mathfrak{b} \neq \mathbf{0}$ and $\mathfrak{b}^{-1} \neq \mathbf{0}$. It follows that $\mathfrak{b} \in \mathcal{J}(\mathfrak{o})$, and $\mathfrak{b} = (\mathfrak{b}^{-1})^{-1} = \mathfrak{a}$. □

Let $\mathfrak{o}$ be a Dedekind domain and $K = \mathsf{q}(\mathfrak{o})$. In what follows we use the monoid-theoretical notions and results of Section 2.1 both for the arithmetic of the monoid $\mathfrak{o}^\bullet$ and the arithmetic of the monoid $\mathcal{J}'(\mathfrak{o})$ of invertible (integral) ideals of $\mathfrak{o}$ (note that $\mathfrak{o} = \mathfrak{o}^{\bullet\circ}$, and $\mathcal{J}'(\mathfrak{o})^\circ = \mathcal{J}'(\mathfrak{o}) \cup \{\mathbf{0}\}$ is the multiplicative semigroup of all ideals of $\mathfrak{o}$). If $\mathfrak{a}, \mathfrak{b} \in \mathcal{J}'(\mathfrak{o})^\circ$ and $a, b \in \mathfrak{o}$, then $\mathfrak{a} \,|\, \mathfrak{b}$ if $\mathfrak{b} = \mathfrak{a}\mathfrak{c}$ for some $\mathfrak{c} \in \mathcal{J}'(\mathfrak{o})^\circ$, and $a \,|\, b$ if $b\mathfrak{o} \subset a\mathfrak{o}$ [equivalently, $b = ac$ for some $c \in \mathfrak{o}$].

Theorem 2.8.3 (Dedekind arithmetic of ideals, I). *Let* $\mathfrak{o}$ *be a Dedekind domain.*

1. *Let* $\mathfrak{a}, \mathfrak{b} \in \mathcal{J}'(\mathfrak{o})$ *and* $a, b \in \mathfrak{o}$.

(a) $\mathfrak{a} \,|\, \mathfrak{b}$ *if and only if* $\mathfrak{b} \subset \mathfrak{a}$.

(b) $a \,|\, b$ *if and only if* $a\mathfrak{o} \,|\, b\mathfrak{o}$.

2. $\mathcal{J}'(\mathfrak{o})$ *is a free monoid with basis* $\mathcal{P}(\mathfrak{o})$ *and unit element* $\mathfrak{o} = \mathbf{1}$. *The quotient group* $\mathsf{q}(\mathcal{J}'(\mathfrak{o})) = \mathcal{J}(\mathfrak{o})$ *is a free abelian group with basis* $\mathcal{P}(\mathfrak{o})$.

For $\mathfrak{p} \in \mathcal{P}(\mathfrak{o})$ we denote by $v_\mathfrak{p} \colon \mathcal{J}(\mathfrak{o})^\circ \to \mathbb{Z} \cup \{\infty\}$ the **$\mathfrak{p}$-adic valuation** of $\mathcal{J}(\mathfrak{o})^\circ$ (see 2.1.4), and for $\mathfrak{a} \in \mathcal{J}(\mathfrak{o})$ we call

$$\mathrm{supp}(\mathfrak{a}) = \{\mathfrak{p} \in \mathcal{P}(\mathfrak{o}) \mid v_\mathfrak{p}(\mathfrak{a}) \neq 0\}$$

the **support** of $\mathfrak{a}$.
Then $v_\mathfrak{p}(\mathbf{0}) = \infty$, $v_\mathfrak{p} \restriction \mathcal{J}(\mathfrak{o}) \colon \mathcal{J}(\mathfrak{o}) \to \mathbb{Z}$ *is a group epimorphism, and if* $\mathfrak{a} \in \mathcal{J}(\mathfrak{o})$, *then* $\mathrm{supp}(\mathfrak{a}) = \emptyset$ *if and only if* $\mathfrak{a} = \mathbf{1}$

3. *If* $\mathfrak{a}, \mathfrak{b} \in \mathcal{J}(\mathfrak{o})$, *then*

- $\mathfrak{a} \in \mathcal{J}'(\mathfrak{o}) \iff v_\mathfrak{p}(\mathfrak{a}) \geq 0$ *for all* $\mathfrak{p} \in \mathcal{P}(\mathfrak{o})$.
- $\mathfrak{b} \subset \mathfrak{a} \iff v_\mathfrak{p}(\mathfrak{a}) \leq v_\mathfrak{p}(\mathfrak{b})$ *for all* $\mathfrak{p} \in \mathcal{P}(\mathfrak{o}) \iff \mathfrak{b} = \mathfrak{a}\mathfrak{c}$ *for some* $\mathfrak{c} \in \mathcal{J}'(\mathfrak{o})$.
- $\mathfrak{a} = \mathfrak{b} \iff v_\mathfrak{p}(\mathfrak{b}) = v_\mathfrak{p}(\mathfrak{a})$ *for all* $\mathfrak{p} \in \mathcal{P}(\mathfrak{o})$;

4. *Let* $\mathfrak{a} \in \mathcal{J}(\mathfrak{o})$.

(a) $\mathfrak{a}$ *has a unique representation in the form* $\mathfrak{a} = \mathfrak{b}\mathfrak{c}^{-1}$, *where* $\mathfrak{b}, \mathfrak{c} \in \mathcal{J}'(\mathfrak{o})$ *and* $\mathfrak{a} + \mathfrak{b} = \mathfrak{o}$.

(b) supp($\mathfrak{a}$) *is finite,*

$$\mathfrak{a} = \prod_{\mathfrak{p}\in\mathcal{P}(\mathfrak{o})} \mathfrak{p}^{v_\mathfrak{p}(\mathfrak{a})} = \prod_{\mathfrak{p}\in\operatorname{supp}(\mathfrak{a})} \mathfrak{p}^{v_\mathfrak{p}(\mathfrak{a})},$$

and for every exponent vector $(n_\mathfrak{p})_{\mathfrak{p}\in\mathcal{P}} \in \mathbb{Z}^{(\mathcal{P}(\mathfrak{o}))}$

$$\mathfrak{a} = \prod_{\mathfrak{p}\in\mathcal{P}(\mathfrak{o})} \mathfrak{p}^{n_\mathfrak{p}} \quad \textit{implies} \quad n_\mathfrak{p} = v_\mathfrak{p}(\mathfrak{a}) \ \textit{ for all } \ \mathfrak{p} \in \mathcal{P}(\mathfrak{o}).$$

(c) $\mathfrak{a}$ *has an (up to the order) unique factorization* $\mathfrak{a} = \mathfrak{p}_1^{e_1} \cdot \ldots \cdot \mathfrak{p}_r^{e_r}$, *where* $r \in \mathbb{N}_0$, $\mathfrak{p}_1, \ldots, \mathfrak{p}_r \in \mathcal{P}(\mathfrak{o})$ *are distinct,* $e_i = v_{\mathfrak{p}_i}(\mathfrak{a}) \in \mathbb{Z}$ *for all* $i \in [1, r]$, *and*

$$\operatorname{supp}(\mathfrak{a}) = \{\mathfrak{p} \in \mathcal{P}(\mathfrak{o}) \mid \mathfrak{a}_\mathfrak{p} \neq \mathfrak{o}_\mathfrak{p}\} = \{\mathfrak{p}_1, \ldots, \mathfrak{p}_r\}.$$

In particular, if $\mathfrak{a} \in \mathcal{J}'(\mathfrak{o})$, *then*

$$\operatorname{supp}(\mathfrak{a}) = \{\mathfrak{p} \in \mathcal{P}(\mathfrak{o}) \mid \mathfrak{p} \supset \mathfrak{a}\} = \{\mathfrak{p} \in \mathcal{P}(\mathfrak{o}) \mid \mathfrak{p} \mid \mathfrak{a}\}.$$

(d) *If* $\mathfrak{p} \in \mathcal{P}(\mathfrak{o})$, *then* $\mathfrak{a} = \mathfrak{p}^{v_\mathfrak{p}(\mathfrak{a})}\mathfrak{c}$, *where* $\mathfrak{c} \in \mathcal{J}(\mathfrak{o})$ *and* $v_\mathfrak{p}(\mathfrak{c}) = 0$.

Proof. 1. Obvious.

2. By 2.6.4 we know that $\mathcal{J}'(\mathfrak{o})$ is a reduced monoid and $\mathsf{q}(\mathcal{J}'(\mathfrak{o}) = \mathcal{J}(\mathfrak{o})$. We must prove that every $\mathfrak{p} \in \mathcal{P}(\mathfrak{o})$ is a prime element of $\mathcal{J}'(\mathfrak{o})$ and every $\mathfrak{a} \in \mathcal{J}'(\mathfrak{o})$ is a product of prime ideals (then $\mathcal{J}'(\mathfrak{o}) = \mathcal{F}(\mathcal{P}(\mathfrak{o})$, and the remaining assertions follow by 2.1.9 and 2.1.4).

Assume now that some ideal $\mathfrak{a} \in \mathcal{J}'(\mathfrak{o})$ is not a product of prime ideals. As $\mathfrak{o}$ is Noetherian, we may assume that $\mathfrak{a}$ is maximal with this property. Then $\mathfrak{a} \neq \mathfrak{o}$ (empty product!) and $\mathfrak{a} \notin \mathcal{P}(\mathfrak{o})$. By Krull's theorem 1.1.5.2 there exists some $\mathfrak{p} \in \mathcal{P}(\mathfrak{o})$ such that $\mathfrak{a} \subsetneq \mathfrak{p}$. Then $\mathfrak{p} \mid \mathfrak{a}$, hence $\mathfrak{a} = \mathfrak{p}\mathfrak{b}$, where $\mathfrak{b} \in \mathcal{J}'(\mathfrak{o})$ and $\mathfrak{a} \subsetneq \mathfrak{b}$. Therefore $\mathfrak{b}$ and thus also $\mathfrak{a}$ is a product of prime ideals, a contradiction.

3. By 2.1.6 and 2.6.3.2.

4. (a) and (b) follow from 2.1.6.

(c) The factorization of $\mathfrak{a}$ is a convenient reformulation of (b), and the subsequent assertions are easily checked.

(d) If $\mathfrak{p} \in \mathcal{P}(\mathfrak{o})$, then 2.1.4 shows that $\mathfrak{a} = \mathfrak{p}^{v_\mathfrak{p}(\mathfrak{a})}\mathfrak{b}^{-1}\mathfrak{d}$ for some $\mathfrak{b}, \mathfrak{d} \in \mathcal{J}'(\mathfrak{o})$ such that $\mathfrak{p} \nmid \mathfrak{b}\mathfrak{d}$. But then $\mathfrak{c} = \mathfrak{b}^{-1}\mathfrak{d} \in \mathcal{J}(\mathfrak{o})$ and $v_\mathfrak{p}(\mathfrak{c}) = -v_\mathfrak{p}(\mathfrak{b}) + v_\mathfrak{p}(\mathfrak{d}) = 0$. □

Theorem 2.8.4 (Dedekind arithmetic of ideals, II). *Let* $\mathfrak{o}$ *be a Dedekind domain.*

1. *Let* $X \subset \mathcal{J}(\mathfrak{o})^\circ$ *and* $\mathfrak{p} \in \mathcal{P}(\mathfrak{o})$. *Then*

$$v_\mathfrak{p}\Big(\sum_{\mathfrak{a}\in X} \mathfrak{a}\Big) = \inf\{v_\mathfrak{p}(\mathfrak{a}) \mid \mathfrak{a} \in X\} \quad \textit{and} \quad v_\mathfrak{p}\Big(\bigcap_{\mathfrak{a}\in X} \mathfrak{a}\Big) = \sup\{v_\mathfrak{p}(\mathfrak{a}) \mid \mathfrak{a} \in X\}.$$

In particular, if $\mathfrak{a}$, $\mathfrak{b} \in \mathcal{I}(\mathfrak{o})^\circ$ *and* $\mathfrak{p} \in \mathcal{P}(\mathfrak{o})$, *then*

$$v_\mathfrak{p}(\mathfrak{a}+\mathfrak{b}) = \min\{v_\mathfrak{p}(\mathfrak{a}), v_\mathfrak{p}(\mathfrak{b})\} \quad \textit{and} \quad v_\mathfrak{p}(\mathfrak{a}\cap\mathfrak{b}) = \max\{v_\mathfrak{p}(\mathfrak{a}), v_\mathfrak{p}(\mathfrak{b})\}.$$

2. Let $\emptyset \neq X \subset \mathcal{I}'(\mathfrak{o})^\circ$. *Then*

$$\gcd(X) = \sum_{\mathfrak{a}\in X} \mathfrak{a} \quad \textit{and} \quad \operatorname{lcm}(X) = \bigcap_{\mathfrak{a}\in X} \mathfrak{a}.$$

In particular, if $\mathfrak{a}$, $\mathfrak{b} \in \mathcal{I}'(\mathfrak{o})$, *then* $\mathfrak{a}+\mathfrak{b} = \mathfrak{o}$ *if and only if* $\min(v_\mathfrak{p}(\mathfrak{a}), v_\mathfrak{p}(\mathfrak{b})) = 0$ *for all* $\mathfrak{p} \in \mathcal{P}(\mathfrak{o})$.

3. If $n \in \mathbb{N}$ *and* $\mathfrak{a}_1, \ldots, \mathfrak{a}_n$, $\mathfrak{b} \in \mathcal{I}(\mathfrak{o})$, *then*

$$\bigcap_{i=1}^{n}(\mathfrak{a}_i + \mathfrak{b}) = \bigcap_{i=1}^{n} \mathfrak{a}_i + \mathfrak{b}.$$

4. (Generalized Chinese remainder theorem) *Let* $n \in \mathbb{N}$, $\mathfrak{a}_1, \ldots, \mathfrak{a}_n \in \mathcal{I}'(\mathfrak{o})$ *and* $a_1, \ldots, a_n \in \mathfrak{o}$ *such that* $a_i \equiv a_j \bmod (\mathfrak{a}_i+\mathfrak{a}_j)$ *for all* $i, j \in [1,n]$. *Then there exists some modulo* $\mathfrak{a}_1 \cap \ldots \cap \mathfrak{a}_n$ *uniquely determined element* $a \in \mathfrak{o}$ *such that* $a \equiv a_i \bmod \mathfrak{a}_i$ *for all* $i \in [1,n]$.

Proof. 1. We use 2.8.3.3. Let

$$\mathfrak{d} = \sum_{\mathfrak{a}\in X} \mathfrak{a} \quad \text{and} \quad \mathfrak{e} = \bigcap_{\mathfrak{a}\in X} \mathfrak{a}.$$

If $\mathfrak{a} \in X$, then $\mathfrak{e} \subset \mathfrak{a} \subset \mathfrak{d}$, hence $v_\mathfrak{p}(\mathfrak{d}) \le v_\mathfrak{p}(\mathfrak{a}) \le v_\mathfrak{p}(\mathfrak{e}$ for all $\mathfrak{p} \in \mathcal{P}(\mathfrak{o})$, and therefore $v_\mathfrak{p}(\mathfrak{d}) \le \inf\{v_\mathfrak{p}(\mathfrak{a}) \mid \mathfrak{a} \in X\}$ and $v_\mathfrak{p}(\mathfrak{e}) \ge \sup\{v_\mathfrak{p}(\mathfrak{a}) \mid \mathfrak{a} \in X\}$ for all $\mathfrak{p} \in \mathcal{P}(\mathfrak{o})$.

To prove equality, assume first that $v_\mathfrak{q}(\mathfrak{d}) < \inf\{v_\mathfrak{q}(\mathfrak{a}) \mid \mathfrak{a} \in X\}$ for some $\mathfrak{q} \in \mathcal{P}(\mathfrak{o})$. Then $v_\mathfrak{q}(\mathfrak{d}\mathfrak{q}) = v_\mathfrak{q}(\mathfrak{d}) + 1 \le \inf\{v_\mathfrak{q}(\mathfrak{a}) \mid \mathfrak{a} \in X\}$ and thus $v_\mathfrak{p}(\mathfrak{d}\mathfrak{q}) \le \inf\{v_\mathfrak{p}(\mathfrak{a}) \mid \mathfrak{a} \in X\}$ for all $\mathfrak{p} \in \mathcal{P}(\mathfrak{o})$. Hence $\mathfrak{a} \subset \mathfrak{d}\mathfrak{q}$ for all $\mathfrak{a} \in X$ and consequently $\mathfrak{d} \subset \mathfrak{d}\mathfrak{q}$, a contradiction.

Assume next that $v_\mathfrak{p}(\mathfrak{e}) > \sup\{v_\mathfrak{q}(\mathfrak{a}) \mid \mathfrak{a} \in X\}$ for some $\mathfrak{q} \in \mathcal{P}(\mathfrak{o})$. Then we obtain $v_\mathfrak{q}(\mathfrak{e}\mathfrak{q}^{-1}) = v_\mathfrak{q}(\mathfrak{e}) - 1 \ge \sup\{v_\mathfrak{q}(\mathfrak{a}) \mid \mathfrak{a} \in X\}$ and thus $v_\mathfrak{p}(\mathfrak{e}\mathfrak{q}^{-1}) \ge \sup\{v_\mathfrak{p}(\mathfrak{a}) \mid \mathfrak{a} \in X\}$ for all $\mathfrak{p} \in X$. Hence $\mathfrak{e}\mathfrak{q}^{-1} \subset \mathfrak{a}$ for all $\mathfrak{a} \in X$, and consequently $\mathfrak{e}\mathfrak{q}^{-1} \subset \mathfrak{e}$, a contradiction.

2. By 2.1.7.1, observing that X is infinite if and only if $\{\mathfrak{a}\mathcal{I}'(\mathfrak{o})^\circ \mid \mathfrak{a} \in X\}$ is infinite (since $\mathcal{I}'(\mathfrak{o})$ is reduced), and this holds if and only if

$$\bigcap_{\mathfrak{a}\in X} \mathfrak{a} = \mathbf{0}.$$

3. By 2.8.3.3 it suffices to prove that

$$v_{\mathfrak{p}}\Big(\bigcap_{i=1}^{n}(\mathfrak{a}_i+\mathfrak{b})\Big)=v_{\mathfrak{p}}\Big(\bigcap_{i=1}^{n}\mathfrak{a}_i+\mathfrak{b}\Big)\quad\text{for all}\ \ \mathfrak{p}\in\mathcal{P}(\mathfrak{o}).$$

Let $\mathfrak{p}\in\mathcal{P}(\mathfrak{o})$. After renumbering we may assume that $v_{\mathfrak{p}}(\mathfrak{a}_1)\geq v_{\mathfrak{p}}(\mathfrak{a}_i)$ and thus $v_{\mathfrak{p}}(\mathfrak{a}_1+\mathfrak{b})=\min\{v_{\mathfrak{p}}(\mathfrak{a}_1),v_{\mathfrak{p}}(\mathfrak{b})\}\geq\min\{v_{\mathfrak{p}}(\mathfrak{a}_i),v_{\mathfrak{p}}(\mathfrak{b})\}=v_{\mathfrak{p}}(\mathfrak{a}_i+\mathfrak{b})$ for all $i\in[1,n]$. Then 1. implies

$$\begin{aligned}v_{\mathfrak{p}}\Big(\bigcap_{i=1}^{n}(\mathfrak{a}_i+\mathfrak{b})\Big)&=\max\{v_{\mathfrak{p}}(\mathfrak{a}_1+\mathfrak{b}),\ldots,v_{\mathfrak{p}}(\mathfrak{a}_n+\mathfrak{b})\}=v_{\mathfrak{p}}(\mathfrak{a}_1+\mathfrak{b})\\&=\min\{v_{\mathfrak{p}}(\mathfrak{a}_1),v_{\mathfrak{p}}(\mathfrak{b})\}\\&=\min\big\{\max\{v_{\mathfrak{p}}(\mathfrak{a}_1),\ldots,v_{\mathfrak{p}}(\mathfrak{a}_n)\},v_{\mathfrak{p}}(\mathfrak{b})\big\}\\&=\min\Big\{v_{\mathfrak{p}}\Big(\bigcap_{i=1}^{n}\mathfrak{a}_i\Big),v_{\mathfrak{p}}(\mathfrak{b})\Big\}=v_{\mathfrak{p}}\Big(\bigcap_{i=1}^{n}\mathfrak{a}_i+\mathfrak{b}\Big).\end{aligned}$$

4. The uniqueness of a modulo $\mathfrak{a}_1\cap\ldots\cap\mathfrak{a}_n$ is obvious. To prove existence, we use induction on n. For $n=1$ there is nothing to do.

$n=2$: Suppose that $a_1-a_2=\alpha_1+\alpha_2$, where $\alpha_1\in\mathfrak{a}_1$ and $\alpha_2\in\mathfrak{a}_2$, and set $a=a_1-\alpha_1=a_2+\alpha_2$. Then $a\equiv a_1 \bmod \mathfrak{a}_1$ and $a\equiv a_2 \bmod \mathfrak{a}_2$.

$n\geq 2$, $n-1\to n$: Let $b\in\mathfrak{o}$ be such that $b\equiv a_i \bmod \mathfrak{a}_i$ for all $i\in[1,n-1]$. Since $a_n-b\equiv a_i-b \bmod (\mathfrak{a}_i+\mathfrak{a}_n)$ for all $i\in[1,n-1]$, it follows by 3. that

$$a_n-b\in\bigcap_{i=1}^{n-1}(\mathfrak{a}_i+\mathfrak{a}_n)=\bigcap_{i=1}^{n-1}\mathfrak{a}_i+\mathfrak{a}_n.$$

As we have just proved this implies the existence of some $a\in\mathfrak{o}$ such that

$$a\equiv b \bmod \bigcap_{i=1}^{n-1}\mathfrak{a}_i\quad\text{and}\quad a\equiv a_n \bmod \mathfrak{a}_n,$$

and we obtain $a\equiv a_i \bmod \mathfrak{a}_i$ for all $i\in[1,n]$. □

Definition and Theorem 2.8.5. *Let $\mathfrak{o}$ be a Dedekind domain and $K=\mathsf{q}(\mathfrak{o})$.* For $\mathfrak{p}\in\mathcal{P}(\mathfrak{o})$ we define the **$\mathfrak{p}$-adic valuation** of K

$$v_{\mathfrak{p}}\colon K\to\mathbb{Z}\cup\{\infty\}\ \text{ by }\ v_{\mathfrak{p}}(a)=v_{\mathfrak{p}}(a\mathfrak{o})\ \text{ for all }\ a\in K.$$

1. *If $\mathfrak{p}\in\mathcal{P}(\mathfrak{o})$, then $v_{\mathfrak{p}}$ is a discrete valuation of K* (see 2.2.1).

2. *Let $\mathfrak{a}\in\mathcal{J}(\mathfrak{o})$. Then $\mathfrak{a}=\{a\in K\mid v_{\mathfrak{p}}(a)\geq v_{\mathfrak{p}}(\mathfrak{a})$ for all $\mathfrak{p}\in\mathcal{P}(\mathfrak{o})\}$, and $v_{\mathfrak{p}}(\mathfrak{a})=\min\{v_{\mathfrak{p}}(a)\mid a\in\mathfrak{a}\}$ for all $\mathfrak{p}\in\mathcal{P}(\mathfrak{o})$.*

If $a, b \in \mathfrak{o}$, *then* $a \equiv b \bmod \mathfrak{a}$ *if and only if* $v_\mathfrak{p}(a-b) \geq v_\mathfrak{p}(\mathfrak{a})$ *for all* $\mathfrak{p} \in \mathcal{P}(\mathfrak{o})$.

Proof. 1. If $x, y \in K$, then $xy\mathfrak{o} = (x\mathfrak{o})(y\mathfrak{o})$ and $(x+y)\mathfrak{o} \subset x\mathfrak{o} + y\mathfrak{o}$. Hence $v_\mathfrak{p}(xy) = v_\mathfrak{p}(xy\mathfrak{o}) = v_\mathfrak{p}(x\mathfrak{o}) + v_\mathfrak{p}(y\mathfrak{o}) = v_\mathfrak{p}(x) + v_\mathfrak{p}(y)$, and by 2.8.3.3, and 2.8.4.1 we obtain

$$\begin{aligned} v_\mathfrak{p}(x+y) &= v_\mathfrak{p}((x+y)\mathfrak{o}) \geq v_\mathfrak{p}(x\mathfrak{o} + y\mathfrak{o}) \\ &= \min\{v_\mathfrak{p}(x\mathfrak{o}), v_\mathfrak{p}(y\mathfrak{o})\} = \min\{v_\mathfrak{p}(x), v_\mathfrak{p}(y)\}. \end{aligned}$$

2. Since

$$\mathfrak{a} = \sum_{a \in \mathfrak{a}} a\mathfrak{o},$$

2.8.4.1 implies $v_\mathfrak{p}(\mathfrak{a}) = \inf\{v_\mathfrak{p}(a\mathfrak{o}) \mid a \in \mathfrak{a}\} = \min\{v_\mathfrak{p}(a) \mid a \in \mathfrak{a}\}$ for all $\mathfrak{p} \in \mathcal{P}(\mathfrak{o})$. If $a \in K$, then $a \in \mathfrak{a}$ if and only if $a\mathfrak{o} \subset \mathfrak{a}$, and by 2.8.3.3 this holds if and only if $v_\mathfrak{p}(\mathfrak{a}) \leq v_\mathfrak{p}(a\mathfrak{o}) = v_\mathfrak{p}(a)$ for all $\mathfrak{p} \in \mathcal{P}(\mathfrak{o})$. □

Theorem 2.8.6 (Factorial Dedekind domains). *Let* $\mathfrak{o}$ *be a domain,* $K = \mathsf{q}(\mathfrak{o})$ *and* $K \neq \mathfrak{o}$.

1. *The following assertions are equivalent*:

(a) $\mathfrak{o}$ *is a factorial Dedekind domain.*

(b) $\mathfrak{o}$ *is a principal ideal domain.*

(c) $\mathfrak{o}$ *is a Dedekind domain, and* $\mathcal{C}(\mathfrak{o}) = \mathbf{1}$.

2. *Let* $\mathfrak{o}$ *be a factorial Dedekind domain,* $\pi \in \mathfrak{o}$ *a prime element and* $\mathfrak{p} = \pi\mathfrak{o}$. *Then* $v_\mathfrak{p} = v_\pi \colon K \to \mathbb{Z} \cup \{\infty\}$ *is the* π*-adic valuation of* K (see 2.1.4).

3. *Every semilocal Dedekind domain is a principal ideal domain.*

Proof. 1. (a) $\Rightarrow$ (b) By 2.8.3.4(c) it suffices to prove that every $\mathfrak{p} \in \mathcal{P}(\mathfrak{o})$ is a principal ideal. Thus let $\mathfrak{p} \in \mathcal{P}(\mathfrak{o})$ and $a \in \mathfrak{p}^\bullet$. Then $a = \pi_1 \cdot \ldots \cdot \pi_r$ for some $r \in \mathbb{N}$ and prime elements $\pi_1, \ldots, \pi_r$. Thus there exists some $i \in [1, r]$ such that $\pi_i \in \mathfrak{p}$, hence $\pi_i\mathfrak{o} \subset \mathfrak{p}$ and therefore $\pi_i\mathfrak{o} = \mathfrak{p}$.

(b) $\Rightarrow$ (a), (c) By 1.1.7.3, $\mathfrak{o}$ is factorial and $\mathcal{P}(\mathfrak{o}) = \max(\mathfrak{o})$. By 2.4.1, $\mathfrak{o}$ is integrally closed. Hence $\mathfrak{o}$ is a factorial Dedekind domain, and $\mathcal{C}(\mathfrak{o}) = \mathbf{1}$.

(c) $\Rightarrow$ (b) If $\mathcal{C}(\mathfrak{o}) = \mathbf{1}$, then every ideal of $\mathfrak{o}$ is principal.

2. Let $x \in K^\times$ and $v_\pi(x) = k \in \mathbb{Z}$. Then $x = \pi^k a^{-1} b$ for some $a, b \in \mathfrak{o}$ such that $\pi \nmid ab$. Hence $ab \notin \mathfrak{p}$ and therefore $v_\mathfrak{p}(a^{-1}b) = -v_\mathfrak{p}(a) + v_\mathfrak{p}(b) = 0$. It follows that $x\mathfrak{o} = \mathfrak{p}^k(a^{-1}b\mathfrak{o})$ and $v_\mathfrak{p}(a^{-1}b\mathfrak{o}) = 0$, hence $v_\mathfrak{p}(x) = v_\mathfrak{p}(x\mathfrak{o}) = k$.

3. By 2.6.6 and 1. □

Theorem 2.8.7 (Quotients of Dedekind domains). *Let $\mathfrak{o}$ be a Dedekind domain and T a submonoid of $\mathfrak{o}^\bullet$.*

1. *$T^{-1}\mathfrak{o}$ is a Dedekind domain.*

2. *Let $\mathfrak{p} \in \mathcal{P}(\mathfrak{o})$ and $\mathfrak{a} \in \mathcal{J}(\mathfrak{o})$.*

(a) *If $T \cap \mathfrak{p} = \emptyset$, then $v_{T^{-1}\mathfrak{p}}(T^{-1}\mathfrak{a}) = v_\mathfrak{p}(\mathfrak{a})$, and $v_{T^{-1}\mathfrak{p}}(z) = v_\mathfrak{p}(z)$ for all $z \in K$. In particular,*

$$\operatorname{supp}(T^{-1}\mathfrak{a}) = \{T^{-1}\mathfrak{p} \mid \mathfrak{p} \in \operatorname{supp}(\mathfrak{a}),\ \mathfrak{p} \cap T = \emptyset\}.$$

(b) *If $\pi \in \mathfrak{p} \setminus \mathfrak{p}^2$, then $\mathfrak{a}_\mathfrak{p} = \mathfrak{a}\mathfrak{o}_\mathfrak{p} = \mathfrak{p}^{v_\mathfrak{p}(\mathfrak{a})}\mathfrak{o}_\mathfrak{p} = \pi^{v_\mathfrak{p}(\mathfrak{a})}\mathfrak{o}_\mathfrak{p}$. In particular:*

- *$\mathcal{P}(\mathfrak{o}_\mathfrak{p}) = \{\mathfrak{p}\mathfrak{o}_\mathfrak{p}\} = \{\pi\mathfrak{o}_\mathfrak{p}\}$, and π is up to associates the unique prime element of $\mathfrak{o}_\mathfrak{p}$.*
- *If $K = \mathsf{q}(\mathfrak{o})$ and $a \in K^\times$, then $a\mathfrak{o}_\mathfrak{p} = \pi^{v_\mathfrak{p}(a)}\mathfrak{o}_\mathfrak{p}$.*

Proof. 1. $T^{-1}\mathfrak{o}$ is Noetherian by 2.7.3.3, integrally closed by 2.7.9.2, and by 2.7.6 we obtain

$$\begin{aligned}\mathcal{P}(T^{-1}\mathfrak{o}) &= \{T^{-1}\mathfrak{p} \mid \mathfrak{p} \in \mathcal{P}(\mathfrak{o}),\ \mathfrak{p} \cap T = \emptyset\} = \{T^{-1}\mathfrak{p} \mid \mathfrak{p} \in \max(\mathfrak{o}),\ \mathfrak{p} \cap T = \emptyset\} \\ &= \max(T^{-1}\mathfrak{o}).\end{aligned}$$

Hence $T^{-1}\mathfrak{o}$ is a Dedekind domain.

2. (a) By 2.6.7.1 and 2.7.3.1,

$$\mathfrak{a} = \prod_{\mathfrak{p} \in \mathcal{P}(\mathfrak{o})} \mathfrak{p}^{v_\mathfrak{p}(\mathfrak{a}} \quad \text{implies} \quad T^{-1}\mathfrak{a} = \prod_{\mathfrak{p} \in \mathcal{P}(\mathfrak{o})} (T^{-1}\mathfrak{p})^{v_\mathfrak{p}(\mathfrak{a})} = \prod_{\substack{\mathfrak{p} \in \mathcal{P}(\mathfrak{o}) \\ \mathfrak{p} \cap T = \emptyset}} (T^{-1}\mathfrak{p})^{v_\mathfrak{p}(\mathfrak{a})}$$

and therefore $v_{T^{-1}\mathfrak{p}}(T^{-1}\mathfrak{a}) = v_\mathfrak{p}(\mathfrak{a})$ for all $\mathfrak{p} \in \mathcal{P}(\mathfrak{o})$ satisfying $\mathfrak{p} \cap T = \emptyset$.

If $z \in K$, then

$$v_{T^{-1}\mathfrak{p}}(z) = v_{T^{-1}\mathfrak{p}}(zT^{-1}\mathfrak{o}) = v_{T^{-1}\mathfrak{p}}(T^{-1}(z\mathfrak{o})) = v_\mathfrak{p}(z\mathfrak{o}) = v_\mathfrak{p}(z).$$

(b) By 1., $\mathfrak{o}_\mathfrak{p}$ is a Dedekind domain, and as $\mathcal{P}(\mathfrak{o}_\mathfrak{p}) = \{\mathfrak{p}\mathfrak{o}_\mathfrak{p}\}$, (a) implies that $\mathfrak{a}\mathfrak{o}_\mathfrak{p} = \mathfrak{a}_\mathfrak{p} = (\mathfrak{p}\mathfrak{o}_\mathfrak{p})^{v_\mathfrak{p}(\mathfrak{a})} = \mathfrak{p}^{v_\mathfrak{p}(\mathfrak{a})}\mathfrak{o}_\mathfrak{p}$. Since $\pi \in \mathfrak{p} \setminus \mathfrak{p}^2$, it follows that $v_\mathfrak{p}(\pi) = 1$ and therefore $\pi\mathfrak{o}_\mathfrak{p} = \mathfrak{p}^{v_\mathfrak{p}(\pi\mathfrak{o}_\mathfrak{p})}\mathfrak{o}_\mathfrak{p} = \mathfrak{p}\mathfrak{o}_\mathfrak{p}$. □

2.9 Ray class groups in Dedekind domains

Let $\mathfrak{o}$ be a Dedekind domain, $K = \mathsf{q}(\mathfrak{o})$ and $\mathfrak{m} \in \mathcal{J}'(\mathfrak{o})$. We shall define the congruence modulo $\mathfrak{m}$ not only for elements of $\mathfrak{o}$ but more generally for

$\mathfrak{m}$-integral elements of K (to be defined below). For this we must first generalize the notion of coprimeness to fractional ideals. This more general congruence together with the congruence regarding signatures to be introduced in 2.9.6 will lead us to the definition of ray classes, a notion which is fundamental for the higher arithmetic of global fields.

Definition 2.9.1. Let $\mathfrak{o}$ be a Dedekind domain, $K = \mathsf{q}(\mathfrak{o})$ and $\mathfrak{m} \in \mathcal{J}'(\mathfrak{o})$.

A fractional ideal $\mathfrak{a} \in \mathcal{J}(\mathfrak{o})$ is called **coprime** to $\mathfrak{m}$ if $v_{\mathfrak{p}}(\mathfrak{a}) = 0$ for all $\mathfrak{p} \in \mathrm{supp}(\mathfrak{m})$. We denote by

- $\mathcal{J}^{\mathfrak{m}}(\mathfrak{o})$ the set of all fractional ideals of $\mathfrak{o}$ coprime to $\mathfrak{m}$;
- $\mathcal{J}'^{\mathfrak{m}}(\mathfrak{o}) = \mathcal{J}^{\mathfrak{m}}(\mathfrak{o}) \cap \mathcal{J}'(\mathfrak{o})$ the set of all (integral) ideals of $\mathfrak{o}$ being to $\mathfrak{m}$;
- $\mathcal{P}^{\mathfrak{m}}(\mathfrak{o}) = \mathcal{P}(\mathfrak{o}) \cap \mathcal{J}^{\mathfrak{m}}(\mathfrak{o}) = \{\mathfrak{p} \in \mathcal{P}(\mathfrak{o}) \mid \mathfrak{p} \nmid \mathfrak{m}\}$ the set of all prime ideals coprime to $\mathfrak{m}$.

By 2.8.4.2 and 2.8.3.2, $\mathcal{J}'^{\mathfrak{m}}(\mathfrak{o}) = \{\mathfrak{a} \in \mathcal{J}'(\mathfrak{o}) \mid \mathfrak{a} + \mathfrak{m} = \mathfrak{o}\}$ is a free monoid with basis $\mathcal{P}^{\mathfrak{m}}(\mathfrak{o})$ and $\mathcal{J}^{\mathfrak{m}}(\mathfrak{o}) = \mathsf{q}(\mathcal{J}'^{\mathfrak{m}}(\mathfrak{o}))$ is a free abelian group with basis $\mathcal{P}^{\mathfrak{m}}(\mathfrak{o})$. Note that for (integral) ideals our notion of coprimeness coincides with that from Section 1.1.

Theorem 2.9.2. *Let $\mathfrak{o}$ be a Dedekind domain, $K = \mathsf{q}(\mathfrak{o})$ and $\mathfrak{m} \in \mathcal{J}'(\mathfrak{o})$. Then every ideal class $C \in \mathcal{C}(\mathfrak{o})$ contains an (integral) ideal which is coprime to $\mathfrak{m}$.*

Proof. Let $C \in \mathcal{C}(\mathfrak{o})$, $\mathfrak{a} \in C$ and $a \in \mathfrak{o}^\bullet$ such that $\mathfrak{a}_0 = a\mathfrak{a} \subset \mathfrak{o}$. For every $\mathfrak{p} \in \mathrm{supp}(\mathfrak{m})$ let $\pi_{\mathfrak{p}} \in \mathfrak{o}$ be such that $v_{\mathfrak{p}}(\pi_{\mathfrak{p}}) = 1$. By the Chinese remainder theorem 1.1.4 there exists some $b \in \mathfrak{o}$ such that $b \equiv \pi_{\mathfrak{p}}^{v_{\mathfrak{p}}(\mathfrak{a}_0)} \mod \mathfrak{p}^{v_{\mathfrak{p}}(\mathfrak{a}_0)+1}$ for all $\mathfrak{p} \in \mathrm{supp}(\mathfrak{m})$, hence $v_{\mathfrak{p}}(b - \pi_{\mathfrak{p}}^{v_{\mathfrak{p}}(\mathfrak{a}_0)}) > v_{\mathfrak{p}}(\mathfrak{a}_0) = v_{\mathfrak{p}}(\pi_{\mathfrak{p}}^{v_{\mathfrak{p}}(\mathfrak{a}_0)})$,

$$v_{\mathfrak{p}}(b) = v_{\mathfrak{p}}(\pi_{\mathfrak{p}}^{v_{\mathfrak{p}}(\mathfrak{a}_0)} + b - \pi_{\mathfrak{p}}^{v_{\mathfrak{p}}(\mathfrak{a}_0)}) = v_{\mathfrak{p}}(\pi_{\mathfrak{p}}^{v_{\mathfrak{p}}(\mathfrak{a}_0)}) = v_{\mathfrak{p}}(\mathfrak{a}_0)$$

and $v_{\mathfrak{p}}(b^{-1}\mathfrak{a}_0) = 0$ for all $\mathfrak{p} \in \mathrm{supp}(\mathfrak{m})$. Hence $b^{-1}\mathfrak{a}_0 \in \mathcal{J}^{\mathfrak{m}}(\mathfrak{o}) = \mathsf{q}(\mathcal{J}'^{\mathfrak{m}}(\mathfrak{o}))$, and $b^{-1}\mathfrak{a}_0 = \mathfrak{c}^{-1}\mathfrak{d}$ for some $\mathfrak{c},\ \mathfrak{d} \in \mathcal{J}'^{\mathfrak{m}}(\mathfrak{o})$. Now let $c \in \mathfrak{o}^\bullet$ be such that $c \equiv 1 \mod \mathfrak{m}$ and $c \equiv 0 \mod \mathfrak{c}$ (again by 1.1.4). Then $c\mathfrak{c}^{-1} \in \mathcal{J}'^{\mathfrak{m}}(\mathfrak{o})$, and $cb^{-1}a\mathfrak{a} = c\mathfrak{c}^{-1}\mathfrak{d} \in C \cap \mathcal{J}'^{\mathfrak{m}}(\mathfrak{o})$ is an ideal in C which is coprime to $\mathfrak{m}$. □

Corollary 2.9.3. *Let $\mathfrak{o}$ be a Dedekind domain and $\mathfrak{a} \in \mathcal{J}(\mathfrak{o})$.*

1. *For every $a \in \mathfrak{a}^\bullet$ there exists some $c \in \mathfrak{a}$ such that $\mathfrak{a} = {}_{\mathfrak{o}}(a, c)$.*

2. *Let $\mathfrak{b} \in \mathcal{J}(\mathfrak{o})$ and $\mathfrak{b} \subset \mathfrak{a}$. Then there exists an $\mathfrak{o}$-isomorphism $\mathfrak{o}/\mathfrak{a}^{-1}\mathfrak{b} \xrightarrow{\sim} \mathfrak{a}/\mathfrak{b}$.*

Proof. 1. Let $a \in \mathfrak{a}^\bullet$ and $\mathfrak{b} \in \mathcal{J}'(\mathfrak{o})$ such that $a\mathfrak{o} = \mathfrak{a}\mathfrak{b}$. By 2.9.2 there exists an ideal $\mathfrak{c} \in [\mathfrak{a}]^{-1}$ such that $\mathfrak{b} + \mathfrak{c} = \mathfrak{o}$. Then $\mathfrak{a}\mathfrak{c} = c\mathfrak{o}$ for some $c \in \mathfrak{a}$, and therefore ${}_{\mathfrak{o}}(a, c) = a\mathfrak{o} + c\mathfrak{o} = \mathfrak{a}\mathfrak{b} + \mathfrak{a}\mathfrak{c} = \mathfrak{a}(\mathfrak{b} + \mathfrak{c}) = \mathfrak{a}$.

2. (due to F. Kainrath) Let $b \in \mathfrak{b}^\bullet$. By 1. there exists some $a \in \mathfrak{a}$ such that $\mathfrak{a} = {}_{\mathfrak{o}}(a, b)$. Then $\mathfrak{a} = a\mathfrak{o} + \mathfrak{b}$ and $\mathfrak{a}/\mathfrak{b} = \mathfrak{o}\alpha$, where $\alpha = a + \mathfrak{b} \in \mathfrak{a}/\mathfrak{b}$. Let $\psi \colon \mathfrak{o} \to \mathfrak{a}/\mathfrak{b}$ be defined $\psi(\lambda) = \lambda\alpha$ for all $\lambda \in \mathfrak{o}$. Then ψ is an $\mathfrak{o}$-epimorphism, and $\operatorname{Ker}(\psi) = \{\lambda \in \mathfrak{o} \mid \lambda\alpha = 0\} = \operatorname{Ann}_{\mathfrak{o}}(\mathfrak{a}/\mathfrak{b}) = \mathfrak{a}^{-1}\mathfrak{b}$.

Therefore ψ induces an $\mathfrak{o}$-isomorphism $\mathfrak{o}/\mathfrak{a}^{-1}\mathfrak{b} \xrightarrow{\sim} \mathfrak{a}/\mathfrak{b}$ as asserted. □

For an ideal $\mathfrak{m} \in \mathcal{J}'(\mathfrak{o})$ we have discussed fractional ideals coprime to $\mathfrak{m}$, and now we go to investigate elements $x \in K$ coprime to $\mathfrak{m}$ in order to define ray class groups modulo $\mathfrak{m}$.

Definitions and Remarks 2.9.4. Let $\mathfrak{o}$ be a Dedekind domain, $K = \mathsf{q}(\mathfrak{o})$ and $\mathfrak{m} \in \mathcal{J}'(\mathfrak{o})$.

An element $z \in K^\times$ is called **coprime** to $\mathfrak{m}$ if the fractional principal ideal $z\mathfrak{o}_K$ is coprime to $\mathfrak{m}$ [equivalently, $v_\mathfrak{p}(z) = 0$ for all $\mathfrak{p} \in \operatorname{supp}(\mathfrak{m})$]. We denote by $K(\mathfrak{m})$ the set of all $z \in K^\times$ which are coprime to $\mathfrak{m}$, and we set

$$\mathfrak{o}(\mathfrak{m}) = \mathfrak{o} \cap K(\mathfrak{m}) = \{b \in \mathfrak{o}^\bullet \mid b\mathfrak{o} + \mathfrak{m} = \mathfrak{o}\} = \{b \in \mathfrak{o}^\bullet \mid b + \mathfrak{m} \in (\mathfrak{o}/\mathfrak{m})^\times\}.$$

$\mathfrak{o}(\mathfrak{m})$ is a saturated submonoid of $\mathfrak{o}^\bullet$, $\mathfrak{o}(\mathfrak{m})^{-1}\mathfrak{o}$ is a subring of K, $K(\mathfrak{m})$ is a subgroup of $K^\times$, $K(\mathfrak{m}) \subset \mathfrak{o}(\mathfrak{m})^{-1}\mathfrak{o}$, $\mathfrak{o}(\mathbf{1}) = \mathfrak{o}^\bullet$ and $K(\mathbf{1}) = K^\times$. As $\mathfrak{o}(\mathfrak{m})$ is saturated, 2.7.1.4 implies $(\mathfrak{o}(\mathfrak{m})^{-1}\mathfrak{o})^\times = \mathsf{q}(\mathfrak{o}(\mathfrak{m})) \subset K(\mathfrak{m})$. The seemingly obvious equality $\mathsf{q}(\mathfrak{o}(\mathfrak{m})) = K(\mathfrak{m})$ will be proved in 2.9.5 below.

The elements of the domain $\mathfrak{o}(\mathfrak{m})^{-1}\mathfrak{o}$ are called **$\mathfrak{m}$-integral**. If $\mathfrak{p} \in \mathcal{P}(\mathfrak{o})$, then $\mathfrak{o}(\mathfrak{p}) = \mathfrak{o} \setminus \mathfrak{p}$ and $\mathfrak{o}_\mathfrak{p}$ is the domain of $\mathfrak{p}$-integral elements of K.

By 2.7.3, $\mathfrak{o}(\mathfrak{m})^{-1}\mathfrak{m}$ is an ideal of $\mathfrak{o}(\mathfrak{m})^{-1}\mathfrak{o}$, $\mathfrak{o}(\mathfrak{m})^{-1}\mathfrak{m} \cap \mathfrak{o} = \mathfrak{m}$, and the embedding $\mathfrak{o} \hookrightarrow \mathfrak{o}(\mathfrak{m})^{-1}\mathfrak{o}$ induces an isomorphism $\mathfrak{o}/\mathfrak{m} \xrightarrow{\sim} \mathfrak{o}(\mathfrak{m})^{-1}\mathfrak{o}/\mathfrak{o}(\mathfrak{m})^{-1}\mathfrak{m}$. We identify: $\mathfrak{o}/\mathfrak{m} = \mathfrak{o}(\mathfrak{m})^{-1}\mathfrak{o}/\mathfrak{o}(\mathfrak{m})^{-1}\mathfrak{m}$. With this identification, the residue class epimorphism $\pi_\mathfrak{m} \colon \mathfrak{o} \to \mathfrak{o}/\mathfrak{m}$, $a \mapsto a + \mathfrak{m}$, extends through quotients to an (equally denoted) ring epimorphism $\pi_\mathfrak{m} \colon \mathfrak{o}(\mathfrak{m})^{-1}\mathfrak{o} \to \mathfrak{o}/\mathfrak{m}$ as follows: If $x = c^{-1}a \in \mathfrak{o}(\mathfrak{m})^{-1}\mathfrak{o}$, where $c \in \mathfrak{o}(\mathfrak{m})$ and $a \in \mathfrak{o}$, then $c + \mathfrak{m} \in (\mathfrak{o}/\mathfrak{m})^\times$, and $\pi_\mathfrak{m}(x) = (c + \mathfrak{m})^{-1}(a + \mathfrak{m}) \in \mathfrak{o}/\mathfrak{m}$.

We define the congruence modulo $\mathfrak{m}$ between $\mathfrak{m}$-integral elements $x, y \in K$ by

$$x \equiv y \mod \mathfrak{m} \quad \text{if and only if} \quad x - y \in \mathfrak{o}(\mathfrak{m})^{-1}\mathfrak{m}.$$

Hence a congruence between $\mathfrak{m}$-integal elements of K is nothing more than a congruence modulo $\mathfrak{o}(\mathfrak{m})^{-1}\mathfrak{m}$ in $\mathfrak{o}(\mathfrak{m})^{-1}\mathfrak{o}$, and since $\mathfrak{o}(\mathfrak{m})^{-1}\mathfrak{m} \cap \mathfrak{o} = \mathfrak{m}$, the above definition coincides with the usual one if $x, y \in \mathfrak{o}$.

Theorem 2.9.5. *Let $\mathfrak{o}$ be a Dedekind domain, $K = \mathsf{q}(\mathfrak{o})$ and $\mathfrak{m} \in \mathcal{J}'(\mathfrak{o})$.*

1. *If $x, y \in \mathfrak{o}(\mathfrak{m})^{-1}\mathfrak{o}$, then it follows that $x \equiv y \mod \mathfrak{m}$ if and only if $v_\mathfrak{p}(x - y) \ge v_\mathfrak{p}(\mathfrak{m})$ for all $\mathfrak{p} \in \operatorname{supp}(\mathfrak{m})$.*

2. $(\mathfrak{o}(\mathfrak{m})^{-1}\mathfrak{o})^\times = \mathsf{q}(\mathfrak{o}(\mathfrak{m})) = K(\mathfrak{m})$, $\pi_\mathfrak{m} \restriction K(\mathfrak{m})\colon K(\mathfrak{m}) \to (\mathfrak{o}/\mathfrak{m})^\times$ *is a group epimorphism, and* $K^{\circ\mathfrak{m}} = \{x \in K(\mathfrak{m}) \mid x \equiv 1 \bmod \mathfrak{m}\}$ *is its kernel.*

The terminology $K^{\circ\mathfrak{m}}$ is motivated by corresponding notions in class field theory and will be elucidated in 3.6.6. We identify: $K(\mathfrak{m})/K^{\circ\mathfrak{m}} = (\mathfrak{o}/\mathfrak{m})^\times$.

3. $K^{\circ\mathfrak{m}} = \{x \in K^\times \mid v_\mathfrak{p}(x-1) \geq v_\mathfrak{p}(\mathfrak{m})$ *for all* $\mathfrak{p} \in \mathrm{supp}(\mathfrak{m})\}$, *and* $K^{\circ\mathfrak{m}} = \mathsf{q}(1+\mathfrak{m})$.

Proof. 1. Let $x, y \in \mathfrak{o}(\mathfrak{m})^{-1}\mathfrak{o}$, say $x = s^{-1}a$ and $y = s^{-1}b$, where $a, b \in \mathfrak{o}$ and $s \in \mathfrak{o}(\mathfrak{m})$. Then $v_\mathfrak{p}(x-y) = -v_\mathfrak{p}(s) + v_\mathfrak{p}(a-b) = v_\mathfrak{p}(a-b)$ for all $\mathfrak{p} \in \mathrm{supp}(\mathfrak{m})$, and therefore

$$\begin{aligned} x \equiv y \mod \mathfrak{m} &\iff s^{-1}(a-b) \in \mathfrak{o}(\mathfrak{m})^{-1}\mathfrak{m} \iff a-b \in \mathfrak{o}(\mathfrak{m})^{-1}\mathfrak{m} \cap \mathfrak{o} = \mathfrak{m} \\ &\iff v_\mathfrak{p}(x-y) = v_\mathfrak{p}(a-b) \geq v_\mathfrak{p}(\mathfrak{m}) \quad \text{for all } \mathfrak{p} \in \mathrm{supp}(\mathfrak{m}). \end{aligned}$$

Clearly, the ring epimorphism $\pi_\mathfrak{m}\colon \mathfrak{o}(\mathfrak{m})^{-1}\mathfrak{o} \to \mathfrak{o}/\mathfrak{m}$ induces a group homomorphism $\pi_\mathfrak{m} \restriction K(\mathfrak{m})\colon K(\mathfrak{m}) \to (\mathfrak{o}/\mathfrak{m})^\times$ with kernel $K^{\circ\mathfrak{m}}$, and if $a \in \mathfrak{o}$ is such that $a+\mathfrak{m} \in (\mathfrak{o}/\mathfrak{m})^\times$, then $a \in \mathfrak{o}(\mathfrak{m}) \subset K(\mathfrak{m})$. Hence $\pi_\mathfrak{m} \restriction K(\mathfrak{m})\colon K(\mathfrak{m}) \to (\mathfrak{o}/\mathfrak{m})^\times$ is an epimorphism.

2. Obviously, $(\mathfrak{o}(\mathfrak{m})^{-1}\mathfrak{o})^\times = \mathsf{q}(\mathfrak{o}(\mathfrak{m})) \subset K(\mathfrak{m})$. Thus let $x \in K(\mathfrak{m})$, say $x\mathfrak{o} = \mathfrak{a}^{-1}\mathfrak{b}$, where $\mathfrak{a}, \mathfrak{b} \in \mathcal{J}'^{\mathfrak{m}}(\mathfrak{o})$ and let $C = [\mathfrak{a}] = [\mathfrak{b}] \in \mathcal{C}(\mathfrak{o})$. By 2.9.2 there exists an ideal $\mathfrak{c} \in C^{-1}$ such that $\mathfrak{c} + \mathfrak{m} = \mathfrak{o}$. Then $\mathfrak{c}\mathfrak{a} = a\mathfrak{o}$ and $\mathfrak{c}\mathfrak{b} = b\mathfrak{o}$ for some $a, b \in \mathfrak{o}(\mathfrak{m})$ and therefore $x\mathfrak{o} = a^{-1}b\mathfrak{o}$. Hence there exists some $u \in \mathfrak{o}^\times$ such that $x = a^{-1}(bu)$, and as $bu \in \mathfrak{o}(\mathfrak{m})$, we eventually obtain $x \in \mathsf{q}(\mathfrak{o}(\mathfrak{m}))$.

3. Since $K^{\circ\mathfrak{m}} \subset \{x \in K^\times \mid v_\mathfrak{p}(x-1) \geq v_\mathfrak{p}(\mathfrak{m})$ for all $\mathfrak{p} \in \mathrm{supp}(\mathfrak{m})\}$ by 1. and obviously $\mathsf{q}(1+\mathfrak{m}) \subset K^{\circ\mathfrak{m}}$, it suffices to prove that

$$\{x \in K^\times \mid v_\mathfrak{p}(x-1) \geq v_\mathfrak{p}(\mathfrak{m}) \ \text{ for all } \ \mathfrak{p} \in \mathrm{supp}(\mathfrak{m})\} \subset \mathsf{q}(1+\mathfrak{m}).$$

Thus let $x \in K^\times$ and $v_\mathfrak{p}(x-1) \geq v_\mathfrak{p}(\mathfrak{m})$ for all $\mathfrak{p} \in \mathrm{supp}(\mathfrak{m})$. If $\mathfrak{p} \in \mathrm{supp}(\mathfrak{m})$, then $v_\mathfrak{p}(x-1) \geq v_\mathfrak{p}(\mathfrak{m}) > 0 = v_\mathfrak{p}(1)$ and $v_\mathfrak{p}(x) = v_\mathfrak{p}((x-1)+1) = v_\mathfrak{p}(1) = 0$. Hence $x \in K(\mathfrak{m}) = \mathsf{q}(\mathfrak{o}(\mathfrak{m}))$ by 2. and thus $x \equiv 1 \bmod \mathfrak{m}$ by 1. Let $x = c^{-1}a$ for some $a, c \in \mathfrak{o}(\mathfrak{m})$, and let $c' \in \mathfrak{o}$ be such that $cc' \equiv 1 \bmod \mathfrak{m}$. Then we obtain $a = cx \equiv c \bmod \mathfrak{m}$, $ac' \equiv cc' \equiv 1 \bmod \mathfrak{m}$, and thus consequently $x = (cc')^{-1}(ac') \in \mathsf{q}(1+\mathfrak{m})$. □

In the theory of algebraic number fields we need a refinement of the concept of class groups by signatures. We introduce this concept for an arbitrary domain.

Definition 2.9.6. Let $\mathfrak{o}$ be a domain and $K = \mathsf{q}(\mathfrak{o})$.

A **real embedding** of K is a field homomorphism $\sigma\colon K \to \mathbb{R}$. Let now $\boldsymbol{\sigma}$ be a finite (possibly empty) set of real embeddings of K. An element $a \in K^\times$ is called **$\boldsymbol{\sigma}$-positive** if $\sigma(a) > 0$ for all $\sigma \in \boldsymbol{\sigma}$. Let $K^{\boldsymbol{\sigma}}$ be the subgroup of all $\boldsymbol{\sigma}$-positive elements of $K^\times$, $\mathcal{H}^{\boldsymbol{\sigma}}(\mathfrak{o}) = \{a\mathfrak{o} \mid a \in K^{\boldsymbol{\sigma}}\}$ and

$$\mathcal{C}^{\boldsymbol{\sigma}}(\mathfrak{o}) = \mathcal{I}(\mathfrak{o})/\mathcal{H}^{\boldsymbol{\sigma}}(\mathfrak{o})$$

the **$\boldsymbol{\sigma}$-ideal class group** of $\mathfrak{o}$. For $\mathfrak{a} \in \mathcal{I}(\mathfrak{o})$ we denote by $[\mathfrak{a}]^{\boldsymbol{\sigma}} = \mathfrak{a}\mathcal{H}^{\boldsymbol{\sigma}}(\mathfrak{o}) \in \mathcal{C}^{\boldsymbol{\sigma}}(\mathfrak{o})$ the **$\boldsymbol{\sigma}$-ideal class** of $\mathfrak{a}$. Note that for $\boldsymbol{\sigma} = \emptyset$ we have $K^{\emptyset} = K^\times$, $\mathcal{H}^{\emptyset}(\mathfrak{o}) = \mathcal{H}(\mathfrak{o})$ and $\mathcal{C}^{\emptyset}(\mathfrak{o}) = \mathcal{C}(\mathfrak{o})$.

The definition of the $\boldsymbol{\sigma}$-ideal class group gives rise to the exact sequences

$$\mathbf{1} \to \mathfrak{o}^\times \cap K^{\boldsymbol{\sigma}} \hookrightarrow K^{\boldsymbol{\sigma}} \xrightarrow{\partial} \mathcal{I}(\mathfrak{o}) \xrightarrow{\pi^{\boldsymbol{\sigma}}} \mathcal{C}^{\boldsymbol{\sigma}}(\mathfrak{o}) \to \mathbf{1}$$

and

$$\mathbf{1} \to K^{\boldsymbol{\sigma}}\mathfrak{o}^\times \hookrightarrow K^\times \xrightarrow{\partial^{\boldsymbol{\sigma}}} \mathcal{C}^{\boldsymbol{\sigma}}(\mathfrak{o}) \xrightarrow{\rho} \mathcal{C}(\mathfrak{o}) \to \mathbf{1},$$

where $\pi^{\boldsymbol{\sigma}}(\mathfrak{a}) = [\mathfrak{a}]^{\boldsymbol{\sigma}}$, $\partial^{\boldsymbol{\sigma}}(a) = [a\mathfrak{o}]^{\boldsymbol{\sigma}}$ and $\rho([\mathfrak{a}]^{\boldsymbol{\sigma}}) = [\mathfrak{a}]$ for all $a \in K^\times$ and $\mathfrak{a} \in \mathcal{I}(\mathfrak{o})$.

Theorem 2.9.7 (Congruences with signatures). *Let $\mathfrak{o}$ be a domain, $K = \mathsf{q}(\mathfrak{o})$, $r \in \mathbb{N}$ and $\boldsymbol{\sigma} = \{\sigma_1, \ldots, \sigma_r\}$ a set of r distinct real embeddings of K.*

1. *The map $s\colon K^\times \to \{\pm 1\}^r$, defined by*

$$s(x) = (\operatorname{sgn}\sigma_1(x), \ldots, \operatorname{sgn}\sigma_r(x)),$$

is an epimorphism, $\operatorname{Ker}(s) = K^{\boldsymbol{\sigma}}$, *and $s(\mathfrak{o}^\bullet) = \{\pm 1\}^r$. In particular, $(K\colon K^{\boldsymbol{\sigma}}) = 2^r$.*

2. *Let $\mathfrak{a}$ be a non-zero ideal of $\mathfrak{o}$, $(\varepsilon_1, \ldots, \varepsilon_r) \in \{\pm 1\}^r$ and $b \in \mathfrak{o}$. Then there exists some $a \in \mathfrak{o}^\bullet$ such that $a \equiv b \bmod \mathfrak{a}$ and $\operatorname{sgn}\sigma_i(a) = \varepsilon_i$ for all $i \in [1, r]$.*

Proof. 1. Obviously, s is a homomorphism, $\operatorname{Ker}(s) = K^{\boldsymbol{\sigma}}$, and we consider the map $\sigma = (\sigma_1, \ldots, \sigma_r)\colon K \to \mathbb{R}^r$. It suffices to prove that $\mathbb{R}\sigma(\mathfrak{o}) = \mathbb{R}^r$, for then it follows that $s(\mathfrak{o}^\bullet) = \{\pm 1\}^r$ and that s is surjective. Since the homomorphisms $\sigma_1 \restriction \mathfrak{o}, \ldots, \sigma_r \restriction \mathfrak{o}$ are distinct, they are linearly independent over $\mathbb{R}$ by 1.5.2, and we suppose that $V = \mathbb{R}\sigma(\mathfrak{o}) \subsetneq \mathbb{R}^r$. Let $l = \alpha_1 X_1 + \ldots + \alpha_r X_r \in \mathbb{R}[X_1, \ldots, X_r]^\bullet$ be a linear form such that $l \restriction V = 0$. Then $(\alpha_1\sigma_1 + \ldots + \alpha_r\sigma_r) \restriction \mathfrak{o} = 0$, which contradicts the linear independence of $\sigma_1 \restriction \mathfrak{o}, \ldots, \sigma_r \restriction \mathfrak{o}$.

2. By 1. there exists some $a_0 \in \mathfrak{o}^\bullet$ such that $\operatorname{sgn}\sigma_i(a_0) = \varepsilon_i$ for all $i \in [1, r]$. Let $c \in \mathfrak{a}^\bullet$, and for $N \in \mathbb{N}$ set $a = b + Na_0c^2$. Then $a \equiv b \bmod \mathfrak{a}$, and for $N \gg 1$ we obtain $a \neq 0$ and $\operatorname{sgn}\sigma_i(a) = \operatorname{sgn}\sigma_i(a_0) = \varepsilon_i$, hence $\varepsilon_i\sigma_i(a) > 0$. $\square$

Definition 2.9.8. Let $\mathfrak{o}$ be a Dedekind domain, $K = \mathsf{q}(\mathfrak{o})$ and $\mathfrak{m} \in \mathcal{J}'(\mathfrak{o})$. Let $\boldsymbol{\sigma}$ be a finite (possibly empty) set of real embeddings of K (see 2.6.5), $K^{\boldsymbol{\sigma}}$ the group of $\boldsymbol{\sigma}$-positive elements of $K^\times$ and

$$K^{\mathfrak{m},\boldsymbol{\sigma}} = \{x \in K(\mathfrak{m}) \cap K^{\boldsymbol{\sigma}} \mid x \equiv 1 \mod \mathfrak{m}\} \subset K(\mathfrak{m}).$$

In particular, for $\boldsymbol{\sigma} = \emptyset$ we obtain $K^{\mathfrak{m},\emptyset} = K^{\circ\mathfrak{m}}$. By definition, $K^{\mathfrak{m},\boldsymbol{\sigma}}$ is a subgroup of $K(\mathfrak{m})$, and therefore $\mathcal{H}^{\mathfrak{m},\boldsymbol{\sigma}}(\mathfrak{o}) = \{a\mathfrak{o} \mid a \in K^{\mathfrak{m},\boldsymbol{\sigma}}\}$ is a subgroup of $\mathcal{J}^{\mathfrak{m}}(\mathfrak{o})$. The factor group

$$\mathcal{C}^{\mathfrak{m},\boldsymbol{\sigma}}(\mathfrak{o}) = \mathcal{J}^{\mathfrak{m}}(\mathfrak{o})/\mathcal{H}^{\mathfrak{m},\boldsymbol{\sigma}}(\mathfrak{o})$$

is called that **$\boldsymbol{\sigma}$-ray class group** of $\mathfrak{o}$ modulo $\mathfrak{m}$. For a fractional ideal $\mathfrak{a} \in \mathcal{J}^{\mathfrak{m}}(\mathfrak{o})$ we denote by $[\mathfrak{a}]^{\mathfrak{m},\boldsymbol{\sigma}} = \mathfrak{a}\mathcal{H}^{\mathfrak{m},\boldsymbol{\sigma}}(\mathfrak{o})$ the **$\boldsymbol{\sigma}$-ray class** modulo $\mathfrak{m}$ of $\mathfrak{a}$. In particular, the **$\boldsymbol{\sigma}$-main ray** $[\mathbf{1}]^{\mathfrak{m},\boldsymbol{\sigma}} = \mathcal{H}^{\mathfrak{m},\boldsymbol{\sigma}}(\mathfrak{o})$ is the unit element of $\mathcal{C}^{\mathfrak{m},\boldsymbol{\sigma}}(\mathfrak{o})$. If $\mathfrak{m} = \mathbf{1}$, then $\mathcal{C}^{\mathbf{1},\boldsymbol{\sigma}}(\mathfrak{o}) = \mathcal{C}^{\boldsymbol{\sigma}}(\mathfrak{o})$ is the $\boldsymbol{\sigma}$-class group of $\mathfrak{o}$ already introduced in 2.9.6.

Theorem 2.9.9. *Let $\mathfrak{o}$ be a Dedekind domain, $K = \mathsf{q}(\mathfrak{o})$, $\mathfrak{m} \in \mathcal{J}'(\mathfrak{o})$ and $\boldsymbol{\sigma}$ a finite set of real embeddings of K.*

1. Let $\mathfrak{c} \in \mathcal{J}'(\mathfrak{o})$. Then every $\boldsymbol{\sigma}$-ray class $C \in \mathcal{C}^{\mathfrak{m},\boldsymbol{\sigma}}(\mathfrak{o})$ contains an (integral) ideal coprime to $\mathfrak{c}$.

2. Let $\pi_{\mathfrak{m}} \colon \mathfrak{o}(\mathfrak{m})^{-1}\mathfrak{o} \to \mathfrak{o}/\mathfrak{m}$ be the residue class epimorphism defined in 2.9.4, and define

$$\pi_{\mathfrak{m}}^{\boldsymbol{\sigma}} \colon K(\mathfrak{m}) \to (\mathfrak{o}/\mathfrak{m})^\times \times \{\pm 1\}^{\boldsymbol{\sigma}} \quad by \quad \pi_{\mathfrak{m}}^{\boldsymbol{\sigma}}(a) = (\pi_{\mathfrak{m}}(a), (\operatorname{sgn} \sigma(a))_{\sigma \in \boldsymbol{\sigma}}).$$

Then $\pi_{\mathfrak{m}}^{\boldsymbol{\sigma}}$ is an epimorphism, and $\operatorname{Ker}(\pi_{\mathfrak{m}}^{\boldsymbol{\sigma}}) = K^{\mathfrak{m},\boldsymbol{\sigma}} \subset K(\mathfrak{m})$. In particular, $\pi_{\mathfrak{m}}^{\emptyset} = \pi_{\mathfrak{m}} \restriction K(\mathfrak{m})$.

3. There is an exact sequence

$$\mathbf{1} \to \mathfrak{o}^\times \cap K^{\boldsymbol{\sigma}}/\mathfrak{o}^\times \cap K^{\mathfrak{m},\boldsymbol{\sigma}} \xrightarrow{\eta} (\mathfrak{o}/\mathfrak{m})^\times \xrightarrow{\theta} \mathcal{C}^{\mathfrak{m},\boldsymbol{\sigma}}(\mathfrak{o}) \xrightarrow{\rho} \mathcal{C}^{\boldsymbol{\sigma}}(\mathfrak{o}) \to \mathbf{1},$$

where

- *$\eta(\varepsilon(\mathfrak{o}^\times \cap K^{\mathfrak{m},\boldsymbol{\sigma}})) = \varepsilon + \mathfrak{m}$ for all $\varepsilon \in \mathfrak{o}^\times \cap K^{\boldsymbol{\sigma}}$;*
- *$\theta(a + \mathfrak{m}) = [a_0\mathfrak{o}]^{\mathfrak{m},\boldsymbol{\sigma}}$ for all $a \in \mathfrak{o}(\mathfrak{m})$, where $a_0 \in \mathfrak{o} \cap K^{\boldsymbol{\sigma}}$ such that $a_0 \equiv a \mod \mathfrak{m}$;*
- *$\rho([\mathfrak{a}]^{\mathfrak{m},\boldsymbol{\sigma}}) = [\mathfrak{a}]^{\boldsymbol{\sigma}}$ for all $\mathfrak{a} \in \mathcal{J}^{\mathfrak{m}}(\mathfrak{o})$.*

Proof. 1. Let $C \in \mathcal{C}^{\mathfrak{m},\boldsymbol{\sigma}}(\mathfrak{o})$ and $\mathfrak{b} \in C \subset \mathcal{J}^{\mathfrak{m}}(\mathfrak{o})$. By 2.9.2 there exists an ideal $\mathfrak{b}_0 \in [\mathfrak{b}]$ such that $\mathfrak{b}_0$ is coprime to $\mathfrak{m}\mathfrak{c}$, and then $\mathfrak{b}_0 = \mathfrak{b}\beta$ for some $\beta \in K(\mathfrak{m})$. By the Chinese remainder theorem 1.1.4 and 2.9.7 there exists some $\beta_1 \in \mathfrak{o}$ such that $\beta\beta_1 \equiv 1 \bmod \mathfrak{m}$, $\beta_1 \equiv 1 \bmod \mathfrak{p}^{v_\mathfrak{p}(\mathfrak{c})}$ for all $\mathfrak{p} \in \mathcal{P}^{\mathfrak{m}}(\mathfrak{o}) \cap \operatorname{supp}(\mathfrak{c})$ and $\sigma(\beta\beta_1) > 0$ for all $\sigma \in \boldsymbol{\sigma}$. Then it follows that β_1 is coprime to $\mathfrak{c}$, and $\mathfrak{b}\beta\beta_1 = \mathfrak{b}_0\beta_1 \in C \cap \mathcal{J}'^{\mathfrak{c}}(\mathfrak{o})$.

2. Apparently, $\pi_\mathfrak{m}^{\boldsymbol{\sigma}}$ is a homomorphism with kernel $K^{\mathfrak{m},\boldsymbol{\sigma}}$, and we must prove that it is surjective. If $a \in \mathfrak{o}(\mathfrak{m})$ and $(\varepsilon_\sigma)_{\sigma\in\boldsymbol{\sigma}} \in \{\pm 1\}^{\boldsymbol{\sigma}}$, then by 2.9.7 there exists some $c \in \mathfrak{o}^\bullet$ such that $c \equiv a \bmod \mathfrak{m}$ and $\operatorname{sgn}\sigma(c) = \varepsilon_\sigma$ for all $\sigma \in \boldsymbol{\sigma}$. Consequently, $c \in \mathfrak{o}(\mathfrak{m}) \subset K(\mathfrak{m})$ and $\pi_\mathfrak{m}^{\boldsymbol{\sigma}}(c) = (a+\mathfrak{m}, (\varepsilon_\sigma)_{\sigma\in\boldsymbol{\sigma}})$.

3. • η is a monomorphism: Since the map $\eta_0 \colon \mathfrak{o}^\times \cap K^{\boldsymbol{\sigma}} \to (\mathfrak{o}/\mathfrak{m})^\times$, $\varepsilon \mapsto \varepsilon + \mathfrak{m}$, is a group homomorphism with kernel $\mathfrak{o}^\times \cap K^{\mathfrak{m},\boldsymbol{\sigma}}$, it induces a monomorphism $\eta \colon \mathfrak{o}^\times \cap K^{\boldsymbol{\sigma}}/\mathfrak{o}^\times \cap K^{\mathfrak{m},\boldsymbol{\sigma}} \to (\mathfrak{o}/\mathfrak{m})^\times$ as asserted.

• ρ is an epimorphism: The map $\rho_0 \colon \mathcal{J}^{\mathfrak{m}}(\mathfrak{o}) \to \mathcal{C}^{\boldsymbol{\sigma}}(\mathfrak{o})$, $\mathfrak{a} \mapsto [\mathfrak{a}]^{\boldsymbol{\sigma}}$, is an epimorphism by 1., and $\operatorname{Ker}(\rho_0) = \mathcal{H}^{\boldsymbol{\sigma}}(\mathfrak{o}) \cap \mathcal{J}^{\mathfrak{m}}(\mathfrak{o}) \supset \mathcal{H}^{\mathfrak{m},\boldsymbol{\sigma}}(\mathfrak{o})$. Hence ρ_0 induces an epimorphism $\rho \colon \mathcal{C}^{\mathfrak{m},\boldsymbol{\sigma}}(\mathfrak{o}) = \mathcal{J}^{\mathfrak{m}}(\mathfrak{o})/\mathcal{H}^{\mathfrak{m},\boldsymbol{\sigma}}(\mathfrak{o}) \to \mathcal{C}^{\boldsymbol{\sigma}}(\mathfrak{o})$ as asserted.

• θ is a well-defined homomorphism, and $\operatorname{Im}(\theta) \subset \operatorname{Ker}(\rho)$: If $a \in \mathfrak{o}(\mathfrak{m})$ and $a_0, a_0' \in \mathfrak{o} \cap K^{\boldsymbol{\sigma}}$ are such that $a_0 \equiv a_0' \equiv a \bmod \mathfrak{m}$, then $a_0^{-1}a_0' \in K^{\mathfrak{m},\boldsymbol{\sigma}}$ and $[a_0'\mathfrak{o}]^{\mathfrak{m},\boldsymbol{\sigma}} = [a_0\mathfrak{o}]^{\mathfrak{m},\boldsymbol{\sigma}}[a_0^{-1}a_0'\mathfrak{o}]^{\mathfrak{m},\boldsymbol{\sigma}} = [a_0\mathfrak{o}]^{\mathfrak{m},\boldsymbol{\sigma}}$. Hence the definition of θ does not depend on the choice of a_0. If $a, b \in \mathfrak{o}(\mathfrak{m})$ and $a_0, b_0 \in \mathfrak{o} \cap K^{\boldsymbol{\sigma}}$ are such that $a_0 \equiv a \bmod \mathfrak{m}$ and $b_0 \equiv b \bmod \mathfrak{m}$, then we obtain $a_0b_0 \in \mathfrak{o} \cap K^{\boldsymbol{\sigma}}$, $a_0b_0 \equiv ab \bmod \mathfrak{m}$, and therefore

$$\theta(a+\mathfrak{m})\theta(b+\mathfrak{m}) = [a_0\mathfrak{o}]^{\mathfrak{m},\boldsymbol{\sigma}}[b_0\mathfrak{o}]^{\mathfrak{m},\boldsymbol{\sigma}} = [a_0b_0\mathfrak{o}]^{\mathfrak{m},\boldsymbol{\sigma}} = \theta(ab+\mathfrak{m}).$$

Hence θ is a homomorphism, and since $\rho([a_0\mathfrak{o}]^{\mathfrak{m},\boldsymbol{\sigma}}) = [a_0\mathfrak{o}]^{\boldsymbol{\sigma}} = [\mathbf{1}]^{\boldsymbol{\sigma}} \in \mathcal{C}^{\boldsymbol{\sigma}}(\mathfrak{o})$, we obtain $\operatorname{Im}(\theta) \subset \operatorname{Ker}(\rho)$.

• $\operatorname{Ker}(\rho) \subset \operatorname{Im}(\theta)$: Let $\mathfrak{a} \in \mathcal{J}^{\mathfrak{m}}(\mathfrak{o})$ be such that $[\mathfrak{a}]^{\mathfrak{m},\boldsymbol{\sigma}} \in \operatorname{Ker}(\rho)$. Then it follows that $[\mathfrak{a}]^{\boldsymbol{\sigma}} = [\mathbf{1}]^{\boldsymbol{\sigma}} \in \mathcal{C}^{\boldsymbol{\sigma}}(\mathfrak{o})$, hence $\mathfrak{a} = a\mathfrak{o}$ for some $a \in K^{\boldsymbol{\sigma}} \cap K(\mathfrak{m})$, and there exists some $a_0 \in K^{\boldsymbol{\sigma}} \cap \mathfrak{o}(\mathfrak{m})$ such that $a \equiv a_0 \bmod \mathfrak{m}$. But then we obtain $a_0^{-1}a \in K(\mathfrak{m})$, $a_0^{-1}a \equiv 1 \bmod \mathfrak{m}$, hence $[a_0^{-1}a]^{\mathfrak{m},\boldsymbol{\sigma}} = [\mathbf{1}]^{\mathfrak{m},\boldsymbol{\sigma}}$ and

$$[\mathfrak{a}]^{\mathfrak{m},\boldsymbol{\sigma}} = [a\mathfrak{o}]^{\mathfrak{m},\boldsymbol{\sigma}} = [a_0\mathfrak{o}]^{\mathfrak{m},\boldsymbol{\sigma}}[a_0^{-1}a\mathfrak{o}]^{\mathfrak{m},\boldsymbol{\sigma}} = [a_0\mathfrak{o}]^{\mathfrak{m},\boldsymbol{\sigma}} = \theta(a_0+\mathfrak{m}) \in \operatorname{Im}(\theta).$$

• $\operatorname{Im}(\eta) \subset \operatorname{Ker}(\theta)$: Obviously $\theta \circ \eta(\varepsilon(\mathfrak{o}^\times \cap K^{\mathfrak{m},\boldsymbol{\sigma}})) = [\varepsilon\mathfrak{o}]^{\mathfrak{m},\boldsymbol{\sigma}} = [\mathbf{1}]^{\mathfrak{m},\boldsymbol{\sigma}}$ for all $\varepsilon \in \mathfrak{o}^\times \cap K^{\boldsymbol{\sigma}}$.

• $\operatorname{Ker}(\theta) \subset \operatorname{Im}(\eta)$: Let $a \in \mathfrak{o}(\mathfrak{m})$ be such that $a + \mathfrak{m} \in \operatorname{Ker}(\theta)$, and let $a_0 \in \mathfrak{o} \cap K^{\boldsymbol{\sigma}}$ such that $a_0 \equiv a \bmod \mathfrak{m}$. Then $\theta(a+\mathfrak{m}) = [a_0\mathfrak{o}]^{\mathfrak{m},\boldsymbol{\sigma}} = [\mathbf{1}]^{\mathfrak{m},\boldsymbol{\sigma}}$ and therefore $a_0\mathfrak{o} = a_1\mathfrak{o}$ for some $a_1 \in K^{\mathfrak{m},\boldsymbol{\sigma}}$. Now we obtain $a_0 = a_1\varepsilon$ for some $\varepsilon \in \mathfrak{o}^\times \cap K^{\boldsymbol{\sigma}}$, and

$$a+\mathfrak{m} = a_0+\mathfrak{m} = (a_1+\mathfrak{m})(\varepsilon+\mathfrak{m}) = \varepsilon+\mathfrak{m} = \eta((\varepsilon+\mathfrak{m})(\mathfrak{o}_\mathfrak{m}^\times \cap K^{\boldsymbol{\sigma}})) \in \operatorname{Im}(\eta). \quad \square$$

2.10 Discrete valuation domains and Dedekind domains

We already introduced discrete valuations in 2.2.1 and used it several times for arithmetical investigations, e.g. in 2.2.2 and 2.8.5. Now we are going to investigate them and the associated ring theory for their own sake.

Definition and Theorem 2.10.1. *Let $\mathfrak{o}$ be a domain and $K = \mathsf{q}(\mathfrak{o}) \ne \mathfrak{o}$.*

1. The following assertions are equivalent:

(a) If $x \in K \setminus \mathfrak{o}$, then $x^{-1} \in \mathfrak{o}$.

(b) If $\mathfrak{a}$ and $\mathfrak{b}$ are $\mathfrak{o}$-submodules of K, then $\mathfrak{a} \subset \mathfrak{b}$ or $\mathfrak{b} \subset \mathfrak{a}$.

(c) If $\mathfrak{a}$ and $\mathfrak{b}$ are ideals of $\mathfrak{o}$, then $\mathfrak{a} \subset \mathfrak{b}$ or $\mathfrak{b} \subset \mathfrak{a}$.

If these conditions are fulfilled, then $\mathfrak{o}$ is called a **valuation domain**.

2. Let $\mathfrak{o}$ be a valuation domain. Then $\mathfrak{o}$ is local, integrally closed, and every finitely generated ideal is principal.

3. The following assertions are equivalent:

(a) $\mathfrak{o}$ is a local Dedekind domain.

(b) $\mathfrak{o}$ is a local principal ideal domain.

(c) $\mathfrak{o}$ is a factorial domain and possesses up to associates only one prime element.

(d) $\mathfrak{o}$ is a Noetherian valuation domain.

If these conditions are fulfilled, then $\mathfrak{o}$ is called a **discrete valuation domain**.

4. Let $\mathfrak{o}$ be a discrete valuation domain and $\pi \in \mathfrak{o}$ a prime element. Then $\mathfrak{p} = \pi\mathfrak{o}$ is the unique maximal ideal of $\mathfrak{o}$, $K = \mathfrak{o}[\pi^{-1}]$, and $\mathfrak{o}$ is a maximal proper subring of K.

Proof. 1. (a) $\Rightarrow$ (b) Let $\mathfrak{a}$ and $\mathfrak{b}$ be $\mathfrak{o}$-submodules of K, $\mathfrak{a} \not\subset \mathfrak{b}$ and $a \in \mathfrak{a} \setminus \mathfrak{b}$. If $b \in \mathfrak{b}^\bullet$, then $b^{-1}a \notin \mathfrak{o}$, hence $a^{-1}b = (b^{-1}a)^{-1} \in \mathfrak{o}$ and $b \in a\mathfrak{o} \subset \mathfrak{a}$. Thus $\mathfrak{b} \subset \mathfrak{a}$.

(b) $\Rightarrow$ (c) Obvious.

(c) $\Rightarrow$ (a) If $x \in K \setminus \mathfrak{o}$, then $x = a^{-1}b$, where $a, b \in \mathfrak{o}^\bullet$ and $b \notin a\mathfrak{o}$. Then $b\mathfrak{o} \not\subset a\mathfrak{o}$, hence $a\mathfrak{o} \subset b\mathfrak{o}$ and $x^{-1} = b^{-1}a \in \mathfrak{o}$.

2. By 1.(c) the set Ω of all proper ideals of $\mathfrak{o}$ is a chain. Hence its union is the largest proper ideal of $\mathfrak{o}$, and thus $\mathfrak{o}$ is local.

If $\mathfrak{a}$ is a finitely generated ideal of $\mathfrak{o}$, say $\mathfrak{a} = (a_1, \ldots, a_n)$, then (using 1.) we may assume that $(a_1) \supset (a_2) \supset \ldots \supset (a_n)$, which implies $\mathfrak{a} = (a_1)$.

Let $x \in K$ be integral over $\mathfrak{o}$, and let $x^n + a_{n-1}x^{n-1} + \ldots + a_0 = 0$ be an integral equation for x over $\mathfrak{o}$. If $x \notin \mathfrak{o}$, then $x^{-1} \in \mathfrak{o}$ and yet

$$x = -\sum_{i=0}^{n-1} a_i (x^{-1})^{n-i} \in \mathfrak{o}.$$

Hence $x \in \mathfrak{o}$, and $\mathfrak{o}$ is integrally closed.

3. (a) $\Leftrightarrow$ (b) By 2.6.6 and 2.8.6.1.

(b) $\Rightarrow$ (c) Let $\pi \in \mathfrak{o}^\bullet$ be such that $\pi\mathfrak{o}$ is the unique maximal ideal of $\mathfrak{o}$. By 1.1.7.3, $\mathfrak{o}$ is factorial and $\mathcal{P}(\mathfrak{o}) = \{\pi\mathfrak{o}\}$. Hence π is up to associates the unique prime element of $\mathfrak{o}$.

(c) $\Rightarrow$ (d) Let $\pi \in \mathfrak{o}^\bullet$ be a prime element of $\mathfrak{o}$ and $x \in K^\times$. Then x has a unique representation $x = \pi^k u$, where $k = v_\pi(x) \in \mathbb{Z}$ and $u \in \mathfrak{o}^\times$. If $x \notin \mathfrak{o}$, then $k < 0$ and $x^{-1} = \pi^{-k}u^{-1} \in \mathfrak{o}$. Hence $\mathfrak{o}$ is a valuation domain.

We prove that $\mathfrak{o}$ is a principal ideal domain (hence Noetherian). Let $\mathfrak{a} \neq \mathbf{0}$ be an ideal of $\mathfrak{o}$, $k = \min\{v_\pi(x) \mid x \in \mathfrak{a}^\bullet\}$ and $a \in \mathfrak{a}$ an element satisfying $v_\pi(a) = k$. Then $a = \pi^k u$, where $u \in \mathfrak{o}^\times$, $a\mathfrak{o} \subset \mathfrak{a}$, and we prove equality. If $x \in \mathfrak{a}^\bullet$, then $x = \pi^l v$, where $l \in \mathbb{Z}$, $l \geq k$ and $v \in \mathfrak{o}^\times$. Consequently, $x = \pi^{l-k}u^{-1}va \in a\mathfrak{o}$.

(d) $\Rightarrow$ (b) By 2., $\mathfrak{o}$ is local. If $\mathfrak{a}$ is an ideal of $\mathfrak{o}$, then by 1.(c) we may assume that $\mathfrak{a} = {}_{\mathfrak{o}}(a_1, \ldots, a_m)$, where $(a_1) \supset (a_2) \supset \ldots \supset (a_m)$, and thus $\mathfrak{a} = (a_1)$.

4. By definition, $\mathfrak{p} = \pi\mathfrak{o}$ is the unique maximal ideal of $\mathfrak{o}$. If $x \in K \setminus \mathfrak{o}$, then $x = \pi^{-n}u = (\pi^{-1})^n u$ for some $n \in \mathbb{N}$ and $u \in \mathfrak{o}^\times$, and therefore $x \in \mathfrak{o}[\pi^{-1}]$. Hence $K = \mathfrak{o}[\pi^{-1}]$.

Now let $\mathfrak{o}'$ be a ring such that $\mathfrak{o} \subsetneq \mathfrak{o}' \subset K$ and $z \in \mathfrak{o}' \setminus \mathfrak{o}$. Then $z = \pi^{-n}u$, where $n \in \mathbb{N}$ and $u \in \mathfrak{o}^\times$. Since $a = \pi^{n-1}u^{-1} \in \mathfrak{o}$ it follows that $\pi^{-1} = az \in \mathfrak{o}'$ and $K = \mathfrak{o}[\pi^{-1}] \subset \mathfrak{o}'$, hence $\mathfrak{o}' = K$. □

We apply discrete valuation domains to characterize Dedekind domains by their localizations in 2.10.2 and thereafter we characterize the overrings of a Dedekind domain in its quotient field in 2.10.3.

Theorem 2.10.2. *Let $\mathfrak{o}$ be a domain and not a field. Then the following assertions are equivalent:*

(a) $\mathfrak{o}$ is a Dedekind domain.

(b) For every $\mathfrak{p} \in \max(\mathfrak{o})$ the localization $\mathfrak{o}_\mathfrak{p}$ is a discrete valuation domain, and every $a \in \mathfrak{o}^\bullet$ lies in only finitely many maximal ideals of $\mathfrak{o}$.

(c) $\mathfrak{o}$ is Noetherian, and for every $\mathfrak{p} \in \max(\mathfrak{o})$ the localization $\mathfrak{o}_\mathfrak{p}$ is a discrete valuation domain.

Proof. (a) $\Rightarrow$ (b) If $\mathfrak{p} \in \mathcal{P}(\mathfrak{o})$, then $\mathfrak{o}_\mathfrak{p}$ is a local Dedekind domain and thus a discrete valuation domain by 2.10.1.3. If $a \in \mathfrak{o}^\bullet$, then a lies in only finitely many maximal ideals of $\mathfrak{o}$ by 2.8.3.4(c).

(b) $\Rightarrow$ (c) Let $\mathbf{0} \neq \mathfrak{a}_0 \subset \mathfrak{a}_1 \subset \mathfrak{a}_2 \subset \ldots$ be an ascending sequence of ideals in $\mathfrak{o}$. For every $\mathfrak{p} \in \max(\mathfrak{o})$, the sequence $\mathfrak{a}_0\mathfrak{o}_\mathfrak{p} \subset \mathfrak{a}_1\mathfrak{o}_\mathfrak{p} \subset \mathfrak{a}_2\mathfrak{o}_\mathfrak{p} \subset \ldots$ becomes stationary since $\mathfrak{o}_\mathfrak{p}$ is Noetherian. Let $n_\mathfrak{p} \in \mathbb{N}_0$ be minimal such that $\mathfrak{a}_n\mathfrak{o}_\mathfrak{p} = a_{n_\mathfrak{p}}\mathfrak{o}_\mathfrak{p}$ for all $n \geq n_\mathfrak{p}$. Since $\mathfrak{a}_0$ lies in only finitely many maximal ideals, we obtain $\mathfrak{a}_0\mathfrak{o}_\mathfrak{p} = \mathfrak{o}_\mathfrak{p}$ and thus $n_\mathfrak{p} = 0$ for almost all $\mathfrak{p} \in \max(\mathfrak{o})$. If $n_0 = \max\{n_\mathfrak{p} \mid \mathfrak{p} \in \max(\mathfrak{o})\}$ and $n \geq n_0$, then $\mathfrak{a}_n\mathfrak{o}_\mathfrak{p} = \mathfrak{a}_{n_0}\mathfrak{p}$ for all $\mathfrak{p} \in \max(\mathfrak{o})$ and thus $\mathfrak{a}_n = \mathfrak{a}_{n_0}$ by 2.7.5. Hence the sequence $(\mathfrak{a}_n)_{n\geq 0}$ becomes stationary, and $\mathfrak{o}$ is Noetherian.

If $\mathfrak{p} \in \max(\mathfrak{o})$, then $\mathfrak{o}_\mathfrak{p}$ is a local Dedekind domain and thus a discrete valuation domain by 2.10.1.3.

(c) $\Rightarrow$ (a) If $\mathfrak{p} \in \max(\mathfrak{o})$, then $\mathfrak{o}_\mathfrak{p}$ is not a field, and thus $\mathfrak{o}$ is not a field. We prove that every non-zero ideal of $\mathfrak{o}$ is invertible.

Thus let $\mathfrak{a} \neq \mathbf{0}$ be an ideal of $\mathfrak{o}$. Then $\mathfrak{a}$ is finitely generated, and $\mathfrak{a}_\mathfrak{p}$ is principal for all $\mathfrak{p} \in \max(\mathfrak{o})$. Hence $\mathfrak{a}$ is invertible 2.7.7. □

Theorem 2.10.3 (Overrings of Dedekind domains). *Let $\mathfrak{o}$ be a Dedekind domain, $K = \mathsf{q}(\mathfrak{o})$ and $\mathfrak{O}$ a domain such that $\mathfrak{o} \subset \mathfrak{O} \subsetneq K$.*

1. *If $\mathfrak{P} \in \mathcal{P}(\mathfrak{O})$ and $\mathfrak{p} = \mathfrak{P} \cap \mathfrak{o}$, then*

$$\mathfrak{p} \in \mathcal{P}(\mathfrak{o}), \quad \mathfrak{o}_\mathfrak{p} = \mathfrak{O}_\mathfrak{P}, \quad \mathfrak{p}\mathfrak{O}_\mathfrak{p} = \mathfrak{P}\mathfrak{O}_\mathfrak{p} = \mathfrak{p}\mathfrak{o}_\mathfrak{p} \text{ and } \mathfrak{P} = \mathfrak{p}\mathfrak{o}_\mathfrak{p} \cap \mathfrak{O}.$$

In particular, $\mathfrak{O}_\mathfrak{P}$ is a discrete valuation domain.

2. *If $\mathfrak{A}$ is an ideal of $\mathfrak{O}$, then $\mathfrak{A} = (\mathfrak{A} \cap \mathfrak{o})\mathfrak{O}$.*

3. *If $\mathfrak{p} \in \mathcal{P}(\mathfrak{o})$, then $\mathfrak{p}\mathfrak{O} \neq \mathfrak{O}$ if and only if $\mathfrak{O} \subset \mathfrak{o}_\mathfrak{p}$, and in this case we have $\mathfrak{p}\mathfrak{O} = \mathfrak{p}\mathfrak{o}_\mathfrak{p} \cap \mathfrak{O} \in \mathcal{P}(\mathfrak{O})$.*

4. *$\mathfrak{O}$ is a Dedekind domain, and*

$$\mathcal{P}(\mathfrak{O}) = \{\mathfrak{p}\mathfrak{O} \mid \mathfrak{p} \in \mathcal{P}(\mathfrak{o}) \text{ and } \mathfrak{p}\mathfrak{O} \neq \mathfrak{O}\}.$$

Proof. 1. If $\mathfrak{P} \in \mathcal{P}(\mathfrak{O})$, then clearly $\mathfrak{p} = \mathfrak{P} \cap \mathfrak{o} \in \mathcal{P}(\mathfrak{o})$ and $\mathfrak{o}_\mathfrak{p} \subset \mathfrak{O}_\mathfrak{P}$. Since $\mathfrak{o}_\mathfrak{p}$ is a discrete valuation domain, 2.10.1.4 implies $\mathfrak{o}_\mathfrak{p} = \mathfrak{O}_\mathfrak{P}$. Hence we immediately obtain $\mathfrak{p}\mathfrak{O}_\mathfrak{p} = \mathfrak{P}\mathfrak{O}_\mathfrak{p} = \mathfrak{p}\mathfrak{o}_\mathfrak{p}$ and $\mathfrak{P} = \mathfrak{P}\mathfrak{O}_\mathfrak{P} \cap \mathfrak{O} = \mathfrak{p}\mathfrak{o}_\mathfrak{p} \cap \mathfrak{O}$.

2. Let $\mathfrak{A}$ be an ideal of $\mathfrak{O}$. Concerning 2.7.5 it suffices to prove that $\mathfrak{A}\mathfrak{O}_\mathfrak{P} = (\mathfrak{A} \cap \mathfrak{o})\mathfrak{O}\mathfrak{O}_\mathfrak{P}$ for all $\mathfrak{P} \in \max(\mathfrak{O})$. Thus let $\mathfrak{P} \in \max(\mathfrak{O})$, $\mathfrak{p} = \mathfrak{P} \cap \mathfrak{o}$ and apply 1. Then $(\mathfrak{A} \cap \mathfrak{o})\mathfrak{O}\mathfrak{O}_\mathfrak{P} = (\mathfrak{A} \cap \mathfrak{o})\mathfrak{o}_\mathfrak{p} \subset \mathfrak{A}\mathfrak{o}_\mathfrak{p} = \mathfrak{A}\mathfrak{O}_\mathfrak{P}$.

As to the reverse inclusion, let $z \in \mathfrak{A}\mathfrak{o}_\mathfrak{p}$, say $z = s^{-1}a$, where $s \in \mathfrak{o} \setminus \mathfrak{p}$ and $a \in \mathfrak{A}$. Since $\mathfrak{A} \subset \mathfrak{O} \subset \mathfrak{o}_\mathfrak{p}$, we obtain $a = t^{-1}c$ for some $t \in \mathfrak{o} \setminus \mathfrak{p}$ and $c \in \mathfrak{o}$. Hence $c = ta \in \mathfrak{A} \cap \mathfrak{o}$ and $z = (st)^{-1}c \in (\mathfrak{A} \cap \mathfrak{o})\mathfrak{o}_\mathfrak{p}$.

3. Let $\mathfrak{p} \in \mathcal{P}(\mathfrak{o})$, and assume first that $\mathfrak{p}\mathfrak{O} \neq \mathfrak{O}$. If $\mathfrak{P} \in \mathcal{P}(\mathfrak{O})$ is such that $\mathfrak{p}\mathfrak{O} \subset \mathfrak{P}$, then $\mathfrak{p} \subset \mathfrak{P} \cap \mathfrak{o} \subsetneq \mathfrak{o}$. Hence $\mathfrak{p} = \mathfrak{P} \cap \mathfrak{O}$, and 1. implies $\mathfrak{O} \subset \mathfrak{O}_{\mathfrak{P}} = \mathfrak{o}_{\mathfrak{p}}$ and $\mathfrak{P} = \mathfrak{p}\mathfrak{o}_{\mathfrak{p}} \cap \mathfrak{O}$. Now it follows by 2. that $\mathfrak{P} = (\mathfrak{P} \cap \mathfrak{o})\mathfrak{O} = (\mathfrak{p}\mathfrak{o}_{\mathfrak{p}} \cap \mathfrak{o})\mathfrak{O} = \mathfrak{p}\mathfrak{O}$.

Conversely, if $\mathfrak{O} \subset \mathfrak{o}_{\mathfrak{p}}$, then $\mathfrak{O}_{\mathfrak{p}} = \mathfrak{o}_{\mathfrak{p}}$, hence $\mathfrak{p}\mathfrak{o}_{\mathfrak{p}} = \mathfrak{p}\mathfrak{O}_{\mathfrak{p}} \neq \mathfrak{O}_{\mathfrak{p}}$, and therefore $\mathfrak{p}\mathfrak{O} \neq \mathfrak{O}$.

4. We prove first that $\mathcal{P}(\mathfrak{O}) = \{\mathfrak{p}\mathfrak{O} \mid \mathfrak{p} \in \mathcal{P}(\mathfrak{o}) \text{ and } \mathfrak{p}\mathfrak{O} \neq \mathfrak{O}\}$. If $\mathfrak{P} \in \mathcal{P}(\mathfrak{O})$, then $\mathfrak{P} = (\mathfrak{P} \cap \mathfrak{o})\mathfrak{O}$ by 2., $\mathfrak{p} = \mathfrak{P} \cap \mathfrak{o} \in \mathcal{P}(\mathfrak{o})$ and $\mathfrak{p}\mathfrak{O} \neq \mathfrak{O}$. Conversely, if $\mathfrak{p} \in \mathcal{P}(\mathfrak{o})$ and $\mathfrak{p}\mathfrak{O} \neq \mathfrak{O}$, then $\mathfrak{p}\mathfrak{O} \in \mathcal{P}(\mathfrak{O})$ by 3.

We show now that every $a \in \mathfrak{O}^\bullet$ lies in only finitely many maximal ideals of $\mathfrak{O}$ (then $\mathfrak{O}$ is a Dedekind domain by 2.10.2). If $a \in \mathfrak{O}^\bullet$, then $a\mathfrak{O} \cap \mathfrak{o} \neq \mathbf{0}$ by 2., and therefore the set $\Omega = \{\mathfrak{p} \in \mathcal{P}(\mathfrak{o}) \mid a \in \mathfrak{p}\}$ is finite. If $\mathfrak{P} \in \mathcal{P}(\mathfrak{O})$ is such that $a \in \mathfrak{P}$, then $a\mathfrak{O} \cap \mathfrak{o} \subset \mathfrak{P} \cap \mathfrak{o}$, hence $\mathfrak{p} = \mathfrak{P} \cap \mathfrak{o} \in \Omega$ and $\mathfrak{P} = \mathfrak{p}\mathfrak{O}$. Consequently there are only finitely many such $\mathfrak{P}$. □

Now we disclose the connection between discrete valuations and discrete valuation domains.

Definition and Theorem 2.10.4. *Let K be a field.*

1. Let $v \colon K \to \mathbb{Z} \cup \{\infty\}$ be a discrete valuation and $\pi \in K$ such that $v(\pi) = 1$. Then $\mathcal{O}_v = \{x \in K \mid v(x) \geq 0\} \subsetneq K$ is a discrete valuation domain, $\mathfrak{p}_v = \pi\mathcal{O}_v = \{x \in K \mid v(x) > 0\}$ is its maximal ideal, $\mathcal{O}_v^\times = \{x \in K \mid v(x) = 0\}$, $\mathcal{J}(\mathcal{O}_v) = \{\pi^n\mathcal{O}_v \mid n \in \mathbb{Z}\}$ and $v = v_\pi = v_{\mathfrak{p}_v}$.

We call $\mathcal{O}_v$ the **valuation domain**, $\mathfrak{p}_v = \pi\mathcal{O}_v$ the **valuation ideal**, $U_v = \mathcal{O}_v^\times$ the **unit group** and $\mathsf{k}_v = \mathsf{k}_{\mathfrak{p}_v} = \mathcal{O}_v/\mathfrak{p}_v$ the **residue class field** of v.

2. Let $\mathfrak{o}$ be a discrete valuation domain with maximal ideal $\mathfrak{p}$, let $K = \mathsf{q}(\mathfrak{o})$ and $v_{\mathfrak{p}} \colon K \to \mathbb{Z} \cup \{\infty\}$ the $\mathfrak{p}$-adic valuation as defined in 2.8.5. Then $\mathfrak{o} = \mathcal{O}_{v_{\mathfrak{p}}}$ and $\mathfrak{p} = \mathfrak{p}_{v_{\mathfrak{p}}}$.

We call $v_{\mathfrak{p}}$ the **defining valuation** of $\mathfrak{o}$.

3. Let $\mathfrak{o}$ and $\mathfrak{o}'$ be discrete valuation domains, $\mathfrak{p}$ and $\mathfrak{p}'$ their maximal ideals and $\mathsf{q}(\mathfrak{o}) = \mathsf{q}(\mathfrak{o}')$. Then the following assertions are equivalent:

(a) $\mathfrak{o} = \mathfrak{o}'$; (b) $\mathfrak{o} \subset \mathfrak{o}'$; (c) $\mathfrak{p} = \mathfrak{p}'$; (d) $\mathfrak{p} \subset \mathfrak{p}'$; (e) $v_{\mathfrak{p}} = v_{\mathfrak{p}'}$.

Proof. 1. If $x, y \in \mathcal{O}_v$, then

$$v(x - y) \geq \min\{v(x), v(y)\} \geq 0 \quad \text{and} \quad v(xy) = v(x) + v(y) \geq 0.$$

Hence $\{1, x - y, xy\} \subset \mathcal{O}_v$ for all $x, y \in \mathcal{O}_v$ and thus $\mathcal{O}_v$ is a subring of K. Moreover, $\mathcal{O}_v \neq K$ since v is surjective. If $x \in K^\times$, then $x \in \mathcal{O}_v^\times$ if and only if $v(x) \geq 0$ and $v(x^{-1}) = -v(x) \geq 0$. Hence $\mathcal{O}_v^\times = \{x \in K \mid v(x) = 0\}$.

Next we prove that $\{\pi^n\mathcal{O}_v \mid n \in \mathbb{Z}\}$ is the set of all fractional ideals of $\mathcal{O}_v$. Thus let $\mathfrak{a} \in \mathcal{I}(\mathcal{O}_v)$, and let $c \in \mathcal{O}_v^\bullet$ be such that $c\mathfrak{a} \subset \mathcal{O}_v$. If $a \in \mathfrak{a}$, then $v(ca) = v(c) + v(a) \geq 0$, hence $v(a) \geq -v(c)$, and there exists some $a \in \mathfrak{a}$ such that $v(a) = \min\{v(x) \mid x \in \mathfrak{a}\}$. We assert that $a\mathcal{O}_v = \mathfrak{a}$. Clearly, $a\mathcal{O}_v \subset \mathfrak{a}$. If $x \in \mathfrak{a}^\bullet$, then $v(a^{-1}x) = -v(a) + v(x) \geq 0$. Hence $a^{-1}x \in \mathcal{O}_v$ and $x \in a\mathcal{O}_v$.

Since $\{\pi^n\mathcal{O}_v \mid n \in \mathbb{Z}\}$ is the set of all fractional ideals of $\mathcal{O}_v$, it follows that $\{\pi^n\mathcal{O}_v \mid n \in \mathbb{N}_0\}$ is the set of all non-zero ideals of $\mathcal{O}_v$. Hence $\mathcal{O}_v$ is a local principal ideal domain with maximal ideal $\mathfrak{p}_v = \pi\mathcal{O}_v$ and thus a discrete valuation domain. If $x \in K^\times$, then $x = \pi^{v(x)}u$ for some $u \in \mathcal{O}_v^\times$. Hence $v = v_\pi$, and 2.8.6.2 implies $v_\pi = v_{\mathfrak{p}_v}$.

2. If $\mathfrak{p} = \pi\mathfrak{o}$, then $v_\mathfrak{p} = v_\pi$ (again by 2.8.6.2), and every $x \in K^\times$ has a unique representation $x = \pi^{v_\mathfrak{p}(x)}u$, where $u \in \mathfrak{o}^\times$. Hence

$$\mathfrak{o} = \{x \in K \mid v_\mathfrak{p}(x) \geq 0\} = \mathcal{O}_{v_\mathfrak{p}} \quad \text{and} \quad \mathfrak{p} = \{x \in K \mid v_\mathfrak{p}(x) > 0\} = \mathfrak{p}_{v_\mathfrak{p}}.$$

3. The equivalence of (a) and (b) follows from 2.10.1.4, the equivalence of (a) and (e) follows from 2., and the implications (a) $\Rightarrow$ (c) $\Rightarrow$ (d) are obvious.

(d) $\Rightarrow$ (a) By 2.10.1.4 it suffices to prove that $\mathfrak{o}' \subset \mathfrak{o}$, and we assume to the contrary that there exists some $x \in \mathfrak{o}' \setminus \mathfrak{o}$. Then $x^{-1} \in \mathfrak{o} \setminus \mathfrak{o}^\times = \mathfrak{p} \subset \mathfrak{p}'$ and therefore $1 = xx^{-1} \in \mathfrak{p}'$, a contradiction. □

Theorem and Definition 2.10.5 (Structure of discrete valuation domains). *Let $\mathfrak{o}$ be a discrete valuation domain, $\mathfrak{p} = \pi\mathfrak{o}$ its maximal ideal, $\mathsf{k}_\mathfrak{p} = \mathfrak{o}/\mathfrak{p}$, $U = \mathfrak{o}^\times$ and $K = \mathsf{q}(\mathfrak{o})$.*

1. *If $n \in \mathbb{Z}$, then $\mathfrak{p}^n = \pi^n\mathfrak{o} = \{x \in K \mid v(x) \geq n\}$,*

$$\bigcap_{n\in\mathbb{N}} \mathfrak{p}^n = \mathbf{0},$$

and for every $k \geq 1$ the map

$$\varphi_{n,k}\colon \mathfrak{o}/\mathfrak{p}^k \to \mathfrak{p}^n/\mathfrak{p}^{n+k}, \quad \textit{defined by}$$
$$\varphi_{n,k}(a + \mathfrak{p}^k) = a\pi^n + \mathfrak{p}^{n+k} \quad \textit{for all} \ \ a \in \mathfrak{o},$$

is a group isomorphism.

In particular, $\varphi_{n,1}\colon \mathfrak{o}/\mathfrak{p} = \mathsf{k}_\mathfrak{p} \to \mathfrak{p}^n/\mathfrak{p}^{n+1}$ is an isomorphism of vector spaces over $\mathsf{k}_\mathfrak{p}$.

2. *For $n \in \mathbb{N}$, let $U^{(n)} = 1 + \mathfrak{p}^n = 1 + \pi^n\mathfrak{o}$.*

Then $\mathfrak{o}^\times = U = U^{(0)} \supset U^{(1)} \supset U^{(2)} \supset \ldots$ is a sequence of subgroups, and

$$\bigcap_{n\in\mathbb{N}} U^{(n)} = \mathbf{1}.$$

The elements of $U^{(1)}$ are called **principal units**, and for $n \in \mathbb{N}$ the elements of $U^{(n)}$ are called **principal units of level** n.

3. The map

$$\psi\colon U/U^{(1)} \to \mathsf{k}_\mathfrak{p}^\times, \ \textit{defined by } \psi(aU^{(1)}) = a+\mathfrak{p} \ \textit{for all } a \in U,$$

is a group isomorphism.

4. If $n \in \mathbb{N}$, then the maps

$$\psi_n\colon U^{(n)}/U^{(n+1)} \to \mathsf{k}_\mathfrak{p}, \ \textit{defined by}$$
$$\psi_n((1+\pi^n a)U^{(n+1)}) = a+\mathfrak{p} \ \textit{for all } a \in \mathfrak{o},$$

and

$$\phi_n\colon U/U^{(n)} \to (\mathfrak{o}/\mathfrak{p}^n)^\times, \ \textit{defined by } \phi_n(uU^{(n)}) = u+\mathfrak{p}^n \ \textit{for all } u \in U,$$

are group isomorphisms.
In particular, $U^{(n)}/U^{(n+1)}$ has a unique structure of a vector space over $\mathsf{k}_\mathfrak{p}$ such that ψ_n is $\mathsf{k}_\mathfrak{p}$-linear.

Proof. Let $v = v_\mathfrak{p} = v_\pi$ (see 2.10.4).

1. If $n \in \mathbb{Z}$, then $\mathfrak{p}^n = (\pi\mathfrak{o})^n = \pi^n\mathfrak{o}$. If $x \in K$, then $x \in \pi^n\mathfrak{o}$ if and only if $\pi^{-n}x \in \mathfrak{o}$, that is, if and only if $0 \le v(\pi^{-n}x) = -n + v(x)$. Hence

$$\mathfrak{p}^n = \pi^n\mathfrak{o} = \{x \in K \mid v(x) \ge n\}, \quad \text{and} \quad \bigcap_{n\in\mathbb{N}} \mathfrak{p}^n = \{x \in K \mid v(x) = \infty\} = \mathbf{0}.$$

If $k \in \mathbb{N}$, then the map $\varphi'_{n,k}\colon \mathfrak{o} \to \mathfrak{p}^n/\mathfrak{p}^{n+k}$, defined by $\varphi'_{n,k}(a) = a\pi^n + \mathfrak{p}^{n+k}$, is a group epimorphism, and

$$\operatorname{Ker}(\varphi'_{n,k}) = \{a \in \mathfrak{o} \mid a\pi_n \in \mathfrak{p}^{n+k}\} = \{a \in \mathfrak{o} \mid v(a) \ge k\} = \mathfrak{p}^k.$$

Hence $\varphi'_{n,k}$ induces an isomorphism $\varphi_{n,k}\colon \mathfrak{o}/\mathfrak{p}^k \to \mathfrak{p}^n/\mathfrak{p}^{n+k}$ as asserted. If $k=1$, then $\mathfrak{p}^n/\mathfrak{p}^{n+1}$ is a vector space over $\mathsf{k}_\mathfrak{p}$ by means of $(\lambda+\mathfrak{p})(a+\mathfrak{p}^{n+1}) = \lambda a + \mathfrak{p}^{n+1}$ for all $\lambda \in \mathfrak{o}$ and $a \in \mathfrak{p}^n$, and $\varphi_{n,1}$ is $\mathsf{k}_\mathfrak{p}$-linear.

2. If $n \in \mathbb{N}$, then obviously $U^{(n+1)} \subset U^{(n)} \subset U$. If $a, b \in U^{(n)}$, say $a = 1+\pi^n u$ and $b = 1+\pi^n v$, where $u, v \in \mathfrak{o}$, then

$$a^{-1}b = \frac{1+\pi^n v}{1+\pi^n u} = 1 + \frac{\pi^n(v-u)}{1+\pi^n u} \in 1+\pi^n\mathfrak{o} = U^{(n)}.$$

Hence $U^{(n)}$ is a subgroup of U, and

$$\bigcap_{n\in\mathbb{N}} U^{(n)} = \bigcap_{n\in\mathbb{N}} (1+\mathfrak{p}^n) = 1 + \bigcap_{n\in\mathbb{N}} \mathfrak{p}^n = \mathbf{1}.$$

3. Since $U = \mathfrak{o} \setminus \mathfrak{p}$, the residue class homomorphism $\rho\colon \mathfrak{o} \to \mathsf{k}_\mathfrak{p}$ induces an epimorphism $\rho \restriction U\colon U \to \mathsf{k}_\mathfrak{p}^\times$ with kernel $1 + \mathfrak{p} = U^{(1)}$ and hence an isomorphism $\psi\colon U/U^{(1)} \to \mathsf{k}_\mathfrak{p}^\times$ as asserted.

4. For $n \in \mathbb{N}$ we define $\psi_n'\colon U^{(n)} \to \mathsf{k}_\mathfrak{p}$ by $\psi_n'(1 + \pi^n a) = a + \mathfrak{p}$ for all $u \in \mathfrak{o}$. If $a,\, b \in \mathfrak{o}$, then

$$\begin{aligned}\psi_n'((1+\pi^n a)(1+\pi^n b)) &= \psi_n'(1 + \pi^n(a+b+\pi^n ab)) = (a+b+\pi^n ab) + \mathfrak{p} \\ &= (a+b) + \mathfrak{p} = \psi_n'(1+\pi^n a) + \psi_n'(1+\pi^n b).\end{aligned}$$

Hence ψ_n' is an epimorphism, and as

$$\mathrm{Ker}(\psi_n') = \{1 + \pi^n a \mid a \in \mathfrak{p}\} = U^{(n+1)},$$

it induces an isomorphism $\psi_n\colon U^{(n)}/U^{(n+1)} \to \mathsf{k}_\mathfrak{p}$.

If $a \in \mathfrak{o}$, then $a + \mathfrak{p}^n \in (\mathfrak{o}/\mathfrak{p}^n)^\times$ if and only if $a \in \mathfrak{o} \setminus \mathfrak{p} = U$. Therefore the map $\phi_n'\colon U \to (\mathfrak{o}/\mathfrak{p}^n)^\times$, defined by $\phi_n'(u) = u + \mathfrak{p}^n$ for all $u \in U$, is an epimorphism, and $\mathrm{Ker}(\phi_n') = \{u \in U \mid u + \mathfrak{p}^n = 1 + \mathfrak{p}^n\} = 1 + \mathfrak{p}^n = U^{(n)}$. Hence ϕ_n' induces an isomorphism $\phi_n\colon U/U^{(n)} \overset{\sim}{\to} (\mathfrak{p}/\mathfrak{p}^n)^\times$ as asserted. □

The following theorem is essentially a result on discrete valuation domains since it deals with a single prime. However, to be more flexible in later applications, we formulate it for Dedekind domains.

Theorem 2.10.6 (π-adic digit expansion). *Let $\mathfrak{o}$ be a Dedekind domain and $a \in \mathfrak{o}$. Let $\mathfrak{p} \in \mathcal{P}(\mathfrak{o})$, $\mathcal{R} \subset \mathfrak{o}$ a set of representatives for $\mathsf{k}_\mathfrak{p}$ such that $0 \in \mathcal{R}$ and $(\pi_n)_{n\ge 0}$ be a sequence in $\mathfrak{o}$ satisfying $v_\mathfrak{p}(\pi_n) = n$ for all $n \ge 0$ (main example: $\pi_n = \pi^n$ for some $\pi \in \mathfrak{p} \setminus \mathfrak{p}^2$).*

1. *For every $k \in \mathbb{N}$ there exists a unique sequence $(a_n)_{n\in[0,k-1]}$ in $\mathcal{R}$ such that*
$$a \equiv \sum_{n=0}^{k-1} a_n \pi_n \mod \mathfrak{p}^k.$$

2. *There exists a unique sequence $(a_n)_{n\ge 0}$ in $\mathcal{R}$ such that*
$$a \equiv \sum_{n=0}^{k-1} a_n \pi_n \mod \mathfrak{p}^k \quad \textit{for all } \ k \ge 0.$$

Proof. 1. By induction on k. For $k = 0$ there is nothing to do.

$k \ge 0,\ k \to k+1$: *Existence*: Let $(a_n)_{n\in[0,k-1]}$ be a sequence in $\mathcal{R}$ such that

$$a \equiv \sum_{n=0}^{k-1} a_n \pi_n \mod \mathfrak{p}^k, \quad \text{and thus } \ c_k = \pi_k^{-1}\Big(a - \sum_{n=0}^{k-1} a_n \pi_n\Big) \in \mathfrak{o}.$$

Let $a_k \in \mathcal{R}$ be such that $c_k \equiv a_k \mod \mathfrak{p}$. Then $c_k\pi_k \equiv a_k\pi_k \mod \mathfrak{p}^{k+1}$ and therefore

$$a = \sum_{n=0}^{k-1} a_n\pi_n + c_k\pi_k \equiv \sum_{n=0}^{k} a_n\pi_n \mod \mathfrak{p}^{k+1}.$$

Uniqueness: If $(a_n)_{n\in[0,k]}$ and $(a'_n)_{n\in[0,k]}$ are sequences in $\mathcal{R}$ such that

$$\sum_{n=0}^{k} a_n\pi_n \equiv \sum_{n=0}^{k} a'_n\pi_n \mod \mathfrak{p}^{k+1}, \quad \text{then} \quad \sum_{n=0}^{k-1} a_n\pi_n \equiv \sum_{n=0}^{k-1} a'_n\pi_n \mod \mathfrak{p}^{k},$$

and thus $a_n = a'_n$ for all $n \in [0, k-1]$ by the induction hypothesis. Hence $a_k\pi_k \equiv a'_k\pi_k \mod \mathfrak{p}^{k+1}$ and therefore $a_k \equiv a'_k \mod \mathfrak{p}$, which implies $a_k = a'_k$, since $a_k,\, a'_k \in \mathcal{R}$.

2. For every $k \in \mathbb{N}$, there exists by 1. a unique sequence $(a_n^{(k)})_{n\in[0,k-1]}$ such that

$$a \equiv \sum_{n=0}^{k-1} a_n^{(k)}\pi_n \mod \mathfrak{p}^k.$$

However, for every $k \in \mathbb{N}$

$$a \equiv \sum_{n=0}^{k} a_n^{(k+1)}\pi_n \mod \mathfrak{p}^{k+1} \quad \text{implies} \quad a \equiv \sum_{n=0}^{k-1} a_n^{(k+1)}\pi_n \mod \mathfrak{p}^k,$$

and therefore $a_n^{(k)} = a_n^{(k+1)}$ for all $n \in [0, k-1]$ by the uniqueness in 1. Hence the sequence $(a_n)_{n\geq 0} = (a_n^{(n+1)})_{n\geq 0}$ is the unique sequence in $\mathcal{R}$ such that 2. holds. □

Theorem 2.10.7. *Let $\mathfrak{o}$ be a Dedekind domain and $K = \mathsf{q}(\mathfrak{o})$.*

1. Let $v\colon K \to \mathbb{Z} \cup \{\infty\}$ be a map such that $v(a) \geq 0$ for all $a \in \mathfrak{o}$. Then v is a discrete valuation if and only if $v = v_\mathfrak{p}$ for some $\mathfrak{p} \in \mathcal{P}(\mathfrak{o})$, and then $\mathcal{O}_v = \mathfrak{o}_\mathfrak{p}$ and $\mathfrak{p}_v = \mathfrak{p}\mathfrak{o}_\mathfrak{p}$.

2. Let $\mathfrak{p} \in \mathcal{P}(\mathfrak{o})$ and $\pi \in \mathfrak{p} \setminus \mathfrak{p}^2$. Then $\mathfrak{o}_\mathfrak{p}$ is the valuation domain and $\mathfrak{p}\mathfrak{o}_\mathfrak{p}$ is the valuation ideal of $v_\mathfrak{p}$. If $n \in \mathbb{Z}$ and $k \in \mathbb{N}$, then $(\mathfrak{p}\mathfrak{o}_\mathfrak{p})^n = \mathfrak{p}^n\mathfrak{o}_\mathfrak{p} = \pi^n\mathfrak{o}_\mathfrak{p}$, $\mathfrak{p}^n/\mathfrak{p}^{n+k} = \mathfrak{p}^n\mathfrak{o}_\mathfrak{p}/\mathfrak{p}^{n+k}\mathfrak{o}_\mathfrak{p}$, $\mathsf{k}_\mathfrak{p} = \mathsf{k}_{\mathfrak{p}\mathfrak{o}_\mathfrak{p}}$, the map

$$\psi_{n,k}\colon \mathfrak{o}/\mathfrak{p}^k \to \mathfrak{p}^n/\mathfrak{p}^{n+k}, \quad \text{defined by} \quad \psi_n(a+\mathfrak{p}) = \pi^n a + \mathfrak{p}^{n+1},$$

is a group isomorphism, and for $k = 1$ it is $\mathsf{k}_\mathfrak{p}$-linear.

Proof. 1. By 2.8.5.1, $v_\mathfrak{p}$ is a discrete valuation of K. Thus let $v\colon K \to \mathbb{Z}\cup\{\infty\}$ be a discrete valuation such that $\mathfrak{o} \subset \mathcal{O}_v$. Then $\mathfrak{p} = \mathfrak{p}_v \cap \mathfrak{o}$ is a prime ideal of $\mathfrak{o}$, and we assert that $\mathfrak{p} \neq \mathbf{0}$. Indeed, if $\mathfrak{p} = \mathbf{0}$, then $v(a) = 0$ for all $a \in \mathfrak{o}^\bullet$ and thus $v \restriction K^\times = 0$, a contradiction. Hence $\mathfrak{p} \in \mathcal{P}(\mathfrak{o})$, $\mathfrak{o}_\mathfrak{p}$ is a discrete valuation domain and $\mathfrak{p}\mathfrak{o}_\mathfrak{p}$ is its maximal ideal. If $x \in \mathfrak{o}_\mathfrak{p}$, then $x = s^{-1}a$ for some $s \in \mathfrak{o}\setminus\mathfrak{p}$ and $a \in \mathfrak{o}$, and therefore $v(x) = v(a) \ge 0$. Hence $\mathfrak{o}_\mathfrak{p} \subset \mathcal{O}_v$, and by 2.10.4.3 we obtain $\mathfrak{o}_\mathfrak{p} = \mathcal{O}_v$, $\mathfrak{p}\mathfrak{o}_\mathfrak{p} = \mathfrak{p}_v$ and $v = v_{\mathfrak{p}_v} = v_\mathfrak{p}$.

2. Using 1. and 2.8.7.2(b), it follows that

$$\mathcal{O}_v = \mathfrak{o}_\mathfrak{p},\ \mathfrak{p}_v = \mathfrak{p}\mathfrak{o}_\mathfrak{p} \ \text{ and } \ (\mathfrak{p}\mathfrak{o}_\mathfrak{p})^n = \mathfrak{p}^n\mathfrak{o}_\mathfrak{p} = \pi^n\mathfrak{o}_\mathfrak{p} \ \text{ for all } \ n \in \mathbb{Z}.$$

In addition, 2.7.6.3 yields $\mathfrak{p}^n/\mathfrak{p}^{n+k} = \mathfrak{p}^n\mathfrak{o}_\mathfrak{p}/\mathfrak{p}^{n+k}\mathfrak{o}_\mathfrak{p}$ for all $k \in \mathbb{N}$, and in particular (for $n = 0$ and $k = 1$) $\mathsf{k}_\mathfrak{p} = \mathsf{k}_{\mathfrak{p}\mathfrak{o}_\mathfrak{p}}$. Regarding this, 2.10.5.1 implies the existence of the isomorphisms $\psi_{n,k}$. □

As an immediate application of 2.10.7 we determine all discrete valuations of $\mathbb{Q}$ and all discrete valuations of a rational function field which are trivial on its field of constants.

Remarks and Definitions 2.10.8.

1. Recall that $\mathcal{P}(\mathbb{Z}) = \{p\mathbb{Z} \mid p \in \mathbb{P}\}$. For a prime p, we call

$$v_p = v_{p\mathbb{Z}} \colon \mathbb{Q} \to \mathbb{Z} \cup \{\infty\}$$

the **p-adic valuation** of $\mathbb{Q}$. If $x = p^n u^{-1} v \in \mathbb{Q}^\times$, where $n,\ u,\ v \in \mathbb{Z}$ and $p \nmid uv$, then $v_p(x) = n$. By 2.10.7.2,

$$\mathcal{O}_{v_p} = \mathbb{Z}_{(p)} = \mathbb{Z}_{p\mathbb{Z}} = \{s^{-1}a \mid s \in \mathbb{Z}\setminus p\mathbb{Z},\ a \in \mathbb{Z}\} \text{ is the valuation domain,}$$

$\mathfrak{p}_{v_p} = p\mathbb{Z}_{(p)}$ is the valuation ideal, and $\mathbb{Z}_{(p)}/p\mathbb{Z}_{(p)} = \mathbb{Z}/p\mathbb{Z} = \mathbb{F}_p$ is the residue class field of v_p. Recall from 2.7.4 that $\mathbb{Z}_{(p)}$ is the domain of p-integral rational numbers.

2. Let F be a field and $F(X)$ the rational function field in X over F. By definition, $F(X) = \mathsf{q}(F[X])$ is the quotient field of the principal domain $F[X]$, and the set $\mathsf{P}(F,X)$ of all monic irreducible polynomials of $F[X]$ is a set of representatives for the prime elements of $F[X]$.

If $p \in \mathsf{P}(F,X)$, then we call $v_p = v_{pF[X]}$ the **p-adic valuation** of $F(X)$. Let $\overline{F}$ be an algebraic closure of F and $\xi_p \in \overline{F}$ such that $p(\xi_p) = 0$. Then it follows that $pF[X] = \{s \in F[X] \mid s(\xi_p) = 0\}$,

$$\begin{aligned} F[X]_{(p)} = F[X]_{pF[X]} &= \{s^{-1}a \mid a, s \in F[X],\ s \notin pF[X]\} \\ &= \{s^{-1}a \mid a, s \in \mathbb{F}[X],\ s(\xi_p) \ne 0\} \text{ is the valuation domain,} \end{aligned}$$

$pF[X]_{(p)}$ is the valuation ideal, and $F[X]_{(p)}/pF[X]_{(p)} = F[X]/pF[X] \cong F(\xi)$ is the residue class field of v_p.

If $T = X^{-1}$, then $F(X) = \mathsf{q}(F[T])$, and there is a discrete valuation v_T of $F(X)$ (the T-adic valuation coming from the prime element T of $F[T]$). We denote it by v_∞ and call it the **degree valuation** with respect to X. This name is justified by the subsequent Theorem 2.10.9.3.

Theorem 2.10.9 (Discrete valuations of rational number and function fields).

1. $\{v_p \mid p \in \mathbb{P}\}$ *is the set of all discrete valuations of* $\mathbb{Q}$.

2. Let $F(X)$ be a rational function field over a field F and $\mathsf{P}(F,X)$ the set of all monic irreducible polynomials in $F[X]$. Then

$$\{v_p \mid p \in \mathsf{P}(F,X) \cup \{v_\infty\}$$

is the set of all discrete valuations $v\colon F(X) \to \mathbb{Z} \cup \{\infty\}$ satisfying $v \restriction F^\times = 0$.

3. If $h = g^{-1}f \in F(X)$, where $f,\, g \in F[X]$ and $g \ne 0$, then $v_\infty(h) = \partial(g) - \partial(f)$.

$$\mathcal{O}_{v_\infty} = \{g^{-1}f \in K(X) \mid f,\, g \in F[X],\ \partial(g) \ge \partial(f)\}$$

is the valuation domain, $\mathfrak{p}_{v_\infty} = X^{-1}\mathcal{O}_{v_\infty}$ is the valuation ideal, and $\mathsf{k}_{v_\infty} = \mathcal{O}_{v_\infty}/X^{-1}\mathcal{O}_{v_\infty} = F[X^{-1}]/X^{-1}F[X^{-1}] = F$ is the residue class field of v_∞.

Proof. 1. Let $v\colon \mathbb{Q} \to \mathbb{Z} \cup \{\infty\}$ be a discrete valuation. Then $\mathbb{Z} \subset \mathcal{O}_v$, and therefore $v = v_p$ for some prime p by 2.10.7.

2. Let $v\colon F(X) \to \mathbb{Z} \cup \{\infty\}$ be a discrete valuation such that $v \restriction F^\times = 0$.

CASE 1: $v(X) \ge 0$. Then $F[X] \subset \mathcal{O}_v$ and therefore $v = v_p$ for some $p \in \mathsf{P}(F,X)$ by 2.10.7.

CASE 2: $v(X) < 0$. If $T = X^{-1}$, then $v(T) > 0$, and as in CASE 1 there exists some $q \in \mathsf{P}(F,T)$ such that $v = v_q$. Since $v(T) > 0$ it follows that $q \mid T$, hence $q = T$, $v = v_\infty$ and $v_\infty(X) = -v_\infty(T) = -1$.

3. Let

$$f = a_nX^n + a_{n-1}X^{n-1} + \ldots + a_0 \in F[X],$$

where $n \in \mathbb{N}$, $a_0, \ldots, a_n \in F$ and $a_n \ne 0$. Since

$$v_\infty(a_nX^n) = v_\infty(a_n) + nv_\infty(X) = -n,\ v(a_i) \in \{0, \infty\},$$

and $v_\infty(a_iX^i) = v_\infty(a_i) - i \ge -i > -n = v_\infty(a_nX^n)$ for all $i \in [0, n-1]$, it follows that $v_\infty(f) = -n = -\partial(f)$. If $h = g^{-1}f \in F(X)^\bullet$ for some f, $g \in F[X]$, then $v_\infty(h) = v_\infty(f) - v_\infty(g) = \partial(g) - \partial(f)$. The remaining assertions follow immediately from the definitions. □

A further important class of discrete valuation domains are domains of formal power series.

Theorem and Definition 2.10.10. *Let R be a commutative ring and X an indeterminate over R.*

By a **formal Laurent series** in X over R we mean a (formal) infinite sum

$$f = \sum_{n \in \mathbb{Z}} a_nX^n, \quad \text{where } a_n \in R, \text{ and } a_{-n} = 0 \text{ for all } n \gg 1,$$

and we call

$$\mathsf{o}(f) = \mathsf{o}_X(f) = \inf\{n \in \mathbb{Z} \mid a_n \ne 0\} \in \mathbb{Z} \cup \{\infty\}$$

the **order** of f. If $\mathsf{o}(f) \geq 0$, then we call f a **formal power series** in X over R; we write it in the form

$$f = \sum_{n=0}^{\infty} a_n X^n \quad \text{and set} \quad \varepsilon(f) = f(0) = a_0.$$

We denote by $R((X))$ the set of all formal Laurent series and by $R[\![X]\!]$ the set of all formal power series in X over R. Addition and multiplication of formal Laurent series are defined by

$$\sum_{n\in\mathbb{Z}} a_n X^n + \sum_{n\in\mathbb{Z}} b_n X^n = \sum_{n\in\mathbb{Z}} (a_n + b_n) X^n$$

and

$$\Bigl(\sum_{n\in\mathbb{Z}} a_n X^n\Bigr)\Bigl(\sum_{n\in\mathbb{Z}} b_n X^n\Bigr) = \sum_{n\in\mathbb{Z}}\Bigl(\sum_{i\in\mathbb{Z}} a_i b_{n-i}\Bigr) X^n.$$

In this way, $R((X))$ is a commutative ring, $R \subset R[\![X]\!] \subset R((X))$, and

$$R((X)) = \bigcup_{n\geq 1} X^{-n} R[\![X]\!].$$

The map $\varepsilon\colon R[\![X]\!] \to R$, $f \mapsto \varepsilon(f)$, is a ring epimorphism with kernel $XR[\![X]\!]$, and induces an isomorphism $R[\![X]\!]/XR[\![X]\!] \stackrel{\sim}{\to} R$. We identify: $R[\![X]\!]/XR[\![X]\!] = R$.

Apparently, the function $\mathsf{o}\colon R((X)) \to \mathbb{Z}\cup\{\infty\}$ has the following properties for all $f,\ g \in R((X))$:

- $\mathsf{o}(f) = \sup\{n \in \mathbb{Z} \mid f \in X^n R[\![X]\!]\}$, and $\mathsf{o}(f) = \infty$ if and only if $f = 0$;
- $\mathsf{o}(f+g) \geq \min\{\mathsf{o}(f), \mathsf{o}(g)\}$, and equality holds if $\mathsf{o}(f) \neq \mathsf{o}(g)$;
- $\mathsf{o}(fg) \geq \mathsf{o}(f) + \mathsf{o}(g)$, and equality holds if R is a domain. In particular, if R is a domain, then $R[\![X]\!]$ is a domain, too.

1. Let $f \in R[\![X]\!]$, say

$$f = \sum_{n=0}^{\infty} a_n X^n.$$

Then $f \in R[\![X]\!]^\times$ if and only if $a_0 \in R^\times$.

2. Let R be a field. Then $R((X))$ is a field, o is a discrete valuation, $\mathcal{O}_{\mathsf{o}} = R[\![X]\!]$ is its valuation domain, $\mathfrak{p}_{\mathsf{o}} = XR[\![X]\!]$ is its valuation ideal and $\mathsf{k}_{\mathsf{o}} = R[\![X]\!]/XR[\![X]\!] = R$ is its residue class field.
The discrete valuation $\mathsf{o} = \mathsf{o}_X\colon R((X)) \to \mathbb{Z} \cup \{\infty\}$ is called the **order valuation** of $R((X))$ with respect to X.

3. If R is a field, then R is relatively algebraically closed in $R((X))$.

Proof. 1. By definition, $f \in R[\![X]\!]^\times$ if and only if there exists a power series $g \in R[\![X]\!]$ such that

$$g = \sum_{n=0}^{\infty} b_n X^n \quad \text{and} \quad fg = \sum_{n=0}^{\infty} \Big(\sum_{i=0}^{n} a_i b_{n-i} \Big) X^n = 1.$$

This holds if and only if there exists a sequence $(b_n)_{n \ge 0}$ in R such that

$$a_0 b_0 = 1 \quad \text{and} \quad b_n = -a_0^{-1}(a_n b_0 + a_{n-1} b_1 + \ldots + a_1 b_{n-1}) \quad \text{for all} \quad n \ge 1.$$

However, this (infinite) system of linear equations for the sequence $(b_n)_{n \ge 0}$ is solvable if and only if $a_0 \in R^\times$.

2. Since $\mathsf{o}(fg) = \mathsf{o}(f) + \mathsf{o}(g)$ for all $f, g \in R((X))$ it follows that $R((X))$ is a domain. If $f \in R((X))^\bullet$, let $n \in \mathbb{Z}$ be maximal such that $X^{-n} f \in R[\![X]\!]$. Then $(X^{-n} f)(0) \ne 0$, hence $X^{-n} f \in R[\![X]\!]^\times \subset R((X))^\times$ by 1. Hence $R((X))$ is a field. The remaining assertions follow from the elementary properties of $\mathsf{o} \colon R((X)) \to \mathbb{Z} \cup \{\infty\}$ stated above.

3. Let R be a field and assume to the contrary that there is some Laurent series $f \in R((X)) \setminus R$ which is algebraic over R, say

$$\sum_{i=0}^{n} b_i f^i = 0 \quad \text{for some} \quad n \ge 1 \ \text{ and } \ b_0, \ldots, b_n \in R, \ b_0 \ne 0.$$

If $\mathsf{o}(f) = d \ge 1$, then $\mathsf{o}(b_i f^i) \ge di \ge 1$ for all $i \in [1, n]$ and therefore

$$0 = \mathsf{o}(-b_0) \ge \min\{\mathsf{o}(b_i f^i) \mid i \in [1, n]\} \ge 1,$$

a contradiction. If $\mathsf{o}(f) \le -1$, then f^{-1} is also algebraic over K, a contradiction since $\mathsf{o}(f^{-1}) \ge 1$. If $\mathsf{o}(f) = 0$, say

$$f = \sum_{n=0}^{\infty} a_n X^n,$$

then $f - a_0$ is algebraic over K, a contradiction since $\mathsf{o}(f - a_0) \ge 1$. □

We close this section with a result concerning free modules over discrete valuation domains which will be used later on.

Theorem 2.10.11. *Let $\mathfrak{o}$ be a discrete valuation domain, $\mathfrak{p} = \pi\mathfrak{o}$ its maximal ideal, $\mathsf{k}_\mathfrak{p} = \mathfrak{o}/\mathfrak{p}$, M a torsion-free $\mathfrak{o}$-module, $n \in \mathbb{N}$ and $u_1, \ldots, u_n \in M$.*

1. If $(u_1 + \pi M, \ldots, u_n + \pi M)$ is linearly independent over $\mathsf{k}_\mathfrak{p}$, then $(u_1, \ldots, u_n)$ is linearly independent over $\mathfrak{o}$.

2. Let $(u_1+\pi M,\ldots,u_n+\pi M)$ be a $\mathsf{k}_\mathfrak{p}$-basis of $M/\pi M$, and suppose that $\operatorname{rk}(M)=n$. *Then $(u_1,\ldots,u_n)$ is an $\mathfrak{o}$-basis of M.*

Proof. 1. Assume to the contrary that $(u_1+\pi M,\ldots,u_n+\pi M)$ is linearly independent over $\mathsf{k}_\mathfrak{p}$, and yet there exist $c_1,\ldots,c_n\in\mathfrak{o}$, not all equal to 0, such that $c_1u_1+\ldots+c_nu_n=0$. If $k=\min\{v_\mathfrak{p}(c_1),\ldots,v_\mathfrak{p}(c_n)\}\in\mathbb{N}_0$, then $\pi^{-k}c_j\notin\mathfrak{p}$ for some $j\in[1,n]$, and

$$\sum_{i=1}^n(\pi^{-k}c_i+\mathfrak{p})(u_i+\pi M)=0\in M/\pi M,\quad\text{a contradiction.}$$

2. Let $K=\mathsf{q}(\mathfrak{o})$ and $U=KM$. Then $(u_1,\ldots,u_n)$ is linearly independent over $\mathfrak{o}$ by 1., and we assert that $M={}_\mathfrak{o}(u_1,\ldots,u_n)$. Thus let $x\in M$. Since $\dim_K(U)=n$, it follows that $(u_1,\ldots,u_n)$ is a K-basis of U, and therefore

$$x=\sum_{i=1}^n c_iu_i,\quad\text{where } c_i\in K.$$

Let $m\in[1,n]$ such that $k=\min\{v_\pi(c_1,\ldots,v_\pi(c_n)\}=v_\pi(c_m)$. If $k\geq 0$, we are done. If $k<0$, then $\pi^{-k}c_i\in\mathfrak{o}$ for all $i\in[1,n]$, $\pi^{-k}c_m\notin\mathfrak{p}$ and

$$0=\pi^{-k}x+\pi M=\sum_{i=1}^n(\pi^{-k}c_i+\mathfrak{p})(u_i+\pi M).$$

This contradicts the linear independence of $(u_1+\pi M,\ldots,u_n+\pi M)$ over $\mathsf{k}_\mathfrak{p}$. □

2.11 Orders in Dedekind domains

Definition 2.11.1. Let $\mathfrak{o}$ be a domain, $K=\mathsf{q}(\mathfrak{o})$ and $\overline{\mathfrak{o}}=\operatorname{cl}_K(\mathfrak{o})$. $\mathfrak{o}$ is called an **order** (in $\overline{\mathfrak{o}}$) if $\mathfrak{o}$ is Noetherian, $\dim(\mathfrak{o})=1$ and $\overline{\mathfrak{o}}$ is a finitely generated $\mathfrak{o}$-module. Then $\overline{\mathfrak{o}}$ and $\mathfrak{f}=(\mathfrak{o}:\overline{\mathfrak{o}})=\{x\in K\mid x\overline{\mathfrak{o}}\subset\mathfrak{o}\}$ are fractional ideals of $\mathfrak{o}$ by 2.6.2.4(a). $\mathfrak{f}$ is called the **conductor** of $\mathfrak{o}$.

Let $\mathfrak{o}$ be an order. Then $\overline{\mathfrak{o}}$ is integrally closed, $\dim(\overline{\mathfrak{o}})=1$ by 2.4.8, and as $\mathfrak{o}$ is Noetherian and $\overline{\mathfrak{o}}$ is a finitely generated $\mathfrak{o}$-module, it follows that every ideal of $\overline{\mathfrak{o}}$ is a finitely generated $\mathfrak{o}$-module and thus a finitely generated ideal. Hence $\overline{\mathfrak{o}}$ is Noetherian and thus a Dedekind domain. In particular, a domain $\mathfrak{o}$ is a Dedekind domain if and only if it is an integrally closed order.

On the other hand, let $\overline{\mathfrak{o}}$ be a Dedekind domain and $\mathfrak{o}$ a subring of $\overline{\mathfrak{o}}$ such that $\mathsf{q}(\mathfrak{o}) = \mathsf{q}(\overline{\mathfrak{o}})$ and $\overline{\mathfrak{o}}$ is a finitely generated $\mathfrak{o}$-module. Then $\mathfrak{o}$ is Noetherian by the Eakin-Nagata theorem (see [46, Theorem 3.7]), and thus $\mathfrak{o}$ is an order in $\overline{\mathfrak{o}}$.

Theorem 2.11.2. *Let* $\mathfrak{o}$ *be an order,* $K = \mathsf{q}(\mathfrak{o})$, $\overline{\mathfrak{o}} = \mathrm{cl}_K(\mathfrak{o})$ *and* $\mathfrak{f} = (\mathfrak{o}:\overline{\mathfrak{o}})$.

1. *$\overline{\mathfrak{o}}$ is a Dedekind domain, and any domain $\mathfrak{o}'$ such that $\mathfrak{o} \subset \mathfrak{o}' \subset \overline{\mathfrak{o}}$ is an order.*

2. *$\mathfrak{f} \in \mathcal{J}'(\overline{\mathfrak{o}})$, and $\mathfrak{f}$ is the largest ideal of $\overline{\mathfrak{o}}$ which lies in $\mathfrak{o}$.*

3. *If $\mathfrak{p} \in \mathcal{P}(\mathfrak{o})$ and $\mathfrak{f} \not\subset \mathfrak{p}$, then $\mathfrak{f}_\mathfrak{p} = \mathfrak{o}_\mathfrak{p} = \overline{\mathfrak{o}}_\mathfrak{p}$ is a discrete valuation domain.*

Proof. 1. We have already seen that $\overline{\mathfrak{o}}$ is a Dedekind domain. Let $\mathfrak{o}'$ be a domain such that $\mathfrak{o} \subset \mathfrak{o}' \subset \overline{\mathfrak{o}}$. Then every ideal of $\mathfrak{o}'$ is an $\mathfrak{o}$-submodule of $\overline{\mathfrak{o}}$, hence finitely generated and thus a finitely generated ideal of $\mathfrak{o}'$. Hence $\mathfrak{o}'$ is Noetherian, and as $\mathfrak{o}'$ is integral over $\mathfrak{o}$, it follows that $\dim(\mathfrak{o}') = 1$. Since $\overline{\mathfrak{o}} = \mathrm{cl}_K(\mathfrak{o}')$ is a finitely generated $\mathfrak{o}$-module, it is all the more a finitely generated $\mathfrak{o}'$-module. Hence $\mathfrak{o}'$ is an order.

2. By definition, $(\mathfrak{f}\overline{\mathfrak{o}})\overline{\mathfrak{o}} = \mathfrak{f}\overline{\mathfrak{o}} \subset \mathfrak{o}$ implies $\mathfrak{f}\overline{\mathfrak{o}} \subset (\mathfrak{o}:\overline{\mathfrak{o}}) = \mathfrak{f}$. Therefore, and since $\mathfrak{f} \subset \mathfrak{f}\overline{\mathfrak{o}} \subset \mathfrak{o}$, it follows that $\mathfrak{f}$ is an ideal of $\overline{\mathfrak{o}}$ which lies in $\mathfrak{o}$. In particular, $\mathfrak{f} \neq \mathbf{0}$ implies $\mathfrak{f} \in \mathcal{J}'(\overline{\mathfrak{o}})$. If $\overline{\mathfrak{a}}$ is an ideal of $\overline{\mathfrak{o}}$ lying in $\mathfrak{o}$, then $\overline{\mathfrak{a}}\,\overline{\mathfrak{o}} = \overline{\mathfrak{a}} \subset \mathfrak{o}$, and consequently $\overline{\mathfrak{a}} \subset (\mathfrak{o}:\overline{\mathfrak{o}}) = \mathfrak{f}$.

3. If $\mathfrak{p} \in \mathcal{P}(\mathfrak{o})$ and $\mathfrak{f} \not\subset \mathfrak{p}$, then $\mathfrak{o}_\mathfrak{p} = \mathfrak{f}_\mathfrak{p} = (\mathfrak{f}\overline{\mathfrak{o}})_\mathfrak{p} = \mathfrak{f}_\mathfrak{p}\overline{\mathfrak{o}}_\mathfrak{p} = \overline{\mathfrak{o}}_\mathfrak{p}$ is a discrete valuation domain. □

Theorem and Definition 2.11.3. *Let* $\mathfrak{o}$ *be an order,* $K = \mathsf{q}(\mathfrak{o})$, $\overline{\mathfrak{o}} = \mathrm{cl}_K(\mathfrak{o})$ *and* $\mathfrak{f} = (\mathfrak{o}:\overline{\mathfrak{o}})$.

1. *Let $\mathfrak{a}$ be an ideal of $\mathfrak{o}$ which is coprime to $\mathfrak{f}$ (i. e., $\mathfrak{a} + \mathfrak{f} = \mathfrak{o}$). Then $\mathfrak{a}\overline{\mathfrak{o}} \in \mathcal{J}'^{\mathfrak{f}}(\overline{\mathfrak{o}})$, $\mathfrak{a}\overline{\mathfrak{o}} \cap \mathfrak{o} = \mathfrak{a}$, $\mathfrak{a}$ is $\mathfrak{o}$-invertible,*

$$\mathfrak{a}^{-1} = (\mathfrak{o}:\mathfrak{a}) = \mathfrak{f}(\overline{\mathfrak{o}}:\mathfrak{a}\overline{\mathfrak{o}}) + \mathfrak{o} = \mathfrak{f}(\mathfrak{a}\overline{\mathfrak{o}})^{-1} + \mathfrak{o} \text{ and } \mathfrak{o}/\mathfrak{a} \cong \overline{\mathfrak{o}}/\mathfrak{a}\overline{\mathfrak{o}}.$$

As in 2.8, we denote by $\mathcal{J}'^{\mathfrak{f}}(\mathfrak{o})$ the set of all ideals of $\mathfrak{o}$ which are coprime to $\mathfrak{f}$ (and thus $\mathfrak{o}$-invertible). Then $\mathcal{J}^{\mathfrak{f}}(\mathfrak{o}) = \mathsf{q}(\mathcal{J}'^{\mathfrak{f}}(\mathfrak{o})) \subset \mathcal{J}(\mathfrak{o})$ and $\mathcal{P}^{\mathfrak{f}}(\mathfrak{o}) = \mathcal{P}(\mathfrak{o}) \cap \mathcal{J}'^{\mathfrak{f}}(\mathfrak{o})$.

2. *If $\overline{\mathfrak{a}} \in \mathcal{J}'^{\mathfrak{f}}(\overline{\mathfrak{o}})$, then $\overline{\mathfrak{a}} \cap \mathfrak{o} + \mathfrak{f} = \mathfrak{o}$ and $\overline{\mathfrak{a}} = (\overline{\mathfrak{a}} \cap \mathfrak{o})\overline{\mathfrak{o}}$.*

3. *Let $c_1, c_2 \in \overline{\mathfrak{o}}$.*

(a) If $c_1c_2 \equiv 1 \bmod \mathfrak{f}$, then $c_1\mathfrak{o} + \mathfrak{f} \in \mathcal{J}(\mathfrak{o})$ and $(c_1\mathfrak{o} + \mathfrak{f})^{-1} = c_2\mathfrak{o} + \mathfrak{f}$.

(b) If $c_1 \equiv c_2 \bmod \mathfrak{f}$ and $c_2 \neq 0$, then $c_1\overline{\mathfrak{o}} \cap \mathfrak{o} = c_1c_2^{-1}(c_2\overline{\mathfrak{o}} \cap \mathfrak{o})$.

Proof. 1. By definition, $\mathfrak{a}\overline{\mathfrak{o}} \in \mathcal{J}'(\overline{\mathfrak{o}})$, and since $1 \in \mathfrak{o} = \mathfrak{a} + \mathfrak{f} \subset \mathfrak{a}\overline{\mathfrak{o}} + \mathfrak{f}$, it follows that $\mathfrak{a}\overline{\mathfrak{o}} + \mathfrak{f} = \overline{\mathfrak{o}}$ and $\mathfrak{a}\overline{\mathfrak{o}} \in \mathcal{J}'^{\mathfrak{f}}(\overline{\mathfrak{o}})$.

In order to prove $\mathfrak{a}\overline{\mathfrak{o}} \cap \mathfrak{o} = \mathfrak{a}$ it suffices to prove that $(\mathfrak{a}\overline{\mathfrak{o}} \cap \mathfrak{o})_\mathfrak{p} = \mathfrak{a}_\mathfrak{p}$ for all $\mathfrak{p} \in \mathcal{P}(\mathfrak{o})$. If $\mathfrak{p} \in \mathcal{P}(\mathfrak{o})$, then $(\mathfrak{a}\overline{\mathfrak{o}} \cap \mathfrak{o})_\mathfrak{p} = \mathfrak{a}_\mathfrak{p}\overline{\mathfrak{o}}_\mathfrak{p} \cap \mathfrak{o}_\mathfrak{p}$. If $\mathfrak{f} \not\subset \mathfrak{p}$, then $\mathfrak{o}_\mathfrak{p} = \overline{\mathfrak{o}}_\mathfrak{p}$ by 2.11.2.3, and $(\mathfrak{a}\overline{\mathfrak{o}} \cap \mathfrak{o})_\mathfrak{p} = \mathfrak{a}_\mathfrak{p}\mathfrak{o}_\mathfrak{p} \cap \mathfrak{o}_\mathfrak{p} = \mathfrak{a}_\mathfrak{p}$. If $\mathfrak{f} \subset \mathfrak{p}$, then $\mathfrak{f}_\mathfrak{p} \subset \mathfrak{p}\mathfrak{o}_\mathfrak{p}$, and since $\mathfrak{o}_\mathfrak{p} = (\mathfrak{a} + \mathfrak{f})_\mathfrak{p} = \mathfrak{a}_\mathfrak{p} + \mathfrak{f}_\mathfrak{p}$, we get $\mathfrak{a}_\mathfrak{p} = \mathfrak{o}_\mathfrak{p}$ and $(\mathfrak{a}\overline{\mathfrak{o}} \cap \mathfrak{o})_\mathfrak{p} = \mathfrak{a}_\mathfrak{p}\overline{\mathfrak{o}}_\mathfrak{p} \cap \mathfrak{o}_\mathfrak{p} = \mathfrak{a}_\mathfrak{p}$.

Since $\mathfrak{a}\,[\,\mathfrak{f}(\overline{\mathfrak{o}} : \mathfrak{a}\overline{\mathfrak{o}}) + \mathfrak{o}] = \mathfrak{a}\,\mathfrak{f}(\overline{\mathfrak{o}} : \mathfrak{a}\overline{\mathfrak{o}}) + \mathfrak{a} = \mathfrak{a}\overline{\mathfrak{o}}\,\mathfrak{f}(\overline{\mathfrak{o}} : \mathfrak{a}\overline{\mathfrak{o}}) + \mathfrak{a} = (\mathfrak{a}\overline{\mathfrak{o}})\mathfrak{f}(\mathfrak{a}\overline{\mathfrak{o}})^{-1} + \mathfrak{a} = \mathfrak{f} + \mathfrak{a} = \mathfrak{o}$, it follows that $\mathfrak{a}$ is $\mathfrak{o}$-invertible and $\mathfrak{a}^{-1} = (\mathfrak{o} : \mathfrak{a}) = \mathfrak{f}(\overline{\mathfrak{o}} : \mathfrak{a}\overline{\mathfrak{o}}) + \mathfrak{o}$.

Since $\mathfrak{a} + \mathfrak{f} = \mathfrak{o}$ and $\mathfrak{a}\overline{\mathfrak{o}} + \mathfrak{f} = \overline{\mathfrak{o}}$ we obtain (observing that $\mathfrak{a} \cap \mathfrak{f} = \mathfrak{a}\mathfrak{f}$ and $\mathfrak{a}\overline{\mathfrak{o}} \cap \mathfrak{f} = \mathfrak{a}\overline{\mathfrak{o}}\mathfrak{f}$ by 1.1.4)

$$\mathfrak{o}/\mathfrak{a} = \mathfrak{a} + \mathfrak{f}/\mathfrak{a} \cong \mathfrak{f}/\mathfrak{a} \cap \mathfrak{f} = \mathfrak{f}/\mathfrak{a}\mathfrak{f} = \mathfrak{f}/\mathfrak{a}\overline{\mathfrak{o}}\mathfrak{f} = \mathfrak{f}/\mathfrak{a}\overline{\mathfrak{o}} \cap \mathfrak{f} \cong \mathfrak{f} + \mathfrak{a}\overline{\mathfrak{o}}/\mathfrak{a}\overline{\mathfrak{o}} = \overline{\mathfrak{o}}/\mathfrak{a}\overline{\mathfrak{o}}.$$

2. Let $\overline{\mathfrak{a}} \in \mathcal{J}'^{\mathfrak{f}}(\overline{\mathfrak{o}})$. As $\mathfrak{f} \subset \mathfrak{o}$, the modular law implies $\overline{\mathfrak{a}} \cap \mathfrak{o} + \mathfrak{f} = (\overline{\mathfrak{a}} + \mathfrak{f}) \cap \mathfrak{o} = \mathfrak{o}$. In order to prove $\overline{\mathfrak{a}} = (\overline{\mathfrak{a}} \cap \mathfrak{o})\overline{\mathfrak{o}}$, it suffices again to prove that $\overline{\mathfrak{a}}_\mathfrak{p} = [(\overline{\mathfrak{a}} \cap \mathfrak{o})\overline{\mathfrak{o}}]_\mathfrak{p}$ for all $\mathfrak{p} \in \mathcal{P}(\mathfrak{o})$. If $\mathfrak{p} \in \mathcal{P}(\mathfrak{o})$, then $[(\overline{\mathfrak{a}} \cap \mathfrak{o})\overline{\mathfrak{o}}]_\mathfrak{p} = (\overline{\mathfrak{a}}_\mathfrak{p} \cap \mathfrak{o}_\mathfrak{p})\overline{\mathfrak{o}}_\mathfrak{p}$. If $\mathfrak{f} \not\subset \mathfrak{p}$, then $\mathfrak{o}_\mathfrak{p} = \overline{\mathfrak{o}}_\mathfrak{p}$ and $(\overline{\mathfrak{a}}_\mathfrak{p} \cap \mathfrak{o}_\mathfrak{p})\overline{\mathfrak{o}}_\mathfrak{p} = \overline{\mathfrak{a}}_\mathfrak{p}$. If $\mathfrak{f} \subset \mathfrak{p}$, then $\mathfrak{f}_\mathfrak{p} \subset \mathfrak{p}\mathfrak{o}_\mathfrak{p}$, and $\mathfrak{o}_\mathfrak{p} = (\overline{\mathfrak{a}} \cap \mathfrak{o} + \mathfrak{f})_\mathfrak{p} = \overline{\mathfrak{a}}_\mathfrak{p} \cap \mathfrak{o}_\mathfrak{p} + \mathfrak{f}_\mathfrak{p}$ implies $\overline{\mathfrak{a}}_\mathfrak{p} \cap \mathfrak{o}_\mathfrak{p} = \mathfrak{o}_\mathfrak{p}$, hence $1 \in \overline{\mathfrak{a}}_\mathfrak{p}$ and $(\overline{\mathfrak{a}}_\mathfrak{p} \cap \mathfrak{o}_\mathfrak{p})\overline{\mathfrak{o}}_\mathfrak{p} = \overline{\mathfrak{o}}_\mathfrak{p} = \overline{\mathfrak{a}}_\mathfrak{p}$.

3. (a) If $c_1c_2 = 1 + b$ for some $b \in \mathfrak{f}$, then $1 = c_1c_2 - b \in c_1c_2\mathfrak{o} + \mathfrak{f}$, hence $c_1c_2\mathfrak{o} + \mathfrak{f} = \mathfrak{o}$, and therefore $c_1c_2\mathfrak{o} + \mathfrak{f}^2 = \mathfrak{o}$ by 1.1.3. Consequently, $(c_1\mathfrak{o} + \mathfrak{f})(c_2\mathfrak{o} + \mathfrak{f}) \supset c_1c_2\mathfrak{o} + \mathfrak{f}^2 = \mathfrak{o}$, hence equality holds, which shows that $c_1\mathfrak{o} + \mathfrak{f} \in \mathcal{J}(\mathfrak{o})$, and $(c_1\mathfrak{o} + \mathfrak{f})^{-1} = c_2\mathfrak{o} + \mathfrak{f}$.

(b) Let $c_1 = c_2 + b$, where $b \in \mathfrak{f}$. If $x = c_1u \in c_1\overline{\mathfrak{o}} \cap \mathfrak{o}$ for some $u \in \overline{\mathfrak{o}}$, then $x = c_1c_2^{-1}(c_2u)$, and $c_2u = c_1u - bu \in c_2\overline{\mathfrak{o}} \cap \mathfrak{o}$. Conversely, if $x = c_1c_2^{-1}(c_2u)$, where $u \in \overline{\mathfrak{o}}$ and $c_2u \in \mathfrak{o}$, then $x = c_1u = c_2u + bu \in c_1\overline{\mathfrak{o}} \cap \mathfrak{o}$. □

Theorem 2.11.4. *Let $\mathfrak{o}$ be an order, $K = \mathsf{q}(\mathfrak{o})$, $\overline{\mathfrak{o}} = \mathrm{cl}_K(\mathfrak{o})$, $\mathfrak{f} = (\mathfrak{o} : \overline{\mathfrak{o}})$, and let $j_{\overline{\mathfrak{o}}/\mathfrak{o}} \colon \mathcal{J}(\mathfrak{o}) \to \mathcal{J}(\overline{\mathfrak{o}})$ be defined by $j_{\overline{\mathfrak{o}}/\mathfrak{o}}(\mathfrak{a}) = \mathfrak{a}\overline{\mathfrak{o}}$.*

1. *$j_{\overline{\mathfrak{o}}/\mathfrak{o}}$ is a group homomorphism, and*

$$\mathrm{Ker}(j_{\overline{\mathfrak{o}}/\mathfrak{o}}) = \{c\mathfrak{o} + \mathfrak{f} \mid c \in \overline{\mathfrak{o}},\ c\overline{\mathfrak{o}} + \mathfrak{f} = \overline{\mathfrak{o}}\}.$$

2. *$j_{\overline{\mathfrak{o}}/\mathfrak{o}} \restriction \mathcal{J}^{\mathfrak{f}}(\mathfrak{o}) \colon \mathcal{J}^{\mathfrak{f}}(\mathfrak{o}) \to \mathcal{J}^{\mathfrak{f}}(\overline{\mathfrak{o}})$ is an isomorphism, $j_{\overline{\mathfrak{o}}/\mathfrak{o}}(\mathcal{P}^{\mathfrak{f}}(\mathfrak{o})) = \mathcal{P}^{\mathfrak{f}}(\overline{\mathfrak{o}})$, and $\mathcal{J}^{\mathfrak{f}}(\mathfrak{o})$ is a free abelian group with basis $\mathcal{P}^{\mathfrak{f}}(\mathfrak{o})$.*

Proof. 1. By 2.6.7.2, $j_{\overline{\mathfrak{o}}/\mathfrak{o}}$ is a group homomorphism. If $c \in \overline{\mathfrak{o}}$ and $c\overline{\mathfrak{o}} + \mathfrak{f} = \overline{\mathfrak{o}}$, then there exists some $c' \in \overline{\mathfrak{o}}$ such that $cc' \equiv 1 \bmod \mathfrak{f}$, hence $c\mathfrak{o} + \mathfrak{f} \in \mathcal{J}(\mathfrak{o})$ by 2.11.3.3(a). Since $(c\mathfrak{o} + \mathfrak{f})\overline{\mathfrak{o}} = c\overline{\mathfrak{o}} + \mathfrak{f} = \overline{\mathfrak{o}}$, we get $c\mathfrak{o} + \mathfrak{f} \in \mathrm{Ker}(j_{\overline{\mathfrak{o}}/\mathfrak{o}})$.

To prove the reverse inclusion, let $\mathfrak{a} \in \mathcal{J}(\mathfrak{o})$ such that $j_{\overline{\mathfrak{o}}/\mathfrak{o}}(\mathfrak{a}) = \mathfrak{a}\overline{\mathfrak{o}} = \overline{\mathfrak{o}}$. Since $\mathfrak{f}\mathfrak{a}^{-1} \subset \mathfrak{f}\mathfrak{a}^{-1}\overline{\mathfrak{o}} = \mathfrak{f}\mathfrak{a}^{-1}\mathfrak{a}\overline{\mathfrak{o}} = \mathfrak{f}\overline{\mathfrak{o}} \subset \mathfrak{o}$, it follows that $\mathfrak{f} \subset (\mathfrak{a}^{-1})^{-1} = \mathfrak{a}$. Suppose that $\mathfrak{f} = \overline{\mathfrak{p}}_1^{e_1} \cdot \ldots \cdot \overline{\mathfrak{p}}_r^{e_r}$, where $r \in \mathbb{N}_0$, $\overline{\mathfrak{p}}_1, \ldots, \overline{\mathfrak{p}}_r \in \mathcal{P}(\overline{\mathfrak{o}})$ are distinct and $e_1, \ldots, e_r \in \mathbb{N}$. We assert that $\mathfrak{a} \not\subset \overline{\mathfrak{p}}_1 \cup \ldots \cup \overline{\mathfrak{p}}_r$. Indeed, otherwise we get

$\mathfrak{a} \subset \overline{\mathfrak{p}}_i$ for some $i \in [1,r]$ by 1.1.6 and thus $\overline{\mathfrak{o}} = \mathfrak{a}\overline{\mathfrak{o}} \subset \overline{\mathfrak{p}}_i$, a contradiction. If $c \in \mathfrak{a}\setminus(\overline{\mathfrak{p}}_1\cup\ldots\cup\overline{\mathfrak{p}}_r)$, then $c\overline{\mathfrak{o}}+\overline{\mathfrak{p}}_i = \overline{\mathfrak{o}}$ for all $i \in [1,r]$, hence $(c\mathfrak{o}+\mathfrak{f})\overline{\mathfrak{o}}) = c\overline{\mathfrak{o}}+\mathfrak{f} = \overline{\mathfrak{o}}$ by 1.1.3, and therefore $\mathfrak{a}_0 = c\mathfrak{o}+\mathfrak{f} \in \mathrm{Ker}(j_{\overline{\mathfrak{o}}/\mathfrak{o}})$. Hence $\mathfrak{a}_0\mathfrak{a}^{-1} \in \mathrm{Ker}(j_{\overline{\mathfrak{o}}/\mathfrak{o}})$ and $\mathfrak{a}_0\mathfrak{a}^{-1} \subset \mathfrak{o}$, since $\mathfrak{a}_0 \subset \mathfrak{a}$. By 2.4.7.4 we obtain $\mathfrak{a}_0\mathfrak{a}^{-1} = \mathfrak{o}$ and $\mathfrak{a} = \mathfrak{a}_0 = c\mathfrak{o}+\mathfrak{f}$.

2. If $\mathfrak{a} \in \mathcal{J}'^{\mathfrak{f}}(\mathfrak{o})$, then $\mathfrak{a}\overline{\mathfrak{o}} \in \mathcal{J}'^{\mathfrak{f}}(\overline{\mathfrak{o}})$ and $\mathfrak{a} = \mathfrak{a}\overline{\mathfrak{o}} \cap \mathfrak{o}$ by 2.11.3.1. Conversely, if $\overline{\mathfrak{a}} \in \mathcal{J}'^{\mathfrak{f}}(\overline{\mathfrak{o}})$, then $\overline{\mathfrak{a}} \cap \mathfrak{o} \in \mathcal{J}'^{\mathfrak{f}}(\mathfrak{o})$ and $\overline{\mathfrak{a}} = (\overline{\mathfrak{a}} \cap \mathfrak{o})\overline{\mathfrak{o}}$ by 2.11.3.2. Hence the restriction $j_{\overline{\mathfrak{o}}/\mathfrak{o}} \restriction \mathcal{J}'^{\mathfrak{f}}(\mathfrak{o})\colon \mathcal{J}'^{\mathfrak{f}}(\mathfrak{o}) \to \mathcal{J}'^{\mathfrak{f}}(\overline{\mathfrak{o}})$ is an isomorphism, and therefore also $j_{\overline{\mathfrak{o}}/\mathfrak{o}} \restriction \mathcal{J}^{\mathfrak{f}}(\mathfrak{o}) = \mathsf{q}(j_{\overline{\mathfrak{o}}/\mathfrak{o}} \restriction \mathcal{J}'^{\mathfrak{f}}(\mathfrak{o}))$ is an isomorphism.

If $\overline{\mathfrak{p}} \in \mathcal{P}^{\mathfrak{f}}(\overline{\mathfrak{o}})$, then $\overline{\mathfrak{p}} \cap \mathfrak{o} \in \mathcal{P}^{\mathfrak{f}}(\mathfrak{o})$, and $\overline{\mathfrak{p}} = (\overline{\mathfrak{p}} \cap \mathfrak{o})\overline{\mathfrak{o}}$ by 2.11.3.2. Conversely, let $\mathfrak{p} \in \mathcal{P}^{\mathfrak{f}}(\mathfrak{o})$. By 2.4.7.3 there exists a prime ideal $\overline{\mathfrak{p}}$ of $\overline{\mathfrak{o}}$ such that $\mathfrak{p}\overline{\mathfrak{o}} \subset \overline{\mathfrak{p}}$. Since $\mathfrak{o} \supsetneq \overline{\mathfrak{p}} \cap \mathfrak{o} \supset \mathfrak{p}\overline{\mathfrak{o}} \cap \mathfrak{o} = \mathfrak{p}$, it follows that $\overline{\mathfrak{p}} \cap \mathfrak{o} = \mathfrak{p} \in \mathcal{P}^{\mathfrak{f}}(\mathfrak{o})$ and thus $\mathfrak{p}\overline{\mathfrak{o}} = (\overline{\mathfrak{p}} \cap \mathfrak{o})\overline{\mathfrak{o}} \in \mathcal{P}^{\mathfrak{f}}(\overline{\mathfrak{o}})$. □

Theorem 2.11.5. *Let $\mathfrak{o}$ be an order, $K = \mathsf{q}(\mathfrak{o})$, $\overline{\mathfrak{o}} = \mathrm{cl}_K(\mathfrak{o})$, $\mathfrak{f} = (\mathfrak{o}:\overline{\mathfrak{o}})$, and let $\boldsymbol{\sigma}$ be a (maybe empty) finite set of real embeddings of K.*

1. *Every $\boldsymbol{\sigma}$-ideal class $C \in \mathcal{C}^{\boldsymbol{\sigma}}(\mathfrak{o})$ contains an ideal $\mathfrak{c} \in \mathcal{J}'^{\mathfrak{f}}(\mathfrak{o})$.*
2. *There is an exact sequence*

$$\mathbf{1} \to \overline{\mathfrak{o}}^\times \cap K^{\boldsymbol{\sigma}}/\mathfrak{o}^\times \cap K^{\boldsymbol{\sigma}} \xrightarrow{\eta} (\overline{\mathfrak{o}}/\mathfrak{f})^\times/(\mathfrak{o}/\mathfrak{f})^\times \xrightarrow{\theta} \mathcal{C}^{\boldsymbol{\sigma}}(\mathfrak{o}) \xrightarrow{\rho} \mathcal{C}^{\boldsymbol{\sigma}}(\overline{\mathfrak{o}}) \to \mathbf{1},$$

where

- *$\eta(\varepsilon(\mathfrak{o}^\times \cap K^{\boldsymbol{\sigma}})) = (\varepsilon+\mathfrak{f})(\mathfrak{o}/\mathfrak{f})^\times$ for all $\varepsilon \in \overline{\mathfrak{o}}^\times \cap K^{\boldsymbol{\sigma}}$;*
- *$\theta((a+\mathfrak{f})(\mathfrak{o}/\mathfrak{f})^\times) = [a_0\overline{\mathfrak{o}} \cap \mathfrak{o}]^{\boldsymbol{\sigma}}$ for all $a \in \overline{\mathfrak{o}}(\mathfrak{f})$, where $a_0 \in \overline{\mathfrak{o}}(\mathfrak{f}) \cap K^{\boldsymbol{\sigma}}$ is such that $a_0 \equiv a \bmod \mathfrak{f}$;*
- *$\rho([\mathfrak{a}]^{\boldsymbol{\sigma}}) = [\mathfrak{a}\overline{\mathfrak{o}}]^{\boldsymbol{\sigma}}$ for all $\mathfrak{a} \in \mathcal{J}(\mathfrak{o})$.*

Proof. We tacitly use 2.11.3 again and again.

1. Let $C \in \mathcal{C}^{\boldsymbol{\sigma}}(\mathfrak{o})$, $\mathfrak{a} \in C \subset \mathcal{J}(\mathfrak{o})$ and $\overline{C} = [\mathfrak{a}\overline{\mathfrak{o}}]^{\boldsymbol{\sigma}} \in \mathcal{C}^{\boldsymbol{\sigma}}(\overline{\mathfrak{o}})$. By 2.9.9.1 there exists an ideal $\overline{\mathfrak{c}} \in \overline{C}$ such that $\overline{\mathfrak{c}}+\mathfrak{f} = \overline{\mathfrak{o}}$. As $[\overline{\mathfrak{c}}]^{\boldsymbol{\sigma}} = [\mathfrak{a}\overline{\mathfrak{o}}]^{\boldsymbol{\sigma}}$ there exists some $y \in K^{\boldsymbol{\sigma}}$ such that $\overline{\mathfrak{c}} = y\mathfrak{a}\overline{\mathfrak{o}}$, and if $\mathfrak{c}_1 = \overline{\mathfrak{c}} \cap \mathfrak{o}$, then $\mathfrak{c}_1 \in \mathcal{J}'^{\mathfrak{f}}(\mathfrak{o})$, $\overline{\mathfrak{c}} = \mathfrak{c}_1\overline{\mathfrak{o}}$ and $y\mathfrak{a}\mathfrak{c}_1^{-1}\overline{\mathfrak{o}} = (y\mathfrak{a}\overline{\mathfrak{o}})\overline{\mathfrak{c}}^{-1} = \overline{\mathfrak{o}}$. Hence $y\mathfrak{a}\mathfrak{c}_1^{-1} \in \mathrm{Ker}(j_{\overline{\mathfrak{o}}/\mathfrak{o}})$, and therefore $y\mathfrak{a}\mathfrak{c}_1^{-1} = c\mathfrak{o}+\mathfrak{f}$ for some $c \in \overline{\mathfrak{o}}$ such that $c\overline{\mathfrak{o}}+\mathfrak{f} = \overline{\mathfrak{o}}$ by 2.11.4.1. By 2.9.7.2 there exists some $c' \in \overline{\mathfrak{o}} \cap K^{\boldsymbol{\sigma}}$ such that $cc' \equiv 1 \bmod \mathfrak{f}$, and we set $\mathfrak{b} = c'y\mathfrak{a}\mathfrak{c}_1^{-1} = cc'\mathfrak{o}+c'\mathfrak{f} \subset \mathfrak{o}$ and $\mathfrak{d} = \mathfrak{c}_1\mathfrak{b} = c'y\mathfrak{a}$. Now $c'y \in K^{\boldsymbol{\sigma}}$ implies $[\mathfrak{d}]^{\boldsymbol{\sigma}} = [\mathfrak{a}]^{\boldsymbol{\sigma}}$, hence $\mathfrak{d} \in C$, and as $\mathfrak{b}+\mathfrak{f} = cc'\mathfrak{o}+\mathfrak{f} = \mathfrak{o}$ and $\mathfrak{c}_1+\mathfrak{f} = \mathfrak{o}$, it follows that $\mathfrak{d}+\mathfrak{f} = \mathfrak{o}$.

2. We show first that the maps η, θ and ρ are well-defined (then it is clear that they are homomorphisms).

If ε, $\varepsilon_1 \in \overline{\mathfrak{o}}^\times \cap K^{\boldsymbol{\sigma}}$ are such that $\varepsilon(\mathfrak{o}^\times \cap K^{\boldsymbol{\sigma}}) = \varepsilon_1(\mathfrak{o}^\times \cap K^{\boldsymbol{\sigma}})$, then $\varepsilon_1 = \varepsilon\varepsilon_0$ for some $\varepsilon_0 \in \mathfrak{o}^\times \cap K^{\boldsymbol{\sigma}}$, and $(\varepsilon_1+\mathfrak{f})(\mathfrak{o}/\mathfrak{f})^\times = (\varepsilon+\mathfrak{f})(\varepsilon_0+\mathfrak{f})(\mathfrak{o}/\mathfrak{f})^\times = (\varepsilon+\mathfrak{f})(\mathfrak{o}/\mathfrak{f})^\times$.

Let $a \in \overline{\mathfrak{o}}(\mathfrak{f})$ and a_0, $a_1 \in \overline{\mathfrak{o}}(\mathfrak{f}) \cap K^{\boldsymbol{\sigma}}$ such that $a_0 \equiv a_1 \equiv a \bmod \mathfrak{f}$. Then we obtain $a_0a_1^{-1} \in K^{\boldsymbol{\sigma}}$, and we assert that $a_0\overline{\mathfrak{o}} \cap \mathfrak{o} = (a_0a_1^{-1})(a_1\overline{\mathfrak{o}} \cap \mathfrak{o})$. Indeed, if $x \in \overline{\mathfrak{o}}$, then $(a_0 - a_1)x \in \mathfrak{o}$, hence $a_0x \in \mathfrak{o}$ if and only if $a_1x \in \mathfrak{o}$, and then $a_0x = (a_0a_1^{-1})a_1x$. Hence $[a_0\overline{\mathfrak{o}} \cap \mathfrak{o}]^\sigma = [a_1\overline{\mathfrak{o}} \cap \mathfrak{o}]^\sigma \in \mathcal{C}^\sigma(\mathfrak{o})$, and therefore there exists a homomorphism $\theta_0 \colon (\overline{\mathfrak{o}}/\mathfrak{f})^\times \to \mathcal{C}^\sigma(\mathfrak{o})$ satisfying $\theta_0(a+\mathfrak{f}) = [a_0\overline{\mathfrak{o}}\cap\mathfrak{o}]^\sigma$ for all $a \in \overline{\mathfrak{o}}(\mathfrak{f})$, provided that $a_0 \in \overline{\mathfrak{o}}(\mathfrak{f}) \cap K^\sigma$ and $a_0 \equiv a \mod \mathfrak{f}$. If additionally $a \in \mathfrak{o}$, then $a_0 \in a + \mathfrak{f} \subset \mathfrak{o}$, hence $a_0\overline{\mathfrak{o}} \cap \mathfrak{o} = a_0\mathfrak{o}$, and therefore we obtain $[a_0\overline{\mathfrak{o}} \cap \mathfrak{o}]^\sigma = [a_0\mathfrak{o}]^\sigma = [1]^\sigma \in \mathcal{C}^\sigma(\mathfrak{o})$. It follows that $(\mathfrak{o}/\mathfrak{f})^\times \subset \mathrm{Ker}(\theta_0)$, and θ_0 induces a homomorphism θ as asserted.

If $\mathfrak{a}$, $\mathfrak{a}_1 \in \mathcal{J}(\mathfrak{o})$ are such that $[\mathfrak{a}]^\sigma = [\mathfrak{a}_1]^\sigma \in \mathcal{C}^\sigma(\mathfrak{o})$, then $\mathfrak{a}_1 = y\mathfrak{a}$ for some $y \in K^\sigma$, and $\mathfrak{a}_1\overline{\mathfrak{o}} = y\mathfrak{a}\overline{\mathfrak{o}}$ implies $[\mathfrak{a}\overline{\mathfrak{o}}]^\sigma = [\mathfrak{a}_1\overline{\mathfrak{o}}]^\sigma \in \mathcal{C}^\sigma(\overline{\mathfrak{o}})$.

- η is injective: Let $\varepsilon \in \overline{\mathfrak{o}}^\times \cap K^\sigma$ be such that $\varepsilon(\mathfrak{o}^\times \cap K^\sigma) \in \mathrm{Ker}(\eta)$. Then $\varepsilon + \mathfrak{f} \in (\mathfrak{o}/\mathfrak{f})^\times$, and there is some $\varepsilon_0 \in \mathfrak{o}$ such that $\varepsilon \equiv \varepsilon_0 \mod \mathfrak{f}$, and $\varepsilon \in \varepsilon_0 + \mathfrak{f} \subset \mathfrak{o}$. Hence $\varepsilon \in \overline{\mathfrak{o}}^\times \cap \mathfrak{o} = \mathfrak{o}^\times$ by 2.4.7.5, and $\varepsilon(\mathfrak{o}^\times \cap K^\sigma) = 1$.
- $\mathrm{Im}(\eta) \subset \mathrm{Ker}(\theta)$: If $\varepsilon \in \overline{\mathfrak{o}}^\times \cap K^\sigma$, then we obtain (with $\varepsilon_0 = \varepsilon$)

$$(\theta\circ\eta)(\varepsilon(\mathfrak{o}^\times \cap K^\sigma)) = \theta((\varepsilon + \mathfrak{f})(\mathfrak{o}/\mathfrak{f})^\times) = [\varepsilon\overline{\mathfrak{o}} \cap \mathfrak{o}]^\sigma = [1]^\sigma \in \mathcal{C}^\sigma(\mathfrak{o})$$

and therefore $\mathrm{Im}(\eta) \subset \mathrm{Ker}(\theta)$.

- $\mathrm{Ker}(\theta) \subset \mathrm{Im}(\eta)$: Let $a \in \overline{\mathfrak{o}}(\mathfrak{f})$ be such that $(a + \mathfrak{f})(\mathfrak{o}/\mathfrak{f})^\times \in \mathrm{Ker}(\theta)$. Let $a_0 \in \overline{\mathfrak{o}}(\mathfrak{f})\cap K^\sigma$ be such that $a \equiv a_0 \mod \mathfrak{f}$. Then $[a_0\overline{\mathfrak{o}}\cap\mathfrak{o}]^\sigma) = [1]^\sigma \in \mathcal{C}^\sigma(\mathfrak{o})$, and by 1. there exists some $c \in \mathfrak{o} \cap K^\sigma$ such that $c\mathfrak{o} + \mathfrak{f} = \mathfrak{o}$ (hence $c + \mathfrak{f} \in (\mathfrak{o}/\mathfrak{f})^\times$) and $a_0\overline{\mathfrak{o}} \cap \mathfrak{o} = c\mathfrak{o}$. Since $a_0\overline{\mathfrak{o}} = (a_0\overline{\mathfrak{o}} \cap \mathfrak{o})\overline{\mathfrak{o}} = c\overline{\mathfrak{o}}$, we obtain $a_0 = c\varepsilon$ for some $\varepsilon \in \overline{\mathfrak{o}}^\times \cap K^\sigma$, and $(a + \mathfrak{f})(\mathfrak{o}/\mathfrak{f})^\times = (a_0 + \mathfrak{f})(\mathfrak{o}/\mathfrak{f})^\times = (\varepsilon + \mathfrak{f})(c + \mathfrak{f})(\mathfrak{o}/\mathfrak{f})^\times = (\varepsilon + \mathfrak{f})(\mathfrak{o}/\mathfrak{f})^\times = \eta(\varepsilon(\mathfrak{o}^\times \cap K^\sigma))$. Hence $(a + \mathfrak{f})(\mathfrak{o}/\mathfrak{f})^\times \in \mathrm{Im}(\eta)$.
- $\mathrm{Im}(\theta) \subset \mathrm{Ker}(\rho)$: If $a \in \overline{\mathfrak{o}}(\mathfrak{f})$ and $a_0 \in \overline{\mathfrak{o}}(\mathfrak{f}) \cap K^\sigma$ is chosen such that $a \equiv a_0 \mod \mathfrak{f}$, then $(\rho\circ\theta)((a + \mathfrak{f})(\mathfrak{o}/\mathfrak{f})^\times) = \rho([a_0\overline{\mathfrak{o}} \cap \mathfrak{o}]^\sigma) = [(a_0\overline{\mathfrak{o}} \cap \mathfrak{o})\overline{\mathfrak{o}}]^\sigma = [a_0\overline{\mathfrak{o}}]^\sigma = [1]^\sigma \in \mathcal{C}^\sigma(\overline{\mathfrak{o}})$, and therefore $\mathrm{Im}(\theta) \subset \mathrm{Ker}(\rho)$.
- $\mathrm{Ker}(\rho) \subset \mathrm{Im}(\theta)$: Let $C \in \mathrm{Ker}(\rho)$. Then $C = [\mathfrak{c}]^\sigma$ for some ideal $\mathfrak{c}$ of $\mathfrak{o}$ such that $\mathfrak{c} + \mathfrak{f} = \mathfrak{o}$ by 1. As $\rho(C) = [\mathfrak{c}\overline{\mathfrak{o}}]^\sigma = [1]^\sigma \in \mathcal{C}^\sigma(\overline{\mathfrak{o}})$, there exists some $c \in \overline{\mathfrak{o}} \cap K^\sigma$ such that $\mathfrak{c}\overline{\mathfrak{o}} = c\overline{\mathfrak{o}}$. Now $\mathfrak{c} + \mathfrak{f} = \mathfrak{o}$ implies $c\overline{\mathfrak{o}} + \mathfrak{f} = \mathfrak{c}\overline{\mathfrak{o}} + \mathfrak{f} = \overline{\mathfrak{o}}$, hence $c \in \overline{\mathfrak{o}}(\mathfrak{f})$, and $C = [\mathfrak{c}]^\sigma = [\mathfrak{c}\overline{\mathfrak{o}} \cap \mathfrak{o}]^\sigma = [c\overline{\mathfrak{o}} \cap \mathfrak{o}]^\sigma = \theta((c + \mathfrak{f})(\mathfrak{o}/\mathfrak{f})^\times) \in \mathrm{Im}(\theta)$.
- ρ is surjective: If $\overline{C} \in \mathcal{C}^\sigma(\overline{\mathfrak{o}})$, then $\overline{C} = [\overline{\mathfrak{c}}]^\sigma$ for some $\overline{\mathfrak{c}} \in \mathcal{J}'^{\mathfrak{f}}(\overline{\mathfrak{o}})$ by 2.9.9.1. Then $\mathfrak{c} = \overline{\mathfrak{c}} \cap \mathfrak{o} \in \mathcal{J}(\mathfrak{o})$, $\overline{\mathfrak{c}} = \mathfrak{c}\overline{\mathfrak{o}}$ and $\overline{C} = \rho(C)$, where $C = [\mathfrak{c}]^\sigma \in \mathcal{C}^\sigma(\mathfrak{o})$. □

Finally we disclose the connection between ray class groups of a Dedekind domain and ideal class groups of suitably associated orders.

Theorem 2.11.6. *Let $\mathfrak{o}$ be an order, $K = \mathsf{q}(\mathfrak{o})$, $\overline{\mathfrak{o}} = \mathrm{cl}_K(\mathfrak{o})$, $\mathfrak{f} = (\mathfrak{o}:\overline{\mathfrak{o}})$, σ a (possibly empty) finite set of real embeddings of K and .*

$$\mathcal{R}^\sigma(\mathfrak{o}) = \{a\overline{\varepsilon} + \mathfrak{f} \in (\overline{\mathfrak{o}}/\mathfrak{f})^\times \mid a \in \mathfrak{o},\ a\mathfrak{o} + \mathfrak{f} = \mathfrak{o},\ \overline{\varepsilon} \in \overline{\mathfrak{o}}^\times \cap K^\sigma\} \subset (\overline{\mathfrak{o}}/\mathfrak{f})^\times.$$

Then there is an exact sequence

$$1 \to \overline{\mathfrak{o}}^\times \cap K^\sigma/\overline{\mathfrak{o}}^\times \cap K^{\mathfrak{f},\sigma} \overset{\eta}{\to} \mathcal{R}^\sigma(\mathfrak{o}) \overset{\theta}{\to} \mathcal{C}^{\mathfrak{f},\sigma}(\overline{\mathfrak{o}}) \overset{\rho}{\to} \mathcal{C}^\sigma(\mathfrak{o}) \to 1,$$

where

- $\eta(\overline{\varepsilon}(\overline{\mathfrak{o}}^\times \cap K^{\mathfrak{f},\boldsymbol{\sigma}}) = \overline{\varepsilon} + \mathfrak{f}$ *for all* $\overline{\varepsilon} \in \overline{\mathfrak{o}}^\times \cap K^{\boldsymbol{\sigma}}$;
- $\theta(a\overline{\varepsilon} + \mathfrak{f}) = [a_0\overline{\mathfrak{o}}]^{\mathfrak{f},\boldsymbol{\sigma}}$ *for all* $a \in \mathfrak{o}$ *satisfying* $a\mathfrak{o} + \mathfrak{f} = \mathfrak{o}$ *and all* $\overline{\varepsilon} \in \overline{\mathfrak{o}}^\times \cap K^{\boldsymbol{\sigma}}$, *where* $a_0 \in \mathfrak{o} \cap K^{\boldsymbol{\sigma}}$ *is such that* $a_0 \equiv a \mod \mathfrak{f}$.
- $\rho([\overline{\mathfrak{a}}]^{\mathfrak{f},\boldsymbol{\sigma}}) = [\overline{\mathfrak{a}} \cap \mathfrak{o}]^{\boldsymbol{\sigma}}$ *for all* $\boldsymbol{\sigma}$*-ray classes* $C = [\overline{\mathfrak{a}}]^{\mathfrak{f},\boldsymbol{\sigma}} \in \mathcal{C}^{\mathfrak{f},\boldsymbol{\sigma}}(\overline{\mathfrak{o}})$ *for some ideal* $\overline{\mathfrak{a}} \in \mathcal{J}'^{\mathfrak{f}}(\overline{\mathfrak{o}})$.

Proof. We show first that the maps η, θ and ρ are well-defined (then it is obvious that they are homomorphisms), and by the way we show that η is injective.

Let $\eta_1 \colon \overline{\mathfrak{o}}^\times \cap K^{\boldsymbol{\sigma}} \to \mathcal{R}^{\boldsymbol{\sigma}}(\mathfrak{o})$ be defined by $\eta_1(\overline{\varepsilon}) = \overline{\varepsilon} + \mathfrak{f}$. Then η_1 is a homomorphism, $\operatorname{Ker}(\eta_1) = \overline{\mathfrak{o}}^\times \cap K^{\mathfrak{f},\boldsymbol{\sigma}}$, and therefore η_1 induces a monomorphism η as asserted.

Suppose that $a,\, a' \in \mathfrak{o}$, $\overline{\varepsilon}, \overline{\varepsilon}' \in \overline{\mathfrak{o}}^\times \cap K^{\boldsymbol{\sigma}}$ and $a_0,\, a_0' \in \mathfrak{o} \cap K^{\boldsymbol{\sigma}}$ are such that $a\mathfrak{o} + \mathfrak{f} = a'\mathfrak{o} + \mathfrak{f} = \mathfrak{o}$, $a\overline{\varepsilon} \equiv a'\overline{\varepsilon}' \mod \mathfrak{f}$, $a_0 \equiv a \mod \mathfrak{f}$ and $a_0' \equiv a' \mod \mathfrak{f}$. Then

$$a_0\overline{\varepsilon} \equiv a\overline{\varepsilon} \equiv a'\overline{\varepsilon}' \equiv a_0'\overline{\varepsilon}' \mod \mathfrak{f} \quad \text{implies} \quad c = (a_0\overline{\varepsilon})(a_0'\overline{\varepsilon}')^{-1} \in K^{\mathfrak{f},\boldsymbol{\sigma}},$$

$a_0\overline{\mathfrak{o}} = a_0\overline{\varepsilon}\,\overline{\mathfrak{o}} = c(a_0'\overline{\varepsilon}'\,\overline{\mathfrak{o}}) = ca_0'\overline{\mathfrak{o}}$ and thus $[a_0\overline{\mathfrak{o}}]^{\mathfrak{f},\boldsymbol{\sigma}} = [a_0'\overline{\mathfrak{o}}]^{\mathfrak{f},\boldsymbol{\sigma}}$. Hence θ is well defined.

Let $j_1 \colon \mathcal{J}^{\mathfrak{f}}(\overline{\mathfrak{o}}) \to \mathcal{C}^{\boldsymbol{\sigma}}(\mathfrak{o})$ be defined by $j_1(\overline{\mathfrak{a}}) = [\overline{\mathfrak{a}} \cap \mathfrak{o}]^{\boldsymbol{\sigma}}$. It suffices to prove that $\mathcal{H}^{\mathfrak{f},\boldsymbol{\sigma}}(\overline{\mathfrak{o}}) \subset \operatorname{Ker}(j_1)$, for then j_1 induces a homomorphism $\rho\colon \mathcal{C}^{\mathfrak{f},\boldsymbol{\sigma}}(\overline{\mathfrak{o}}) \to \mathcal{C}^{\boldsymbol{\sigma}}(\mathfrak{o})$ such that $\rho([\overline{\mathfrak{a}}]^{\mathfrak{f},\boldsymbol{\sigma}}) = [\overline{\mathfrak{a}} \cap \mathfrak{o}]^{\boldsymbol{\sigma}}$ for all $\boldsymbol{\sigma}$-ray classes $C = [\overline{\mathfrak{a}}]^{\mathfrak{f},\boldsymbol{\sigma}} \in \mathcal{C}^{\mathfrak{f},\boldsymbol{\sigma}}(\overline{\mathfrak{o}})$, where $\mathfrak{a} \in \mathcal{J}'^{\mathfrak{f}}(\overline{\mathfrak{o}})$. Thus let $\overline{\mathfrak{a}} = c\overline{\mathfrak{o}} \in \mathcal{H}^{\mathfrak{f},\boldsymbol{\sigma}}(\overline{\mathfrak{o}})$, where $c \in K^{\mathfrak{f},\boldsymbol{\sigma}}$. By 2.9.5.2 it follows that $c = c_1c_2^{-1}$, where $c_1,\, c_2 \in \overline{\mathfrak{o}}(\mathfrak{f})$, by 2.11.3.3(b) we obtain $c_1\overline{\mathfrak{o}} \cap \mathfrak{o} = c(c_2\overline{\mathfrak{o}} \cap \mathfrak{o})$ and consequently $c\mathfrak{o} = (c_1\overline{\mathfrak{o}} \cap \mathfrak{o})(c_2\overline{\mathfrak{o}} \cap \mathfrak{o})^{-1} \in \mathcal{J}^{\mathfrak{f}}(\mathfrak{o})$. As $j_{\overline{\mathfrak{o}}/\mathfrak{o}}(c\mathfrak{o}) = c\overline{\mathfrak{o}}$, we get $j_1(\mathfrak{a}) = [c\mathfrak{o}]^{\boldsymbol{\sigma}} = [1]^{\boldsymbol{\sigma}} \in \mathcal{C}^{\boldsymbol{\sigma}}(\mathfrak{o})$.

- $\operatorname{Im}(\eta) \subset \operatorname{Ker}(\theta)$: If $\overline{\varepsilon} \in \overline{\mathfrak{o}}^\times \cap K^{\boldsymbol{\sigma}}$, then $\theta \circ \eta(\overline{\varepsilon}(\overline{\mathfrak{o}}^\times \cap K^{\mathfrak{f},\boldsymbol{\sigma}})) = \theta(\overline{\varepsilon} + \mathfrak{f}) = [1]^{\mathfrak{f},\boldsymbol{\sigma}}$, and therefore $\operatorname{Im}(\eta) \subset \operatorname{Ker}(\theta)$.
- $\operatorname{Ker}(\theta) \subset \operatorname{Im}(\eta)$: Let $a \in \mathfrak{o}$ and $\overline{\varepsilon} \in \overline{\mathfrak{o}}^\times \cap K^{\boldsymbol{\sigma}}$ be such that $a\mathfrak{o} + \mathfrak{f} = \mathfrak{o}$, $a\overline{\varepsilon} + \mathfrak{f} \in \operatorname{Ker}(\theta)$, and let $a_0 \in \mathfrak{o} \cap K^{\boldsymbol{\sigma}}$ be such that $a_0 \equiv a \mod \mathfrak{f}$. Then we obtain $\theta(a\overline{\varepsilon}+\mathfrak{f}) = [a_0\overline{\mathfrak{o}}]^{\mathfrak{f},\boldsymbol{\sigma}} = [1]^{\mathfrak{f},\boldsymbol{\sigma}} \in \mathcal{C}^{\mathfrak{f},\boldsymbol{\sigma}}(\overline{\mathfrak{o}})$, hence $a_0 = c\overline{\varepsilon}_0$ for some $c \in K^{\mathfrak{f},\boldsymbol{\sigma}}$ and $\overline{\varepsilon}_0 \in \overline{\mathfrak{o}}^\times$. Consequently,

$$a\overline{\varepsilon} + \mathfrak{f} = a_0\overline{\varepsilon} + \mathfrak{f} = c\overline{\varepsilon}\,\overline{\varepsilon}_0 + \mathfrak{f} = \overline{\varepsilon}\,\overline{\varepsilon}_0 + \mathfrak{f} = \eta(\overline{\varepsilon}\,\overline{\varepsilon}_0(\overline{\mathfrak{o}}_{\mathfrak{f}}^\times \cap K^{\boldsymbol{\sigma}})) \in \operatorname{Im}(\eta).$$

- $\operatorname{Im}(\theta) \subset \operatorname{Ker}(\rho)$: Let $a\overline{\varepsilon} + \mathfrak{f} \in \mathcal{R}^{\boldsymbol{\sigma}}(\mathfrak{o})$, where $a \in \mathfrak{o}$, $a\mathfrak{o} + \mathfrak{f} = \mathfrak{o}$ and $\overline{\varepsilon} \in \overline{\mathfrak{o}}^\times \cap K^{\boldsymbol{\sigma}}$. If $a_0 \in \mathfrak{o} \cap K^{\boldsymbol{\sigma}}$ and $a_0 \equiv a \mod \mathfrak{f}$, then

$$\rho\circ\theta(a\overline{\varepsilon} + \mathfrak{f}) = \rho([a_0\overline{\mathfrak{o}}]^{\mathfrak{f},\boldsymbol{\sigma}}) = [a_0\overline{\mathfrak{o}}]^{\boldsymbol{\sigma}} = [1]^{\boldsymbol{\sigma}},$$

and therefore $\operatorname{Im}(\theta) \subset \operatorname{Ker}(\rho)$.

- $\operatorname{Ker}(\rho) \subset \operatorname{Im}(\theta)$: Let $\overline{C} \in \operatorname{Ker}(\rho)$. By 2.9.9.1 there exists an ideal $\overline{\mathfrak{a}} \in \mathcal{J}'^{\mathfrak{f}}(\overline{\mathfrak{o}})$ such that $\overline{C} = [\overline{\mathfrak{a}}]^{\mathfrak{f},\boldsymbol{\sigma}}$, and then $\theta(\overline{C}) = [\overline{\mathfrak{a}} \cap \mathfrak{o}]^{\boldsymbol{\sigma}} = [1]^{\boldsymbol{\sigma}} \in \mathcal{C}^{\boldsymbol{\sigma}}(\mathfrak{o})$. Hence there exists some $c \in \mathfrak{o} \cap K^{\boldsymbol{\sigma}}$ such that $\overline{\mathfrak{a}} \cap \mathfrak{o} = c\mathfrak{o}$. It follows that $c\mathfrak{o} + \mathfrak{f} = \overline{\mathfrak{a}} \cap \mathfrak{o} + \mathfrak{f} = \mathfrak{o}$ and $\overline{C} = [\overline{\mathfrak{a}}]^{\mathfrak{f},\boldsymbol{\sigma}} = [c\overline{\mathfrak{o}}]^{\mathfrak{f},\boldsymbol{\sigma}} = \theta(a + \mathfrak{f}) \in \operatorname{Im}(\theta)$.
- ρ is surjective: Let $C \in \mathcal{C}^{\boldsymbol{\sigma}}(\mathfrak{o})$. By 2.11.5.1, there exists some $\mathfrak{c} \in \mathcal{J}'^{\mathfrak{f}}(\mathfrak{o})$ such that $C = [\mathfrak{c}]^{\boldsymbol{\sigma}} = \rho([\mathfrak{c}\overline{\mathfrak{o}}]^{\mathfrak{f},\boldsymbol{\sigma}}) \in \operatorname{Im}(\rho)$. □

2.12 Extensions of Dedekind domains 1: General theory

Definition 2.12.1. Let $\mathfrak{o}$ and $\mathfrak{O}$ be Dedekind domains such that $\mathfrak{o} \subset \mathfrak{O}$. Let $\mathfrak{P} \in \mathcal{P}(\mathfrak{O})$ and $\mathfrak{p} \in \mathcal{P}(\mathfrak{o})$ such that $\mathfrak{P}$ lies above $\mathfrak{p}$ (i. e., $\mathfrak{P} \cap \mathfrak{o} = \mathfrak{p}$). Then we call $\mathfrak{P}$ a **prime divisor** of $\mathfrak{p}$ in $\mathfrak{O}$ and we write $\mathfrak{P} \,|\, \mathfrak{p}$.

Let $\mathfrak{P} \in \mathcal{P}(\mathfrak{O})$ and $\mathfrak{p} = \mathfrak{P} \cap \mathfrak{o} \in \mathcal{P}(\mathfrak{o})$. Then $\mathfrak{p}\mathfrak{O} \in \mathcal{J}'(\mathfrak{O})$, $\mathfrak{p}\mathfrak{O} \subset \mathfrak{P}$, and we call

$$e(\mathfrak{P}/\mathfrak{p}) = v_{\mathfrak{P}}(\mathfrak{p}\mathfrak{O}) \in \mathbb{N}$$

the **ramification index** of $\mathfrak{P}$ over $\mathfrak{p}$.

As in 2.5.3 we assume that $\mathsf{k}_{\mathfrak{p}} \subset \mathsf{k}_{\mathfrak{P}}$ and denote by

$$f(\mathfrak{P}/\mathfrak{p}) = [\mathsf{k}_{\mathfrak{P}} : \mathsf{k}_{\mathfrak{p}}] = \dim_{\mathsf{k}_{\mathfrak{p}}} \mathsf{k}_{\mathfrak{P}} \in \mathbb{N} \cup \{\infty\}$$

the (residue class) degree of $\mathfrak{P}$ over $\mathfrak{p}$. If $e(\mathfrak{P}/\mathfrak{p}) = 1$ and $\mathsf{k}_{\mathfrak{P}}/\mathsf{k}_{\mathfrak{p}}$ is separable, then $\mathfrak{P}/\mathfrak{p}$ is called **unramified**; otherwise it is called **ramified**. If $\mathrm{char}(\mathsf{k}_{\mathfrak{p}}) \nmid e(\mathfrak{P}/\mathfrak{p})$ and $\mathsf{k}_{\mathfrak{P}}/\mathsf{k}_{\mathfrak{p}}$ is separable, then $\mathfrak{P}/\mathfrak{p}$ is called **tamely ramified**; otherwise it is called **wildly ramified**.

Let $\mathfrak{o}$ be a Dedekind domain, T a submonoid of $\mathfrak{o}^\bullet$, $\mathfrak{p} \in \mathcal{P}(\mathfrak{o})$ and $\mathfrak{p} \cap T = \emptyset$. Then $T^{-1}\mathfrak{o}$ is a Dedekind domain, $T^{-1}\mathfrak{p} = \mathfrak{p}T^{-1}\mathfrak{o} \in \mathcal{P}(T^{-1}\mathfrak{o})$, $T^{-1}\mathfrak{p} \cap \mathfrak{o} = \mathfrak{p}$ and $\mathsf{k}_{T^{-1}\mathfrak{p}} = \mathsf{k}_{\mathfrak{p}}$. Consequently, $e(T^{-1}\mathfrak{p}/\mathfrak{p}) = f(T^{-1}\mathfrak{p}/\mathfrak{p}) = 1$.

Theorem 2.12.2. *Let $\mathfrak{o}$ and $\mathfrak{O}$ be Dedekind domains such that $\mathfrak{o} \subset \mathfrak{O}$ and $K = \mathsf{q}(\mathfrak{o}) \subset \mathsf{q}(\mathfrak{O})$. Let $\mathfrak{p} \in \mathcal{P}(\mathfrak{o})$ and $\mathfrak{P} \in \mathcal{P}(\mathfrak{O})$.*

1. *$\mathfrak{p}\mathfrak{O}_{\mathfrak{p}} \cap \mathfrak{O} = \mathfrak{p}\mathfrak{O}$.*

2. *$\mathfrak{P} \cap \mathfrak{o} = \mathfrak{p}$ if and only if $\mathfrak{p} \subset \mathfrak{P}$, and if $K \cap \mathfrak{O} = \mathfrak{o}$, then $\mathfrak{P} \cap K = \mathfrak{P} \cap \mathfrak{o}$.*

3. *Suppose that $\mathfrak{P} \cap \mathfrak{o} = \mathfrak{p}$ and $e = e(\mathfrak{P}/\mathfrak{p})$.*

(a) *$v_{\mathfrak{P}}(\mathfrak{a}\mathfrak{O}) = e v_{\mathfrak{p}}(\mathfrak{a})$ for all $\mathfrak{a} \in \mathcal{J}(\mathfrak{o})$, and $v_{\mathfrak{P}}(x) = e v_{\mathfrak{p}}(x)$ for all $x \in K$.*

(b) *If $k \in \mathbb{N}$, then $\mathfrak{P}^k \cap \mathfrak{o} = \mathfrak{p}^l$, where $l = \lceil k/e \rceil$.*

4. *Let $\mathfrak{o}'$ be a Dedekind domain, $\mathfrak{o} \subset \mathfrak{o}' \subset \mathfrak{O}$, $\mathfrak{p} = \mathfrak{P} \cap \mathfrak{o}$ and $\mathfrak{p}' = \mathfrak{P} \cap \mathfrak{o}'$. Then*

$$e(\mathfrak{P}/\mathfrak{p}) = e(\mathfrak{P}/\mathfrak{p}')e(\mathfrak{p}'/\mathfrak{p}) \text{ and } f(\mathfrak{P}/\mathfrak{p}) = f(\mathfrak{P}/\mathfrak{p}')f(\mathfrak{p}'/\mathfrak{p}).$$

Additionally, $\mathfrak{P}/\mathfrak{p}$ is unramified [tamely ramified] if and only if both $\mathfrak{P}/\mathfrak{p}'$ and $\mathfrak{p}'/\mathfrak{p}$ are unramified [tamely ramified].

5. *Let T be a submonoid of $\mathfrak{o}^\bullet$, $T \cap \mathfrak{p} = \emptyset$ and $\mathfrak{p} \subset \mathfrak{P}$. Then*

$$f(T^{-1}\mathfrak{P}/T^{-1}\mathfrak{p}) = f(\mathfrak{P}/\mathfrak{p}) \quad \text{and} \quad e(T^{-1}\mathfrak{P}/T^{-1}\mathfrak{p}) = e(\mathfrak{P}/\mathfrak{p}).$$

Proof. 1. Obviously, $\mathfrak{p}\mathfrak{O} \subset \mathfrak{p}\mathfrak{O}_\mathfrak{p} \cap \mathfrak{O}$. If $x \in \mathfrak{p}\mathfrak{O}_\mathfrak{p} \cap \mathfrak{O} = (\mathfrak{p}\mathfrak{O})_\mathfrak{p} \cap \mathfrak{O}$, then $x = s^{-1}a$, where $s \in \mathfrak{o} \setminus \mathfrak{p}$ and $a \in \mathfrak{p}\mathfrak{O}$. Let $s' \in \mathfrak{o}$ be such that $s's \equiv 1 \mod \mathfrak{p}$, say $s's = 1 + c$, where $c \in \mathfrak{p}$. Then $x = (s's - c)x = s'a - cx \in \mathfrak{p}\mathfrak{O}$.

2. If $\mathfrak{p} \subset \mathfrak{P}$, then $\mathfrak{p} \subset \mathfrak{P} \cap \mathfrak{o} \subsetneq \mathfrak{o}$ and thus $\mathfrak{p} = \mathfrak{P} \cap \mathfrak{o}$. The converse is obvious. If $\mathfrak{O} \cap K = \mathfrak{o}$, then $\mathfrak{P} \cap K = \mathfrak{P} \cap \mathfrak{O} \cap K = \mathfrak{P} \cap \mathfrak{o}$.

3. (a) Let first $\mathfrak{a} \in \mathcal{J}'(\mathfrak{o})$ and $v_\mathfrak{p}(\mathfrak{a}) = m \in \mathbb{N}_0$. Then $\mathfrak{a} = \mathfrak{p}^m\mathfrak{c}$, where $\mathfrak{c} \in \mathcal{J}'(\mathfrak{o})$ and $\mathfrak{c}+\mathfrak{p} = \mathfrak{o}$. As $1 \in \mathfrak{c}+\mathfrak{p} \subset \mathfrak{c}\mathfrak{O}+\mathfrak{P}$, it follows that $\mathfrak{c}\mathfrak{O}+\mathfrak{P} = \mathfrak{O}$ and $\mathfrak{a}\mathfrak{O} = (\mathfrak{p}^m\mathfrak{O})(\mathfrak{c}\mathfrak{O})$. By definition, $\mathfrak{p}\mathfrak{O} = \mathfrak{P}^e\mathfrak{C}$, where $\mathfrak{C} \in \mathcal{J}'(\mathfrak{O})$ and $\mathfrak{P}+\mathfrak{C} = \mathfrak{O}$. Hence $\mathfrak{a}\mathfrak{O} = \mathfrak{P}^{me}\mathfrak{C}^m(\mathfrak{c}\mathfrak{O})$, and since $\mathfrak{P}+\mathfrak{C}^m(\mathfrak{c}\mathfrak{O}) = \mathfrak{O}$ by 1.1.3, it follows that $v_\mathfrak{P}(\mathfrak{a}\mathfrak{O}) = me = ev_\mathfrak{p}(\mathfrak{a})$.

If $\mathfrak{a} \in \mathcal{J}(\mathfrak{o})$ is arbitrary, then $\mathfrak{a} = \mathfrak{b}^{-1}\mathfrak{c}$, where $\mathfrak{b},\, \mathfrak{c} \in \mathcal{J}'(\mathfrak{o})$. Hence it follows that $\mathfrak{a}\mathfrak{O} = (\mathfrak{b}\mathfrak{O})^{-1}(\mathfrak{a}\mathfrak{O})$, and

$$v_\mathfrak{P}(\mathfrak{a}\mathfrak{O}) = -v_\mathfrak{P}(\mathfrak{b}\mathfrak{O}) + v_\mathfrak{P}(\mathfrak{a}\mathfrak{O}) = -ev_\mathfrak{p}(\mathfrak{b}) + ev_\mathfrak{p}(\mathfrak{c}) = ev_\mathfrak{p}(\mathfrak{a}).$$

If $x \in K$, then $x\mathfrak{O} = (x\mathfrak{o})\mathfrak{O}$ and $v_\mathfrak{P}(x) = v_\mathfrak{P}(x\mathfrak{O}) = ev_\mathfrak{p}(x\mathfrak{o}) = ev_\mathfrak{p}(x)$.

(b) If $\mathfrak{q} \in \mathcal{P}(\mathfrak{o}) \setminus \{\mathfrak{p}\}$, then $1 \in \mathfrak{p}^k + \mathfrak{q} \subset \mathfrak{P}^k \cap \mathfrak{o} + \mathfrak{q}$, hence $v_\mathfrak{q}(\mathfrak{P}^k \cap \mathfrak{o}) = 0$, and $\mathfrak{P}^k \cap \mathfrak{o} = \mathfrak{p}^l$, where $l \in \mathbb{N}_0$ is minimal such that $\mathfrak{p}^l \subset \mathfrak{P}^k$. If $m \in \mathbb{N}_0$, then

$$\mathfrak{p}^m \subset \mathfrak{P}^k \iff \mathfrak{p}^m\mathfrak{O} = \mathfrak{P}^{em} \subset \mathfrak{P}^k \iff em \ge k \iff m \ge \lfloor \frac{e}{k} \rfloor,$$

it follows that $l = \lfloor e/k \rfloor$.

4. Since $\mathfrak{p}' \cap \mathfrak{o} = \mathfrak{p}$ and $\mathsf{k}_\mathfrak{p} \subset \mathsf{k}_{\mathfrak{p}'} \subset \mathsf{k}_\mathfrak{P}$, it follows that

$$f(\mathfrak{p}'/\mathfrak{p})f(\mathfrak{P}/\mathfrak{p}') = [\mathsf{k}_{\mathfrak{p}'} : \mathsf{k}_\mathfrak{p}]\,[\mathsf{k}_\mathfrak{P} : \mathsf{k}_{\mathfrak{p}'}] = [\mathsf{k}_\mathfrak{P} : \mathsf{k}_\mathfrak{p}] = f(\mathfrak{P}/\mathfrak{p}),$$

and $\mathsf{k}_\mathfrak{P}/\mathsf{k}_\mathfrak{p}$ is separable if and only if both $\mathsf{k}_\mathfrak{P}/\mathsf{k}_{\mathfrak{p}'}$ and $\mathsf{k}_{\mathfrak{p}'}/\mathsf{k}_\mathfrak{p}$ are separable. By 2., $e(\mathfrak{P}/\mathfrak{p}) = v_\mathfrak{P}(\mathfrak{p}\mathfrak{O}) = v_\mathfrak{P}(\mathfrak{p}\mathfrak{o}'\mathfrak{O}) = e(\mathfrak{P}/\mathfrak{p}')v_{\mathfrak{p}'}(\mathfrak{p}\mathfrak{o}') = e(\mathfrak{P}/\mathfrak{p}')e(\mathfrak{p}'/\mathfrak{p})$. Hence $\mathfrak{P}/\mathfrak{p}$ is unramified [tamely ramified] if and only if both $\mathfrak{P}/\mathfrak{p}'$ and $\mathfrak{p}'/\mathfrak{p}$ are unramified [tamely ramified].

5. By 4., as $f(T^{-1}\mathfrak{P}/\mathfrak{P}) = f(T^{-1}\mathfrak{p}/\mathfrak{p}) = e(T^{-1}\mathfrak{P}/\mathfrak{P}) = e(T^{-1}\mathfrak{p}/\mathfrak{p}) = 1$. □

The behavior of Dedekind domains in finite and in particular Galois field extensions is fundamental for the theory of algebraic number fields and algebraic function fields. We are now going to derive the basics of this theory. A full insight requires valuation theory and will only be obtained in Chapter 5 when we have the appropriate methods at our disposal.

Theorem 2.12.3 (Extensions of Dedekind domains). *Let $\mathfrak{o}$ be a Dedekind domain, $K = \mathsf{q}(\mathfrak{o})$, L/K a finite field extension, $[L\!:\!K] = n$ and $\mathfrak{O} = \mathrm{cl}_L(\mathfrak{o})$.*

> *1. $\mathfrak{O}$ is a Dedekind domain, $L = \mathsf{q}(\mathfrak{O})$, $\mathfrak{O} \cap K = \mathfrak{o}$, and the ideal extension $j_{\mathfrak{O}/\mathfrak{o}} \colon \mathcal{J}(\mathfrak{o}) \to \mathcal{J}(\mathfrak{O})$ is a monomorphism. If L/K is separable, then $\mathfrak{O}$ is a finitely generated $\mathfrak{o}$-module.*
>
> *2. Let L/K be separable and $\mathfrak{O}'$ a subring of $\mathfrak{O}$ such that $\mathfrak{o} \subset \mathfrak{O}'$ and $\mathsf{q}(\mathfrak{O}') = L$. Then $\mathfrak{O}'$ is an order in $\mathfrak{O}$, and there exists some $q \in \mathfrak{o}^\bullet$ satisfying $q\mathfrak{O} \subset \mathfrak{O}'$.*

Proof. 1. By 2.5.1.1 it follows that $\mathfrak{O}$ is an integrally closed domain, $L = \mathsf{q}(\mathfrak{O})$, $\mathfrak{O} \cap K = \mathfrak{o}$ and by 2.4.8 we obtain $\dim(\mathfrak{O}) = 1$. If L/K is separable, then $\mathfrak{O}$ is Noetherian, hence a Dedekind domain, and a finitely generated $\mathfrak{o}$-module by 2.5.1.5.

Assume now that $\mathrm{char}(K) = p > 0$ and L/K is inseparable. Let $\overline{K} \supset L$ be an algebraic closure of K, and let L_0 be the separable closure of K in L. Then $\mathfrak{O}_0 = \mathrm{cl}_{L_0}(\mathfrak{o})$ is a Dedekind domain and $\mathfrak{O} = \mathrm{cl}_L(\mathfrak{O}_0)$. Since $[L:K] < \infty$, there exists some p-power $q = p^e$ such that $L \subset L_0^{1/q} \subset \overline{K}$ (see 1.4.7.2), and we assert that $\mathfrak{O}_0^{1/q} = \mathrm{cl}_{L_0^{1/q}}(\mathfrak{o})$. Indeed, if $x \in \mathrm{cl}_{L_0^{1/q}}(\mathfrak{o})$ and $x^n + a_{n-1}x^{n-1} + \ldots + a_1x + a_0 = 0$ is an integral equation for x over $\mathfrak{o}$, then $x^q \in L_0$, and $(x^q)^n + a_{n-1}^q(x^q)^{n-1} + \ldots + a_1^q x^q + a_0^q = 0$ is an integral equation for x^q over $\mathfrak{o}$. Hence $x^q \in \mathfrak{O}_0$ and $x \in \mathfrak{O}_0^{1/q}$. Conversely, if $x \in \mathfrak{O}_0^{1/q}$, then $x^q \in \mathfrak{O}_0$, hence x is integral over $\mathfrak{O}_0$ and thus over $\mathfrak{o}$.

Now $\mathfrak{O}_0^{1/q} \cap L = \mathrm{cl}_{L_0^{1/q}}(\mathfrak{o}) \cap L = \mathrm{cl}_L(\mathfrak{o}) = \mathfrak{O} \subset \mathfrak{O}_0^{1/q}$. Since the assignment $x \mapsto x^{1/q}$ defines an isomorphism $\mathfrak{O}_0 \to \mathfrak{O}_0^{1/q}$, it follows that $\mathfrak{O}_0^{1/q}$ is a Dedekind domain. We shall prove that every non-zero ideal of $\mathfrak{O}$ is invertible (then $\mathfrak{O}$ is a Dedekind domain). Thus let $\mathfrak{a} \neq \mathbf{0}$ be an ideal of $\mathfrak{O}$. Then $\widetilde{\mathfrak{a}} = \mathfrak{a}\mathfrak{O}_0^{1/q}$ is an invertible ideal of $\mathfrak{O}_0^{1/q}$. Hence there exist some $n \in \mathbb{N}$, $a_1, \ldots, a_n \in \widetilde{\mathfrak{a}}$ and $x_1, \ldots, x_n \in L_0^{1/q}$ such that $x_i\widetilde{\mathfrak{a}} \subset \mathfrak{O}_0^{1/q}$ for all $i \in [1,n]$ and $1 = a_1x_1 + \ldots + a_nx_n$. For $i \in [1,n]$ we obtain

$$a_i = \sum_{j=1}^{k} s_{i,j}^{1/q} a_{i,j} \quad \text{for some} \quad k \in \mathbb{N}\,, \ s_{i,j} \in \mathfrak{O}_0 \ \text{ and } \ a_{i,j} \in \mathfrak{a}\,.$$

With this it follows that

$$1 = \Bigl(\sum_{i=1}^{n} a_i x_i\Bigr)^q = \sum_{i=1}^{n}\sum_{j=1}^{k} s_{i,j}a_{i,j}^q x_i^q = \sum_{i=1}^{n}\sum_{j=1}^{k} a_{i,j}(s_{i,j}a_{i,j}^{q-1}x_i^q).$$

If $i \in [1,n]$ and $j \in [1,k]$, then

$$s_{i,j}a_{i,j}^{q-1}x_i^q\mathfrak{a} \subset s_{i,j}x_i^q\mathfrak{a}^q \subset s_{i,j}(x_i\widetilde{\mathfrak{a}})^q \subset \mathfrak{O}_0 \subset \mathfrak{O},$$

and $s_{i,j}a_{i,j}^{q-1}x_i^q \in L$ implies $s_{i,j}a_{i,j}^{q-1}x_i^q \in \mathfrak{a}^{-1}$. Hence $1 \in \mathfrak{a}\mathfrak{a}^{-1}$ and $\mathfrak{a}$ is invertible.

It remains to prove that $j_{\mathfrak{O}/\mathfrak{o}}$ is injective. If $\mathfrak{a} \in \mathrm{Ker}(j_{\mathfrak{O}/\mathfrak{o}})$, then $\mathfrak{a}\mathfrak{O} = \mathfrak{O}$, hence $\mathfrak{a} \subset \mathfrak{O} \cap K = \mathfrak{o}$ and therefore $\mathfrak{a} = \mathfrak{o}$ by 2.4.7.4.

2. By 1., $\mathfrak{O}$ is a finitely generated $\mathfrak{o}$-module. Every ideal of $\mathfrak{O}'$ is an $\mathfrak{o}$-submodule of $\mathfrak{O}$, hence a finitely generated $\mathfrak{o}$-module and thus a finitely generated ideal of $\mathfrak{O}'$. Therefore $\mathfrak{O}'$ is Noetherian. As $\mathfrak{O} = \mathrm{cl}_K(\mathfrak{o}) = \mathrm{cl}_K(\mathfrak{O}')$ is a finitely generated $\mathfrak{o}$-module, it is a finitely generated $\mathfrak{O}'$-module, and $\dim(\mathfrak{O}') = \dim(\mathfrak{O}) = 1$ by 2.4.8. Hence $\mathfrak{O}'$ is an order, 2.11.2.2 implies $\mathfrak{F} = (\mathfrak{O}' : \mathfrak{O}) \in \mathcal{I}'(\mathfrak{O})$, and 2.4.7.1 implies $\mathfrak{F} \cap \mathfrak{o} \neq \mathbf{0}$. If $q \in \mathfrak{F} \cap \mathfrak{o}^\bullet$, then $q\mathfrak{O} \subset \mathfrak{O}'$. $\square$

Theorem 2.12.4 (Factorization of prime ideals in finite extensions). *Let $\mathfrak{o}$ be a Dedekind domain, $K = \mathsf{q}(\mathfrak{o})$, L/K a finite field extension, $[L\colon K] = n$, $\mathfrak{O} = \mathrm{cl}_L(\mathfrak{o})$ and $\mathfrak{p} \in \mathcal{P}(\mathfrak{o})$. Then*

$$\mathfrak{p}\mathfrak{O} = \mathfrak{P}_1^{e_1} \cdot \ldots \cdot \mathfrak{P}_r^{e_r},$$

where $r \in \mathbb{N}$, $\mathfrak{P}_1, \ldots, \mathfrak{P}_r \in \mathcal{P}(\mathfrak{O})$ are distinct, $e_i = e(\mathfrak{P}_i/\mathfrak{p})$ for all $i \in [1, r]$, and $\{\mathfrak{P}_1, \ldots, \mathfrak{P}_r\} = \{\mathfrak{P} \in \mathcal{P}(\mathfrak{O}) \mid \mathfrak{P} \supset \mathfrak{p}\}$. If $f_i = f(\mathfrak{P}_i/\mathfrak{p}) = \dim_{\mathsf{k}_\mathfrak{p}} \mathsf{k}_{\mathfrak{P}_i}$ for all $i \in [1, r]$, then

$$\sum_{i=1}^{r} e_i f_i \le n, \tag{$\dagger$}$$

and equality holds if and only if $\mathfrak{O}_\mathfrak{p}$ is a finitely generated $\mathfrak{o}_\mathfrak{p}$-module (in particular, if L/K is separable).
If $\mathfrak{o}$ is a discrete valuation domain, then $\mathfrak{O}$ is a semilocal principal ideal domain.

Proof. By 2.12.3 it follows that $\mathfrak{O}$ is a Dedekind domain and $j_{\mathfrak{O}/\mathfrak{o}}$ is injective. Hence $\mathfrak{p}\mathfrak{O} \ne \mathfrak{O}$, and by 2.8.3.4 there exists a factorization $\mathfrak{p}\mathfrak{O} = \mathfrak{P}_1^{e_1} \cdot \ldots \cdot \mathfrak{P}_r^{e_r}$, where $r \in \mathbb{N}$, $\mathfrak{P}_1, \ldots, \mathfrak{P}_r \in \mathcal{P}(\mathfrak{O})$ are distinct, $e_i = v_{\mathfrak{P}_i}(\mathfrak{p}\mathfrak{O}) = e(\mathfrak{P}_i/\mathfrak{p}) \in \mathbb{N}$ for all $i \in [1, r]$, and $\{\mathfrak{P}_1, \ldots, \mathfrak{P}_r\} = \{\mathfrak{P} \in \mathcal{P}(\mathfrak{O}) \mid \mathfrak{P} \supset \mathfrak{p}\}$.

Concerning 2.7.9.1, $\mathfrak{O}_\mathfrak{p} = \mathrm{cl}_L(\mathfrak{o}_\mathfrak{p})$,

$$\{\mathfrak{P}' \in \mathcal{P}(\mathfrak{O}_\mathfrak{p}) \mid \mathfrak{P}' \cap \mathfrak{o}_\mathfrak{p} = \mathfrak{p}\mathfrak{o}_\mathfrak{p}\} = \{\mathfrak{P}_1\mathfrak{O}_\mathfrak{p}, \ldots, \mathfrak{P}_r\mathfrak{O}_\mathfrak{p}\},$$

$$(\mathfrak{p}\mathfrak{o}_\mathfrak{p})\mathfrak{O}_\mathfrak{p} = \prod_{i=1}^{r} (\mathfrak{P}_i\mathfrak{O}_\mathfrak{p})^{e_i} \in \mathcal{J}'(\mathfrak{O}_\mathfrak{p}),$$

and $e_i = e(\mathfrak{P}_i\mathfrak{O}_\mathfrak{p}/\mathfrak{p}\mathfrak{o}_\mathfrak{p})$ for all $i \in [1, r]$. Since $\mathsf{k}_\mathfrak{p} = \mathsf{k}_{\mathfrak{p}\mathfrak{o}_\mathfrak{p}}$ and $\mathsf{k}_{\mathfrak{P}_i} = \mathsf{k}_{\mathfrak{P}_i\mathfrak{O}_\mathfrak{p}}$, it follows that $f_i = f(\mathfrak{P}_i\mathfrak{O}_\mathfrak{p}/\mathfrak{p}\mathfrak{o}_\mathfrak{p})$ for all $i \in [1, r]$.

By these arguments, for the proof of ($\dagger$) we may assume that $\mathfrak{o}$ is a discrete valuation domain with maximal ideal $\mathfrak{p} = \pi\mathfrak{o}$. Then $\mathfrak{O}$ is a Dedekind domain, and as $\mathcal{P}(\mathfrak{O}) = \{\mathfrak{P}_1, \ldots, \mathfrak{P}_r\}$, 2.8.6.3 implies that $\mathfrak{O}$ is a semilocal principal ideal domain. If $i \in [1, r]$, then $\pi\mathfrak{O} = \mathfrak{P}_1^{e_1} \cdot \ldots \cdot \mathfrak{P}_r^{e_r} \subset \mathfrak{P}_i^{e_i}$, and $\mathfrak{O}/\mathfrak{P}_i^{e_i}$ is a vector space over $\mathsf{k}_\mathfrak{p}$. The isomorphism

$$\mathfrak{O}/\pi\mathfrak{O} \ \overset{\sim}{\to}\ \prod_{i=1}^{r} \mathfrak{O}/\mathfrak{P}_i^{e_i},$$

given by the Chinese remainder theorem 1.1.4, is $\mathsf{k}_\mathfrak{p}$-linear, and therefore

$$\dim_{\mathsf{k}_\mathfrak{p}} \mathfrak{O}/\pi\mathfrak{O} = \sum_{i=1}^{r} \dim_{\mathsf{k}_\mathfrak{p}} \mathfrak{O}/\mathfrak{P}_i^{e_i}.$$

Next we prove:

> **A.** $\dim_{\mathsf{k}_\mathfrak{p}} \mathfrak{O}/\pi\mathfrak{O} \le n$, and equality holds if and only if $\mathfrak{O}$ is a finitely generated $\mathfrak{o}$-module.

Proof of **A.** We tacitly use 2.10.11 several times. If $m \in \mathbb{N}$ and $u_1, \ldots, u_m \in \mathfrak{O}$ are such that $(u_1 + \pi\mathfrak{O}, \ldots, u_m + \pi\mathfrak{O})$ is linearly independent over $\mathsf{k}_\mathfrak{p}$, then $(u_1, \ldots, u_m)$ is linearly independent over $\mathfrak{o}$ and thus over K. Hence $\dim_{\mathsf{k}_\mathfrak{p}}(\mathfrak{O}/\mathfrak{p}\mathfrak{O}) \le n$.

If $\dim_{\mathsf{k}_\mathfrak{p}}(\mathfrak{O}/\pi\mathfrak{O}) = n$, $u_1, \ldots, u_n \in \mathfrak{O}$ and $(u_1 + \pi\mathfrak{O}, \ldots, u_n + \pi\mathfrak{O})$ is a $\mathsf{k}_\mathfrak{p}$-basis of $\mathfrak{O}/\pi\mathfrak{O}$, then $(u_1, \ldots, u_n)$ is an $\mathfrak{o}$-basis of $\mathfrak{O}$, since $\mathrm{rk}_\mathfrak{o}(\mathfrak{O}) = n$. Hence $\mathfrak{O}$ is a finitely generated $\mathfrak{o}$-module.

Conversely, if $\mathfrak{O}$ is a finitely generated $\mathfrak{o}$-module, then $\mathfrak{O}$ is $\mathfrak{o}$-free by 2.3.5.2, and as $K\mathfrak{O} = L$, we obtain $n = \mathrm{rk}(\mathfrak{O}) = \dim_{\mathsf{k}_\mathfrak{p}}(\mathfrak{O}/\pi\mathfrak{O})$. □[**A.**]

If L/K is separable, then $\mathfrak{O}$ is a finitely generated $\mathfrak{o}$-module by 2.5.1.5.

If $i \in [1, r]$, then **A** implies $\dim_{\mathsf{k}_\mathfrak{p}}(\mathfrak{O}/\mathfrak{P}_i^{e_i}) < \infty$, and it remains to prove that

$$\dim_{\mathsf{k}_\mathfrak{p}} \mathfrak{O}/\mathfrak{P}_i^{e_i} = e_i f_i.$$

For that purpose we consider the descending chain of vector spaces over $\mathsf{k}_\mathfrak{p}$

$$\mathfrak{O}/\mathfrak{P}_i^{e_i} \supset \mathfrak{P}_i/\mathfrak{P}_i^{e_i} \supset \ldots \supset \mathfrak{P}_i^{e_i-1}/\mathfrak{P}_i^{e_i} \supset \mathbf{0}.$$

As $(\mathfrak{P}_i^\nu/\mathfrak{P}_i^{e_i})/(\mathfrak{P}_i^{\nu+1}/\mathfrak{P}_i^{e_i}) \cong \mathfrak{P}_i^\nu/\mathfrak{P}_i^{\nu+1} \cong \mathfrak{O}/\mathfrak{P}_i = \mathsf{k}_{\mathfrak{P}_i}$ for all $\nu \in [0, e_i - 1]$ (see 2.10.7.2), it follows that

$$\dim_{\mathsf{k}_\mathfrak{p}} \mathfrak{O}/\mathfrak{P}_i^{e_i} = \sum_{\nu=0}^{e_i-1} \dim_{\mathsf{k}_\mathfrak{p}} (\mathfrak{P}_i^\nu/\mathfrak{P}_i^{e_i})/(\mathfrak{P}_i^{\nu+1}/\mathfrak{P}_i^{e_i}) = e_i \dim_{\mathsf{k}_\mathfrak{p}} \mathsf{k}_{\mathfrak{P}_i} = e_i f_i. \quad \square$$

Definition 2.12.5. Let $\mathfrak{o}$ be a Dedekind domain, $K = \mathsf{q}(\mathfrak{o})$, L/K a finite field extension, $[L\colon K] = n$, $\mathfrak{O} = \mathrm{cl}_L(\mathfrak{o})$, $\mathfrak{p} \in \mathcal{P}(\mathfrak{o})$ and $\mathfrak{p}\mathfrak{O} = \mathfrak{P}_1^{e_1} \cdot \ldots \cdot \mathfrak{P}_r^{e_r}$, where $r \in \mathbb{N}$, $\mathfrak{P}_1, \ldots, \mathfrak{P}_r \in \mathcal{P}(\mathfrak{O})$ are distinct and $e_1, \ldots, e_r \in \mathbb{N}$. Then $\mathfrak{P}_1, \ldots, \mathfrak{P}_r$ are the prime divisors of $\mathfrak{p}$ in $\mathfrak{O}$, we write

$$\mathfrak{p}\mathfrak{O} = \prod_{\mathfrak{P} \mid \mathfrak{p}} \mathfrak{P}^{e(\mathfrak{P}/\mathfrak{p})} \quad \text{and we suppose that} \quad \sum_{\mathfrak{P} \mid \mathfrak{p}} e(\mathfrak{P}/\mathfrak{p}) f(\mathfrak{P}/\mathfrak{p}) = n \quad \text{(see 2.12.4).}$$

We call $\mathfrak{p}$ in L

- **unramified** if $\mathfrak{P}/\mathfrak{p}$ is unramified for all prime divisors $\mathfrak{P}$ of $\mathfrak{p}$ in $\mathfrak{O}$, and **ramified** otherwise;
- **tamely ramified** if $\mathfrak{P}/\mathfrak{p}$ is tamely ramified for all prime divisors $\mathfrak{P}$ of $\mathfrak{p}$ in $\mathfrak{O}$, and **wildly ramified** otherwise;
- **undecomposed** if $r = 1$, and **decomposed** otherwise;
- **fully decomposed** if $r = n$ [equivalently $e(\mathfrak{P}/\mathfrak{p}) = f(\mathfrak{P}/\mathfrak{p}) = 1$ for all prime divisors $\mathfrak{P}$ of $\mathfrak{p}$ in $\mathfrak{O}$].

If $\mathfrak{p}$ is undecomposed in L and $\mathfrak{P}$ is the prime divisor of $\mathfrak{p}$ in $\mathfrak{O}$, then we call $\mathfrak{p}$ in L

- **fully ramified** if $e(\mathfrak{P}/\mathfrak{p}) = n$ [and then $f(\mathfrak{P}/\mathfrak{p}) = 1$];
- **inert** if $f(\mathfrak{P}/\mathfrak{p}) = n$ [and then $e(\mathfrak{P}/\mathfrak{p}) = 1$].

In particular, if $L = K$ (and thus $\mathfrak{o} = \mathfrak{O}$, $\mathfrak{p} = \mathfrak{P}$), then $\mathfrak{p}$ is at the same time unramified, tamely ramified, undecomposed, fully decomposed, fully ramified and inert.

Theorem 2.12.6 (Kummer's factorization law). *Let $\mathfrak{o}$ be a Dedekind domain, $K = \mathsf{q}(\mathfrak{o})$ and $\mathfrak{p} \in \mathcal{P}(\mathfrak{o})$. Let $\pi \colon \mathfrak{o}_\mathfrak{p}[X] \to \mathfrak{o}_\mathfrak{p}[X]/\mathfrak{p}\mathfrak{o}_\mathfrak{p}[X] = \mathsf{k}_\mathfrak{p}[X]$ be the residue class homomorphism, and for $h \in \mathfrak{o}_\mathfrak{p}[X]$ we denote by $\overline{h}$ the residue class polynomial of h modulo $\mathfrak{p}$.*

Let L/K be a finite separable field extension, $\mathfrak{O} = \mathrm{cl}_L(\mathfrak{o})$ and $a \in \mathfrak{O}$ such that $L = K(a)$ and $\mathfrak{O}_\mathfrak{p} = \mathfrak{o}_\mathfrak{p}[a]$. Let $g \in \mathfrak{o}[X]$ be the minimal polynomial of a over K, and let $\overline{g} = \overline{g}_1^{e_1} \cdot \ldots \cdot \overline{g}_r^{e_r}$, where $r \in \mathbb{N}$, $g_1, \ldots, g_r \in \mathfrak{o}[X]$ are monic such that $\overline{g}_1, \ldots, \overline{g}_r \in \mathsf{k}_\mathfrak{p}[X]$ are distinct and irreducible, and $e_1, \ldots, e_r \in \mathbb{N}$. Let A be an algebraically closed extension field of $\mathsf{k}_\mathfrak{p}$, and for $i \in [1, r]$ let $\alpha_i \in A$ be such that $\overline{g}_i(\alpha_i) = 0$.

1. *$\mathfrak{p}\mathfrak{O} = \mathfrak{P}_1^{e_1} \cdot \ldots \cdot \mathfrak{P}_r^{e_r}$, where $\mathfrak{P}_1, \ldots, \mathfrak{P}_r \in \mathcal{P}(\mathfrak{O})$ are distinct. If $i \in [1, r]$, then $\mathfrak{P}_i = \mathfrak{p}\mathfrak{O} + g_i(a)\mathfrak{O}$, $f(\mathfrak{P}_i/\mathfrak{p}) = \partial(g_i)$, and there is a $\mathsf{k}_\mathfrak{p}$-isomorphism $\phi_i \colon \mathsf{k}_{\mathfrak{P}_i} \overset{\sim}{\to} \mathsf{k}_\mathfrak{p}(\alpha_i)$ satisfying $\phi_i(a + \mathfrak{P}_i) = \alpha_i$.*
   We identify: $\mathsf{k}_{\mathfrak{P}_i} = \mathsf{k}_\mathfrak{p}(\alpha_i)$.

2. *For every monic polynomial $q \in \mathfrak{o}[X] \setminus \mathfrak{o}$ we have*

$$\sum_{\substack{i=1 \\ v_{\mathfrak{P}_i}(q(a))>0}}^{r} f(\mathfrak{P}_i/\mathfrak{p}) \le \partial(q).$$

Proof. 1. Let $i \in [1, r]$. If $x \in \mathfrak{O} \subset \mathfrak{o}_\mathfrak{p}[a]$, then there exists some polynomial $h \in \mathfrak{o}_\mathfrak{p}[X]$ such that $x = h(a)$, and then $\overline{h}(\alpha_i)$ is independent of the choice of h. Indeed, if $h_1, h_2 \in \mathfrak{o}_\mathfrak{p}[X]$ and $h_1(a) = h_2(a)$, then $g \mid (h_1 - h_2)$ in $K[X]$ (and thus in $\mathfrak{o}_\mathfrak{p}[X]$ by 2.2.5.4). If $h_1 - h_2 = gh$, where $h \in \mathfrak{o}_\mathfrak{p}[X]$, then $\overline{h}_1(\alpha_i) - \overline{h}_2(\alpha_i) = \overline{g}(\alpha_i)\overline{h}(\alpha_i) = 0$.

We define $\Phi_i \colon \mathfrak{O} \to A$ by $\Phi_i(x) = \overline{h}(\alpha_i)$. Then Φ_i is the unique ring homomorphism such that $\Phi_i \restriction \mathfrak{o} \colon \mathfrak{o} \to \mathsf{k}_\mathfrak{p}$ is the residue class epimorphism and $\Phi_i(a) = \alpha_i$. Therefore $\mathrm{Im}(\Phi_i) = \mathsf{k}_\mathfrak{p}(\alpha_i)$, and if $\mathfrak{P}_i = \mathrm{Ker}(\Phi_i) \in \mathcal{P}(\mathfrak{O})$, then $\mathfrak{P}_i \cap \mathfrak{o} = \mathfrak{p}$, and Φ_i induces a $\mathsf{k}_\mathfrak{p}$-isomorphism $\phi_i \colon \mathsf{k}_{\mathfrak{P}_i} = \mathfrak{O}/\mathfrak{P}_i \to \mathsf{k}_\mathfrak{p}(\alpha_i)$ satisfying $\phi_i(a + \mathfrak{P}_i) = \alpha_i$. We identify: $\mathsf{k}_{\mathfrak{P}_i} = \mathsf{k}_\mathfrak{p}(\alpha_i)$.

It follows that $f(\mathfrak{P}_i/\mathfrak{p}) = [\mathsf{k}_{\mathfrak{P}_i} : \mathsf{k}_\mathfrak{p}] = [\mathsf{k}_\mathfrak{p}(\alpha_i) : \mathsf{k}_\mathfrak{p}] = \partial(g_i)$, and it remains to prove the following two assertions:

A. $\mathfrak{P}_i = \mathfrak{p}\mathfrak{O} + g_i(a)\mathfrak{O}$ for all $i \in [1, r]$.

B. $\mathfrak{p}S = \mathfrak{P}_1^{e_1} \cdot \ldots \cdot \mathfrak{P}_r^{e_r}$.

Proof of **A.** Let $i \in [1, r]$. Since $\Phi_i \restriction \mathfrak{o} = \pi \restriction \mathfrak{o}$ and $\Phi_i(g_i(a)) = \overline{g}_i(\alpha_i) = 0$ we obtain $\mathfrak{p}\mathfrak{O} + g_i(a)\mathfrak{O} \subset \mathfrak{P}_i$. For the converse, let $x = h(a) \in \mathfrak{P}_i$, where $h \in \mathfrak{o}_\mathfrak{p}[X]$. Then $\Phi_i(x) = \overline{h}(\alpha_i) = 0$, hence $\overline{g}_i \mid \overline{h}$ in $\mathsf{k}_\mathfrak{p}[X]$, and there exists some polynomial $h_1 \in \mathfrak{o}[X]$ such that $\overline{h} = \overline{g}_i\overline{h}_1 \in \mathsf{k}_\mathfrak{p}[X] = \mathfrak{o}_\mathfrak{p}[X]/\mathfrak{p}\mathfrak{o}_\mathfrak{p}[X]$. It follows that $h - g_ih_1 \in \mathfrak{p}\mathfrak{o}_\mathfrak{p}[X]$, $h(a) - g_i(a)h_1(a) \in \mathfrak{p}\mathfrak{o}_\mathfrak{p}[a] \cap \mathfrak{O} = \mathfrak{p}\mathfrak{O}_\mathfrak{p} \cap \mathfrak{O} = \mathfrak{p}\mathfrak{O}$ by 2.12.2.1, and consequently

$$x = h(a) = [h(a) - g_i(a)h_1(a)] + g_i(a)h_1(a) \in \mathfrak{p}\mathfrak{O} + g_i(a)\mathfrak{O}. \qquad \square[\mathbf{A}.]$$

Proof of **B.** Obviously, $\{\mathfrak{P}_1, \ldots, \mathfrak{P}_r\} \subset \{\mathfrak{P} \in \mathcal{P}(\mathfrak{O}) \mid \mathfrak{P} \cap \mathfrak{o} = \mathfrak{p}\}$, and we assert that equality holds. Thus let $\mathfrak{P} \in \mathcal{P}(\mathfrak{O})$ be such that $\mathfrak{P} \cap \mathfrak{o} = \mathfrak{p}$. Since $\mathfrak{O}_\mathfrak{p} = \mathfrak{o}_\mathfrak{p}[a]$, we get $\mathsf{k}_\mathfrak{P} = \mathfrak{O}_\mathfrak{p}/\mathfrak{P}\mathfrak{O}_\mathfrak{p} = \mathsf{k}_\mathfrak{p}[a + \mathfrak{P}]$, and $g(a) = 0$ implies $\overline{g}(a + \mathfrak{P}) = 0$. Hence there is some $i \in [1, r]$ such that $\overline{g}_i(a + \mathfrak{P}) = 0$, and there exists a $\mathsf{k}_\mathfrak{p}$-isomorphism $\psi_i \colon \mathsf{k}_\mathfrak{P} \overset{\sim}{\to} \mathsf{k}_\mathfrak{p}(\alpha_i)$. If $\Psi_i \colon \mathfrak{O} \to \mathsf{k}_\mathfrak{p}(\alpha_i)$ is defined by $\Psi_i(x) = \psi_i(x + \mathfrak{P})$, then $\Psi_i \restriction \mathfrak{o}$ is the residue class epimorphism and $\Psi_i(a) = \alpha_i$. Hence $\Psi_i = \Phi_i$ and $\mathfrak{P} = \mathrm{Ker}(\Psi_i) = \mathfrak{P}_i$.

Now let $\mathfrak{p}\mathfrak{O} = \mathfrak{P}_1^{d_1} \cdot \ldots \cdot \mathfrak{P}_r^{d_r}$, where $d_1, \ldots, d_r \in \mathbb{N}$. Since $\overline{g} = \overline{g}_1^{e_1} \cdot \ldots \cdot \overline{g}_r^{e_r}$, it follows that

$$g_1(a)^{e_1} \cdot \ldots \cdot g_r(a)^{e_r} = (g_1^{e_1} \cdot \ldots \cdot g_r^{e_r} - g)(a) \in \mathfrak{O} \cap \mathfrak{p}\mathfrak{o}_\mathfrak{p}[a] = \mathfrak{O} \cap \mathfrak{p}\mathfrak{O}_\mathfrak{p} = \mathfrak{p}\mathfrak{O},$$

and

$$\prod_{i=1}^r \mathfrak{P}_i^{e_i} = \prod_{i=1}^r (\mathfrak{p}\mathfrak{O} + g_i(a)\mathfrak{O})^{e_i} \subset \mathfrak{p}\mathfrak{O} + \prod_{i=1}^r g_i(a)^{e_i}\mathfrak{O} = \mathfrak{p}\mathfrak{O} = \prod_{i=1}^r \mathfrak{P}_i^{d_i}.$$

Hence $e_i \geq d_i$ for all $i \in [1, r]$, and by 2.12.4 we obtain

$$[L\!:\!K] = \sum_{i=1}^r d_i f(\mathfrak{P}_i/\mathfrak{p}) \leq \sum_{i=1}^r e_i\, \partial(g_i) = \partial(g) = [L\!:\!K].$$

Consequently $e_i = d_i$ for all $i \in [1, r]$. $\square$[**B.**]

2. Let $q \in \mathfrak{o}[X]$ be monic and $i \in [1, r]$ be such that $v_{\mathfrak{P}_i}(q(a)) > 0$. Since $\Phi_i(q(a)) = \overline{q}(\alpha_i) = 0$, it follows that $\overline{g}_i \mid \overline{q}$. Since $\overline{g}_1, \ldots, \overline{g}_r$ are distinct irreducible polynomials and q is monic, we obtain

$$\partial(q) = \partial(\overline{q}) \geq \sum_{\substack{i=1 \\ v_{\mathfrak{P}_i}(q(a))>0}}^r \partial(\overline{g}_i) = \sum_{\substack{i=1 \\ v_{\mathfrak{P}_i}(q(a))>0}}^r f(\mathfrak{P}_i/\mathfrak{p}). \qquad \square$$

Let assumptions be as in 2.12.4 and assume that L/K is separable. Later, in 5.10.1, we shall define the **discriminant** $\mathfrak{d}_{\mathfrak{O}/\mathfrak{o}} \in \mathcal{J}'(\mathfrak{o})$, and we shall prove that a prime ideal $\mathfrak{p} \in \mathcal{P}(\mathfrak{o})$ is ramified in L if and only if $v_\mathfrak{p}(\mathfrak{d}_{\mathfrak{O}/\mathfrak{o}}) > 0$. Hence only finitely many prime ideals are ramified in L. In the next theorem we shall prove this weaker qualitative result, and that in contrast in a purely inseparable extension all prime ideals are ramified.

Theorem 2.12.7. *Let $\mathfrak{o}$ be a Dedekind domain, $K = \mathsf{q}(\mathfrak{o})$, L/K a finite field extension and $\mathfrak{O} = \mathrm{cl}_L(\mathfrak{o})$.*

1. *If L/K is separable, then only finitely many prime ideals of $\mathfrak{o}$ are ramified in L.*

2. *Assume that $\mathrm{char}(K) = p > 0$. Let L/K be purely inseparable, $[L:K] = p^n$ and $\mathfrak{p} \in \mathcal{P}(\mathfrak{o})$. Then $\mathfrak{O} = \{x \in L \mid x^{p^n} \in K\}$, $\mathfrak{P} = \{x \in L \mid x^{p^n} \in \mathfrak{p}\} \in \mathcal{P}(\mathfrak{O})$, $\mathfrak{P}$ is the only prime ideal of $\mathfrak{O}$ lying above $\mathfrak{p}$, and $e(\mathfrak{P}/\mathfrak{p}) = p^r$ for some $r \in [0,n]$. If $\mathsf{k}_\mathfrak{p}$ is perfect, then $f(\mathfrak{P}/\mathfrak{p}) = 1$.*

Proof. 1. Let $a \in \mathfrak{O}$ be such that $L = K(a)$. By 2.12.3.2, $\mathfrak{o}[a]$ is an order, and there exists some $q \in \mathfrak{o}^\bullet$ such that $q\mathfrak{O} \subset \mathfrak{o}[a]$. Let $g \in \mathfrak{o}[X]$ be the minimal polynomial of a over K, $\Delta = \mathsf{D}(g) \in \mathfrak{o}^\bullet$ its discriminant and $\mathfrak{p} \in \mathcal{P}(\mathfrak{o})$ such that $v_\mathfrak{p}(\Delta q) = 0$. Then $\mathfrak{O}_\mathfrak{p} = (q\mathfrak{O})_\mathfrak{p} \subset \mathfrak{o}[a]_\mathfrak{p} \subset \mathfrak{O}_\mathfrak{p}$ and thus $\mathfrak{O}_\mathfrak{p} = \mathfrak{o}[a]_\mathfrak{p} = \mathfrak{o}_\mathfrak{p}[a]$.

For $h \in \mathfrak{o}_\mathfrak{p}[X]$ let $\overline{h} \in \mathsf{k}_\mathfrak{p}[X]$ be the residue class polynomial. Since $\overline{\Delta} = \mathsf{D}(\overline{g}) \neq 0$ (see 1.6.5.4), $\overline{g}$ is a product of distinct irreducible monic polynomials, and therefore $\mathfrak{p}$ is unramified in L by 2.12.6.1. Since $v_\mathfrak{p}(\Delta q) = 0$ for almost all $\mathfrak{p} \in \mathcal{P}(\mathfrak{o})$, it follows that only finitely many $\mathfrak{p} \in \mathcal{P}(\mathfrak{o})$ are ramified in L.

2. If $z \in L$ and $z^{p^n} = c \in \mathfrak{o}$, then $z^{p^n} - c = 0$ is an integral equation for z over $\mathfrak{o}$, which implies $z \in \mathfrak{O}$. Conversely, if $z \in \mathfrak{O}$ and $r \in [0,n]$ is minimal such that $z^{p^r} \in K$, then $X^{p^r} - z^{p^r} \in K[X]$ is the minimal polynomial of z over K, which implies $z^{p^r} \in \mathfrak{o}$ by 2.5.1.3 and therefore $z^{p^n} = (z^{p^r})^{p^{n-r}} \in \mathfrak{o}$.

Let $\mathfrak{P}_0 = \{z \in L \mid z^{p^n} \in \mathfrak{p}\}$ and $\mathfrak{P} \in \mathcal{P}(\mathfrak{o})$ such that $\mathfrak{P} \cap K = \mathfrak{p}$. Then $\mathfrak{P}_0$ is a proper ideal of $\mathfrak{O}$, and if $z \in \mathfrak{P}$, then $z^{p^n} \in K \cap \mathfrak{P} = \mathfrak{p}$ and thus $z \in \mathfrak{P}_0$. Hence $\mathfrak{P} \subset \mathfrak{P}_0$ and therefore $\mathfrak{P} = \mathfrak{P}_0$. In particular, $\mathfrak{P}$ is the unique prime divisor of $\mathfrak{p}$ in $\mathfrak{O}$. If $\pi \in \mathfrak{P} \setminus \mathfrak{P}^2$, then $\pi^{p^n} \in \mathfrak{p}$, and $v_\mathfrak{P}(\pi^{p^n}) = p^n = e(\mathfrak{P}/\mathfrak{p})v_\mathfrak{p}(\pi^{p^n})$, which implies $e(\mathfrak{P}/\mathfrak{p}) \mid p^n$ and therefore $e(\mathfrak{P}/\mathfrak{p}) = p^r$ for some $r \in [0,n]$.

Let finally $\mathsf{k}_\mathfrak{p}$ be perfect and $\xi = x + \mathfrak{P} \in \mathsf{k}_\mathfrak{P}$ for some $x \in \mathfrak{O}$. Since $x^{p^n} \in \mathfrak{o}$, we obtain $\xi^{p^n} \in \mathsf{k}_\mathfrak{p}$ and therefore $\xi \in \mathsf{k}_\mathfrak{p}$, since $\mathrm{char}(\mathsf{k}_\mathfrak{p}) = p$. Hence it follows that $\mathsf{k}_\mathfrak{P} = \mathsf{k}_\mathfrak{p}$ and $f(\mathfrak{P}/\mathfrak{p}) = 1$. □

Theorem 2.12.8 (Eisenstein's factorization law). *Let $\mathfrak{o}$ be a Dedekind domain, $K = \mathsf{q}(\mathfrak{o})$, L/K a finite field extension, $[L:K] = n$, $\mathfrak{O} = \mathrm{cl}_L(\mathfrak{o})$, $\mathfrak{p} \in \mathcal{P}(\mathfrak{o})$ and $\mathfrak{P} \in \mathcal{P}(\mathfrak{O})$ such that $\mathfrak{P} \cap \mathfrak{o} = \mathfrak{p}$.*

1. *Let $g \in \mathfrak{o}[X]$ be a $\mathfrak{p}$-Eisenstein polynomial, and let $L = K(a)$ such that $g(a) = 0$. Then g is irreducible in $K[X]$, $\mathfrak{O}_\mathfrak{p} = \mathfrak{o}_\mathfrak{p}[a]$, $\mathfrak{p}$ is fully ramified in L, and $v_\mathfrak{P}(a) = 1$.*

2. *Let $\mathfrak{p}$ be fully ramified in L, $a \in \mathfrak{O}$ such that $v_\mathfrak{P}(a) = 1$, and let $g \in \mathfrak{o}[X]$ be the minimal polynomial of a over K. Then g is a $\mathfrak{p}$-Eisenstein polynomial, and $\mathfrak{O}_\mathfrak{p} = \mathfrak{o}_\mathfrak{p}[a]$.*

Proof. 1. Let $g = X^n + c_{n-1}X^{n-1} + \ldots + c_1X + c_0$, where $c_0, \ldots, c_{n-1} \in \mathfrak{p}$ and $c_0 \notin \mathfrak{p}^2$. By 1.1.11 and 2.5.2 g is irreducible in $K[X]$. Let $\mathfrak{P} \in \mathcal{P}(\mathfrak{O})$ be a prime divisor of $\mathfrak{p}$. For $i \in [0, n-1]$ we obtain

$$v_{\mathfrak{P}}(c_ia^i) = e(\mathfrak{P}/\mathfrak{p})v_{\mathfrak{p}}(c_i) + iv_{\mathfrak{P}}(a) \geq e(\mathfrak{P}/\mathfrak{p}) \geq 1.$$

Since $a^n = -c_{n-1}a^{n-1} - \ldots - c_1a - c_0$, we get $v_{\mathfrak{P}}(a^n) = nv_{\mathfrak{P}}(a) \geq 1$ and thus $v_{\mathfrak{P}}(a) \geq 1$. If $i \in [1, n-1]$, then $v_{\mathfrak{P}}(c_ia^i) \geq e(\mathfrak{P}/\mathfrak{p}) + i > e(\mathfrak{P}/\mathfrak{p}) = v_{\mathfrak{P}}(a_0)$, and therefore it follows that $nv_{\mathfrak{P}}(a) = v_{\mathfrak{P}}(a^n) = v_{\mathfrak{P}}(a_0) = e(\mathfrak{P}/\mathfrak{p}) \leq n \leq nv_{\mathfrak{P}}(a)$. Hence $v_{\mathfrak{P}}(a) = 1$ and $e(\mathfrak{P}/\mathfrak{p}) = n$.

We assume now that contrary to our assertion $\mathfrak{o}_{\mathfrak{p}}[a] \subsetneq \mathfrak{O}_{\mathfrak{p}}$. By 2.12.3.2 $\mathfrak{o}[a]$ is an order in $\mathfrak{O}$, and there exists some $q \in \mathfrak{o}^\bullet$ such that $q\mathfrak{O} \subset \mathfrak{o}[a]$. Let $\pi \in \mathfrak{p} \setminus \mathfrak{p}^2$ and $v_{\mathfrak{p}}(q) = N \geq 0$. Therefore $q = \pi^N u$ for some $u \in \mathfrak{O}_{\mathfrak{p}}^\times$, and $q\mathfrak{O}_{\mathfrak{p}} = \pi^N\mathfrak{O}_{\mathfrak{p}} \subset \mathfrak{o}_{\mathfrak{p}}[a] \subsetneq \mathfrak{O}_{\mathfrak{p}}$. Hence $N \geq 1$, and there exists some $x \in \mathfrak{O}_{\mathfrak{p}} \setminus \mathfrak{o}_{\mathfrak{p}}[a]$ such that $\pi x \in \mathfrak{o}_{\mathfrak{p}}[a]$. We set $\pi x = b_0 + b_1a + \ldots + b_{n-1}a^{n-1}$, where $b_0, \ldots, b_{n-1} \in \mathfrak{o}_{\mathfrak{p}}$ and $b_i \notin \pi\mathfrak{o}_{\mathfrak{p}}$ for at least one $i \in [0, n-1]$. Let $k \in [0, n-1]$ be minimal such that $b_k \notin \pi\mathfrak{o}_{\mathfrak{p}}$. Since

$$\frac{a^n}{\pi} = -\sum_{j=0}^{n-1} \frac{c_j}{\pi}a^j \in \mathfrak{o}_{\mathfrak{p}}[a] \subset \mathfrak{O}_{\mathfrak{p}}$$

and

$$\pi x a^{n-k-1} = \sum_{\nu=0}^{n} b_\nu a^{\nu+n-k-1} = b_ka^{n-1} + \sum_{\nu=0}^{k-1} b_\nu a^{\nu+n-k-1} + a^n \sum_{\nu=k+1}^{n} b_\nu a^{\nu-k-1},$$

it follows that

$$z = \frac{b_ka^{n-1}}{\pi} = xa^{n-k-1} - \sum_{\nu=0}^{k-1} \frac{b_\nu}{\pi}a^{\nu+n-k-1} - \frac{a^n}{\pi}\sum_{\nu=k+1}^{n-1} b_\nu a^{\nu-k-1} \in \mathfrak{O}_{\mathfrak{p}}.$$

Observing $\mathrm{cl}_L(\mathfrak{o}_{\mathfrak{p}}) = \mathfrak{O}_{\mathfrak{p}}$ and using 1.6.2.1 and 2.5.1.3 we get $\mathsf{N}_{L/K}(z) \in \mathfrak{o}_{\mathfrak{p}}$, but

$$\begin{aligned} v_{\mathfrak{p}}(\mathsf{N}_{L/K}(z)) &= v_{\mathfrak{p}}\Big(\Big(\frac{b_k}{\pi}\Big)^n \mathsf{N}_{L/K}(a)^{n-1}\Big) = n[v_{\mathfrak{p}}(b_k) - v_{\mathfrak{p}}(\pi)] + (n-1)v_{\mathfrak{p}}(c_0) \\ &= -n + (n-1) = -1, \quad \text{a contradiction.} \end{aligned}$$

2. We prove first that $(1, a, \ldots, a^{n-1})$ is linearly independent over K. Assume to the contrary that $c_0 + c_1a + \ldots + c_{n-1}a^{n-1} = 0$ for some $c_0, \ldots, c_{n-1} \in K$, not all equal to 0. If $i \in [0, n-1]$ such that $c_i \neq 0$. Then $v_{\mathfrak{P}}(c_ia^i) = nv_{\mathfrak{p}}(c_i) + i \equiv i \bmod n$, hence all these values are distinct, and consequently

$$v_{\mathfrak{P}}\Big(\sum_{i=0}^{n-1} c_ia^i\Big) = \min\{v_{\mathfrak{P}}(c_ia^i) \mid i \in [0, n-1]\} < \infty, \quad \text{a contradiction.}$$

Since $(1, a, \ldots, a^{n-1})$ is linearly independent over K, it follows that $L = K(a)$, and we denote by $g = X^n + c_{n-1}X^{n-1} + \ldots + c_1 X + c_0 \in \mathfrak{o}[X]$ the minimal polynomial of a over K. Then $a^n = -c_0 - c_1 a - \ldots - c_{n-1}a^{n-1}$, and again, as

$$v_{\mathfrak{P}}(c_i a^i) = n v_{\mathfrak{p}}(c_i) + i \equiv i \mod n \quad \text{for all } i \in [0, n-1] \text{ such that } c_i \neq 0,$$

it follows that $n = v_{\mathfrak{P}}(a^n) = \min\{n v_{\mathfrak{p}}(c_i) + i \mid i \in [0, n-1]\}$ and therefore $v_{\mathfrak{p}}(c_0) = 1$. We assume that g is not a $\mathfrak{p}$-Eisenstein polynomial, and let $i \in [1, n-1]$ be minimal such that $v_{\mathfrak{p}}(c_i) = 0$. Then

$$i = v_{\mathfrak{P}}(c_i a^i) = v_{\mathfrak{P}}\Big(-a^n - \sum_{j=0}^{i-1} c_j a^j - \sum_{j=i+1}^{n-1} c_j a^j\Big)$$

but

$$v_{\mathfrak{P}}(a^n) = n > i \quad \text{and} \quad v_{\mathfrak{P}}(c_j a^j) \geq \begin{cases} n v_{\mathfrak{p}}(c_j) + j \geq n > i & \text{if } j \in [0, i-1], \\ n v_{\mathfrak{p}}(c_j) + j \geq j > i & \text{if } j \in [i+1, n-1], \end{cases}$$

a contradiction. Now $\mathfrak{O}_{\mathfrak{p}} = \mathfrak{o}_{\mathfrak{p}}[a]$ follows from 1. □

Let $\mathfrak{o}$ be a Dedekind domain, $K = \mathsf{q}(\mathfrak{o})$, L/K a finite separable field extension and $\mathfrak{O} = \mathrm{cl}_L(\mathfrak{o})$. An order $\mathfrak{O}'$ in $\mathfrak{O}$ is called an **$\mathfrak{o}$-order** in $\mathfrak{O}$ if $\mathfrak{o} \subset \mathfrak{O}'$, and then its conductor $\mathfrak{F} = (\mathfrak{O}' : \mathfrak{O}) \in \mathcal{J}'(\mathfrak{O})$ is called an **$\mathfrak{o}$-conductor ideal** (of $\mathfrak{O}$). By 2.12.3.2 every domain $\mathfrak{O}'$ such that $\mathfrak{o} \subset \mathfrak{O}' \subset \mathfrak{O}$ and $\mathsf{q}(\mathfrak{O}') = L$ is an $\mathfrak{o}$-order in $\mathfrak{O}$. In general, it is hopeless to obtain an explicit description of all $\mathfrak{o}$-orders in $\mathfrak{O}$. Only in the quadratic case this is easy and will be done in 3.8.2. Nevertheless it is possible to characterize all $\mathfrak{o}$-conductor ideals of $\mathfrak{O}$.

Theorem 2.12.9. *Let $\mathfrak{o}$ be a Dedekind domain, $K = \mathsf{q}(\mathfrak{o})$, L/K a finite separable field extension and $\mathfrak{O} = \mathrm{cl}_L(\mathfrak{o})$.*

1. *Let $\mathfrak{O}'$ be an $\mathfrak{o}$-order in $\mathfrak{O}$, $\mathfrak{F} = (\mathfrak{O}' : \mathfrak{O})$ and $\mathfrak{p} \in \mathcal{P}(\mathfrak{o})$. Then $\mathfrak{O}'_{\mathfrak{p}}$ is an $\mathfrak{o}_{\mathfrak{p}}$-order in $\mathfrak{O}_{\mathfrak{p}}$, and $(\mathfrak{O}'_{\mathfrak{p}} : \mathfrak{O}_{\mathfrak{p}}) = \mathfrak{F}_{\mathfrak{p}}$.*

2. *Let $(\mathfrak{O}'^{(\mathfrak{p})})_{\mathfrak{p} \in \mathcal{P}(\mathfrak{o})}$ be a family of $\mathfrak{o}_{\mathfrak{p}}$-orders in $\mathfrak{O}_{\mathfrak{p}}$, $\mathfrak{O}'^{(\mathfrak{p})} = \mathfrak{O}_{\mathfrak{p}}$ for almost all $\mathfrak{p} \in \mathcal{P}(\mathfrak{o})$, and*

$$\mathfrak{O}' = \bigcap_{\mathfrak{p} \in \mathcal{P}(\mathfrak{o})} \mathfrak{O}'^{(\mathfrak{p})}.$$

Then $\mathfrak{O}'$ is an $\mathfrak{o}$-order in $\mathfrak{O}$ such that $\mathfrak{O}'_{\mathfrak{p}} = \mathfrak{O}'^{(\mathfrak{p})}$ for all $\mathfrak{p} \in \mathcal{P}(\mathfrak{o})$, and

$$(\mathfrak{O}' : \mathfrak{O}) = \bigcap_{\mathfrak{p} \in \mathcal{P}(\mathfrak{o})} (\mathfrak{O}'^{(\mathfrak{p})} : \mathfrak{O}_{\mathfrak{p}}).$$

3. Let $\mathfrak{F} \in \mathcal{I}'(\mathfrak{O})$. Then $\mathfrak{o}+\mathfrak{F}$ is an $\mathfrak{o}$-order in $\mathfrak{O}$, and the following assertions are equivalent:

(a) $\mathfrak{F} = (\mathfrak{o}+\mathfrak{F} : \mathfrak{O})$.

(b) $\mathfrak{F}$ *is an $\mathfrak{o}$-conductor ideal of* $\mathfrak{O}$.

(c) $\mathfrak{F}_\mathfrak{p}$ *is an $\mathfrak{o}_\mathfrak{p}$-conductor ideal of $\mathfrak{O}_\mathfrak{p}$ for all $\mathfrak{p} \in \mathcal{P}(\mathfrak{o})$.*

Proof. We tacitly use 2.12.3.2 again and again.

1. Since $\mathfrak{O}_\mathfrak{p} = \mathrm{cl}_L(\mathfrak{o}_\mathfrak{p})$ and $\mathsf{q}(\mathfrak{O}'_\mathfrak{p}) = \mathsf{q}(\mathfrak{O}') = L$, it follows that $\mathfrak{O}'_\mathfrak{p}$ is an $\mathfrak{o}_\mathfrak{p}$-order in $\mathfrak{O}_\mathfrak{p}$, and as $\mathfrak{O}'_\mathfrak{p}$ is a finitely generated $\mathfrak{o}_\mathfrak{p}$-module, we obtain $(\mathfrak{O}'_\mathfrak{p} : \mathfrak{O}_\mathfrak{p}) = \mathfrak{F}_\mathfrak{p}$ by 2.7.1.2.

2. Obviously, $\mathfrak{O}'$ is a domain satisfying $\mathfrak{o} \subset \mathfrak{O}' \subset \mathfrak{O}_\mathfrak{p}$ for all $\mathfrak{p} \in \mathcal{P}(\mathfrak{o})$. For all $\mathfrak{p}, \mathfrak{q} \in \mathcal{P}(\mathfrak{o})$ we obtain

$$(\mathfrak{O}'^{(\mathfrak{p})})_\mathfrak{q} = \begin{cases} \mathfrak{O}'^{(\mathfrak{p})} & \text{if } \mathfrak{p} = \mathfrak{q}, \\ L & \text{if } \mathfrak{p} \neq \mathfrak{q}, \end{cases} \quad \text{since} \quad (\mathfrak{o}_\mathfrak{p})_\mathfrak{q} = \begin{cases} \mathfrak{o}_\mathfrak{p} & \text{if } \mathfrak{p} = \mathfrak{q}, \\ K & \text{if } \mathfrak{p} \neq \mathfrak{q}. \end{cases}$$

As $\mathfrak{O}'^{(\mathfrak{p})} = \mathfrak{O}_\mathfrak{p}$ for all but finitely many $\mathfrak{p} \in \mathcal{P}(\mathfrak{o})$, it follows that

$$\mathfrak{O}'_\mathfrak{p} = \bigcap_{\mathfrak{q} \in \mathcal{P}(\mathfrak{o})} (\mathfrak{O}'^{(\mathfrak{q})})_\mathfrak{p} = \mathfrak{O}'^{(\mathfrak{p})} \quad \text{for all} \quad \mathfrak{p} \in \mathcal{P}(\mathfrak{o}),$$

$$\mathfrak{O}' \subset \bigcap_{\mathfrak{p} \in \mathcal{P}(\mathfrak{o})} \mathfrak{O}_\mathfrak{p} = \mathfrak{O} \quad \text{and} \quad (\mathfrak{O}' : \mathfrak{O}) = \Big(\bigcap_{\mathfrak{p} \in \mathcal{P}(\mathfrak{o})} \mathfrak{O}'^{(\mathfrak{p})} : \mathfrak{O} \Big) = \bigcap_{\mathfrak{p} \in \mathcal{P}(\mathfrak{o})} (\mathfrak{O}'^{(\mathfrak{p})} : \mathfrak{O}_\mathfrak{p}).$$

It remains to prove that $L = \mathsf{q}(\mathfrak{O}')$. Let $\{\mathfrak{p}_1, \ldots, \mathfrak{p}_m\} = \{\mathfrak{p} \in \mathcal{P}(\mathfrak{o}) \mid \mathfrak{O}'_\mathfrak{p} \neq \mathfrak{O}_\mathfrak{p}\}$ and $z \in L$. Let $q, q_1, \ldots, q_m \in \mathfrak{o}^\bullet$ be such that $qz \in \mathfrak{O}$ and $q_j \mathfrak{O} \subset q_j \mathfrak{O}_{\mathfrak{p}_j} \subset \mathfrak{O}'_{\mathfrak{p}_j}$ for all $j \in [1, m]$. Then we get $qq_1 \cdot \ldots \cdot q_m z \in \mathfrak{O}'_{\mathfrak{p}_1} \cap \ldots \cap \mathfrak{O}'_{\mathfrak{p}_m} = \mathfrak{O}'$ and $z \in \mathsf{q}(\mathfrak{O}')$.

3. Obviously, $\mathfrak{o} + \mathfrak{F}$ is a subring of $\mathfrak{O}$. If $c \in \mathfrak{F}^\bullet$, then $\mathfrak{O}c \subset \mathfrak{F} \subset \mathfrak{o} + \mathfrak{F}$ and therefore $L = \mathsf{q}(\mathfrak{O}) = \mathsf{q}(\mathfrak{o} + \mathfrak{F})$. Hence $\mathfrak{o} + \mathfrak{F}$ is an $\mathfrak{o}$-order in $\mathfrak{O}$.

(a) $\Rightarrow$ (b) By definition, since $\mathfrak{o} + \mathfrak{F}$ is an $\mathfrak{o}$-order in $\mathfrak{O}$.

(b) $\Rightarrow$ (c) By 1.

(c) $\Rightarrow$ (b) For $\mathfrak{p} \in \mathcal{P}(\mathfrak{o})$ let $\mathfrak{O}'^{(\mathfrak{p})}$ be an $\mathfrak{o}_\mathfrak{p}$-order in $\mathfrak{O}_\mathfrak{p}$ with conductor $\mathfrak{F}_\mathfrak{p} = (\mathfrak{O}'^{(\mathfrak{p})} : \mathfrak{O}_\mathfrak{p})$. For almost all $\mathfrak{p} \in \mathcal{P}(\mathfrak{o})$ we have $\mathfrak{F}_\mathfrak{p} = \mathfrak{O}_\mathfrak{p}$, hence $\mathfrak{O}'^{(\mathfrak{p})} = \mathfrak{O}_\mathfrak{p}$, and therefore 2. implies that

$$\mathfrak{O}' = \bigcap_{\mathfrak{p} \in \mathcal{P}(\mathfrak{o})} \mathfrak{O}'^{(\mathfrak{p})} \quad \text{is an } \mathfrak{o}\text{-order in } \mathfrak{O}, \quad \text{and} \quad (\mathfrak{O}' : \mathfrak{O})$$
$$= \bigcap_{\mathfrak{p} \in \mathcal{P}(\mathfrak{o})} (\mathfrak{O}'^{(\mathfrak{p})} : \mathfrak{O}_\mathfrak{p}) = \bigcap_{\mathfrak{p} \in \mathcal{P}(\mathfrak{o})} \mathfrak{F}_\mathfrak{p} = \mathfrak{F}.$$

(b) $\Rightarrow$ (a) Let $\mathfrak{O}'$ be an $\mathfrak{o}$-order in $\mathfrak{O}$ such that $\mathfrak{F} = (\mathfrak{O}' : \mathfrak{O})$. Since $\mathfrak{F} \subset \mathfrak{o} + \mathfrak{F} \subset \mathfrak{O}'$, it follows that $\mathfrak{F} \subset (\mathfrak{o} + \mathfrak{F} : \mathfrak{O}) \subset (\mathfrak{O}' : \mathfrak{O}) = \mathfrak{F}$, and consequently $\mathfrak{F} = (\mathfrak{o} + \mathfrak{F} : \mathfrak{O})$. $\square$

Theorem 2.12.10 (Characterization of conductor ideals). *Let $\mathfrak{o}$ be a Dedekind domain, $K = \mathsf{q}(\mathfrak{o})$, L/K a finite separable field extension and $\mathfrak{O} = \mathrm{cl}_L(\mathfrak{o})$.*

1. Let $\mathfrak{F} \in \mathcal{J}'(\mathfrak{O})$, and set

$$\mathfrak{F} = \prod_{\mathfrak{p}\in\mathcal{P}(\mathfrak{o})} \mathfrak{F}^{(\mathfrak{p})}, \quad \textit{where } \mathfrak{F}^{(\mathfrak{p})} = \prod_{\mathfrak{P}\,|\,\mathfrak{p}} \mathfrak{P}^{v_{\mathfrak{P}}(\mathfrak{F})} \in \mathcal{J}'(\mathfrak{O}) \quad \textit{for all } \mathfrak{p} \in \mathcal{P}(\mathfrak{o}).$$

Then $\mathfrak{F}$ is an $\mathfrak{o}$-conductor ideal of $\mathfrak{O}$ if and only if for all $\mathfrak{p} \in \mathcal{P}(\mathfrak{o})$ its $\mathfrak{p}$-component $\mathfrak{F}^{(\mathfrak{p})}$ is an $\mathfrak{o}$-conductor ideal of $\mathfrak{O}$.

2. Let $\mathfrak{p} \in \mathcal{P}(\mathfrak{o})$, and suppose that $\mathfrak{p}\mathfrak{O} = \mathfrak{P}_1^{e_1} \cdot\ldots\cdot \mathfrak{P}_r^{e_r}$, where $r \in \mathbb{N}$, $\mathfrak{P}_1, \ldots, \mathfrak{P}_r \in \mathcal{P}(\mathfrak{O})$ are distinct, $e_i = e(\mathfrak{P}_i/\mathfrak{p})$ and $f_i = f(\mathfrak{P}_i/\mathfrak{p})$ for all $i \in [1,r]$. Let $k_1, \ldots, k_r \in \mathbb{N}_0$ and $\mathfrak{F} = \mathfrak{P}_1^{k_1} \cdot\ldots\cdot \mathfrak{P}_r^{k_r} \in \mathcal{J}'(\mathfrak{O})$. Then the following assertions are equivalent:

(a) $\mathfrak{F}$ is an $\mathfrak{o}$-conductor ideal.

(b) For all $i \in [1,r]$ satisfying $f_i = 1$ and $k_i \equiv 1 \bmod e_i$, there exists some $j \in [1,r] \setminus \{i\}$ such that $k_j e_i > (k_i - 1)e_j$.

3. If $[L\colon K] = 2$, then an ideal $\mathfrak{F} \in \mathcal{J}'(\mathfrak{O})$ is an $\mathfrak{o}$-conductor ideal if and only if $\mathfrak{F} = \mathfrak{f}\mathfrak{O}$ for some ideal $\mathfrak{f} \in \mathcal{J}'(\mathfrak{o})$.

Proof. 1. We apply 2.12.9.3. If $\mathfrak{p}$, $\mathfrak{q} \in \mathcal{P}(\mathfrak{o})$, then

$$(\mathfrak{F}^{(\mathfrak{p})})_{\mathfrak{q}} = \begin{cases} \mathfrak{F}_{\mathfrak{p}} & \text{if} \quad \mathfrak{p} = \mathfrak{q}, \\ \mathfrak{O}_{\mathfrak{q}} & \text{if} \quad \mathfrak{p} \neq \mathfrak{q}. \end{cases}$$

Let first $\mathfrak{F}$ be an $\mathfrak{o}$-conductor ideal of $\mathfrak{O}$ and $\mathfrak{p} \in \mathcal{P}(\mathfrak{o})$. Then $\mathfrak{F}_{\mathfrak{p}} = (\mathfrak{F}^{(\mathfrak{p})})_{\mathfrak{p}}$ is an $\mathfrak{o}_{\mathfrak{p}}$-conductor ideal of $\mathfrak{O}_{\mathfrak{p}}$, and therefore $(\mathfrak{F}^{(\mathfrak{p})})_{\mathfrak{q}}$ is an $\mathfrak{o}_{\mathfrak{q}}$-conductor ideal of $\mathfrak{O}_{\mathfrak{q}}$ for all $\mathfrak{q} \in \mathcal{P}(\mathfrak{o})$. Hence $\mathfrak{F}^{(\mathfrak{p})}$ is an $\mathfrak{o}$-conductor ideal of $\mathfrak{O}$. Conversely, suppose that $\mathfrak{F}^{(\mathfrak{p})}$ is an $\mathfrak{o}$-conductor ideal of $\mathfrak{O}$ for all $\mathfrak{p} \in \mathcal{P}(\mathfrak{o})$. Then $\mathfrak{F}_{\mathfrak{p}} = (\mathfrak{F}^{(\mathfrak{p})})_{\mathfrak{p}}$ is an $\mathfrak{o}_{\mathfrak{p}}$-conductor ideal of $\mathfrak{O}_{\mathfrak{p}}$ for all $\mathfrak{p} \in \mathcal{P}(\mathfrak{o})$, and thus $\mathfrak{F}$ is an $\mathfrak{O}$-conductor ideal of $\mathfrak{O}$.

2. If $\mathfrak{q} \in \mathcal{P}(\mathfrak{o}) \setminus \{\mathfrak{p}\}$, then $\mathfrak{F}_{\mathfrak{q}} = \mathfrak{O}_{\mathfrak{q}}$. Hence $\mathfrak{F}$ is an $\mathfrak{o}$-conductor ideal of $\mathfrak{O}$ if and only if $\mathfrak{F}_{\mathfrak{p}}$ is an $\mathfrak{o}_{\mathfrak{p}}$-conductor ideal of $\mathfrak{O}_{\mathfrak{p}}$. Since $\mathfrak{F}_{\mathfrak{p}} = \mathfrak{P}_{1\mathfrak{p}}^{k_1} \cdot\ldots\cdot \mathfrak{P}_{r\mathfrak{p}}^{k_r}$, $e_i = e(\mathfrak{P}_{i\mathfrak{p}}/\mathfrak{p}\mathfrak{o}_{\mathfrak{p}})$ and $f_i = f(\mathfrak{P}_{i\mathfrak{p}}/\mathfrak{p}\mathfrak{o}_{\mathfrak{p}})$ for all $i \in [1,r]$, we may assume that $\mathfrak{o}$ is a discrete valuation domain with maximal ideal $\mathfrak{p} = \pi\mathfrak{o}$. Then $\mathfrak{O}$ is a semilocal principal ideal domain by 2.12.4. By 2.12.9.3, $\mathfrak{F}$ is an $\mathfrak{o}$-conductor ideal if and only if $\mathfrak{F} = (\mathfrak{o} + \mathfrak{F} : \mathfrak{O})$, and this holds if and only if $\mathfrak{P}_i^{-1}\mathfrak{F} \not\subset \mathfrak{o} + \mathfrak{F}$ for all $i \in [1,r]$ such that $k_i \ge 1$. Hence it suffices to prove the following assertion.

A. Let $i \in [1,r]$, $\mathfrak{P} = \mathfrak{P}_i = \Pi\mathfrak{O}$, $e = e_i$, $f = f_i$, $k = k_i \ge 1$ and

$$\mathfrak{Q} = \mathfrak{P}_1^{k_1} \cdot\ldots\cdot \mathfrak{P}_{i-1}^{k_{i-1}} \mathfrak{P}_{i+1}^{k_{i+1}} \cdot\ldots\cdot \mathfrak{P}_r^{k_r}.$$

Then $\mathfrak{P}^{-1}\mathfrak{F} = \mathfrak{P}^{k-1}\mathfrak{Q} \subset \mathfrak{o} + \mathfrak{F}$ if and only if $f = 1$, $k \equiv 1 \bmod e$ and $k_j e \le (k-1)e_j$ for all $j \in [1,r] \setminus \{i\}$.

Proof of **A.** If $\mathfrak{P}^{k-1}\mathfrak{O} \subset \mathfrak{o} + \mathfrak{F}$, then

$$\mathfrak{P}^{k-1}/\mathfrak{P}^k = (\mathfrak{P}^{k-1}\mathfrak{O} + \mathfrak{P}^k)/\mathfrak{P}^k \subset (\mathfrak{o} + \mathfrak{F} + \mathfrak{P}^k)/\mathfrak{P}^k = (\mathfrak{o} + \mathfrak{P}^k)/\mathfrak{P}^k,$$

and we prove first:

A′. $\mathfrak{P}^{k-1}/\mathfrak{P}^k \subset (\mathfrak{o} + \mathfrak{P}^k)/\mathfrak{P}^k$ if and only if $f = 1$ and $k \equiv 1 \bmod e$.

Proof of **A′.** Let $\mathcal{R} \subset \mathfrak{o}$ be a set of representatives for $\mathsf{k}_\mathfrak{p}$ such that $0 \in \mathcal{R}$ and $\{\omega_1 = 1, \omega_2, \ldots, \omega_f\} \subset \mathfrak{O}$ a set of representatives for a $\mathsf{k}_\mathfrak{p}$-basis of $\mathsf{k}_\mathfrak{P}$. Then

$$\Big\{\sum_{j=1}^{f} a_j\omega_j \,\Big|\, a_j \in \mathcal{R}\Big\}$$

is a set of representatives for $\mathsf{k}_\mathfrak{P}$ in $\mathfrak{O}$. We set $k-1 = me+r$, where $m \in \mathbb{N}_0$ and $r \in [0, e-1]$. Then every $n \in [0, k-1]$ has a unique representation $n = \mu e + \rho$, where $\mu \in [0, m]$ and $\rho \in [0, r]$, and then $v_\mathfrak{P}(\pi^\mu \Pi^\rho) = \mu e + \rho = n$. By 2.10.6.1 every $a \in \mathfrak{O}$ has a representation

$$a \equiv \sum_{\mu=0}^{m}\sum_{\rho=0}^{r}\sum_{j=1}^{f} a_{\mu e+\rho,j}\,\omega_j \pi^\mu \Pi^\rho \bmod \mathfrak{P}^k \quad \text{with uniquely determined} \quad a_{\mu e+\rho,j} \in \mathcal{R}. \tag{$*$}$$

If a satisfies $(*)$, then $a + \mathfrak{P}^k \in \mathfrak{o} + \mathfrak{P}^k$ if and only if $a_{\mu e+\rho,j} = 0$ for all $\rho \in [1, r]$ and $j \in [2, f]$, that is, if

$$a \equiv \sum_{\mu=0}^{m} a_{\mu e,1}\pi^\mu \bmod \mathfrak{P}^k.$$

If a satisfies $(*)$, then $a \in \mathfrak{P}^{k-1}$ if and only if $a_{\mu e+\rho,j} = 0$ for all $\mu \in [0, m-1]$ and $\rho \in [0, r-1]$, that is, if

$$a \equiv \sum_{j=1}^{f} a_{me+r,j}\,\pi^m \Pi^r \bmod \mathfrak{P}^k.$$

Due to these representations, $\mathfrak{P}^{k-1}/\mathfrak{P}^k \subset (\mathfrak{o} + \mathfrak{P}^k)/\mathfrak{P}^k$ holds if and only if $f = 1$ and $r = 0$, i.e., $k - 1 \equiv 0 \bmod e$. □[**A′.**]

To complete the proof of **A**, we show:

A″. Suppose that $\mathfrak{P}^{k-1}/\mathfrak{P}^k \subset (\mathfrak{o} + \mathfrak{P}^k)/\mathfrak{P}^k$. Then $\mathfrak{P}^{k-1}\mathfrak{O} \subset \mathfrak{o} + \mathfrak{F}$ if and only if $k_j e \le (k-1)e_j$ for all $j \in [1, r] \setminus \{i\}$.

Proof of **A″.** If $\mathfrak{P}^{k-1}/\mathfrak{P}^k \subset (\mathfrak{o} + \mathfrak{P}^k)/\mathfrak{P}^k$, then $\mathfrak{P}^{k-1} \subset \mathfrak{o} + \mathfrak{P}^k$, $f = 1$ and $k - 1 = me$ for some $m \in \mathbb{N}$ by **A′.** Thus, if $a \in \mathfrak{P}^{k-1} \setminus \mathfrak{P}^k$, then we obtain $a \equiv a_0\pi^m \bmod \mathfrak{P}^k$ for some $a_0 \in \mathcal{R}^\bullet$.

Assume first that $\mathfrak{P}^{k-1}\mathfrak{O} \subset \mathfrak{o} + \mathfrak{F} = \mathfrak{o} + \mathfrak{P}^k\mathfrak{O}$, $j \in [1, r] \setminus \{i\}$ and $\mathfrak{O} = Q\mathfrak{O}$ for some $Q \in \mathfrak{O}$. Then $\mathfrak{P}^{k-1}\mathfrak{O} = \Pi^{k-1}Q\mathfrak{O}$, and $\Pi^{k-1}Q \equiv a \bmod \mathfrak{P}^k\mathfrak{O}$ for some $a \in \mathfrak{o}$. On the other hand, since $\Pi^{k-1}Q \in \mathfrak{P}^{k-1} \setminus \mathfrak{P}^k$, it follows that $\Pi^{k-1}Q \equiv a_0\pi^m \mod \mathfrak{P}^k$ for some $a_0 \in \mathcal{R}^\bullet$, and $a \equiv a_0\pi^m \mod \mathfrak{P}^k \cap \mathfrak{o} = \mathfrak{p}^{m+1}$ by 2.12.2.3(b). Hence $v_\mathfrak{p}(a - a_0\pi^m) \geq m+1$, and $v_\mathfrak{p}(a_0\pi^m) = m$ implies $v_\mathfrak{p}(a) = m$ and $v_{\mathfrak{P}_j}(a) = e_jm$. Since $a - \Pi^kQ \in \mathfrak{P}^k\mathfrak{O}$, it follows that

$$e_jm = v_{\mathfrak{P}_j}(a) \geq \min\{v_{\mathfrak{P}_j}(a - \Pi^kQ), v_{\mathfrak{P}_j}(\Pi^kQ)\} \geq k_j,$$

and therefore $(k-1)e_j \geq k_je$.

Assume now that $k_je \leq (k-1)e_j$, hence $k_j \leq e_jm$ for all $j \in [1,r] \setminus \{i\}$, and let $c \in \mathfrak{P}^{k-1}\mathfrak{O} \subset \mathfrak{P}^{k-1}$. Then $c \in a_0\pi^m + \mathfrak{P}^k$ for some $a_0 \in \mathcal{R}$, and for all $j \in [1,r] \setminus \{i\}$ we get

$$v_{\mathfrak{P}_j}(c - a_0\pi^m) \geq \min\{v_{\mathfrak{P}_j}(c), v_{\mathfrak{P}_j}(a_0\pi^m)\} \geq \min\{k_j, v_{\mathfrak{P}_j}(a_0\pi^m)\} \geq k_j,$$

since $v_{\mathfrak{P}_j}(a_0\pi^m) \in \{e_jm, \infty\}$ and $e_jm \geq k_j$. Therefore $c - a_0\pi^m \in \mathfrak{O}$ and $c \in \mathfrak{o} + \mathfrak{F}$. □[**A″.**]

3. By 1. we must prove: If $\mathfrak{p} \in \mathcal{P}(\mathfrak{o})$ and

$$\mathfrak{F} = \prod_{\mathfrak{P} \mid \mathfrak{p}} \mathfrak{P}^{k_\mathfrak{P}}, \quad \text{where} \quad k_\mathfrak{P} \in \mathbb{N}_0,$$

then $\mathfrak{F}$ is an $\mathfrak{o}$-conductor ideal if and only if $\mathfrak{F} = \mathfrak{p}^k\mathfrak{O}$ for some $k \in \mathbb{N}$.

Let $\mathfrak{p} \in \mathcal{P}(\mathfrak{o})$. Then $\mathfrak{p}\mathfrak{O} = \mathfrak{P}_1^{e_1} \cdot \ldots \cdot \mathfrak{P}_r^{e_r}$, where $r \in \mathbb{N}$, $\mathfrak{P}_1, \ldots, \mathfrak{P}_r \in \mathcal{P}(\mathfrak{O})$ are distinct, $e_i = e(\mathfrak{P}_i/\mathfrak{p})$, $f_i = f(\mathfrak{P}_i/\mathfrak{p})$ and $e_1f_1 + \ldots + e_rf_r = 2$. Hence there are 3 possibilities:

- $r = 1$, $f_1 = 1$, $e_1 = 2$: Then $\mathfrak{p}\mathfrak{O} = \mathfrak{P}^2$ ($\mathfrak{p}$ is ramified) and $\mathfrak{F} = \mathfrak{P}^k$ for some $k \in \mathbb{N}$. Hence $\mathfrak{F}$ is an $\mathfrak{o}$-conductor ideal if and only if $k \equiv 0 \mod 2$, that is $k = 2m$ and $\mathfrak{F} = \mathfrak{p}^m\mathfrak{O}$.

- $r = 1$, $f_1 = 2$, $e_1 = 1$: Then $\mathfrak{p}\mathfrak{O} = \mathfrak{P} \in \mathcal{P}(\mathfrak{O})$ ($\mathfrak{p}$ is inert). Hence $\mathfrak{P}^k = \mathfrak{p}^k\mathfrak{O}$ is an $\mathfrak{o}$-conductor ideal for every $k \in \mathbb{N}$.

- $r = 2$, $e_1 = e_2 = f_1 = f_2 = 1$: Then $\mathfrak{p}\mathfrak{O} = \mathfrak{P}_1\mathfrak{P}_2$ ($\mathfrak{p}$ is decomposed). If $k_1, k_2 \in \mathbb{N}_0$, then $\mathfrak{F} = \mathfrak{P}_1^{k_1}\mathfrak{P}_2^{k_2}$ is an $\mathfrak{o}$-conductor ideal if and only if $k_1 = k_2$, that is $\mathfrak{F} = \mathfrak{p}^{k_1}\mathfrak{O}$. □

2.13 Extensions of Dedekind domains 2: Galois extensions

We refine the results of the previous section in the case of finite Galois extensions. To do so, we make the following general assumption:

G. Let $\mathfrak{o}$ be a Dedekind domain, $K = \mathsf{q}(\mathfrak{o})$, L/K a finite Galois extension, $n = [L\colon K]$, $G = \mathrm{Gal}(L/K)$ and $\mathfrak{O} = \mathrm{cl}_L(\mathfrak{o})$.

If $\mathfrak{p} \in \mathcal{P}(\mathfrak{o})$, $\mathfrak{P} \in \mathcal{P}(\mathfrak{O})$ and $\mathfrak{P} \cap \mathfrak{o} = \mathfrak{p}$, then we denote by $e(\mathfrak{P}/\mathfrak{p}) = v_{\mathfrak{P}}(\mathfrak{p}\mathfrak{O})$ the ramification index, by $f(\mathfrak{P}/\mathfrak{p}) = [\mathsf{k}_{\mathfrak{P}} : \mathsf{k}_{\mathfrak{p}}]$ the residue class degree, by $G_{\mathfrak{P}} = \{\sigma \in G \mid \sigma\mathfrak{P} = \mathfrak{P}\}$ the decomposition group and by $D_{\mathfrak{P}} = L^{G_{\mathfrak{P}}}$ the decomposition field of $\mathfrak{P}$ over K. We recall the following results from 2.5.3 and 2.5.4:

- If $\sigma \in G$, then $\sigma\mathfrak{P} \in \mathcal{P}(\mathfrak{O})$, $\sigma\mathfrak{P} \cap \mathfrak{o} = \mathfrak{p}$, $G_{\sigma\mathfrak{P}} = \sigma G_{\mathfrak{P}}\sigma^{-1}$, σ induces a $\mathsf{k}_{\mathfrak{p}}$-isomorphism $\sigma_{\mathfrak{P}} \colon \mathsf{k}_{\mathfrak{P}} \to \mathsf{k}_{\sigma\mathfrak{P}}$ satisfying $\sigma_{\mathfrak{P}}(a + \mathfrak{P}) = \sigma(a) + \sigma\mathfrak{P}$ for all $a \in \mathfrak{O}$, and $f(\mathfrak{P}/\mathfrak{p}) = f(\sigma\mathfrak{P}/\mathfrak{p})$.

- If $\sigma \in G_{\mathfrak{P}}$, then $\sigma_{\mathfrak{P}} \in \mathrm{Gal}(\mathsf{k}_{\mathfrak{P}}/\mathsf{k}_{\mathfrak{p}})$, $\mathsf{k}_{\mathfrak{P}}/\mathsf{k}_{\mathfrak{p}}$ is normal, and there is an epimorphism $\rho_{\mathfrak{P}} \colon G_{\mathfrak{P}} \to \mathrm{Gal}(\mathsf{k}_{\mathfrak{P}}/\mathsf{k}_{\mathfrak{p}})$, defined by $\rho_{\mathfrak{P}}(\sigma) = \sigma_{\mathfrak{P}}$.

- If $(G{:}G_{\mathfrak{P}}) = r$ and $G/G_{\mathfrak{P}} = \{\sigma_1 G_{\mathfrak{P}}, \ldots, \sigma_r G_{\mathfrak{P}}\}$, then $\sigma_1\mathfrak{P}, \ldots, \sigma_r\mathfrak{P}$ are the distinct prime divisors of $\mathfrak{p}$ in $\mathfrak{O}$.

Theorem 2.13.1. *Assume* **G**. *Let* $\mathfrak{p} \in \mathcal{P}(\mathfrak{o})$, $\mathfrak{P} \in \mathcal{P}(\mathfrak{O})$ *such that* $\mathfrak{P} \cap \mathfrak{o} = \mathfrak{p}$, $e = e(\mathfrak{P}/\mathfrak{p})$ *and* $f = f(\mathfrak{P}/\mathfrak{p})$.

1. *Let* $\sigma \in G$.

(a) *If* $\mathfrak{A} \in \mathcal{J}(\mathfrak{O})$, *then* $\sigma\mathfrak{A} \in \mathcal{J}(\mathfrak{O})$ *and* $v_{\sigma\mathfrak{P}}(\mathfrak{A}) = v_{\mathfrak{P}}(\sigma^{-1}\mathfrak{A})$.

(b) $v_{\sigma\mathfrak{P}}(x) = v_{\mathfrak{P}}(\sigma^{-1}(x))$ *for all* $x \in L$.

(c) $e(\sigma\mathfrak{P}/\mathfrak{p}) = e$ *and* $f(\sigma\mathfrak{P}/\mathfrak{p}) = f$.

2. *Let* $(G{:}G_{\mathfrak{P}}) = r$ *and* $G/G_{\mathfrak{P}} = \{\sigma_1 G_{\mathfrak{P}}, \ldots, \sigma_r G_{\mathfrak{P}}\}$. *Then*

$$\mathfrak{p}\mathfrak{O} = (\sigma_1\mathfrak{P} \cdot \ldots \cdot \sigma_r\mathfrak{P})^e, \quad |G_{\mathfrak{P}}| = ef \quad \textit{and} \quad efr = n.$$

Proof. 1.(a) As $\sigma \restriction \mathfrak{O} \colon \mathfrak{O} \to \mathfrak{O}$ is a ring isomorphism, it follows that $\sigma\mathfrak{A} \in \mathcal{J}(\mathfrak{O})$ for all $\mathfrak{A} \in \mathcal{J}(\mathfrak{O})$. Assume first that $\mathfrak{A} \in \mathcal{J}'(\mathfrak{O})$ and $\mathfrak{A} = \mathfrak{P}^m\mathfrak{C}$, where $m = v_{\mathfrak{P}}(\mathfrak{A})$ and $\mathfrak{C} \in \mathcal{J}'(\mathfrak{O})$ such that $\mathfrak{P} + \mathfrak{C} = \mathfrak{O}$. Then we obtain $\sigma\mathfrak{A} = (\sigma\mathfrak{P})^m(\sigma\mathfrak{C})$ and $\sigma\mathfrak{C} + \sigma\mathfrak{P} = \sigma(\mathfrak{P} + \mathfrak{C}) = \mathfrak{O}$, hence $v_{\sigma\mathfrak{P}}(\sigma\mathfrak{A}) = v_{\mathfrak{P}}(\mathfrak{A})$. If $\mathfrak{A} \in \mathcal{J}(\mathfrak{O})$ is arbitrary, say $\mathfrak{A} = \mathfrak{B}^{-1}\mathfrak{C}$, where $\mathfrak{B}$, $\mathfrak{C} \in \mathcal{J}'(\mathfrak{O})$, then $\sigma\mathfrak{A} = (\sigma\mathfrak{B})^{-1}\sigma\mathfrak{C}$, and we obtain $v_{\sigma\mathfrak{P}}(\sigma\mathfrak{A}) = -v_{\sigma\mathfrak{P}}(\sigma\mathfrak{B}) + v_{\sigma\mathfrak{P}}(\sigma\mathfrak{C}) = -v_{\mathfrak{P}}(\mathfrak{B}) + v_{\mathfrak{P}}(\mathfrak{C}) = v_{\mathfrak{P}}(\mathfrak{A})$.

It follows that $v_{\sigma\mathfrak{P}}(\mathfrak{A}) = v_{\mathfrak{P}}(\sigma^{-1}\mathfrak{A})$ for all $\mathfrak{A} \in \mathcal{J}(\mathfrak{O})$.

(b) If $x \in L$, then $v_{\sigma\mathfrak{P}}(x) = v_{\sigma\mathfrak{P}}(x\mathfrak{O}) = v_{\mathfrak{P}}(\sigma^{-1}x\mathfrak{O}) = v_{\mathfrak{P}}(\sigma^{-1}x)$.

(c) $e(\sigma\mathfrak{P}/\mathfrak{p}) = v_{\sigma\mathfrak{P}}(\mathfrak{p}\mathfrak{O}) = v_{\mathfrak{P}}(\sigma^{-1}(\mathfrak{p}\mathfrak{O})) = v_{\mathfrak{P}}(\mathfrak{p}\mathfrak{O}) = e(\mathfrak{P}/\mathfrak{p})$. As $\sigma_{\mathfrak{P}} \colon \mathsf{k}_{\mathfrak{P}} \to \mathsf{k}_{\sigma\mathfrak{P}}$ is a $\mathsf{k}_{\mathfrak{p}}$-isomorphism, we obtain $f(\mathfrak{P}/\mathfrak{p}) = [\mathsf{k}_{\mathfrak{P}} : \mathsf{k}_{\mathfrak{p}}] = [\mathsf{k}_{\sigma\mathfrak{P}} : \mathsf{k}_{\mathfrak{p}}] = f(\sigma\mathfrak{P}/\mathfrak{p})$.

2. By 2.12.4 and 1.(c) it follows that

$$\mathfrak{p}\mathfrak{O} = \prod_{i=1}^{r} (\sigma_i\mathfrak{P})^{e(\sigma_i\mathfrak{P}/\mathfrak{p})} = \prod_{i=1}^{r} (\sigma_i\mathfrak{P})^e, \quad n = efr \text{ and } |G_{\mathfrak{P}}| = \frac{n}{(G{:}G_{\mathfrak{P}})} = \frac{n}{r} = ef.$$

□

Definition and Theorem 2.13.2. *Assume* **G**. *Let* $\mathfrak{P} \in \mathcal{P}(\mathfrak{O})$, $\mathfrak{p} = \mathfrak{P} \cap \mathfrak{o}$, *and let* $\rho_{\mathfrak{P}} \colon G_{\mathfrak{P}} \to \mathrm{Gal}(\mathsf{k}_{\mathfrak{P}}/\mathsf{k}_{\mathfrak{p}})$ *be defined by* $\rho_{\mathfrak{P}}(\sigma) = \sigma_{\mathfrak{P}}$. The group

$$\begin{aligned} I_{\mathfrak{P}} &= \mathrm{Ker}(\rho_{\mathfrak{P}}) = \{\sigma \in \mathrm{G}_{\mathfrak{P}} \mid \sigma_{\mathfrak{P}} = \mathrm{id}_{\mathsf{k}_{\mathfrak{P}}}\} \\ &= \{\sigma \in G_{\mathfrak{P}} \mid \sigma a \equiv a \mod \mathfrak{P} \text{ for all } a \in \mathfrak{O}\} \end{aligned}$$

is called the **inertia group**, and its fixed field $T_{\mathfrak{P}} = L^{I_{\mathfrak{P}}}$ is called the **inertia field** of $\mathfrak{P}$ over K. Then $K \subset D_{\mathfrak{P}} \subset T_{\mathfrak{P}} \subset L$, and as $I_{\mathfrak{P}}$ is a normal subgroup of $G_{\mathfrak{P}}$, the field extension $T_{\mathfrak{P}}/D_{\mathfrak{P}}$ is Galois, and $\mathrm{Gal}(T_{\mathfrak{P}}/D_{\mathfrak{P}}) = G_{\mathfrak{P}}/I_{\mathfrak{P}}$ (as usual, we identify $\sigma \restriction D_{\mathfrak{P}}$ with $\sigma I_{\mathfrak{P}}$. By definition, there is an isomorphism $G_{\mathfrak{P}}/I_{\mathfrak{P}} \overset{\sim}{\to} \mathrm{Gal}(\mathsf{k}_{\mathfrak{P}}/\mathsf{k}_{\mathfrak{p}})$, given by $\sigma I_{\mathfrak{P}} \mapsto \sigma_{\mathfrak{P}}$.

We extend the series of subgroups $G \supset G_{\mathfrak{P}} \supset I_{\mathfrak{P}}$ and define

$$\begin{aligned} G_{\mathfrak{P},i} &= \{\sigma \in G \mid \sigma a \equiv a \mod \mathfrak{P}^{i+1} \text{ for all } a \in \mathfrak{O}\,\} \\ &= \{\sigma \in G \mid v_{\mathfrak{P}}(\sigma a - a) \ge i+1 \text{ for all } a \in \mathfrak{O}\,\} \quad \text{for all } \ i \ge 0. \end{aligned}$$

Then $(G_{\mathfrak{P},i})_{i\ge 0}$ is a sequence of subgroups (see below). The group $G_{\mathfrak{P},i}$ is called the ***i*-th ramification group** and its fixed field $V_{\mathfrak{P},i} = L^{G_{\mathfrak{P},i}}$ is called the ***i*-th ramification field** of $\mathfrak{P}$ over K. By definition, $G_{\mathfrak{P},0} = I_{\mathfrak{P}}$ and $V_{\mathfrak{P},0} = T_{\mathfrak{P}}$.

The sequence $G \supset G_{\mathfrak{P}} \supset G_{\mathfrak{P},0} \supset G_{\mathfrak{P},1} \supset \ldots$ is called the **Hilbert subgroup series** for $\mathfrak{P}$ over K.

The items of the Hilbert subgroup series are invariant under localization as follows:

If $\mathfrak{P} \in \mathcal{P}(\mathfrak{O})$ and $\mathfrak{p} = \mathfrak{P} \cap \mathfrak{o}$, then $\mathfrak{P}\mathfrak{O}_{\mathfrak{p}} \in \mathcal{P}(\mathfrak{O}_{\mathfrak{p}})$, $\mathfrak{P}\mathfrak{O}_{\mathfrak{p}} \cap \mathfrak{o}_{\mathfrak{p}} = \mathfrak{p}\mathfrak{o}_{\mathfrak{p}}$ and $v_{\mathfrak{P}} = v_{\mathfrak{P}\mathfrak{O}_{\mathfrak{p}}}$; if $\sigma \in G$, then $\sigma\mathfrak{P} = \mathfrak{P}$ if and only if $\sigma\mathfrak{P}\mathfrak{O}_{\mathfrak{p}} = \mathfrak{P}\mathfrak{O}_{\mathfrak{p}}$, hence $G_{\mathfrak{P}} = G_{\mathfrak{P}\mathfrak{O}_{\mathfrak{p}}}$ and consequently $D_{\mathfrak{P}} = D_{\mathfrak{P}\mathfrak{O}_{\mathfrak{p}}}$; if $\sigma \in G$ and $i \ge 0$, then evidently $v_{\mathfrak{P}}(\sigma a - a) \ge i+1$ for all $a \in \mathfrak{O}$ if and only if $v_{\mathfrak{P}\mathfrak{O}_{\mathfrak{p}}}(\sigma a - a) \ge i+1$ for all $a \in \mathfrak{O}_{\mathfrak{p}}$. Hence $G_{\mathfrak{P},i} = G_{\mathfrak{P}\mathfrak{O}_{\mathfrak{p}},i}$, and consequently $V_{\mathfrak{P},i} = V_{\mathfrak{P}\mathfrak{O}_{\mathfrak{p}},i}$.

1. $G \supset G_{\mathfrak{P}} \supset I_{\mathfrak{P}} = G_{\mathfrak{P},0} \supset G_{\mathfrak{P},1} \supset G_{\mathfrak{P},2} \supset \ldots$ *is a sequence of subgroups and* $G_{\mathfrak{P},i} = \mathbf{1}$ *for all* $i \gg 1$.

2. $K \subset D_{\mathfrak{P}} \subset T_{\mathfrak{P}} = V_{\mathfrak{P},0} \subset V_{\mathfrak{P},1} \subset \ldots,\ V_{\mathfrak{P},i} = L$ *for all* $i \gg 1$, $I_{\mathfrak{P}}$ *is a normal subgroup of* $G_{\mathfrak{P}}$, $T_{\mathfrak{P}}/D_{\mathfrak{P}}$ *is Galois, and* $\mathrm{Gal}(T_{\mathfrak{P}}/D_{\mathfrak{P}}) = G_{\mathfrak{P}}/I_{\mathfrak{P}}$ *(as usual, for* $\sigma \in G_{\mathfrak{P}}$ *we identify* $\sigma I_{\mathfrak{P}}$ *with* $\sigma \restriction D_{\mathfrak{P}}$ *), and there is an isomorphism*

$$\mathrm{Gal}(T_{\mathfrak{P}}/D_{\mathfrak{P}}) = G_{\mathfrak{P}}/I_{\mathfrak{P}} \overset{\sim}{\to} \mathrm{Gal}(\mathsf{k}_{\mathfrak{P}}/\mathsf{k}_{\mathfrak{p}}, \quad \textit{given by} \quad \sigma I_{\mathfrak{P}} \mapsto \sigma_{\mathfrak{P}}.$$

3. If $\tau \in G$ and $i \geq 0$, then $G_{\tau\mathfrak{P},i} = \tau G_{\mathfrak{P},i}\tau^{-1}$, $V_{\tau\mathfrak{P},i} = \tau V_{\mathfrak{P},i}$, and $G_{\mathfrak{P},i}$ is a normal subgroup of $G_{\mathfrak{P}}$.

Proof. 1. Clearly, $G_{\mathfrak{P},i+1} \subset G_{\mathfrak{P},i}$ for all $i \geq 0$, and if $\sigma \in I_{\mathfrak{P}}$, then we get $\sigma a \equiv a \mod \mathfrak{P}$ for all $a \in \mathfrak{O}$, hence $\sigma\mathfrak{P} = \mathfrak{P}$ and $\sigma \in G_{\mathfrak{P}}$.

If $i \geq 0$ and $\sigma, \tau \in G_{\mathfrak{P},i}$, then $\sigma\tau a - a = \sigma(\tau a - a) + (\tau a - a) \in \mathfrak{P}^{i+1}$ for all $a \in \mathfrak{O}$, hence $\sigma\tau \in G_{\mathfrak{P},i}$, and $\sigma^{-1}a - a = -\sigma^{-1}(\sigma a - a) \in \sigma^{-1}\mathfrak{P}^{i+1} = \mathfrak{P}^{i+1}$ for all $a \in \mathfrak{O}$ (since $G_{\mathfrak{P},i} \subset G_{\mathfrak{P}}$), hence $\sigma^{-1} \in G_{\mathfrak{P},i}$. Therefore

$$G \supset G_{\mathfrak{P}} \supset I_{\mathfrak{P}} = G_{\mathfrak{P},0} \supset G_{\mathfrak{P},1} \supset G_{\mathfrak{P},2} \supset \ldots$$

is a series of subgroups as asserted.

If $\sigma \in G_{\mathfrak{P},i}$ for all $i \geq 0$, then $\sigma a - a \in \mathfrak{P}^{i+1}$ for all $a \in \mathfrak{O}$ and all $i \geq 0$, hence $\sigma a = a$ for all $a \in \mathfrak{O}$ and therefore $\sigma = \mathrm{id}_L$. Consequently, $G_i = \mathbf{1}$ for all $i \gg 1$.

2. Obvious by 1. and the definition of $I_{\mathfrak{P}}$.

3. Let $\sigma, \tau \in G$. Then:

$$\begin{aligned} \sigma \in G_{\tau\mathfrak{P},i} &\iff v_{\tau\mathfrak{P}}(\sigma a - a) = v_{\mathfrak{P}}(\tau^{-1}\sigma\tau\tau^{-1}a - \tau^{-1}a) \geq i+1 \text{ for all } a \in \mathfrak{O} \\ &\iff v_{\mathfrak{P}}(\tau^{-1}\sigma\tau a - a) \geq i+1 \text{ for all } a \in \tau^{-1}\mathfrak{O} = \mathfrak{O} \\ &\iff \tau^{-1}\sigma\tau \in G_{\mathfrak{P},i}. \end{aligned}$$

Hence $G_{\tau\mathfrak{P},i} = \tau G_{\mathfrak{P},i}\tau^{-1}$, and therefore $V_{\tau\mathfrak{P},i} = L^{G_{\tau\mathfrak{P},i}} = \tau V_{\mathfrak{P},i}$. If $\tau \in G_{\mathfrak{P}}$ and $i \geq 0$, then $G_{\mathfrak{P},i} = G_{\tau\mathfrak{P},i} = \tau G_{\mathfrak{P},i}\tau^{-1}$, and therefore $G_{\mathfrak{P},i}$ is a normal subgroup of $G_{\mathfrak{P}}$. □

Theorem 2.13.3. *Assume* **G**. *Let $\mathfrak{P} \in \mathcal{P}(\mathfrak{O})$, $\mathfrak{p} = \mathfrak{P} \cap \mathfrak{o}$, and let K' be an intermediate field of L/K. Let $G \supset G_{\mathfrak{P}} \supset I_{\mathfrak{P}} = G_{\mathfrak{P},0} \supset G_{\mathfrak{P},1} \supset \ldots$ be the Hilbert subgroup series for $\mathfrak{P}$ over K and $G' = \mathrm{Gal}(L/K')$.*

1. $G'_{\mathfrak{P}} = G_{\mathfrak{P}} \cap G'$ is the decomposition group, $I'_{\mathfrak{P}} = I_{\mathfrak{P}} \cap G'$ is the inertia group and $G'_{\mathfrak{P},i} = G_{\mathfrak{P},i} \cap G'$ is the i-th ramification group of $\mathfrak{P}$ over K' for all $i \geq 0$.

2. Let K'/K be Galois, $\overline{G} = G/G' = \mathrm{Gal}(K'/K)$ and $\mathfrak{p}' = \mathfrak{P} \cap K'$. Let $\overline{G}_{\mathfrak{p}'}$ be the decomposition group and $\overline{I}_{\mathfrak{p}'}$ the inertia group of $\mathfrak{p}'$ over K. Suppose that $\mathfrak{g}_{\mathfrak{P}/\mathfrak{p}} = \mathrm{Gal}(\mathsf{k}_{\mathfrak{P}}/\mathsf{k}_{\mathfrak{p}})$, $\mathfrak{g}_{\mathfrak{P}/\mathfrak{p}'} = \mathrm{Gal}(\mathsf{k}_{\mathfrak{P}}/\mathsf{k}_{\mathfrak{p}'})$ and $\mathfrak{g}_{\mathfrak{p}'/\mathfrak{p}} = \mathrm{Gal}(\mathsf{k}_{\mathfrak{p}'}/\mathsf{k}_{\mathfrak{p}})$. Then the natural injections, restrictions and residue class maps fit into the following commutative diagram with exact rows and columns.

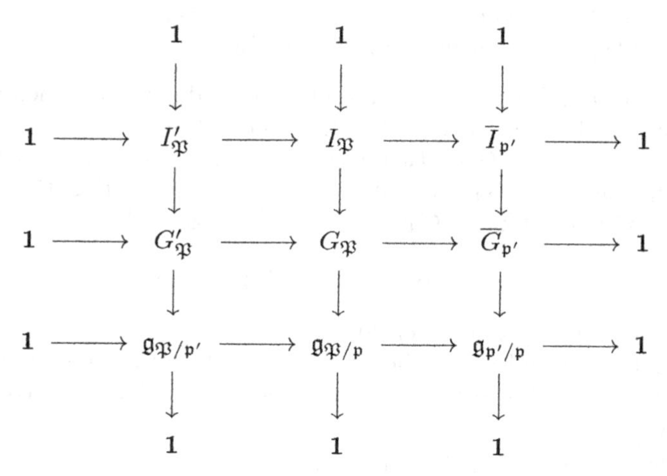

Proof. 1. Obvious by the definitions.

2. The columns of the diagram are exact by the very definition of the inertia groups in 2.13.2, and the bottom row is exact by Galois theory.

It is not all obvious and must be proved that the first two rows are well defined. If $\sigma \in G_{\mathfrak{P}}$, then $\sigma\mathfrak{p}' = \sigma(\mathfrak{P} \cap K') = \sigma\mathfrak{P} \cap \sigma K' = \mathfrak{P} \cap K' = \mathfrak{p}'$, and therefore $\sigma \restriction K' \in \overline{G}_{\mathfrak{p}'}$. Also, $\sigma \restriction K' = \mathrm{id}_{K'}$ if and only if $\sigma \in G_{\mathfrak{P}} \cap G' = G_{\mathfrak{P}'}$. If $\sigma \in I_{\mathfrak{P}}$, then $\sigma a \equiv a \bmod \mathfrak{P}$ for all $a \in \mathfrak{O}$, hence $\sigma a \equiv a \bmod \mathfrak{p}'$ for all $a \in \mathfrak{O} \cap K' = \mathrm{cl}_{K'}(\mathfrak{o})$ and thus $\sigma \restriction K' = \overline{I}_{\mathfrak{p}'}$. Consequently, the restriction $\sigma \mapsto \sigma \restriction K'$ defines homomorphisms $G_{\mathfrak{P}} \to \overline{G}_{\mathfrak{p}'}$ resp. $I_{\mathfrak{P}} \to \overline{I}_{\mathfrak{p}'}$ with kernels $G'_{\mathfrak{P}}$ resp. $I'_{\mathfrak{P}}$.

The left squares and the right upper square are obviously commutative, and the right bottom square commutes since $(\sigma \restriction K')_{\mathfrak{p}'} = \sigma_{\mathfrak{P}} \restriction \mathsf{k}_{\mathfrak{p}'}$ for all $\sigma \in G_{\mathfrak{P}}$.

As to the exactness of the second row, we must prove that the restriction map $G_{\mathfrak{P}} \to \overline{G}_{\mathfrak{p}'}$ is surjective. Let $\tau \in \overline{G}_{\mathfrak{p}'}$ and $\sigma \in G$ such that $\sigma \restriction K' = \tau$. Then $\sigma\mathfrak{P} \cap K' = \mathfrak{p}' = \mathfrak{P} \cap K'$, and thus there exists some $\rho \in G'$ such that $\mathfrak{P} = \rho\sigma\mathfrak{P}$. Hence $\rho\sigma \in G_{\mathfrak{P}}$ and $\rho\sigma \restriction K' = \tau$.

Now the surjectivity of the restriction map $I_{\mathfrak{P}} \to \overline{I}_{\mathfrak{p}'}$ and therewith the exactness of the first row follow by a simple diagram chasing as follows. Let $\tau \in \overline{I}_{\mathfrak{p}'}$ and $\sigma \in G_{\mathfrak{P}}$ such that $\sigma \restriction K' = \tau$. Then $\sigma_{\mathfrak{P}} \restriction \mathsf{k}_{\mathfrak{p}'} = \tau_{\mathfrak{p}'} = \mathrm{id}_{\mathsf{k}_{\mathfrak{p}'}}$, hence $\sigma_{\mathfrak{P}} \in \mathfrak{g}_{\mathfrak{P}/\mathfrak{p}'}$, and therefore $\sigma_{\mathfrak{P}} = \sigma'_{\mathfrak{P}}$ for some $\sigma' \in G'_{\mathfrak{P}}$. Since $(\sigma'^{-1}\sigma)_{\mathfrak{P}} = \mathrm{id}_{\mathsf{k}_{\mathfrak{P}}}$ and $\sigma' \restriction K' = \mathrm{id}_{K'}$, we get $\sigma'^{-1} \circ \sigma \in I_{\mathfrak{P}}$ and $\sigma'^{-1} \circ \sigma \restriction K' = \sigma \restriction K' = \tau$. □

Next we show how the decomposition group and the inertia group determine the factorization type of prime ideals in extension fields. After that, in 2.13.8, we investigate the structure of the higher ramification groups.

Theorem 2.13.4. *Assume* **G**. *Let* $\mathfrak{P} \in \mathcal{P}(\mathfrak{O})$, $\mathfrak{p} = \mathfrak{P} \cap \mathfrak{o}$, *and suppose that* $\mathsf{k}_{\mathfrak{P}}/\mathsf{k}_{\mathfrak{p}}$ *is separable. Let* $D_{\mathfrak{P}} = L^{G_{\mathfrak{P}}}$ *be the decomposition field and* $T_{\mathfrak{P}} = G^{I_{\mathfrak{P}}}$

the inertia field of $\mathfrak{P}$ *over* K. *Let* $e = e(\mathfrak{P}/\mathfrak{p})$, $f = f(\mathfrak{P}/\mathfrak{p})$, $\mathfrak{P}_D = \mathfrak{P} \cap D_{\mathfrak{P}}$, $\mathfrak{P}_T = \mathfrak{P} \cap T_{\mathfrak{P}}$ *and* r *the number of prime divisors of* $\mathfrak{p}$ *in* L.

1. *We have* $[L:T_{\mathfrak{P}}] = |I_{\mathfrak{P}}| = e$, $[T_{\mathfrak{P}}:D_{\mathfrak{P}}] = (G_{\mathfrak{P}}:I_{\mathfrak{P}}) = f$ *and* $[D_{\mathfrak{P}}:K] = (G:G_{\mathfrak{P}}) = r$.

2. $\mathfrak{P}_D$ *is undecomposed in* L *and inert in* $T_{\mathfrak{P}}$, *and* $\mathfrak{P}_T$ *is fully ramified in* L. *In particular,* $e(\mathfrak{P}/\mathfrak{P}_T) = e$, $f(\mathfrak{P}/\mathfrak{P}_T) = 1$, $e(\mathfrak{P}_T/\mathfrak{P}_D) = 1$, $f(\mathfrak{P}_T/\mathfrak{P}_D) = f$, $e(\mathfrak{P}_D/\mathfrak{p}) = f(\mathfrak{P}_D/\mathfrak{p}) = 1$, $\mathsf{k}_{\mathfrak{P}_D} = \mathsf{k}_{\mathfrak{p}}$ *and* $\mathsf{k}_{\mathfrak{P}} = \mathsf{k}_{\mathfrak{P}_T}$.

$$
\begin{array}{ccccccc}
K & \xrightarrow{\;r\;} & D_{\mathfrak{P}} & \xrightarrow{\;f\;} & T_{\mathfrak{P}} & \xrightarrow{\;e\;} & L \\
\mathfrak{p} & \xrightarrow[e=f=1]{} & \mathfrak{P}_D & \xrightarrow[e=1]{} & \mathfrak{P}_T & \xrightarrow[f=1]{} & \mathfrak{P}
\end{array}
$$

3. $T_{\mathfrak{P}}$ *(resp.* $D_{\mathfrak{P}}$ *) is the largest intermediate field* K' *of* L/K *such that* $e(\mathfrak{P} \cap K'/\mathfrak{p}) = 1$ *(resp.* $e(\mathfrak{P} \cap K'/\mathfrak{p}) = f(\mathfrak{P} \cap K'/\mathfrak{p}) = 1$ *).*

4. *Let* K' *be an intermediate field of* L/K. *Then* $\mathfrak{p}$ *is fully decomposed (resp. unramified) in* K' *if and only if* $K' \subset \sigma D_{\mathfrak{P}}$ *(resp.* $K' \subset \sigma T_{\mathfrak{P}}$ *) for all* $\sigma \in G$. *In particular,* $\mathfrak{P}$ *is fully decomposed (resp. unramified) in* L *if and only if* $D_{\mathfrak{P}} = L$ *(resp* $T_{\mathfrak{P}} = L$ *).*

Proof. From 2.13.1.2 we obtain $\mathfrak{p}\mathfrak{O} = (\mathfrak{P}\mathfrak{P}_2 \cdot \ldots \cdot \mathfrak{P}_r)^e$, where $r = (G:G_{\mathfrak{P}})$ and $\mathfrak{P}_2, \ldots, \mathfrak{P}_r \in \mathcal{P}(\mathfrak{O}) \setminus \{\mathfrak{P}\}$ are distinct, $|G_{\mathfrak{P}}| = ef$ and $n = ref$.

1. Clearly, $[D_{\mathfrak{P}}:K] = (G:G_{\mathfrak{P}}) = r$, and as $\mathsf{k}_{\mathfrak{P}}/\mathsf{k}_{\mathfrak{p}}$ is separable, we obtain

$$
f = [\mathsf{k}_{\mathfrak{P}}:\mathsf{k}_{\mathfrak{p}}] = |\mathrm{Gal}(\mathsf{k}_{\mathfrak{P}}/\mathsf{k}_{\mathfrak{p}})| = (G_{\mathfrak{P}}:I_{\mathfrak{P}}) = |\mathrm{Gal}(T_{\mathfrak{P}}/D_{\mathfrak{P}})| = [T_{\mathfrak{P}}:D_{\mathfrak{P}}]
$$

and

$$
[L:T_{\mathfrak{P}}] = \frac{n}{[T_{\mathfrak{P}}:D_{\mathfrak{P}}]\,[D_{\mathfrak{P}}:K]} = \frac{n}{fr} = e.
$$

2. By 2.13.3.1 the group $G_{\mathfrak{P}} = \mathrm{Gal}(L/D_{\mathfrak{P}})$ itself is the decomposition group of $\mathfrak{P}$ over $D_{\mathfrak{P}}$. Hence $\mathfrak{P}$ is the only prime divisor of $\mathfrak{P}_D$ in L, and as

$$
|G_{\mathfrak{P}}| = e(\mathfrak{P}/\mathfrak{P}_D)f(\mathfrak{P}/\mathfrak{P}_D) = ef = e(\mathfrak{P}/\mathfrak{P}_D)e(\mathfrak{P}_D/\mathfrak{p})f(\mathfrak{P}/\mathfrak{P}_D)f(\mathfrak{P}_D/\mathfrak{p}),
$$

it follows that $e(\mathfrak{P}_D/\mathfrak{p}) = f(\mathfrak{P}_D/\mathfrak{p}) = 1$, $f = f(\mathfrak{P}/\mathfrak{P}_D)$ and $e = e(\mathfrak{P}/\mathfrak{P}_D)$.

Alike, $I_{\mathfrak{P}} = \mathrm{Gal}(L/T_{\mathfrak{P}})$ is the inertia group of $\mathfrak{P}$ over $T_{\mathfrak{P}}$. Hence

$$
e = |I_{\mathfrak{P}}| = e(\mathfrak{P}/\mathfrak{P}_T),\ f(\mathfrak{P}/\mathfrak{P}_T) = 1,\ e(\mathfrak{P}_T/\mathfrak{P}_D) = 1 \text{ and } f(\mathfrak{P}_T/\mathfrak{P}_D) = f.
$$

Eventually $f(\mathfrak{P}_D/\mathfrak{p}) = 1$ implies $\mathsf{k}_{\mathfrak{p}} = \mathsf{k}_{\mathfrak{P}_D}$, and $f(\mathfrak{P}/\mathfrak{P}_T) = 1$ implies $\mathsf{k}_{\mathfrak{P}} = \mathsf{k}_{\mathfrak{P}_T}$.

3. Let K' be an intermediate field of L/K, $G' = \mathrm{Gal}(L/K')$, and let $\mathfrak{p}' = \mathfrak{P} \cap K'$. Recall that $G_{\mathfrak{P}} \cap G'$ is the decomposition group and $I_{\mathfrak{P}} \cap G'$ is the inertia group of $\mathfrak{P}$ over K'.

If $f(\mathfrak{p}'/\mathfrak{p}) = e(\mathfrak{p}'/\mathfrak{p}) = 1$, then $f(\mathfrak{P}/\mathfrak{p}') = f$ and $e(\mathfrak{P}/\mathfrak{p}') = e$. Hence it follows that $|G_{\mathfrak{P}} \cap G'| = ef = |G_{\mathfrak{P}}|$, which implies $G_{\mathfrak{P}} \subset G'$ and $K' \subset D_{\mathfrak{P}}$.

If $e(\mathfrak{p}'/\mathfrak{p}) = 1$, then $e(\mathfrak{P}/\mathfrak{p}') = e$. Hence $|I_{\mathfrak{P}} \cap G'| = e = |I_{\mathfrak{P}}|$, which implies $I_{\mathfrak{P}} \subset G'$ and $K' \subset T_{\mathfrak{P}}$.

4. $\mathfrak{p}$ is fully decomposed (resp. unramified) in K' if and only if

$$e(\sigma\mathfrak{P}\cap K'/\mathfrak{p}) = f(\sigma\mathfrak{P}\cap K'/\mathfrak{p}) = 1 \quad (\text{resp. } e(\sigma\mathfrak{P}\cap K'/\mathfrak{p}) = 1) \quad \text{for all } \sigma \in G,$$

and by 3. this holds if and only if $K' \subset \sigma D_{\mathfrak{P}}$ (resp. $K' \subset \sigma T_{\mathfrak{P}}$) for all $\sigma \in G$. □

In the following theorem we apply our results to determine the factorization of prime ideals in a compositum of fields.

Theorem 2.13.5. *Assume* **G**, *let* $\mathfrak{p} \in \mathcal{P}(\mathfrak{o})$, *and assume that* $\mathsf{k}_{\mathfrak{p}}$ *is perfect.*

1. *Let* $K_1, \ldots, K_n$ *be intermediate fields of* L/K, *and suppose that* $K' = K_1 \cdot \ldots \cdot K_n$. *Then* $\mathfrak{p}$ *is fully decomposed (resp. unramified) in* K' *if and only if* $\mathfrak{p}$ *is fully decomposed (resp. unramified) in* K_i *for all* $i \in [1, n]$.

2. *Let* K' *be an intermediate field of* L/K *such that* L *is the normal closure of* K'/K. *Then* $\mathfrak{p}$ *is fully decomposed (resp. unramified) in* L *if and only if* $\mathfrak{p}$ *is fully decomposed (resp. unramified) in* K'.

3. *Let* K_1 *and* K' *be intermediate fields of* L/K, $K_1' = K_1K'$ *and* $\mathfrak{o}' = \mathrm{cl}_{K'}(\mathfrak{o})$. *Let* $\mathfrak{p} \in \mathcal{P}(\mathfrak{o})$ *be unramified in* K_1 *and* $\mathfrak{p}' \in \mathcal{P}(\mathfrak{o}')$ *such that* $\mathfrak{p}' \cap K = \mathfrak{p}$. *Then* $\mathfrak{p}'$ *is unramified in* K_1'.

$$\begin{array}{ccccc} K_1 & \longrightarrow & K_1' & \longrightarrow & L \\ \uparrow & & \uparrow & & \\ K & \longrightarrow & K' & & \end{array}$$

Proof. Let $\mathfrak{P} \in \mathcal{P}(\mathfrak{O})$ such that $\mathfrak{P} \cap \mathfrak{o} = \mathfrak{p}$.

1. By 2.13.4.4, $\mathfrak{p}$ is fully decomposed (resp. unramified) in K' if and only if $K' \subset \sigma D_{\mathfrak{P}}$ (resp. $K' \subset \sigma T_{\mathfrak{P}}$) for all $\sigma \in \mathrm{Gal}(L/K)$. Similarly, if $i \in [1, n]$, then $\mathfrak{p}$ is fully decomposed (resp. unramified) in K_i if and only if $K_i \subset \sigma D_{\mathfrak{P}}$ (resp. $K_i \subset \sigma T_{\mathfrak{P}}$) for all $\sigma \in \mathrm{Gal}(L/K)$. However, if L' is any intermediate field of L/K, then $K' \subset L'$ if and only if $K_i \subset L'$ for all $i \in [1, n]$. Hence the assertion follows.

2. Let $G = \{\sigma_1, \ldots, \sigma_n\}$ and $K_i = \sigma K'$. Then $L = K_1 \cdot \ldots \cdot K_n$, and $\mathfrak{p}$ is fully decomposed (resp. unramified) in K' if and only if it is fully decomposed (resp. unramified) in K_i for all $i \in [1, n]$. Hence the assertion follows by 1.

3. We must prove: If $\mathfrak{P} \in \mathcal{P}(\mathfrak{O})$ is such that $\mathfrak{P} \cap K_1/\mathfrak{P} \cap K$ is unramified, then $\mathfrak{P} \cap K_1'/\mathfrak{P} \cap K'$ is unramified, too. We apply 2.13.4.3.

Let $\mathfrak{P} \in \mathcal{P}(\mathfrak{O})$, and let $\mathfrak{P} \cap K_1/\mathfrak{P} \cap K$ be unramified. Then $K_1 \subset T_{\mathfrak{P}}$ and therefore $K_1' \subset T_{\mathfrak{P}}K'$. If $G' = \mathrm{Gal}(L/K')$, then $I_{\mathfrak{P}}' = I_{\mathfrak{P}} \cap G'$ is the inertia group and therefore $T_{\mathfrak{P}}' = T_{\mathfrak{P}}K'$ is the inertia field of $\mathfrak{P}$ over K'. Hence $\mathfrak{P} \cap K_1'/\mathfrak{P} \cap K'$ is unramified. □

Corollary and Definition 2.13.6. *Assume* **G**. *Let L/K be abelian, $\mathfrak{p} \in \mathcal{P}(\mathfrak{o})$ such that $\mathsf{k}_{\mathfrak{p}}$ is perfect and $\mathfrak{P} \in \mathcal{P}(\mathfrak{O})$ such that $\mathfrak{P} \cap \mathfrak{o} = \mathfrak{p}$.*
The decomposition group $G_{\mathfrak{P}}$, the inertia group $I_{\mathfrak{P}}$ and the ramification groups $G_{\mathfrak{P},i}$ for $i \geq 0$ only depend on $\mathfrak{p}$ and not on the individual prime divisor $\mathfrak{P}$ of $\mathfrak{p}$. Thus we set $G_{\mathfrak{p}} = G_{\mathfrak{P}}$, $I_{\mathfrak{p}} = I_{\mathfrak{P}}$, $G_{\mathfrak{p},i} = G_{\mathfrak{P},i}$ for all $i \geq 0$, and accordingly $D_{\mathfrak{p}} = D_{\mathfrak{P}}$, $T_{\mathfrak{p}} = T_{\mathfrak{P}}$, etc. Correspondingly we call these groups resp. fields the decomposition group, the inertia group, the ramification groups resp. the decomposition field, the inertia field and the ramification fields of $\mathfrak{p}$ in L, and we call $G \supset G_{\mathfrak{p}} \supset G_{\mathfrak{p},0} \supset G_{\mathfrak{p},1} \supset \ldots$ the Hilbert subgroup series for $\mathfrak{p}$ in L.

1. *$\mathfrak{p}$ is fully decomposed in $D_{\mathfrak{p}}$, the prime divisors of $\mathfrak{p}$ in $D_{\mathfrak{p}}$ are inert in $T_{\mathfrak{p}}$, and the prime divisors of $\mathfrak{p}$ in $T_{\mathfrak{p}}$ are fully ramified in L.*

2. *$D_{\mathfrak{p}}$ is the largest intermediate field of L/K in which $\mathfrak{p}$ is fully decomposed, and $T_{\mathfrak{p}}$ is the largest intermediate field of L/K in which $\mathfrak{p}$ is unramified.*

Proof. Obvious by 2.13.4. □

We return to the general case and study the factorization of a prime ideal in intermediate fields of a finite Galois extension. For this, we need the concept of double cosets. Let G_1 and G_2 be subgroups of a group G. If $g, g' \in G$, then the double cosets G_1gG_2 and $G_1g'G_2$ are either equal or disjoint, and we denote by $G_1\backslash G/G_2 = \{G_1gG_2 \mid g \in G\}$ the **double coset space**.

Theorem 2.13.7. *Assume* **G**. *Let $\mathfrak{P} \in \mathcal{P}(\mathfrak{O})$ and $\mathfrak{P} \cap \mathfrak{o} = \mathfrak{p}$. Let K' be an intermediate field of L/K, $G' = \mathrm{Gal}(L/K')$, $\mathfrak{o}' = \mathfrak{O} \cap K'$, $|G'\backslash G/G_{\mathfrak{P}}| = s$ and $G'\backslash G/G_{\mathfrak{P}} = \{G'\sigma_1G_{\mathfrak{P}}, \ldots, G'\sigma_sG_{\mathfrak{P}}\}$.*

Then $\sigma_1(\mathfrak{P}) \cap K', \ldots, \sigma_s(\mathfrak{P}) \cap K' \in \mathcal{P}(\mathfrak{o}')$ are the distinct prime divisors of $\mathfrak{p}$ in $\mathfrak{o}'$.

Proof. Let $\theta\colon G \to \{\mathfrak{q} \in \mathcal{P}(\mathfrak{o}') \mid \mathfrak{q} \cap K = \mathfrak{p}\}$ be defined by $\theta(\sigma) = \sigma(\mathfrak{P}) \cap K'$. Then θ is surjective, and if σ, $\tau \in G$, then we obtain:

$$\begin{aligned} \sigma\mathfrak{P} \cap K' = \tau\mathfrak{P} \cap K' &\iff \sigma\mathfrak{P} = \rho\tau\mathfrak{P} \text{ for some } \rho \in G' \\ &\iff \sigma^{-1}\rho\tau \in G_{\mathfrak{P}} \text{ for some } \rho \in G' \\ &\iff \tau \in \rho^{-1}\sigma G_{\mathfrak{P}} \text{ for some } \rho \in G' \\ &\iff \tau \in G'\sigma G_{\mathfrak{P}} \iff G'\sigma G_{\mathfrak{P}} = G'\tau G_{\mathfrak{P}}. \end{aligned}$$

Hence θ induces a bijective map $\theta^*\colon G'\backslash G/G_{\mathfrak{P}} \to \{\mathfrak{q} \in \mathcal{P}(\mathfrak{o}') \mid \mathfrak{q} \cap K = \mathfrak{p}\}$ satisfying $\theta^*(G'\sigma G_{\mathfrak{P}}) = \sigma\mathfrak{P} \cap K'$ for all $\sigma \in G$. □

We continue with an investigation of the structure of the higher ramification groups.

Theorem 2.13.8. *Assume* **G**. *Let* $\mathfrak{P} \in \mathcal{P}(\mathfrak{O})$ *and* $\mathfrak{p} = \mathfrak{P} \cap \mathfrak{o}$ *such that* $\mathsf{k}_{\mathfrak{P}}/\mathsf{k}_{\mathfrak{p}}$ *is separable. Let* $U_{\mathfrak{P}} = U_{\mathfrak{P}}^{(0)} = \mathfrak{O}_{\mathfrak{P}}^{\times}$, *and for* $i \ge 1$ *let* $U_{\mathfrak{P}}^{(i)} = 1 + \mathfrak{P}^i\mathfrak{O}_{\mathfrak{P}}$ *be the group of principal units of level* i *in* $\mathfrak{O}_{\mathfrak{P}}$. *Let* $G_{\mathfrak{P}}$ *be the decomposition group,* $I_{\mathfrak{P}}$ *the inertia group and* $G_{\mathfrak{P},i}$ *for* $i \ge 0$ *the* i*-th ramification group of* $\mathfrak{P}$ *over* K.

1. *Let* $\pi \in L$ *be such that* $v_{\mathfrak{P}}(\pi) = 1$.

(a) *If* $\sigma \in I_{\mathfrak{P}}$, *then* $v_{\mathfrak{P}}(\sigma\pi - \pi) = \inf\{v_{\mathfrak{P}}(\sigma a - a) \mid a \in \mathfrak{O}\}$.

(b) *If* $i \ge 0$, *then*

$$G_{\mathfrak{P},i} = \Big\{\sigma \in I_{\mathfrak{P}} \;\Big|\; \frac{\sigma\pi}{\pi} \in U_{\mathfrak{P}}^{(i)}\Big\} = \{\sigma \in I_{\mathfrak{P}} \mid v_{\mathfrak{P}}(\sigma\pi - \pi) \ge i+1\},$$

and

$$\begin{aligned} \phi_i\colon G_{\mathfrak{P},i}/G_{\mathfrak{P},i+1} &\to U_{\mathfrak{P}}^{(i)}/U_{\mathfrak{P}}^{(i+1)}, \\ \textit{defined by } \phi_i(\sigma G_{\mathfrak{P},i+1}) &= \frac{\sigma\pi}{\pi}\,U_{\mathfrak{P}}^{(i+1)}, \end{aligned}$$

is a monomorphism which does not depend on π.

2. *The group* $I_{\mathfrak{P}}/G_{\mathfrak{P},1}$ *is cyclic.*

(a) *If* $\mathrm{char}(\mathsf{k}_{\mathfrak{p}}) = 0$, *then* $I_{\mathfrak{P}}$ *is cyclic, and* $G_{\mathfrak{P},1} = \mathbf{1}$.

(b) *If* $\mathrm{char}(\mathsf{k}_{\mathfrak{p}}) = p > 0$, *then* $p \nmid (I_{\mathfrak{P}} : G_{\mathfrak{P},1})$, *and* $G_{\mathfrak{P},i}/G_{\mathfrak{P},i+1}$ *is an elementary abelian* p*-group for all* $i \ge 1$.

3. *Let* $\mathsf{k}_{\mathfrak{p}}$ *be finite. Then* $G_{\mathfrak{P}}/I_{\mathfrak{P}}$ *is cyclic and* $G_{\mathfrak{P}}$ *is solvable.*

Proof. 1. (a) Let $\sigma \in I_{\mathfrak{P}}$. It suffices to prove that $v_{\mathfrak{P}}(\sigma\pi - \pi) \le v_{\mathfrak{p}}(\sigma a - a)$ for all $a \in \mathfrak{O}$. Let $T_{\mathfrak{P}} = L^{I_{\mathfrak{P}}}$ be the inertia field of $\mathfrak{P}$ over K, $\mathfrak{O}_T = \mathfrak{O} \cap T_{\mathfrak{P}}$ and $\mathfrak{P}_T = \mathfrak{P} \cap T_{\mathfrak{P}}$. Then 2.13.4.2 implies $\mathsf{k}_{\mathfrak{P}} = \mathsf{k}_{\mathfrak{P}_T}$, and therefore $\mathsf{k}_{\mathfrak{P}}$ has a set of representatives $\mathcal{R}$ in $\mathfrak{O}_T$ such that $0 \in \mathcal{R}$. Now let $a \in \mathfrak{O}$. By 2.10.6.2 there exists a sequence $(r_n)_{n\ge 0}$ in $\mathcal{R}$ such that $a \equiv r_0 + r_1\pi + \ldots + r_i\pi^i \mod \mathfrak{P}^{i+1}\mathfrak{O}_{\mathfrak{P}}$ for all $i \ge 0$, and therefore

$$\begin{aligned}\sigma(a) - a &\equiv \sum_{j=0}^{i} r_j\sigma\pi^j - \sum_{j=0}^{i} r_j\pi^j \\ &\equiv (\sigma\pi - \pi)\sum_{j=1}^{i} r_j \sum_{\nu=1}^{j} (\sigma\pi)^{j-\nu}\pi^{\nu-1} \mod \mathfrak{P}^{i+1}\mathfrak{O}_{\mathfrak{P}}.\end{aligned}$$

For all $i \ge 0$ this implies $\sigma a - a = (\sigma\pi - \pi)b_i + \pi^{i+1}c_i$ for some $b_i,\, c_i \in \mathfrak{O}_{\mathfrak{P}}$, hence $v_{\mathfrak{P}}(\sigma a - a) \ge \min\{v_{\mathfrak{P}}(\sigma\pi - \pi), i+1\}$, and therefore eventually $v_{\mathfrak{P}}(\sigma a - a) \ge v_{\mathfrak{P}}(\sigma\pi - \pi)$.

(b) Let $i \ge 0$. If $\sigma \in I_{\mathfrak{P}}$, then $v_{\mathfrak{P}}(\sigma\pi - \pi) = \inf\{v_{\mathfrak{P}}(\sigma a - a) \mid a \in \mathfrak{O}\}$ by 1., and therefore

$$\sigma \in G_{\mathfrak{P},i} \iff v_{\mathfrak{P}}(\sigma\pi - \pi) = v_{\mathfrak{P}}\Bigl(\frac{\sigma\pi}{\pi} - 1\Bigr) + 1 \ge i+1 \iff \frac{\sigma\pi}{\pi} \in U_{\mathfrak{P}}^{(i)}.$$

It follows that

$$G_{\mathfrak{P},i} = \Bigl\{\sigma \in I_{\mathfrak{P}} \Bigm| \frac{\sigma\pi}{\pi} \in U_{\mathfrak{P}}^{(i)}\Bigr\}.$$

If $\sigma \in G_{\mathfrak{P},i}$ and $u \in U_{\mathfrak{P}}$, then

$$\frac{\sigma(\pi u)}{\pi u} = \frac{\sigma\pi}{\pi}\,\frac{\sigma u}{u} \quad \text{and} \quad v_{\mathfrak{P}}\Bigl(\frac{\sigma u}{u} - 1\Bigr) = v_{\mathfrak{P}}(\sigma u - u) \ge i+1,$$

hence

$$\frac{\sigma(\pi u)}{\pi u} \in \frac{\sigma\pi}{\pi}\,U_{\mathfrak{P}}^{(i+1)}.$$

Therefore the map

$$\phi_i'\colon G_{\mathfrak{P},i} \to U_{\mathfrak{P}}^{(i)}/U_{\mathfrak{P}}^{(i+1)}, \quad \text{defined by} \quad \phi_i'(\sigma) = \frac{\sigma\pi}{\pi}\,U_{\mathfrak{P}}^{(i+1)}$$

does not depend on π. If $\sigma,\, \tau \in G$, then

$$\phi_i'(\sigma\tau) = \frac{\sigma\tau\pi}{\pi}U_{\mathfrak{P}}^{(i+1)} = \Bigl(\frac{\sigma(\tau\pi)}{\tau\pi}U_{\mathfrak{P}}^{(i+1)}\Bigr)\Bigl(\frac{\tau\pi}{\pi}U_{\mathfrak{P}}^{(i+1)}\Bigr) = \phi_i'(\sigma)\phi_i'(\tau).$$

Hence ϕ_i' is a homomorphism. By definition, $\mathrm{Ker}(\phi_i') = G_{\mathfrak{P},i+1}$, and ϕ_i' induces a monomorphism

$$\phi_i\colon G_{\mathfrak{P},i}/G_{\mathfrak{P},i+1} \to U_{\mathfrak{P}}^{(i)}/U_{\mathfrak{P}}^{(i+1)}$$
$$\text{such that } \phi_i(\sigma G_{\mathfrak{P},i+1}) = \frac{\sigma\pi}{\pi}\,U_{\mathfrak{P}}^{(i+1)} \quad \text{for all } \sigma \in G_{\mathfrak{P},i}.$$

2. By 1. we get a monomorphism $\phi_0\colon I_{\mathfrak{P}}/G_{\mathfrak{P},1} = G_{\mathfrak{P},0}/G_{\mathfrak{P},1} \to U_{\mathfrak{P}}/U_{\mathfrak{P}}^{(1)}$, and by 2.10.5.3 we obtain $U_{\mathfrak{P}}/U_{\mathfrak{P}}^{(1)} \cong \mathsf{k}_{\mathfrak{P}}^{\times}$. By 1.2.1, $I_{\mathfrak{P}}/G_{\mathfrak{P},1}$ is cyclic, and $p \nmid (I_{\mathfrak{P}}\!:\!G_{\mathfrak{P},1})$ if $\mathrm{char}(\mathsf{k}_{\mathfrak{p}}) = p > 0$.

If $i \geq 1$, let $\phi_i\colon G_{\mathfrak{P},i}/G_{\mathfrak{P},i+1} \to U_{\mathfrak{P}}^{(i)}/U_{\mathfrak{P}}^{(i+1)}$ be this monomorphism given by 1. By 2.10.5.4 we obtain $U_{\mathfrak{P}}^{(i)}/U_{\mathfrak{P}}^{(i+1)} \cong \mathsf{k}_{\mathfrak{P}}$. As a consequence, $G_{\mathfrak{P},i}/G_{\mathfrak{P},i+1}$ is isomorph to a finite subgroup of the additive group $\mathsf{k}_{\mathfrak{P}}$, hence it is trivial if $\mathrm{char}(\mathsf{k}_{\mathfrak{p}}) = 0$, and it is an elementary abelian p-group if $\mathrm{char}(\mathsf{k}_{\mathfrak{p}}) = p > 0$. In particular, if $\mathrm{char}(\mathsf{k}_{\mathfrak{p}}) = 0$, then $G_{\mathfrak{P},i} = G_{\mathfrak{P},i+1}$ for all $i \geq 1$, hence $G_{\mathfrak{P},1} = \mathbf{1}$ by 2.13.2.1, and $I_{\mathfrak{P}}$ is cyclic.

3. By 2.13.2, $I_{\mathfrak{P}}$ is a normal subgroup of $G_{\mathfrak{P}}$, $G_{\mathfrak{P}}/I_{\mathfrak{P}} \cong \mathrm{Gal}(\mathsf{k}_{\mathfrak{P}}/\mathsf{k}_{\mathfrak{p}})$ is cyclic, and the subgroup series $G_{\mathfrak{P}} = \mathrm{Gal}(L/D_{\mathfrak{P}}) \supset I_{\mathfrak{P}} = G_{\mathfrak{P},0} \supset G_{\mathfrak{P},1} \supset \ldots$ shows that $G_{\mathfrak{P}}$ is solvable. □

We close this section with an example which demonstrates our theory in the case of a non-abelian field extension of degree 6 (for more details see [23]).

Example 2.13.9. Let $\mathfrak{o}$ be a Dedekind domain, $K = \mathsf{q}(\mathfrak{o})$ and L/K a Galois extension with Galois group $G = \mathfrak{S}_3 = \langle \sigma, \tau \rangle$, where $\sigma = (123)$ and $\tau = (12)$. Then the groups G, $A = \langle \sigma \rangle$, $H = \langle \tau \rangle$, $H' = \langle \sigma\tau \rangle$, $H'' = \langle \sigma^2\tau \rangle$ and $\mathbf{1}$ are all subgroups of G, and we consider their fixed fields

$$M = L^H,\ M' = L^{H'},\ M'' = L^{H''} \text{ and } L_0 = L^A.$$

The fields M, M', M'' are conjugate and we consider M as a representative. Note that $[M\!:\!K] = 3$ and $[L_0\!:\!K] = 2$. The following two diagrams illustrate the lattices of intermediate fields of L/K and corresponding subgroups of G.

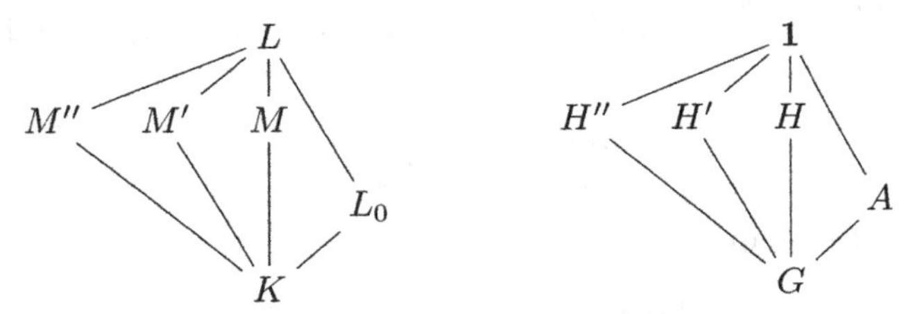

For prime ideals $\mathfrak{p} \in \mathcal{P}(\mathfrak{o})$ and $\mathfrak{P} \in \mathcal{P}(\mathfrak{O})$ such that $\mathfrak{P} \cap K = \mathfrak{p}$ and $\mathsf{k}_{\mathfrak{P}}/\mathsf{k}_{\mathfrak{p}}$ is separable the table below lists all possible Hilbert subgroup series

$$G \supset G_{\mathfrak{P}} \supset I_{\mathfrak{P}} = G_{\mathfrak{P},0} \supset G_{\mathfrak{P},1}$$

and the corresponding behavior of the prime divisors of $\mathfrak{p}$ in the actual intermediate fields. The factorization type of $\mathfrak{p}$ in the (non-trivial) intermediate fields follows using 2.13.3.1, 2.13.4 and 2.13.7.

If $\mathrm{char}(\mathsf{k}_{\mathfrak{p}}) = p > 0$, then we get the following restrictions from 2.13.8: line 1 cannot occur; the lines 3 and 6 are only possible for $p \neq 3$; the lines 2 and 5 are only possible for $p = 3$; line 9 is only possible for $p \neq 2$; line 8 is only possible for $p = 2$; if $\mathsf{k}_{\mathfrak{p}}$ is finite, then line 4 cannot occur.

	G	$G_{\mathfrak{P}}$	$I_{\mathfrak{P}}$	$G_{\mathfrak{P},1}$	L	L_0	M
1	G	G	G	G	$\mathfrak{P}^6$	$\mathfrak{Q}^2$	$\mathfrak{R}^3$
2	G	G	A	A	$\mathfrak{P}^3$	$\mathfrak{Q}$	$\mathfrak{R}^3$
3	G	G	A	1	$\mathfrak{P}^3$	$\mathfrak{Q}$	$\mathfrak{R}^3$
4	G	G	1	1	$\mathfrak{P}$	$\mathfrak{Q}$	$\mathfrak{R}$
5	G	A	A	A	$(\mathfrak{P}\mathfrak{P}')^3$	$\mathfrak{Q}\mathfrak{Q}'$	$\mathfrak{R}^3$
6	G	A	A	1	$(\mathfrak{P}\mathfrak{P}')^3$	$\mathfrak{Q}\mathfrak{Q}'$	$\mathfrak{R}^3$
7	G	A	1	1	$\mathfrak{P}\mathfrak{P}'$	$\mathfrak{Q}\mathfrak{Q}'$	$\mathfrak{R}$
8	G	H	H	H	$(\mathfrak{P}\mathfrak{P}'\mathfrak{P}'')^2$	$\mathfrak{Q}^2$	$\mathfrak{R}\mathfrak{R}'^2$
9	G	H	H	1	$(\mathfrak{P}\mathfrak{P}'\mathfrak{P}'')^2$	$\mathfrak{Q}^2$	$\mathfrak{R}\mathfrak{R}'^2$
10	G	H	1	1	$\mathfrak{P}\mathfrak{P}'\mathfrak{P}''$	$\mathfrak{Q}$	$\mathfrak{R}\mathfrak{R}'$
11	G	1	1	1	$\mathfrak{P}\mathfrak{P}'\mathfrak{P}''\mathfrak{P}_1\mathfrak{P}_1'\mathfrak{P}_1''$	$\mathfrak{Q}\mathfrak{Q}'$	$\mathfrak{R}\mathfrak{R}'\mathfrak{R}''$

2.14 Ideal norms and Frobenius automorphisms

A Dedekind domain $\mathfrak{o}$ is said to have **finite residue class rings** if $\mathfrak{o}/\mathfrak{a}$ is finite for all $\mathfrak{a} \in \mathcal{J}'(\mathfrak{o})$. If this is the case, then we call $\mathfrak{N}(\mathfrak{a}) = |\mathfrak{o}/\mathfrak{a}|$ the **absolute norm** of $\mathfrak{a}$, we set $\varphi_{\mathfrak{o}}(\mathfrak{a}) = \varphi(\mathfrak{a}) = |(\mathfrak{o}/\mathfrak{a})^\times|$ and we call $\varphi_{\mathfrak{o}} \colon \mathcal{J}'(\mathfrak{o}) \to \mathbb{N}$ the **Euler phi function** for $\mathfrak{o}$.

$\mathbb{Z}$ is a Dedekind domain with finite residue class rings. If $a \in \mathbb{Z}^\bullet$, then $\mathfrak{N}(a\mathbb{Z}) = |a|$ and $\varphi(a\mathbb{Z}) = \varphi(|a|)$ (the usual Euler phi function).

Theorem 2.14.1. *Let $\mathfrak{o}$ be a Dedekind domain such that $\mathsf{k}_{\mathfrak{p}} = \mathfrak{o}/\mathfrak{p}$ is finite for all $\mathfrak{p} \in \mathcal{P}(\mathfrak{o})$. Then $\mathfrak{o}$ has finite residue class rings $\mathfrak{a} = \mathfrak{p}_1^{e_1} \cdot \ldots \cdot \mathfrak{p}_r^{e_r} \in \mathcal{J}'(\mathfrak{o})$, where $r \in \mathbb{N}_0$, $\mathfrak{p}_1, \ldots, \mathfrak{p}_r \in \mathcal{P}(\mathfrak{o})$ are distinct and $e_1, \ldots, e_r \in \mathbb{N}$, then*

$$\mathfrak{N}(\mathfrak{a}) = \prod_{i=1}^r \mathfrak{N}(\mathfrak{p}_i)^{e_i} \text{ and } \varphi(\mathfrak{a}) = \prod_{i=1}^r \mathfrak{N}(\mathfrak{p}_i)^{e_i-1}(\mathfrak{N}(\mathfrak{p}_i)-1) = \mathfrak{N}(\mathfrak{a}) \prod_{i=1}^r \Big(1 - \frac{1}{\mathfrak{N}(\mathfrak{p}_i)}\Big).$$

In particular, it follows that $\mathfrak{N}(\mathfrak{a}\mathfrak{b}) = \mathfrak{N}(\mathfrak{a})\mathfrak{N}(\mathfrak{b})$ for all $\mathfrak{a}, \mathfrak{b} \in \mathcal{J}'(\mathfrak{o})$, and $\varphi_{\mathfrak{o}}(\mathfrak{a}\mathfrak{b}) = \varphi_{\mathfrak{o}}(\mathfrak{a})\varphi_{\mathfrak{o}}(\mathfrak{b})$ for all $\mathfrak{a}, \mathfrak{b} \in \mathcal{J}'(\mathfrak{o})$ such that $\mathfrak{a} + \mathfrak{b} = \mathfrak{o}$.

Proof. Let first $\mathfrak{p} \in \mathcal{P}(\mathfrak{o})$, $|\mathsf{k}_{\mathfrak{p}}| = \mathfrak{N}(\mathfrak{p}) < \infty$, $\pi \in \mathfrak{p} \setminus \mathfrak{p}^2$ and $e \in \mathbb{N}$. Let $\mathcal{R}$ be a set of representatives for $\mathsf{k}_{\mathfrak{p}}$ in $\mathfrak{o}$ such that $0 \in \mathfrak{R}$. By 2.10.6.1 every $a \in \mathfrak{o}$ has a unique representation $a \equiv a_0 + a_1\pi + \ldots + a_{e-1}\pi^{e-1} \bmod \mathfrak{p}^e$ with $a_0, \ldots, a_{e-1} \in \mathcal{R}$. Hence $|\mathfrak{o}/\mathfrak{p}^e| = \mathfrak{N}(\mathfrak{p})^e < \infty$, and since $a + \mathfrak{p}^e \in (\mathfrak{o}/\mathfrak{p}^e)^\times$ if and only if $a_0 \notin \mathfrak{p}$, we obtain $|(\mathfrak{o}(\mathfrak{p}^e)^\times| = (\mathfrak{N}(\mathfrak{p}) - 1)\mathfrak{N}(\mathfrak{p})^{e-1}$.

Let now $\mathfrak{a} \in \mathcal{J}'(\mathfrak{o})$ be as in the theorem. By the Chinese remainder theorem 1.1.4 there exist isomorphisms

$$\mathfrak{o}/\mathfrak{a} \stackrel{\sim}{\to} \prod_{i=1}^r \mathfrak{o}/\mathfrak{p}_i^{e_i} \quad \text{and} \quad (\mathfrak{o}/\mathfrak{a})^\times \stackrel{\sim}{\to} \prod_{i=1}^r (\mathfrak{o}/\mathfrak{p}_i^{e_i})^\times.$$

Hence $\mathfrak{o}/\mathfrak{a}$ is finite,

$$\mathfrak{N}(\mathfrak{a}) = |\mathfrak{o}/\mathfrak{a}| = \prod_{i=1}^{r} \mathfrak{N}(\mathfrak{p}_i)^{e_i}$$

and

$$\varphi(\mathfrak{a}) = |(\mathfrak{o}/\mathfrak{a})^\times| = \prod_{i=1}^{r} \mathfrak{N}(\mathfrak{p}_i)^{e_i-1}(\mathfrak{N}(\mathfrak{p}_i)-1) = \mathfrak{N}(\mathfrak{a}) \prod_{i=1}^{r}\Bigl(1-\frac{1}{\mathfrak{N}(\mathfrak{p}_i)}\Bigr). \quad \square$$

By 2.14.1, $\mathfrak{N}\colon \mathcal{J}'(\mathfrak{o}) \to \mathbb{N}$ is a monoid homomorphism. It uniquely extends to an (equally denoted) group homomorphism $\mathfrak{N} = \mathfrak{N}_\mathfrak{o}\colon \mathcal{J}(\mathfrak{o}) \to \mathbb{Q}_{>0}$, which we call the **absolute norm** (of $\mathfrak{o}$).

Next we define the ideal norm for extensions of Dedekind domains, which is a relative counterpart to the absolute norm. For this we restrict to separable field extensions in order to avoid pathological behavior.

Definition and Theorem 2.14.2. *Let $\mathfrak{o}$ be a Dedekind domain, $K = \mathsf{q}(\mathfrak{o})$, L/K a finite separable field extension, $[L:K] = n$ and $\mathfrak{O} = \mathrm{cl}_L(\mathfrak{o})$.*
The **ideal norm** $\mathcal{N}_{\mathfrak{O}/\mathfrak{o}}\colon \mathcal{J}(\mathfrak{O}) \to \mathcal{J}(\mathfrak{o})$ is the unique group homomorphism satisfying $\mathcal{N}_{\mathfrak{O}/\mathfrak{o}}(\mathfrak{P}) = \mathfrak{p}^{f(\mathfrak{P}/\mathfrak{p})}$ for all $\mathfrak{P} \in \mathcal{P}(\mathfrak{O})$ and $\mathfrak{p} = \mathfrak{P} \cap \mathfrak{o}$.

1. *Let K' be an intermediate field of L/K and $\mathfrak{o}' = \mathrm{cl}_{K'}(\mathfrak{o})$. Then $\mathcal{N}_{\mathfrak{O}/\mathfrak{o}} = \mathcal{N}_{\mathfrak{o}'/\mathfrak{o}} \circ \mathcal{N}_{\mathfrak{O}/\mathfrak{o}'}\colon \mathcal{J}(\mathfrak{O}) \to \mathcal{J}(\mathfrak{o})$.*
2. *If $\mathfrak{a} \in \mathcal{J}(\mathfrak{o})$, then $\mathcal{N}_{\mathfrak{O}/\mathfrak{o}}(\mathfrak{a}\mathfrak{O}) = \mathfrak{a}^n$.*
3. *Let L/K be Galois, $G = \mathrm{Gal}(L/K)$ and $\mathfrak{A} \in \mathcal{J}(\mathfrak{O})$. Then*

$$\mathcal{N}_{\mathfrak{O}/\mathfrak{o}}(\mathfrak{A})\mathfrak{O} = \prod_{\sigma\in G} \sigma\mathfrak{A}.$$

4. *If $a \in L^\times$, then $\mathcal{N}_{\mathfrak{O}/\mathfrak{o}}(a\mathfrak{O}) = \mathsf{N}_{L/K}(a)\mathfrak{o}$.*

Proof. 1. It suffices to prove that $\mathcal{N}_{\mathfrak{O}/\mathfrak{o}}(\mathfrak{P}) = \mathcal{N}_{\mathfrak{o}'/\mathfrak{o}} \circ \mathcal{N}_{\mathfrak{O}/\mathfrak{o}'}(\mathfrak{P})$ for $\mathfrak{P} \in \mathcal{P}(\mathfrak{O})$. If $\mathfrak{P} \in \mathcal{P}(\mathfrak{O})$, $\mathfrak{p}' = \mathfrak{P} \cap \mathfrak{o}'$ and $\mathfrak{p} = \mathfrak{P} \cap \mathfrak{o}$, then $f(\mathfrak{P}/\mathfrak{p}) = f(\mathfrak{P}/\mathfrak{p}')f(\mathfrak{p}'/\mathfrak{p})$, and consequently

$$\mathcal{N}_{\mathfrak{o}'/\mathfrak{o}} \circ \mathcal{N}_{\mathfrak{O}/\mathfrak{o}'}(\mathfrak{P}) = \mathcal{N}_{\mathfrak{o}'/\mathfrak{o}}(\mathfrak{p}'^{f(\mathfrak{P}/\mathfrak{p}')}) = \mathfrak{p}^{f(\mathfrak{P}/\mathfrak{p}')f(\mathfrak{p}'/\mathfrak{p})} = \mathfrak{p}^{f(\mathfrak{P}/\mathfrak{p})} = \mathcal{N}_{\mathfrak{O}/\mathfrak{o}}(\mathfrak{P}).$$

2. It suffices to prove the assertion for $\mathfrak{a} = \mathfrak{p} \in \mathcal{P}(\mathfrak{o})$. If $\mathfrak{p} \in \mathcal{P}(\mathfrak{o})$, then

$$\mathfrak{p}\mathfrak{O} = \prod_{\mathfrak{P}\,|\,\mathfrak{p}} \mathfrak{P}^{e(\mathfrak{P}/\mathfrak{p})} \quad\text{and}\quad \mathcal{N}_{\mathfrak{O}/\mathfrak{o}}(\mathfrak{p}\mathfrak{O}) = \prod_{\mathfrak{P}\,|\,\mathfrak{p}} \mathfrak{p}^{e(\mathfrak{P}/\mathfrak{p})f(\mathfrak{P}/\mathfrak{p})} = \mathfrak{p}^n \quad\text{by 2.12.4.}$$

3. It suffices to prove the assertion for $\mathfrak{A} = \mathfrak{P} \in \mathcal{P}(\mathfrak{O})$. Let $\mathfrak{P} \in \mathcal{P}(\mathfrak{O})$, $\mathfrak{p} = \mathfrak{P} \cap \mathfrak{o}$, $e = e(\mathfrak{P}/\mathfrak{p})$, $f = f(\mathfrak{P}/\mathfrak{p})$, $(G : G_{\mathfrak{P}}) = r$, and suppose that $G/G_{\mathfrak{P}} = \{\sigma_1 G_{\mathfrak{P}}, \ldots, \sigma_r G_{\mathfrak{P}}\}$. Then 2.13.1.2 implies $|G_{\mathfrak{P}}| = ef$ and

$$\prod_{\sigma \in G} \sigma\mathfrak{P} = \prod_{i=1}^{r} \prod_{\tau \in G_{\mathfrak{P}}} \sigma_i\tau\mathfrak{P} = \prod_{i=1}^{r} (\sigma_i\mathfrak{P})^{ef} = (\mathfrak{p}\mathfrak{O})^f = \mathcal{N}_{\mathfrak{O}/\mathfrak{o}}(\mathfrak{P})\mathfrak{O}.$$

4. Let L' be a Galois closure of L/K. Let $\mathfrak{O}' = \mathrm{cl}_{L'}(\mathfrak{O}) = \mathrm{cl}_{L'}(\mathfrak{o})$, $G = \mathrm{Gal}(L'/K)$ and $d = [L' : L]$. Then L'/K is Galois, and

$$\begin{aligned}\mathcal{N}_{\mathfrak{O}'/\mathfrak{o}}(a\mathfrak{O}')\mathfrak{O}' &= \prod_{\sigma \in G} \sigma(a\mathfrak{O}') = \prod_{\sigma \in G} \sigma(a)\mathfrak{O}' = \mathsf{N}_{L'/K}(a)\mathfrak{O}' = \mathsf{N}_{L/K} \circ \mathsf{N}_{L'/L}(a)\mathfrak{O}' \\ &= \mathsf{N}_{L/K}(a^d)\mathfrak{O}' = [\mathsf{N}_{L/K}(a)\mathfrak{o}]^d\mathfrak{O}'.\end{aligned}$$

On the other hand, $\mathcal{N}_{\mathfrak{O}'/\mathfrak{o}}(a\mathfrak{O}')\mathfrak{O}' = \mathcal{N}_{\mathfrak{O}/\mathfrak{o}} \circ \mathcal{N}_{\mathfrak{O}'/\mathfrak{O}}(a\mathfrak{O}')\mathfrak{O}' = \mathcal{N}_{\mathfrak{O}/\mathfrak{o}}(a\mathfrak{O})^d\mathfrak{O}'$. Since the ideal extension $j_{\mathfrak{O}'/\mathfrak{o}} \colon \mathcal{J}(\mathfrak{o}) \to \mathcal{J}(\mathfrak{O}')$ is injective, we get $[\mathsf{N}_{L/K}(a)\mathfrak{o}]^d = \mathcal{N}_{\mathfrak{O}/\mathfrak{o}}(a\mathfrak{O})^d$, and consequently $\mathsf{N}_{L/K}(a)\mathfrak{o} = \mathcal{N}_{\mathfrak{O}/\mathfrak{o}}(a\mathfrak{O})$, since $\mathcal{J}(\mathfrak{o})$ is a free abelian group. □

Corollary 2.14.3. *Let $\mathfrak{o}$ be a Dedekind domain with finite residue class rings. Let $K = \mathsf{q}(\mathfrak{o})$, L/K a finite separable field extension and $\mathfrak{O} = \mathrm{cl}_L(\mathfrak{o})$. Then $\mathfrak{O}$ has finite residue class rings, too, and $\mathfrak{N}_{\mathfrak{O}}(\mathfrak{A}) = \mathfrak{N}_{\mathfrak{o}}(\mathcal{N}_{\mathfrak{O}/\mathfrak{o}}(\mathfrak{A}))$ for all $\mathfrak{A} \in \mathcal{J}(\mathfrak{O})$.*

Proof. If $\mathfrak{P} \in \mathcal{P}(\mathfrak{O})$ and $\mathfrak{p} = \mathfrak{P} \cap \mathfrak{o}$, then $[\mathsf{k}_{\mathfrak{P}} : \mathsf{k}_{\mathfrak{p}}] = f(\mathfrak{P}/\mathfrak{p})$, and as $\mathsf{k}_{\mathfrak{p}}$ is finite, we obtain $|\mathsf{k}_{\mathfrak{P}}| = |\mathsf{k}_{\mathfrak{p}}|^{f(\mathfrak{P}/\mathfrak{p})} < \infty$. Hence $\mathfrak{O}$ has finite residue class rings, and $\mathfrak{N}_{\mathfrak{O}}(\mathfrak{P}) = \mathfrak{N}_{\mathfrak{o}}(\mathfrak{p})^{f(\mathfrak{P}/\mathfrak{p})} = \mathfrak{N}_{\mathfrak{o}}(\mathfrak{p}^{f(\mathfrak{P}/\mathfrak{p})}) = \mathfrak{N}_{\mathfrak{o}}(\mathcal{N}_{\mathfrak{O}/\mathfrak{o}}(\mathfrak{P}))$. As $\mathcal{J}(\mathfrak{O})$ is a free abelian group with basis $\mathcal{P}(\mathfrak{O})$ and the maps $\mathfrak{N}_{\mathfrak{O}} \colon \mathcal{P}(\mathfrak{O}) \to \mathbb{Q}_{>0}$ and $\mathfrak{N}_{\mathfrak{o}} \circ \mathcal{N}_{\mathfrak{O}/\mathfrak{o}} \colon \mathcal{P}(\mathfrak{O}) \to \mathbb{Q}_{>0}$ are group homomorphisms, it follows that $\mathfrak{N}_{\mathfrak{O}}(\mathfrak{A}) = \mathfrak{N}_{\mathfrak{o}}(\mathcal{N}_{\mathfrak{O}/\mathfrak{o}}(\mathfrak{A}))$ for all $\mathfrak{A} \in \mathcal{J}(\mathfrak{O})$. □

Theorem 2.14.4. *Let $\mathfrak{o}$ be a Dedekind domain, $K = \mathsf{q}(\mathfrak{o})$, L/K a finite separable field extension, $[L:K] = n$ and $\mathfrak{O} = \mathrm{cl}_L(\mathfrak{o})$.*

1. If $\mathfrak{A} \in \mathcal{J}(\mathfrak{O})$ and $a \in L^\times$, then

$$v_{\mathfrak{p}}(\mathcal{N}_{\mathfrak{O}/\mathfrak{o}}(\mathfrak{A})) = \sum_{\substack{\mathfrak{P} \in \mathcal{P}(\mathfrak{O}) \\ \mathfrak{P} \cap \mathfrak{o} = \mathfrak{p}}} f(\mathfrak{P}/\mathfrak{p}) v_{\mathfrak{P}}(\mathfrak{A}) \quad \textit{and}$$

$$v_{\mathfrak{p}}(\mathsf{N}_{L/K}(a)) = \sum_{\substack{\mathfrak{P} \in \mathcal{P}(\mathfrak{O}) \\ \mathfrak{P} \cap \mathfrak{o} = \mathfrak{p}}} f(\mathfrak{P}/\mathfrak{p}) v_{\mathfrak{P}}(a).$$

2. *If* $\mathfrak{m} \in \mathcal{J}'(\mathfrak{o})$ *and* $\mathfrak{A} \in \mathcal{J}^{\mathfrak{m}\mathfrak{O}}(\mathfrak{O})$, *then* $\mathcal{N}_{\mathfrak{O}/\mathfrak{o}}(\mathfrak{A}) \in \mathcal{J}^{\mathfrak{m}}(\mathfrak{o})$.

3. *If* $\mathfrak{A} \in \mathcal{J}(\mathfrak{O})$ *and* $\mathfrak{p} \in \mathcal{P}(\mathfrak{o})$, *then*

$$\mathcal{N}_{\mathfrak{O}/\mathfrak{o}}(\mathfrak{A})\mathfrak{o}_\mathfrak{p} = \mathcal{N}_{\mathfrak{O}_\mathfrak{p}/\mathfrak{o}_\mathfrak{p}}(\mathfrak{A}_\mathfrak{p}) \quad \text{and} \quad \mathcal{N}_{\mathfrak{O}/\mathfrak{o}}(\mathfrak{A}) = \sum_{a\in\mathfrak{A}} \mathsf{N}_{L/K}(a)\mathfrak{o}.$$

Proof. 1. Let $\mathfrak{A} \in \mathcal{J}(\mathfrak{O})$. By definition,

$$\mathcal{N}_{\mathfrak{O}/\mathfrak{o}}(\mathfrak{A}) = \prod_{\mathfrak{P}\in\mathcal{P}(\mathfrak{O})} \mathcal{N}_{\mathfrak{O}/\mathfrak{o}}(\mathfrak{P})^{v_\mathfrak{P}(\mathfrak{A})} = \prod_{\mathfrak{p}\in\mathcal{P}(\mathfrak{o}} \prod_{\substack{\mathfrak{P}\in\mathcal{P}(\mathfrak{O}) \\ \mathfrak{P}\cap\mathfrak{o}=\mathfrak{p}}} \mathfrak{p}^{f(\mathfrak{P}/\mathfrak{p})v_\mathfrak{P}(\mathfrak{A})}$$

and therefore

$$v_\mathfrak{p}(\mathcal{N}_{\mathfrak{O}/\mathfrak{o}}(\mathfrak{A})) = \sum_{\substack{\mathfrak{P}\in\mathcal{P}(\mathfrak{O}) \\ \mathfrak{P}\cap\mathfrak{o}=\mathfrak{p}}} f(\mathfrak{P}/\mathfrak{p})v_\mathfrak{P}(\mathfrak{A}).$$

If $a \in L$, then (using 2.14.2.4) we obtain

$$\begin{aligned} v_\mathfrak{p}(\mathsf{N}_{L/K}(a)) = v_\mathfrak{p}(\mathsf{N}_{L/K}(a)\mathfrak{o}) = v_\mathfrak{p}(\mathcal{N}_{\mathfrak{O}/\mathfrak{o}}(a\mathfrak{O})) &= \sum_{\substack{\mathfrak{P}\in\mathcal{P}(\mathfrak{O}) \\ \mathfrak{P}\cap\mathfrak{o}=\mathfrak{p}}} f(\mathfrak{P}/\mathfrak{p})v_\mathfrak{P}(a\mathfrak{O}) \\ &= \sum_{\substack{\mathfrak{P}\in\mathcal{P}(\mathfrak{O}) \\ \mathfrak{P}\cap\mathfrak{o}=\mathfrak{p}}} f(\mathfrak{P}/\mathfrak{p})v_\mathfrak{P}(a). \end{aligned}$$

2. Let $\mathfrak{m} \in \mathcal{J}'(\mathfrak{o})$, $\mathfrak{A} \in \mathcal{J}^{\mathfrak{m}\mathfrak{O}}(\mathfrak{O})$ and $\mathfrak{p} \in \operatorname{supp}(\mathfrak{m})$. If $\mathfrak{P} \in \mathcal{P}(\mathfrak{O})$ and $\mathfrak{P}\cap\mathfrak{o} = \mathfrak{p}$, then $v_\mathfrak{P}(\mathfrak{m}\mathfrak{O}) = e(\mathfrak{P}/\mathfrak{p})v_\mathfrak{p}(\mathfrak{m}) > 0$, hence $v_\mathfrak{P}(\mathfrak{A}) = 0$, and therefore $v_\mathfrak{p}(\mathcal{N}_{\mathfrak{O}/\mathfrak{o}}(\mathfrak{A})) = 0$ by (a). Consequently, $\mathcal{N}_{\mathfrak{O}/\mathfrak{o}}(\mathfrak{A}) \in \mathcal{J}^{\mathfrak{m}}(\mathfrak{o})$.

3. Let $\mathfrak{A} \in \mathcal{J}(\mathfrak{O})$ and $\mathfrak{p} \in \mathcal{P}(\mathfrak{o})$. By 2.7.8.3 we get

$$\{\mathfrak{P}' \in \mathcal{P}(\mathfrak{O}_\mathfrak{p}) \mid \mathfrak{P}'\cap\mathfrak{o}_\mathfrak{p} = \mathfrak{p}\mathfrak{o}_\mathfrak{p}\} = \{\mathfrak{P}_\mathfrak{p} \mid \mathfrak{P} \in \mathcal{P}(\mathfrak{O}) \mid \mathfrak{P}\cap\mathfrak{o} = \mathfrak{p}\},$$

and if $\mathfrak{P} \in \mathcal{P}(\mathfrak{O})$ and $\mathfrak{P}\cap\mathfrak{o} = \mathfrak{p}$, then $v_{\mathfrak{P}_\mathfrak{p}}(\mathfrak{A}_\mathfrak{p})f(\mathfrak{P}_\mathfrak{p}/\mathfrak{p}\mathfrak{o}_\mathfrak{p}) = v_\mathfrak{P}(\mathfrak{A})f(\mathfrak{P}/\mathfrak{p})$ by 2.8.7.2(a). Hence

$$\begin{aligned} \mathcal{N}_{\mathfrak{O}_\mathfrak{p}/\mathfrak{o}_\mathfrak{p}}(\mathfrak{A}_\mathfrak{p}) &= \prod_{\substack{\mathfrak{P}'\in\mathcal{P}(\mathfrak{O}_\mathfrak{p}) \\ \mathfrak{P}'\cap\mathfrak{o}_\mathfrak{p}=\mathfrak{p}\mathfrak{o}_\mathfrak{p}}} \mathfrak{P}'^{v_{\mathfrak{P}'}(\mathfrak{A}_\mathfrak{p})f(\mathfrak{P}'/\mathfrak{p}\mathfrak{o}_\mathfrak{p})} \\ &= \prod_{\substack{\mathfrak{P}\in\mathcal{P}(\mathfrak{O}) \\ \mathfrak{P}\cap\mathfrak{o}=\mathfrak{p}}} \mathfrak{P}_\mathfrak{p}^{v_\mathfrak{P}(\mathfrak{A})f(\mathfrak{P}/\mathfrak{p})} = \mathcal{N}_{\mathfrak{O}/\mathfrak{o}}(\mathfrak{A})\mathfrak{o}_\mathfrak{p}. \end{aligned}$$

As to the second equality, it suffices to prove that

$$v_\mathfrak{p}(\mathcal{N}_{\mathfrak{O}/\mathfrak{o}}(\mathfrak{A})) = v_\mathfrak{p}\Big(\sum_{a\in\mathfrak{A}} \mathsf{N}_{L/K}(a)\mathfrak{o}\Big) \quad \text{for all } \mathfrak{p} \in \mathcal{P}(\mathfrak{o}).$$

Let $\mathfrak{p} \in \mathcal{P}(\mathfrak{o})$, $\{\mathfrak{P}_1, \ldots, \mathfrak{P}_r\} = \{\mathfrak{P} \in \mathcal{P}(\mathfrak{O}) \mid \mathfrak{P}\cap\mathfrak{o} = \mathfrak{p}\}$. We show first:

A. There exists some $a_0 \in \mathfrak{A}$ such that $v_{\mathfrak{P}_i}(\mathfrak{A}) = v_{\mathfrak{P}_i}(a_0)$ for all $i \in [1, r]$.

Proof of **A.** Let first $\mathfrak{A} \subset \mathfrak{O}$. For $i \in [1, r]$ let $a_i \in \mathfrak{A}$ be such that $v_{\mathfrak{P}_i}(\mathfrak{A}) = v_{\mathfrak{P}_i}(a_i)$ (see 2.8.5.2) and $e \in \mathbb{N}$, $e > \max\{v_{\mathfrak{P}_i}(\mathfrak{A}) \mid i \in [1, r]\}$. By the Chinese remainder theorem 1.1.4 there exists some $a_0 \in \mathfrak{O}$ such that $a_0 \equiv a_i \mod \mathfrak{P}_i^e$ for all $i \in [1, r]$ and $a_0 \equiv 0 \mod \mathfrak{P}^{v_{\mathfrak{P}}(\mathfrak{A})}$ for all $\mathfrak{P} \in \operatorname{supp}(\mathfrak{A}) \setminus \{\mathfrak{P}_1, \ldots, \mathfrak{P}_r\}$. Then $a_0 \in \mathfrak{A}$ and $v_{\mathfrak{P}_i}(a_0) = v_{\mathfrak{P}_i}(a_i) = v_{\mathfrak{P}_i}(\mathfrak{A})$ for all $i \in [1, r]$.

Let now $\mathfrak{A} \in \mathcal{J}(\mathfrak{O})$ be arbitrary. Then $\mathfrak{A}$ is a finitely generated $\mathfrak{o}$-module by 2.5.1.5, by 2.5.1.2 there exists some $q \in \mathfrak{o}^\bullet$ such that $q\mathfrak{A} \subset \mathfrak{O}$, and as above there exists some $a_0' \in q\mathfrak{A}$ such that $v_{\mathfrak{P}_i}(a_0') = v_{\mathfrak{P}_i}(q\mathfrak{A})$ for all $i \in [1, r]$. Therefore we obtain $a_0 = q^{-1}a_0' \in \mathfrak{A}$ and $v_{\mathfrak{P}_i}(\mathfrak{A}) = v_{\mathfrak{P}_i}(a_0)$ for all $i \in [1, r]$. $\square$[**A.**]

Let $a_0 \in \mathfrak{A}$ be as in **A**. If $a \in \mathfrak{A}$, then $v_{\mathfrak{P}_i}(a) \ge v_{\mathfrak{P}_i}(a_0)$ for all $i \in [1, r]$, and

$$v_{\mathfrak{p}}(\mathsf{N}_{L/K}(a)) = \sum_{i=1}^{r} f(\mathfrak{P}_i/\mathfrak{p}) v_{\mathfrak{P}_i}(a) \ge \sum_{i=1}^{r} f(\mathfrak{P}_i/\mathfrak{p}) v_{\mathfrak{P}_i}(a_0) = v_{\mathfrak{p}}(\mathsf{N}_{L/K}(a_0)).$$

Hence

$$\begin{aligned} v_{\mathfrak{p}}(\mathcal{N}_{\mathfrak{O}/\mathfrak{o}}(\mathfrak{A})) &= \sum_{i=1}^{r} f(\mathfrak{P}_i/\mathfrak{p}) v_{\mathfrak{P}_i}(\mathfrak{A}) = \sum_{i=1}^{r} f(\mathfrak{P}_i/\mathfrak{p}) v_{\mathfrak{P}_i}(a_0) = v_{\mathfrak{p}}(\mathsf{N}_{L/K}(a_0)) \\ &= \min\{v_{\mathfrak{p}}(\mathsf{N}_{L/K}(a)) \mid a \in \mathfrak{A}\} = v_{\mathfrak{p}}\Big(\sum_{a \in \mathfrak{A}} \mathsf{N}_{L/K}(a)\mathfrak{o}\Big). \qquad \square \end{aligned}$$

Definition 2.14.5 (Frobenius automorphism)**.** Let $\mathfrak{o}$ be a Dedekind domain with finite residue class rings, $K = \mathsf{q}(\mathfrak{o})$, L/K a finite Galois extension, $G = \operatorname{Gal}(L/K)$, $\mathfrak{O} = \operatorname{cl}_L(\mathfrak{o})$, $\mathfrak{P} \in \mathcal{P}(\mathfrak{O})$ and $\mathfrak{p} = \mathfrak{P} \cap \mathfrak{o}$. An automorphism $\varphi \in G$ is called a **Frobenius automorphism** of $\mathfrak{P}$ (over K) if

$$\varphi(x) \equiv x^{\mathfrak{N}(\mathfrak{p})} \mod \mathfrak{P} \quad \text{for all} \quad x \in \mathfrak{O}.$$

This concept is invariant under localization as follows: If $\mathfrak{P} \in \mathcal{P}(\mathfrak{O})$ and $\mathfrak{p} = \mathfrak{P} \cap \mathfrak{o}$, then $\mathfrak{P}\mathfrak{O}_\mathfrak{p} \in \mathcal{P}(\mathfrak{O}_\mathfrak{p})$, $\mathfrak{P}\mathfrak{O}_\mathfrak{p} \cap \mathfrak{o}_\mathfrak{p} = \mathfrak{p}\mathfrak{o}_\mathfrak{p}$ and $\mathfrak{N}(\mathfrak{p}) = \mathfrak{N}(\mathfrak{p}\mathfrak{o}_\mathfrak{p})$. If $\varphi \in G$, then it evidently follows that $\varphi(x) \equiv x^{\mathfrak{N}(\mathfrak{p})} \mod \mathfrak{P}$ for all $x \in \mathfrak{O}$ if and only if $\varphi(x) \equiv x^{\mathfrak{N}(\mathfrak{p}\mathfrak{o}_\mathfrak{p})} \mod \mathfrak{P}\mathfrak{O}_\mathfrak{p}$ for all $x \in \mathfrak{O}_\mathfrak{p}$. Hence φ is a Frobenius automorphism of $\mathfrak{P}$ over K if and only if φ is a Frobenius automorphism of $\mathfrak{P}\mathfrak{O}_\mathfrak{p}$ over K.

Theorem 2.14.6 (Existence and uniqueness of Frobenius automorphisms)**.** *Let $\mathfrak{o}$ be a Dedekind domain with finite residue class rings, $K = \mathsf{q}(\mathfrak{o})$, L/K a finite Galois extension, $G = \operatorname{Gal}(L/K)$, $\mathfrak{O} = \operatorname{cl}_L(\mathfrak{o})$, $\mathfrak{P} \in \mathcal{P}(\mathfrak{O})$ and $\mathfrak{p} = \mathfrak{P} \cap \mathfrak{o}$. Let $D_\mathfrak{P} = L^{G_\mathfrak{P}}$ be the decomposition field and $T_\mathfrak{P} = L^{I_\mathfrak{P}}$ the inertia field of $\mathfrak{P}$ over K.*

1. An automorphism $\varphi \in G$ *is a Frobenius automorphism of* $\mathfrak{P}$ *over* K *if and only if* $\varphi \in G_{\mathfrak{P}}$, *and the induced residue class automorphism* $\varphi_{\mathfrak{P}} \in \mathrm{Gal}(\mathsf{k}_{\mathfrak{P}}/\mathsf{k}_{\mathfrak{p}})$ (as defined in 2.5.3) *satisfies* $\varphi_{\mathfrak{P}}(\xi) = \xi^{\mathfrak{N}(\mathfrak{p})}$ *for all* $\xi \in \mathsf{k}_{\mathfrak{P}}$.

2. There exists a Frobenius automorphism of $\mathfrak{P}$ *over* K, *and if* φ *is a Frobenius automorphism of* $\mathfrak{P}$ *over* K, *then:*

(a) $\mathrm{Gal}(T_{\mathfrak{P}}/D_{\mathfrak{P}}) = \langle \varphi \restriction T_{\mathfrak{P}} \rangle$, *the coset* $\varphi I_{\mathfrak{P}} \subset G_{\mathfrak{P}}$ *is the set of all Frobenius automorphisms of* $\mathfrak{P}$ *over* K, *and* φ *is also a Frobenius automorphism of* $\mathfrak{P}$ *over* $D_{\mathfrak{P}}$.

(b) If $\sigma \in G$, *then* $\sigma\varphi\sigma^{-1}$ *is a Frobenius automorphism of* $\sigma\mathfrak{P}$ *over* K.

3. Let φ *be a Frobenius automorphism of* $\mathfrak{P}$ *over* K, M *an intermediate field of* L/K *and* $\mathfrak{q} = \mathfrak{P} \cap M$.

(a) $\varphi^{f(\mathfrak{q}/\mathfrak{p})}$ *is a Frobenius automorphism of* $\mathfrak{P}$ *over* M.

(b) If M/K *is Galois, then* $\varphi \restriction M$ *is a Frobenius automorphism of* $\mathfrak{q}$ *over* K.

Proof. 1. If φ is a Frobenius automorphism of $\mathfrak{P}$ over K and $x \in \mathfrak{P}$, then $\varphi(x) \equiv x^{\mathfrak{N}(\mathfrak{p})} \equiv 0 \bmod \mathfrak{P}$ implies $\varphi(x) \in \mathfrak{P}$, hence $\varphi(\mathfrak{P}) \subset \mathfrak{P}$, consequently $\varphi(\mathfrak{P}) = \mathfrak{P}$ and $\varphi \in G_{\mathfrak{P}}$. If $\varphi \in G_{\mathfrak{P}}$, then $\varphi_{\mathfrak{P}} \in \mathrm{Gal}(\mathsf{k}_{\mathfrak{P}}/\mathsf{k}_{\mathfrak{p}})$, and the congruence $\varphi(x) \equiv x^{\mathfrak{N}(\mathfrak{p})} \bmod \mathfrak{P}$ for all $x \in \mathfrak{O}$ is equivalent to $\varphi(\xi) = \xi^{\mathfrak{N}(\mathfrak{p})}$ for all $\xi \in \mathsf{k}_{\mathfrak{P}}$.

2. Since $|\mathsf{k}_{\mathfrak{p}}| = \mathfrak{N}(\mathfrak{p})$, 1.7.1.2 implies $\mathrm{Gal}(\mathsf{k}_{\mathfrak{P}}/\mathsf{k}_{\mathfrak{p}}) = \langle \phi \rangle$, if $\phi(\xi) = \xi^{\mathfrak{N}(\mathfrak{p})}$ for all $\xi \in \mathsf{k}_{\mathfrak{P}}$. By 2.5.4.2 and 2.13.2 the map $\rho_{\mathfrak{P}} \colon G_{\mathfrak{P}} \to \mathrm{Gal}(\mathsf{k}_{\mathfrak{P}}/\mathsf{k}_{\mathfrak{p}})$, $\sigma \mapsto \sigma_{\mathfrak{P}}$, is an epimorphism, and $\mathrm{Ker}(\rho_{\mathfrak{P}}) = I_{\mathfrak{P}}$. By 1., an automorphism $\varphi \in G_{\mathfrak{P}}$ is a Frobenius automorphism of $\mathfrak{P}$ over K if and only if $\rho_{\mathfrak{P}}(\varphi) = \phi$. Hence there exists a Frobenius automorphism φ of $\mathfrak{P}$ over K, and the coset $\rho_{\mathfrak{P}}^{-1}(\phi) = \varphi I_{\mathfrak{P}} \in G_{\mathfrak{P}}/I_{\mathfrak{P}}$ is the set of all Frobenius automorphisms of $\mathfrak{P}$ over K.

Let $\varphi \in G_{\mathfrak{P}}$ be a Frobenius automorphism of $\mathfrak{P}$ over K. By 2.13.2.2 $\rho_{\mathfrak{P}}$ induces an isomorphism $\rho_{\mathfrak{P}}^* \colon G_{\mathfrak{P}}/I_{\mathfrak{P}} \xrightarrow{\sim} \mathrm{Gal}(\mathsf{k}_{\mathfrak{P}}/\mathsf{k}_{\mathfrak{p}})$, and as $G_{\mathfrak{P}}/I_{\mathfrak{P}} = \mathrm{Gal}(T_{\mathfrak{P}}/D_{\mathfrak{P}})$, it follows that $G_{\mathfrak{P}}/I_{\mathfrak{P}} = \mathrm{Gal}(T_{\mathfrak{P}}/D_{\mathfrak{P}}) = \langle \varphi I_{\mathfrak{P}} \rangle = \langle \varphi \restriction T_{\mathfrak{P}} \rangle$. In particular, $\varphi \restriction T_{\mathfrak{P}}$ is an arithmetically distinguished generator of the cyclic Galois group of $T_{\mathfrak{P}}/D_{\mathfrak{P}}$. If $\mathfrak{P}_D = \mathfrak{P} \cap D_{\mathfrak{P}}$, then $\mathsf{k}_{\mathfrak{P}_D} = \mathsf{k}_{\mathfrak{p}}$, and by definition φ is a Frobenius automorphism of $\mathfrak{P}$ over $D_{\mathfrak{P}}$.

If $\sigma \in G$, then $\sigma\mathfrak{O} = \mathfrak{O}$, hence

$$\sigma\varphi\sigma^{-1}(x) - x^{\mathfrak{N}(\mathfrak{p})} = \sigma\big(\varphi(\sigma^{-1}x) - (\sigma^{-1}x)^{\mathfrak{N}(\mathfrak{p})}\big) \in \sigma\mathfrak{P} \text{ for all } x \in \mathfrak{O},$$

and therefore $\sigma\varphi\sigma^{-1}$ is a Frobenius automorphism of $\sigma\mathfrak{P}$ over K.

3. (a) If $x \in \mathfrak{O}$, then $\varphi^{f(\mathfrak{q}/\mathfrak{p})}(x) \equiv x^{\mathfrak{N}(\mathfrak{p})f(\mathfrak{q}/\mathfrak{p})} \equiv x^{\mathfrak{N}(\mathfrak{q})} \bmod \mathfrak{P}$, and therefore $\varphi^{f(\mathfrak{q}/\mathfrak{p}}$ is a Frobenius automorphism of $\mathfrak{P}$ over M.

(b) If $x \in \mathfrak{O} \cap M$, then $(\varphi \restriction M)(x) - x^{\mathfrak{N}(\mathfrak{p})} \in \mathfrak{P} \cap M = \mathfrak{q}$, and therefore $\varphi \restriction M$ is a Frobenius automorphism of $\mathfrak{q}$ over K. □

Definitions and Remarks 2.14.7 (Frobenius symbol and Artin symbol). Let $\mathfrak{o}$ be a Dedekind domain with finite residue class rings, $K = \mathsf{q}(\mathfrak{o})$, L/K a finite Galois extension, $G = \mathrm{Gal}(L/K)$ and $\mathfrak{O} = \mathrm{cl}_L(\mathfrak{o})$.

1. Let $\mathfrak{P} \in \mathcal{P}(\mathfrak{O})$ and $\mathfrak{p} = \mathfrak{P} \cap \mathfrak{o}$ such that $\mathfrak{P}/\mathfrak{p}$ is unramified. Then $I_{\mathfrak{P}} = \mathbf{1}$, $T_{\mathfrak{P}} = L$, the map $\varphi_{\mathfrak{P}} \colon G_{\mathfrak{P}} \to \mathrm{Gal}(\mathsf{k}_{\mathfrak{P}}/\mathsf{k}_{\mathfrak{p}})$, $\sigma \mapsto \sigma_{\mathfrak{P}}$, is an isomorphism, and there exists a unique Frobenius automorphism of $\mathfrak{P}$ over K which we denote by

$$(\mathfrak{P}, L/K) \quad \text{or} \quad \Big(\frac{L/K}{\mathfrak{P}}\Big).$$

By definition, $G_{\mathfrak{P}} = \langle (\mathfrak{P}, L/K) \rangle$, $\mathrm{ord}(\mathfrak{P}, L/K) = |G_{\mathfrak{P}}| = f(\mathfrak{P}/\mathfrak{p})$, and if $\sigma, \sigma' \in G_{\mathfrak{P}}$, then $\sigma = \sigma'$ if and only if $\sigma_{\mathfrak{P}} = \sigma'_{\mathfrak{P}}$, i. e., $\sigma(x) \equiv \sigma'(x) \bmod \mathfrak{P}$ for all $x \in \mathfrak{O}$.

2. Let $\mathfrak{p} \in \mathcal{P}(\mathfrak{o})$ be unramified in L. Then we define the **Frobenius symbol** of $\mathfrak{p}$ for L/K by

$$[\mathfrak{p}, L/K] = \Big[\frac{L/K}{\mathfrak{p}}\Big] = \Big\{\Big(\frac{L/K}{\mathfrak{P}}\Big) \;\Big|\; \mathfrak{P} \in \mathcal{P}(\mathfrak{O}),\ \mathfrak{P} \cap \mathfrak{o} = \mathfrak{p}\Big\} \subset G.$$

If $\mathfrak{P} \in \mathcal{P}(\mathfrak{O})$ and $\mathfrak{P} \cap \mathfrak{o} = \mathfrak{p}$, then

$$\Big[\frac{L/K}{\mathfrak{p}}\Big] = \Big\{\Big(\frac{L/K}{\sigma\mathfrak{P}}\Big) \;\Big|\; \sigma \in G\Big\} = \Big\{\sigma\Big(\frac{L/K}{\mathfrak{P}}\Big)\sigma^{-1} \;\Big|\; \sigma \in G\Big\} \subset G$$

is a conjugacy class in G. Note that $[\mathfrak{p}, L/K] = \mathbf{1}$ holds if and only if $(\mathfrak{P}, L/K) = \mathrm{id}_L$ for some (and then for all) prime divisors $\mathfrak{P}$ of $\mathfrak{p}$ in L, and this holds if and only if $\mathfrak{p}$ is fully decomposed in L.

3. Let L/K be abelian, $\mathfrak{P} \in \mathcal{P}(\mathfrak{O})$ and $\mathfrak{p} = \mathfrak{P} \cap \mathfrak{o}$. Then $(\mathfrak{P}, L/K)$ only depends on $\mathfrak{p}$, we define

$$(\mathfrak{p}, L/K) = \Big(\frac{L/K}{\mathfrak{p}}\Big) = \Big(\frac{L/K}{\mathfrak{P}}\Big) \in G,$$

and we call $(\mathfrak{p}, L/K)$ the **Frobenius automorphism** of $\mathfrak{p}$ in K. In this case, $[\mathfrak{p}, L/K] = \{(\mathfrak{p}, L/K)\}$, and since the Frobenius automorphism is invariant under localization, we obtain $(\mathfrak{p}, L/K) = (\mathfrak{p}\mathfrak{o}_{\mathfrak{p}}, L/K)$.

4. Again let L/K be an arbitrary finite Galois field extension, and let $\mathfrak{m} \in \mathcal{J}'(\mathfrak{o})$ be the product of the (by 2.12.7) finitely many in L ramified prime ideals of $\mathfrak{o}$. For a fractional ideal $\mathfrak{b} \in \mathcal{J}^{\mathfrak{m}}(\mathfrak{o})$ we define the **Artin symbol** of $\mathfrak{b}$ for L/K by

$$(\mathfrak{b}, L/K) = \Big(\frac{L/K}{\mathfrak{b}}\Big) = \prod_{\mathfrak{p} \in \mathcal{P}^{\mathfrak{m}}(\mathfrak{o})} \Big(\frac{L/K}{\mathfrak{p}}\Big)^{v_{\mathfrak{p}}(\mathfrak{b})}.$$

The Artin symbol defines a group homomorphism $(\,\cdot\,, L/K) \colon \mathcal{J}^{\mathfrak{m}}(\mathfrak{o}) \to G$, called the **Artin map**. This is the fundamental map of global class field theory and will be studied in the forthcoming volume on class field theory.

Theorem 2.14.8 (Functorial properties of the Artin symbol). *Let $\mathfrak{o}$ be a Dedekind domain with finite residue class rings and $K = \mathsf{q}(\mathfrak{o})$. Let L'/K be a finite field extension, and let L and K' be intermediate fields of L'/K such that L'/K' and L/K are abelian. Let $\mathfrak{O} = \mathrm{cl}_L(\mathfrak{o})$, $\mathfrak{o}' = \mathrm{cl}_{K'}(\mathfrak{o})$, $\mathfrak{O}' = \mathrm{cl}_{L'}(\mathfrak{o})$, let $\mathfrak{m} \in \mathcal{J}'(\mathfrak{o})$ be the product of all in L' ramified prime ideals of $\mathfrak{o}$, $\mathfrak{a} \in \mathcal{J}^{\mathfrak{m}}(\mathfrak{o})$ and $\mathfrak{b} \in \mathcal{J}^{\mathfrak{m}\mathfrak{o}'}(\mathfrak{o}')$.*

1. $(\mathfrak{b}, L'/K') \restriction L = (\mathcal{N}_{\mathfrak{o}'/\mathfrak{o}}(\mathfrak{b}), L/K)$.

2. *Assume that* $K' \subset L$.

(a) $(\mathfrak{b}, L/K') = (\mathcal{N}_{\mathfrak{o}'/\mathfrak{o}}(\mathfrak{b}), L/K)$.

(b) *If K'/K is abelian, then* $(\mathfrak{a}, L/K) \restriction K' = (\mathfrak{a}, K'/K)$.

(c) $(\mathcal{N}_{\mathfrak{O}/\mathfrak{o}}(\mathfrak{A}), L/K) = \mathbf{1}$ *for all* $\mathfrak{A} \in \mathcal{J}^{\mathfrak{m}\mathfrak{O}}(\mathfrak{O})$.

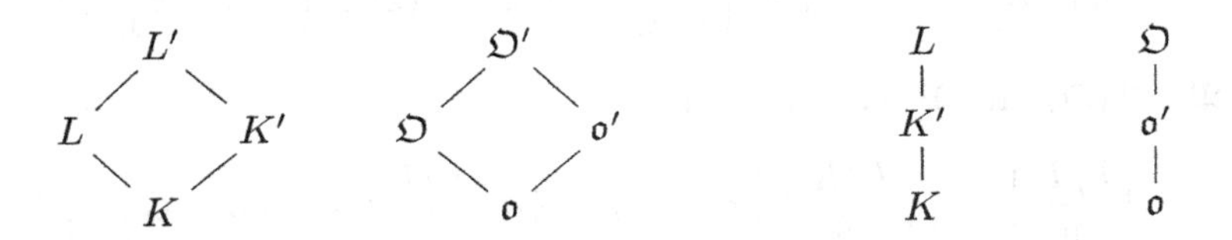

Proof. 1. It suffices to prove the assertion for a prime ideal $\mathfrak{b} = \mathfrak{p}' \in \mathcal{P}^{\mathfrak{m}\mathfrak{o}'}(\mathfrak{o}')$. Thus let $\mathfrak{p}' \in \mathcal{P}^{\mathfrak{m}\mathfrak{o}'}(\mathfrak{o}')$, $\mathfrak{p} = \mathfrak{p}' \cap \mathfrak{o}$, $\mathfrak{P}' \in \mathcal{P}(\mathfrak{O}')$, $\mathfrak{p}' = \mathfrak{P}' \cap \mathfrak{o}'$, $\mathfrak{P} = \mathfrak{P}' \cap \mathfrak{O}$, and let $G_{\mathfrak{P}} \subset \mathrm{Gal}(L/K)$ be the decomposition group of $\mathfrak{P}$ over K. If $x \in \mathfrak{O}$, then

$$\Big(\frac{L'/K'}{\mathfrak{p}'}\Big)(x) - x^{\mathfrak{N}(\mathfrak{p}')} \in \mathfrak{P}' \cap \mathfrak{O} = \mathfrak{P}.$$

In particular,

$$x \in \mathfrak{P} \quad \text{implies} \quad \Big(\frac{L'/K'}{\mathfrak{p}'}\Big)(x) \in \mathfrak{P}, \quad \text{and therefore} \quad \Big(\frac{L'/K'}{\mathfrak{p}'}\Big) \restriction L \in G_{\mathfrak{P}}.$$

On the other hand, it follows by definition that

$$\Big(\frac{L/K}{\mathcal{N}_{\mathfrak{o}'/\mathfrak{o}}(\mathfrak{p}')}\Big) = \Big(\frac{L/K}{\mathfrak{p}}\Big)^{f(\mathfrak{p}'/\mathfrak{p})} \in G_{\mathfrak{P}},$$

and for all $x \in \mathfrak{O}$ we obtain

$$\Big(\frac{L'/K'}{\mathfrak{p}'}\Big)(x) \equiv x^{\mathfrak{N}(\mathfrak{p}')} \equiv x^{\mathfrak{N}(\mathfrak{p})^{f(\mathfrak{p}'/\mathfrak{p})}} \equiv \Big(\frac{L/K}{\mathfrak{p}}\Big)^{f(\mathfrak{p}'/\mathfrak{p})}(x)$$
$$\equiv \Big(\frac{L/K}{\mathcal{N}_{\mathfrak{o}'/\mathfrak{o}}(\mathfrak{p}')}\Big)(x) \bmod \mathfrak{P},$$

which implies

$$\Big(\frac{L'/K'}{\mathfrak{p}'}\Big) \restriction L = \Big(\frac{L/K}{\mathcal{N}_{\mathfrak{o}'/\mathfrak{o}}(\mathfrak{p}')}\Big).$$

2. (a) By 1., applied with $L = L'$.

(b) It suffices to prove the assertion for $\mathfrak{a} = \mathfrak{p} \in \mathcal{P}^{\mathfrak{m}}(\mathfrak{o})$. But therefore it holds by 2.14.6.3(b).

(c) By 1., applied with $K' = L$ and $\mathfrak{o}' = \mathfrak{O}$. □

The following final result shows that the Frobenius automorphism controls the factorization type of unramified prime ideals in finite field extensions. We introduce the following terminology.

Let $\mathfrak{o}$ be a Dedekind domain, $K = \mathsf{q}(\mathfrak{o})$, K'/K a finite separable field extension, $\mathfrak{o}' = \mathrm{cl}_{K'}(\mathfrak{o})$, and let $\mathfrak{p} \in \mathcal{P}(\mathfrak{o})$ be unramified in L. We say that $\mathfrak{p}$ has **factorization type** $(f_1, \ldots, f_r)$ in L if $\mathfrak{p}\mathfrak{o}' = \mathfrak{p}_1' \cdot \ldots \cdot \mathfrak{p}_r'$ such that $\mathfrak{p}_i' \in \mathcal{P}(\mathfrak{o}')$ and $f(\mathfrak{p}_i'/\mathfrak{p}) = f_i$ for all $i \in [1, r]$.

We refer to 1.5.7 for the definiton of the cycle type of a permutation.

Theorem 2.14.9. *Let $\mathfrak{o}$ be a Dedekind domain with finite residue class rings and $K = \mathsf{q}(\mathfrak{o})$. Let L/K be a finite Galois extension, $G = \mathrm{Gal}(L/K)$, and let $\mathfrak{O} = \mathrm{cl}_L(\mathfrak{o})$. Let $\mathfrak{p} \in \mathcal{P}(\mathfrak{o})$ be unramified in L, $\mathfrak{P} \in \mathcal{P}(\mathfrak{O})$ such that $\mathfrak{P} \cap \mathfrak{o} = \mathfrak{p}$ and $\varphi \in G_{\mathfrak{P}}$ the Frobenius automorphism of $\mathfrak{P}$ over K. Let $a \in L$, $[K(a):K] = n$, and let $A = \{a = a_1, \ldots, a_n\}$ be the set of all conjugates of a in L. Then $\varphi_a = \varphi \restriction A \in \mathfrak{S}(A)$, and the cycle type of φ_a coincides with the factorization type of $\mathfrak{p}$ in $K(a)$.*

In particular, $\mathfrak{p}$ is fully decomposed in $K(a)$ if and only if $\varphi \restriction K(a) = \mathrm{id}_{K(a)}$.

Proof. Let $K' = K(a)$, $H = \mathrm{Gal}(L/K')$, $\mathfrak{o}' = \mathfrak{O} \cap K' = \mathrm{cl}_{K'}(\mathfrak{o})$, and let $\mathfrak{p}' = \mathfrak{P} \cap \mathfrak{o}'$. Clearly, $\varphi_a = \varphi \restriction A$ is a permutation of A, and G acts transitively on A. By 2.5.3.1 it follows that $\{\sigma\mathfrak{P} \mid \sigma \in G\}$ is the set of all prime divisors of $\mathfrak{p}$ in L, and for $\sigma \in G$ we set $\mathfrak{q}_\sigma = \sigma^{-1}\mathfrak{P} \cap K'$. Then $\{\mathfrak{q}_\sigma \mid \sigma \in G\}$ is the set of all prime divisors of $\mathfrak{p}$ in K', and it suffices to prove:

A. If $\sigma, \sigma' \in G$, then $\mathfrak{q}_\sigma = \mathfrak{q}_{\sigma'}$ if and only if $\sigma(a)$ and $\sigma'(a)$ lie in the same cycle of φ_a.

B. If $\sigma \in G$ and $m \in \mathbb{Z}$, then $\varphi^m\sigma(a) = \sigma(a)$ if and only if $f(\mathfrak{q}_\sigma/\mathfrak{p}) \mid m$.

Indeed, by **A** the assignment $\mathfrak{q}_\sigma \mapsto \sigma(a)$ induces a bijective map from the set of prime divisors of $\mathfrak{p}$ in K' onto the cycles of φ_a, and by **B** the residue class degree $f(\mathfrak{q}_\sigma/\mathfrak{p})$ equals the order of the cycle of $\sigma(a)$.

Proof of **A.** Let σ, $\sigma' \in G$. Then we obtain:

$$\begin{aligned}
\mathfrak{q}_\sigma = \mathfrak{q}_{\sigma'} &\iff \sigma^{-1}\mathfrak{P} \cap K' = \sigma'^{-1}\mathfrak{P} \cap K' \\
&\iff \tau\sigma^{-1}\mathfrak{P} = \sigma'^{-1}\mathfrak{P} \ \text{ for some } \ \tau \in H \\
&\iff \sigma'\tau\sigma^{-1} \in G_{\mathfrak{P}} = \langle\varphi\rangle \quad \text{for some } \ \tau \in H \\
&\iff \sigma'\tau = \varphi^m\sigma \ \text{ for some } \ \tau \in H \ \text{ and } \ m \in \mathbb{Z} \\
&\iff \sigma'(a) = \varphi^m\sigma(a) \ \text{ for some } \ m \in \mathbb{Z} \\
&\iff \sigma(a) \ \text{ and } \ \sigma'(a) \ \text{ belong to the same cycle of } \ \varphi_a. \qquad \square[\mathbf{A}.]
\end{aligned}$$

Proof of **B.** Let $\sigma \in G$ and $m \in \mathbb{Z}$. Let $G_{\sigma^{-1}\mathfrak{P}}$ be the decomposition group of $\sigma^{-1}\mathfrak{P}$ and $\varphi_{\sigma^{-1}\mathfrak{P}}$ the Frobenius automorphism of $\sigma^{-1}\mathfrak{P}$ over K. By 2.14.6 it follows that $\varphi_{\sigma^{-1}\mathfrak{P}} = \sigma^{-1}\varphi\sigma$, and $\varphi_{\sigma^{-1}\mathfrak{P}}^{f(\mathfrak{q}_\sigma/\mathfrak{p})}$ is the Frobenius automorphism of $\sigma^{-1}\mathfrak{P}$ over K'. Hence $G_{\sigma^{-1}\mathfrak{P}} \cap H = \langle\varphi_{\sigma^{-1}\mathfrak{P}}^{f(\mathfrak{q}_\sigma/\mathfrak{p})}\rangle$ is the decomposition group of $\sigma^{-1}\mathfrak{P}$ over K'. Now we obtain:

$$\begin{aligned}
\varphi^m\sigma a = \sigma a &\iff \sigma^{-1}\varphi^m\sigma = (\sigma^{-1}\varphi\sigma)^m = \varphi_{\sigma^{-1}\mathfrak{P}}^m \in H \cap G_{\sigma^{-1}\mathfrak{P}} = \langle\varphi_{\sigma^{-1}\mathfrak{P}}^{f(\mathfrak{q}_\sigma/\mathfrak{p})}\rangle \\
&\iff f(\mathfrak{q}_\sigma/\mathfrak{p}) \mid m. \qquad \square
\end{aligned}$$

2.15 Exercises for Chapter 2

1. Let $\mathfrak{o}$ be a domain and T a submonoid of $\mathfrak{o}^\bullet$. Then

$$\widetilde{T} = (T^{-1}\mathfrak{o})^\times \cap \mathfrak{o} = \bigcup_{\substack{\mathfrak{p}\in\max(\mathfrak{o}) \\ \mathfrak{p}\cap T=\emptyset}} \mathfrak{p}$$

is the smallest saturated submonoid of $\mathfrak{o}^\bullet$ containing T, and $T^{-1}\mathfrak{o} = \widetilde{T}^{-1}\mathfrak{o}$.

2. Let $\mathfrak{o}$ be a domain. A function $f\colon \mathfrak{o}^\bullet \to \mathbb{N}$ is called a **quasinorm** if it has the following property:

> If x, $y \in \mathfrak{o}^\bullet$, $y \nmid x$ and $f(y) \le f(x)$, then there exist a, $b \in \mathfrak{o}$ such that $ax - by \in \mathfrak{o}^\bullet$ and $f(ax - by) < f(y)$.

a. If $\mathfrak{o}$ possesses a quasinorm, then $\mathfrak{o}$ is a principal ideal domain.

b. Let $\mathfrak{o}$ be a principal ideal domain. If $a = up_1 \cdot \ldots \cdot p_r \in \mathfrak{o}^\bullet$, where $u \in \mathfrak{o}^\times$, $r \in \mathbb{N}_0$ and $p_1, \ldots, p_r \in \mathfrak{o}$ are prime elements, then we call $\ell(a) = r$ the **length** of a. The map $f\colon \mathfrak{o}^\bullet \to \mathbb{N}_0$, defined by $f(a) = 2^{\ell(a)}$, is a quasinorm.

3. Projective modules (see [50, Sec. 1.3]). Let R be a (not necessarily commutative) ring. An R-module P is called **projective** if every epimorphism $p\colon M \to P$ of R-modules splits (see 1.1.1).

a. For an R-module P the following assertions are equivalent:

(a) P is projective.

(b) P is a direct summand of a free module.

(c) There exists a family $(a_i)_{i\in I}$ in P and a family $(f_i\colon P\to R)_{i\in I}$ of R-homomorphisms such the set $\{i\in I\mid f_i(a)\neq 0\}$ is finite for all $a\in P$ and

$$a=\sum_{i\in I} f_i(a)a_i.$$

b. Let $\mathfrak{o}$ be a domain and $\mathfrak{a}$ a fractional ideal of $\mathfrak{o}$. Then $\mathfrak{a}$ is invertible if and only if $\mathfrak{a}$ is a projective $\mathfrak{o}$-module.

4. A ring is called **Artinian** if it satisfies the descending chain condition for ideals (that is, every sequence $\mathfrak{a}_0\supset\mathfrak{a}_1\supset\mathfrak{a}_2\supset\ldots$ becomes stationary).

A domain $\mathfrak{o}$ which is not a field is a Dedekind domain if and only if it is Noetherian, integrally closed, and for every non-zero ideal $\mathfrak{a}$ of $\mathfrak{o}$ the ring $\mathfrak{o}/\mathfrak{a}$ is Artinian.

5. Let $\mathfrak{o}$ be a domain and $K=\mathsf{q}(\mathfrak{o})$. Then $\mathfrak{o}\,(\!(X)\!)$ is a subring of $K\,(\!(X)\!)$, and if

$$f=\sum_{n\in\mathbb{Z}} a_nX^n\in\mathfrak{o}\,(\!(X)\!)^\bullet\quad\text{and}\quad d=\min\{n\in\mathbb{Z}\mid a_n\neq 0\}$$

then $\mathrm{ld}(f)=a_d$ is called the **leading coefficient** of f. Set $\mathrm{ld}(0)=0$.

a. If $f,\,g\in\mathfrak{o}\,(\!(X)\!)$, then $\mathrm{ld}(fg)=\mathrm{ld}(f)\mathrm{ld}(g)$.

b. If $\mathfrak{a}$ is an ideal of $\mathfrak{o}\,(\!(X)\!)$, then $\mathrm{ld}(\mathfrak{a})=\{\mathrm{ld}(f)\mid f\in\mathfrak{a}\}$ is an ideal of $\mathfrak{o}$, and if $\mathrm{ld}(\mathfrak{a})={}_{\mathfrak{o}}(\mathrm{ld}(f_1),\ldots,\mathrm{ld}(f_r)$, then $\mathfrak{a}={}_{\mathfrak{o}\,(\!(X)\!)}(f_1,\ldots,f_r)$.

c. If $\mathfrak{o}$ is a Dedekind domain, then $\mathfrak{o}\,(\!(X)\!)$ is a Dedekind domain, too. Hint: Use **b.** and Exercise 4.

6. Let $\mathfrak{o}$ be an integrally closed domain, $n\in\mathbb{N}$, $n\geq 2$ and $a,\,b\in\mathfrak{o}$ such that $a^{n-1}b\in(a^n,b^n)$. Then (a,b) is invertible. Hint: Induction on n. If $a^{n-1}b=a^nx+b^ny$, then $z=by/a$ is integral over $\mathfrak{o}$, and if $n=2$, then $(a,b)(y,1-z)=(a)$.

7. Let $\mathfrak{o}$ be a Dedekind domain.

a. For every $n\in\mathbb{N}$ there are only finitely many ideals $\mathfrak{a}$ of $\mathfrak{o}$ such that $|\mathfrak{o}/\mathfrak{a}|\leq n$.

b. If $f,\,g\in\mathfrak{o}[X]$, then $\mathsf{c}(fg)=\mathsf{c}(f)\mathsf{c}(g)$ (where $\mathsf{c}(\cdot)$ denotes the content ideal, see 1.6.3).

8. Let $\mathfrak{o}$ be a factorial domain and $\boldsymbol{X}$ a family of indeterminates over $\mathfrak{o}$. Then $\mathfrak{o}[\boldsymbol{X}]$ is factorial.

9. Let R and S be commutative rings such that $R\subset S$ and S is integral over R. Let $\mathfrak{a}$ be an ideal of R and $\alpha\in\mathfrak{a}S$. Then there exist $a_0,\ldots,a_{n-1}\in\mathfrak{a}$ such that $\alpha^n+a_{n-1}\alpha^{n-1}+\ldots+a_1\alpha+a_0=0$.

10. Let $\mathfrak{o}$ be an integrally closed domain, $K = \mathsf{q}(\mathfrak{o})$, L/K a finite field extension and $\mathfrak{O} = \mathrm{cl}_L(\mathfrak{o})$.

a. For every $\mathfrak{p} \in \max(\mathfrak{o})$ there are only finitely many $\mathfrak{P} \in \max(\mathfrak{O})$ lying above $\mathfrak{p}$.

b. Let $\alpha \in \mathfrak{O}$ and $g = X^n + a_{n-1}X^{n-1} + \ldots + a_1X + a_0 \in \mathfrak{o}[X]$ the minimal polynomial of α over K.

(i) Let $\mathfrak{p} \in \mathcal{P}(\mathfrak{o})$. If $\alpha \in \mathfrak{p}\mathfrak{O}$, then $a_0, \ldots, a_{n-1} \in \mathfrak{p}$. Hint: Use Exercise 9.

(ii) $\alpha \in \mathfrak{O}^\times \iff a_0 \in \mathfrak{o}^\times \iff \mathsf{N}_{L/K}(\alpha) \in \mathfrak{o}^\times$.

11. Let $m,\, n \in \mathbb{Z}$ be odd integers, p an odd prime, $p \nmid n$, and suppose that $f = X^5 + pmX + pn$. Then $\mathrm{Gal}_{\mathbb{Q}}(f) = \mathfrak{S}_5$. Hint: Factorize f modulo 2.

12. Let $\mathfrak{o}$ be a domain and not a field.

a. If every prime ideal of $\mathfrak{o}$ is invertible, then $\mathfrak{o}$ is a Dedekind domain.

b. If $\mathfrak{o}$ is Noetherian and every maximal ideal of $\mathfrak{o}$ is invertible, then $\mathfrak{o}$ is a Dedekind domain.

13. Let $\mathfrak{o}$ be a domain which is not a field, and suppose that every non-zero ideal is a product of prime ideals. Then $\mathfrak{o}$ is a Dedekind domain. Hint: Prove first that every non-zero prime ideal is maximal. For this, let $\mathfrak{p}$ be a non-zero non-maximal prime ideal, $a \in \mathfrak{o} \setminus \mathfrak{p}$ such that $\mathfrak{p} + a\mathfrak{o} \neq \mathfrak{o}$. Consider the prime factorization of $\mathfrak{p} + a\mathfrak{o}$ and of $\mathfrak{p} + a^2\mathfrak{o}$ in the residue class domain $\mathfrak{o}/\mathfrak{p}$ (see [29, Ch. VIII, §6]).

14. Let $\mathfrak{o}$ be a domain. A map $\nu\colon \mathfrak{o} \to \mathbb{N}_0$ is called a **Euclidean norm** if:

- $\nu(a) = 0$ if and only if $a = 0$.
- For every $a \in \mathfrak{o}^\bullet$ and $b \in \mathfrak{o}$ there exist $q,\, r \in \mathfrak{o}$ such that $b = aq + r$ and $\nu(r) < \nu(a)$.

A domain with a Euclidean norm is called a **Euclidean domain**. The map $a \mapsto |a|$ is a Euclidean norm of $\mathbb{Z}$. If K is a field, then $f \mapsto 2^{\partial(f)}$ is a Euclidean norm of $K[X]$ (where $2^{-\infty} = 0$).

a. Every Euclidean domain is a principal ideal domain.

b. For a squarefree integer $D \in \mathbb{Z} \setminus \{1\}$ let $\mathbb{Z}[\sqrt{D}] = \{a + b\sqrt{D} \mid a,\, b \in \mathbb{Z}\}$. $\mathbb{Z}[\sqrt{D}]$ is a domain, $\mathsf{q}(\mathbb{Z}[\sqrt{D}]) = \mathbb{Q}(\sqrt{D}) = \{u + v\sqrt{D} \mid u,\, v \in \mathbb{Q}\}$, and the norm

$$\mathsf{N} = \mathsf{N}_{\mathbb{Q}(\sqrt{D})/\mathbb{Q}}\colon \mathbb{Q}(\sqrt{D}) \to \mathbb{Q}, \quad \text{defined by} \quad \mathsf{N}(u + v\sqrt{D}) = u^2 - v^2D,$$

satisfies $\mathsf{N}(\alpha\beta) = \mathsf{N}(\alpha)\mathsf{N}(\beta)$ for all $\alpha,\, \beta \in \mathbb{Q}(\sqrt{D})$. The map $\alpha \mapsto |\mathsf{N}(\alpha)|$ is a Euclidean norm of $\mathbb{Z}[\sqrt{D}]$ (and then we call $\mathbb{Z}[\sqrt{D}]$ norm-Euclidean) if and only if for every $\xi \in \mathbb{Q}(\sqrt{D}\,)$ there exists some $\alpha \in \mathbb{Z}[\sqrt{D}\,]$ such that $|\mathsf{N}_{K/\mathbb{Q}}(\xi-\alpha)| < 1$.

c. If $D \in \{-1,-2,2,3,6,7\}$, then $\mathbb{Z}[\sqrt{D}]$ is norm-Euclidean.

15. Integral elements in polynomial rings (see [18, II. § 10]). Let R and S be commutative rings such that $R \subset S$ and $R' = \mathrm{cl}_S(R)$.

a. Let $g,\, h \in S[X]$ be such that $gh \in R[X]$. If h is monic, then $g \in R'[X]$ (use double induction on ∂f and ∂h).

b. $R'[X] = \mathrm{cl}_{S[X]}(R[X])$.

c. Let $\mathfrak{o}$ be an integrally closed domain and $\boldsymbol{X}$ a family of indeterminates over $\mathfrak{o}$. Then $\mathfrak{o}[\boldsymbol{X}]$ is integrally closed.

16. Let L/K be a field extension.

a. If $[L\!:\!K] < \infty$, then $L[\![X]\!] = \mathrm{cl}_{L[\![X]\!]}(K[\![X]\!])$.

b. Let $\mathrm{char}(K) = p > 0$. Then $L[\![X]\!]$ is integral over $K[\![X]\!]$ if and only if $L^{p^m} \subset K$ for some $m \in \mathbb{N}$.

17. Let $\mathfrak{o}$ be an order, $K = \mathsf{q}(\mathfrak{o})$, $\overline{\mathfrak{o}} = \mathrm{cl}_K(\mathfrak{o})$ and $\mathfrak{f} = (\mathfrak{o}\!:\!\overline{\mathfrak{o}})$.

a. For every non-zero ideal $\mathfrak{a}$ of $\mathfrak{o}$ there is a (natural) isomorphism

$$\mathfrak{o}/\mathfrak{a} \overset{\sim}{\rightarrow} \bigoplus_{\mathfrak{p}\in\mathcal{P}(\mathfrak{o})} \mathfrak{o}_\mathfrak{p}/\mathfrak{a}\mathfrak{o}_\mathfrak{p} \cong \bigoplus_{\substack{\mathfrak{p}\in\mathcal{P}(\mathfrak{o})\\ \mathfrak{p}\supset\mathfrak{a}}} \mathfrak{o}_\mathfrak{p}/\mathfrak{a}\mathfrak{o}_\mathfrak{p}.$$

b. There is a (natural) exact sequence

$$\mathbf{1} \to \mathfrak{o}^\times \to \overline{\mathfrak{o}}^\times \to \bigoplus_{\mathfrak{p}\in\mathcal{P}(\mathfrak{o})} \overline{\mathfrak{o}}_\mathfrak{p}^\times/\mathfrak{o}_\mathfrak{p}^\times \to \mathcal{C}(\mathfrak{o}) \to \mathcal{C}(\overline{\mathfrak{o}}) \to \mathbf{1},$$

and

$$\bigoplus_{\mathfrak{p}\in\mathcal{P}(\mathfrak{o})} \overline{\mathfrak{o}}_\mathfrak{p}^\times/\mathfrak{o}_\mathfrak{p}^\times \cong \bigoplus_{\substack{\mathfrak{p}\in\mathcal{P}(\mathfrak{o})\\ \mathfrak{p}\supset\mathfrak{f}}} \overline{\mathfrak{o}}_\mathfrak{p}^\times/\mathfrak{o}_\mathfrak{p}^\times$$

18. Generalize Exercise 2.13.9 to the case of a Galois field extension L/K with dihedral Galois group of degree $2p$, i. e.,

$$\mathrm{Gal}(L/K) = D_{2p} = \langle \sigma, \tau \mid \sigma^p = \tau^2 = 1,\ \sigma\tau = \tau\sigma^{-1}\rangle.$$

19. Let $\mathfrak{o}$ be a domain, not a field, and $K = \mathsf{q}(\mathfrak{o})$.

a. The following assertions are equivalent:

(a) Every non-zero finitely generated ideal of $\mathfrak{o}$ is invertible.

(b) Every non-zero ideal of $\mathfrak{o}$ which is generated by two elements is invertible.

(c) For every $\mathfrak{p} \in \mathcal{P}(\mathfrak{o})$ the local domain $\mathfrak{o}_\mathfrak{p}$ is a valuation domain.

If these conditions are fulfilled, then $\mathfrak{o}$ is called a **Prüfer domain**. In particular, $\mathfrak{o}$ is a Dedekind domain if and only if it is a Noetherian Prüfer domain. A domain in which every finitely generated ideal is principal is called a **Bezout domain**. Every valuation domain is a Bezout domain.

b. Let $\mathfrak{o}$ be a Dedekind domain, L/K a (not necessarily finite) algebraic field extension and $\mathfrak{O} = \mathrm{cl}_L(\mathfrak{o})$. Then $\mathfrak{O}$ is a Prüfer domain. Note that this also holds if $\mathfrak{o}$ is merely a Prüfer domain (see [18, IV. (22.3)]). The domain $\overline{\mathbb{Z}} = \mathrm{cl}_{\mathbb{Z}}(\overline{\mathbb{Q}})$ of all algebraic integers is a Bezout domain but not a Dedekind domain.

20. Let $\sigma = (\mathbb{Q} \hookrightarrow \mathbb{R})$, $\boldsymbol{\sigma} = \{\sigma\}$, $m \in \mathbb{N}$ and $\mathcal{C}^{m\mathbb{Z},\boldsymbol{\sigma}}(\mathbb{Z})$ the $\boldsymbol{\sigma}$-ray class group module $m\mathbb{Z}$. Then there is an isomorphism $\mathcal{C}^{m\mathbb{Z},\boldsymbol{\sigma}}(\mathbb{Z}) \overset{\sim}{\to} (\mathbb{Z}/m\mathbb{Z})^\times$, given by $[a\mathbb{Z}]^{m\mathbb{Z},\boldsymbol{\sigma}} \mapsto a + m\mathbb{Z}$ for all $a \in \mathbb{Z}$ such that $(a,m) = 1$.

3

Algebraic Number Fields: Elementary and Geometric Methods

This chapter contains those parts of the classical theory of algebraic number fields which allow a satisfactory treatment by ideal-theoretic and geometric methods. Apart from the books [52], [50], [40], [62], [36] and [43] I mention [2], [15] and [7] as standard references. A wealth of explicit material and additional exercises are contained in [56] and [45].

Integral bases, discriminants and decomposition laws for quadratic, cubic and cyclotomic number fields are discussed in the Sections 3.1 and 3.2. Special diligence was devoted to the various concepts of characters and their arithmetical relevance (abstract group characters, Dirichlet characters, residue class characters, quadratic characters). Section 3.3 deals with the basics of the theory of abelian number fields using Dirichlet characters. In Section 3.4 quadratic characters are investigated in detail. In particular, we consider Gauss sums and prove the quadratic reciprocity law. The finiteness theorems for units and ideal classes are proved in Section 3.5, and asymptotic results concerning the distribution of ideals in ray classes are achieved in Section 3.6.

Section 3.7 contains a short outline (without proofs) of the main results of class field theory. In particular, there we connect the description of abelian number fields using Dirichlet characters with those using class field theory. For an exhaustive treatment of abelian number fields I refer to [61]; an elementary introduction containing illustrative examples is in [6, Ch. 2].

Special emphasis is devoted to the theory of quadratic number fields and quadratic orders. In particular, I mention the ideal-theoretic version of Gauss's genus theory for non-principal quadratic orders in Section 2.9. For an exhaustive presentation of the theory of quadratic orders I refer to [22] and [47].

As already announced, the theory of differents and discriminants is postponed to Chapter 5, since then we have valuation-theoretic methods at our disposal.

3.1 Complete modules, integral bases and discriminants

Recall from 1.4.10 that we denoted by $\overline{\mathbb{Q}}$ the field of all algbraic numbers, and from 2.4.5 that we denoted by $\overline{\mathbb{Z}} = \mathrm{cl}_{\overline{\mathbb{Q}}}(\mathbb{Z})$ the domain of all algebraic integers.

Definition and Remarks 3.1.1. An **algebraic number field** K is a subfield of $\mathbb{C}$ which is finite over $\mathbb{Q}$ (then it is algebraic over $\mathbb{Q}$ and thus a subfield of $\overline{\mathbb{Q}}$). By a **basis** of K we mean a $\mathbb{Q}$-basis, and we call $[K:\mathbb{Q}]$ the **degree** of K. A **quadratic** [**cubic, quartic, ...**] **number field** is an algebraic number field of degree 2 [3, 4, ...]. An algebraic number field K is called **Galois** [**abelian, cyclic, ...**] if the field extension $K/\mathbb{Q}$ is Galois [abelian, cyclic, ...].

Let K be an algebraic number field and $[K:\mathbb{Q}] = n$. Every field homomorphism $K \to \mathbb{C}$ is a $\mathbb{Q}$-homomorphism; let $\mathrm{Hom}(K,\mathbb{C}) = \mathrm{Hom}^{\mathbb{Q}}(K,\mathbb{C})$ be the set of all field homomorphisms $K \to \mathbb{C}$. Then $|\mathrm{Hom}(K,\mathbb{C})| = n$, and we assume that $\mathrm{Hom}(K,\mathbb{C}) = \{\sigma_1,\ldots,\sigma_n\}$, where $n = r_1 + 2r_2$ such that $\sigma_\nu(K) \subset \mathbb{R}$ for all $\nu \in [1,r_1]$ and $\sigma_{r_1+r_2+\nu}(x) = \overline{\sigma_{r_1+\nu}(x)}$ for all $x \in K$ and all $\nu \in [1,r_2]$ (we denote this henceforth by $\sigma_{r_1+r_2+\nu} = \overline{\sigma_{r_1+\nu}} \ne \sigma_{r_1+\nu}$). We call (r_1,r_2) the **signature** of K, $\sigma_1,\ldots,\sigma_{r_1}$ the **real embeddings** and $(\sigma_{r_1+1},\overline{\sigma_{r_1+1}}),\ldots,(\sigma_{r_1+r_2},\overline{\sigma_{r_1+r_2}})$ the pairs of conjugate **complex embeddings** of K (recall that this terminology is consistent with 2.6.5). An algebraic number $x \in K$ is called **totally positive**, $x \gg 0$ if $\sigma_i(x) > 0$ for all $i \in [1,r_1]$. If $r_1 > 0$, then we tacitly assume that $K \subset \mathbb{R}$.

An algebraic number field K of degree n is called **totally real** if $r_1 = n$ (and then $r_2 = 0$) and **totally imaginary** if $r_1 = 0$ (and then $2r_2 = n$). In this terminology, $\mathbb{Q}$ is a totally real algebraic number field of degree 1. If K is a Galois algebraic number field, then $\mathrm{Hom}(K,\mathbb{C}) = \mathrm{Gal}(K/\mathbb{Q})$, and thus K is either totally real or totally imaginary, and then we simply say that K is **real** or that K is **imaginary**. In particular, a quadratic number field is either real ($r_1 = 2$, $r_2 = 0$) or it is imaginary ($r_1 = 0$, $r_2 = 1$). A cubic number field is either totally real ($r_1 = 3$, $r_2 = 0$) or it satisfies $r_1 = r_2 = 1$ (and then it is called **simply real**). For an algebraic number field K we denote by $\mathfrak{o}_K = \mathrm{cl}_K(\mathbb{Z}) = \overline{\mathbb{Z}} \cap K$ the domain of all algebraic integers in K. By 2.12.3 and 2.14.3 (applied for the field extension $K/\mathbb{Q}$) it follows that $\mathfrak{o}_K$ is a Dedekind domain with finite residue class rings. In particular, $\mathfrak{o}_\mathbb{Q} = \mathbb{Z}$. By 2.5.1, $\mathfrak{o}_K$ is a finitely generated (additive) abelian group and contains a basis of K. Hence $\mathfrak{o}_K$ is a free abelian group of rank $[K:\mathbb{Q}]$ by 2.3.5.2.

The Dedekind domain $\mathfrak{o}_K$ is called the **ring of integers** or the **maximal order** of K, and its unit group $U_K = \mathfrak{o}_K^\times$ is (by abuse of language) called the **unit group** of K. A ($\mathbb{Z}$-)basis of $\mathfrak{o}_K$ is called an **integral basis** of K.

In this chapter we apply the results of Chapter 2 to the maximal order $\mathfrak{o}_K$ of an algebraic number field, and for this we adjust our terminology according to the following catalogue (recall that all fractional ideals of $\mathfrak{o}_K$ are invertible).

We denote by

- $\mathcal{P}_K = \mathcal{P}(\mathfrak{o}_K)$ the set of all maximal ideals of $\mathfrak{o}_K$;
- $\mathcal{J}_K = \mathcal{J}(\mathfrak{o}_K)$ the group of all fractional ideals of $\mathfrak{o}_K$;
- $\mathcal{J}'_K = \mathcal{J}'(\mathfrak{o}_K)$ the monoid of all non-zero (integral) ideals of $\mathfrak{o}_K$;
- $\mathcal{H}_K = \mathcal{H}(\mathfrak{o}_K)$ the group of all fractional principal ideals of $\mathfrak{o}_K$;
- $\mathcal{H}'_K = \mathcal{H}'(\mathfrak{o}_K)$ the monoid of all non-zero (integral) principal ideals of $\mathfrak{o}_K$;
- $\mathcal{C}_K = \mathcal{J}_K/\mathcal{H}_K = \mathcal{C}(\mathfrak{o}_K)$ the ideal class group or Picard group of $\mathfrak{o}_K$

(see 2.6.1, 2.6.5 and 2.8.1).

We call $\mathcal{C}_K$ die **(ideal) class group** and $h_K = |\mathcal{C}_K|$ the **class number** of K. We shall see in 3.5.7 that it is finite. By 2.8.6 we know that $\mathfrak{o}_K$ is factorial if and only if $h_K = 1$, and it is a traditional belief in algebraic number theory that the class group measures the deviation of $\mathfrak{o}_K$ from being factorial. For recent quantitative results which support this belief the interested reader should consult [17]. Even more is true. In 2.1.1 we pointed out that the monoid $\mathcal{H}'_K = \mathcal{H}'(\mathfrak{o}_K)$ of all non-zero (integral) ideals of $\mathfrak{o}_K$ has the same factorization properties as the domain $\mathfrak{o}_K$ itself, and in 4.4.5 we shall prove that the monoid $\mathcal{H}'_K$ is (up to isomorphisms) uniquely determined by the ideal class group $\mathcal{C}_K$.

For $K = \mathbb{Q}$, the situation is special. The map $\mathbb{Q}_{>0} \to \mathcal{J}_\mathbb{Q}$, $a \mapsto a\mathbb{Z}$, is a group isomorphism, and we shall occasionally make the following identifications:

$\mathcal{J}_\mathbb{Q} = \mathcal{H}_\mathbb{Q} = \mathbb{Q}^+ = \mathbb{Q}_{>0}$, $\mathcal{C}_\mathbb{Q} = \mathbf{1}$, $\mathcal{P}_\mathbb{Q} = \mathbb{P}$, $\mathcal{J}'_\mathbb{Q} = \mathbb{N}$ and $\mathfrak{N}(a) = \mathfrak{N}_\mathbb{Z}(a) = a$ for all $a \in \mathcal{J}_\mathbb{Q} = \mathbb{Q}^+$.

If L/K is a finite extension of algebraic number fields, then $\mathfrak{o}_L = \mathrm{cl}_L(\mathfrak{o}_K)$, and we call $\mathcal{N}_{L/K} = \mathcal{N}_{\mathfrak{o}_L/\mathfrak{o}_K} : \mathcal{J}_L \to \mathcal{J}_K$ the **ideal norm** for L/K. It has the following properties (see Section 2.14):

- If $a \in L^\times$, then $\mathcal{N}_{L/K}(a\mathfrak{o}_L) = \mathsf{N}_{L/K}(a)\mathfrak{o}_K$.
- If $\mathfrak{A} \in \mathcal{J}_L$, then $\mathfrak{N}(\mathfrak{A}) = \mathfrak{N}(\mathcal{N}_{L/K}(\mathfrak{A}))$.
- If $a \in \mathbb{Q}^+$, then $\mathfrak{N}(a) = a$.
- If $\mathfrak{a} \in \mathcal{J}_K$, then $\mathcal{N}_{K/\mathbb{Q}}(\mathfrak{a}) = \mathfrak{N}(\mathfrak{a}) \in \mathbb{Q}^+ = \mathcal{J}_Q$.

For $m \in \mathbb{N}$, we denote by $\mathbb{Z}(m) = \{a \in \mathbb{Z} \mid (a,m) = 1\}$ the monoid of all integers coprime to m and by $\mathbb{Q}(m) = \mathsf{q}(\mathbb{Z}(m))$ the multiplicative group of all rational numbers coprime to m (see 2.9.4). Recall that the residue class homomorphism $\pi_m\colon \mathbb{Z}(m) \to (\mathbb{Z}/m\mathbb{Z})^\times$ extends by virtue of quotients to an (equally denoted) group homomorphism $\pi_m\colon \mathbb{Q}(m) \to (\mathbb{Z}/m\mathbb{Z})^\times$ with kernel

$$\mathbb{Q}^{\circ m} = \mathsf{q}(1+m\mathbb{Z}) = \{a^{-1}b \mid a,\, b \in \mathbb{Z}(m),\ a \equiv b \bmod m\}.$$

Subsequently, in 3.6.6, we shall introduce the groups

$$\mathbb{Q}^+(m) = \mathbb{Q}(m) \cap \mathbb{Q}^+, \quad \mathbb{Q}^m = \mathbb{Q}^{\circ m} \cap \mathbb{Q}^+ = \{a^{-1}b \mid a,\, b \in \mathbb{N},\ a \equiv b \equiv 1 \bmod m\}$$
$$\text{and } \ \mathbb{Q}^m_\pm = \{a \in \mathbb{Q}^m \mid a \equiv \pm 1 \bmod m\},$$

and in Section 3.7 we will see the role of these groups for the class field theory of the m-th cyclotomic field $\mathbb{Q}^{(m)}$.

Definitions and Remarks 3.1.2. Let K be an algebraic number field of degree $[K\colon\mathbb{Q}] = n$.

1. A **complete module** of K is a finitely generated (additive) subgroup $\mathfrak{a}$ of K which contains a basis of K [equivalently, $\mathbb{Q}\mathfrak{a} = K$]. By 2.3.5.2, a subgroup $\mathfrak{a}$ of K is a complete module of K if and only if it is a free abelian group of rank n, and then every ($\mathbb{Z}$-)basis of $\mathfrak{a}$ is a basis of K.

If $\mathfrak{a}$ is a complete module of K and $\lambda \in K^\times$, then $\lambda\mathfrak{a}$ is a complete module of K as well. Two complete modules $\mathfrak{a}$ and $\mathfrak{b}$ of K are called **equivalent**, $\mathfrak{a} \sim \mathfrak{b}$ if $\mathfrak{b} = \lambda\mathfrak{a}$ for some $\lambda \in K^\times$.

An **order** of K is an order in $\mathfrak{o}_K$ as defined in 2.11.1. Recall that a subring $\mathfrak{o}$ of K is an order of K if and only if $\mathfrak{o}$ is Noetherian, $\dim(\mathfrak{o}) = 1$, $K = \mathsf{q}(\mathfrak{o})$ and $\mathfrak{o}_K = \mathrm{cl}_K(\mathfrak{o})$. In particular, $\mathfrak{o}_K$ is an order of K.

2. Let $\mathfrak{a}$ be a complete module of K, and let $(u_1,\ldots,u_n)$ and $(v_1,\ldots,v_n)$ be bases of $\mathfrak{a}$. Then $(v_1,\ldots,v_n) = (u_1,\ldots,u_n)A$ for some matrix $A \in \mathsf{GL}_n(\mathbb{Z})$, and since $|\det(A)| = 1$, we get $d_{K/\mathbb{Q}}(u_1,\ldots,u_n) = d_{K/\mathbb{Q}}(v_1,\ldots,v_n)$ by 1.6.9.3. We call

$$\Delta(\mathfrak{a}) = d_{K/\mathbb{Q}}(u_1,\ldots,u_n) = \det(\sigma_\nu(u_i))^2_{\nu,\, i\in[1,n]}.$$

the **discriminant** of $\mathfrak{a}$ and $\mathsf{d}_K = \Delta(\mathfrak{o}_K)$ the **discriminant** of K. If $\mathfrak{a} \subset \mathfrak{o}_K$, then 2.5.1.3 implies $\Delta(\mathfrak{a}) \in \mathbb{Z}$, hence in particular $\mathsf{d}_K \in \mathbb{Z}$. Note that $\mathfrak{o}_\mathbb{Q} = \mathbb{Z}$, (1) is an integral basis of $\mathbb{Q}$ and $\mathsf{d}_\mathbb{Q} = 1$.

3. If $\mathfrak{a}$ is a complete module of K, then we call $\mathcal{R}(\mathfrak{a}) = \{x \in K \mid x\mathfrak{a} \subset \mathfrak{a}\}$ the **ring of multipliers** of $\mathfrak{a}$.

The following comprehensive Theorem 3.1.3 gathers elementary properties of the notions just introduced and their mutual dependences.

Theorem 3.1.3. *Let K be an algebraic number field.*

1. Let $\mathfrak{a}$ be a complete module of K and $\mathfrak{b}$ a finitely generated subgroup of K. Then $\mathfrak{b}$ is a complete module of K if and only if there exists some $q \in K^\times$ such that $q\mathfrak{a} \subset \mathfrak{b}$, and then there is even some $q \in \mathbb{N}$ such that $q\mathfrak{a} \subset \mathfrak{b}$.

2. For a subset $\mathfrak{o}$ of K the following assertions are equivalent:

(a) $\mathfrak{o}$ is an order of K.

(b) $\mathfrak{o}$ is a subring and a complete module of K.

(c) $\mathfrak{o} = \mathcal{R}(\mathfrak{a})$ for some complete module $\mathfrak{a}$ of K.

3. If $\mathfrak{a}$ and $\mathfrak{b}$ are complete modules of K and $\mathfrak{a} \subset \mathfrak{b}$, then

$$\Delta(\mathfrak{a}) = \Delta(\mathfrak{b})(\mathfrak{b}:\mathfrak{a})^2.$$

4. Let $\mathfrak{a}$ be a complete module of K.

(a) If $\lambda \in K^\times$, then $\lambda\mathfrak{a}$ is a complete module of K, $\mathcal{R}(\lambda\mathfrak{a}) = \mathcal{R}(\mathfrak{a})$, and $\Delta(\lambda\mathfrak{a}) = \mathsf{N}_{K/\mathbb{Q}}(\lambda)^2\Delta(\mathfrak{a})$.

(b) Let $\mathfrak{b}$ be a complete module of K and $\lambda \in K^\times$ such that $\mathfrak{a} \subset \mathfrak{b}$ and $\lambda\mathfrak{a} \subset \mathfrak{b}$. Then $(\mathfrak{b}:\lambda\mathfrak{a}) = |\mathsf{N}_{K/\mathbb{Q}}(\lambda)|(\mathfrak{b}:\mathfrak{a})$.

(c) $\mathfrak{a} \cap \mathbb{N} \neq \emptyset$, and for every $a \in K$ there exists some $a_0 \in a + \mathfrak{a}$ such that $a_0 \gg 0$.

5. Let $\mathfrak{o}$ be an order of K.

(a) An (additive) subgroup $\mathfrak{a}$ of K is a fractional ideal of $\mathfrak{o}$ if and only if $\mathfrak{a}$ is a complete module of K and $\mathfrak{o} \subset \mathcal{R}(\mathfrak{a})$.

(b) If $\mathfrak{a}$ is a non-zero ideal of $\mathfrak{o}$, then $\mathfrak{o}/\mathfrak{a}$ is finite.

6. Let $\mathfrak{a}$ and $\mathfrak{b}$ be complete modules of K. Then $\mathfrak{a} \sim \mathfrak{b}$ if and only if $\mathcal{R}(\mathfrak{a}) = \mathcal{R}(\mathfrak{b})$ and $\mathfrak{a}$ and $\mathfrak{b}$ are isomorphic $\mathcal{R}(\mathfrak{a})$-modules.

Proof. 1. Let $(u_1, \ldots, u_n)$ be a basis of $\mathfrak{a}$. If $q \in K^\times$ and $q\mathfrak{a} \subset \mathfrak{b}$, then $(qu_1, \ldots, qu_n)$ is a basis of K in $\mathfrak{b}$, and thus $\mathfrak{b}$ is a complete module of K. Conversely let $\mathfrak{b}$ be a complete module of K and $(v_1, \ldots, v_n)$ a basis of $\mathfrak{b}$. Then there exists some matrix $A \in \mathsf{GL}_n(\mathbb{Q})$ such that $(u_1, \ldots, u_n) = (v_1, \ldots, v_n)A$. If $q \in \mathbb{N}$ is such that $qA \in \mathsf{M}_n(\mathbb{Z})$, then $qu_i \in \mathfrak{b}$ for all $i \in [1, n]$ and thus $q\mathfrak{a} \subset \mathfrak{b}$.

2. (a) $\Rightarrow$ (b) By definition, $\mathfrak{o}$ is a subring of K and if $q \in (\mathfrak{o}:\mathfrak{o}_K)^\bullet$, then $q\mathfrak{o}_K \subset \mathfrak{o}$. If $(u_1, \ldots, u_n)$ is an integral basis of K, then $(qu_1, \ldots, qu_n)$ is a basis of K in $\mathfrak{o}$. Hence $\mathfrak{o}$ is a complete module of K.

(b) $\Rightarrow$ (c) Observe that $\mathfrak{o} = \mathcal{R}(\mathfrak{o})$.

(c) $\Rightarrow$ (a) Let $\mathfrak{a}$ be a complete module of K and $\mathfrak{o} = \mathcal{R}(\mathfrak{a})$. If $x, y \in \mathfrak{o}$, then $(x-y)\mathfrak{a} \subset x\mathfrak{a} + y\mathfrak{a} \subset \mathfrak{a}$ and $xy\mathfrak{a} \subset x\mathfrak{a} \subset \mathfrak{a}$, hence $\{x-y, xy\} \subset \mathfrak{o}$, and as $1 \in \mathfrak{o}$, it follows that $\mathfrak{o}$ is a subring of K. If $x \in \mathfrak{o}$, then $x\mathfrak{a} \subset \mathfrak{a}$, and therefore $x \in \mathrm{cl}_K(\mathbb{Z}) = \mathfrak{o}_K$ by 2.4.2. Hence $\mathfrak{o} \subset \mathfrak{o}_K$, every ideal of $\mathfrak{o}$ is a finitely generated abelian group by 2.3.5 and $\mathfrak{o}_K = \mathrm{cl}_K(\mathfrak{o})$. Therefore $\mathfrak{o}$ is Noetherian, $\dim(\mathfrak{o}) = \dim(\mathfrak{o}_K) = 1$ by 2.4.8, and apparently $\mathfrak{o}_K$ is a finitely generated $\mathfrak{o}$-module. It remains to prove that $K = \mathsf{q}(\mathfrak{o})$. If $\lambda \in K^\times$, then $\lambda\mathfrak{a}$ is a complete module of K, and by 1. there exists some $q \in \mathbb{N}$ such that $q\lambda\mathfrak{a} \subset \mathfrak{a}$. Thus we obtain $q\lambda \in \mathcal{R}(\mathfrak{a}) = \mathfrak{o}$ and $\lambda \in q^{-1}\mathfrak{o} \subset \mathsf{q}(\mathfrak{o})$.

3. Let $\mathfrak{a}$ and $\mathfrak{b}$ be complete modules of K such that $\mathfrak{a} \subset \mathfrak{b}$, $(u_1, \ldots, u_n)$ a basis of $\mathfrak{a}$, $(v_1, \ldots, v_n)$ a basis of $\mathfrak{b}$ and $(u_1, \ldots, u_n) = (v_1, \ldots, v_n)A$ for some $A \in \mathsf{M}_n(\mathbb{Z})$. Then 2.3.6.3 implies that $(\mathfrak{b}:\mathfrak{a}) = |\det(A)|$, and by 1.6.9.3 we obtain

$$\Delta(\mathfrak{a}) = d_{K/\mathbb{Q}}(u_1, \ldots, u_n) = \det(A)^2 d_{K/\mathbb{Q}}(v_1, \ldots, v_n) = (\mathfrak{b}:\mathfrak{a})^2\Delta(\mathfrak{b}).$$

4. (a) Obviously, $\lambda\mathfrak{a}$ is a complete module of K, and $\mathcal{R}(\mathfrak{a}) = \mathcal{R}(\lambda\mathfrak{a})$.

If $(u_1, \ldots, u_n)$ is a basis of $\mathfrak{a}$, then $(\lambda u_1, \ldots, \lambda u_n)$ is a basis of $\lambda\mathfrak{a}$, and by 1.6.9.1 we obtain

$$\Delta(\lambda\mathfrak{a}) = d_{K/\mathbb{Q}}(\lambda u_1, \ldots, \lambda u_n) = \mathsf{N}_{K/\mathbb{Q}}(\lambda)^2 d_{K/\mathbb{Q}}(u_1, \ldots, u_n) = \mathsf{N}_{K/\mathbb{Q}}(\lambda)^2\Delta(\mathfrak{a}).$$

(b) Using 3. and (a), we obtain

$$(\mathfrak{b}:\lambda\mathfrak{a})^2 = \frac{\Delta(\lambda\mathfrak{a})}{\Delta(\mathfrak{b})} = \frac{|\mathsf{N}_{K/\mathbb{Q}}(\lambda)|^2\Delta(\mathfrak{a})}{\Delta(\mathfrak{b})} = |\mathsf{N}_{K/\mathbb{Q}}(\lambda)|^2(\mathfrak{b}:\mathfrak{a})^2.$$

(c) By 1. there exists some $q \in \mathbb{N}$ such that $q\mathfrak{o}_K \subset \mathfrak{a}$. Now 2.4.7.1 implies that $\mathfrak{o}_K \cap \mathbb{Z} \neq \emptyset$, and thus $\mathfrak{o}_K \cap \mathbb{N} \neq \emptyset$. If $c \in \mathfrak{o}_K \cap \mathbb{N}$, then $mqc \in \mathfrak{a} \cap \mathbb{N}$ for all $m \in \mathbb{N}$, and if $a \in \mathfrak{a}$, then $a_0 = a + mqc \gg 0$ for all $m \gg 1$.

5.(a) If $\mathfrak{a}$ is a fractional ideal of $\mathfrak{o}$, then $\mathfrak{o}\mathfrak{a} \subset \mathfrak{a}$, hence $\mathfrak{o} \subset \mathcal{R}(\mathfrak{a})$, and there exists some $c \in \mathfrak{o}^\bullet$ such that $c\mathfrak{o} \subset \mathfrak{a}$. Hence $\mathfrak{a}$ is a complete module of K by 1.

As to the converse, let $\mathfrak{a}$ be a complete module of K and $\mathfrak{o} \subset \mathcal{R}(\mathfrak{a})$. Since $\mathfrak{o}\mathfrak{a} \subset \mathcal{R}(\mathfrak{a})\mathfrak{a} = \mathfrak{a}$, it follows that $\mathfrak{a}$ is an $\mathfrak{o}$-module. Obviously, $\mathfrak{a} \neq \mathbf{0}$, and by 1. there exists some $q \in K^\times$ such that $q\mathfrak{a} \subset \mathfrak{o}$. Hence $\mathfrak{a}$ is a fractional ideal of $\mathfrak{o}$.

(b) Let $\mathfrak{f} = (\mathfrak{o} : \mathfrak{o}_K) \in \mathcal{I}'_K$ be the conductor of $\mathfrak{o}$. By 4.(c) there exist integers $a \in \mathfrak{a} \cap \mathbb{N}$ and $f \in \mathfrak{f} \cap \mathbb{N}$. Then $fa\mathfrak{o}_K \subset a\mathfrak{o} \subset \mathfrak{a} \subset \mathfrak{o} \subset \mathfrak{o}_K$ and therefore we obtain $(\mathfrak{o}:\mathfrak{a}) \le (\mathfrak{o}_K : f\mathfrak{o}_K) = f^n < \infty$, where $n = [K:\mathbb{Q}] = \mathrm{rk}(\mathfrak{o}_K)$.

6. If $\mathfrak{a} \sim \mathfrak{b}$, then $\mathfrak{b} = \lambda\mathfrak{a}$ for some $\lambda \in K^\times$, hence $\mathcal{R}(\mathfrak{a}) = \mathcal{R}(\mathfrak{b})$ by 4.(a), and the assignment $a \mapsto \lambda a$ defines an $\mathcal{R}(\mathfrak{a})$-isomorphism $\mathfrak{a} \to \mathfrak{b}$.

Conversely, let $\mathcal{R}(\mathfrak{a}) = \mathcal{R}(\mathfrak{b})$ and $\varphi\colon \mathfrak{a} \to \mathfrak{b}$ be an $\mathcal{R}(\mathfrak{a})$-isomorphism. By 1. there exists some $q \in K^\times$ such that $qa,\, qb \in \mathcal{R}(\mathfrak{a})$, and then it follows that $\varphi(qab) = qa\varphi(b) = qb\varphi(a)$. Hence $a^{-1}\varphi(a) = b^{-1}\varphi(b) = \lambda \in K^\times$ does not depend on a, and thus $\varphi(a) = \lambda a$ for all $a \in \mathfrak{a}$ and $\mathfrak{b} = \lambda\mathfrak{a}$. □

Corollary 3.1.4. *Let K be an algebraic number field, $a \in K^\times$ and $\mathfrak{a} \in \mathcal{I}_K$. Then*

$$\mathfrak{N}(a\mathfrak{a}) = |\mathsf{N}_{K/\mathbb{Q}}(a)|\,\mathfrak{N}(\mathfrak{a}) \quad \textit{and} \quad \Delta(\mathfrak{a}) = \mathfrak{N}(\mathfrak{a})^2\,\mathsf{d}_K.$$

Proof. By 2.14.3 and the above remarks concerning the ideal theory of $\mathbb{Z}$ we get $\mathfrak{N}(a\mathfrak{a}) = \mathfrak{N}(a\mathfrak{o}_K)\mathfrak{N}(\mathfrak{a})$ and $\mathfrak{N}(a\mathfrak{o}_K) = \mathfrak{N}_{\mathbb{Z}}(\mathsf{N}_{K/\mathbb{Q}}(a\mathfrak{o}_K)) = |\mathsf{N}_{K/\mathbb{Q}}(a)|$.

For the proof of the second formula let $\lambda \in \mathfrak{o}_K^\bullet$ be such that $\lambda\mathfrak{a} \subset \mathfrak{o}_K$. Then $\Delta(\lambda\mathfrak{a}) = \mathsf{N}_{K/\mathbb{Q}}(\lambda)^2\Delta(\mathfrak{a})$, and by 3.1.3.3 and 2.14.2.4 we obtain

$$\Delta(\lambda\mathfrak{a}) = (\mathfrak{o}_K : \lambda\mathfrak{a})^2\mathsf{d}_K = \mathfrak{N}(\lambda\mathfrak{a})^2\mathsf{d}_K = \mathfrak{N}(\lambda\mathfrak{o}_K)^2\mathfrak{N}(\mathfrak{a})^2\mathsf{d}_K = \mathsf{N}_{K/\mathbb{Q}}(\lambda)^2\mathfrak{N}(\mathfrak{a})^2\mathsf{d}_K.$$

Hence $\Delta(\mathfrak{a}) = \mathfrak{N}(\mathfrak{a})^2\mathsf{d}_K$. □

The following result on bases of complete modules of algebraic number fields comprises an algorithm for their construction.

Theorem 3.1.5 (Basis theorem for complete modules). *Let K be an algebraic number field, $[K:\mathbb{Q}] = n$, $\mathfrak{a}$ a complete module of K, $\mathfrak{a} \subset \mathfrak{o}_K$, $(v_1,\ldots,v_n) \in \mathfrak{a}^n$ a basis of K and $d = |d_{K/\mathbb{Q}}(v_1,\ldots,v_n)|$. Then $d \in \mathbb{N}$, and we set $d = d_0^2d_1$, where $d_0, d_1 \in \mathbb{N}$ and d_1 is squarefree. For $i \in [1,n]$ let $b_{i,i} \in \mathbb{N}$ be minimal such there exist integers $b_{1,i},\ldots,b_{i-1,i} \in \mathbb{Z}$ satisfying*

$$u_i = \frac{1}{d_0}\sum_{j=1}^{i} b_{j,i}v_j \in \mathfrak{a}.$$

Then $(u_1,\ldots,u_n)$ is a basis of $\mathfrak{a}$. In particular, $\mathfrak{a}$ has a basis $(u_1,\ldots,u_n)$ such that $u_1 = \min(\mathfrak{a}\cap\mathbb{N})$, and every order of K has a basis $(u_1,\ldots,u_n)$ such that $u_1 = 1$.

Proof. We consider the complete module $\mathfrak{a}_0 = \mathbb{Z}v_1 + \ldots + \mathbb{Z}v_n \subset \mathfrak{a} \subset \mathfrak{o}_K$. By 3.1.3.3 and 2.5.1.3, $d = |d_{K/\mathbb{Q}}(v_1,\ldots,v_n)| = |\Delta(\mathfrak{a}_0)| = |\Delta(\mathfrak{a})|\,(\mathfrak{a}:\mathfrak{a}_0)^2 \in \mathbb{N}$, hence $(\mathfrak{a}:\mathfrak{a}_0)^2 \mid d$, and therefore $(\mathfrak{a}:\mathfrak{a}_0) \mid d_0$ and $d_0\mathfrak{a} \subset \mathfrak{a}_0$. From the definition of the numbers $b_{i,j}$ we get the matrix equation

$$(u_1,\ldots,u_n) = (v_1,\ldots,v_n)B, \quad \text{where}$$

$$B = \frac{1}{d_0}\begin{pmatrix} b_{1,1} & b_{1,2} & \ldots & . & b_{n,1} \\ 0 & b_{2,2} & \ldots & . & b_{n,2} \\ 0 & 0 & \ldots & . & . \\ . & . & \ddots & . & . \\ 0 & 0 & \ldots & 0 & b_{n,n} \end{pmatrix} \in \mathsf{GL}_n(\mathbb{Q})\,.$$

Hence $(u_1,\ldots,u_n)$ is a basis of K, $\mathbb{Z}u_1 + \ldots + \mathbb{Z}u_n \subset \mathfrak{a}$, and we show that equality holds. Therefore it suffices to prove the following assertion for all $i \in [0,n]$:

> **A.** If $c_1,\ldots,c_i \in \mathbb{Z}$ are such that $x = d_0^{-1}(c_1v_1 + \ldots + c_iv_i) \in \mathfrak{a}$, then $x \in \mathbb{Z}u_1 + \ldots + \mathbb{Z}u_i$.

Indeed, if $x \in \mathfrak{a}$, then $d_0x \in \mathfrak{a}_0$, hence $x = d_0^{-1}(c_1v_1 + \ldots + c_nv_n)$ for some $c_1,\ldots,c_n \in \mathbb{Z}$, and then **A** (applied with $i = n$) implies $x \in \mathbb{Z}u_1 + \ldots \mathbb{Z}u_n$.

Proof of **A.** By induction on i. For $i=0$ there is nothing to do.

$i \ge 1$, $i-1 \to i$: Let $c_1,\ldots,c_i \in \mathbb{Z}$ and $x = d_0^{-1}(c_1v_1+\ldots+c_iv_i) \in \mathfrak{a}$. If $c_i = kb_{i,i}+r$, where $k \in \mathbb{Z}$ and $r \in [0, b_{i,i}-1]$, then

$$x - ku_i = \frac{1}{d_0}\sum_{j=1}^{i}(c_j - kb_{j,i})v_j \in \mathfrak{a} \quad \text{and} \quad c_i - kb_{i,i} = r \in [0, b_{i,i}-1]\,.$$

The minimality of $b_{i,i}$ implies $c_i - kb_{i,i} = 0$, hence $x - ku_i \in \mathbb{Z}u_1+\ldots+\mathbb{Z}u_{i-1}$ by the induction hypothesis, and $x \in \mathbb{Z}u_1+\ldots+\mathbb{Z}u_i$.

In particular, it is possible to choose $(v_1,\ldots,v_n)$ such that $v_1 = \min(\mathfrak{a}\cap\mathbb{N})$, and then we obtain $u_1 = v_1$. □

Theorem 3.1.6. *Let K be an algebraic number field, (r_1, r_2) its signature and $n = r_1+2r_2 = [K:\mathbb{Q}]$. Then*

$$\mathsf{d}_K \equiv 0 \text{ or } 1 \bmod 4 \quad \text{and} \quad \operatorname{sgn}(\mathsf{d}_K) = (-1)^{r_2}.$$

Proof. Let $(u_1,\ldots,u_n)$ be an integral basis of K and

$$\operatorname{Hom}(K,\mathbb{C}) = \{\sigma_1,\ldots,\sigma_n\},$$

where $\sigma_i(K) \subset \mathbb{R}$ for $i \in [1, r_1]$ and $\sigma_{r_1+r_2+i} = \overline{\sigma_{r_1+i}} \ne \sigma_{r_1+i}$ for $i \in [1, r_2]$. Let

$$L = \prod_{i=1}^{r_1+r_2} \sigma_i(K) \subset \mathbb{C}$$

be the compositum of the conjugate fields of K. By definition,

$$\mathsf{d}_K = \det(\sigma_\nu(u_i))^2_{\nu,i\in[1,n]} = (A-B)^2 = (A+B)^2 - 4AB,$$

where

$$A = \sum_{\substack{\pi\in\mathfrak{S}_n\\ \operatorname{sgn}(\pi)=1}} \prod_{\nu=1}^{n} \sigma_{\pi(\nu)}(u_\nu) \in \mathfrak{o}_L \quad \text{and} \quad B = \sum_{\substack{\pi\in\mathfrak{S}_n\\ \operatorname{sgn}(\pi)=-1}} \prod_{\nu=1}^{n} \sigma_{\pi(\nu)}(u_\nu) \in \mathfrak{o}_L.$$

$L/\mathbb{Q}$ is Galois, and for $\tau \in G = \operatorname{Gal}(L/\mathbb{Q})$ we define $\mathsf{s}_\tau \in \mathfrak{S}_n$ by $\tau\sigma_i = \sigma_{\mathsf{s}_\tau(i)}$ for all $i \in [1,n]$. Then

$$\tau(A) = \sum_{\substack{\pi\in\mathfrak{S}_n\\ \operatorname{sgn}(\pi)=1}} \prod_{\nu=1}^{n} \sigma_{\mathsf{s}_\tau\circ\pi(\nu)}(u_\nu), \qquad \tau(B) = \sum_{\substack{\pi\in\mathfrak{S}_n\\ \operatorname{sgn}(\pi)=-1}} \prod_{\nu=1}^{n} \sigma_{\mathsf{s}_\tau\circ\pi(\nu)}(u_\nu),$$

and consequently

$$(\tau(A),\tau(B)) = \begin{cases} (\tau(A),\tau(B) & \text{if } \mathsf{s}_\tau \text{ is even,} \\ (\tau(B),\tau(A) & \text{if } \mathsf{s}_\tau \text{ is odd.} \end{cases}$$

It follows that $\tau(A+B) = A+B$ and $\tau(AB) = AB$ for all $\tau \in G$, hence

$$\{A+B, AB\} \subset \mathfrak{o}_L \cap \mathbb{Q} = \mathbb{Z} \ \text{ and } \ \mathsf{d}_K = (A+B)^2 - 4AB \equiv 0 \text{ or } 1 \bmod 4.$$

The matrix $\big(\overline{\sigma_\nu(u_i)}\big)_{\nu,i\in[1,n]}$ arises from the matrix $(\sigma_\nu(u_i))_{\nu,i\in[1,n]}$ by interchanging r_2 columns, and therefore

$$\overline{\det(\sigma_\nu(u_i))_{\nu,i\in[1,n]}} = \det\big(\overline{\sigma_\nu(u_i)}\big)_{\nu,i\in[1,n]} = (-1)^{r_2}\det(\sigma_\nu(u_i))_{\nu,i\in[1,n]}$$

If $d = \det(\sigma_\nu(u_i))_{\nu,i\in[1,n]}$, then $\overline{d} = (-1)^{r_2}\overline{d}$, and therefore it follows that $(-1)^{r_2}\mathsf{d}_K = (-1)^{r_2}d^2 = d\overline{d} = |d|^2 > 0$, hence $\operatorname{sgn}(\mathsf{d}_K) = (-1)^{r_2}$. □

Recall that a polynomial $f = X^n + c_{n-1}X^{n-1} + \ldots + c_1X + c_0 \in \mathbb{Z}[X] \setminus \mathbb{Z}$ is a p-Eisenstein polynomial for some prime p if $p \mid c_i$ for all $i \in [0, n-1]$ and $p^2 \nmid c_0$. Every p-Eisenstein polynomial is irreducible in $\mathbb{Z}[X]$ and thus in $\mathbb{Q}[X]$ (see 1.1.11 and 1.1.12). Eisenstein polynomials are a useful tool for the calculation of integral bases and discriminants. We have already seen in 2.12.8 that they are basic for the characterization of ramification, and we shall meet them again in Section 5.7. We proceed with a simple criterion for integral bases.

Theorem 3.1.7. *Let K be an algebraic number field, $[K:\mathbb{Q}] = n$, $a \in \mathfrak{o}_K$, $K = \mathbb{Q}(a)$, and let $f \in \mathbb{Z}[X]$ be the minimal polynomial of a over $\mathbb{Q}$.*

1. $\Delta(\mathbb{Z}[a]) = d_{K/\mathbb{Q}}(1, a, \ldots, a^{n-1}) = \mathsf{D}(f) = \mathsf{d}_K(\mathfrak{o}_K : \mathbb{Z}[a])^2$. In particular:

- *If p is a prime such that $p^2 \nmid \mathsf{D}(f)$, then $p \nmid (\mathfrak{o}_K : \mathbb{Z}[a])$.*
- *If $\mathsf{D}(f)$ is squarefree, then $\mathsf{d}_K = \mathsf{D}(f)$, and $(1, a, \ldots, a^{n-1})$ is an integral basis of K.*

2. Let p be a prime. Then we obtain $p \nmid (\mathfrak{o}_K : \mathbb{Z}[a])$ if and only if $\mathfrak{o}_{K,(p)} = \mathbb{Z}_{(p)}[a]$. In particular, if f is a p-Eisenstein polynomial, then $p \nmid (\mathfrak{o}_K : \mathbb{Z}[a])$.

Proof. 1. Since $(1, a, \ldots, a^{n-1})$ is a basis of $\mathbb{Z}[a]$, 1.6.9.1 and 3.1.3.3 imply that

$$\Delta(\mathbb{Z}[a]) = d_{K/\mathbb{Q}}(1, a, \ldots, a^{n-1}) = \mathsf{D}(f) = \Delta(\mathfrak{o}_K)(\mathfrak{o}_K : \mathbb{Z}[a])^2 = \mathsf{d}_K(\mathfrak{o}_K : \mathbb{Z}[a])^2.$$

2. Obviously, $\mathbb{Z}_{(p)}[a] = \mathbb{Z}[a]_{(p)} \subset \mathfrak{o}_{K,(p)}$, and let $q = (\mathfrak{o}_K : \mathbb{Z}[a])$, hence $q\mathfrak{o}_K \subset \mathbb{Z}[a]$. If $p \nmid q$, then $q \in \mathbb{Z}_{(p)}^\times$, hence $\mathfrak{o}_{K,(p)} = q\mathfrak{o}_{K,(p)} \subset \mathbb{Z}[a]_{(p)}$ and therefore $\mathbb{Z}_{(p)}[a] = \mathfrak{o}_{K,(p)}$.

Conversely, if $\mathbb{Z}[a]_{(p)} = \mathfrak{o}_{K,(p)}$, then there exists some $t \in \mathbb{N}$ such that $p \nmid t$ and $t\mathfrak{o}_K \subset \mathbb{Z}[a]$ (see 2.7.1.1(d)). As $(\mathfrak{o}_K : \mathbb{Z}[a]) \mid (\mathfrak{o}_K : t\mathfrak{o}_K) = t^n$ by 2.3.6.3, it follows that $p \nmid (\mathfrak{o}_K : \mathbb{Z}[a])$.

If f is a p-Eisenstein polynomial, then $\mathbb{Z}_{(p)}[a] = \mathfrak{o}_{K,(p)}$ by 2.12.8.1. □

In the sequel, we will tacitly apply 3.1.7 again and again to determine integral bases and discriminants of quadratic, cubic and cyclotomic number fields.

Theorem 3.1.8. *Let K be a quadratic number field.*

1. *There exists a uniquely determined squarefree integer $D \in \mathbb{Z}$ such that $K = \mathbb{Q}(\sqrt{D})$,* called the **radicand** *of K. If D is the radicand of K and $a \in \mathbb{Q}^\times$, then $K = \mathbb{Q}(\sqrt{a})$ if and only if D is the squarefree kernel of a.*

2. *Let D be the radicand of K and*

$$\omega_D = \begin{cases} \sqrt{D} & \text{if } \ D \not\equiv 1 \mod 4, \\ (1+\sqrt{D})/2 & \text{if } \ D \equiv 1 \mod 4. \end{cases}$$

Then $(1, \omega_D)$ is an integral basis of K, $\mathfrak{o}_K = \mathbb{Z}[\omega_D]$,

$$\mathsf{d}_K = \begin{cases} 4D & \text{if } \ D \not\equiv 1 \mod 4, \\ D & \text{if } \ D \equiv 1 \mod 4, \end{cases}$$

and $K = \mathbb{Q}(\sqrt{\mathsf{d}_K})$. In particular, K is uniquely determined by d_K.

Proof. 1. Obvious (see Exercise 3.a of Chapter 1).

2. Let $N = (\mathfrak{o}_K : \mathbb{Z}[\sqrt{D}])$ and $f = X^2 - D \in \mathbb{Z}[X]$ (the minimal polynomial of $\sqrt{D}$). If $p \in \mathbb{P}$ and $p \mid D$, then f is a p-Eisenstein polynomial, hence it follows that $p \nmid N$ and $(N, D) = 1$. Since

$$\mathsf{d}_K N^2 = \mathsf{D}(f) = -\mathsf{N}_{K/\mathbb{Q}}(f'(\sqrt{D})) = -\mathsf{N}_{K/\mathbb{Q}}(2\sqrt{D}) = 4D$$

(by 3.1.7.1 and 1.6.9.2), it follows that $N \in \{1, 2\}$, and $N = 1$ if $2 \mid D$. If $N = 1$, then $(1, \sqrt{D})$ is an integral basis of K, and $\mathsf{d}_K = 4D$.

If $D \equiv 3 \bmod 4$, then $f_1 = f(X+1) = X^2 + 2X + 1 - D \in \mathbb{Z}[X]$ is a 2-Eisenstein polynomial, and since $f_1(\sqrt{D} - 1) = 0$, it follows that

$$2 \nmid (\mathfrak{o}_K : \mathbb{Z}[\sqrt{D} - 1]) = (\mathfrak{o}_K : \mathbb{Z}[\sqrt{D}]) = N \quad \text{and thus} \quad N = 1.$$

If $D \equiv 1 \bmod 4$, then $D = 1 + 4d$, where $d \in \mathbb{Z}$, and we consider the polynomial $g = X^2 - X - d \in \mathbb{Z}[X]$. If $\omega_D = (1 + \sqrt{D})/2$, then $g(\omega_D) = 0$, hence $\omega_D \in \mathfrak{o}_K$, and since $\mathbb{Z}[\sqrt{D}] \subsetneq \mathbb{Z}[\omega_D] \subset \mathfrak{o}_K$ it follows that $N = 2$, $\mathsf{d}_K = D$, $(1, \omega_D)$ is an integral basis of K, and $\mathsf{d}_K = D$. □

We already considered quadratic orders in Exercise 14 of Chapter 2. There we invited the ambitious reader to prove that $\mathbb{Z}[\sqrt{D}]$ is Euclidean (and thus a principal ideal domain) if $D \in \{-1, -2, 2, 3, 5, 6, 7\}$. In Exercise 2 of the current chapter we assert in addition that $\mathbb{Z}[\omega_D]$ is also norm-Euclidean if $D \in \{-3, -7, -11, 13, 17, 21, 29\}$, and that $\mathbb{Z}[\omega_D]$ is not even Euclidean if

$D < -11$. Overall there are 16 positive radicands D for which $\mathbb{Z}[\omega_D]$ is norm-Euclidean, and there exist positive radicands D for which $\mathbb{Z}[\omega_D]$ is Euclidean but not norm-Euclidean.

There are exactly 4 more negative radicands for which $\mathbb{Z}[\omega_D]$ is a principal ideal domain, namely -19, -43, -67, -163 (for a readable proof of this famous fact we refer to [9, Ch. 3, §12]). In contrast, an old conjecture of Gauss states that there exist infinitely many positive radicands D for which $\mathbb{Z}[\omega_D]$ is a principal ideal domain.

That a quadratic number field is uniquely determined by its discriminant, is a specialty of quadratic number fields. In 3.2.7 we shall construct three non-isomorphic cubic number fields with the same discriminant.

Next we determine discriminants and integral bases of pure cubic number fields. If $m \in \mathbb{N} \setminus \mathbb{N}^3$, then the polynomial $X^3 - m$ is irreducible in $\mathbb{Q}[X]$ by 1.7.9, hence $[\mathbb{Q}(\sqrt[3]{m}):\mathbb{Q}] = 3$, and we call $\mathbb{Q}(\sqrt[3]{m})$ a **pure cubic number field**. If $m \in \mathbb{N} \setminus \mathbb{N}^3$, then m has a unique factorization $m = ab^2q^3$, where $a,\, b,\, q \in \mathbb{N}$, $ab \ne 1$ and ab is squarefree, given as follows: For a prime p, let $v_p(m) = 3\alpha_p + \beta_p$, where $\alpha_p \in \mathbb{N}_0$ and $\beta_p \in \{0,1,2\}$. Then

$$a = \prod_{\substack{p \in \mathbb{P} \\ \beta_p = 1}} p, \quad b = \prod_{\substack{p \in \mathbb{P} \\ \beta_p = 2}} p, \quad q = \prod_{p \in \mathbb{P}} p^{\alpha_p}, \quad \mathbb{Q}(\sqrt[3]{m}\,) = \mathbb{Q}(\sqrt[3]{ab^2}\,) = \mathbb{Q}(\sqrt[3]{a^2b}\,),$$

and by interchanging a and b if necessary, we may assume that $3 \nmid b$.

Theorem 3.1.9. *Let $m = ab^2$, where $a,\, b \in \mathbb{N}$ are such that $3 \nmid b$, $ab \ne 1$ and ab is squarefree. Let $K = \mathbb{Q}(\sqrt[3]{m}\,)$ and $\theta = \sqrt[3]{m}$.*

1. *If $m \not\equiv \pm 1 \bmod 9$, then $\mathsf{d}_K = -27a^2b^2$, and*

$$\Big(1,\, \theta,\, \frac{\theta^2}{b}\Big) \quad \textit{is an integral basis of } K.$$

2. *If $e \in \{\pm 1\}$ and $m \equiv e \bmod 9$, then $\mathsf{d}_K = -3a^2b^2$, and*

$$\Big(1,\, \frac{\theta^2}{b},\, \frac{1 + e\theta + \theta^2}{3}\Big) \quad \textit{is an integral basis of } K.$$

3. *$(\mathfrak{o}_K : \mathbb{Z}[\theta]) = db$ and $(\mathfrak{o}_K : \mathbb{Z}[b^{-1}\theta^2]) = da$, where*

$$d = \begin{cases} 1 & \textit{if} \quad m \not\equiv \pm 1 \bmod 9, \\ 3 & \textit{if} \quad m \equiv \pm 1 \bmod 9. \end{cases}$$

Proof. We show first:

A. $\mathsf{d}_K = -3^N a^2 b^2$, where $N \in \{1,3\}$, and $N = 3$ if $3 \,|\, m$.

Proof of **A.** If $g = X^3 - ab^2$ and $g_1 = X^3 - a^2b$, then $g(\theta) = g_1(b^{-1}\theta^2) = 0$, and by 1.6.3 and 3.1.7, we get $\Delta(\mathbb{Z}[\theta]) = \mathsf{D}(g) = -27a^2b^4 = \mathsf{d}_K(\mathfrak{o}_K : \mathbb{Z}[\theta])^2$ and $\Delta(\mathbb{Z}[b^{-1}\theta^2]) = \mathsf{D}(g_1) = -27a^4b^2 = \mathsf{d}_K(\mathfrak{o}_K : \mathbb{Z}[b^{-1}\theta^2])^2$. Hence it follows that $\mathsf{d}_K < 0$ and $\mathsf{d}_K \mid 27a^2b^2$.

If $p \in \mathbb{P}$ and $p \mid a$, then g is a p-Eisenstein polynomial, hence $p \nmid (\mathfrak{o}_K : \mathbb{Z}[\theta])$, and therefore $(a, (\mathfrak{o}_K : \mathbb{Z}[\theta])) = 1$. It follows that

$$3a^2 \mid \mathsf{d}_K \ \text{ if } \ 3 \nmid a, \quad \text{and} \quad 27a^2 \mid \mathsf{d}_K \ \text{ if } \ 3 \mid a.$$

If $p \in \mathbb{P}$ and $p \mid b$, then g_1 is a p-Eisenstein polynomial, hence $p \nmid (\mathfrak{o}_K : \mathbb{Z}[b^{-1}\theta^2])$, and therefore $(b, (\mathfrak{o}_K : \mathbb{Z}[b^{-1}\theta^2])) = 1$. It follows that

$$3b^2 \mid \mathsf{d}_K \ \text{ if } \ 3 \nmid b, \quad \text{and} \quad 27b^2 \mid \mathsf{d}_K \ \text{ if } \ 3 \mid b.$$

Since $(a, b) = 1$, we obtain $3a^2b^2 \mid \mathsf{d}_K \mid 27a^2b^2$, and $\mathsf{d}_K = -3^N a^2b^2$ for some $N \in \{1, 3\}$. If $3 \mid m$, then $3 \mid a$ or $3 \mid b$, hence $27a^2b^2 \mid \mathsf{d}_K$ and consequently $\mathsf{d}_K = -27a^2b^2$. $\square$[**A.**]

1. Suppose that $m \not\equiv \pm 1 \bmod 9$. Since $9 \nmid m$, we get $m^3 \not\equiv m \bmod 9$, and $h = (X + m)^3 - m = X^3 + 3X^2m + 3Xm^2 + m^3 - m$ is a 3-Eisenstein polynomial satisfying $h(\theta - m) = 0$ and $h'(\theta - m) = 3\theta^2$. Hence it follows that $3 \nmid (\mathfrak{o}_K : \mathbb{Z}[\theta - m]) = (\mathfrak{o}_K : \mathbb{Z}[\theta])$ and

$$\mathsf{d}_K(\mathfrak{o}_K : \mathbb{Z}[\theta])^2 = \mathsf{D}(h) = -\mathsf{N}_{K/\mathbb{Q}}(h'(\theta - m)) = -27a^2b^4.$$

Together with **A** we obtain $\mathsf{d}_K = -27a^2b^2$ and $(\mathfrak{o}_K : \mathbb{Z}[\theta]) = b$. Since

$$\mathbb{Z}[\theta] \subset \mathbb{Z} + \mathbb{Z}\theta + \mathbb{Z}b^{-1}\theta^2 \subset \mathfrak{o}_K \quad \text{and} \quad (\mathbb{Z} + \mathbb{Z}\theta + \mathbb{Z}b^{-1}\theta^2 : \mathbb{Z}[\theta]) = b,$$

it follows that $\mathfrak{o}_K = \mathbb{Z} + \mathbb{Z}\theta + \mathbb{Z}b^{-1}\theta^2$.

2. Let $e \in \{\pm 1\}$ be such that $m \equiv e \bmod 9$. We calculate the minimal polynomial of $\omega = (1 + e\theta + \theta^2)/3$. As

$$\begin{aligned} 3\omega \begin{pmatrix} 1 & \theta & \theta^2 \end{pmatrix} &= \begin{pmatrix} 3\omega & m + \theta + e\theta^2 & em + m\theta + \theta^2 \end{pmatrix} \\ &= \begin{pmatrix} 1 & \theta & \theta^2 \end{pmatrix} \begin{pmatrix} 1 & m & em \\ e & 1 & m \\ 1 & e & 1 \end{pmatrix}, \end{aligned}$$

it follows that 3ω is a zero of the polynomial

$$\det \begin{pmatrix} X - 1 & -m & -em \\ -e & X - 1 & -m \\ -1 & -e & X - 1 \end{pmatrix} = X^3 - 3X^2 + 3(1 - em)X - (m - e)^2,$$

and therefore ω is a zero of the polynomial

$$h = X^3 - X^2 + \frac{1 - em}{3} X - \frac{(m - e)^2}{27} \in \mathbb{Z}[X],$$

which implies $\omega \in \mathfrak{o}_K$. Since

$$(1 \quad \theta \quad \theta^2) = (1 \quad b^{-1}\theta^2 \quad \omega)\begin{pmatrix} 1 & -e & 0 \\ 0 & -be & b \\ 0 & 3e & 0 \end{pmatrix} \quad \text{and} \quad \det\begin{pmatrix} 1 & -e & 0 \\ 0 & -be & b \\ 0 & 3e & 0 \end{pmatrix} = -3be,$$

it follows that $(\mathbb{Z} + \mathbb{Z}b^{-1}\theta^2 + \mathbb{Z}\omega : \mathbb{Z}[\theta]) = 3b$ and

$$-27a^2b^4 = \Delta(\mathbb{Z}[\theta]) = \Delta(\mathbb{Z} + \mathbb{Z}b^{-1}\theta^2 + \mathbb{Z}\omega) \cdot 9b^2.$$

Hence $\Delta(\mathbb{Z}+\mathbb{Z}b^{-1}\theta^2+\mathbb{Z}\omega) = -3a^2b^2 = \mathsf{d}_K(\mathfrak{o}_K : \mathbb{Z}+\mathbb{Z}b^{-1}\theta^2+\mathbb{Z}\omega)$. Using **A**, we get $\mathsf{d}_K = -3a^2b^2$, and $(1, b^{-1}\theta^2, \omega)$ is an integral basis of K.

3. By 1. and 2.

$$-27a^2b^2 = d^2\mathsf{d}_K, \quad \text{where} \quad d = \begin{cases} 1 & \text{if} \quad m \not\equiv \pm 1 \mod 9, \\ 3 & \text{if} \quad m \equiv \pm 1 \mod 9, \end{cases}$$

and from the proof of **A** we get

$$(\mathfrak{o}_K : \mathbb{Z}[\theta])^2 = \frac{-27a^2b^4}{\mathsf{d}_K} = (bd)^2 \quad \text{and} \quad (\mathfrak{o}_K : \mathbb{Z}[b^{-1}\theta^2])^2 = \frac{-27a^4b^2}{\mathsf{d}_K} = (ad)^2.$$

Hence $(\mathfrak{o}_K : \mathbb{Z}[\theta]) = db$ and $(\mathfrak{o}_K : \mathbb{Z}[b^{-1}\theta^2]) = da$. □

As a prepartation for the treatment of cyclotomic fields, we prove the following result on composita.

Theorem 3.1.10. *Let K and L be Galois algebraic number fields, and let $N = KL$ be their compositum. Suppose that $[K:\mathbb{Q}] = n$, $[L:\mathbb{Q}] = m$, $K \cap L = \mathbb{Q}$ and $(\mathsf{d}_K, \mathsf{d}_L) = 1$. Let $(u_1, \ldots, u_n)$ be an integral basis of K and $(v_1, \ldots, v_m)$ an integral basis of L. Then $[N:K] = mn$, $(u_iv_j)_{(i,j)\in[1,n]\times[1,m]}$ is an integral basis of N, and $\mathsf{d}_N = \mathsf{d}_K^m\mathsf{d}_L^n$.*

Proof. Let $\mathrm{Gal}(N/L) = \{\sigma_1, \ldots, \sigma_n\}$ and $\mathrm{Gal}(N/K) = \{\tau_1, \ldots, \tau_m\}$. Then N is Galois by 1.5.6.2, and there are isomorphisms

$$\begin{cases} \mathrm{Gal}(N/L) & \stackrel{\sim}{\to} & \mathrm{Gal}(K/\mathbb{Q}) \\ \sigma & \mapsto & \sigma\restriction K \end{cases}, \qquad \begin{cases} \mathrm{Gal}(N/K) & \stackrel{\sim}{\to} & \mathrm{Gal}(L/\mathbb{Q}) \\ \tau & \mapsto & \tau\restriction L \end{cases}$$

and

$$\begin{cases} \mathrm{Gal}(N/K) & \stackrel{\sim}{\to} & \mathrm{Gal}(K/\mathbb{Q}) \times \mathrm{Gal}(L/\mathbb{Q}) \\ \sigma & \mapsto & (\sigma\restriction K, \sigma\restriction L) \end{cases}.$$

Hence $[N:\mathbb{Q}] = mn$, and as $N = {\scriptstyle\mathbb{Q}}(\{u_iv_j \mid (i,j) \in [1,n] \times [1,m]\})$, it follows that $(u_iv_j)_{(i,j)\in[1,n]\times[1,m]}$ is a basis of N lying in $\mathfrak{o}_N$.

Let $\alpha \in \mathfrak{o}_N$. Then

$$\alpha = \sum_{i=1}^{n}\sum_{j=1}^{m} a_{i,j}u_iv_j, \quad \text{where } a_{i,j} \in \mathbb{Q} \text{ for all } (i,j) \in [1,n]\times[1,m],$$

and we must prove that $a_{i,j} \in \mathbb{Z}$ for all $(i,j) \in [1,n]\times[1,m]$. For $j \in [1,m]$, we set

$$b_j = \sum_{i=1}^{n} a_{i,j}u_i \in K, \quad \text{hence} \quad \alpha = \sum_{j=1}^{m} b_j v_j, \quad \text{and} \quad \tau_\mu(\alpha) = \sum_{j=1}^{m} b_j \tau_\mu(v_j) \in \mathfrak{o}_N$$

for all $\mu \in [1,m]$. Using the matrix $T = (\tau_\mu(v_j))_{j,\mu\in[1,m]} \in \mathsf{M}_m(\mathfrak{o}_N)$ and its adjoint $T^\# \in \mathsf{M}_m(\mathfrak{o}_N)$ we obtain

$$(\tau_1\alpha,\ldots,\tau_m\alpha)T^\# = (b_1,\ldots,b_m)TT^\# = (b_1,\ldots,b_m)\,\det(T) \in \mathfrak{o}_N^m.$$

As $\mathrm{Gal}(L/\mathbb{Q}) = \{\tau_1 \restriction L,\ldots,\tau_m \restriction L)$, we get $\det(T)^2 = d_{L/\mathbb{Q}}(v_1,\ldots,v_m) = \mathsf{d}_L$ and therefore $b_j\mathsf{d}_L \in \mathfrak{o}_N \cap K = \mathfrak{o}_K$ for all $j \in [1,m]$. Since

$$b_j\mathsf{d}_L = \sum_{i=1}^{n} a_{i,j}\mathsf{d}_L u_i \in \mathfrak{o}_K \quad \text{for all} \quad j \in [1,m]$$

and $(u_1,\ldots,u_n)$ is an integral basis of K, it follows that $a_{i,j}\mathsf{d}_L \in \mathbb{Z}$ for all $(i,j) \in [1,n]\times[1,m]$.

By interchanging the role of L and K, we obtain $a_{i,j}\mathsf{d}_K \in \mathbb{Z}$, and since $(\mathsf{d}_K,\mathsf{d}_L) = 1$, it follows that $a_{i,j} \in \mathbb{Z}$ for all $(i,j) \in [1,n]\times[1,m]$.

To calculate d_N, we use the tensor product of matrices. Let R be a commutative ring, $A = (a_{i,\nu})_{i,\nu\in[1,n]} \in \mathsf{M}_n(R)$, $B = (b_{j,\mu})_{j,\mu\in[1,m]} \in \mathsf{M}_m(R)$, and $A\otimes B = \big(a_{i,\nu}b_{j,\mu}\big)_{(i,j),(\nu,\mu)\in[1,n]\times[1,m]} \in \mathsf{M}_{mn}(R)$.

Then

$$\begin{aligned} A\otimes B &= \begin{pmatrix} a_{1,1}B & a_{1,2}B & \ldots & a_{1,n}B \\ \cdot & \cdot & \ldots & \cdot \\ \cdot & \cdot & \ldots & \cdot \\ a_{n,1}B & a_{n,2}B & \ldots & a_{n,n}B \end{pmatrix} \\ &= \begin{pmatrix} a_{1,1}I_m & a_{1,2}I_m & \ldots & a_{1,n}I_m \\ \cdot & \cdot & \ldots & \cdot \\ \cdot & \cdot & \ldots & \cdot \\ a_{n,1}I_m & a_{n,2}I_m & \ldots & a_{n,n}I_m \end{pmatrix} \begin{pmatrix} B & \mathbf{0} & \ldots & \mathbf{0} \\ \mathbf{0} & B & \ldots & \mathbf{0} \\ \mathbf{0} & \mathbf{0} & \ddots & \mathbf{0} \\ \mathbf{0} & \mathbf{0} & \ldots & B \end{pmatrix}, \end{aligned}$$

and consequently $\det(A\otimes B) = \det(A)^m \det(B)^n$. Now it follows that

$$\begin{aligned} \mathsf{d}_N &= \det\big(\sigma_\nu\tau_\mu(u_iv_j)\big)^2_{(\nu,\mu),\,(i,j)\in[1,n]\times[1,m]} \\ &= \det\big((\sigma_\nu(u_i))_{\nu,i\in[1,n]} \otimes (\tau_\mu(v_j))_{\mu,j\in[1,m]}\big)^2 \\ &= [\det(\sigma_\nu u_i)^m_{\nu,i\in[1,n]}]^2\,[\det(\tau_\mu v_j)^n_{\mu,j\in[1,m]}]^2 = \mathsf{d}_K^m\mathsf{d}_L^n. \end{aligned}$$

□

The assumption $K \cap L = \mathbb{Q}$ in Theorem 3.1.10 is redundant; it follows from the assumption $(\mathsf{d}_K,\mathsf{d}_L) = 1$ using the theory of relative discriminants to be developed in Chapter 5. Indeed, suppose that $\mathbb{Q} \ne M = K\cap L$. Then 5.9.8.2 implies $1 < |\mathsf{d}_M| \mid (\mathsf{d}_K,\mathsf{k}_L)$.

Theorem 3.1.11. *Let $n \in \mathbb{N}$, $n \geq 3$, $\zeta_n \in \mathbb{C}$ a primitive n-th root of unity and $\mathbb{Q}^{(n)} = \mathbb{Q}(\zeta_n)$. Then $\mathfrak{o}_{\mathbb{Q}^{(n)}} = \mathbb{Z}[\zeta_n]$, $(1, \zeta_n, \zeta_n^2, \ldots, \zeta_n^{\varphi(n)-1})$ is an integral basis of $\mathbb{Q}^{(n)}$, and*

$$\mathsf{d}_{\mathbb{Q}^{(n)}} = (-1)^{\varphi(n)/2} n^{\varphi(n)} \Big[\prod_{p \mid n} p^{\varphi(n)/(p-1)}\Big]^{-1}.$$

If p is a prime and $e \in \mathbb{N}$, then

$$\mathsf{d}_{\mathbb{Q}^{(p^e)}} = (-1)^{p^{e-1}(p-1)/2} p^{p^{e-1}(ep-e-1)} \quad \text{and} \quad \mathsf{N}_{\mathbb{Q}^{(p^e)}/\mathbb{Q}}(\zeta_{p^e} - 1) = p.$$

If n is composite, then $\mathsf{N}_{\mathbb{Q}^{(n)}/\mathbb{Q}}(\zeta_n - 1) = 1$.

Proof. We assume first that $n = p^e \geq 3$ for some prime p and $e \in \mathbb{N}$, and we set $\zeta = \zeta_{p^e}$. By 1.7.5, $[\mathbb{Q}^{(p^e)} : \mathbb{Q}] = \varphi(p^e) = p^{e-1}(p-1) \equiv 0 \mod 2$, $(1, \zeta, \ldots, \zeta^{\varphi(p^e)-1})$ is a basis of $\mathbb{Z}[\zeta] = \mathbb{Z}[\zeta - 1]$ and thus also of $\mathbb{Q}^{(p^e)}$. The polynomial

$$\Phi_{p^e} = \frac{X^{p^e} - 1}{X^{p^{e-1}} - 1} = \sum_{j=0}^{p-1} X^{p^{e-1}j} \in \mathbb{Z}[X]$$

is the minimal polynomial of ζ, $\Psi = \Phi_{p^e}(X+1) \in \mathbb{Z}[X]$ is the minimal polynomial of $\zeta - 1$, and $\mathsf{N}_{\mathbb{Q}^{(p^e)}/\mathbb{Q}}(\zeta - 1) = (-1)^{\varphi(p^e)}\Psi(0) = \Phi_{p^e}(1) = p$ by 1.6.2.1.

If $\overline{\Psi} = \Psi + p\mathbb{Z}[X] \in \mathbb{F}_p[X]$ denotes the residue class polynomial of Ψ, then

$$\overline{\Psi} = \frac{(X+1)^{p^e} - 1}{(X+1)^{p^{e-1}} - 1} = \frac{X^{p^e}}{X^{p^{e-1}}} = X^{p^{e-1}(p-1)}.$$

Therefore, and since $\Psi(0) = p$, it follows that Ψ is a p-Eisenstein polynomial, and by 3.1.7.2(b) we get $p \nmid (\mathfrak{o}_{\mathbb{Q}^{(p^e)}} : \mathbb{Z}[\zeta])$.

If $\xi = \zeta_p = \zeta^{p^{e-1}}$ and $\Phi_p = X^{p-1} + X^{p-2} + \ldots + X + 1$, then

$$\Phi'_{p^e}(\zeta) = \sum_{j=1}^{p-1} p^{e-1} j \zeta^{p^{e-1}j-1} = p^{e-1}\zeta^{-1} \sum_{j=1}^{p-1} j\xi^j = p^{e-1}\zeta^{-1}\Phi'_p(\xi),$$

Since $\mathsf{N}_{\mathbb{Q}^{(p^e)}/\mathbb{Q}}(\zeta) = \Phi_{p^e}(0) = 1$ and $[\mathbb{Q}^{(p^e)} : \mathbb{Q}^{(p)}] = p^{e-1}$, we obtain

$$\mathsf{N}_{\mathbb{Q}^{(p^e)}/\mathbb{Q}}(\Phi'_{p^e}(\zeta) = p^{(e-1)p^{e-1}(p-1)} \mathsf{N}_{\mathbb{Q}^{(p)}/\mathbb{Q}}(\Phi'_p(\xi))^{p^{e-1}}.$$

As $X^p - 1 = (X-1)\Phi_p$, it follows that $pX^{p-1} = (X-1)\Phi'_p + \Phi_p$, hence

$$\Phi'_p(\xi) = \frac{p\xi^{-1}}{\xi - 1}, \quad \text{and} \quad \mathsf{N}_{\mathbb{Q}^{(p)}/\mathbb{Q}}(\Phi'_p(\xi)) = \frac{p^{p-1}\mathsf{N}_{\mathbb{Q}^{(p)}/\mathbb{Q}}(\xi)^{-1}}{\mathsf{N}_{\mathbb{Q}^{(p)}/\mathbb{Q}}(\xi - 1)} = \frac{p^{p-1}}{p} = p^{p-2}.$$

Since $(-1)^{\varphi(p^e)(\varphi(p^e)-1)/2} = (-1)^{p^{e-1}(p-1)/2}$, we obtain

$$\begin{aligned}\Delta(\mathbb{Z}[\zeta]) &= (-1)^{p^{e-1}(p-1)/2}\mathsf{N}_{\mathbb{Q}^{(p^e)}/\mathbb{Q}}(\Phi'_{p^e}(\zeta)) \\ &= (-1)^{p^{e-1}(p-1)/2}p^{(e-1)p^{e-1}(p-1)+(p-2)p^{e-1}} \\ &= (-1)^{p^{e-1}(p-1)/2}p^{p^{e-1}(ep-e-1)} = \mathsf{d}_{\mathbb{Q}^{(p^e)}}(\mathfrak{o}_{\mathbb{Q}^{(p^e)}}:\mathbb{Z}[\zeta])^2,\end{aligned}$$

and as $p \nmid (\mathfrak{o}_{\mathbb{Q}^{(p^e)}}:\mathbb{Z}[\zeta])$, it follows that

$$\begin{aligned}\mathsf{d}_{\mathbb{Q}^{(p^e)}} &= \Delta(\mathbb{Z}[\zeta]) = (-1)^{p^{e-1}(p-1)/2}p^{p^{e-1}(ep-e-1)} \\ &= (-1)^{\varphi(p^e)/2}p^{e\varphi(p^e)-\varphi(p^e)/(p-1)}.\end{aligned}$$

Now we do the general case by induction on the number of prime divisors of n. Suppose that $n = p^e m$, where $p \in \mathbb{P}$, $e, m \in \mathbb{N}$, $m \geq 2$ and $p \nmid m$. If $p^e = 2 \nmid m$, then $\mathbb{Q}^{(2m)} = \mathbb{Q}^{(m)}$, $\varphi(2m) = \varphi(m)$ and

$$\begin{aligned}&(-1)^{\varphi(2m)/2}(2m)^{\varphi(2m)}\Big[\prod_{p\,|\,2m} p^{\varphi(2m)/(p-1)}\Big]^{-1} \\ &\quad = (-1)^{\varphi(m)/2}m^{\varphi(m)}\Big[\prod_{p\,|\,m} p^{\varphi(m)/(p-1)}\Big]^{-1} \\ &\quad = \mathsf{d}_{\mathbb{Q}^{(m)}} = \mathsf{d}_{\mathbb{Q}^{(2m)}}.\end{aligned}$$

Thus assume that $p^e \geq 3$, $m \geq 3$, $n = p^e m$, and set $\zeta = \zeta_n$. Then ζ^m is a primitive p^e-th and ζ^{p^e} is a primitive p^m-th root of unity, $\mathbb{Q}^{(n)} = \mathbb{Q}^{(p^e)}\mathbb{Q}^{(m)}$ and $\mathbb{Q}^{(p^e)} \cap \mathbb{Q}^{(m)} = \mathbb{Q}$ by 1.7.7. By the induction hypothesis we know that $\mathsf{d}_{\mathbb{Q}^{(p^e)}}$ is a power of p, $p \nmid \mathsf{d}_{\mathbb{Q}^{(m)}}$, $(1, \zeta^m, \ldots, \zeta^{m(\varphi(p^e)-1)})$ is an integral basis of $\mathbb{Q}^{(p^e)}$ and $(1, \zeta^{p^e}, \ldots, \zeta^{p^e(\varphi(m)-1)})$ is an integral basis of $\mathbb{Q}^{(m)}$. Since $\mathbb{Q}^{(n)}$ is totally imaginary and $\varphi(n) \equiv 0 \mod 4$, 3.1.6 implies $\mathsf{d}_{\mathbb{Q}^{(n)}} > 0$, and by 3.1.10 we obtain

$$\begin{aligned}\mathsf{d}_{\mathbb{Q}^{(n)}} &= |\mathsf{d}_{\mathbb{Q}^{(p^e)}}|^{\varphi(m)}|\mathsf{d}_{\mathbb{Q}^{(m)}}|^{\varphi(p^e)} \\ &= \Big[p^{e\varphi(p^e)-\varphi(p^e)/(p-1)}\Big]^{\varphi(m)}\Big[m^{\varphi(m)}\prod_{p\,|\,m} p^{-\varphi(m)/(p-1)}\Big]^{\varphi(p^e)} \\ &= n^{\varphi(n)}\prod_{p\,|\,n} p^{-\varphi(n)/(p-1)} = (-1)^{\varphi(n)/2}n^{\varphi(n)}\Big[\prod_{p\,|\,n} p^{\varphi(n)/(p-1)}\Big]^{-1},\end{aligned}$$

and that $\{\zeta^{\nu m+\mu p^{e-1}} \mid \nu \in [0, \varphi(p^e)-1],\ \mu \in [0, \varphi(m)-1]\}$ is an integral basis of $\mathbb{Q}^{(n)}$. Hence $\mathfrak{o}_{\mathbb{Q}^{(n)}} = \mathbb{Z}[\zeta]$, and therefore $(1, \zeta, \ldots, \zeta^{\varphi(n)-1})$ is an integral basis of $\mathbb{Q}^{(n)}$.

It remains to calculate $\mathsf{N}_{\mathbb{Q}^{(n)}/\mathbb{Q}}(\zeta_n - 1) = (-1)^{\varphi(n)}\Phi_n(1) = \Phi_n(1)$ for composite n (observe that $\Phi_n(X+1)$ is the minimal polynomial of $\zeta_n - 1$ and

$\varphi(n) \equiv 0 \bmod 2$). Regardless of whether n is composite or not, we obtain, using 1.7.5 and 1.7.2,

$$\Phi_n = \prod_{1 \le d \,|\, n} (X^{n/d} - 1)^{\mu(d)} = \prod_{1 \le d \,|\, n} \Big(\frac{X^{n/d} - 1}{X - 1}\Big)^{\mu(d)} \prod_{1 \le d \,|\, n} (X - 1)^{\mu(d)}$$

$$= \prod_{1 \le d \,|\, n} \Big(\sum_{j=0}^{n/d-1} X^j\Big)^{\mu(d)}, \quad \text{since} \quad \sum_{1 \le d \,|\, n} \mu(d) = 0,$$

and therefore

$$\Phi_n(1) = \prod_{1 \le d \,|\, n} \Big(\frac{n}{d}\Big)^{\mu(d)} = \begin{cases} p & \text{if } n \text{ is a } p-\text{power}, \\ 1 & \text{if } n \text{ is composite.} \end{cases} \qquad \square$$

3.2 Factorization of primes in algebraic number fields

Let K be an algebraic number field and $K \ne \mathbb{Q}$. If $\mathfrak{p} \in \mathcal{P}_K$ and $\mathfrak{p} \cap \mathbb{Z} = p\mathbb{Z}$ for a prime p, then we say that $\mathfrak{p}$ lies above p or that $\mathfrak{p}$ is a prime divisor of p, and we write $\mathfrak{p} \,|\, p$. The residue class field $\mathsf{k}_\mathfrak{p} = \mathfrak{o}_K/\mathfrak{p}$ is a vector space over $\mathbb{F}_p$. We call $f_\mathfrak{p} = f(\mathfrak{p}/p\mathbb{Z}) = \dim_{\mathbb{F}_p} \mathsf{k}_\mathfrak{p}$ the **degree** and $e_\mathfrak{p} = e(\mathfrak{p}/p\mathbb{Z}) = \mathsf{v}_\mathfrak{p}(p)$ the **ramification index** of $\mathfrak{p}$. By definition, $\mathfrak{N}(\mathfrak{p}) = |\mathsf{k}_\mathfrak{p}| = p^{f_\mathfrak{p}}$. A prime ideal $\mathfrak{p} \in \mathcal{P}_K$ is called **inert** if $f_\mathfrak{p} > 1$, **unramified** if $e_\mathfrak{p} = 1$, and **ramified** if $e_\mathfrak{p} > 1$. If L/K is a finite field extension, $\mathfrak{P} \in \mathcal{P}_L$, $\mathfrak{p} \in \mathcal{P}_K$ and $\mathfrak{P} \,|\, \mathfrak{p} \,|\, p$, then $e_\mathfrak{P} = e(\mathfrak{P}/\mathfrak{p})e_\mathfrak{p}$ and $f_\mathfrak{P} = f(\mathfrak{P}/\mathfrak{p})f_\mathfrak{p}$.

If p is a prime, then $p\mathfrak{o}_K = \mathfrak{p}_1^{e_1} \cdot \ldots \cdot \mathfrak{p}_r^{e_r}$, where $r \in \mathbb{N}$, $\mathfrak{p}_1, \ldots, \mathfrak{p}_r$ are the distinct prime divisors of p in K and $e_i = e_{\mathfrak{p}_i}$ for all $i \in [1, r]$. If $f_i = f_{\mathfrak{p}_i}$ for all $i \in [1, r]$, then

$$\sum_{i=1}^{r} e_i f_i = [L\!:\!K].$$

We refer to the equation $p\mathfrak{o}_K = \mathfrak{p}_1^{e_1} \cdot \ldots \cdot \mathfrak{p}_r^{e_r}$ as the **factorization** of p in K. We call p **unramified** in K if $e_i = 1$ for all $i \in [1, r]$ and **ramified** otherwise. We have seen in 2.12.7 that only finitely many primes are ramified in K, and in 5.9.7 we shall prove that a prime is ramified in K if and only if it divides d_K. In K, a prime p is called **decomposed** if $r > 1$, **fully decomposed** if $r = n$, and **undecomposed** if $r = 1$. If p is undecomposed in K and $\mathfrak{p} \in \mathcal{P}_K$ is the unique prime divisor of p in K, then p is called **inert** resp. **fully ramified** in K if $f_\mathfrak{p} = n$ resp. $e_\mathfrak{p} = n$.

If K is a Galois number field, p is a prime and $r_p(K)$ is the number of prime divisors of p in K, then $e_p(K)f_p(K)r_p(K) = n$, where $e_p(K) = e_\mathfrak{p}$ and $f_p(K) = f_\mathfrak{p}$ for all prime divisors $\mathfrak{p}$ of p in K.

If $K/\mathbb{Q}$ is abelian and p is an in K unramified prime, then

$$(p, K) = \Big(\frac{K/\mathbb{Q}}{p\mathbb{Z}}\Big) \in \mathrm{Gal}(K/\mathbb{Q})$$

is called the **Frobenius automorphism** of p in K (see 2.14.7.**3**).

More generally, if $a \in \mathbb{Q}_{>0}$ and $v_p(a) = 0$ for all in K ramified primes p, then

$$(a, K) = \Big(\frac{K/\mathbb{Q}}{a\mathbb{Z}}\Big) = \prod_{p \mid a} (p, K)^{v_p(a)} \in \mathrm{Gal}(K/\mathbb{Q})$$

is the **Artin symbol** of a in K as introduced in 2.14.7.**4** in the setting of Dedekind domains. Note that here the product denotes the composition of automorphisms.

A **factorization law** for K is a criterion to determine the factorization type of a given prime p in K. In the current section, we shall prove factorization laws for quadratic, pure cubic and cyclotomic number fields, and in Section 3.3 we will do the same for all subfields of cyclotomic number fields. By the famous Kronecker-Weber theorem every abelian number field is contained in some cyclotomic number field (see 3.7.9). Hence our investigations in 3.3 comprise all abelian number fields.

More generally, a **factorization law** for a finite extension L/K of algebraic number fields is a criterion to determine the prime factorization of $\mathfrak{p}\mathfrak{o}_L$ for a given prime ideal $\mathfrak{p} \in \mathcal{P}_K$. Global class field theory provides a factorization law for all abelian extensions L/K of algebraic number fields. A tentative formulation of this theory will be given in 3.7.7. For a thorough presentation we refer to the forthcoming volume on class field theory.

We start with a reformulation of what we know from the abstract theory of Dedekind domains.

Theorem 3.2.1. *Let $K = \mathbb{Q}(a)$ be an algebraic number field, where $a \in \mathfrak{o}_K$. Let $g \in \mathbb{Z}[X]$ be the minimal polynomial of a and p a prime. For a polynomial $h \in \mathbb{Z}[X]$ we denote by $\overline{h} = h + p\mathbb{Z}[X] \in \mathbb{F}_p[X]$ the residue class polynomial of h.*

1. *Assume that $p \nmid (\mathfrak{o}_K : \mathbb{Z}[a])$, and let $g_1, \ldots, g_r \in \mathbb{Z}[X]$ be monic such that the residue class polynomials $\overline{g}_1, \ldots, \overline{g}_r \in \mathbb{F}_p[X]$ are distinct and irreducible, and let $\overline{g} = \overline{g}_1^{e_1} \cdot \ldots \cdot \overline{g}_r^{e_r}$, where $e_1, \ldots, e_r \in \mathbb{N}$. Then $p\mathfrak{o}_K = \mathfrak{p}_1^{e_1} \cdot \ldots \cdot \mathfrak{p}_r^{e_r}$, where $\mathfrak{p}_1, \ldots, \mathfrak{p}_r \in \mathcal{P}_K$ are distinct, $\mathfrak{p}_i = p\mathfrak{o}_K + g_i(a)\mathfrak{o}_K$ and $f_{\mathfrak{p}_i} = \partial(g_i)$ for all $i \in [1, r]$.*

2. *If g is a p-Eisenstein polynomial, then $p \nmid (\mathfrak{o}_K : \mathbb{Z}[a])$, p is fully ramified in K, and if $\mathfrak{p} \in \mathcal{P}_K$ is the prime divisor of p in K, then $v_{\mathfrak{p}}(a) = 1$.*

Proof. By 2.12.6, 2.12.8 and 3.1.7. □

In order to formulate the factorization law for quadratic number fields we need some preparations from elementary number theory.

Let $m \in \mathbb{N}$ and $a \in \mathbb{Z}$. We call a a **quadratic residue** modulo m if there exists some $x \in \mathbb{Z}$ such that $x^2 \equiv a \bmod m$ [equivalently, $a + m\mathbb{Z}$ is a square in $\mathbb{Z}/m\mathbb{Z}$]. Otherwise we call a a **quadratic non-residue** modulo m. By the Chinese remainder theorem 1.1.4 a is a quadratic residue modulo m if and only if a is a quadratic residue modulo $p^{v_p(m)}$ for all primes p dividing m.

Theorem 3.2.2.

1. *Let $p \neq 2$ be a prime, $e \in \mathbb{N}$ and $a \in \mathbb{Z} \setminus p\mathbb{Z}$. Then the following assertions are equivalent:*

(a) a is a quadratic residue modulo p^e.

(b) a is a quadratic residue modulo p.

(c) $a^{(p-1)/2} \equiv 1 \bmod p$.

If a is a quadratic non-residue modulo p, then it follows that $a^{(p-1)/2} \equiv -1 \bmod p$.

2. *Let $e \in \mathbb{N}$, $e \geq 3$ and $a \in \mathbb{Z} \setminus 2\mathbb{Z}$. Then the following assertions are equivalent:*

(a) a is a quadratic residue modulo 2^e.

(b) a is a quadratic residue modulo 8.

(c) $a \equiv 1 \bmod 8$.

Proof. 1. By 1.7.3.2, $(\mathbb{Z}/p^e\mathbb{Z})^\times$ is a cyclic group of order $\varphi(p^e) = p^{e-1}(p-1)$, and $\mathrm{ord}_{p^e}(1+p) = p^{e-1}$. Therefore $a+p^e\mathbb{Z}$ is a square in $(\mathbb{Z}/p^e\mathbb{Z})^\times$ if and only if $a^{p^{e-1}(p-1)/2} \equiv 1 \bmod p^e$, and this holds if and only if $a^{(p-1)/2} \equiv 1 \bmod p$. Therefore (a), (b) and (c) are equivalent.

If $a^{(p-1)/2} \not\equiv 1 \bmod p$, then $a^{(p-1)/2} \equiv -1 \bmod p$, since $a^{p-1} \equiv 1 \bmod p$ and the polynomial $X^2 - 1 \in \mathbb{F}_p[X]$ has only the zeros $\pm 1 + p\mathbb{Z}$.

2. (a) $\Rightarrow$ (b) $\Rightarrow$ (c) Obvious.

(c) $\Rightarrow$ (a) By 1.7.3.3 we obtain $(\mathbb{Z}/2^e\mathbb{Z})^\times = \langle -1 + 2^e\mathbb{Z}, 5 + 2^e\mathbb{Z}\rangle$, and $\{a + 2^e\mathbb{Z} \mid a \equiv 1 \bmod 8\} = \langle 5^2 + 2^e\mathbb{Z}\rangle = (\mathbb{Z}/2^e\mathbb{Z})^{\times 2}$. □

Definition and Remarks 3.2.3. For an odd prime p and $a \in \mathbb{Z} \setminus p\mathbb{Z}$ we define the **Legendre symbol** by

$$\left(\frac{a}{p}\right) = \begin{cases} 1 & \text{if } a \text{ is a quadratic residue modulo } p, \\ -1 & \text{otherwise.} \end{cases}$$

From 3.2.2 we obtain Euler's criterion

$$\Big(\frac{a}{p}\Big) \equiv a^{(p-1)/2} \bmod p \quad \text{and} \quad \Big(\frac{-1}{p}\Big) = (-1)^{(p-1)/2}.$$

If $c \in \mathbb{Z} \setminus p\mathbb{Z}$, then

$$\Big(\frac{ac}{p}\Big) = \Big(\frac{a}{p}\Big)\Big(\frac{c}{p}\Big), \quad \Big(\frac{ac^2}{p}\Big) = \Big(\frac{a}{p}\Big) \quad \text{and} \quad \Big(\frac{c^2}{p}\Big) = 1.$$

If $a \in \mathbb{Z} \setminus 2\mathbb{Z}$, we define

$$\Big(\frac{a}{2}\Big) = \begin{cases} 1 & \text{if} \quad a \equiv \pm 1 \bmod 8, \\ -1 & \text{if} \quad a \equiv \pm 5 \bmod 8. \end{cases}$$

If p is a prime and $a \in p\mathbb{Z}$ we define

$$\Big(\frac{a}{p}\Big) = 0.$$

With this enlarged domain of definition, we call

$$\Big(\frac{a}{p}\Big) \quad \text{the } \textbf{extended Legendre symbol}.$$

After these preparations we turn to quadratic number fields. If K is a quadratic number field, then there are the following possibilities for the factorization of a prime p in K:

- $p\mathfrak{o}_K = \mathfrak{p}_+\mathfrak{p}_-$, where $\mathfrak{p}_\pm \in \mathcal{P}_K$, $\mathfrak{p}_+ \neq \mathfrak{p}_-$ and $\mathfrak{N}(\mathfrak{p}_\pm) = p$ (p is decomposed);
- $p\mathfrak{o}_K \in \mathcal{P}_K$ and $\mathfrak{N}(p\mathfrak{o}_K) = p^2$ (p is inert);
- $p\mathfrak{o}_K = \mathfrak{p}^2$, where $\mathfrak{p} \in \mathcal{P}_K$ and $\mathfrak{N}(\mathfrak{p}) = p$ (p is ramified).

In the subsequent factorization law for quadratic number fields we do not give the explicit form of the prime factors. This will be done more generally for arbitrary quadratic orders in 3.8.5.

Theorem 3.2.4 (Factorization law for quadratic number fields). *Let K be a quadratic number field and p a prime. Then p is in K*

decomposed if $\Big(\frac{\mathsf{d}_K}{p}\Big) = 1$, *inert if* $\Big(\frac{\mathsf{d}_K}{p}\Big) = -1$, *and ramified if* $\Big(\frac{\mathsf{d}_K}{p}\Big) = 0$.

If $p \nmid \mathsf{d}_K$, then the Frobenius automorphism (p, K) is given by

$$(p, K)(\sqrt{\mathsf{d}_K}) = \Big(\frac{\mathsf{d}_K}{p}\Big)\sqrt{\mathsf{d}_K}.$$

Proof. Let D be the radicand of K. Then $\mathsf{d}_K = D$ if $D \equiv 1 \bmod 4$, and $\mathsf{d}_K = 4D$ otherwise (see 3.1.8). Therefore

$$\Big(\frac{\mathsf{d}_K}{p}\Big) = \Big(\frac{D}{p}\Big) \text{ if } p \neq 2, \quad \text{and} \quad \Big(\frac{\mathsf{d}_K}{2}\Big) = \begin{cases} 0 & \text{if } D \equiv 2 \text{ or } 3 \bmod 4, \\ 1 & \text{if } D \equiv 1 \bmod 8, \\ -1 & \text{if } D \equiv 5 \bmod 8. \end{cases}$$

We determine the factorization type of p by means of 3.2.1. For $g \in \mathbb{Z}[X]$ we denote by $\overline{g} = g + p\mathbb{Z}[X] \in \mathbb{F}_p[X]$ the residue class polynomial of g.

$\boldsymbol{p \neq 2}$ **:** Then $p \nmid (\mathfrak{o}_K : \mathbb{Z}[\sqrt{D}])$, and $g = X^2 - D \in \mathbb{Z}[X]$ is the minimal polynomial of $\sqrt{D}$. If $p \mid D$, then $\overline{g} = X^2$ and p is ramified. If $p \nmid D$ and D is a quadratic residue modulo p, then $\overline{D} = \alpha^2$ for some $\alpha \in \mathbb{F}_p^\times$, hence $\overline{g} = (X - \alpha)(X + \alpha)$, and p is decomposed. If $p \nmid D$ and D is a quadratic non-residue modulo p, then $\overline{g} \in \mathbb{F}_p[X]$ is irreducible, and p is inert.

$\boldsymbol{p = 2}$ **:** If $D \equiv 2$ or $3 \bmod 4$, then $\mathfrak{o}_K = \mathbb{Z}[\sqrt{D}]$, $g = X^2 - D$ is the minimal polynomial of $\sqrt{D}$, and $\overline{g} = (X - \overline{D})^2 \in \mathbb{F}_2[X]$. Therefore 2 is ramified in K.

Assume now that $D \equiv 1 \bmod 4$, $D = 4d + 1$, where $d \in \mathbb{Z}$, and let $\omega = (1 + \sqrt{D})/2$. Then $g = X^2 - X - d \in \mathbb{Z}[X]$ is the minimal polynomial of ω, and $\mathfrak{o}_K = \mathbb{Z}[\omega]$. If $D \equiv 1 \bmod 8$, then $\overline{g} = X(X + 1) \in \mathbb{F}_2[X]$, and 2 is decomposed in K. If $D \equiv 5 \bmod 8$, then $\overline{g} = X^2 + X + 1 \in \mathbb{F}_2[X]$ is irreducible, and 2 is inert in K.

Now let p be unramified in K and $G = \mathrm{Gal}(K/\mathbb{Q}) = \{1, \tau\}$, where $\tau(\sqrt{D}) = -\sqrt{D}$, and let G_p be the decomposition group of p in K (see 2.13.6). Then

$$G_p = \langle (p, K) \rangle = \begin{cases} 1, & \text{and consequently } (p, K) = 1 \text{ if } p \text{ is decomposed in } K, \\ G, & \text{and consequently } (p, K) = \tau \text{ if } p \text{ is inert in } K. \end{cases}$$

Overall it follows that

$$(p, K)(\sqrt{\mathsf{d}_K}) = \Big(\frac{\sqrt{d_K}}{p}\Big)\sqrt{\mathsf{d}_K}. \qquad \square$$

Next we consider pure cubic fields $K = \mathbb{Q}(\sqrt[3]{m})$, where $m \in \mathbb{N} \setminus \mathbb{N}^3$. For the factorization of a prime p in K we have the following possibilities:

- $p\mathfrak{o}_K = \mathfrak{p}\mathfrak{q}\mathfrak{r}$, where $\mathfrak{p}$, $\mathfrak{q}$, $\mathfrak{r} \in \mathcal{P}_K$ are distinct and $f_\mathfrak{p} = f_\mathfrak{q} = f_\mathfrak{r} = 1$ (p is fully decomposed);

- $p\mathfrak{o}_K = \mathfrak{p}\mathfrak{q}$, where $\mathfrak{p}$, $\mathfrak{q} \in \mathcal{P}_K$, $f_\mathfrak{p} = 2$ and $f_\mathfrak{q} = 1$ (p is unramified and decomposed, but not fully decomposed);

- $p\mathfrak{o}_K = \mathfrak{p} \in \mathcal{P}_K$ and $f_\mathfrak{p} = 3$ (p is inert);

- $p\mathfrak{o}_K = \mathfrak{p}^3$, where $\mathfrak{p} \in \mathcal{P}_K$ and $f_\mathfrak{p} = 1$ (p is fully ramified);
- $p\mathfrak{o}_K = \mathfrak{p}^2\mathfrak{q}$, where $\mathfrak{p},\ \mathfrak{q} \in \mathcal{P}_K$, $\mathfrak{p} \ne \mathfrak{q}$ and $f_\mathfrak{p} = f_\mathfrak{q} = 1$ (p is ramified, but not fully ramified).

Let p be a prime and $a \in \mathbb{Z} \setminus p\mathbb{Z}$. We call a a cubic residue modulo p if there exists some $x \in \mathbb{Z}$ such that $x^3 \equiv a \mod p$ [equivalently, $a+p\mathbb{Z} \in \mathbb{F}_p^{\times 3}$].

If $p \not\equiv 1 \mod 3$, then $3 \nmid |\mathbb{F}_p^{\times 3}| = p-1$, and therefore every $a \in \mathbb{Z} \setminus p\mathbb{Z}$ is a cubic residue modulo p.

Theorem 3.2.5 (Factorization law for pure cubic fields). *Let $m = ab^2$, where $a,\ b \in \mathbb{N}$, $ab \ne 1$, ab is squarefree and $3 \nmid b$. Let $K = \mathbb{Q}(\sqrt[3]{m})$ and $\theta = \sqrt[3]{m}$. Then a prime p factorizes in K as follows:*

1. *If $p \,|\, ab$, then p is fully ramified.*
2. *Assume that $p \ne 3$ and $p \nmid ab$.*

   (a) If $p \equiv -1 \mod 3$, then p is unramified and decomposed, but not fully decomposed.

   (b) If $p \equiv 1 \mod 3$, then p is fully decomposed if m is a cubic residue modulo p, and p is inert otherwise.
3. *If $m \not\equiv \pm 1 \mod 9$, then 3 is fully ramified.*
4. *Assume that $m \equiv \pm 1 \mod 9$.*

   (a) If $a^2 \not\equiv b^2 \mod 9$, then 3 is fully ramified.

   (b) If $a^2 \equiv b^2 \mod 9$, then 3 is ramified, but not fully ramified.

Proof. Let p be a prime, and for $h \in \mathbb{Z}[X]$ we denote by $\overline{h} = h+p\mathbb{Z}[X] \in \mathbb{F}_p[X]$ the residue class polynomial of h. We tacitly use 3.2.1.

1. If $p \,|\, a$, then $g = X^3 - ab^2 \in \mathbb{Z}[X]$ is the minimal polynomial of θ, it is a p-Eisenstein polynomial and $\overline{g} = X^3$. Hence $p\mathfrak{o}_K = \mathfrak{p}^3$ for some $\mathfrak{p} \in \mathcal{P}_K$.

If $p \,|\, b$, then $g_1 = X^3 - a^2b \in \mathbb{Z}[X]$ is the minimal polynomial of $b^{-1}\theta^2$, it is a p-Eisenstein polynomial and $\overline{g}_1 = X^3$. Hence $p\mathfrak{o}_K = \mathfrak{p}^3$ for some $\mathfrak{p} \in \mathcal{P}_K$.

2. $g = X^3 - m$ is the minimal polynomial of θ, $\overline{g} = X^3 - \overline{m} \in \mathbb{F}_p[X]$, and $p \nmid (\mathfrak{o}_K : \mathbb{Z}[\theta])$ by 3.1.9.3.

(a) If $p \equiv -1 \mod 3$, then there exits a unique $\alpha \in \mathbb{F}_p$ such that $\alpha^3 = \overline{m}$, and therefore $X^3 - \overline{m} = (X-\alpha)\psi$ for some irreducible polynomial $\psi \in \mathbb{F}_p[X]$. Hence $p\mathfrak{o}_K = \mathfrak{p}\mathfrak{q}$, where $\mathfrak{p},\ \mathfrak{q} \in \mathcal{P}_K$ and $\mathfrak{p} \ne \mathfrak{q}$.

(b) If $p \equiv 1 \mod 3$ and m is a cubic residue modulo p, then $X^3 - \overline{m} \in \mathbb{F}_p[X]$ splits (since $\mu_3(\mathbb{F}_p) = 3$), and therefore $p\mathfrak{o}_K = \mathfrak{p}\mathfrak{q}\mathfrak{r}$ with distinct $\mathfrak{p},\ \mathfrak{q},\ \mathfrak{r} \in \mathcal{P}_K$. If m is not a cubic residue modulo p, then $X^3 - \overline{m} \in \mathbb{F}_p[X]$ is irreducible, and $p\mathfrak{o}_K \in \mathcal{P}_K$.

3. Let $m \not\equiv \pm 1 \mod 9$. If $3 \mid a$, then 3 is fully ramified by 1. Thus suppose that $3 \nmid a$. The polynomial $h = (X+m)^3 - m = X^3 + 3X^2m + 3Xm^2 + m^3 - m$ is the minimal polynomial of $\theta - m$, and since $9 \nmid m^3 - m$ it is an Eisenstein polynomial. Moreover, $3 \nmid (\mathfrak{o}_K : \mathbb{Z}[\theta]) = (\mathfrak{o}_K : \mathbb{Z}[\theta - m])$ by 3.1.9.3, and therefore $3\mathfrak{o}_K = \mathfrak{p}^3$ for some $\mathfrak{p} \in \mathcal{P}_K$.

4. Since $3 \nmid ab$, we get $a^2 \equiv b^2 \mod 3$ and then $r = v_3(a^2 - b^2) \geq 1$. Then $\alpha = \theta - a \in \mathfrak{o}_K$, $h = (X+a)^3 - m = X^3 + 3aX^2 + 3a^2X + a(a^2 - b^2) \in \mathbb{Z}[X]$ is the minimal polynomial of α and

$$h_1 = \frac{1}{27}\, h(3X) = X^3 + aX^2 + \frac{a^2}{3}X + \frac{a(a^2 - b^2)}{27} \in \mathbb{Q}[X] \setminus \mathbb{Z}[X]$$

is the minimal polynomial of $\alpha/3$. Hence $\alpha \notin 3\mathfrak{o}_K$. However,

$$\alpha^3 + 3a\alpha^2 + 3a^2\alpha + a(a^2 - b^2) = \alpha^3 + 3a\alpha\theta + a(a^2 - b^2) = 0 \qquad (\dagger)$$

implies $\alpha^3 \in 3\mathfrak{o}_K$. Therefore there exists some $\mathfrak{p} \in \mathcal{P}_K$ such that $v_\mathfrak{p}(\alpha) < v_\mathfrak{p}(3)$ and $v_\mathfrak{p}(\alpha^3) = 3v_\mathfrak{p}(\alpha) \geq v_\mathfrak{p}(3)$. This is only possible if either $3\mathfrak{o}_K = \mathfrak{p}^3$, or $3\mathfrak{o}_K = \mathfrak{p}^2\mathfrak{q}$, where $q \in \mathcal{P}_K \setminus \{\mathfrak{p}\}$. Hence it suffices to prove that $3\mathfrak{o}_K = \mathfrak{p}^3$ implies $r = 1$ and $3\mathfrak{o}_K = \mathfrak{p}^2\mathfrak{q}$ implies $r \geq 2$.

First let $3\mathfrak{o}_K = \mathfrak{p}^3$, and set $s = v_\mathfrak{p}(\alpha)$. Then $s \in \{1, 2\}$, $v_\mathfrak{p}(\alpha^3) = 3s$ and $v_\mathfrak{p}(3a\alpha\theta) = 3 + s$ (as $v_\mathfrak{p}(\theta) = 0$). Since $3s \neq 3 + s$, ($\dagger$) implies

$$v_\mathfrak{p}(a(a^2 - b^2)) = v_\mathfrak{p}(a^2 - b^2) = 3r = \min\{3 + s, 3s\}.$$

It follows that $3s = 3r < 3 + s \leq 5$, and consequently $r = s = 1$.

Now let $3\mathfrak{o}_K = \mathfrak{p}^2\mathfrak{q}$, where $\mathfrak{q} \in \mathcal{P}_K \setminus \{\mathfrak{p}\}$. In this case we obtain $v_\mathfrak{q}(\alpha) \geq 1$, $v_\mathfrak{q}(\alpha^3) \geq 3$, $v_\mathfrak{q}(3a\alpha\theta) \geq 2$, and ($\dagger$) implies

$$r = v_3(a^2 - b^2) = v_3(a(a^2 - b^2)) = v_\mathfrak{q}(a(a^2 - b^2)) \geq 2. \qquad \square$$

Next we consider cyclotomic number fields and refer to 1.7.5 and 3.1.11 for basic definitions and results. For $n \in \mathbb{N}$, $n \geq 2$, recall from 3.1.1 the residue class homomorphism $\pi_n \colon \mathbb{Q}(n) \to (\mathbb{Z}/n\mathbb{Z})^\times$: If $a = b^{-1}c \in \mathbb{Q}(n)$, where $b, c \in \mathbb{Z}$ and $(b, n) = (c, n) = 1$, then $\pi_n(a) = (b + n\mathbb{Z})^{-1}(c + n\mathbb{Z}) = b'c + n\mathbb{Z}$ if $b' \in \mathbb{Z}$ and $bb' \equiv 1 \mod n$.

If $a \in \mathbb{Q}(n)$ and $\xi \in \mu_n(\mathbb{C})$, then $\xi^a = \xi^{\pi_n(a)} = \xi^k$, where $k \in \mathbb{Z}$ is such that $bk \equiv c \mod n$ [equivalently, $\pi_n(a) = k + n\mathbb{Z}$].

Theorem 3.2.6 (Factorization law for cyclotomic number fields). *Let $n \geq 3$, ζ_n be a primitive n-th root of unity, $\mathbb{Q}^{(n)} = \mathbb{Q}(\zeta_n)$ and $G = \mathrm{Gal}(\mathbb{Q}^{(n)}/\mathbb{Q})$.*

1. Let p be a prime and $n = p^e m$, where $e \in \mathbb{N}_0$, $m \in \mathbb{N}$ and $p \nmid m$. Let $f \in \mathbb{N}$ be minimal such that $p^f \equiv 1 \mod m$. Then $\varphi(m) = fr$, where $r \in \mathbb{N}$, and

$$p\mathfrak{o}_{\mathbb{Q}^{(n)}} = (\mathfrak{P}_1 \cdot \ldots \cdot \mathfrak{P}_r)^{\varphi(p^e)},$$

where $\mathfrak{P}_1, \ldots, \mathfrak{P}_r \in \mathcal{P}_{\mathbb{Q}^{(n)}}$ are distinct, and $f = f_p(\mathbb{Q}^{(n)})$.

In particular:

- *p is unramified in $\mathbb{Q}^{(n)}$ if and only if $p \nmid n$ or $p = 2$ and $4 \nmid n$.*
- *p is fully decomposed in $\mathbb{Q}^{(n)}$ if and only if $p \equiv 1 \bmod n$.*
- *If $n = p^e$ then p is the only ramified prime in $\mathbb{Q}^{(p^e)}$, and it is fully ramified.*

2. For every $a \in \mathbb{Q}(n)$ let $\sigma_n(a) \in G$ be such that $\sigma_n(a)(\xi) = \xi^a$ for all $\xi \in \mu_n(\mathbb{C})$. In particular, $\sigma_n(-1)(z) = \overline{z}$ for all $z \in \mathbb{Q}^{(n)}]$. The map $\sigma_n \colon \mathbb{Q}(n) \to G$ is an epimorphism and induces an (equally denoted) isomorphism $\sigma_n \colon (\mathbb{Z}/n\mathbb{Z})^\times \overset{\sim}{\to} G$.

3. If q is a prime and $q \nmid n$, then $\sigma_n(q)$ is the Frobenius automorphism of q in $\mathbb{Q}^{(n)}$, and if $a \in \mathbb{Q}^+(n)$, then

$$\sigma_n(a) = (a, \mathbb{Q}^{(n)}/\mathbb{Q}) = \prod_{q \in \mathbb{P}} (q, \mathbb{Q}^{(n)}/\mathbb{Q})^{v_q(a)}).$$

In particular, $\sigma_n = (\,\cdot\,, \mathbb{Q}^{(n)}/\mathbb{Q}) \colon \mathbb{Q}^+(n) \to \mathrm{Gal}(\mathbb{Q}^{(n)}/\mathbb{Q})$ is the Artin map and induces the (equally denoted) Artin isomorphism

$$(\,\cdot\,, \mathbb{Q}^{(n)}/\mathbb{Q}) \colon (\mathbb{Z}/n\mathbb{Z})^\times \to \mathrm{Gal}(\mathbb{Q}^{(n)}/\mathbb{Q}).$$

Proof. 1. By definition, $f = \mathrm{ord}_m(p)$, hence $f \mid \varphi(m)$, and we set $\varphi(m) = fr$, where $r \in \mathbb{N}$. In the sequel we use 3.2.1. Let $\overline{\Phi}_n = \Phi_n + p\mathbb{Z}[X] \in \mathbb{F}_p[X]$ be the residue class polynomial of the cyclotomic polynomial Φ_n. Since $\mathfrak{o}_K = \mathbb{Z}[\zeta]$ it suffices to prove that $\overline{\Phi}_n = \overline{\Phi}_m^{\varphi(p^e)}$, and $\overline{\Phi}_m = \psi_1 \cdot \ldots \cdot \psi_r$, where $\psi_1, \ldots, \psi_r$ are distinct monic irreducible polynomials in $\mathbb{F}_p[X]$ of degree f. Since

$$\Phi_n = \prod_{1 \le d \mid n} (X^{n/d} - 1)^{\mu(d)} = \prod_{1 \le d \mid m} \prod_{j=0}^{e} (X^{mp^{e-j}/d} - 1)^{\mu(d)\mu(p^j)}$$
$$= \prod_{1 \le d \mid m} \Bigl(\frac{X^{mp^e/d} - 1}{X^{mp^{e-1}/d} - 1} \Bigr)^{\mu(d)},$$

it follows that

$$\overline{\Phi}_n = \prod_{1 \le d \mid m} \Bigl(\frac{(X^{m/d} - 1)^{p^e}}{(X^{m/d} - 1)^{p^{e-1}}} \Bigr)^{\mu(d)} = \prod_{1 \le d \mid m} (X^{m/d} - \overline{1})^{\varphi(p^e)\mu(d)} = \overline{\Phi}_m^{\varphi(p^e)}.$$

The polynomial

$$X^m - \overline{1} = \prod_{1 \le d \mid m} \overline{\Phi}_d \in \mathbb{F}_p[X]$$

is separable and its zeros in an algebraic closure $\overline{\mathbb{F}}_p$ of $\mathbb{F}_p$ are primitive m-th roots of unity. Hence the splitting field E of $X^m - \overline{1}$ is the smallest subfield of $\overline{\mathbb{F}}_p$ satisfying $|E^\times| \equiv 0 \bmod m$. Therefore $E = \mathbb{F}_{p^f} = \mathbb{F}_p(\xi)$ for every primitive m-th root of unity $\xi \in \overline{\mathbb{F}}_p$, and $f = [\mathbb{F}_p(\xi) : \mathbb{F}_p]$ is the degree of every monic irreducible factor of $\overline{\Phi}_m$.

2. By 1.7.5, there exists an isomorphism $\theta\colon G \to (\mathbb{Z}/n\mathbb{Z})^\times$ such that $\sigma(\xi) = \xi^{\theta(\sigma)}$ for all $\sigma \in G$ and $\xi \in \mu_n(\mathbb{C})$. Hence $\sigma_n = \theta^{-1} \circ \pi_n\colon \mathbb{Q}(n) \to G$ is an epimorphism, and $\sigma_n(a)(\xi) = \xi^a$ for all $\xi \in \mu_n(\mathbb{C})$. Now it is clear that the epimorphism $\sigma_n\colon \mathbb{Q}(n) \to G$ induces the equally denoted isomorphism $\sigma_n = \theta^{-1}\colon (\mathbb{Z}/n\mathbb{Z})^\times \overset{\sim}{\to} G$. By definition, $\sigma_n(-1)(\zeta) = \zeta^{-1} = \overline{\zeta}$, and therefore $\sigma_n(-1)(z) = \overline{z}$ for all $z \in \mathbb{Q}^{(n)}$.

3. Let q be a prime, $q \nmid n$ and $\mathfrak{q} \in P_{\mathbb{Q}^{(n)}}$ such that $\mathfrak{q}|q$. Then it follows that $|\mathsf{k}_\mathfrak{q}| = q^{f_\mathfrak{q}} \equiv 1 \bmod n$, and therefore $\mathsf{k}_\mathfrak{q}$ contains n distinct n-th roots of unity. Consequently, if $\xi,\, \xi' \in \mu_n(\mathbb{C})$ and $\xi \equiv \xi' \bmod \mathfrak{q}$, then $\xi = \xi'$. By definition, $(q, \mathbb{Q}^{(n)})(\zeta) \equiv \zeta^q \bmod \mathfrak{q}$, which implies $(q, \mathbb{Q}^{(n)})(\zeta) = \zeta^q = \sigma_n(q)(\zeta)$. If $a \in \mathbb{Q}^+(n)$, then

$$(a, \mathbb{Q}^{(n)})(\zeta) = \prod_{q \in \mathbb{P}} (q, \mathbb{Q}^{(n)})^{v_q(a)}(\zeta) = \zeta^S, \quad \text{where} \quad S = \prod_{q \in \mathbb{P}} p^{v_q(a)} = a$$

(observe that the product is taken in G and means composition of automorphisms). Hence $(a, \mathbb{Q}^{(n)})(\zeta) = \sigma_n(a)(\zeta)$ and therefore $(a, \mathbb{Q}^{(n)}) = \sigma_n(a)$. If $\iota \in G$ is the complex conjugation, then $\iota(\zeta) = \zeta^{-1} = \sigma_n(-1)(\zeta)$ and therefore $\iota = \sigma_n(-1)$. □

Example 3.2.7. Three cubic number fields with equal discriminants. We consider the polynomials

$$g_1 = X^3 - 18X - 6, \quad g_2 = X^3 - 36X - 78, \quad g_3 = X^3 - 54X - 150.$$

They are 2- and 3-Eisenstein polynomials, hence irreducible in $\mathbb{Q}[X]$, and for $i \in [1,3]$ we set $K_i = \mathbb{Q}(a_i)$, where $g_i(a_i) = 0$. For a cubic polynomial $g = X^3 + pX + q$ we have $\mathsf{D}(g) = -4p^3 - 27q^2$, which implies

$$\mathsf{D}(g_1) = \mathsf{D}(g_2) = \mathsf{D}(g_3) = 22356 = 2^2 \cdot 3^5 \cdot 23.$$

Let $i \in [1,3]$. Then $a_i \in \mathfrak{o}_{K_i}$, and if $N_i = (\mathfrak{o}_{K_i} : \mathbb{Z}[a_i])$, then it follows from 3.1.7.2 that $\mathsf{D}(g_i) = 2^2 \cdot 3^5 \cdot 23 = \mathsf{d}_{K_i} N_i^2$ and $(6, N_i) = 1$. Consequently, $N_i = 1$, $(1, a_i, a_i^2)$ is an integral basis of K_i, and $\mathsf{d}_{K_i} = 2^2 \cdot 3^5 \cdot 23$.

To prove that K_1, K_2, K_3 are distinct, we consider the factorization of the primes 5 and 11 using 3.2.1.

$p = 5$: For $i \in [1,3]$ set $\overline{g}_i = g_i + 5\mathbb{Z}[X] \in \mathbb{F}_5[X]$. Then

$$\overline{g}_1 = X^3 + 2X - \overline{1}, \quad \overline{g}_2 = X^3 - X + \overline{2}, \quad \overline{g}_3 = X^3 + X = X(X - \overline{2})(X + \overline{2}).$$

$\overline{g}_1$ and $\overline{g}_2$ have no zero in $\mathbb{F}_5$ and are thus irreducible in $\mathbb{F}_5[X]$. Hence 5 is inert in both K_1 and K_2, and 5 is fully decomposed in K_3.

$p = 11$: For $i \in [1,3]$ set $\overline{g}_i = g_i + 11\mathbb{Z}[X] \in \mathbb{F}_{11}[X]$. Then

$$\overline{g}_1 = X^3 + \overline{4}X + \overline{5} = (X + \overline{1})(X + \overline{2})(X - \overline{3}) \quad \text{and} \quad \overline{g}_2 = X^3 - \overline{3}X - \overline{1}.$$

$\overline{g}_2$ has no zero in $\mathbb{F}_{11}$ and is thus irreducible in $\mathbb{F}_{11}[X]$. Hence 11 is fully decomposed in K_1 and inert in K_2. Consequently, K_1, K_2 and K_3 are distinct.

We proceed with a norm-based criterion to determine the factorization of a prime in an algebraic number field.

Theorem 3.2.8. *Let K be an algebraic number field, $[K:\mathbb{Q}] = n$ and $a \in \mathfrak{o}_K$ such that $K = \mathbb{Q}(a)$. Let p be a prime, $R_p \subset \mathbb{Z}$ a set of representatives for $\mathbb{F}_p$ such that $0 \in R_p$ (e. g., $R_p = [0, p-1]$), and assume that $p \nmid (\mathfrak{o}_K : \mathbb{Z}[a])$.*

> *1. Let $f,\, k \in [1, n]$, and suppose that p has k distinct prime divisors of degree f in K. Then there exist k distinct f-tuples $(c_0^{(i)]}, \ldots, c_{f-1}^{(i)}) \in R_p^f$ such that*
>
> $$v_p(\mathsf{N}_{K/\mathbb{Q}}(a^f + c_{f-1}^{(i)}a^{f-1} + \ldots + c_1^{(i)}a + c_0^{(i)})) \ge f \quad \textit{for all} \ \ i \in [1, k],$$
>
> *and with them we obtain*
>
> $$\mathfrak{p}_i = p\mathfrak{o}_K + (a^f + c_{f-1}^{(i)}a^{f-1} + \ldots + c_1^{(i)}a + c_0^{(i)})\mathfrak{o}_K \in \mathcal{P}_K, \quad \mathfrak{p}_i \,|\, p, \quad f_{\mathfrak{p}_i} = f,$$
>
> *and*
>
> $$v_{\mathfrak{p}_i}(a^f + c_{f-1}^{(i)}a^{f-1} + \ldots + c_1^{(i)}a + c_0^{(i)}) \ge 1 \qquad \textit{for all} \ \ i \in [1, k].$$
>
> *2. Let $k \in [1, n]$, $a_0, \ldots, a_{k-1} \in \mathbb{Z}$,*
>
> $$y = a^k + c_{k-1}a^{k-1} + \ldots + c_1 a + c_0 \neq 0$$
>
> *and $m = v_p(\mathsf{N}_{K/\mathbb{Q}}(y)) > 0$. Then*
>
> $$\sum_{\substack{\mathfrak{p} \in \mathcal{P}_K \\ \mathfrak{p} \,|\, p, \ v_{\mathfrak{p}}(y) > 0}} f_{\mathfrak{p}} \le k \quad \textit{and} \quad m = \sum_{\substack{\mathfrak{p} \in \mathcal{P}_K \\ \mathfrak{p} \,|\, p}} f_{\mathfrak{p}} v_{\mathfrak{p}}(y).$$

Proof. For a polynomial $h \in \mathbb{Z}[X]$ we denote by $\overline{h} = h + p\mathbb{Z}[X] \in \mathbb{F}_p[X]$ the residue class polynomial of h. In the sequel we use 3.2.1.

1. Let $g \in \mathbb{Z}[X]$ be the minimal polynomial of a, let $g_1, \ldots, g_r \in \mathbb{Z}[X]$ be monic polynomials such that $\overline{g}_1, \ldots, \overline{g}_r \in \mathbb{F}_p[X]$ are the distinct irreducible monic polynomials dividing $\overline{g}$, and let $k \in [1, r]$ be such that for all $i \in [1, k]$ we have $\partial(g_i) = f$, $\mathfrak{p}_i = p\mathfrak{o}_K + g_i(a)\mathfrak{o}_K$ and $g_i = X^f + c_{f-1}^{(i)}X^{f-1} + \ldots + c_1^{(i)}X + c_0^{(i)}$, where $c_0^{(i)}, \ldots, c_{f-1}^{(i)} \in R_p$. Then the k f-tuples $(c_0^{(i)}, \ldots, c_{f-1}^{(i)})$ are distinct, and for all $i \in [1, k]$ we have $g_i(a) \in \mathfrak{p}_i$, hence $\mathsf{N}_{K/\mathbb{Q}}(g_i(a)) \in \mathcal{N}_{K/\mathbb{Q}}(\mathfrak{p}_i) = p^f\mathbb{Z}$ by 2.14.4.3, and therefore $v_p(\mathsf{N}_{K/\mathbb{Q}}(g_i(a))) \ge f$.

2. By 2.12.6.2 we obtain

$$\sum_{\substack{\mathfrak{p} \in \mathcal{P}_K \\ \mathfrak{p} \,|\, p, \ v_{\mathfrak{p}}(y) > 0}} f_{\mathfrak{p}} \le k.$$

Since

$$y\mathfrak{o}_K = \prod_{\mathfrak{p}\,|\,p} \mathfrak{p}^{v_\mathfrak{p}(y)}\,\mathfrak{a} \quad \text{for some} \quad \mathfrak{a} \in \mathcal{J}'_K \ \text{ such that } \ p \nmid \mathfrak{N}(\mathfrak{a}),$$

it follows that

$$v_p(\mathsf{N}_{K/\mathbb{Q}}(y)) = v_p(\mathfrak{N}(y\mathfrak{o}_K)) = v_p\Big(\prod_{\mathfrak{p}\,|\,p} \mathfrak{N}(\mathfrak{p}^{v_\mathfrak{p}(y)})\Big) = \sum_{\mathfrak{p}\,|\,p} f_\mathfrak{p} v_\mathfrak{p}(y). \qquad \square$$

Example 3.2.9. We determine the factorizations of the primes 2, 3, 5 and 7 in the field $K = \mathbb{Q}(\theta)$, where $\theta^5 = 2$. We use 3.1.7, 3.2.1 and 3.2.8.

The polynomial $g = X^5 - 2$ is the minimal polynomial of θ, and it is a 2-Eisenstein polynomial. Hence $2\mathfrak{o}_K = \mathfrak{p}_2^5$, where $\mathfrak{p}_2 = \theta\mathfrak{o}_K \in \mathcal{P}_K$, $f_{\mathfrak{p}_2} = 1$, and $2 \nmid (\mathfrak{o}_K : \mathbb{Z}[\theta])$.

The polynomial

$$h = (X+2)^5 - 2 = X^5 + 10X^4 + 40X^3 + 80X^2 + 80X + 30$$

is the minimal polynomial of $\theta - 2$, and it is a 5-Eisenstein polynomial. Hence $5\mathfrak{o}_K = \mathfrak{p}_5^5$, where $\mathfrak{p}_5 \in \mathcal{P}_K$, $f_{\mathfrak{p}_5} = 1$, and $5 \nmid (\mathfrak{o}_K : \mathbb{Z}[\theta - 2]) = (\mathfrak{o}_K : \mathbb{Z}[\theta])$.

Since $\Delta(\mathsf{Z}[\theta]) = \mathsf{D}(g) = \mathsf{N}_{K/\mathbb{Q}}(g'(\theta)) = \mathsf{N}_{K/\mathbb{Q}}(5\theta^4) = 5^5 \cdot 2^4$ it follows that at most the primes 2 and 5 divide $(\mathfrak{o}_K : \mathbb{Z}[\theta])$. Since they do not, we obtain $\mathfrak{o}_K = \mathbb{Z}[\theta]$, $\mathsf{d}_K = 5^5 \cdot 2^4$, and $(1, \theta, \theta^2, \theta^3, \theta^4)$ is an integral basis of K.

The set $R_3 = \{0, \pm 1\}$ is a set of representatives for $\mathbb{F}_3$. Since $\mathsf{N}_{K/\mathbb{Q}}(\theta) = 2$, $\mathsf{N}_{K/\mathbb{Q}}(\theta + 1) = 3$ and $\mathsf{N}_{K/\mathbb{Q}}(\theta - 1) = 1$, we see that 3 has in K at most one prime divisor of degree 1, and since $\mathsf{N}_{K/\mathbb{Q}}(\theta+1) = \mathfrak{N}((\theta+1)\mathfrak{o}_K) = 3$, it follows that $\mathfrak{p}_3$ is the only prime divisor of degree 1 of 3 in K.

Among the 9 numbers $\mathsf{N}_{K/\mathbb{Q}}(\theta^2 + c_1\theta + c_0)$, where $c_0, c_1 \in R_3$, none is divisible by 3^2. Hence 3 has in K no prime divisor of degree 2, and consequently $3\mathfrak{o}_K = \mathfrak{p}_3\mathfrak{p}'_3$ for some $\mathfrak{p}'_3 \in \mathcal{P}_K$ of degree 4.

The set $R_7 = \{0, \pm 1, \pm 2, \pm 3\}$ is a set of representatives for $\mathbb{F}_7$, and exactly one of the 7 numbers $\mathsf{N}_{K/\mathbb{Q}}(\theta + c)$ for $c \in R_7$ is divisible by 7, it is given by $\mathsf{N}_{K/\mathbb{Q}}(\theta + 3) = 5 \cdot 7^2$, and therefore 7 has in K at most one prime divisor of degree 1. By 3.2.8.2 we obtain

$$\sum_{\substack{\mathfrak{p}\,|\,7 \\ v_\mathfrak{p}(\theta+3)>0}} f_\mathfrak{p} \le 1,$$

and therefore there is no $\mathfrak{p} \in \mathcal{P}_K$ such that $f_\mathfrak{p} \ge 2$ and $v_\mathfrak{p}(\theta + 3) > 0$. Hence $(\theta + 3)\mathfrak{o}_K = \mathfrak{p}_5\mathfrak{p}_7^2$, and $\mathfrak{p}_7$ is the only prime divisor of degree 1 dividing 7.

Among the 49 numbers $\mathsf{N}_{K/\mathbb{Q}}(\theta^2+c_1\theta+c_0)$, where $c_0, c_1 \in R_7$, exactly 2 are divisible by 7^2, namely $\mathsf{N}_{K/\mathbb{Q}}(\theta^2+3\theta) = 2\cdot 5\cdot 7^2$ and $\mathsf{N}_{K/\mathbb{Q}}(\theta^2+2\theta-3) = 5\cdot 7^2$. But these two numbers yield no prime divisor of 7 of degree 2, since

$$(\theta^2 + 3\theta)\mathfrak{o}_K = \theta(\theta + 3)\mathfrak{o}_K = \mathfrak{p}_2\mathfrak{p}_5\mathfrak{p}_7^2 \quad \text{and}$$
$$(\theta^2 + 2\theta - 3)\mathfrak{o}_K = (\theta + 3)(\theta - 1)\mathfrak{o}_K = \mathfrak{p}_5\mathfrak{p}_7^2.$$

Consequently, $7\mathfrak{o}_K = \mathfrak{p}_7\mathfrak{p}'_7$ for some $\mathfrak{p}'_7 \in \mathcal{P}_K$ of degree 4.

3.3 Dirichlet characters and abelian number fields

Let G be a (multiplicative) abelian group and $\mathbb{T} = \{z \in \mathbb{C} \mid |z| = 1\} \subset \mathbb{C}^\times$. The abelian group $\mathsf{X}(G) = \mathrm{Hom}(G, \mathbb{T})$ (endowed with pointwise multiplication) is called the **character group** or **dual group** of G; its elements are called **(linear) characters** of G. The unit element of $\mathsf{X}(G)$ is the **unit character** 1_G, defined by $1_G(g) = 1$ for all $g \in G$. If $\chi \in \mathsf{X}(G)$, then the conjugate character $\overline{\chi} = \chi^{-1} \colon G \to \mathbb{T}$ is the inverse of χ in $\mathsf{X}(G)$. For a subset H of G we set

$$H^\perp = \{\chi \in \mathsf{X}(G) \mid \chi \restriction H = 1_H\}$$

Apparently, $H^\perp = \langle H \rangle^\perp$ is a subgroup of $\mathsf{X}(G)$.

If $|G| = n < \infty$ and $\mu_n = \mu_n(\mathbb{C}) = \langle \mathrm{e}^{2\pi \mathrm{i}/n} \rangle \subset \mathbb{T}$, then $\mathsf{X}(G) = \mathrm{Hom}(G, \mu_n)$.

Theorem 3.3.1 (Main theorem on characters of finite abelian groups). *Let G be a (multiplicative) finite abelian group and $n = |G|$.*

1. Let $G = \langle g \rangle$ be cyclic, $\mu_n = \langle \zeta \rangle$ for some primitive n-th root of unity ζ, and let $\chi \colon G \to \mathbb{T}$ be defined by $\chi(g^k) = \zeta^k$ for all $k \in \mathbb{Z}$. Then $\mathsf{X}(G) = \langle \chi \rangle$ and $|\mathsf{X}(G)| = n$.

2. Let $k \in \mathbb{N}$, and let $G = G_1 \cdot \ldots \cdot G_k$ be the (inner) direct product of subgroups $G_1, \ldots, G_k$ of G. Then the map

$$\Phi \colon \mathsf{X}(G) \to \mathsf{X}(G_1) \times \ldots \times \mathsf{X}(G_k),$$
$$\text{defined by } \Phi(\chi) = (\chi \restriction G_1, \ldots, \chi \restriction G_k).$$

is an isomorphism. In particular, G and $\mathsf{X}(G)$ are isomorphic.

3. Let H be a subgroup of G and $\pi \colon G \to G/H$ the residue class homomorphism. Then there exist isomorphisms

$$\Phi \colon \mathsf{X}(G/H) \xrightarrow{\sim} H^\perp \quad \text{and} \quad \Psi \colon \mathsf{X}(G)/H^\perp \xrightarrow{\sim} \mathsf{X}(H)$$

such that $\Phi(\varphi) = \varphi \circ \pi$ for all $\varphi \in \mathsf{X}(G/H)$ and $\Psi(\chi H^\perp) = \chi \restriction H$ for all $\chi \in \mathsf{X}(G)$. Moreover, for every $a \in G \setminus \{1\}$ there exists a character $\varphi \in \mathsf{X}(G)$ such that $\varphi(a) \neq 1$.

4. If $g \in G$ and $\mathrm{ord}(g) = f$, then

$$(1 - X^f)^{n/f} = \prod_{\chi \in \mathsf{X}(G)} (1 - \chi(g)X) \in \mathbb{C}[X].$$

5. *If* $a \in G$ *and* $\psi \in \mathsf{X}(G)$, *then*

$$\sum_{g \in G} \psi(g) = \begin{cases} n & \text{if } \psi = 1_G, \\ 0 & \text{if } \psi \neq 1_G, \end{cases} \qquad \text{and} \qquad \sum_{\chi \in \mathsf{X}(G)} \chi(a) = \begin{cases} n & \text{if } a = 1, \\ 0 & \text{if } a \neq 1. \end{cases}$$

Proof. 1. It is clear that $\chi \in \mathsf{X}(G)$ and $\operatorname{ord}(\chi) = n$. If $\varphi \in \mathsf{X}(G)$ and $\varphi(g) = \zeta^k$ for some $k \in \mathbb{N}$, then $(\chi^{-k}\varphi)(g) = 1$, hence $\chi^{-k}\varphi = 1_G$ and $\varphi = \chi^k \in \langle \chi \rangle$. Hence $\mathsf{X}(G) = \langle \chi \rangle$ is a cyclic group of order n.

2. Obviously, Φ is a monomorphism. Let $(\chi_1, \ldots, \chi_k) \in \mathsf{X}(G_1) \times \ldots \times \mathsf{X}(G_k)$ and define $\chi \colon G \to \mathbb{T}$ by $\chi(g) = \chi_1(g_1) \cdot \ldots \cdot \chi_k(g_k)$ if $g = g_1 \cdot \ldots \cdot g_k \in G$ (where $g_i \in G_i$ for all $i \in [1, k]$). Then $\chi \in \mathsf{X}(G)$ and $\Phi(\chi) = (\chi_1, \ldots, \chi_k)$.

By 2.3.7, G is an inner direct product of cyclic subgroups $C_1, \ldots, C_k$, and therefore $\mathsf{X}(G) = \mathsf{X}(C_1 \cdot \ldots \cdot C_k) \cong \mathsf{X}(C_1) \times \ldots \times \mathsf{X}(C_k) \cong C_1 \times \ldots \times C_k \cong G$.

3. If $\varphi \in \mathsf{X}(G/H)$, then $\varphi \circ \pi \in \mathsf{X}(G)$, $\varphi \circ \pi \restriction H = 1_H$, and thus $\varphi \circ \pi \in H^\perp$. Hence $\Phi \colon \mathsf{X}(G/H) \to H^\perp$, defined by $\Phi(\varphi) = \varphi \circ \pi$, is a homomorphism. For every $\chi \in H^\perp$ there exists a unique character $\varphi \in \mathsf{X}(G/H)$ such that $\varphi \circ \pi = \chi$; it is given by $\varphi(gH) = \chi(g)$ for all $g \in G$. Hence Φ is an isomorphism.

Let $\psi \colon \mathsf{X}(G) \to \mathsf{X}(H)$ be defined by $\psi(\chi) = \chi \restriction H$. Then ψ is a homomorphism, and since $\operatorname{Ker}(\psi) = H^\perp$, it follows that ψ induces a monomorphism $\Psi \colon \mathsf{X}(G)/H^\perp \to \mathsf{X}(H)$ satisfying $\Psi(\chi H^\perp) = \chi \restriction H$ for all $\chi \in \mathsf{X}(G)$. Since

$$|\operatorname{Im}(\Psi)| = \frac{|\mathsf{X}(G)|}{|H^\perp|} = \frac{|\mathsf{X}(G)|}{|\mathsf{X}(G/H)|} = \frac{|G|}{|G/H|} = |H| = |\mathsf{X}(H)|,$$

Ψ is surjective and thus it is an isomorphism.

If $a \in G \setminus \{1\}$, then $1 < |\langle a \rangle| = |\mathsf{X}(\langle a \rangle)| = |\mathsf{X}(G)/\langle a \rangle^\perp|$, hence $\langle a \rangle^\perp \neq \mathsf{X}(G)$, and therefore there exists some $\varphi \in \mathsf{X}(G)$ such that $\varphi(a) \neq 1$.

4. By 3., the map $\theta \colon \mathsf{X}(G) \to \mathsf{X}(\langle g \rangle)$, defined by $\theta(\chi) = \chi \restriction \langle g \rangle$, is an epimorphism, and $|\operatorname{Ker}(\theta)| = |\langle g \rangle^\perp| = |\mathsf{X}(G/\langle g \rangle)|$. If $\psi \in \mathsf{X}(H)$, then

$$|\{\chi \in \mathsf{X}(G) \mid \chi \restriction \langle g \rangle = \psi\}| = |\theta^{-1}(\psi)| = |\operatorname{Ker}(\theta)| = |\mathsf{X}(G/\langle g \rangle)| = |G/\langle g \rangle| = \frac{n}{f}.$$

If $\xi = \mathrm{e}^{2\pi \mathrm{i}/f}$, then $\{\psi(g) \mid \psi \in \mathsf{X}(H)\} = \{\xi^j \mid j \in [0, f-1]\}$, and we get

$$\prod_{\chi \in \mathsf{X}(G)} (1 - \chi(g)X) = \prod_{\psi \in \mathsf{X}(\langle g \rangle)} \prod_{\substack{\chi \in \mathsf{X}(G) \\ \chi \restriction \langle g \rangle = \psi}} (1 - \chi(g)X) = \prod_{\psi \in \mathsf{X}(\langle g \rangle)} (1 - \psi(g)X)^{n/f}$$

$$= \prod_{j=0}^{f-1} (1 - \xi^j X)^{n/f} = (1 - X^f)^{n/f}.$$

5. If $\psi = 1_G$ or $a = 1$, there is nothing to do. If $\psi \neq 1_G$ and $b \in G$ is such that $\psi(b) \neq 1$, then

$$\psi(b) \sum_{g \in G} \psi(g) = \sum_{g \in G} \psi(bg) = \sum_{g \in G} \psi(g), \quad \text{which implies} \quad \sum_{g \in G} \psi(g) = 0.$$

If $a \neq 1$, then by 3. there exists some $\varphi \in \mathsf{X}(G)$ such that $\varphi(a) \neq 1$, and we obtain

$$\varphi(a) \sum_{\chi \in \mathsf{X}(G)} \chi(a) = \sum_{\chi \in \mathsf{X}(G)} \varphi\chi(a) = \sum_{\chi \in \mathsf{X}(G)} \chi(a), \quad \text{which implies} \quad \sum_{\chi \in \mathsf{X}(G)} \chi(a) = 0.$$

□

Theorem 3.3.2. *Let G be a (multiplicative) abelian group.*

1. For $x \in G$ we define $\widehat{x}\colon \mathsf{X}(G) \to \mathbb{T}$ by means of $\widehat{x}(\chi) = \chi(x)$ for all $\chi \in \mathsf{X}(G)$. Then $\widehat{x} \in \mathsf{X}(\mathsf{X}(G))$, and the map

$$u_G\colon G \to \mathsf{X}(\mathsf{X}(G)) \quad \textit{defined by} \quad u_G(x) = \widehat{x},$$

is an isomorphism. We identify: $G = \mathsf{X}(\mathsf{X}(G))$.
Consequently, $x = \widehat{x}$, $x(\chi) = \chi(x)$ for all $x \in G$ and $\chi \in \mathsf{X}(G)$, and if $Y \subset \mathsf{X}(G)$, then $Y^\perp = \langle Y\rangle^\perp = \{x \in G \mid \chi(x) = 0 \text{ for all } \chi \in Y\}$.

2. $(H^\perp)^\perp = H$ for every subgroup H of G.

Proof. 1. If $x \in G$, then clearly $\widehat{x} \in \mathsf{XX}(G)$, and u_G is a homomorphism. If $x \in \operatorname{Ker}(u_G)$ then $\widehat{x}(\chi) = \chi(x) = 1$ for all $\chi \in \mathsf{X}(G)$, and thus $x = 1$ by 3.3.1.3. Hence u_G is a monomorphism, and since $|\mathsf{X}(\mathsf{X}(G))| = |G|$ by 3.3.1.2, u_G is an isomorphism.

2. Let H be a subgroup of G. If $x \in H$, then $x(\chi) = \chi(x) = 1$ for all $\chi \in H^\perp$ and thus $x \in (H^\perp)^\perp$. Hence $H \subset (H^\perp)^\perp$, and equality holds, since by 3.3.1.(2. and 3.) $|(H^\perp)^\perp| = |\mathsf{X}(\mathsf{X}(G)/H^\perp)| = |\mathsf{X}(G)/H^\perp| = |\mathsf{X}(H)| = |H|$. □

Definition and Theorem 3.3.3. For a homomorphism $\varphi\colon G \to G'$ of finite abelian groups we define the **dual homomorphism**

$$\mathsf{X}(\varphi)\colon \mathsf{X}(G') \to \mathsf{X}(G) \quad \text{by} \quad \mathsf{X}(\varphi)(\chi) = \chi{\circ}\varphi.$$

1. Let $\mathbf{1} \to H \overset{i}{\to} G \overset{\pi}{\to} G_1 \to \mathbf{1}$ be an exact sequence of finite abelian groups. Then the dual sequence

$$\mathbf{1} \;\to\; \mathsf{X}(G_1) \;\overset{\mathsf{X}(\pi)}{\to}\; \mathsf{X}(G) \;\overset{\mathsf{X}(i)}{\to}\; \mathsf{X}(H) \;\to\; \mathbf{1}$$

is exact, too.

2. Let $K \subset H \subset G$ be finite abelian groups, and let $\pi\colon H \to H/K$ be the residue class epimorphism. Then there is an isomorphism

$$\Phi\colon \mathsf{X}(H/K) \overset{\sim}{\to} K^\perp/H^\perp \quad \textit{satisfying}$$
$$\Phi(\varphi) = (\varphi{\circ}\pi)H^\perp \quad \textit{for all} \quad \varphi \in \mathsf{X}(H/K).$$

Proof. 1. We may assume that H is a subgroup of G, $G_1 = G/H$, π is the residue class epimorphism and $i = (H \hookrightarrow G)$. Then the sequence is exact by 3.3.1.3.

2. The natural exact sequence $\mathbf{1} \to H/K \to G/K \to G/H \to \mathbf{1}$ induces by 1. the following commutative diagram with exact rows and vertical isomorphisms given by 3.3.1.3:

$$\begin{array}{ccccccccc} \mathbf{1} & \longrightarrow & \mathsf{X}(G/H) & \longrightarrow & \mathsf{X}(G/K) & \longrightarrow & \mathsf{X}(H/K) & \longrightarrow & \mathbf{1} \\ & & \Phi_H \big\downarrow & & \Phi_K \big\downarrow & & & & \\ \mathbf{1} & \longrightarrow & H^\perp & \longrightarrow & K^\perp & \longrightarrow & K^\perp/H^\perp & \longrightarrow & \mathbf{1}. \end{array}$$

Hence there exists a unique isomorphism $\Phi\colon \mathsf{X}(H/K) \stackrel{\sim}{\to} K^\perp/H^\perp$ which makes the diagram commutative, and this has the desired property. □

Remarks and Definitions 3.3.4. For $m \in \mathbb{N}$, we set $\mathsf{X}(m) = \mathsf{X}((\mathbb{Z}/m\mathbb{Z})^\times)$. A character $\chi \in \mathsf{X}(m)$ is called a **Dirichlet character** modulo m. For a Dirichlet character $\chi \in \mathsf{X}(m)$ we consider the (equally denoted) function $\chi\colon \mathbb{Z} \to \mathbb{C}$, defined by

$$\chi(a) = \begin{cases} \chi(a + m\mathbb{Z}) & \text{if } (a, m) = 1, \\ 0 & \text{if } (a, m) \ne 1. \end{cases}$$

It has the following properties (for all $a,\ b \in \mathbb{Z}$):

D1. $\chi(ab) = \chi(a)\chi(b)$

D2. $\chi(a) = \chi(b)$ if $a \equiv b \bmod m$.

D3. $\chi(a) = 0$ if and only if $(a, m) > 1$.

Conversely, every function $\chi\colon \mathbb{Z} \to \mathbb{C}$ satisfying **D1**, **D2**, and **D3**, is induced by a Dirichlet character modulo m, and therefore it is itself called a Dirichlet character modulo m. Under this wording, a map $\chi\colon \mathbb{Z} \to \mathbb{C}$ is a Dirichlet character modulo m if and only if $\chi \restriction \mathbb{Z}(m) \to \mathbb{T}$ is a monoid homomorphism satisfying $\chi \restriction (1 + m\mathbb{Z}) = 1$, and $\chi \restriction \mathbb{Z} \setminus \mathbb{Z}(m) = 0$.

If $\chi\colon \mathbb{Z} \to \mathbb{T}$ is a Dirichlet character modulo m, then χ has a unique extension to an (equally denoted) map

$$\chi\colon \mathbb{Q} \to \mathbb{C} \text{ such that } \chi \restriction \mathbb{Q}(m)\colon \mathbb{Q}(m) \to \mathbb{T} \text{ is a group homomorphism}$$
$$\text{with kernel } \mathbb{Q}^{\circ m} = \mathsf{q}(1 + m\mathbb{Z}) \text{ and } \chi \restriction \mathbb{Q} \setminus \mathbb{Q}(m) = 0.$$

By a convenient abuse of language we shall call all these maps χ **Dirichlet characters** modulo m, regardless of whether they are defined on $(\mathbb{Z}/m\mathbb{Z})^\times$, on $\mathbb{Z}$ or on $\mathbb{Q}$.

Again let $m \in \mathbb{N}$. We denote by $\mathbf{1}_m \in \mathsf{X}(m)$ the unit character modulo m. By definition, if $a \in \mathbb{Z}$, then

$$\mathbf{1}_m(a) = \begin{cases} 1 & \text{if} \quad (a,m) = 1, \\ 0 & \text{if} \quad (a,m) \neq 1. \end{cases}$$

Clearly, $\mathbf{1}_1 = \mathrm{id}_{\mathbb{Z}}$, $\mathsf{X}(1) = \{\mathbf{1}\}$ and $\mathsf{X}(2) = \{\mathbf{1}_2\}$. For $\chi \in \mathsf{X}(m)$ and $a \in \mathbb{Z}$, we set $\overline{\chi}(a) = \overline{\chi(a)}$, and if $(a,m) = 1$, then $\overline{\chi}(a) = \chi(a)^{-1}$.

Lemma 3.3.5. *Let $m \in \mathbb{N}$ and $\chi \in \mathsf{X}(m)$.*

1. *If $c \in \mathbb{Z}$, then*

$$\sum_{t \in (\mathbb{Z}/m\mathbb{Z})^\times} \chi(t) = \sum_{j=c}^{c+m-1} \chi(j) = \begin{cases} \varphi(m) & \text{if } \chi = \mathbf{1}_m, \\ 0 & \text{if } \chi \neq \mathbf{1}_m. \end{cases}$$

2. *Let $a, b \in \mathbb{Z}$ and $(a,m) = (b,m) = 1$. Then*

$$\sum_{\chi \in \mathsf{X}(m)} \overline{\chi}(a)\chi(b) = \begin{cases} \varphi(m) & \text{if } a \equiv b \mod m, \\ 0 & \text{if } a \not\equiv b \mod m. \end{cases}$$

3. *If $\chi \neq \mathbf{1}_m$, $a, b \in \mathbb{Z}$ and $a < b$, then*

$$\Big| \sum_{n=a+1}^{b} \chi(n) \Big| < m.$$

Proof. 1. Obvious by 3.3.1.5 and the very definition of a Dirichlet character.

2. If $\overline{a} = a + m\mathbb{Z}$, $\overline{b} = b + m\mathbb{Z} \in (\mathbb{Z}/m\mathbb{Z})^\times$, then 3.3.1.5 implies

$$\sum_{\chi \in \mathsf{X}(m)} \overline{\chi}(a)\chi(b) = \sum_{\chi \in \mathsf{X}(m)} \chi(\overline{a}^{-1}\overline{b}) = \begin{cases} \varphi(m) & \text{if } \overline{a} = \overline{b}, \\ 0 & \text{if } \overline{a} \neq \overline{b}. \end{cases}$$

3. Let $b = a + km + r$, where $k \in \mathbb{N}_0$ and $r \in [0, m-1]$. Then

$$\sum_{n=a+1}^{b} \chi(n) = \sum_{j=0}^{k-1} \sum_{i=1}^{m} \chi(a + jm + i) + \sum_{i=1}^{r} \chi(a + km + i) = \sum_{i=1}^{r} \chi(a + km + i)$$

by 1., and therefore

$$\Big| \sum_{n=a+1}^{b} \chi(n) \Big| \leq \sum_{i=1}^{r} |\chi(a + km + i)| \leq r < m. \qquad \square$$

We proceed with a comparison of Dirichlet characters for distinct moduli of definition. Let $m,\, d \in \mathbb{N}$ and $d \,|\, m$. Then $\mathbb{Z}(m) \subset \mathbb{Z}(d)$, the residue class epimorphism $\mathbb{Z}/m\mathbb{Z} \to \mathbb{Z}/d\mathbb{Z}$ induces a group homomorphism $\pi_{m,d} \colon (\mathbb{Z}/m\mathbb{Z})^\times \to (\mathbb{Z}/d\mathbb{Z})^\times$, and we assert that $\pi_{m,d}$ is surjective. Indeed, if $a_0 \in \mathbb{Z}(d)$, there exists some $a \in \mathbb{Z}$ such that $a \equiv a_0 \bmod d$ and $(a,p) = 1$ for all primes p such that $p \,|\, m$ and $p \nmid d$; then it follows that $(a,m) = 1$ and $\pi_{m,d}(a + m\mathbb{Z}) = a_0 + d\mathbb{Z}$. By definition,

$$\operatorname{Ker}(\pi_{m,d}) = \{a + m\mathbb{Z} \mid a \in \mathbb{Z}(m),\ a \equiv 1 \bmod d\} \subset (\mathbb{Z}/m\mathbb{Z})^\times.$$

By 3.3.3.1 the epimorphism $\pi_{m,d} \colon (\mathbb{Z}/m\mathbb{Z})^\times \to (\mathbb{Z}/d\mathbb{Z})^\times$ induces an exact sequence

$$\mathbf{1} \to \mathsf{X}(d) \overset{\widehat{\pi}_{m,d}}{\to} \mathsf{X}(m) \overset{\rho}{\to} \mathsf{X}(\operatorname{Ker}(\pi_{m,d})) \to \mathbf{1},$$

where $\widehat{\pi}_{m,d} = \mathsf{X}(\pi_{m,d})$ and $\rho(\chi) = \chi \restriction \operatorname{Ker}(\pi_{m,d})$. If $\chi = \widehat{\pi}_{m,d}(\chi_1) = \chi_1 \circ \pi_{m,d}$ for some $\chi_1 \in \mathsf{X}(d)$, then we say that χ is **induced** by χ_1. Hence χ is induced by χ_1 if and only if $\chi(a) = \chi_1(a)$ for all $a \in \mathbb{Z}(m)$. Let

$$\mathsf{X}(m)^{(d)} = \operatorname{Im}(\widehat{\pi}_{m,d}) = \operatorname{Ker}(\rho) = \operatorname{Ker}(\pi_{m,d})^\perp \cong \mathsf{X}(d)$$

be the group of all characters $\chi \in \mathsf{X}(m)$ which are induced by some $\chi_1 \in \mathsf{X}(d)$. If $\chi \in \mathsf{X}(m)^{(d)}$, then we call d a **modulus of definition** for χ. By definition, d is a modulus of definition for χ if and only if $\chi \restriction (1 + d\mathbb{Z}) \cap \mathbb{Z}(m) = 1$. Consequently, if d is a modulus of definition for χ, then every $d' \in \mathbb{N}$ satisfying $d \,|\, d' \,|\, m$ is also a modulus of definition for χ.

The smallest modulus of definition for a character $\chi \in \mathsf{X}(m)$ is called the **conductor** of χ and is denoted by $f(\chi)$. If $f(\chi) = m$ then χ is called **primitive**. By definition, a character $\chi \in \mathsf{X}(m)$ is primitive if and only if for every proper divisor d of m there exists some $a \in \mathbb{Z}$ such that $(a,m) = 1$, $a \equiv 1 \bmod d$ and $\chi(a) \neq 1$. If $\chi \in \mathsf{X}(m)$, then $\widehat{\pi}_{m,f(\chi)}$ is injective, and therefore there exists a unique primitive character $\chi' \in \mathsf{X}(f(\chi))$ such that χ is induced by χ'. We call χ' the **associated primitive character** of χ.

By definition, $1_m' = 1_1$ for all $m \in \mathbb{N}$, and if p is a prime, then every character $\chi \in \mathsf{X}(p) \setminus \{1_p\}$ is primitive.

Theorem 3.3.6. *Let $m \in \mathbb{N}$*

1. *Let $\chi \in \mathsf{X}(m)$, let d_1 and d_2 be moduli of definition for χ and $d = (d_1, d_2)$. Then d is a modulus of definition for χ, too. In particular, $f(\chi)$ is the* gcd *of all moduli of definition for χ.*
2. *If $d \in \mathbb{N}$ and $d \,|\, m$, then $\mathsf{X}(m)^{(d)} = \{\chi \in \mathsf{X}(m) \mid f(\chi) \,|\, d\}$.*
3. *Let $m = m_1 \cdot \ldots \cdot m_r$, where $r, m_1, \ldots, m_r \in \mathbb{N}$ and $(m_i, m_j) = 1$ for all $i,\, j \in [1,r]$ such that $i \neq j$. Then $\mathsf{X}(m)$ is an (inner) direct product*

$$\mathsf{X}(m) = \mathsf{X}(m)^{(m_1)} \cdot \ldots \cdot \mathsf{X}(m)^{(m_r)} \cong \prod_{i=1}^{r} \mathsf{X}(m_i).$$

Every $\chi \in \mathsf{X}(m)$ *has a unique factorization* $\chi = \chi_1 \cdot \ldots \cdot \chi_r$ *such that* $\chi_i \in \mathsf{X}(m)^{(m_i)}$ *for all* $i \in [1,r]$, *and* $f(\chi) = f(\chi_1) \cdot \ldots \cdot f(\chi_r)$. *Explicitly, the characters* $\chi_1, \ldots, \chi_r$ *are obtained as follows:*

For $a \in \mathbb{Z}$ *let* $a(i) \in \mathbb{Z}$ *be such that* $a(i) \equiv a \bmod m_i$ *and* $a(i) \equiv 1 \bmod m_j$ *for all* $j \in \{1,r\} \setminus \{i\}$. *Then it follows that* $a \equiv a(1) \cdot \ldots \cdot a(r) \bmod m$ *and* $\chi_i(a) = \chi(a(i))$ *for all* $i \in [1,r]$.

Proof. 1. Let $a \in \mathbb{Z}(m)$ and $a \equiv 1 \bmod d$. By 2.8.4.4 there exists some modulo d uniquely determined $c_1 \in \mathbb{Z}$ such that $c_1 \equiv a \bmod d_1$ and $c_1 \equiv 1 \bmod d_2$. Since $\pi_{m,d}$ is surjective, there exists some $c \in \mathbb{Z}(m)$ such that $c \equiv c_1 \bmod d$. Then it follows that $c \equiv a \bmod d_1$, $c \equiv 1 \bmod d_2$, and as d_1 and d_2 are moduli of definition of χ, we obtain $\chi(a) = \chi(c) = \chi(1) = 1$.

2. Let $\chi \in \mathsf{X}(m)$ and $d \in \mathbb{N}$ such that $d \mid m$. If $f(\chi) \mid d$, then obviously d is a modulus of definition for χ. Conversely, if d is a modulus of definition for χ, then $(d, f(\chi))$ is also a modulus of definition for χ, and since $(d, f(\chi)) \mid f(\chi)$, it follows that $f(\chi) = (d, f(\chi)) \mid d$.

3. By the Chinese remainder theorem 1.1.4, $(\mathbb{Z}/m\mathbb{Z})^\times = \Lambda_1 \cdot \ldots \cdot \Lambda_r$, where $\Lambda_i \cong (\mathbb{Z}(m_i\mathbb{Z})^\times$ for $i \in [1,r]$ is given as follows. For $a \in \mathbb{Z}(m)$ let $a(i) \in \mathbb{Z}$ be such that $a(i) \equiv a \bmod m_i$ and $a(i) \equiv 1 \bmod m_j$ for all $j \in [1,r] \setminus \{i\}$. Then $\Lambda_i = \{a(i) + m\mathbb{Z} \mid a \in \mathbb{Z}(m)\} \subset (\mathbb{Z}/m\mathbb{Z})^\times$ for all $i \in [1,r]$, and

$$a + m\mathbb{Z} = \prod_{i=1}^{r} (a(i) + m\mathbb{Z}) \quad \text{for all} \quad a \in \mathbb{Z}(m).$$

By 3.3.1.2 there is an isomorphism

$$\Psi \colon \mathsf{X}(m) \overset{\sim}{\to} \mathsf{X}(\Lambda_1) \times \ldots \times \mathsf{X}(\Lambda_r) \quad \text{such that} \quad \Psi(\chi) = (\chi \restriction \Lambda_1, \ldots, \chi \restriction \Lambda_r),$$

and therefore $\mathsf{X}(m) = \mathsf{X}_1 \cdot \ldots \cdot \mathsf{X}_r$, where for all $i \in [1,r]$

$$\mathsf{X}_i = \{\chi \in \mathsf{X}(m) \mid \chi \restriction \Lambda_j = 1 \text{ for all } j \in [1,r] \setminus \{i\}\} = \mathsf{X}(m)^{(m_i)} \cong \mathsf{X}(m_i).$$

Explicitly, if $\chi \in \mathsf{X}(m)$ and $a \in \mathbb{Z}(m)$, $a \equiv a(1) \cdot \ldots \cdot a(r) \bmod m$ as above, then $\chi = \chi_1 \cdot \ldots \cdot \chi_r \in \mathsf{X}(m)$, where $\chi_i \in \mathsf{X}_i = \mathsf{X}(m)^{(m_i)}$ is given by $\chi_i(a) = \chi(a(i))$.

It remains to prove that $f(\chi) = f(\chi_1) \cdot \ldots \cdot f(\chi_r)$. For $i \in [1,r]$ let $f_i \in \mathbb{N}$ such that $f_i \mid m_i$, and set $f = f_1 \cdot \ldots \cdot f_r$. Then it suffices to prove that f is a modulus of definition for χ if and only if f_i is a modulus of definition for χ_i for all $i \in [1,r]$.

Assume first that f_i is a modulus of definition for χ_i for all $i \in [1,r]$, and let $a \in \mathbb{Z}$ be such that $a \equiv 1 \bmod f$. Then $a(i) \equiv 1 \bmod f_i$ for all $i \in [1,r]$, and therefore $\chi(a) = \chi_1(a(1)) \cdot \ldots \cdot \chi_r(a(r)) = 1$. Conversely, let f be a modulus of definition for χ, and let $i \in [1,r]$ and $a \in \mathbb{Z}$ be such that $a \equiv 1 \bmod f_i$. Let $a' \in \mathbb{Z}$ be such that $a' \equiv a \bmod m_i$ and $a' \equiv 1 \bmod m_j$ for all $j \in [1,r] \setminus \{i\}$. Then $a' \equiv 1 \bmod f$ and $1 = \chi(a') = \chi_1(a') \cdot \ldots \cdot \chi_r(a') = \chi_i(a') = \chi_i(a)$. $\square$

Let $m = p^e d$, where p is a prime, $e \in \mathbb{N}_0$, $d \in \mathbb{N}$, $p \nmid d$, and then we set $\mathsf{X}(m)_p = \mathsf{X}(m)^{(p^e)}$. By 3.3.6.3, every $\chi \in \mathsf{X}(m)$ has a unique factorization $\chi = \chi_p \chi_1$, where $\chi_p \in \mathsf{X}(m)_p$ and $\chi_1 \in \mathsf{X}(m)^{(d)}$. We call χ_p the **p-component** of χ. If $f(\chi_p) = p^c$ for some $c \in [0, e]$ and $f(\chi_1) = d_1 \mid d$, then $f(\chi) = p^c d_1$. Moreover, if χ' is the associated primitive character of χ, then $(\chi')_p = \chi'_p$ is the associated primitive character of χ_p. Note that $\chi'_p \neq 1_m$ if and only if $p \mid f(\chi)$, and this holds if and only if $\chi'(p) = 0$. The situation is illustrated by the following diagram:

$$\begin{array}{ccccc} \chi: & \mathsf{X}(m) & \xrightarrow{\pi} & \mathsf{X}(f) & \xrightarrow{\chi'} & \mathbb{T} \\ & \| & & \| & & \| \\ (\chi_p, \chi_1): & \mathsf{X}(m)_p \cdot \mathsf{X}(m)^{(d)} & \xrightarrow{\pi} & \mathsf{X}(f)_p \cdot \mathsf{X}(f)^{(d_1)} & \xrightarrow{(\chi'_p, \chi'_1)} & \mathbb{T} \end{array}$$

In particular, if $m = p_1^{e_1} \cdot \ldots \cdot p_r^{m_r}$, where $r \in \mathbb{N}$, $p_1, \ldots, p_r$ are distinct primes and $e_1, \ldots, e_r \in \mathbb{N}$, then $\chi = \chi_{p_1} \cdot \ldots \cdot \chi_{p_r}$. Consequently, every $\chi \in \mathsf{X}(m)$ is the product of its p-components.

For a subgroup Z of $\mathsf{X}(m)$ we denote by $Z' = \{\chi' \mid \chi \in Z\}$ the set of associated primitive characters and by $Z_p = \{\chi_p \mid \chi \in Z\} \subset \mathsf{X}(m)_p$ the set of p-components of characters in Z.

Note that $\mathsf{X}(\mathsf{X}(m)) = (\mathbb{Z}/m\mathbb{Z})^\times$ by 3.3.2.2. If C is a subgroup of $(\mathbb{Z}/m\mathbb{Z})^\times$, then

$$C^\perp = \{\chi \in \mathsf{X}(m) \mid \chi(a) = 1 \text{ for all } a \in \mathbb{Z}(m) \text{ with } a + m\mathbb{Z} \in C\} \subset \mathsf{X}(m),$$

and if Z is a subgroup of $\mathsf{X}(m)$, then

$$Z^\perp = \{a + m\mathbb{Z} \mid a \in \mathbb{Z}(m) \text{ and } \chi(a) = 1 \text{ for all } \chi \in Z\} \subset (\mathbb{Z}/m\mathbb{Z})^\times.$$

From 3.3.1.3 we get an isomorphism

$$(\mathbb{Z}/m\mathbb{Z})^\times / Z^\perp = \mathsf{X}(\mathsf{X}(m))/Z^\perp \xrightarrow{\sim} \mathsf{X}(Z),$$
$$(a + m\mathbb{Z})Z^\perp \mapsto [\chi \mapsto \chi(a) \ \text{ for all } \ \chi \in Z],$$

and we identify: $\mathsf{X}(Z) = (\mathbb{Z}/m\mathbb{Z})^\times / Z^\perp$.

Example 3.3.7. We discuss $\mathsf{X}(12)$.

We set $1_{12} = 1$, and for $a \in \mathbb{Z}$ we set $\overline{a} = a + 12\mathbb{Z}$. $(\mathbb{Z}/12\mathbb{Z})^\times = \{\pm\overline{1}, \pm\overline{5}\}$, $\mathsf{X}(12) = \mathsf{X}(12)^{(4)} \cdot \mathsf{X}(12)^{(3)} = \{1, \chi, \psi, \chi\psi\}$, where the characters are defined by the following table:

	χ	π	$\chi\psi$
$\overline{1}$	1	1	1
$-\overline{1}$	-1	-1	1
$\overline{5}$	1	-1	-1
$-\overline{5}$	-1	1	-1

Hence $\mathsf{X}(12)^{(4)} = \{1, \chi\}$ and $\mathsf{X}(12)^{(3)} = \{1, \psi\}$.

If $Z = \langle \chi\psi \rangle = \{1, \chi\psi\} \subset \mathsf{X}(12)$, then $Z_2 = \langle \chi \rangle$, $Z_3 = \langle \psi \rangle$ and $Z^\perp = \langle -\overline{1} \rangle$.

Now we are ready to describe the connection between Dirichlet characters and subfields of cyclotomic fields.

Definition and Theorem 3.3.8. *Let* $m \in \mathbb{N}$, $m \geq 3$, $\mathbb{Q}^{(m)} = \mathbb{Q}(\mu_m) \subset \mathbb{C}$ *the* m*-th cyclotomic field and* $G_m = \mathrm{Gal}(\mathbb{Q}^{(m)}/\mathbb{Q})$.
Recall the epimorphism $\sigma_m \colon \mathbb{Q}(m) \to G_m$ defined in 3.2.6: If $a \in \mathbb{Q}(m)$, then $\sigma_m(a)(\xi) = \xi^a$ for all $\xi \in \mu_m$, and

$$\sigma_m \restriction \mathbb{Q}^+(m) = (\,\cdot\,, \mathbb{Q}^{(m)}/\mathbb{Q}) \colon \mathbb{Q}^+(m) \to G_m \text{ is the Artin map.}$$

By definition, $\sigma_m(-1)$ is the complex conjugation, and σ_m induces the (equally denoted) Artin isomorphism $\sigma_m = (\,\cdot\,, \mathbb{Q}^{(m)}/\mathbb{Q}) \colon (\mathbb{Z}/m\mathbb{Z})^\times \overset{\sim}{\to} G_m$.

For a subgroup Z of $\mathsf{X}(m)$ we consider the group

$$\sigma_m(Z^\perp) = \{\sigma_m(a) \mid a \in \mathbb{Z}(m) \text{ and } \chi(a) = 1 \text{ for all } \chi \in Z\} \subset G_m,$$

and we define

$$\mathbb{Q}_Z = (\mathbb{Q}^{(m)})^{\sigma_m(Z^\perp)} = (\mathbb{Q}^{(m)})^{(Z^\perp, \mathbb{Q}^{(m)}/\mathbb{Q})} \subset \mathbb{Q}^{(m)}.$$

For $\chi \in \mathsf{X}(m)$ we set $\mathbb{Q}_\chi = \mathbb{Q}_{\langle\chi\rangle} = (\mathbb{Q}^{(m)})^{\sigma_m(\mathrm{Ker}(\chi))}$. We call $\mathbb{Q}_Z$ resp. $\mathbb{Q}_\chi$ the **associated subfield** of Z resp. χ inside $\mathbb{Q}^{(m)}$.

A character $\chi \in \mathsf{X}(m)$ is called **even** if $\chi(-1) = 1$, and it is called **odd** if $\chi(-1) = -1$. For a subgroup Z of $\mathsf{X}(m)$ we denote by Z^+ the set of all even and by Z^- the set of all odd characters in Z. Then $Z^+ = Z \cap \mathsf{X}(m)^+$ is a subgroup of Z, and $(Z : Z^+) \leq 2$.

1. *Let Z be a subgroup of* $\mathsf{X}(m)$. *Then* $\mathrm{Gal}(\mathbb{Q}^{(m)}/\mathbb{Q}_Z) = \sigma_m(Z^\perp)$, *there is an isomorphism*

$$\theta_Z \colon (\mathbb{Z}/m\mathbb{Z})^\times/Z^\perp \to \mathrm{Gal}(\mathbb{Q}_Z/\mathbb{Q}), \quad \textit{given by}$$
$$\theta_Z((a + m\mathbb{Z})Z^\perp) = \sigma_m(a) \restriction \mathbb{Q}_Z,$$

and $[\mathbb{Q}_Z : \mathbb{Q}] = |Z|$.

2. *Let $\mathfrak{X}(m)$ be the set of all subgroups of $\mathsf{X}(m)$ and $\mathfrak{K}(m)$ the set of all subfields of $\mathbb{Q}^{(m)}$. Then the maps*

$$\begin{cases} \mathfrak{X}(m) & \to & \mathfrak{K}(m) \\ Z & \mapsto & \mathbb{Q}_Z \end{cases} \quad \textit{and} \quad \begin{cases} \mathfrak{K}(m) & \to & \mathfrak{X}(m) \\ L & \mapsto & [\sigma_m^{-1}\mathrm{Gal}(\mathbb{Q}^{(m)}/L)]^\perp \end{cases}$$

are mutually inverse inclusion-preserving bijections. In particular:

- *If K is a subfield of $\mathbb{Q}^{(m)}$, and Z is a subgroup of $\mathsf{X}(m)$, then*

$$K = \mathbb{Q}_Z \iff Z = \{\chi \in \mathsf{X}(m) \mid \chi \restriction \sigma_m^{-1}\mathrm{Gal}(\mathbb{Q}^{(m)}/K) = 1\}.$$

- *If Z and Z' are subgroups of* $\mathsf{X}(m)$, *then*

$$\mathbb{Q}_{ZZ'} = \mathbb{Q}_Z\mathbb{Q}_{Z'}, \quad \mathbb{Q}_{Z\cap Z'} = \mathbb{Q}_Z \cap \mathbb{Q}_{Z'} \quad \textit{and} \quad \mathbb{Q}_Z \;=\; \prod_{\chi\in Z} \mathbb{Q}_\chi.$$

If Z is a subgroup of $\mathsf{X}(m)$ and $K = \mathbb{Q}_Z \subset \mathbb{Q}^{(m)}$, then we call Z the **character group** of K modulo m.

3. *If Z is a subgroup of* $\mathsf{X}(m)$, *then* $\mathbb{Q}_{Z^+} = \mathbb{Q}_Z \cap \mathbb{R}$.

4. *Let $d \in \mathbb{N}$ and $d \mid m$.*

(a) $\mathbb{Q}^{(d)} = (\mathbb{Q}^{(m)})^{\sigma_m(\mathrm{Ker}(\pi_{m,d}))} = \mathbb{Q}_{\mathsf{X}(m)^{(d)}}$.

(b) *d is a modulus of definition for a character $\chi \in \mathsf{X}(m)$ if and only if $\mathbb{Q}_\chi \subset \mathbb{Q}^{(d)}$. In particular,* $f(\chi) = \min\{d \in \mathbb{N} \mid \mathbb{Q}(\chi) \subset \mathbb{Q}^{(d)}\}$.

(c) *If Z_1 is a subgroup of $\mathsf{X}(d)$ and $Z = \widehat{\pi}_{m,d}(Z_1) \subset \mathsf{X}(m)$, then* $\mathbb{Q}_{Z_1} = \mathbb{Q}_Z$.

(d) *Let $K \subset \mathbb{Q}^{(d)} \subset \mathbb{Q}^{(m)}$ be a subfield, and let $Z_1 \subset \mathsf{X}(d)$ and $Z \subset \mathsf{X}(m)$ be subgroups such that $K = \mathbb{Q}_{Z_1} = \mathbb{Q}_Z$. Then we get* $Z = \widehat{\pi}_{m,d}(Z_1) \cong Z_1$.

Proof. 1. By 1.5.4.2, $\mathbb{Q}_Z = (\mathbb{Q}^{(m)})^{\sigma_m(Z^\perp)}$ implies $\sigma_m(Z^\perp) = \mathrm{Gal}(\mathbb{Q}^{(m)}/\mathbb{Q}_Z)$, and σ_m induces an isomorphism

$$\theta_Z \colon (\mathbb{Z}/m\mathbb{Z})^\times/Z^\perp \to G_m/\mathrm{Gal}(\mathbb{Q}^{(m)}/\mathbb{Q}_Z) = \mathrm{Gal}(\mathbb{Q}_Z/\mathbb{Q}),$$

given by $\theta_Z((a+m\mathbb{Z})Z^\perp) = \sigma_m(a)\restriction\mathbb{Q}_Z$ for all $a \in \mathbb{Z}(m)$. In particular, using 3.3.1.3 we obtain

$$[\mathbb{Q}_Z:\mathbb{Q}] = |(\mathbb{Z}/m\mathbb{Z})^\times/Z^\perp| = |\mathsf{X}(\mathsf{X}(m))/Z^\perp| = |\mathsf{X}(Z)| = |Z|.$$

2. If $Z \in \mathfrak{X}(m)$, then

$$[\sigma_m^{-1}\mathrm{Gal}(\mathbb{Q}^{(m)}/\mathbb{Q}_Z)]^\perp = [\sigma_m^{-1}(\sigma_m(Z^\perp)]^\perp = Z^{\perp\perp} = Z.$$

If $L \in \mathfrak{K}(m)$ and

$$H = \mathrm{Gal}(\mathbb{Q}^{(m)}/L), \text{ then } \mathbb{Q}_{(\sigma_m H)^\perp} = (\mathbb{Q}^{(m)})^{\sigma_m[\sigma_m^{-1}H^{\perp\perp}]} = (\mathbb{Q}^{(m)})^H = L.$$

If $Z' \subset Z \subset \mathsf{X}(m)$ are subgroups, then $Z'^\perp \supset Z^\perp$, and consequently $\sigma_m(Z'^\perp) \supset \sigma_m(Z^\perp)$ and $\mathbb{Q}_{Z'} \subset \mathbb{Q}_Z$.

3. Since $\mathbb{Q}^{(m)} \cap \mathbb{R} = (\mathbb{Q}^{(m)})^{\langle\sigma_m(-1)\rangle} = (\mathbb{Q}^{(m)})^{\sigma_m(\mathsf{X}(m)^{+\perp})} = \mathbb{Q}_{\mathsf{X}(m)^+}$, it follows that $\mathbb{Q}_Z \cap \mathbb{R} = \mathbb{Q}_Z \cap \mathbb{Q}_{\mathsf{X}(m)^+} = \mathbb{Q}_{Z^+}$.

4. (a) Since

$$\mathrm{Gal}(\mathbb{Q}^{(m)}/\mathbb{Q}^{(d)}) = \{\sigma_m(a) \mid a \in \mathbb{Z}(m),\ a \equiv 1 \bmod d\} = \sigma_m(\mathrm{Ker}(\pi_{m,d})$$

and $\mathrm{Ker}(\pi_{m,d})^\perp = \mathsf{X}(m)^{(d)}$, it follows that $\mathrm{Ker}(\pi_{m,d}) = [(\mathsf{X}(m)^{(d)}]^\perp$, and consequently $\mathbb{Q}^{(d)} = (\mathbb{Q}^{(m)})^{\sigma_m([\mathsf{X}(m)^{(d)}]^\perp)} = \mathbb{Q}_{\mathsf{X}(m)^{(d)}}$.

(b) Let $\chi \in \mathsf{X}(m)$. Then $\text{Ker}(\pi_{m,d}) \subset \text{Ker}(\chi)$ if and only if

$$\mathbb{Q}_\chi = (\mathbb{Q}^{(m)})^{\sigma_m(\text{Ker}(\chi))} \subset (\mathbb{Q}^{(m)})^{\sigma_m(\text{Ker}(\pi_{m,d})} = \mathbb{Q}^{(d)},$$

and therefore d is a modulus of definition for χ if and only if $\mathbb{Q}_\chi \subset \mathbb{Q}^{(d)}$.

(c) By definition, $Z \subset \text{Im}(\widehat{\pi}_{m,d}) = \mathsf{X}_m^{(d)}$ implies $\mathbb{Q}_Z \subset \mathbb{Q}_{\mathsf{X}(m)^{(d)}} = \mathbb{Q}^{(d)}$. Moreover, $\mathbb{Q}_{Z_1} = (\mathbb{Q}^{(d)})^{\sigma_d(Z_1^\perp)}$ and $\mathbb{Q}_Z = (\mathbb{Q}^{(m)})^{\sigma_m(\widehat{\pi}_{m,d}(Z_1)^\perp)}$. Hence we must prove that $\sigma_m(\widehat{\pi}_{m,d}(Z_1)^\perp) \restriction \mathbb{Q}^{(d)} = \sigma_d(Z_1^\perp)$.

Let $a + d\mathbb{Z} = \pi_{m,d}(a + m\mathbb{Z})$, where $a \in \mathbb{Z}(m)$. Then $a + d\mathbb{Z} \in \widehat{\pi}_{m,d}(Z_1)^\perp$ if and only if $\chi(a) = 1$ for all $\chi \in Z_1$, i.e., if and only if $a \in Z_1^\perp$. If $\xi \in \mu_d$, then $\sigma_m(a)(\xi) = \xi^d = \sigma_d(a)(\xi)$, and consequently $\sigma_m(a) \restriction \mathbb{Q}^{(d)} = \sigma_d(a)$.

(d) By (c) and 2. Observe that $\widehat{\pi}_{m,d}(Z_1) \cong Z_1$, since $\widehat{\pi}_{m,d}$ is a monomorphism. □

Theorem 3.3.9. *Let $m \in \mathbb{N}$, $m \ge 3$, $K = \mathbb{Q}_Z \subset \mathbb{Q}^{(m)}$ for some subgroup Z of $\mathsf{X}(m)$, and let p be a prime. Then $e_p(K) = |Z_p|$, and p is unramified in K if and only if $p \nmid f(\chi)$ for all $\chi \in Z$.*

Proof. Let $m = p^e d$, where $e \in \mathbb{N}_0$, $d \in \mathbb{N}$ and $p \nmid d$. Then $K\mathbb{Q}^{(d)} \subset \mathbb{Q}^{(m)}$, and $K\mathbb{Q}^{(d)} = \mathbb{Q}_Z\mathbb{Q}_{\mathsf{X}(m)^{(d)}} = \mathbb{Q}_{Z\mathsf{X}(m)^{(d)}}$. If $\chi \in Z$, then $\chi = \chi_p\chi_1$, where $\chi_p \in Z_p$ and $\chi_1 \in \mathsf{X}(m)^{(d)}$. Hence it follows that $Z\mathsf{X}(m)^{(d)} = Z_p\mathsf{X}(m)^{(d)}$ and consequently $K\mathbb{Q}^{(d)} = \mathbb{Q}_{Z_p\mathsf{X}(m)^{(d)}} = \mathbb{Q}_{Z_p}\mathbb{Q}^{(d)}$.

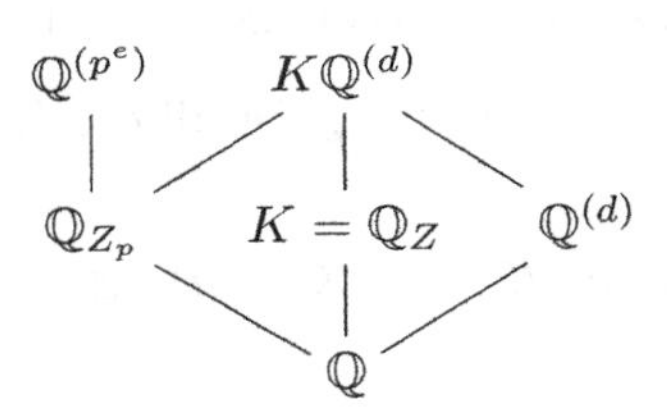

Since $\mathbb{Q}_{Z_p} \subset \mathbb{Q}_{\mathsf{X}(m)_p} = \mathbb{Q}^{(p^e)}$, p is fully ramified in $\mathbb{Q}_{Z_p}$. As p is unramified in $\mathbb{Q}^{(d)}$, 2.13.5.3 implies that all prime divisors of p in K are unramified in $K\mathbb{Q}^{(d)}$, and all prime divisors of p in $\mathbb{Q}_{Z_p}$ are unramified in $\mathbb{Q}_{Z_p}\mathbb{Q}^{(d)} = K\mathbb{Q}^{(d)}$. Hence $e_p(K) = e_p(K\mathbb{Q}^{(d)}) = e_p(\mathbb{Q}_{Z_p}\mathbb{Q}^{(d)}) = e_p(\mathbb{Q}_{Z_p}) = [\mathbb{Q}_{Z_p} : \mathbb{Q}] = |Z_p|$.

In particular, p is unramified in K if and only if $Z_p = \mathbf{1}$, and this holds if and only if $p \nmid f(\chi)$ for all $\chi \in Z$. □

Theorem 3.3.10. *Let $m \in \mathbb{N}$, $m \ge 3$, and let Z be a subgroup of $\mathsf{X}(m)$. Let $K = \mathbb{Q}_Z \subset \mathbb{Q}^{(m)}$, and let p be a prime. Set*

$$Y = \{\chi \in Z \mid \chi'(p) \ne 0\} \quad \textit{and} \quad W = \{\chi \in Z \mid \chi'(p) = 1\}$$

(where χ' denotes the associated primitive character of χ). Then $W \subset Y \subset Z$ are subgroups, $e_p(K) = (Z : Y)$, $f_p(K) = (Y : W)$ and $r_p(K) = |W|$. Let G_p be the decomposition group and I_p the inertia group of p in K. Then $\mathsf{X}(G_p) \cong Z/W$, $\mathsf{X}(I_p) \cong Z/Y$ and Y/W is a cyclic group.

Proof. Let D_p be the decomposition field and T_p the inertia field of p in K (see 2.13.6). Then T_p is the largest subfield of K in which p is unramified. If $\chi \in Z$, then $\chi'(p) \neq 0$ if and only if $p \nmid f(\chi)$. Hence $Y = \{\chi \in Z \mid p \nmid f(\chi)\}$, and by 3.3.9 it follows that $T_p = \mathbb{Q}_Y$.

If $d = \mathrm{lcm}(\{f(\chi) \mid \chi \in Y\})$, then we get $Y \subset \mathsf{X}(m)^{(d)} = \widehat{\pi}_{m,d}(\mathsf{X}(d))$ and $T_p \subset \mathbb{Q}^{(d)}$. In what follows we identify $\mathsf{X}(m)^{(d)}$ with $\mathsf{X}(d)$ by means of $\widehat{\pi}_{m,d}$. Then $T_p = \mathbb{Q}_Y$ (built inside $\mathbb{Q}^{(d)}$), and there exists a subgroup W of Y such that $D_p = \mathbb{Q}_W$. Explicitly,

$$Y = \{\chi \in \mathsf{X}(d) \mid \chi \restriction \sigma_d^{-1}\mathrm{Gal}(\mathbb{Q}^{(d)}/T_p) = 1\}.$$

Let $\varphi_p \in \mathrm{Gal}(T_p/W_p)$ be the Frobenius automorphism. Using 2.14.6 and 3.2.6.3 we obtain $\varphi_p = \sigma_d(p) \restriction T_p$ and $\mathrm{Gal}(T_p/D_p) = \langle \varphi_p \rangle$. Hence

$$\begin{aligned} W = \{\chi \in \mathsf{X}(d) \mid \chi \restriction \sigma_d^{-1}\mathrm{Gal}(\mathbb{Q}^{(d)}/D_p) = 1\} &= \{\chi \in Y \mid \chi \restriction \sigma_d^{-1}(\langle \varphi_p \rangle) = 1\} \\ &= \{\chi \in Y \mid \chi(p) = 1\} \end{aligned}$$

From now on we view again all fields as subfields of $\mathbb{Q}^{(m)}$. Using 3.3.8 and 3.3.3 again and again, we get isomorphisms

$$\begin{aligned} G_p &= \mathrm{Gal}(\mathbb{Q}^{(m)}/D_p)/\mathrm{Gal}(\mathbb{Q}^{(m}/K) \cong W^\perp/Z^\perp \\ &\cong \mathsf{X}(Z/W), \text{hence} \quad \mathsf{X}(G_p) \cong Z/W \end{aligned}$$

and

$$\begin{aligned} I_p &= \mathrm{Gal}(\mathbb{Q}^{(m)}/T_p)/\mathrm{Gal}(\mathbb{Q}^{(m}/K) \cong Y^\perp/Z^\perp \\ &\cong \mathsf{X}(Z/Y), \text{hence} \quad \mathsf{X}(I_p) \cong Z/Y. \end{aligned}$$

Moreover, $G_p/I_p \cong (W^\perp/Z^\perp)/(Y^\perp/Z^\perp) \cong W^\perp/Y^\perp \cong \mathsf{X}(Y/W) \cong Y/W$, and therefore Y/W is cyclic by 2.13.8.3.

By 2.13.4.1, we get $f_p(K) = (G_p\!:\!I_p) = (Y\!:\!W)$, $e_p(K) = |I_p| = (Z\!:\!Y)$ and

$$r_p(K) = (G\!:\!G_p) = \frac{|Z|}{|Z/W|} = |W|. \qquad \square$$

3.4 Quadratic characters and quadratic reciprocity

Let $m \in \mathbb{N}$. A (Dirichlet) character $\chi \in \mathsf{X}(m)$ is called

- **real** if $\chi(\mathbb{Z}) \subset \{0, \pm 1\}$,
- **quadratic** if $\mathrm{ord}(\chi) = 2$ [equivalently, χ is real and $\chi \neq 1_m$].

If p is an odd prime, then the extended Legendre symbol $\varphi_p\colon \mathbb{Z} \to \{0, \pm 1\}$ (see 3.2.3) is the unique quadratic character modulo p, and it is primitive.

Theorem 3.4.1.

1. *Let p be an odd prime, $p^* = (-1)^{(p-1)/2}p$ and $n \in \mathbb{N}$. Then $\mathbb{Q}(\sqrt{p^*}\,)$ is the unique quadratic subfield of $\mathbb{Q}^{(p^n)}$.*

2. *If K is any quadratic number field, then $K \subset \mathbb{Q}^{(|\mathsf{d}_K|)}$.*

Proof. We tacitly use 3.2.4 and 3.1.8.

1. $\mathbb{Q}^{(p^n)}/\mathbb{Q}$ is a cyclic field extension of degree $p^{n-1}(p-1)$, and thus it has a unique quadratic subfield K. By 3.2.6.1 p is the only prime which is ramified $\mathbb{Q}^{(p^n)}$. Hence at most p is ramified in K, and therefore $\mathsf{d}_K = p^*$.

2. Let $\mathsf{d}_K = 2^e d$, where $e \in \{0, 2, 3\}$, $d \in \mathbb{Z}$ and $|d| = p_1 \cdot \ldots \cdot p_r$ for some $r \in \mathbb{N}_0$ and distinct odd primes $p_1, \ldots, p_r$. For all $i \in [1, r]$, we set $p_i^* = (-1)^{p_i-1)/2}p_i$ and obtain $(-1)^{(d-1)/2}d = p_1^* \cdot \ldots \cdot p_r^* \equiv 1 \bmod 4$.

CASE 1: $e = 0$. Then $\mathsf{d}_K = p_1^* \cdot \ldots \cdot p_r^*$ and $|\mathsf{d}_K| = p_1 \cdot \ldots \cdot p_r$. Consequently, $\mathbb{Q}^{(|\mathsf{d}_K|)} = \mathbb{Q}^{(p_1)} \cdot \ldots \cdot \mathbb{Q}^{(p_r)} \supset \mathbb{Q}(\sqrt{p_1^*}, \ldots, \sqrt{p_r^*}\,) \supset \mathbb{Q}(\sqrt{\mathsf{d}_K}\,)$.

CASE 2: $e = 2$. Then $\mathsf{d}_K = -4p_1^* \cdot \ldots \cdot p_r^*$ and $|\mathsf{d}_K| = 4p_1 \cdot \ldots \cdot p_r$. Consequently,

$$\mathbb{Q}^{(|\mathsf{d}_K|)} = \mathbb{Q}^{(4)}\mathbb{Q}^{(p_1)} \cdot \ldots \cdot \mathbb{Q}^{(p_r)} \supset \mathbb{Q}(\sqrt{-1}, \sqrt{p_1^*}, \ldots, \sqrt{p_r^*}) \supset \mathbb{Q}(\sqrt{\mathsf{d}_K}\,).$$

CASE 3: $e = 3$. Then $\mathsf{d}_K = 8(-1)^{(d-1)/2}p_1^* \cdot \ldots \cdot p_r^*$ and therefore $|\mathsf{d}_K| = 8p_1 \cdot \ldots \cdot p_r$. Consequently,

$$\mathbb{Q}^{(|\mathsf{d}_K|)} = \mathbb{Q}^{(8)}\mathbb{Q}^{(p_1)} \cdot \ldots \cdot \mathbb{Q}^{(p_r)} \supset \mathbb{Q}(\sqrt{2}, \sqrt{-2}, \sqrt{p_1^*}, \ldots, \sqrt{p_r^*}\,) \supset \mathbb{Q}(\sqrt{\mathsf{d}_K}\,). \ \square$$

The following Theorem 3.4.2 contains the famous quadratic reciprocity law of Gauss. Here we present a proof which depends on the embedding of a quadratic number field into a cyclotomic number field and a comparison of the factorization laws using the Artin symbol. Certainly, this proof is neither the shortest nor the simplest one, but it gives an idea of why the quadratic reciprocity law can be viewed as an origin of algebraic number theory. This will even become more evident in the forthcoming volume on class field theory; there the quadratic reciprocity law will appear as a special case of the general reciprocity law of class field theory.

Later, after 3.4.8, we shall give a quick and simple proof of the quadratic reciprocity law using Gauss sums.

Theorem 3.4.2 (Quadratic Reciprocity Law).

1. *Let p be an odd prime. Then*

$$\Big(\frac{-1}{p}\Big) = (-1)^{(p-1)/2} = \begin{cases} 1 & \text{if } \ p \equiv 1 \bmod 4, \\ -1 & \text{if } \ p \equiv -1 \bmod 4, \end{cases}$$

and

$$\left(\frac{2}{p}\right) = (-1)^{(p^2-1)/8} = \begin{cases} 1 & \textit{if} \quad p \equiv \pm 1 \bmod 8, \\ -1 & \textit{if} \quad p \equiv \pm 5 \bmod 8. \end{cases}$$

2. Let p and q be distinct odd primes and $p^* = (-1)^{(p-1)/2}p$. Then

$$\left(\frac{p^*}{q}\right) = \left(\frac{q}{p}\right) \quad \textit{and}$$

$$\left(\frac{p}{q}\right)\left(\frac{q}{p}\right) = (-1)^{\frac{p-1}{2}\frac{q-1}{2}} = \begin{cases} -1 & \textit{if} \ p \equiv q \equiv -1 \bmod 4, \\ 1 & \textit{otherwise.} \end{cases}$$

Proof. 1. We have already seen in 3.2.3 that

$$\left(\frac{-1}{p}\right) = (-1)^{(p-1)/2} = \begin{cases} 1 & \text{if} \quad p \equiv 1 \bmod 4, \\ -1 & \text{if} \quad p \equiv -1 \bmod 4 \end{cases}$$

To prove the second formula, we consider the cyclotomic field $\mathbb{Q}^{(8)} = \mathbb{Q}(\zeta_8)$ that $\zeta_8 = (1+\mathrm{i})/\sqrt{2}$, $\sqrt{2} = \zeta_8 + \zeta_8^{-1} \in \mathbb{Q}^{(8)}$, and $(p, \mathbb{Q}^{(8)}/\mathbb{Q})(\sqrt{2}) = \zeta_8^p + \zeta_8^{-p}$ by 3.2.6.3. Now we use 2.14.9 and obtain:

$$\left(\frac{2}{p}\right) = 1 \iff p \ \text{ is decomposed in } \ \mathbb{Q}(\sqrt{2}\,) \iff (p, \mathbb{Q}^{(8)}) \restriction \mathbb{Q}(\sqrt{2}) = 1$$
$$\iff (p, \mathbb{Q}^{(8)})(\sqrt{2}) = \sqrt{2} \iff \zeta_8^p + \zeta_8^{-p} = \zeta_8 + \zeta_8^{-1}$$
$$\iff p \equiv \pm 1 \bmod 8.$$

Hence it follows that

$$\left(\frac{2}{p}\right) = (-1)^{(p^2-1)/8} = \begin{cases} 1 & \text{if} \quad p \equiv \pm 1 \bmod 8, \\ -1 & \text{if} \quad p \equiv \pm 5 \bmod 8. \end{cases}$$

2. Let D_q be the decomposition field of q in $\mathbb{Q}^{(p)}$, and let $f \in \mathbb{N}$ be minimal such that $q^f \equiv 1 \bmod p$. Then $f = f_q(\mathbb{Q}^{(p)}) = [\mathbb{Q}^{(p)} : D_q]$, and by 3.4.1.1 we obtain $\mathbb{Q}(\sqrt{p^*}\,) \subset \mathbb{Q}^{(p)}$. By 2.13.4.4 it follows that q is (fully) decomposed in $\mathbb{Q}(\sqrt{p^*}\,)$ if and only if $\mathbb{Q}(\sqrt{p^*}\,) \subset D_q$, and as $\mathbb{Q}^{(p)}/\mathbb{Q}$ is cyclic, this holds if and only if $f \mid (p-1)/2$. Now we obtain:

$$\left(\frac{p^*}{q}\right) = 1 \iff q \ \text{ is decomposed in } \ \mathbb{Q}(\sqrt{p^*}\,) \iff f \,\Big|\, \frac{p-1}{2}$$
$$\iff q^{(p-1)/2} \equiv 1 \bmod p \iff \left(\frac{q}{p}\right) = 1.$$

Hence it follows that

$$\left(\frac{q}{p}\right) = \left(\frac{p^*}{q}\right) = \left(\frac{-1}{q}\right)^{(p-1)/2}\left(\frac{q}{p}\right)$$

and

$$\Big(\frac{p}{q}\Big)\Big(\frac{q}{p}\Big) = (-1)^{\frac{p-1}{2}\frac{q-1}{2}} = \begin{cases} -1 & \text{if } \ p \equiv q \equiv -1 \bmod 4, \\ 1 & \text{otherwise.} \end{cases}$$ □

In order to interpret the Artin symbol for composite modules in the context of quadratic number fields and to obtain a corresponding reciprocity law we first extend the Legendre symbol to the Kronecker symbol, and then we consider quadratic characters and Gauss sums in this context.

Let $m,\, n \in \mathbb{Z}^\bullet$, and set $m = 2^\beta m_1$, where $\beta \in \mathbb{N}_0$ and $m_1 \in \mathbb{Z}$. Assume that $|m_1| = p_1 \cdot \ldots \cdot p_r$ for some $r \in \mathbb{N}_0$ and (not necessarily distinct) odd primes $p_1, \ldots, p_r$. If $(n,m) = 1$, then we define the **Kronecker symbol** using Legendre symbols by

$$\Big(\frac{n}{m}\Big) = \varepsilon(m,n)(-1)^{\beta(n^2-1)/8}\prod_{i=1}^{r}\Big(\frac{n}{p_i}\Big), \quad \text{where}$$

$$\varepsilon(m,n) = \begin{cases} -1 & \text{if } \ m < 0 \text{ and } n < 0\,, \\ 1 & \text{otherwise.} \end{cases}$$

$$\text{If } \ (m,n) > 1, \quad \text{we set } \ \Big(\frac{n}{m}\Big) = 0.$$

If $m \in \mathbb{N}$ is odd, then the Kronecker symbol is called the **Jacobi symbol**, and using the elementary properties of the Legendre symbol we obtain for all $n \in \mathbb{Z}$:

$$\Big(\frac{n}{m}\Big) = \prod_{i=1}^{r}\Big(\frac{n}{p_i}\Big).$$

In particular: If n is a quadratic residue modulo m, then

$$\Big(\frac{n}{m}\Big) = 1$$

(the converse holds if m is a prime, but not in general), and

$$\Big(\frac{n}{m}\Big) = \Big(\frac{n'}{m}\Big) \quad \text{if } \ n' \in \mathbb{Z} \ \text{ and } \ n \equiv n' \bmod m.$$

Theorem 3.4.3 (Properties of the Kronecker symbol)**.** *Let m and n be non-zero coprime integers.*

1. *If m and n are odd, then*

$$\Big(\frac{-1}{m}\Big) = (-1)^{(m-1)/2} \quad \textit{and}$$

$$\Big(\frac{n}{2}\Big) = (-1)^{(n^2-1)/8} = \Big(\frac{2}{|n|}\Big) = \Big(\frac{2}{n}\Big).$$

2. If $m = m'm''$ and $n = n'n''$, where $m', m'', n', n'' \in \mathbb{Z}$, then

$$\Big(\frac{n}{m}\Big) = \Big(\frac{n}{m'}\Big)\Big(\frac{n}{m''}\Big) = \Big(\frac{n'}{m}\Big)\Big(\frac{n''}{m}\Big) \quad \textit{and} \quad \Big(\frac{n}{-1}\Big) = \operatorname{sgn}(n).$$

3. If $m = 2^\beta m_1$ and $n = 2^\delta n_1$, where $\beta, \delta \in \mathbb{N}_0$, $m_1, n_1 \in \mathbb{Z}$ and $2 \nmid m_1 n_1$, then

$$\Big(\frac{n}{m}\Big)\Big(\frac{m}{n}\Big) = \varepsilon(m,n)\,(-1)^{\frac{m_1-1}{2}\,\frac{n_1-1}{2}}.$$

Proof. We will again and again use the following simple argument: If $n \in \mathbb{N}$ is odd and $n = n_1 \cdot \ldots \cdot n_k$ for some $k \ge 1$ and $n_1, \ldots, n_k \in \mathbb{N}$, then

$$\sum_{i=1}^k \frac{n_i - 1}{2} \equiv \frac{n-1}{2} \mod 2 \quad \text{and} \quad \sum_{i=1}^k \frac{n_i^2 - 1}{8} \equiv \frac{n^2-1}{8} \mod 2.$$

1. Let m and n be odd. Then

$$\begin{aligned}\Big(\frac{-1}{m}\Big) &= \operatorname{sgn}(m)\prod_{i=1}^r \Big(\frac{-1}{p_i}\Big) = \operatorname{sgn}(m)\prod_{i=1}^r (-1)^{p_i-1)/2} \\ &= \operatorname{sgn}(m)(-1)^{(|m|-1)/2} = (-1)^{(m-1)/2},\end{aligned}$$

and if $|n| = q_1 \cdot \ldots \cdot q_s$ for some $s \in \mathbb{N}_0$ and odd primes $q_1, \ldots, q_s$, then

$$\Big(\frac{2}{n}\Big) = \Big(\frac{2}{|n|}\Big) = \prod_{j=1}^s \Big(\frac{2}{q_j}\Big) = \prod_{j=1}^s (-1)^{(q_j^2-1)/8} = (-1)^{(n^2-1)/8} = \Big(\frac{n}{2}\Big),$$

where the last equality holds by the very definition.

2. Obvious by the definition.

3. By symmetry, and since $(m,n) = 1$, we may assume that $\delta = 0$, $n = n_1$, and we set $|m| = p_1 \cdot \ldots \cdot p_r$ and $|n| = q_1 \cdot \ldots \cdot q_s$, where $r, s \in \mathbb{N}_0$ and $p_1, \ldots, p_r, q_1, \ldots, q_s$ are odd primes. Then we obtain, using 3.4.2,

$$\begin{aligned}\Big(\frac{n}{m}\Big)\Big(\frac{m}{n}\Big) &= \varepsilon(m,n)(-1)^{\beta(n^2-1)/8}\prod_{i=1}^r \Big(\frac{n}{p_i}\Big)\,\varepsilon(n,m)\prod_{j=1}^s \Big(\frac{m}{q_j}\Big) \\ &= (-1)^{\beta(n^2-1)/8}\prod_{i=1}^r \Big[\Big(\frac{\operatorname{sgn}(n)}{p_i}\Big)\prod_{j=1}^s \Big(\frac{q_j}{p_i}\Big)\Big]\prod_{j=1}^s \Big[\Big(\frac{\operatorname{sgn}(m)}{q_j}\Big)\Big(\frac{2}{q_j}\Big)^\beta \prod_{i=1}^r \Big(\frac{p_i}{q_j}\Big)\Big] \\ &= (-1)^{\beta(n^2-1)/8}\prod_{i=1}^r \operatorname{sgn}(n)^{(p_i-1)/2}\prod_{j=1}^s \Big[\operatorname{sgn}(m)^{(q_j-1)2}(-1)^{\beta(q_j^2-1)/8}\Big] \\ &\quad \cdot \prod_{i=1}^r\prod_{j=1}^s \Big(\frac{q_j}{p_i}\Big)\Big(\frac{p_i}{q_j}\Big)\end{aligned}$$

$$= \operatorname{sgn}(n)^{(|m_1|-1)/2}\operatorname{sgn}(m)^{(|n|-1)/2}\prod_{i=1}^{r}\prod_{j=1}^{s}(-1)^{\frac{p_i-1}{2}\frac{q_j-1}{2}}$$

$$= \operatorname{sgn}(n)^{(|m_1|-1)/2}\operatorname{sgn}(m)^{(|n|-1)/2}(-1)^{\frac{|m_1|-1}{2}\frac{|n|-1}{2}}.$$

If $m > 0$, then

$$\Big(\frac{n}{m}\Big)\Big(\frac{m}{n}\Big) = \Big[\operatorname{sgn}(n)(-1)^{(|n|-1)/2}\Big]^{(m_1-1)/2} = (-1)^{\frac{m_1-1}{2}\frac{n-1}{2}}.$$

If $n > 0$, then

$$\Big(\frac{n}{m}\Big)\Big(\frac{m}{n}\Big) = \Big[\operatorname{sgn}(m)(-1)^{(|m_1|-1)/2}\Big]^{(n-1)/2} = (-1)^{\frac{m_1-1}{2}\frac{n-1}{2}}.$$

If $m < 0$ and $n < 0$, then

$$\Big(\frac{n}{m}\Big)\Big(\frac{m}{n}\Big) = (-1)^S,$$

where

$$S = \frac{m_1-1}{2} + \frac{n-1}{2} + \frac{-m_1-1}{2}\,\frac{-n_1-1}{2} \equiv 1 + \frac{m_1-1}{2}\,\frac{n_1-1}{2} \mod 2,$$

and therefore

$$\Big(\frac{n}{m}\Big)\Big(\frac{m}{n}\Big) = \varepsilon(m,n)(-1)^{\frac{m_1-1}{2}\frac{n-1}{2}}. \qquad \square$$

Definition 3.4.4. An integer $d \in \mathbb{Z}$ is called a **quadratic discriminant** if d is not a square and $d \equiv 0$ or $1 \bmod 4$. Then $K = \mathbb{Q}(\sqrt{d})$, and if $d = \mathsf{d}_K$, then d is called a **fundamental discriminant**.

It follows from 3.1.8.2 that a discriminant d is a fundamental discriminant if and only if

- either d is squarefree and $d \equiv 1 \bmod 4$,
- or $d = 4D$ for some squarefree integer $D \not\equiv 1 \bmod 4$.

Every quadratic discriminant d has a unique factorization $d = d_0 f^2$, where d_0 is a fundamental discriminant and $f \in \mathbb{N}$. This factorization is obtained as follows: If d^* is the squarefree kernel of d, then $d_0 = d$ if $d \equiv 1 \bmod 4$, and $d_0 = 4d$ otherwise. We call d_0 the **associated fundamental discriminant** and f the **conductor** of d. If $K = \mathbb{Q}(\sqrt{d})$, then $d_0 = \mathsf{d}_K$.

For a quadratic discriminant d, we define the **associated quadratic character** $\chi_d \colon \mathbb{Z} \to \{\pm 1, 0\}$ with the aid of the Kronecker symbol by

$$\chi_d(a) = \Big(\frac{d}{a}\Big).$$

If K is a quadratic number field K, then we call

$$\chi_K = \chi_{\mathsf{d}_K} = \Big(\frac{\mathsf{d}_K}{\cdot}\Big) \quad \text{the **character** of } K.$$

The following Theorem 3.4.5 shows that the quadratic character actually controls the Artin symbol for composite modules.

Theorem 3.4.5 (Quadratic characters). *Let $d \in \mathbb{Z}$ be a quadratic discriminant.*

1. $\chi_d \in \mathsf{X}(|d|)$ is a quadratic character modulo $|d|$. If $a \in \mathbb{Z}(|d|)$, then $\chi_d(a)$ is given as follows:

(a) If $d \equiv 1 \bmod 4$, then

$$\chi_d(a) = \Big(\frac{a}{|d|}\Big).$$

(b) If $d = 2^e D$, where $e \in \mathbb{N}$, $e \geq 2$, $D \in \mathbb{Z}$ and $2 \nmid D$, then

$$\chi_d(a) = (-1)^{\frac{a-1}{2}\frac{D-1}{2}} \Big(\frac{a}{|d|}\Big).$$

2. Let K be a quadratic number field. Then $K \subset \mathbb{Q}^{(|\mathsf{d}_K|)}$, and for $a \in \mathbb{Z}(|\mathsf{d}_K|)$ let $\sigma_a \in \mathrm{Gal}(\mathbb{Q}^{(|\mathsf{d}_K|)}/\mathbb{Q})$ be defined by $\sigma_a(\xi) = \xi^a$ for all $\xi \in \mu_{|\mathsf{d}_K|}(K)$. Then

$$\chi_K(a) = \frac{\sigma_a(\sqrt{\mathsf{d}_K}\,)}{\sqrt{\mathsf{d}_K}} = \varepsilon(a, \mathsf{d}_K)\frac{(a, K)(\sqrt{\mathsf{d}_K}\,)}{\sqrt{\mathsf{d}_K}},$$

where

$$\varepsilon(a, \mathsf{d}_K) = \begin{cases} -1 & \text{if } a < 0 \text{ and } \mathsf{d}_K < 0, \\ 1 & \text{otherwise.} \end{cases}$$

$\chi_K \in \mathsf{X}(|\mathsf{d}_K|)$ is primitive, $\chi_K(-1) = \mathrm{sgn}(\mathsf{d}_K)$, and $K = \mathbb{Q}_{\chi_K}$ according to 3.3.8.

Proof. 1. Let $a \in \mathbb{Z}(|d|)$. We tacitly use 3.4.3.

(a) Let $d \equiv 1 \bmod 4$. Then

$$\chi_d(a) = \Big(\frac{d}{a}\Big) = \Big(\frac{a}{d}\Big)\varepsilon(a, d) = \Big(\frac{a}{|d|}\Big)\Big(\frac{a}{\mathrm{sgn}(d)}\Big)\varepsilon(a, d) = \Big(\frac{a}{|d|}\Big).$$

If $a,\, b \in \mathbb{Z}$, then $\chi_d(ab) = \chi_d(a)\chi_d(b)$; if $a \equiv b \bmod |d|$, then $\chi_d(a) = \chi_d(b)$; and if $a \equiv 1 \bmod |d|$, then $\chi_d(a) = 1$. Hence $\chi_d \in \mathsf{X}(|d|)$.

(b) By definition

$$\begin{aligned}\chi_d(a) &= \Big(\frac{d}{a}\Big) = (-1)^{\frac{a-1}{2}\frac{D-1}{2}} \Big(\frac{a}{d}\Big)\varepsilon(a, d) = (-1)^{\frac{a-1}{2}\frac{D-1}{2}} \Big(\frac{a}{d}\Big)\Big(\frac{a}{\mathrm{sgn}(d)}\Big) \\ &= (-1)^{\frac{a-1}{2}\frac{D-1}{2}} \Big(\frac{a}{|d|}\Big).\end{aligned}$$

2. We use 3.2.6.3 and 3.2.4. We know that $K \subset \mathbb{Q}^{(|\mathsf{d}_K|)}$, and for a prime p such that $p \nmid \mathsf{d}_K$ we obtain

$$\chi_K(p) = \Big(\frac{\mathsf{d}_K}{p}\Big) = \frac{(p,K)(\sqrt{\mathsf{d}_K}\,)}{\sqrt{\mathsf{d}_K}} = \frac{(p,\mathbb{Q}^{(m)})(\sqrt{\mathsf{d}_K}\,)}{\sqrt{\mathsf{d}_K}} = \frac{\sigma_p(\sqrt{\mathsf{d}_K}\,)}{\sqrt{\mathsf{d}_K}},$$

and

$$\chi_K(-1) = \Big(\frac{\mathsf{d}_K}{-1}\Big) = \operatorname{sgn}(\mathsf{d}_K) = \frac{\overline{\sqrt{\mathsf{d}_K}}}{\sqrt{\mathsf{d}_K}} = \frac{\sigma_{-1}(\sqrt{\mathsf{d}_K})}{\sqrt{\mathsf{d}_K}}.$$

Assume now that $a \in \mathbb{Z}(|\mathsf{d}_K|)$ and $|a| = p_1 \cdot \ldots \cdot p_r$ for some $r \in \mathbb{N}_0$ and primes $p_1, \ldots, p_r$. If $a > 0$, then

$$\begin{aligned}\chi_K(a) &= \Big(\frac{\mathsf{d}_K}{a}\Big) = \prod_{i=1}^{r}\Big(\frac{\mathsf{d}_K}{p_i}\Big) = \prod_{i=1}^{r}\frac{\sigma_{p_i}(\sqrt{\mathsf{d}_K}\,)}{\sqrt{\mathsf{d}_K}} = \prod_{i=1}^{r}\frac{\sigma_{p_i}\circ(\sigma_{p_{i-1}}\circ\ldots\circ\sigma_{p_1})(\sqrt{d})}{(\sigma_{p_{i-1}}\circ\ldots\circ\sigma_{p_1})(\sqrt{d})}\\ &= \frac{\sigma_{p_r}\circ\sigma_{p_{r-1}}\circ\ldots\circ\sigma_{p_1}(\sqrt{d})}{\sqrt{d}} = \frac{\sigma_a(\sqrt{\mathsf{d}_K}\,)}{\sqrt{\mathsf{d}_K}} = \frac{(a,\mathbb{Q}^{(m)})(\sqrt{\mathsf{d}_K}\,)}{\sqrt{\mathsf{d}_K}}\\ &= \frac{(a,K)(\sqrt{\mathsf{d}_K}\,)}{\sqrt{\mathsf{d}_K}},\end{aligned}$$

and

$$\begin{aligned}\chi_K(-a) &= \Big(\frac{\mathsf{d}_K}{-a}\Big) = \Big(\frac{\mathsf{d}_K}{-1}\Big)\Big(\frac{\mathsf{d}_K}{a}\Big) = \operatorname{sgn}(\mathsf{d}_K)\chi_K(a)\\ &= \frac{\sigma_{-1}(\sqrt{\mathsf{d}_K}\,)}{\sqrt{\mathsf{d}_K}}\,\frac{\sigma_a(\sqrt{\mathsf{d}_K}\,)}{\sqrt{\mathsf{d}_K}} = \frac{\sigma_{-1}\circ\sigma_a(\sqrt{\mathsf{d}_K}\,)}{\sigma_a(\sqrt{\mathsf{d}_K}\,)}\,\frac{\sigma_a(\sqrt{\mathsf{d}_K}\,)}{\sqrt{\mathsf{d}_K}} = \frac{\sigma_{-a}(\sqrt{\mathsf{d}_K}\,)}{\sqrt{\mathsf{d}_K}}.\end{aligned}$$

Recall that the isomorphism $\sigma_{|\mathsf{d}_K|} \colon (\mathbb{Z}/\mathsf{d}_K\mathbb{Z})^\times \stackrel{\sim}{\to} \operatorname{Gal}(\mathbb{Q}^{(|\mathsf{d}_K|)}/\mathbb{Q})$ is defined by $\sigma_{|\mathsf{d}_K|}(a + \mathsf{d}_K\mathbb{Z}) = \sigma_a$ for all $a \in \mathbb{Z}(|\mathsf{d}_K|)$. Therefore we obtain

$$\begin{aligned}\sigma_{|\mathsf{d}_K|}(\langle\chi_K\rangle^\perp) &= \{\sigma_a \mid a \in \mathbb{Z}(|\mathsf{d}_K|),\ \chi_K(a) = 1\}\\ &= \{\sigma_a \mid a \in \mathbb{Z}(|\mathsf{d}_K|),\ \sigma_a \restriction K = 1\} \ = \operatorname{Gal}(\mathbb{Q}^{(|\mathsf{d}_K|}/K),\end{aligned}$$

and consequently $K = \mathbb{Q}_{\chi_K}$.

It remains to prove that χ_K is primitive. Let $\mathsf{d}_K = 2^eD$, where $e \in \{0,2,3\}$ and $D \in \mathbb{Z}$ is odd and squarefree. Let $k \in \mathbb{N}$ be such that $k \mid \mathsf{d}_K$ and $k < |\mathsf{d}_K|$. We must prove that there exists some $a \in \mathbb{Z}(|\mathsf{d}_K|)$ such that $a \equiv 1 \bmod k$ and $\chi_K(a) = -1$. We apply the Chinese remainder theorem 1.1.4.

CASE 1: $D \nmid k$. Then there exists some prime p such that $p \mid D$ and $p \nmid k$. Let $a \in \mathbb{Z}(|\mathsf{d}_K|)$ be such that

$$a \equiv 1 \bmod \frac{|\mathsf{d}_K|}{p} \quad\text{and}\quad \Big(\frac{a}{p}\Big) = -1.$$

Then $a \equiv 1 \bmod k$ and

$$\chi_K(a) = \Big(\frac{a}{|\mathsf{d}_K|}\Big) = \Big(\frac{a}{|\mathsf{d}_K|/p}\Big)\Big(\frac{a}{p}\Big) = -1.$$

CASE 2: $D \mid k$. Then $e \in \{2,3\}$, and $k = 2^c|D|$ for some $c \in [1, e-1]$.

CASE 2a: $e = 3$: Let $a \in \mathbb{Z}$ be such that $a \equiv 5 \mod 8$ and $a \equiv 1 \mod |D|$. Then $a \equiv 1 \mod k$, and

$$\chi_K(a) = (-1)^{\frac{a-1}{2}\frac{D-1}{2}}\Big(\frac{a}{|\mathsf{d}_K|}\Big) = \Big(\frac{a}{2|D|}\Big)$$
$$= (-1)^{(a^2-1)/8}\Big(\frac{a}{|D|}\Big) = (-1)^{(a^2-1)/8} = -1.$$

CASE 2b: $e = 2$: Observe that $D \equiv 3 \mod 4$. If $a \in \mathbb{Z}$, $a \equiv 3 \mod 4$ and $a \equiv 1 \mod |D|$, then $a \equiv 1 \mod k$, and

$$\chi_K(a) = (-1)^{\frac{a-1}{2}\frac{D-1}{2}}\Big(\frac{a}{|\mathsf{d}_K|}\Big) = -\Big(\frac{a}{|D|}\Big) = -1. \qquad \square$$

As already announced, we now introduce Gauss sums. Using them, we shall obtain a short proof of the quadratic reciprocity law. The deeper arithmetical relevance of Gauss sums however comes from their role in the analytic theory of algebraic number fields (see 4.2.9).

Definition 3.4.6. Let $m \in \mathbb{N}$ and $\zeta_m = \mathsf{e}^{2\pi\mathsf{i}/m}$. For a character $\chi \in \mathsf{X}(m)$ and $a \in \mathbb{Z}$ we define the **Gauss sum** with parameters χ and a by

$$\tau(\chi, a) = \sum_{t\in(\mathbb{Z}/m\mathbb{Z})^\times} \chi(t)\zeta_m^{at}, \quad \text{and we set } \tau(\chi) = \tau(\chi,1) = \sum_{t\in(\mathbb{Z}/m\mathbb{Z})^\times} \chi(t)\zeta_m^{t}.$$

By definition, if $d = \mathrm{lcm}\{m, \varphi(m)\}$, then $\tau(\chi, a) \in \mathbb{Z}[\zeta_d] = \mathfrak{o}_{\mathbb{Q}^{(d)}}$. Obviously, $\tau(\chi,a)$ depends only on the residue class $a + m\mathbb{Z}$, and for every $c \in \mathbb{Z}$ we obtain

$$\tau(\chi, a) = \sum_{j=c}^{c+m-1} \chi(j)\zeta_m^{aj}.$$

In particular, if $\chi \neq 1_m$ and $a \equiv 0 \mod m$, then $\tau(\chi, a) = 0$ by 3.3.5.1.

Theorem 3.4.7. *Let $m \in \mathbb{N}$, $a \in \mathbb{Z}$ and $\chi \in \mathsf{X}(m)$.*

1. $\overline{\tau(\chi, a)} = \chi(-1)\,\tau(\overline{\chi}, a)$.
2. *If either $(a, m) = 1$ or χ is primitive, then $\tau(\chi, a) = \overline{\chi}(a)\,\tau(\chi)$.*
3. *If χ is primitive, then $|\tau(\chi)| = \sqrt{m}$, and $\tau(\chi)\tau(\overline{\chi}) = \chi(-1)m$.*

Proof. 1. By definition,

$$\overline{\tau(\chi, a)} = \sum_{t\in(\mathbb{Z}/m\mathbb{Z})^\times} \overline{\chi}(t)\zeta_m^{-at} \underset{[s=-t]}{=} \sum_{s\in(\mathbb{Z}/m\mathbb{Z})^\times} \overline{\chi}(-s)\zeta_m^{as} = \chi(-1) \sum_{s\in(\mathbb{Z}/m\mathbb{Z})^\times} \overline{\chi}(s)\zeta_m^{as}$$
$$= \chi(-1)\tau(\overline{\chi}, a)\,.$$

2. If $(a,m)=1$, then

$$\chi(a)\,\tau(\chi,a)=\sum_{t\in(\mathbb{Z}/m\mathbb{Z})^\times}\chi(at)\zeta_m^{at}\underset{[s=at]}{=}\sum_{s\in(\mathbb{Z}/m\mathbb{Z})^\times}\chi(s)\zeta_m^s=\tau(\chi)$$

and therefore $\tau(\chi,a)=\overline{\chi}(a)\tau(\chi)$.

Now let χ be primitive and $(a,m)=d>1$. Since $\overline{\chi}(a)=0$, we must prove that $\tau(\chi,a)=0$. Let $a=db$ and $m=dk$, where $b\in\mathbb{Z}$, $k\in\mathbb{N}$, and let $c\in\mathbb{Z}(m)$ be such that $c\equiv 1 \mod k$ and $\chi(c)\neq 1$. If $t\in(\mathbb{Z}/m\mathbb{Z})^\times$, then

$$\zeta_m^{at}=\zeta_m^{dbt}=\zeta_k^{bt}=\zeta_k^{btc}=\zeta_m^{dbtc}=\zeta_m^{atc}$$

and

$$\begin{aligned}\chi(c)\,\tau(\chi,a)&=\sum_{t\in(\mathbb{Z}/m\mathbb{Z})^\times}\chi(ct)\zeta_m^{at}\\&=\sum_{t\in(\mathbb{Z}/m\mathbb{Z})^\times}\chi(ct)\zeta_m^{atc}\underset{[s=ct]}{=}\sum_{s\in(\mathbb{Z}/m\mathbb{Z})^\times}\chi(s)\zeta_m^{as}=\tau(\chi,a),\end{aligned}$$

which implies that $\tau(\chi,a)=0$.

3. Let χ be primitive. Then

$$\begin{aligned}|\tau(\chi)|^2&=\tau(\chi)\overline{\tau(\chi)}=\tau(\chi)\sum_{\nu=0}^{m-1}\overline{\chi}(\nu)\zeta_m^{-\nu}=\sum_{\nu=0}^{m-1}\tau(\chi,\nu)\zeta_m^{-\nu}=\sum_{\nu=0}^{m-1}\sum_{\mu=0}^{m-1}\chi(\mu)\zeta_m^{\mu\nu-\nu}\\&=\sum_{\mu=0}^{m-1}\chi(\mu)\sum_{\nu=0}^{m-1}\zeta_m^{(\mu-1)\nu}=m,\quad\text{since}\quad\sum_{\nu=0}^{m-1}\zeta_m^{(\mu-1)\nu}=\begin{cases}0&\text{if }\ \mu\neq1,\\m&\text{if }\ \mu=1.\end{cases}\end{aligned}$$

Hence $|\tau(\chi)|=\sqrt{m}$, and by 1. we obtain

$$m=|\tau(\chi)|^2=\tau(\chi)\overline{\tau(\chi)}=\chi(-1)\tau(\chi)\tau(\overline{\chi})\ \text{and}\ \tau(\chi)\tau(\overline{\chi})=\chi(-1)m.\quad\square$$

Theorem 3.4.8. *Let p be an odd prime, $m\in\mathbb{N}$, $p\nmid m$ and $\varphi\in\mathsf{X}(m)$ a primitive real character modulo m. Then*

$$\varphi(p)=\Big(\frac{\varphi(-1)m}{p}\Big).$$

Proof. We calculate $\tau(\varphi)\in\mathbb{Z}[\zeta_m]$ modulo $p\mathbb{Z}[\zeta_m]$. Since $\varphi^p=\varphi=\overline{\varphi}$ and $\mathbb{Z}[\zeta_m]/p\mathbb{Z}[\zeta_m]$ is a ring of characteristic p, we get

$$\begin{aligned}\tau(\varphi)^p&=\Big(\sum_{t\in(\mathbb{Z}/m\mathbb{Z})^\times}\varphi(t)\zeta_m^t\Big)^p\\&\equiv\sum_{t\in(\mathbb{Z}/m\mathbb{Z})^\times}\varphi(t)\zeta_m^{pt}\equiv\tau(\varphi,p)\equiv\varphi(p)\tau(\varphi)\ \text{mod}\ p\mathbb{Z}[\zeta_m],\end{aligned}$$

and therefore

$$[\tau(\varphi)^2]^{(p-1)/2}\tau(\varphi)^2 = \tau(\varphi)^{p+1} \equiv \varphi(p)\tau(\varphi)^2 \mod p\mathbb{Z}[\zeta_m]\,.$$

As $\tau(\varphi)^2 = \varphi(-1)m$ and $p \nmid m$ we may cancel $\tau(\varphi)^2$ in this congruence and obtain, using Euler's criterion (see 3.2.3)

$$\Big(\frac{\varphi(-1)m}{p}\Big) \equiv \big[\tau(\varphi)^2\big]^{(p-1)/2} \equiv \varphi(p) \mod p\,, \quad \text{and thus} \quad \Big(\frac{\varphi(-1)m}{p}\Big) = \varphi(p). \quad \square$$

Now we use 3.4.8 to give the promised simple proof of the quadratic reciprocity law.

1. Let p and q be distinct odd primes. The quadratic character $\varphi_p \in \mathsf{X}(p)$ is primitive, and by 3.4.8 it follows that

$$\Big(\frac{q}{p}\Big) = \varphi_p(q) = \Big(\frac{\varphi_p(-1)p}{q}\Big) = \Big(\frac{(-1)^{(p-1)/2}p}{q}\Big) = \Big(\frac{p^*}{q}\Big).$$

2. Let $\chi\colon (\mathbb{Z}/8\mathbb{Z})^\times \to \{\pm 1\}$ be defined by

$$\chi(a+8\mathbb{Z}) = (-1)^{(a^2-1)/8} \quad \text{for all} \quad a \in \mathbb{Z}(2).$$

Then $\chi \in \mathsf{X}(8)$ is a primitive quadratic character modulo 8, $\chi(-1) = 1$, and by 3.4.8 it follows that

$$\Big(\frac{2}{p}\Big) = \Big(\frac{\chi(-1)8}{p}\Big) = \chi(p) = (-1)^{(p^2-1)/8}.$$

3.5 The finiteness results for algebraic number fields

Among others, we prove in this section the two finiteness results for an algebraic number field K with signature (r_1, r_2):

> *The ideal class group* $\mathcal{C}_K$ *is finite* (see 3.5.7), *and the unit group* $U_K = \mathfrak{o}_K^\times$ *is the direct product of the finite group* $\mu(K)$ *of roots of unity in* K *and a free abelian group of rank* $r_1 + r_2 - 1$ (see 3.5.8).

The groups $\mathcal{C}_K$ and U_K fit into the exact sequence

$$\mathbf{1} \to U_K \hookrightarrow K^\times \xrightarrow{\partial} \mathcal{J}_K \to \mathcal{C}_K \to \mathbf{1}, \qquad (*)$$

where $\partial(x) = x\mathfrak{o}_K$ for all $x \in K^\times$. Note that $\mathrm{Im}(\partial) = \mathcal{H}_K \subset \mathcal{J}_K$ and $\mathcal{J}_K$ is the free abelian group generated by the set $\mathcal{P}_K$ of non-zero prime ideals of $\mathfrak{o}_K$. In particular, $\mathcal{I}_K = \mathsf{q}(\mathcal{I}'_K)$, $\partial(\mathfrak{o}_K) \subset \mathcal{I}'_K$, and $\mathcal{I}'_K$ is a free (and thus a reduced factorial) monoid. In this regard, the finiteness theorems together with the exact sequence $(*)$ describe the deviation from unique factorization in an algebraic number field.

We start with some geometric preliminaries. Let V be a real vector space and $\dim_{\mathbb{R}}(V) = n < \infty$. A subset Γ of V is called a **lattice** (in V) if Γ is a free abelian group with an $\mathbb{R}$-linearly independent basis. If Γ is a lattice in V, then $\mathrm{rk}(\Gamma) = \dim_{\mathbb{R}}(\mathbb{R}\Gamma) \le n$, every basis of Γ is linearly independent over $\mathbb{R}$, and by 2.3.6 every subgroup Γ' of Γ is also a lattice (called a **sublattice**).

Let Γ be a lattice in V, $\mathrm{rk}(\Gamma) = m$ and $(v_1, \ldots, v_m)$ a basis of Γ. Then the set

$$\mathcal{G} = \Big\{ \sum_{j=0}^{m} x_j v_j \,\Big|\, x_1, \ldots, x_m \in [0,1) \Big\} \subset \mathbb{R}\Gamma$$

is called a **fundamental parallelotope** of Γ,

$$\mathbb{R}\Gamma = \biguplus_{x \in \mathcal{G}} (x + \Gamma) = \biguplus_{u \in \Gamma} (u + \mathcal{G}),$$

and $\mathcal{G}$ is a set of representatives for $\mathbb{R}\Gamma/\Gamma$ in $\mathbb{R}\Gamma$.

If V is an Euclidean vector space, then $\mathcal{G}$ possesses an elementary m-dimensional volume

$$\mathrm{vol}(\Gamma) = \mathrm{vol}_m(\mathcal{G}) = \sqrt{\det(\langle v_i, v_j \rangle)_{i,j \in [1,m]}},$$

which does not depend on $(v_1, \ldots, v_m)$.

In particular, if $V = \mathbb{R}^n$ and $(e_{m+1}, \ldots, e_n)$ is an orthonormal basis of the orthogonal complement $(\mathbb{R}\Gamma)^\perp$, then $\mathrm{vol}(\Gamma) = |\det(v_1, \ldots, v_m, e_{m+1}, \ldots, e_n)|$.

Theorem 3.5.1. *Let V be a real vector space, $n = \dim_{\mathbb{R}}(V) \in \mathbb{N}$ and Γ a subgroup of V.*

1. *The following assertions are equivalent:*

   (a) *Γ is a lattice.*

   (b) *0 is not an accumulation point of Γ.*

   (c) *Γ is a discrete subset of V (with respect to the canonical topology).*

2. *Let Γ be a lattice. Then $\mathrm{rk}(\Gamma) = n$ if and only if there exists a bounded set of representatives for V/Γ in V.*

3. *Let Γ be a lattice and Γ' a sublattice of Γ. Then $(\Gamma\colon\Gamma') < \infty$ if and only if $\mathbb{R}\Gamma = \mathbb{R}\Gamma'$, and if V is an Euclidean vector space, then*

$$(\Gamma\colon\Gamma') = \frac{\mathrm{vol}(\Gamma')}{\mathrm{vol}(\Gamma)},$$

Proof. 1. (a) $\Rightarrow$ (b) Let $\mathrm{rk}(\Gamma) = m \in [0, n]$, $(u_1, \ldots, u_m)$ a basis of Γ, and let $u_{m+1}, \ldots, u_n \in V$ be such that $\boldsymbol{u} = (u_1, \ldots, u_m, u_{m+1}, \ldots, u_n)$ is an $\mathbb{R}$-basis of V. For any $c_1, \ldots, c_n \in \mathbb{R}$ we set $\|c_1u_1 + \ldots + c_nu_n\|_{\boldsymbol{u}} = \max\{|c_1|, \ldots, |c_n|\}$. Then $\|\cdot\|_{\boldsymbol{u}}$ is a norm of V, $U = \{x \in V \mid \|x\|_{\boldsymbol{u}} < 1\}$ is a neighbourhood of 0 in V, and $\Gamma \cap U = \{0\}$. Hence 0 is not an accumulation point of Γ.

(b) $\Rightarrow$ (c) We assume to the contrary that Γ has an accumulation point $c \in V$. Let $(x_n)_{n \geq 0}$ be a sequence in $\Gamma \setminus \{c\}$ such that $(x_n)_{n\geq 0} \to c$. Then $(x_{n+1} - x_n)_{n\geq 0}$ is a sequence in Γ and $(x_{n+1} - x_n)_{n\geq 0} \to 0$. As 0 is not an accumulation point of Γ, the sequence $(x_n)_{n\geq 0}$ is ultimately constant and thus $x_n = c$ for all $n \gg 1$, a contradiction.

(c) $\Rightarrow$ (a) Let $V_0 = \mathbb{R}\Gamma$, $m = \dim_{\mathbb{R}}(V_0)$, $(u_1, \ldots, u_m) \in \Gamma^m$ an $\mathbb{R}$-basis of V_0, and set $\Gamma_0 = \mathbb{Z}u_1 + \ldots + \mathbb{Z}u_m$. Then Γ_0 is a lattice in V_0, $\Gamma_0 \subset \Gamma$, and

$$\mathcal{G}_0 = \{\lambda_1u_1 + \ldots + \lambda_mu_m \mid \lambda_1, \ldots, \lambda_m \in [0, 1)\}$$

is a fundamental parallelotope of Γ_0. Hence

$$V_0 = \biguplus_{x\in\mathcal{G}_0} (x + \Gamma_0), \quad \Gamma = \biguplus_{x\in\mathcal{G}_0} (x + \Gamma_0) \cap \Gamma = \biguplus_{u\in\Gamma\cap\mathcal{G}_0} (u + \Gamma_0)$$

and $\Gamma/\Gamma_0 = \{u + \Gamma \mid u \in \Gamma\cap\mathcal{G}_0\}$. Since $\Gamma\cap\mathcal{G}_0$ is a discrete and bounded subset of V_0, it is finite, and therefore $d = |\Gamma\cap\mathcal{G}_0| = (\Gamma\!:\!\Gamma_0) < \infty$. We obtain $d\Gamma \subset \Gamma_0$, hence $\Gamma \subset d^{-1}\Gamma_0$, and since $d^{-1}\Gamma_0$ is free with basis $d^{-1}u_1, \ldots, d^{-1}u_m$, it follows by 2.3.5 that Γ is a free abelian group of rank $k \leq m$. If $(v_1, \ldots, v_k)$ is a ($\mathbb{Z}$-)basis of Γ, then $V_0 = \mathbb{R}\Gamma = \mathbb{R}v_1 + \ldots \mathbb{R}v_k$, hence $k = m$, $(v_1, \ldots, v_m)$ is linearly independent over $\mathbb{R}$, and Γ is a lattice.

2. If $\mathrm{rk}(\Gamma) = n$, then every fundamental parallelotope is a bounded set of representatives for V/Γ in V. As to the converse, let M be a bounded set of representatives for V/Γ in V. Then $V_0 = \mathbb{R}\Gamma \subset V$, and we prove equality. Let $v \in V$. For every $k \in \mathbb{N}$ there exist $u_k \in \Gamma$ and $m_k \in M$ such that $kv = u_k + m_k$ and therefore $v = k^{-1}u_k + k^{-1}m_k$. Since M is bounded it follows that $(k^{-1}m_k)_{k\geq 1} \to 0$, hence $(k^{-1}u_k)_{k\geq 1} \to v$, and since $k^{-1}u_k \in \mathbb{R}\Gamma = V_0$ for all $k \geq 0$ and V_0 is closed in V, we eventually obtain $v \in V_0$.

3. By 2.3.6.1 and since $\mathbb{R}\Gamma' \subset \mathbb{R}\Gamma$, we obtain:

$$\begin{aligned}(\Gamma\!:\!\Gamma') < \infty &\iff \mathrm{rk}(\Gamma) = \mathrm{rk}(\Gamma') \iff \dim_{\mathbb{R}}(\mathbb{R}\Gamma) = \dim_{\mathbb{R}}(\mathbb{R}\Gamma') \\ &\iff \mathbb{R}\Gamma' = \mathbb{R}\Gamma.\end{aligned}$$

Assume now that $\mathbb{R}\Gamma = \mathbb{R}\Gamma'$ and $m = \dim_{\mathbb{R}}(\mathbb{R}\Gamma) = \mathrm{rk}(\Gamma)$. By 2.3.6.1 there exist a basis $(u_1, \ldots, u_m)$ of Γ and $e_1, \ldots, e_m \in \mathbb{N}$ such that $(e_1u_1, \ldots, e_mu_m)$ is a basis of Γ', and then $(\Gamma : \Gamma') = e_1 \cdot \ldots \cdot e_m$. If V is a Euclidean vector space, then

$$\frac{\mathrm{vol}(\Gamma')}{\mathrm{vol}(\Gamma)} = \frac{\sqrt{\det(\langle e_iu_i, e_ju_j\rangle)_{i,j\in[1,m]}}}{\sqrt{\det(\langle u_i, u_j\rangle)_{i,j\in[1,m]}}} = \prod_{j=1}^{m} e_j = (\Gamma\!:\!\Gamma'). \qquad \square$$

Corollary 3.5.2. *Let W be a (multiplicative) subgroup of $\mathbb{R}_{>0}$. Then the following assertions are equivalent:*

(a) *W is discrete.*

(b) *W is cyclic.*

(c) *1 is not an accumulation point of W.*

If these conditions are fulfilled and $\rho = \max\{x \in W \mid x < 1\}$), then $W = \langle \rho \rangle$.

Proof. Since $\log\colon \mathbb{R}_{>0} \to \mathbb{R}$ is a topological isomorphism, we obtain:

- W is discrete if and only if $\log(W)$ is discrete;
- W is cyclic if and only if $\log(W)$ is a lattice in $\mathbb{R}$;
- 1 is an accumulation point of W if and only if 0 is an accumulation point of $\log(W)$.

Hence the equivalence of (a), (b) and (c) follows from 3.5.1. □

We proceed with Minkowski's convex body theorem and start with some preliminaries concerning the geometry of $\mathbb{R}^n$. Every compact subset X of $\mathbb{R}^n$ has a finite Lebesgue measure $\lambda(X)$. A subset X of $\mathbb{R}^n$ is called **symmetric** if $X = -X$. A closed subset X of $\mathbb{R}^n$ is convex if and only if

$$\frac{1}{2}(X+X) = \Big\{\frac{1}{2}(\boldsymbol{x}+\boldsymbol{y}) \;\Big|\; \boldsymbol{x},\, \boldsymbol{y} \in X\Big\} \subset X.$$

Theorem 3.5.3 (Minkowski's convex body theorem). *Let $n \in \mathbb{N}$ and Γ a lattice in $\mathbb{R}^n$ such that $\operatorname{rk}(\Gamma) = n$. Let X be a compact, convex and symmetric subset of $\mathbb{R}^n$, and suppose that $\lambda(X) \geq 2^n \operatorname{vol}(\Gamma)$. Then $X \cap \Gamma \supsetneq \mathbf{0}$.*

Proof. Assume first that $\lambda(X) > 2^n \operatorname{vol}(\Gamma)$. Then it suffices to prove that there exist $v_1,\, v_2 \in \Gamma$ such that

$$v_1 \neq v_2 \quad \text{and} \quad \Big(\frac{1}{2}X + v_1\Big) \cap \Big(\frac{1}{2}X + v_2\Big) \neq \emptyset. \qquad (\dagger)$$

Indeed, then there exist $x_1,\, x_2 \in X$ such that

$$\frac{1}{2}x_1 + v_1 = \frac{1}{2}x_2 + v_2, \quad \text{and then} \quad \mathbf{0} \neq v_1 - v_2 = \frac{1}{2}\,[x_2 + (-x_1)] \in X \cap \Gamma.$$

Assume to the contrary that

$$\Big(\frac{1}{2}X + v\Big)_{v \in \Gamma}$$

is a family of pairwise disjoint sets, and let $\mathcal{G}$ be a fundamental parallelotope of Γ. Then

$$\mathbb{R}^n = \biguplus_{v\in\Gamma} (\mathcal{G} - v), \quad \text{hence} \quad \frac{1}{2}X = \biguplus_{v\in\Gamma} (\mathcal{G} - v) \cap \frac{1}{2}X,$$

and therefore

$$\begin{aligned}\lambda(\mathcal{G}) = \mathrm{vol}(\Gamma) < \frac{1}{2^n}\lambda(X) = \lambda\Big(\frac{1}{2}X\Big) &= \sum_{v\in\Gamma} \lambda\Big((\mathcal{G}-v)\cap\frac{1}{2}X\Big) \\ &= \sum_{v\in\Gamma} \lambda\Big(\mathcal{G}\cap\big(\frac{1}{2}X+v\big)\Big) \le \lambda(\mathcal{G}), \quad \text{a contradiction.}\end{aligned}$$

Now assume that $\lambda(X) = 2^n \,\mathrm{vol}(\Gamma)$. If $\varepsilon \in \mathbb{R}_{>0}$, then the set $X_\varepsilon = (1+\varepsilon)X$ is compact, convex, symmetric, and as $\lambda(X_\varepsilon) = (1+\varepsilon)^n\lambda(X) > 2^n \,\mathrm{vol}(\Gamma)$, it follows that $X_\varepsilon \cap \Gamma \supsetneq \mathbf{0}$. As Γ is discrete and X_ε is bounded, the set $X_\varepsilon \cap \Gamma$ is finite. If $X \cap \Gamma = \mathbf{0}$, then there exists some $\varepsilon \in \mathbb{R}_{>0}$ such that $X_\varepsilon \cap \Gamma = \mathbf{0}$, a contradiction. □

Let K be an algebraic number field, (r_1, r_2) its signature, and let $[K:\mathbb{Q}] = n = r_1 + 2r_2$. Let $\sigma_1, \ldots, \sigma_{r_1}$ be the real embeddings and $(\sigma_{r_1+1}, \overline{\sigma_{r_1+1}}), \ldots, (\sigma_{r_1+r_2}, \overline{\sigma_{r_1+r_2}})$ the pairs of conjugate complex embeddings of K. Then $\mathrm{Hom}(K,\mathbb{C}) = \{\sigma_1, \ldots, \sigma_n\}$, and we define the **field embedding** $\varphi\colon K \to \mathbb{R}^n$ by

$$\begin{aligned}\varphi(x) = \big(\sigma_1(x), \ldots, \sigma_{r_1}(x), \Im\sigma_{r_1+1}(x), \ldots, \Im\sigma_{r_1+r_2}(x),& \\ \Re\sigma_{r_1+1}(x), \ldots, \Re\sigma_{r_1+r_2}(x)\big)^{\mathrm{t}}&.\end{aligned}$$

Obviously, φ is a $\mathbb{Q}$-linear monomorphism.

Theorem 3.5.4. *Let K be an algebraic number field, $[K:\mathbb{Q}] = n = r_1 + 2r_2$, and let $\mathrm{Hom}(K,\mathbb{C}) = \{\sigma_1, \ldots, \sigma_n\}$, where $\sigma_i(K) \subset \mathbb{R}$ for all $i \in [1, r_1]$ and $\sigma_{r_1+r_2+i} = \overline{\sigma_{r_1+i}} \ne \sigma_{r_1+i}$ for all $i \in [1, r_2]$. Let $\varphi\colon K \to \mathbb{R}^n$ be the field embedding and $\mathfrak{a}$ a complete module of K.*

1. *$\varphi(\mathfrak{a})$ is a lattice in $\mathbb{R}^n$ of rank n, and* $\mathrm{vol}(\varphi(\mathfrak{a})) = 2^{-r_2}\sqrt{|\Delta(\mathfrak{a})|}$.

2. *For every $C \in \mathbb{R}_{>0}$, the set*

$$\mathfrak{a}_C = \{\alpha \in \mathfrak{a} \mid |\sigma_i(\alpha)| \le C \textit{ for all } i \in [1, r_1+r_2]\}$$

is finite.

3. *There exists some $\alpha \in \mathfrak{o}_K$ such that $K = \mathbb{Q}(\alpha)$ and*

$$|\sigma_i(\alpha)| \le 2^{n-1}\sqrt{|\mathsf{d}_K|} + \frac{1}{2} \quad \textit{for all} \quad i \in [1, n].$$

Proof. 1. Let $(u_1, \ldots, u_n)$ be a basis of $\mathfrak{a}$. For $j \in [1,n]$ we consider the column

$$(\sigma_1(u_j), \ldots, \sigma_n(u_j))^{\mathsf{t}} = (S_j, T_j, \overline{T}_j)^{\mathsf{t}} \in \mathbb{C}^n,$$

where $S_j = (\sigma_1(u_j), \ldots, \sigma_{r_1}(u_j))^{\mathsf{t}}$ and $T_j = (\sigma_{r_1+1}(u_j), \ldots, \sigma_{r_1+r_2}(u_j))^{\mathsf{t}}$. It follows that

$$\varphi(u_j) = \begin{pmatrix} S_j \\ \frac{1}{2\mathrm{i}}\,(T_j - \overline{T}_j) \\ \frac{1}{2}\,(T_j + \overline{T}_j) \end{pmatrix} = B\begin{pmatrix} S_j \\ T_j \\ \overline{T}_j, \end{pmatrix} \quad \text{where} \quad B = \begin{pmatrix} I & \mathbf{0} & \mathbf{0} \\ \mathbf{0} & \frac{1}{2\mathrm{i}}I & -\frac{1}{2\mathrm{i}}I \\ \mathbf{0} & \frac{1}{2}I & \frac{1}{2}I \end{pmatrix}.$$

Since $|\det(B)| = 2^{-r_2}$, we obtain

$$|\det(\varphi(u_1), \ldots, \varphi(u_n))| = 2^{-r_2}|\det(\sigma_i(u_j))_{i,j\in[1,n]}| = 2^{-r_2}\sqrt{|\Delta(\mathfrak{a})|} \ne 0.$$

Hence $(\varphi(u_1), \ldots, \varphi(u_n))$ is linearly independent over $\mathbb{R}$, $\varphi(\mathfrak{a})$ is a lattice in $\mathbb{R}^n$ of rank n, and $\mathrm{vol}(\varphi(\mathfrak{a})) = 2^{-r_2}\sqrt{|\Delta(\mathfrak{a})|}$.

2. If $C \in \mathbb{R}_{>0}$, then $\varphi(\mathfrak{a}_C)$ is bounded and discrete by the very definition, and thus it is finite.

3. The assertion obviously holds for $K = \mathbb{Q}$. Thus assume that $n > 1$ and set $M = 2^{n-1}\sqrt{|\mathsf{d}_K|}$. Then the set

$$X = [-M, M] \times \Big[-\frac{1}{2}, \frac{1}{2}\Big]^{n-1} \subset \mathbb{R}^n$$

is symmetric, compact, convex, and

$$2^n\,\mathrm{vol}(\varphi(\mathfrak{o}_K)) = 2^{n-r_2}\sqrt{|\mathsf{d}_K|} \le 2M = \lambda(X).$$

By 3.5.3 that there exists some $\alpha \in \mathfrak{o}_K^\bullet$ such that $\varphi(\alpha) \in X$. We shall prove that

$$K = \mathbb{Q}(\alpha), \quad \text{and} \quad |\sigma_i(\alpha)| \le 2^{n-1}\sqrt{|\mathsf{d}_K|} + \frac{1}{2} \quad \text{for all} \quad i \in [1, r_1 + r_2].$$

We set $m = [K : \mathbb{Q}(\alpha)] = |\{j \in [1,n] \mid \sigma_j(\alpha) = \sigma_1(\alpha)\}|$.

CASE 1: $r_1 > 0$. Then we obtain

$$|\sigma_1(\alpha)| \le M = 2^{n-1}\sqrt{|\mathsf{d}_K|}, \qquad |\sigma_i(\alpha)| \le \frac{1}{2} \quad \text{for all} \quad i \in [2, r_1],$$

and

$$|\sigma_{r_1+i}(\alpha)|^2 = \Im\sigma_{r_1+i}(\alpha)^2 + \Re\sigma_{r_1+i}(\alpha)^2 \le \frac{1}{2} \quad \text{for all} \quad i \in [1, r_2].$$

Since

$$1 \le |\mathsf{N}_{K/\mathbb{Q}}(\alpha)| = |\sigma_1(\alpha)| \prod_{i=2}^{r_1} |\sigma_i(\alpha)| \prod_{i=1}^{r_2} |\sigma_{r_1+i}(\alpha)|^2 < |\sigma_1(\alpha)|,$$

it follows that $|\sigma_i(\alpha)| < |\sigma_1(\alpha)|$ for all $i \in [2, n]$ and thus $m = 1$.

CASE 2: $r_1 = 0$. Then we obtain

$$|\Im\sigma_1(\alpha)| \le M = 2^{n-1}\sqrt{|\mathsf{d}_K|}, \qquad |\Im\sigma_i(\alpha| \le \frac{1}{2} \ \text{ for all } \ i \in [2, r_2] \cup [r_2+2, 2r_2]$$

$$\Im\sigma_{r_2+1}(\alpha) = -\Im\sigma_1(\alpha), \quad |\Re\sigma_i(\alpha)| \le \frac{1}{2} \ \text{ for all } \ i \in [1, r_2],$$

$$|\sigma_1(\alpha)| \le 2^{n-1}\sqrt{|\mathsf{d}_K|} + \frac{1}{2} \quad \text{and} \quad |\sigma_i(\alpha)|^2 = \Im\sigma_i(\alpha)^2 + \Re\sigma_i(\alpha)^2 \le \frac{1}{2}$$
$$\text{for all } \ i \in [2, r_2].$$

Since

$$\begin{aligned} 1 \le |\mathsf{N}_{K/\mathbb{Q}}(\alpha)| &= |\sigma_1(\alpha)|^2 \prod_{i=2}^{r_2} |\sigma_i(\alpha)|^2 \\ &\le \Big(\frac{1}{2}\Big)^{r_2-1} |\sigma_1(\alpha)|^2 \le |\sigma_1(\alpha)|^2 \le \Im\sigma_1(\alpha)^2 + \frac{1}{4}, \end{aligned}$$

it follows that

$$|\Im\sigma_1(\alpha)| \ge \frac{\sqrt{3}}{2} > \frac{1}{2},$$

hence $\sigma_j(\alpha) \ne \sigma_1(\alpha)$ for all $j \in [2, n]$ and $m = 1$. □

Theorem 3.5.5. A. *Let r_1, $r_2 \in \mathbb{N}_0$ and $n = r_1 + 2r_2 \in \mathbb{N}$. For $a \in \mathbb{R}_{>0}$ we set*

$$U_{r_1,r_2}(a) = \Big\{(x_1, \ldots, x_{r_1}, y_1, z_1, \ldots, y_{r_2}, z_{r_2}) \in \mathbb{R}^n \;\Big|\; \sum_{i=1}^{r_1} |x_i| + 2\sum_{i=1}^{r_2} \sqrt{y_i^2 + z_i^2} \le a\Big\},$$

and for $\mathbf{c} = (c_1, \ldots, c_{r_1+r_2}) \in \mathbb{R}_{>0}^{r_1+r_2}$ we set

$$W(\mathbf{c}) = \Big\{(x_1, \ldots, x_{r_1}, y_1, z_1, \ldots, y_{r_2}, z_{r_2}) \in \mathbb{R}^n \;\Big|\; |x_i| \le c_i \ \text{ for } \ i \in [1, r_1]$$
$$\text{and } \ \sqrt{y_i^2 + z_i^2} \le c_{r_1+i} \ \text{ for } \ i \in [1, r_2]\Big\}.$$

Then $U_{r_1,r_2}(a)$ and $W(\mathbf{c})$ are symmetric compact and convex sets satisfying

$$\lambda(U_{r_1,r_2}(a)) = 2^{r_1}\Big(\frac{\pi}{2}\Big)^{r_2}\frac{a^n}{n!} \qquad \text{and} \qquad \lambda(W(\mathbf{c})) = 2^{r_1}\pi^{r_2}\prod_{i=1}^{r_1} c_i \prod_{i=1}^{r_2} c_{r_1+i}^2.$$

B. *Let K be an algebraic number field, $[K : \mathbb{Q}] = n = r_1 + 2r_2$, and suppose that $\mathrm{Hom}(K, \mathbb{C}) = \{\sigma_1, \ldots, \sigma_n\}$, $\sigma_i(K) \subset \mathbb{R}$ for $i \in [1, r_1]$ and $\sigma_{r_1+r_2+i} = \overline{\sigma_{r_1+i}} \ne \sigma_{r_1+i}$ for $i \in [1, r_2]$. Let $\varphi\colon K \to \mathbb{R}^n$ be the field embedding and $\mathfrak{a}$ a complete module of K.*

1. *Let* $\mathbf{c} = (c_1, \ldots, c_{r_1+r_2}) \in \mathbb{R}_{>0}^{r_1+r_2}$ *be such that*

$$\prod_{i=1}^{r_1} c_i \prod_{i=1}^{r_2} c_{r_1+i}^2 = \Big(\frac{2}{\pi}\Big)^{r_2} \sqrt{|\Delta(\mathfrak{a})|}\,.$$

Then there exists some $\alpha \in \mathfrak{a}^\bullet$ *such that*

$$|\sigma_i(\alpha)| \le c_i \quad \textit{for all} \quad i \in [1, r_1 + r_2], \qquad \textit{and}$$
$$|\mathsf{N}_{K/\mathbb{Q}}(\alpha)| \le \Big(\frac{2}{\pi}\Big)^{r_2} \sqrt{|\Delta(\mathfrak{a})|}.$$

2. *There exists some* $\alpha \in \mathfrak{a}^\bullet$ *such that*

$$|\mathsf{N}_{K/\mathbb{Q}}(\alpha)| \le \frac{n!}{n^n}\Big(\frac{4}{\pi}\Big)^{r_2} \sqrt{|\Delta(\mathfrak{a})|}\,.$$

3. *If* $\mathfrak{a} \in \mathfrak{J}_K$, *then there exists some* $\alpha \in \mathfrak{a}^\bullet$ *such that*

$$|\mathsf{N}_{K/\mathbb{Q}}(\alpha)| \le \min\{B_K, M_K\}\,\mathfrak{N}(\mathfrak{a}),$$

where

$$B_K = \Big(\frac{2}{\pi}\Big)^{r_2} \sqrt{|\mathsf{d}_K|} \quad \textit{and} \quad M_K = \frac{n!}{n^n}\Big(\frac{4}{\pi}\Big)^{r_2} \sqrt{|\mathsf{d}_K|},$$

B_K is called the **Blichfeld constant**, and M_K is called the **Minkowski constant**.

Proof. **A.** Let $a,\ c_1, \ldots, c_{r_1+r_2} \in \mathbb{R}_{>0}$. The sets $U_{r_1,r_2}(a)$ and $W(\mathbf{c})$ are apparently symmetric, compact and convex. By the very definition,

$$W(\mathbf{c}) = \prod_{i=1}^{r_1}[-c_i, c_i] \times \prod_{i=1}^{r_2}\{(y,z) \in \mathbb{R}^2 \mid y^2 + z^2 \le c_{r_1+i}^2\},$$

which implies that

$$\lambda(W(\mathbf{c})) = \prod_{i=1}^{r_1}(2c_i)\prod_{i=1}^{r_2}(c_{r_1+i}^2\pi) = 2^{r_1}\pi^{r_2}\prod_{i=1}^{r_1} c_i \prod_{i=1}^{r_2} c_{r_1+i}^2.$$

For the computation of $\lambda(U_{r_1,r_2}(a))$ we use induction on r_1 and r_2. Since

$$U_{1,0}(a) = \{x \in \mathbb{R} \mid |x| \le a\} \quad \text{and} \quad U_{0,1}(a) = \{(y,z) \in \mathbb{R}^2 \mid 2\sqrt{y^2 + z^2} \le a\},$$

it follows that

$$\lambda(U_{1,0}(a)) = 2a \quad \text{and} \quad \lambda(U_{0,1}(a)) = \frac{a^2\pi}{4}.$$

If $r_1 \geq 1$ and $r_2 \geq 0$, then

$$U_{r_1,r_2}(a) = \{(x_1,\ldots,x_{r_1},y_1,z_1,\ldots,y_{r_2},z_{r_2}) \in \mathbb{R}^n \mid |x_{r_1}| \leq a \text{ and } (x_1,\ldots,x_{r_1-1},y_1,z_1,\ldots,y_{r_2},z_{r_2}) \in U_{r_1-1,r_2}(a-|x_1|)\},$$

and therefore

$$\lambda(U_{r_1,r_2}(a)) = \int_{-a}^{a} \lambda(U_{r_1-1,r_2}(a-|x_r|))dx_r = \int_{-a}^{a} 2^{r_1-1}\Big(\frac{\pi}{2}\Big)^{r_2}\frac{(a-|x_r|)^{n-1}}{(n-1)!}dx_r$$
$$= 2^{r_1}\Big(\frac{\pi}{2}\Big)^{r_2}\frac{1}{(n-1)!}\int_0^a (a-x_r)^{n-1}dx_r = 2^{r_1}\Big(\frac{\pi}{2}\Big)^{r_2}\frac{a^n}{n!}.$$

If $r_1 \geq 0$ and $r_2 \geq 1$, then

$$U_{r_1,r_2}(a) = \Big\{(x_1,\ldots,x_{r_1},y_1,z_1,\ldots,y_{r_2},z_{r_2}) \in \mathbb{R}^n \;\Big|\; 2\sqrt{y_r^2+z_r^2} \leq a \text{ and } (x_1,\ldots,x_{r_1},y_1,z_1,\ldots,y_{r_2-1},z_{r_2-1}) \in U_{r_1,r_2-1}\big(a-2\sqrt{y_{r_2}^2+z_{r_2}^2}\,\big)\Big\},$$

and therefore

$$\lambda(U_{r_1,r_2}(a)) = \int_{2\sqrt{y_{r_2}^2+z_{r_2}^2}\leq a} U_{r_1,r_2-1}\big(a-2\sqrt{y_{r_2}^2+z_{r_2}^2}\,\big)d(y_{r_2},z_{r_2})$$
$$= \int_{2\sqrt{y_{r_2}^2+z_{r_2}^2}\leq a} 2^{r_1}\Big(\frac{\pi}{2}\Big)^{r_2-1}\frac{\big(a-2\sqrt{y_{r_2}^2+z_{r_2}^2}\,\big)^{n-2}}{(n-2)!}$$
$$[\; y_{r_2} = t\cos\varphi,\; z_{r_2} = t\sin\varphi\,]$$
$$= 2^{r_1}\Big(\frac{\pi}{2}\Big)^{r_2-1}\int_{\varphi=0}^{2\pi}\int_{t=0}^{a/2}\frac{(a-2t)^{n-2}}{(n-2)!}\,t\,d\varphi\,dt$$
$$= 2^{r_1}\Big(\frac{\pi}{2}\Big)^{r_2-1}\frac{2\pi}{(n-2)!}\int_0^{a/2}(a-2t)^{n-2}t\,dt = 2^{r_1}\Big(\frac{\pi}{2}\Big)^{r_2}\frac{a^n}{n!}.$$

B. 1. By 3.5.4.1 and **A** we obtain

$$2^n\mathrm{vol}(\varphi(\mathfrak{a})) = 2^{r_1+r_2}\sqrt{|\Delta(\mathfrak{a})|} = 2^{r_1}\pi^{r_2}\prod_{i=1}^{r_1}c_i\prod_{i=1}^{r_2}c_{r_1+i}^2 = \lambda(W(\boldsymbol{c})),$$

and 3.5.3 implies $W(\boldsymbol{c})\cap\varphi(\mathfrak{a}) \supsetneq \mathbf{0}$. Hence there exists some $\alpha \in \mathfrak{a}^\bullet$ satisfying $|\sigma_i(\alpha)| \leq c_i$ for all $i \in [1, r_1+r_2]$, and

$$|\mathsf{N}_{K/\mathbb{Q}}(\alpha)| \leq \prod_{i=1}^{r_1}c_i\prod_{i=1}^{r_2}c_{r_1+i}^2 = \Big(\frac{2}{\pi}\Big)^{r_2}\sqrt{|\Delta(\mathfrak{a})|}\,.$$

2. Let $a \in \mathbb{R}_{>0}$ be such that

$$a^n = n!\Big(\frac{4}{\pi}\Big)^{r_2}\sqrt{|\Delta(\mathfrak{a})|}.$$

By 3.5.4.1 and **A** we obtain

$$2^n \text{vol}(\varphi(\mathfrak{a})) = 2^{r_1+r_2}\sqrt{|\Delta(\mathfrak{a})|} = 2^{r_1}\Big(\frac{\pi}{2}\Big)^{r_2}\frac{a^n}{n!} = \lambda(U_{r_1,r_2}(a)),$$

and 3.5.3 implies $U_{r_1,r_2}(a) \cap \varphi(\mathfrak{a}) \supsetneqq \mathbf{0}$. Hence there exists some $\alpha \in \mathfrak{a}^\bullet$ satisfying

$$\sum_{i=1}^{r_1}|\sigma_i(\alpha)| + 2\sum_{i=1}^{r_2}|\sigma_{r_1+i}(\alpha)| \le a.$$

Using the inequality between arithmetic and geometric mean, we obtain

$$\begin{aligned}\sqrt[n]{|\mathsf{N}_{K/\mathbb{Q}}(\alpha)|} &= \sqrt[n]{\prod_{i=1}^{r_1}|\sigma_i(\alpha)|\prod_{i=1}^{r_2}|\sigma_{r_1+i}(\alpha)|^2} \\ &\le \frac{1}{n}\Big(\sum_{i=1}^{r_1}|\sigma_i(\alpha)| + 2\sum_{i=1}^{r_2}|\sigma_{r_1+i}(\alpha)|\Big) \le \frac{a}{n},\end{aligned}$$

and consequently

$$|\mathsf{N}_{K/\mathbb{Q}}(\alpha)| \le \frac{a^n}{n^n} = \frac{n!}{n^n}\Big(\frac{4}{\pi}\Big)^{r_2}\sqrt{|\Delta(\mathfrak{a})|}.$$

3. Let $\mathfrak{a} \in \mathcal{J}_K$. Then $\sqrt{|\Delta(\mathfrak{a})|} = \mathfrak{N}(\mathfrak{a})\sqrt{|\mathsf{d}_K|}$ by 3.1.4, and therefore the bounds in 1. and 2. are given by $B_K\mathfrak{N}(\mathfrak{a})$ and $M_K\mathfrak{N}(\mathfrak{a})$. □

Theorem 3.5.6 (Discriminant theorem of Minkowski and Hermite).

1. Let K be an algebraic number field, (r_1, r_2) its signature and $[K:\mathbb{Q}] = n \ge 2$. Then

$$|\mathsf{d}_K| \ge \Big(\frac{\pi^{r_2}n^n}{4n!}\Big)^2 > 1,$$

2. For every $C \in \mathbb{R}_{>0}$ there are only finitely many algebraic number fields K satisfying $|\mathsf{d}_K| \le C$.

Proof. 1. By 3.5.5.**B**.3 there exists some $\alpha \in \mathfrak{o}_K^\bullet$ such that $|\mathsf{N}_{K/\mathbb{Q}}(\alpha)| \le M_K$, and $|\mathsf{N}_{K/\mathbb{Q}}(\alpha)| \ge 1$ implies

$$|\mathsf{d}_K| \ge \Big[\frac{n^n}{n!}\Big(\frac{\pi}{4}\Big)^{r_2}\Big]^2 \ge \Big(\frac{\pi}{4}\Big)^n\Big(\frac{n^n}{n!}\Big)^2 = \Phi(n).$$

Since

$$\frac{\Phi(n+1)}{\Phi(n)} = \frac{\pi}{4}\Big(1+\frac{1}{n}\Big)^{2n} \ge \pi \quad \text{for all } n \ge 1,$$

the sequence $(\Phi(n))_{n\geq 2}$ is monotonically increasing. Hence it follows that $\Phi(n) \geq \Phi(2) = \pi^2/4 > 2$ for all $n \geq 2$, and therefore

$$|\mathsf{d}_K| \geq \Big(\frac{\pi^{r_2}\, n^n}{4n!}\Big)^2 \geq \Phi(n) > 1.$$

2. Let $C \in \mathbb{R}_{>0}$. As $(\Phi(n))_{n\geq 2} \to \infty$, it follows that $|\mathsf{d}_K| > C$ for all $n \gg 1$. Hence it remains to prove:

For every $n \in \mathbb{N}$ and $M \in \mathbb{R}_{>0}$ there exist only finitely many algebraic number fields $K \subset \mathbb{C}$ such that $[K:\mathbb{Q}] = n$ and $|\mathsf{d}_K| \leq M$.

If K is an algebraic number field of degree n and $|\mathsf{d}_K| \leq M$, then 3.5.4.3 shows that $K = \mathbb{Q}(\alpha)$ for some $\alpha \in \mathfrak{o}_K$ such that $|\alpha'| \leq 2^{n-1}\sqrt{M} + 1$ for all conjugates α' of α in $\mathbb{C}$. Therefore it suffices to prove that for every $n \in \mathbb{N}$ and $B \in \mathbb{R}_{>0}$ there exist only finitely many algebraic integers $\alpha \in \mathbb{C}$ of degree n for which all of its conjugates $\alpha' \in \mathbb{C}$ satisfy $|\alpha'| \leq B$. To prove this, we show that there exist only finitely many monic polynomials $f \in \mathbb{Z}[X]$ such that $\partial(f) = n$ and $|\alpha| \leq B$ for all zeros α of f in $\mathbb{C}$. Let

$$f = X^n + a_1X^{n-1} + \ldots + a_{n-1}X + a_n = (X - \alpha_1) \cdot \ldots \cdot (X - \alpha_n) \in \mathbb{Z}[X],$$

where $\alpha_1, \ldots, \alpha_n \in \mathbb{C}$ and $|\alpha_\nu| \leq B$ for all $\nu \in [1, n]$. Then we obtain

$$|a_i| = \Big|\sum_{1\leq \nu_1 < \ldots < \nu_i \leq n} \alpha_{\nu_1} \cdot \ldots \cdot \alpha_{\nu_i}\Big| \leq \binom{n}{i} B^i \quad \text{for all} \quad i \in [1, n],$$

and evidently there are only finitely many polynomials with this property. □

In 5.9.8 we shall prove that a prime p is ramified in K if and only if $p \,|\, \mathsf{d}_K$. As $|\mathsf{d}_K| > 1$, at least one prime is ramified in K.

Theorem 3.5.7 (Finiteness of the class number). *Let K be an algebraic number field.*

1. *Let $\mathfrak{a}$ be a complete module of K and $N \in \mathbb{N}$. Then there exist only finitely many subgroups A of K containing $\mathfrak{a}$ and satisfying $(A : \mathfrak{a}) \leq N$; and there exist only finitely many subgroups B of $\mathfrak{a}$ satisfying $(\mathfrak{a}:B) \leq N$.*

2. *Let $\mathfrak{o}$ be an order of K. For every $N \in \mathbb{N}$ there exist only finitely many ideals $\mathfrak{a}$ of $\mathfrak{o}$ such that $\mathfrak{N}(\mathfrak{a}) \leq N$.*

3. *Let $\mathfrak{o}$ be an order of K. Then there are only finitely many pairwise not equivalent complete modules $\mathfrak{a}$ of K such that $\mathcal{R}(\mathfrak{a}) = \mathfrak{o}$.*

4. $h_K = |\mathcal{C}_K| < \infty$.

Proof. 1. As $\mathfrak{a}$ is a free abelian group of rank $[K:\mathbb{Q}]$, it follows that for all $n \in \mathbb{N}$ the groups $\mathfrak{a}/n\mathfrak{a}$ and $n^{-1}\mathfrak{a}/\mathfrak{a}$ are finite.

Let A and B be subgroups of K such that $B \subset \mathfrak{a} \subset A$, $(A:\mathfrak{a}) \le N$ and $(\mathfrak{a}:B) \le N$. Then $(A:\mathfrak{a}) \mid N!$, $(\mathfrak{a}:B) \mid N!$, and $N!\mathfrak{a} \subset B \subset \mathfrak{a} \subset A \subset N!^{-1}\mathfrak{a}$. Since $N!^{-1}\mathfrak{a}/\mathfrak{a}$ and $\mathfrak{a}/N!\mathfrak{a}$ are finite, the assertion follows.

2. By 1., applied with $\mathfrak{a} = \mathfrak{o}$.

3. By 1. it suffices to prove that for every complete module $\mathfrak{a}$ of K with $\mathcal{R}(\mathfrak{a}) = \mathfrak{o}$ there exists a complete module $\mathfrak{a}'$ of K such that $\mathfrak{a}' \sim \mathfrak{a}$, $\mathfrak{a}' \supset \mathfrak{o}$ and $(\mathfrak{a}':\mathfrak{o}) \le C$ for some constant $C \in \mathbb{R}_{>0}$ which only depends on $\mathfrak{o}$.

Thus let $\mathfrak{a}$ be a complete module of K such that $\mathcal{R}(\mathfrak{a}) = \mathfrak{o}$. By 3.5.5.**B**.1 there exists some $\alpha \in \mathfrak{a}^\bullet$ such that

$$|\mathsf{N}_{K/\mathbb{Q}}(\alpha)| \le \Big(\frac{2}{\pi}\Big)^{r_2} \sqrt{|\Delta(\mathfrak{a})|}.$$

Then $\mathfrak{o}\alpha \subset \mathfrak{a}$, hence $\mathfrak{o} \subset \mathfrak{a}' = \alpha^{-1}\mathfrak{a}$ and $\alpha^{-1}\mathfrak{a} \sim \mathfrak{a}$. By 3.1.3 it follows that

$$(\mathfrak{a}':\mathfrak{o}) = \frac{\sqrt{|\Delta(\mathfrak{o})|}}{\sqrt{|\Delta(\alpha^{-1}\mathfrak{a})|}} = |\mathsf{N}_{K/\mathbb{Q}}(\alpha)| \frac{\sqrt{|\Delta(\mathfrak{o})|}}{\sqrt{|\Delta(\mathfrak{a})|}} \le \Big(\frac{2}{\pi}\Big)^{r_2} \sqrt{|\Delta(\mathfrak{o})|} = C \quad \text{(say)}.$$

4. By 3.1.3.5, a subgroup $\mathfrak{a}$ of K is a complete module of K satisfying $\mathcal{R}(\mathfrak{a}) = \mathfrak{o}_K$ if and only if $\mathfrak{a} \in \mathfrak{J}_K$, and two fractional ideals $\mathfrak{a}, \mathfrak{b} \in \mathfrak{J}_K$ are equivalent if and only if $[\mathfrak{a}] = [\mathfrak{b}] \in \mathcal{C}(\mathfrak{o}_K)$. Therefore $\mathcal{C}_K = \mathcal{C}(\mathfrak{o}_K)$ is finite by 3. □

Theorem 3.5.8 (Dirichlet's unit theorem). *Let K be an algebraic number field, (r_1, r_2) its signature, $[K:\mathbb{Q}] = n = r_1 + 2r_2$ and $r = r_1 + r_2$. Let $\mathfrak{o}$ be an order of K. Then $\mathfrak{o}^\times = \{\alpha \in \mathfrak{o} \mid |\mathsf{N}_{K/\mathbb{Q}}(\alpha)| = 1\} = \mathfrak{o} \cap U_K$, the group $\mu(\mathfrak{o})$ of roots of unity in $\mathfrak{o}$ is a finite cyclic group, and $\mathfrak{o}^\times \cong \mu(\mathfrak{o}) \times \mathbb{Z}^{r-1}$.*

Explicitly, $\mathfrak{o}^\times = \langle \zeta, \varepsilon_1, \ldots, \varepsilon_{r-1}\rangle$, where ζ is a root of unity in K, and $\langle \varepsilon_1, \ldots, \varepsilon_{r-1}\rangle$ is a free abelian group with basis $(\varepsilon_1, \ldots, \varepsilon_{r-1})$.

Proof. Since $\mathfrak{o}^\times = \{\alpha \in \mathfrak{o} \mid \alpha\mathfrak{o} = \mathfrak{o}\}$ and $(\mathfrak{o}:\alpha\mathfrak{o}) = |\mathsf{N}_{K/\mathbb{Q}}(\alpha)|$ for all $\alpha \in \mathfrak{o}$ by 3.1.3.4(a), it follows that $\mathfrak{o}^\times = \{\alpha \in \mathfrak{o} \mid |\mathsf{N}_{K/\mathbb{Q}}(\alpha)| = 1\} = \mathfrak{o} \cap U_K$.

Let $\mathrm{Hom}(K, \mathbb{C}) = \{\sigma_1, \ldots, \sigma_n\}$, and assume that $\sigma_i(K) \subset \mathbb{R}$, $l_i = 1$ for $i \in [1, r_1]$, and $\sigma_{r_1+r_2+i} = \overline{\sigma_{r_1+i}} \ne \sigma_{r_1+i}$, $l_{r_1+i} = 2$ for $i \in [1, r_2]$. We consider the hyperplane

$$H = \Big\{(x_1, \ldots, x_r) \in \mathbb{R}^r \ \Big|\ \sum_{i=1}^{r} x_i = 0\Big\} \subset \mathbb{R}^r$$

and the logarithmic embedding $\ell = (\ell_1, \ldots, \ell_r)\colon K^\times \to \mathbb{R}^r$, defined by

$$\ell_i(x) = l_i \log|\sigma_i(x)| \text{ for all } x \in K^\times \text{ and } i \in [1, r].$$

Then $\ell\colon K^\times \to \mathbb{R}^r$ is a homomorphism, $\dim_{\mathbb{R}}(H) = r-1$,

$$\log|\mathsf{N}_{K/\mathbb{Q}}(\alpha)| = \sum_{i=1}^r \ell_i(\alpha) \quad \text{for all} \quad \alpha \in K^\times, \qquad \text{and therefore} \quad \ell(\mathfrak{o}^\times) \subset H.$$

Let $\|\cdot\|$ be the maximum norm on $\mathbb{R}^r$. If $C \in \mathbb{R}_{>0}$ and $\alpha \in K^\times$, then $\|\ell(\alpha)\| \le C$ implies $|\sigma_i(\alpha)| \le \mathrm{e}^{C/l_i}$ for all $i \in [1,r]$. Therefore it follows by 3.5.4.2 that the set $\{\alpha \in \mathfrak{o}^\bullet \mid \|\ell(\alpha)\| \le C\}$ is finite. Consequently, $\ell(\mathfrak{o}^\times)$ is a discrete subgroup of H, hence a lattice in H, and $\mathrm{rk}(\ell(\mathfrak{o}^\times)) = s \in [0, r-1]$. Let $\varepsilon_1, \ldots, \varepsilon_s \in \mathfrak{o}^\times$ be such that $(\ell(\varepsilon_1), \ldots, \ell(\varepsilon_s))$ is a basis of $\ell(\mathfrak{o}^\times)$, and let $j\colon \ell(\mathfrak{o}^\times) \to \mathfrak{o}^\times$ be the unique homomorphism satisfying $j(\ell(\varepsilon_i)) = \varepsilon_i$ for all $i \in [1,s]$. Then $\ell \circ j = \mathrm{id}_{\ell(\mathfrak{o}^\times)}$, $j\colon \ell(\mathfrak{o}^\times) \to \langle \varepsilon_1, \ldots, \varepsilon_s\rangle$ is an isomorphism, and we obtain $\mathfrak{o}^\times = \mathrm{Ker}(\ell \restriction \mathfrak{o}^\times) \cdot \langle \varepsilon_1, \ldots, \varepsilon_s\rangle \cong \mathrm{Ker}(\ell \restriction \mathfrak{o}^\times) \times \mathbb{Z}^s$.

$\mathrm{Ker}(\ell \restriction \mathfrak{o}^\times) = \{\varepsilon \in \mathfrak{o}^\times \mid \ell(\varepsilon) = \mathbf{0}\}$ is a finite subgroup of $\mathfrak{o}^\times$, hence cyclic by 1.2.1 and contained in $\mu(\mathfrak{o})$. Conversely, if $\zeta \in \mu(\mathfrak{o})$, then $|\sigma_i(\zeta)| = 1$ for all $i \in [1,r]$, hence $\ell(\zeta) = \mathbf{0}$ and $\zeta \in \mathrm{Ker}(\ell \restriction \mathfrak{o}^\times)$. Consequently, $\mathrm{Ker}(\ell \restriction \mathfrak{o}^\times) = \mu(\mathfrak{o})$.

It remains to prove that $s = r-1$, and therefore we must show that the lattice $\ell(\mathfrak{o}^\times)$ contains $r-1$ linearly independent vectors. We need the following two tools.

I. For every $k \in [1,r]$ there exists some $\varepsilon \in \mathfrak{o}^\times$ such that $|\sigma_i(\varepsilon)| < 1$ for all $i \in [1,r] \setminus \{k\}$.

II. Let $C = (c_{i,j})_{i,j \in [1,r]} \in \mathsf{M}_r(\mathbb{R})$ be a matrix satisfying

$$\sum_{i=1}^r c_{i,j} = 0 \text{ for all } \; j \in [1,r], \quad \text{and}$$

$$c_{i,j} \begin{cases} > 0 & \text{if } \; i = j, \\ < 0 & \text{if } \; i \ne j \end{cases} \quad \text{for all} \;\; i,\, j \in [1,r].$$

Then C has rank $r-1$.

We assume now first that **I** and **II** hold and finish the proof. By **I** there exist $\varepsilon_1, \ldots, \varepsilon_r \in \mathfrak{o}^\times$ such that $|\sigma_i(\varepsilon_j)| < 1$ for all $i,\, j \in [1,r]$ with $i \ne j$. Then the matrix $(\ell_i(\varepsilon_j))_{i,j \in [1,r]}$ has rank $r-1$ by **II**, and therefore $\dim_{\mathbb{R}} \langle \ell(\varepsilon_1), \ldots, \ell(\varepsilon_r)\rangle = r-1$.

Proof of **I.** Let $k \in [1,r]$. Starting with an arbitrary $\alpha_1 \in \mathfrak{o}^\times$, 3.5.5.**B**.1 implies the existence of a sequence $(\alpha_n)_{n \ge 1}$ in $\mathfrak{o}^\bullet$ such that, for all $n \ge 1$,

$$|\sigma_i(\alpha_{n+1})| < |\sigma_i(\alpha_n)| \;\; \text{for all} \;\; i \in [1,r] \setminus \{k\}$$

and

$$|\mathsf{N}_{K/\mathbb{Q}}(\alpha_{n+1})| \le B = \Big(\frac{2}{\pi}\Big)^{r_2} \sqrt{|\Delta(\mathfrak{o})|}.$$

By 3.5.7.2 and 3.1.3.4(a) there exist only finitely many principal ideals $\alpha\mathfrak{o}$ of $\mathfrak{o}$ satisfying $(\mathfrak{o} : \alpha\mathfrak{o}) = |\mathsf{N}_{K/\mathbb{Q}}(\alpha)| \le B$. Therefore there exist $h,\, n \in \mathbb{N}$

such that $h < n$ and $\alpha_h\mathfrak{o} = \alpha_n\mathfrak{o}$, and if $\varepsilon = \alpha_h^{-1}\alpha_n$, then $|\sigma_i(\varepsilon)| < 1$ for all $i \in [1,r] \setminus \{k\}$, □[**I.**]

Proof of **II.** We show that the first $r-1$ rows of C are linearly independent. We assume the contrary. Then there exist $\lambda_1, \ldots, \lambda_{r-1} \in \mathbb{R}$ and $k \in [1, r-1]$ such that

$$\sum_{i=1}^{r-1} \lambda_i c_{i,j} = 0 \quad \text{for all} \quad j \in [1,r] \quad \text{and} \quad \lambda_k = \max\{\lambda_1, \ldots, \lambda_{r-1}\} > 0.$$

It follows that

$$0 = \sum_{i=1}^{r-1} \lambda_k^{-1}\lambda_i c_{i,k} \ge \sum_{i=1}^{r-1} c_{i,k} > \sum_{i=1}^{r} c_{i,k} = 0, \qquad \text{a contradiction.}$$ □

Corollary 3.5.9. *Let K be an algebraic number field. Then $K^\times = \mu(K) \cdot T$ for some free abelian group T which is isomorphic to $K^\times/\mu(K)$.*

Proof. We consider the exact sequence

$$\mathbf{1} \to U_K/\mu(K) \to K^\times/\mu(K) \xrightarrow{\partial} \mathcal{H}_K \to \mathbf{1},$$

where $\partial(x\mu(K)) = x\mathfrak{o}_K$ for all $x \in K^\times$. As $\mathcal{H}_K$ is a subgroup of the free abelian group $\mathcal{I}_K$, it is itself free by 2.3.4. By 1.1.2 the exact sequence splits, and

$$K^\times/\mu(K) = T_0 \cdot U_K/\mu(K), \quad \text{where} \quad T_0 \cong \mathcal{H}_K.$$

By 3.5.8 the group $U_K/\mu(K)$ is free, hence $K^\times/\mu(K) = T_0 \cdot U_K/\mu(K)$ is a free abelian group, the exact sequence $\mathbf{1} \to \mu(K) \to K^\times \to K^\times/\mu(K) \to \mathbf{1}$ splits, and therefore $K^\times = \mu(K) \cdot T$ for some free abelian subgroup T of $K^\times$ such that $T \cong K^\times/\mu(K)$. □

Definition and Theorem 3.5.10 (Regulator). *Let K be an algebraic number field, (r_1, r_2) its signature, $[K:\mathbb{Q}] = n = r_1 + 2r_2$ and $r = r_1 + r_2$. Assume that* $\operatorname{Hom}(K,\mathbb{C}) = \{\sigma_1, \ldots, \sigma_n\}$, *where $\sigma_i(K) \subset \mathbb{R}$, $l_i = 1$ for $i \in [1, r_1]$, and $\sigma_{r_1+r_2+i} = \overline{\sigma_{r_1+i}} \ne \sigma_{r_1+i}$, $l_{r_1+i} = 2$ for $i \in [1, r_2]$. Let $\ell = (\ell_1, \ldots, \ell_r)\colon K^\times \to \mathbb{R}^r$ be the logarithmic embedding of K, defined by $\ell_i(x) = l_i \log|\sigma_i(x)|$ for all $i \in [1,r]$, and set*

$$H = \{(x_1, \ldots, x_r) \in \mathbb{R}^r \mid x_1 + \ldots + x_r = 0\} \subset \mathbb{R}^r.$$

We call $r-1$ the **unit rank** of K. If $\mu(K) = \langle\zeta\rangle$, then by Dirichlet's theorem 3.5.8 there exist $\varepsilon_1, \ldots, \varepsilon_{r-1} \in U_K$ such that

$$U_K = \langle \zeta, \varepsilon_1, \ldots, \varepsilon_{r-1}\rangle = \mu(K) \cdot \langle \varepsilon_1, \ldots, \varepsilon_{r-1}\rangle \cong \mu(K) \times \mathbb{Z}^{r-1},$$

and every such $(r-1)$-tuple $(\varepsilon_1,\ldots,\varepsilon_{r-1}) \in U_K^{r-1}$ is called a system of **fundamental units** of K. We denote by $w_K = |\mu(K)|$ the number of roots of unity in K.

Let $s \in [1, r-1]$. An s-tuple $(\eta_1,\ldots,\eta_s) \in U_K^s$ is called a system of **independent units** of K if, for all $(b_1,\ldots,b_s) \in \mathbb{Z}^s$, $\eta_1^{b_1}\cdot\ldots\cdot\eta_s^{b_s} = 1$ implies $b_1 = \ldots = b_s = 0$. An s-tuple $(\eta_1,\ldots,\eta_s) \in U_K^s$ is a system of independent units if and only if $(\ell(\eta_1),\ldots,\ell(\eta_s))$ is linearly independent.

Let U be a subgroup of U_K. Then $\ell(U)$ is a sublattice of $\ell(U_K)$, and by 2.3.6 there exists some $s \in [0, r-1]$ and there exist $\eta_1,\ldots,\eta_s \in U$ such that $(\ell(\eta_1),\ldots,\ell(\eta_s))$ is a basis of $\ell(U)$ (and then $(\eta_1,\ldots,\eta_s)$ is a system of independent units). The epimorphism $\ell \restriction U \colon U \to \ell(U)$ splits, and since $\mathrm{Ker}(\ell \restriction U) = \mu(U) = \mu(K) \cap U$, it follows that $U = \mu(U)\cdot\langle\eta_1,\ldots,\eta_s\rangle$.

The epimorphism $\ell^*\colon U_K/U \to \ell(U_K)/\ell(U)$, defined by $\ell^*(\varepsilon U) = \ell(\varepsilon) + \ell(U)$ for all $\varepsilon \in U$, has the kernel $\mu(K)U/U \cong \mu(K)/\mu(K)\cap U = \mu(K)/\mu(U)$, and yields an exact sequence

$$\mathbf{1} \to \mu(K)/\mu(U) \to U_K/U \to \ell(U_K)/\ell(U) \to \mathbf{0}. \qquad (*)$$

Since $\mu(K)/\mu(U)$ is finite, it follows that U_K/U is finite if and only if $\ell(U_K)/\ell(U)$ is finite. This holds if and only if $s = \mathrm{rk}(\ell(U)) = r-1$, and then we call

$$\mathsf{R}(U) = \mathsf{R}(\eta_1,\ldots,\eta_{r-1}) = \frac{1}{\sqrt{r}}\,\mathrm{vol}(\ell(U))$$

the **regulator** of U or of $(\eta_1,\ldots,\eta_{r-1})$. In particular,

$$\mathsf{R}_K = \mathsf{R}(U_K) = \mathsf{R}(\varepsilon_1,\ldots,\varepsilon_r)$$

is called the **regulator** of K.

If U is a subgroup of U_K and $(U_K\!:\!U) < \infty$, then $(*)$ together with 3.5.1.3 implies

$$\begin{aligned}(U_K\!:\!U) &= (\mu(K)\!:\!\mu(U))(\ell(U_K)\!:\!\ell(U)) = (\mu(K)\!:\!\mu(U))\frac{\mathrm{vol}(\ell(U))}{\mathrm{vol}(\ell(U_K))}\\ &= (\mu(K)\!:\!\mu(U))\,\frac{\mathsf{R}(U)}{\mathsf{R}_K}.\end{aligned}$$

Let $\mathfrak{o}$ be an order of K. Then $\mathfrak{o}^\times = \langle\zeta,\eta_1,\ldots,\eta_{r-1}\rangle \cong \mu(\mathfrak{o})\times\mathbb{Z}^{r-1}$, where $\mu(\mathfrak{o}) = \mu(K)\cap\mathfrak{o} = \langle\zeta\rangle$ (see 3.5.8). In particular, $(\eta_1,\ldots,\eta_{r-1})$ is a system of independent units of K, and we call $\mathsf{R}(\mathfrak{o}) = \mathsf{R}(\eta_1,\ldots,\eta_{r-1})$ the **regulator** of $\mathfrak{o}$.

1. Let $(\eta_1,\ldots,\eta_{r-1})$ be a system of independent units of K. If λ is any minor of order $r-1$ of the matrix

$$\Lambda(\eta_1,\ldots,\eta_{r-1}) = \begin{pmatrix} l_1\log|\sigma_1(\eta_1)| & \ldots & l_1\log|\sigma_1(\eta_{r-1})| \\ \cdot & \ldots & \cdot \\ \cdot & \ldots & \cdot \\ l_r\log|\sigma_r(\eta_1)| & \ldots & l_r\log|\sigma_r(\eta_{r-1})| \end{pmatrix} \in \mathsf{M}_{r,r-1}(\mathbb{R}),$$

then $|\lambda| = \mathsf{R}(\eta_1,\ldots,\eta_{r-1})$.

2. Let $U' \subset U \subset U_K$ be subgroups such that $(U_K\!:\!U') < \infty$. Then

$$(U\!:\!U') = \frac{|\mu(U)|}{|\mu(U')|}\,\frac{\mathsf{R}(U')}{\mathsf{R}(U)}.$$

In particular, if $\mathfrak{o}$ is an order of K, then

$$(U_K\!:\!\mathfrak{o}^\times) = \frac{|\mu(K)|}{|\mu(\mathfrak{o})|}\,\frac{\mathsf{R}(\mathfrak{o})}{\mathsf{R}_K}.$$

Proof. 1. Let $U = \langle \eta_1, \ldots, \eta_{r-1}\rangle$. Then $\ell(U)$ is a lattice in H with basis $(\ell(\eta_1), \ldots, \ell(\eta_{r-1}))$, $H = \mathbb{R}\ell(U)$, and

$$\mathbb{R}^r = H \perp \mathbb{R}\boldsymbol{e}, \quad \text{where} \quad \boldsymbol{e} = \frac{1}{\sqrt{r}}(1, \ldots, 1) \in \mathbb{R}^r.$$

By definition,

$$\mathrm{vol}(\ell(U)) = |\det(\ell(\eta_1), \ldots, \ell(\eta_{r-1}), \boldsymbol{e})| = \frac{1}{\sqrt{r}}\left|\det\begin{pmatrix} \ell_1(\eta_1) & \ldots & \ell_1(\eta_{r-1}) & 1 \\ \cdot & \ldots & \cdot & \cdot \\ \cdot & \ldots & \cdot & \cdot \\ \ell_r(\eta_1) & \ldots & \ell_r(\eta_{r-1}) & 1 \end{pmatrix}\right|.$$

Now let $i \in [1, r]$. If in the matrix above we replace the i-th row with the sum of all rows, then we obtain

$$\mathrm{vol}(\ell(U)) = \frac{1}{\sqrt{r}}\left|\det\begin{pmatrix} \ell_1(\eta_1) & \ldots\ldots & \ell_1(\eta_{r-1} & 1 \\ \cdot & \ldots\ldots & \cdot & \cdot \\ \ell_{i-1}(\eta_1) & \ldots\ldots & \ell_{i-1}(\eta_{r-1} & 1 \\ 0 & \ldots\ldots & 0 & r \\ \ell_{i+1}(\eta_1) & \ldots\ldots & \ell_{i+1}(\eta_{r-1} & 1 \\ \cdot & \ldots\ldots & \cdot & \cdot \\ \ell_r(\eta_1) & \ldots\ldots & \ell_r(\eta_{r-1} & 1 \end{pmatrix}\right| = \sqrt{r}\,|\det(\Lambda_i)|$$

where $\Lambda_i \in \mathsf{M}_{r-1}(\mathbb{R})$ is the submatrix of $\Lambda(\eta_1, \ldots, \eta_{r-1})$ which is formed by deleting the i-th row. Hence $\det(\Lambda_i)$ is the i-th minor of order $r-1$ of $\Lambda(\eta_1, \ldots, \eta_{r-1})$, and $\mathsf{R}(\eta_1, \ldots, \eta_{r-1}) = \mathsf{R}(U) = |\det(\Lambda_i)|$.

2. The logarithmic embedding induces an exact sequence

$$\mathbf{1} \to \mu(U)/\mu(U') \to U/U' \to \ell(U)/\ell(U') \to \mathbf{0},$$

and together with 3.5.1.3 it follows that

$$(U\!:\!U') = \frac{|\mu(U)|}{|\mu(U')|}\,(\ell(U)\!:\!\ell(U')) = \frac{|\mu(U)|}{|\mu(U')|}\,\frac{\mathrm{vol}(\ell(U'))}{\mathrm{vol}(\ell(U))} = \frac{|\mu(U)|}{|\mu(U')|}\,\frac{\mathsf{R}(U')}{\mathsf{R}(U)}. \quad \square$$

Remarks 3.5.11 (Fields with unit rank 0 or 1). Let K be an algebraic number field, (r_1, r_2) its signature and $s = r_1 + r_2 - 1$ its unit rank.

$\boldsymbol{s = 0}$ **:** Then either $K = \mathbb{Q}$ $(r_1 = 1,\ r_2 = 0)$ or K is an imaginary quadratic number field $(r_1 = 0,\ r_2 = 1)$. In these cases $U_K = \mu(K)$, and $|\mu(K)| = m$ if m is maximal such that $\mathbb{Q}^{(m)} \subset K$. Clearly, $\mu(\mathbb{Q}) = \{\pm 1\}$, and if K is imaginary quadratic, then

$$|\mu(K)| = \begin{cases} 6 & \text{if } \mathsf{d}_K = -3, \\ 4 & \text{if } \mathsf{d}_K = -4, \\ 2 & \text{if } \mathsf{d}_K < -4. \end{cases}$$

$\boldsymbol{s = 1}$ **:** Then either K is real quadratic $(r_1 = 2,\ r_2 = 0)$, or K is simply real cubic $(r_1 = r_2 = 1)$, or K is totally imaginary quartic $(r_1 = 0,\ r_2 = 2)$. In these cases $U_K = \mu(K) \cdot \langle \varepsilon \rangle = \langle \zeta, \varepsilon \rangle$ for some root of unity $\zeta \in \mu(K)$ and some unit $\varepsilon \in U_K$.

If $r_1 > 0$, we may assume that $K \subset \mathbb{R}$, $\mu_K = \{\pm 1\}$. If $\varepsilon' \in U_K$ is such that $U_K = \langle -1, \varepsilon \rangle = \langle -1, \varepsilon' \rangle$, then $\varepsilon' \in \{\pm\varepsilon, \pm\varepsilon^{-1}\}$. Hence there is a uniquely determined $\varepsilon_0 \in U_K$ satisfying $\varepsilon_0 > 1$ and $U_K = \langle -1, \varepsilon_0 \rangle$. We call this ε_0 the **fundamental unit** of K. Note that $\{\varepsilon \in U_K \mid \varepsilon > 1\} = \{\varepsilon_0^n \mid n \in \mathbb{N}\}$, $\varepsilon_0 = \min\{\varepsilon \in U_K \mid \varepsilon > 1\}$ and $\mathsf{R}_K = \log \varepsilon_0$.

If $r_2 = 2$, then $[K:\mathbb{Q}] = 4$, and we assert that $|\mu(K)| \in \{2, 4, 6, 8, 10, 12\}$. Indeed, $|\mu(K)| = m$ if m is maximal such that $\mathbb{Q}^{(m)} \subset K$. If $\mathbb{Q}^{(m)} \subset K$, then $\varphi(m) \mid 4$, it is easily checked that this holds if and only if $m \in \{1, 2, 3, 4, 5, 6, 8, 10, 12\}$. However, $\mathbb{Q}^{(1)} = \mathbb{Q}^{(2)} = \mathbb{Q}$, $\mathbb{Q}^{(3)} = \mathbb{Q}^{(6)}$ and $\mathbb{Q}^{(5)} = \mathbb{Q}^{(10)}$.

3.6 Class groups of algebraic number fields

Let K be an algebraic number field. In 3.5.7.4 we have seen that the (ideal) class group $\mathcal{C}_K = \mathcal{C}(\mathfrak{o}_K)$ is finite, and by 2.9.9.1 we know that every ideal class contains ideals coprime to a given one. In this section we investigate class groups and their generalizations more closely. We start with a criterion which allows us to calculate the class number.

Theorem 3.6.1. *Let K be an algebraic number field, (r_1, r_2) is its signature, and $[K:\mathbb{Q}] = n = r_1 + 2r_2$. Then every ideal class $C \in \mathcal{C}_K$ contains an ideal $\mathfrak{c} \in \mathcal{J}'_K$ satisfying*

$$\mathfrak{N}(\mathfrak{c}) \le M_K = \frac{n!}{n^n} \Big(\frac{4}{\pi}\Big)^{r_2} \sqrt{|\mathsf{d}_K|}.$$

Proof. Let $\mathfrak{a} \in \mathcal{I}'_K$ be an ideal in C^{-1}. By 3.5.5.**B**.2 there exists some $\alpha \in \mathfrak{a}$ such that $|\mathsf{N}_{K/\mathbb{Q}}(\alpha)| \le M_K \mathfrak{N}(\mathfrak{a})$. Then $\mathfrak{c} = \alpha\mathfrak{a}^{-1} \in C \cap \mathcal{I}'_K$, and from 3.1.4 we get $\mathfrak{N}(\mathfrak{c}) = |\mathsf{N}_{K/\mathbb{Q}}(\alpha)|\, \mathfrak{N}(\mathfrak{a})^{-1} \le M_K$. □

We proceed with a few examples to show how the Minkowski bound can be used to calculate the class number. For quadratic number fields (and more generally for quadratic orders) we shall prove in 3.8.8 that the class number equals the number of equivalenc classes of quadratic irraionals. The calculation ot the latter can efficiently be accomplished using the reduction theory of quadratic irrationals. We refer to [22] and [47].

Example 3.6.2. Let K be an algebraic number field. If $M_K < 2$, then $h_K = 1$ since the $\mathbf{1} = \mathfrak{o}_K$ is the only ideal of $\mathfrak{o}_K$ with absolute norm 1.

a. If K is a real-quadratic number field, then $r_1 = 2,\ r_2 = 0$ and since $M_K = \sqrt{\mathsf{d}_K}/2$, it follows that $h_K = 1$ if $\mathsf{d}_K \in \{5, 8, 13\}$. Using other methods, we have seen quadratic number fields with class number 1 in the discussion after 3.1.8.

b. Let $K = \mathbb{Q}(x)$, where $x \in \mathbb{R}$ and $x^3 = x + 1$. Note that the polynomial $g = X^3 - X - 1$ is irreducible and has a unique real zero. Hence K is a simply real cubic number field, $r_1 = r_2 = 1$, and

$$M_K = \frac{3!}{3^3}\,\frac{4}{\pi}\,\sqrt{23} < 2 \qquad \text{implies} \qquad h_K = 1.$$

Example 3.6.3. Let $K = \mathbb{Q}(\sqrt{10})$. Then $r_1 = 1,\ r_2 = 0,\ \mathfrak{o}_K = \mathbb{Z}[\sqrt{10}\,]$ and $\mathsf{d}_K = 40$, hence $M_K = \sqrt{40}/2 \approx 3.16$. By 3.2.4 we obtain $2\mathfrak{o}_K = \mathfrak{z}^2$ and $3\mathfrak{o}_K = \mathfrak{p}\mathfrak{p}'$, where $\mathfrak{z},\ \mathfrak{p},\ \mathfrak{p}' \in \mathcal{P}_K,\ \ \mathfrak{p} \ne \mathfrak{p}',\ \ \mathfrak{N}(\mathfrak{z}) = 2$ and $\mathfrak{N}(\mathfrak{p}) = \mathfrak{N}(\mathfrak{p}') = 3$. We assert that $\pm 2 \notin \mathsf{N}_{K/\mathbb{Q}}(\mathfrak{o}_K)$. Indeed, otherwise there exist $x,\ y \in \mathbb{Z}$ such that $\pm 2 = \mathsf{N}_{K/\mathbb{Q}}(x + y\sqrt{10}\,) = x^2 - 10y^2 \equiv x^2 \bmod 5$, which is impossible. Hence $\mathfrak{z}$ is not a principal ideal. Since $-6 = \mathsf{N}_{K/\mathbb{Q}}(2 \pm \sqrt{10}\,)$ and $6 = \mathfrak{N}(\mathfrak{z}\mathfrak{p}) = \mathfrak{N}(\mathfrak{z}\mathfrak{p}')$, it follows that $\mathfrak{z}\mathfrak{p}$ and $\mathfrak{z}\mathfrak{p}'$ are principal ideals, and therefore $\mathcal{C}_K = \{[\mathbf{1}],\ [\mathfrak{p}]\}$, $[\mathfrak{p}]^2 = [\mathbf{1}],\ \ [\mathfrak{z}] = [\mathfrak{p}] = [\mathfrak{p}']$ and $h_K = 2$.

Example 3.6.4. Let $K = \mathbb{Q}(\sqrt{-14}\,)$. Then $r_1 = 0,\ \ r_2 = 1,\ \ \mathfrak{o}_K = \mathbb{Z}[\sqrt{-14}\,]$ and $\mathsf{d}_K = 56$, hence

$$M_K = \frac{2!}{2^2}\,\frac{4}{\pi}\,\sqrt{56} \approx 4.76.$$

By 3.2.4, $2\mathfrak{o}_K = \mathfrak{z}^2$ and $3\mathfrak{o}_K = \mathfrak{p}\mathfrak{p}'$, where $\mathfrak{z},\ \mathfrak{p},\ \mathfrak{p}' \in \mathcal{P}_K,\ \mathfrak{p} \ne \mathfrak{p}',\ \mathfrak{N}(\mathfrak{z}) = 2$ and $\mathfrak{N}(\mathfrak{p}) = \mathfrak{N}(\mathfrak{p}') = 3$. If $x,\ y \in \mathbb{Z}^\bullet$, then $\mathsf{N}_{K/\mathbb{Q}}(x + y\sqrt{-14}) = x^2 + 14y^2 \ge 15$, and thus $\mathfrak{z},\ \mathfrak{p},\ \mathfrak{p}',\ \mathfrak{p}^2,\ \mathfrak{p}'^2$ are not principal ideals.

Since $\mathfrak{N}(\mathfrak{p}^4) = 81 = \mathsf{N}_{K/\mathbb{Q}}(5 + 2\sqrt{-14})$, the ideals $\mathfrak{p}^4$ and $\mathfrak{p}'^4$ are principal, and since $\mathfrak{N}(\mathfrak{z}\mathfrak{p}^2) = \mathfrak{N}(\mathfrak{z}\mathfrak{p}'^2) = \mathsf{N}_{K/\mathbb{Q}}(2 \pm \sqrt{-14})$, the ideals $\mathfrak{z}\mathfrak{p}^2$ and $\mathfrak{z}\mathfrak{p}'^2$ are principal. Hence $\mathcal{C}_K = \{[\mathbf{1}],\ [\mathfrak{p}], [\mathfrak{p}]^2,\ [\mathfrak{p}]^3\}$, $[\mathfrak{p}]^4 = [\mathbf{1}]$, $[\mathfrak{p}'] = [\mathfrak{p}]^{-\mathbf{1}}$, $[\mathfrak{z}] = [\mathfrak{p}]^{\mathbf{2}}$, $h_K = 4$ and $\mathcal{C}_K$ is cyclic.

Example 3.6.5. Let $K = \mathbb{Q}(\theta)$, where $\theta^3 = m \in \{2, 5, 7\}$. Then $r_1 = r_2 = 1$, $\mathfrak{o}_K = \mathbb{Z}[\theta]$ and $\mathsf{d}_K = -27m^2$ by 3.1.9, hence

$$M_K = \frac{3!}{3^3}\,\frac{4}{\pi}\,3m\sqrt{3} \approx \begin{cases} 2.94 & \text{if } m = 2, \\ 7.35 & \text{if } m = 5, \\ 10.29 & \text{if } m = 7. \end{cases}$$

If $\alpha = a + b\theta + c\theta^2 \in \mathfrak{o}_K$, where $a,\, b,\, c \in \mathbb{Z}$, then

$$\begin{pmatrix} \alpha \\ \alpha\theta \\ \alpha\theta^2 \end{pmatrix} = \begin{pmatrix} a & b & c \\ mc & a & b \\ mb & mc & a \end{pmatrix} \begin{pmatrix} 1 \\ \theta \\ \theta^2 \end{pmatrix}$$

and

$$\mathsf{N}_{K/\mathbb{Q}}(\alpha) = \det \begin{pmatrix} a & b & c \\ mc & a & b \\ mb & mc & a \end{pmatrix} = a^3 + mb^3 + m^2c^3 - 3mabc.$$

In the sequel we use 3.2.5.

$\boldsymbol{m = 2}$**:** $2\mathfrak{o}_K = \mathfrak{p}_2^3$ and $\mathfrak{p}_2 = \theta\mathfrak{o}_K$. Hence $h_K = 1$.

$\boldsymbol{m = 5}$**:** $2\mathfrak{o}_K = \mathfrak{p}_2\mathfrak{p}_2'$ $(f_{\mathfrak{p}_2} = 1,\ f_{\mathfrak{p}_2'} = 2)$, $3\mathfrak{o}_K = \mathfrak{p}_3^3$, $5\mathfrak{o}_K = \mathfrak{p}_5^3$, $\mathfrak{p}_5 = \theta\mathfrak{o}_K$ and $7\mathfrak{o}_K \in \mathcal{P}_K$. Since $\mathsf{N}_{K/\mathbb{Q}}(1+\theta) = 6$, it follows that $\mathfrak{p}_2\mathfrak{p}_3 = (1+\theta)\mathfrak{o}_K$, and since $\mathsf{N}_{K/\mathbb{Q}}(-2+\theta) = -3$, we obtain $\mathfrak{p}_3 = (-2+\theta)\mathfrak{o}_K$.

As $(-2+\theta)(3+2\theta+\theta^2)) = -(1+\theta)$, it follows that $\mathfrak{p}_2 = (3+2\theta+\theta^2)\mathfrak{o}_K$, and as $(3+2\theta+\theta^2)(1+\theta-\theta^2) = -2$, it follows that $\mathfrak{p}_2' = (1+\theta-\theta^2)\mathfrak{o}_K$. Hence all ideals $\mathfrak{a} \in \mathcal{J}_K'$ with $\mathfrak{N}(\mathfrak{a}) \le M_K$ are principal and thus $h_K = 1$.

$\boldsymbol{m = 7}$**:** $2\mathfrak{o}_K = \mathfrak{p}_2\mathfrak{p}_2'$ $(f_{\mathfrak{p}_2} = 1,\ f_{\mathfrak{p}_2'} = 2)$, $3\mathfrak{o}_K = \mathfrak{p}_3^3$, $5\mathfrak{o}_K = \mathfrak{p}_5\mathfrak{p}_5'$ $(f_{\mathfrak{p}_5} = 1,\ f_{\mathfrak{p}_5'} = 2)$ and $7\mathfrak{o}_K = \mathfrak{p}_7^3$. Hence we obtain the following list of all ideals $\mathfrak{a} \in \mathcal{J}_K'$ satisfying $\mathfrak{N}(\mathfrak{a}) \le 10$:

$\mathfrak{a}$	$\mathfrak{p}_2$	$\mathfrak{p}_2^2$	$\mathfrak{p}_2^3$	$\mathfrak{p}_2\mathfrak{p}_2'$	$\mathfrak{p}_2\mathfrak{p}_3$	$\mathfrak{p}_2\mathfrak{p}_5$	$\mathfrak{p}_2'$	$\mathfrak{p}_3$	$\mathfrak{p}_5$	$\mathfrak{p}_7$
$\mathfrak{N}(\mathfrak{a})$	2	4	8	8	6	10	4	3	5	7

By definition, $\mathfrak{p}_7 = \theta\mathfrak{o}_K$. Since $\mathsf{N}_{K/\mathbb{Q}}(1+\theta) = 8$ and $2 \nmid 1+\theta$, we obtain $\mathfrak{p}_2^3 = (1+\theta)\mathfrak{o}_K$. Since $\mathsf{N}_{K/\mathbb{Q}}(-1+\theta) = 6$ and $\mathsf{N}_{K/\mathbb{Q}}(2+\theta) = 15$, it follows that $\mathfrak{p}_2\mathfrak{p}_3 = (1+\theta)\mathfrak{o}_K$ and $\mathfrak{p}_2\mathfrak{p}_5 = (2+\theta)\mathfrak{o}_K$. Hence the ideals $\mathfrak{p}_2^3$, $\mathfrak{p}_2\mathfrak{p}_2'$, $\mathfrak{p}_2\mathfrak{p}_3$, $\mathfrak{p}_3\mathfrak{p}_5$ and $\mathfrak{p}_7$ are principal, and therefore the following relations hold in $\mathcal{C}_K$: $[\mathfrak{p}_2]^3 = [\mathbf{1}]$, $[\mathfrak{p}_2'] = [\mathfrak{p}_3] = [\mathfrak{p}_2]^{-1}$ and $[\mathfrak{p}_5] = [\mathfrak{p}_3]^{-1} = [\mathfrak{p}_2]$. It follows that $\mathcal{C}_K = \langle[\mathfrak{p}_2]\rangle$, and we assert that $\mathfrak{p}_2$ is not principal. Indeed, assume to the contrary that $\mathfrak{p}_2 = \alpha\mathfrak{o}_K$, where $\alpha = a + b\theta + c\theta^2$ for some $a,\, b,\, c \in \mathbb{Z}$. Then $\pm 2 = \mathsf{N}(\alpha) = a^3 + 7b^3 + 49c^3 - 21abc \equiv a^3 \bmod 7$, a contradiction. Overall we obtain $h_K = 3$ and $\mathcal{C}_K = \{[\mathbf{1}],\, [\mathfrak{p}_2],\, [\mathfrak{p}_2^2]\}$.

Next we adjust and amend the notions of generalized ideal class groups and ray class groups as introduced in 2.6.5 and 2.9.8 for algebraic number fields.

Definition 3.6.6. Let K be an algebraic number field with signature (r_1, r_2), $[K:\mathbb{Q}] = n = r_1 + 2r_2$ and $\mathfrak{m} \in \mathcal{J}'_K$. Let $\boldsymbol{\sigma}(K)$ be the set of all real embeddings of K, hence $|\boldsymbol{\sigma}(K)| = r_1$, and let $\boldsymbol{\sigma} \subset \boldsymbol{\sigma}(K)$. Recall that a number $x \in K^\times$ is called $\boldsymbol{\sigma}$-positive if $\sigma(x) > 0$ for all $\sigma \in \boldsymbol{\sigma}$, and that $K^{\boldsymbol{\sigma}}$ denotes the set of all $\boldsymbol{\sigma}$-positive elements of $K^\times$. In particular, $K^{\emptyset} = K^\times$, and $x \in K$ is totally positive, $x \gg 0$, if and only if x is $\boldsymbol{\sigma}(K)$-positive.

Recall that $K(\mathfrak{m})$ denotes the set of all $x \in K$ which are coprime to $\mathfrak{m}$, and

$$\begin{aligned} K^{\circ\mathfrak{m}} &= \{x \in K(\mathfrak{m}) \mid x \equiv 1 \bmod \mathfrak{m}\} \\ &= \{x \in K^\times \mid v_{\mathfrak{p}}(x-1) \ge v_{\mathfrak{p}}(\mathfrak{m})\} = \mathsf{q}(1+\mathfrak{m}) \end{aligned}$$

by 2.9.5. In particular, $K^{\circ 1} = K^\times$.

In addition, we consider the following groups:

- $K^+ = K^{\boldsymbol{\sigma}(K)} = \{x \in K \mid x \gg 0\}$, the group of all totally positive numbers,
- $K^{\mathfrak{m},\boldsymbol{\sigma}} = K^{\circ\mathfrak{m}} \cap K^{\boldsymbol{\sigma}}$, and
- $K^{\mathfrak{m}} = K^{\mathfrak{m},\boldsymbol{\sigma}(K)} = \{x \in K(\mathfrak{m}) \mid x \gg 0 \text{ and } x \equiv 1 \bmod \mathfrak{m}\} = K^{\circ\mathfrak{m}} \cap K^+$.
  In particular, $K^{1,\boldsymbol{\sigma}} = K^{\boldsymbol{\sigma}}$, $K^1 = K^+$, and $K^{\mathfrak{m},\emptyset} = K^{\circ\mathfrak{m}}$.
- $U_K^{\mathfrak{m},\boldsymbol{\sigma}} = U_K \cap K^{\mathfrak{m},\boldsymbol{\sigma}}$.
  Note that $(U_K : U_K^{\mathfrak{m},\boldsymbol{\sigma}}) < \infty$. Indeed, if $m = |(\mathfrak{o}_K/\mathfrak{m})^\times|$ and $\varepsilon \in U_K$, then $\varepsilon^{2m} \in U_K^{\mathfrak{m},\boldsymbol{\sigma}}$. Hence $U_K/U_K^{\mathfrak{m},\boldsymbol{\sigma}}$ is a finitely generated torsion group and thus finite.
- $U_K^{\mathfrak{m}} = U_K \cap K^{\mathfrak{m}}$, $U_K^{\circ\mathfrak{m}} = U_K \cap K^{\circ\mathfrak{m}}$, and $U_K^1 = U_K^+ = U_K \cap K^+$.
- $\mathcal{H}_K = \mathcal{H}(\mathfrak{o}_K)$, the group of fractional principal ideals;
- $\mathcal{J}_K^{\mathfrak{m}} = \mathcal{J}^{\mathfrak{m}}(\mathfrak{o}_K)$, the group of all fractional ideals of $\mathfrak{o}_K$ which are coprime to $\mathfrak{m}$;
- $\mathcal{H}_K^{\mathfrak{m},\boldsymbol{\sigma}} = \mathcal{H}^{\mathfrak{m},\boldsymbol{\sigma}}(\mathfrak{o}_K) = \{a\mathfrak{o}_K \mid a \in K^{\mathfrak{m},\boldsymbol{\sigma}}\}$;
- $\mathcal{H}_K^{\mathfrak{m}} = \mathcal{H}_K^{\mathfrak{m},\boldsymbol{\sigma}(K)} = \{a\mathfrak{o}_K \mid a \gg 0 \text{ and } a \equiv 1 \bmod \mathfrak{m}\}$, the **principal ray** modulo $\mathfrak{m}$;
- $\mathcal{H}_K^{\circ\mathfrak{m}} = \mathcal{H}_K^{\mathfrak{m},\emptyset} = \{a\mathfrak{o}_K \mid a \equiv 1 \bmod \mathfrak{m}\}$, the **large principal ray** modulo $\mathfrak{m}$;
- $\mathcal{C}_K^{\mathfrak{m},\boldsymbol{\sigma}} = \mathcal{C}^{\mathfrak{m},\boldsymbol{\sigma}}(\mathfrak{o}_K) = \mathcal{J}_K^{\mathfrak{m}}/\mathcal{H}_K^{\mathfrak{m},\boldsymbol{\sigma}}$, the **$\boldsymbol{\sigma}$-ray class group** modulo $\mathfrak{m}$,
- $\mathcal{C}_K^{\mathfrak{m}} = \mathcal{C}_K^{\mathfrak{m},\boldsymbol{\sigma}(K)} = \mathcal{J}_K^{\mathfrak{m}}/\mathcal{H}_K^{\mathfrak{m}}$, the **ray class group** modulo $\mathfrak{m}$, and
- $\mathcal{C}_K^{\circ\mathfrak{m}} = \mathcal{C}_K^{\mathfrak{m},\emptyset} = \mathcal{J}_K^{\mathfrak{m}}/\mathcal{H}_K^{\circ\mathfrak{m}}$, the **small ray class group** modulo $\mathfrak{m}$.

Ray class groups are of central importance in class field theory. What we do here is merely a preliminary consideration. For a fractional ideal $\mathfrak{a} \in \mathcal{J}_K^{\mathfrak{m}}$ we denote by $[\mathfrak{a}]^{\mathfrak{m},\boldsymbol{\sigma}} = \mathfrak{a}\mathcal{H}_K^{\mathfrak{m},\boldsymbol{\sigma}} \in \mathcal{C}_K^{\mathfrak{m},\boldsymbol{\sigma}}$ the **$\boldsymbol{\sigma}$-ray class**, by $[\mathfrak{a}]^{\mathfrak{m}} = \mathfrak{a}\mathcal{H}_K^{\mathfrak{m}} \in \mathcal{C}_K^{\mathfrak{m}}$ the **ray class** and by $[\mathfrak{a}]^{\circ\mathfrak{m}} = \mathfrak{a}\mathcal{H}_K^{\circ\mathfrak{m}} \in \mathcal{C}_K^{\circ\mathfrak{m}}$ the **small ray class** modulo $\mathfrak{m}$ of $\mathfrak{a}$. In the special case $\mathfrak{m} = \mathbf{1}$ we obtain $\mathcal{H}_K^{\mathbf{1}} = \mathcal{H}_K^{+} = \{\alpha\mathfrak{o}_K \mid \alpha \in K^+\}$ and $\mathcal{H}_K^{\circ\mathbf{1}} = \mathcal{H}_K$. The class group $\mathcal{C}_K^{\mathbf{1}} = \mathcal{C}_K^{+} = \mathcal{J}_K/\mathcal{H}_K^{+}$ is called the **narrow class group** and its order $h_K^{+} = |\mathcal{C}_K^{+}|$ is called the **narrow class number** of K. For $\mathfrak{a} \in \mathcal{J}_K$ we denote by $[\mathfrak{a}]^{+} = \mathfrak{a}\mathcal{H}_K^{+}$ the **narrow ideal class** of $\mathfrak{a}$. Finally, $\mathcal{C}_K^{\circ\mathbf{1}} = \mathcal{C}_K$ is the (ordinary) class group of K.

We finally discuss ray class groups in $\mathbb{Q}$. Apparently $\mathbb{Q}^+ = \mathbb{Q}_{>0} = \mathsf{q}(\mathbb{N})$, and we continue the discussion made at the beginning of Section 3.1. There we identified $\mathcal{J}_{\mathbb{Q}} = \mathcal{H}_{\mathbb{Q}} = \mathbb{Q}^+$ and $\mathcal{J}'_{\mathbb{Q}} = \mathbb{N}$. The inclusion $\mathbb{Q} \hookrightarrow \mathbb{R}$ is the unique real embedding of $\mathbb{Q}$. In 3.1.1 we identified $\mathbb{N} = \mathcal{J}'_{\mathbb{Q}}$, for $m \in \mathbb{N}$ we set $\mathbb{Z}(m) = \{a \in \mathbb{Z} \mid (a, m) = 1\}$,

$$\mathbb{Q}(m) = \mathsf{q}(\mathbb{Z}(m)) = \{a^{-1}b \mid a \in \mathbb{N},\ b \in \mathbb{Z},\ a \equiv 1 \bmod m,\ (b, m) = 1\},$$

and we extended the residue class epimorphism $\pi_m \colon \mathbb{Z}(m) \to (\mathbb{Z}/m\mathbb{Z})^\times$ by virtue of quotients to a unique (equally denoted) group epimorphism $\pi_m \colon \mathbb{Q}(m) \to (\mathbb{Z}/m\mathbb{Z})^\times$ with kernel

$$\begin{aligned}\mathbb{Q}^{\circ m} = \mathsf{q}(1 + m\mathbb{Z}) &= \{x \in \mathbb{Q}(m) \mid x \equiv 1 \bmod m\} \\ &= \{a^{-1}b \mid a \in \mathbb{N},\ b \in \mathbb{Z},\ a \equiv b \equiv 1 \bmod m\} \quad \text{(see 2.9.5.2)}.\end{aligned}$$

By virtue of the identification $\mathcal{J}_{\mathbb{Q}} = \mathbb{Q}^+$ we obtain

$$\mathcal{J}_{\mathbb{Q}}^m = \mathbb{Q}^+(m) = \mathbb{Q}(m) \cap \mathbb{Q}^+, \quad \mathcal{H}_{\mathbb{Q}}^m = \mathbb{Q}^m = \mathbb{Q}^{\circ m} \cap \mathbb{Q}^+,$$

and

$$\mathcal{H}_{\mathbb{Q}}^{\circ m} = \mathbb{Q}_{\pm}^m = \{a \in \mathbb{Q}(m) \cap \mathbb{Q}^+ \mid a \equiv \pm 1 \bmod m\}.$$

The residue class epimorphism $\pi_m \colon \mathbb{Q}(m) \to (\mathbb{Z}/m\mathbb{Z})^\times$ satisfies

$$\pi_m(\mathbb{Q}^+(m)) = (\mathbb{Z}/m\mathbb{Z})^\times, \quad \pi_m(\mathbb{Q}^m) = 1 + m\mathbb{Z} \quad \text{and} \quad \pi_m(\mathbb{Q}_{\pm}^m) = \pm 1 + m\mathbb{Z}.$$

Hence π_m induces isomorphisms

$$\begin{aligned}\mathcal{C}_{\mathbb{Q}}^m &= \mathbb{Q}^+(m)/\mathbb{Q}^m \overset{\sim}{\to} (\mathbb{Z}/m\mathbb{Z})^\times \quad \text{and} \\ \mathcal{C}_{\mathbb{Q}}^{\circ m} &= \mathbb{Q}^+(m)/\mathbb{Q}_{\pm}^m \overset{\sim}{\to} (\mathbb{Z}/m\mathbb{Z})^\times/\{\pm 1 + m\mathbb{Z}\}.\end{aligned}$$

We identify: $\mathcal{C}_{\mathbb{Q}}^m = (\mathbb{Z}/m\mathbb{Z})^\times$ and $\mathcal{C}_{\mathbb{Q}}^{\circ m} = (\mathbb{Z}/m\mathbb{Z})^\times/\{\pm 1 + m\mathbb{Z}\}$.

The following Theorem 3.6.7 relates the order of the various generalized class groups with the ordinary class number of K.

Theorem 3.6.7. *Let K be an algebraic number field, (r_1, r_2) its signature and $\mathfrak{m} \in \mathcal{J}'_K$.*

1. *Let σ be a set of real embeddings of K. Then*

$$|\mathcal{C}_K^{\mathfrak{m},\sigma}| = \frac{2^{|\sigma|} h_K\, \varphi_K(\mathfrak{m})}{(U_K : U_K^{\mathfrak{m},\sigma})}, \quad \textit{where} \quad \varphi_K(\mathfrak{m}) = |(\mathfrak{o}_K/\mathfrak{m})^\times|.$$

In particular,

$$|\mathcal{C}_K^{\mathfrak{m}}| = \frac{2^{r_1} h_K\, \varphi_K(\mathfrak{m})}{(U_K : U_K^{\mathfrak{m}})}, \quad |\mathcal{C}_K^{\circ\mathfrak{m}}| = \frac{h_K\, \varphi_K(\mathfrak{m})}{(U_K : U_K^{\circ\mathfrak{m}})} \quad \textit{and} \quad |\mathcal{C}_K^{+}| = \frac{2^{r_1} h_K}{(U_K : U_K^{+})}.$$

2. *There are (natural) exact sequences*

$$\mathbf{1} \to \mathcal{H}_K^{\circ\mathfrak{m}}/\mathcal{H}_K^{\mathfrak{m}} \to \mathcal{C}_K^{\mathfrak{m}} \to \mathcal{C}_K^{\circ\mathfrak{m}} \to \mathbf{1}$$

and

$$\mathbf{1} \to U_K^{\circ\mathfrak{m}}/U_K^{\mathfrak{m}} \to K^{\circ\mathfrak{m}}/K^{\mathfrak{m}} \to \mathcal{H}_K^{\circ\mathfrak{m}}/\mathcal{H}_K^{\mathfrak{m}} \to \mathbf{1}.$$

In particular,

$$\frac{|\mathcal{C}_K^{\mathfrak{m}}|}{|\mathcal{C}_K^{\circ\mathfrak{m}}|} = (\mathcal{H}_K^{\circ\mathfrak{m}} : \mathcal{H}_K^{\mathfrak{m}}) = \frac{2^{r_1}}{(U_K^{\circ\mathfrak{m}} : U_K^{\mathfrak{m}})}.$$

Proof. 1. By 2.9.2 the map $\rho\colon \mathcal{J}_K^{\mathfrak{m}} \to \mathcal{J}_K/\mathcal{H}_K = \mathcal{C}_K$, defined by $\rho(\mathfrak{a}) = [\mathfrak{a}]$ for all $\mathfrak{a} \in \mathcal{J}_K^{\mathfrak{m}}$, is an epimorphism with kernel $\mathcal{H}_K(\mathfrak{m})$, and therefore

$$|\mathcal{C}_K^{\mathfrak{m},\sigma}| = (\mathcal{J}_K^{\mathfrak{m}} : \mathcal{H}_K^{\mathfrak{m},\sigma}) = (\mathcal{J}_K^{\mathfrak{m}} : \mathcal{H}_K(\mathfrak{m}))(\mathcal{H}_K(\mathfrak{m}) : \mathcal{H}_K^{\mathfrak{m},\sigma}) = h_K\, (\mathcal{H}_K(\mathfrak{m}) : \mathcal{H}_K^{\mathfrak{m},\sigma}).$$

The map $\theta\colon K(\mathfrak{m}) \to \mathcal{H}_K(\mathfrak{m})/\mathcal{H}_K^{\mathfrak{m},\sigma}$, defined by $\theta(a) = (a\mathfrak{o}_K)\mathcal{H}_K^{\mathfrak{m},\sigma}$, is an epimorphism with kernel $U_K K^{\mathfrak{m},\sigma}$. Since $U_K K^{\mathfrak{m},\sigma}/K^{\mathfrak{m},\sigma} \cong U_K/U_K^{\mathfrak{m},\sigma}$, we obtain

$$(\mathcal{H}_K(\mathfrak{m}) : \mathcal{H}_K^{\mathfrak{m},\sigma}) = (K(\mathfrak{m}) : U_K K^{\mathfrak{m},\sigma}) = \frac{(K(\mathfrak{m}) : K^{\mathfrak{m},\sigma})}{(U_K K^{\mathfrak{m},\sigma} : K^{\mathfrak{m},\sigma})} = \frac{(K(\mathfrak{m}) : K^{\mathfrak{m},\sigma})}{(U_K : U_K^{\mathfrak{m},\sigma})}.$$

By 2.9.9.2, there exists an epimorphism $\pi_{\mathfrak{m}}^{\sigma}\colon K(\mathfrak{m}) \to (\mathfrak{o}_K/\mathfrak{m})^\times \times \{\pm 1\}^\sigma$ with kernel $K^{\mathfrak{m}\sigma}$, and therefore $(K(\mathfrak{m}) : K^{\mathfrak{m},\sigma}) = 2^{|\sigma|}\varphi_K(\mathfrak{m})$. Putting all together, we obtain

$$|\mathcal{C}_K^{\mathfrak{m},\sigma}| = h_K\, \frac{(K(\mathfrak{m}) : K^{\mathfrak{m},\sigma})}{(U_K : U_K^{\mathfrak{m},\sigma})} = h_K\, \frac{2^{|\sigma|}\varphi_K(\mathfrak{m})}{(U_K : U_K^{\mathfrak{m},\sigma})}.$$

2. The exact sequences come from the very definitions, and together with 1. the final formula follows. □

Next we derive an asymptotic formula for the number of ideals in a given ray class. This formula is the arithmetical basis for the analytic class number formula to be derived in 4.2.5. We start with an asymptotic estimation of the number of lattice points in expanding domains.

Let $n \in \mathbb{N}$, $\|\cdot\|$ the maximum norm and λ the Lebesgue measure on $\mathbb{R}^n$. A map $f\colon [0,1]^{n-1} \to \mathbb{R}^n$ is called a **Lipschitz map** if there exists some $C \in \mathbb{R}_{>0}$ such that $\|f(\boldsymbol{x}) - f(\boldsymbol{y})\| \le C\|\boldsymbol{x} - \boldsymbol{y}\|$ for all $\boldsymbol{x}, \boldsymbol{y} \in [0,1]^{n-1}$. Every such C is called a **Lipschitz constant** of f. Note that every C^1-map $f\colon [0,1]^{n-1} \to \mathbb{R}^n$ is a Lipschitz map, and $\sup\{\|df(\boldsymbol{x})\| \mid \boldsymbol{x} \in [0,1[^{n-1}\}$ is a Lipschitz constant of f.

A subset B of $\mathbb{R}^n$ is called $(n-1)$-**Lipschitz parametrizable** if there exist finitely many Lipschitz maps $f_1, \ldots, f_k\colon [0,1]^{n-1} \to \mathbb{R}^n$ such that

$$B \subset \bigcup_{i=1}^{k} f_i([0,1]^{n-1}).$$

If a subset D of $\mathbb{R}^n$ has an $(n-1)$-Lipschitz parametrizable boundary ∂D, then $\lambda(\partial D) = 0$ and D is (even Jordan) measurable.

Theorem 3.6.8. *Let D be a subset of $\mathbb{R}^n$ with an $(n-1)$-Lipschitz parametrizable boundary ∂D, Γ a lattice of rank n in $\mathbb{R}^n$ and $\boldsymbol{y} \in \mathbb{R}^n$. Then*

$$|tD \cap (\boldsymbol{y}+\Gamma)| = \frac{\lambda(D)}{\mathrm{vol}(\Gamma)}\, t^n + O(t^{n-1}) \quad \textit{for all } t \in \mathbb{R}_{\ge 1}, \ \textit{where } tD = \{t\boldsymbol{x} \mid \boldsymbol{x} \in D\}.$$

Proof. For a subset J of $\mathbb{R}^n$ we denote by $\mathrm{diam}(J) = \sup\{\|\boldsymbol{x}-\boldsymbol{x}'\| \mid \boldsymbol{x}, \boldsymbol{x}' \in J\}$ its diameter. Let $\mathcal{G}$ be a fundamental parallelotope of Γ, and for $\boldsymbol{v} \in \Gamma$ let $\mathcal{G}_{\boldsymbol{v}} = \boldsymbol{v} + \mathcal{G}$. We prove first:

A. There exists a function $\theta_\Gamma\colon \mathbb{R}_{\ge 0} \to \mathbb{R}_{>0}$ such that

$$|\{\boldsymbol{v} \in \Gamma \mid \mathcal{G}_{\boldsymbol{v}} \cap J \ne \emptyset\}| \le \theta_\Gamma(\mathrm{diam}(J)) \quad \text{for every subset } J \text{ of } \mathbb{R}^n.$$

Proof of **A.** We assume first that $\Gamma = \mathbb{Z}^n$ and denote by $\mathcal{D} = [0,1)^n$ the fundamental parallelotope of $\mathbb{Z}^n$. Let $J \subset \mathbb{R}^n$, $\mathrm{diam}(J) = \delta \in \mathbb{R}_{\ge 0}$, $\boldsymbol{a} = (a_1, \ldots, a_n) \in J$ and consider the cube

$$Q = \prod_{i=1}^{n} [a_i - \delta, a_i + \delta].$$

Then $J \subset Q$, and

$$\begin{aligned} |\{\boldsymbol{v} \in \mathbb{Z}^n \mid \mathcal{D}_{\boldsymbol{v}} \cap J \ne \emptyset\}| &\le |\{\boldsymbol{v} \in \mathbb{Z}^n \mid \mathcal{D}_{\boldsymbol{v}} \cap Q \ne \emptyset\}| \\ &= \prod_{i=1}^{n} \big|\mathbb{Z} \cap [a_i - \delta, a_i + \delta]\big| = (2\delta + 1)^n. \end{aligned}$$

Now let Γ be arbitrary and $(\boldsymbol{u}_1, \ldots, \boldsymbol{u}_n)$ a basis of Γ such that

$$\mathcal{G} = \{c_1\boldsymbol{u}_1 + \ldots + c_n\boldsymbol{u}_n \mid c_1, \ldots, c_n \in [0,1)\}.$$

Let $\varphi\colon \mathbb{R}^n \to \mathbb{R}^n$ be the $\mathbb{R}$-linear map satisfying $\varphi(\boldsymbol{u}_i) = \boldsymbol{e}_i$ (the i-th unit vector) for all $i \in [1,n]$. Then $\varphi(\Gamma) = \mathbb{Z}^n$, $\varphi(\mathcal{G}) = \mathcal{D}$, and φ is a Lipschitz map with Lipschitz constant $\|\varphi\|$. If $J \subset \mathbb{R}^n$ and $\operatorname{diam}(J) = \delta$, then we obtain $\operatorname{diam}(\varphi(J)) \le \|\varphi\|\delta$, and

$$|\{\boldsymbol{v} \in \Gamma \mid \mathcal{G}_{\boldsymbol{v}} \cap J \ne \emptyset\}| = |\{\boldsymbol{x} \in \mathbb{Z} \mid \mathcal{D}_{\varphi(\boldsymbol{v})} \cap \varphi(J) \ne \emptyset\}| \le (2\|\varphi\|\delta + 1)^n.$$

Hence the function $\theta(x) = (2\|\varphi\|x + 1)^n$ fulfills our requirements. □[**A**]

For $t \in \mathbb{R}_{\ge 1}$, we set

$$n(t) = |tD \cap (\boldsymbol{y} + \Gamma)| = |\{\boldsymbol{v} \in \Gamma \mid \boldsymbol{y} + \boldsymbol{v} \in tD\}|,$$

$$m(t) = |\{\boldsymbol{v} \in \Gamma \mid \boldsymbol{y} + \mathcal{G}_{\boldsymbol{v}} \subset tD\}| \quad \text{and} \quad M(t) = |\{\boldsymbol{v} \in \Gamma \mid (\boldsymbol{y} + \mathcal{G}_{\boldsymbol{v}}) \cap tD \ne \emptyset\}|.$$

Since $m(t) \le n(t) \le M(t)$ and $m(t)\operatorname{vol}(\Gamma) \le \lambda(tD) = t^n\lambda(D) \le M(t)\operatorname{vol}(\Gamma)$, it follows that

$$-(M(t) - m(t)) \le m(t) - n(t) \le \frac{\lambda(D)}{\operatorname{vol}(\Gamma)}\,t^n - n(t) \le M(t) - m(t),$$

and therefore we must show that $M(t) - m(t) = O(t^{n-1})$. Since $\partial(D)$ is $(n-1)$-Lipschitz parametrizable and

$$\begin{aligned} M(t) - m(t) &\le |\{\boldsymbol{v} \in \Gamma \mid (\boldsymbol{y} + \mathcal{G}_{\boldsymbol{v}}) \cap \partial(tD) \ne \emptyset\}| \\ &= |\{\boldsymbol{v} \in \Gamma \mid \mathcal{G}_{\boldsymbol{v}} \cap (-\boldsymbol{y} + t\partial(D)) \ne \emptyset\}|, \end{aligned}$$

it suffices to prove: If $f\colon [0,1]^{n-1} \to \mathbb{R}^n$ is a Lipschitz map, then

$$|\{\boldsymbol{v} \in \Gamma \mid \mathcal{G}_{\boldsymbol{v}} \cap [-\boldsymbol{y} + tf([0,1]^{n-1})] \ne \emptyset\}| = O(t^{n-1}) \quad \text{for all} \quad t \ge 1.$$

Thus let $f\colon [0,1]^{n-1} \to \mathbb{R}^n$ be a Lipschitz map with Lipschitz constant $C > 0$, let $t \ge 1$ and $s = \lceil t \rceil$. Then $[0,1]^{n-1}$ is the union of s^{n-1} (small) cubes of edge length s^{-1}. If W is such a small cube, then

$$\operatorname{diam}(-\boldsymbol{y} + tf(W)) = t\operatorname{diam}(f(W)) \le ts^{-1}C \le C$$

and therefore $|\{\boldsymbol{v} \in \Gamma \mid \mathcal{G}_{\boldsymbol{v}} \cap (-\boldsymbol{y} + tf(W)) \ne \emptyset\}| \le \theta_\Gamma(C)$. Summing up over the small cubes, we obtain

$$\big|\{\boldsymbol{v} \in \Gamma \mid \mathcal{G}_{\boldsymbol{v}} \cap \big[-\boldsymbol{y} + tf([0,1]^{n-1})\big] \ne \emptyset\}\big| \le \theta_\Gamma(C)s^{n-1} \ll t^{n-1}. \qquad \square$$

Now let K be an algebraic number field, (r_1, r_2) its signature, and set $[K:\mathbb{Q}] = n = r_1 + 2r_2$ and $r = r_1 + r_2$. Let $\operatorname{Hom}(K,\mathbb{C}) = \{\sigma_1, \ldots, \sigma_n\}$, and assume that $\sigma_i(K) \subset \mathbb{R}$ for $i \in [1, r_1]$, and $\sigma_{r_1+r_2+i} = \overline{\sigma_{r_1+i}} \ne \sigma_{r_1+i}$ for $i \in [1, r_2]$. We consider the **Minkowski space** $K_\mathbb{R} = \mathbb{R}^{r_1} \times \mathbb{C}^{r_2}$ and the **Minkowski embedding**

$$\psi\colon K \to K_\mathbb{R} = \mathbb{R}^{r_1} \times \mathbb{C}^{r_2}, \quad \text{defined by} \quad \psi(x) = (\sigma_1(x), \ldots, \sigma_r(x)).$$

The Minkowski space, equipped with the scalar product

$$\langle (x_1,\ldots,x_{r_1},w_1,\ldots,w_{r_2}),(x_1',\ldots,x_{r_1}',w_1',\ldots,w_{r_2}')\rangle = \sum_{i=1}^{r_1} x_ix_i' + \sum_{i=1}^{r_2} \Re(w_i\overline{w_i'}\,),$$

is an n-dimensional Euclidean vector space, and the map $\Phi\colon \mathbb{R}^n \to K_{\mathbb{R}}$, defined by

$$\Phi(x_1,\ldots,x_{r_1},y_1,\ldots,y_{r_2},z_1,\ldots,z_{r_2}) = (x_1,\ldots,x_{r_1},z_1+\mathrm{i}y_{r_1},\ldots,z_{r_2}+\mathrm{i}y_{r_2}),$$

is an isometry. If $\varphi\colon K \to \mathbb{R}^n$ is the field embedding as defined in 3.5.4, then $\psi = \Phi\circ\varphi\colon K \to K_{\mathbb{R}}$ is the Minkowski embedding. As Φ is an isometry, we obtain:

If $\mathfrak{a} \in \mathcal{J}_K$ is a fractional ideal, then $\psi(\mathfrak{a}) \subset K_{\mathbb{R}}$ is a lattice of rank n, and $\mathrm{vol}(\psi(\mathfrak{a})) = 2^{-r_2}\sqrt{|\Delta(\mathfrak{a})|} = 2^{-r_2}\mathfrak{N}(\mathfrak{a})\sqrt{|\mathsf{d}_K|}$ by 3.5.4.1.

Theorem 3.6.9 (Ideal asymptotic). *Let K be an algebraic number field, (r_1,r_2) its signature and $n=[K:\mathbb{Q}]=r_1+2r_2$. Let $w_K=|\mu(K)|$ be the number of roots of unity in K and $\boldsymbol{\sigma}$ a set of real embeddings of K. Let $\mathfrak{m}\in\mathcal{J}_K'$ be an ideal, $C\in\mathcal{C}_K^{\mathfrak{m},\boldsymbol{\sigma}}$ a $\boldsymbol{\sigma}$-ray class modulo $\mathfrak{m}$ and S a finite subset of $\mathcal{P}_K\setminus\mathrm{supp}(\mathfrak{m})$. Then*

$$\begin{aligned}\mathsf{a}(C,t,S) &= |\{\mathfrak{a}\in C\cap\mathcal{J}_K' \mid \mathfrak{p}\nmid\mathfrak{a} \text{ for all } \mathfrak{p}\in S \text{ and } \mathfrak{N}(\mathfrak{a})\le t\}| \\ &= \rho t\prod_{\mathfrak{p}\in S}\Big(1-\frac{1}{\mathfrak{N}(\mathfrak{p})}\Big) + O(t^{1-1/n}) \quad \text{for all } t\ge 1,\end{aligned}$$

where

$$\rho = \frac{2^{r_1-|\boldsymbol{\sigma}|}(2\pi)^{r_2}\,\mathsf{R}_K(U_K:U_K^{\mathfrak{m},\boldsymbol{\sigma}})}{w_K\,\mathfrak{N}(\mathfrak{m})\sqrt{|\mathsf{d}_K|}}.$$

Proof. **1.** We suppose first that $S=\emptyset$ and set $\mathsf{a}(C,t)=\mathsf{a}(C,t,\emptyset)$. Let $\mathrm{Hom}(K,\mathbb{C})=\{\sigma_1,\ldots,\sigma_n\}$ be such that $\sigma_i(K)\subset\mathbb{R}$, $l_i=1$ for $i\in[1,[r_1]$, and $\sigma_{r_1+r_2+i}=\overline{\sigma_{r_1+i}}\ne\sigma_{r_1+i}$, $l_{r_1+i}=2$ for $i\in[1,r_2]$. Set $r=r_1+r_2$, and let $\psi\colon K\to K_{\mathbb{R}}=\mathbb{R}^{r_1}\times\mathbb{C}^{r_2}$ be the Minkowski embedding (as defined above). Suppose that $\boldsymbol{\sigma}=\{\sigma_1,\ldots,\sigma_s\}$, where $s=|\boldsymbol{\sigma}|\in[0,r_1]$, and consider the group

$$K_{\mathbb{R}}^{\boldsymbol{\sigma}} = \mathbb{R}_{>0}^s\times(\mathbb{R}^\times)^{r_1-s}\times(\mathbb{C}^\times)^{r_2}\subset K_{\mathbb{R}},$$

equipped with component-wise multiplication.

Let $t\in\mathbb{R}_{\ge1}$, and fix an ideal $\mathfrak{b}\in C^{-1}\cap\mathcal{J}_K'(\mathfrak{m})$. If $\mathfrak{a}\in C\cap\mathcal{J}_K'$ and $\mathfrak{N}(\mathfrak{a})\le t$, then $\mathfrak{a}\mathfrak{b}=a\mathfrak{o}$ for some $a\in\mathfrak{b}\cap K^{\mathfrak{m},\boldsymbol{\sigma}}$, and $\mathfrak{N}(a\mathfrak{o})=|\mathsf{N}_{K/\mathbb{Q}}(a)|\le t\mathfrak{N}(\mathfrak{b})$. There the coset $aU_K^{\mathfrak{m},\boldsymbol{\sigma}}$ is uniquely determined by $\mathfrak{a}$. Conversely, if $a\in\mathfrak{b}\cap K^{\mathfrak{m},\boldsymbol{\sigma}}$ and $|\mathsf{N}_{K/\mathbb{Q}}(a)|\le t\mathfrak{N}(\mathfrak{b})$, then $\mathfrak{a}=a\mathfrak{b}^{-1}\in C$ and $\mathfrak{N}(\mathfrak{a})\le t$. Hence we obtain

$$\mathsf{a}(C,t) = |\{aU_K^{\mathfrak{m},\boldsymbol{\sigma}} \mid a\in K^{\mathfrak{m},\boldsymbol{\sigma}}\cap\mathfrak{b},\ \ |\mathsf{N}_{K/\mathbb{Q}}(a)|\le\mathfrak{N}(\mathfrak{b})t\,\}|. \tag{1}$$

Since $\mathfrak{b} + \mathfrak{m} = \mathfrak{o}_K$, there exists some $a_0 \in \mathfrak{b}$ such that $a_0 \equiv 1 \bmod \mathfrak{m}$, and we obtain $K^{\mathfrak{m},\boldsymbol{\sigma}} \cap \mathfrak{b} = \{a \in K^{\boldsymbol{\sigma}} \cap \mathfrak{o}_K \mid a \equiv a_0 \bmod \mathfrak{b}\mathfrak{m}\}$. Since $(U_K : U_K^{\mathfrak{m},\boldsymbol{\sigma}}) < \infty$, 3.5.10 implies $U_K^{\mathfrak{m},\boldsymbol{\sigma}} = \mu(U_K^{\mathfrak{m},\boldsymbol{\sigma}}) \cdot V_0$, where $\mu(U_K^{\mathfrak{m},\boldsymbol{\sigma}}) = \mu(K) \cap U_K^{\mathfrak{m},\boldsymbol{\sigma}}$ and $V_0 = \langle \eta_1, \ldots, \eta_{r-1} \rangle$ is a free subgroup of $U_K^{\mathfrak{m},\boldsymbol{\sigma}}$ of rank $r-1$. Hence every coset $aU_K^{\mathfrak{m},\boldsymbol{\sigma}}$ is the disjoint union of $w_K^{\mathfrak{m},\boldsymbol{\sigma}} = |\mu(U_K^{\mathfrak{m},\boldsymbol{\sigma}})|$ cosets $a'V_0$, and we obtain

$$\mathsf{a}(C,t) = \frac{1}{w_K^{\mathfrak{m},\boldsymbol{\sigma}}} \left| \{aV_0 \mid a \in K^{\boldsymbol{\sigma}} \cap \mathfrak{o}_K,\ a \equiv a_0 \bmod \mathfrak{b}\mathfrak{m},\ |\mathsf{N}_{K/\mathbb{Q}}(a)| \le \mathfrak{N}(\mathfrak{b})t\} \right|. \tag{2}$$

Now we shift the calculation of $\mathsf{a}(C,t)$ to the Minkowski space. For a vector $\boldsymbol{\xi} = (\xi_1, \ldots, \xi_r) \in K_{\mathbb{R}}^{\boldsymbol{\sigma}} = \mathbb{R}_{>0}^s \times (\mathbb{R}^\times)^{r_1 - s} \times (\mathbb{C}^\times)^{r_2}$ and $a \in K$ we define

$$a\boldsymbol{\xi} = (\sigma_1(a)\xi_1, \ldots, \sigma_r(a)\xi_r) \quad \text{and} \quad \mathsf{N}(\boldsymbol{\xi}) = \prod_{i=1}^r |\xi_i|^{l_i}.$$

Then it follows that $\mathsf{N}(a\boldsymbol{\xi}) = |\mathsf{N}_{K/\mathbb{Q}}(a)|\mathsf{N}(\boldsymbol{\xi})$ and $\mathsf{N}(\psi(a)) = |\mathsf{N}_{K/\mathbb{Q}}(a)|$ for all $\boldsymbol{\xi} \in K_{\mathbb{R}}^{\boldsymbol{\sigma}}$ and $a \in K$. Set $V = \psi(V_0) \subset K_{\mathbb{R}}^{\boldsymbol{\sigma}}$, $\boldsymbol{\alpha} = \psi(a_0) \in K_{\mathbb{R}}$, and let D be a set of representatives for $K_{\mathbb{R}}^{\boldsymbol{\sigma}}/V$ in $K_{\mathbb{R}}^{\boldsymbol{\sigma}}$. Then $\Gamma = \psi(\mathfrak{b}\mathfrak{m})$ is a lattice in $K_{\mathbb{R}}$ of rank r, $\operatorname{vol}(\Gamma) = 2^{-r_2}\mathfrak{N}(\mathfrak{b})\mathfrak{N}(\mathfrak{m})\sqrt{|\mathsf{d}_K|}$, and

$$\mathsf{a}(C,t) = \frac{1}{w_K^{\mathfrak{m},\boldsymbol{\sigma}}} \left| \{\boldsymbol{\xi} \in D \cap (\boldsymbol{\alpha} + \Gamma) \mid \mathsf{N}(\boldsymbol{\xi}) \le \mathfrak{N}(\mathfrak{b})t\} \right|. \tag{3}$$

We must find a suitable set of representatives D for $K_{\mathbb{R}}^{\boldsymbol{\sigma}}/V$. For this, we use a modified logarithmic embedding $\mathcal{L}\colon K_{\mathbb{R}}^{\boldsymbol{\sigma}} \to \mathbb{R}^r$, defined by

$$\mathcal{L}(\boldsymbol{\xi}) = \Big(l_1 \log \frac{|\xi_1|}{\mathsf{N}(\boldsymbol{\xi})^{1/n}}, \ldots, l_r \log \frac{|\xi_r|}{\mathsf{N}(\boldsymbol{\xi})^{1/n}} \Big) \quad \text{for all} \quad \boldsymbol{\xi} = (\xi_1, \ldots, \xi_r) \in K_{\mathbb{R}}^{\boldsymbol{\sigma}}.$$

Since $l_1 + \ldots + l_r = n$, it follows that

$$\sum_{i=1}^r l_i \log \frac{|\xi_i|}{\mathsf{N}(\boldsymbol{\xi})^{1/n}} = \log \prod_{i=1}^r |\xi_i|^{l_i} - \sum_{i=1}^r \frac{l_i}{n} \log \mathsf{N}(\boldsymbol{\xi}) = 0 \quad \text{for all} \quad \boldsymbol{\xi} \in K_{\mathbb{R}},$$

which implies $\mathcal{L}(K_{\mathbb{R}}^{\boldsymbol{\sigma}}) \subset H = \{(x_1, \ldots, x_r) \in \mathbb{R}^r \mid x_1 + \ldots + x_r = 0\}$.

Note that $\mathcal{L} \circ \psi \restriction V_0 = \ell \restriction V_0$, where $\ell\colon K^\times \to \mathbb{R}^r$ is the logarithmic embedding used in the proof of Dirichlet's unit theorem 3.5.8. For $j \in [1, r-1]$ we set

$$\boldsymbol{v}_j = \mathcal{L} \circ \psi(\eta_j) = \ell(\eta_j) = (l_1 \log |\sigma_1(\eta_j)|, \ldots, l_r \log |\sigma_r(\eta_j)|) \in H.$$

Then $\mathcal{L}(V) = \ell(V_0)$ is a lattice in H of rank $r-1$, and if $(\boldsymbol{v}_1, \ldots, \boldsymbol{v}_{r-1})$ is a basis of $\mathcal{L}(V)$, then its fundamental parallelotope

$$F = \Big\{ \sum_{j=1}^{r-1} c_j \boldsymbol{v}_j \;\Big|\; c_1, \ldots, c_{r-1} \in [1,0) \Big\}$$

is a set of representatives for $H/\ell(V)$ in H. Hence $D = \mathcal{L}^{-1}(F) \subset K_{\mathbb{R}}^{\boldsymbol{\sigma}}$ is a set of representatives for $K_{\mathbb{R}}^{\boldsymbol{\sigma}}/V$, and we use it to calculate $\mathsf{a}(C,t)$. We set

$$D(t) = \{\boldsymbol{\xi} \in D \mid \mathsf{N}(\boldsymbol{\xi}) \le t\} = \{\boldsymbol{\xi} \in D \mid \mathsf{N}(t^{1/n}\boldsymbol{\xi}) \le 1\}$$
$$= t^{1/n}\{\boldsymbol{\xi} \in t^{-1/n}D \mid \mathsf{N}(\boldsymbol{\xi}) \le 1\}$$

If $t \in \mathbb{R}_{>0}$ and $\boldsymbol{\xi} = (\xi_1, \ldots, \xi_r) \in D$, then

$$\mathcal{L}(t\boldsymbol{\xi}) = \Big(l_i \log \frac{|t\xi_i|}{\mathsf{N}(t\boldsymbol{\xi})}\Big)_{i\in[1,r]} = \Big(l_i \log \frac{|\xi_i|}{\mathsf{N}(\boldsymbol{\xi})}\Big)_{i\in[1,r]} = \mathcal{L}(\boldsymbol{\xi}) \in F,$$

and therefore $tD = D$. It follows that $D(t) = t^{1/n}D(1)$, and

$$\mathsf{a}(C,t) = \frac{1}{w_K^{\mathfrak{m},\boldsymbol{\sigma}}}\big|(\mathfrak{N}(\mathfrak{b})t)^{1/n}\,D(1) \cap (\boldsymbol{\alpha} + \Lambda)\big|. \tag{4}$$

$D(1)$ consists of all $\boldsymbol{\xi} = (\xi_1, \ldots, \xi_r) \in K_{\mathbb{R}}^{\boldsymbol{\sigma}}$ satisfying

$$\mathsf{N}(\boldsymbol{\xi}) = \prod_{i=1}^{r} |\xi_i|^{l_i} \le 1$$

and

$$\log \frac{|\xi_i|}{\mathsf{N}(\xi)^{1/n}} = \sum_{j=1}^{r-1} c_j \log|\sigma_i(\eta_j)|, \quad \text{where } c_1, \ldots, c_{r-1} \in [0,1), \quad \text{for all } i \in [1,r].$$

In particular, if $\boldsymbol{\xi} = (\xi_1, \ldots, \xi_r) \in D(1)$, then $|\xi_i| \le \mathsf{e}^{(r-1)B}$ for all $i \in [1,r]$, where $B = \max\{|\sigma_i(\eta_j)| \mid i \in [1,r],\ j \in [1, r-1]\}$. Hence $D(1)$ is bounded. Its boundary $\partial D(1)$ consists of $\mathbf{0}$ and all $\boldsymbol{\xi} = (\xi_1, \ldots, \xi_r) \in K_{\mathbb{R}}^{\boldsymbol{\sigma}}$ satisfying

$$\log|\xi_i| = \sum_{j=1}^{r-1} c_j \log|\sigma_i(\eta_j)| \quad \text{for some} \quad (c_1, \ldots, c_{r-1}) \in \{0,1\}^{r-1}.$$

Hence $\partial D(1)$ is $(n-1)$-Lipschitz parametrizable, and by 3.6.8 we obtain

$$\mathsf{a}(C,t) = \frac{1}{w_K^{\mathfrak{m},\boldsymbol{\sigma}}}\,|(\mathsf{N}(\mathfrak{b})t)^{1/n}D(1) \cap (\boldsymbol{\alpha} + \Gamma)| = \frac{1}{w_K^{\mathfrak{m},\boldsymbol{\sigma}}}\Big[\frac{\lambda(D(1)}{\mathrm{vol}(\Gamma)}\,\mathfrak{N}(\mathfrak{b})t + O(t^{1-1/n})\Big]$$
$$= \frac{2^{r_2}\lambda(D(1))}{w_K^{\mathfrak{m},\boldsymbol{\sigma}}\mathfrak{N}(\mathfrak{m})\sqrt{|\mathsf{d}_K|}}\,t + O(t^{1-1/n}).$$

For the calculation of $\lambda(D(1))$ we use polar coordinates in the imaginary conjugates. For $\boldsymbol{\xi} = (\xi_1, \ldots, \xi_r) \in D(1)$ we set $\xi_i = \rho_i > 0$ for $i \in [1,s]$, $\xi_i = \varepsilon_i\rho_i$ for $i \in [s+1, r_1]$, where $\rho_i = |\xi_i|$ and $\varepsilon_i \in \{\pm 1\}$, and $\xi_i = \rho_i\mathsf{e}^{\mathsf{i}\varphi_i}$ for $i \in [r_1+1, r_2]$, where $\rho_i = |\xi_i|$ and $\varphi_i \in [0, 2\pi)$.

Let Ω be the set of all $(\rho_1, \ldots, \rho_r) \in \mathbb{R}_{>0}^r$ satisfying

$$\prod_{i=1}^{r} \rho_i^{l_i} \le 1$$

and, for all $i \in [1, r]$,

$$\log \rho_i - \frac{1}{n} \log \prod_{\nu=1}^{r} \rho_\nu^{l_\nu} = \sum_{j=1}^{r-1} c_j \log |\sigma_i(\eta_j)|, \text{ for some } c_1, \ldots, c_{r-1} \in [0, 1).$$

Then $D(1)$ consists of all r-tuples

$$(\rho_1, \ldots, \rho_s, \varepsilon_{s+1}\rho_{s+1}, \ldots, \varepsilon_{r_1}\rho_{r_1}, \rho_{r_1+1}\mathrm{e}^{\mathrm{i}\varphi_1}, \ldots, \rho_r\mathrm{e}^{\mathrm{i}\varphi_r}),$$

where $(\rho_1, \ldots, \rho_r) \in \Omega$ and $\varphi_i \in [0, 2\pi)$ for all $i \in [r_1+1, r]$, and we obtain

$$\lambda(D(1)) = 2^{r_1-s}(2\pi)^{r_2} \int_\Omega \rho_{r_1+1} \cdot \ldots \cdot \rho_r d(\rho_{r_1+1}, \ldots, \rho_r).$$

To calculate the integral, let $S = (0, 1] \times [0, 1)^{r-1}$, and let $\Phi \colon S \to \Omega$ be defined by $\Phi(u, c_1, \ldots, c_{r-1}) = (\rho_1, \ldots, \rho_r)$, where

$$\rho_i = \rho_i(u, c_1, \ldots, c_{r-1}) = u^{1/n} \exp\Bigl\{\sum_{j=1}^{r-1} c_j \log |\sigma_i(\eta_j)|\Bigr\} \quad \text{for all} \quad i \in [1, r].$$

Then $\Phi \restriction S^\circ$ is a C^1-map,

$$u = \prod_{i=1}^{r} \rho_i^{l_i} \quad \text{and}$$

$$\bigl|\det(\log |\sigma_i(\eta_j)|)_{i,\,j \in [1,[r-1]}\bigr| = \frac{1}{l_1 \cdot \ldots \cdot l_{r-1}} \mathsf{R}(\eta_1, \ldots, \eta_{r-1}) \neq 0.$$

Hence Φ is bijective. By the transformation theorem we obtain

$$\lambda(D(1)) =$$
$$2^{r_1-s}(2\pi)^{r_2} \int_{[0,1]^r} \rho_{r_1+1} \cdot \ldots \cdot \rho_r \det \mathsf{J}(\Phi)(u, c_1, \ldots, c_{r-1})\, d(u, c_1, \ldots, c_{r-1}).$$

If $i \in [1, r]$, then

$$\frac{\partial \rho_i}{\partial u} = \frac{1}{n} u^{1/n-1} \exp\Bigl\{\sum_{j=1}^{r-1} c_j \log |\sigma_i(\eta_j)|\Bigr\} = \frac{\rho_i}{nu}$$

and

$$\frac{\partial \rho_i}{\partial c_j} = u^{1/n} \exp\Bigl\{\sum_{j=1}^{r-1} c_j \log |\sigma_i(\eta_j)|\Bigr\} \log |\sigma_i(\eta_j)| = \rho_i \log |\sigma_i(\eta_j)|$$

for all $j \in [1, r-1]$. Hence

$$\det \mathsf{J}(\Phi)(u, c_1, \ldots, c_{r-1}) = \det \begin{pmatrix} \frac{\rho_1}{nu} & \rho_1 \log|\sigma_1(\eta_1)| & \ldots & \rho_1 \log|\sigma_1(\eta_{r-1})| \\ \cdot & \cdot & \ldots & \cdot \\ \cdot & \cdot & \ldots & \cdot \\ \cdot & \cdot & \ldots & \cdot \\ \frac{\rho_r}{nu} & \rho_r \log|\sigma_r(\eta_1)| & \ldots & \rho_r \log|\sigma_r(\eta_{r-1})| \end{pmatrix}$$

$$= \frac{\rho_1 \cdot \ldots \cdot \rho_r}{2^{r_2} nu} \det \begin{pmatrix} l_1 & l_1 \log|\sigma_1(\eta_1)| & \ldots & l_1 \log|\sigma_1(\eta_{r-1})| \\ \cdot & \cdot & \ldots & \cdot \\ \cdot & \cdot & \ldots & \cdot \\ \cdot & \cdot & \ldots & \cdot \\ l_r & l_r \log|\sigma_r(\eta_1)| & \ldots & l_r \log|\sigma_r(\eta_{r-1})| \end{pmatrix}$$

$$= \frac{\rho_1 \cdot \ldots \cdot \rho_r}{2^{r_2} nu} \det \begin{pmatrix} l_1 & l_1 \log|\sigma_1(\eta_1)| & \ldots & l_1 \log|\sigma_1(\eta_{r-1})| \\ \cdot & \cdot & \ldots & \cdot \\ \cdot & \cdot & \ldots & \cdot \\ l_{r-1} & l_{r-1} \log|\sigma_{r-1}(\eta_1)| & \ldots & l_{r-1} \log|\sigma_{r-1}(\eta_{r-1})| \\ n & 0 & \ldots & 0 \end{pmatrix}$$

$$= \frac{\rho_1 \cdot \ldots \cdot \rho_r}{2^{r_2} u} \mathsf{R}(\eta_1, \ldots, \eta_{r-1}) = \frac{2^{-r_2} \mathsf{R}(U_K^{\mathfrak{m}\boldsymbol{\sigma}})}{\rho_{r_1+1} \cdot \ldots \cdot \rho_r},$$

and we obtain

$$\lambda(D(1)) = 2^{r_1 - s}(2\pi)^{r_2} \int_{[0,1]^r} 2^{-r_2} \mathsf{R}(U_K^{\mathfrak{m},\boldsymbol{\sigma}}) d(\rho_1, \ldots, \rho_r) = 2^{r_1 - s} \pi^{r_2} \mathsf{R}(U_K^{\mathfrak{m},\boldsymbol{\sigma}}).$$

By 3.5.10.2 it follows that

$$\mathsf{R}(U_K^{\mathfrak{m},\boldsymbol{\sigma}}) = \frac{\mathsf{R}_K(U_K : U_K^{\mathfrak{m},\boldsymbol{\sigma}}) w_K^{\mathfrak{m},\boldsymbol{\sigma}}}{w_K},$$

and consequently

$$\mathsf{a}(C, t) = \rho t + O(t^{1-1/n}), \quad \text{where} \quad \rho = \frac{2^{r_1 - s}(2\pi)^{r_2} \, \mathsf{R}_K(U_K : U_K^{\mathfrak{m},\boldsymbol{\sigma}})}{w_K \, \mathfrak{N}(\mathfrak{m}) \sqrt{|\mathsf{d}_K|}}.$$

2. Now we can do the general case. If $\mathfrak{s} \in \mathcal{I}_K'$, then the special case just done shows that

$$|\{\mathfrak{a} \in C \cap \mathcal{I}_K' \mid \mathfrak{s} \,|\, \mathfrak{a} \text{ and } \mathfrak{N}(\mathfrak{a}) \le t\}| = |\{\mathfrak{b} \in C([\mathfrak{s}]^{\mathfrak{m},\boldsymbol{\sigma}})^{-1} \mid \mathfrak{N}(\mathfrak{b}) \le t\mathfrak{N}(\mathfrak{s})^{-1}\}|$$
$$= \rho \mathfrak{N}(\mathfrak{b})^{-1} t \; + \; O(t^{1-1/n}) \quad \text{for all } t \ge 1.$$

Let $|S| = k \geq 1$ and $S = \{\mathfrak{p}_1, \ldots, \mathfrak{p}_k\}$. By the inclusion-exclusion principle we obtain

$$\begin{aligned}
\mathsf{a}(C,t,S) &= \bigl|\{\mathfrak{a} \in C \cap \mathcal{J}'_K \mid \mathfrak{N}(\mathfrak{a}) \leq t\}\bigr| \\
&\quad + \sum_{i=1}^{k} (-1)^i \sum_{1 \leq \nu_1 < \ldots < \nu_i \leq k} \bigl|\{\mathfrak{a} \in C \cap \mathcal{J}'_K \mid \mathfrak{p}_{\nu_1} \cdot \ldots \cdot \mathfrak{p}_{\nu_i} \mid \mathfrak{a},\ \mathfrak{N}(\mathfrak{a}) \leq t\}\bigr| \\
&= \rho t + \sum_{i=1}^{k} (-1)^i \sum_{1 \leq \nu_1 < \ldots < \nu_i \leq k} \rho t\, \mathfrak{N}(\mathfrak{p}_{\nu_1} \cdot \ldots \cdot \mathfrak{p}_{\nu_i})^{-1} + O(t^{1-1/n}) \\
&= \rho t \sum_{i=0}^{k} (-1)^i \sum_{1 \leq \nu_1 < \ldots < \nu_i \leq k} \frac{1}{\mathfrak{N}(\mathfrak{p}_{\nu_1}) \cdot \ldots \cdot \mathfrak{N}(\mathfrak{p}_{\nu_i})} + O(t^{1-1/n}) \\
&= \rho t \prod_{i=1}^{k} \Bigl(1 - \frac{1}{\mathfrak{N}(\mathfrak{p}_i)}\Bigr) + O(t^{1-1/n}) \quad \text{for all } t \geq 1. \qquad \square
\end{aligned}$$

Corollary 3.6.10. *Let K be an algebraic number field, $n = [K:\mathbb{Q}] = r_1 + 2r_2$, where (r_1, r_2) is its signature, $w_K = |\mu(K)|$ the number of roots of unity in K and $C \in \mathcal{C}_K$ an (ordinary) ideal class. Then*

$$|\{\mathfrak{a} \in C \cap \mathcal{J}'_K \mid \mathfrak{N}(\mathfrak{a}) \leq t\}| = \frac{2^{r_1} (2\pi)^{r_2}\, \mathsf{R}_K}{w_K \sqrt{|\mathsf{d}_K|}}\, t + O(t^{1-1/n}) \quad \textit{for } t \geq 1.$$

and

$$|\{\mathfrak{a} \in \mathcal{J}'_K \mid \mathfrak{N}(\mathfrak{a}) \leq t\}| = \frac{2^{r_1} (2\pi)^{r_2}\, \mathsf{R}_K h_K}{w_K \sqrt{|\mathsf{d}_K|}}\, t + O(t^{1-1/n}).$$

Proof. By 3.6.9, applied with $\mathfrak{m} = \mathbf{1}$ and $\boldsymbol{\sigma} = \emptyset$. $\square$

3.7 The main theorems of classical class field theory

Throughout this section, let K be an algebraic number field. All algebraic field extensions of K are tacitly assumed to be contained $\overline{\mathbb{Q}}$.

Definition 3.7.1. Let L/K be a finite Galois extension and $G = \mathrm{Gal}(L/K)$. We denote by $\mathrm{Split}(L/K)$ the set of all in L fully decomposed and by $\mathrm{Ram}(L/K)$ the (finite) set of all in L ramified prime ideals $\mathfrak{p} \in \mathcal{P}_K$ (see 2.12.7.1). Let $\mathfrak{m} \in \mathcal{J}'_K$ be an ideal such that $\mathrm{Ram}(L/K) \subset \mathrm{supp}(\mathfrak{m})$, and let $\mathcal{P}_K^{\mathfrak{m}} = \mathcal{P}^{\mathfrak{m}}(\mathfrak{o}_K)$ be the set of all $\mathfrak{p} \in \mathcal{P}_K$ not dividing $\mathfrak{m}$ (see 2.9.1). Recall that $\mathcal{N}_{L/K} = \mathcal{N}_{\mathfrak{o}_L/\mathfrak{o}_K} \colon \mathcal{J}_L \to \mathcal{J}_K$ denotes the ideal norm, and that $\mathcal{N}_{L/K}(\mathcal{J}_L^{\mathfrak{m}\mathfrak{o}_L}) \subset \mathcal{J}_K^{\mathfrak{m}}$ by 2.14.4.2.

For $\mathfrak{a} \in \mathcal{J}_K^{\mathfrak{m}}$ let $(\mathfrak{a}, L/K) \in G$ be the Artin symbol and $(\,\cdot\,, L/K)\colon \mathcal{J}_K^{\mathfrak{m}} \to G$ the **Artin map**. Then $\mathcal{N}_{L/K}(\mathcal{J}_L^{\mathfrak{m}\mathfrak{o}_L}) \subset \operatorname{Ker}(\,\cdot\,, L/K)$ by 2.14.8.2(c), and a prime ideal $\mathfrak{p} \in \mathcal{P}_K^{\mathfrak{m}}$ is fully decomposed in L if and only if $(\mathfrak{p}, L/K) = 1$ (see 2.14.7.**2.**).

We are now going to quote without proofs the main theorems of classical class field theory due to Hilbert, Weber, Takagi, Hasse and Artin. For a detailed exposistion we refer to the forthcoming volume of class field theory.

Theorem and Definition 3.7.2 (Main theorem of class field theory).

1. (Artin reciprocity law) *Let L/K be a finite abelian field extension. Then there exists an ideal $\mathfrak{m} \in \mathcal{J}'_K$ such that* $\operatorname{Ram}(L/K) \subset \operatorname{supp}(\mathfrak{m})$ *and the Artin map $(\,\cdot\,, L/K)\colon \mathcal{J}_K^{\mathfrak{m}} \to G$ is an epimorphism with kernel*

$$\mathcal{N}_{L/K}^{\mathfrak{m}} = \mathcal{N}_{L/K}(\mathcal{J}_L^{\mathfrak{m}\mathfrak{o}_L})\mathcal{H}_K^{\mathfrak{m}}.$$

In particular, $\mathcal{P}_K^{\mathfrak{m}} \cap \operatorname{Split}(L/K) = \mathcal{P}_K^{\mathfrak{m}} \cap \mathcal{N}_{L/K}^{\mathfrak{m}}$.

2. (Existence theorem) *Let $\mathfrak{m} \in \mathcal{J}'_K$ be an ideal and $\mathcal{G}$ a subgroup of $\mathcal{J}_K^{\mathfrak{m}}$ such that $\mathcal{H}_K^{\mathfrak{m}} \subset \mathcal{G}$. Then there exists a unique over K finite abelian extension field $K[\mathcal{G}]$ such that $\mathcal{G} = \mathcal{N}_{K[\mathcal{G}]/K}^{\mathfrak{m}}$. Furthermore,* $\mathcal{P}_K \cap \mathcal{G} = \mathcal{P}_K^{\mathfrak{m}} \cap \operatorname{Split}(K[\mathcal{G}]/K)$ *and* $\operatorname{Ram}(K[\mathcal{G}]/K) \subset \operatorname{supp}(\mathfrak{m})$.

We call $K[\mathcal{G}]$ the **class field** to $\mathcal{G}$. The Artin map

$$(\,\cdot\,, K[\mathcal{G}]/K)\colon \mathcal{J}_K^{\mathfrak{m}} \to \operatorname{Gal}(K[\mathcal{G}]/K)$$

induces an (equally denoted) isomorphism

$$(\,\cdot\,, K[\mathcal{G}]/K)\colon \mathcal{J}_K^{\mathfrak{m}}/\mathcal{G} \xrightarrow{\sim} \operatorname{Gal}(K[\mathcal{G}]/K),$$

called the **Artin isomorphism**. The class field $K[\mathfrak{m}] = K[\mathcal{H}_K^{\mathfrak{m}}]$ to $\mathcal{H}_K^{\mathfrak{m}}$ is called the **ray class field** modulo $\mathfrak{m}$ (of K).

The assignment $\mathcal{G} \mapsto K[\mathcal{G}]$ defines an inclusion-reversing bijection from the set of all subgroups $\mathcal{G}$ of $\mathcal{J}_K^{\mathfrak{m}}$ containing $\mathcal{H}_K^{\mathfrak{m}}$ onto the set of all intermediate fields of $K[\mathfrak{m}]/K$.

3. Let L/K be a finite abelian field extension and $\mathfrak{m} \in \mathcal{J}'_K$ the product of all in L ramified prime ideals. Then there exists some $m \in \mathbb{N}$ such that $L \subset K[\mathfrak{m}^m]$.

Our definition of class fields is due to Takagi and Artin. Classically, Weber used the set $\operatorname{Split}(L/K)$ of all fully decomposed prime ideals as a basis for his definition as follows. If $\mathfrak{m} \in \mathcal{J}'_K$ and $\mathcal{G}$ is a subgroup of $\mathcal{J}_K^{\mathfrak{m}}$ such that $\mathcal{H}_K^{\mathfrak{m}} \subset \mathcal{G}$, then an over K Galois extension field L of K is a class field to $\mathcal{G}$ (according to Weber) if $\mathcal{P}_K \cap \mathcal{G} = \mathcal{P}_K^{\mathfrak{m}} \cap \operatorname{Split}(L/K)$.

Next, in 3.7.3, we quote a classical result due to M. Bauer by which a finite Galois extension field L of K is uniquely determined by the set $\mathrm{Split}(L/K)$. We show as a consequence that the two definitions of class fields coincide. We shall prove Bauer's theorem in 4.3.5 using analytic methods.

Theorem 3.7.3 (Splitting Theorem). *Let L/K and L'/K be finite Galois field extensions. If $\mathrm{Split}(L/K)$ and $\mathrm{Split}(L'/K)$ only differ by a finite set, then $L = L'$.*

Proof. See 4.3.5. □

We proceed with a series of simple (but important) consequences of the main results 3.7.2 and 3.7.3 which we phrase as corollaries and equip them with proofs.

Corollary 3.7.4. *Let $\mathfrak{m} \in \mathcal{I}'_K$, $\mathcal{G}$ be a subgroup of $\mathcal{I}_K^{\mathfrak{m}}$ such that $\mathcal{H}_K^{\mathfrak{m}} \subset \mathcal{G}$, and let L/K be a finite abelian field extension such that $\mathcal{P}_K^{\mathfrak{m}} \cap \mathrm{Split}(L/K) = \mathcal{P}_K \cap \mathcal{G}$. Then $L = K[\mathcal{G}]$.*

Proof. Since $\mathcal{P}_K^{\mathfrak{m}} \cap \mathrm{Split}(L/K) = \mathcal{P}_K \cap \mathcal{G} = \mathcal{P}_K^{\mathfrak{m}} \cap \mathrm{Split}(K_{\mathcal{G}}/K)$, the sets $\mathrm{Split}(L/K)$ and $\mathrm{Split}(K_{\mathcal{G}}/K)$ only differ by a finite set, and thus $L = K[\mathcal{G}]$ by 3.7.3. □

Corollary 3.7.5. *Suppose that $\mathfrak{m} \in \mathcal{I}'_K$; let $K[\mathfrak{m}]$ be the ray class field modulo $\mathfrak{m}$, $\mathcal{G}$ a subgroup of $\mathcal{I}_K^{\mathfrak{m}}$ such that $\mathcal{H}_K^{\mathfrak{m}} \subset \mathcal{G}$, and let $L = K[\mathfrak{m}]^{(\mathcal{G},L/K)}$ be the fixed field of $(\mathcal{G}, L/K)$. Then L is the class field to $\mathcal{G}$.*

Proof. Let $G_{\mathfrak{m}} = \mathrm{Gal}(K[\mathfrak{m}]/K)$ and $G = \mathrm{Gal}(L/K) = G_{\mathfrak{m}}/(\mathcal{G}, L/K)$. We must prove that $\mathcal{G}$ is the kernel of the Artin map $(\,\cdot\,, L/K)\colon \mathcal{I}_K^{\mathfrak{m}} \to G$. If $\mathfrak{a} \in \mathcal{I}_K^{\mathfrak{m}}$, then 2.14.8.2(b) implies

$$(\mathfrak{a}, L/K) = (\mathfrak{a}, K[\mathfrak{m}]/K)\restriction L = 1 \iff (\mathfrak{a}, K[\mathfrak{m}]/K) \in (\mathcal{G}, L/K) \iff \mathfrak{a} \in \mathcal{G}.$$

□

Corollary 3.7.6. *Let $\mathfrak{m}, \mathfrak{m}_0 \in \mathcal{I}'_K$ and $\mathfrak{m}_0 \mid \mathfrak{m}$. Then $\mathcal{H}_K^{\mathfrak{m}} \subset \mathcal{H}_K^{\mathfrak{m}_0} \cap \mathcal{I}_K^{\mathfrak{m}} \subset \mathcal{I}_K^{\mathfrak{m}}$, and $K[\mathfrak{m}_0]$ is the class field to $\mathcal{H}_K^{\mathfrak{m}_0} \cap \mathcal{I}_K^{\mathfrak{m}}$. In particular, $K[\mathfrak{m}_0] \subset K[\mathfrak{m}]$.*

Proof. It follows from the very definition that $\mathcal{H}_K^{\mathfrak{m}} \subset \mathcal{H}_K^{\mathfrak{m}_0} \cap \mathcal{I}_K^{\mathfrak{m}} \subset \mathcal{I}_K^{\mathfrak{m}}$. If L is the class field to $\mathcal{H}_K^{\mathfrak{m}_0} \cap \mathcal{I}_K^{\mathfrak{m}}$, then $L \subset K[\mathfrak{m}]$ by 3.7.5, and

$$\mathrm{Split}(L/K) \cap \mathcal{P}_K^{\mathfrak{m}} = \mathcal{P}_K \cap \mathcal{H}_K^{\mathfrak{m}_0} \cap \mathcal{H}_K^{\mathfrak{m}} = \mathcal{P}_K^{\mathfrak{m}} \cap \mathcal{H}_K^{\mathfrak{m}_0} = \mathrm{Split}(K[\mathfrak{m}_0]/K) \cap \mathcal{P}_K^{\mathfrak{m}}.$$

Hence $\mathrm{Split}(L/K)$ and $\mathrm{Split}(K[\mathfrak{m}_0]/K)$ only differ by a finite set, and by 3.7.3 it follows that $K[\mathfrak{m}_0] = L \subset K[\mathfrak{m}]$. □

Corollary 3.7.7 (Factorization law of class field theory). *Let $\mathfrak{m} \in \mathcal{I}'_K$, $\mathcal{G}$ a subgroup of $\mathcal{I}_K^{\mathfrak{m}}$ such that $\mathcal{H}_K^{\mathfrak{m}} \subset \mathcal{G}$. Let $\mathfrak{p} \in \mathcal{P}_K^{\mathfrak{m}}$ and $f \in \mathbb{N}$ minimal such that $\mathfrak{p}^f \in \mathcal{G}$. Then $f = f(\mathfrak{P}/\mathfrak{p})$ for all $\mathfrak{P} \in \mathcal{P}_{K[\mathcal{G}]}$ lying above $\mathfrak{p}$.*

Proof. Let $G = \mathrm{Gal}(K[\mathcal{G}]/K)$ and $\mathfrak{P} \in \mathcal{P}_{K[\mathcal{G}]}$ such that $\mathfrak{P} \mid \mathfrak{p}$. By definition (see 2.14.7.1), $(\mathfrak{p}, K[\mathcal{G}]/K) \in G$ is the Frobenius automorphism of $\mathfrak{P}$ over K and $f(\mathfrak{P}/\mathfrak{p})$ is its order. By means of the Artin isomorphism $(\,\cdot\,, L/K)\colon \mathcal{I}_K^{\mathfrak{m}}/\mathcal{G} \stackrel{\sim}{\to} G$ we see that $f(\mathfrak{P}/\mathfrak{p})$ is the order of $\mathfrak{p}\mathcal{G}$ in $\mathcal{I}_K^{\mathfrak{m}}/\mathcal{G}$. □

The ray class field $K[\mathbf{1}]$ is called the **large Hilbert class field** and the class field $\mathsf{H}_K = K[\mathcal{H}_K]$ is called the **Hilbert class field** of K. Since $\mathcal{H}_K \supset \mathcal{H}_K^{1}$, it follows that $K[\mathbf{1}] \supset \mathsf{H}_K$. Furthermore we obtain $\mathrm{Gal}(K[\mathbf{1}]/K) \cong \mathcal{C}_K^{+}$, $\mathrm{Split}(K[\mathbf{1}]/K) = \mathcal{P}_K \cap \mathcal{H}_K^{+}$, $\mathrm{Gal}(\mathsf{H}_K/K) \cong \mathcal{C}_K$, and $\mathrm{Split}(\mathsf{H}_K/K) = \mathcal{P}_K \cap \mathcal{H}_K$.

Corollary 3.7.8. *The large Hilbert class field $K[\mathbf{1}]$ is the largest over K abelian extension field L satisfying $\mathrm{Ram}(L/K) = \emptyset$.*

Proof. By 3.7.2.2 we have $\mathrm{Ram}(K[\mathbf{1}]/K) = \emptyset$. Conversely, if L/K is any finite abelian field extension such that $\mathrm{Ram}(L/K) = \emptyset$, then $L \subset K[\mathbf{1}]$ by 3.7.2.3. □

We close this section with a discussion of the class field theory over $\mathbb{Q}$.

Definition and Theorem 3.7.9. *Let $m \in \mathbb{N}$, $m \geq 3$, and let $\mathbb{Q}^{(m)}$ be the m-th cyclotomic field.*
We tacitly use the identifications made in 3.6.6 and denote by $\mathbb{Q}[m]$ the ray class field modulo m of $\mathbb{Q}$.

1. $\mathbb{Q}[m] = \mathbb{Q}^{(m)}$.

2. (Kronecker-Weber) *Every abelian number field is contained in a cyclotomic field.*

Proof. 1. By 3.7.2 and 3.2.6.1 we obtain

$$\mathrm{Split}(\mathbb{Q}[m]/\mathbb{Q}) = \mathbb{Q}^m \cap \mathbb{P} = \{p \in \mathbb{P} \mid p \equiv 1 \bmod m\} = \mathrm{Split}(\mathbb{Q}^{(m)}/\mathbb{Q}),$$

and therefore 3.7.3 implies $\mathbb{Q}[m] = \mathbb{Q}^{(m)}$.

2. By 1. and 3.7.2.3. □

The above and all that follows also hold (by trivial reason) for $m \in \{1, 2\}$ where $\mathbb{Q}^{(m)} = \mathbb{Q}$. We disregard these cases in the sequel and assume that $m \geq 3$.

Recall that the Artin map $(\,\cdot\,, \mathbb{Q}^{(m)}/\mathbb{Q})\colon \mathbb{Q}^{+}(m) \to \mathrm{Gal}(\mathbb{Q}^{(m)}/\mathbb{Q})$ was already studied in 3.2.6.3, and there we proved:
If $a \in \mathbb{Q}^{+}(m)$ and ζ is an m-th root of unity, then

$$(a, \mathbb{Q}^{(m)}/\mathbb{Q})(\zeta) = \sigma_a(\zeta) = \zeta^a.$$

For every subgroup $\mathcal{G}$ of $\mathbb{Q}^+(m)$ containing $\mathbb{Q}^m$ we denote by $\mathbb{Q}[\mathcal{G}]$ the class field to $\mathcal{G}$. By 3.7.5 it is the fixed field of $(\mathcal{G}, \mathbb{Q}^{(m)}/\mathbb{Q})$, and by 3.7.4 it is the unique subfield of $\mathbb{Q}^{(m)}$ satisfying $\mathrm{Split}(\mathbb{Q}[\mathcal{G}]/\mathbb{Q}) = \mathbb{P} \cap \mathcal{G}$. In particular:

- $\mathbb{Q}[\mathbb{Q}^m_\pm] = \mathbb{Q}^{(m)+} = \mathbb{Q}(\zeta_m + \zeta_m^{-1}) = \mathbb{Q}^{(m)} \cap \mathbb{R}$ is the greatest real subfield of $\mathbb{Q}^{(m)}$, and $\mathrm{Split}(\mathbb{Q}^{(m)+}/\mathbb{Q}) = \{p \in \mathbb{P} \mid p \equiv \pm 1 \bmod m\}$.

- $\mathbb{Q}[\mathbb{Q}^+(m)^2\mathbb{Q}^m]$ is the compositum of all quadratic subfields of $\mathbb{Q}^{(m)}$, and a prime p lies in $\mathrm{Split}(\mathbb{Q}[\mathbb{Q}^+(m)^2\mathbb{Q}^m]/\mathbb{Q})$ if and only if it is a quadratic residue moduo m.

Now we identify $\mathbb{Q}^+(m)/\mathbb{Q}^m = (\mathbb{Z}/m\mathbb{Z})^\times$. Then 3.2.6.3 shows that the Artin isomorphism is given by

$$\sigma_m = (\,\cdot\,, \mathbb{Q}^{(m)}/K) \colon (\mathbb{Z}/m\mathbb{Z})^\times \xrightarrow{\sim} \mathrm{Gal}(\mathbb{Q}^{(m)}/\mathbb{Q}),$$

where $\sigma_m(a + m\mathbb{Z})(\zeta) = \zeta^a$ for all $a \in \mathbb{Z}(m)$ and all m-th roots of unity ζ. Hence 3.7.2.2 and 3.7.5 merely comprise the well-known fact that the assignment

$$\Gamma \mapsto \mathbb{Q}[\Gamma] = (\mathbb{Q}^{(m)})^{(\Gamma, \mathbb{Q}^{(m)}/\mathbb{Q})}$$

defines an inclusion-reversing bijection between the subgroups Γ of $(\mathbb{Z}/m\mathbb{Z})^\times$ and the subfields of $\mathbb{Q}^{(m)}$.

In 3.3.8 we established an inclusion-preserving bijection between the subgroups Z of the character group $\mathsf{X}(m)$ of $(\mathbb{Z}/m\mathbb{Z})^\times$ and the subfields of $\mathbb{Q}^{(m)}$ by the assignment

$$Z \mapsto \mathbb{Q}_Z = (\mathbb{Q}^{(m)})^{\sigma_m(Z^\perp)}.$$

Theorem 3.7.10. *Let $m \in \mathbb{N}$, $m \ge 3$. Let Γ be a subgroup of $(\mathbb{Z}/m\mathbb{Z})^\times$ and Z a subgroup of $\mathsf{X}(m)$. Then*

$$\mathbb{Q}[\Gamma] = \mathbb{Q}_Z \quad \textit{if and only if} \quad \Gamma = Z^\perp.$$

Proof. Since $\mathbb{Q}[\Gamma] = (\mathbb{Q}^{(m)})^{(\Gamma, \mathbb{Q}^{(m)}/\mathbb{Q})}$ and $\mathbb{Q}_Z = (\mathbb{Q}^{(m)})^{\sigma_m(Z^\perp)}$, the fields coincide if and only if $(\Gamma, \mathbb{Q}^{(m)}/\mathbb{Q}) = \sigma_m(\Gamma) = \sigma_m(Z^\perp)$, and since $\sigma_m \colon (\mathbb{Z}/m\mathbb{Z})^\times \to \mathrm{Gal}(\mathbb{Q}^{(m)}/\mathbb{Q})$ is an isomorphism, the assertion follows. □

3.8 Arithmetic of quadratic orders

Before we start with the arithmetic theory of quadratic orders, we present the fundamentals of the theory of quadratic irrationals which are an important tool for these investigations. For a thorough analysis of quadratic irrationals together with the connections to the theory of binary quadratic forms we refer to [22].

Throughout this section, for $x_1, \ldots, x_n \in \mathbb{C}$ we set

$$\langle x_1, \ldots, x_n \rangle = \mathbb{Z}x_1 + \ldots + \mathbb{Z}x_n.$$

Definitions and Remarks 3.8.1. Recall that we normalized $\sqrt{d}$ to $\sqrt{d} > 0$ if $d > 0$, and $\Im(\sqrt{d}) > 0$ if $d < 0$.

1. A complex number ξ is called a **quadratic irrational** if $\mathbb{Q}(\xi)$ is a quadratic number field. Then $\mathbb{Q}(\xi) = \mathbb{Q}(\sqrt{d})$ for some $d \in \mathbb{Q}^\times \setminus \mathbb{Q}^{\times 2}$, and $\xi = u + v\sqrt{d}$ with uniquely determined $u, v \in \mathbb{Q}$. If $d \in \mathbb{Q}^\times \setminus \mathbb{Q}^{\times 2}$, then $\mathrm{Gal}(\mathbb{Q}(\sqrt{d})/\mathbb{Q}) = \{\mathrm{id}_{\mathbb{Q}(\sqrt{d})}, \tau\}$, where $\tau(\sqrt{d}) = -\sqrt{d}$. If $\xi = u + v\sqrt{d}$ with $u, v \in \mathbb{Q}$, then we call $\xi' = \tau(\xi) = u - v\sqrt{d}$ the **conjugate** of ξ, and then $\xi\xi' = \mathsf{N}_{\mathbb{Q}(\sqrt{d})/\mathbb{Q}}(\xi)$. For a subset X of $\mathbb{Q}(\xi)$, we set $X' = \{x' \mid x \in X\}$.

2. Let ξ be a quadratic irrational. Then elementary arguments regarding the normalization of $\sqrt{d}$ show that there exists a unique triple $(a, b, c) \in \mathbb{Z}^3$ such that

$$a \neq 0, \quad (a, b, c) = 1, \quad d = b^2 - 4ac \text{ is not a square and } \xi = \frac{b + \sqrt{d}}{2a}. \quad (*)$$

We call (a, b, c) the **type** and d the **discriminant** of ξ. Note that d is a quadratic discriminant as defined in 3.4.4.

3. We define a group operation $\mathsf{GL}_2(\mathbb{Z}) \times (\mathbb{C} \setminus \mathbb{Q}) \to \mathbb{C} \setminus \mathbb{Q}$ by

$$\begin{pmatrix} \alpha & \beta \\ \gamma & \delta \end{pmatrix} z = \frac{\alpha z + \beta}{\gamma z + \delta} \quad \text{for all} \quad \begin{pmatrix} \alpha & \beta \\ \gamma & \delta \end{pmatrix} \in \mathsf{GL}_2(\mathbb{Q}) \text{ and } z \in \mathbb{C} \setminus \mathbb{Q}.$$

For matrices A, $A_1 \in \mathsf{GL}_2(\mathbb{Z})$ and $z \in \mathbb{C} \setminus \mathbb{Q}$ it is easily checked that

$$A_1(Az) = (A_1A)z, \quad \text{and if } A = \begin{pmatrix} \alpha & \beta \\ \gamma & \delta \end{pmatrix}, \text{ then } \Im(Az) = \frac{\det(A)}{|\gamma z + \delta|^2} \Im(z).$$

Hence $\mathsf{SL}_2(\mathbb{Z})$ also operates on the upper half plane $\mathcal{H}^+ = \{z \in \mathbb{C} \mid \Im(z) > 0\}$. If $I \in \mathsf{SL}_2(\mathbb{Z})$ denotes the unit matrix, then $(-I)z = z$ for all $z \in \mathbb{C} \setminus \mathbb{Q}$. Hence also the **modular group** $\mathsf{PSL}_2(\mathbb{Z}) = \mathsf{SL}_2(\mathbb{Z})/\{\pm I\}$ operates on $\mathcal{H}^+$.

3. Two irrational numbers z, $z_1 \in \mathbb{C} \setminus \mathbb{Q}$ are called **equivalent**, $z \sim z_1$ if $z_1 = Az$ for some $A \in \mathsf{GL}_2(\mathbb{Z})$. $\sim$ is an equivalence relation on $\mathbb{C} \setminus \mathbb{Q}$.

If ξ is a quadratic irrational of type (a, b, c) with discriminant d and

$$\xi_1 = \begin{pmatrix} \alpha & \beta \\ \gamma & \delta \end{pmatrix} \xi,$$

then a tedious but simple calculation shows that ξ_1 is a quadratic irrational of type (a_1, b_1, c_1) with discriminant d, where

$$\begin{aligned} a_1 &= a\delta^2 + b\gamma\delta + c\gamma^2, \\ b_1 &= 2a\beta\delta + b(\alpha\delta + \beta\gamma) + 2c\alpha\gamma, \\ c_1 &= a\beta^2 + b\alpha\beta + c\alpha^2, \qquad \text{and} \quad d = b_1^2 - 4a_1c_1. \end{aligned}$$

Now we turn to the theory of quadratic orders.

Definition and Theorem 3.8.2 (Classification of quadratic orders). *Let d be a quadratic discriminant and $K = \mathbb{Q}(\sqrt{d})$. Then $d = 4D + \sigma_d$, where $D \in \mathbb{Z}$,*

$$\sigma_d = \begin{cases} 0 & \text{if } d \equiv 0 \bmod 4, \\ 1 & \text{if } d \equiv 1 \bmod 4, \end{cases} \qquad \text{we set} \qquad \omega_d = \frac{\sigma_d + \sqrt{d}}{2} \quad \text{and} \quad \mathcal{O}_d = \mathbb{Z}[\omega_d].$$

We call ω_d the **basis number** and $\mathcal{O}_d$ the **quadratic order with discriminant** d, and we set $U_d = \mathcal{O}_d^\times$. If $K = \mathbb{Q}(\sqrt{d})$, then $d = \mathsf{d}_K f^2$, where d_K is the associated fundamental discriminant and f is the conductor of d (see 3.4.4). In particular, $\mathfrak{o}_K = \mathcal{O}_{\mathsf{d}_K} = \mathbb{Z}[\omega_{\mathsf{d}_K}]$ by 3.1.8, and $U_K = U_{\mathsf{d}_K}$.

1. *Let f be the conductor of d. Then*

$$\mathcal{O}_d = \Bigl\{\frac{u + v\sqrt{d}}{2} \Bigm| u, v \in \mathbb{Z},\ u \equiv vd \bmod 2\Bigr\} = \langle 1, \omega_d\rangle.$$

$\mathcal{O}_d$ is an order of K,

$$\Delta(\mathcal{O}_d) = d, (\mathfrak{o}_K : \mathcal{O}_d) = f, \text{ and } (\mathcal{O}_d : \mathfrak{o}_K) = f\mathfrak{o}_K.$$

If $m \in \mathbb{N}$, then $\mathcal{O}_{dm^2} = \mathbb{Z}[m\omega_d] = \langle 1, m\omega_d\rangle$ and $(\mathcal{O}_d : \mathcal{O}_{dm^2}) = m$. In particular, $\mathcal{O}_d = \mathbb{Z} + f\mathfrak{o}_K$.

2. *Let K be a quadratic number field, $\mathfrak{o}$ an order of K with index $(\mathfrak{o}_K : \mathfrak{o}) = f$. Then $\mathfrak{o} = \mathcal{O}_{\mathsf{d}_K f^2}$.*

Proof. For the straightforward proofs we refer to [22, Theorem 5.1.7] □

Theorem 3.8.3 (Unit groups of quadratic orders). *Let d be a quadratic discriminant. Then*

$$U_d = \Bigl\{\frac{u + v\sqrt{d}}{2} \Bigm| u, v \in \mathbb{Z},\ |u^2 - dv^2| = 4\Bigr\}.$$

1. *If $d < 0$, then $U_d = \mu(\mathcal{O}_d)$ is the group of roots of unity of $\mathcal{O}_d$, and*

$$|U_d| = \begin{cases} 6 & \text{if } d = -3, \\ 4 & \text{if } d = -4, \\ 2 & \text{if } d < -4. \end{cases}$$

2. *If $d > 0$, then $U_d = \langle -1, \varepsilon_d\rangle \cong \mathbb{Z}/2\mathbb{Z} \times \mathbb{Z}$, where*

$$\varepsilon_d = \min\{\varepsilon \in U_d \mid \varepsilon > 1\},$$

and

$$\begin{aligned}\{\varepsilon \in U_d \mid \varepsilon \geq 1\} &= \{\varepsilon_d^k \mid k \in \mathbb{N}_0\} \\ &= \Bigl\{\frac{u + v\sqrt{d}}{2} \Bigm| u, v \in \mathbb{N}_0,\ |u^2 - dv^2| = 4\Bigr\}.\end{aligned}$$

ε_d is called the **fundamental unit** of $\mathcal{O}_d$. Note that $\varepsilon_K = \varepsilon_{\mathfrak{o}_K}$ is the fundamental unit of K by 3.5.11.

Proof. Let $K = \mathbb{Q}(\sqrt{d})$. By 3.5.8.1 and 3.8.2 it follows that

$$U_d = \{\alpha \in \mathcal{O}_d \mid |\mathsf{N}_{K/\mathbb{Q}}(\alpha)| = 1\} = \Big\{\frac{u+v\sqrt{d}}{2} \;\Big|\; u,\, v \in \mathbb{Z},\; |u^2-dv^2| = 4\Big\}.$$

Now 3.5.8.2 implies $U_d = \mu(\mathcal{O}_d)$ if $d<0$, and $U_d = \mu(\mathcal{O}_d)\cdot\langle\varepsilon_0\rangle$ if $d>0$.

If $d<0$, then $|u^2-dv^2| = u^2+|d|v^2$ for all $(u,v) \in \mathbb{Z}^2$, and the Diophantine equation $|u^2-dv^2| = 4$ has the asserted number of solutions.

If $d>0$, then $\mathcal{O}_d \subset \mathbb{R}$ implies $\mu(\mathcal{O}_d) = \{\pm1\}$. If $U_d = \langle -1, \varepsilon_0\rangle$ and $\varepsilon_1 \in U_d$ is any unit of $\mathcal{O}_d$, then $U_d = \langle -1, \varepsilon_1\rangle$ if and only if $\varepsilon_1 \in \{\pm\varepsilon_0, \pm\varepsilon_0^{-1}\}$, and exactly one of these four units is greater that 1. Hence there exists a unique $\varepsilon_d \in \mathbb{R}_{>1}$ such that $U_d = \langle -1, \varepsilon_d\rangle$. Then $U_d \cap \mathbb{R}_{>1} = \{\varepsilon_d^n \mid n \in \mathbb{N}\}$, and $\varepsilon_d = \min\{\varepsilon \in U_d \mid \varepsilon > 1\}$.

Assume now that

$$\varepsilon = \frac{u+v\sqrt{d}}{2}, \quad \text{where } u,\, v \in \mathbb{Z} \text{ and } |u^2-dv^2| = 4.$$

Then

$$\max\{\pm\varepsilon, \pm\varepsilon^{-1}\} = \max\Big\{\frac{u \pm v\sqrt{d}}{2}, \frac{-u \pm v\sqrt{d}}{2}\Big\} = \frac{|u|+|v|\sqrt{d}}{2},$$

and therefore

$$\{\varepsilon \in U_d \mid \varepsilon \ge 1\} = \Big\{\frac{u+v\sqrt{d}}{2} \;\Big|\; u,\, v \in \mathbb{N}_0,\; |u^2-dv^2| = 4\Big\}.$$

Since $U_K \cap \mathcal{O}_d = \langle -1, \varepsilon_K\rangle \cap \mathcal{O}_d = \langle -1, \varepsilon_d\rangle$, it follows that $\varepsilon_d = \varepsilon_K^m$ and $(U_K : U_d) = m$, where $m \in \mathbb{N}$ is minimal such that $\varepsilon_K^m \in \mathcal{O}_d$. □

Beyond U_d we consider the group $U_d^+ = U_d \cap K^+$ of totally positive units. By definition, $U_d = U_d^+$ if $d<0$. If $d>0$, then

$$U_d^+ = \begin{cases} \langle\varepsilon_d\rangle & \text{if } \mathsf{N}_{K/\mathbb{Q}}(\varepsilon_d) = 1, \\ \langle\varepsilon_d^2\rangle & \text{if } \mathsf{N}_{K/\mathbb{Q}}(\varepsilon_d) = -1, \end{cases} \quad \text{and} \quad (U_d : U_d^+) = \begin{cases} 2 & \text{if } \mathsf{N}_{K/\mathbb{Q}}(\varepsilon_d) = 1 \\ 4 & \text{if } \mathsf{N}_{K/\mathbb{Q}}(\varepsilon_d) = -1. \end{cases}$$

In any case, $(U_d^+ : U_d^{+2}) = 2$. Note that $\mathsf{N}_{K/\mathbb{Q}}(\varepsilon_d) = -1$ holds if and only if the **Pell minus equation** $x^2 - dy^2 = -4$ has a solution.

We proceed with the ideal theory of quadratic orders. Recall that quadratic orders are one-dimensional Noetherian domains and that all residue class rings of non-zero ideals are finite (see 2.11.1 and 3.1.3.5(b)).

Definition and Theorem 3.8.4 (Structure of ideals in $\mathcal{O}_d$). *Let d be a quadratic discriminant.*

For a quadratic irrational ξ of type (a,b,c) and discriminant d we set

$$I(\xi) = a\langle 1, \xi\rangle = \Big\langle a, \frac{b+\sqrt{d}}{2}\Big\rangle \subset \mathcal{O}_d.$$

An ideal $\mathfrak{a}$ of $\mathcal{O}_d$ is called

- $\mathcal{O}_d$-**primitive** if $e^{-1}\mathfrak{a} \not\subset \mathcal{O}_d$ for all $e \in \mathbb{N}_{\ge 2}$;
- $\mathcal{O}_d$-**regular** if it is $\mathcal{O}_d$-primitive and $\mathcal{O}_d$-invertible.

1. Let $a,\, b,\, c,\, e \in \mathbb{Z}$ and $d = b^2 - 4ac$. Then

$$\mathfrak{a} = e\Big\langle a, \frac{b+\sqrt{d}}{2}\Big\rangle$$

is an ideal of $\mathcal{O}_d$.

2. Let $\mathfrak{a} \ne \mathbf{0}$ be an ideal of $\mathcal{O}_d$. Then there exist $a,\, e \in \mathbb{N}$ and $b,\, c \in \mathbb{Z}$ such that $d = b^2 - 4ac$ and

$$\mathfrak{a} = e\Big\langle a, \frac{b+\sqrt{d}}{2}\Big\rangle.$$

There the numbers a, e and the residue class $b + 2a\mathbb{Z} \in \mathbb{Z}/2a\mathbb{Z}$ are uniquely determined by $\mathfrak{a}$. If $g = (a, b, c) \in \mathbb{N}$ and $d_0 = g^{-2}d$, then the following assertions hold:

(a) $\mathfrak{a} \cap \mathbb{Z} = \mathbb{Z}ae$, $ae^2 = (\mathcal{O}_d : \mathfrak{a}) \equiv 0 \bmod g$, and $e^{-1}\mathfrak{a}$ is $\mathcal{O}_d$-primitive.

(b) d_0 is a quadratic discriminant, $\mathfrak{a}\mathfrak{a}' = gae^2\mathcal{O}_{d_0}$ and $\mathcal{R}(\mathfrak{a}) = \mathcal{O}_{d_0}$.

(c) The ideal $\mathfrak{a}$ is

- *$\mathcal{O}_d$-primitive if and only if $e = 1$;*
- *$\mathcal{O}_d$-invertible if and only if $\mathcal{R}(\mathfrak{a}) = \mathcal{O}_d$, and this holds if and only if $g = 1$;*
- *$\mathcal{O}_d$-regular if and only if $e = g = 1$, and then $\mathfrak{a}\mathfrak{a}' = m\mathcal{O}_d$, where $m = (\mathcal{O}_d : \mathfrak{a})$.*

3. Let $\mathfrak{a}$ be an $\mathcal{O}_d$-invertible fractional ideal of $\mathcal{O}_d$. Then $\mathfrak{a}\mathfrak{a}' = a\mathcal{O}_d$ for some $a \in \mathbb{Q}^+$, and there exists some $r \in \mathbb{Q}_{>0}$ such that $r\mathfrak{a}$ is an $\mathcal{O}_d$-regular ideal.

4. If ξ is a quadratic irrational with discriminant d, then $I(\xi)$ is an $\mathcal{O}_d$-regular ideal, and every $\mathcal{O}_d$-regular ideal is of this form.

Proof. Elementary. We refer to [22, Sect. 5.4]. □

Theorem 3.8.5 (Prime ideals in quadratic orders). *Let d be a quadratic discriminant, f its conductor, χ_d the quadratic character associated with d* (see 3.4.4) *and p a prime.*

1. $\chi_d(p) \neq -1$ *if and only if* d *is a quadratic residue modulo* $4p$.

2. Let $\chi_d(p) = 1$ *and* $b \in \mathbb{Z}$ *such that* $d \equiv b^2 \bmod 4p$. *Then it follows that* $p\mathcal{O}_d = \mathfrak{p}\mathfrak{p}'$, *where*

$$\mathfrak{p} = \Big\langle p, \frac{b+\sqrt{d}}{2}\Big\rangle, \quad \mathfrak{p}' = \Big\langle p, \frac{b-\sqrt{d}}{2}\Big\rangle \quad \text{and} \quad \mathfrak{p} \neq \mathfrak{p}'.$$

$\mathfrak{p}$ *and* $\mathfrak{p}'$ *are* $\mathcal{O}_d$*-regular prime ideals, and they are the only prime ideals of* $\mathcal{O}_d$ *containing* p.

3. If $\chi_d(p) = -1$, *then* $p\,\mathcal{O}_d$ *is the only prime ideal of* $\mathcal{O}_d$ *containing* p.

4. If $\chi_d(p) = 0$, *then there exists a unique prime ideal* $\mathfrak{p}$ *of* $\mathcal{O}_d$ *such that* $p \in \mathfrak{p}$, *namely*:

- *If* $p \neq 2$, *then*

$$\mathfrak{p} = \begin{cases} \langle p, \sqrt{D}\rangle & \text{if } \ d = 4D \equiv 0 \bmod 4, \\ \Big\langle p, \dfrac{p+\sqrt{d}}{2}\Big\rangle & \text{if } \ d \equiv 1 \bmod 4. \end{cases}$$

- *If* $p = 2$ *and* $d = 4D$ *with* $D \in \mathbb{Z}$, *then*

$$\mathfrak{p} = \begin{cases} \langle 2, \sqrt{D}\rangle & \text{if } \ 2 \mid D, \\ \langle 2, 1+\sqrt{D}\rangle & \text{if } \ 2 \nmid D. \end{cases}$$

In any case, $\mathfrak{p}$ *has the following properties*:

- *If* $p \nmid f$, *then* $\mathfrak{p}$ *is* $\mathcal{O}_d$*-regular, and* $\mathfrak{p}^2 = p\,\mathcal{O}_d$.
- *If* $p \mid f$, *then* $\mathfrak{p}$ *is not* $\mathcal{O}_d$*-invertible, and* $\mathfrak{p}^2 = p\,\mathfrak{p}$.

Proof. 1. If $p \neq 2$, then $d \equiv 0$ or $1 \bmod 4$, and thus d is a quadratic residue modulo 4. Hence d is a quadratic residue modulo $4p$ if and only if it is a quadratic residue modulo p, and by definition this is equivalent to $\chi_d(p) \neq -1$.

Thus assume that $p = 2$. If $2 \mid d$, then $\chi_d(2) = 0$ and $d \equiv 4$ or $0 \bmod 8$. Hence d is a quadratic residue modulo 8. If $2 \nmid d$, then $\chi_d(2) \neq 0$, $d \equiv 1 \bmod 4$, and d is a quadratic residue modulo 8 if and only if $d \equiv 1 \bmod 8$, which by definition is equivalent to $\chi_d(2) = 1$.

In the sequel we use 3.8.4 again and again.

2. Let $c \in \mathbb{Z}$ be such that $d = b^2 - 4pc$. As $p \nmid d$, it follows that $(p, b) = 1$, hence $(p, b, c) = 1$, and therefore $\mathfrak{p}$ and $\mathfrak{p}'$ are $\mathcal{O}_d$-regular. We obtain

$$\begin{aligned}\mathfrak{p}\mathfrak{p}' &= \Big\langle p^2, p\frac{b+\sqrt{d}}{2}, p\frac{b-\sqrt{d}}{2}, p\frac{b^2-d}{4}\Big\rangle = p\Big\langle p, b, \frac{b+\sqrt{d}}{2}, c\Big\rangle \\ &= p\Big\langle 1, \frac{b+\sqrt{d}}{2}\Big\rangle = p\mathcal{O}_d.\end{aligned}$$

Since $b \not\equiv -b \bmod 2p$, we obtain $\mathfrak p \ne \mathfrak p'$, and since $(\mathcal O_d : \mathfrak p) = (\mathcal O_d : \mathfrak p') = p$, it follows that $\mathfrak p$ and $\mathfrak p'$ are prime ideals. If $\mathfrak q$ is a prime ideal of $\mathcal O_d$ containing p, then $p\mathcal O_d = \mathfrak p \mathfrak p' \subset \mathfrak q$, hence $\mathfrak p \subset \mathfrak q$ or $\mathfrak p' \subset \mathfrak q$, and therefore $\mathfrak q \in \{\mathfrak p, \mathfrak p'\}$.

3. It suffices to prove that p is a prime element in $\mathcal O_d$. Thus let $x,\, y \in \mathcal O_d$ such that $p \,|\, xy$. Then $p^2 = \mathsf N_{K/\mathbb Q}(p) \,|\, \mathsf N_{K/\mathbb Q}(x)\mathsf N_{K/\mathbb Q}(y)$, and we may assume that $p \,|\, \mathsf N_{K/\mathbb Q}(x)$. If $x = (u + v\sqrt d)/2$, where $u,\, v \in \mathbb Z$, then $p \,|\, u^2 - v^2 d$, and as $\chi_d(p) = -1$, it follows that $u \equiv v \equiv 0 \bmod p$ and consequently $p \,|\, x$.

4. CASE 1: $d = 4D$, where $D \in \mathbb Z$ and $p \,|\, D$, say $D = pD'$ and $D' \in \mathbb Z$. Then $\mathfrak p = \mathbb Z p \oplus \mathbb Z\sqrt D$ is an $\mathcal O_d$-primitive ideal satisfying $(\mathcal O_d : \mathfrak p) = p$. Hence $\mathfrak p$ is a prime ideal, and it is $\mathcal O_d$-regular if and only if $p \nmid D'$. We obtain

$$\mathfrak p^2 = \langle p^2, p\sqrt D, D\rangle = p\,\langle p, \sqrt D, D'\rangle = \begin{cases} p\,\langle p, \sqrt D\rangle = p\,\mathfrak p & \text{if } \ p \,|\, D', \\ p\,\langle 1, \sqrt D\rangle = p\,\mathcal O_d & \text{if } \ p \nmid D', \end{cases}$$

and it remains to prove that $p \,|\, D'$ if and only if $p \,|\, f$.

Assume first that $p \,|\, D'$. Then $p^2 \,|\, d$ if $p \ne 2$, $\ 2^4 \,|\, d$ if $p = 2$, and thus $p \,|\, f$. If $p \nmid D'$, then $\mathsf v_p(d) = 1$ if $p \ne 2$, $\ \mathsf v_2(d) = 3$ if $p = 2$, and thus $p \nmid f$.

CASE 2: $d \equiv 1 \bmod 4$ and $p \,|\, d$, say $d = pd'$, where $d' \in \mathbb Z$ and $d' \equiv p \bmod 4$. As above,

$$\mathfrak p = \Big\langle p, \frac{p + \sqrt d}{2}\Big\rangle$$

is an $\mathcal O_d$-primitive prime ideal which is $\mathcal O_d$-regular if and only if $p \nmid d'$, which holds if and only if $p \nmid f$. Now we obtain

$$\begin{aligned} \mathfrak p^2 &= \Big\langle p^2, \frac{p^2 + p\sqrt d}{2}, \frac{p^2 + d + 2p\sqrt d}{4}\Big\rangle = \Big\langle p^2, \frac{p^2 + p\sqrt d}{2}, \frac{-p^2 + d}{4}\Big\rangle \\ &= p\,\Big\langle p, \frac{p + \sqrt d}{2}, \frac{-p + d'}{4}\Big\rangle \ = \ \begin{cases} p\,\mathfrak p & \text{if } \ p \,|\, d', \\ p\,\mathcal O_d & \text{if } \ p \nmid d'. \end{cases} \end{aligned}$$

CASE 3: $p = 2$ and $d = 4D$, where $D \in \mathbb Z$ and $2 \nmid D$. As above, it follows that $\mathfrak p = \langle 2, 1 + \sqrt D\rangle$ is an $\mathcal O_d$-primitive prime ideal which is $\mathcal O_d$-regular if and only if $(2, 2, (D-1)/2) = 1$. However,

$$(2, 2, (D-1)/2) = 1 \iff D \equiv 3 \bmod 4 \iff 2 \nmid f.$$

Finally we obtain

$$\begin{aligned} \mathfrak p^2 &= \langle 4, 2 + 2\sqrt D, 1 + D + 2\sqrt D\rangle = \langle 4, 2 + 2\sqrt D, D - 1\rangle \\ &= 2\Big\langle 2, 1 + \sqrt D, \frac{-1 + D}{2}\Big\rangle \ = \ \begin{cases} 2\,\mathfrak p & \text{if } \ 2 \,|\, f, \\ 2\,\mathcal O_d & \text{if } \ 2 \nmid f. \end{cases} \end{aligned}$$

In any case, we assert that $\mathfrak p$ is the only prime ideal of $\mathcal O_d$ containing p. Indeed, let $\mathfrak q$ be any prime ideal containing p. As $\mathfrak p^2 = p\mathfrak p$ or $\mathfrak p^2 = p\mathcal O_d$, it follows that $\mathfrak p^2 \subset \mathfrak q$, hence $\mathfrak p \subset \mathfrak q$ and therefore $\mathfrak p = \mathfrak q$. □

Next we investigate class groups of quadratic orders.

Definitions and Remarks 3.8.6.

Let d be a quadratic discriminant and $K = \mathbb{Q}(\sqrt{d})$. If $d < 0$, then we have $\boldsymbol{\sigma}(K) = \emptyset$ and consequently $K^+ = K^\times$. If $d > 0$, then $|\boldsymbol{\sigma}(K)| = 2$, $K^+ = \{a \in K \mid a > 0 \text{ and } a' > 0\}$ and $(K\!:\!K^+) = 4$ by 2.9.7.1. Corresponding to 2.6.5 and 2.9.8 we define

- $\mathcal{H}_d = \mathcal{H}(\mathcal{O}_d) = \{a\mathcal{O}_d \mid a \in K^\times\}, \quad \mathcal{C}_d = \mathcal{I}(\mathcal{O}_d)/\mathcal{H}_d \quad \text{and} \quad h_d = |\mathcal{C}_d|,$
- $\mathcal{H}_d^+ = \mathcal{H}^+(\mathcal{O}_d) = \{a\mathcal{O}_d \mid a \in K^+\}, \quad \mathcal{C}_d^+ = \mathcal{I}(\mathcal{O}_d)/\mathcal{H}_d^+ \quad \text{and} \quad h_d^+ = |\mathcal{C}_d^+|.$

We call $\mathcal{C}_d$ the **class group** and $\mathcal{C}_d^+$ the **narrow class group**, h_d die **class number** and h_d^+ the **narrow class number** of $\mathcal{O}_d$. For an invertible ideal $\mathfrak{a} \in \mathcal{I}(\mathcal{O}_d)$, we denote by $[\mathfrak{a}] = \mathfrak{a}\mathcal{H}_d$ the ideal class and by $[\mathfrak{a}]^+ = \mathfrak{a}\mathcal{H}_d^+$ the narrow ideal class of $\mathfrak{a}$. Observe that $\mathcal{H}_d^+ = \{a\mathcal{O}_d \mid a \in K^\times,\ \mathsf{N}_{K/\mathbb{Q}}(a) > 0\}$, and by 3.8.4.(3. and 4.) it follows that

$$\begin{aligned}\mathcal{C}_d^{(+)} &= \{[\mathfrak{a}]^{(+)} \mid \mathfrak{a} \text{ is an } \mathcal{O}_d\text{-regular ideal of } \mathcal{O}_d\} \\ &= \{[I(\xi)]^{(+)} \mid \xi \text{ is a quadratic irrational of discriminant } d\}.\end{aligned}$$

Note that $[\mathcal{O}_d]^{(+)}$ is the unit element of $\mathcal{C}_d^{(+)}$. If $K = \mathbb{Q}(\sqrt{d})$, then $h_K^{(+)} = h_{\mathsf{d}_K}^{(+)}$.

In the following Theorem 3.8.7 we connect the numbers h_K, h_d and h_d^+.

Theorem 3.8.7 (Class groups of quadratic orders). *Let d be a quadratic discriminant, f its conductor and $K = \mathbb{Q}(\sqrt{d})$.*

1. *There are (natural) exact sequences*

$$1 \to \mathcal{H}_d/\mathcal{H}_d^+ \to \mathcal{C}_d^+ \xrightarrow{\pi} \mathcal{C}_d \to 1,$$

where $\pi([\mathfrak{a}]^+) = \pi([\mathfrak{a}])$ for all $\mathfrak{a} \in \mathcal{I}(\mathcal{O}_d)$,

$$1 \to U_d/U_d^+ \to K^\times/K^+ \xrightarrow{\rho} \mathcal{H}_d/\mathcal{H}_d^+ \to 1,$$

where $\rho(aK^+) = (a\mathcal{O}_d)\mathcal{H}_d^+$ for all $a \in K^\times$, and

$$\frac{|\mathcal{H}_d|}{|\mathcal{H}_d^+|} = \frac{h_d^+}{h_d} = \begin{cases} 2 & \text{if } d > 0 \text{ and } \mathsf{N}_{K/\mathbb{Q}}(\varepsilon_d) = 1, \\ 1 & \text{otherwise.}\end{cases}$$

2. *There is a (natural) exact sequence*

$$1 \to U_K/U_d \to (\mathfrak{o}_K/f\mathfrak{o}_K)^\times/(\mathcal{O}_d/f\mathfrak{o}_K)^\times \to \mathcal{C}_d \xrightarrow{\iota} \mathcal{C}_K \to 1,$$

where $\iota([\mathfrak{a}]) = [\mathfrak{a}\mathfrak{o}_K]$ for all $\mathfrak{a} \in \mathcal{I}(\mathcal{O}_d)$, and

$$h_d = \frac{h_K f}{(U_K\!:\!U_d)} \prod_{p \mid f}\Big(1 - \frac{\chi_K(p)}{p}\Big).$$

Proof. 1. The exact sequences come from the very definitions, and they also show that

$$\frac{h_d^+}{h_d} = \frac{|\mathcal{C}_d^+|}{|\mathcal{C}_d|} = \frac{|\mathcal{H}_d|}{|\mathcal{H}_d^+|} = \frac{(K^\times : K^+)}{(U_d : U_d^+)}.$$

Since

$$(K^\times : K^+) = \begin{cases} 4 & \text{if } d > 0, \\ 1 & \text{if } d < 0 \end{cases} \quad \text{and} \quad (U_d : U_d^+) = \begin{cases} 4 & \text{if } d > 0, \ \mathsf{N}_{K/\mathbb{Q}}(\varepsilon_d) = -1, \\ 2 & \text{if } d > 0, \ \mathsf{N}_{K/\mathbb{Q}}(\varepsilon_d) = 1, \\ 1 & \text{if } d < 0, \end{cases}$$

the assertion follows.

2. The exact sequence comes from 2.11.5, applied with $\boldsymbol{\sigma} = \emptyset$. Using 3.8.2.1, we obtain $\mathcal{O}_d / f\mathfrak{o}_K = \mathbb{Z} + f\mathfrak{o}_K / f\mathfrak{o}_K \cong \mathbb{Z}/f\mathfrak{o}_K \cap \mathbb{Z}$, and $\mathfrak{o}_K = \mathbb{Z} + \mathbb{Z}\omega_{\mathsf{d}_K}$ implies $f\mathfrak{o}_K \cap \mathbb{Z} = (f\mathbb{Z} + f\mathbb{Z}\omega_{\mathsf{d}_K}) \cap \mathbb{Z} = f\mathbb{Z}$. Hence $(\mathcal{O}_d / f\mathfrak{o}_K)^\times \cong (\mathbb{Z}/f\mathbb{Z})^\times$, and

$$h_d = \frac{h_K}{(U_K : U_d)} \, \frac{|(\mathfrak{o}_K / f\mathfrak{o}_K)^\times|}{|(\mathbb{Z}/f\mathbb{Z})^\times|}.$$

By 2.14.1 we obtain

$$\begin{aligned} |(\mathfrak{o}_K / f\mathfrak{o}_K)^\times| &= \mathfrak{N}(f\mathfrak{o}_K) \prod_{p \mid f} \prod_{\substack{\mathfrak{p} \in \mathcal{P}_K \\ \mathfrak{p} \mid p}} \Bigl(1 - \frac{1}{\mathfrak{N}(\mathfrak{p})}\Bigr) \\ &= f^2 \prod_{\substack{p \mid f \\ \chi_K(p) = 1}} \Bigl(1 - \frac{1}{p}\Bigr)^2 \prod_{\substack{p \mid f \\ \chi_K(p) = -1}} \Bigl(1 - \frac{1}{p^2}\Bigr) \prod_{\substack{p \mid f \\ \chi_K(p) = 0}} \Bigl(1 - \frac{1}{p}\Bigr), \end{aligned}$$

and since

$$(\mathbb{Z}/f\mathbb{Z})^\times = \varphi(f) = f \prod_{p \mid f} \Bigl(1 - \frac{1}{p}\Bigr),$$

it follows that

$$\frac{|(\mathfrak{o}_K / f\mathfrak{o}_K)^\times|}{|(\mathbb{Z}/f\mathbb{Z})^\times|} = f \prod_{p \mid f} \Bigl(1 - \frac{\chi_K(p)}{p}\Bigr) \quad \text{and} \quad h_d = \frac{h_K f}{(U_K : U_d)} \prod_{p \mid f} \Bigl(1 - \frac{\chi_K(p)}{p}\Bigr).$$

□

We proceed with a description of the class group using quadratic irrationals.

Theorem 3.8.8. *Let ξ be a quadratic irrational of type (a, b, c) and ξ_1 a quadratic irrational of type (a_1, b_1, c_1).*

1. Assume that

$$\xi_1 = A\xi, \quad \textit{where} \quad A = \begin{pmatrix} \alpha & \beta \\ \gamma & \delta \end{pmatrix} \in \mathsf{GL}_2(\mathbb{Z}).$$

Then

$$\mathsf{N}_{\mathbb{Q}(\xi)/\mathbb{Q}}(\gamma\xi+\delta) = \frac{a_1}{a}\,\det(A), \quad \text{and} \quad I(\xi) = \frac{a}{a_1}\,(\gamma\xi+\delta)\,I(\xi_1).$$

2. If $\lambda \in K^\times$ *and* $I(\xi) = \lambda I(\xi_1)$*, then* $\xi_1 = A\xi$ *for some matrix*

$$A = \begin{pmatrix} \alpha & \beta \\ \gamma & \delta \end{pmatrix} \in \mathsf{GL}_2(\mathbb{Z}) \quad \text{such that} \quad \lambda = \frac{a}{a_1}(\gamma\xi+\delta)\,.$$

3. $\xi \sim \xi_1$ *if and only if* $[I(\xi)] = [I(\xi_1)] \in \mathcal{C}_d$*. In particular,* h_d *is the number of equivalence classes of quadratic irrationals with discriminant* d*.*

Proof. 1. Since

$$\xi_1 = \frac{\alpha\xi+\beta}{\gamma\xi+\delta} \quad \text{and} \quad \begin{pmatrix} \alpha\xi+\beta \\ \gamma\xi+\delta \end{pmatrix} = A\begin{pmatrix} \xi \\ 1 \end{pmatrix},$$

it follows that

$$I(\xi_1) = a_1\Big(\mathbb{Z} + \mathbb{Z}\,\frac{\alpha\xi+\beta}{\gamma\xi+\delta}\Big) = \frac{a_1}{\gamma\xi+\delta}\,(\mathbb{Z}(\alpha\xi+\beta) + \mathbb{Z}(\gamma\xi+\delta)) = \frac{a_1}{\gamma\xi+\delta}\,(\mathbb{Z}+\mathbb{Z}\xi)$$
$$= \frac{a_1}{a(\gamma\xi+\delta)}\,I(\xi), \quad \text{and thus} \quad I(\xi) = \frac{a}{a_1}\,(\gamma\xi+\delta)\,I(\xi_1).$$

Since

$$\frac{\sqrt{d}}{a_1} = \xi_1 - \xi_1' = \frac{\alpha\xi+\beta}{\gamma\xi+\delta} - \frac{\alpha\xi'+\beta}{\gamma\xi'+\delta} = \frac{(\alpha\delta-\beta\gamma)(\xi-\xi')}{\mathsf{N}_{\mathbb{Q}(\xi)/\mathbb{Q}}(\gamma\xi+\delta)} = \frac{\det(A)}{\mathsf{N}_{\mathbb{Q}(\xi)/\mathbb{Q}}(\gamma\xi+\delta)}\,\frac{\sqrt{d}}{a},$$

we obtain

$$\mathsf{N}_{\mathbb{Q}(\xi)/\mathbb{Q}}(\gamma\xi+\delta) = \frac{a_1}{a}\,\det(A).$$

2. If $\lambda \in K^\times$ and $I(\xi) = \lambda I(\xi_1)$, then $\mathbb{Z}a + \mathbb{Z}a\xi = \mathbb{Z}\lambda a_1 + \mathbb{Z}\lambda a_1\xi_1$, which implies

$$\begin{pmatrix} \lambda a_1\xi_1 \\ \lambda a_1 \end{pmatrix} = \begin{pmatrix} \alpha & \beta \\ \gamma & \delta \end{pmatrix}\begin{pmatrix} a\xi \\ a \end{pmatrix} \quad \text{for some matrix} \quad A = \begin{pmatrix} \alpha & \beta \\ \gamma & \delta \end{pmatrix} \in \mathsf{GL}_2(\mathbb{Z}).$$

From this, we obtain

$$\lambda = \frac{\lambda a_1}{a_1} = \frac{a}{a_1}\,(\gamma\xi+\delta) \quad \text{and} \quad \xi_1 = \frac{\lambda a_1\xi_1}{\lambda a_1} = \frac{\alpha a\xi+\beta a}{\gamma a\xi+\delta a} = A\xi\,.$$

3. Obvious by 1. and 2. $\square$

3.9 Genus theory of quadratic orders

Throughout this section, let d be a quadratic discriminant and $K = \mathbb{Q}(\sqrt{d})$.

The determination of the structure of the class group $\mathcal{C}_d$ is a wide open problem. Precise results are only known for the 2-rank. This is the content of the famous Gauss's genus theory which we now develop in the context of ideal theory.

Theorem and Definition 3.9.1.

1. Let $\mathfrak{a}$ be an $\mathcal{O}_d$-invertible fractional ideal of $\mathcal{O}_d$. Then $\mathfrak{a} = \mathfrak{a}'$ if and only if $\mathfrak{a}^2 = a\mathcal{O}_d$ for some $a \in \mathbb{Q}^+$.

If these conditions are fulfilled, then $\mathfrak{a}$ is called **ambiguous**.

Every ambiguous $\mathcal{O}_d$-invertible fractional ideal $\mathfrak{a}$ of $\mathcal{O}_d$ has a factorization $\mathfrak{a} = a\mathfrak{a}_0$, where $a \in \mathbb{Q}^+$ and $\mathfrak{a}_0$ is an $\mathcal{O}_d$-regular ambiguous ideal.

2. For a (narrow) ideal class $C \in \mathcal{C}_d^{(+)}$ we set $C' = \{\mathfrak{a}' \mid \mathfrak{a} \in C\}$; obviously, $CC' = [\mathcal{O}_d]^{(+)} \in \mathcal{C}_d^{(+)}$.
If $C \in \mathcal{C}_K^{(+)}$, then $C = C'$ if and only if $C^2 = [\mathcal{O}_d]^{(+)}$.

If these conditions are fulfilled, then C is called **ambiguous**. We denote by $\mathcal{C}_d^{(+)}[2]$ the elementary abelian 2-group of all (narrow) ambiguous classes. *There is a (natural) exact sequence*

$$1 \to \mathcal{H}_d/\mathcal{H}_d^+ \to \mathcal{C}_d^+[2] \to \mathcal{C}_d[2] \to 1,$$

$$\frac{|\mathcal{C}_d^+[2]|}{|\mathcal{C}_d[2]|} = \begin{cases} 2 & \text{if } d > 0 \text{ and } \mathsf{N}_{K/\mathbb{Q}}(\varepsilon_d) = 1, \\ 1 & \text{otherwise,} \end{cases}$$

and there is a (natural) isomorphism $\mathcal{C}_d^{(+)}/(\mathcal{C}_d^{(+)})^2 \cong \mathcal{C}_d^{(+)}[2]$.
The group $(\mathcal{C}_d^{(+)})^2$ is called the **(narrow) principal genus** and the group $\mathcal{C}_d^{(+)}/(\mathcal{C}_d^{(+)})^2$ is called the **(narrow) genus group** of $\mathcal{O}_d$.

Proof. 1. Obvious by 3.8.4.3.

2. If $C \in \mathcal{C}_d^{(+)}$, then $CC' = [\mathcal{O}_d]^{(+)}$ by 1., hence $C' = C^{-1}$, and (a) and (b) are equivalent. The remaining assertions follow by 3.8.7.1. □

Theorem 3.9.2. *Every narrow ambiguous class $C \in \mathcal{C}_d^+$ contains an $\mathcal{O}_d$-regular ambiguous ideal.*

Proof. Let $C = [\mathfrak{a}]^+ \in \mathcal{C}_d^+$, where $\mathfrak{a} \in \mathcal{I}'(\mathcal{O}_d)$. Then $[\mathfrak{a}']^+ = [\mathfrak{a}]^+$ and thus $\mathfrak{a}' = c\mathfrak{a}$ for some $c \in K^+$. It follows that $\mathfrak{a} = (\mathfrak{a}')' = c'\mathfrak{a}' = cc'\mathfrak{a}$. Since $cc' \in \mathbb{Q}^+$, we obtain $cc' = 1$, and if $b = c+1$, then $b \in K^+$ and

$$\frac{b}{b'} = \frac{(c+1)c}{(c'+1)c} = \frac{c^2+c}{1+c} = c.$$

Now $(b\mathfrak{a})' = b'\mathfrak{a}' = b'c\mathfrak{a} = b\mathfrak{a}$, hence $b\mathfrak{a}$ is an ambiguous invertible ideal, and there exists some $r \in \mathbb{Q}^+$ such that $rb\mathfrak{a} \in C$ is an ambiguous $\mathcal{O}_d$-regular ideal. □

In the following Theorem 3.9.3 we determine the structure of ambigous $\mathcal{O}_d$-regular ideals and determine $|\mathcal{C}_d^+[2]|$.

Theorem 3.9.3 (Structure of ambiguous ideals).
I. *A subgroup $\mathfrak{a}$ of $\mathcal{O}_d$ is an ambiguous $\mathcal{O}_d$-regular ideal if and only if it is of one of the following two types:*

$$\mathfrak{a} = \langle a, \sqrt{D}\rangle, \quad \text{where } d = 4D \equiv 0 \bmod 4,\ a \in \mathbb{N},\ a \,|\, D \ \text{ and } \ \Big(a, \frac{D}{a}\Big) = 1; \tag{a}$$

$$\mathfrak{a} = \Big\langle a, \frac{a+\sqrt{d}}{2}\Big\rangle, \quad \text{where } \ a \in \mathbb{N},\ 4a \,|\, a^2 - d \ \text{ and } \ \Big(a, \frac{a^2-d}{4a}\Big) = 1. \tag{b}$$

II. *Suppose that $d = \varepsilon 2^e p_1^{e_1} \cdot \ldots \cdot p_r^{e_r}$, where $\varepsilon \in \{\pm 1\}$, $r, e \in \mathbb{N}_0$, $e_1, \ldots, e_r \in \mathbb{N}$, and $p_1, \ldots, p_r$ are distinct odd primes. For $j \in [1, r]$ we set $q_j = p_j^{e_j}$, and*

$$\mathfrak{q}_j = \begin{cases} \Big\langle q_j, \dfrac{q_j+\sqrt{d}}{2}\Big\rangle & \text{if } d \equiv 1 \bmod 4, \\ \langle q_j, \sqrt{D}\rangle & \text{if } d = 4D \equiv 0 \bmod 4. \end{cases}$$

If $d \equiv 1 \bmod 4$ or $d \equiv 4 \bmod 16$, we set $\mathfrak{q}_0 = \mathfrak{q}_0^* = \mathfrak{q}_0^{**} = \mathbf{1}$.
If $d = 4D \equiv 12 \bmod 16$, we set $\mathfrak{q}_0 = \langle 2, 1+\sqrt{D}\rangle$ and $\mathfrak{q}_0^* = \mathfrak{q}_0^{**} = \mathbf{1}$.
If $d = 4D \equiv 0 \bmod 8$ and $e = v_2(d) \ge 3$, we set $\mathfrak{q}_0 = \langle 2^{e-2}, \sqrt{D}\rangle$, and

$$\begin{cases} \mathfrak{q}_0^* = \mathfrak{q}_0^{**} = \mathbf{1} & \text{if } d \equiv 16 \bmod 32, \\ \mathfrak{q}_0^* = [4, 2+\sqrt{D}], \quad \mathfrak{q}_0^{**} = [2^{e-2}, 2^{e-3}+\sqrt{D}] & \text{if } d \equiv 0 \bmod 32. \end{cases}$$

We set

$$\tau(d) = \begin{cases} r & \text{if } d \equiv 1 \bmod 4 \ \text{ or } \ d \equiv 4 \bmod 16, \\ r+1 & \text{if } d \equiv 8 \text{ or } 12 \bmod 16 \ \text{ or } \ d \equiv 16 \bmod 32, \\ r+2 & \text{if } d \equiv 0 \bmod 32. \end{cases}$$

Then $2^{\tau(d)}$ is the number of $\mathcal{O}_d$-regular ambiguous ideals of $\mathcal{O}_d$. These are in a unique way given by

$$\mathfrak{a} = \mathfrak{q}_0^{\delta_0} (\mathfrak{q}_0^*)^{\delta_0^*} (\mathfrak{q}_0^{**})^{\delta_0^{**}} \prod_{j=1}^r \mathfrak{q}_j^{\delta_j},$$

where $\delta_0,\ \delta_0^*,\ \delta_0^{**},\ \delta_1, \ldots, \delta_r \in \{0,1\}$, $\delta_0 = \delta_0^* = \delta_0^{**} = 0$ if $\tau(d) = r$, $\delta_0^* = \delta_0^{**} = 0$ if $\tau(d) = r+1$, and $\delta_0^*\delta_0^{**} = 0$ if $\tau(d) = r+2$.

III. (Main theorem of genus theory)

$$|\mathcal{C}_d^+[2]| = |\{C \in \mathcal{C}_d \mid C^2 = [\mathcal{O}_d]^+\}| = 2^{\tau(d)-1}.$$

If $d > 0$ and and either $d \equiv 1 \bmod 4$ or $d \equiv 4 \bmod 16$, then $\mathsf{N}_{K/\mathbb{Q}}(\varepsilon_d) = -1$.

Proof. **I.** By 3.8.4 a subgroup $\mathfrak{a}$ of $\mathcal{O}_d$ is an $\mathcal{O}_d$-regular ideal if and only if

$$\mathfrak{a} = \Big\langle a, \frac{b+\sqrt{d}}{2} \Big\rangle,$$

where $a \in \mathbb{N}$, $b \in [0, 2a-1]$, and if $c \in \mathbb{Z}$ such that $d = b^2 - 4ac$, then $(a,b,c) = 1$. Moreover,

$$\mathfrak{a} \text{ is ambiguous} \iff \mathfrak{a} = \mathfrak{a}' \iff b \equiv -b \bmod 2a \iff b \in \{0, a\}$$

If $b = 0$, then $d = 4D \equiv 0 \bmod 4$, and

$$\mathfrak{a} = \langle a, \sqrt{D} \rangle \text{ is } \mathcal{O}_d\text{-regular ambiguous} \iff a \,|\, D \text{ and } \Big(a, \frac{D}{a}\Big) = 1.$$

If $b = a$, then

$$\mathfrak{a} = \Big\langle a, \frac{a+\sqrt{d}}{2} \Big\rangle \text{ is } \mathcal{O}_d\text{-regular ambiguous} \iff 4a \,|\, a^2 - d \text{ and } \Big(a, \frac{a^2-d}{4a}\Big) = 1.$$

This proves **I** and shows that $\mathfrak{q}_0,\ \mathfrak{q}_0^*,\ \mathfrak{q}_0^{**}, \mathfrak{q}_1, \ldots, \mathfrak{q}_r$ are $\mathcal{O}_d$-regular ambiguous ideals of $\mathcal{O}_d$.

II. Before we start with the actual proof, we provide a series of useful identities between ideals which can easily be proved by checking them term by term.

Let $c = q_1^{\delta_1} \cdot \ldots \cdot q_r^{\delta_r}$, where $\delta_1, \ldots, \delta_r \in \{0,1\}$.

(1) We obtain

$$\prod_{j=1}^r \mathfrak{q}_j^{\delta_j} = \begin{cases} \Big\langle c, \dfrac{c+\sqrt{d}}{2} \Big\rangle & \text{if } d \equiv 1 \bmod 4, \\ \langle c, \sqrt{D} \rangle & \text{if } d = 4D \equiv 0 \bmod 4. \end{cases}$$

We prove this by induction on r, and for this we must show: If $c = c'c''$, where $c', c'' \in \mathbb{N}$ and $(c', c'') = 1$, then

$$(*) \qquad \Big\langle c, \frac{c+\sqrt{d}}{2}\Big\rangle = \Big\langle c', \frac{c'+\sqrt{d}}{2}\Big\rangle \Big\langle c'', \frac{c''+\sqrt{d}}{2}\Big\rangle \quad \text{if } d \equiv 1 \mod 4$$

and

$$(**) \qquad \langle c, \sqrt{D}\rangle = \langle c', \sqrt{D}\rangle \langle c'', \sqrt{D}\rangle \quad \text{if } d = 4D \equiv 0 \mod 4.$$

Exemplifying we shall prove $(*)$ in the subsequent Lemma 3.9.5.

(2) Let $d = 4D \equiv 12 \mod 16$. Then $\langle 2, 1+\sqrt{D}\rangle\langle c, \sqrt{D}\rangle = \langle 2c, \sqrt{D}\rangle$.

(3) Let $v_2(d) = e \geq 3$ and $d = 4D$. Then

$$\langle 2^{e-2}c, \sqrt{D}\rangle = \langle 2^{e-2}, \sqrt{D}\rangle\langle c, \sqrt{D}\rangle.$$

(4) Let $v_2(d) = e \geq 5$ and $d = 4D$. Then

$$\langle 4c, 2c+\sqrt{D}\rangle = \langle 4, 2+\sqrt{D}\rangle\langle c, \sqrt{D}\rangle$$

and

$$\langle 2^{e-2}c, 2^{e-3}c+\sqrt{D}\rangle = \langle 2^{e-2}, 2^{e-3}+\sqrt{D}\rangle\langle c, \sqrt{D}\rangle.$$

(5) $\mathfrak{q}_j^2 = q_j\mathcal{O}_d$ for all $j \in [1, r]$, and

$$\mathfrak{q}_0^2 = \begin{cases} 2\mathcal{O}_d & \text{if } d \equiv 12 \mod 16, \\ 2^{e-2}\mathcal{O}_d & \text{if } d \equiv 0 \mod 8. \end{cases}$$

If $d \equiv 0 \mod 32$, then $\mathfrak{q}_0^{*2} = 4\mathcal{O}_d$, $\mathfrak{q}_0^{**2} = 2^{e-2}\mathcal{O}_d$, $\mathfrak{q}_0\mathfrak{q}_0^* = 2\mathfrak{q}_0^{**}$, $\mathfrak{q}_0\mathfrak{q}_0^{**} = 2^{e-3}\mathfrak{q}_0^*$ and $\mathfrak{q}_0^*\mathfrak{q}_0^{**} = 2\mathfrak{q}_0$.

Now we are ready for the actual proof of **II** and discuss the various cases in detail.

CASE 1: $d \equiv 1 \mod 4$. Apparently, there are no $\mathcal{O}_d$-regular ambiguous ideals of type (b), and we consider $\mathcal{O}_d$-regular ambiguous ideals of type (a). If $a \in \mathbb{N}$, then

$$4a \mid a^2 - d \text{ and } \Big(a, \frac{a^2-d}{4a}\Big) = 1 \iff a \mid d \text{ and } \Big(a, \frac{d}{a}\Big) = 1$$

$$\iff a = \prod_{j=1}^{r} q_j^{\delta_j} \text{ for some } \delta_1, \ldots, \delta_r \in \{0, 1\}.$$

If this is the case, then **(1)** shows that

$$\mathfrak{a} = \Big\langle a, \frac{a+\sqrt{d}}{2}\Big\rangle = \prod_{j=1}^{r} \mathfrak{q}_j^{\delta_j}.$$

CASE 2: $d = 4D = 2^e d_0$, where $e \ge 2$ and $d_0 \equiv 1 \bmod 2$.

(a) We first consider $\mathcal{O}_d$-regular ambiguous ideals of type (a). If $a \in \mathbb{N}$, then

$$a \mid D \ \text{ and } \ \Bigl(a, \frac{D}{a}\Bigr) = 1 \iff a = q_0^{\delta_0} c, \ \text{ where } \ c = \prod_{j=1}^r q_j^{\delta_j} \ \text{ and } \ q_0 = 2^{e-2},$$

$$\text{for some } \ \delta_0, \delta_1, \ldots, \delta_r \in \{0,1\} \ \text{ such that } \ \delta_0 = 0 \ \text{ if } \ e = 2.$$

If this is the case, then

$$\mathfrak{a} = \langle 2^{(e-2)\delta_0} c, \sqrt{D}\rangle = \langle 2^{(e-2)\delta_0}, \sqrt{D}\rangle\langle c, \sqrt{D}\rangle = \mathfrak{q}_0^{\delta_0} \prod_{j=1}^r \mathfrak{q}_j^{\delta_j} \quad \text{by } (\mathbf{1}) \text{ and } (\mathbf{3}).$$

(b) Now we consider $\mathcal{O}_d$-regular ambiguous ideals of type (b). If $a \in \mathbb{N}$, then $4a \mid a^2 - d$ if and only if $a = 2c$, where $1 \le c \mid D$. Thus assume that $a = 2c$ and $1 \le c \mid D$. Then

$$\Bigl(a, \frac{a^2-d}{4a}\Bigr) = \Bigl(2c, \frac{c^2-D}{2c}\Bigr) \equiv 0 \bmod 2 \ \text{ if } \ D \equiv 1 \bmod 4.$$

Hence if $d = 4D \equiv 4 \bmod 16$, there are no $\mathcal{O}_d$-regular ambiguous ideals of type (b).

b1) $D \equiv 3 \bmod 4$ ($d \equiv 12 \bmod 16$). Then

$$\Bigl(2c, \frac{c^2-D}{2c}\Bigr) = \Bigl(c, \frac{c^2-D}{c}\Bigr) = \Bigl(c, \frac{D}{c}\Bigr),$$

and consequently

$$4a \mid a^2 - d \ \text{ and } \ \Bigl(a, \frac{a^2-d}{4a}\Bigr) = 1 \iff a = 2c, \ \text{ where } \ c = \prod_{j=1}^r q_j^{\delta_j}$$

$$\text{for some } \ \delta_1, \ldots, \delta_r \in \{0,1\}.$$

If this is the case, then

$$\mathfrak{a} = \Bigl\langle a, \frac{a+\sqrt{d}}{2}\Bigr\rangle = \langle 2c, c+\sqrt{D}\rangle = \langle 2, 1+\sqrt{D}\rangle\langle c, \sqrt{D}\rangle = \mathfrak{q}_0 \prod_{j=1}^r \mathfrak{q}_j^{\delta_j}$$

by (**1**) and (**2**). Together with **(a)** it follows that in the case $d \equiv 12 \bmod 16$ all $\mathcal{O}_d$-regular ideals are of the form

$$\mathfrak{a} = \mathfrak{q}_0^{\delta_0} \prod_{i=1}^r \mathfrak{q}_j^{\delta_j}, \ \text{ where } \ \delta_0, \delta_1, \ldots, \delta_r \in \{0,1\}, \ \ a = 2^{\delta_0} a_1 \ \text{ and } \ a_1 \mid D.$$

b2) $D \equiv 0 \bmod 2$ ($e \ge 3$). If $a \in \mathbb{N}$, then $4a \mid a^2 - d$ if and only if $a = 4a_1$ such that $16a_1 \mid 16a_1^2 - 2^e d_0$. This is impossible if $e = 3$, and equivalent to $a_1 \mid 2^{e-4} d_0$ if $e \ge 4$.

Thus let $e \geq 4$, $a = 4a_1$ and $c \,|\, 2^{e-4}d_0$. If $e = 4$, then a_1 is odd, and

$$\Big(a, \frac{a^2-d}{4a}\Big) = \Big(4a_1, \frac{16a_1^2 - 2^e d_0}{16a_1}\Big) = \Big(4a_1, a_1 - \frac{2^{e-4}d_0}{a_1}\Big) \equiv 0 \mod 2.$$

Consequently, if $e \in \{3,4\}$, there are no $\mathcal{O}_d$-regular ambiguous ideals of type (b).

Assume finally that $e \geq 5$, $a = 4a_1$ and $a_1 \,|\, 2^{e-4}d_0$. Then

$$\Big(a, \frac{a^2-d}{4a}\Big) = \Big(4a_1, a_1 - \frac{2^{e-4}d_0}{a_1}\Big) = 1$$

if and only if either

$$a_1 \,|\, d_0 \quad \text{and} \quad \Big(a_1, \frac{d_0}{a_1}\Big) = 1,$$

or

$$a_1 = 2^{e-4}a_2, \quad \text{where } \ a_2 \in \mathbb{N} \ \text{ is such that } \ a_2 \,|\, d_0 \ \text{ and } \ \Big(a_2, \frac{d_0}{a_2}\Big) = 1.$$

Hence we obtain

$$4a \,|\, a^2 - d \ \text{ and } \ \Big(a, \frac{a^2-d}{4a}\Big) = 1 \iff a = 4c \ \text{ or } \ a = 2^{e-2}c, \ \text{ where } \ c = \prod_{j=1}^{r} q_j^{\delta_j}$$

$$\text{for some } \ \delta_1, \ldots, \delta_r \in \{0,1\}.$$

Using (**1**) and (**4**), it follows that either

$$\mathfrak{a} = \Big\langle a, \frac{a+\sqrt{d}}{2}\Big\rangle = \langle 4c, 2c+\sqrt{D}\rangle = \langle 4, 2+\sqrt{D}\rangle\langle c, \sqrt{D}\rangle = \mathfrak{q}_0^* \prod_{j=1}^{r} \mathfrak{q}_j^{\delta_j}$$

or

$$\mathfrak{a} = \Big\langle a, \frac{a+\sqrt{d}}{2}\Big\rangle = \langle 2^{e-2}c, 2^{e-3}c+\sqrt{D}\rangle$$

$$= \langle 2^{e-2}, 2^{e-3}+\sqrt{D}\rangle\langle c, \sqrt{D}\rangle = \mathfrak{q}_0^{**} \prod_{j=1}^{r} \mathfrak{q}_j^{\delta_j}.$$

Summarizing all cases, **II** follows (the uniqueness of the exponents is obvious).

III. We continue to use the relations (**1**) to (**5**) given above.

Let $\mathcal{A} = \langle \mathfrak{q}_1, \ldots, \mathfrak{q}_r, \mathfrak{q}_0, \mathfrak{q}_0^* \rangle$ be the group of all $\mathcal{O}_d$-invertible ambiguous ideals of $\mathcal{O}_d$ generated by $\{\mathfrak{q}_1, \ldots, \mathfrak{q}_r, \mathfrak{q}_0, \mathfrak{q}_0^*\}$ (note that we disregarded $\mathfrak{q}_0^{**}$ if $d \equiv 0 \mod 32$).

> **A.** For every $\mathcal{O}_d$-invertible fractional ambiguous ideal $\mathfrak{a}$ of $\mathcal{O}_d$ there exists some $r \in \mathbb{Q}^+$ such that $r\mathfrak{a} \in \mathcal{A}$.

Proof of **A.** Let $\mathfrak{a}$ be an $\mathcal{O}_d$-invertible fractional ideal of $\mathcal{O}_d$. By 3.8.4.3 there exists some $r_0 \in \mathbb{Q}^+$ such that $r_0\mathfrak{a}$ is $\mathcal{O}_d$-regular, and then $r_0\mathfrak{a} = \mathfrak{b}(\mathfrak{q}_0^{**})^\delta$ for some $\mathfrak{b} \in \mathcal{A}$ by **II**. If $\delta = 0$, we set $r = r_0$. If $\delta = 1$, then $2r_0\mathfrak{a} = \mathfrak{b}\mathfrak{q}_0\mathfrak{q}_0^* \in \mathcal{A}$ by (**5**), and we set $r = 2r_0$. □[**A.**]

B. $\{a\mathcal{O}_d \mid a \in \mathbb{Q}^+\} \cap \mathcal{A} = \mathcal{A}^2$, and $\dim_{\mathbb{F}_2} \mathcal{A}/\mathcal{A}^2 = \tau(d)$.

Proof of **B.** By (**5**) we obtain $\mathcal{A}^2 \subset \{a\mathcal{O}_d \mid a \in \mathbb{Q}^+\} \cap \mathcal{A}$, and obviously $\dim_{\mathbb{F}_2} \mathcal{A}/\mathcal{A}^2 \le \tau(d)$. Thus suppose that either $\mathcal{A}^2 \subsetneq \{a\mathcal{O}_d \mid a \in \mathbb{Q}^+\} \cap \mathcal{A}$ or $\dim_{\mathbb{F}_2} \mathcal{A}/\mathcal{A}^2 < \tau(d)$. In either case we obtain a non-trivial relation of the form

$$\mathfrak{q}_0^{\delta_0}(\mathfrak{q}_0^*)^{\delta_0^*} \prod_{j=1}^r \mathfrak{q}_j^{\delta_j} = a\mathcal{O}_d \quad \text{for some } a \in \mathbb{Q}^+,$$

where $\delta_0, \delta_0^*, \delta_1, \ldots, \delta_r \in \{0,1\}$ such that $\delta_0 = \delta_0^* = 0$ if $d \equiv 1 \bmod 4$ or $d \equiv 4 \bmod 16$, and $\delta_0^* = 0$ if $d \not\equiv 0 \bmod 32$. However, if $c = q_1^{\delta_1} \cdot \ldots \cdot q_r^{\delta_r}$ and $\mathfrak{c} = \mathfrak{q}_1^{\delta_1} \cdot \ldots \cdot \mathfrak{q}_r^{\delta_r}$, then in the course of proving **II** we have seen that

$$\mathfrak{c} = \begin{cases} \left\langle c, \dfrac{c+\sqrt{d}}{2} \right\rangle & \text{if } d \equiv 1 \bmod 4, \\ \langle c, \sqrt{D} \rangle & \text{if } d = 4D \equiv 0 \bmod 4, \end{cases}$$

$$\mathfrak{q}_0\mathfrak{c} = \begin{cases} \langle 2c, c+\sqrt{D} \rangle & \text{if } d = 4D \equiv 12 \bmod 16, \\ \langle 2^{e-2}c, \sqrt{D} \rangle & \text{if } d = 4D \equiv 0 \bmod 8 \end{cases}$$

and $\mathfrak{q}_0^*\mathfrak{c} = \langle 4c, 2c+\sqrt{D} \rangle$ if $d = 4D \equiv 0 \bmod 32$. But neither of these ideals is of the form $a\mathcal{O}_d$ for some $a \in \mathbb{Q}^+$. □[**B.**]

We define

$$\nu_d \colon \mathcal{A} \to \mathcal{C}_d^+[2] \quad \text{by} \quad \nu_d(\mathfrak{a}) = [\mathfrak{a}]^+.$$

Apparently ν_d is a homomorphism, $\mathcal{A}^2 \subset \mathrm{Ker}(\nu_d)$, and ν_d is surjective by **A.** We shall prove:

C. There exists an epimorphism $\rho \colon \mathrm{Ker}(\nu_d) \to U_d^+/U_d^{+2}$ such that $\mathrm{Ker}(\rho) = \mathcal{A}^2$.

From **C** we conclude (using **B**)

$$|\mathcal{C}_d^+[2]| = (\mathcal{A}\colon\mathrm{Ker}(\nu_d)) = \frac{(\mathcal{A}\colon\mathcal{A}^2)}{(\mathrm{Ker}(\nu_d)\colon\mathcal{A}^2)} = \frac{(\mathcal{A}\colon\mathcal{A}^2)}{(U_d^+\colon U_d^{+2})} = \frac{2^{\tau(d)}}{2} = 2^{\tau(d)-1}$$

Proof of **C.** Let $\mathfrak{a} \in \mathrm{Ker}(\nu_d)$. Then $\mathfrak{a} = a\mathcal{O}_d$ for some $a \in K^+$. If $\mathfrak{a} = \mathfrak{a}'$, then $a' = au$ for some $u \in U_d^+$, and we assert that the coset uU_d^{+2} is uniquely determined by $\mathfrak{a}$. Indeed, if $\mathfrak{a} = a_1\mathcal{O}_d$ for some $a_1 \in K^+$ and $a_1' = a_1u_1$ with $u_1 \in U_d^+$, then $a_1 = a\varepsilon$ with $\varepsilon \in U_d^+$, and $u_1^{-1}u = a_1a_1'^{-1}a^{-1}a' = \varepsilon\varepsilon'^{-1} = \varepsilon^2$. We define

$$\rho \colon \mathrm{Ker}(\nu_d) \to U_d^+/U_d^{+2} \quad \text{by} \quad \rho(\mathfrak{a}) = uU_d^{+2} = a^{-1}a'U_d^{+2}.$$

If $\mathfrak{a}, \mathfrak{a}_1 \in \mathrm{Ker}(\nu_d)$, $\mathfrak{a} = a\mathcal{O}_d$, $\mathfrak{a}_1 = a_1\mathcal{O}_d$, $u = a^{-1}a'$ and $u_1 = a_1^{-1}a_1'$ are such that $a, a_1 \in K^+$ and $u, u_1 \in U_d^+$, then $\mathfrak{a}\mathfrak{a}_1 = aa_1\mathcal{O}_d$, $uu_1 = (aa_1)^{-1}(aa_1)'$, and $\rho(\mathfrak{a}\mathfrak{a}_1) = uu_1U_d^+ = \rho(\mathfrak{a})\rho(\mathfrak{a}_1)$. Hence ρ is a homomorphism, and apparently $\mathcal{A}^2 \subset \mathrm{Ker}(\rho)$.

Next we prove that ρ is surjective. Thus let $\varepsilon \in U_d^+$ and $a = 1 + \varepsilon'$. Since $a \in K^+$ and $a' = a\varepsilon$, the principal ideal $a\mathcal{O}_d$ is ambiguous, and therefore $\mathfrak{a} = ra\mathcal{O}_d \in \mathcal{A}$ for some $r \in \mathbb{Q}^+$ by **B**. Hence $\mathfrak{a} \in \mathrm{Ker}(\nu_d)$, and as $ra = (ra)'\varepsilon$, we obtain $\varepsilon U_d^2 = \rho(\mathfrak{a})$.

It remains to prove that $\mathrm{Ker}(\rho) \subset \mathcal{A}^2$. If $\mathfrak{a} \in \mathrm{Ker}(\rho)$, then $\mathfrak{a} = a\mathcal{O}_d \in \mathcal{A}$ for some $a \in K^+$ such that $a^{-1}a' = u^2$ for some $u \in U_d^+$. Then it follows that $(au)' = a'u^{-1} = au$, hence $a_0 = au \in \mathbb{Q}^+$ and $\mathfrak{a} = a_0\mathcal{O}_d \subset \mathcal{A}^2$ by **B**. □[**C.**]

Assume finally that $d \equiv 1 \bmod 4$ or $d \equiv 4 \bmod 16$ and yet $\mathsf{N}_{K/\mathbb{Q}}(\varepsilon_d) = 1$. Then $\tau(d) = 1$ and therefore $1 = |\mathcal{C}_d^+[2]| = 2|\mathcal{C}_d[2]|$ by 3.9.1.2, a contradiction. □

Corollary 3.9.4.

1. *If $d < 0$, then $h_d = h_d^+$ is odd if and only if*
   - *either $d \in \{-p^r, -4p^r\}$ for some prime $p \equiv 3 \bmod 4$ and an odd exponent $r \in \mathbb{N}$,*
   - *or $d \in \{-4, -8, -16\}$.*
2. *If $d > 0$, then h_d^+ is odd if and only if*
   - *either $d \in \{p^r, 4p^r\}$ for some prime $p \equiv 1 \bmod 4$ and an odd exponent $r \in \mathbb{N}$,*
   - *or $d = 8$.*

In all these cases $\mathsf{N}_{K/\mathbb{Q}}(\varepsilon_d) = -1$.

Proof. Obvious by 3.9.3.**III**, since h_d^+ is odd if and only if $\mathcal{C}_d^+[2] = \mathbf{1}$. □

We close with the lemma promised in the course of the proof of Theorem 3.9.3.

Lemma 3.9.5. *Let $d, c', c'' \in \mathbb{N}$, $d \equiv 1 \bmod 4$, $(c', c'') = 1$ and $c = c'c'' \mid d$. Then*

$$\Big\langle c, \frac{c+\sqrt{d}}{2}\Big\rangle = \Big\langle c', \frac{c'+\sqrt{d}}{2}\Big\rangle\Big\langle c''\frac{c''+\sqrt{d}}{2}\Big\rangle$$
$$= \Big\langle c, \frac{c+c'\sqrt{d}}{2}, \frac{c+c''\sqrt{d}}{2}, \frac{c+d+(c'+c'')\sqrt{d}}{4}\Big\rangle$$

Proof. $\subset$: Let u', $u'' \in \mathbb{Z}$ be such that $u'c' + u''c'' = 1$, and set $u' + u'' = 2v$. Then

$$\frac{c+\sqrt{d}}{2} = -vc + u'\,\frac{c+c'\sqrt{d}}{2} + u''\,\frac{c+c''\sqrt{d}}{2} \in \left\langle c, \frac{c'+\sqrt{d}}{2}, \frac{c''+\sqrt{d}}{2}\right\rangle.$$

$\supset$: Since

$$\frac{c+c'\sqrt{d}}{2} = c\,\frac{1-c'}{2} + \frac{c+\sqrt{d}}{2}\,c' \quad\text{and}\quad \frac{c+c''\sqrt{d}}{2} = c\,\frac{1-c''}{2} + \frac{c+\sqrt{d}}{2}\,c'',$$

both numbers are contained in the left-hand side. For the proof of the remaining containment we observe first that

$$1 + c^{-1}d - c' - c'' \equiv 0 \mod 4. \qquad (\dagger)$$

Indeed, since $c' \equiv c'' \equiv 1 \mod 2$, we obtain $c'' \equiv \varepsilon c' \mod 4$ for some $\varepsilon \in \{\pm 1\}$, hence $c = c'c'' \equiv \varepsilon \mod 4$ and $c^{-1}d \equiv \varepsilon \mod 4$. Consequently, it follows that

$$1 + c^{-1}d - c' - c'' \equiv 1 + \varepsilon - c'(1+\varepsilon) \equiv (1+\varepsilon)(1-c') \equiv 0 \mod 4.$$

By means of $(\dagger)$ we finally obtain

$$c\,\frac{1+c^{-1}d-c'-c''}{4} + \frac{c+\sqrt{d}}{2}\,\frac{c'+c''}{2} = \frac{c+d+(c'+c'')\sqrt{d}}{4} \in \left\langle c, \frac{c+\sqrt{d}}{2}\right\rangle.$$

□

3.10 Exercises for Chapter 3

1. Let p and q be odd primes.

a. If $p \equiv 1 \mod 8$, then $h_{-4p} \equiv 0 \mod 4$.

b. Assume that $p \not\equiv q \mod 4$. Then $h_{-pq} \equiv 0 \mod 4$ if and only if $\left(\frac{q}{p}\right) = 1$.

2. Let d be a quadratic discriminant.

a. d has a unique factorization $d = d_1 \cdot \ldots \cdot d_r$, where $d_1, \ldots, d_r$ are distinct discriminants such that either $|d_i|$ is a prime or $d_i \in \{-4, \pm 8\}$ for all $i \in [1, r]$ (the d_i's are called **prime discriminants**).

b. (Compare Exercise 14 of Chapter 2 and [50, Sec. 3.4]).
If $d \in \{13, 17, 21, 29\}$, then $\mathcal{O}_d$ is norm-Euclidean.
If $d < 0$, then $\mathcal{O}_d$ is norm-Euclidean if and only if $d \in \{-1, -2, -3, -7, -11\}$, and if $d < -11$, then $\mathcal{O}_d$ is not even Euclidean.

3. Let K be a totally complex Galois number field, $G = \mathrm{Gal}(K/\mathbb{Q})$, $\iota \in G$ the complex conjugation and $K_0 = K^{\langle\iota\rangle}$ the fixed field of ι. Then the following assertions are equivalent:

(a) $K_0/\mathbb{Q}$ is Galois.

(b) K_0 is totally real.

(c) ι lies in the center of G.

(d) $(U_K : U_{K_0}) < \infty$.

If these conditions are fulfilled then K is called a **CM-field**.

4. Let K be a CM-field (see Exercise 3) and K_0 its maximal totally real subfield. Then $\delta_K = (U_K : U_{K_0}\mu_K) \le 2$. Hint: Consider the epimorphism $\lambda\colon U_K \to \mu_K/\mu_K^2$, defined by $\lambda(\varepsilon) = \varepsilon^{-1}\overline{\varepsilon}\mu_K^2$.

a. Let $K = \mathbb{Q}^{(p^n)}$ for some odd prime power p^n. Then $\delta_K = 1$. Hint: Assume $\delta_K = 2$. Then there exists some $u \in U_{K_0}$ such that $K = K_0(\sqrt{u})$, and a prime divisor $\mathfrak{p}$ of p in K_0 is both ramified and unramified in K.

b. Let $K = \mathbb{Q}^{(2^n)}$ for some $n \ge 2$. Then $\delta_K = 1$. Hint: Assume $\delta_K = 2$. Then there is some $\varepsilon \in U_{K_0}\mu_K$ such that $\varepsilon^{-1}\overline{\varepsilon} = \zeta$, a primitive 2^n-th root of unity, $\mathsf{N}_{K/\mathbb{Q}(i)}(\varepsilon^{-1}\overline{\varepsilon}) = \pm 1$ but $\mathsf{N}_{K/\mathbb{Q}(i)}(\zeta) = \pm i$.

c. Let $m \in \mathbb{N} \setminus \{1\}$ be not a prime power and $K = \mathbb{Q}^{(m)}$. Then $\delta_K = 2$. Hint: Consider $\lambda(1-\zeta)$ for a primitive m-th root of unity.

d. Let $[K:\mathbb{Q}] = 4$, ε_0 be the fundamental unit of K_0 and $\mathsf{N}_{K_0/\mathbb{Q}}(\varepsilon_0) = -1$. Then $\delta_K = 1$.

e. Let $K = \mathbb{Q}(\sqrt{-p_1}, \sqrt{-p_2})$, where p_1, p_2 are primes such that $p_1 \equiv p_2 \equiv -1 \bmod 4$. Then $\delta_K = 2$.

5. Let $K = \mathbb{Q}(\sqrt{a}, \sqrt{b})$, where $a, b \in \mathbb{Q}$ and $[K:\mathbb{Q}] = 4$.

a. There exist distinct squarefree integers m, n, k such that $K = \mathbb{Q}(\sqrt{m}, \sqrt{n}, \sqrt{k})$ and one of the following three alternatives hold:

(α) $m \equiv 3 \bmod 4$, $n \equiv k \equiv 2 \bmod 4$.

(β) $m \equiv 1 \bmod 4$, $n \equiv k \equiv 2$ or $3 \bmod 4$.

(γ) $m \equiv n \equiv k \equiv 1 \bmod 4$.

b. An integral basis of K and the discriminant d_K in the various cases above are as follows:

$$\Big(1,\ \sqrt{m},\ \sqrt{n},\ \frac{\sqrt{n}+\sqrt{k}}{2}\Big), \quad \text{and} \quad \mathsf{d}_K = 64mnk \quad \text{in case } (\alpha);$$

$$\Big(1,\ \frac{1+\sqrt{m}}{2},\ \sqrt{n},\ \frac{\sqrt{n}+\sqrt{k}}{2}\Big), \quad \text{and} \quad \mathsf{d}_K = 16mnk \quad \text{in case } (\beta);$$

$$\Big(1,\ \frac{1+\sqrt{m}}{2},\ \frac{1+\sqrt{n}}{2},\ \frac{(1+\sqrt{m})(1+\sqrt{k})}{4}\Big), \quad \text{and} \quad \mathsf{d}_K = mnk \quad \text{in case } (\gamma).$$

Hint: $(1, \sqrt{m}, \sqrt{n}, \sqrt{k})$ is a basis of K. If $\alpha \in K$ and M is a quadratic subfield of K, then $\alpha \in \mathfrak{o}_K$ if and only if $\mathsf{N}_{K/M}(\alpha) \in \mathfrak{o}_M$ and $\mathsf{Tr}_{K/M}(\alpha) \in \mathfrak{o}_M$.

6. There exists some $C \in \mathbb{R}_{>0}$ such that $[K:\mathbb{Q}] \le C \log|\mathsf{d}_K|$ for all algebraic number fields $K \ne \mathbb{Q}$.

7. Let $g = X^3 + X^2 - 2X + 8,\ \ g(t) = 0,\ \ K = \mathbb{Q}(t)$ and $u = 4/t$.

a. f is irreducible in $\mathbb{Q}[X]$.

b. $\mathsf{d}_K = -503$, and $(1, t, u)$ is an integral basis of K.

c. $2\mathfrak{o}_K$ is a product of 3 principal prime ideals, $5\mathfrak{o}_K$ is a product of 2 principal prime ideals, and $3\mathfrak{o}_K$, $7\mathfrak{o}_K$ and $11\mathfrak{o}_K$ are prime ideals.

d. $h_K = 1$.

8. Let $g = X^3 - 7X - 7,\ \ g(t) = 0$ and $K = \mathbb{Q}(t)$. The polynomial g is irreducible in $\mathbb{Q}[X]$. Determine an integral basis of K, calculate d_K and determine the factorizations of 2, 3, 5 and 7 in K.
Do the same for $g = X^3 - X - 4$.

9. Let $K = \mathbb{Q}(\sqrt{7}, \sqrt{-7}\,)$. Determine the character group of this abelian number field, formulate the decomposition law for all primes using Legendre symbols and determine the corresponding Hilbert subgroup series.

10. Let $m = 2^e p_1^{e_1} \cdot \ldots \cdot p_r^{e_r}$, where $e \in \mathbb{N}_0$, $e_1, \ldots, e_r \in \mathbb{N}$ and $p_1, \ldots, p_r$ are distinct odd primes. Determine all quadratic characters $\chi \in \mathsf{X}(m)$, their conductors $f(\chi)$ and the associated quadratic number fields $\mathbb{Q}_\chi$.

11. Determine all subgroups Z of $\mathsf{X}(15)$ and the associated subfields $\mathbb{Q}_Z$ of $\mathbb{Q}^{(15)}$. Among them, there is a relative quadratic field extension L/K in which all prime ideals of K are unramified.

12. Let $a, b \in \mathbb{Z}^\bullet$ and $a \equiv 0$ or $1 \bmod 4$. Then

$$\Bigl(\frac{a}{|a|-1}\Bigr) = \operatorname{sgn}(a) \ \text{ if } a \ne 1, \quad \text{and} \quad \Bigl(\frac{a}{b}\Bigr) = \operatorname{sgn}(a)\Bigl(\frac{a}{b'}\Bigr) \ \text{ if } b + b' \equiv 0 \bmod |a|.$$

13. Let $K \subset \mathbb{R}$ be a simply real cubic fields $(r_1 = r_2 = 1)$, $\varepsilon \in U_K$ and $\varepsilon > 1$.

a. (Artin) $|\mathsf{d}_K| < 4\varepsilon^3 + 24$.
Hint: Set $\varepsilon = \rho^2$, and let $\rho^{-1}\mathsf{e}^{\mathsf{i}\theta}$, $\rho^{-1}\mathsf{e}^{-\mathsf{i}\theta}$ be the conjugates of ε, where $\rho \in \mathbb{R}_{>0}$ and $\theta \in (0, \pi)$. Then $|\mathsf{d}_K| \le |\Delta(\mathbb{Z}[\omega])| = D^2$, where $D = 4(x \sin\theta - \cos 2\theta)$ and $2x = \rho^3 + \rho^{-3}$. Fix x, let $\theta_0 \in (0, \pi)$ be such that $x \sin\theta - \cos 2\theta$ has its maximum in θ_0, and set $c = \cos\theta_0$. Then $cx - 2c^2 + 1 = 0$, $c < -(2\rho^3)^{-1}$ and obtain the estimate $D^2 \le 4(\rho^6 + 6 - 4c^2 - 4c^4 + \rho^{-6}) < 4\varepsilon^3 + 24$. For details see [15, V.8.(3.2)].

b. If $4\varepsilon^{3/2} + 24 < |\mathsf{d}_K|$, then ε is the fundamental unit of K.

14. Let $l \in \mathbb{N}$ be such that $l \ge 2$ and $4l^3 + 27$ is squarefree. There exists a unique $x \in \mathbb{R}$ such that $x^3 - lx^2 - 1 = 0$. The field $K = \mathbb{Q}(x)$ is a simply real cubic number field, $\mathsf{d}_K = -(4l^3 + 27)$, and x is the fundamental unit of K. Hint: Exercise 13.b.

15. Let $a \in \mathbb{N} \setminus \mathbb{N}^3$, $K = \mathbb{Q}(\sqrt[3]{a})$, $r \in \mathbb{Z}$ and $|a - r^3| = 1$. Then $\sqrt[3]{a} + r \in U_K$. Determine the fundamental unit of K for $a \in \{2, 3, 7\}$. Hint: Exercise 13.b.

16. Let $K_0 = \mathbb{Q}(\sqrt{-6})$ and $K = K_0(\sqrt{2})$. Calculate h_K and h_{K_0} (note that $h_K < h_{K_0}$).

17. Calculate h_K for $K = \mathbb{Q}(\sqrt[3]{3})$ and $K = \mathbb{Q}(\sqrt[3]{6})$.

18. Two quadratic irrationals ξ, ξ_1 are called **properly equivalent** if $\xi_1 = A\xi$ for some $A \in \mathsf{SL}_2(\mathbb{Z})$.

a. For a quadratic discriminant d the following assertions are equivalent:

(a) Any two equivalent quadratic irrationals of discriminant d are properly equivalent.

(b) Some [every] quadratic irrational ξ of discriminant d is properly equivalent to $-\xi$.

(c) Pell's minus equation $x^2 - dy^2 = -4$ has integral solutions.

b. For $i \in \{1,2\}$ let ξ_i be a quadratic irrational of type (a_i, b_i, c_i) such that $a_1a_2 > 0$. Then the quadratic irrationals ξ_1 and ξ_2 are properly equivalent if and only if $[I(\xi)]^+ = [I(\xi_1)]^+$.

19. Let $d > 0$ be a quadratic discriminant such that $\mathsf{N}_{\mathbb{Q}(\sqrt{d})/\mathbb{Q}}(\varepsilon_d) = 1$. Then there are $2^{\mu(d)-2}$ ambiguous classes $C \in \mathcal{C}_d$ without an ambiguous ideal. Discuss the case $d = 34$: Determine ε_d, h_d, h_d^+ and representing ideals for all (narrow) ambiguous classes.

20. Let $D \in \mathbb{N}$ be squarefree and $D = n^2 + r$, where $r \in \mathbb{Z}$, $r \mid 4n$ and $|r| < n$. Then the quadratic number field $K = \mathbb{Q}(\sqrt{D}\,)$ is called a **Richaud-Degert field**. Its fundamental unit is given by

$$\varepsilon_K = \begin{cases} n + \sqrt{D} & \text{if } r \in \{\pm 1\} \text{ and } D \neq 5, \\ (n + \sqrt{D})/2 & \text{if } r \in \{\pm 4\}, \\ (2n^2 + r + 2n\sqrt{D})/r & \text{if } r \notin \{\pm, \pm 4\}. \end{cases}$$

4

Elementary Analytic Theory

Analytical methods are indispensable for an advanced understanding of the arithmetic of number fields. Most topics of this chapter are contained in the already mentioned books on algebraic number theory [40], [52] and [50]. For the quantitative results about quadratic number fields we refer to [28], and for more results concerning the distribution of primes we refer to [34]. As to the arithmetical relevance of the class group we refer to [17]; there one can also find an abstract introduction to analytical methods.

After some preliminaries on Dirichlet series and Euler products we introduce zeta and L functions of algebraic number fields. Section 4.3 is devoted to density results for prime ideals. We discuss the connection between Dirichlet and natural density, prove Bauer's theorem and finally the prime ideal theorem using a slight version of the Tauber theorem of Ikehara and Delange. In the following Section 4.4 we consider several density results assuming the existence theorem of class field theory. By this means we prove the Čebotarev density theorem, Dirichlet's theorem for ray classes and the prime ideal theorem for non-principal orders. In particular, we show that each ideal class of an algebraic number field contains infinitely many prime ideals and as a consequence we deduce that the ideal class group completely determines the factorization properties of the principal order.

4.1 Euler products and Dirichlet series

For $\vartheta \in \mathbb{R}$, we set $\mathcal{H}_\vartheta = \{s \in \mathbb{C} \mid \Re(s) > \vartheta\}$ and $\overline{\mathcal{H}}_\vartheta = \{s \in \mathbb{C} \mid \Re(s) \geq \vartheta\}$. We slightly generalize and amend the notion of multiplicative functions introduced in 1.7.2.

Definition 4.1.1. Let $D = \mathcal{F}(P)$ be a free monoid (see 2.1.9). A function $f\colon D \to \mathbb{C}$ is called

- **multiplicative** if $f(1) = 1$ and $f(ab) = f(a)f(b)$ for all $a,\, b \in D$ such that $(a, b) = 1$;

- **completely multiplicative** if $f(1) = 1$ and $f(ab) = f(a)f(b)$ holds for all $a, b \in D$.

Theorem 4.1.2 (Euler products). *Let $D = \mathcal{F}(P)$ be a free monoid with a denumerable basis P and $f\colon D \to \mathbb{C}$ a function.*

1. Let f be multiplicative. Then

$$\sum_{a\in D} |f(a)| = \prod_{p\in P}\Big(\sum_{n=0}^{\infty} |f(p^n)|\Big) \in (0,\infty],$$

and if this expression is finite, then

$$\sum_{a\in D} f(a) = \prod_{p\in P}\Big(\sum_{n=0}^{\infty} f(p^n)\Big).$$

2. Let f be completely multiplicative and $|f(p)| < 1$ for all $p \in P$. Then

$$\sum_{p\in P}\sum_{n=1}^{\infty} \frac{|f(p^n)|}{n} \le \sum_{a\in D} |f(a)| = \prod_{p\in P} \frac{1}{1-|f(p)|} \in (0,\infty],$$

and if this expression is finite, then

$$\exp\Big\{\sum_{p\in P}\sum_{n=1}^{\infty} \frac{f(p^n)}{n}\Big\} = \sum_{a\in D} f(a) = \prod_{p\in P} \frac{1}{1-f(p)} \ne 0.$$

Proof. 1. Observe that

$$\prod_{p\in P}\Big(\sum_{n=0}^{\infty} |f(p^n)|\Big) = \sup\Big\{\prod_{p\in F}\Big(\sum_{n=0}^{\infty} |f(p^n)|\Big) \,\Big|\, F \in \mathbb{P}_{\text{fin}}(P)\Big\}.$$

For every finite subset E of D, we denote by E^* the (finite) set of all $p \in P$ dividing some $a \in E$ and obtain

$$\sum_{a\in E} |f(a)| \le \prod_{p\in E^*}\sum_{n=0}^{\infty} |f(p^n)| = \sum_{a\in\mathcal{F}(E^*)} |f(a)| \le \sum_{a\in D} |f(a)|.$$

Taking the supremum over all finite subsets E of D yields

$$\sum_{a\in D} |f(a)| = \prod_{p\in P}\Big(\sum_{n=0}^{\infty} |f(p^n)|\Big).$$

Assume that this expression is finite and let $\varepsilon > 0$. Then there exists a finite subset E of D such that

$$\sum_{a\in D\setminus E} |f(a)| < \varepsilon.$$

For every finite subset E' of P satisfying $E \subset \mathcal{F}(E')$ we obtain (by multiplication of the absolutely convergent series)

$$\Big|\sum_{a\in D} f(a) - \prod_{p\in E'}\Big(\sum_{n=0}^{\infty} f(p^n)\Big)\Big| = \Big|\sum_{a\in D\setminus\mathcal{F}(E')} f(a)\Big| \le \sum_{a\in D\setminus\mathcal{F}(E')} |f(a)|$$
$$\le \sum_{a\in D\setminus E} |f(a)| < \varepsilon,$$

and therefore

$$\sum_{a\in D} f(a) = \prod_{p\in P}\Big(\sum_{n=0}^{\infty} f(p^n)\Big).$$

2. By 1. it follows that

$$\sum_{a\in D} |f(a)| = \prod_{p\in P}\Big(\sum_{n=0}^{\infty} |f(p^n)|\Big)$$
$$= \prod_{p\in P}\Big(\sum_{n=0}^{\infty} |f(p)|^n\Big) = \prod_{p\in P}\big(1-|f(p)|\big)^{-1} \in (0,\infty]\,.$$

If this expression is finite, then

$$\prod_{p\in\mathbb{P}}\sum_{n=1}^{\infty}\frac{|f(p^n)|}{n} \le \sum_{a\in D}|f(a)| < \infty,$$

and consequently

$$\exp\Big\{\sum_{p\in P}\sum_{n=1}^{\infty}\frac{f(p^n)}{n}\Big\} = \prod_{p\in P}\exp\Big\{\sum_{n=1}^{\infty}\frac{f(p^n)}{n}\Big\} = \prod_{p\in P}\exp\Big\{\log\frac{1}{1-f(p)}\Big\}$$
$$= \prod_{p\in P}\frac{1}{1-f(p)} = \prod_{p\in P}\Big(\sum_{n=0}^{\infty} f(p)^n\Big) = \sum_{a\in D} f(a) \ne 0. \qquad \square$$

The most classical example for the application of 4.1.2 is $D = \mathbb{N} = \mathcal{F}(\mathbb{P})$ and the completely multiplicative function $f(n) = n^{-s}$ for $s \in \mathbb{C}$. Suppose that $\sigma = \Re(s) > 1$. Then

$$\sum_{n=1}^{\infty}\Big|\frac{1}{n^s}\Big| = \sum_{n=1}^{\infty}\frac{1}{n^\sigma} < \infty \quad\text{and therefore}\quad \zeta(s) = \sum_{n=1}^{\infty}\frac{1}{n^s} = \prod_{p\in\mathbb{P}}\frac{1}{1-p^{-s}},$$

where $\zeta\colon \mathcal{H}_1 \to \mathbb{C}$ is the classical **Riemann zeta function** whose subtle analytic properties reflect the distribution of prime numbers. For details we refer to books on analytic number theory and in particular to [30].

More elementary but probably less common is the case of polynomials over a finite field. In the following Theorem 4.1.3 we apply 4.1.2 to count the number of irreducible monic polynomials of a fixed degree over a finite field.

Theorem 4.1.3. *Let q be a prime power and $\mathsf{P}(\mathbb{F}_q, X)$ the set of all irreducible monic polynomials in $\mathbb{F}_q[X]$. For $n \in \mathbb{N}$, let*

$$B_n(q) = |\{p \in \mathsf{P}(\mathbb{F}_q, X) \mid \partial(p) = n\}|$$

be the number of irreducible monic polynomials of degree n in $\mathbb{F}_q[X]$. Then

$$q^n = \sum_{1 \le d \,|\, n} dB_d(q) \quad \textit{and} \quad B_n(q) = \frac{1}{n} \sum_{1 \le d \,|\, n} \mu(d) q^{n/d},$$

where μ denotes the Möbius function (see 1.7.2). *In particular,*

$$B_n(q) = \frac{q^n}{n} + O\Big(\frac{q^{n/2}}{n}\Big) \qquad \text{(Prime number theorem for polynomials).}$$

Proof. The monoid $\mathsf{D}(\mathbb{F}_q, X)$ of all monic polynomials in $\mathbb{F}_q[X]$ is the free monoid with basis $\mathsf{P}(\mathbb{F}_q, X)$. For $s \in \mathcal{H}_1$, the function $\mathsf{D}(\mathbb{F}_q, X) \to \mathbb{C}$, defined by $g \mapsto q^{-s\partial(g)}$, is completely multiplicative, and if $\sigma = \Re(s)$, then

$$\sum_{g \in \mathsf{D}(\mathbb{F}_q, X)} |q^{-s\partial(g)}| = \sum_{n=0}^{\infty} A_n(q) q^{-\sigma n} = \sum_{n=0}^{\infty} q^{n(1-\sigma)} = \frac{1}{1 - q^{1-\sigma}} < \infty,$$

where $A_n(q) = |\{g \in \mathsf{D}(\mathbb{F}_q, X) \mid \partial(g) = n\}| = q^n$.

For $s \in \mathcal{H}_1$ we consider the zeta function of $\mathbb{F}_q[X]$, defined by

$$Z(s) = \sum_{g \in \mathsf{D}(\mathbb{F}, X)} q^{-s\partial(g)} = \sum_{n=1}^{\infty} A_n(q) q^{-ns} = \sum_{n=0}^{\infty} q^{(1-s)n} = \frac{1}{1 - q^{1-s}}.$$

By 4.1.2 we obtain

$$Z(s) = \prod_{p \in \mathsf{P}(\mathbb{F}, X)} \frac{1}{1 - q^{-s\partial(p)}} = \prod_{d=1}^{\infty} \Big(\frac{1}{1 - q^{-sd}}\Big)^{B_d(q)}.$$

If $t = q^{-s}$, then $|t| < 1$ and

$$1 - qt = \prod_{d=1}^{\infty} (1 - t^d)^{B_d(q)}.$$

Observing locally uniform absolute convergence in $|t| < q^{-1}$, logarithmic differentiation ($\varphi \mapsto \varphi'/\varphi$) yields

$$\frac{-q}{1 - qt} = \sum_{d=1}^{\infty} \frac{-dB_d(q) t^{d-1}}{1 - t^d}$$

and therefore

$$\sum_{n=1}^{\infty} q^n t^n = \frac{qt}{1-qt} = \sum_{d=1}^{\infty} dB_d(q)\frac{t^d}{1-t^d} = \sum_{d=1}^{\infty} dB_d(q)\sum_{\nu=0}^{\infty} t^{d(\nu+1)}$$

$$\underset{d(\nu+1)=n}{=} \sum_{n=1}^{\infty}\Big(\sum_{1\le d\,|\,n} dB_d(q)\Big)t^n.$$

Now we compare coefficients. We obtain

$$q^n = \sum_{1\le d\,|\,n} dB_d(q) \quad \text{for all } \ n \ge 1,$$

and by the additive version of 1.7.2.2 it follows that

$$B_n(q) = \frac{1}{n}\sum_{1\le d\,|\,n} \mu(d)q^{n/d} \quad \text{for all } \ n \ge 1. \qquad \square$$

We proceed with a series of analytic preparations for the number field case.

Definition and Theorem 4.1.4 (Dirichlet series and integrals). An infinite series

$$\mathsf{D}(s) = \sum_{n=1}^{\infty} \frac{a_n}{n^s}$$

built with coefficients $a_n \in \mathbb{C}$ and $s \in \mathbb{C}$ is called a **Dirichlet series**, and

$$c = \inf\{\sigma \in \mathbb{R} \mid \mathsf{D}(\sigma) \ \text{ is absolutely convergent}\,\} \in \mathbb{R} \cup \{\pm\infty\}$$

is called its **abscissa of absolute convergence**.

An improper (Riemann) integral

$$\mathsf{D}^*(s) = \int_1^{\infty} \frac{g(\xi)}{\xi^s}\,d\xi$$

built with a piece-wise smooth function $g\colon [1,\infty) \to \mathbb{C}$ and $s \in \mathbb{C}$ is called a **Dirichlet integral**, and

$$c = \inf\{\sigma \in \mathbb{R} \mid D^*(\sigma) \ \text{ is absolutely convergent}\,\} \in \mathbb{R} \cup \{\pm\infty\}$$

is called its **abscissa of absolute convergence**.

Let $c < \infty$ be the abscissa of absolute convergence of a Dirichlet series $\mathsf{D}(s)$ resp. a Dirichlet integral $\mathsf{D}^*(s)$ as above. Then both $\mathsf{D}(s)$ and $\mathsf{D}^*(s)$ are locally uniformly and absolutely convergent in the half-plane $\mathcal{H}_c$, and there they define holomorphic functions $f,\ f^*\colon \mathcal{H}_c \to \mathbb{C}$ satisfying

$$f^{(k)}(s) = (-1)^k \sum_{n=1}^{\infty} \frac{a_n(\log n)^k}{n^s} \quad \textit{and} \quad f^{*(k)}(s) = (-1)^k \int_1^{\infty} \frac{g(\xi)(\log \xi)^k}{\xi^s}\,ds$$

for all $k \geq 0$ and $s \in \mathcal{H}_c$.

To avoid awkward wording, we will often not distinguish between a Dirichlet series (a Dirichlet integral) and the function which it defines in its domain of absolute convergence.

Proof. It suffices to prove absolute and uniform convergence of $\mathsf{D}(s)$ and $\mathsf{D}^*(s)$ in every closed half-plane $\overline{\mathcal{H}}_{c'}$ with $c' > c$. Let $c' > c$ and $\sigma \in (c, c')$. Then the series $\mathsf{D}(\sigma)$ and the integral $\mathsf{D}^*(\sigma)$ are absolutely convergent, and for all $s \in \overline{\mathcal{H}}_{c'}$ we obtain

$$\sum_{n=1}^{\infty} \left| \frac{a_n}{n^s} \right| \leq \sum_{n=1}^{\infty} \frac{|a_n|}{n^\sigma} < \infty \quad \text{and} \quad \int_1^\infty \left| \frac{g(\xi)}{\xi^s} \right| d\xi \leq \int_1^\infty \frac{|g(\xi)|}{\xi^\sigma} \, d\xi < \infty.$$

From these estimates the assertions follow by the Weierstrass criteria for series and integrals (see [54, Ch. 3] for series and [12, II.13.8.6] for integrals). □

Theorem 4.1.5 (Abel's summation formula). *Let $(a_n)_{n \geq 1}$ be a sequence of complex numbers and*

$$A(x) = \sum_{n=1}^{\lfloor x \rfloor} a_n \quad \textit{for } x \in \mathbb{R}_{\geq 0}.$$

1. *If $(b_n)_{n \geq 1}$ is a sequence of complex numbers and $N \in \mathbb{N}$, then*

$$\sum_{n=1}^{N} a_n b_n = A(N) b_N + \sum_{n=1}^{N-1} A(n)(b_n - b_{n+1}).$$

2. *If $b \colon \mathbb{R}_{\geq 1} \to \mathbb{C}$ is continuously differentiable and $X \in \mathbb{R}_{\geq 1}$, then*

$$\sum_{n=1}^{\lfloor X \rfloor} a_n b(n) = A(X) b(X) - \int_1^X A(\xi) b'(\xi) \, d\xi.$$

Proof. 1. Since $A(0) = 0$, we obtain

$$\sum_{n=1}^{N} a_n b_n = \sum_{n=1}^{N} [A(n) - A(n-1)] b_n = \sum_{n=1}^{N} A(n) b_n - \sum_{n=1}^{N-1} A(n) b_{n+1}$$
$$= A(N) b_N + \sum_{n=1}^{N-1} A(n)(b_n - b_{n+1}) \, .$$

2. With $N = \lfloor X \rfloor$, 1. implies

$$\sum_{n=1}^{N} a_n b(n) = A(N)b(N) - \sum_{n=1}^{N-1} A(n)[b(n+1) - b(n)]$$

$$= A(N)b(X) - A(N)[b(X) - b(N)] - \sum_{n=1}^{N-1} A(n)[b(n+1) - b(n)]$$

$$= A(X)b(X) - \int_N^X A(\xi)b'(\xi)\,d\xi - \sum_{n=1}^{N-1} \int_n^{n+1} A(\xi)b'(\xi)\,d\xi$$

$$= A(X)b(X) - \int_1^X A(\xi)b'(\xi)\,d\xi\,.$$

□

For a non-empty subset D of $\mathbb{C}$ we denote by $\mathcal{O}(D)$ the set of all complex functions which are holomorphic in some open neighborhood of D.

For holomorphic functions $f,\, g \in \mathcal{O}(\mathcal{H}_1)$ we set

$$f \sim g \quad \text{if} \quad f - g \in \mathcal{O}(\mathcal{H}_1 \cup \{1\}) \quad \big[\text{then there exists } \lim_{s\to 1+} (f(s) - g(s)) \in \mathbb{C}\big].$$

Obviously, $\sim$ is an equivalence relation on $\mathcal{O}(\mathcal{H}_1)$, and $f \sim 0$ means that f has a holomorphic extension to a neighborhood of 1.

Theorem 4.1.6. *Let $c \in \mathbb{R}$ and $f \in \mathcal{O}(\mathcal{H}_c \cup \{c\})$. Suppose that there exists a sequence $(a_n)_{n\ge 1}$ in $\mathbb{R}_{\ge 0}$ such that*

$$f(s) = \sum_{n=1}^{\infty} \frac{a_n}{n^s} \quad \textit{for all} \quad s \in \mathcal{H}_c,$$

and let c_0 be the abscissa of absolute convergence of the Dirichlet series defining f. Then $c_0 < c$.

Proof. If $a = 1 + c$, then there exists some $\rho > 1$ such that f is holomorphic in the disc $B(a;\rho) = \{s \in \mathbb{C} \mid |s - a| < \rho\}$, and 4.1.4 implies

$$f(s) = \sum_{k=0}^{\infty} \frac{f^{(k)}(a)}{k!}(s-a)^k = \sum_{k=0}^{\infty} \frac{(-1)^k}{k!}\Big(\sum_{n=1}^{\infty} \frac{a_n (\log n)^k}{n^a}\Big)(s-a)^k$$

for all $s \in B(a;\rho)$.

Now let $\varepsilon \in (0, \rho - 1)$. Then $|(c - \varepsilon) - a| = \varepsilon + 1 < \rho$ and

$$f(c-\varepsilon) = \sum_{k=0}^{\infty}\Big(\sum_{n=1}^{\infty} \frac{a_n(\log n)^k}{n^a}\Big)\frac{(1+\varepsilon)^k}{k!} = \sum_{n=1}^{\infty}\Big(\sum_{k=0}^{\infty} \frac{(1+\varepsilon)^k(\log n)^k}{k!}\Big)\frac{a_n}{n^a}$$

$$= \sum_{n=1}^{\infty} \frac{\mathrm{e}^{(1+\varepsilon)\log n} a_n}{n^a} = \sum_{n=1}^{\infty} \frac{a_n}{n^{c-\varepsilon}} < \infty,$$

which implies $c_0 \le c - \varepsilon < c$. □

Theorem 4.1.7. *Let $\theta \in \mathbb{R}_{\geq 0}$, $(a_n)_{n\geq 1}$ be a sequence in $\mathbb{C}$ and*

$$A(x) = \sum_{n=1}^{\lfloor x \rfloor} a_n.$$

1. Assume that for every $\varepsilon \in \mathbb{R}_{>0}$ we have $A(x) \ll x^{\theta+\varepsilon}$ for all $x \in \mathbb{R}_{\geq 1}$. Then the Dirichlet series

$$f(s) = \sum_{n=1}^{\infty} \frac{a_n}{n^s}$$

converges locally uniformly in the half-plane $\mathcal{H}_\theta$. The function $f\colon \mathcal{H}_\theta \to \mathbb{C}$ is holomorphic, and

$$\sum_{n=1}^{\infty} \frac{a_n}{n^s} = s \int_1^\infty \frac{A(x)}{x^{s+1}}\, dx \quad \text{for all } s \in \mathcal{H}_\theta.$$

2. Let θ be the abscissa of absolute convergence of the Dirichlet series in 1. Then $A(x) \ll x^{\theta+\varepsilon}$ for every $\varepsilon \in \mathbb{R}_{>0}$.

Proof. 1. Let $s \in \mathcal{H}_\theta$, $\sigma = \Re(s)$, $0 < \varepsilon < \sigma - \theta$ and and $x \in \mathbb{R}_{>1}$. Then

$$\frac{A(x)}{x^s} \ll x^{\theta+\varepsilon-\sigma}, \quad \text{hence} \quad \lim_{x\to\infty} \frac{A(x)}{x^s} = 0,$$

and

$$\int_1^\infty \left| \frac{A(\xi)}{\xi^{s+1}} \right| d\xi \ll \int_1^\infty \frac{d\xi}{\xi^{\sigma+1-\theta-\varepsilon}} = \frac{1}{\sigma-\theta-\varepsilon} < \infty.$$

As ε was arbitrary, the Dirichlet integral

$$\int_1^\infty \frac{A(\xi)}{\xi^{s+1}}\, d\xi$$

has an abscissa of absolute convergence $c \leq \theta$, and thus it converges locally uniformly in $\mathcal{H}_\theta$. For $N \in \mathbb{N}$, 4.1.5.2 implies

$$\sum_{n=1}^{N} \frac{a_n}{n^s} = \frac{A(N)}{N^s} + s \int_1^N \frac{A(\xi)}{\xi^{s+1}}\, d\xi.$$

Since locally uniformly in $\mathcal{H}_\theta$ we have

$$\left(\frac{A(N)}{N^s}\right)_{N\geq 1} \to 0 \quad \text{and} \quad \left(s \int_1^N \frac{A(\xi)}{\xi^{s+1}}\, d\xi\right)_{N\geq 1} \to s \int_1^\infty \frac{A(\xi)}{\xi^{s+1}}\, d\xi,$$

it follows that the Dirichlet series defining f converges locally uniformly in $\mathcal{H}_\theta$, and thus $f \in \mathcal{O}(\mathcal{H}_\theta)$ by [12, I. 9.12.1].

2. If $x \in \mathbb{R}_{>1}$ and $\varepsilon \in \mathbb{R}_{>0}$, then

$$\left|\frac{A(x)}{x^{\theta+\varepsilon}}\right| \le \sum_{n=1}^{\lfloor x\rfloor} \frac{|a_n|}{x^{\theta+\varepsilon}} \le \sum_{n=1}^{\lfloor x\rfloor} \frac{|a_n|}{n^{\theta+\varepsilon}} \le \sum_{n=1}^{\infty} \frac{|a_n|}{n^{\theta+\varepsilon}} < \infty,$$

and consequently $A(x) \ll x^{\theta+\varepsilon}$. □

Theorem 4.1.8 (Convergence theorem for Dirichlet series). *Let $(a_n)_{n\ge1}$ be a sequence in $\mathbb{C}$, $\rho \in \mathbb{C}$, $\vartheta \in (0,1)$, and for $x \in \mathbb{R}_{\ge1}$ we assume that*

$$A(x) = \sum_{n\le x} a_n = \rho x + O(x^{\vartheta}).$$

Then the Dirichlet series

$$F(s) = \sum_{n=1}^{\infty} \frac{a_n}{n^s}$$

converges locally uniformly in $\mathcal{H}_1$. The function $F(s)$ has a holomorphic extension to $\mathcal{H}_\vartheta \setminus \{1\}$ with a simple pole in 1 and residuum ρ, and if $s \in \mathcal{H}_1$, then

$$F(s) = \frac{\rho}{s-1} + g(s), \quad \textit{where} \quad g(s) = \rho + s\int_1^{\infty} \frac{A(\xi)-\rho\xi}{\xi^{s+1}}\,d\xi.$$

The integral defining $g(s)$ converges absolutely and locally uniformly in the half-plane $\mathcal{H}_\vartheta$. In particular, $g\colon \mathcal{H}_\vartheta \to \mathbb{C}$ is holomorphic, and

$$F(s) \sim \frac{\rho}{s-1}.$$

Proof. Since $A(x) \ll x$ for $x \in \mathbb{R}_{\ge1}$, the Dirichlet series converges locally uniformly in $\mathcal{H}_1$ by 4.1.7.1. Let $M \in \mathbb{R}_{>0}$ be such that $|A(x) - \rho x| \le Mx^{\vartheta}$ for all $x \ge 1$. By 4.1.5.2 we obtain, for $X \in \mathbb{R}_{\ge1}$ and $s \in \mathcal{H}_1$,

$$\sum_{n=1}^{\lfloor X\rfloor} \frac{a_n}{n^s} = \frac{A(X)}{X^s} + s\int_1^X \frac{A(\xi)}{\xi^{s+1}}\,d\xi = \frac{A(X)}{X^s} + s\rho\int_1^X \frac{d\xi}{\xi^s} + s\int_1^X \frac{A(\xi)-\rho\xi}{\xi^{s+1}}\,d\xi.$$

If $s \in \mathcal{H}_1$, then

$$\int_1^{\infty} \frac{d\xi}{\xi^s} = \frac{1}{s-1},$$

and if $\Re(s) \ge \sigma > \vartheta$, then

$$\int_1^{\infty} \left|\frac{A(\xi)-\rho\xi}{\xi^{s+1}}\right| d\xi \le M\int_1^{\infty} \xi^{\vartheta-\sigma-1}\,d\xi = \frac{M}{\sigma-\vartheta}.$$

Hence the integral defining $g(s)$ converges absolutely and locally uniformly in the half-plane $\mathcal{H}_\vartheta$. For $s \in \mathcal{H}_1$ it follows that

$$F(s) = \sum_{n=1}^{\infty} \frac{a_n}{n^s} = \frac{\rho s}{s-1} + s\int_1^{\infty} \frac{A(\xi)-\rho\xi}{\xi^s}\,d\xi = \frac{\rho}{s-1} + g(s).$$ □

4.2 Dirichlet L functions

Let K be an algebraic number field, $[K:\mathbb{Q}] = N$, (r_1, r_2) its signature, d_K its discriminant, R_K its regulator, h_K its class number and $w_K = |\mu(K)|$ the number of roots of unity in K.

We tacitly use the notation of Chapter 3.

Theorem 4.2.1 (Dominance of degree 1 primes). *Let $\mathcal{P}_K^1$ be the set of all prime ideals $\mathfrak{p} \in \mathcal{P}_K$ of degree 1, and let $a\colon \mathcal{P}_K \to \mathbb{C}$ be a bounded function. Then the Dirichlet series*

$$\Phi(s) = \sum_{\mathfrak{p}\in\mathcal{P}_K} \frac{a(\mathfrak{p})}{\mathfrak{N}(\mathfrak{p})^s} = \sum_{n\geq 1} \frac{a_n}{n^s}, \quad \text{where} \quad a_n = \sum_{\substack{\mathfrak{p}\in\mathcal{P}_K \\ \mathfrak{N}(\mathfrak{p})=n}} a(\mathfrak{p})$$

converges absolutely and locally uniformly in $\mathcal{H}_1$, and the Dirichlet series

$$\Phi_1(s) = \sum_{\mathfrak{p}\in\mathcal{P}_K\setminus\mathcal{P}_K^1} \frac{a(\mathfrak{p})}{\mathfrak{N}(\mathfrak{p})^s}$$

converges absolutely and locally uniformly in $\mathcal{H}_{1/2}$. In particular, $\Phi\colon \mathcal{H}_1 \to \mathbb{C}$ and $\Phi_1\colon \mathcal{H}_{1/2} \to \mathbb{C}$ are holomorphic functions, and

$$\sum_{\mathfrak{p}\in\mathcal{P}_K} \frac{a(\mathfrak{p})}{\mathfrak{N}(\mathfrak{p})^s} = \sum_{\mathfrak{p}\in\mathcal{P}_K^1} \frac{a(\mathfrak{p})}{\mathfrak{N}(\mathfrak{p})^s} + \Phi_1(s) \sim \sum_{\mathfrak{p}\in\mathcal{P}_K^1} \frac{a(\mathfrak{p})}{\mathfrak{N}(\mathfrak{p})^s} \quad \text{for } s \in \mathcal{H}_1.$$

Proof. Let $A \in \mathbb{R}_{>0}$ be such that $|a(\mathfrak{p})| \leq A < \infty$ for all $\mathfrak{p} \in \mathcal{P}_K$, and note that $|\{\mathfrak{p} \in \mathcal{P}_K \mid \mathfrak{p}\,|\,p\}| \leq N$ for all $p \in \mathbb{P}$. If $s \in \mathcal{H}_1$ and $\sigma = \Re(s)$, then

$$\sum_{\mathfrak{p}\in\mathcal{P}_K} \left|\frac{a(\mathfrak{p})}{\mathfrak{N}(\mathfrak{p})^s}\right| \leq \sum_{p\in\mathbb{P}} \sum_{\substack{\mathfrak{p}\in\mathcal{P}_K \\ \mathfrak{p}\,|\,p}} \frac{A}{p^{f(\mathfrak{p}/p)\sigma}} \leq AN \sum_{p\in\mathbb{P}} \frac{1}{p^\sigma} < \infty.$$

If $s \in \mathcal{H}_{1/2}$ and $\sigma = \Re(s)$, then

$$\left|\sum_{\mathfrak{p}\in\mathcal{P}_K\setminus\mathcal{P}_K^1} \frac{a(\mathfrak{p})}{\mathfrak{N}(\mathfrak{p})^s}\right| \leq A \sum_{\substack{\mathfrak{p}\in\mathcal{P}_K \\ f(\mathfrak{p})\geq 2}} \frac{1}{\mathfrak{N}(\mathfrak{p})^\sigma} \leq AN \sum_{p\in\mathbb{P}} \frac{1}{p^{2\sigma}} < \infty.$$

Now the asssertions follow by 4.1.4. $\square$

Definition 4.2.2. Let $\mathfrak{m} \in \mathcal{J}'_K$ and $\mathcal{C}_K^{\mathfrak{m}} = \mathcal{J}_K^{\mathfrak{m}}/\mathcal{H}_K^{\mathfrak{m}}$ be the ray class group modulo $\mathfrak{m}$ (see 3.6.6). A character $\chi \in \mathsf{X}(\mathcal{C}_K^{\mathfrak{m}})$ is called a **ray class character** modulo $\mathfrak{m}$ or a **(generalized) Dirichlet character** modulo $\mathfrak{m}$.

Let $1_{\mathfrak{m}} \in \mathsf{X}(\mathcal{C}_K^{\mathfrak{m}})$ be the unit character. Every ray class character $\chi \in \mathsf{X}(\mathcal{C}_K^{\mathfrak{m}})$ induces an (equally denoted) homomorphism $\chi\colon \mathcal{J}_K^{\mathfrak{m}} \to \mathbb{T}$, defined by $\chi(\mathfrak{a}) = \chi([\mathfrak{a}]^{\mathfrak{m}})$ for all $\mathfrak{a} \in \mathcal{J}_K^{\mathfrak{m}}$; it satisfies $\chi \restriction \mathcal{H}_K^{\mathfrak{m}} = 1$, and we set $\chi(\mathfrak{a}) = 0$ for all $\mathfrak{a} \in \mathcal{J}_K \setminus \mathcal{J}_K^{\mathfrak{m}}$. Conversely, every homomorphism $\chi\colon \mathcal{J}_K^{\mathfrak{m}} \to \mathbb{T}$ satisfying $\chi \restriction \mathcal{H}_K^{\mathfrak{m}} = 1$ is induced by a ray class character modulo $\mathfrak{m}$ and is (just to avoid awkward wording) itself called a **ray class character** modulo $\mathfrak{m}$.

For a ray class $C \in \mathcal{C}_K^{\mathfrak{m}}$ and $s \in \mathcal{H}_1$ we call

$$\zeta(s, C) = \sum_{\mathfrak{a} \in C \cap \mathcal{J}'_K} \frac{1}{\mathfrak{N}(\mathfrak{a})^s}$$

the **partial zeta function** of C. For a ray class character $\chi \in \mathsf{X}(\mathcal{C}_K^{\mathfrak{m}})$ and $s \in \mathcal{H}_1$ we call

$$L(s, \chi) = \sum_{\mathfrak{a} \in \mathcal{J}'_K} \frac{\chi(\mathfrak{a})}{\mathfrak{N}(\mathfrak{a})^s} = \sum_{C \in \mathcal{C}_K^{\mathfrak{m}}} \chi(C)\zeta(s, C)$$

the **(generalized) Dirichlet L function** built with χ, and we call

$$\zeta_K(s) = L(s, 1_1) = \sum_{\mathfrak{a} \in \mathcal{J}'_K} \frac{1}{\mathfrak{N}(\mathfrak{a})^s} = \sum_{C \in \mathcal{C}_K^{+}} \zeta(s, C)$$

the **Dedekind zeta function** of K.

If $K = \mathbb{Q}$ and $\mathfrak{m} = m\mathbb{Z}$ for some $m \in \mathbb{N}$, then in 3.6.6 we identified $\mathcal{J}'_{\mathbb{Q}} = \mathbb{N}$ (by identifying the ideal $a\mathbb{Z}$ with $|a|$ for all $a \in \mathbb{Z}$) and $\mathcal{C}_{\mathbb{Q}}^{\mathfrak{m}} = (\mathbb{Z}/m\mathbb{Z})^{\times}$ (by identifying the ray class $[a\mathbb{Z}]^m$ with the residue class $|a| + m\mathbb{Z} \in (\mathbb{Z}/m\mathbb{Z})^{\times}$). By this identification a ray class character modulo $m\mathbb{Z}$ is just a Dirichlet character modulo m as defined in 3.3.4, and $\mathsf{X}(\mathcal{C}_{\mathbb{Q}}^{\mathfrak{m}}) = \mathsf{X}((\mathbb{Z}/m\mathbb{Z})^{\times}) = \mathsf{X}(m)$. The resulting Dirichlet L function

$$L(s, \chi) = \sum_{a\mathbb{Z} \in \mathcal{J}'_{\mathbb{Q}}} \frac{\chi(a\mathbb{Z})}{\mathfrak{N}(a\mathbb{Z})^s} = \sum_{n \geq 1} \frac{\chi(n)}{n^s}, \quad \text{defined for } s \in \mathcal{H}_1,$$

is called the **(ordinary) Dirichlet L function** built with the character $\chi \in \mathsf{X}(m)$. The function

$$\zeta(s) = L(s, 1) = \zeta_{\mathbb{Q}}(s)$$

is the classical Riemann zeta function.

Theorem 4.2.3. *Let $\mathfrak{m} \in \mathcal{J}'_K$, $C \in \mathcal{C}_K^{\mathfrak{m}}$ a ray class and $\chi \in \mathsf{X}(\mathcal{C}_K^{\mathfrak{m}})$ a ray class character modulo $\mathfrak{m}$.*

1. The Dirichlet series

$$\sum_{\mathfrak{p}\in\mathcal{P}_K}\frac{1}{\mathfrak{N}(\mathfrak{p})^s},\qquad L(s,\chi)=\sum_{\mathfrak{a}\in\mathcal{J}'_K}\frac{\chi(\mathfrak{a})}{\mathfrak{N}(\mathfrak{a})^s},\qquad \zeta(s,C)=\sum_{\mathfrak{a}\in C\cap\mathcal{J}'_K}\frac{1}{\mathfrak{N}(\mathfrak{a})^s}$$

and

$$\zeta_K(s)=\sum_{\mathfrak{a}\in\mathcal{J}'_K}\frac{1}{\mathfrak{N}(\mathfrak{a})^s}$$

converge absolutely and locally uniformly in $\mathcal{H}_1$, and there they define holomorphic functions satisfying

$$L(s,\chi)=\prod_{\mathfrak{p}\in\mathcal{P}_K}\frac{1}{1-\chi(\mathfrak{p})\mathfrak{N}(\mathfrak{p})^{-s}}\neq 0,\ \ \zeta_K(s)=\prod_{\mathfrak{p}\in\mathcal{P}_K}\frac{1}{1-\mathfrak{N}(\mathfrak{p})^{-s}}\neq 0$$

and

$$L(s,1_{\mathfrak{m}})=\zeta_K(s)\prod_{\substack{\mathfrak{p}\in\mathcal{P}_K\\ \mathfrak{p}\,|\,\mathfrak{m}}}(1-\mathfrak{N}(\mathfrak{p})^{-s}).$$

2. $L(\,\cdot\,,\chi)$ has a holomorphic logarithm $\log L(\,\cdot\,,\chi)$ in $\mathcal{H}_1$ satisfying

$$\log L(s,\chi)\ =\ \sum_{\mathfrak{p}\in\mathcal{P}_K}\sum_{k=1}^{\infty}\frac{\chi(\mathfrak{p})^k}{k\mathfrak{N}(\mathfrak{p})^{sk}}\ =\ \sum_{\mathfrak{p}\in\mathcal{P}_K}\frac{\chi(\mathfrak{p})}{\mathfrak{N}(\mathfrak{p})^s}+G(s,\chi)$$

for some holomorphic function $G(\,\cdot\,,\chi)\colon\mathcal{H}_{1/2}\to\mathbb{C}$. In particular,

$$\log L(s,\chi)\sim\sum_{\mathfrak{p}\in\mathcal{P}_K}\frac{\chi(\mathfrak{p})}{\mathfrak{N}(\mathfrak{p})^s}\quad \textit{and}$$

$$\log L(s,1_{\mathfrak{m}})\sim\log\zeta_K(s)\sim\sum_{\mathfrak{p}\in\mathcal{P}_K}\frac{1}{\mathfrak{N}(\mathfrak{p})^s}.$$

Proof. 1. This follows by 4.2.1 and 4.1.2 since for $s\in\mathbb{C}$ the function

$$\mathcal{J}'_K\to\mathbb{C},\quad\text{defined by}\quad \mathfrak{a}\mapsto\frac{\chi(\mathfrak{a})}{\mathfrak{N}(\mathfrak{a})^s},$$

is completely multiplicative).

2. As the function $L(\,\cdot\,,\chi)$ is holomorphic and zero-free in the simply connected region $\mathcal{H}_1$, it has there a holomorphic logarithm, and since

$$\exp\Big\{\sum_{\mathfrak{p}\in\mathcal{P}_K}\sum_{k=1}^{\infty}\frac{\chi(\mathfrak{p})^k}{k\mathfrak{N}(\mathfrak{p})^{sk}}\Big\}=\prod_{p\in\mathcal{P}_K}\frac{1}{1-\chi(\mathfrak{p})\mathfrak{N}(\mathfrak{p})^{-s}}$$

$$=\prod_{\mathfrak{p}\in\mathcal{P}_K}\sum_{k=1}^{\infty}\frac{\chi(\mathfrak{p})^k}{k\mathfrak{N}(\mathfrak{p})^{sk}}\quad\text{for}\ \ s\in\mathcal{H}_1$$

by 4.1.2, the series

$$\log L(s,\chi) = \sum_{\mathfrak{p}\in\mathcal{P}_K}\sum_{k=1}^{\infty}\frac{\chi(\mathfrak{p})^k}{k\mathfrak{N}(\mathfrak{p})^{sk}}$$

defines a holomorphic logarithm of $L(\,\cdot\,,\chi)$ in $\mathcal{H}_1$. If $N=[K:\mathbb{Q}]$, $s\in\mathcal{H}_{1/2}$ and $\sigma=\Re(s)$, then

$$\sum_{\mathfrak{p}\in\mathcal{P}_K}\sum_{k=2}^{\infty}\Big|\frac{\chi(\mathfrak{p})^k}{k\mathfrak{N}(\mathfrak{p})^{sk}}\Big| \le \sum_{p\in\mathbb{P}} N\sum_{k=2}^{\infty}\frac{1}{p^{\sigma k}} \le N\sum_{p\in\mathbb{P}}\frac{1}{p^{2\sigma}}\,\frac{1}{1-p^{-\sigma}}$$
$$\le \frac{N}{1-2^{-\sigma}}\sum_{p\in\mathbb{P}}\frac{1}{p^{2\sigma}}<\infty,$$

and therefore the function

$$G(s,\chi) = \sum_{\mathfrak{p}\in\mathcal{P}_K}\sum_{k=2}^{\infty}\frac{\chi(\mathfrak{p})^k}{k\mathfrak{N}(\mathfrak{p})^{sk}}$$

is holomorphic in $\mathcal{H}_{1/2}$ by 4.1.4. □

Theorem 4.2.4. *Let $\mathfrak{m}\in\mathcal{J}_K'$ be an ideal, $C\in\mathcal{C}_K^{\mathfrak{m}}$ a ray class and $\chi\in\mathsf{X}(\mathcal{C}_K^{\mathfrak{m}})$ a ray class character modulo $\mathfrak{m}$.*

1. *The partial zeta function $\zeta(\,\cdot\,,C)$ has a holomorphic extension to $\mathcal{H}_{1-1/N}\setminus\{1\}$ with a simple pole in 1 and residuum*

$$\rho(\mathfrak{m}) = \frac{(2\pi)^{r_2}\mathsf{R}_K(U_K:U_K^{\mathfrak{m}})}{w_K\mathfrak{N}(\mathfrak{m})\sqrt{|\mathsf{d}_K|}}.$$

If $s\in\mathcal{H}_{1-1/N}\setminus\{1\}$, then

$$\zeta(s,C) = \frac{\rho(\mathfrak{m})}{s-1} + g(s,C),$$

for some holomorphic function $g(\,\cdot\,,C)\colon \mathcal{H}_{1-1/N}\to\mathbb{C}$.

2. *$L(\,\cdot\,,\chi)$ has a holomorphic extension to $\mathcal{H}_{1-1/N}\setminus\{1\}$, and there it satisfies*

$$L(s,\chi) = \frac{\rho^*(\mathfrak{m})}{s-1}\varepsilon(\chi)+G(s,\chi), \quad \textit{where } \varepsilon(\chi)=\begin{cases}0 & \textit{if } \chi\ne 1_{\mathfrak{m}},\\ 1 & \textit{if } \chi=1_{\mathfrak{m}},\end{cases}$$

$G(\,\cdot\,,\chi)\colon H_{1-1/N}\to\mathbb{C}$ is a holomorphic function, and

$$\rho^*(\mathfrak{m}) = \rho(K)\prod_{\substack{\mathfrak{p}\in\mathcal{P}_K\\ \mathfrak{p}\,|\,\mathfrak{m}}}\Big(1-\frac{1}{\mathfrak{N}(\mathfrak{p})}\Big), \quad \textit{where } \rho(K)=\frac{2^{r_1}(2\pi)^{r_2}h_K\mathsf{R}_K}{w_K\sqrt{|\mathsf{d}_K|}}.$$

In particular:

- *If* $\chi \neq 1_{\mathfrak{m}}$, *then* $L(\,\cdot\,,\chi)$ *has a holomorphic extension to* $\mathcal{H}_{1-1/N}$.
- $L(\,\cdot\,,1_{\mathfrak{m}})$ *has a simple pole in* 1 *with residuum* $\rho^*(\mathfrak{m})$.

3. Let $m(\chi)$ *be the order of* $L(\,\cdot\,,\chi)$ *in* 1 *(that is,* $m(1_{\mathfrak{m}})=-1$ *and* $m(\chi)\geq 0$ *if* $\chi\neq 1_{\mathfrak{m}}$ *). Then*

$$\log L(s,\chi)\ \sim\ -m(\chi)\log\frac{1}{s-1}.$$

In fact, $L(1,\chi)\neq 0$ and thus $m(\chi)=0$ holds for all $\chi\neq 1_{\mathfrak{m}}$. We shall prove this for $K=\mathbb{Q}$ in 4.2.8, and for general K in 4.4.1.2 using the (hitherto unproved) existence of a ray class field modulo $\mathfrak{m}$.

Proof. 1. By definition,

$$\zeta(s,C)=\sum_{\mathfrak{a}\in C\cap\mathcal{J}'_K}\frac{1}{\mathfrak{N}(\mathfrak{a})^s}=\sum_{n=1}^{\infty}\frac{a_n}{n^s},\quad\text{where}\quad a_n=|\{\mathfrak{a}\in C\cap\mathcal{J}'_K\mid\mathfrak{N}(\mathfrak{a})=n\}|.$$

For $x\in\mathbb{R}_{\geq 1}$, 3.6.9 (applied with $\boldsymbol{\sigma}=\boldsymbol{\sigma}(K)$) shows that

$$A(x)=\sum_{n\leq x}a_n=|\{\mathfrak{a}\in C\cap\mathcal{J}'_K\mid\mathfrak{N}(\mathfrak{a})\leq x\}|=\rho(\mathfrak{m})x+O(x^{1-1/N}),$$

and the assertion follows by 4.1.8.

2. By 1., using 3.3.1.5, it follows that

$$\begin{aligned}L(s,\chi)&=\sum_{C\in\mathcal{C}_K^{\mathfrak{m}}}\chi(C)\zeta(s,\chi)=\sum_{C\in\mathcal{C}_K^{\mathfrak{m}}}\chi(C)\Big[\frac{\rho(\mathfrak{m})}{s-1}+g(s,C)\Big]\\&=\frac{\rho(\mathfrak{m})|\mathcal{C}_K^{\mathfrak{m}}|}{s-1}\,\varepsilon(\chi)+G(s,\chi)\end{aligned}$$

for some holomorphic function $G(\,\cdot\,,\chi)\colon H_{1-1/N}\to\mathbb{C}$. By 3.6.7 and 2.14.1 we obtain

$$\begin{aligned}\rho^*(\mathfrak{m})=\rho(\mathfrak{m})|\mathcal{C}_K^{\mathfrak{m}}|&=\frac{2\pi)^{r_2}\mathsf{R}_K(U_K\!:\!U_K^{\mathfrak{m}})}{w_K\mathfrak{N}(\mathfrak{m})\sqrt{|\mathsf{d}_K|}}\,\frac{2^{r_1}h_K\varphi_K(\mathfrak{m})}{(U_K\!:\!U_K^{\mathfrak{m}})}=\frac{2^{r_1}(2\pi)^{r_2}h_K\mathsf{R}_K}{w_K\sqrt{|\mathsf{d}_K|}}\,\frac{\varphi_K(\mathfrak{m})}{\mathfrak{N}(\mathfrak{m})}\\&=\rho(K)\prod_{\substack{\mathfrak{p}\in\mathcal{P}_K\\ \mathfrak{p}\,|\,\mathfrak{m}}}\Big(1-\frac{1}{\mathfrak{N}(\mathfrak{p})}\Big).\end{aligned}$$

In particular, if $\chi\neq 1_{\mathfrak{m}}$, then $L(\,\cdot\,,\chi)$ has a holomorphic extension to $\mathcal{H}_{1-1/N}$, while $L(s,1_{\mathfrak{m}})$ has a simple pole in 1 with residuum $\rho^*(\mathfrak{m})$.

3. By the definition of $m(\chi)$, we obtain $L(s,\chi)=(s-1)^{m(\chi)}g(s,\chi)$ for some holomorphic function $g(\,\cdot\,,\chi)\colon\mathcal{H}_{1-1/N}\to\mathbb{C}$ satisfying $g(1,\chi)\neq 0$. Hence there exists an open disc $U=B_\varepsilon(1)\subset\mathcal{H}_{1-1/N}$ such that $g(\,\cdot\,,\chi)$ is holomorphic and

zero-free in the simply connected region $\mathcal{H}_1 \cup U$, and thus it possesses there a holomorphic logarithm. If $s \in \mathcal{H}_1 \cup U$, then

$$\exp\Big\{\log L(s,\chi) + m(\chi)\log\frac{1}{s-1}\Big\} = L(s,\chi)(s-1)^{-m(\chi)} = g(s,\chi),$$

and therefore the function $\log g(\,\cdot\,\chi)\colon \mathcal{H}_1 \cup U \to \mathbb{C}$, defined by

$$\log g(s,\chi) = \log L(s,\chi) + m(\chi)\log\frac{1}{s-1},$$

is a holomorphic logarithm of $g(\,\cdot\,,\chi)$ in $\mathcal{H}_1 \cup U$. It follows that

$$\log L(s,\chi) \sim -m(\chi)\log\frac{1}{s-1}. \qquad \square$$

Corollary 4.2.5. *The Dedekind zeta function $\zeta_K(s)$ has*

- *a holomorphic extension to $\mathcal{H}_{1-1/N} \setminus \{1\}$ with a simple pole in* 1 *and residuum*

$$\rho(K) = \frac{2^{r_1}(2\pi)^{r_2}h_K\mathsf{R}_K}{w_K\sqrt{|\mathsf{d}_K|}},$$

hence

$$\zeta_K(s) \sim \frac{\rho(K)}{s-1} \qquad \textit{and} \qquad \lim_{s\to 1}(s-1)\zeta_K(s) = \rho(K);$$

- *a holomorphic logarithm $\log\zeta_K(s)$ in $\mathcal{H}_1$ satisfying*

$$\log\zeta_K(s) \sim \sum_{\mathfrak{p}\in\mathcal{P}_K}\frac{1}{\mathfrak{N}(\mathfrak{p})^s} \sim \log\frac{1}{s-1}.$$

Proof. By 4.2.3.2 and 4.2.4. $\square$

The first assertion in 4.2.5 is called the **analytic class number formula**. Its significance lies in the fact that it connects the arithmetic invariants of an algebraic number field with the analytic behavior of the Dedekind zeta function.

In the following theorem we consider the class number formula for subfields of cyclotomic fields (that is, by the Kronecker-Weber theorem, for abelian number fields). We use the notation and results given in 3.3.8, 3.3.9, and 3.3.10.

Theorem 4.2.6. *Let $m \in \mathbb{N}$, $m \geq 3$ and K be a subfield of $\mathbb{Q}^{(m)}$. Let $Z \subset \mathsf{X}(m)$ be the character group of K modulo m, and for $\chi \in Z$ we denote by χ' the associated primitive character. Then*

$$\zeta_K(s) = \prod_{\chi \in Z} L(1, \chi') \quad \text{for all } s \in \mathcal{H}_1, \quad \text{and} \quad \mathsf{R}_K h_K = \kappa_K \prod_{\substack{\chi \in Z \\ \chi \neq \mathbf{1}_m}} L(1, \chi'),$$

where

$$\kappa_K = \frac{\sqrt{|\mathsf{d}_K|}}{2^{n-1}}, \text{ if } K \text{ is real, and } \kappa_K = \frac{w_K \sqrt{|\mathsf{d}_K|}}{(2\pi)^{n/2}}, \text{ if } K \text{ is imaginary.}$$

Proof. Since $\mathbf{1}'_m = \mathbf{1}_1 = 1$, we get $L(s, \mathbf{1}'_m) = \zeta(s)$ for all $s \in \mathcal{H}_1$, and therefore

$$\lim_{s \to 1} (s-1) \prod_{\chi \in Z} L(s, \chi') = \lim_{s \to 1} (s-1)\zeta(s) \prod_{\substack{\chi \in Z \\ \chi \neq \mathbf{1}_m}} L(s, \chi') = \prod_{\substack{\chi \in Z \\ \chi \neq \mathbf{1}_m}} L(1, \chi').$$

By 4.2.5 we obtain

$$\lim_{s \to 1} (s-1)\zeta_K(s) = \frac{h_K \mathsf{R}_K}{\kappa_K},$$

and thus it suffices to prove the product formula for $\zeta_K(s)$. Since

$$\zeta_K(s) = \prod_{p \in \mathbb{P}} \prod_{\substack{\mathfrak{p} \in \mathcal{P}_K \\ \mathfrak{p} \mid p}} \frac{1}{1 - \mathfrak{N}(\mathfrak{p})^{-s}} \quad \text{and} \quad \prod_{\chi \in Z} L(1, \chi') = \prod_{p \in \mathbb{P}} \prod_{\chi \in Z} \frac{1}{1 - \chi'(p) p^{-s}},$$

it suffices to prove that

$$\prod_{\substack{\mathfrak{p} \in \mathcal{P}_K \\ \mathfrak{p} \mid p}} \frac{1}{1 - \mathfrak{N}(\mathfrak{p})^{-s}} = \prod_{\chi \in Z} \frac{1}{1 - \chi'(p) p^{-s}} \quad \text{for all } p \in \mathbb{P}.$$

Let $p \in \mathbb{P}$, and assume that and $p\mathfrak{o}_K = (\mathfrak{p}_1 \cdot \ldots \cdot \mathfrak{p}_r)^e$, where $e = e_p(K)$, $r = r_p(K)$, $\mathfrak{p}_1, \ldots, \mathfrak{p}_r \in \mathcal{P}_K$ are distinct, and $f = f_p(K) = f_{\mathfrak{p}_i}$ for all $i \in [1, r]$. If

$$Y = \{\chi \in Z \mid \chi'(p) \neq 0\} \quad \text{and} \quad W = \{\chi \in Z \mid \chi'(p) = 1\},$$

then $W \subset Y \subset Z$ are subgroups, $(Z\!:\!Y) = e$, Y/W is cyclic, $(Y\!:\!W) = f$ and $|W| = r$. Let $\chi_0 \in Y$ be such that $Y/W = \langle \chi_0 W \rangle$ (see 3.3.10). Then $\zeta = \chi_0(p)$ is a primitive f-th root of unity, every $\chi \in Y$ has a unique factorization $\chi = \chi_0^j \chi_1$, where $j \in [0, f-1]$ and $\chi_1 \in W$, and then $\chi'(p) = \zeta^j$. Hence

$$\prod_{\chi \in Z} \frac{1}{1 - \chi'(p) p^{-s}} = \prod_{\chi \in W} \prod_{j=0}^{f-1} \frac{1}{1 - \zeta^j p^{-s}} = \Big(\frac{1}{1 - p^{-sf}} \Big)^r = \prod_{\substack{\mathfrak{p} \in \mathcal{P}_K \\ \mathfrak{p} \mid p}} \frac{1}{1 - \mathfrak{N}(\mathfrak{p})^{-s}}.$$

$\square$

There is a comprehensive literature on the class numbers of abelian and in particular of cyclotomic and quadratic number fields. We cannot go into further detail here and refer to [24], [61] and [22].

We close this section with explicit bounds and formulas for ordinary Dirichlet L functions.

Theorem 4.2.7. *Let $m \in \mathbb{N}$ and $\chi \in \mathsf{X}(m)$ a Dirichlet character.*

1. *If $s \in \mathcal{H}_1$, then*

$$L(s,\chi) = \sum_{n=1}^{\infty} \frac{\chi(n)}{n^s} = \prod_{p\in\mathbb{P}} \frac{1}{1-\chi(p)p^{-s}} = \prod_{\substack{p\in\mathbb{P}\\ p\nmid m}} \frac{1}{1-\chi(p)p^{-s}} \neq 0,$$

$$\log L(s,\chi) = \sum_{p\in\mathbb{P}} \sum_{k=1}^{\infty} \frac{\chi(p)^k}{kp^{sk}} \sim \sum_{p\in\mathbb{P}} \frac{\chi(p)}{p^s},$$

$$\log L(s,1_m) \sim \log\zeta(s) \sim \log\frac{1}{s-1},$$

and

$$\zeta(s) = \sum_{n=1}^{\infty} \frac{1}{n^s} = \prod_{p\in\mathbb{P}} \frac{1}{1-p^{-s}} = L(s,1_m) \prod_{\substack{p\in\mathbb{P}\\ p\,|\,m}} \frac{1}{1-p^{-s}}.$$

2. *The Riemann zeta function has a holomorphic extension to $\mathcal{H}_0 \setminus \{1\}$ with a simple pole in 1 and residuum 1. If $s \in \mathcal{H}_0 \setminus \{1\}$, then*

$$\zeta(s) = \frac{1}{s-1} + \sum_{n=1}^{\infty} F_n(s), \quad \textit{where} \quad F_n(s) = \int_n^{n+1} \Big(\frac{1}{n^s} - \frac{1}{\xi^s}\Big)d\xi.$$

The functions $F_n\colon \mathcal{H}_0 \to \mathbb{C}$ are holomorphic for all $n \geq 1$, the series converges locally uniformly and absolutely in $\mathcal{H}_0$, and

$$\Big|\zeta(s) - \frac{1}{s-1}\Big| \leq \frac{|s|}{\Re(s)} \quad \textit{for all} \quad s \in \mathcal{H}_0.$$

3. *Let $m \geq 3$, $\chi \neq 1_m$, and for $x \in \mathbb{R}_{\geq 1}$ let*

$$A(x) = \sum_{n\leq x} \chi(x).$$

If $s \in \mathcal{H}_0$ and $\sigma = \Re(s)$, then

$$L(s,\chi) = \sum_{n=1}^{\infty} \frac{\chi(n)}{n^s} = s\int_1^{\infty} \frac{A(x)}{x^{s+1}}\,dx, \quad |L(s,\chi| \leq \frac{m|s|}{\sigma},$$

and the Dirichlet series converges locally uniformly in $\mathcal{H}_0$.

Proof. 1. Obvious by 4.2.3 and 4.2.4.3.

2. If $s \in \mathcal{H}_1$, then

$$\zeta(s) - \frac{1}{s-1} = \sum_{n=1}^{\infty} \frac{1}{n^s} - \int_1^{\infty} \frac{d\xi}{\xi^s} = \sum_{n=1}^{\infty} \int_n^{n+1} \Big(\frac{1}{n^s} - \frac{1}{\xi^s}\Big) d\xi = \sum_{n=1}^{\infty} F_n(s),$$

and obviously F_n is holomorphic in $\mathcal{H}_0$ for all $n \geq 1$. If $n \in \mathbb{N}$, $\xi \in [n, n+1]$, $s \in \mathcal{H}_0$ and $\sigma = \Re(s)$, then

$$\Big|\frac{1}{n^s} - \frac{1}{\xi^s}\Big| = \Big|\int_n^{\xi} \frac{s}{t^{s+1}}\,dt\Big| \leq |s| \int_n^{n+1} \frac{dt}{t^{\sigma+1}}, \quad \text{which implies}$$

$$|F_n(s)| \leq |s| \int_n^{n+1} \frac{dt}{t^{\sigma+1}},$$

and if $\sigma \geq \sigma_0 > 0$ and $|s| \leq M$, then

$$\sum_{n=1}^{\infty} |F_n(s)| \leq |s| \int_1^{\infty} \frac{dt}{t^{\sigma+1}} = \frac{|s|}{\sigma} \leq \frac{M}{\sigma_0} < \infty.$$

Hence the series of the functions F_n converges locally uniformly and absolutely to a holomorphic function in $\mathcal{H}_0$, and by the identity theorem for holomorphic functions we obtain, for all $s \in \mathcal{H}_0$,

$$\zeta(s) - \frac{1}{s-1} = \sum_{n=1}^{\infty} F_n(s), \quad \text{and thus also } \Big|\zeta(s) - \frac{1}{s-1}\Big| \leq \frac{|s|}{\Re(s)}.$$

3. By 3.3.5 it follows that $A(m) = 0$ and $|A(x)| < m$ for all $x \in \mathbb{R}_{>0}$. Hence 4.1.7.1 implies that

$$L(s,\chi) = \sum_{n=1}^{\infty} \frac{\chi(n)}{n^s} = s \int_1^{\infty} \frac{A(\xi)}{\xi^{s+1}} d\xi \quad \text{for all } s \in \mathcal{H}_0,$$

and the Dirichlet series converges locally uniformly in $\mathcal{H}_0$. From this, we obtain the estimate

$$|L(s,\chi)| \leq m|s| \int_1^{\infty} \frac{d\xi}{\xi^{\sigma+1}} = \frac{m|s|}{\sigma}. \qquad \square$$

In the following Theorem 4.2.8 we prove that $L(1,\chi) \neq 0$. This result is fundamental for Dirichlet's theorem on primes in arithmetic progressions (see 4.3.11.2), and we offer two different proofs of this important fact.

The same result holds for (generalized) Dirichlet L functions in algebraic number fields, and both proofs have their counterparts there. One of them uses class field theory and the other uses the analytic continuation of the L functions. For a thorough exposition we must refer to the forthcoming volume on class field theory.

In the after next section however, in 4.4.1.2 we shall give a proof of $L(1,\chi) \neq 0$ for a generalized Dirichlet L function assuming the existence of ray class fields as formulated in 3.7.2, and afterwards we shall draw some striking consequences for the theory of algebraic number fields.

Theorem 4.2.8. *Let $m \in \mathbb{N}$. Then*

$$L(1,\chi) \neq 0 \quad \textit{and} \quad \log L(s,\chi) \sim 0 \quad \textit{for all} \;\; \chi \in \mathsf{X}(m) \setminus \{1_m\}.$$

Proof. By 4.2.4.3 it suffices to prove that $L(1,\chi) \neq 0$ for all $\chi \in \mathsf{X}(m) \setminus \{1_m\}$.

First proof. We use the arithmetic of the field $K = \mathbb{Q}^{(m)} = \mathbb{Q}(\zeta)$, where ζ is a primitive m-th root of unity (see 1.7.5, 3.1.11 and 3.2.6). For $s \in \mathcal{H}_1$, 4.2.7.1 implies

$$\prod_{\chi \in \mathsf{X}(m)} L(s,\chi) = \prod_{\substack{p \in \mathbb{P} \\ p \nmid m}} \prod_{\chi \in \mathsf{X}(m)} (1 - \chi(p)p^{-s})^{-1}.$$

Let $p \in \mathbb{P}$, $p \nmid m$ and $f = \mathrm{ord}_m(p) = \mathrm{ord}_{(\mathbb{Z}/m\mathbb{Z})^\times}(p + m\mathbb{Z})$. Then $f \mid \varphi(m)$, say $\varphi(m) = df$. Then $p\mathfrak{o}_K = \mathfrak{p}_1 \cdot \ldots \cdot \mathfrak{p}_d$, where $\mathfrak{p}_1, \ldots, \mathfrak{p}_d \in \mathcal{P}_K$ are distinct and $\mathfrak{N}(\mathfrak{p}_i) = p^f$ for all $i \in [1,d]$. Now 3.3.1.4 implies

$$\prod_{\chi \in \mathsf{X}(m)} (1 - \chi(p)p^{-s}) = (1 - p^{-sf})^d = \prod_{i=1}^{d} (1 - \mathfrak{N}(\mathfrak{p}_i)^{-s}).$$

It follows that

$$\prod_{\chi \in \mathsf{X}(m)} L(s,\chi) = \prod_{\substack{p \in \mathbb{P} \\ p \nmid m}} \prod_{\substack{\mathfrak{p} \in \mathcal{P}_K \\ \mathfrak{p} \mid p}} (1 - \mathfrak{N}(\mathfrak{p})^{-s})^{-1} = \prod_{\mathfrak{p} \in \mathcal{P}_K} (1 - \mathfrak{N}(\mathfrak{p})^{-s})^{-1} \prod_{\substack{\mathfrak{p} \in \mathcal{P}_K \\ \mathfrak{p} \mid m}} (1 - \mathfrak{N}(\mathfrak{p})^{-s})$$

$$= \zeta_K(s) \prod_{\substack{\mathfrak{p} \in \mathcal{P}_K \\ \mathfrak{p} \mid m}} (1 - \mathfrak{N}(\mathfrak{p})^{-s}),$$

and since

$$L(s, 1_m) = \zeta(s) \prod_{\substack{p \in \mathbb{P} \\ p \mid m}} (1 - p^{-s}),$$

we ultimately obtain

$$\prod_{\substack{\chi \in \mathsf{X}(m) \\ \chi \neq 1_m}} L(s,\chi) = \frac{\zeta_K(s)}{\zeta(s)} \prod_{\substack{p \in \mathbb{P} \\ p \mid m}} \Big[(1 - p^{-s})^{-1} \prod_{\substack{\mathfrak{p} \in \mathcal{P}_K \\ \mathfrak{p} \mid p}} (1 - \mathfrak{N}(\mathfrak{p})^{-s}) \Big] \quad \text{for all} \;\; s \in \mathcal{H}_1. \qquad (\dagger)$$

By 4.2.5 both functions ζ_K and ζ are holomorphic in $\mathcal{H}_{1-1/\varphi(m)}$ with a simple pole in $s = 1$, hence their quotient is holomorphic and non-zero in 1. By 4.2.7.3 the left-hand side of $(\dagger)$ is holomorphic in $\mathcal{H}_0$, and by the identity theorem

(†) holds for all $s \in \mathcal{H}_{1-1/\varphi(m)}$. The right-hand side is non-zero in 1, and thus $L(1,\chi) \neq 1$ for all $\chi \in \mathsf{X}(m) \setminus \{1_m\}$.

Second proof. Let $\chi \in \mathsf{X}(m) \setminus \{1_m\}$.

CASE 1: $\chi^2 = 1_m$. We assume that $L(1,\chi) = 0$. Then $L(\,\cdot\,,1_m)L(\,\cdot\,,\chi)$ is holomorphic in $\mathcal{H}_0$, and for $s \in \mathcal{H}_1$, 4.2.7.1 implies

$$L(s,1_m)L(s,\chi) = \exp\Bigl\{\sum_{p\in\mathbb{P}}\sum_{k=1}^{\infty}\frac{1+\chi(p)^k}{kp^{sk}}\Bigr\} = \mathrm{e}^{g(s)},$$

where $g\colon \mathcal{H}_1 \to \mathbb{C}$, is defined by

$$g(s) = \sum_{p\in\mathbb{P}}\sum_{k=1}^{\infty}\frac{1+\chi(p)^k}{kp^{sk}} = \sum_{n=1}^{\infty}\frac{A_n}{n^s}$$

and

$$A_n = \frac{1+\chi(p)^k}{k} \quad \text{if } n = p^k \text{ is a prime power, and } A_n = 0 \text{ otherwise.}$$

For $s \in \mathbb{R}_{>0}$ we obtain

$$g(s) \geq \sum_{\substack{p\in\mathbb{P}\\ p\nmid m}}\sum_{k=1}^{\infty}\frac{2}{2kp^{2sk}} \geq \sum_{\substack{p\in\mathbb{P}\\ p\nmid m}}\frac{1}{p^{2s}} \sim \log L(2s,1_m) \sim \frac{1}{2s-1} \quad \text{by 4.2.7.1.}$$

Hence the Dirichlet series defining $g(s)$ diverges for $s = 1/2$, and thus its abscissa of absolute convergence satisfies $c \geq 1/2$.

For real $s > c$ and $i \geq 0$ exponentiation of the absolute convergent series yields

$$g(s)^i = \sum_{n=0}^{\infty}\frac{A_n^{(i)}}{n^s} \quad \text{with some coefficients } A_n^{(i)} \in \mathbb{R}_{\geq 0}.$$

By a rearrangement of the absolute convergent double series it follows that, for all real $s > c$,

$$\mathrm{e}^{g(s)} = \sum_{i=0}^{\infty}\frac{1}{i!}\sum_{n=1}^{\infty}\frac{A_n^{(i)}}{n^s} = \sum_{n=1}^{\infty}\frac{\overline{A}_n}{n^s} \quad \text{where } \overline{A}_n \in \mathbb{R}_{\geq 0}.$$

Hence the Dirichlet series defining $\mathrm{e}^{g(s)}$ has an abscissa of absolute convergence $\overline{c} \leq c$. But the Dirichlet defining $g(s)$ is a subseries of those defining $\mathrm{e}^{g(s)}$, and thus $\overline{c} = c \geq 1/2$. As $\mathrm{e}^g \in \mathcal{O}(\mathcal{H}_0)$, this contradicts 4.1.6.

CASE 2: $\chi^2 \neq 1_m$. We assume that $L(1,\chi) = 0$ and consider the function $g\colon \mathcal{H}_0 \setminus \{1\} \to \mathbb{C}$, defined by

$$g(s) = L(s,1_m)^3 L(s,\chi)^4 L(s,\chi^2).$$

Since the order of g in 1 satisfies $\operatorname{ord}(g;1) \geq -3+4 = 1$, it follows that $g \in \mathcal{O}(\mathcal{H}_0)$ and $g(1) = 0$. Now 4.2.7.1 implies

$$g(s) = \exp\Big\{\sum_{p\in P}\sum_{k=1}^{\infty} \frac{3+4\chi(p^k)+\chi^2(p^k)}{kp^{sk}}\Big\} \quad \text{for all } \ s \in \mathcal{H}_1,$$

and therefore

$$|g(s)| = \exp\Big\{\sum_{p\in P}\sum_{k=1}^{\infty} \frac{3+4\Re\{\chi(p^k)\}+\Re\{\chi(p^{2k})\}}{kp^{sk}}\Big\} \quad \text{for all real } \ s > 1.$$

If $p \in \mathbb{P}$, $k \in \mathbb{N}$ and $\Re\{\chi(p)\} = \cos\theta$, then

$$\begin{aligned}3+4\Re\{\chi(p^k)\}+\Re\{\chi(p^{2k})\} &= 3+4\cos k\theta+\cos 2k\theta = 2+4\cos k\theta+2\cos^2 k\theta \\ &= 2(1+\cos k\theta)^2 \geq 0.\end{aligned}$$

It follows that $|g(s)| \geq 1$ for all $s \in \mathbb{R}_{>1}$, a contradiction. □

We close this section with the derivation of explicit formulas for $L(1,\chi)$. In doing so, we may restrict ourselves to primitive Dirichlet characters. Indeed, let $m \in \mathbb{N}$, $\chi \in \mathsf{X}(m)$, $f = f(\chi)$, and let $\chi' \in \mathsf{X}(f)$ be the associated primitive Dirichlet character of χ (see 3.3.4). Then

$$L(s,\chi) = L(s,\chi')\prod_{p\,|\,m}(1-\chi'(p)p^{-s}).$$

We use Gauss sums as defined in 3.4.6.

Theorem 4.2.9 (Summation of L series). *Let $m \in \mathbb{N}$, $m \geq 3$, and let $\chi \in \mathsf{X}(m)$ be a primitive Dirichlet character modulo m (hence $\chi \neq 1_m$). Then*

$$L(1,\chi) = -\frac{\tau(\chi)}{m}\sum_{j=1}^{m-1}\overline{\chi}(j)\log\sin\frac{\pi j}{m} \quad \textit{if } \ \chi(-1) = 1,$$

and

$$L(1,\chi) = \frac{\pi \mathrm{i}\tau(\chi)}{m^2}\sum_{j=1}^{m-1}\overline{\chi}(j)j \quad \textit{if } \ \chi(-1) = -1.$$

Proof. Let $\zeta_m = \mathrm{e}^{2\pi\mathrm{i}/m}$, and use 1.7.4.1, 3.4.6, 3.4.7.2 and 3.3.5. If $s \in \mathbb{R}_{>1}$, then

$$\begin{aligned}L(s,\chi) &= \sum_{n=1}^{\infty}\frac{\chi(n)}{n^s} = \sum_{k=1}^{m-1}\chi(k)\sum_{\substack{n=1\\ n\equiv k \bmod m}}^{\infty}\frac{1}{n^s} = \sum_{k=1}^{m-1}\chi(k)\sum_{n=1}^{\infty}\Big(\frac{1}{m}\sum_{j=0}^{m-1}\zeta_m^{\pm(n-k)j}\Big)\frac{1}{n^s} \\ &= \frac{1}{m}\sum_{j=0}^{m-1}\sum_{k=1}^{m-1}\chi(k)\zeta_m^{\mp kj}\sum_{n=1}^{\infty}\frac{\zeta_m^{\pm jn}}{n^s} = \frac{1}{m}\sum_{j=1}^{m-1}\tau(\chi,\mp j)\sum_{n=1}^{\infty}\frac{\zeta_m^{\pm jn}}{n^s},\end{aligned}$$

since $\tau(\chi,0)=0$. By 1.7.4 we obtain

$$\sum_{n=1}^{\lfloor x\rfloor}\zeta_m^{\pm jn}\ll 1 \quad\text{for}\quad x\in\mathbb{R}_{>1},$$

and thus 4.1.7 implies the locally uniform convergence of the latter Dirichlet series in $\mathcal{H}_0$. Therefore we obtain

$$\lim_{s\to 1}\sum_{n=1}^{\infty}\frac{\zeta_m^{\pm jn}}{n^s}=\sum_{n=1}^{\infty}\frac{\zeta_m^{\pm j}}{n}=-\log(1-\zeta_m^{\pm j}.$$

Here log is the principal branch of the complex logarithm: If $z=re^{i\varphi}$, where $r\in\mathbb{R}_{>0}$ and $\varphi\in(-\pi,\pi)$, then $\log z=\log r+\mathrm{i}\varphi$. Observe that

$$\log(1-x)=\sum_{n=1}^{\infty}\frac{x^n}{n}\qquad\text{if } x\in\mathbb{C},\ |x|\le 1\ \text{and}\ x\ne 1\quad\text{(see [35, Satz 243]).}$$

If $j\in[1,m-1]$, then $\tau(\chi,\mp j)=\chi(\mp 1)\tau(\chi)\overline{\chi}(j)$, and thus

$$L(1,\chi)=\lim_{s\to 1+}L(s,\chi)=\frac{-\tau(\chi)}{m}\sum_{j=1}^{m-1}\chi(\mp 1)\overline{\chi}(j)\log(1-\zeta_m^{\pm j}).$$

We add the (+)- and the (−)-equation and obtain

$$2L(1,\chi)=\frac{-\tau(\chi)}{m}\sum_{j=1}^{m-1}\overline{\chi}(j)\left[\log(1-\zeta_m^{-j})+\chi(-1)\log(1-\zeta_m^{j})\right].$$

If $j\in[1,m-1]$, then

$$1-\zeta_m^j=-\zeta_{2m}^j(\zeta_{2m}^j-\zeta_{2m}^{-j})=-\zeta_{2m}^j\Big(2\mathrm{i}\sin\frac{\pi j}{m}\Big)=\mathrm{e}^{\pi\mathrm{i}(\frac{j}{m}-\frac12)}\Big(2\sin\frac{\pi j}{m}\Big),$$

hence

$$1-\zeta_m^{-j}=\overline{1-\zeta_m^j}=\mathrm{e}^{-\pi\mathrm{i}(\frac{j}{m}-\frac12)}\Big(2\sin\frac{\pi j}{m}\Big),$$

and

$$\log(1-\zeta_m^{\pm j})=\log\Big(2\sin\frac{\pi j}{m}\Big)\pm\pi\mathrm{i}\Big(\frac{j}{m}-\frac12\Big),\qquad\text{since}\quad\Big|\frac{j}{m}-\frac12\Big|<1.$$

If $\chi(-1)=1$, then

$$\log(1-\zeta_m^{-j})+\log(1-\zeta_m^{j})=2\log\Big(\sin\frac{\pi j}{m}\Big)+2\log 2,$$

and therefore

$$L(1,\chi)=-\frac{\tau(\chi)}{m}\sum_{j=1}^{m-1}\overline{\chi}(j)\log\sin\frac{\pi j}{m},\qquad\text{since}\quad\sum_{j=1}^{m-1}\overline{\chi}(j)=0.$$

If $\chi(-1) = -1$, then

$$\log(1-\zeta_m^{-j}) - \log(-1\zeta_m^j) = -2\pi\mathrm{i}\Big(\frac{j}{m} - \frac{1}{2}\Big),$$

and therefore

$$L(1,\chi) = \frac{\pi\mathrm{i}\tau(\chi)}{m^2}\sum_{j=1}^{m-1}\overline{\chi}(j)j, \quad \text{again since} \quad \sum_{j=1}^{m-1}\overline{\chi}(j) = 0. \qquad \square$$

4.3 Density of prime ideals

With the view to applications in more general situations and in particular for function fields we derive the elementary properties of densities within the abstract scope of Dirichlet systems.

Definition 4.3.1. A **Dirichlet system** is a pair $(P, |\cdot|)$, consisting of a denumerable set P and a function $|\cdot|\colon P \to \mathbb{N}$ such that

$$A_n = |\{p \in P \mid |p| \le n\}| \ll n \text{ for all } n \ge 1,$$

$$\sum_{p\in P}\frac{1}{|p|^\sigma} < \infty \quad \text{for all } \sigma \in \mathbb{R}_{>1}, \quad \text{and} \quad \lim_{\sigma\to 1+}\sum_{p\in P}\frac{1}{|p|^\sigma} = \infty.$$

For a subset T of P we define the **natural density** $\delta(T)$ by

$$\delta(T) = \lim_{N\to\infty}\frac{|\{p\in T \mid |p| \le N\}|}{|\{p \in P \mid |p| \le N\}|} \in [0,1]$$

and the **Dirichlet density** by

$$d(T) = \lim_{\sigma\to 1+}\frac{\sum_{p\in T}|p|^{-\sigma}}{\sum_{p\in P}|p|^{-\sigma}} \in [0,1],$$

provided that the corresponding limits exist. Obviously, $\delta(P) = d(P) = 1$, and if T is a finite subset of P, then $\delta(T) = d(T) = 0$. For subsets T_1 and T_2 of P we define containment and the equality up to a set of Dirichlet density 0 by

- $T_1 \dot{\subset} T_2$ if $T_1 \subset T_2 \cup T$ for some subset T of P such that $d(T) = 0$;
- $T_1 \doteq T_2$ if $d(T_1 \setminus T_2) = d(T_2 \setminus T_1) = 0$.

Lemma 4.3.2. *Let $(P, |\cdot|)$ be a Dirichlet system and $T_1, T_2 \subset P$.*

1. $T_1 \dot\subset T_2$ if and only if $T_1 \doteq T_1 \cap T_2$.

2. $T_1 \doteq T_2$ if and only if $T_1 \dot\subset T_2$ and $T_2 \dot\subset T_1$. In particular, if T_1 and T_2 only differ by a finite set, then $T_1 \doteq T_2$.

3. Suppose that $d(T_1)$ and $d(T_2)$ exist.

(a) If $d(T_1 \cap T_2)$ exists, then

$$d(T_1 \cup T_2) = d(T_1) + d(T_2) - d(T_1 \cap T_2) \le d(T_1) + d(T_2).$$

(b) If $T_1 \dot\subset T_2$, then $d(T_1) \le d(T_2)$.

(c) If $d(T_1) = d(T_2) = 0$, then $d(T_1 \cup T_2) = 0$.

4. Suppose that $d(T_1)$ exists.

(a) $d(P \setminus T_1) = 1 - d(T_1)$.

(b) If $T_2 \dot\subset T_1$ and $d(T_1) = 0$, then $d(T_2) = 0$.

(c) If $T_2 \doteq T_1$, then $d(T_2) = d(T_1)$.

Proof. Straightforward. □

Theorem 4.3.3. *Let $(P, |\cdot|)$ be a Dirichlet system, and let T be a subset of P with natural density $\delta(T)$. Then T has a Dirichlet density $d(T) = \delta(T)$.*

Proof. For $n \in \mathbb{N}$, let $b_n = |\{p \in T \mid |p| = n\}|$, $a_n = |\{p \in P \mid |p| = n\}|$,

$$B_n = \sum_{\nu=1}^{n} b_\nu = |\{p \in T \mid |p| \le n\}| \text{ and } A_n = \sum_{\nu=1}^{n} a_\nu = |\{p \in P \mid |p| \le n\}|.$$

Then $B_n \le A_n \ll n$ for all $n \in \mathbb{N}$, and

$$\delta = \delta(T) = \lim_{n\to\infty} \frac{B_n}{A_n}.$$

For $n \in \mathbb{N}$ we set $B_n = \delta A_n + \varepsilon_n A_n$, where $\varepsilon_n \in \mathbb{R}$, and $(\varepsilon_n)_{n\ge 1} \to 0$. If $\sigma \in \mathbb{R}_{>1}$, then

$$\lim_{N\to\infty} \sum_{n=1}^{N} a_n n^{-\sigma} = \sum_{p\in P} |p|^{-\sigma} < \infty, \qquad \lim_{N\to\infty} \sum_{n=1}^{N} b_n n^{-\sigma} = \sum_{p\in T} |p|^{-\sigma} < \infty,$$

and we must prove that

$$\delta = \lim_{\sigma\to 1+} \frac{\sum_{p\in T} |p|^{-\sigma}}{\sum_{p\in P} |p|^{-\sigma}} = \lim_{\sigma\to 1+} \lim_{N\to\infty} \frac{\sum_{n=1}^{N} b_n n^{-\sigma}}{\sum_{n=1}^{N} a_n n^{-\sigma}}.$$

We show that for every $\varepsilon \in \mathbb{R}_{>0}$ there exists some $\tau \in \mathbb{R}_{>0}$ such that

$$\Big|\lim_{N\to\infty} \frac{\sum_{n=1}^N b_n n^{-\sigma}}{\sum_{n=1}^N a_n n^{-\sigma}} - \delta\Big| \le \varepsilon \quad \text{for all} \quad \sigma \in (1, 1+\tau).$$

Let $\varepsilon \in \mathbb{R}_{>0}$ and $\sigma \in \mathbb{R}_{>1}$. For $n \in \mathbb{N}$ we set $\Delta_n(\sigma) = n^{-\sigma} - (n+1)^{-\sigma}$. For $N \in \mathbb{N}$, Abel's summation formula 4.1.5 implies

$$\sum_{n=1}^N b_n n^{-\sigma} = B_N N^{-\sigma} + \sum_{n=1}^{N-1} B_n \Delta_n(\sigma) = B_N N^{-\sigma} + \delta \sum_{n=1}^{N-1} A_n \Delta_n(\sigma) + \sum_{n=1}^{N-1} \varepsilon_n A_n \Delta_n(\sigma)$$

and

$$\sum_{n=1}^N a_n n^{-\sigma} = A_N N^{-\sigma} + \sum_{n=1}^{N-1} A_n \Delta_n(\sigma).$$

From this and $A_N \ll N$ we deduce the existence of the limits

$$\lim_{N\to\infty} \sum_{n=1}^{N-1} A_n \Delta_n(\sigma) = \lim_{N\to\infty} \Big(\sum_{n=1}^N a_n n^{-\sigma} - A_N N^{-\sigma}\Big) = \sum_{n=1}^\infty a_n n^{-\sigma} = \sum_{p\in P} |p|^{-\sigma} \in \mathbb{R}_{>0}$$

and

$$\lim_{N\to\infty} \sum_{n=1}^{N-1} \varepsilon_n A_n \Delta_n(\sigma).$$

Let $m \in \mathbb{N}$ be such that $|\varepsilon_n| \le \varepsilon/2$ for all $n > m$, and let $N > m$. Then

$$\begin{aligned}
\Big|\frac{\sum_{n=1}^N b_n n^{-\sigma}}{\sum_{n=1}^N a_n n^{-\sigma}} - \delta\Big| \le{}& \Big|\frac{B_N N^{-\sigma} + \delta \sum_{n=1}^{N-1} A_n \Delta_n(\sigma)}{A_N N^{-\sigma} + \sum_{n=1}^{N-1} A_n \Delta_n(\sigma)} - \delta\Big| \\
&+ \Big|\frac{\sum_{n=1}^{N-1} \varepsilon_n A_n \Delta_n(\sigma)}{A_N N^{-\sigma} + \sum_{n=1}^{N-1} A_n \Delta_n(\sigma)} - \frac{\sum_{n=1}^{N-1} \varepsilon_n A_n \Delta_n(\sigma)}{\sum_{n=1}^{N-1} A_n \Delta_n(\sigma)}\Big| \\
&+ \Big|\frac{\sum_{n=1}^{m} \varepsilon_n A_n \Delta_n(\sigma)}{\sum_{n=1}^{N-1} A_n \Delta_n(\sigma)}\Big| + \Big|\frac{\sum_{n=m+1}^{N-1} \varepsilon_n A_n \Delta_n(\sigma)}{\sum_{n=1}^{N-1} A_n \Delta_n(\sigma)}\Big|.
\end{aligned}$$

As $N \to \infty$, the four summands above behave as follows:

$$\lim_{N\to\infty} \Big|\frac{B_N N^{-\sigma} + \delta \sum_{n=1}^{N-1} A_n \Delta_n(\sigma)}{A_N N^{-\sigma} + \sum_{n=1}^{N-1} A_n \Delta_n(\sigma)} - \delta\Big| = \lim_{N\to\infty} \Big|\frac{B_N N^{-\sigma}}{A_N N^{-\sigma} + \sum_{n=1}^{N-1} A_n \Delta_n(\sigma)}\Big| = 0,$$

$$\lim_{N\to\infty} \Big|\frac{\sum_{n=1}^{N-1} \varepsilon_n A_n \Delta_n(\sigma)}{A_N N^{-\sigma} + \sum_{n=1}^{N-1} A_n \Delta_n(\sigma)} - \frac{\sum_{n=1}^{N-1} \varepsilon_n A_n \Delta_n(\sigma)}{\sum_{n=1}^{N-1} A_n \Delta_n(\sigma)}\Big| = 0,$$

$$\lim_{N\to\infty} \Big|\frac{\sum_{n=1}^{m} \varepsilon_n A_n \Delta_n(\sigma)}{\sum_{n=1}^{N-1} A_n \Delta_n(\sigma)}\Big| = \Big|\frac{\sum_{n=1}^{m} \varepsilon_n A_n \Delta_n(\sigma)}{\sum_{n=1}^{\infty} A_n \Delta_n(\sigma)}\Big|$$

and

$$\Big|\frac{\sum_{n=m+1}^{N-1} \varepsilon_n A_n \Delta_n(\sigma)}{\sum_{n=1}^{N-1} A_n \Delta_n(\sigma)}\Big| \le \frac{\varepsilon}{2} \Big|\frac{\sum_{n=m+1}^{N-1} A_n \Delta_n(\sigma)}{\sum_{n=1}^{N-1} A_n \Delta_n(\sigma)}\Big| \le \frac{\varepsilon}{2}.$$

Therewith we now obtain

$$\lim_{N\to\infty}\left|\frac{\sum_{n=1}^N b_n n^{-\sigma}}{\sum_{n=1}^N a_n n^{-\sigma}} - \delta\right| \le \frac{\varepsilon}{2} + \left|\frac{\sum_{n=1}^m \varepsilon_n A_n \Delta_n(\sigma)}{\sum_{p\in P}|p|^{-\sigma}}\right|.$$

Since

$$\lim_{\sigma\to 1+}\sum_{p\in P}|p|^{-\sigma} = \infty \quad \text{and} \quad \lim_{\sigma\to 1+}\Delta_n(\sigma) = \frac{1}{n(n+1)},$$

there exists some $\tau \in \mathbb{R}_{>0}$ such that

$$\left|\frac{\sum_{n=1}^m \varepsilon_n A_n \Delta_n(\sigma)}{\sum_{p\in P}|p|^{-\sigma}}\right| \le \frac{\varepsilon}{2} \quad \text{for all} \quad \sigma \in (1, 1+\tau).$$

This finally implies that

$$\left|\lim_{N\to\infty}\frac{\sum_{n=1}^N b_n n^{-\sigma}}{\sum_{n=1}^N a_n n^{-\sigma}} - \delta\right| \le \varepsilon \quad \text{for all} \quad \sigma \in (1, 1+\tau). \qquad \square$$

Definition and Remarks 4.3.4. Let K be an algebraic number field. Then $(\mathcal{P}_K, \mathfrak{N})$ is a Dirichlet system. Indeed, by 3.6.10 it follows that

$$|\{\mathfrak{p} \in \mathcal{P}_K \mid \mathfrak{N}(\mathfrak{p}) \le n\}| \le |\{\mathfrak{a} \in \mathcal{J}_K' \mid \mathfrak{N}(\mathfrak{a}) \le n\}| \ll n \quad \text{for all} \quad n \in \mathbb{N},$$

and 4.2.5 shows that

$$\sum_{\mathfrak{p}\in\mathcal{P}_K}\frac{1}{\mathfrak{N}(\mathfrak{p})^s} = \log\frac{1}{s-1} + g(s), \quad \text{where} \quad g \in \mathcal{O}(\mathcal{H}_1) \cup \{1\}).$$

In particular,

$$\lim_{\sigma\to 1+}\sum_{\mathfrak{p}\in\mathcal{P}_K}\mathfrak{N}(\mathfrak{p})^{-\sigma} = \infty \quad \text{and} \quad \lim_{\sigma\to 1+}\frac{\sum_{\mathfrak{p}\in\mathcal{P}_K}\mathfrak{N}(\mathfrak{p})^{-\sigma}}{\log\frac{1}{\sigma-1}} = 1.$$

Consequently, for every subset T of $\mathcal{P}_K$ its Dirichlet density is given by

$$d(T) = \lim_{\sigma\to 1+}\frac{\sum_{\mathfrak{p}\in\mathcal{P}_K}\mathfrak{N}(\mathfrak{p})^{-\sigma}}{\log\frac{1}{\sigma-1}},$$

provided this limit exists.

A subset T of $\mathcal{P}_K$ is called **regular** with density $c \in [0,1]$ if

$$\sum_{\mathfrak{p}\in T}\frac{1}{\mathfrak{N}(\mathfrak{p})^s} \sim c\log\frac{1}{s-1} \quad \text{for} \quad s \in \mathcal{H}_1.$$

Obviously, $\mathcal{P}_K$ is regular with density 1, and if a set T is regular with density c, then $c = d(T)$ is its Dirichlet density.

In the sequel we tacitly use the notation and results of 4.3.1, 4.3.2 and 4.3.4 for algebraic number fields.

As a first striking application of the concept of Dirichlet densities we prove in 4.3.5 an ancient theorem of Bauer asserting that a Galois extension of algebraic number fields is uniquely determined by the set of fully decomposed primes. In modern times Bauer's result was the starting point for a comprehensive theory. Some of these further developments will be discussed in the forthcoming volume on class field theory. For a thorough presentation of this theory however we refer to [34].

Recall from 3.7.1 that for a finite extension L/K of algebraic number fields we denote by $\mathrm{Split}(L/K)$ the set of all in L fully decomposed prime ideals $\mathfrak{p} \in \mathcal{P}_K$.

Theorem 4.3.5. *Let K be an algebraic number field and $\mathcal{P}_K^1$ be the set of prime ideals of degree 1 in $\mathfrak{o}_K$.*

1. *If T is any subset of $\mathcal{P}_K$, then $T \doteq T \cap \mathcal{P}_K^1$.*

2. *Let L/K be a finite Galois extension. Then $\mathrm{Split}(L/K)$ is a regular subset of $\mathcal{P}_K$ with density*

$$d(\mathrm{Split}(L/K)) = \frac{1}{[L:K]}.$$

3. (Splitting Theorem of M. Bauer, weak form) *Let L/K and L'/K be finite Galois extensions (inside a fixed algebraic closure). Then*

- *$L \subset L'$ if and only if $\mathrm{Split}(L'/K) \mathrel{\dot\subset} \mathrm{Split}(L/K)$, and*
- *$L = L'$ if and only if $\mathrm{Split}(L/K) \doteq \mathrm{Split}(L'/K)$.*

4. *Let L/K be a finite field extension such that $\mathcal{P}_K \doteq \mathrm{Split}(L/K)$. Then $L = K$.*

Proof. 1. $T \setminus (T \cap \mathcal{P}_K^1) \subset \mathcal{P}_K \setminus \mathcal{P}_K^1$, and 4.2.1 implies $d(\mathcal{P}_K \setminus \mathcal{P}_K^1) = 0$. Hence $d(T \setminus (T \cap \mathcal{P}_K^1)) = 0$ and $T \doteq T \cap \mathcal{P}_K^1$.

2. If $\mathfrak{P} \in \mathcal{P}_L$ and $\mathfrak{p} = \mathfrak{P} \cap K$, then $\mathfrak{p} \in \mathcal{P}_K^1 \cap \mathrm{Split}(L/K)$ if and only if $\mathfrak{P} \in \mathcal{P}_L^1$ and $\mathfrak{P}/\mathfrak{p}$ is unramified. Thus let $\mathcal{P}_L'$ be the set of all $\mathfrak{P} \in \mathcal{P}_L^1$ such that $\mathfrak{P}/\mathfrak{P} \cap K$ is unramified. Then $\mathcal{P}_L^1 \setminus \mathcal{P}_L'$ is finite, and the map $\rho\colon \mathcal{P}_L' \to \mathrm{Split}(L/K) \cap \mathcal{P}_K^1$, defined by $\rho(\mathfrak{P}) = \mathfrak{P} \cap K$, is surjective. If $\mathfrak{p} \in \mathrm{Split}(L/K) \cap \mathcal{P}_K^1$, then we have $|\rho^{-1}(\mathfrak{p})| = [L:K]$ and $\mathfrak{N}(\mathfrak{P}) = \mathfrak{N}(\mathfrak{p})$ for all $\mathfrak{P} \in \rho^{-1}(\mathfrak{p})$. It follows that

$$\sum_{\mathfrak{p} \in \mathrm{Split}(L/K)} \frac{1}{\mathfrak{N}(\mathfrak{p})^s} \sim \sum_{\mathfrak{p} \in \mathrm{Split}(L/K) \cap \mathcal{P}_K^1} \frac{1}{\mathfrak{N}(\mathfrak{p})^s} = \frac{1}{[L:K]} \sum_{\mathfrak{P} \in \mathcal{P}_L'} \sim \frac{1}{[L:K]} \sum_{\mathfrak{P} \in \mathcal{P}_L^1} \frac{1}{\mathfrak{N}(\mathfrak{P})^s}$$

$$\sim \frac{1}{[L:K]} \log \frac{1}{s-1} \quad \text{for } s \in \mathcal{H}_1,$$

and therefore $\mathrm{Split}(L/K)$ is regular with density $1/[L:K]$.

3. It suffices to prove the first statement. It follows from the very definition that $L \subset L'$ implies $\mathrm{Split}(L'/K) \subset \mathrm{Split}(L/K)$. Thus let $\mathrm{Split}(L'/K) \,\dot{\subset}\, \mathrm{Split}(L/K)$. Then $\mathrm{Split}(L'/K) \cap \mathrm{Split}(L/K) \doteq \mathrm{Split}(L'/K)$, and $\mathrm{Split}(LL'/K) = \mathrm{Split}(L/K) \cap \mathrm{Split}(L'/K)$ by 2.13.5.1.

Hence it follows that

$$\frac{1}{[LL':K]} = d(\mathrm{Split}(LL'/K)) = d(\mathrm{Split}(L'/K)) = \frac{1}{[L':K]},$$

and consequently $L \subset LL' = L'$.

4. Let N be a Galois closure of L/K. Then $\mathrm{Split}(L/K) = \mathrm{Split}(N/K)$ by 2.13.5.2,

$$1 = d(\mathrm{Split}(L/K)) = d(\mathrm{Split}(N/K)) = \frac{1}{[N:K]} \quad \text{by 2.},$$

and consequently $N = K = L$. □

Remark 4.3.6. Note that 4.3.5 implies 3.7.3, and thus Weber's definition of class fields coincides with Takagi's. In particular, we have obtained a complete proof of 3.7.4 and thus of the fact that the cyclotomic field $\mathbb{Q}^{(m)}$ is the ray class field modulo m of $\mathbb{Q}$.

The usual way to proceed from density statements to asymptotic results is the application of Tauber theorems. Here we present a special case of the Tauber theorem due to Ikehara and Delange. For more general versions of that theorem we refer to [10], [49, Ch. III, §3] or [22, Theorem 8.2.5].

Before however, we recall (without proofs) the definition and the main properties of the Gamma function.

Definition and Theorem 4.3.7. For $s \in \mathcal{H}_0$ the Gamma function is defined by the absolutely convergent improper integral

$$\Gamma(s) = \int_0^\infty t^{s-1} \mathrm{e}^{-t} dt.$$

1. The Gamma function has an extension to a zero-free holomorphic function in $\mathbb{C} \setminus (-\mathbb{N}_0)$. If $n \in \mathbb{N}_0$, then Γ has a simple pole in $-n$, and

$$\frac{d}{dz}\Big[\frac{1}{\Gamma(z)}\Big]_{z=-n} = \mathrm{Res}\{\Gamma; -n\} = \frac{(-1)^n}{n!}.$$

2. For all $s \in \mathbb{C}$ we have

$$\Gamma(s+1) = s\Gamma(s), \quad \Gamma(s)\Gamma(1-s) = \frac{\pi}{\sin \pi s}$$

and

$$\Gamma(s)\Gamma\big(s+\frac{1}{2}\big) = 2^{1-2s}\sqrt{\pi}\,\Gamma(2s).$$

3. $\Gamma(k+1) = k!$ for all $k \in \mathbb{N}_0$, and $\Gamma\big(\frac{1}{2}\big) = \sqrt{\pi}$.

Proof. [55, Ch. A.2] □

Theorem 4.3.8 (Ikehara - Delange). *Let $(a_n)_{n\geq 1}$ be a sequence in $\mathbb{R}_{\geq 0}$ and $c \in \mathbb{R}_{>0}$ be such that*

$$\sum_{n=1}^{\infty} \frac{a_n}{n^s} \sim c \log \frac{1}{s-1} \quad \textit{for} \ \ s \in \mathcal{H}_1.$$

Then

$$A(x) = \sum_{n=1}^{\lfloor x \rfloor} a_n = (c + o(1))\frac{x}{\log x} \quad \textit{for} \ \ x \to \infty.$$

Before we start with the real proof the theorem, we prepend an analytical lemma.

Lemma 4.3.9. *Let $c \in \mathbb{R}_{>0}$, and let $B\colon \mathbb{R}_{\geq 1} \to \mathbb{R}$ be defined by*

$$B(x) = 0 \ \ \textit{if} \ \ x \in [1, \mathsf{e}] \quad \textit{and} \quad B(x) = \frac{cx}{\log x} \ \ \textit{if} \ \ x \geq \mathsf{e}.$$

Then there exists an entire function $H \in \mathcal{O}(\mathbb{C})$ such that

$$\int_1^{\infty} \frac{B(x)}{x^{s+2}} dx = c \log \frac{1}{s} + H(s) \quad \textit{for all} \ \ s \in \mathcal{H}_0.$$

Proof of Lemma 4.3.9. If $s \in \mathcal{H}_0$ and $\Re(s) \geq \sigma > 0$, then

$$\frac{B(x)}{x^{s+2}} \ll \frac{1}{x^{\sigma+1}},$$

and thus the integral converges absolutely and locally uniformly in $\mathcal{H}_0$. We define

$$b\colon \mathbb{R}_{\geq 1} \to \mathbb{R} \quad \text{by} \quad b(u) = \frac{d}{dz}\Big[\frac{u^{z-1}}{\Gamma(z)}\Big]_{z=1},$$

and we assert there exist entire functions H_1, $H_2 \in \mathcal{O}(\mathbb{C})$ such that, for all $s \in \mathcal{H}_0$,

$$\int_1^{\infty} b(u)\mathsf{e}^{-su} du = \frac{1}{s} \log \frac{1}{s} + H_1(s) \quad \text{and} \quad \int_1^{\infty} \frac{\mathsf{e}^{-su}}{u} du = \log \frac{1}{s} + H_2(s). \quad \textbf{(A)}$$

The second equation in (**A**) implies Lemma 4.3.9, since

$$\int_1^\infty \frac{B(x)}{x^{s+2}}\,dx = c\int_e^\infty \frac{dx}{x^{s+1}\log x} \underset{[x=\mathrm{e}^u]}{=} c\int_1^\infty \frac{\mathrm{e}^{-su}}{u}\,du = c\log\frac{1}{s} + cH_2(s),$$

and thus it suffices to prove (**A**).

Proof of (**A**). For $z \in \mathcal{H}_0$ and $s \in \mathbb{R}_{>0}$ the definition of the Gamma function yields

$$\Gamma(z) = \int_0^\infty t^{z-1}\mathrm{e}^{-t}dt \underset{[t=su]}{=} s^z\int_0^\infty u^{z-1}\mathrm{e}^{-su}du. \tag{$*$}$$

If $u \in \mathbb{R}_{\geq 0}$, $x = \Re(z)$, $s \in \mathcal{H}_0$ and $\sigma = \Re(s)$, then $|u^{z-1}\mathrm{e}^{-su}| = u^{x-1}\mathrm{e}^{-\sigma u}$, and

$$\int_0^\infty u^{x-1}\mathrm{e}^{-\sigma u}du = \int_0^1 u^{x-1}\mathrm{e}^{-\sigma u}du + \int_1^\infty u^{x-1}\mathrm{e}^{-\sigma u}du < \infty,$$

which implies

$$\Big(s \mapsto \int_0^\infty u^{z-1}\mathrm{e}^{-su}du\Big) \in \mathcal{O}(\mathcal{H}_0).$$

By the identity theorem for holomorphic functions ($*$) holds for all $s \in \mathcal{H}_0$, and thus

$$\frac{1}{s^z} = \int_0^\infty \frac{u^{z-1}}{\Gamma(z)}\mathrm{e}^{-su}du \quad \text{for all } \ s,\, z \in \mathcal{H}_0.$$

We will differentiate the integral with respect to z. To do so, fix $s \in \mathcal{H}_0$ and set $\sigma = \Re(s)$. For every bounded region R in $\mathcal{H}_0$ there exist constants $M, x_1, x_2 \in \mathbb{R}_{>0}$ such that

$$\left|\frac{u^{z-1}}{\Gamma(z)}\mathrm{e}^{-su}\right| \leq \begin{cases} M\mathrm{e}^{-\sigma u}u^{x_1} & \text{for } u \in (0,1), \\ M\mathrm{e}^{-\sigma u}u^{x_2} & \text{for } u \in (1,\infty). \end{cases}$$

As these bounds are integrable, we may interchange differentiation and integration. This yields

$$\begin{aligned} \frac{1}{s}\log\frac{1}{s} = \frac{d}{dz}\Big[\frac{1}{s^z}\Big]_{z=1} &= \int_0^\infty \frac{d}{dz}\Big[\frac{u^{z-1}}{\Gamma(z)}\Big]_{z=1}\mathrm{e}^{-su}du \\ &= \int_0^1 b(u)\mathrm{e}^{-su}du + \int_1^\infty b(u)\mathrm{e}^{-su}du, \end{aligned}$$

and consequently

$$\int_1^\infty b(u)\mathrm{e}^{-su} = \frac{1}{s}\log\frac{1}{s} + H_1(s), \quad \text{where} \quad H_1 \in \mathcal{O}(\mathbb{C}).$$

To evaluate the second integral in (**A**), observe that

$$b(u) = \Big[\frac{u^{z-1}\log u}{\Gamma(z)}\Big]_{z=1} + \Big[u^{z-1}\frac{d}{dz}\Big(\frac{1}{\Gamma(z)}\Big)\Big]_{z=1} = \log u + \Big(\frac{1}{\Gamma}\Big)'(1),$$

and therefore

$$b'(u) = \frac{1}{u} \quad \text{and} \quad \lim_{X\to\infty} b(X)\mathrm{e}^{-sX} = 0.$$

Partial integration yields

$$\int_1^\infty \frac{\mathrm{e}^{-su}}{u}\,du = \int_1^\infty \mathrm{e}^{-su} b'(u)du = \Big[\mathrm{e}^{-su} b(u)\Big]_1^\infty + s\int_1^\infty b(u)\mathrm{e}^{-su}du$$
$$= -b(1)\mathrm{e}^{-s} + s\int_1^\infty b(u)\mathrm{e}^{-su}du = \log\frac{1}{s} + sH_1(s) - b(1)\mathrm{e}^{-s},$$

and we obtain

$$\int_{\mathrm{e}}^\infty \frac{dx}{x^{s+1}\log x} \underset{x=\mathrm{e}^u}{=} \int_1^\infty \frac{\mathrm{e}^{-su}}{u}du = \log\frac{1}{s} + H_2(s),$$

where $H_2(s) = sH_1(s) - b(1)\mathrm{e}^{-s}$, which completes the proof of (**A**). □

Proof of Theorem 4.3.8. Let $f_0\colon \mathcal{H}_1 \to \mathbb{C}$ be defined by

$$f_0(s) = \sum_{n=1}^\infty \frac{a_n}{n^s}.$$

By assumption, 1 is the abscissa of absolute convergence of the Dirichlet series defining $f_0(s)$. Hence $f_0 \in \mathcal{O}(\mathcal{H}_1)$, and if $s \in \mathcal{H}_1$, then our assumption and 4.1.7 imply

$$f_0(s) = \sum_{n=1}^\infty \frac{a_n}{n^s} = s\int_1^\infty \frac{A(x)}{x^{s+1}}dx = c\log\frac{1}{s-1} + f_1(s)$$

for some function $f_1 \in \mathcal{O}(\mathcal{H}_1 \cup \{1\})$. Let $L \in (0,\infty]$ be such that f_1 is holomorphic in $\Omega = \mathcal{H}_1 \cup \{1+it \mid t \in \mathbb{R},\ |t| < L\}$, and let $B\colon \mathbb{R}_{\ge 1} \to \mathbb{R}$ be as in 4.3.9. We define

$$F\colon \mathbb{R}_{\ge 1} \to \mathbb{R} \ \text{ by } \ F(x) = A(x) - B(x) \ \text{ and } \ f\colon \mathcal{H}_1 \to \mathbb{C} \ \text{ by } \ f(s) = \int_1^\infty \frac{F(x)}{x^{s+1}}\,dx.$$

If $x \in \mathbb{R}_{\ge 1}$, then $F(x) \ll x^{1+\varepsilon}$ for every $\varepsilon \in \mathbb{R}_{>0}$, and

$$\int_1^\infty \frac{F(x)}{x^{s+1}}\,dx \ll \int_1^\infty \frac{dx}{x^{\sigma-\varepsilon}} < \infty \quad \text{for all } \sigma > 1+\varepsilon.$$

Hence 1 is the abscissa of absolute convergence of the Dirichlet integral defining f, and thus $f \in \mathcal{O}(\mathcal{H}_1)$. For $s \in \mathcal{H}_1$ we obtain (using 4.3.9)

$$f(s) = \int_1^\infty \frac{F(x)}{x^{s+1}}\,dx = \int_1^\infty \frac{A(x)}{x^{s+1}}\,dx - \int_1^\infty \frac{B(x)}{x^{s+1}}\,dx$$
$$= \Big[\frac{c}{s}\log\frac{1}{s-1} + \frac{f_1(s)}{s}\Big] - \Big[c\log\frac{1}{s-1} + H(s-1)\Big]$$
$$= c\Big(\frac{1}{s} - 1\Big)\log\frac{1}{s-1} + f_2(s) \qquad \text{for some function } f_2 \in \mathcal{O}(\mathcal{H}_1),$$

and consequently

$$f'(1+s) = c\Big[\frac{1}{s+1} - \frac{1}{(s+1)^2}\log\frac{1}{s} + f_2'(1+s)\Big] \ll \log\frac{1}{|s|} \quad \text{for } s \in \mathcal{H}_0.$$

Next we define $h\colon \Omega \to \mathbb{C}$ by

$$h(s) = \frac{1}{c}\int_0^{1/2} f''(s+u)du = \frac{1}{c}\Big[f'\Big(s+\frac{1}{2}\Big) - f'(s)\Big] \text{ if } s \neq 1, \text{ and } h(1) = 0.$$

Then h is holomorphic in $\Omega \setminus \{1\}$, and for $s \in \mathcal{H}_0$ we obtain

$$h(1+s) \ll \log\frac{1}{|s|} \quad \text{and} \quad h(1+s) = \frac{1}{c}\int_0^{1/2}\int_1^\infty \frac{F(x)(\log x)^2}{x^{2+s+u}}dx\,du.$$

Let the auxiliary function $v\colon \mathbb{R}_{\geq 1} \to \mathbb{R}$ be defined by

$$v(x) = \frac{1}{cx}\int_0^{1/2}\frac{du}{x^u} = \frac{1}{cx\log x}\Big(1 - \frac{1}{\sqrt{x}}\Big).$$

There exists some $x_0 \in \mathbb{R}_{>1}$ such that v is monotonically decreasing in $[x_0, \infty)$, and

$$\lim_{x\to\infty}\frac{v(x\alpha)}{v(x)} = \frac{1}{\alpha} \quad \text{for all } \alpha \in \mathbb{R}_{>1}.$$

For $x \in \mathbb{R}_{\geq 1}$ we set $I(x) = (\log x)^2 A(x)v(x)$ and $J(x) = (\log x)^2 B(x)v(x)$. Then

$$I(x) = J(x) + (\log x)^2 F(x)v(x), \quad A(x)J(x) = B(x)I(x) \text{ for all } x \geq \mathsf{e},$$

$$\lim_{x\to\infty} J(x) = 1, \text{ and as } \lim_{x\to\infty}\frac{A(x)\log x}{cx} = \lim_{x\to\infty}\frac{I(x)\log x}{(\log x)^2 v(x)x} = \lim_{x\to\infty} I(x),$$

we must prove that

$$\lim_{x\to\infty} I(x) = 1.$$

For λ, η, $\varepsilon \in \mathbb{R}_{>0}$ such that $0 < 2\lambda < L$ we consider the integral

$$\Phi_{\varepsilon,\lambda}(\eta) = \int_{-2\lambda}^{2\lambda} h(1+\varepsilon+\mathsf{i}y)\Big(1 - \frac{|y|}{2\lambda}\Big)\eta^{-\mathsf{i}y}dy$$

and evaluate it for $\varepsilon \to 0+$. For $y \in [-2\lambda, 2\lambda]$ and $\varepsilon \in (0,1)$ we have

$$h(1+\varepsilon+\mathsf{i}y)\Big(1 - \frac{|y|}{2\lambda}\Big)\eta^{-\mathsf{i}y} \ll \log\frac{1}{|y|}, \quad \text{since } h(1+s) \ll \log\frac{1}{|s|} \text{ for } s \in \mathcal{H}_0.$$

Hence dominated convergence implies

$$\lim_{\varepsilon\to 0+}\Phi_{\varepsilon,\lambda}(\eta)=\int_{-2\lambda}^{2\lambda}\lim_{\varepsilon\to 0+}h(1+\varepsilon+\mathrm{i}y)\Big(1-\frac{|y|}{2\lambda}\Big)\eta^{-\mathrm{i}y}dy$$
$$=\int_{-2\lambda}^{2\lambda}h(1+\mathrm{i}y)\Big(1-\frac{|y|}{2\lambda}\Big)\eta^{-\mathrm{i}y}dy.$$

By a standard calculation we obtain

$$\int_{-2\lambda}^{2\lambda}\Big(1-\frac{|y|}{2\lambda}\Big)(\eta x)^{-\mathrm{i}y}\,dy=2\int_0^{2\lambda}\Big(1-\frac{y}{2\lambda}\Big)\cos\big[y\log(\eta x)\big]\,dy=\frac{2\sin^2[\lambda\log(\eta x)]}{\lambda\,[\log(\eta x)]^2},$$

and we evaluate $\Phi_{\varepsilon,\lambda}(\eta)$ using Fubini's Theorem as follows.

$$\Phi_{\varepsilon,\lambda}(\eta)=\frac{1}{c}\int_{-2\lambda}^{2\lambda}\Big\{\int_0^{1/2}f''(1+\varepsilon+\mathrm{i}y+u)du\Big\}\Big(1-\frac{|y|}{2\lambda}\Big)\eta^{-\mathrm{i}y}\,dy$$
$$=\frac{1}{c}\int_{-2\lambda}^{2\lambda}\Big\{\int_0^{1/2}\Big[\int_1^\infty\frac{(\log x)^2F(x)}{x^{2+\varepsilon+\mathrm{i}y+u}}\,dx\Big]du\Big\}\Big(1-\frac{|y|}{2\lambda}\Big)\eta^{-\mathrm{i}y}\,dy$$
$$=\frac{1}{c}\int_0^{1/2}\Big\{\int_1^\infty\frac{(\log x)^2F(x)}{x^{2+\varepsilon+u}}\Big[\int_{-2\lambda}^{2\lambda}\Big(1-\frac{|y|}{2\lambda}\Big)(\eta x)^{-\mathrm{i}y}\,dy\Big]\,dx\Big\}du$$
$$=\frac{1}{c}\int_0^{1/2}\Big\{\int_1^\infty\frac{(\log x)^2F(x)}{x^{2+\varepsilon+u}}\Big[\frac{2\sin^2\big[\lambda\log(\eta x)\big]}{\lambda\big[\log(\eta x)\big]^2}\Big]\,dx\Big\}du$$
$$=\int_1^\infty\frac{(\log x)^2F(x)}{x^{1+\varepsilon}}\,\frac{2\sin^2\big[\lambda\log(\eta x)\big]}{\lambda\big[\log(\eta x)\big]^2}\Big\{\frac{1}{cx}\int_0^{1/2}\frac{du}{x^u}\Big\}dx$$
$$=2\int_1^\infty\frac{(\log x)^2F(x)v(x)\,\sin^2[\lambda\log(\eta x)]}{x^{1+\varepsilon}\lambda\,[\log(\eta x)]^2}\,dx\,.$$

Our next task is to consider the integral

$$I_\lambda(\eta)=\int_1^\infty\frac{I(x)\,\sin^2[\lambda\log(\eta x)]}{\lambda x[\log(\eta x)]^2}\,dx$$

and to prove that

$$\lim_{\eta\to 0+}I_\lambda(\eta)=\pi. \tag{††}$$

First we must verify that $I_\lambda(\eta)$ is finite. Obviously, the integrand is non-negative and measurable. For every $\varepsilon\in\mathbb{R}_{>0}$ we obtain

$$\int_1^\infty\frac{I(x)\sin^2[\lambda\log(\eta x)]}{\lambda x^{1+\varepsilon}[\log(\eta x)]^2}dx=\int_1^\infty\frac{J(x)\sin^2[\lambda\log(\eta x)]}{\lambda x^{1+\varepsilon}[\log(\eta x)]^2}dx+\frac{1}{2}\Phi_{\varepsilon,\lambda}(\eta),$$

and since J is bounded, it follows that

$$\int_1^\infty\frac{J(x)\sin^2[\lambda\log(\eta x)]}{\lambda x^{1+\varepsilon}[\log(\eta x)]^2}dx\ll\int_1^\infty\frac{\sin^2[\lambda\log(\eta x)]}{\lambda x[\log(\eta x)]^2}dx$$
$$\underset{[\,t=\lambda\log(\eta x)\,]}{=}\int_{\lambda\log\eta}^\infty\frac{\sin^2 t}{t^2}\ll 1.$$

Now we apply monotone convergence to obtain

$$\begin{aligned} I_\lambda(\eta) &= \int_1^\infty \frac{I(x)\,\sin^2[\lambda\log(\eta x)]}{\lambda x[\log(\eta x)]^2}\,dx = \lim_{\varepsilon\to0+}\int_1^\infty \frac{I(x)\,\sin^2[\lambda\log(\eta x)]}{\lambda x^{1+\varepsilon}[\log(\eta x)]^2}\,dx \\ &= \lim_{\varepsilon\to0+}\int_1^\infty \frac{J(x)\,\sin^2[\lambda\log(\eta x)]}{\lambda x^{1+\varepsilon}[\log(\eta x)]^2}\,dx \;+\; \lim_{\varepsilon\to0+}\frac{1}{2}\,\Phi_{\varepsilon,\lambda}(\eta) \\ &= \int_1^\infty \frac{J(x)\,\sin^2[\lambda\log(\eta x)]}{\lambda x[\log(\eta x)]^2}\,dx \;+\; \frac{1}{2}\int_{-2\lambda}^{2\lambda} h(1+\mathrm{i}y)\Big(1-\frac{|y|}{2\lambda}\Big)\eta^{-iy}\,dy\,. \end{aligned}$$

We evaluate these two integrals for $\eta\to0+$.

As to the first integral, we use the substitution $t=\lambda\log(\eta x)$ and obtain

$$\int_1^\infty \frac{J(x)\,\sin^2[\lambda\log(\eta x)]}{\lambda x[\log(\eta x)]^2}\,dx = \int_{-\infty}^\infty J_\eta(t)\,\frac{\sin^2 t}{t^2}\,dt,$$

where

$$J_\eta(t)=\begin{cases} J(\eta^{-1}\mathrm{e}^{t/\lambda}) & \text{if } t\ge\lambda\log\eta, \\ 0 & \text{if } t<\lambda\log\eta, \end{cases} \quad \text{and} \quad \lim_{\eta\to0+}J_\eta(t)=\lim_{x\to\infty}J(x)=1.$$

Together with J the functions J_η are bounded, and dominated convergence yields

$$\begin{aligned} \lim_{\eta\to0+}\int_1^\infty J(x)\,\frac{\sin^2[\lambda\log(\eta x)]}{\lambda x[\log(\eta x)]^2}\,dx &= \lim_{\eta\to0+}\int_{-\infty}^\infty J_\eta(t)\,\frac{\sin^2 t}{t^2}\,dt \\ &= \int_{-\infty}^\infty \frac{\sin^2 t}{t^2}\,dt=\pi. \end{aligned}$$

To evaluate the second integral, we set

$$\int_{-2\lambda}^{2\lambda} h(1+\mathrm{i}y)\Big(1-\frac{|y|}{2\lambda}\Big)\eta^{-\mathrm{i}y}\,dy = \int_{-\infty}^\infty g_\lambda(y)\mathrm{e}^{-\mathrm{i}y\log\eta}\,dy,$$

where

$$g_\lambda(y)=\begin{cases} h(1+\mathrm{i}y)\big(1-\frac{|y|}{2\lambda}\big) & \text{if } 0\le|y|\le2\lambda, \\ 0 & \text{if } |y|>2\lambda. \end{cases}$$

Then the Riemann-Lebesgue lemma implies

$$\lim_{\eta\to0+}\int_{-2\lambda}^{2\lambda} h(1+\mathrm{i}y)\Big(1-\frac{|y|}{2\lambda}\Big)\eta^{-\mathrm{i}y}\,dy = \lim_{t\to\infty}\int_{-\infty}^\infty g_\lambda(y)\mathrm{e}^{\mathrm{i}yt}\,dy=0,$$

and overall we obtain $\lim_{\eta\to0} I_\lambda(\eta)=\pi$. □[††]

Now let $\lambda,\,\theta,\,\eta\in\mathbb{R}_{>0}$ be such that $0<2\lambda<L$ and $\eta^{-1}\mathrm{e}^{-\theta}\ge x_0$. If $0<a<b$, then

$$\int_a^b \frac{\sin^2[\lambda\log(\eta x)]}{\lambda x[\log(\eta x)]^2}\,dx \underset{[t=\lambda\log(\eta x)]}{=} \int_{\lambda\log(\eta a)}^{\lambda\log(\eta b)} \frac{\sin^2 t}{t^2}\,dt.$$

For $x \in [\eta^{-1}\mathrm{e}^{-\theta}, \eta^{-1}\mathrm{e}^{\theta}]$, the monotonicity of A and v imply the inequalities

$$I(x) \geq [\log(\eta^{-1}\mathrm{e}^{-\theta})]^2 A(\eta^{-1}\mathrm{e}^{-\theta})\, v(\eta^{-1}\mathrm{e}^{\theta}) = I(\eta^{-1}\mathrm{e}^{-\theta})\, \frac{v(\eta^{-1}\mathrm{e}^{\theta})}{v(\eta^{-1}\mathrm{e}^{-\theta})},$$
$$I(x) \leq [\log(\eta^{-1}\mathrm{e}^{\theta})]^2 A(\eta^{-1}\mathrm{e}^{\theta})\, v(\eta^{-1}\mathrm{e}^{-\theta}) = I(\eta^{-1}\mathrm{e}^{\theta})\, \frac{v(\eta^{-1}\mathrm{e}^{-\theta})}{v(\eta^{-1}\mathrm{e}^{\theta})},$$

and consequently

$$I_\lambda(\eta) \geq \int_{\eta^{-1}\mathrm{e}^{-\theta}}^{\eta^{-1}\mathrm{e}^{\theta}} \frac{I(x)\,\sin^2[\lambda\log(\eta x)]}{\lambda x[\log(\eta x)]^2}\,dx \;\geq\; I(\eta^{-1}\mathrm{e}^{-\theta})\,\frac{v(\eta^{-1}\mathrm{e}^{\theta})}{v(\eta^{-1}\mathrm{e}^{-\theta})}\int_{-\lambda\theta}^{\lambda\theta}\frac{\sin^2 t}{t^2}\,dt \tag{$'$}$$

and

$$\int_{\eta^{-1}\mathrm{e}^{-\theta}}^{\eta^{-1}\mathrm{e}^{\theta}} \frac{I(x)\,\sin^2[\lambda\log(\eta x)]}{\lambda x[\log(\eta x)]^2}\,dx \;\leq\; I(\eta^{-1}\mathrm{e}^{\theta})\,\frac{v(\eta^{-1}\mathrm{e}^{-\theta})}{v(\eta^{-1}\mathrm{e}^{\theta})}\int_{-\lambda\theta}^{\lambda\theta}\frac{\sin^2 t}{t^2}\,dt. \tag{$''$}$$

By ($'$) we obtain

$$I(\eta^{-1}\mathrm{e}^{-\theta}) \leq \frac{v(\eta^{-1}\mathrm{e}^{-\theta})}{v(\eta^{-1}\mathrm{e}^{\theta})}\Big(\int_{-\lambda\theta}^{\lambda\theta}\frac{\sin^2 t}{t^2}\,dt\Big)^{-1} I_\lambda(\eta).$$

Since

$$\lim_{\eta\to 0}\frac{v(\eta^{-1}\mathrm{e}^{-\theta})}{v(\eta^{-1}\mathrm{e}^{\theta})} = \mathrm{e}^{2\theta} \quad\text{and}\quad \lim_{\eta\to 0} I_\lambda(\eta) = \pi,$$

it follows that

$$\limsup_{x\to\infty} I(x) = \limsup_{\eta\to 0} I(\eta^{-1}\mathrm{e}^{-\theta}) \leq \mathrm{e}^{2\theta}\pi\Big(\int_{-\lambda\theta}^{\lambda\theta}\frac{\sin^2 t}{t^2}\,dt\Big)^{-1},$$

and therefore

$$\limsup_{x\to\infty} I(x) \leq \lim_{\theta\to 0}\Big[\lim_{\lambda\to\infty}\Big\{\mathrm{e}^{2\theta}\pi\Big(\int_{-\lambda\theta}^{\lambda\theta}\frac{\sin^2 t}{t^2}\,dt\Big)^{-1}\Big\}\Big] = \lim_{\theta\to 0}\mathrm{e}^{2\theta} = 1.$$

In particular, $M = \sup\{I(x) \mid x \in \mathbb{R}_{\geq 1}\} < \infty$, and ($''$) implies

$$I(\eta^{-1}\mathrm{e}^{\theta}) \geq \frac{v(\eta^{-1}\mathrm{e}^{\theta})}{v(\eta^{-1}\mathrm{e}^{-\theta})}\Big(\int_{-\lambda\theta}^{\lambda\theta}\frac{\sin^2 t}{t^2}\,dt\Big)^{-1}\big[I_\lambda(\eta) - T\big],$$

where the bound for $I(x)$ and again the substitution $t = \lambda\log(\eta x)$ show that

$$T = \Big\{\int_1^{\eta^{-1}\mathrm{e}^{-\theta}} + \int_{\eta^{-1}\mathrm{e}^{\theta}}^{\infty}\Big\}\frac{I(x)\,\sin^2[\lambda\log(\eta x)]}{\lambda x[\log(\eta x)]^2}\,dx \leq M\Big\{\int_{\lambda\log\eta}^{-\lambda\theta} + \int_{\lambda\theta}^{\infty}\Big\}\frac{\sin^2 t}{t^2}\,dt$$
$$= M\Big\{\int_{\lambda\theta}^{-\lambda\log\eta} + \int_{\lambda\theta}^{\infty}\Big\}\frac{\sin^2 t}{t^2}\,dt \leq 2M\int_{\lambda\theta}^{\infty}\frac{\sin^2 t}{t^2}\,dt.$$

It follows that

$$\liminf_{x\to\infty} I(x) = \liminf_{\eta\to 0} I(\eta^{-1}\mathrm{e}^{-\theta}) \ge \mathrm{e}^{2\theta}\Big(\int_{-\lambda\theta}^{\lambda\theta}\frac{\sin^2 t}{t^2}\,dt\Big)^{-1}\Big[\pi - 2M\int_{\lambda\theta}^{\infty}\frac{\sin^2 t}{t^2}\,dt\Big],$$

and therefore

$$\begin{aligned}\liminf_{x\to\infty} I(x) &\ge \lim_{\theta\to 0}\Big[\lim_{\lambda\to\infty}\Big\{\mathrm{e}^{2\theta}\Big(\int_{-\lambda\theta}^{\lambda\theta}\frac{\sin^2 t}{t^2}\,dt\Big)^{-1}\Big[\pi - 2M\int_{\lambda\theta}^{\infty}\frac{\sin^2 t}{t^2}\,dt\Big]\Big\}\Big]\\ &= \lim_{\theta\to 0}\mathrm{e}^{2\theta} = 1.\end{aligned}$$

Eventually

$$\limsup_{x\to\infty} I(x) \le 1 \quad\text{and}\quad \liminf_{x\to\infty} I(x) \ge 1 \quad\text{implies}\quad \lim_{x\to\infty} I(x) = 1. \qquad \square$$

Theorem 4.3.10 (Prime ideal theorem). *Let K be an algebraic number field and T a regular subset of $\mathcal{P}_K$ with density $d(T) = c \in [0,1]$. Then*

$$\pi_T(x) = |\{\mathfrak{p}\in T \mid \mathfrak{N}(\mathfrak{p}) \le x\}| = (c+o(1))\frac{x}{\log x} \quad \textit{for}\ \ x\to\infty.$$

In particular,

$$\pi_K(x) = |\{\mathfrak{p}\in \mathcal{P}_K \mid \mathfrak{N}(\mathfrak{p}) \le x\}| \sim \frac{x}{\log x} \quad \textit{for}\ \ x\to\infty.$$

Proof. For $s\in\mathcal{H}_1$, our assumption reads

$$\sum_{\mathfrak{p}\in T}\frac{1}{\mathfrak{N}(\mathfrak{p})^s} = \sum_{n=1}^{\infty}\frac{a_n}{n^s} \sim c\log\frac{1}{s-1}, \quad \text{where}\quad a_n = |\{\mathfrak{p}\in T \mid \mathfrak{N}(\mathfrak{p}) = n\}|.$$

If $c>0$, then 4.3.8 implies

$$\pi_T(x) = \sum_{1\le n\le x} a_n = (c+o(1))\frac{x}{\log x} \quad \text{for}\ \ x\to\infty.$$

Thus suppose that $c=0$. For every $n\in\mathbb{N}$, 4.3.5.2 implies that there exists a subset T_n of $\mathcal{P}_K$ such that $d(T_n) = 1/n$, hence

$$d(T) \le d(T\cup T_n) = \frac{1}{n} \quad\text{and thus}\quad \pi_T(x) \le \Big(\frac{1}{n}+o(1)\Big)\frac{x}{\log x} \quad \text{for}\ \ x\to\infty.$$

As $n\to\infty$, we obtain

$$\pi_T(x) = o(1)\frac{x}{\log x} \quad \text{for}\ \ x\to\infty. \qquad \square$$

In the subsequent Theorem 4.3.11 we emphasize that the classical results of rational number theory, specifically the prime number theorem and Dirichlet's theorem on primes in arithmetic progressions, are already covered by our investigations.

Theorem 4.3.11.

1. (Prime number theorem)

$$\pi(x) = |\{p \in \mathbb{P} \mid p \leq x\}| \sim \frac{x}{\log x} \quad \textit{for } x \to \infty.$$

2. (Dirichlet's theorem on primes in arithmetic progressions) *Let* $k,\ m \in \mathbb{N}$ *and* $(m,k) = 1$. *Then the set*

$$\mathbb{P} \cap (k + m\mathbb{Z}) = \{p \in \mathbb{P} \mid p \equiv k \mod m\}$$

has Dirichlet density $\varphi(m)^{-1}$, *and*

$$|\{p \in \mathbb{P} \mid p \leq x,\ p \equiv k \mod m\}| \sim \frac{1}{\varphi(m)} \frac{x}{\log x} \quad \textit{for } x \to \infty.$$

Proof. It suffices to prove 2. If p is prime such that $p \nmid m$, then 3.3.5.2, implies

$$\sum_{\chi \in \mathsf{X}(m)} \chi(p)\overline{\chi}(k) = \begin{cases} \varphi(m) & \text{if } p \equiv k \mod m, \\ 0 & \text{if } p \not\equiv k \mod m. \end{cases}$$

For $s \in \mathcal{H}_1$ we obtain, using 4.2.7.1 and 4.2.8,

$$\sum_{\substack{p \in \mathbb{P} \\ p \equiv k \bmod m}} \frac{1}{p^s} = \frac{1}{\varphi(m)} \sum_{p \in \mathbb{P}} \frac{1}{p^s} \sum_{\chi \in \mathsf{X}(m)} \chi(p)\overline{\chi}(k) = \frac{1}{\varphi(m)} \sum_{\chi \in \mathsf{X}(m)} \overline{\chi}(k) \sum_{p \in \mathbb{P}} \frac{\chi(p)}{p^s}$$

$$\sim \frac{1}{\varphi(m)} \sum_{\chi \in \mathsf{X}(m)} \overline{\chi}(k) \log L(s,\chi) \sim \frac{1}{\varphi(m)} \log L(s, 1_m) \sim \frac{1}{\varphi(m)} \log \frac{1}{s-1}.$$

Hence the set $\mathbb{P} \cap (k + m\mathbb{Z})$ has Dirichlet density $\varphi(m)^{-1}$ by 4.3.4, and the asymptotic behavior follows by 4.3.8. □

4.4 Density results using class field theory

In this section, we continue to use all assumptions of Section 4.2 *and we use the (hitherto unproved) results of* Section 3.7.

In the early foundations of class field theory, analytic methods played a central role (see e.g. [25], [26] [33] or [40]). In the forthcoming volume on class field theory we shall dispense with analytic methods for its foundation. Nevertheless, we now present the analytic proof of the norm index inequality, usually called the *first inequality* in class field theory. The reason is twofold. On the one hand, this proof together with the existence of ray class fields shows the non-vanishing of Dirichlet L series, and on the other hand it gives a good impression of what analytic methods could do for class field theory.

Theorem 4.4.1 (Norm index inequality). *Let L/K be a finite Galois extension of algebraic number fields and $\mathfrak{m} \in \mathcal{I}'_K$ such that* $\mathrm{Ram}(L/K) \subset \mathrm{supp}(\mathfrak{m})$. *As in 3.7.2 we set $\mathcal{N}^{\mathfrak{m}}_{L/K} = \mathcal{N}_{L/K}(\mathcal{I}_L^{\mathfrak{m}o_L})\mathcal{H}_K^{\mathfrak{m}} \subset \mathcal{I}_K^{\mathfrak{m}}$ and $\mathcal{G} = \mathcal{N}^{\mathfrak{m}}_{L/K}/\mathcal{H}_K^{\mathfrak{m}} \subset \mathcal{C}_K^{\mathfrak{m}}$.*

1. $(\mathcal{I}_K^{\mathfrak{m}} : \mathcal{N}^{\mathfrak{m}}_{L/K}) \le [L:K]$, and if $\chi \in \mathsf{X}(\mathcal{C}_K^{\mathfrak{m}}) \setminus \{1_{\mathfrak{m}}\}$ is a ray class character such that $\chi \restriction \mathcal{N}^{\mathfrak{m}}_{L/K} = 1$, then $L(1,\chi) \ne 0$.

2. Let $L = K[\mathfrak{m}]$ be a ray class field modulo $\mathfrak{m}$. Then $L(1,\chi) \ne 0$ (and thus $m(\chi) = 0$) for all $\chi \in \mathsf{X}(\mathcal{C}_K^{\mathfrak{m}}) \setminus \{1_{\mathfrak{m}}\}$.

Proof. 1. Let $h_{\mathfrak{m}} = |\mathcal{C}_K^{\mathfrak{m}}/\mathcal{G}| = (\mathcal{I}_K^{\mathfrak{m}} : \mathcal{N}^{\mathfrak{m}}_{L/K})$, and observe that, for all $C \in \mathcal{C}_K^{\mathfrak{m}}$, 3.3.1.2 implies

$$\sum_{\substack{\chi \in \mathsf{X}(\mathcal{C}_K^{\mathfrak{m}}) \\ \chi \restriction \mathcal{G} = 1}} \chi(C) = \sum_{\psi \in \mathsf{X}(\mathcal{C}_K^{\mathfrak{m}}/\mathcal{G})} \psi(C\mathcal{G}) = \begin{cases} h_{\mathfrak{m}} & \text{if } C \in \mathcal{G}, \\ 0 & \text{otherwise.} \end{cases}$$

For $\chi \in \mathsf{X}(\mathcal{C}_K^{\mathfrak{m}})$ and $s \in \mathcal{H}_1$, 4.2.4.3 and 4.2.3.2 imply

$$\log L(s,\chi) \sim -m(\chi) \log \frac{1}{s-1} \sim \sum_{\mathfrak{p} \in \mathcal{P}_K} \frac{\chi(\mathfrak{p})}{\mathfrak{N}(\mathfrak{p})^s} = \sum_{C \in \mathcal{C}_K^{\mathfrak{m}}} \chi(C) \sum_{\mathfrak{p} \in C \cap \mathcal{P}_K} \frac{1}{\mathfrak{N}(\mathfrak{p})^s},$$

where $m(\chi)$ is the order of the meromorphic function $L(\,\cdot\,,\chi)$ in 1. Hence (again by 3.3.1.2)

$$\sum_{\mathfrak{p} \in \mathcal{P}_K \cap \mathcal{N}^{\mathfrak{m}}_{L/K}} \frac{1}{\mathfrak{N}(\mathfrak{p})^s} = \sum_{C \in \mathcal{G}} \sum_{\mathfrak{p} \in C \cap \mathcal{P}_K} \frac{1}{\mathfrak{N}(\mathfrak{p})^s} = \sum_{C \in \mathcal{C}_K^{\mathfrak{m}}} \frac{1}{h_{\mathfrak{m}}} \sum_{\substack{\chi \in \mathsf{X}(\mathcal{C}_K^{\mathfrak{m}}) \\ \chi \restriction \mathcal{G} = 1}} \chi(C) \sum_{\mathfrak{p} \in C \cap \mathcal{P}_K} \frac{1}{\mathfrak{N}(\mathfrak{p})^s}$$

$$\sim \frac{1}{h_{\mathfrak{m}}} \sum_{\substack{\chi \in \mathsf{X}(\mathcal{C}_K^{\mathfrak{m}}) \\ \chi \restriction \mathcal{G} = 1}} -m(\chi) \log \frac{1}{s-1} = \frac{1}{h_{\mathfrak{m}}} \Big[1 - \sum_{\substack{\chi \in \mathsf{X}(\mathcal{C}_K^{\mathfrak{m}}) \setminus \{1_{\mathfrak{m}}\} \\ \chi \restriction \mathcal{G} = 1}} m(\chi) \Big] \log \frac{1}{s-1}.$$

If $\mathfrak{p} \in \mathcal{P}_K \cap \mathcal{N}_{L/K}(\mathcal{I}_L^{\mathfrak{m}o_L})$, $\mathfrak{P} \in \mathcal{P}_L$ and $\mathfrak{P} \,|\, \mathfrak{p}$, then $\mathfrak{p} = \mathcal{N}_{L/K}(\mathfrak{P})$, hence $f(\mathfrak{P}/\mathfrak{p}) = 1$ and therefore $\mathfrak{p} \in \mathrm{Split}(L/K)$. Conversely, if $\mathfrak{p} \in \mathrm{Split}(L/K) \cap \mathcal{I}_K^{\mathfrak{m}}$, then apparently $\mathfrak{p} \in \mathcal{P}_K \cap \mathcal{N}_{L/K}(\mathcal{I}_L^{\mathfrak{m}o_L})$. It follows that

$$\mathcal{P}_K \cap \mathcal{N}^{\mathfrak{m}}_{L/K} \supset \mathcal{P}_K \cap \mathcal{N}_{L/K}(\mathcal{I}_L^{\mathfrak{m}o_L}) = \mathrm{Split}(L/K) \cap \mathcal{I}_K^{\mathfrak{m}} \doteq \mathrm{Split}(L/K)$$

and

$$\frac{1}{h_{\mathfrak{m}}} \Big[1 - \sum_{\substack{\chi \in \mathsf{X}(\mathcal{C}_K^{\mathfrak{m}}) \setminus \{1_{\mathfrak{m}}\} \\ \chi \restriction \mathcal{G} = 1}} m(\chi) \Big] \log \frac{1}{s-1} \sim \sum_{\mathfrak{p} \in \mathrm{Split}(L/K) \cap \mathcal{I}_K^{\mathfrak{m}}} \frac{1}{\mathfrak{N}(\mathfrak{p})^s} + \Phi(s),$$

where

$$\Phi(s) = \sum_{\mathfrak{p} \in \mathcal{P}_K \cap \mathcal{N}^{\mathfrak{m}}_{L/K} \setminus \mathrm{Split}(L/K)} \frac{1}{\mathfrak{N}(\mathfrak{p})^s}$$

and

$$\sum_{\mathfrak{p}\in\mathrm{Split}(L/K)\cap\mathcal{J}_K^{\mathfrak{m}}} \frac{1}{\mathfrak{N}(\mathfrak{p})^s} \sim \sum_{\mathfrak{p}\in\mathrm{Split}(L/K)} \frac{1}{\mathfrak{N}(\mathfrak{p})^s} \sim \frac{1}{[L:K]}\log\frac{1}{s-1} \quad \text{by 4.3.5.3.}$$

Overall we obtain

$$\Big[1 - \sum_{\substack{\chi\in\mathsf{X}(\mathcal{C}_K^{\mathfrak{m}})\setminus\{1_{\mathfrak{m}}\}\\ \chi\restriction\mathcal{G}=1}} m(\chi) - \frac{h_{\mathfrak{m}}}{[L:K]}\Big]\log\frac{1}{s-1} = h_{\mathfrak{m}}\Phi(s) + G(s)$$

for some function $G \in \mathcal{O}(\mathcal{H}_1 \cup \{1\})$, and consequently

$$\Big[1 - \sum_{\substack{\chi\in\mathsf{X}(\mathcal{C}_K^{\mathfrak{m}})\setminus\{1_{\mathfrak{m}}\}\\ \chi\restriction\mathcal{G}=1}} m(\chi) - \frac{h_{\mathfrak{m}}}{[L:K]}\Big] = \lim_{\sigma\to1+}\frac{h_{\mathfrak{m}}\Phi(\sigma)+G(\sigma)}{\log\frac{1}{\sigma-1}}$$

$$= \lim_{\sigma\to1+}\frac{h_{\mathfrak{m}}\Phi(\sigma)}{\log\frac{1}{\sigma-1}} \ge 0.$$

From this we infer $h_{\mathfrak{m}} \le [L:K]$. Moreover, if $\chi \in \mathsf{X}(\mathcal{C}_K^{\mathfrak{m}}) \setminus \{1_{\mathfrak{m}}\}$ and $\chi\restriction\mathcal{G} = 1$, then $m(\chi) = 0$ and thus $L(1,\chi) \ne 0$.

Remark. The Artin reciprocity law for L/K (see 3.7.2) implies that actually $\Phi = 0$ and $h_{\mathfrak{m}} = [L:K]$.

2. If $L = K[\mathfrak{m}]$, then $\mathcal{N}_{L/K}^{\mathfrak{m}} = \mathcal{H}_K^{\mathfrak{m}}$ by 3.7.2, and thus the assertion follows from 1. □

For the following results we need the definitions of the various types of ray class groups given in 3.6.6.

Theorem 4.4.2 (Dirichlet's theorem for ray classes). *Let K be an algebraic number field. Let $\mathfrak{m} \in \mathcal{J}_K'$, and suppose that $L(1,\chi) \ne 0$ for all ray class characters $\chi \in \mathsf{X}(\mathcal{C}_K^{\mathfrak{m}}) \setminus \{1_{\mathfrak{m}}\}$.*

1. If $C \in \mathcal{C}_K^{\mathfrak{m}}$ is a ray class modulo $\mathfrak{m}$, then the set $C \cap \mathcal{P}_K$ is regular with Dirichlet density

$$d(C\cap\mathcal{P}_K) = \frac{1}{|\mathcal{C}_K^{\mathfrak{m}}|}, \quad \textit{and} \quad |\{\mathfrak{p}\in C\cap\mathcal{P}_K \mid \mathfrak{N}(\mathfrak{p})\le x\}| \sim \frac{1}{|\mathcal{C}_K^{\mathfrak{m}}|}\,\frac{x}{\log x}.$$

2. If $C \in \mathcal{C}_K^{\circ\mathfrak{m}}$ is a small ray class modulo $\mathfrak{m}$, then the set $C \cap \mathcal{P}_K$ is regular with Dirichlet density

$$d(C\cap\mathcal{P}_K) = \frac{1}{|\mathcal{C}_K^{\circ\mathfrak{m}}|}, \quad \textit{and} \quad |\{\mathfrak{p}\in C\cap\mathcal{P}_K \mid \mathfrak{N}(\mathfrak{p})\le x\}| \sim \frac{1}{|\mathcal{C}_K^{\circ\mathfrak{m}}|}\,\frac{x}{\log x}.$$

Proof. 1. We mimic the proof of 4.3.11.2. Let $C \in \mathcal{C}_K^{\mathfrak{m}}$ and $s \in \mathcal{H}_1$. Then 3.3.1.5, 4.2.3.2 and 4.2.4.3 imply

$$\sum_{\mathfrak{p} \in C \cap \mathcal{P}_K} \frac{1}{\mathfrak{N}(\mathfrak{p})^s} = \frac{1}{|\mathcal{C}_K^{\mathfrak{m}}|} \sum_{\mathfrak{p} \in \mathcal{P}_K} \sum_{\chi \in \mathsf{X}(\mathcal{C}_K^{\mathfrak{m}})} \chi(\mathfrak{p})\overline{\chi}(C) \sim \frac{1}{|\mathcal{C}_K^{\mathfrak{m}}|} \sum_{\chi \in \mathsf{X}(\mathcal{C}_K^{\mathfrak{m}})} \overline{\chi}(C) \log L(s,\chi)$$
$$\sim \frac{1}{|\mathcal{C}_K^{\mathfrak{m}}|} \sum_{\chi \in \mathsf{X}(\mathcal{C}_K^{\mathfrak{m}})} \overline{\chi}(C) \Big[-m(\chi) \log \frac{1}{s-1}\Big] = \frac{1}{|\mathcal{C}_K^{\mathfrak{m}}|} \log \frac{1}{s-1},$$

since $m(1_{\mathfrak{m}}) = -1$ and $m(\chi) = 0$ for all $\chi \in \mathsf{X}(\mathcal{C}_K^{\mathfrak{m}}) \setminus \{1_{\mathfrak{m}}$. Now the assertions follow by 4.3.4 and 4.3.10.

2. Every class $C \in \mathcal{C}_K^{\circ\mathfrak{m}}$ is the (disjoint) union of $|\mathcal{C}_K^{\mathfrak{m}}|/|\mathcal{C}_K^{\circ\mathfrak{m}}|$ classes from $\mathcal{C}_K^{\mathfrak{m}}$. □

Corollary 4.4.3 (Dirichlet's theorem for ideal classes). *Let K be an algebraic number field, and suppose that $L(1,\chi) \neq 0$ for all $\chi \in \mathsf{X}(\mathcal{C}_K^+) \setminus \{1\}$.*

1. *If $C \in \mathcal{C}_K^+$ is a narrow ideal class, then the set $C \cap \mathcal{P}_K$ is regular with Dirichlet density*

$$d(C \cap \mathcal{P}_K) = \frac{1}{h_K^+}, \quad \text{and} \quad |\{\mathfrak{p} \in C \cap \mathcal{P}_K \mid \mathfrak{N}(\mathfrak{p}) \le x\}| \sim \frac{1}{h_K^+} \frac{x}{\log x}.$$

2. *If $C \in \mathcal{C}_K$ is an ordinary ideal class, then the set $C \cap \mathcal{P}_K$ is regular with Dirichlet density*

$$d(C \cap \mathcal{P}_K) = \frac{1}{h_K}, \quad \text{and} \quad |\{\mathfrak{p} \in C \cap \mathcal{P}_K \mid \mathfrak{N}(\mathfrak{p}) \le x\}| \sim \frac{1}{h_K} \frac{x}{\log x}.$$

Proof. Obvious by 4.4.2 with $\mathfrak{m} = \mathbf{1}$. □

The following Corollary 4.4.4 presents the consequences of Dirichlet's theorem 4.4.3 for the analytic theory of orders in algebraic number fields. We use the notation and results of Section 2.11.

Corollary 4.4.4 (Prime ideal theorem for orders). *Let $\mathfrak{o}$ be an order in an algebraic number field K and $C \in \mathcal{C}(\mathfrak{o})$ an ideal class of $\mathfrak{o}$. Then*

$$|\{\mathfrak{p} \in C \cap \mathcal{P}(\mathfrak{o}) \mid (\mathfrak{o}:\mathfrak{p}) \le x\}| \sim \frac{1}{|\mathcal{C}(\mathfrak{o})|} \frac{x}{\log x} \quad \text{for} \quad x \to \infty,$$

Proof. Let $\mathfrak{f} = (\mathfrak{o}:\mathfrak{o}_K)$. By 2.11.6 there is an epimorphism

$$\rho\colon \mathcal{C}_K^{\circ\mathfrak{f}} \to \mathcal{C}(\mathfrak{o}), \quad \text{given by} \quad \rho([\overline{\mathfrak{a}}]^{\circ\mathfrak{f}}) = [\overline{\mathfrak{a}} \cap \mathfrak{o}] \ \text{for all} \ \overline{\mathfrak{a}} \in \mathcal{J}_K^{\mathfrak{f}},$$

and therefore

$$|\mathcal{C}(\mathfrak{o})| = \frac{|\mathcal{C}_K^{\circ\mathfrak{f}}|}{|\mathrm{Ker}(\rho)|} = \frac{|\mathcal{C}_K^{\circ\mathfrak{f}}|}{|\rho^{-1}(C)|} \quad \text{for every class} \ C \in \mathcal{C}(\mathfrak{o}).$$

From 2.11.3.1 and 2.11.4.2 it follows that the assignment $\overline{\mathfrak{p}} \mapsto \overline{\mathfrak{p}} \cap \mathfrak{o}$ defines a bijective map $\mathcal{I}_K^{\mathsf{f}} \to \mathcal{I}^{\mathsf{f}}(\mathfrak{o})$ satisfying $\mathfrak{N}(\overline{\mathfrak{p}}) = (\mathfrak{o}_K : \overline{\mathfrak{p}}) = (\mathfrak{o} : \overline{\mathfrak{p}} \cap \mathfrak{o})$ for all $\overline{\mathfrak{p}} \in \mathcal{P}_K^{\mathsf{f}}$. Therefore we obtain, for every class $C \in \mathcal{C}(\mathfrak{o})$,

$$\begin{aligned} |\{\mathfrak{p} \in C \cap \mathcal{P}^{\mathsf{f}}(\mathfrak{o}) \mid (\mathfrak{o}:\mathfrak{p}) \le x\}| &= \sum_{\overline{C} \in \rho^{-1}(C)} |\{\overline{\mathfrak{p}} \in \overline{C} \cap \mathcal{P}_K^{\mathsf{f}} \mid \mathfrak{N}(\overline{\mathfrak{p}}) \le x\}| \\ &\sim \sum_{\overline{C} \in \rho^{-1}(C)} |\{\overline{\mathfrak{p}} \in \overline{C} \cap \mathcal{P}_K \mid \mathfrak{N}(\overline{\mathfrak{p}}) \le x\}| \\ &\sim \frac{|\rho^{-1}(C)|}{|\mathcal{C}_K^{\mathrm{of}}|} \frac{x}{\log x} = \frac{1}{|\mathcal{C}(\mathfrak{o})|} \frac{x}{\log x} \quad \text{for} \quad x \to \infty, \end{aligned}$$

and since $C \cap \mathcal{P}(\mathfrak{o}) \doteq C \cap \mathcal{P}^{\mathsf{f}}(\mathfrak{o})$, the assertion follows. □

In 2.1.1 we outlined why the monoid of all non-zero principal ideals of a domain has the same factorization properties as the domain itself, and we announced in 3.1.1 that the class group $\mathcal{C}_K$ determines the monoid $\mathcal{H}_K'$ of non-zero principal ideals of $\mathfrak{o}_K$ up to isomorphisms. Now we have the necessary tools at our disposal to honor these announcements. Note that the group $\mathcal{H}_K$ of fractional principal ideals of $\mathfrak{o}_K$ is the quotient monoid of $\mathcal{H}_K'$.

Theorem 4.4.5 (The class group determines the arithmetic). *Let K and L be algebraic number fields. Then $\mathcal{H}_K' \cong \mathcal{H}_L'$ if and only if $\mathcal{C}_K \cong \mathcal{C}_L$.*

Proof. **I.** Let first $\varphi\colon \mathcal{H}_K' \xrightarrow{\sim} \mathcal{H}_L'$ be a monoid isomorphism. Then φ has a unique extension to an (equally denoted) isomorphism $\varphi\colon \mathcal{H}_K \xrightarrow{\sim} \mathcal{H}_L$ of the quotient groups, and this regards inclusions. Indeed, if $a,\ b \in K^\times$, then:

$$\begin{aligned} a\mathfrak{o}_K \subset b\mathfrak{o}_K &\iff (a\mathfrak{o}_K)(b\mathfrak{o}_K)^{-1} \subset \mathfrak{o}_K \\ &\iff \varphi((a\mathfrak{o}_K)(b\mathfrak{o}_K)^{-1}) = \varphi(a\mathfrak{o}_K)\varphi((b\mathfrak{o}_K)^{-1}) \subset \mathfrak{o}_L \\ &\iff \varphi(a\mathfrak{o}_K) \subset \varphi(b\mathfrak{o}_K). \end{aligned}$$

It suffices to prove that there exists a group isomorphism $\phi\colon \mathcal{I}_K \xrightarrow{\sim} \mathcal{I}_L$ such that $\phi \restriction \mathcal{H}_K = \varphi$, for then ϕ induces an isomorphism $\phi^*\colon \mathcal{C}_K \xrightarrow{\sim} \mathcal{C}_L$.

We define $\phi\colon \mathcal{I}_K \to \mathcal{I}_L$ by $\displaystyle \phi(\mathfrak{a}) = \sum_{a \in \mathfrak{a}} \varphi(a\mathfrak{o}_K) \in \mathcal{I}_L$ for all $\mathfrak{a} \in \mathcal{I}_K$,

and we prove first:

A. If $\mathfrak{a} \in \mathcal{I}_K$ and $\mathfrak{a} = {}_{\mathfrak{o}_K}(X)$ for some subset X of $\mathfrak{a}$, then

$$\phi(\mathfrak{a}) = \sum_{a \in X} \varphi(a\mathfrak{o}_K) = \bigcap_{\substack{c \in K^\times \\ X \subset c\mathfrak{o}_K}} \varphi(c\mathfrak{o}_K),$$

Proof of **A.** Let $\mathfrak{a} = {}_{\mathfrak{o}_K}(X)$, observe that $\varphi\colon \mathcal{H}_K \to \mathcal{H}_L$ is an isomorphism, and set

$$\mathfrak{b} = \sum_{a\in X} \varphi(a\mathfrak{o}_K) = (\mathfrak{b}^{-1})^{-1} = \bigcap_{\substack{d\in L^\times \\ \mathfrak{b}\subset d\mathfrak{o}_L}} d\mathfrak{o}_L$$

(see 2.8.2 and 2.6.1). For $d \in L^\times$ let $c \in K^\times$ be such that $d\mathfrak{o}_L = \varphi(c\mathfrak{o}_K)$. Then

$$\begin{aligned}\mathfrak{b} \subset d\mathfrak{o}_L &\iff \varphi(a\mathfrak{o}_K) \subset \varphi(c\mathfrak{o}_K) \ \text{ for all } \ a \in X \\ &\iff a\mathfrak{o}_K \subset c\mathfrak{o}_K \ \text{ for all } \ a \in X \iff X \subset c\mathfrak{o}_K.\end{aligned}$$

Hence we obtain

$$\mathfrak{b} = \bigcap_{\substack{d\in L^\times \\ \mathfrak{b}\subset d\mathfrak{o}_L}} d\mathfrak{o}_L = \bigcap_{\substack{c\in K^\times \\ X\subset c\mathfrak{o}_K}} \varphi(c\mathfrak{o}_K) = \bigcap_{\substack{c\in K^\times \\ \mathfrak{a}\subset c\mathfrak{o}_K}} \varphi(c\mathfrak{o}_K) = \phi(\mathfrak{a}),$$

independent of X. □[**A.**]

B. If $\mathfrak{a} \in \mathcal{J}_K$ and $a \in K^\times$, then $a\mathfrak{o}_K \subset \mathfrak{a}$ holds if and only if $\varphi(a\mathfrak{o}_K) \subset \phi(\mathfrak{a})$. In particular, ϕ is injective.

Proof of **B.** Obviously, $a\mathfrak{o}_K \subset \mathfrak{a}$ implies $\varphi(a\mathfrak{o}_K) \subset \phi(\mathfrak{a})$. Thus assume that $\mathfrak{a} \in \mathcal{J}_K$, $a \in K^\times$ and

$$\varphi(a\mathfrak{o}_K) \subset \phi(\mathfrak{a}) = \bigcap_{\substack{c\in K^\times \\ \mathfrak{a}\subset c\mathfrak{o}_K}} \varphi(c\mathfrak{o}_K).$$

Then $\varphi(a\mathfrak{o}_K) \subset \varphi(c\mathfrak{o}_K)$ and thus $a\mathfrak{o}_K \subset c\mathfrak{o}_K)$ for all $c \in \mathfrak{o}_K$ such that $\mathfrak{a} \subset c\mathfrak{o}_K$. Hence it follows that

$$\phi(\mathfrak{a}) = \bigcap_{\substack{c\in K^\times \\ \mathfrak{a}\subset c\mathfrak{o}_K}} c\mathfrak{o}_K \supset a\mathfrak{o}_K.$$

Now it follows that ϕ is injective, since

$$\mathfrak{a} = \bigcap_{\substack{a\in K^\times \\ \varphi(a\mathfrak{o}_K)\subset\phi(\mathfrak{a})}} a\mathfrak{o}_K \quad \text{for all } \ \mathfrak{a} \in \mathcal{J}_K.$$ □[**B.**]

If $\mathfrak{b} \in \mathcal{J}_L$, then

$$\mathfrak{a} = \sum_{b\in\mathfrak{b}} \varphi^{-1}(b\mathfrak{o}_L) \in \mathcal{J}_K, \quad \text{and} \quad \phi(\mathfrak{a}) = \mathfrak{b} \ \text{ by } \mathbf{A}.$$

Hence ϕ is bijective. It remains to prove that ϕ is a homomorphism. Thus let $\mathfrak{a}$, $\mathfrak{b} \in \mathcal{K}_K$. Then

$$\begin{aligned}\phi(\mathfrak{a}\mathfrak{b}) &= \sum_{a\in\mathfrak{a}}\sum_{b\in\mathfrak{b}} \varphi(ab\mathfrak{o}_K) = \sum_{a\in\mathfrak{a}}\sum_{b\in\mathfrak{b}} \varphi(a\mathfrak{o}_K)\varphi(b\mathfrak{o}_K) = \Big(\sum_{a\in\mathfrak{a}} \varphi(a\mathfrak{o}_K)\Big)\Big(\sum_{b\in\mathfrak{b}} \varphi(b\mathfrak{o}_K)\Big) \\ &= \phi(\mathfrak{a})\phi(\mathfrak{b}).\end{aligned}$$

II. Let $\theta\colon \mathcal{C}_K \to \mathcal{C}'_K$ be an isomorphism. For every $C \in \mathcal{C}_K$, the sets $C \cap \mathcal{P}_K$ and $\theta(C) \cap \mathcal{P}_L$ are denumerable by 4.4.3.2 and therefore there exists a bijective map $\phi_C\colon C \cap \mathcal{P}_K \to \theta(C) \cap \mathcal{P}_L$. Let $\phi\colon \mathcal{P}_K \to \mathcal{P}_L$ be the unique bijective map satisfying $\phi \restriction C \cap \mathcal{P}_K = \phi_C$ for all $C \in \mathcal{C}_K$. Since $\mathcal{J}'_K$ is the free monoid with basis $\mathcal{P}_K$ there exists a unique monoid isomorphism $\Phi\colon \mathcal{J}'_K \to \mathcal{J}'_L$ such that $\Phi \restriction \mathcal{P}_K = \phi$. Then $[\Phi(\mathfrak{p})] = \theta([\mathfrak{p}])$ for all $\mathfrak{p} \in \mathcal{P}_K$, and we assert that $\Phi(\mathcal{H}'_K) = \mathcal{H}'_L$.

Thus let $\mathfrak{a} = \mathfrak{p}_1 \cdot \ldots \cdot \mathfrak{p}_m \in \mathcal{J}'_K$, where $m \in \mathbb{N}$ and $\mathfrak{p}_1, \ldots, \mathfrak{p}_m \in \mathcal{P}_K$. It follows that $[\mathfrak{a}] = [\mathfrak{p}_1] \cdot \ldots \cdot [\mathfrak{p}_m]$ and therefore

$$\theta([\mathfrak{a}]) = \prod_{j=1}^m \theta([\mathfrak{p}_j]) = \prod_{j=1}^m [\Phi(\mathfrak{p}_j)] = \Big[\prod_{j=1}^m \Phi(\mathfrak{p}_j)\Big] = [\Phi(\mathfrak{a})].$$

Hence $[\mathfrak{a}] = [\mathbf{1}]$ if and only if $[\Phi(\mathfrak{a})] = [\mathbf{1}]$, that is, $\mathfrak{a} \in \mathcal{H}'_K$ if and only if $\Phi(\mathfrak{a}) \in \mathcal{H}'_L$. □

We close this section with a proof of the Čebotarev density theorem using class field theory.

Theorem 4.4.6 (Čebotarev's density theorem). *Let L/K be a finite Galois extension of algebraic number fields, $G = \mathrm{Gal}(L/K)$, $\sigma \in G$ and $\mathcal{P}(L/K, \sigma)$ the set of all $\mathfrak{p} \in \mathcal{P}_K$ which are unramified in L and have there a prime divisor $\mathfrak{P}$ satisfying $(\mathfrak{P}, L/K) = \sigma$. Then the set $\mathcal{P}(L/K, \sigma)$ is regular with Dirichlet density*

$$d(\mathcal{P}(L/K,\sigma)) = \frac{|\Gamma_\sigma|}{[L:K]}, \quad \text{where } \Gamma_\sigma \text{ denotes the conjugacy class of } \sigma \text{ in } G.$$

Proof. We assume first that L/K is cyclic and $G = \langle\sigma\rangle$ (then $\Gamma_\sigma = \{\sigma\}$), and we use 3.7.2. Let $\mathfrak{m} \in \mathcal{J}'_K$ be an ideal and $\mathcal{G} \subset \mathcal{J}_K(\mathfrak{m})$ a subgroup satisfying $\mathcal{H}_K^\mathfrak{m} \subset \mathcal{G}$ such that L is the class field to $\mathcal{G}$. If $\theta_\mathcal{G}\colon \mathcal{J}_K(\mathfrak{m})/\mathcal{G} \to G$ denotes the Artin isomorphism, then $\mathcal{P}(L/K,\sigma) = \theta_\mathcal{G}^{-1}(\sigma) \cap \mathcal{P}_K$, and $\theta_\mathcal{G}^{-1}(\sigma) \in \mathcal{J}_K(\mathfrak{m})/\mathcal{G}$ is the disjoint union of $(\mathcal{G} : \mathcal{H}_K^\mathfrak{m})$ ray classes modulo $\mathfrak{m}$. By 4.4.2, the set $\mathcal{P}(L/K,\sigma)$ is regular with Dirichlet density

$$d(\mathcal{P}(L/K,\sigma)) = \frac{(\mathcal{G}:\mathcal{H}_K^\mathfrak{m})}{|\mathcal{C}_K^\mathfrak{m}|} = \frac{1}{(\mathcal{J}_K(\mathfrak{m}):\mathcal{G})} = \frac{1}{[L:K]}.$$

Now we can do the general case. Let $\sigma \in G$, $D = L^{\langle\sigma\rangle}$ be the fixed field of σ and $f = \mathrm{ord}(\sigma) = [L:D]$. By the special case already done, the subset $\mathcal{P}(L/D,\sigma)$ of $\mathcal{P}_D$ is regular with Dirichlet density $1/f$. Now we consider the sets

$$\mathcal{P}^*(L/K,\sigma) = \{\mathfrak{P} \in \mathcal{P}_L \mid \mathfrak{P} \cap K \in \mathcal{P}(L/K,\sigma) \text{ and } (\mathfrak{P}, L/K) = \sigma\}$$

and

$$\mathcal{P}'(L/D,\sigma) = \{\mathfrak{q} \in \mathcal{P}(L/D,\sigma) \mid \mathfrak{q} \cap K \text{ is fully decomposed in } D\}.$$

If $\mathfrak{q} \in \mathcal{P}(L/D,\sigma)\setminus\mathcal{P}'(L/D,\sigma)$, then $e(\mathfrak{q}/\mathfrak{q}\cap K) \geq 2$ or $f(\mathfrak{q}/\mathfrak{q}\cap K) \geq 2$, and therefore $\mathcal{P}(L/K,\sigma) \doteq \mathcal{P}'(L/K,\sigma)$. Hence $\mathcal{P}'(L/K,\sigma)$ is regular with Dirichlet density $1/f$.

If $\mathfrak{P} \in \mathcal{P}^*(L/K,\sigma)$, then $G_{\mathfrak{P}} = \langle\sigma\rangle$ is the decomposition group and D is the decomposition field of $\mathfrak{P}$ over K. Consequently, if $\mathfrak{q} = \mathfrak{P} \cap D$, then $\mathfrak{q} \in \mathcal{P}(L/D,\sigma)$, $\mathfrak{q}$ is inert in L and $\mathfrak{q}\cap K$ is fully decomposed in D. Hence the assignment $\mathfrak{P} \mapsto \mathfrak{P}\cap D$ defines a bijective map $\mathcal{P}^*(L/K,\sigma) \to \mathcal{P}'(L/D,\sigma)$, and the assignment $\mathfrak{q} \mapsto \mathfrak{q}\cap K$ defines a surjective map $\mathcal{P}'(L/D,\sigma) \to \mathcal{P}(L/K,\sigma)$.

Hence the map $\rho\colon \mathcal{P}^*(L/K,\sigma) \to \mathcal{P}(L/K,\sigma)$, defined by $\rho(\mathfrak{P}) = \mathfrak{P}\cap K$, is surjective, and we consider its fibers. Let $\mathfrak{p} \in \mathcal{P}(L/K,\sigma)$ and $\mathfrak{P}_0 \in \rho^{-1}(\mathfrak{p})$. If $\mathfrak{P} \in \mathcal{P}_L$, then

$$\begin{aligned}\mathfrak{P} \in \rho^{-1}(\mathfrak{p}) &\iff \mathfrak{P}\cap K = \mathfrak{p} \text{ and } (\mathfrak{P}, L/K) = \sigma \\ &\iff \mathfrak{P} = \tau\mathfrak{P}_0 \text{ for some } \tau \in G \text{ and } \sigma = \tau\sigma\tau^{-1}.\end{aligned}$$

Let $\mathsf{Z}(\sigma) = \{\tau \in G \mid \tau\sigma\tau^{-1} = \sigma\}$ be the centralizer of σ. Then we have $|G| = |\Gamma_\sigma|\,|\mathsf{Z}(\sigma)|$, and

$$|\rho^{-1}(\mathfrak{p})| = |\{\tau\mathfrak{P}_0 \mid \tau \in G,\ \sigma = \tau\sigma\tau^{-1}\}| = (\mathsf{Z}(\sigma)\,{:}\,G_{\mathfrak{P}_0}) = (\mathsf{Z}(\sigma)\,{:}\,\langle\sigma\rangle) = \frac{|G|}{|\Gamma_\sigma|f}.$$

Now it follows that, for $s \in \mathcal{H}_1$,

$$\frac{|G|}{|\Gamma_\sigma|f}\sum_{\mathfrak{p}\in\mathcal{P}(L/K,\sigma)}\frac{1}{\mathfrak{N}(\mathfrak{p})^s} = \sum_{\mathfrak{P}\in\mathcal{P}^*(L/K,\sigma)}\frac{1}{\mathfrak{N}(\mathfrak{P})^{fs}} = \sum_{\mathfrak{q}\in\mathcal{P}'(L/D,\sigma)}\frac{1}{\mathfrak{N}(\mathfrak{q})^s} \sim \frac{1}{f}\log\frac{1}{s-1},$$

and therefore

$$\sum_{\mathfrak{p}\in\mathcal{P}(L/K,\sigma)}\frac{1}{\mathfrak{N}(\mathfrak{p})^s} \sim \frac{|\Gamma_\sigma|}{|G|}\log\frac{1}{s-1}, \quad \text{which implies} \quad d(\mathcal{P}(L/K,\sigma) = \frac{|\Gamma_\sigma|}{[L\,{:}\,K]}.$$

□

Using the Frobenius symbol (see 2.14.7.2), we can reformulate Čebotarev's density theorem as follows.

Corollary 4.4.7. *Let L/K be a finite Galois extension of algebraic number fields, $G = \mathrm{Gal}(L/K)$, and Γ a conjugacy class in G. Then the set*

$$\Big\{\mathfrak{p} \in \mathcal{P}_K \;\Big|\; \mathfrak{p} \text{ unramified in } L \text{ and } \Big[\frac{L/K}{\mathfrak{p}}\Big] = C\Big\} \quad \text{has density} \quad \frac{|\Gamma|}{[L\,{:}\,K]}.$$

Proof. Obvious by 4.4.6. □

We proceed with some simple consequences of Theorem 4.4.6. For a proof in a more general context and more subtle applications we refer the reader to the forthcoming volume on class field theory.

For a finite extension L/K of algebraic number fields we define the **Kronecker set** $\mathrm{Kron}(L/K)$ to be the set of all $\mathfrak{p} \in \mathcal{P}_K$ which are unramified in L and have at least one prime divisor $\mathfrak{P} \in \mathcal{P}_L$ such that $f(\mathfrak{P}/\mathfrak{p}) = 1$.

If L/K is Galois and $G = \mathrm{Gal}(L/K)$, then $\mathrm{Kron}(L/K) = \mathrm{Split}(L/K)$ consists of all $\mathfrak{p} \in \mathcal{P}_K$ such that $(\mathfrak{P}, L/K) = \mathrm{id}_L \in G$ for one (and then for all) $\mathfrak{P} \in \mathcal{P}_L$ lying above $\mathfrak{p}$ (see 2.5.3.1 and 2.14.7.**2**). Hence 4.4.6 implies

$$d(\mathrm{Split}(L/K)) = \frac{1}{[L:K]},$$

which is a fresh proof of the weak form of Bauer's theorem 4.3.5.3.

Theorem 4.4.8. *Let L/K be a finite extension of algebraic number fields.*

1. Let N/K be a finite Galois field extension such that $L \subset N$, $G = \mathrm{Gal}(N/K)$ and $H = \mathrm{Gal}(N/L)$. For $\sigma \in G$ let Γ_σ be the conjugacy class of σ in G. Then

$$\mathrm{Kron}(L/K) = \biguplus_{\substack{\sigma \in G \\ H \cap \Gamma_\sigma \neq \emptyset}} \mathcal{P}(N/K, \sigma), \text{ and } \ d(\mathrm{Kron}(L/K)) \geq \frac{1}{[L:K]},$$

with equality if and only if L/K is Galois.

2. L/K is Galois if and only if $\mathrm{Kron}(L/K) \doteq \mathrm{Split}(L/K)$.

Proof. 1. Let $\mathfrak{p} \in \mathcal{P}_K$ be unramified in N, $\mathfrak{P} \in \mathcal{P}_N$ such that $\mathfrak{P} \cap K = \mathfrak{p}$ and $\sigma = (\mathfrak{P}, N/K)$. By 2.13.4.3 it follows that $\mathfrak{p} \in \mathrm{Kron}(L/K)$ if and only if L is contained in the decomposition field of some prime divisor of $\mathfrak{p}$ in L, and this holds if and only if $H \supset \langle \tau\sigma\tau^{-1} \rangle$ for some $\tau \in G$. Consequently, $\mathfrak{p} \in \mathrm{Kron}(L/K)$ if and only if $H \cap \Gamma_\sigma \neq \emptyset$. Therefore $\mathrm{Kron}(L/K)$ is the disjoint union of the sets $\mathcal{P}(L/K, \sigma)$, where $\sigma \in G$ and $\Gamma_\sigma \cap H \neq \emptyset$. By 4.4.6 we obtain

$$d(\mathrm{Kron}(L/K) = \sum_{\substack{\sigma \in G \\ H \cap \Gamma_\sigma \neq \emptyset}} d(\mathcal{P}(L/K, \sigma)) = \sum_{\substack{\sigma \in G \\ H \cap \Gamma_\sigma \neq \emptyset}} \frac{|\Gamma_\sigma|}{|G|} = \frac{1}{|G|} \Big| \bigcup_{\substack{\sigma \in G \\ H \cap \Gamma_\sigma \neq \emptyset}} \Gamma_\sigma \Big|$$

$$\geq \frac{|H|}{|G|} = \frac{1}{[L:K]}, \quad \text{since} \quad H \subset \bigcup_{\substack{\sigma \in G \\ H \cap \Gamma_\sigma \neq \emptyset}} \Gamma_\sigma.$$

Equality holds if and only if, for all $\sigma \in G$, $H \cap \Gamma_\sigma \neq \emptyset$ implies $\Gamma_\sigma \subset H$, that is, if and only if H is a normal subgroup of G and L/K is Galois.

2. If L/K is Galois, then we already know that $\mathrm{Kron}(L/K) = \mathrm{Split}(L/K)$. Thus assume that $\mathrm{Kron}(L/K) \doteq \mathrm{Split}(L/K)$, and let N be a Galois closure of L/K. Then

$$\frac{1}{[N:K]} = d(\mathrm{Split}(N/K)) = d(\mathrm{Split}(L/K)) \geq d(\mathrm{Kron}(L/K)) \geq \frac{1}{[L:K]},$$

hence $[N:K] \leq [L:K]$ and thus $N = L$. □

Theorem 4.4.9 (Theorem of Bauer, strong form). *Let L/K and M/K be finite extensions of algebraic number fields (inside a fixed algebraic closure of K), and let L/K be Galois. Then*

$$\mathrm{Kron}(M/K) \dot{\subset} \mathrm{Kron}(L/K) \quad \text{if and only if} \quad L \subset M.$$

Proof. Clearly, $L \subset M$ implies $\mathrm{Kron}(M/K) \subset \mathrm{Kron}(L/K)$. Now let N/K be a finite Galois extension containing L and M, $G = \mathrm{Gal}(N/K)$, and for $\sigma \in G$ let Γ_σ be the conjugacy class of σ. Let $H = \mathrm{Gal}(N/L)$, $U = \mathrm{Gal}(N/M)$, and suppose that $\mathrm{Kron}(M/K) \dot{\subset} \mathrm{Kron}(L/K)$. Then 4.4.8.1 implies

$$\bigcup_{\substack{\sigma\in G \\ \Gamma_\sigma\cap U\neq\emptyset}} \mathcal{P}(N/K,\sigma) \dot{\subset} \bigcup_{\substack{\tau\in G \\ \Gamma_\tau\cap H\neq\emptyset}} \mathcal{P}(N/K,\tau).$$

Let $\sigma \in U$. Then $d(\mathcal{P}(N/K,\sigma)) > 0$, and there exists some $\tau \in G$ such that $\Gamma_\tau \cap H \neq \emptyset$ and $\mathcal{P}(N/K,\sigma) \cap \mathcal{P}(N/K,\tau) \neq \emptyset$. If $\mathfrak{p} \in \mathcal{P}(N/K,\sigma) \cap \mathcal{P}(N/K,\tau)$, then $\mathfrak{p}$ has prime divisors $\mathfrak{P}$ and $\mathfrak{Q}$ in N such that $(\mathfrak{P}, N/K) = \sigma$ and $(\mathfrak{Q}, N/K) = \tau$.

Since $\mathfrak{P} \cap K = \mathfrak{Q} \cap K$, there exists some $\nu \in G$ that satisfies $\mathfrak{P} = \nu\mathfrak{Q}$, and consequently $\sigma = \nu\tau\nu^{-1} \in H$. Hence $U \subset H$ and therefore $L \subset M$. □

4.5 Exercises for Chapter 4

1. The Gamma function satisfies the identities

$$\Gamma\Big(\frac{s}{2}\Big)\Gamma\Big(\frac{1+s}{2}\Big) = 2^{1-s}\sqrt{\pi}\Gamma(s) \quad \text{and} \quad \Gamma\Big(\frac{1-s}{2}\Big)\Gamma\Big(\frac{1+s}{2}\Big) = \frac{\pi}{\cos\frac{\pi s}{2}}.$$

2. For $n \in \mathbb{N}$ let $\mu(n)$ be the Möbious function, $\mathsf{d}(n)$ the number of positive divisors of n and $\nu(n)$ the number of primes dividing n, that is,

$$\mathsf{d}(n) = \sum_{1\le d\,|\,n} 1 \quad \text{and} \quad \nu(n) = \sum_{\substack{p\in\mathbb{P} \\ p|n}} 1.$$

If $s \in \mathcal{H}_1$, then

$$\frac{1}{\zeta(s)} = \sum_{n=1}^{\infty} \frac{\mu(n)}{n^s}, \qquad \frac{\zeta(s)^2}{\zeta(2s)} = \sum_{n=1}^{\infty} \frac{2^{\nu(n)}}{n^s}, \qquad \frac{\zeta(s)^3}{\zeta(2s)} = \sum_{n=1}^{\infty} \frac{d(n^2)}{n^s}$$

and

$$\sum_{n=1}^{\infty} \frac{\nu(n)}{n^s} = \zeta(s) \sum_{p\in\mathbb{P}} \frac{1}{p^s}.$$

3. Let $f\colon \mathbb{N} \to \mathbb{C}$ be a completely multiplicative function and

$$F(s) = \sum_{n=1}^{\infty} \frac{f(n)}{n^s} \quad \text{for} \ \ s \in \mathcal{H}_c,$$

where $c \in \mathbb{R}_{>0}$ is the abscissa of absolute convergence of the defining Dirichlet series.

a. Let $\Lambda(n)$ be the von Mangoldt function, defined by

$$\Lambda(n) = \begin{cases} \log p & \text{if} \ \ n = p^m \ \ \text{for some prime} \ \ p \ \ \text{and} \ \ m \in \mathbb{N}, \\ 0 & \text{otherwise.} \end{cases}$$

Then

$$\frac{F'(s)}{F(s)} = -\sum_{n=1}^{\infty} \frac{f(n)\Lambda(n)}{n^s} \quad \text{for} \ \ s \in \mathcal{H}_c.$$

b. Let $\lambda(n)$ be the Liouville function, defined by

$$\lambda(n) = (-1)^S, \quad \text{where} \quad S = \sum_{p \in \mathbb{P}} v_p(n).$$

If $n \in \mathbb{N}$, then

$$\sum_{1 \le d \,|\, n} \lambda(d) = \begin{cases} 1 & \text{if} \ \ n \ \ \text{is a square,} \\ 0 & \text{otherwise.} \end{cases}$$

If $f(\mathbb{N}) \subset \{0,1\}$, then

$$F(s) \ne 0 \quad \text{and} \quad \frac{F(2s)}{F(s)} = \sum_{n=1}^{\infty} \frac{f(n)\lambda(n)}{n^s} \quad \text{for all} \ \ s \in \mathcal{H}_c.$$

4. Let K_1, K_2 be quadratic number fields, $K_1 \ne K_2$, $K = K_1K_2$, and let K_3 be the third quadratic subfield of K.

a. $L(s, \chi_{K_3}) = L(s, \chi_{K_1})L(s, \chi_{K_2})$, and $\zeta_K(s) = \zeta(s)^2\zeta_{K_1}(s)\zeta_{K_2}(s)\zeta_{K_3}(s)$.

b. For $i \in \{1,2\}$ let $K_i = \mathbb{Q}(\sqrt{-p_i}\,)$ for some prime $p_i \equiv -1 \bmod 4$. Then $\mathsf{d}_K = (p_1p_2)^2$, and $h_K = h_{K_1}h_{K_2}h_{K_3}$ is odd. Hint: Ch. 3, Ex. 4.e.

5. $h_5 = h_{20} = 1$, $h_{80} = 2$, $h_{5\cdot 4^n} = 4$ for all $n \ge 3$, and $h_{5^{2n+1}} = 1$ for all $n \ge 0$.

6. Let $K = \mathbb{Q}_Z \subset \mathbb{Q}^{(m)}$ be an abelian number field defined by its character group $Z \subset \mathsf{X}(m)$. Then

$$\zeta_K(s) = \zeta(s)J(s) \prod_{\chi \in Z \setminus \{1_m\}} L(s, \chi), \quad \text{where}$$

$$J(s) = \prod_{\substack{p \in \mathbb{P} \\ p | m}} \Big[(1 - p^{-s}) \prod_{\substack{\mathfrak{p} \in \mathcal{P}_K \\ \mathfrak{p} | p}} (1 - \mathfrak{N}(\mathfrak{p})^{-s})^{-1}\Big].$$

7. Let K be an algebraic number field, $T \subset \mathcal{P}_K$ and $r \in [0,1] \cap \mathbb{Q}$. We say that T has **polar densitiy** r if

$$r = \frac{m}{n} \quad \text{for some } m, n \in \mathbb{N} \text{ such that } (s-1)^m \prod_{\mathfrak{p} \in T} \Big(\frac{1}{1 - \mathfrak{N}(\mathfrak{p})^{-s}}\Big)^n \in \mathcal{O}(\mathcal{H}_1 \cup \{1\}).$$

This definition only depends on r and not on m and n. If T has polar density r, then it also has Dirichlet density $d(T) = r$.

8. Let $K \subset L \subset N$ be algebraic number fields, $\alpha \in \mathcal{O}_L$ such that $L = K(\alpha)$, $g \in \mathfrak{o}_K[X]$ the minimal polynomial of α over K and N a splitting field of g over K.

a. For all but finitely many prime ideals $\mathfrak{p} \in \mathcal{P}_K$ the following assertions are equivalent:

(a) There exists some $b \in \mathfrak{o}_K$ such that $g(b) \equiv 0 \bmod \mathfrak{p}$.

(b) There exists some prime ideal $\mathfrak{P} \in \mathcal{P}_L$ lying above $\mathfrak{p}$ such that $f(\mathfrak{P}/\mathfrak{p}) = 1$.

(c) There exists a prime ideal $\mathfrak{Q} \in \mathcal{P}_N$ such that $(\mathfrak{Q}, N/K) \restriction L = \mathrm{id}_L$.

b. There are infinitely many prime ideals $\mathfrak{p} \in \mathcal{P}_K$ such that g has in K no zero modulo $\mathfrak{p}$ [that is, $g(x) \not\equiv 0 \bmod \mathfrak{p}$ for all $x \in \mathfrak{o}_K$]. Hint: A finite group cannot be the union of the conjugates of a proper subgroup.

9. Let $K \subset L$ be algebraic number fields, $n = [L\!:\!K]$, $\alpha \in \mathcal{O}_L$ such that $L = K(\alpha)$ and $g \in \mathfrak{o}_K[X]$ the minimal polynomial of α over K.

a. For all but finitely many prime ideals $\mathfrak{p} \in \mathcal{P}_K$ the following assertions are equivalent:

(a) The residue class polynomial $\overline{g} \in \mathsf{k}_\mathfrak{p}[X]$ is irreducible.

(b) $\mathfrak{p}$ is inert in L.

(c) There exists some $\mathfrak{P} \in \mathcal{P}_L$ lying above $\mathfrak{p}$ such that $f(\mathfrak{P}/\mathfrak{p}) = n$.

b. If n is a prime, then infinitely many $\mathfrak{p} \in \mathcal{P}_K$ are inert in L.

10. Let q be a prime power, and let $\mathsf{D}(\mathbb{F}_q, X)$ be the free monoid of all monic polynomials in $\mathbb{F}_q[X]$ with basis $\mathsf{P}(\mathbb{F}_q, X)$. For $n \in \mathbb{N}$ let b_n be the number of squarefree polynomials in $\mathsf{D}(\mathbb{F}_q, X)$ of degree n. Then

$$b_n = \begin{cases} q & \text{if } n = 1, \\ q^n - q^{n-1} & \text{if } n \ge 2. \end{cases}$$

Hint: Use 4.1.3. For $f \in \mathsf{D}(\mathbb{F}_q, X)$ set $\delta(f) = 1$ if f is squarefree, and $\delta(f) = 0$ otherwise. Use the identity

$$\sum_{n=0}^{\infty} b_n t^n = \sum_{f \in \mathsf{D}(\mathbb{F}_q, X)} \delta(f) t^{\partial(f)} = \prod_{p \in \mathsf{P}(\mathbb{F}_q, X)} (1 + t^{\partial(p)}) = \frac{Z(2t)}{Z(t)}.$$

11. Let L/K be a cyclic extension of algebraic number fields, $[L\!:\!K]=n$, $d\,|\,n$ and for $\mathfrak{p}\in\mathcal{P}_K$ let $\sigma_\mathfrak{p}=(\mathfrak{p},L/K)$. Let $\mathcal{A}$ be the set of all $\mathfrak{p}\in\mathcal{P}_K$ such that $\operatorname{ord}(\sigma_\mathfrak{p})=d$ and $\mathcal{B}$ the set of all $\mathfrak{p}\in\mathcal{P}_K$ having exacty d prime divisors in L. Calculate the Dirichlet densities of $\mathcal{A}$ and $\mathcal{B}$.

12. Let K be an algebraic number field and $a\in\mathfrak{o}_K$. For every $m\in\mathbb{N}$ there exist infinitely many $\mathfrak{p}\in\mathcal{P}_K$ such that a is an m-th power residue modulo $\mathfrak{p}$.

13. Let $n\in\mathbb{N}$, and let $a_1,\ldots,a_n\in\mathbb{Z}$ be multiplicatively independent modulo $\mathbb{Q}^{\times 2}$. For any $\varepsilon_1,\ldots,\varepsilon_n\in\{\pm1\}$ there exist infinitely many primes p such that

$$\Big(\frac{a_i}{p}\Big)=\varepsilon_i \quad \text{for all} \quad i\in[1,n].$$

Hint: Consider the Dirichlet series

$$\sum_{p\in\mathbb{P}}\Big[1+\varepsilon_i\Big(\frac{a_i}{p}\Big)\Big]\frac{1}{p^s}.$$

14. For a real number $a\in(0,1]$ and $s\in\mathcal{H}_1$ the **Hurwitz zeta function** is defined by the normally convergent Dirichlet series

$$\zeta(s,a)=\sum_{n=0}^{\infty}\frac{1}{(n+a)^s}.$$

$\zeta(\,\cdot\,,a)$ is holomorphic in $\mathcal{H}_1$, and there

$$\frac{d^k}{ds^k}\zeta(s,a)=(-1)^k\sum_{n=0}^{\infty}\frac{[\log(n+a)]^k}{(n+a)^s} \quad \text{for all} \quad k\in\mathbb{N}_0.$$

b. If $m\in\mathbb{N}$, $\chi\in\mathsf{X}(m)$ and $s\in\mathcal{H}_1$, then

$$L(s,\chi)=\frac{1}{m^s}\sum_{j=1}^{m}\chi(j)\zeta\Big(s,\frac{j}{m}\Big).$$

c. If $s\in\mathcal{H}_1$, then

$$\Gamma(s)\zeta(s,a)=\int_0^\infty\frac{x^{s-1}\mathsf{e}^{-ax}}{1-\mathsf{e}^{-x}}\,dx.$$

15. Let $D=\mathcal{F}(P)$ be a free monoid. For functions $f,\,g\colon D\to\mathbb{C}$ we define the **Dirichlet convolution** $f*g\colon D\to\mathbb{C}$ by

$$f*g(a)=\sum_{\substack{(b,c)\in D^2\\ bc=a}}f(b)g(c).$$

The set $Z=\mathbb{C}^D$ of all functions $f\colon D\to\mathbb{C}$, equipped with pointwise addition and the Dirichlet convolution as multiplication, is a domain with unit element $\varepsilon\in Z$, defined by

$$\varepsilon(a)=\begin{cases}1 & \text{if } a=1,\\ 0 & \text{if } a\neq 1.\end{cases}$$

$f \in Z$ is (Dirichlet-)invertible if and only if $f(1) \neq 0$. In this case we denote by $f^- \in Z$ its Dirichlet inverse satisfying $f * f^- = \varepsilon$. Every multiplicative function is Dirichlet invertible.

16. Let $D = \mathcal{F}(P)$ and $Z = \mathbb{C}^D$ as in Exercise 15. Let $\iota \in Z$ be defined by $\iota(a) = 1$ for all $a \in D$, and let $\mu \in Z$ be the unique multiplicative function satisfying $\mu(1) = 1$, $\mu(p) = -1$ for all $p \in P$ and $\mu(p^n) = 0$ for all $p \in P$ and $n \geq 2$ (the abstract Möbius function, see 1.7.2).

a. $\mu * \iota = \varepsilon$, if $f \in Z$, then $Sf = \iota * f$ is the summatory function of f, and $f = \mu * Sf$ is the Möbius inversion formula.

b. If $f,\ g \in Z$, then $f * g$ is multiplicative if and only if both f and g are multiplicative.

c. Let $f \in Z$ be multiplicative. Then its Dirichlet inverse f^- is also multiplicative, and f is completely multiplicative if and only if $f^-(a) = \mu(a)f(a)$ for all $a \in D$.

17. Let $f,\ g \colon \mathbb{N} \to \mathbb{C}$, and let

$$F(s) = \sum_{n=1}^{\infty} f(n)n^{-s} \quad \text{and} \quad G(s) = \sum_{n=1}^{\infty} g(n)n^{-s}$$

be Dirichlet series which converge absolutely in some half-plane $\mathcal{H}_c$. Then

$$F(s)G(s) = \sum_{n=1}^{\infty} (f * g)(n)n^{-s} \quad \text{for all } s \in \mathbb{C}.$$

If $f(1) \neq 0$, $F(s) \neq 0$ for $s \in \mathcal{H}_c$ and $f' \colon \mathbb{N} \to \mathbb{C}$ is defined by $f'(n) = f(n)\log n$, then $F(s) = \mathsf{e}^{H(s)}$ for all $s \in \mathcal{H}_c$, where

$$H(s) = \log f(1) + \sum_{n=2}^{\infty} \frac{(f' * f^-)(n)}{\log n} n^{-s}.$$

18. Let K be an algebraic number field. For an ideal $\mathfrak{a} \in \mathcal{J}'_K$ and $m \in \mathbb{N}$ let $d_m(\mathfrak{a}) = |\{(\mathfrak{a}_1, \ldots, \mathfrak{a}_m) \in (\mathcal{J}'_K)^m \mid \mathfrak{a} = \mathfrak{a}_1 \cdot \ldots \cdot \mathfrak{a}_m\}|$. Then

$$\zeta_K(s)^m = \sum_{\mathfrak{a} \in \mathcal{J}'_K} d_m(\mathfrak{a})\mathfrak{N}(\mathfrak{a})^{-s} \quad \text{for all } \ s \in \mathcal{H}_1.$$

19. Let $m \in \mathbb{N}$, $m \geq 3$ and $\chi \in \mathsf{X}(m) \setminus \{1_m\}$. Then

$$|L(1, \chi)| < 2 + \log m.$$

20. Let K be a quadratic number field and χ_K its character (see 3.4.4). Then

$$\zeta_K(s) = \zeta(s)L(s, \chi_K) \quad \text{for } \ s \in \mathcal{H}_1, \quad \text{and} \quad L(1, \chi_K) > 0.$$

5

Valuation Theory

At least since the works of Hensel (p-adic numbers) and Hasse (local-global principles) valuation theory has played a central role in algebraic number theory, and following E. Artin [1], several authors built the theory from its beginning on a valuation-theoretic basis (see e. g. [27] or [63]). Although this is advantageous for several reasons, I decided to start with ideal-theoretic, geometric and elementary analytic methods. This corresponds to historical development, and at first glance these methods are closer to the subject. It is our hope that the aforegoing familiarity with the basic ideal theory will promote the appreciation of the valuation-theoretic point of view.

Although valuation-theoretic aspects are addressed in almost every book on algebraic number theory, my main references for the present chapter are [1], [13], [5], [42, §§ 23ff.], [52] and [4].

After preliminaries on valued fields and their completions we specialize to non-Archimedian absolute values from Section 5.3 on. Section 5.4 contains various versions of Hensel's lemma together with some applications, in particular criteria for local powers and the existence of distinguished residue systems. In Section 5.5 we investigate extensions of absolute values, first in the perfect and then in the general case with a focus on $\mathfrak{p}$-adic absolute values, Galois actions and extensions of Laurent series fields. In particular, we determine all absolute values of an algebraic number field and prove the product formula. The subsequent sections deal with special properties of unramified and ramified extensions of local fields, and the existence and uniqueness of Frobenius automorphisms. Section 8 contains Krasner's lemma and among others the classification of local fields, the theory of cyclotomic extensions of $\mathfrak{p}$-adic fields and the p-adic structure of principal units.

Section 5.9 is of central importance to the theory of algebraic numbers and algebraic functions. As already several times announced, it addresses the theory of differents and discriminants. We first develop the theory in the frame of Dedekind domains, and afterwards we specialize to $\mathfrak{p}$-adic and algebraic number fields. Only then is the classical theory of algebraic number fields complete. Finally, Section 5.11 contains the theory of higher ramification including a proof of the Hasse-Arf theorem for general residue class fields using Tsen's theorem.

5.1 Absolute values

Definition 5.1.1. An **absolute value** of a field K is a map $|\cdot|\colon K \to \mathbb{R}_{\geq 0}$ with the following properties:

A1. $|0| = 0$, $|x| > 0$ for all $x \in K^\times$, and $0 < |x| < 1$ for some $x \in K$.

A2. $|xy| = |x|\,|y|$ for all $x, y \in K$.

A3. There exists some $c \in \mathbb{R}_{\geq 1}$ such that $|x+y| \leq c \max\{|x|, |y|\}$ for all $x, y \in K$.

Let $|\cdot|\colon K \to \mathbb{R}_{\geq 0}$ be an absolute value. Then $|\cdot| \restriction K^\times \colon K^\times \to \mathbb{R}_{>0}$ is a group homomorphism, $|K^\times|$ is a subgroup of $\mathbb{R}_{>0}$, $|x^n| = |x|^n$ for all $x \in K^\times$ and $n \in \mathbb{Z}$, $|z| = 1$ for all roots of unity $z \in \mu(K)$, $|-x| = |x|$ for all $x \in K$, and there exists some $x \in K$ such that $|x| > 1$.

An absolute value $|\cdot|\colon K \to \mathbb{R}_{\geq 0}$ is called

- **non-Archimedean** if $|x+y| \leq \max\{|x|, |y|\}$ for all $x, y \in K$; otherwise it is called **Archimedean**;
- **discrete** if $|K^\times|$ is a discrete subgroup of $\mathbb{R}_{>0}$. By 3.5.2 this holds if and only if $|K^\times| = \langle \rho \rangle$ for some $\rho \in (0, 1)$.

If $|\cdot|\colon K \to \mathbb{R}_{\geq 0}$ is an absolute value and $\lambda \in \mathbb{R}_{>0}$, then $|\cdot|^\lambda\colon K \to \mathbb{R}_{\geq 0}$ is an absolute value, too. Two absolute values $|\cdot|_1, |\cdot|_2\colon K \to \mathbb{R}_{\geq 0}$ are called **equivalent** if $|\cdot|_1 = |\cdot|_2^\lambda$ for some $\lambda \in \mathbb{R}_{>0}$. Equivalence of absolute values is an equivalence relation. If $|\cdot|_1, |\cdot|_2$ are equivalent absolute values of a field K and $|\cdot|_1$ is Archimedean [non-Archimedean, discrete], then $|\cdot|_2$ is also Archimedean [non-Archimedean, discrete].

If $|\cdot|$ is an absolute value of a field K, then the pair $(K, |\cdot|)$ is called a **valued field**. If there is no danger of confusion, we will dispense with mentioning $|\cdot|$ explicitly: Then we call $K = (K, |\cdot|)$ a valued field and $|\cdot|$ its absolute value. A valued field is called **Archimedean** [**non-Archimedean, discrete**] if its absolute value has this property.

Let $K = (K, |\cdot|)$ and $K' = (K', |\cdot|')$ be valued fields. A field homomorphism $f\colon K \to K'$ is called **value-preserving** or a **homomorphism of valued fields** if $|\varphi(x)|' = |x|$ for all $x \in K$. We sometimes write $\varphi\colon (K, |\cdot|) \to (K', |\cdot|')$ to highlight the involved absolute values.

Theorem 5.1.2. *Let K be a field.*

1. *Let $|\cdot|\colon K \to \mathbb{R}_{\geq 0}$ be a map with the properties* **A1** *and* **A2** *of* 5.1.1, *and let $c \in \mathbb{R}_{\geq 1}$. Then the following assertions are equivalent:*

   (a) $|x+y| \leq c \max\{|x|, |y|\}$ *for all $x, y \in K$.*

   (b) *For all $x \in K$, $|x| \leq 1$ implies $|x+1| \leq c$.*

2. For an absolute value $|\cdot|\colon K \to \mathbb{R}_{\ge 0}$ *the following assertions are equivalent:*

(a) $|\cdot|$ *fulfills the* **triangle inequality** $|x+y| \le |x| + |y|$ *for all* $x,\, y \in K$.

(b) $|x+y| \le 2\max\{|x|, |y|\}$ *for all* $x,\, y \in K$.

In particular:

- *Every non-Archimedean absolute value of* K *fulfills the triangle inequality.*
- *Every absolute value of* K *is equivalent to an absolute value which fulfills the triangle inequality.*

3. Let $|\cdot|\colon K \to \mathbb{R}_{\ge 0}$ *be an absolute value which fulfills the triangle inequality. Then* $\big||x| - |y|\big| \le |x \pm y| \le |x| + |y|$ *for all* $x,\, y \in K$, *and the map*

$$d\colon K \times K \to \mathbb{R}_{\ge 0}, \quad \textit{defined by} \;\; d(x,y) = |x-y|,$$

is a metric.

Proof. 1. (a) $\Rightarrow$ (b) If $x \in K$ and $|x| \le 1$, then $|x+1| \le c\max\{|x|, |1|\} = c$.

(b) $\Rightarrow$ (a) Let $x,\, y \in K$ and $|y| \le |x|$. If $x = 0$, there is nothing to do. If $x \ne 0$, then $|x^{-1}y| = |x|^{-1}|y| \le 1$, hence $|1 + x^{-1}y| \le c$ and consequently $|x+y| = |x|\,|1+x^{-1}y| \le c|x|$.

2. (a) $\Rightarrow$ (b) Obvious.

(b) $\Rightarrow$ (a) For $r \in \mathbb{N}$ and $x_1, \ldots, x_{2^r} \in K$, a simple induction on r shows that

$$\Big|\sum_{i=1}^{2^r} x_i\Big| \le 2^r \max\{|x_1|, \ldots, |x_{2^r}|\}.$$

Now let $n \in \mathbb{N}$ and $x_1, \ldots, x_n \in K$. Let $r \in \mathbb{N}$ such that $2^{r-1} \le n < 2^r$, and set $x_i = 0$ for all $i \in [n+1, 2^r]$. Then it follows that

$$\Big|\sum_{i=1}^{n} x_i\Big| = \Big|\sum_{i=1}^{2^r} x_i\Big| \le 2^r \max\{|x_1|, \ldots, |x_{2^r}|\} \le 2n\max\{|x_1|, \ldots, |x_n|\}.$$

In particular, $|nx| \le 2n|x|$ for all $n \in \mathbb{N}$ and $x \in K$. If $x,\, y \in K$ and $n \in \mathbb{N}$, then

$$\begin{aligned} |x+y|^n &= \Big|\sum_{i=0}^{n} \binom{n}{i} x^i y^{n-i}\Big| \le 2(n+1)\max\Big\{\Big|\binom{n}{i} x^i y^{n-i}\Big| \;\Big|\; i \in [0,n]\Big\} \\ &\le 2(n+1)\max\Big\{2\binom{n}{i}|x|^i|y|^{n-i} \;\Big|\; i \in [0,n]\Big\} \le 4(n+1)\sum_{i=0}^{n}\binom{n}{i}|x|^i|y|^{n-i} \\ &= 4(n+1)(|x|+|y|)^n, \quad \text{and therefore} \quad |x+y| \le \sqrt[n]{4(n+1)}\,(|x|+|y|). \end{aligned}$$

As $n \to \infty$, it follows that $|x+y| \le |x| + |y|$.

Clearly every non-Archimedean absolute value fulfills the triangle inequality. Let $|\cdot|\colon K \to \mathbb{R}_{\geq 0}$ be any absolute value and $c \in \mathbb{R}_{\geq 1}$ such that $|x+y| \leq c\max\{|x|, |y|\}$ for all $x,\, y \in K$. If $\lambda \in \mathbb{R}_{>0}$ is such that $c^\lambda \leq 2$, then $|x+y|^\lambda \leq 2\max\{|x|^\lambda, |y|^\lambda\}$ for all $x,\, y \in K$. Hence $|\cdot|^\lambda$ is an absolute value which is equivalent to $|\cdot|$ and fulfills the triangle inequality.

3. If $x,\, y \in K$, then

$$|x-y| \leq |x| + |-y| = |x| + |y| \quad \text{and} \quad |x| = |(x\pm y) \mp y| \leq |x \pm y| + |y|.$$

It follows that $|x| - |y| \leq |x \pm y|$, alike $|y| - |x| \leq |x \pm y|$, and consequently $\big||x| - |y|\big| \leq |x \pm y|$.

If $x,\, y,\, z \in K$, then $d(x,z) = |x-z| \leq |x-y| + |y-z| = d(x,y) + d(y,z)$. Apparently, $d(x,y) = d(y,x)$, and $d(x,y) = 0$ if and only if $x = y$. Hence d is a metric. □

Definitions, Remarks and Examples 5.1.3.

1. Let K be a field, K_0 a subfield of K, $|\cdot|\colon K \to \mathbb{R}_{\geq 0}$ an absolute value and $|\cdot|_0 = |\cdot| \restriction K_0$. Then $|\cdot|_0$ is an absolute value if and only if there exists some $x \in K_0^\times$ such that $|x| \neq 1$. In this case we call $|\cdot|$ an **extension** of $|\cdot|_0$ to K and $|\cdot|_0$ the **restriction** of $|\cdot|$ to K_0.

If $|\cdot|_0$ is an absolute value of K_0 and $|\cdot|_1$ and $|\cdot|_2$ are distinct extensions of $|\cdot|_0$ to K, then $|\cdot|_1$ and $|\cdot|_2$ are not equivalent. Indeed, if $|\cdot|_1 = |\cdot|_2^\lambda$ for some $\lambda \in \mathbb{R}_{>0}$, then $|\cdot|_0 = |\cdot|_1 \restriction K_0 = |\cdot|_2 \restriction K_0$ implies $\lambda = 1$.

2. The ordinary absolute value of $\mathbb{C}$ (which we henceforth denote by $|\cdot|_\infty$) is an Archimedean absolute value and fulfills the triangle inequality. We denote the restriction of $|\cdot|_\infty$ to $\mathbb{R}$ and to $\mathbb{Q}$ likewise with by $|\cdot|_\infty$. Note that the absolute value $|\cdot|_\infty^2$ of $\mathbb{C}$ (which plays an essential role in the sequel) does not satisfy the triangle inequality.

3. Let K be an algebraic number field and (r_1, r_2) be its signature. Let $[K:\mathbb{Q}] = n = r_1 + 2r_2$ and $\text{Hom}(K,\mathbb{C}) = \{\sigma_1, \ldots, \sigma_n\}$ such that $\sigma_1, \ldots, \sigma_{r_1}$ are the real embeddings and $(\sigma_{r_1+1}, \overline{\sigma_{r_1+1}}), \ldots, (\sigma_{r_1+r_2}, \overline{\sigma_{r_1+r_2}})$ the pairs of conjugate complex embeddings of K (see Section 3.1). For $j \in [1, r_1+r_2]$ we define

$$|\cdot|_{\infty,j}\colon K \to \mathbb{R}_{\geq 0} \ \text{by} \ |a|_{\infty,j} = |\sigma_j(a)|_\infty^{l_j}, \ \text{where} \ l_j = \begin{cases} 1 & \text{if } j \in [1, r_1], \\ 2 & \text{if } j \in [r_1+1, r_2]. \end{cases}$$

Then $|\cdot|_{\infty,1}, \ldots, |\cdot|_{\infty,r_1+r_2}$ are Archimedean absolute values of K satisfying $|\cdot|_{\infty,j} \restriction \mathbb{Q} = |\cdot|_\infty^{l_j}$ for all $j \in [1, r_1+r_2]$. They are called the **normalized Archimedean absolute values** of K.

4. Let K be a field, $v\colon K \to \mathbb{Z} \cup \{\infty\}$ a discrete valuation (see 2.2.1) and $\rho \in (0,1)$. Then it is easily checked that

$$|\cdot|_{v,\rho}\colon K \to \mathbb{R}_{\geq 0}, \quad \text{defined by} \quad |a|_{v,\rho} = \rho^{v(a)} \quad (\text{where } \rho^\infty = 0)$$

is a discrete absolute value of K. We call $|\cdot|_{v,\rho}$ an **absolute value belonging to** v. Any two absolute values belonging to v are equivalent. More precisely, if $|\cdot|_v$ is an absolute value belonging to v, then another absolute value belongs to v if and only if it is equivalent to $|\cdot|_v$.

Let $\mathfrak{o}$ be a discrete valuation domain, $K = \mathsf{q}(\mathfrak{o})$ and v a discrete valuation defining $\mathfrak{o}$ (see 2.10.4.2). Any absolute value $|\cdot|\colon K \to \mathbb{R}_{\geq 0}$ belonging to v is called an **absolute value defining** $\mathfrak{o}$; then

$$\mathfrak{o} = \{x \in K \mid v(x) \geq 0\} = \{x \in K \mid |x| \leq 1\}.$$

5. Let $\mathfrak{o}$ be a Dedekind domain, $K = \mathsf{q}(\mathfrak{o})$, $\mathfrak{p} \in \mathcal{P}(\mathfrak{o})$, $v_\mathfrak{p}\colon K \to \mathbb{Z} \cup \{\infty\}$ the $\mathfrak{p}$-adic valuation of K (see 2.8.5) and $\rho \in (0,1)$. Then the absolute value $|\cdot|_{\mathfrak{p},\rho} = |\cdot|_{v_\mathfrak{p},\rho}$ (belonging to $v_\mathfrak{p}$) is called a **$\mathfrak{p}$-adic absolute value** of K. If $q = |\mathsf{k}_\mathfrak{p}| < \infty$, then $|\cdot|_\mathfrak{p} = |\cdot|_{\mathfrak{p},1/q}$ is called the **normalized $\mathfrak{p}$-adic absolute value** of K; it satisfies $|a|_\mathfrak{p} = |\mathsf{k}_\mathfrak{p}|^{-v_\mathfrak{p}(a)}$ for all $a \in K$ (where $|\mathsf{k}_\mathfrak{p}|^{-\infty} = 0$).

If $v\colon K \to \mathbb{Z} \cup \{\infty\}$ is a discrete valuation, then we call (K, v) also a **discrete valued field**, and then every absolute value belonging to v is a $\mathfrak{p}_v$-adic absolute value of K.

6. Let $K = (K, |\cdot|)$ be a discrete valued field and $\rho \in (0,1)$ such that $|K^\times| = \langle\rho\rangle$. We define

$$v\colon K \to \mathbb{Z} \cup \{\infty\} \text{ by } v(a) = \frac{\log|a|}{\log\rho} \quad \Big(\text{where } \log 0 = -\infty \text{ and } \frac{\log 0}{\log\rho} = \infty\Big).$$

Then v is a discrete valuation of K, and $|\cdot| = |\cdot|_{v,\rho}$ is an absolute value belonging to v. We call v the **discrete valuation belonging to** $|\cdot|$.

Definition and Theorem 5.1.4. *Let p be a prime, $v_p = v_{p\mathbb{Z}}\colon \mathbb{Q} \to \mathbb{Z} \cup \{\infty\}$ the p-adic valuation of $\mathbb{Q}$ and*

$$|\cdot|_p = |\cdot|_{p\mathbb{Z},1/p}\colon \mathbb{Q} \to \mathbb{R}_{\geq 0}$$

the normalized p-adic absolute value of $\mathbb{Q}$, given by

$$|a|_p = p^{-v_p(a)} \text{ for all } a \in \mathbb{Q}, \quad (\text{where } p^{-\infty} = 0)$$

(see 2.10.9.1 and the examples **4**, **5** and **6** in 5.1.3). $|\cdot|_p\colon \mathbb{Q} \to \mathbb{R}_{\geq 0}$ *is a discrete absolute value,* $|\mathbb{Q}^\times|_p = \langle p^{-1}\rangle = \langle p\rangle \subset \mathbb{R}_{>0}$, *and* $v_p\colon \mathbb{Q} \to \mathbb{Z} \cup \{\infty\}$ *is the corresponding discrete valuation. In particular,* $(\mathbb{Q}, v_p)$ *is a discrete valued field.*

$|\cdot|_p\colon \mathbb{Q} \to \langle p\rangle \cup \{0\} \subset \mathbb{R}_{\geq 0}$ is called the **p-adic absolute value.**

The family of all p-adic absolute values of $\mathbb{Q}$ together with the Archimedian absolute value satisfies the **product formula**

$$\prod_{p \in \mathbb{P} \cup \{\infty\}} |a|_p = 1 \quad \text{for all } a \in \mathbb{Q}^\times.$$

Proof. It suffices to prove the product formula. If $a \in \mathbb{Q}^\times$, then $|a|_p = p^{-v_p(a)}$ for all primes p, and consequently

$$\prod_{p \in \mathbb{P} \cup \{\infty\}} |a|_p = \prod_{p \in \mathbb{P}} p^{-v_p(a)} \, |a|_\infty = \prod_{p \in \mathbb{P}} p^{-v_p(a)} \prod_{p \in \mathbb{P}} p^{v_p(a)} = 1. \qquad \square$$

The product formula shows that the system of p-adic absolute values together with the Archimedian absolute value of $\mathbb{Q}$ form a complete system. We shall meet an analogous phenomenon and see its striking consequences in higher number fields and in function fields.

Theorem 5.1.5. *Let K be a field, and let $|\cdot|_1,\ |\cdot|_2\colon K \to \mathbb{R}_{\geq 0}$ be absolute values of K. Then the following assertions are equivalent:*

(a) *$|\cdot|_1$ and $|\cdot|_2$ are equivalent absolute values.*

(b) *$\{x \in K \mid |x|_1 \leq 1\} \subset \{x \in K \mid |x|_2 \leq 1\}$.*

(c) *$\{x \in K \mid |x|_1 < 1\} \subset \{x \in K \mid |x|_2 < 1\}$.*

Proof. (a) $\Rightarrow$ (b) Obvious.

(b) $\Rightarrow$ (c) We assume to the contrary that there exists some $x \in K$ such that $|x|_1 < 1$ and $|x|_2 \geq 1$. Then $|x|_2 = 1$, and we pick some $a \in K$ such that $|a|_2 > 1$. For $n \gg 1$ we obtain $|x^n a|_1 = |x|_1^n |a|_1 < 1$ and $|x^n a|_2 = |a|_2 > 1$, a contradiction.

(c) $\Rightarrow$ (a) Let $a \in K^\times$ be such that $|a|_1 < 1$. Then also $|a|_2 < 1$, and there exists some $\lambda \in \mathbb{R}_{>0}$ such that $|a|_2 = |a|_1^\lambda$. We assert that $|x|_2 = |x|_1^\lambda$ for all $x \in K^\times$. Let $x \in K^\times$ and $s \in \mathbb{R}$ such that $|x|_1 = |a|_1^s$.

Let $r \in \mathbb{Q}$ be such that $r < s$, and set $r = m/n$ for some $m \in \mathbb{Z}$ and $n \in \mathbb{N}$. Then we obtain $|a|_1^r > |a|_1^s = |x|_1$, hence $|a|_1^m > |x|_1^n$ and therefore $|x^n a^{-m}|_1 = |x|_1^n |a|_1^{-m} < 1$. It follows that $|x^n a^{-m}|_2 = |x|_2^n |a|_2^{-m} < 1$, that is, $|x|_2^n < |a|_2^m$, and therefore $|x|_2 < |a_2|^r$. Overall we obtain

$$|x|_2 \leq \inf\{|a|_2^r \mid r \in \mathbb{Q},\ r < s\} = |a|_2^s.$$

Alike it follows that $|x|_2 \geq \sup\{|a|_2^r \mid r \in \mathbb{Q},\ r > s\} = |a|_2^s$, and consequently $|x|_2 = |a|_2^s = |a|_1^{s\lambda} = |x|_1^\lambda$. $\square$

Theorem 5.1.6 (Weak approximation theorem). *Let K be a field, $r \geq 2$, and let $|\cdot|_1, \ldots, |\cdot|_r$ be pairwise not equivalent absolute values of K.*

1. *There exists some $z \in K$ such that $|z|_1 > 1$, and $|z|_i < 1$ for all $i \in [2, r]$.*

2. *Let $(x_1, \ldots, x_r) \in K^r$.*

(a) *For every* $\varepsilon \in \mathbb{R}_{>0}$ *there exists some* $x \in K$ *such that*

$$|x - x_i|_i < \varepsilon \textit{ for all } i \in [1, r].$$

(b) *Suppose that* $[1,r] = I \cup J$, *where* $I,\ J \neq \emptyset$ *and* $I \cap J = \emptyset$. *Then there exists some* $x \in K$ *such that* $|x|_i < 1$ *for all* $i \in I$, *and* $|x|_i > 1$ *for all* $i \in J$.

(c) *Let* $\varepsilon \in \mathbb{R}_{>0}$ *and* $s \in [0,r]$. *For* $i \in [1,s]$ *let* $|\cdot|_i$ *be discrete,* $m_i \in \mathbb{Z}$, *and let* $v_i \colon K \to \mathbb{Z} \cup \{\infty\}$ *be the discrete valuation belonging to* $|\cdot|_i$. *Then there exists some* $x \in K$ *such that* $v_i(x - x_i) = m_i$ *for all* $i \in [1,s]$ *and* $|x - x_i|_i < \varepsilon$ *for all* $i \in [s+1, r]$.

Proof. 1. By induction on r.

$r = 2$: By 5.1.5 there exist $\alpha,\ \beta \in K$ such that $|\alpha|_1 < 1$, $|\alpha|_2 \geq 1$, $|\beta|_2 < 1$ and $|\beta|_1 \geq 1$. If $z = \alpha^{-1}\beta$, then $|z|_1 > 1$ and $|z|_2 < 1$.

$r \geq 3$, $r-1 \to r$: By the induction hypothesis and the case $r = 2$, there exist $x,\ y \in K$ such that $|x|_1 > 1$ and $|x|_i < 1$ for all $i \in [2, r-1]$, $|y|_1 > 1$ and $|y|_r < 1$.

CASE 1: $|x|_r \leq 1$. For all $n \in \mathbb{N}$ we have $|x^n y|_1 = |x|_1^n |y|_1 > 1$ and $|x^n y|_r = |x|_r^n |y|_r < 1$, and if $n \gg 1$, then $|x^n y|_i = |x|_i^n |y|_i < 1$ for all $i \in [2, r-1]$. Hence $z = x^n y$ has for $n \gg 1$ the required properties.

CASE 2: $|x|_r > 1$. If $n \gg 1$, then

$$\left|\frac{x^n y}{1 + x^n}\right|_1 = \frac{|y|_1}{|1 + x^{-n}|_1} \geq \frac{|y|_1}{1 + |x|_1^{-n}} > 1, \quad \left|\frac{x^n y}{1 + x^n}\right|_r = \frac{|y|_r}{|1 + x^{-n}|_r} \leq \frac{|y|_r}{|1 - |x|_r^{-n}|} < 1,$$

and

$$\left|\frac{x^n y}{1 + x^n}\right|_i \leq \frac{|x|_i^n\, |y|_i}{|1 - |x|_i^n|} < 1 \quad \text{for all} \quad i \in [2, r-1].$$

Hence

$$z = \frac{x^n y}{1 + x^n} \quad \text{has for} \quad n \gg 1 \quad \text{the required properties.}$$

2. (a) Let $\varepsilon \in \mathbb{R}_{>0}$. For every $i \in [1,r]$ there exists by 1. some $z_i \in K$ such that $|z_i|_i > 1$ and $|z_i|_j < 1$ for all $j \in [1,r] \setminus \{i\}$. For $n \in \mathbb{N}$ and $i \in [1,r]$ let

$$y_i = \frac{z_i^n}{1 + z_i^n}.$$

For $n \gg 1$ and $i,\ j \in [1,r]$ such that $i \neq j$ we obtain

$$|y_i - 1|_i = \frac{1}{|1 + z_i^n|_i} \leq \frac{1}{|z_i|_i^n - 1} < \frac{\varepsilon}{r|x_i|_i}$$

and

$$|y_i|_j = \frac{|z_i|_j^n}{|1 + z_i^n|_j} = \frac{1}{|1 + z_i^{-n}|_j} \leq \frac{1}{|z_i|_j^{-n} - 1} < \frac{\varepsilon}{r|x_j|}.$$

If

$$x = \sum_{j=1}^{r} y_j x_j \in K,$$

then

$$|x - x_i|_i \le |y_i - 1|_i \, |x_i|_i + \sum_{\substack{j=1 \\ j \ne i}}^{r} |y_j|_i \, |x_j|_i < \varepsilon \quad \text{for all} \quad i \in [1, r].$$

(b) Let $\varepsilon \in (0, 1)$. For $i \in I$ let $x_i \in K$ be such that $|x_i|_i < 1 - \varepsilon$, and for $i \in J$ let $x_i \in K$ be such that $|x_i|_i > 1 + \varepsilon$. By (a), there exists some $x \in K$ such that $|x - x_i|_i < \varepsilon$ for all $i \in [1, r]$. If $i \in I$, then

$$|x|_i = |(x - x_i) + x_i|_i \le |x - x_i|_i + |x_i|_i < 1,$$

and if $i \in J$, then

$$|x|_i = |x_i + (x - x_i)|_i \ge |x_i|_i - |x - x_i|_i > 1.$$

(c) For $i \in [1, s]$ let $|K_i^\times|_i = \langle \rho_i \rangle$, where $\rho_i \in (0, 1)$, and let $z_i \in K$ such that $v_i(z_i) = m_i$ (and thus $|z_i|_i = \rho_i^{m_i}$). For $i \in [s + 1, r]$ let $z_i = 0$. By 1. there exist x', $x'' \in K$ such that

- $|x' - x_i|_i < \rho_i^{m_i}$ and $|x'' - z_i|_i < \rho_i^{m_i}$ for $i \in [1, s]$, and
- $|x' - x_i|_i < \varepsilon/2$ and $|x''|_i < \varepsilon/2$ for $i \in [s + 1, r]$.

We set $x = x' + x''$ and observe that $x - x_i = (x_- x_i) + (x'' - z_i) + z_i$.
If $i \in [s + 1, r]$, then $|x - x_i|_i = |(x' - x_i) + x''| \le |x' - x_i|_i + |x''|_i < \varepsilon$.
If $i \in [1, s]$, then $v_i(x' - x_i) > m_i$, $v_i(x'' - z_i) > m_i$, and since $v_i(z_i) = m_i$, it follows that $v(x - x_i) = m_i$. □

Theorem 5.1.7. *Let K be a field, F its prime ring and $|\cdot| \colon K \to \mathbb{R}_{\ge 0}$ an absolute value.*

1. The following assertions are equivalent:

(a) $|\cdot|$ is non-Archimedean.

(b) $|x| \le 1$ for all $x \in F$.

(c) $|F|$ is bounded.

If $\operatorname{char}(K) > 0$, then every absolute value of K is non-Archimedean.

2. Let $|\cdot|$ be non-Archimedean.

(a) If x, $y \in K$ and $|x| \ne |y|$, then $|x \pm y| = \max\{|x|, |y|\}$.

(b) *Let* $n \geq 2$ *and* $x_1, \ldots, x_n \in K$ *such that* $x_1 + \ldots + x_n = 0$. *Then there exist* $i, j \in [1, n]$ *such that* $i \neq j$ *and*

$$|x_i| = |x_j| = \max\{|x_1|, \ldots, |x_n|\}.$$

(c) *Let* $n \in \mathbb{N}$, $x, a_0, \ldots, a_{n-1} \in K$ *and*

$$x^n + a_{n-1}x^{n-1} + \ldots + a_1 x + a_0 = 0.$$

If $|a_i| \leq 1$ *for all* $i \in [0, n-1]$, *then* $|x| \leq 1$.

(d) *Let* K_0 *be a subfield of* K *such that* K/K_0 *is algebraic. Then* $|\cdot| \restriction K_0$ *is an absolute value of* K_0.

Proof. 1. (a) $\Rightarrow$ (b) If $x \in F$, then $x = \pm n 1_K$ for some $n \in \mathbb{N}_0$, and therefore $|x| = |1_K + \ldots + 1_K| \leq |1_K| = 1$.

(b) $\Rightarrow$ (c) Obvious.

(c) $\Rightarrow$ (a) By 5.1.2.2 there exists some $\lambda \in \mathbb{R}_{>0}$ such that $|\cdot|^\lambda$ fulfills the triangle inequality. Now $|F|$ is bounded if and only if $|F|^\lambda$ is bounded, and $|\cdot|$ is non-Archimedian if and only if $|\cdot|^\lambda$ is non-Archimedian. Hence we may assume that $|\cdot|$ fulfills the triangle inequality.

Let $B \in \mathbb{R}$ such that $|z| \leq B$ for all $z \in F$. If $x, y \in K$ and $n \in \mathbb{N}$, then

$$|x+y|^n = |(x+y)^n| = \Bigl|\sum_{i=0}^{n} \binom{n}{i} x^i y^{n-i}\Bigr| \leq \sum_{i=0}^{n} \Bigl|\binom{n}{i} 1_K\Bigr|\, |x|^i |y|^{n-i}$$
$$\leq (n+1)B \max\{|x|, |y|\}^n, \ \text{hence } |x+y| \leq \sqrt[n]{(n+1)B}\, \max\{|x|, |y|\}.$$

As $(\sqrt[n]{(n+1)B})_{n \geq 1} \to 1$, it follows that $|x| + |y| \leq \max\{|x|, |y|\}$.

If $\text{char}(K) > 0$, then every absolute value of K is non-Archimedian by (c).

2. (a) If $|x| < |y|$, then

$$|y| = |(x+y) + (-x)| \leq \max\{|x+y|, |x|\} \leq \max\{|x|, |y|\}) = |y|,$$

hence $|y| = \max\{|x+y|, |x|\}$, and consequently $|y| = |x+y|$.

(b) Assume the contrary, and suppose that $|x_1| > |x_i|$ for all $i \in [2, n]$. Then $|x_2 + \cdots + x_n| \leq \max\{|x_2|, \ldots, |x_n|\} < |x_1|$ and therefore

$$0 = |x_1 + (x_2 + \ldots + x_n)| = |x_1|$$

by (a), a contradiction.

(c) If $|x| > 1$, then $|x^n| > |x^i| \geq |a_i x^i|$ for all $i \in [0, n-1]$, which contradicts (b).

(d) Assume to the contrary that $|a| \leq 1$ for all $a \in K_0$. If $x \in K$, then there exist some $n \in \mathbb{N}$ and $a_0, \ldots, a_{n-1} \in K_0$ such that $x^n + a_{n-1}x^{n-1} + \ldots + a_0 = 0$. By (c) we obtain $|x| \leq 1$ for all $x \in K$, a contradiction. □

Theorem 5.1.8. *Let $K = (K, |\cdot|)$ be a non-Archimedean valued field,*

$$\mathcal{O} = \{x \in K \mid |x| \le 1\} \quad \text{and} \quad \mathfrak{P} = \{x \in K \mid |x| < 1\}.$$

1. $\mathcal{O}$ is a valuation domain, $\mathfrak{p}_K$ is its maximal ideal, $K = \mathsf{q}(\mathcal{O})$ and $\mathcal{O}^\times = \{x \in K \mid |x| = 1\}$. $\mathcal{O}$ is Noetherian (and thus a discrete valuation domain) if and only if $|\cdot|$ is discrete.

2. Let $\mathfrak{o}$ be a Dedekind domain, $K = \mathsf{q}(\mathfrak{o})$ and $|x| \le 1$ for all $x \in \mathfrak{o}$. Then there exists a unique prime ideal $\mathfrak{p} \in \mathcal{P}(\mathfrak{o})$ such that $|\cdot|$ is a $\mathfrak{p}$-adic absolute value of K.

Proof. 1. If $x,\, y \in \mathcal{O}$, then $|x-y| \le \max\{|x|, |y|\} \le 1$ and $|xy| = |x|\,|y| \le 1$. Hence $\{1, x-y, xy\} \subset \mathcal{O}$ for all $x,\, y \in \mathcal{O}$, thus $\mathcal{O}$ is a proper subring of K, and

$$\begin{aligned}\mathcal{O}^\times &= \{x \in K^\times \mid x \in \mathcal{O} \text{ and } x^{-1} \in \mathcal{O}\} = \{x \in K^\times \mid |x| \le 1 \text{ and } |x^{-1}| \le 1\} \\ &= \{x \in K \mid |x| = 1\} = \mathcal{O} \setminus \mathfrak{P}.\end{aligned}$$

If $x \in K \setminus \mathcal{O}$, then $|x| > 1$, hence $|x^{-1}| < 1$ and thus $x^{-1} \in \mathcal{O}$. Therefore $\mathcal{O}$ is a valuation domain and $K = \mathsf{q}(\mathcal{O})$.

If $\mathcal{O}$ is Noetherian, then it is a discrete valuation domain by 2.10.1.3, and if π is a prime element of $\mathcal{O}$, then $K^\times = \langle \pi \rangle \cdot \mathcal{O}_K^\times$, hence $|K^\times| = \langle |\pi| \rangle \subset \mathbb{R}^\times$, and therefore $|\cdot|$ is discrete. Conversely, if $|\cdot|$ is discrete, then $|\cdot| = |\cdot|_{v,\rho}$ for some discrete valuation $v\colon K \to \mathbb{Z} \cup \{\infty\}$ and $\rho \in (0,1)$ by 5.1.3.**6**, and then $\mathcal{O} = \{x \in K \mid |x| \le 1\} = \{x \in K \mid v(x) \ge 0\} = \mathcal{O}_v$ is a discrete valuation domain.

2. The ideal $\mathfrak{p} = \mathfrak{P} \cap \mathfrak{o}$ is a prime ideal of $\mathfrak{o}$, and we assert that $\mathfrak{p} \ne \mathbf{0}$. Indeed, if $\mathfrak{p} = \mathbf{0}$, then $|x| = 1$ for all $x \in \mathfrak{o}^\bullet$ and thus $|x| = 1$ for all $x \in K^\times$, a contradiction. Hence $\mathfrak{p} \in \mathcal{P}(\mathfrak{o})$.

If $x \in \mathfrak{o}_\mathfrak{p}$, then $x = s^{-1}a$ for some $a \in \mathfrak{o}$ and $s \in \mathfrak{o} \setminus \mathfrak{p} \subset \mathcal{O} \setminus \mathfrak{P}$. As $|s| = 1$, it follows that $|x| = |s^{-1}a| \le 1$, and thus $x \in \mathcal{O}$. Hence $\mathfrak{o}_\mathfrak{p} \subset \mathcal{O}$, and thus equality holds since $\mathfrak{o}_\mathfrak{p}$ is a discrete valuation domain and thus a maximal proper subring (see 2.10.7.2 and 2.10.1.4). In particular, if $|\cdot|_\mathfrak{p}$ is any $\mathfrak{p}$-adic absolute value of K, then $|x|_\mathfrak{p} \le 1$ if and only $|x| \le 1$. Hence $|\cdot|_\mathfrak{p}$ and $|\cdot|$ are equivalent by 5.1.5, and $|\cdot|$ is itself a $\mathfrak{p}$-adic absolute value.

The uniqueness of $\mathfrak{p}$ follows since $\mathfrak{p} = \{x \in \mathfrak{o} \mid |x| < 1\}$. □

Theorem 5.1.9. *Let $|\cdot|\colon \mathbb{Q} \to \mathbb{R}_{\ge 0}$ be an absolute value.*

1. If $|\cdot|$ is non-Archimedean, then $|\cdot|$ is equivalent to the p-adic absolute value $|\cdot|_p$ for some prime p.

2. If $|\cdot|$ is Archimedean, then $|\cdot|$ is equivalent to $|\cdot|_\infty$.

Proof. 1. Let $|\cdot|$ be non-Archimedian. Then $|x| \le 1$ for all $x \in \mathbb{Z}$ by 5.1.7.1, and thus the assertion follows by 5.1.8.2 with $\mathfrak{o} = \mathbb{Z}$.

2. Let $|\cdot|$ be Archimedian. By 5.1.2.3 we may assume that $|\cdot|$ satisfies the triangle inequality, and by 5.1.7.1 there exists some $m \in \mathbb{N}$ such that $|m| > 1$. For arbitrary $k,\, n \in \mathbb{N}$ with $n \ge 2$ we consider the n-adic digit expansion of m^k, given by

$$m^k = a_0 + a_1 n + \ldots + a_s n^s, \quad \text{where} \quad s \in \mathbb{N}_0\,,\ a_0, \ldots, a_s \in [0, n-1] \ \text{and} \ a_s \ne 0\,.$$

Then $n^s \le m^k$, hence $s \log n \le k \log m$, and $|a_i| = |1 + \ldots + 1| \le a_i < n$ for all $i \in [0, s]$. It follows that

$$\begin{aligned} |m|^k = |m^k| \le \sum_{i=0}^{s} |a_i|\,|n|^i &< (s+1)n \max\{1, |n|^s\} \\ &\le \Big(\frac{k \log m}{\log n} + 1\Big) n \max\big\{1,\ |n|^{(k \log m)/\log n}\big\}, \end{aligned}$$

hence

$$|m| \le \sqrt[k]{kn\Big(\frac{\log m}{\log n} + \frac{1}{k}\Big)} \max\big\{1,\ |n|^{\log m/\log n}\big\},$$

and consequently,

$$\begin{aligned} |m| &\le \lim_{k\to\infty} \sqrt[k]{kn\Big(\frac{\log m}{\log n} + \frac{1}{k}\Big)} \max\big\{1,\ |n|^{\log m/\log n}\big\} \\ &= |n|^{\log m/\log n}, \ \text{since} \ |m| > 1. \end{aligned}$$

It follows that

$$|n| > 1 \quad \text{and} \quad \frac{\log |m|}{\log m} \le \frac{\log |n|}{\log n}.$$

Now we can interchange the roles of m and n. We obtain

$$\frac{\log |m|}{\log m} = \frac{\log |n|}{\log n} \ \text{ for all } \ m,\, n \in \mathbb{N}_{\ge 2}, \quad \text{and we set} \ \lambda = \frac{\log |m|}{\log m} \in \mathbb{R}_{>0}.$$

It follows that $|n| = n^\lambda = |n|_\infty^\lambda$ for all $n \in \mathbb{N}$ and thus also for all $x \in \mathbb{Q}$. Hence $|\cdot| = |\cdot|_\infty^\lambda$ is equivalent to $|\cdot|_\infty$. □

Corollary 5.1.10. *Let $(K, |\cdot|)$ be an Archimedian valued field. Then $\mathbb{Q} \subset K$, and $|\cdot|\restriction\mathbb{Q} = |\cdot|_\infty^\lambda$ for some $\lambda \in \mathbb{R}_{>0}$. In particular, $|\cdot|$ is not discrete.*

Proof. By 5.1.7.1 we obtain $\mathrm{char}(K) = 0$, hence $\mathbb{Q} \subset K$, and 5.1.9 implies $|\cdot|\restriction\mathbb{Q} = |\cdot|_\infty^\lambda$ for some $\lambda \in \mathbb{R}_{>0}$. In particular, $\mathbb{Q}_{>0}^\lambda \subset |K^\times|$, and thus $|K^\times|$ is not discrete. □

5.2 Topology and completion of valued fields

Remarks and Definitions 5.2.1. Let $K = (K, |\cdot|)$ be a valued field. For $a \in K$ and $\varepsilon \in \mathbb{R}_{>0}$ we consider the open ε-ball around a, defined by

$$B_\varepsilon(a) = B_\varepsilon(a, |\cdot|) = \{x \in K \mid |x - a| < \varepsilon\} = a + B_\varepsilon(0, |\cdot|).$$

Let $\lambda \in \mathbb{R}_{>0}$ such that $|\cdot|^\lambda$ fulfills the triangle inequality. Then the map

$$K \times K \to \mathbb{R}_{\geq 0}, \quad \text{defined by} \quad (x, y) \mapsto |x - y|^\lambda,$$

is a metric on K, and $\{B_\varepsilon(a, |\cdot|^\lambda) \mid \varepsilon \in \mathbb{R}_{>0}\} = \{B_\varepsilon(a, |\cdot|) \mid \varepsilon \in \mathbb{R}_{>0}\}$ is for every $a \in K$ a fundamental system of neighborhoods of a in the topology $\mathcal{T}$ induced by this metric. This shows that the topology $\mathcal{T}$ does not depend on λ but only on the equivalence class of $|\cdot|$. We call $\mathcal{T}$ the **$|\cdot|$-topology** or the **valuation topology** of the valued field $(K, |\cdot|)$. In all topological considerations we may assume that $|\cdot|$ satisfies the triangle inequality (otherwise we replace $|\cdot|$ with $|\cdot|^\lambda$ for a suitable $\lambda \in \mathbb{R}_{>0}$). We make use of the notions and results of the theory of metric spaces. In particular, we apply the concepts of convergent and Cauchy sequences, of completeness, and of (uniform) continuity of mappings. We adapt these concepts for the theory of valued fields and for that we assume now that the absolute value in question satisfies the triangle inequality.

A sequence $(x_n)_{n\geq 0}$ in K is called **convergent** if there exists some $x \in K$ such that $(|x - x_n|)_{n\geq} \to 0$ (in $\mathbb{R}$). In this case, x is uniquely determined and is called the **limit point** of $(x_n)_{n\geq 0}$. We write $(x_n)_{n\geq 0} \to x$, and more precisely (if necessary)

$$|\cdot|\text{-}\lim_{n\to\infty} x_n = x \quad \text{or} \quad (x_n)_{n\geq 0} \xrightarrow{|\cdot|} x.$$

A sequence $(x_n)_{n\geq 0}$ in K is called

- a **Cauchy sequence** if for every $\varepsilon \in \mathbb{R}_{>0}$ there exists some $n_0 \geq 0$ such that $|x_n - x_m| < \varepsilon$ for all $n \geq m \geq n_0$;
- a **zero sequence** if $(x_n)_{n\geq 0} \to 0$ [equivalently, $(|x_n|)_{n\geq 0} \to 0$];
- **bounded** if the sequence $(|x_n|)_{n\geq 0}$ is bounded.

An element $x \in K$ is called a **cluster point** of a sequence $(x_n)_{n\geq 0}$ if it is the limit point of a subsequence [equivalently, for every $\varepsilon \in \mathbb{R}_{>0}$ there are infinitely many $n \geq 0$ satisfying $|x - x_n| < \varepsilon$].

A map $f\colon K \to K'$ between valued fields $K = (K, |\cdot|)$ and $K' = (K', |\cdot|')$ is called

- **continuous** in a point $a \in K$ if $(x_n)_{n\geq 0} \to a$ implies $(f(x_n))_{n\geq 0} \to f(a)$ for every sequence $(x_n)_{n\geq 0}$ in K [equivalently, for every $\varepsilon \in \mathbb{R}_{>0}$ there exists some $\delta \in \mathbb{R}_{>0}$ such that $|x-a| < \delta$ implies $|f(x)-f(a)|' < \varepsilon$ for all $x \in K$];

- **uniformly continuous** in a subset A of K if for every $\varepsilon \in \mathbb{R}_{>0}$ there exists some $\delta \in \mathbb{R}_{>0}$ such that $|x-y| < \delta$ implies $|f(x)-f(y)|' < \varepsilon$ for all $x,\ y \in K$.

The following assertions concerning properties of a valued field $K = (K, |\cdot|)$ or of a map $f\colon K \to K'$ between valued fields $K = (K, |\cdot|)$ and $K' = (K', |\cdot|')$ are proved literally as in the elementary analysis.

a. Let $(x_n)_{n\geq 0}$ and $(y_n)_{n\geq 0}$ be sequences in K and $x,\ y \in K$ such that $(x_n)_{n\geq 0} \to x$ and $(y_n)_{n\geq 0} \to y$. Then:

- $(x_n \pm y_n)_{n\geq 0} \to x \pm y$ and $(x_n y_n)_{n\geq 0} \to xy$.

- If $x \neq 0$, then $x_n \neq 0$ for all $n \gg 1$, and $(x_n^{-1})_{n\gg 1} \to x^{-1}$.

- $(|x_n|)_{n\geq 0} \to |x|$.

In particular, the maps

$$\begin{cases} K\times K & \to & K \\ (x,y) & \mapsto & x+y \end{cases}, \quad \begin{cases} K\times K & \to & K \\ (x,y) & \mapsto & xy \end{cases} \quad \text{and} \quad \begin{cases} K^{\times} & \to & K \\ x & \mapsto & x^{-1} \end{cases}$$

are continuous, and K, equipped with the $|\cdot|$-topology, is a topological field. For a thorough study of topological groups and fields we refer to the forthcoming volume on the foundations and main theorems of class field theory.

b. If $(x_n)_{n\geq 0}$ is a zero sequence and $(y_n)_{n\geq 0}$ is bounded, then $(x_n y_n)_{n\geq 0}$ is a zero sequence.

c. Every convergent sequence is a Cauchy sequence, and every Cauchy sequence is bounded.

d. Let $(x_n)_{n\geq 0}$ and $(y_n)_{n\geq 0}$ be Cauchy sequences. Then $(x_n \pm y_n)_{n\geq 0}$ and $(x_n y_n)_{n\geq 0}$ are Cauchy sequences, too . If 0 is not a cluster point of $(x_n)_{n\geq 0}$, then $x_n \neq 0$ for all $n \gg 1$, and $(x_n^{-1})_{n\gg 1}$ is a Cauchy sequence.

e. Every cluster point of a Cauchy sequence is a limit point.

f. If $(x_n)_{n\geq 0}$ is a Cauchy sequence, then the sequence $(|x_n|)_{n\geq 0}$ is a Cauchy sequence in $\mathbb{R}$ (and thus convergent).

g. The map $|\cdot|\colon K \to \mathbb{R}_{\geq 0}$ is uniformly continuous.

h. Let $f\colon K \to K'$ be uniformly continuous. If $(x_n)_{n\geq 0}$ is a Cauchy sequence in K, then $(f(x_n))_{n\geq 0}$ is a Cauchy sequence in K'.

i. A field homomorphism $\varphi\colon K \to K'$ is uniformly continuous if and only if it is continuous in 0.

j. Let $\varphi\colon K \to K'$ be a value-preserving homomorphism. Then its image $(\varphi(K), |\cdot|' \restriction \varphi(K))$ is a valued field, and $\varphi\colon K \to \varphi(K)$ is an isomorphism of valued fields. In particular, φ is a topological isomorphism [i. e., it is both an isomorphism and a homeomorphism].

Theorem 5.2.2. *Two absolute values of a field K are equivalent if and only if their valuation topologies coincide.*

Proof. Equivalent absolute values have the same valuation topology by the very definiton of the valuation topology. Thus let $|\cdot|_1$ and $|\cdot|_2$ be absolute values of a field K which are not equivalent. By 5.1.5 there exists some $x \in K$ such that $|x|_1 < 1$ and $|x|_2 \geq 1$. Then it follows that $(x^n)_{n\geq 0} \overset{|\cdot|_1}{\to} 0$, whereas $|x^n|_2 \geq 1$ for all $n \geq 0$. Hence the two valuation topologies are distinct. □

Definitions and Remarks 5.2.3. Let $K = (K, |\cdot|)$ be a valued field.

1. K is called **complete** if every Cauchy sequence in K is convergent. A subset A of K is called **complete** if every Cauchy sequence in A has a limit point in A. Every complete subset of K is closed, and if K is complete, then every closed subset of K is complete.

2. A valued field $K' = (K', |\cdot|')$ is called a **completion** of $(K, |\cdot|)$ if K' is complete, K is a dense subfield of K', and $|\cdot|' \restriction K = |\cdot|$.

3. If $(K', |\cdot|')$ is a completion of $(K, |\cdot|)$ and $\lambda \in \mathbb{R}_{>0}$, then $(K', |\cdot|'^{\lambda})$ is a completion of $(K, |\cdot|^{\lambda})$.

If $(K', |\cdot|')$ is a completion of $(K, |\cdot|)$, then we usually write again $|\cdot|$ instead of $|\cdot|'$ and call $K' = (K', |\cdot|)$ a completion of K.

4. $(\mathbb{R}, |\cdot|_\infty)$ is a completion of $(\mathbb{Q}, |\cdot|_\infty)$.

Theorem 5.2.4 (Completion theorem). *Let $K = (K, |\cdot|)$ be a valued field.*

> *1. K has a completion.*
>
> *2. Let $K^* = (K^*, |\cdot|^*)$ be a complete valued field, $f\colon K \to K^*$ a value-preserving homomorphism and $K' = (K', |\cdot|')$ a completion of K. Then there exists a unique value-preserving homomorphism $f'\colon K' \to K^*$ such that $f' \restriction K = f$.*
>
> *3. Let $K_1 = (K_1, |\cdot|_1)$ be another valued field and $\varphi\colon K \to K_1$ an isomorphism of valued fields. Let K' be a completion of K and K_1' a completion of K_1. Then there exists a unique isomorphism of valued fields $\varphi'\colon K' \to K_1'$ such that $\varphi' \restriction K = \varphi$. In particular: If K' and K'' are completions of K, then there exists a unique isomorphism of valued fields $\Phi\colon K' \to K''$ such that $\Phi \restriction K = \mathrm{id}_K$.*
>
> *4. Let K be complete, K_0 a subfield of K and $|\cdot|_0 = |\cdot| \restriction K_0$ an absolute value of K_0. Then the (topological) closure $\overline{K_0}$ of K_0 in K is a completion of the valued field $K_0 = (K_0, |\cdot| \restriction K_0)$.*

Proof. We may assume that all involved absolute values fulfill the triangle inequality (see 5.2.1).

1. We show:

A. There exists a valued field $K^* = (K^*, |\cdot|^*)$ and a homomorphism of valued fields $f\colon K \to K^*$ such that the following conditions are fulfilled:

- For every Cauchy sequence $(x_n)_{n\geq 0}$ in K the sequence $(f(x_n))_{n\geq 0}$ is convergent.
- For every $x^* \in K^*$ there exists a sequence $(x_n)_{n\geq 0}$ in K such that $(f(x_n))_{n\geq 0} \to x^*$.

Proof of **A.** Let CF be the set of all Cauchy sequences in K, NF the set of all zero sequences in K, and let $\mathsf{c}\colon K \to \mathsf{CF}$ be defined by $\mathsf{c}(x) = (x)_{n\geq 0}$. As we have mentioned in 5.2.1.**d**, CF is a ring under the component-wise addition and multiplication of sequences, $\mathsf{c}\colon K \to \mathsf{CF}$ is a ring monomorphism, and $\mathsf{NF} = \mathsf{CF} \setminus \mathsf{CF}^\times$ is a maximal ideal of CF. We set $K^* = \mathsf{CF}/\mathsf{NF}$, and we define $f\colon K \to K^*$ by $f(x) = \mathsf{c}(x) + \mathsf{NF}$. Then f is a field homomorphism.

If $(x_n)_{n\geq 0} \in \mathsf{CF}$, then $(|x_n|)_{n\geq 0}$ is a convergent sequence in $\mathbb{R}$, and if $(y_n)_{n\geq 0} \in \mathsf{NF}$, then $\big||x_n + y_n| - |x_n|\big| \leq |y_n|$ for all $n \geq 0$, which implies that

$$\lim_{n\to\infty} |x_n| = \lim_{n\to\infty} |x_n + y_n| \quad \text{only depends on the coset} \quad (x_n)_{n\geq 0} + \mathsf{NF} \in K^*.$$

We define $|\cdot|^*\colon K^* \to \mathbb{R}_{\geq 0}$ by

$$\big|(x_n)_{n\geq 0} + \mathsf{NF}\big|^* = \lim_{n\to\infty} |x_n|.$$

It is easily checked that $|\cdot|^*$ is an absolute value of K^*. To finish the proof of **A** we must show: If $(x_n)_{n\geq 0} \in \mathsf{CF}$, then $(f(x_n))_{n\geq 0} \to (x_n)_{n\geq 0} + \mathsf{NF}$ in K^*.

Let $(x_n)_{n\geq 0} \in \mathsf{CF}$ and $\varepsilon \in \mathbb{R}_{>0}$. There exists some $k_0 \geq 0$ such that $|x_k - x_n| \leq \varepsilon$ for all $n \geq k \geq k_0$, and for $k \geq k_0$ we then obtain

$$\big|f(x_k) - [(x_n)_{n\geq 0} + \mathsf{NF}]\big|^* = \big|(x_k - x_n)_{n\geq 0} + \mathsf{NF}\big|^* = \lim_{n\to\infty} |x_k - x_n| \leq \varepsilon. \qquad \square[\mathbf{A}.]$$

Using the exchange principle, **A** implies: There exists an extension field K' of K and an isomorphism $f'\colon K' \to K^*$ such that $f' \restriction K = f$. Now we define $|\cdot|'\colon K' \to \mathbb{R}_{\geq 0}$ by $|x|' = |f'(x)|^*$ for all $x \in K'$. Then $|\cdot|'$ is an absolute value of K', $|\cdot|' \restriction K = |\cdot|$, every Cauchy sequence of K is convergent in K', and every $x \in K'$ is the limit point of a sequence in K. Thus it remains to prove that $K' = (K', |\cdot|')$ is complete.

Let $(x'_n)_{n\geq 0}$ be a Cauchy sequence in K', and for $n \geq 0$ let $(x_{n,k})_{k\geq 0}$ be a sequence in K such that $(x_{n,k})_{k\geq 0} \to x'_n$ in K'. For every $n \geq 1$ let $k_n \geq 0$ be such that $|x_{n,k_n} - x'_n|' < 1/n$. Then we obtain, for all $m \geq n \geq 1$,

$$\begin{aligned} |x_{n,k_n} - x_{m,k_m}| &= |x_{n,k_n} - x_{m,k_m}|' \\ &\leq |x_{n,k_n} - x'_n|' + |x'_n - x'_m|' + |x_{m,k_m} - x'_m|' \\ &< |x'_n - x'_m|' + \frac{1}{n} + \frac{1}{m}. \end{aligned}$$

Hence $(x_{n,k_n})_{n\geq 1} \in \mathsf{CF}$, and thus there exists some $x' \in K'$ such that $(x_{n,k_n})_{n\geq 0} \to x'$. Since $|x'-x'_n|' \leq |x'-x_{n,k_n}|' + |x_{n,k_n} - x'_n|'$ for all $n \geq 0$, it follows that $(x'_n)_{n\geq 0} \to x'$.

2. *Uniqueness*: Let $f' \colon K' \to K^*$ be a homomorphism of valued fields such that $f' \restriction K = f$. If $x' \in K'$, then there exists a sequence $(x_n)_{n\geq 0}$ in K such that $(x_n)_{n\geq 0} \to x'$. It follows that $|f(x_n) - f'(x')|^* = |f'(x_n - x')|^* = |x_n - x'|'$, and therefore $f'(x') = |\cdot|^*\text{-}\lim_{n\to\infty} f'(x)$.

Existence: For $x' \in K'$ let $(x_n)_{n\geq 0}$ be a sequence in K such that $(x_n)_{n\geq 0} \to x'$. For all $n \geq m \geq 0$ we obtain

$$\begin{aligned} |f(x_n) - f(x_m)|^* &= |f(x_m - x_n)|^* = |x_m - x_n| \\ &= |x_m - x_n|' \leq |x_m - x'|' + |x_n - x'|', \end{aligned}$$

hence $(f(x_n))_{n\geq 0}$ is a Cauchy sequence in K^*, and therefore $(f(x_n))_{n\geq 0} \to x^*$ for some $x^* \in K^*$. We assert that x^* only depends on x'. Indeed, if $(y_n)_{n\geq 0}$ is another sequence in K with limit point x', then $(x_n - y_n)_{n\geq 0} \to 0$, and as $|f(x_n) - f(y_n)|^* = |x_n - y_n|$, it follows that $(f(x_n) - f(y_n))_{n\geq 0} \to 0$ and $(f(y_n))_{n\geq 0} \to x^*$.

Define $f' \colon K' \to K^*$ by $f'(x') = |\cdot|^*\text{-}\lim_{n\to\infty} f(x_n)$. Then $f' \restriction K = f$, and since

$$|f'(x')|^* = \lim_{n\to\infty} |f(x_n)|^* = \lim_{n\to\infty} |x_n| = \lim_{n\to\infty} |x_n|' = |x'|,$$

it follows that f' is value-preserving. If $x', x'' \in K'$, let $(x'_n)_{n\geq 0}$, $(x''_n)_{n\geq 0}$ be sequences in K such that $(x'_n)_{n\geq 0} \to x'$ and $(x''_n)_{n\geq 0} \to x''$. Then apparently

$$\begin{aligned} &(x'_n \mathbin{+\!\!.} x''_n)_{n\geq 0} \to x' \mathbin{+\!\!.} x'', \ (f(x'_n))_{n\geq 0} \to f(x'), \ (f(x''_n))_{n\geq 0} \to f(x''), \\ &(f(x'_n \mathbin{+\!\!.} x''_n))_{n\geq 0} \to f(x' \mathbin{+\!\!.} x''), \\ &\text{and also } (f(x'_n \mathbin{+\!\!.} x''_n))_{n\geq 0} = (f(x'_n) \mathbin{+\!\!.} f(x''_n))_{n\geq 0} \to f(x') \mathbin{+\!\!.} f(x''). \end{aligned}$$

Hence it follows that $f(x' \mathbin{+\!\!.} x'') = f(x') \mathbin{+\!\!.} f(x'')$, and f is a homomorphism.

3. By 2., there exist homomorphisms of valued fields $\Phi \colon K' \to K'_1$ such that $\Phi \restriction K = \varphi$ and $\Phi_1 \colon K'_1 \to K'$ such that $\Phi_1 \restriction K_1 = \varphi^{-1}$. Then $\Phi_1 \circ \Phi \colon K' \to K'$ is a homomorphism of valued fields satisfying $\Phi_1 \circ \Phi \restriction K = \mathrm{id}_{K'} \restriction K = \mathrm{id}_K$ and the uniqueness in 2. implies $\Phi_1 \circ \Phi = \mathrm{id}_{K'}$. Alike we obtain $\Phi \circ \Phi_1 = \mathrm{id}_{K'_1}$, and thus Φ is an isomorphism of valued fields.

4. Being closed in K, $\overline{K}_0$ is complete, and K_0 is dense in $\overline{K}_0$. Hence it suffices to prove that $\overline{K}_0$ is a subfield of K. If $x, y \in \overline{K}_0$, then there exist sequences $(x_n)_{n\geq 0}$ and $(y_n)_{n\geq 0}$ in K_0 such that $(x_n)_{n\geq 0} \to x$ and $(y_n)_{n\geq 0} \to y$. Then $(x_n + y_n)_{n\geq 0} \to x + y$, $(x_n y_n)_{n\geq 0} \to xy$, and if $x \neq 0$, then $(x_n^{-1})_{n\gg 1} \to x^{-1}$. Hence $x + y \in \overline{K}_0$, $xy \in \overline{K}_0$, and if $x \neq 0$, then $x^{-1} \in \overline{K}_0$. Consequently, $\overline{K}_0$ is a subfield of K. □

Corollary 5.2.5. *Let K be an algebraic number field and (r_1, r_2) its signature. Let $\sigma_1, \ldots, \sigma_{r_1}$ be the real embeddings and $(\sigma_{r_1+1}, \overline{\sigma_{r_1+1}}), \ldots, (\sigma_{r_1+r_2}, \overline{\sigma_{r_1+r_2}})$ the pairs of conjugate complex embeddings of K* (see 3.1.1). *For $j \in [1, r_1]$ we set $l_j = 1$ and $\mathbb{K}_j = \mathbb{R}$, and for $j \in [r_1+1, r_1+r_2]$ we set $l_j = 2$ and $\mathbb{K}_j = \mathbb{C}$.*

For $j \in [1, r_1 + r_2]$ let $|\cdot|_{\infty,j} = |\cdot|_\infty^{l_j} \circ \sigma_j$ be the normalized Archimedean absolute value of K (as in 5.1.3.**3**) *and $\widehat{K}_j$ a completion of $(K, |\cdot|_{\infty,j})$. Then the embedding $\sigma_j \colon K \to \mathbb{K}_j$ has a unique extension to an isomorphism of valued fields $\widehat{\sigma}_j \colon \widehat{K}_j \xrightarrow{\sim} (\mathbb{K}_j, |\cdot|_\infty^{l_j})$.*
We call $|\cdot|_{\infty,j}$ the absolute value and $\widehat{K}_j$ the completion corresponding to the embedding σ_j.

Proof. Let $j \in [1, r_1+r_2]$ and note that $|\cdot|_{\infty,j} \restriction \mathbb{Q} = |\cdot|_\infty^{l_j}$. Let $\widehat{K}_j = (\widehat{K}_j, |\cdot|_{\infty,j})$ be a completion of $(K, |\cdot|_{\infty,j})$, $\nu_j = (K \hookrightarrow \widehat{K}_j)$ the inclusion and $\widetilde{\mathbb{Q}}$ the (topological) closure of $\mathbb{Q}$ in $\widehat{K}_j$,

$$\begin{array}{ccccc} \sigma_j : K & \xrightarrow{\nu_j} & \widehat{K}_j & \xrightarrow{\widehat{\sigma}_j} & \mathbb{K}_j \\ \uparrow & & \uparrow & & \uparrow \\ \mathbb{Q} & \longrightarrow & \widetilde{\mathbb{Q}} & \xrightarrow{\cong} & \mathbb{R}. \end{array}$$

By definition, the embedding $\sigma_j \colon (K, |\cdot|_{\infty,j}) \to (\mathbb{K}_j, |\cdot|_\infty^{l_j})$ is a value-preserving homomorphism, and $(\mathbb{K}_j, |\cdot|_\infty^{l_j})$ is complete. By 5.2.4.2, $\sigma_j \colon K \to \mathbb{K}_j$ has a unique extension to a value-preserving homomorphism $\widehat{\sigma}_j \colon \widehat{K}_j \to (\mathbb{K}_j, |\cdot|_\infty^{l_j})$. Since $\widehat{\sigma}_j \restriction \mathbb{Q} = \sigma_j \restriction \mathbb{Q} = \mathrm{id}_\mathbb{Q}$ and $(\widetilde{\mathbb{Q}}, |\cdot|_{\infty,j} \restriction \widetilde{\mathbb{Q}})$ is a completion of $(\mathbb{Q}, |\cdot|_\infty^{l_j})$ by 5.2.4.4, it follows by 5.2.4.3 that $\widehat{\sigma}_j \restriction \widetilde{\mathbb{Q}} \colon (\widetilde{\mathbb{Q}}, |\cdot|_{\infty,j} \restriction \widetilde{\mathbb{Q}}) \to (\mathbb{R}, |\cdot|_\infty^{l_j})$ is an isomorphism of valued fields. Since $\mathbb{R} = \widehat{\sigma}_j(\widetilde{\mathbb{Q}}) \subset \widehat{\sigma}_j(\widehat{K}_j) \subset \mathbb{K}_j$ and $\sigma_j(K) \subset \widehat{\sigma}_j(K_j)$, we conclude that $\widehat{\sigma}_j(\widehat{K}_j) = \mathbb{K}_j$ and that $\widehat{\sigma}_j$ is an isomorphism of valued fields. $\square$

Remark and Example 5.2.6. Let assumptions be as in 5.2.5.

For fixed $j \in [1, r_1 + r_2]$, we can identify $\widetilde{\mathbb{Q}}$ with $\mathbb{R}$ and $\widehat{K}_j$ with $\mathbb{K}_j$ by means of $\widehat{\sigma}_j$ (so that $\mathbb{R} \subset \widehat{K}_j$), and sometimes this will be convenient. However, as soon as different embeddings come into play, one must be careful. For example, if $\sigma_1, \sigma_2 \colon K \to \mathbb{R}$ are distinct real embeddings and $\widehat{K}_1$, $\widehat{K}_2$ are the corresponding completions, then $\widehat{K}_1 \cong \mathbb{R}$ and $\widehat{K}_2 \cong \mathbb{R}$, but a simultaneous identification $\widehat{K}_1 = \mathbb{R} = \widehat{K}_2$ would cause confusion.

As an example, consider $K = \mathbb{Q}(\sqrt[4]{2})$. There are 2 real embeddings $\sigma_1, \sigma_2 \colon K \to \mathbb{R}$, defined by $\sigma_1(\sqrt[4]{2}) = \sqrt[4]{2}$ (hence $\sigma_1 = (K \hookrightarrow \mathbb{R})$, and $\sigma_2(\sqrt[4]{2}) = -\sqrt[4]{2}$; and one pair of complex conjgate embeddings $(\sigma_3, \overline{\sigma}_3)$, given by $\sigma_3(\sqrt[4]{2}) = \mathrm{i}\sqrt[4]{2}$. The corresponding absolute values are given as follows for q_0, a_1, a_2, $a_3 \in \mathbb{Q}$:

$$\begin{aligned} |a_0 + a_1\sqrt[4]{2} + a_2\sqrt{2} + a_3\sqrt[4]{8}\,|_{\infty,1} &= |a_0 + a_1\sqrt[4]{2} + a_2\sqrt{2} + a_3\sqrt[4]{8}\,|_\infty, \\ |a_0 + a_1\sqrt[4]{2} + a_2\sqrt{2} + a_3\sqrt[4]{8}\,|_{\infty,2} &= |a_0 - a_1\sqrt[4]{2} + a_2\sqrt{2} - a_3\sqrt[4]{8}\,|_\infty, \\ |a_0 + a_1\sqrt[4]{2} + a_2\sqrt{2} + a_3\sqrt[4]{8}\,|_{\infty,3} &= |a_0 + \mathrm{i}a_1\sqrt[4]{2} - a_2\sqrt{2} - \mathrm{i}a_3\sqrt[4]{8}\,|_\infty^2. \end{aligned}$$

For $i \in \{1,2\}$, $\sigma_i\colon K \to \mathbb{R}$ extends to $\widehat{\sigma}_i\colon \widehat{K}_i \stackrel{\sim}{\to} \mathbb{R}$, and $\sigma_3\colon K \to \mathbb{C}$ extends to $\widehat{\sigma}_3\colon \widehat{K}_3 \to \mathbb{C}$. In particular, $\widehat{\sigma}_2^{-1} \circ \widehat{\sigma}_1\colon \widehat{K}_1 \to \widehat{K}_2$ is a field isomorphism, but it is not value-preserving.

In the following results we assume again that the absolute values in question satisfy the triangle inequality (see 5.2.1). In all relevant applications this assumption is automatically fulfilled.

Definitions and Remarks 5.2.7. Let $(K, |\cdot|)$ be a valued field such that $|\cdot|$ fulfills the triangle inequality, and let V be a vector space over K.

1. A ($|\cdot|$-compatible) **norm** of V is a map $\|\cdot\|\colon V \to \mathbb{R}_{\ge 0}$ having the following properties for all u, $v \in V$ and $\lambda \in K$:

(N1) $\|u\| = 0$ if and only if $u = 0$.

(N2) $\|u+v\| \le \|u\| + \|v\|$.

(N3) $\|\lambda u\| = |\lambda|\,\|u\|$.

2. Let $\|\cdot\|\colon V \to \mathbb{R}_{\ge 0}$ be a norm. The map $V \times V \to \mathbb{R}_{\ge 0}$, $(u,v) \mapsto \|u-v\|$, is a metric. Hence $V = (V, \|\cdot\|)$ is a metric space, and as for K itself we adapt the basic concepts from the theory of metric spaces.

For $a \in V$ and $\varepsilon \in \mathbb{R}_{>0}$, the open ε-ball around a is given by

$$B_\varepsilon(a) = B_\varepsilon(a, \|\cdot\|) = \{u \in V \mid \|u-a\| < \varepsilon\} = a + B_\varepsilon(0, \|\cdot\|),$$

and $\{B_\varepsilon(a) \mid \varepsilon \in \mathbb{R}_{>0}\}$ is a fundamental system of neighborhoods of a.

A sequence $(u_n)_{n\ge 0}$ in V is called **convergent** if there exists some $u \in V$ such that $(\|u_n - u\|)_{n \ge 0} \to 0$. If this is the case, then u is uniquely determined, and we write

$$(u_n)_{n\ge 0} \to u \quad \text{or (more precisely if necessary)} \quad \|\cdot\|\text{-}\lim_{n\to\infty} u_n = u.$$

A sequence $(u_n)_{n \ge 0}$ in V is called a **Cauchy sequence** if for every $\varepsilon \in \mathbb{R}_{>0}$ there exists some $n_0 \ge 0$ such that $\|u_n - u_m\| < \varepsilon$ for all $m \ge n \ge n_0$. Every convergent sequence in V is a Cauchy sequence, and $V = (V, \|\cdot\|)$ is called **complete** if every Cauchy sequence is convergent.

3. Two norms $\|\cdot\|_1$ and $\|\cdot\|_2$ of V are called **equivalent** if there exist C_1, $C_2 \in \mathbb{R}_{>0}$ such that $\|u\|_2 \le C_1 \|u\|_1$ and $\|u\|_1 \le C_2 \|u\|_2$ for all $u \in V$. Equivalent norms induce the same topology (and thus the same notions of convergence, Cauchy sequences and completeness).

4. Let $\boldsymbol{u} = (u_1, \ldots, u_n)$ be a K-basis of V, and define the $\boldsymbol{u}$-norm

$$\|\cdot\|_{\boldsymbol{u}}\colon V \to \mathbb{R}_{\ge 0} \quad \text{by} \quad \|c_1u_1 + \ldots + c_nu_n\|_{\boldsymbol{u}} = \max\{|c_1|, \ldots, |c_n|\}.$$

Apparently $\|\cdot\|_{\boldsymbol{u}}$ is a $|\cdot|$-compatible norm of V.

5. Let $\|\cdot\|_0 \colon K^n \to \mathbb{R}_{\geq 0}$ be the maximum norm of K^n, defined by

$$\|(c_1,\ldots,c_n)\|_0 = \max\{|c_1|,\ldots,|c_n|\}.$$

$\|\cdot\|_0$ is the $\boldsymbol{u}$-norm with respect to the canonical basis $\boldsymbol{u} = (\boldsymbol{e}_1,\ldots,\boldsymbol{e}_n)$ of K^n; it is a $|\cdot|$-compatible norm of K^n, and if $\boldsymbol{c} = (c_1,\ldots,c_n) \in K^n$, then

$$B_\varepsilon(\boldsymbol{c}) = \prod_{i=1}^{n} B_\varepsilon(c_i) \quad \text{for all} \quad \varepsilon \in \mathbb{R}_{>0}.$$

Consequently $\|\cdot\|_0$ induces the product topology on K^n, and for every sequence $(\boldsymbol{x}_k)_{k\geq 0} = (x_{k,1},\ldots,x_{k,n})_{n\geq 0}$ in K^n we obtain:

- $(\boldsymbol{x}_k)_{k\geq 0} \to \boldsymbol{x} = (x_1,\ldots,x_n) \in K^n$ if and only if $(x_{k,i})_{k\geq 0} \to x_i$ for all $i \in [1,n]$;
- $(\boldsymbol{x}_k)_{k\geq 0}$ is a Cauchy sequence if and only if $(x_{k,i})_{k\geq 0}$ is a Cauchy sequence for all $i \in [1,n]$.

In particular, if K is complete, then K^n is complete with respect to $\|\cdot\|_0$.

6. Again let V be any vector space over K, $\dim_K(V) = n \in \mathbb{N}$, let $\boldsymbol{u} = (u_1,\ldots,u_n)$ a basis of V and $\|\cdot\|_{\boldsymbol{u}}$ the $\boldsymbol{u}$-norm defined in **4.** above. Let $\Phi \colon K^n \to V$ be the isomorphism defined by $\Phi(c_1,\ldots,c_n) = c_1u_1 + \ldots + c_nu_n$. Then $\|\Phi(\boldsymbol{x})\|_{\boldsymbol{u}} = \|\boldsymbol{x}\|_0$ for all $\boldsymbol{x} \in K^n$, and Φ is a topological isomorphism. In particular, if K is complete, then $(V, \|\cdot\|_{\boldsymbol{u}})$ is complete.

Theorem 5.2.8 (Norm equivalence theorem). *Let $(K, |\cdot|)$ be a complete valued field such that $|\cdot|$ fulfills the triangle inequality, and let V be a finite-dimensional vector space over K. Then any two norms of V are equivalent, and V is complete with respect to each of them.*

Proof. Let $\dim_K(V) = n$ and $\boldsymbol{u} = (u_1,\ldots,u_n)$ be a K-basis of V. By 5.2.7.6 it suffices to prove that every $|\cdot|$-compatible norm of V is equivalent to $\|\cdot\|_{\boldsymbol{u}}$. We proceed by induction on n. Let $\|\cdot\|$ be a $|\cdot|$-compatible norm of V.

$n = 1$: Let $V = Ku_1$ and $u = c_1u_1 \in V$, where $c_i \in K$. Then it follows that $\|u\| = |c_1|\,\|u_1\| = \|u\|_{\boldsymbol{u}}\|u_1\|$, and therefore $\|\cdot\|_{\boldsymbol{u}}$ and $\|\cdot\|$ are equivalent.

$n \geq 2$, $n-1 \to n$. If $a = c_1u_1 + \ldots + c_nu_n \in V$, where $c_1,\ldots,c_n \in K$, then

$$\|a\| \leq \sum_{i=1}^{n} |c_i|\,\|u_i\| \leq \Big(\sum_{i=1}^{n} \|u_i\|\Big)\|a\|_{\boldsymbol{u}},$$

and it remains to prove:

There exists some $C \in \mathbb{R}_{>0}$ such that $\|a\|_{\boldsymbol{u}} \leq C\,\|a\|$ for all $a \in V$.

We assume the contrary. Then for every $k \geq 1$ these exists some $a_k \in V$ such that $\|a_k\|_{\boldsymbol{u}} > k\|a_k\|$, say $a_k = a_{k,1}u_1 + \ldots + a_{k,n}u_n$, where $a_{k,i} \in K$, and there exists some $i \in [1,n]$ such that $\|a_k\|_{\boldsymbol{u}} = |a_{k,i}|$ for infinitely many $n \geq 1$. By renumbering (if necessary) and proceeding with a suitable subsequence we may assume that $i = n$ and $\|a_k\|_{\boldsymbol{u}} = |a_{k,n}| > k\|a_k\|$ for all $k \geq 1$. We set $b_k = a_{k,n}^{-1}a_k = b_{k,1}u_1 + \ldots + b_{k,n-1}u_{n-1} + u_n$, where $b_{k,i} = a_{k,n}^{-1}a_{k,i}$ for all $i \in [1,n]$. Then $\|b_k\|_{\boldsymbol{u}} = |a_{k,n}|^{-1}\|a_k\|_{\boldsymbol{u}} = 1 > k\|b_k\|$, hence

$$\|b_k\| < \frac{1}{k} \quad \text{for all} \quad k \geq 1, \qquad \text{and therefore} \qquad (b_k)_{k\geq 1} \to 0.$$

Let $V_0 = Ku_1 + \ldots + Ku_{n-1} \subset V$, and let $\pi\colon V \to V_0$ be the unique K-homomorphism satisfying $\pi \restriction V_0 = \mathrm{id}_{V_0}$ and $\pi(u_n) = 0$. Obviously, $\|\cdot\| \restriction V_0$ is a $|\cdot|$-compatible norm, and the induction hypothesis implies that $(V_0, \|\cdot\| \restriction V_0)$ is complete.

For all $m \geq k \geq 1$ we obtain (observing that $b_m - b_k \in V_0$)

$$\|\pi(b_k) - \pi(b_m)\| = \|\pi(b_k - b_m)\| = \|b_k - b_m\| \leq \|b_k\| + \|b_m\| < \frac{1}{k} + \frac{1}{m}.$$

Hence $(\pi(b_k))_{k\geq 1}$ is a Cauchy sequence in V_0 and therefore $(\pi(b_k))_{k\geq 1} \to b^*$ for some $b^* \in V_0$. Since $b_k = \pi(b_k) + u_n$ for all $k \geq 1$, it ultimately follows that $(b_k)_{k\geq 1} \to b^* + u_n$, a contradiction to $(b_k)_{k\geq 1} \to 0$. □

Corollary 5.2.9. *Let $(K, |\cdot|)$ be a complete valued field such that $|\cdot|$ satisfies the triangle equality, L/K a field extension and $|\cdot|_L$ an absolute value of L such that $|\cdot|_L \restriction K = |\cdot|$.*

1. *Assume that $[L{:}K] < \infty$. Then $|\cdot|_L$ is the only extension of $|\cdot|$ to L, and $(L, |\cdot|_L)$ is complete. If $(u_1, \ldots, u_n)$ is a K-basis of L and K^n is equipped with the product topology, then the K-isomorphism $\Phi\colon K^n \to L$, defined by $\Phi(c_1, \ldots, c_n) = c_1u_1 + \ldots + c_nu_n$, is a homeomorphism.*

2. *Let $\mathcal{O}$ be a closed subring of K such that $K = \mathsf{q}(\mathcal{O})$.*

(a) *Let M be a free $\mathcal{O}$-submodule of L, $\mathrm{rk}_{\mathcal{O}}(M) = n \in \mathbb{N}$ and $(u_1 \ldots, u_n)$ an $\mathcal{O}$-basis of M. Then M is closed in L, and if $\mathcal{O}^n$ is equipped with the product topology, then the $\mathcal{O}$-isomorphism $\phi_0\colon \mathcal{O}^n \to M$, defined by $\phi_0(c_1, \ldots, c_n) = c_1u_1 + \ldots + c_nu_n$, is a homeomorphism.*

(b) *Let $\mathcal{O}_0$ be a dense subring of $\mathcal{O}$ and M_0 a finitely generated $\mathcal{O}_0$-submodule of L. Then M_0 is dense in $\mathcal{O}M_0$, and if $\mathcal{O}$ is a principal ideal domain, then $\mathcal{O}M_0$ is complete and closed in L.*

Proof. 1. Let $|\cdot|'$ be another absolute value of L such that $|\cdot|' \restriction K = |\cdot|$. Then $|\cdot|_L$ and $|\cdot|'$ are $|\cdot|$-compatible norms of L. By 5.2.8 $|\cdot|_L$ and $|\cdot|'$ are equivalent norms on L, and $(L, |\cdot|_L)$ is complete. Since $|\cdot|_L$ and $|\cdot|'$ induce the same topology, they are equivalent absolute values by 5.2.2, and thus 5.1.3.**1** implies $|\cdot|_L = |\cdot|'$. If L is equipped with the $\boldsymbol{u}$-norm, then Φ is a topological isomorphism by 5.2.7.**6**. However, by 5.2.8 the $\boldsymbol{u}$-norm induces the $|\cdot|_L$-topology.

2. Being a closed subring of K, $\mathcal{O}$ is complete.

(a) $KM = Ku_1 + \ldots + Ku_n$ is a K-subspace of L, $(u_1, \ldots, u_n)$ is a K-basis of KM, and $|\cdot| \restriction KM$ is a $|\cdot|$-compatible norm of KM. By 5.2.8, KM is complete, and the K-isomorphism

$$\Phi\colon K^n \to KM, \text{ defined by } \quad \Phi(c_1, \ldots, c_n) = c_1u_1 + \ldots + c_nu_n,$$

is a homeomorphism by 5.2.7.**6**. Since $M = \Phi(\mathcal{O}^n)$, $\phi_0 = \phi \restriction \mathcal{O}^n$, and $\mathcal{O}^n$ is complete in the product topology, it follows that M is complete and therefore closed in L.

(b) Let $z \in \mathcal{O}M_0$, say $z = c_1x_1 + \ldots + c_kx_k$, where $k \in \mathbb{N}$, $c_1, \ldots, c_k \in \mathcal{O}$ and $x_1, \ldots, x_k \in M_0$. For every $i \in [1, k]$ there exists a sequence $(c_{i,n})_{n \ge 0}$ in $\mathcal{O}_0$ such that $(x_{i,n})_{n \ge 0} \to x_i$. It follows that $z_n = c_{1,n}x_1 + \ldots + c_{k,n}x_k \in M_0$ for all $n \ge 0$, and $(z_n)_{n \ge 0} \to z$. Hence M_0 is dense in $\mathcal{O}M_0$. If $\mathcal{O}$ is a principal ideal domain, then $\mathcal{O}M_0$ is a finitely generated torsion-free $\mathcal{O}$-module, hence free by 2.3.5, and thus $\mathcal{O}M_0$ is complete and closed in L by (a). □

The following theorem of Ostrowski shows that $\mathbb{R}$ and $\mathbb{C}$ are essentially the only complete Archimedian valued field. Recall that $|\cdot|_\infty$ denotes the ordinary absolute on $\mathbb{C}$ and its subfields. For a map $\Phi\colon X \to \mathbb{C}$ of a set X into $\mathbb{C}$ we denote by $\overline{\Phi}\colon X \to \mathbb{C}$ the conjugate complex map, defined by $\overline{\Phi}(x) = \overline{\Phi(x)}$ for all $x \in X$.

Theorem 5.2.10 (Theorem of Ostrowski). *Let $(K, |\cdot|)$ be a complete Archimedian valued field. Then there exists an isomorphism of valued fields*

$$\Phi\colon (K, |\cdot|) \to (\mathbb{K}, |\cdot|_\infty^\lambda), \quad \textit{where} \quad \mathbb{K} \in \{\mathbb{R}, \mathbb{C}\} \quad \textit{and} \quad \lambda \in \mathbb{R}_{>0}.$$

If $\Phi'\colon (K, |\cdot|) \to (\mathbb{K}', |\cdot|_\infty^{\lambda'})$ is another isomorphism of valued fields, where $\mathbb{K}' \in \{\mathbb{R}, \mathbb{C}\}$ and $\lambda' \in \mathbb{R}_{>0}$, then $\mathbb{K}' = \mathbb{K}$, $\lambda' = \lambda$ and $\Phi' \in \{\Phi, \overline{\Phi}\}$.

Proof. We may assume that $|\cdot|$ satisfies the triangle inequality (otherwise we replace $|\cdot|$ with $|\cdot|^\kappa$ for a suitable exponent $\kappa \in \mathbb{R}_{>0}$).

Uniqueness: Let $\Phi\colon (K, |\cdot|) \to (\mathbb{K}, |\cdot|_\infty^\lambda)$ and $\Phi'\colon (K, |\cdot|) \to (\mathbb{K}', |\cdot|_\infty^{\lambda'})$ be isomorphisms of valued fields, where $\mathbb{K}, \mathbb{K}' \in \{\mathbb{R}, \mathbb{C}\}$ and $\lambda, \lambda' \in \mathbb{R}_{>0}$. Then

$$\Phi' \circ \Phi^{-1}\colon (\mathbb{K}, |\cdot|^\lambda) \to (\mathbb{K}', |\cdot|^{\lambda'})$$

is an isomorphism of valued fields, hence $\mathbb{K} = \mathbb{K}'$, $\lambda = \lambda'$, and $\Phi' \circ \Phi^{-1}$ is a continuous automorphism of $\mathbb{K}$. Since $\Phi' \circ \Phi^{-1} \restriction \mathbb{Q} = \mathrm{id}_\mathbb{Q}$; it follows that

$\Phi'\circ\Phi^{-1}\restriction\mathbb{R}=\mathrm{id}_{\mathbb{R}}$, and consequently either $\Phi'\circ\Phi=\mathrm{id}_{\mathbb{K}}$, or $\mathbb{K}=\mathbb{C}$ and $\Phi'\circ\Phi$ is the complex conjugation. In any case we obtain $\Phi'\in\{\Phi,\overline{\Phi}\}$.

Existence: By 5.1.10 it follows that $\mathbb{Q}\subset K$ and $|\cdot|\restriction\mathbb{Q}=|\cdot|_\infty^\lambda$ for some $\lambda\in\mathbb{R}_{>0}$. Since $(\mathbb{R},|\cdot|_\infty^\lambda)$ is a completion of $(\mathbb{Q},|\cdot|_\infty^\lambda)$, there exists a homomorphism of valued fields $\Psi\colon(\mathbb{R},|\cdot|_\infty^\lambda)\to(K,|\cdot|)$ by 5.2.4.2. Using the exchange principle, we may assume that $\mathbb{R}\subset K$ and $|\cdot|\restriction\mathbb{R}=|\cdot|_\infty^\lambda$. If now $\mathbb{R}=K$, then $\Phi=\Psi^{-1}\colon(K,|\cdot|)\to(\mathbb{R},|\cdot|_\infty^\lambda)$ is an isomorphism of valued fields, and we are done.

Thus assume that $\mathbb{R}\subsetneq K$. We prove the following assertion **A** which is the crucial point of our argument.

> **A.** For every $\xi\in K$ there exists some polynomial $g\in\mathbb{R}[X]$ such that $\partial(g)=2$ and $g(\xi)=0$.

Proof of **A.** Let $\xi\in K$. We assert that there exists some $z\in\mathbb{C}$ such that ξ is a zero of the polynomial $g=X^2-(z+\overline{z})X+z\overline{z}\in\mathbb{R}[X]$. We assume the contrary and define $f\colon\mathbb{C}\to\mathbb{R}_{\ge 0}$ by $f(z)=|\xi^2-(z+\overline{z})\xi+z\overline{z}|$. Then apparently f is continuous, and for all $z\in\mathbb{C}$ we obtain $f(z)>0$ and

$$f(z)\ge|z\overline{z}|\left[1-\frac{|\xi|^2}{|z\overline{z}|}-|\xi|\left|\frac{z+\overline{z}}{z\overline{z}}\right|\right]=|z|_\infty^{2\lambda}\left[1-\frac{|\xi|^2}{|z|_\infty^{2\lambda}}-|\xi|\left|\frac{1}{z}+\frac{1}{\overline{z}}\right|_\infty^\lambda\right].$$

Thus it follows that

$$\lim_{z\to\infty}f(z)=\infty,\quad\text{and therefore there exists}\quad m=\min f(\mathbb{C})\in\mathbb{R}_{>0}.$$

The set $S=\{z\in\mathbb{C}\mid f(z)=m\}$ is closed and bounded, hence compact, and thus there exists some $z_0\in S$ such that $|z_0|_\infty\ge|z|_\infty$ for all $z\in S$. Let $\varepsilon\in(0,m^{1/\lambda})$ and consider the polynomial

$$g_\varepsilon=X^2-(z_0+\overline{z}_0)X+z_0\overline{z}_0+\varepsilon=(X-z_1)(X-z_2)\in\mathbb{R}[X],$$

where z_1, $z_2\in\mathbb{C}$ and $|z_1|_\infty\ge|z_2|_\infty$. As $|z_1|_\infty^2\ge|z_1z_2|_\infty=z_0\overline{z}_0+\varepsilon>|z_0|_\infty^2$, it follows that $z_1\notin S$ and therefore $f(z_1)>m$.

For $n\in\mathbb{N}$ we consider the polynomial $G_n=(g_\varepsilon-\varepsilon)^n-(-\varepsilon)^n\in\mathbb{R}[X]$. Then $\partial(G_n)=2n,\ G_n(z_1)=0,$

$$G_n=\prod_{i=1}^{2n}(X-\alpha_i),\quad\text{where}\quad\alpha_1=z_1,\alpha_2,\ldots,\alpha_{2n}\in\mathbb{C},\quad\text{and}$$

$$G_n=\overline{G}_n=\prod_{i=1}^{2n}(X-\overline{\alpha}_i).$$

Hence it follows that

$$|G_n(\xi)|^2=\prod_{i=1}^{2n}|(\xi-\alpha_i)(\xi-\overline{\alpha}_i)|=\prod_{i=1}^{2n}|\xi^2-(\alpha_i+\overline{\alpha}_i)\xi+\alpha_i\overline{\alpha}_i|$$
$$=\prod_{i=1}^{2n}f(\alpha_i)\ge f(z_1)m^{2n-1},$$

but also

$$|G_n(\xi)| \le |g_\varepsilon(\xi) - \varepsilon|^n + |\varepsilon|^n = |\xi^2 - (z_0 + \overline{z}_0)\xi + z_0\overline{z}_0|^n + \varepsilon^n$$
$$= f(z_0)^n + |\varepsilon|^n = m^n + \varepsilon^{\lambda n}.$$

These two inequalities imply

$$\frac{f(z_1)}{m} \le \frac{|G_n(\xi)|^2}{m^{2n}} \le \frac{(m^n + \varepsilon^{\lambda n})^2}{m^{2n}} = \Big[1 + \Big(\frac{\varepsilon^\lambda}{m}\Big)^n\Big]^2, \quad \text{hence}$$
$$f(z_1) \le m\Big[1 + \Big(\frac{\varepsilon^\lambda}{m}\Big)^n\Big]^2,$$

and as $n \to \infty$ we obtain $f(z_1) \le m$, a contradiction. □[**A.**]

By means of **A** we can now finish the proof. In particular, **A** implies that $K/\mathbb{R}$ is algebraic. Since $\mathbb{C}$ is the algebraic closure of $\mathbb{R}$, there exists an $\mathbb{R}$-homomorphism $\Phi\colon K \to \mathbb{C}$, and as $K \supsetneq \mathbb{R}$, Φ is an isomorphism. Then $|\cdot|' = |\cdot|_\infty^\lambda \circ \Phi$ is an absolute value of K, $\Phi\colon (K, |\cdot|') \to (\mathbb{C}, |\cdot|_\infty^\lambda)$ is an $\mathbb{R}$-isomorphism of valued fields, hence $|\cdot|\restriction\mathbb{R} = |\cdot|'\restriction\mathbb{R} = |\cdot|_\infty^\lambda$, and consequently $|\cdot| = |\cdot|'$ by 5.2.9.1. □

5.3 Non-Archimedian valued fields 1

We start by fixing the notation and conventions to be used in the sequel.

Definitions, Conventions and Remarks 5.3.1. Let $K = (K, |\cdot|)$ be a non-Archimedian valued field. We set

$$\mathcal{O}_K = \big\{x \in K \bigm| |x| \le 1\big\}, \quad \mathfrak{p}_K = \big\{x \in K \bigm| |x| < 1\big\} \quad \text{and}$$
$$U_K = \big\{x \in K \bigm| |x| = 1\big\}.$$

In 5.1.8.1 we proved that $\mathcal{O}_K$ is a valuation domain with maximal ideal $\mathfrak{p}_K$ and unit group $\mathcal{O}_K^\times = U_K$, and that $\mathcal{O}_K$ is Noetherian (and then a discrete valuation domain) if and only if $|\cdot|$ is discrete. As $|\cdot|\colon K \to \mathbb{R}$ is continuous, it follows that $\mathcal{O}_K$ and U_K are closed and $\mathfrak{p}_K$ is open in K. We call $\mathcal{O}_K$ the **valuation domain**, $\mathfrak{p}_K$ the **valuation ideal**, $U_K = \mathcal{O}_K^\times$ the **unit group** and $\mathsf{k}_K = \mathsf{k}_{\mathfrak{p}_K} = \mathcal{O}_K/\mathfrak{p}_K$ the **residue class field** of the valued field K.

Let $|\cdot|$ be discrete. Then we denote by $v_K\colon K \to \mathbb{Z} \cup \{\infty\}$ the discrete valuation belonging to $|\cdot|$ and call it the discrete valuation of $(K, |\cdot|)$ (see 5.1.3.**6.**). Then we have $\mathcal{O}_K = \mathcal{O}_{v_K}$, $\mathfrak{p}_K = \mathfrak{p}_{v_K}$, $U_K = U_{v_K}$ and $\mathsf{k}_K = \mathsf{k}_{v_K}$. If $\pi \in K$, then $v_K(\pi) = 1$ if and only if $\mathfrak{p}_K = \pi\mathcal{O}_K$, and this holds if and only if π is a prime element of $\mathcal{O}_K$. Moreover, if $\rho = |\pi|$, then $|K^\times| = \langle\rho\rangle$, $|\cdot| = |\cdot|_{v_K,\rho}$, $|x| = \rho^{v_K(x)}$ for all $x \in K$, and we obtain

$$\mathcal{O}_K = \{x \in K \mid v_K(x) \geq 0\}, \quad \mathfrak{p}_K = \{x \in K \mid v_K(x) \geq 1\},$$

and $U_K = \{x \in K \mid v_K(x) = 0\}$. We set $U_K^{(0)} = U_K$ and $U_K^{(n)} = 1 + \mathfrak{p}_K^n$ for all $n \geq 1$. By 2.10.5 there exist for all $n \in \mathbb{N}$ (natural) isomorphisms

$$U_K/U_K^{(1)} \stackrel{\sim}{\to} \mathsf{k}_K^\times, \quad U_K^{(n)}/U_K^{(n+1)} \stackrel{\sim}{\to} \mathsf{k}_K \quad \text{and} \quad U_K/U_K^{(n)} \stackrel{\sim}{\to} (\mathcal{O}_K/\mathfrak{p}_K^n)^\times.$$

The elements of $U_K^{(1)}$ are called **principal units**, and the elements of $U_K^{(n)}$ for $n \geq 1$ are called **principal units of level** n of K.

If $K' = (K', |\cdot|')$ is another non-Archimedian valued field and $\varphi\colon K \to K'$ is an isomorphism of valued fields, then $\varphi(\mathcal{O}_K) = \mathcal{O}_{K'}$, $\varphi(\mathfrak{p}_K) = \mathfrak{p}_{K'}$, and φ induces an isomorpnism $\overline{\varphi}\colon \mathsf{k}_K \stackrel{\sim}{\to} \mathsf{k}_{K'}$ by means of $\overline{\varphi}(x + \mathfrak{p}_K) = \varphi(x) + \mathfrak{p}_{K'}$ for all $x \in \mathcal{O}_K$.

We proceed with the properties of convergence and the structure of completions for non-Archimedian and in particular discrete valued fields.

Theorem 5.3.2. *Let $K = (K, |\cdot|)$ be a non-Archimedian valued field.*

1. *Let $(x_n)_{n\geq 0}$ be a sequence in K.*

(a) *If $x \in K^\times$ and $(x_n)_{n\geq 0} \to x$, then $|x_n| = |x|$ for all $n \gg 1$.*

(b) *$(x_n)_{n\geq 0}$ is a Cauchy sequence if and only if $(x_{n+1} - x_n)_{n\geq 0} \to 0$.*

2. *Let K be complete and $(x_n)_{n\geq 0}$ a zero sequence in K. Then*

$$\sum_{n\geq 0} x_n \ \text{ converges in } K, \text{ and } \ \Big|\sum_{n=0}^{\infty} x_n\Big| \leq \sup\{|x_n| \mid n \geq 0\}.$$

3. *Let $|\cdot|$ be discrete and π a prime element of $\mathcal{O}_K$.*

(a) *The sets $\mathcal{O}_K$, $\mathfrak{p}_K^n$ (for all $n \in \mathbb{Z}$) and $U_K^{(n)}$ (for all $n \in \mathbb{N}_0$) are both open and closed in K, $\{\mathfrak{p}_K^n \mid n \in \mathbb{N}\}$ is a fundamental system of neighborhoods of 0, and $\{U_K^{(n)} \mid n \in \mathbb{N}\}$ is a fundamental system of neighborhoods of 1 in K.*

(b) *Let $(x_n)_{n\geq 0}$ be a sequence in K and $x \in K$.*

i. *$(x_n)_{n\geq 0}$ is bounded if and only if the sequence $(v_K(x_n))_{n\geq 0}$ is bounded from below.*

ii. *$(x_n)_{n\geq 0} \to x$ if and only if $(v_K(x_n - x))_{n\geq 0} \to \infty$.*

iii. *If $(x_n)_{n\geq 0} \to x$ and $x \neq 0$, then $v_K(x_n) = v_K(x)$ for all $n \gg 1$.*

4. Let $|\cdot|$ be discrete and $|\mathsf{k}_K| = q < \infty$. Then $\mathcal{O}_K$ has finite residue class rings, and if $n \in \mathbb{N}$, then

$$\mathfrak{N}(\mathfrak{p}_K^n) = (\mathcal{O}_K : \mathfrak{p}_K^n) = q^n \quad \textit{and}$$
$$(U_K : U_K^{(n)}) = |(\mathcal{O}_K/\mathfrak{p}_K^n)^\times| = q^{n-1}(q-1).$$

Proof. 1. (a) If $(x_n)_{n\geq 0} \to x \in K^\times$ and $n \gg 1$, then $|x_n - x| < |x|$ and therefore $|x_n| = |(x_n - x) + x| = |x|$.

(b) If $(x_n)_{n\geq 0}$ is a Cauchy sequence, then obviously $(x_{n+1} - x_n)_{n\geq 0} \to 0$. Thus assume that $(x_{n+1} - x_n)_{n\geq 0} \to 0$, and let $\varepsilon \in \mathbb{R}_{>0}$. Then there exists some $n_0 \geq 0$ such that $|x_{n+1} - x_n| < \varepsilon$ for all $n \geq n_0$, and for all $m \geq n \geq n_0$ it follows that

$$|x_m - x_n| = \Bigl|\sum_{i=n}^{m-1}(x_{i+1} - x_i)\Bigr| \leq \max\{|x_{i+1} - x_i| \mid i \in [n, m-1]\} < \varepsilon.$$

2. For $n \in \mathbb{N}_0$ we set $s_n = x_0 + x_1 + \ldots + x_n$. Then

$$(x_n)_{n\geq 0} = (s_n - s_{n-1})_{n\geq 1}$$

is a zero sequence. Hence $(s_n)_{n\geq 0}$ is a Cauchy sequence by 1. and thus convergent. If $N \geq 0$, then

$$\Bigl|\sum_{n=0}^{N} x_n\Bigr| \leq \max\{|x_0|, \ldots, |x_N|\} \leq \sup\{|x_n| \mid n \geq 0\},$$

and therefore

$$\Bigl|\sum_{n=0}^{\infty} x_n\Bigr| = \lim_{N\to\infty}\Bigl|\sum_{n=0}^{N} x_n\Bigr| \leq \sup\{|x_n| \mid n \geq 0\}.$$

3. Let $\rho \in (0,1)$ be such that $|K^\times| = \langle\rho\rangle$.

(a) By 5.3.1, $\mathcal{O}_K$ and U_K are closed in K. Since

$$\mathcal{O}_K = \{x \in K \mid |x| < \rho^{-1}\} \quad \text{and} \quad U_K = \{x \in K \mid \rho < |x| < \rho^{-1}\},$$

it follows that $\mathcal{O}_K$ and U_K are also open in K. If $\pi \in \mathcal{O}_K$ is a prime element, then $\mathfrak{p}_K^n = \pi^n\mathcal{O}_K$ for all $n \in \mathbb{Z}$ and $U_K^{(n)} = 1 + \mathfrak{p}_K^n$ for all $n \in \mathbb{N}$. Since the assignments $x \mapsto \pi^n x$ and $x \mapsto 1+x$ define homeomorphisms $K \to K$, the sets $\mathfrak{p}_K^n$ for all $n \in \mathbb{Z}$ and $U_K^{(n)}$ for all $n \in \mathbb{N}$ are also both open and closed. Since $\mathfrak{p}_K^n = \{x \in K \mid |x| \leq \rho^n\}$, $U_K^{(n)} = \{x \in K \mid |x - 1| \leq \rho^n\}$ and $(\rho^n)_{n\geq 1} \to 0$, it follows that $\{\mathfrak{p}_K^n \mid n \in \mathbb{N}\}$ is a fundamental system of neighborhoods of 0 and $\{U_K^{(n)} \mid n \in \mathbb{N}\}$ is a fundamental system of neighborhoods of 1.

(b) Observe that for all $x \in K$

$$v_K(x) = \frac{\log|x|}{\log \rho} \quad \text{and} \quad |x| = \rho^{v_K(x)} \quad (\text{where } \log 0 = -\infty \text{ and } \rho^{-\infty} = 0).$$

Hence (i) and (ii) follow from the very definition, and (iii) follows from 1.(a).

4. By 2.14.1 and 2.10.5.4. □

Theorem 5.3.3. *Let $K = (K, |\cdot|)$ be a non-Archimedian valued field, and let $\widehat{K} = (\widehat{K}, |\cdot|)$ be a completion of K.*

1. $\widehat{K}$ is a non-Archimedian valued field, and $|\widehat{K}| = |K| \subset \mathbb{R}_{\geq 0}$.

2. $\mathcal{O}_K = \mathcal{O}_{\widehat{K}} \cap K$ is dense in $\mathcal{O}_{\widehat{K}}$, and $\mathfrak{p}_K = \mathfrak{p}_{\widehat{K}} \cap K$ is dense in $\mathfrak{p}_{\widehat{K}}$. The embedding $K \hookrightarrow \widehat{K}$ induces isomorphisms of additive groups $K/\mathcal{O}_K \overset{\sim}{\to} \widehat{K}/\mathcal{O}_{\widehat{K}}$ and $K/\mathfrak{p}_K \overset{\sim}{\to} \widehat{K}/\mathfrak{p}_{\widehat{K}}$, and the embedding $\mathcal{O}_K \hookrightarrow \mathcal{O}_{\widehat{K}}$ induces an isomorphism $\mathsf{k}_K \overset{\sim}{\to} \mathsf{k}_{\widehat{K}}$.
We identify: $\widehat{K}/\mathcal{O}_{\widehat{K}} = K/\mathcal{O}_K$, $\widehat{K}/\mathfrak{p}_{\widehat{K}} = K/\mathfrak{p}_K$ and $\mathsf{k}_{\widehat{K}} = \mathsf{k}_K$.

3. Let $(K, |\cdot|)$ be discrete and $\pi \in \mathcal{O}_K$ a prime element.

(a) $(\widehat{K}, |\cdot|)$ is discrete, $v_{\widehat{K}} \restriction K = v_K \colon K \to \mathbb{Z} \cup \{\infty\}$, and π is a prime element of $\mathcal{O}_{\widehat{K}}$.
In the sequel we write v_K instead of $v_{\widehat{K}}$.

(b) If $n \in \mathbb{Z}$, then $\mathfrak{p}_{\widehat{K}}^n = \pi^n \mathcal{O}_{\widehat{K}} = \mathfrak{p}_K \mathcal{O}_{\widehat{K}}$, and $\mathfrak{p}_K^n = \mathfrak{p}_{\widehat{K}}^n \cap K$ is dense in $\mathfrak{p}_{\widehat{K}}^n$.

(c) For $n \in \mathbb{Z}$ and $k \in \mathbb{N}$ the embedding $K \hookrightarrow \widehat{K}$ induces group isomorphisms $K/\mathfrak{p}_K^n \overset{\sim}{\to} \widehat{K}/\mathfrak{p}_{\widehat{K}}^n$ and $\mathfrak{p}_K^n/\mathfrak{p}_K^{n+k} \overset{\sim}{\to} \mathfrak{p}_{\widehat{K}}^n/\mathfrak{p}_{\widehat{K}}^{n+k}$, and a ring isomorphism $\mathcal{O}_K/\mathfrak{p}_K^k \overset{\sim}{\to} \mathcal{O}_{\widehat{K}}/\mathfrak{p}_{\widehat{K}}^k$.
We identify:
$K/\mathfrak{p}_K^n = \widehat{K}/\mathfrak{p}_{\widehat{K}}^n$, $\mathfrak{p}_K^n/\mathfrak{p}_K^{n+k} = \mathfrak{p}_{\widehat{K}}^n/\mathfrak{p}_{\widehat{K}}^{n+k}$ and $\mathcal{O}_K/\mathfrak{p}_K^k = \mathcal{O}_{\widehat{K}}/\mathfrak{p}_{\widehat{K}}^k$.

Proof. 1. By 5.1.7.1 it follows that $\widehat{K}$ is non-Archimedian, and it suffices to prove that $|\widehat{K}^\times| \subset |K^\times|$. If $x \in \widehat{K}^\times$, then there exists a sequence $(x_n)_{n\geq 0}$ in K such that $(x_n)_{n\geq 0} \to x$, and 5.3.2.1(a) implies that $|x| = |x_n| \in |K|$ for all $n \gg 1$.

2. By the very definition $\mathcal{O}_K = \mathcal{O}_{\widehat{K}} \cap K$ and $\mathfrak{p}_K = \mathfrak{p}_{\widehat{K}} \cap \mathcal{O}_K$. Hence the embedding $K \hookrightarrow \widehat{K}$ induces group monomorphisms $j \colon K/\mathcal{O}_K \to \widehat{K}/\mathcal{O}_{\widehat{K}}$ and $j_1 \colon K/\mathfrak{p}_K \to \widehat{K}/\mathfrak{p}_{\widehat{K}}$, and the embedding $\mathcal{O}_K \hookrightarrow \mathcal{O}_{\widehat{K}}$ induces a field homomorphism $j_0 \colon \mathsf{k}_K \to \mathsf{k}_{\widehat{K}}$. Let $x \in \widehat{K}^\times$ and $(x_n)_{n\geq 0}$ be a sequence in K such that $(x_n)_{n\geq 0} \to x$. Then $|x_n| = |x|$ for all $n \gg 1$, hence $x \in \mathcal{O}_{\widehat{K}}$ implies $x_n \in \mathcal{O}_K$, and $x \in \mathfrak{p}_{\widehat{K}}$ implies $x_n \in \mathfrak{p}_K$ for all $n \gg 1$. Therefore $\mathcal{O}_K$ is dense

in $\mathcal{O}_{\widehat{K}}$ and $\mathfrak{p}_K$ is dense in $\mathfrak{p}_{\widehat{K}}$. Moreover, if $n \gg 1$, then $|x_n - x| < 1$, hence $x + \mathcal{O}_{\widehat{K}} = j(x_n + \mathcal{O}_K)$ and $x + \mathfrak{p}_{\widehat{K}} = j_1(x_n + \mathfrak{p}_K)$. Consequently j, j_1 and j_0 are isomorphisms.

3. (a) Since $|K| = |\widehat{K}|$, it follows that K is discrete, and $v_{\widehat{K}} \restriction K = v_K$. We write v_K instead of $v_{\widehat{K}}$, and as $v_K(\pi) = 1$, it follows that π is a prime element of $\mathcal{O}_{\widehat{K}}$.

(b) If $n \in \mathbb{Z}$, then 2.10.5.1 implies $\mathfrak{p}_K^n = \pi^n \mathcal{O}_K = \{x \in K \mid v_K(x) \geq n\}$ and $\mathfrak{p}_{\widehat{K}}^n = \pi^n \mathcal{O}_{\widehat{K}} = \{x \in \widehat{K} \mid v_K(x) \geq n\}$. Hence $\mathfrak{p}_{\widehat{K}}^n = \mathfrak{p}_K^n \mathcal{O}_{\widehat{K}}$, $\mathfrak{p}_K^n = \mathfrak{p}_{\widehat{K}}^n \cap K$, and since $\mathcal{O}_K$ is dense in $\mathcal{O}_{\widehat{K}}$, it follows that $\mathfrak{p}_K^n$ is dense in $\mathfrak{p}_{\widehat{K}}^n$.

(c) Let $n \in \mathbb{Z}$ and $k \in \mathbb{N}$. By (b) the embeddings $K \hookrightarrow \widehat{K}$, $\mathfrak{p}_K^n \hookrightarrow \mathfrak{p}_{\widehat{K}}^n$ and $\mathcal{O}_K \hookrightarrow \mathcal{O}_{\widehat{K}}$ induce monomorphisms $j_n \colon K/\mathfrak{p}_K^n \to \widehat{K}/\mathfrak{p}_{\widehat{K}}^n$, $j_{n,k} \colon \mathfrak{p}_K^n/\mathfrak{p}_K^{n+k} \to \mathfrak{p}_{\widehat{K}}^n/\mathfrak{p}_{\widehat{K}}^{n+k}$ and $j_k' \colon \mathcal{O}_K/\mathfrak{p}_K^k \to \mathcal{O}_{\widehat{K}}/\mathfrak{p}_{\widehat{K}}^{n+k}$. To prove that they are surjective, let $x \in \widehat{K}$ and $(x_i)_{i\geq 0}$ be a sequence in K such that $(x_i)_{i\geq 0} \to x$. If $i \gg 1$, then $v_K(x - x_i) \geq |n| + k \geq \max\{k, n\}$, and therefore $x + \mathfrak{p}_{\widehat{K}}^n = j_n(x_i + \mathfrak{p}_K^n)$; if $x \in \mathfrak{p}_{\widehat{K}}^n$, then $x + \mathfrak{p}_{\widehat{K}}^{n+k} = j_{n,k}(x_i + \mathfrak{p}_K^{n+k})$; and if $x \in \mathcal{O}_{\widehat{K}}$, then $x + \mathfrak{p}_{\widehat{K}}^k = j_k'(x_i + \mathfrak{p}_K^k)$. □

Finally we apply our results to the completion of $\mathfrak{p}$-adic valued quotient fields of Dedekind domains. These are relevant examples in the theory of algebraic numbers and algebraic functions.

Theorem and Definition 5.3.4. *Let $\mathfrak{o}$ be a Dedekind domain and $K = \mathsf{q}(\mathfrak{o})$. Let $\mathfrak{p} \in \mathcal{P}(\mathfrak{o})$, $\pi \in \mathfrak{p} \setminus \mathfrak{p}^2$ and $v_\mathfrak{p} \colon K \to \mathbb{Z} \cup \{\infty\}$ the $\mathfrak{p}$-adic valuation (see 2.8.5).*

By 2.10.7.2, $v_\mathfrak{p}$ is a discrete valuation of K with valuation domain $\mathfrak{o}_\mathfrak{p}$, valuation ideal $\mathfrak{p}\mathfrak{o}_\mathfrak{p}$ and residue class field $\mathsf{k}_\mathfrak{p} = \mathfrak{o}/\mathfrak{p} = \mathfrak{o}_\mathfrak{p}/\mathfrak{p}\mathfrak{o}_\mathfrak{p} = \mathsf{k}_{\mathfrak{p}\mathfrak{o}_\mathfrak{p}}$. Moreover, if $n \in \mathbb{Z}$ and $k \in \mathbb{N}$, then $(\mathfrak{p}\mathfrak{o}_\mathfrak{p})^n = \mathfrak{p}^n\mathfrak{o}_\mathfrak{p}$ and $\mathfrak{p}^n/\mathfrak{p}^{n+k} = \mathfrak{p}^n\mathfrak{o}_\mathfrak{p}/\mathfrak{p}^{n+k}\mathfrak{o}_\mathfrak{p}$. If $|\cdot|_\mathfrak{p} \colon K \to \mathbb{R}_{\geq 0}$ is a $\mathfrak{p}$-adic absolute value of K (see 5.1.3.5), then $(K, |\cdot|_\mathfrak{p})$ is a discrete valued field.

We denote by $K_\mathfrak{p} = (K_\mathfrak{p}, |\cdot|_\mathfrak{p})$ a completion of $(K, |\cdot|_\mathfrak{p})$. We call $K_\mathfrak{p}$ the **$\mathfrak{p}$-adic completion** of K, and we rephrase and amend 5.3.3 in the present context.

$K_\mathfrak{p}$ is a discrete valued field, we denote by $\widehat{\mathfrak{o}}_\mathfrak{p}$ its valuation domain and by $\widehat{\mathfrak{p}}$ its valuation ideal. Then π is a prime element of $\widehat{\mathfrak{o}}_\mathfrak{p}$, $v_{\widehat{\mathfrak{p}}} \restriction K = v_\mathfrak{p}$, and we write again $v_\mathfrak{p}$ instead of $v_{\widehat{\mathfrak{p}}}$; $\mathfrak{o}_\mathfrak{p} = \widehat{\mathfrak{o}}_\mathfrak{p} \cap K$ is dense in $\widehat{\mathfrak{o}}_\mathfrak{p}$. If $n \in \mathbb{Z}$, then $\widehat{\mathfrak{p}}^n = \pi^n \widehat{\mathfrak{o}}_\mathfrak{p}$, and $\mathfrak{p}^n\mathfrak{o}_\mathfrak{p} = \widehat{\mathfrak{p}}^n \cap K$ is dense in $\widehat{\mathfrak{p}}^n$. For $n \in \mathbb{Z}$ and $k \in \mathbb{N}$ we restate the identifications made in 5.3.3.3 by means of natural isomorphisms, and we combine them with the identifications already made in 2.10.7.2:

$$K/\mathfrak{p}^n\mathfrak{o}_\mathfrak{p} = K_\mathfrak{p}/\widehat{\mathfrak{p}}^n, \quad \mathfrak{p}^n/\mathfrak{p}^{n+k} = \mathfrak{p}^n\mathfrak{o}_\mathfrak{p}/\mathfrak{p}^{n+k}\mathfrak{o}_\mathfrak{p} = \widehat{\mathfrak{p}}^n/\widehat{\mathfrak{p}}^{n+k},$$
$$\mathfrak{o}/\mathfrak{p}^k = \mathfrak{o}_\mathfrak{p}/\mathfrak{p}^k\mathfrak{o}_\mathfrak{p} = \widehat{\mathfrak{o}}_\mathfrak{p}/\widehat{\mathfrak{p}}^k,$$

and in particular, $\mathsf{k}_\mathfrak{p} = \mathfrak{o}/\mathfrak{p} = \mathsf{k}_{\mathfrak{p}\mathfrak{o}_\mathfrak{p}} = \mathfrak{o}_\mathfrak{p}/\mathfrak{p}\mathfrak{o}_\mathfrak{p} = \widehat{\mathfrak{o}}_\mathfrak{p}/\widehat{\mathfrak{p}} = \mathsf{k}_{\widehat{\mathfrak{p}}}$.

After these preliminary adaptations we proceed with the very results which are special for the $\mathfrak{p}$-adic case.

The ideal extension $\iota\colon \mathcal{J}(\mathfrak{o}) \to \mathcal{J}(\widehat{\mathfrak{o}}_\mathfrak{p})$, *defined by* $\iota(\mathfrak{a}) = \mathfrak{a}\widehat{\mathfrak{o}}_\mathfrak{p}$, *is a group epimorphism. If* $\mathfrak{a} \in \mathcal{J}(\mathfrak{o})$, *then*

$$\mathfrak{a} \text{ is dense in } \mathfrak{a}\,\widehat{\mathfrak{o}}_\mathfrak{p},\quad \mathfrak{a}\widehat{\mathfrak{o}}_\mathfrak{p} \cap K = \mathfrak{a}_\mathfrak{p},\quad \text{and}\quad v_\mathfrak{p}(\mathfrak{a}) = v_\mathfrak{p}(\mathfrak{a}_\mathfrak{p}) = v_\mathfrak{p}(\mathfrak{a}\widehat{\mathfrak{o}}_\mathfrak{p}).$$

In particular, $\mathfrak{o}$ *is dense in* $\widehat{\mathfrak{o}}_\mathfrak{p}$, *and for all* $n \in \mathbb{Z}$, $\mathfrak{p}^n$ *is dense in* $\widehat{\mathfrak{p}}^n$.

Proof. By 2.6.7.2 the ideal extension ι is a group homomorphism. If $\mathfrak{a} \in \mathcal{J}(\mathfrak{o})$ and $v_\mathfrak{p}(\mathfrak{a}) = m \in \mathbb{Z}$, then $\mathfrak{a}_\mathfrak{p} = \mathfrak{a}\mathfrak{o}_\mathfrak{p} = \pi^m\mathfrak{o}_\mathfrak{p}$ by 2.8.7.2(b). Hence $\mathfrak{a}\widehat{\mathfrak{o}}_\mathfrak{p} = \pi^m\widehat{\mathfrak{o}}_\mathfrak{p}$, $v_\mathfrak{p}(\mathfrak{a}\widehat{\mathfrak{o}}_\mathfrak{p}) = m = v_\mathfrak{p}(\mathfrak{a})$ and $\iota(\mathcal{J}(\mathfrak{o})) = \{\pi^n\widehat{\mathfrak{o}}_\mathfrak{p} \mid n \in \mathbb{Z}\} = \mathcal{J}(\widehat{\mathfrak{o}}_\mathfrak{p})$. Since $\mathfrak{a}_\mathfrak{p} = \pi^m\mathfrak{o}_\mathfrak{p}$ is dense in $\mathfrak{a}\widehat{\mathfrak{o}}_\mathfrak{p}$, it suffices to prove that $\mathfrak{a}$ is dense in $\mathfrak{a}_\mathfrak{p}$.

Let $s^{-1}a \in \mathfrak{a}_\mathfrak{p}$, where $a \in \mathfrak{a}$ and $s \in \mathfrak{o}\setminus\mathfrak{p}$. If $n \in \mathbb{N}$, then $s+\mathfrak{p}^n \in (\mathfrak{o}/\mathfrak{p}^n)^\times$, and thus there exists some $s_n \in \mathfrak{o}$ such that $ss_n \equiv 1 \mod \mathfrak{p}^n$. Then $s_na \in \mathfrak{a}$, and

$$v_\mathfrak{p}(s_na - s^{-1}a) = v_\mathfrak{p}(s^{-1}a) + v_\mathfrak{p}(ss_n - 1) \ge v_\mathfrak{p}(s^{-1}a) + n \quad \text{for all } n \ge 1.$$

Hence $(s_na)_{n\ge1} \to s^{-1}a$, and therefore $\mathfrak{a}$ is dense in $\mathfrak{a}_\mathfrak{p}$. □

The following result on series expansions of elements in a complete valued field shows the strong analogy with Weierstrass's theory of complex functions.

Theorem 5.3.5. *Let* K *be a complete discrete valued field and* $(\pi_n)_{n\ge0}$ *a sequence in* K *such that* $v_K(\pi_n) = n$ *for all* $n \ge 0$ *(most important example:* $\pi_n = \pi^n$ *for some prime element* π *of* $\mathcal{O}_K$*). Let* $\mathcal{R} \subset \mathcal{O}_K$ *be a set of representatives for* k_K *such that* $0 \in \mathcal{R}$.

1. *For every* $x \in \mathcal{O}_K$ *there exists a unique sequence* $(a_n)_{n\ge0}$ *in* $\mathcal{R}$ *such that*

$$x = \sum_{n=0}^{\infty} a_n\pi_n.$$

What is more, $v_K(x) = \inf\{n \in \mathbb{N}_0 \mid a_n \ne 0\} \in \mathbb{N}_0 \cup \{\infty\}$, *and if we endow* $\mathcal{R}$ *with the discrete topology and* $\mathcal{R}^{\mathbb{N}_0}$ *with the product topology, then the map*

$$\Phi\colon \mathcal{R}^{\mathbb{N}_0} \to \mathcal{O}_K, \quad \text{defined by} \quad \Phi\big((a_n)_{n\ge0}\big) = \sum_{n=0}^{\infty} a_n\pi_n\,,$$

is a homeomorphism.

2. *If* k_K *is finite, then* $\mathcal{O}_K$ *is compact, and so are the sets* $\mathfrak{p}_K^n$ *for all* $n \in \mathbb{Z}$ *and* $U_K^{(n)}$ *for all* $n \ge 0$.

3. If $x \in K^\times$ and $v_K(x) = d \in \mathbb{Z}$, then there exists a unique sequence $(a_n)_{n\ge d}$ in $\mathcal{R}$ such that $a_d \ne 0$ and

$$x = \sum_{n=d}^{\infty} a_n \pi_n.$$

Proof. Since $\mathcal{R}$ is a set of representatives for k_K, it follows that for every $a \in \mathcal{O}_K$ there exists a unique $a_0 \in \mathcal{R}$ satisfying $a \equiv a_0 \mod \mathfrak{p}_K$.

1. If $(a_n)_{n\ge 0}$ is any sequence in $\mathcal{R}$, then $(a_n\pi_n)_{n\ge 0} \to 0$, and by 5.3.2.2 the series

$$\sum_{n\ge 0} a_n \pi_n \quad \text{converges.}$$

If $x \in \mathcal{O}_K$ and $(a_n)_{n\ge 0}$ is a sequence in $\mathcal{R}$, then

$$x = \sum_{n=0}^{\infty} a_n\pi_n \ \text{ if and only if } \ x \equiv \sum_{n=0}^{N} a_n\pi_n \mod \mathfrak{p}_K^{N+1} \ \text{ for all } \ N \ge 0.$$

Hence the existence and uniqueness of the series expansion of x follows using 2.10.6.2, and therefore Φ is bijective. If in the above series $m \ge 0$ is minimal such that $a_m \ne 0$, then

$$x = \sum_{n=0}^{\infty} a_n\pi_n = a_m\pi_m(1+\pi t), \quad \text{where } \ t \in \mathcal{O}_K, \quad \text{and thus } \ v_K(x) = m.$$

For $\boldsymbol{a} = (a_n)_{n\ge 0} \in \mathcal{R}^{\mathbb{N}_0}$ and $N \in \mathbb{N}_0$, we set

$$B_N(\boldsymbol{a}) = \{(b_n)_{n\ge 0} \in \mathcal{R}^{\mathbb{N}_0} \mid b_n = a_n \text{ for all } n \in [0,N]\}.$$

Then $\{B_N \mid N \in \mathbb{N}_0\}$ is a fundamental system of neighborhoods of $\boldsymbol{a}$ in $\mathcal{R}^{\mathbb{N}_0}$, and $\Phi(B_N) = \Phi(\boldsymbol{a}) + \mathfrak{p}_K^{N+1}$ for all $N \ge 0$. Hence Φ is a homeomorphism.

2. If k_K is finite, then $\mathcal{R}$ is finite and $\mathcal{R}^{\mathbb{N}_0}$ is compact by Tychonoff's theorem. Hence $\mathcal{O}_K$ is compact and so are the sets $\mathfrak{p}_K^n = \pi^n\mathcal{O}_K$ for all $n \in \mathbb{Z}$. For $n \in \mathbb{N}_0$, $U_K^{(n)}$ is closed in $\mathcal{O}_K$ by 5.3.2.3(a), and thus it is compact.

3. If $x \in K^\times$ and $d = v_K(x)$, then the assertion follows by 1., applied for $\pi^{-d}x$. □

Before we proceed with the general theory, we discuss two basic structures: The field of p-adic numbers in 5.3.6, and the field of formal Laurent series in 5.3.7.

Remarks and Definitions 5.3.6. Let p be a prime and $|\cdot|_p \colon \mathbb{Q} \to \mathbb{R}_{\ge 0}$ the p-adic absolute value. We recall and amend our knowledge from 5.1.4 and 2.10.8.

$|\cdot|_p$ is a discrete absolute value, the p-adic valuation $v_p \colon \mathbb{Q} \to \mathbb{Z} \cup \{\infty\}$ is the corresponding discrete valuation, the domain $\mathbb{Z}_{(p)}$ of p-integral rational numbers is its valuation domain, $p\mathbb{Z}_{(p)}$ is the valuation ideal and the prime field $\mathbb{F}_p = \mathbb{Z}/p\mathbb{Z} = \mathbb{Z}_{(p)}/p\mathbb{Z}_{(p)}$ is its residue class field.

We fix a completion $\mathbb{Q}_p = (\mathbb{Q}_p, |\cdot|_p)$ of $(\mathbb{Q}, |\cdot|_p)$ and call it the **p-adic number field**. It is a complete discrete valued field, and by 5.2.4.3 it is unique up to unique isomorphisms of valued fields. Its valuation

$$v_p = v_{\mathbb{Q}_p} \colon \mathbb{Q}_p \to \mathbb{Z} \cup \{\infty\}$$

is (likewise) called the **p-adic valuation**, its valuation domain

$$\mathbb{Z}_p = \mathcal{O}_{\mathbb{Q}_p} = \{x \in \mathbb{Q}_p \mid |x|_p \le 1\} = \{x \in \mathbb{Q}_p \mid v_p(x) \ge 0\}$$

is called the **domain of p-adic integers**, and its unit group

$$U_p = U_p^{(0)} = \mathbb{Z}_p^\times = \{x \in \mathbb{Q}_p \mid v_p(x) = 0\}$$

is called the **group of p-adic units**. For $n \in \mathbb{N}$ we denote by $U_p^{(n)} = 1 + p^n\mathbb{Z}_p$ the group of p-adic principal units of level n. The structure of these groups will be investigated in detail in 5.8.11.

By 2.8.7.2 and 5.3.4 it follows that $\mathbb{Z} \subset \mathbb{Z}_{(p)} \subset \mathbb{Z}_p$, and $\mathbb{Z}$ is dense in $\mathbb{Z}_p$. If $a \in \mathbb{Q}^\times$, then $a\mathbb{Z}_{(p)} = p^{v_p(a)}\mathbb{Z}_{(p)} \subset p^{v_p(a)}\mathbb{Z}_p = a\mathbb{Z}_p$, and $a\mathbb{Z}$ is dense in $a\mathbb{Z}_p$. If $a \in \mathbb{Z}$, then $\mathbb{Z}_p/a\mathbb{Z}_p = \mathbb{Z}_p/p^{v_p(a)}\mathbb{Z}_p = \mathbb{Z}/p^{v_p(a)}\mathbb{Z}$. Furthermore, if $n \in \mathbb{Z}$ and $k \in \mathbb{N}$, then $p^n\mathbb{Z}/p^{n+k}\mathbb{Z} = p^n\mathbb{Z}_{(p)}/p^{n+k}\mathbb{Z}_{(p)} = p^n\mathbb{Z}_p/p^{n+k}\mathbb{Z}_p$.

Now we adapt 5.3.5 for $\mathbb{Q}_p$. For every $a \in \mathbb{Z}_p$, there exists a unique sequence $(a_n)_{n\ge 0}$ in $[0, p-1]$, called the **p-adic digit sequence** of a, such that

$$a = \sum_{n=0}^{\infty} a_n p^n,$$

and then $v_p(a) = \inf\{n \in \mathbb{N}_0 \mid a_n \ne 0\} \in \mathbb{N}_0 \cup \{\infty\}$. The p-adic digit sequence $(a_n)_{n\ge 0}$ is uniquely determined by the system of congruences

$$a \equiv \sum_{n=0}^{N} a_n p^n \mod p^{N+1} \quad \text{for all } \ N \in \mathbb{N}_0.$$

In particular,

$$-1 = \sum_{n=0}^{\infty} (p-1)p^n \quad \text{and} \quad \frac{1}{1-p} = \sum_{n=0}^{\infty} p^n.$$

Next we investigate the fields of formal Laurent series from a valuation-theoretic point of view. They will play an important role in the arithmetic of algebraic function fields in Chapter 6.

Theorem 5.3.7. *Let $F((X))$ be the field of formal Laurent series in X over a field F and $\mathsf{o}_X \colon F((X)) \to \mathbb{Z} \cup \{\infty\}$ the order valuation with respect to X.* Recall from 2.10.10 that $F((X)) = (F((X)), \mathsf{o}_X)$ is a discrete valued field with valuation domain $\mathcal{O}_{F((X))} = F[\![X]\!]$, valuation ideal $\mathfrak{p}_{F((X))} = XF[\![X]\!]$ and residue class field $\mathsf{k}_{F((X))} = F$.

1. *Let $d \in \mathbb{Z}$, $f \in F((X))$ and $(f^{(k)})_{k \ge 0}$ a sequence in $F((X))$ such that*

$$f = \sum_{n=d}^{\infty} a_n X^n \quad \text{and} \quad f^{(k)} = \sum_{n=d}^{\infty} a_n^{(k)} X^n \quad \text{for all} \quad k \ge 0.$$

Then $(f^{(k)})_{k \ge 0} \to f$ if and only if for every $n \ge d$ there exists some $k_n \ge 0$ such that $a_n^{(k)} = a_n$ for all $k \ge k_n$.

2. *$(F((X)), \mathsf{o}_X)$ is complete, and $F(X)$ is dense in $F((X))$.*

3. *Let $t \in F[\![X]\!]$ and $\mathsf{o}_X(t) = m \ge 1$.*

(a) *Let $(a_n)_{n \in \mathbb{Z}}$ be a sequence in F such that $a_{-n} = 0$ for all $n \gg 1$, and let $d = \inf\{n \in \mathbb{Z} \mid a_n \ne 0\} \in \mathbb{Z} \cup \{\infty\}$. Then*

$$\sum_{n \in \mathbb{Z}} a_n t^n \quad \text{converges in} \quad F((X)),$$

$$\text{and} \quad \mathsf{o}_X\Big(\sum_{n \in \mathbb{Z}} a_n t^n\Big) = md \in \mathbb{Z} \cup \{\infty\},$$

In particular (for $t = X$), every formal Laurent series is in fact a convergent series in $F((X))$.

(b) *Let $F((t))$ be the set of all Laurent series*

$$\sum_{n \in \mathbb{Z}} a_n t^n,$$

where $(a_n)_{n \in \mathbb{Z}} \in F^{\mathbb{Z}}$ and $a_{-n} = 0$ for all $n \gg 1$. Then $F((t))$ is a subfield of $F((X))$, and the map

$$\Phi \colon F((X)) \to F((t)), \quad \text{defined by} \quad \Phi\Big(\sum_{n \in \mathbb{Z}} a_n X^n\Big) = \sum_{n \in \mathbb{Z}} a_n t^n$$

for all sequences $(a_n)_{n \in \mathbb{Z}} \in F^{\mathbb{Z}}$ such that $a_{-n} = 0$ for all $n \gg 1$, is a field isomorphism. In particular, $F((t))$ is the field of formal Laurent series in t over F, $\mathsf{o}_t = \mathsf{o}_X \circ \Phi^{-1} \colon F((t)) \to \mathbb{Z} \cup \{\infty\}$ is the order valuation of $F((t))$ with respect to t, and

$$\mathsf{o}_X \restriction F((t)) = m\mathsf{o}_t \colon F((t)) \to \mathbb{Z} \cup \{\infty\}.$$

(c) *If $m = 1$, then $F((X)) = F((t))$ and $\mathsf{o}_X = \mathsf{o}_t$.*

Proof. 1. Assume first that $(f^{(k)})_{k\ge 0} \to f$, and let $n \ge d$. Then there exists some $k_0 \ge 0$ such that $\mathsf{o}(f^{(k)} - f) > n$ and thus $a_n^{(k)} = a_n$ for all $k \ge k_0$.

Suppose conversely that for every $n \ge d$ there exists some $k_n \ge 0$ such that $a_n = a_n^{(k)}$ for all $k \ge k_n$. Let $m \ge d$ and $k^* = \max\{k_n \mid n \in [d,m]\}$. If $k \ge k^*$ and $n \in [d,m]$, then $k^* \ge k_n$ and thus $a_n^{(k)} = a_n$. Hence $\mathsf{o}(f^{(k)} - f) \ge m$.

2. For the proof of completeness let

$$\Bigl(f^{(k)} = \sum_{n\in\mathbb{Z}} a_n^{(k)} X^n\Bigr)_{k\ge 0} \quad \text{be a Cauchy sequence in } F((X)).$$

As $(f^{(k)})_{k\ge 0}$ is bounded, there exists some $d \in \mathbb{Z}$ such that $\mathsf{o}(f^{(k)}) \ge d$ for all $k \ge 0$, and as $(f^{(k)})_{k\ge 0}$ is a Cauchy sequence, $(f^{(k+1)} - f^{(k)})_{k\ge 0} \to 0$. For $n \ge d$ let $k_n \ge 0$ be such that $\mathsf{o}(f^{(k+1)} - f^{(k)}) \ge n+1$ for all $k \ge k_n$. If $k \ge k_n$, then $a_i^{(k+1)} - a_i^{(k)} = 0$ for all $i \le n$, and in particular $a_n^{(k)} = a_n^{(k_n)}$ for all $k \ge k_n$. If $a_n = a_n^{(k_n)}$ for all $n \ge d$, then 1. implies

$$(f^{(k)})_{k\ge 0} \to f = \sum_{n=d}^{\infty} a_n X^n.$$

Clearly, $F[X] \subset F[\![X]\!]$ implies $F(X) = \mathsf{q}(F[X]) \subset \mathsf{q}(F[\![X]\!]) = F((X))$, and if $d \in \mathbb{Z}$, then

$$f = \sum_{n=d}^{\infty} a_n X^n = \lim_{N\to\infty} \sum_{n=d}^{N} a_n X^n.$$

Hence $F[X]$ and thus all the more $F(X)$ is dense in $F((X))$.

3. Let $(a_n)_{n\in\mathbb{Z}}$ be a sequence in F such that $a_{-n} = 0$ for all $n \gg 1$,

$$y = \sum_{n\in\mathbb{Z}} a_n t^n \in F((X)) \quad \text{and} \quad d = \inf\{n \in \mathbb{Z} \mid a_n \ne 0\} \in \mathbb{Z} \cup \{\infty\}.$$

(a) Since $(a_n t^n)_{n\ge 0} \to 0$, the series converges, and as $y = a_d t^d + t^{d+1} u$ for some $a_d \in F^\times$ and $u \in F[\![X]\!]$, it follows that

$$\mathsf{o}_X(t^{d+1}u) \ge (d+1)m > dm = \mathsf{o}_X(a_d t^d),$$

and therefore $\mathsf{o}_X(y) = dm$.

(b) Obviously, $F((t))$ is a subring of $F((X))$, and Φ is a ring isomorphism. Hence $F((t))$ is a subfield of $F((X))$. The remaining assertions follow immediately from the definitions.

(c) By 5.3.5.3, since F is the residue class field of $(F((X)), \mathsf{o}_X)$ and t is a prime element of $F[\![X]\!]$. □

As an application we discuss the differential calculus and the theory of logarithms and exponentials in formal power series.

Definition and Theorem 5.3.8. *Let $F((X))$ be the field of formal Laurent series in X over a field F.*

A. For a formal Laurent series

$$f = \sum_{n\in\mathbb{Z}} a_n X^n \quad \text{we define its derivative by} \quad f' = \frac{df}{dX} = \sum_{n\in\mathbb{Z}} na_n X^{n-1}.$$

1. If $f,\, g \in F((X))$ and $c \in F$, then $(cf)' = cf'$, $(f+g)' = f'+g'$,

$$(fg)' = f'g + fg', \quad \Big(\frac{f}{g}\Big)' = \frac{f'g - fg'}{g^2} \quad \text{if } g \neq 0,$$

and if $(f_n)_{n\geq 0}$ is a sequence in $F((X))$ such that $(f_n)_{n\geq 0} \to f$, then $(f_n')_{n\geq 0} \to f'$.

2. Let $f \in F((X))$. Then $f' = 0$ if and only if either $f \in K$ or $\mathrm{char}(F) = p > 0$ and $f \in F((X^p))$.

B. *Assume that $\mathrm{char}(F) = 0$. The power series*

$$L(X) = \sum_{n=1}^{\infty} \frac{(-1)^{n-1}}{n} X^n \in F[\![X]\!] \quad \text{and} \quad E(X) = \sum_{n=0}^{\infty} \frac{1}{n!} X^n \in F[\![X]\!].$$

satisfy the functional equations

$$L(X+Y+XY) = L(X) + L(Y), \quad E(X+Y) = E(X)E(Y)$$

and

$$E(L(X)) - 1 = L(E(X) - 1) = X.$$

We define the **logarithm** $\mathrm{Log}\colon 1+XF[\![X]\!] \to XF[\![X]\!]$ and the **exponential** $\mathrm{Exp}\colon XF[\![X]\!] \to 1+XF[\![X]\!]$ by

$$\mathrm{Log}(1+x) = L(x) \quad \text{and} \quad \mathrm{Exp}(x) = E(x) \quad \text{for all } x \in XF[\![X]\!].$$

The maps Log *and* Exp *are mutually inverse group isomorphisms (notice that $1+XF[\![X]\!] = U^{(1)}_{F((X))}$).*

Proof. 1. The rules $(cf)' = cf'$, $(f+g)' = f'+g'$ and $(fg)' = f'g + fg'$ are easily checked. If $g \neq 0$, then

$$f' = \Big(\frac{f}{g}g\Big)' = \Big(\frac{f}{g}\Big)'g + \frac{f}{g}g' \quad \text{implies} \quad \Big(\frac{f}{g}\Big)' = \frac{f'g - fg'}{g^2}.$$

By definition, $\mathsf{o}(f') \geq \mathsf{o}(f) + 1$ for all $f \in F((X))$, where $\mathsf{o} = \mathsf{o}_X$ is the order valuation. Hence the assignment $f \mapsto f'$ is continuous and thus it commutes with limits.

2. Obvious (as for polynomials).

3. By differentiation in $F[\![X,Y]\!] = F[\![X]\!][\![Y]\!] = F[\![Y]\!][\![X]\!]$. We use the standard notation of partial derivatives. Obviously, $L(0) = 0$, $E(0) = 1$,

$$L'(X) = \sum_{n=1}^{\infty}(-1)^{n-1}X^{n-1} = \frac{1}{1+X} \quad \text{and}$$

$$E'(X) = \sum_{n=1}^{\infty}\frac{1}{(n-1)!}X^{n-1} = E(X).$$

If $G(X,Y) = L(X+Y+XY) - L(X) - L(Y)$, then $G(0,0) = 0$,

$$\frac{\partial G(X,Y)}{\partial X} = \frac{1+Y}{1+X+Y+XY} - \frac{1}{1+X} = 0, \quad \text{and alike} \quad \frac{\partial G(X,Y)}{\partial Y} = 0,$$

hence $G(X,Y) = 0$ and $L(X+Y+XY) = L(X) + L(Y)$. If

$$H(X,Y) = \frac{E(X+Y)}{E(X)E(Y)}, \quad \text{then} \quad H(0,0) = 1 \quad \text{and}$$

$$\frac{\partial H(X,Y)}{\partial X} = \frac{\partial H(X,Y)}{\partial Y} = 0,$$

hence $H(X,Y) = 1$ and $E(X+Y) = E(X)E(Y)$. If

$$A(X) = \frac{E(L(X))}{1+X}, \quad \text{then} \quad A(0) = 1 \quad \text{and} \quad A'(X) = 0,$$

hence $A(X) = 1$ and $E(L(X)) = 1 + X$. If

$$B(X) = L(E(X)-1) - X, \quad \text{then} \quad B(0) = 0 \quad \text{and} \quad B'(X) = 0,$$

hence $B(X) = 0$ and $L(E(X)-1) = X$.

From these calculation rules it is obvious that Log and Exp are mutually inverse isomorphisms as asserted. □

We close this section with a characterization of the most important class of complete discrete valued fields.

Theorem and Definition 5.3.9. *Let $K = (K, |\cdot|)$ be a non-Archimedean valued field. Then the following assertions are equivalent:*

(a) K is locally compact.

(b) $\mathcal{O}_K$ is compact.

(c) k_K is finite, K is complete, and $|\cdot|$ is discrete.

If these conditions are fulfilled, then K is called a **(non-Archimedean) local field**.

Proof. (a) $\Rightarrow$(b) Let U be a compact neighborhood of 0, $\varepsilon \in \mathbb{R}_{>0}$ and $B_\varepsilon(0) \subset U$. If $a \in K^\times$ and $|a| < \varepsilon$, then $A = \{x \in K \mid |x| \le a\} \subset B_\varepsilon(0) \subset U$, and as A is closed, it is compact. Hence $\mathcal{O}_K = a^{-1}A$ is compact.

(b) $\Rightarrow$(c) We show first that the set $M(r) = \{x \in K \mid |x| \le r\}$ is compact for every $r \in \mathbb{R}_{>0}$. If $r \in \mathbb{R}_{>0}$ and $a \in K^\times$ is such that $|a|r \le 1$, then $aM(r) \subset \mathcal{O}_K$ is closed and thus compact. Hence $M(r)$ is compact as well.

If $\mathcal{R} \subset \mathcal{O}_K$ is a set of representatives for k_K, then $\{a + \mathfrak{p}_K \mid a \in \mathcal{R}\}$ is an open covering of $\mathcal{O}_K$ consisting of pairwise disjoint sets. Hence $\mathcal{R}$ and therefore k_K is finite.

Let $(x_n)_{n\ge0}$ be a Cauchy sequence in K. As it is bounded, there exists some $r \in \mathbb{R}_{>0}$ such that $|x_n| \le r$ for all $n \ge 0$. The set $M(r) = \{x \in K \mid |x| \le r\}$ is compact, hence $(x_n)_{n\ge0}$ has a cluster point $x \in M(r)$, and consequently $(x_n)_{n\ge0} \to x$.

We assume finally that $|K^\times|$ is not discrete and that α is a cluster point of $|K^\times|$. Then there exists a sequence $(x_n)_{n\ge0}$ in K such that $(|x_n|)_{n\ge0} \to \alpha$ and $|x_n| \ne \alpha$ for all $n \ge 0$. Let $r \in \mathbb{R}_{>0}$ be such that $|x_n| \le r$ for all $n \ge 0$. As $M(r) = \{x \in K \mid |x| \le r\}$ is compact, the sequence $(x_n)_{n\ge0}$ has a cluster point and thus a convergent subsequence in $M(r)$. We may assume that already $(x_n)_{n\ge0}$ converges, say $(x_n)_{n\ge0} \to x$. Then $(|x_n|)_{n\ge0} \to |x|$, hence $|x| = \alpha$ and $x \ne 0$. But now 5.3.2.1.(a) implies $|x_n| = \alpha$ for all $n \gg 1$, a contradiction.

(c) $\Rightarrow$(a) By 5.3.5.2, $\mathcal{O}_K$ is compact, and if $a \in K$, then $a + \mathcal{O}_K$ is a compact neighborhood of a. □

Definition and Remark 5.3.10. In general, a **local field** is a locally compact topological field. If K is a local field, then the additive group K has an up to a constant factor uniquely determined Haar measure μ, and its module $m_\mu \colon K \to \mathbb{R}_{\ge0}$, defined by $\mu(xC) = m_\mu(x)\mu(C)$ for every $x \in K$ and every compact subset C of K, turns out to be a normalized absolute value as defined in 5.1.3.**3** and 5.1.3.**5**. We refer to [62] for a foundation of the theory on this basis.

In 5.2.10 we have proved that (up to isomorphisms) $\mathbb{R}$ and $\mathbb{C}$ are the only Archimedian local fields, and in 5.8.3 we shall prove that (up to isomorphisms) every non-Archimedian local field is either a finite extension of some $\mathbb{Q}_p$ or a field of formal Laurent series over a finite field.

5.4 Hensel's lemma, generalizations and applications

We tacitly continue to use the notation introduced in 5.3.1 and 5.3.4.

Let $K = (K, |\cdot|)$ be a complete non-Archimedian valued field, $f \in \mathcal{O}_K[X]$ and $\overline{f} = f + \mathfrak{p}_K\mathcal{O}_K[X] \in \mathsf{k}_K[X]$ the residue class polynomial of f. Hensel's lemma in its crude form asserts that under suitable conditions it is possible to deduce the factorization properties of f from those of $\overline{f}$. In the present section we discuss various statements of this kind and give a series of first applications (criteria for power residues, existence of roots of unity and of suitable residue systems). In the further course of this volume we shall encounter this powerful concept (which traces back to the work of K. Hensel) again and again.

Definition and Theorem 5.4.1. *Let $K = (K, |\cdot|)$ be a non-Archimedean valued field. For a polynomial*

$$f = \sum_{\nu \geq 0} a_\nu X^\nu \in K[X] \quad \textit{we define} \quad |f| = \max\{|a_\nu| \mid \nu \geq 0\}.$$

Then $|\cdot|\colon K[X] \to \mathbb{R}_{\geq 0}$ has a unique extension to an (equally denoted) non-Archimedean absolute value $|\cdot|\colon K(X) \to \mathbb{R}_{\geq 0}$.

Let $d \in \mathbb{N}_0$ and $K[X]_d = \{f \in K[X] \mid \partial(f) \leq d\}$. Then the map

$$\theta\colon K^{d+1} \to K[X]_d, \quad \textit{defined by} \quad \theta(a_0, \ldots, a_d) = \sum_{i=0}^{d} a_i X^i,$$

is a topological isomorphism, and $\mathcal{O}_K[X]_d = K[X]_d \cap \mathcal{O}_K[X] = \theta(\mathcal{O}_K^{d+1})$ is closed in $K[X]_d$. If $(K, |\cdot|)$ is complete, then $K[X]_d$ is complete, too.

Proof. We prove first:

A. $|f + g| \leq \max\{|f|, |g|\}$ and $|fg| = |f|\,|g|$ for all $f, g \in K[X]$.

Proof of **A.** Let

$$f = \sum_{\nu \geq 0} a_\nu X^\nu \quad \text{and} \quad g = \sum_{\nu \geq 0} b_\nu X^\nu,$$

hence

$$f + g = \sum_{\nu \geq 0} (a_\nu + b_\nu) X^\nu \quad \text{and} \quad fg = \sum_{\nu \geq 0} \Big(\sum_{\substack{i,j \geq 0 \\ i+j=\nu}} a_i b_j \Big) X^\nu.$$

Then

$$\begin{aligned} |f + g| &= \max\{|a_\nu + b_\nu| \mid \nu \geq 0\} \leq \max\{\max\{|a_\nu|, |b_\nu|\} \mid \nu \geq 0\} \\ &= \max\{\max\{|a_\nu| \mid \nu \geq 0\}, \max\{|b_\nu| \mid \nu \geq 0\}\} = \max\{|f|, |g|\}. \end{aligned}$$

As to the product we obtain for all $\nu \geq 0$

$$\begin{aligned} \Big| \sum_{\substack{i,j \geq 0 \\ i+j=\nu}} a_i b_j \Big| &\leq \max\{|a_i|\,|b_j| \mid i, j \geq 0,\ i + j = \nu\} \\ &\leq \max\{|a_i| \mid i \geq 0\} \max\{|b_j| \mid j \geq 0\} = |f|\,|g|, \end{aligned}$$

and therefore $|fg| \le |f|\,|g|$. To prove the reverse inequality, let $k,\, l \ge 0$ be maximal such that $|f| = |a_k|$ and $|g| = |b_l|$. Then $|a_k| > |a_i|$ for all $i > k$ and $|b_l| > |b_j|$ for all $j > l$. Let $i,\, j \ge 0$ be such that $i+j = k+l$ and $(i,j) \ne (k,l)$. Then it follows that either $i > k$ or $j > l$, hence $|a_i| < |a_k|,\ \ |b_j| \le |b_l|$ in the first case, $|a_i| \le |a_k|,\ \ |b_j| < |b_l|$ in the second case, and consequently $|a_i|\,|b_j| < |a_k|\,|b_l|$ in any case. Using this, we eventually obtain

$$|f|\,|g| = |a_k|\,|b_l| = \Bigl| \sum_{\substack{i,j\ge 0 \\ i+j=k+l}} a_i b_j \Bigr| \le |fg|. \qquad \square[\mathbf{A}.]$$

If $f \in K[X]$, then $|f| = 0$ if and only if $f = 0$. For a rational function $h = g^{-1}f \in K(X)$, where $f,\, g \in K[X]$ and $g \ne 0$, we define $|h| = |g|^{-1}|f|$. Using **A**, it is easily checked that $|\cdot|\colon K(X) \to \mathbb{R}_{\ge 0}$ is a well-defined non-Archimedian absolute value of $K(X)$. Its uniqueness is obvious.

By definition, θ an isomorphism of vector spaces over K, and it is a homeomorphism, since the maximum norm induces the product topology. Since $\mathcal{O}_K$ is closed in K, it follows that $\mathcal{O}_K^{d+1}$ is closed in K^{d+1}, and thus $\mathcal{O}_K[X]_d$ is closed in $K[X]_d$. If K is complete, then K^{d+1} and thus also $K[X]_{d+1}$ is complete (recall that a sequence in K^{d+1} is convergent resp. a Cauchy sequence if and only if all component sequences have this property). $\square$

Theorem 5.4.2 (Hensel-Newton). *Let $K = (K,\, |\cdot|)$ be a complete non-Archimedian valued field, $f \in \mathcal{O}_K[X]$ and $a_0 \in \mathcal{O}_K$ such that $|f(a_0)| < |f'(a_0)|^2$. Then there exists a unique $a \in \mathcal{O}_K$ such that*

$$f(a) = 0 \quad \text{and} \quad |a - a_0| \le \frac{|f(a_0)|}{|f'(a_0)|}.$$

Let the sequence $(a_j)_{j\ge 0}$ be recursively defined by

$$a_{j+1} = a_j - \frac{f(a_j)}{f'(a_j)} \quad \text{if } f'(a_j) \ne 0, \quad \text{and} \quad a_{j+1} = 0 \quad \text{if } f'(a_j) = 0.$$

Then

$$a = \lim_{j\to\infty} a_j \quad \text{and} \quad |f'(a)| = |f'(a_0)|.$$

Proof. For a polynomial $F \in \mathcal{O}_K[X]$ we define the sequence of formal derivatives $(D^\nu F)_{\nu\ge 0}$ in $\mathcal{O}_K[X]$ by means of the identity

$$F(X+Y) = \sum_{\nu\ge 0} (D^\nu F)(X) Y^\nu\,.$$

Obviously, $D^0F = F,\ \ D^1F = F'$ and $D^\nu F = 0$ for all $\nu > \partial(F)$. We prove first the following assertions for all $j \ge 0$:

A. $a_j \in \mathcal{O}_K$, $|f(a_{j+1})| \le |f(a_j)| \le |f(a_0)|$, $|f(a_j)| < |f'(a_j)|^2$, $|f'(a_j)| = |f'(a_0)|$,

$$|f(a_{j+1})|\,|f'(a_0)|^2 \le |f(a_j)|\,|f(a_0)| \quad \text{and} \quad |f(a_j)| \le |f(a_0)|\Big(\frac{|f(a_0)|}{|f'(a_0)|^2}\Big)^j.$$

Proof of **A.** By induction on j.

By assumption, $a_0 \in \mathcal{O}_K$ and $|f(a_0)| < |f'(a_0)|^2$. We assume now that $j \ge 0$, $a_j \in \mathcal{O}_K$, $|f(a_j)| \le |f(a_0)|$, $|f(a_j)| < |f'(a_j)|^2$, $|f'(a_j)| = |f'(a_0)| > 0$ and

$$|f(a_j)| \le |f(a_0)|\Big(\frac{|f(a_0)|}{|f'(a_0)|^2}\Big)^j.$$

Then

$$a_{j+1} = a_j - \frac{f(a_j)}{f'(a_j)},$$

and $f'(a_0) \in \mathcal{O}_K$ implies

$$\frac{|f(a_j)|}{|f'(a_j)|} \le \frac{|f(a_0)|}{|f'(a_0)|} < |f'(a_0)| = |f'(a_j)| \le 1, \quad \text{hence}$$
$$\frac{f(a_j)}{f'(a_j)} \in \mathcal{O}_K \quad \text{and} \quad a_{j+1} \in \mathcal{O}_K.$$

From

$$f(a_{j+1}) = f\Big(a_j - \frac{f(a_j)}{f'(a_j)}\Big) = f(a_j) - f'(a_j)\frac{f(a_j)}{f'(a_j)} + \sum_{\nu\ge2}(D^\nu f)(a_j)\Big(-\frac{f(a_j)}{f'(a_j)}\Big)^\nu$$
$$= \sum_{\nu\ge2}(D^\nu f)(a_j)\Big(-\frac{f(a_j)}{f'(a_j)}\Big)^\nu$$

we derive

$$|f(a_{j+1})| \le \max\Big\{|(D^\nu f)(a_j)|\Big(\frac{|f(a_j)|}{|f'(a_j)|}\Big)^\nu \,\Big|\, \nu \ge 2\Big\} \le \frac{|f(a_j)|^2}{|f'(a_j)|^2},$$

hence $|f(a_{j+1})|\,|f'(a_0)|^2 = |f(a_{j+1})|\,|f'(a_j)|^2 \le |f(a_j)|^2 \le |f(a_j)|\,|f(a_0)|$ and

$$|f(a_{j+1})| \le |f(a_j)|\frac{|f(a_0)|}{|f'(a_0)|^2} \le |f(a_j)|.$$

From

$$f'(a_{j+1}) = f'\Big(a_j - \frac{f(a_j)}{f'(a_j)}\Big) = f'(a_j) + \sum_{\nu\ge1}(D^\nu f')(a_j)\Big(-\frac{f(a_j)}{f'(a_j)}\Big)^\nu$$

we derive

$$|f'(a_{j+1}) - f'(a_j)| \le \max\Big\{|(D^\nu f')(a_j)|\Big(\frac{|f(a_j)|}{|f'(a_j)|}\Big)^\nu \,\Big|\, \nu \ge 1\Big\}$$
$$\le \frac{|f(a_j)|}{|f'(a_j)|} < |f'(a_j)|,$$

and $|f'(a_{j+1})| = \big|[f'(a_{j+1}) - f'(a_j)] + f'(a_j)\big| = |f'(a_j)| = |f'(a_0)|$. With this it follows that $|f(a_{j+1})| \le |f(a_j)| < |f'(a_j)|^2 = |f'(a_{j+1})|^2$, and

$$|f(a_{j+1})| \le |f(a_j)| \frac{|f(a_0)|}{|f'(a_0)|^2} \le |f(a_0)| \Big(\frac{|f(a_0)|}{|f'(a_0)|^2}\Big)^{j+1}. \qquad \Box[\mathbf{A}.]$$

By means of **A** we obtain

$$\lim_{j\to\infty} |f(a_j)| = 0 \quad \text{and} \quad \lim_{j\to\infty} |a_{j+1} - a_j| = \lim_{j\to\infty} \frac{|f(a_j)|}{|f'(a_j)|} = \lim_{j\to\infty} \frac{|f(a_j)|}{|f'(a_0)|} = 0.$$

Hence $(a_j)_{j\ge0}$ is a Cauchy sequence, $(a_j)_{j\ge0} \to a$ for some $a \in \mathcal{O}_K$, $(f(a_j))_{j\ge0} \to f(a)$, and consequently $f(a) = 0$. If $n > j \ge 0$, then

$$|a_n - a_j| = \Big|\sum_{\nu=j}^{n-1} (a_{\nu+1} - a_\nu)\Big| \le \sup\Big\{\frac{|f(a_\nu)|}{|f'(a_\nu)|} \;\Big|\; \nu \in [j, n-1]\Big\} \le \frac{|f(a_0)|}{|f'(a_0)|},$$

hence

$$|a - a_j| = \lim_{n\to\infty} |a_n - a_j| \le \frac{|f(a_0)|}{|f'(a_0)|} \quad \text{and} \quad |f'(a)| = \lim_{n\to\infty} |f'(a_n)| = |f'(a_0)|.$$

To prove the uniqueness of a, we assume that there is some $b \in \mathcal{O}_K$ such that $b \ne a$, $f(b) = 0$ and

$$|b - a_0| \le \frac{|f(a_0)|}{|f'(a_0)|}, \quad \text{hence as well} \quad |b-a| = |(b-a_0)+(a_0-a)| \le \frac{|f(a_0)|}{|f'(a_0)|} < 1.$$

Since $b = a + (b - a)$ and $f(a) = 0$, it follows that

$$0 = f(b) = \sum_{\nu\ge0} (D^\nu f)(a)(b-a)^\nu = f'(a)(b-a) + \sum_{\nu\ge2} (D^\nu f)(a)(b-a)^\nu,$$

hence

$$f'(a) = -\sum_{\nu\ge2} (D^\nu f)(a)(b-a)^{\nu-1}$$

and therefore

$$|f'(a_0)| = |f'(a)| \le \max_{\nu\ge2} |(D^\nu f)(a)|\,|b-a|^{\nu-1} \le |b-a| \le \frac{|f(a_0)|}{|f'(a_0)|},$$

which implies the contradiction $|f'(a_0)|^2 \le |f(a_0)|$. $\Box$

We proceed with a first series of applications.

Theorem 5.4.3. *Let $K = (K, |\cdot|)$ be a complete non-Archimedean valued field. For a polynomial $f \in \mathcal{O}_K[X]$ let $\overline{f} = f + \mathfrak{p}_K\mathcal{O}_K[X] \in \mathsf{k}_K[X]$ be the residue class polynomial (if $a \in \mathcal{O}_K$, then $\overline{a} = a + \mathfrak{p}_K \in \mathsf{k}_K$ is the residue class of a).*

1. Let $f \in \mathcal{O}_K[X]$ and $\alpha \in \mathsf{k}_K$ a simple zero of $\overline{f}$. Then there exists a unique $a \in \mathcal{O}_K$ such that $f(a) = 0$ and $\overline{a} = \alpha$.

2. Let $m \in \mathbb{N}$ be such that $\mathrm{char}(\mathsf{k}_K) \nmid m$.

(a) Let $a_0 \in U_K$ and $\alpha \in \mathsf{k}_K$ be such that $\alpha^m = \overline{a}_0$. Then there exists a unique $a \in U_K$ satisfying $a^m = a_0$ and $\overline{a} = \alpha$. In particular, $U_K^{(1)} = (U_K^{(1)})^m$.

(b) Let $\xi \in \mathsf{k}_K$ be a primitive m-th root of unity. Then there exists a unique primitive m-th root of unity $\zeta \in \mathcal{O}_K$ such that $\overline{\zeta} = \xi$, and the residue class homomorphism $a \mapsto \overline{a}$ induces an isomorphism $\mu_m(K) \overset{\sim}{\rightarrow} \mu_m(\mathsf{k}_K)$. In particular, if $|\mathsf{k}_K| = q < \infty$, then K contains a primitive $(q-1)$-th root of unity, and the residue class homomorphism induces an isomorphism $\mu_{q-1}(K) \overset{\sim}{\rightarrow} \mathsf{k}_K^\times = \mu_{q-1}(\mathsf{k}_K)$.

3. Let $\mathrm{char}(\mathsf{k}_K) = p > 0$ and let $\mu'(K)$ resp. $\mu'(\mathsf{k}_K)$ be the group of roots of unity of order coprime to p in K resp. k_K. Then the residue class homomorphism induces an isomorphism $\mu'(K) \overset{\sim}{\rightarrow} \mu'(\mathsf{k}_K)$. In particular, if $|\mathsf{k}_K| = q < \infty$, then $|\mu'(K)| = q - 1$.

Proof. 1. Let $a_0 \in \mathcal{O}_K$ be such that $\overline{a}_0 = \alpha$. Then $\overline{f(a_0)} = \overline{f}(\alpha) = 0$ and $\overline{f'(a_0)} = \overline{f}'(\alpha) \neq 0$, hence $|f(a_0)| < 1$ and $|f'(a_0)| = 1$. By 5.4.2 there exists a unique $a \in \mathcal{O}_K$ satisfying $f(a) = 0$ and $|a - a_0| < 1$, that is, $\overline{a} = \alpha$.

2.(a) Let $f = X^m - a_0 \in \mathcal{O}_K[X]$. Then $\overline{f} = X^m - \overline{a}_0 \in \mathsf{k}_K[X]$ is separable and thus α is a simple zero of $\overline{f}$. By 1. there exists a unique $a \in \mathcal{O}_K$ such that $f(a) = 0$ and $\overline{a} = \alpha$, which implies $a \in U_K$ and $a^m = a_0$. In particular, if $a_0 \in U_K^{(1)}$, then $\overline{a}_0 = \overline{1}$, and with $\alpha = \overline{1}$ it follows that $\overline{a} = \overline{1}$, $a \in U_K^{(1)}$ and $a_0 \in (U_K^{(1)})^m$.

(b) By (a) there exists some $\zeta \in K$ such that $\zeta^m = 1$ and $\overline{\zeta} = \xi$. If $\zeta^d = 1$ for some $d \in [1, m-1]$, then $\xi^d = 1$, a contradiction. Hence ζ is a primitive m-th root of unity, and the residue class map induces an isomorphism $\mu_m(K) = \langle \zeta \rangle \overset{\sim}{\rightarrow} \langle \xi \rangle = \mu_m(\mathsf{k}_K)$.

3. By 2.(b), since

$$\mu'(K) = \bigcup_{\substack{m \in \mathbb{N} \\ p \nmid m}} \mu_m(K) \quad \text{and} \quad \mu'(\mathsf{k}_K) = \bigcup_{\substack{m \in \mathbb{N} \\ p \nmid m}} \mu_m(\mathsf{k}_K). \qquad \square$$

Theorem 5.4.4 (Hensel-Ore). *Let $K = (K, |\cdot|)$ be a complete non-Archimedean valued field, $f \in \mathcal{O}_K[X]^\bullet$, $\delta \in (0,1)$ and $\omega \in (0,1]$.*

1. Let $G, H \in \mathcal{O}_K[X]$ be such that G is monic, $|\mathsf{R}(G,H)| = \omega$ and $|f - GH| = \delta < \omega^2$. Then there exist polynomials $g, h \in \mathcal{O}_K[X]$ such that g is monic, $\partial(g) = \partial(G)$,

$$f = gh, \quad |g - G| \le \omega^{-1}\delta \quad \textit{and} \quad |h - H| \le \omega^{-1}\delta.$$

2. *Let f be monic, $r \ge 2$, and let $F_1, \ldots, F_r \in \mathcal{O}_K[X]$ be monic.*

   (a) *Suppose that $|f - F_1 \cdot \ldots \cdot F_r| = \delta < |\mathsf{D}(f)|$. Then there exist monic polynomials $f_1, \ldots, f_r \in \mathcal{O}_K[X]$ such that*

$$f = f_1 \cdot \ldots \cdot f_r \quad \text{and} \quad |f_i - F_i| < |\mathsf{D}(F_i)| = |\mathsf{D}(f_i)| \text{ for all } i \in [1, r].$$

   (b) *Let $|\cdot|$ be discrete, and $f \equiv F_1 \cdot \ldots \cdot F_r \mod \mathfrak{p}_K^{v_K(\mathsf{D}(f))+1}\mathcal{O}_K[X]$. Then there exist monic polynomials $f_1, \ldots, f_r \in \mathcal{O}_K[X]$ such that*

$$f = f_1 \cdot \ldots \cdot f_r \quad \text{and} \quad f_i \equiv F_i \mod \mathfrak{p}_K^{v_K(\mathsf{D}(f_i))+1}\mathcal{O}_K[X] \text{ for all } i \in [1, r].$$

Proof. For $\eta \in (0,1)$ set $\mathfrak{I}_\eta = \{x \in \mathcal{O}_K \mid |x| \le \eta\} \in \mathfrak{I}'(\mathcal{O}_K)$, and for a polynomial $P \in \mathcal{O}_K[X]$ let $\overline{P} = P + \mathfrak{I}_\delta\mathcal{O}_K[X] \in \mathcal{O}_K[X]/\mathfrak{I}_\delta\mathcal{O}_K[X] = (\mathcal{O}_K/\mathfrak{I}_\delta)[X]$ be the residue class polynomial modulo $\mathfrak{I}_\delta$.

1. Let $\delta = \omega^2\rho$, where $\rho \in (0,1)$. Since $|f - GH| = \delta < 1$, it follows that $\overline{f} = \overline{G}\,\overline{H}$. If $G = 1$, we set $g = 1$ and $h = f$. Thus assume that $\partial(G) = d \in \mathbb{N}$, $\partial(H) = m \in \mathbb{N}_0$, $\partial(f) = r \in \mathbb{N}_0$,

$$G = X^d + \ldots, \quad H = c_m X^m + c_{m-1}X^{m-1} + \ldots + c_1 X + c_0 \text{ and } f = aX^r + \ldots,$$

where $a, c_0, \ldots, c_m \in \mathcal{O}_K$ and $ac_m \ne 0$. Now $\delta < \omega^2 \le \omega$ implies $\overline{\mathsf{R}(G,H)} \ne \overline{0}$, hence $\overline{H} \ne \overline{0}$ by 1.6.5.2, and we set $e = \partial(\overline{H}) \in [0, m]$. Then we obtain

$$\partial(\overline{f}\,) = \partial(\overline{G}\,) + \partial(\overline{H}\,) = e + d \le r = \partial(f).$$

We construct two sequences $(g_n)_{n\ge0}$, $(h_n)_{n\ge0}$ in $\mathcal{O}_K[X]$ such that $g_0 = G$, $\overline{h}_0 = \overline{H}$ and the following assertions hold for all $n \ge 0$:

- $g_n = X^d + \ldots$, $h_n = aX^{r-d} + \ldots$, and $|\mathsf{R}(g_n, h_n)| = \omega$;
- $|f - g_n h_n| \le \rho^n \delta$;
- $|g_{n+1} - g_n| \le \rho^n \omega^{-1}\delta$ and $|h_{n+1} - h_n| \le \rho^n \omega^{-1}\delta$;
- $|g_n - G| \le \omega^{-1}\delta$ and $|h_n - H| \le \omega^{-1}\delta$.

We first finish the proof using these two sequences. Since $(g_n)_{n\ge0}$ is a Cauchy sequence in $\mathcal{O}_K[X]_d$, and $(h_n)_{n\ge0}$ is a Cauchy sequence in $\mathcal{O}_K[X]_{r-d}$, there exist polynomials $g \in \mathcal{O}_K[X]_d$ and $h \in \mathcal{O}_K[X]_{n-d}$ such that $(g_n)_{n\ge0} \to g$ and $(h_n)_{n\ge0} \to h$. Hence $(g_n h_n)_{n\ge0} \to gh$, and as $(g_n h_n)_{n\ge0} \to f$ by construction, it follows that $f = gh$. As $g_n = X^d + \ldots$, $h_n = aX^{r-d} + \ldots$, $|g_n - G| \le \omega^{-1}\delta$ and $|h_n - H| \le \omega^{-1}\delta$ for all $n \ge 0$, we obtain $g = X^d + \ldots$, $h = aX^{r-d} + \ldots$, $|g - G| \le \omega^{-1}\delta$ and $|h - H| \le \omega^{-1}\delta$.

Now we construct recursively the crucial sequences $(g_n)_{n\ge 0}$ and $(h_n)_{n\ge 0}$. We set $g_0 = G$ and

$$h_0 = \begin{cases} aX^e + c_{e-1}X^{e-1} + \ldots + c_1X + c_0 & \text{if} \quad d+e=r, \\ aX^{r-d} + c_eX^e + \ldots + c_1X + c_0 & \text{if} \quad d+e<r. \end{cases}$$

Since $\overline{f} = \overline{a}X^r + \ldots = \overline{G}\,\overline{H} = (X^d+\ldots)(\overline{c}_eX^e+\ldots) = \overline{c}_eX^{d+e}+\ldots$, we obtain

$$\overline{a} = \begin{cases} \overline{0} & \text{if} \quad d+e<r, \\ \overline{c}_e & \text{if} \quad d+e=r. \end{cases}$$

In any case, $\overline{h}_0 = \overline{H}$, and thus $|h_0 - H| \le \delta \le \omega^{-1}\delta$, $|g_0 - G| \le \delta \le \omega^{-1}\delta$ and $\overline{f} - \overline{g}_0\overline{h}_0 = \overline{f} - \overline{G}\,\overline{H} = \overline{0}$, hence $|f - g_0h_0| \le \delta$.

Now let $n \ge 0$, and suppose that we have already constructed polynomials $g_n,\, h_n \in \mathcal{O}_K[X]$ such that $g_n = X^d+\ldots$, $h_n = aX^{r-d}+\ldots$, $|\mathsf{R}(g_n,h_n)| = \omega$ and $|f - g_nh_n| \le \rho^n\delta$. By 1.6.4.1 and since $\partial(f-g_nh_n) < r$, there exist polynomials $U,\,V \in \mathsf{c}(f-g_nh_n)\mathcal{O}_K[X]$ such that $\partial(U) < d$, $\partial(V) < r-d$ and $g_nV + h_nU = (f-g_nh_n)\mathsf{R}(g_n,h_n)$. We set

$$g_{n+1} = g_n + \mathsf{R}(g_n,h_n)^{-1}U \quad \text{and} \quad h_{n+1} = h_n + \mathsf{R}(g_n,h_n)^{-1}V.$$

With this we obtain $g_{n+1} = X^d+\ldots \in \mathcal{O}_K[X]$, $h_{n+1} = aX^{r-d}+\ldots \in \mathcal{O}_K[X]$,

$$|g_{n+1} - g_n| = \frac{|U|}{|\mathsf{R}(g_n,h_n)|} \le \frac{|f-g_nh_n|}{\omega} \le \rho^n\omega^{-1}\delta \le \rho\omega,$$

hence $g_{n+1} \equiv g_n \mod \mathfrak{J}_{\rho\omega}\mathcal{O}_K[X]$, and

$$|h_{n+1} - h_n| = \frac{|V|}{|\mathsf{R}(g_n,h_n)|} \le \frac{|f-g_nh_n|}{\omega} \le \rho^n\omega^{-1}\delta \le \rho\omega,$$

hence $h_{n+1} \equiv h_n \mod \mathfrak{J}_{\rho\omega}\mathcal{O}_K[X]$. Now it follows that

$$\mathsf{R}(g_n,h_n) \equiv \mathsf{R}(g_{n+1},h_{n+1}) \mod \mathfrak{J}_{\rho\omega}, \quad |\mathsf{R}(g_{n+1},h_{n+1}) - \mathsf{R}(g_n,h_n)| \le \rho\omega < \omega,$$

and consequently

$$|\mathsf{R}(g_{n+1},h_{n+1})| = |[R(g_{n+1},h_{n+1}) - \mathsf{R}(g_n,h_n)] + \mathsf{R}(g_n,h_n)| = \omega.$$

By the very definition it follows that

$$f - g_{n+1}h_{n+1} = f - g_nh_n - \frac{h_nU + g_nV}{\mathsf{R}(g_n,h_n)} - \frac{UV}{\mathsf{R}(g_n,h_n)^2} = -\frac{UV}{\mathsf{R}(g_n,h_n)^2}$$

and therefore

$$|f - g_{n+1}h_{n+1}| \le \frac{|UV|}{|\mathsf{R}(g_n, h_n)|^2} = \frac{|U|}{|\mathsf{R}(g_n, h_n)|}\,\frac{|V|}{|\mathsf{R}(g_n, h_n)|}$$
$$\le \rho^{2n}\omega^{-2}\delta^2 = \rho^{2n+1}\delta \le \rho^{n+1}\delta.$$

Finally,

$$|g_{n+1} - G| \le \max\{|g_{n+1} - g_n|, |g_n - G|\} \le \max\{\rho^n\omega^{-1}\delta, \omega^{-1}\delta\} = \omega^{-1}\delta,$$

and alike $|h_{n+1} - H| \le \omega^{-1}\delta$.

2.(a) By induction on r.

$r = 2$: We set $\omega = |\mathsf{R}(F_1, F_2)|$. As $|f - F_1F_2| = \delta < |\mathsf{D}(f)|$, we get $\overline{f} = \overline{F_1}\,\overline{F_2}$ and $\mathsf{D}(\overline{f}) = \mathsf{D}(\overline{F_1}\,\overline{F_2}) \ne \overline{0}$. Hence 1.6.5.4 implies

$$\overline{\mathsf{D}(f)} = \mathsf{D}(\overline{f}) = \mathsf{D}(\overline{F_1}\,\overline{F_2}) = \overline{\mathsf{D}(F_1F_2)},\ |\mathsf{D}(f) - \mathsf{D}(F_1F_2)| \le \delta < |\mathsf{D}(f)|$$

and

$$|\mathsf{D}(F_1F_2)| = |\mathsf{D}(f) - [\mathsf{D}(f) - \mathsf{D}(F_1F_2)]| = |\mathsf{D}(f)|.$$

Since $\mathsf{D}(F_1F_2) = \mathsf{D}(F_1)\mathsf{D}(F_2)\mathsf{R}(F_1, F_2)^2$ (see 1.6.6), it follows that

$$\delta < |\mathsf{D}(f)| = |\mathsf{D}(F_1)|\,|\mathsf{D}(F_2)|\,|\mathsf{R}(F_1, F_2)|^2 \le \omega^2,$$

and we apply 1. There exist polynomials $f_1, f_2 \in \mathcal{O}_K[X]$ such that f_1 is monic, $f = f_1f_2$ and $|f_i - F_i| \le \omega^{-1}\delta$ for $i \in \{1,2\}$. Then f_2 is monic, too, and it remains to prove that $|f_i - F_i| < |\mathsf{D}(F_i)| = |\mathsf{D}(f_i)|$ for $i \in \{1,2\}$.

Let $i \in \{1,2\}$ and $\eta = |f_i - F_i|$. Then it follows that $F_i \equiv f_i \mod \mathfrak{I}_\eta\mathcal{O}_K[X]$, hence $\mathsf{D}(F_i) \equiv \mathsf{D}(f_i) \mod \mathfrak{I}_\eta$,

$$|\mathsf{D}(F_i) - \mathsf{D}(f_i)| \le \eta \le \frac{\delta}{\omega} < \frac{|\mathsf{D}(f)|}{|\mathsf{R}(F_1, F_2)|} = |\mathsf{D}(F_1)|\,|\mathsf{D}(F_2)|\,|\mathsf{R}(F_1, F_2)| \le |\mathsf{D}(F_i)|,$$

and $|\mathsf{D}(f_i)| = |\mathsf{D}(F_i) - [\mathsf{D}(F_i) - \mathsf{D}(f_i)]| = |\mathsf{D}(F_i)|$.

$r \ge 3$, $r-1 \to r$: By the case $r = 2$, there exist monic polynomials $f_1, h \in \mathcal{O}_K[X]$ such that

$$f = f_1h,\ |f_1 - F_1| < |\mathsf{D}(f_1)| = |\mathsf{D}(F_1)| \text{ and } |h - F_2 \cdot \ldots \cdot F_r| < |\mathsf{D}(h)|.$$

The induction hypothesis yields monic polynomials $f_2, \ldots, f_r \in \mathcal{O}_K[X]$ such that $h = f_2 \cdot \ldots \cdot f_r$ and $|f_i - F_i| < |\mathsf{D}(F_i)| = |\mathsf{D}(f_i)|$ for all $i \in [2, r]$.

(b) Let $\pi \in \mathcal{O}_K$ be a prime element. Then $f - F_1 \cdot \ldots \cdot F_r = \pi^{v_K(\mathsf{D}(f))}G$, where $G \in \pi\mathcal{O}_K[X]$, and $|f - F_1 \cdot \ldots \cdot F_r| = |\mathsf{D}(f)|\,|G| < |\mathsf{D}(f)|$. By (a) there exist polynomials $f_1, \ldots, f_r \in \mathcal{O}_K[X]$ such that $f = f_1 \cdot \ldots \cdot f_r$ and $|f_i - F_i| < |\mathsf{D}(f_i)|$ for all $i \in [1, r]$. Then $f_i - F_i = \pi^{v_K(\mathsf{D}(f_i))}G_i$, where $|G_i| < 1$ and consequently $G_i \in \pi\mathcal{O}_K[X]$. For all $i \in [1, r]$ it follows that $f_i \equiv F_i \mod \mathfrak{p}_K^{v_K(\mathsf{D}(f_i))+1}\mathcal{O}_K[X]$. $\square$

Theorem 5.4.5 (Hensel's lemma, usual form). *Let $K = (K, |\cdot|)$ be a complete non-Archimedian valued field and $f \in \mathcal{O}_K[X]$. For $P \in \mathcal{O}_K[X]$ let $\overline{P} \in \mathsf{k}_K[X]$ be the residue class polynomial (in particular, $\overline{a} = a + \mathfrak{p}_K \in \mathsf{k}_K$ for all $a \in \mathcal{O}_K$).*

1. *Let $\varphi, \psi \in \mathsf{k}_K[X]^\bullet$ be coprime polynomials such that $\overline{f} = \varphi\psi$. Then there exist polynomials $g, h \in \mathcal{O}_K[X]$ such that $f = gh$, $\partial(g) = \partial(\varphi)$, $\overline{g} = \varphi$, $\overline{h} = \psi$, and if φ is monic, then g is monic, too.*

2. *Suppose that f is monic, $\overline{f}$ is separable and $\overline{f} = \varphi_1 \cdot \ldots \cdot \varphi_r$ for some $r \in \mathbb{N}$ and monic polynomials $\varphi_1, \ldots, \varphi_r \in \mathsf{k}_K[X]$. Then there exist monic polynomials $f_1, \ldots, f_r \in \mathcal{O}_K[X]$ such that $f = f_1 \cdot \ldots \cdot f_r$ and $\overline{f}_i = \varphi_i$ for all $i \in [1, r]$.*

Proof. 1. Let first φ be monic, and let $F, G \in \mathcal{O}_K[X]$ be any polynomials such that $\overline{F} = \varphi$, $\overline{G} = \psi$, F is monic and $\partial(F) = \partial(\varphi)$. Then we obtain $\overline{\mathsf{R}(G, H))} = \mathsf{R}(\varphi, \psi) \neq 0$ by 1.6.7.1 and 1.6.5.3, hence

$$|\mathsf{R}(G, H)| = 1 > |f - GH|,$$

and by 5.4.4.1 there exist polynomials $g, h \in \mathcal{O}_K[X]$ such that g is monic, $\partial(g) = \partial(\varphi)$, $\overline{g} = \varphi$, $\overline{h} = \psi$ and $f = gh$.

Now let φ be arbitrary, and let $a \in \mathcal{O}_K^\times$ be such that $\overline{a}^{-1}\varphi$ is monic. Then $\overline{f} = (\overline{a}^{-1}\varphi)(\overline{a}\psi)$, and there exist polynomials $g_1, h_1 \in \mathcal{O}_K[X]$ such that $\partial(g_1) = \partial(\varphi)$, $\partial(h_1) = \partial(\psi)$, $\overline{g}_1 = \overline{a}^{-1}\varphi$, $\overline{h}_1 = \overline{a}\psi$ and $f = g_1 h_1$. The polynomials $g = a^{-1}g_1$ and $h = ah_1$ fulfill our requirements.

2. By induction on r. For $r = 1$ there is nothing to do.

$r \geq 2$, $r - 1 \to r$: As $\overline{f}$ is separable, $\varphi_1, \ldots, \varphi_r$ are pairwise coprime, and thus φ_1 and $\varphi_2 \cdot \ldots \cdot \varphi_r$ are coprime. By 1. there exist monic polynomials $f_1, g \in \mathcal{O}_K[X]$ such that $f = f_1 g$, $\overline{f}_1 = \varphi_1$ and $\overline{g} = \varphi_2 \cdot \ldots \cdot \varphi_r$. Applying the induction hypothesis for g yields the assertion. □

Theorem 5.4.6. *Let $K = (K, |\cdot|)$ be a complete non-Archimedian valued field and $f \in K[X]$ an irreducible polynomial. If $a \in K^\times$ is the leading coefficient of f, then $|f| = \max\{|a|, |f(0)|\}$. In particular, if f is monic and $f(0) \in \mathcal{O}_K$, then $f \in \mathcal{O}_K[X]$.*

Proof. For a polynomial $P \in \mathcal{O}_K[X]$ let $\overline{P} \in \mathsf{k}_K[X]$ be the residue class polynomial. By definition, $|f| \geq \max\{|a|, |f(0)|\}$, and we assume to the contrary that $|f| > \max\{|a|, |f(0)|\}$. We shall prove that f is reducible. Let $c \in K^\times$ be such that $|f| = |c|$, and let

$$f_0 = c^{-1}f = a_0 + a_1X + \ldots + a_nX^n, \text{ where } a_0 = c^{-1}f(0) \text{ and } a_n = c^{-1}a.$$

Then $1 = |f_0| > \max\{|a_n|, |a_0|\}$, and if $r = \min\{\nu \in [0, n] \mid |a_\nu| = 1\}$, then $0 < r < n$ and $\overline{f} = X^r(\overline{a}_r + \overline{a}_{r+1}X + \ldots + \overline{a}_nX^{n-r})$. Therefore f_0 and thus also f is reducible by 5.4.5.1.

In particular, if f is monic and $f(0) \in \mathcal{O}_K$, then $|f| \le 1$ and consequently $f \in \mathcal{O}_K[X]$. □

If K is a discrete valued field, $m \in \mathbb{N}$ and $a \in U_K^m$, say $a = x^m$ for some $x \in U_K$. Then $a \equiv x^m \mod \mathfrak{p}_K^k$ for every $k \ge 1$, and in this case we say that a is an **m-th power residue** modulo $\mathfrak{p}_K^k$. The following Theorem 5.4.7 shows that for sufficiently large k the converse is true. This sheds new light on and is a far-reaching generalization of 3.2.2, where we showed a weaker result for quadratic residues.

Theorem 5.4.7 (Power residues vs. local powers). *Let $K = (K, |\cdot|)$ be a complete discrete valued field, $m \in \mathbb{N}$ and $a \in U_K$. Then a is an m-th power in U_K if and only if a is an m-th power residue modulo $\mathfrak{p}_K^{2v_K(m)+1}$.*

In particular: If p is a prime, $a \in \mathbb{Z}$, $p \nmid a$ and $m \in \mathbb{N}$, then $a \in \mathbb{Q}_p^{\times m}$ if and only if a is an m-th power residue modulo $p^{2v_p(m)+1}$.

Proof. If $a \in U_K$, then $a \in K^{\times m}$ if and only if $a \in U_K^m$, and if $a = x^m$ for some $x \in U_K$, then $a \equiv x^m \mod \mathfrak{p}_K^k$ for all $k \in \mathbb{N}$. Thus suppose that conversely there exists some $x \in U_K$ such that $a \equiv x^m \mod \mathfrak{p}_K^{2v_\mathfrak{p}(m)+1}$ and consider the polynomial $f = X^m - b \in \mathcal{O}_K[X]$. If $\pi \in \mathcal{O}_K$ is a prime element, then

$$|f'(x)|^2 = |mx^{m-1}|^2 = |m|^2 = |\pi|^{2v_\mathfrak{p}(m)} > |\pi|^{2v_\mathfrak{p}(m)+1} \ge |x^m - a| = |f(x)|.$$

By 5.4.2, there exists some $c \in \mathcal{O}_K$ such that $c^m = a$, and then obviously $c \in U_K$. □

Theorem 5.4.8. *Let $K = (K, |\cdot|)$ be a complete non-Archimedean valued field, and let $|\cdot|_0$ be any discrete absolute value of K. Then $|\cdot|_0$ is equivalent to $|\cdot|$.*

In particular: If $K = F((X))$ is the field of formal Laurent series over a field F (see 2.10.10), *then the order valuation o_X is the only discrete valuation of K.*

Proof. For a polynomial $P \in \mathcal{O}_K[X]$, we denote by $\overline{P} \in \mathsf{k}_K[X]$ its residue class polynomial. Assume to the contrary that $|\cdot|_0$ is not equivalent to $|\cdot|$. Let $|K^\times|_0 = \langle \rho \rangle$, where $\rho \in (0,1)$ and $\pi \in K$ such that $|\pi|_0 = \rho$. By the weak approximation theorem 5.1.6.2(a) there exists some $d \in K$ such that $|d - \pi|_0 < \rho$ and $|d - 1| < 1$, hence $|d|_0 = |(d-\pi) + \pi|_0 = |\pi|_0 = \rho$ and $\overline{d} = \overline{1} \in \mathsf{k}_K$. Let $n \ge 2$ be such that $\mathrm{char}(\mathsf{k}_K) \nmid n$, and consider the polynomial $g = X^n - d \in \mathcal{O}_K[X]$. Then $\overline{g} = X^n - \overline{1}$, and $\overline{d}$ is a simple zero of $\overline{g}$ in k_K. By 5.4.3.1 there exists some $x \in \mathcal{O}_K$ such that $x^n = d$, and then it follows that $\rho = |d|_0 = |x|_0^n \in |K^\times|_0^n$, a contradiction. □

Definition and Theorem 5.4.9. *Let K be a complete discrete valued field, and let $\rho \colon \mathcal{O}_K \to \mathsf{k}_K$ the residue class homomorphism.*

Note that a subset $\mathcal{R}$ of $\mathcal{O}_K$ is a set of representatives for k_K if and only if the residue class map $\rho{\restriction}\mathcal{R} \colon \mathcal{R} \to \mathsf{k}_K$ is bijective. A subset $\mathcal{R}$ of $\mathcal{O}_K$ is called

- a **multiplicative residue system** if $0 \in \mathcal{R}$, $\mathcal{R}^\bullet$ is a subgroup of U_K, and $\rho\restriction\mathcal{R}^\bullet\colon \mathcal{R}^\bullet \to \mathsf{k}_K^\times$ is a group isomorphism;

- a **coefficient field** (of K) if $\mathcal{R}$ is a subfield of $\mathcal{O}_K$, and $\rho\restriction\mathcal{R}\colon \mathcal{R} \to \mathsf{k}_K$ is a field isomorphism.

Every coefficient field is a multiplicative residue system, and every multiplicative residue system is a set of representatives for k_K. If $|\mathsf{k}_K| = q < \infty$, then 5.4.3.2(b) shows that $\mathcal{R} = \mu_{q-1}(K) \cup \{0\}$ is a multiplicative residue system.

Let $\mathcal{R} \subset \mathcal{O}_K$ be a multiplicative residue system and $\pi \in \mathcal{O}_K$ a prime element (then $\langle\pi\rangle \cong \mathbb{Z}$ and $\mathcal{R}^\bullet \cong \mathsf{k}_K^\times$). If we endow $\langle\pi\rangle$ and $\mathcal{R}^\bullet$ with the discrete topology, then $U_K = \mathcal{R}^\bullet{\cdot}U_K^{(1)}$ and $K^\times = \langle\pi\rangle{\cdot}\mathcal{R}^\bullet{\cdot}U_K^{(1)}$ are topological inner direct products (this means, the maps $\mathcal{R}^\bullet \times U_K^{(1)} \to U_K$, defined by $(\rho, u) \mapsto \rho u$ and $\langle\pi\rangle \times \mathcal{R}^\bullet \times U \to K^\times$, defined by $(\pi^m, \rho, u) \mapsto \pi^m \rho u$, are topological isomorphisms).

Proof. If $\sigma = (\rho\restriction\mathcal{R})^{-1}\colon \mathsf{k}_K \to \mathcal{R}$, then $\sigma\restriction\mathsf{k}_K^\times\colon \mathsf{k}_K^\times \to \mathcal{R}^\bullet \hookrightarrow U_K$ is a homomorphism satisfying $\sigma\circ\rho\restriction U_K = \mathrm{id}_{U_K}$. Therefore the epimorphism $\rho\restriction U_K\colon U_K \to \mathsf{k}_K^\times$ splits (see 1.1.2), we obtain

$$U_K = \mathrm{Im}(\sigma\restriction\mathsf{k}_K^\times){\cdot}\mathrm{Ker}(\rho\restriction U_K) = \mathcal{R}^\bullet{\cdot}U_K^{(1)},$$

and obviously $K^\times = \langle\pi\rangle{\cdot}U_K$ (see 2.1.6.1).

In order to prove that the products are also topological, we must show: For every sequence $(x_n = \pi^{k_n} r_n u_n)_{n\ge0}$ in $K^\times$ and $x = \pi^k r u \in K^\times$ (where $k_n, k \in \mathbb{Z}$, $r_n, r \in \mathcal{R}$ and $u_n, u \in U_K^{(1)}$) it follows that $(x_n)_{n\ge0} \to x$ if and only if $k_n = k$ and $r_n = r$ for all $n \gg 1$, and $(u_n)_{n\ge0} \to u$. Obviously, if $k_n = k$ and $r_n = r$ for all $n \gg 1$ and $(u_n)_{n\ge0} \to u$, then $(x_n)_{n\ge0} \to x$. Assume conversely that $(x_n)_{n\ge0} \to x$. Then $k_n = v_K(x_n) = v_K(x) = k$ for all $n \gg 1$ by 5.3.2.3(b) and $(r^{-1}r_n u_n)_{n\ge0} \to u$. As $U_K^{(1)}$ is open in $K^\times$, we infer $r^{-1}r_n \in U_K^{(1)}$ and thus $r = r_n$ for all $n \gg 1$, and finally $(u_n)_{n\gg0} \to u$. □

Theorem 5.4.10 (Existence of multiplicative residue systems). *Let K be a complete discrete valued field, π a prime element of $\mathcal{O}_K$ and $\rho\colon \mathcal{O}_K \to \mathsf{k}_K$ the residue class epimorphism.*

1. *Let K_0 be a coefficient field of K, $\mathsf{k}_K((X))$ the field of formal Laurent series, and let $\Phi\colon K \to \mathsf{k}_K((X))$ be defined by*

$$\Phi\Big(\sum_{n\in\mathbb{Z}} a_n\pi^n\Big) = \sum_{n\in\mathbb{Z}} \rho(a_n)X^n$$

for all sequences $(a_n)_{n\in\mathbb{Z}}$ in K_0 satisfying $a_{-n} = 0$ for $n \gg 1$. Then Φ is an isomorphism, $\Phi\restriction K_0 = \rho\restriction K_0\colon K_0 \overset{\sim}{\to} \mathsf{k}_K$, $\Phi(\pi) = X$ and $\mathsf{o}_X\circ\Phi = v_K$.

We identify: $K_0 = \mathsf{k}_K \subset K$, and $K = \mathsf{k}_K((\pi))$, the field of formal Laurent series in π over $K_0 = \mathsf{k}_K$.

2. If $\mathrm{char}(\mathsf{k}_K) = 0$*, then K possesses a coefficient field.*

3. Suppose that $\mathrm{char}(\mathsf{k}_K) = p > 0$*, and let k_K be perfect. Then K possesses a unique multiplicative residue system $\mathcal{R}$, and if* $\mathrm{char}(K) = p$*, then $\mathcal{R}$ is even a coefficient field.*

In particular, if $\mathrm{char}(K) = \mathrm{char}(\mathsf{k}_K)$ *and k_K is perfect, then K possesses a coefficient field K_0. If we identify $K_0 = \mathsf{k}_K$ by 1., then $K = \mathsf{k}_K((\pi))$ and $v_K = \mathsf{o}_\pi$ for every prime element π of $\mathcal{O}_K$.*

Proof. 1. By 5.3.5.3 for every $x \in K$ there exists a unique sequence $(a_n)_{n\in\mathbb{Z}}$ in K_0 satisfying $a_{-n} = 0$ for all $n \gg 1$ such that

$$x = \sum_{n\in\mathbb{Z}} a_n \pi^n.$$

Since $\rho\colon K_0 \to \mathsf{k}_K$ is an isomorphism, it follows that Φ is an isomorphism. By definition, $\Phi \restriction K_0 = \rho$, $\Phi(\pi) = X$, $v_K \circ \Phi^{-1}\colon \mathsf{k}_K((X)) \to \mathbb{Z} \cup \{\infty\}$ is a discrete valuation, and $v_K \circ \Phi^{-1} = \mathsf{o}_X$ by 5.4.8. Hence $v_K = \mathsf{o}_X \circ \Phi$.

2. Suppose that $\mathrm{char}(\mathsf{k}_K) = 0$. Then $\mathrm{char}(K) = 0$, too, we may assume that $\mathbb{Q} \subset \mathcal{O}_K$ and $\mathbb{Q} \subset \mathsf{k}_K$. We use 1.8.2. Let $\boldsymbol{\tau} = (\tau_\lambda)_{\lambda\in\Lambda}$ be a transcendence basis of $\mathsf{k}_K/\mathbb{Q}$, and set $\mathsf{k}_1 = \mathbb{Q}(\boldsymbol{\tau})$. Then $\mathsf{k}_K/\mathsf{k}_1$ is algebraic. For $\lambda \in \Lambda$ let $t_\lambda \in \mathcal{O}_K$ be such that $\rho(t_\lambda) = \tau_\lambda$, set $\boldsymbol{t} = (t_\lambda)_{\lambda\in\Lambda}$ and $K_1 = \mathbb{Q}(\boldsymbol{t})$. Then $\boldsymbol{t}$ is algebraic independent over $\mathbb{Q}$, and $\rho \restriction \mathbb{Q}[\boldsymbol{t}]\colon \mathbb{Q}[\boldsymbol{t}] \to \mathbb{Q}[\boldsymbol{\tau}]$ is an isomorphism. Hence $\mathbb{Q}[\boldsymbol{t}]^\bullet \subset \mathcal{O}_K^\times$, $K_1 = \mathbb{Q}(\boldsymbol{t}) \subset \mathcal{O}_K$, and $\rho \restriction K_1\colon K_1 \overset{\sim}{\to} \mathsf{k}_1$ is an isomorphism.

Let K_0 be the relative algebraic closure of K_1 in K. As $K_1 \subset \mathcal{O}_K$ and $\mathcal{O}_K$ is integrally closed, we get $K_0 \subset \mathcal{O}_K$, and $\rho \restriction K_0\colon K_0 \to \mathsf{k}_K$ is a monomorphism. We shall prove that $\rho(K_0) = \mathsf{k}_K$, and then we are done. Let $\alpha \in \mathsf{k}_K$ and $g \in K_1[X]$ such that $\rho(g) \in \mathsf{k}_1[X]$ is the minimal polynomial of α over k_1. Then α is a simple zero of $\rho(g)$, and by 5.4.3.1 there exists a unique $a \in \mathcal{O}_K$ such that $g(a) = 0$ and $\rho(a) = \alpha$. Since a is algebraic over K_1, it follows that $a \in K_0$.

3. For $a \in \mathcal{O}_K$ we set $\overline{a} = a + \mathfrak{p}_K \in \mathsf{k}_K$. We shall prove:

I. There exists a unique map $\sigma\colon \mathsf{k}_K \to \mathcal{O}_K$ such that $\rho\circ\sigma = \mathrm{id}_{\mathsf{k}_K}$, and additionally $\sigma(\alpha^p) = \sigma(\alpha)^p$ for all $\alpha \in \mathsf{k}_K$.

II. Let $\sigma\colon \mathsf{k}_K \to \mathcal{O}_K$ be as in **I.** Then $\sigma(0) = 0$, $\sigma(\alpha\beta) = \sigma(\alpha)\sigma(\beta)$ for all $\alpha, \beta \in \mathsf{k}_K$, and if $\mathrm{char}(K) = p$, then $\sigma(\alpha+\beta) = \sigma(\alpha) + \sigma(\beta)$ for all $\alpha, \beta \in \mathsf{k}_K$, too.

Proof of **I.** We prove first:

$(*)$ If $b, c \in \mathcal{O}_K$ and $b \equiv c \bmod \mathfrak{p}_K$, then $b^{p^n} \equiv c^{p^n} \bmod \mathfrak{p}_K^{n+1}$ for all $n \ge 0$.

We prove $(*)$ by induction on n. For $n=0$ there is nothing to do. Thus suppose that $n \ge 0$ and $c^{p^n} = b^{p^n} + q$ for some $q \in \mathfrak{p}_K^{n+1}$. Then $c^{p^{n+1}} = b^{p^{n+1}} + pb^{p^n(p-1)}q + q^2u$ for some $u \in \mathcal{O}_K$. Due to $q^2u \in \mathfrak{p}_K^{2(n+1)} \subset \mathfrak{p}_K^{n+2}$ and $pb^{p^n(p-1)}q \in p\mathfrak{p}_K^{n+1} \subset \mathfrak{p}_K^{n+2}$, it follows that $c^{p^{n+1}} \equiv b^{p^{n+1}} \mod \mathfrak{p}_K^{n+2}$. $\square[(*)]$

Uniqueness of σ: Let $\sigma, \sigma' \colon \mathsf{k}_K \to \mathcal{O}_K$ such that $\rho \circ \sigma = \rho \circ \sigma' = \mathrm{id}_{\mathsf{k}_K}$, $\sigma(\alpha^p) = \sigma(\alpha)^p$ and $\sigma'(\alpha^p) = \sigma'(\alpha)^p$ for all $\alpha \in \mathsf{k}_K$. If $\alpha \in \mathsf{k}_K$ and $n \in \mathbb{N}$, then $\rho \circ \sigma(\alpha^{1/p^n}) = \alpha^{1/p^n} = \rho \circ \sigma'(\alpha^{1/p^n})$, hence $\sigma(\alpha^{1/p^n}) \equiv \sigma'(\alpha^{1/p^n}) \mod \mathfrak{p}_K$ and therefore $\sigma(\alpha^{1/p^n})^{p^n} \equiv \sigma'(\alpha^{1/p^n})^{p^n} \mod \mathfrak{p}_K^{n+1}$ by $(*)$. By a simple induction on n it follows that $\sigma(\alpha) = \sigma(\alpha^{1/p^n})^{p^n}$ and $\sigma'(\alpha) = \sigma'(\alpha^{1/p^n})^{p^n}$, hence $\sigma(\alpha) \equiv \sigma'(\alpha) \mod \mathfrak{p}_K^{n+1}$ for all $n \ge 0$, and consequently $\sigma(\alpha) = \sigma'(\alpha)$.

Construction of σ: For $\alpha \in \mathsf{k}_K$ and $n \ge 0$ let $a_n \in \mathcal{O}_K$ be such that $\overline{a}_n = \alpha^{1/p^n}$. Then $\overline{a_n^{p^n}} = \alpha = \overline{a}_0$, hence $a_n^{p^n} \in a_0 + \mathfrak{p}_K$, $a_{n+1}^p \equiv a_n \mod \mathfrak{p}_K$, and $(*)$ implies $a_{n+1}^{p^{n+1}} \equiv a_n^{p^n} \mod \mathfrak{p}_K^{n+1}$ for all $n \ge 0$.

Hence $(a_{n+1}^{p^{n+1}} - a_n^{p^n})_{n\ge 0} \to 0$, and $(a_n^{p^n})_{n \ge 0}$ is a Cauchy sequence in $a_0 + \mathfrak{p}_K$. Since $a_0 + \mathfrak{p}_K$ is closed in K, there exists some $a \in a_0 + \mathfrak{p}_K$ such that $(a_n^{p^n})_{n\ge 0} \to a$, we set $\sigma(\alpha) = a \in a_0 + \mathfrak{p}_K$, and then $\rho \circ \sigma(\alpha) = \overline{a} = \overline{a}_0 = \alpha$.

We assert that $a = \sigma(\alpha)$ does not depend on the sequence $(a_n)_{n \ge 0}$. To this end let $(a_n')_{n\ge 0}$ be another sequence in $\mathcal{O}_K$ satisfying $\overline{a_n'} = \alpha^{1/p^n}$ for all $n \ge 0$. Then we obtain $a_n' \equiv a_n \mod \mathfrak{p}_K$, hence $a_n'^{p^n} \equiv a_n^{p^n} \mod \mathfrak{p}_K^{n+1}$ for all $n \ge 0$ by $(*)$, $(a_n'^{p^n} - a_n^{p^n})_{n \ge 0} \to 0$ and $(a_n'^{p^n})_{n\ge 0} \to a$.

It remains to prove that $\sigma(\alpha^p) = \sigma(\alpha)^p$ for all $\alpha \in \mathsf{k}_K$. Let $\alpha \in \mathsf{k}_K$ and $(a_n)_{n \ge 0}$ be a sequence in $\mathcal{O}_K$ such that $\overline{a}_n = \alpha^{1/p^n}$ for all $n \ge 0$. Then $(a_n^{p^n})_{n \ge 0} \to \sigma(\alpha)$ and therefore $(a_n^{p^{n+1}})_{n\ge 0} \to \sigma(\alpha)^p$. But since $\overline{a_n^p} = (\alpha^p)^{1/p^n}$ for all $n \ge 0$, we also obtain $(a_n^{p^{n+1}})_{n \ge 0} = ((a_n^p)^{p^n})_{n\ge 0} \to \sigma(\alpha^p)$, and therefore $\sigma(\alpha)^p = \sigma(\alpha^p)$. $\square$[**I.**]

Proof of **II.** If in **I** we use the constant sequence $(0)_{n \ge 0}$ to construct $\sigma(0)$, then $\sigma(0) = 0$. Let now $\alpha, \beta \in \mathsf{k}_K$, $a = \sigma(\alpha)$, $b = \sigma(\beta)$, and let $(a_n)_{n\ge 0}$ and $(b_n)_{n \ge 0}$ be sequences in $\mathcal{O}_K$ such that $\overline{a}_n = \alpha^{1/p^n}$ and $\overline{b}_n = \beta^{1/p^n}$ for all $n \ge 0$. Then it follows that $(a_n^{p^n})_{n\ge 0} \to \sigma(\alpha)$, $(b_n^{p^n})_{n \ge 0} \to \sigma(\beta)$ and consequently $((a_nb_n)^{p^n})_{n\ge 0} \to \sigma(\alpha)\sigma(\beta)$. Since $\overline{a_nb_n} = (\alpha\beta)^{1/p^n}$, we also obtain $((a_nb_n)^{p^n})_{n \ge 0} \to \sigma(\alpha\beta)$ and $\sigma(\alpha\beta) = \sigma(\alpha)\sigma(\beta)$.

If $\mathrm{char}(K) = p$, then $((a_n + b_n)^{p^n})_{n\ge 0} = (a_n^{p^n} + b_n^{p^n})_{n \ge 0} \to \sigma(\alpha) + \sigma(\beta)$. On the other hand, since $\overline{a_n + b_n} = \alpha^{1/p^n} + \beta^{1/p^n} = (\alpha+\beta)^{1/p^n}$ for all $n \ge 0$, we obtain $((a_n + b_n)^{p^n})_{n\ge 0} \to \sigma(\alpha+\beta)$, and $\sigma(\alpha+\beta) = \sigma(\alpha) + \sigma(\beta)$. $\square$[**II.**]

Finally, we prove the existence and uniqueness of $\mathcal{R}$. Thus let $\sigma \colon \mathsf{k}_K \to \mathcal{O}_K$ be as in **I**. Then $\mathcal{R} = \sigma(\mathsf{k}_K) \subset \mathcal{O}_K$ is a set of representatives for k_K. From **II** we obtain $0 \in \mathcal{R}$, $\mathcal{R}^\bullet$ is a subgroup of U_K, and since $\rho \restriction \mathcal{R}^\bullet = (\sigma \restriction \mathsf{k}_K^\times)^{-1}$, it follows that $\rho \restriction \mathcal{R}^\bullet \colon \mathcal{R}^\bullet \xrightarrow{\sim} \mathsf{k}_K^\times$ is an isomorphism. Hence $\mathcal{R}$ is a multiplicative residue system. If $\mathrm{char}(K) = p$, then $\mathcal{R} + \mathcal{R} = \mathcal{R}$, hence $\mathcal{R}$ is a subfield of $\mathcal{O}_K$, $\rho \restriction \mathcal{R} = \sigma^{-1} \colon \mathcal{R} \xrightarrow{\sim} \mathsf{k}_K$ is an isomorphism, and $\mathcal{R}$ is a coefficient field.

Let finally $\mathcal{R}' \subset \mathcal{O}_K$ be another multiplicative residue system, and $\sigma' = (\rho \restriction \mathcal{R}')^{-1} \colon \mathsf{k}_K \to \mathcal{R}'$. Then $\sigma' = \sigma$ by **I** and $\mathcal{R}' = \sigma'(\mathsf{k}_K) = \sigma(\mathsf{k}_K) = \mathcal{R}$. $\square$

5.5 Extension of absolute values

In this section we investigate the behavior of absolute values in algebraic field extensions. In 5.5.1 we introduce additional notations and conventions which we will tacitly use together with those from 5.3.1 in the sequel.

Definitions, Conventions and Remarks 5.5.1. Let L/K be a field extension and $|\cdot|\colon L \to \mathbb{R}_{\geq 0}$ a non-Archimedian absolute value such that $|\cdot|\restriction K\colon K \to \mathbb{R}_{\geq 0}$ is an absolute value (by 5.1.7.2(d) this is the case if L/K is algebraic, and then $|\cdot|$ is an extension of $|\cdot|\restriction K$).

1. By definition, we have $\mathcal{O}_L = \{x \in L \mid |x| \leq 1\}$, $\mathfrak{p}_L = \{x \in L \mid |x| < 1\}$, $\mathsf{k}_L = \mathcal{O}_L/\mathfrak{p}_L = \mathsf{k}_{\mathfrak{p}_L}$, $\mathcal{O}_K = \mathcal{O}_L \cap K$, $\mathfrak{p}_K = \mathfrak{p}_L \cap K = \mathfrak{p}_L \cap \mathcal{O}_K$, and also $\mathsf{k}_K = \mathcal{O}_K/\mathfrak{p}_K$. The embedding $\mathcal{O}_K \hookrightarrow \mathcal{O}_L$ induces a monomorphism $\mathsf{k}_K \to \mathsf{k}_L$, and we identify. Then $\mathsf{k}_K \subset \mathsf{k}_L$, we call $\mathsf{k}_L/\mathsf{k}_K$ the **residue class extension** and $f(L/K) = [\mathsf{k}_L : \mathsf{k}_K] \in \mathbb{N} \cup \{\infty\}$ the **residue class degree** of L/K. The index $e(L/K) = (|L^\times| : |K^\times|) \in \mathbb{N} \cup \{\infty\}$ is called the **ramification index** of L/K. For an intermediate field M of L/K we obtain $f(L/K) = f(M/K)f(L/M)$ and $e(L/K) = e(M/K)e(L/M)$.

If $(L, |\cdot|)$ is discrete, then $(K, |\cdot|)$ is also discrete. Conversely, if $(K, |\cdot|)$ is discrete and $e(L/K) < \infty$, then $(L, |\cdot|)$ is discrete (indeed, if $|K^\times| = \langle\rho\rangle$, where $\rho \in (0,1)$, then $|L^\times| = \langle\rho^{1/e(L/K)}\rangle$).

If $\widehat{K}$ is a completion of K and $\widehat{L}$ is a completion of L, then the inclusion $K \hookrightarrow \widehat{L}$ has a unique extension to a value-preserving field homomorphism $\widehat{K} \to \widehat{L}$ by 5.2.4.1. We identify and assume that $\widehat{K} \subset \widehat{L}$. By 5.3.3 we obtain $|K| = |\widehat{K}|$, $|L| = |\widehat{L}|$, $\mathsf{k}_K = \mathsf{k}_{\widehat{K}}$, $\mathsf{k}_L = \mathsf{k}_{\widehat{L}}$, and consequently $e(L/K) = e(\widehat{L}/\widehat{K})$ and $f(L/K) = f(\widehat{L}/\widehat{K})$.

For a polynomial $P \in \mathcal{O}_L[X]$ we denote by $\overline{P} = P + \mathfrak{p}_L\mathcal{O}_L[X] \in \mathsf{k}_L[X]$ the residue class polynomial. If $P \in \mathcal{O}_K[X]$, then $P + \mathfrak{p}_K\mathcal{O}_K[X]) = \overline{P}$, since $\mathsf{k}_K \subset \mathsf{k}_L$. In particular, if $a \in \mathcal{O}_K$, then $\overline{a} = a + \mathfrak{p}_K \in \mathsf{k}_K \subset \mathsf{k}_L$.

2. Let $(L, |\cdot|)$ be discrete, $\rho \in (0,1)$ such that $|L^\times| = \langle\rho\rangle$ and $e = e(L/K)$. Then $e \in \mathbb{N}$ and $|K^\times| = \langle\rho^e\rangle$. Let $v_L\colon L \to \mathbb{Z} \cup \{\infty\}$ and $v_K\colon K \to \mathbb{Z} \cup \{\infty\}$ be the corresponding discrete valuations (see 5.1.3.**6**). If $x \in K$, then

$$v_L(x) = \frac{\log|x|}{\log\rho} = e\,\frac{\log|x|}{\log\rho^e} = ev_K(x), \quad \text{and thus} \quad v_L\restriction K = e(L/K)v_K.$$

Let $\mathfrak{p}_K = \pi_K\mathcal{O}_K$ and $\mathfrak{p}_L = \pi_L\mathcal{O}_L$ with prime elements $\pi_K \in \mathcal{O}_K$ and $\pi_L \in \mathcal{O}_L$. Then (according to 5.1.3.**5**) $|\cdot|\colon L \to \mathbb{R}_{\geq 0}$ is a $\mathfrak{p}_L$-adic and $|\cdot|\restriction K\colon K \to \mathbb{R}_{\geq 0}$ is a $\mathfrak{p}_K$-adic absolute value. Corresponding to 2.12.1, $f(L/K) = f(\mathfrak{p}_L/\mathfrak{p}_K)$ and $e(L/K) = v_L(\pi_K) = v_{\mathfrak{p}_L}(\mathfrak{p}_K\mathcal{O}_L) = e(\mathfrak{p}_L/\mathfrak{p}_K)$. In particular, $e(L/K) = 1$ if and only if every prime element of $\mathcal{O}_K$ is also a prime element of $\mathcal{O}_L$.

Theorem 5.5.2. *Let L/K be a field extension and $|\cdot|\colon L \to \mathbb{R}_{\geq 0}$ a non-Archimedean absolute value such that $|\cdot|\restriction K\colon K \to \mathbb{R}_{\geq 0}$ is an absolute value.*

1. *Let $e,\, f \in \mathbb{N}$, $u_1, \ldots, u_f \in \mathcal{O}_L$ and $a_1, \ldots, a_e \in L^\times$ be such that the residue classes $\overline{u}_1, \ldots, \overline{u}_f \in \mathsf{k}_L$ are linearly independent over k_K, and the cosets $|a_1|\,|K^\times|, \ldots, |a_e|\,|K^\times| \in |L^\times|/|K^\times|$ are distinct. Then $\{u_i a_j \mid i \in [1,f],\ j \in [1,e]\}$ is linearly independent over K, and $e(L/K)f(L/K) \leq [L\!:\!K]$.*

2. *Let $(L, |\cdot|)$ be discrete, $e = e(L/K)$, $f = f(L/K) < \infty$, and let $\pi_K \in \mathcal{O}_K$ and $\pi_L \in \mathcal{O}_L$ be prime elements. Let $u_1, \ldots, u_f \in \mathcal{O}_L$ be such that $(\overline{u}_1, \ldots, \overline{u}_f)$ is a k_K-basis of k_L, let $\mathcal{R} \subset \mathcal{O}_K$ be a set of representatives for k_K such that $0 \in \mathcal{R}$, and set*

$$\mathcal{R}' = \{a_1u_1 + \ldots + a_fu_f \mid a_1, \ldots, a_f \in \mathcal{R}\} \subset \mathcal{O}_L.$$

   (a) *$\mathcal{R}'$ is a set of representatives for k_L, $0 \in \mathcal{R}'$, and if $a \in \mathcal{R}'$, then there exists a unique f-tuple $(a_1, \ldots, a_f) \in \mathcal{R}^f$ such that $a = a_1u_1 + \ldots + a_fu_f$.*

   (b) *For every $a \in \mathcal{O}_L$ there exists a uniquely determined family $(a_{i,l,j})_{i\in[1,f],\, l\in[0,e-1],\, j\geq 0}$ in $\mathcal{R}$ such that*

$$a \equiv \sum_{i=1}^{f}\sum_{l=0}^{e-1}\sum_{j=0}^{k} a_{i,l,j}u_i\pi_L^l\pi_K^j \mod \mathfrak{p}_L^{e(k+1)} \quad \textit{for all } k \geq 0.$$

Proof. 1. We prove first:

A. If $c_1, \ldots, c_f \in K$, then $|c_1u_1 + \ldots + c_fu_f| = \max\{|c_1|, \ldots, |c_f|\}$.

Proof of **A.** Let $c_1, \ldots, c_f \in K$, and assume that $|c_1| = \max\{|c_1|, \ldots, |c_f|\}$. Since $\overline{u}_i \neq \overline{0}$, we have $|u_i| = 1$ for all $i \in [1,f]$, hence

$$|c_1u_1 + \ldots + c_fu_f| \leq \max\{|c_1|\,|u_1|, \ldots, |c_f|\,|u_f|\} \leq |c_1|,$$

and we assume to the contrary that $|c_1u_1 + \ldots + c_fu_f| < |c_1|$. Then it follows that $|u_1 + c_1^{-1}c_2u_2 + \ldots + c_1^{-1}c_fu_f| < 1$. Since $|c_1^{-1}c_i| \leq 1$ and thus $c_1^{-1}c_i \in \mathcal{O}_K$ for all $i \in [2,f]$, we get $\overline{0} = \overline{u}_1 + \overline{a_1^{-1}a_2}\overline{u}_2 + \ldots + \overline{a_1^{-1}a_f}\overline{u}_f$, which contradicts the linear independence of $\overline{u}_1, \ldots, \overline{u}_f$. □[**A.**]

Assume now that $c_{i,j} \in K$ (for $i \in [1,f]$ and $j \in [1,e]$) are such that

$$0 = \sum_{i=1}^{f}\sum_{j=1}^{e} c_{i,j}u_ia_j = \sum_{j=1}^{e} b_ja_j, \quad \text{where } b_j = \sum_{i=1}^{f} c_{i,j}u_i \quad \text{for all } j \in [1,e].$$

By **A** we obtain $|b_j| = \max\{|c_{1,j}|, \ldots, |c_{f,j}|\} \in |K|$ for all $j \in [1,e]$. As the cosets $|a_1|\,|K^\times|, \ldots, |a_e|\,|K^\times| \in |L^\times|/|K^\times|$ are distinct, the non-zero elements of the sequence $(|a_j|\,|b_j| = |b_ja_j|)_{j\in[1,e]}$ are pairwise distinct. Hence

$$0 = \Big|\sum_{j=1}^{e} b_ja_j\Big| = \max\{|b_ja_j| \mid j \in [1,e]\}, \text{ and therefore } b_j = 0 \text{ for all } j \in [1,e].$$

Consequently, $c_{i,j} = 0$ for all $i \in [1,f]$ and $j \in [1,e]$.

For the proof of $e(L/K)f(L/K) \le [L:K]$ let $e, f \in \mathbb{N}$, $e \le e(L/K)$ and $f \le f(L/K)$. Then there exist $u_1, \ldots, u_f \in \mathcal{O}_L$ and $a_1, \ldots, a_e \in L^\times$ such that $\overline{u}_1, \ldots, \overline{u}_r \in \mathsf{k}_L$ are linearly independent over k_K, and the cosets $|a_1|\,|K^\times|, \ldots, |a_s|\,|K^\times|$ in $|L^\times|/|K^\times|$ are distinct. It follows that $ef \le [L:K]$, and (since e and f were arbitrary) $e(L/K)f(L/K) \le [L:K]$.

2.(a) We must prove that for every $\alpha \in \mathsf{k}_L$ there exists uniquely determined elements $a_1, \ldots, a_f \in \mathcal{R}$ such that $\overline{a_1u_1 + \ldots + a_fu_f} = \alpha$. If $\alpha \in \mathsf{k}_L$, then there exist $\alpha_1, \ldots, \alpha_f \in \mathsf{k}_K$ such that $\alpha = \alpha_1\overline{u}_1 + \ldots + \alpha_f\overline{u}_f$. For every $i \in [1, f]$ there exists some $a_i \in \mathcal{R}$ such that $\overline{a}_i = \alpha_i$, and then $\overline{a_1u_1 + \ldots + a_fu_f} = \alpha$. To prove uniqueness, let likewise $b_1, \ldots, b_f \in \mathcal{R}$ be such that $\overline{b_1u_1 + \ldots + b_fu_f} = \alpha$. Then $\alpha = \overline{b}_1\overline{u}_1 + \ldots + \overline{b}_f\overline{u}_f$, hence $\overline{b}_i = \alpha_i = \overline{a}_i$ and thus $b_i = a_i$ for all $i \in [1, f]$.

(b) If $k \in \mathbb{N}_0$ and $m \in [0, e(k+1)-1]$, then $m = ej + l$ with uniquely determined $j \in [0, k]$ and $l \in [0, e-1]$, and then $v_L(\pi_L^l\pi_k^j) = ej + l = m$. Now 2.10.6.2 implies that for every $a \in \mathcal{O}_L$ there exists a unique family $(a_{l,j})_{l\in[0,e-1],\, j\ge 0}$ in $\mathcal{R}'$ such that

$$a \equiv \sum_{l=0}^{e-1}\sum_{j=0}^{k} a_{l,j}\pi_L^l\pi_K^j \mod \mathfrak{p}_L^{e(k+1)} \quad \text{for all } k \ge 0.$$

By (a), every coefficient $a_{l,j} \in \mathcal{R}'$ has a unique representation

$$a_{l,j} = \sum_{i=1}^{f} a_{i,l,j}u_i, \quad \text{where } a_{i,l,j} \in \mathcal{R},$$

which implies the existence and uniqueness of the asserted expansion. □

Theorem 5.5.3. (Extension of non-Archimedian values, complete case)
Let $K = (K, |\cdot|)$ be a complete non-Archimedian valued field and L/K an algebraic field extension. Then $|\cdot|$ has a unique extension to an absolute value of L. In particular, if $\overline{K}$ denotes an algebraic closure of K, then $|\cdot|$ has a unique extension to an absolute value $|\cdot|^*\colon \overline{K} \to \mathbb{R}_{\ge 0}$, and if L is an intermediate field of $\overline{K}/K$, then $|\cdot|^* \upharpoonright L$ is the unique extension of $|\cdot|$ to an absolute value of L.

We fix an algebraic closure $\overline{K}$ of K, and we assume that all over K algebraic extension fields of K are contained in $\overline{K}$. For every algebraic extension L/K we denote the unique absolute value of L extending $|\cdot|$ again by $|\cdot|$ (instead of $|\cdot|^* \upharpoonright L$). In particular, if p is a prime and $\overline{\mathbb{Q}}_p$ is an algebraic closure of $\mathbb{Q}_p$, then we call the extension $|\cdot|_p\colon \overline{\mathbb{Q}}_p \to \mathbb{R}_\ge$ again the p-adic absolute value. Consequently $\overline{\mathbb{Q}}_p = (\overline{\mathbb{Q}}_p, |\cdot|_p)$ is an algebraically closed non-Archimedian valued field. However, $\overline{\mathbb{Q}}_p$ is not complete (see [5, Ch. 8.3]).

1. *If $\sigma \in \mathrm{Hom}^K(L, \overline{K})$, then $|\sigma x| = |x|$ for all $x \in L$, and if $[L:K] < \infty$ then maps $\mathsf{N}_{L/K}\colon L \to K$ and $\mathsf{Tr}_{L/K}\colon L \to K$ are continuous.*

2. Suppose that $[L\colon K] = n < \infty$. *Then*

$$|x| = \sqrt[n]{|\mathsf{N}_{L/K}(x)|} \quad \textit{for all} \quad x \in L,$$

and $L = (L, |\cdot|)$ *is complete. If* K *is discrete, then* L *is discrete, and if* K *is a local field, then* L *is a local field, too.*

3. $\mathrm{cl}_L(\mathcal{O}_K) = \mathcal{O}_L$, $\mathfrak{p}_L$ *is the only prime ideal of* $\mathcal{O}_L$ *lying above* $\mathfrak{p}_K$, *and the residue class extension* $\mathsf{k}_L/\mathsf{k}_K$ *is algebraic.*

4. If $\sigma \in \mathrm{Gal}(L/K)$, *then* $\sigma\mathcal{O}_L = \mathcal{O}_L$, $\sigma\mathfrak{p}_L = \mathfrak{p}_L$, *and there exists a unique automorphism* $\overline{\sigma} \in \mathrm{Gal}(\mathsf{k}_L/\mathsf{k}_K)$ *satisfying* $\overline{\sigma}(\overline{a}) = \overline{\sigma(a)}$ *for all* $a \in \mathcal{O}_L$.
$\overline{\sigma}$ is called the **residue class automorphism** induced by σ.

Proof. We start with the finite case.

A. Let L/K be a field extension, $[L\colon K] = n < \infty$, and define

$$|\cdot|_L \colon L \to \mathbb{R}_{\geq 0} \quad \text{by} \quad |x|_L = \sqrt[n]{|\mathsf{N}_{L/K}(x)|} \quad \text{for all} \quad x \in L,$$

Then $|\cdot|_L$ is an absolute value of L, $|\cdot|_L \restriction K = |\cdot|$, $(L, |\cdot|_L)$ is complete, and $|\cdot|_L$ is the only extension of $|\cdot|$ to L. If $|\cdot|$ is discrete, then $|\cdot|_L$ is also discrete, and if $(K, |\cdot|)$ is a local field, then $(L, |\cdot|_L)$ is a local field, too.

Proof of **A.** We show that $|\cdot|_L \colon L \to \mathbb{R}_{\geq 0}$ satisfies the conditions **A1**, **A2** and **A3** of 5.1.1. **A1** and **A2** are obvious. To prove **A3**, we use 5.1.2.1 and prove that $|z|_L \leq 1$ implies $|1 + z|_L \leq 1$ for all $z \in L$. Let $z \in L$, $|z|_L \leq 1$, and let $f \in K[X]$ be the minimal polynomial of z over K (then $f(X-1)$ is the minimal polynomial of $1+z$ over K). Using 1.6.2.1 it follows that $1 \geq |z|_L^n = |\mathsf{N}_{L/K}(z)| = |f(0)|^{[L:K(z)]}$, hence $|f(0)| \leq 1$, and therefore $f \in \mathcal{O}_K[X]$ by 5.4.6. Now it follows that

$$|1+z|^n = |\mathsf{N}_{L/K}(1+z)| = |f(-1)|^{[L:K(x)]} \leq 1, \quad \text{and consequently} \quad |1+z| \leq 1.$$

Hence $|\cdot|_L$ is an absolute value of L. Obviously, $|\cdot|_L \restriction K = |\cdot|$, and if $|\cdot|$ is discrete, then so is $|\cdot|_L$. By 5.2.9.1 it follows that $|\cdot|_L$ is the only absolute value of L which extends $|\cdot|$ and $(L, |\cdot|_L)$ is complete. By 5.5.2.1 we obtain $f(L/K) = [\mathsf{k}_L : \mathsf{k}_K] \leq n$. Hence if k_K is finite, then so is k_L. Consequently if $(K, |\cdot|)$ is a local field, then so is $(L, |\cdot|_L)$. $\square$[**A.**]

Now let L/K be an arbitrary algebraic field extension. For every over K finite intermediate field M of L/K let $|\cdot|_M$ be the unique absolute value of M extending $|\cdot|$ according to **A**. We define $\|\cdot\| \colon L \to \mathbb{R}_{\geq 0}$ by $\|x\| = |x|_{K(x)}$ for all $x \in L$, and we assert that $\|\cdot\|$ is the only extension of $|\cdot|$ to an absolute value of L. First note that if $x \in L$ and K' is any over K finite intermediate field of L/K such that $x \in K'$, then $|\cdot|_{K'} \restriction K(x) = |\cdot|_{K(x)}$, and therefore $\|x\| = |x|_{K'}$.

We show that $\|\cdot\|$ is an absolute value (see 5.1.1). **A1** is obvious. As to **A2** and **A3**, let $x, y \in L$ and $K' = K(x,y)$. Then $\|xy\| = |xy|_{K'} = |x|_{K'}|y|_{K'} = \|x\|\,\|y\|$ and $\|x+y\| = |x+y|_{K'} \le \max\{|x|_{K'}, |y|_{K'}\} = \max\{\|x\|, \|y\|\}$. Hence $\|\cdot\|$ is an absolute value of L. Clearly, $\|\cdot\| \restriction K = |\cdot|$, and $\|\cdot\|$ is non-Archimedian. If $|\cdot|^*\colon L \to \mathbb{R}_{\ge 0}$ is any absolute value of L satisfying $|\cdot|^* \restriction K = |\cdot|$ and $x \in L$, then $|\cdot|^* \restriction K(x)$ is an absolute value of $K(x)$ which extends $|\cdot|$, hence $|\cdot|^* \restriction K(x) = |\cdot|_{K(x)}$ and thus $|x|^* = \|x\|$.

Now we are well-prepared to prove the assertions 1. to 4.

1. Let $\sigma \in \mathrm{Hom}^K(L, \overline{K})$ and define $|\cdot|_\sigma\colon L \to \mathbb{R}_{\ge 0}$ by $|x|_\sigma = |\sigma x|$ for all $x \in L$. Then $|\cdot|_\sigma$ is an absolute value of L satisfying $|\cdot|_\sigma \restriction K = |\cdot|$, and consequently $|\cdot|_\sigma = |\cdot|\colon L \to \mathbb{R}_{\ge 0}$. In particular, $\sigma\colon L \to \overline{K}$ is continuous. If $[L:K] < \infty$ and $q = [L:K]_{\mathsf{i}}$, then

$$\mathsf{N}_{L/K}(x) = \prod_{\sigma \in \mathrm{Hom}^K(L,\overline{K})} \sigma(x)^q \quad \text{and}$$

$$\mathsf{Tr}_{L/K}(x) = q \sum_{\sigma \in \mathrm{Hom}^K(L,\overline{K})} \sigma(x) \quad \text{for all } x \in L$$

by 1.6.2.2, and therefore $\mathsf{N}_{L/K}\colon L \to K$ and $\mathsf{Tr}_{L/K}\colon L \to K$ are continuous.

2. By **A**.

3. Since $\mathcal{O}_L$ is integrally closed, we obtain $\mathrm{cl}_L(\mathcal{O}_K) \subset \mathrm{cl}(\mathcal{O}_L) = \mathcal{O}_L$. Thus assume that $x \in \mathcal{O}_L$, $m = [K(x):K]$, and let $f \in K[X]$ be the minimal polynomial of x over K. Since $|x| = \sqrt[m]{|\mathsf{N}_{K(x)/K}(x)|} = \sqrt[m]{|f(0)|} \le 1$ by 1.6.2.1, it follows that $|f(0)| \le 1$, hence $f \in \mathcal{O}_K[X]$ by 5.4.6, and consequently $x \in \mathrm{cl}_L(\mathcal{O}_K)$.

By definition, $\mathfrak{p}_K = \mathfrak{p}_L \cap \mathcal{O}_K \subset \mathfrak{p}_L$. As to uniqueness, let $\mathfrak{P} \in \mathcal{P}(\mathcal{O}_L)$ be such that $\mathfrak{P} \cap \mathcal{O}_K = \mathfrak{p}_K$, $x \in \mathfrak{P}$ and

$$f = X^d + a_{d-1}X^{d-1} + \ldots + a_1X + a_0 \in \mathcal{O}_K[X]$$

the minimal polynomial of x over K. Then

$$a_0 = -x(a_1 + \ldots + a_{d-1}x^{d-2} + x^{d-1}) \in \mathfrak{P} \cap \mathcal{O}_K = \mathfrak{p}_K,$$

and $|a_0| = |\mathsf{N}_{K(x)/K}(x)| = |x|^{[K(x):K]} < 1$. Hence $|x| < 1$, which implies $x \in \mathfrak{p}_L$, and consequently $\mathfrak{P} \subset \mathfrak{p}_L$. Now equality follows from 2.4.7.3.

To prove that $\mathsf{k}_L/\mathsf{k}_K$ is algebraic, let $\alpha \in \mathsf{k}_L$, $a \in \mathcal{O}_L$ such that $\alpha = \overline{a}$ and $g \in \mathcal{O}_K[X]$ be the minimal polynomial of a over K. Then $\overline{g} \in \mathsf{k}_K[X]$ is monic and $\overline{g}(\alpha) = \overline{g(\alpha)} = 0$. Hence α is algebraic over k_K.

4. Let $\sigma \in \mathrm{Gal}(L/K)$. Since $|\sigma x| = |x|$ for all $x \in L$, we get $\sigma\mathcal{O}_L = \mathcal{O}_L$, $\sigma\mathfrak{p}_L = \mathfrak{p}_L$, and there exists a unique automorphism $\overline{\sigma}\colon \mathsf{k}_L \to \mathsf{k}_L$ such that $\overline{\sigma}(\overline{a}) = \overline{\sigma a}$ for all $a \in \mathcal{O}_L$. As $\sigma \restriction \mathcal{O}_K = \mathrm{id}_{\mathcal{O}_K}$, it follows that $\overline{\sigma} \restriction \mathsf{k}_L = \mathrm{id}_{\mathsf{k}_L}$ and thus $\overline{\sigma} \in \mathrm{Gal}(\mathsf{k}_L/\mathsf{k}_K)$. □

Corollary 5.5.4. *Let K be a non-Archimedian local field. Then every finite Galois field extension L/K is solvable.*

Proof. Let L/K be a finite Galois field extension. Then $\sigma\mathfrak{p}_L = \mathfrak{p}_L$ for all $\sigma \in \mathrm{Gal}(L/K)$ by 5.5.3.4. Hence $\mathrm{Gal}(L/K)$ is the decomposition group of $\mathfrak{p}_L$ over K (see 2.5.3), and therefore L/K is solvable by 2.13.8.3. □

Although an infinite algebraic extension of a complete non-Archimedian valued field need not be complete, its completion is special as the following corollary shows.

Corollary 5.5.5. *Let $K = (K, |\cdot|)$ be a complete non-Archimedian valued field, L/K an algebraic field extension, $(\widehat{L}, |\cdot|)$ a completion of $(L, |\cdot|)$ and $\widetilde{L}$ the relative algebraic closure of L in $\widehat{L}$. Then $\widetilde{L}/L$ is purely inseparable.*

Proof. Let $\overline{K}$ be an algebraic closure of K such that $\widetilde{L} \subset \overline{K}$. If now $\varphi \in \mathrm{Hom}^L(\widetilde{L}, \overline{K})$, then φ is valuation-preserving and thus continuous by 5.5.3.1, $\varphi \restriction L = \mathrm{id}_L$, and as L is dense in $\widetilde{L}$, it follows that $\varphi = \mathrm{id}_{\widetilde{L}}$. Hence $[\widetilde{L}:L]_s = |\mathrm{Hom}^L(\widetilde{L}, \overline{K})| = 1$, and therefore $\widetilde{L}/L$ is purely inseparable by 1.4.7.1. □

In our next result we amend Theorem 5.5.3 in the case of discrete absolute values.

Theorem 5.5.6 (Extension of discrete absolute values, complete case). *Let $K = (K, |\cdot|)$ be a complete discrete valued field and L/K an algebraic field extension such that $(L, |\cdot|)$ is discrete. Let $\pi_L \in \mathcal{O}_L$ be a prime element, $e = e(L/K)$, and suppose that $f = f(L/K) < \infty$* (note that $e < \infty$ by 5.5.1.1).

1. *Let $u_1, \ldots, u_f \in \mathcal{O}_L$ be such that $(\overline{u}_1, \ldots \overline{u}_f)$ is a k_K-basis of k_L. Then $\mathcal{O}_L$ is a free $\mathcal{O}_K$-module with basis*

$$\{u_i\pi_L^j \mid i \in [1, f],\ j \in [0, e-1]\}.$$

In particular, $[L:K] = ef < \infty$, and $(L, |\cdot|)$ is complete.

2. *$v_K \circ \mathsf{N}_{L/K} = f v_L$ and $v_L \restriction K = e v_K$.*

3. *If $k \in \mathbb{N}_0$, then $\mathsf{N}_{L/K}(\mathfrak{p}_L^k) \subset \mathfrak{p}_K^{fk}$ and $\mathsf{N}_{L/K}(U_L^{(ke)}) \subset U_K^{(k)}$.*

4. *Let $\mathsf{k}_L/\mathsf{k}_K$ be separable. Then there exists some $a_0 \in \mathcal{O}_L$ such that $\mathcal{O}_L = \mathcal{O}_K[a]$ for all $a \in \mathcal{O}_L$ satisfying $a \equiv a_0 \bmod \mathfrak{p}_L^2$.*

Proof. 1. By definition, $|L^\times|/|K^\times| = \{|\pi_L|^j|K^\times| \mid j \in [0, e-1]\}$, and therefore $\{u_i\pi_L^j \mid i \in [1, f],\ j \in [0, e-1]\}$ is linearly independent over K by 5.5.2.1. Hence

$$A = \sum_{i=1}^{f}\sum_{j=0}^{e-1} u_i\pi_L^j\mathcal{O}_K \subset \mathcal{O}_L$$

is a free $\mathcal{O}_K$-module of rank ef, and A is closed in L by 5.2.9.2(a). We assert that $A = \mathcal{O}_L$. Indeed, let $a \in \mathcal{O}_L$, let $\mathcal{R} \subset \mathcal{O}_K$ be a set of representatives for k_K such that $0 \in \mathcal{R}$, and let $\pi_K \in \mathcal{O}_K$ be a prime element. By 5.5.2.2(b) there exists a family $(a_{i,l,j})_{i\in[1,f],\,l\in[0,e-1],\,j\geq 0}$ in $\mathcal{R}$ such that

$$a \equiv \sum_{i=1}^{f}\sum_{l=0}^{e-1}\sum_{j=0}^{k} a_{i,l,j}u_i\pi_L^l\pi_K^j \mod \mathfrak{p}_L^{e(k+1)} \quad \text{for all } k \geq 0.$$

By construction,

$$a_k = \sum_{i=1}^{f}\sum_{l=0}^{e-1}\sum_{j=0}^{k} a_{i,l,j}u_i\pi_L^l\pi_K^j \in A \quad \text{for all } k \geq 0,$$

and therefore $a = \lim_{k\to\infty} a_k \in A$.

Consequently, since $\mathcal{O}_L$ is a free $\mathcal{O}_K$-module of rank ef and $L = K\mathcal{O}_L$, it follows that $[L\colon K] = ef < \infty$, and $(L, |\cdot|)$ is complete by 5.5.3.2.

2. By 5.5.1.**2.**, $v_L \restriction K = ev_K$. If $|K^\times| = \langle\rho\rangle$, where $\rho \in (0,1)$, then $|L^\times| = \langle\rho^{1/e}\rangle$, and if $x \in L^\times$, then

$$v_K(\mathsf{N}_{L/K}(x)) = \frac{\log|\mathsf{N}_{L/K}(x)|}{\log\rho} = \frac{n\log|x|}{e\log\rho^{1/e}} = f\,\frac{\log|x|}{\log\rho^{1/e}} = fv_L(x).$$

Hence $v_K\circ\mathsf{N}_{L/K} = fv_L\colon L \to \mathbb{Z}\cup\{\infty\}$.

3. If $b \in \mathfrak{p}_L^k$, then $v_K(\mathsf{N}_{L/K}(b)) = fv_L(b) \geq fk$, and $\mathsf{N}_{L/K}(b) \in \mathfrak{p}_K^{fk}$.

Let $\overline{K}$ be an algebraic closure of K and $q = [L\colon K]_{\mathsf{i}}$. If $u \in U_L$, then

$$|\mathsf{N}_{L/K}(u)| = \prod_{\sigma\in\mathrm{Hom}_K(L,\overline{K})} |\sigma(u)|^q = 1 \quad \text{and therefore } \mathsf{N}_{L/K}(u) \in U_K.$$

Now let $k \geq 1$ and $u \in U_L^{(ke)}$. Then $u = 1 + a$, where $a \in \mathfrak{p}_L^{ke}$ and thus $|a| \leq |\pi_L|^{ke}$. Then

$$\mathsf{N}_{L/K}(u) = \prod_{\sigma\in\mathrm{Hom}_K(L,\overline{K})} (1+\sigma(a))^q = 1 + b,$$

where $b \in \overline{K}$ is a sum of products of the form $b_j = \sigma_1(a)\cdot\ldots\cdot\sigma_r(a)$ for some $r \in \mathbb{N}$ and $\sigma_1, \ldots, \sigma_r \in \mathrm{Hom}_K(L, \overline{K})$. For each such product b_j we obtain

$$|b_j| = \prod_{i=1}^{r} |\sigma_i(a)| = |a|^r \leq |\pi_L|^{ker} \leq |\pi_L|^{ke}.$$

Hence $|b| \leq |\pi_L|^{ke}$, $ev_K(b) = v_L(b) \geq ke$, $v_K(b) \geq k$ and $\mathsf{N}_{L/K}(u) \in U_K^{(k)}$.

4. By 1.5.1 there exists some $b \in \mathcal{O}_L$ such that $\mathsf{k}_L = \mathsf{k}_K(\overline{b})$. Let $g \in \mathcal{O}_K[X]$ be a monic polynomial such that $\overline{g} \in \mathsf{k}_K[X]$ is the minimal polynomial of $\overline{b}$ over k_K. Then $\overline{g(b)} = \overline{g}(\overline{b}) = \overline{0}$, hence $v_L(g(b)) \geq 1$, we set

$$a_0 = \begin{cases} b & \text{if } \ v_L(g(b)) = 1, \\ b + \pi_L & \text{if } \ v_L(g(b)) \geq 2, \end{cases}$$

and we assert that $v_L(g(a_0)) = 1$. This is clear if $v_L(g(b)) = 1$. If $v_L(g(b)) \geq 2$, then $g(a_0) = g(b) + \pi_L g'(b) + \pi_L^2 y$ for some $y \in \mathcal{O}_L$, and since $\overline{g}$ is separable, it follows that $\overline{g'(b)} = \overline{g}'(\overline{b}) \neq \overline{0}$, hence $v_L(g'(b)) = 0$ and $v_L(g(a_0)) = 1$.

Let $a \in \mathcal{O}_L$ be such that $a \equiv a_0 \mod \mathfrak{p}_L^2$. Then $g(a) \equiv g(a_0) \mod \mathfrak{p}_L^2$ and therefore $v_L(g(a)) = 1$. By 1., applied with $(u_1, \ldots, u_f) = (1, a, \ldots, a^{f-1})$ and $\pi_L = g(a)$, it follows that

$$\mathcal{O}_L = \sum_{i=1}^{f}\sum_{j=0}^{e-1} \mathcal{O}_K a^{i-1} g(a)^j \subset \mathcal{O}_K[a] \subset \mathcal{O}_L \quad \text{and therefore} \quad \mathcal{O}_L = \mathcal{O}_K[a]. \ \square$$

In 5.5.3 we proved the uniqueness of the extension of an absolute value using completeness. A careful study of the process however shows that an essentially weaker condition, namely the validity of Hensel's lemma, is not only sufficient but also necessary to guarantee unique extensions of absolute values. We refer to [52, Ch. II, § 6] for an exposition of this theory. Here we proceed with the extension of absolute values in the non-complete case.

Theorem 5.5.7 (Extension of absolute values, general case). *Let $K = (K, |\cdot|)$ be a valued field, L/K a finite field extension, $\widehat{K} = (\widehat{K}, |\cdot|)$ a completion of K and $\widehat{K}^* = (\widehat{K}^*, |\cdot|)$ an algebraic closure of $\widehat{K}$.*

1. Let $|\cdot|_L$ be an absolute value of L such that $|\cdot|_L \restriction K = |\cdot|$. Then $(L, |\cdot|_L)$ has a completion $\widehat{L}$ such that $\widehat{K} \subset \widehat{L}$, and every such completion satisfies $\widehat{L} = \widehat{K}L$, $[\widehat{L}:\widehat{K}] \leq [L:K]$ and $[\widehat{L}:\widehat{K}]_{\mathsf{i}} \mid [L:K]_{\mathsf{i}}$.

2. Let $\tau \in \mathrm{Hom}^K(L, \widehat{K}^*)$. Then $|\cdot|_\tau = |\cdot| \circ \tau \colon L \to \mathbb{R}_{\geq 0}$ is an absolute value of L such that $|\cdot|_\tau \restriction K = |\cdot|$. If $\widehat{L}_\tau$ is a completion of $(L, |\cdot|_\tau)$ satisfying $\widehat{K} \subset \widehat{L}_\tau$, then $\widehat{K}\tau(L)$ is a completion of $(\tau(L), |\cdot|)$, and τ has a unique extension to a $\widehat{K}$-isomorphism of valued fields $\widehat{\tau} \colon (\widehat{L}_\tau, |\cdot|_\tau) \overset{\sim}{\to} (\widehat{K}\tau(L), |\cdot|)$.

3. If $\tau, \tau' \in \mathrm{Hom}^K(L, \widehat{K}^*)$, then $|\cdot|_\tau = |\cdot|_{\tau'}$ if and only if there exists an automorphism $\sigma \in \mathrm{Gal}(\widehat{K}^*/\widehat{K})$ such that $\tau' = \sigma \circ \tau$.
In this case, τ and τ' are called $\widehat{K}$-**conjugate**.

4. Let $\{\tau_1, \ldots, \tau_r\} \subset \mathrm{Hom}_K(L, \widehat{K}^*)$ be a maximal set of pairwise not $\widehat{K}$-conjugate K-homomorphisms. Then $|\cdot|_{\tau_1}, \ldots, |\cdot|_{\tau_r} \colon L \to \mathbb{R}_{\geq 0}$ are precisely the distinct extensions of $|\cdot|$ to L, and they have the following properties:

(a) *If* $l = [L\colon K]_{\mathsf{i}}$ *and* $l_i = [\widehat{L}_{\tau_i}\colon \widehat{K}]_{\mathsf{i}}$ *for all* $i \in [1, r]$, *then*

$$[L\colon K] = \sum_{i=1}^{r} \frac{l}{l_i}\, [\widehat{L}_{\tau_i}\colon \widehat{K}], \qquad [L\colon K]_{\mathsf{s}} = \sum_{i=1}^{r} \frac{l}{l_i}\, [\widehat{L}_{\tau_i}\colon \widehat{K}]_{\mathsf{s}},$$

and, for all $x \in L$,

$$\mathsf{N}_{L/K}(x) = \prod_{i=1}^{r} \mathsf{N}_{\widehat{L}_{\tau_i}/\widehat{K}}(x)^{l/l_i} \quad \text{and} \quad \mathsf{Tr}_{L/K}(x) = \sum_{i=1}^{r} \frac{l}{l_i} \mathsf{Tr}_{\widehat{L}_{\tau_i}/\widehat{K}}(x).$$

(b) *The image of* L *under the diagonal embedding*

$$\delta\colon L \to \widehat{L}_{\tau_1} \times \ldots \times \widehat{L}_{\tau_r}$$

is dense with respect to the product topology.
In particular: For every r*-tuple* $(z_1, \ldots, z_r) \in \widehat{L}_{\tau_1} \times \ldots \times \widehat{L}_{\tau_r}$ *there exists a sequence* $(x_n)_{n\geq 0}$ *in* L *such that*

$$|\cdot|_{\tau_i}\text{-}\lim_{n\to\infty} x_n = z_i \quad \text{for all } i \in [1, r].$$

5. *Let* L/K *be separable,* $L = K(\alpha)$, $f \in K[X]$ *the minimal polynomial of* α *over* K *and* $f = f_1 \cdot \ldots \cdot f_r$, *where* $f_1, \ldots, f_r \in \widehat{K}[X]$ *are monic and irreducible. For* $i \in [1, r]$ *let* $\alpha_i \in \widehat{K}^*$ *be such that* $f_i(\alpha_i) = 0$, *and let* $\tau_i \in \mathrm{Hom}^K(L, \widehat{K}^*)$ *be the unique* K*-homomorphism satisfying* $\tau_i(\alpha) = \alpha_i$. *Then* $\{\tau_1, \ldots, \tau_r\}$ *is a maximal set of pairwise not* $\widehat{K}$*-conjugate* K*-homomorphisms, and if* $i \in [1, r]$, *then* $\widehat{L}_{\tau_i} = \widehat{K}(\widehat{\tau}_i^{-1}(\alpha_i))$, *where* $\widehat{\tau}_i\colon \widehat{L}_{\tau_i} \xrightarrow{\sim} \widehat{K}\tau_i(L) \hookrightarrow \widehat{K}^*$ *is the isomorphism given in* 2.. *In particular, if* $L \subset \widehat{K}^*$, *then* $\widehat{K}(\alpha)$ *is a completion of* $(L, |\cdot|)$.

Using the formalism of tensor products, this structure can be described by means of a natural identification

$$L \otimes_K \widehat{K} = \bigoplus_{i=1}^{r} \widehat{L}_{\tau_i}, \quad \alpha \otimes 1 = (\alpha, \ldots, \alpha) \in \prod_{i=1}^{r} \widehat{L}_{\tau_i}.$$

For details we refer to the forthcoming volume on class field theory.

Proof. We illustrate the situation by the following diagram:

$$\begin{array}{ccccccc}
\tau: L & \longrightarrow & \widehat{L}_\tau & \xrightarrow{\widehat{\tau}} & \widehat{K}\tau L & \longrightarrow & \widehat{K}^* \\
\uparrow & & \uparrow & & \uparrow & & \\
K & \longrightarrow & \overline{K} & \xrightarrow{=} & \widehat{K}. & &
\end{array}$$

1. Let $\widehat{L} = (\widehat{L}, |\cdot|)$ be a completion of $(L, |\cdot|_L)$ and $\overline{K}$ the (topological) closure of K in $\widehat{L}$. Then $(\overline{K}, |\cdot|_L \restriction \overline{K})$ is a completion of K, and there exists a K-isomorphism of valued fields $(\widehat{K}, |\cdot|) \stackrel{\sim}{\to} (\overline{K}, |\cdot|_L \restriction \overline{K})$. We identify: $\widehat{K} = \overline{K} \subset \widehat{L}$. It follows that $L \subset \widehat{K}L \subset \widehat{L}$ and $[\widehat{K}L : \widehat{K}] \le [L:K]$. Now L is dense in $\widehat{K}L$, and $\widehat{K}L$ is complete by 5.2.9.1. Hence $\widehat{K}L$ is closed in $\widehat{L}$, thus $\widehat{K}L = \widehat{L}$, and $[\widehat{L}:\widehat{K}] \le [L:K]$.

If L_0 is the separable closure of K in L, then $\widehat{K}L_0/\widehat{K}$ is separable by 1.4.5.1, and therefore $[\widehat{L}:\widehat{K}]_{\rm i} \mid [\widehat{L}:\widehat{K}L_0] = [\widehat{K}L:\widehat{K}L_0] \mid [L:L_0] = [L:K]_{\rm i}$.

2. Obviously, $|\cdot|_\tau \colon L \to \mathbb{R}_{\ge 0}$ is an absolute value and $|\cdot|_\tau \restriction K = |\cdot|$. By 1., $(L, |\cdot|_\tau)$ has a completion $\widehat{L}_\tau$ such that $\widehat{K} \subset \widehat{L}_\tau$, and by definition $\widehat{K}\tau(L) \subset \widehat{K}^*$. Since $[\widehat{K}\tau(L) : \widehat{K}] \le [\tau(L) : K] = [L : K] < \infty$, it follows that $\widehat{K}\tau(L)$ is complete. If $\overline{\tau(L)}$ is the (topological) closure of $\tau(L)$ in $\widehat{K}\tau(L)$, then $\overline{\tau(L)}$ is a completion of $\tau(L)$, and as $\widehat{K} = \overline{K} \subset \overline{\tau(L)}$, it follows that $\overline{\tau(L)} = \widehat{K}\tau(L)$. Therefore τ has a unique extension to an isomorphism of valued fields $\widehat{\tau} \colon (\widehat{L}_\tau, |\cdot|_\tau) \stackrel{\sim}{\to} (\widehat{K}\tau(L), |\cdot|)$, and $\widehat{\tau} \restriction K = {\rm id}_K$ implies $\widehat{\tau} \restriction \widehat{K} = {\rm id}_{\widehat{K}}$.

3. Let τ, $\tau' \in {\rm Hom}_K(L, \widehat{K}^*)$.

If $|\cdot|_\tau = |\cdot|_{\tau'}$, then $|\tau' \circ \tau^{-1}(z)| = |\tau^{-1}(z)|_{\tau'} = |\tau^{-1}(z)|_\tau = |z|$ for all $z \in \tau(L)$. Hence $\tau' \circ \tau^{-1} \colon (\tau(L), |\cdot|) \to (\tau'(L), |\cdot|)$ is a K-isomorphism of valued fields and has a unique extension to a $\widehat{K}$-isomorphism of valued fields $\tau^* \colon \widehat{K}\tau(L) \to \widehat{K}\tau'(L) \hookrightarrow \widehat{K}^*$. By 1.2.7.3 there exists some $\sigma \in {\rm Gal}(\widehat{K}^*/\widehat{K})$ such that $\sigma \restriction \widehat{K}\tau(L) = \tau^*$. Then it follows that $\sigma \restriction \tau(L) = \tau^* \restriction \tau(L) = \tau' \circ \tau^{-1}$ and therefore $\tau' = \sigma \circ \tau$.

If $\sigma \in {\rm Gal}(\widehat{K}^*/\widehat{K})$ and $\tau' = \sigma \circ \tau$, then

$$|x|_{\tau'} = |\tau'(x)| = |\sigma\tau(x)| = |\tau(x)| = |x|_\tau$$

for all $x \in L$ by 5.5.3.1 and therefore $|\cdot|_\tau = |\cdot|_{\tau'}$.

4. By 3. it follows that $|\cdot|_{\tau_1}, \ldots, |\cdot|_{\tau_r}$ are distinct extensions of $|\cdot|$ to an absolute value of L. Thus let $|\cdot|_L$ be any absolute value of L such that $|\cdot|_L \restriction K = |\cdot|$, and let $(\widehat{L}, |\cdot|_L)$ be a completion of $(L, |\cdot|_L)$ such that $\widehat{K} \subset \widehat{L}$. Since $[\widehat{L}:\widehat{K}] \le [L:K]$ (by 1.), there exists some $\widehat{\tau} \in {\rm Hom}^{\widehat{K}}(\widehat{L}, \widehat{K}^*)$, and then $|\cdot| \circ \widehat{\tau} \colon \widehat{L} \to \mathbb{R}_{\ge 0}$ is an absolute value satisfying $|\cdot| \circ \widehat{\tau} \restriction \widehat{K} = |\cdot| = |\cdot|_L \restriction \widehat{K}$. Hence 5.2.9.1 implies $|\cdot| \circ \widehat{\tau} = |\cdot|_L \colon \widehat{L} \to \mathbb{R}_{\ge 0}$, and therefore $|\cdot|_L = |\cdot|_\tau$, where $\tau = \widehat{\tau} \restriction L \in {\rm Hom}^K(L, \widehat{K}^*)$. If $i \in [1, r]$ is such that τ is $\widehat{K}$-conjugate to τ_i, then $|\cdot|_L = |\cdot|_{\tau_i}$.

For the proof of the remaining assertions in 4. we show first:

B. ${\rm Hom}^K(L, \widehat{K}^*) = \{\sigma \circ \tau_i \mid i \in [1, r],\ \sigma \in {\rm Hom}^{\widehat{K}}(\widehat{K}\tau_i(L), \widehat{K}^*)\}$, and if $\sigma \circ \tau_i = \sigma' \circ \tau_j$ for some i, $j \in [1, r]$, $\sigma \in {\rm Hom}^{\widehat{K}}(\widehat{K}\tau_i(L), \widehat{K}^*)$ and $\sigma' \in {\rm Hom}^{\widehat{K}}(\widehat{K}\tau_j(L), \widehat{K}^*)$, then $i = j$ and $\sigma = \sigma'$.

Proof of **B.** If $i \in [1, r]$ and $\sigma \in {\rm Hom}^{\widehat{K}}(\widehat{K}\tau_i(L), \widehat{K}^*)$, then apparently $\sigma \circ \tau_i \in {\rm Hom}^K(L, \widehat{K}^*)$. Conversely, if $\varphi \in {\rm Hom}^K(L, \widehat{K}^*)$, let $i \in [1, r]$ be such that φ is $\widehat{K}$-conjugate to τ_i. If $\sigma_1 \in \mathrm{Gal}(\widehat{K}^*/\widehat{K})$ is such that $\varphi = \sigma_1 \circ \tau_i$, then $\sigma = \sigma_1 \restriction \widehat{K}\tau_i(L) \in \mathrm{Hom}^{\widehat{K}}(\widehat{K}\tau_i(L), \widehat{K}^*)$ and $\varphi = \sigma \circ \tau_i$.

Now let $i,\ j \in [1,r]$, $\sigma \in \mathrm{Hom}^{\widehat{K}}(\widehat{K}\tau_i(L), \widehat{K}^*)$ and $\sigma' \in \mathrm{Hom}^{\widehat{K}}(\widehat{K}\tau_j(L), \widehat{K}^*)$ be such that $\sigma\circ\tau_i = \sigma'\circ\tau_j$. If $\sigma_1,\ \sigma_1' \in \mathrm{Gal}(\widehat{K}^*/\widehat{K})$ are such that $\sigma_1 \restriction \widehat{K}\tau_i(L) = \sigma$ and $\sigma_1' \restriction \widehat{K}\tau_j(L) = \sigma'$, then $\sigma_1 \circ \tau_i = \sigma_1' \circ \tau_j$, hence $\tau_j = (\sigma_1'^{-1} \circ \sigma_1) \circ \tau_i$, and τ_i and τ_j are $\widehat{K}$-conjugate. Hence $i = j$, $\sigma_1 \restriction \tau_i(L) = \sigma_1' \restriction \tau_i(L) \colon \tau_i(L) \to \widehat{K}^*$, and as σ_1 and σ_1' are value-preserving, we finally get $\sigma = \sigma_1 \restriction \widehat{K}\tau_i(L) = \sigma_1' \restriction \widehat{K}\tau_i(L) = \sigma'$. $\square$[**B.**]

(a) For $i \in [1,r]$ there is an isomorphism $\widehat{\tau}_i \colon \widehat{L}_{\tau_i} \overset{\sim}{\to} \widehat{K}\tau_i(L)$ by 2., and therefore the map $\mathrm{Hom}^{\widehat{K}}(\widehat{K}\tau_i(L), \widehat{K}^*) \to \mathrm{Hom}^{\widehat{K}}(\widehat{L}_{\tau_i}, \widehat{K}^*)$, $\sigma \mapsto \widehat{\sigma} = \sigma \circ \widehat{\tau}_i$, is bijective. Hence

$$\frac{[L\colon K]}{l} = [L\colon K]_{\mathsf{s}} = |\mathrm{Hom}_K(L, \widehat{K}^*)| = \sum_{i=1}^{r} |\mathrm{Hom}^{\widehat{K}}(\widehat{K}\tau_i(L), \widehat{K}^*)|$$
$$= \sum_{i=1}^{r} |\mathrm{Hom}^{\widehat{K}}(\widehat{L}_{\tau_i}, \widehat{K}^*)| = \sum_{i=1}^{r} [\widehat{L}_{\tau_i}, \widehat{K}]_{\mathsf{s}} = \sum_{i=1}^{r} \frac{[\widehat{L}_{\tau_i} \colon \widehat{K}]}{l_i}.$$

If $x \in L$, then 1.6.2.2 implies

$$\mathsf{N}_{L/K}(x) = \prod_{\tau \in \mathrm{Hom}^K(L,\widehat{K}^*)} \tau(x)^l = \prod_{i=1}^{r} \prod_{\sigma \in \mathrm{Hom}^{\widehat{K}}(\widehat{K}\tau_i(L),\widehat{K}^*)} \sigma\circ\tau_i(x)^l$$
$$= \prod_{i=1}^{r} \prod_{\widehat{\sigma} \in \mathrm{Hom}^{\widehat{K}}(\widehat{L}_{\tau_i},\widehat{K}^*)} \widehat{\sigma}(x)^l = \prod_{i=1}^{r} \mathsf{N}_{\widehat{L}_{\tau_i}/\widehat{K}}(x)^{l/l_i},$$

and alike for the trace.

(b) We must prove: For all $\varepsilon \in \mathbb{R}_{>0}$ and $(z_1, \ldots, z_r) \in \widehat{L}_{\tau_1} \times \ldots \times \widehat{L}_{\tau_r}$ there exists some $x \in L$ such that $|x - z_i|_{\tau_i} < \varepsilon$.

Thus let $\varepsilon \in \mathbb{R}_{>0}$ and $(z_1, \ldots, z_r) \in \widehat{L}_{\tau_1} \times \ldots \times \widehat{L}_{\tau_r}$. If $i \in [1,r]$, then there exists some $x_i \in L$ such that $|x_i - z_i|_{\tau_i} < \varepsilon/2$. By the weak approximation theorem 5.1.6.2(a) there exists some $x \in L$ such that $|x - x_i|_{\tau_i} < \varepsilon/2$ and consequently $|x - z_i|_{\tau_i} \le |x - x_i|_{\tau_i} + |x_i - z_i|_{\tau_i} < \varepsilon$ for all $i \in [1,r]$.

5. Let $f = f_1 \cdot \ldots \cdot f_r = (X - \alpha_1) \cdot \ldots \cdot (X - \alpha_n)$, where $\alpha_1, \ldots, \alpha_n \in \widehat{K}^*$. For $\nu \in [1,n]$ let $\tau_\nu \in \mathrm{Hom}^K(L, \widehat{K}^*)$ be the unique K-homomorphism satisfying $\tau_\nu(\alpha) = \alpha_\nu$ and thus $\tau_\nu(L) = K(\alpha_\nu)$. By 2., $\widehat{K}(\alpha_\nu) = \widehat{K}\tau_\nu(L)$ is a completion of $\tau_\nu(L)$, and τ_ν has a unique extension to a $\widehat{K}$-isomorphism of valued fields $\widehat{\tau}_\nu \colon \widehat{L}_{\tau_\nu} \to \widehat{K}(\alpha_\nu)$. Since $\widehat{\tau}_\nu(\alpha) = \alpha_\nu$, it follows that $\widehat{L}_{\tau_\nu} = \widehat{K}(\tau_\nu^{-1}(\alpha_\nu))$. If $\nu,\ \mu \in [1,n]$, then τ_ν and τ_μ are $\widehat{K}$-conjugate if and only if there exists some $i \in [1,r]$ such that $f_i(\alpha_\nu) = f_i(\alpha_\mu) = 0$.

If $L \subset \widehat{K}^*$, we may assume that $\alpha = \alpha_1$ and $\tau_1 = 1$. $\square$

Theorem 5.5.8 (Extension of absolute values to Galois field extensions). *Let $K = (K, |\cdot|)$ be a valued field, $\widehat{K}$ a completion of K, L/K a finite Galois extension and $G = \mathrm{Gal}(L/K)$.*

For an absolute value $\|\cdot\|\colon L \to \mathbb{R}_{\geq 0}$ and $\sigma \in G$ we define

$$^{\sigma}\|\cdot\| = \|\cdot\| \circ \sigma^{-1}\colon L \to \mathbb{R}_{\geq 0}.$$

Then $^{\sigma}\|\cdot\|\colon L \to \mathbb{R}_{\geq 0}$ is an absolute value of L, $^{\sigma}\|\cdot\| \restriction K = \|\cdot\| \restriction K$, and $^{\sigma}\|\sigma x\| = \|x\|$ for all $x \in L$. Hence $\sigma\colon (L, \|\cdot\|) \xrightarrow{\sim} (L, {}^{\sigma}\|\cdot\|)$ is a K-isomorphism of valued fields. If $\widehat{L}$ is a completion of $(L, \|\cdot\|)$ such that $\widehat{K} \subset \widehat{L}$ and $^{\sigma}\widehat{L}$ is a completion of $(L, {}^{\sigma}\|\cdot\|)$ such that $\widehat{K} \subset {}^{\sigma}\widehat{L}$, then σ has a unique extension to a $\widehat{K}$-isomorphism of valued fields $\widehat{\sigma}\colon \widehat{L} \xrightarrow{\sim} {}^{\sigma}\widehat{L}$. If $\sigma, \tau \in G$, then $^{\sigma\tau}\|\cdot\| = \|\cdot\| \circ (\sigma\tau)^{-1} = (\|\cdot\| \circ \tau^{-1}) \circ \sigma^{-1} = {}^{\sigma}({}^{\tau}\|\cdot\|)$. Hence the assignment $(\sigma, \|\cdot\|) \mapsto {}^{\sigma}\|\cdot\|$ defines an operation of G on the set of all absolute values of L.

Let $|\cdot|_1, \ldots, |\cdot|_r\colon L \to \mathbb{R}_{\geq 0}$ be the distinct extensions of $|\cdot|$ to an absolute value of L, and for $i \in [1, r]$ let $\widehat{L}_i$ be a completion of $(L, |\cdot|_i)$ such that $\widehat{K} \subset \widehat{L}_i$.

1. *G operates transitively on $\{|\cdot|_1, \ldots, |\cdot|_r\}$. If $\sigma \in G$, $i, j \in [1, r]$ and $^{\sigma}|\cdot|_i = |\cdot|_j$, then $\sigma\colon (L, |\cdot|_i) \xrightarrow{\sim} (L, |\cdot|_j)$ is a K-isomorphism of valued fields and has a unique extension to a $\widehat{K}$-isomorphism of valued fields $\widehat{\sigma}\colon \widehat{L}_i \xrightarrow{\sim} \widehat{L}_j$.*

2. *Let $i \in [1, r]$ and $G_i = \{\sigma \in G \mid {}^{\sigma}|\cdot|_i = |\cdot|_i\}$ be the isotropy group of $|\cdot|_i$. Then $(G{:}G_i) = r$, and $G/G_i = \{\tau_1 G_i, \ldots, \tau_r G_i\}$ implies $\{|\cdot|_1, \ldots, |\cdot|_r\} = \{{}^{\tau_1}|\cdot|_i, \ldots, {}^{\tau_r}|\cdot|_i\}$.*
$\widehat{L}_i/\widehat{K}$ is Galois, and the map

$$\rho_i\colon G_i \to \mathrm{Gal}(\widehat{L}_i/\widehat{K}), \quad \textit{defined by} \ \ \rho_i(\sigma) = \widehat{\sigma} \ \ \textit{for all} \ \ \sigma \in G_i,$$

is an isomorphism. We identify: $G_i = \mathrm{Gal}(\widehat{L}_i/\widehat{K})$.
In particular, $[\widehat{L}_i{:}\widehat{K}] \mid [L{:}K]$.

Proof. 1. By definition, G operates on $\{|\cdot|_1, \ldots, |\cdot|_r\}$, and we assume to the contrary that this operation it not transitive. Then $r \geq 2$, and after renumbering we may assume that $\{{}^{\sigma}|\cdot|_1 \mid \sigma \in G\} \cap \{{}^{\sigma}|\cdot|_2 \mid \sigma \in G\} = \emptyset$. By 5.1.6.2(b) there exists some $x \in L$ such that $^{\sigma}|x|_1 < 1$ and $^{\sigma}|x|_2 \geq 1$ for all $\sigma \in G$. Then

$$|\mathsf{N}_{L/K}(x)| = |\mathsf{N}_{L/K}(x)|_1 = \prod_{\sigma \in G} |\sigma^{-1}x|_1 = \prod_{\sigma \in G} {}^{\sigma}|x|_1 < 1$$

and

$$|\mathsf{N}_{L/K}(x)| = |\mathsf{N}_{L/K}(x)|_2 = \prod_{\sigma \in G} |\sigma^{-1}x|_2 = \prod_{\sigma \in G} {}^{\sigma}|x|_2 \geq 1,$$

a contradiction.

2. Let $i \in [1, r]$. As G operates transitively, elementary group theory implies $(G{:}G_i) = r$, and if $G/G_i = \{\tau_1 G_i, \ldots, \tau_r G_i\}$, then $^{\tau_1}|\cdot|_i, \ldots, {}^{\tau_r}|\cdot|_i$ are the

distinct extensions of $|\cdot|$. If $\sigma \in G_i$, then $\widehat{\sigma} \in \mathrm{Gal}(\widehat{L}_i/\widehat{K})$, and as $\widehat{\sigma} \upharpoonright L = \sigma$, the map $\rho_i \colon G_i \to \mathrm{Gal}(\widehat{L}_i/\widehat{K})$, defined by $\rho_i(\sigma) = \widehat{\sigma}$, is a monomorphism. Hence $|G_i| \le |\mathrm{Gal}(\widehat{L}_i/\widehat{K})| \le [\widehat{L}_i \colon \widehat{K}]$, but 5.5.7.4(a) shows that

$$[L\colon K] = \sum_{j=1}^{r} [\widehat{L}_j \colon \widehat{K}] \ge \sum_{j=1}^{r} |\mathrm{Gal}(\widehat{L}_j/\widehat{K})| \ge \sum_{j=1}^{r} |G_j| = r|G_i| = |G| = [L\colon K].$$

Therefore $|G_j| = |\mathrm{Gal}(\widehat{L}_j : \widehat{K})| = [\widehat{L}_j : \widehat{K}]$ for all $j \in [1,r]$. In particular, it follows that $\widehat{L}_i/\widehat{K}$ is Galois and ρ_i is an isomorphism. $\square$

The following Theorem 5.5.9 connects the extension theory of absolute values with the extension theory of Dedekind domains as developed in 2.12 and 2.13.

Theorem 5.5.9 (Extension of $\mathfrak{p}$-adic absolute values). *Let $\mathfrak{o}$ be a Dedekind domain, $K = \mathsf{q}(\mathfrak{o})$, L/K a finite field extension, and $l = [L : K]_{\mathsf{i}}$. Let $\mathfrak{O} = \mathrm{cl}_L(\mathfrak{o})$, $\mathfrak{p} \in \mathcal{P}(\mathfrak{o})$, and assume that $\mathfrak{p}\mathfrak{O} = \mathfrak{P}_1^{e_1} \cdot \ldots \cdot \mathfrak{P}_r^{e_r}$, where $r \in \mathbb{N}$, $e_1, \ldots, e_r \in \mathbb{N}$, and $\mathfrak{P}_1, \ldots, \mathfrak{P}_r \in \mathcal{P}(\mathfrak{O})$ are distinct. Let $K_\mathfrak{p}$ be a completion of $(K, |\cdot|_{\mathfrak{p},\rho})$ for some $\rho \in (0,1)$, $\widehat{\mathfrak{o}}_\mathfrak{p}$ its valuation domain and $\widehat{\mathfrak{p}}$ its valuation ideal. For $i \in [1,r]$ let $L_{\mathfrak{P}_i}$ be a completion of $(L, |\cdot|_{\mathfrak{P}_i,\rho^{1/e_i}})$ such that $K_\mathfrak{p} \subset L_{\mathfrak{P}_i}$, $\widehat{\mathfrak{O}}_{\mathfrak{P}_i}$ its valuation domain, $\widehat{\mathfrak{P}}_i$ its valuation ideal and $l_i = [L_{\mathfrak{P}_i} : K_\mathfrak{p}]_{\mathsf{i}}$.*

1. *$|\cdot|_{\mathfrak{P}_1,\rho^{1/e_1}}, \ldots, |\cdot|_{\mathfrak{P}_r,\rho^{1/e_r}}$ are the distinct extensions of $|\cdot|_{\mathfrak{p},\rho}$ to an absolute value of L.*

2. *Let $i \in [1,r]$ and $f_i = f(\mathfrak{P}_i/\mathfrak{p})$. Then*

$$f_i = f(L_{\mathfrak{P}_i}/K_\mathfrak{p}),\ e_i = e(\mathfrak{P}_i/\mathfrak{p}) = e(L_{\mathfrak{P}_i}/K_\mathfrak{p}),$$

$$[L_{\mathfrak{P}_i} : K_\mathfrak{p}] = e_i f_i,\ v_{\mathfrak{P}_i} \upharpoonright K = e_i v_\mathfrak{p},$$

$$[L\colon K]_{\mathsf{s}} = \sum_{i=1}^{r} [L_{\mathfrak{P}_i} : K_\mathfrak{p}]_{\mathsf{s}} \ \text{ and } \ v_\mathfrak{p} \circ \mathsf{N}_{L/K} = \sum_{i=1}^{r} \frac{l}{l_i} f_i\, v_{\mathfrak{P}_i} \colon L \to \mathbb{Z} \cup \{\infty\}.$$

In particular, if L/K is separable, then

$$[L\colon K] = \sum_{i=1}^{r} [L_{\mathfrak{P}_i} : K_\mathfrak{p}] = \sum_{i=1}^{n} e_i f_i,$$

$$\mathsf{N}_{L/K}(x) = \prod_{i=1}^{s} \mathsf{N}_{\widehat{L}_{\mathfrak{P}_i}/\widehat{K}_\mathfrak{p}}(x) \ \text{ and } \ \mathsf{Tr}_{L/K}(x) = \sum_{i=1}^{s} \mathsf{Tr}_{\widehat{L}_{\mathfrak{P}_i}/\widehat{K}_\mathfrak{p}}(x)$$

for all $x \in L$.

3. Assume that $\mathfrak{N}(\mathfrak{p}) = q < \infty$.
Let $|\cdot|_{\mathfrak{p}} = |\cdot|_{\mathfrak{p},1/q}$ *and* $|\cdot|_{\mathfrak{P}_i} = |\cdot|_{\mathfrak{P}_i,1/q^{f_i}}$ *for* $i \in [1,r]$ *be the normalized absolute values. Then*

$$|\mathsf{N}_{L/K}(x)|_{\mathfrak{p}} = \prod_{i=1}^{r} |x|_{\mathfrak{P}_i}^{l/l_i} \quad \textit{for all } x \in L.$$

4. Let L/K *be Galois,* $G = \mathrm{Gal}(L/K)$ *and* $i \in [1,r]$.

(a) If $\sigma \in G$, *then* ${}^{\sigma}|\cdot|_{\mathfrak{P}_i,\rho^{1/e_i}} = |\cdot|_{\sigma\mathfrak{P}_i,\rho_i^{1/e_i}}$, *and the isotropy group of* $|\cdot|_{\mathfrak{P}_i,\rho^{1/e_i}}$ *is just the decomposition group of* $\mathfrak{P}_i$ *over* K (see 2.5.3).

(b) Let $G_{\mathfrak{P}_i}$ *be the decomposition group and* $(G_{\mathfrak{P}_i,j})_{j\ge 0}$ *the sequence of the ramification groups of* $\mathfrak{P}_i$ *over* K (see 2.13.2). *Then* $G_{\mathfrak{P}_i} = \mathrm{Gal}(L_{\mathfrak{P}_i}/K_{\mathfrak{p}})$, *and* $(G_{\mathfrak{P}_i,j})_{j\ge 0}$ *is just the sequence of the ramification groups of* $\widehat{\mathfrak{P}}_i$ *over* $K_{\mathfrak{p}}$.

5. Let L/K *be Galois, and suppose that* $\mathfrak{p}$ *is unramified in* L, $\mathsf{k}_{\mathfrak{p}}$ *is finite and* $\mathfrak{P} \in \{\mathfrak{P}_1, \ldots, \mathfrak{P}_r\}$. *For an automorphism* $\sigma \in G_{\mathfrak{P}}$ *the following assertions are equivalent:*

(a) σ *is a Frobenius automorphism of* $\mathfrak{P}$ *over* K (as defined in 2.14.5 for the Dedekind domains $\mathfrak{o} \subset \mathfrak{O}$).

(b) σ *is a Frobenius automorphism of* $\widehat{\mathfrak{P}}$ *over* $K_{\mathfrak{p}}$ (as defined in 2.14.5 for the Dedekind domains $\widehat{\mathfrak{o}}_{\mathfrak{p}} \subset \widehat{\mathfrak{O}}_{\mathfrak{P}}$).

Proof. 1. If $i \in [1,r]$, then we get $e_i = e(\mathfrak{P}_i/\mathfrak{p}) = e(\widehat{\mathfrak{P}}_i/\widehat{\mathfrak{p}}) = e(L_{\mathfrak{P}_i}/K_{\mathfrak{p}})$ and $v_{\mathfrak{P}_i} \restriction K = e_i v_{\mathfrak{p}}$ by 2.12.2.3(a) and 5.5.1. If $x \in K$, then

$$|x|_{\mathfrak{P}_i,\rho^{1/e_i}} = (\rho^{1/e_i})^{v_{\mathfrak{P}_i}(x)} = \rho^{v_{\mathfrak{p}}(x)} = |x|_{\mathfrak{p},\rho},$$

and consequently $|\cdot|_{\mathfrak{P}_i,\rho^{1/e_i}} \restriction K = |\cdot|_{\mathfrak{p},\rho}$.

Let $|\cdot|\colon L \to \mathbb{R}_{\ge 0}$ be any absolute value of L such that $|\cdot| \restriction K = |\cdot|_{\mathfrak{p},\rho}$. Then $|\cdot|$ is non-Archimedian, and if $\mathcal{O}$ is the valuation domain of $(L, |\cdot|)$, then

$$\mathfrak{o} \subset \{x \in K \mid |x|_{\mathfrak{p},\rho} \le 1\} \subset \{x \in L \mid |x| \le 1\} = \mathcal{O}.$$

Hence $\mathfrak{O} = \mathrm{cl}_L(\mathfrak{o}) \subset \mathrm{cl}_L(\mathcal{O}) = \mathcal{O}$, and therefore $|\cdot| = |\cdot|_{\mathfrak{P},\rho_1}$ for some prime ideal $\mathfrak{P} \in \mathcal{P}(\mathfrak{O})$ and $\rho_1 \in (0,1)$ by 5.1.8.2. Since $\mathfrak{P} = \{x \in \mathfrak{O} \mid |x| < 1\}$ and $\mathfrak{P} \cap \mathfrak{o} = \{x \in \mathfrak{o} \mid |x|_{\mathfrak{p},\rho} < 1\} = \mathfrak{p}$, there exists some $i \in [1,r]$ such that $\mathfrak{P} = \mathfrak{P}_i$. If $x \in K$, then $|x|_{\mathfrak{p},\rho} = |x|_{\mathfrak{P}_i,\rho_1} = |x|_{\mathfrak{P}_i,\rho^{1/e_i}}$, and therefore $\rho_1 = \rho^{1/e_i}$.

2. If $i \in [1,r]$, then $f_i = f(\mathfrak{P}_i/\mathfrak{p}) = f(L_{\mathfrak{P}_i}/K_{\mathfrak{p}})$ by 5.5.1.1. The assertions concerning e_i were already proved in 1., and the equality $e_i f_i = [L_{\mathfrak{P}_i} : K_{\mathfrak{p}}]$ follows from 5.5.6.1. By 5.5.7.4 we obtain

$$[L\colon K]_{\mathsf{s}} = \sum_{i=1}^{r} [L_{\mathfrak{P}_i}\colon K_{\mathfrak{p}}]_{\mathsf{s}},$$

$$\mathsf{N}_{L/K}(x) = \prod_{i=1}^{s} \mathsf{N}_{\widehat{L}_{\mathfrak{P}_i}/\widehat{K}_{\mathfrak{p}}}(x)^{l/l_i} \quad \text{and}$$

$$\mathsf{Tr}_{L/K}(x) = \frac{l}{l_i}\sum_{i=1}^{s} \mathsf{Tr}_{\widehat{L}_{\mathfrak{P}_i}/\widehat{K}_{\mathfrak{p}}}(x) \ \text{ for all } \ x \in L.$$

Hence 5.5.6.2 implies

$$v_{\mathfrak{p}}(\mathsf{N}_{L/K}(x)) = \sum_{i=1}^{s} \frac{l}{l_i}\, v_{\mathfrak{p}}(\mathsf{N}_{\widehat{L}_{\mathfrak{P}_i}/\widehat{K}_{\mathfrak{p}}}(x)) = \sum_{i=1}^{r} \frac{l}{l_i}\, f_i v_{\mathfrak{P}_i}(x) \quad \text{for all } \ x \in L.$$

3. Observe that $\mathfrak{N}(\mathfrak{p}) = q$ implies $\mathfrak{N}(\mathfrak{P}_i) = q^{f_i}$ for all $i \in [1, r]$. Hence $|\cdot|_{\mathfrak{p}} = |\cdot|_{\mathfrak{p},1/q}$ and $|\cdot|_{\mathfrak{P}_i} = |\cdot|_{\mathfrak{P}_i,1/q^{f_i}}$ are the normalized $\mathfrak{p}$-adic and $\mathfrak{P}_i$-adic absolute values. If $x \in L$, then 2. implies

$$|\mathsf{N}_{L/K}(x)|_{\mathfrak{p}} = q^{-v_{\mathfrak{p}}(\mathsf{N}_{L/K}(x))} = \prod_{i=1}^{r} \bigl(q^{-f_i v_{\mathfrak{P}_i}(x)}\bigr)^{l/l_i} = \prod_{i=1}^{r} |x|_{\mathfrak{P}_i}^{l/l_i}.$$

4.(a) Let $\sigma \in G$. By definition and 2.13.1.1 we obtain, for all $x \in L$,

$$^{\sigma}|x|_{\mathfrak{P}_i,\rho^{1/e_i}} = |\sigma^{-1}x|_{\mathfrak{P}_i,\rho^{1/e_i}} = \rho^{v_{\mathfrak{P}_i}(\sigma^{-1}x)/e_i} = \rho^{v_{\sigma\mathfrak{P}_i}(x)/e_i} = |x|_{\sigma\mathfrak{P}_i,\rho^{1/e_i}},$$

and therefore $^{\sigma}|\cdot|_{\mathfrak{P}_i,\rho^{1/e_i}} = |\cdot|_{\sigma\mathfrak{P}_i,\rho^{1/e_i}}$. Moreover, $^{\sigma}|\cdot|_{\mathfrak{P}_i,\rho^{1/e_i}} = |\cdot|_{\mathfrak{P}_i,\rho^{1/e_i}}$ if and only if $\mathfrak{P}_i = \sigma\mathfrak{P}_i$, and therefore the isotropy group of $|\cdot|_{\mathfrak{P}_i,\rho^{1/e_i}}$ coincides with the decomposition group of $\mathfrak{P}_i$ over K.

(b) By (a), $G_{\mathfrak{P}_i}$ is the isotropy group of $|\cdot|_{\mathfrak{P}_i,\rho^{1/e_i}}$, and by 5.5.8.2 it follows that $G_{\mathfrak{P}_i} = \mathrm{Gal}(L_{\mathfrak{P}_i}/K_{\mathfrak{p}})$. By definition (see 2.13.2) we have, for all $j \ge 0$,

$$G_{\mathfrak{P}_i,j} = \{\sigma \in G \mid \mathsf{v}_{\mathfrak{P}_i}(\sigma a - a) \ge j+1 \text{ for all } \ a \in \mathfrak{O}\} \subset G_{\mathfrak{P}_i},$$

and

$$G_{\widehat{\mathfrak{P}}_i,j} = \{\sigma \in G_{\mathfrak{P}_i} \mid \mathsf{v}_{\mathfrak{P}_i}(\sigma a - a) \ge j+1 \text{ for all } a \in \widehat{\mathfrak{O}}_{\mathfrak{P}_i}\}.$$

Apparently $\mathfrak{O} \subset \widehat{\mathfrak{O}}_{\mathfrak{P}_i}$ implies $G_{\widehat{\mathfrak{P}}_i,j} \subset G_{\mathfrak{P}_i,j}$. To prove equality, let $\sigma \in G_{\mathfrak{P}_i,j}$ and $a \in \widehat{\mathfrak{O}}_{\mathfrak{P}_i}$. Since $\mathfrak{O}$ is dense in $\widehat{\mathfrak{O}}_{\mathfrak{P}_i}$ (see 5.3.4), there exists a sequence $(a_n)_{n\ge 0}$ in $\mathfrak{O}$ such that $(a_n)_{n\ge 0} \to a$ in $L_{\mathfrak{P}_i}$. As σ is continuous, it follows that $(\sigma a_n - a_n)_{n\ge 0} \to \sigma a - a$, and $\mathsf{v}_{\mathfrak{P}_i}(\sigma a_n - a_n) \ge j+1$ for all $n \ge 0$ implies $\mathsf{v}_{\mathfrak{P}_i}(\sigma a - a) \ge j+1$. As this holds for all $a \in \widehat{\mathfrak{O}}_{\mathfrak{P}_i}$, we get $\sigma \in G_{\widehat{\mathfrak{P}}_i,j}$.

5. Recall the identifications $k_{\widehat{\mathfrak{P}}} = \widehat{\mathfrak{O}}_{\mathfrak{P}}/\widehat{\mathfrak{P}} = \mathfrak{O}/\mathfrak{P} = \mathsf{k}_{\mathfrak{P}}$ made in 5.3.4 and $G_{\mathfrak{P}} = \mathrm{Gal}(L_{\mathfrak{P}}/K_{\mathfrak{p}})$ made in 5.5.8.2. For $\sigma \in G_{\mathfrak{P}}$ let $\sigma_{\mathfrak{P}} \in \mathrm{Gal}(\mathsf{k}_{\mathfrak{P}}/\mathsf{k}_{\mathfrak{p}})$ be the residue class automorphism induced by σ (see 2.5.3). Observing these identifications, 2.14.6.1 shows that both (a) and (b) hold if and only if $\sigma_{\mathfrak{P}}(\xi) = \xi^{|\mathsf{k}_{\mathfrak{p}}|}$ for all $\xi \in \mathsf{k}_{\mathfrak{p}}$. □

We close this section with an explicit description of all absolute values of an algebraic number field. In particular, we prove the product formula for the normalized absolute values as we already did for $\mathbb{Q}$ in 5.1.4.

Theorem 5.5.10 (The absolute values of an algebraic number field). *Let K be an algebraic number field, (r_1, r_2) its signature, $[K:\mathbb{Q}] = n = r_1 + 2r_2$, and suppose that* $\mathrm{Hom}(K, \mathbb{C}) = \{\sigma_1, \ldots, \sigma_n\}$. *Let $\sigma_1, \ldots, \sigma_{r_1}$ be the real embeddings and $(\sigma_{r_1+1}, \overline{\sigma_{r_1+1}}), \ldots, (\sigma_{r_1+r_2}, \overline{\sigma_{r_1+r_2}})$ the pairs of conjugate complex embeddings of K* (see 3.1.1). *For $j \in [1, r_1]$ we set $l_j = 1$ and $\mathbb{K}_j = \mathbb{R}$, and for $j \in [r_1+1, r_1+r_2]$ we set $l_j = 2$ and $\mathbb{K}_j = \mathbb{C}$. For every $j \in [1, r_1+r_2]$ let $|\cdot|_{\infty,j} = |\cdot|_\infty^{l_j} \circ \sigma_j$ be the normalized Archimedean absolute value of K* (as in 5.1.3.**3**).

1. *For every Archimedean absolute value $|\cdot|$ of K there is a unique $j \in [1, r_1+r_2]$ such that $|\cdot|$ is equivalent to $|\cdot|_{\infty,j}$.*
2. *If $x \in K$, then*

$$|\mathsf{N}_{K/\mathbb{Q}}(x)|_\infty = \prod_{j=1}^{r_1+r_2} |x|_{j,\infty}.$$

3. *If $K = \mathbb{Q}(\alpha)$ and $f \in \mathbb{Q}[X]$ is the minimal polynomial of α, then f splits in $\mathbb{R}[X]$ in a product of r_1 linear factors and r_2 irreducible factors of degree 2.*
4. *For every non-Archimedean absolute value $|\cdot|$ of K there exists a unique prime ideal $\mathfrak{p} \in \mathcal{P}_K$ such that $|\cdot|$ is equivalent to the normalized $\mathfrak{p}$-adic absolute value $|\cdot|_\mathfrak{p}$, and*

$$\prod_{\mathfrak{p} \in \mathcal{P}_K} |x|_\mathfrak{p} \prod_{j=1}^{r_1+r_2} |x|_{\infty,j} = 1 \quad \textit{for all } x \in K^\times.$$

Proof. 1. *Existence*: Let $|\cdot|$ be an Archimedean absolute value of K and $\widehat{K}$ a completion of $(K, |\cdot|)$. By 5.2.10 there exist some $\lambda \in \mathbb{R}_{>0}$ and an isomorphism of valued fields $\Phi\colon (\widehat{K}, |\cdot|) \to (\mathbb{K}, |\cdot|_\infty^\lambda)$, where $\mathbb{K} \in \{\mathbb{R}, \mathbb{C}\}$. As $\Phi \restriction K \in \mathrm{Hom}(K, \mathbb{C})$, we obtain $\Phi \restriction K \in \{\sigma_j, \overline{\sigma}_j\}$ for some $j \in [1, r_1+r_2]$. But then $|\cdot| = |\cdot|_\infty^\lambda \circ \sigma_j = |\cdot|_{\infty,j}^\lambda$ is equivalent to $|\cdot|_{\infty,j}$.

Uniqueness: We must prove that the absolute values $|\cdot|_{\infty,1}, \ldots, |\cdot|_{\infty,r_1+r_2}$ are mutually not equivalent. We assume to the contrary that there exist some $i, j \in [1, r_1+r_2]$ and $\lambda \in \mathbb{R}_{>0}$ such that $i \ne j$ and $|\cdot|_{\infty,i} = |\cdot|_{\infty,j}^\lambda$. For $a \in \mathbb{N}$ this implies $a^{l_i} = a^{l_j\lambda}$, hence $l_i = l_j\lambda$, and therefore we obtain $|\sigma_i(x)|_\infty = |x|_{\infty,i}^{1/l_i} = |x|_{\infty,j}^{\lambda/l_i} = |x|_{\infty,j}^{1/l_j} = |\sigma_j(x)|_\infty$ for all $x \in K$. Since $i \ne j$, there exists some $z \in K$ such that $\sigma_i(z) \ne \sigma_j(z)$ and $\sigma_i(z) \ne \overline{\sigma_j(z)}$. But then there exists some $g \in \mathbb{N}$ such that $|\sigma_i(z) + g|_\infty \ne |\sigma_j(z) + g|_\infty$, and consequently $|\sigma_i(z+g)|_\infty \ne |\sigma_j(z+g)|_\infty$, a contradiction.

2. By 1.6.2.2, if $x \in K$, then

$$|\mathsf{N}_{K/\mathbb{Q}}(x)|_\infty = \prod_{i=1}^{n} |\sigma_i(x)|_\infty = \prod_{j=1}^{r_1+r_2} |\sigma_j(x)|_\infty^{l_j} = \prod_{j=1}^{r_1+r_2} |x|_{\infty,j}.$$

3. By 5.5.7.5.

4. Let $|\cdot|$ be a non-Archimedian absolute value of K and $\mathcal{O}$ its valuation domain. As $\mathbb{Z} \subset \mathcal{O}$ and $\mathcal{O}$ is integrally closed, we get $\mathfrak{o}_K = \mathrm{cl}_K(\mathbb{Z}) \subset \mathcal{O}$. By 5.1.8.2 there exists a unique $\mathfrak{p} \in \mathcal{P}_K$ such that $|\cdot|$ is a $\mathfrak{p}$-adic absolute value and thus equivalent to $|\cdot|_\mathfrak{p}$.

If $a \in K^\times$, then using 5.1.4, 5.5.9.3 and 2. we get

$$1 = \prod_{p \in \mathbb{P} \cup \{\infty\}} |\mathsf{N}_{K/\mathbb{Q}}(a)|_p = \prod_{p \in \mathbb{P}} \prod_{\substack{\mathfrak{p} \in \mathcal{P}_K \\ \mathfrak{p} \mid p}} |a|_\mathfrak{p} \prod_{j=1}^{r_1+r_2} |a|_{\infty,j} = \prod_{\mathfrak{p} \in \mathcal{P}_K} |a|_\mathfrak{p} \prod_{j=1}^{r_1+r_2} |a|_{\infty,j}.$$

□

The product formula is one evidence for the singularization of normalized absolute values. A more serious one is their appearance as the module of a Haar measure on the completions of K (see 5.3.10).

5.6 Unramified field extensions

We continue to use the notation from 5.3.1 *and* 5.5.1.

Definitions and Remarks 5.6.1. Let $L = (L, |\cdot|)$ be a non-Archimedian valued field and K a subfield of L such that L/K is algebraic. Then $|\cdot| \restriction K$ is an absolute value of K by 5.1.7.2(d). We set $f = f(L/K) = [\mathsf{k}_L : \mathsf{k}_K]$ and $e = e(L/K) = (|L^\times| : |K^\times|)$.

1. Let $[L:K] = n < \infty$ (then $ef \le n$ by 5.5.2.1). The field extension L/K is called

- **unramified** if $f = n$ and $\mathsf{k}_L/\mathsf{k}_K$ is separable (then $e = 1$); otherwise L/K is called **ramified**;

- **fully ramified** if $e = n$ (then $f = 1$).

In particular, the trivial field extension K/K is both unramified and fully ramified.

If M is an intermediate field of L/K, then L/K is unramified [fully ramified] if and only if both L/M and M/K are unramified [fully ramified].

Let $\widehat{K}$ be a completion of K and $\widehat{L}$ a completion of L such that $\widehat{K} \subset \widehat{L}$. Then $e = e(\mathfrak{p}_L/\mathfrak{p}_K) = e(\widehat{L}/\widehat{K})$, $f = f(\mathfrak{p}_L/\mathfrak{p}_K) = f(\widehat{L}/\widehat{K})$, $\mathsf{k}_K = \mathsf{k}_{\widehat{K}}$ and $\mathsf{k}_L = \mathsf{k}_{\widehat{L}}$. Hence $\mathfrak{p}_L/\mathfrak{p}_K$ is unramified if and only if L/K is unramified, and this holds if and only if $\widehat{L}/\widehat{K}$ is unramified.

2. Let L/K be arbitrary, and let $\mathcal{E}(L/K)$ be the set of all over K finite intermediate fields of L/K. The field extension L/K is called **unramified** [**fully ramified**] if E/K is unramified [fully ramified] for all $E \in \mathcal{E}(L/K)$. Note that this new definition is consistent with the old one if $[L\!:\!K] < \infty$.

Instead of saying that L/K is unramified [fully ramified] we also say that L is unramified [fully ramified] over K.

3. Let $(L', |\cdot|')$ be another non-Archimedian valued field, $\varphi\colon L \to L'$ a value-preserving isomorphism and $K' = \varphi(K)$. Then $|\cdot|' \restriction K'$ is an absolute value of K', $e(L'/K') = e(L/K)$ and $f(L'/K') = f(L/K)$. In particular, L'/K' is (un)ramified resp. fully ramified if and only if L/K has the same property.

For complete discrete valued fields the concepts simplify considerably as the following Theorem 5.6.2 shows.

Theorem 5.6.2. *Let $K = (K, |\cdot|)$ be a complete discrete valued field and L/K an algebraic field extension.*

1. *L/K is unramified if and only if $e(L/K) = 1$ and $\mathsf{k}_L/\mathsf{k}_K$ is separable.*

2. *L/K is fully ramified if and only if $f(L/K) = 1$.*

3. *Let M be an intermediate field of L/K. Then L/K is unramified [fully ramified] if and only if both L/M and M/K are unramified [fully ramified].*

Proof. Let $\mathcal{E}(L/K)$ be the set of all over K finite intermediate fields of L/K. Then

$$|L^\times| = \bigcup_{E \in \mathcal{E}(L/K)} |E^\times| \quad \text{and} \quad \mathsf{k}_L = \bigcup_{E \in \mathcal{E}(L/K)} \mathsf{k}_E.$$

The extension $\mathsf{k}_L/\mathsf{k}_K$ is separable if and only if $\mathsf{k}_E/\mathsf{k}_K$ is separable for all $E \in \mathcal{E}(L/K)$, and $[E\!:\!K] = e(E/K)f(E/K)$ for all $E \in \mathcal{E}(L/K)$. With these observations in mind, the assertions follow easily. □

Theorem 5.6.3 (Characterization of unramified extensions). *Let K be a complete non-Archimedian valued field and L/K a finite field extension. For $P \in \mathcal{O}_L[X]$ we denote by $\overline{P} \in \mathsf{k}_L[X]$ the residue class polynomial.*

1. *Let L/K be unramified, $a \in \mathcal{O}_L$ such that $\mathsf{k}_L = \mathsf{k}_K(\overline{a})$, and let $g \in \mathcal{O}_K[X]$ be the minimal polynomial of a over K. Then its residue class polynomial $\overline{g} \in \mathsf{k}_K[X]$ is the minimal polynomial of $\overline{a}$ over k_K, $L = K(a)$, L/K is separable and $\mathcal{O}_L = \mathcal{O}_K[a]$.*

2. *Let $L = K(a)$, and let $g \in \mathcal{O}_K[X]$ be a monic polynomial such that $g(a) = 0$ and the residue class polynomial $\overline{g} \in \mathsf{k}_K[X]$ is separable. Then L/K is unramified, and $\mathsf{k}_L = \mathsf{k}_K(\overline{a})$.*

Proof. 1. Let $\varphi \in \mathsf{k}_K[X]$ be the minimal polynomial of $\overline{a}$ over k_K. As $g(a) = 0$, it follows that $\overline{g}(\overline{a}) = \overline{g(a)} = \overline{0}$, hence $\varphi \,|\, \overline{g}$, and

$$[L\!:\!K] \geq [K(a)\!:\!K] = \partial(g) = \partial(\overline{g}) \geq \partial(\varphi) = [\mathsf{k}_L\!:\!\mathsf{k}_K] = f(L/K) = [L\!:\!K].$$

Hence we obtain $\overline{g} = \varphi$ and $K(a) = L$. Since $\mathsf{k}_L/\mathsf{k}_K$ is separable, φ is separable, and therefore g and L/K are separable.

Let $\mathcal{R} \subset \mathcal{O}_K$ be a set of representatives for k_K such that $0 \in \mathcal{R}$, let $\pi \in \mathcal{O}_K$ be a prime element and $d = [L:K]$. Then π is a prime element of $\mathcal{O}_L$, $(\overline{1}, \overline{a}, \ldots, \overline{a}^{d-1})$ is k_K-basis of k_L, and by 5.5.2.2(a) the set

$$\mathcal{R}' = \{c_0 + c_1 a + \ldots + c_{d-1}a^{d-1} \mid c_0, \ldots, c_{d-1} \in \mathcal{R}\} \subset \mathcal{O}_L$$

is a set of representatives for k_L such that $0 \in \mathcal{R}'$. By 5.3.5.1 every $x \in \mathcal{O}_L$ has a representation

$$x = \sum_{n=0}^{\infty}\Big(\sum_{i=0}^{d-1} c_{n,i}a^i\Big)\pi^n = \sum_{i=0}^{d-1}\Big(\sum_{n=0}^{\infty} c_{n,i}\pi^n\Big)a^i, \quad \text{where } c_{n,i} \in \mathcal{R},$$

which shows that $x \in \mathcal{O}_K[a]$. Hence $\mathcal{O}_L = \mathcal{O}_K[a]$.

2. Let $g_1 \in K[X]$ be the minimal polynomial of a over K. Since $g \in \mathcal{O}_K[X]$, it follows that a is integral over $\mathcal{O}_K$, hence $g_1 \in \mathcal{O}_K[X]$ by 2.5.1.3 and $g_1 \,|\, g$ in $\mathcal{O}_K[X]$. Hence $\overline{g}_1 \,|\, \overline{g}$ in $\mathsf{k}_K[X]$, and $\overline{g}_1$ is separable. By 5.4.5.2 $\overline{g}_1$ is irreducible, thus it is the minimal polynomial of $\overline{a}$ over k_K and $\mathsf{k}_K(\overline{a})/\mathsf{k}_K$ is separable. Since

$$[L\!:\!K] = \partial(g_1) = \partial(\overline{g}_1) = [\mathsf{k}_K(\overline{a})\!:\!\mathsf{k}_K] \leq [\mathsf{k}_L\!:\!\mathsf{k}_K] = f(L/K) \leq [L\!:\!K],$$

it follows that $\mathsf{k}_L = \mathsf{k}_K(\overline{a})$ and $f(L/K) = [L\!:\!K]$, whence L/K is unramified. □

Theorem 5.6.4 (Shifting theorem for unramified field extensions). *Let K be a complete non-Archimedian valued field, $\overline{K}$ an algebraic closure of K, and let L and K' be intermediate fields of $\overline{K}/K$.*

1. *If L/K is unramified, then L/K is separable, and LK'/K' is unramified, too.*

2. *Let L be the compositum of a family $(L_i)_{i \in I}$ of over K unramified intermediate fields of L/K. Then L/K is unramified.*

Proof. 1. Let L/K be unramified. If $a \in L$, then $K(a)/K$ is a finite unramified field extension, and therefore it is separable by 5.6.3.1. Thus a is separable over K, and consequently L/K is separable.

Assume now first that $[L:K] < \infty$. Let $a \in \mathcal{O}_L$ be such that $\mathsf{k}_L = \mathsf{k}_K(\overline{a})$, and let $g \in \mathcal{O}_K[X]$ be the minimal polynomial of a over K. By 5.6.3.1, $L = K(a)$, $\overline{g} \in \mathsf{k}_K[X]$ is the minimal polynomial of $\overline{a}$ over k_K, and therefore $\overline{g}$ is separable. Since $LK' = K'(a)$, $g \in \mathcal{O}_{K'}[X]$, $g(a) = 0$ and $\overline{g} \in \mathsf{k}_{K'}[X]$ is separable, it follows that LK'/K' is unramified by 5.6.3.2.

Now let L/K be arbitrary, and let M be an over K' finite intermediate field of LK'/K'. As $LK' = K'(L)$, it follows that $M \subset K'(B)$ for some finite subset B of L, and we set $M_0 = K(B)$. Then M_0/K is finite and unramified, hence M_0K'/K' is unramified, and since $M \subset M_0K'$, it follows that M/K' is unramified.

2. If M is an over K finite intermediate field of L/K, then there exists a finite subset J of I and for every $j \in J$ an over K finite intermediate field M_j of L_j/K such that M is contained in the compositum of the finite family $(M_j)_{j\in J}$. Thus it suffices to prove:

> $(*)$ If $n \in \mathbb{N}$ and $L_1, \ldots, L_n$ are over K finite and unramified intermediate fields of $\overline{K}/K$, then their compositum is unramified over K.

We prove $(*)$ by induction on n. It is obviously sufficient to do the case $n = 2$. Then L_1L_2/L_1 is unramified by 1., and as L_1/K is unramified, it follows that L_1L_2/K is unramified, too. □

Theorem and Definition 5.6.5. *Let K be a complete non-Archimedian valued field and L/K an algebraic field extension.*

1. *If L is an algebraic closure of K, then k_L is an algebraic closure of k_K.*

2. *Let $\mathcal{Z}_\mathsf{u}(L/K)$ be the set of all over K finite unramified intermediate fields of L/K and $\mathcal{Z}_\mathsf{s}(\mathsf{k}_L/\mathsf{k}_K)$ the set of all over k_K finite separable intermediate fields of $\mathsf{k}_L/\mathsf{k}_K$. Then the map*

$$\mathcal{Z}_\mathsf{u}(L/K) \to \mathcal{Z}_\mathsf{s}(\mathsf{k}_L/\mathsf{k}_K), \quad M \mapsto \mathsf{k}_M$$

is an inclusion-preserving bijection. If $M \in \mathcal{Z}_\mathsf{u}(L/K)$, then M/K is Galois if and only if $\mathsf{k}_M/\mathsf{k}_K$ is Galois.

3. *There exists a largest over K unramified intermediate field T of L/K. Its residue class field k_T is the separable closure of k_K in k_L.* The field T is called the **inertia field** of L/K. If $\overline{K}$ is an algebraic closure of K, then we denote by K_∞ the inertia field of $\overline{K}/K$ and call it the **maximal unramified extension** of K (in $\overline{K}$). *In particular, k_{K_∞} is a separable closure of k_K.*

4. *Let L/K be normal and T the inertia field of L/K. Then $\mathsf{k}_L/\mathsf{k}_K$ is also normal, and the field extensions T/K and $\mathsf{k}_T/\mathsf{k}_K$ are Galois.* The Galois group $\mathrm{Gal}(L/T)$ is called the **inertia group** of L/K.

5. *Let L/K be a finite Galois extension, $G = \mathrm{Gal}(L/K)$ and T the inertia field of L/K. For $\sigma \in G$ let $\overline{\sigma} \in \mathrm{Gal}(\mathsf{k}_L/\mathsf{k}_K)$ be the residue class automorphism induced by σ* (see 5.5.3.4). *Then the map $\rho_{L/K}\colon G \to \mathrm{Gal}(\mathsf{k}_T/\mathsf{k}_K)$, definied by $\rho_{L/K}(\sigma) = \overline{\sigma} \restriction \mathsf{k}_T$ for all $\sigma \in G$, is an epimorphism, $\mathrm{Ker}(\rho_{L/K}) = \mathrm{Gal}(L/T)$, and $\rho_{L/K}$ induces an isomorphism $\rho_{T/K}\colon \mathrm{Gal}(T/K) \overset{\sim}{\to} \mathrm{Gal}(\mathsf{k}_T/\mathsf{k}_K)$ such that $\overline{\tau} = \rho_{T/K}(\tau)$ is the residue class automorphism induced by τ for every $\tau \in \mathrm{Gal}(T/K)$.*
If $(K, |\cdot|)$ is discrete, then $\mathrm{Gal}(L/T)$ is the inertia group and T is the inertia field of $\mathfrak{p}_L$ over K as defined in 2.13.2.

Proof. For a polynomial $f \in \mathcal{O}_L[X]$ we denote by $\overline{f} \in \mathsf{k}_L[X]$ its residue class polynomial. In particular, if $a \in \mathcal{O}_K$, then $\overline{a} = a + \mathfrak{p}_L \in \mathsf{k}_L$.

1. By 5.5.3.3 the residue class extension $\mathsf{k}_L/\mathsf{k}_K$ is algebraic. Let L be an algebraic closure of K, $\varphi \in \mathsf{k}_L[X] \setminus \mathsf{k}_L$ and $f \in \mathcal{O}_L[X]$ such that $\overline{f} = \varphi$. Then f splits in $L[X]$, and as $\mathcal{O}_L$ is integrally closed, f splits in $\mathcal{O}_L[X]$. Hence φ splits in $\mathsf{k}_L[X]$, and thus k_L is algebraically closed.

2. It suffices to prove: For every $\mathsf{m} \in \mathcal{Z}_{\mathsf{s}}(\mathsf{k}_L/\mathsf{k}_K)$ there exists an intermediate field $M \in \mathcal{Z}_{\mathsf{u}}(L/K)$ with the following properties:

- $\mathsf{k}_M = \mathsf{m}$.
- If $E \in \mathcal{Z}_{\mathsf{u}}(L/K)$ is such that $\mathsf{m} \subset \mathsf{k}_E$, then $M \subset E$.
- M/K is Galois if and only if m/k_K is Galois.

Let $\mathsf{m} = \mathsf{k}_K(\alpha) \in \mathcal{Z}_s(\mathsf{k}_L/\mathsf{k}_K)$ and $g \in \mathcal{O}_K[X]$ a monic polynomial such that $\overline{g} \in \mathsf{k}_K[X]$ is the minimal polynomial of α over k_K. Then $\overline{g}$ is separable and irreducible, hence g is separable and irreducible in $\mathcal{O}_K[X]$ (and thus irreducible in $K[X]$ by 2.5.2). Let $a_0 \in \mathcal{O}_L$ be such that $\overline{a}_0 = \alpha$, and set $M_0 = K(a_0)$. Then we obtain $M_0 \in \mathcal{Z}_{\mathsf{u}}(L/K)$ by 5.6.3.2, $a_0 \in \mathcal{O}_L \cap M_0 = \mathcal{O}_{M_0}$, $g \in \mathcal{O}_{M_0}[X]$, and by 5.4.3.1 there exists a unique $a \in \mathcal{O}_{M_0}$ such that $g(a) = 0$ and $\overline{a} = \alpha$. We set $M = K(a) \in \mathcal{Z}_{\mathsf{u}}(L/K)$. Then $\overline{a} = \alpha \in \mathsf{k}_M$ implies $\mathsf{m} \subset \mathsf{k}_M$, and equality follows since $[\mathsf{m}:\mathsf{k}_K] = \partial(\overline{g}) = \partial(g) = [M:K] = [\mathsf{k}_M:\mathsf{k}_K]$.

Now let $E \in \mathcal{Z}_{\mathsf{u}}(L/K)$ be such that $\mathsf{m} \subset \mathsf{k}_E$, let $b \in \mathcal{O}_E$ be such that $\overline{b} = \alpha$ and $E_0 = K(b)$. Then $b \in \mathcal{O}_L \cap E_0 = \mathcal{O}_{E_0}$, $g \in \mathcal{O}_{E_0}[X]$, and by 5.4.3.1 there exists a unique $a_0 \in \mathcal{O}_{E_0}$ such that $g(a_0) = 0$ and $\overline{a}_0 = \alpha$. If $E_1 = ME_0$, then $b \in \mathcal{O}_{E_1}$, $g \in \mathcal{O}_{E_1}[X]$, and by 5.4.3.1 there exists a unique $a_1 \in \mathcal{O}_{E_1}$ such that $g(a_1) = 0$ and $\overline{a}_1 = \alpha$. It follows that $a = a_1 = a_0$, and as a consequence we obtain $M = K(a) = K(a_0) \subset E_0 = K(b) \subset E$.

If M/K is Galois, then g splits in $M[X]$ and thus in $\mathcal{O}_M[X]$ since $\mathcal{O}_M$ is integrally closed. Hence $\overline{g}$ splits in $\mathsf{k}_M[X] = \mathsf{m}[X]$, and therefore m is a splitting field of $\overline{g}$. Consequently, m/k_K is Galois.

If m/k_K is Galois, then $\overline{g}$ splits in $\mathsf{m}[X] = \mathsf{k}_M[X]$, hence g splits in $\mathcal{O}_M[X]$ by 5.4.5.2, M is a splitting field of g, and therefore M/K is Galois.

3. Let T be the compositum of all fields in $\mathcal{Z}_{\mathsf{u}}(L/K)$. Then T/K is unramified by 5.6.4.2. If M is any over K unramified intermediate field of L/K and

$x \in M$, then $K(x) \in \mathcal{Z}_{\mathsf{u}}(L/K)$ and thus $K(x) \subset T$. Hence T is the largest over K unramified intermediate field of L/K. In particular, T/K and $\mathsf{k}_T/\mathsf{k}_K$ are separable, and we must prove that k_T is the separable closure of k_K in k_L.

Thus let $\alpha \in \mathsf{k}_L$ be separable over k_K. By 2. there exists an intermediate field $M \in \mathcal{Z}_{\mathsf{u}}(L/K)$ such that $\mathsf{k}_M = \mathsf{k}_K(\alpha)$, and $M \subset T$ implies $\alpha \in \mathsf{k}_M \subset \mathsf{k}_T$.

4. Let $\varphi \in \mathsf{k}_K[X]$ be irreducible and $\alpha \in \mathsf{k}_L$ such that $\varphi(\alpha) = 0$. Let $a \in \mathcal{O}_L$ be such that $\alpha = \overline{a}$, and let $g \in \mathcal{O}_K[X]$ be the minimal polynomial of a over K. Since $\overline{g}(\alpha) = \overline{g(a)} = \overline{0}$, we get $\varphi \,|\, \overline{g}$. As L/K is normal, g splits in $L[X]$, hence in $\mathcal{O}_L[X]$, and therefore $\overline{g}$ and thus φ splits in $\mathsf{k}_L[X]$. Hence $\mathsf{k}_L/\mathsf{k}_K$ is normal, and $\mathsf{k}_T/\mathsf{k}_K$ is Galois by 1.4.8.4.

Since T/K is separable by 5.6.4.1, it suffices to prove that T/K is normal. Let $\overline{L}$ be an algebraic closure of L, $\varphi \in \mathrm{Hom}^K(T,\overline{L})$ and $\overline{\varphi} \in \mathrm{Hom}^K(L,\overline{L})$ and extension of φ. Then $\varphi(L) = L$, and $\varphi \restriction L\colon L \to L$ is a K-isomorphism of valued fields by 5.5.3.1. Hence $\varphi(T)/K$ is unramified, $\varphi(T) \subset T$, and thus equality holds. Therefore T/K is normal. Applied to $L = \overline{K}$, it follows that k_{K_∞} is a separable closure of k_K.

5. T/K is Galois by 4., and evidently $\rho_{L/K}$ is a homomorphism. If $\sigma \in G$, then $\rho_{L/K}(\sigma) = \overline{\sigma} \restriction \mathsf{k}_T = \overline{\sigma \restriction T}$, and therefore $\mathrm{Gal}(L/T) \subset \mathrm{Ker}(\rho_{L/K})$.

Since k_T is the separable closure of k_K in k_L, $[T:K] = [\mathsf{k}_T:\mathsf{k}_K]$, and the restriction $\overline{\sigma} \mapsto \overline{\sigma} \restriction \mathsf{k}_T$ defines an isomorphism $\mathrm{Gal}(\mathsf{k}_L/\mathsf{k}_K) \overset{\sim}{\to} \mathrm{Gal}(\mathsf{k}_T/\mathsf{k}_K)$. We identify: $\mathrm{Gal}(\mathsf{k}_L/\mathsf{k}_K) = \mathrm{Gal}(\mathsf{k}_T/\mathsf{k}_K)$. Now it follows by 2.5.4.2 that $\rho_{L/K}$ is an epimorphism. Since

$$|\mathrm{Ker}(\rho_{L/K})| = \frac{|G|}{|\mathrm{Gal}(\mathsf{k}_T/\mathsf{k}_K)|} = \frac{[L:K]}{[T:K]} = [L:T] = |\mathrm{Gal}(L/T)|,$$

we obtain $\mathrm{Ker}(\rho_{L/K}) = \mathrm{Gal}(L/T)$.

We identify $G/\mathrm{Gal}(L/T) = \mathrm{Gal}(T/K)$ by means of $\sigma\mathrm{Gal}(L/T) = \sigma \restriction T$. Then $\rho_{L/K}$ induces an isomorphism

$$\rho_{T/K}\colon \mathrm{Gal}(T/K) = G/\mathrm{Gal}(L/K) \overset{\sim}{\to} \mathrm{Gal}(\mathsf{k}_T/\mathsf{k}_K)$$

as asserted. □

The assertion in 5.6.5.5 remains true for infinite field extensions if it is stated appropriately using the infinite Galois theory. We shall do this in the announced forthcoming volume on class field theory.

We continue with supplementary statements and consequences of 5.6.5 in special situations.

Corollary 5.6.6. *Let $K = (K, |\cdot|)$ be a complete discrete valued field, L/K a finite field extension and T the inertia field of L/K. Then*

$$[L:K] = e(L/K)f(L/K), [T:K] = [\mathsf{k}_L:\mathsf{k}_K]_{\mathsf{s}}, [L:T] = e(L/K)[\mathsf{k}_L:\mathsf{k}_K]_{\mathsf{i}},$$

and if $\mathsf{k}_L/\mathsf{k}_K$ is separable, then L/T is fully ramified.

Proof. The equality $[L\!:\!K] = e(L/K)f(L/K)$ holds by 5.5.6.1. As k_T is the separable closure of k_K in k_L (by 5.6.5.3) and $e(T/K) = 1$ it follows that

$$[T\!:\!K] = f(T/K) = [\mathsf{k}_T\!:\!\mathsf{k}_K] = [\mathsf{k}_L\!:\!\mathsf{k}_K]_\mathsf{s}$$

and

$$[L\!:\!T] = \frac{[L\!:\!K]}{[T\!:\!K]} = \frac{e(L/K)f(L/K)}{[\mathsf{k}_L\!:\!\mathsf{k}_K]_\mathsf{s}} = e(L/K)[\mathsf{k}_L\!:\!\mathsf{k}_K]_\mathsf{i}.$$

If $\mathsf{k}_L/\mathsf{k}_K$ is separable, then $[L\!:\!T] = e(L/K) = e(L/T)$, and therefore L/T is fully ramified. □

Definition and Theorem 5.6.7. *Let K be a complete non-Archimedian valued field with finite residue class field $\mathsf{k} = \mathsf{k}_K = \mathbb{F}_q$ for some prime power q. Let $\overline{K}$ be an algebraic closure of K and K_∞ the maximal unramified extension of K in $\overline{K}$* (see 5.6.5.3). *Then $\overline{\mathsf{k}} = \mathsf{k}_{\overline{K}} = \mathsf{k}_{K_\infty}$ is an algebraic closure of k, and for $x \in \mathcal{O}_{K_\infty}$ we set $\overline{x} = x + \mathfrak{p}_{K_\infty} \in \overline{\mathsf{k}}$. Let $\phi_\mathsf{k} \in \mathrm{Gal}(\overline{\mathsf{k}}/\mathsf{k})$ be the Frobenius automorphism over k, defined by $\phi_\mathsf{k}(\xi) = \xi^q$ for all $\xi \in \overline{\mathsf{k}}$, and for $n \in \mathbb{N}$ let k_n be the unique intermediate field of $\overline{\mathsf{k}}/\mathsf{k}$ such that $[\mathsf{k}_n\!:\!\mathsf{k}] = n$* (see 1.7.1).

1. *For every $n \in \mathbb{N}$ there exists a unique over K unramified intermediate field K_n of $\overline{K}/K$ such that $[K_n\!:\!K] = n$. It has the following properties:*

(a) *$K_n \subset K_\infty$, $\mathsf{k}_{K_n} = \mathsf{k}_n$ and K_n/K is cyclic.*
For $\sigma \in \mathrm{Gal}(K_n/K)$ we denote by $\overline{\sigma} \in \mathrm{Gal}(\mathsf{k}_n/\mathsf{k})$ the residue class automorphism defined by $\overline{\sigma}(\overline{x}) = \overline{\sigma(x)}$ for all $x \in \mathcal{O}_{K_n}$.
Then $\sigma \mapsto \overline{\sigma}$ defines an isomorphism $\mathrm{Gal}(K_n/K) \overset{\sim}{\to} \mathrm{Gal}(\mathsf{k}_n/\mathsf{k})$, *and there exists a unique automorphism $\varphi_{K_n/K} \in \mathrm{Gal}(K_n/K)$ satisfying $\varphi_{K_n/K}(x) \equiv x^q \bmod \mathfrak{p}_{K_n}$ for all $x \in \mathcal{O}_{K_n}$ [i. e., $\overline{\varphi_{K_n/K}} = \phi_\mathsf{k} \restriction \mathsf{k}_n$].*
$\varphi_{K_n/K}$ is called the **Frobenius automorphism** of K_n/K.
If $(K, |\cdot|)$ is discrete, then $\varphi_{K_n/K} = (\mathfrak{p}_{K_n}, K_n/K)$ is the Frobenius automorphism of $\mathfrak{p}_{K_n}$ over K (as defined in 2.14.5).

(b) *$K_n = K^{(q^n-1)} = K(\zeta)$ for some primitive (q^n-1)-th root of unity ζ* (see 1.7.5.1), *$\mathsf{k}_{K_n}^\times = \langle\overline{\zeta}\rangle$, $\mathrm{Gal}(K_n/K) = \langle\varphi_{K_n/K}\rangle$, and $\varphi_{K_n/K}(\xi) = \xi^q$ for every $\xi \in \mu_{q^n-1}(K_n)$.*

(c) *If $m, n \in \mathbb{N}$, then $K_m \subset K_n$ if and only if $m \mid n$, and*

$$K_\infty = \bigcup_{n\in\mathbb{N}} K_n,$$

For $\sigma \in \mathrm{Gal}(K_\infty/K)$ we denote by $\overline{\sigma} \in \mathrm{Gal}(\overline{\mathsf{k}}/\mathsf{k})$ the residue class automorphism defined by $\overline{\sigma}(\overline{x}) = \overline{\sigma(x)}$ for all $x \in \mathcal{O}_{K_\infty}$.
If $\sigma \in \mathrm{Gal}(K_\infty/K)$ and $n \in \mathbb{N}$, then $\overline{\sigma} \restriction \mathsf{k}_n = \overline{\sigma \restriction K_n}$.

2. There exists a unique automorphism $\varphi_K \in \mathrm{Gal}(K_\infty/K)$ *such that* $\varphi_K \upharpoonright K_n = \varphi_{K_n/K}$ *for all* $n \in \mathbb{N}$ [*equivalently,* $\phi_{\mathsf{k}} = \overline{\varphi_K}$].
It satisfies $\varphi_K(x) \equiv x^q \mod \mathfrak{p}_{K_\infty}$ *for all* $x \in \mathcal{O}_{K_\infty}$, *and* $\varphi_{K_n} = \varphi_K^n$ *for all* $n \in \mathbb{N}$.
φ_K is called the **Frobenius automorphism** over K.

3. Let $\mu' = \{\xi \in \mu(\overline{K}) \mid (\mathrm{ord}(\xi), q) = 1\}$ *be the group of all roots of unity in* $\overline{K}$ *of an order coprime to* q. *Then we have* $K_\infty = K(\mu')$, $\varphi_K(\xi) = \xi^q$ *for all* $\xi \in \mu'$, *and* $|\mu' \cap K_n| = q^n - 1$ *for all* $n \in \mathbb{N}$.

4. Let $m \in \mathbb{N}$ *be such that* $(q, m) = 1$, *let* $f \in \mathbb{N}$ *be minimal such that* $q^f \equiv 1 \mod m$, *and let* $\zeta_m \in \overline{K}$ *be a primitive* m*-th root of unity.*
Then $K(\zeta_m)/K$ *is unramified,* $[K(\zeta_m){:}K] = f$, $\mathcal{O}_{K(\zeta_m)} = \mathcal{O}_K[\zeta_m]$, *and* $\mathrm{Gal}(K(\zeta_m)/K) = \langle \varphi_K \upharpoonright K(\zeta_m)\rangle$.

Proof. As k_K is perfect, 5.6.5.3 implies $\mathsf{k}_{K_\infty} = \overline{\mathsf{k}}_K$.

1. If L is an over K unramified intermediate field of $\overline{K}/K$, then $L \subset K_\infty$ by the very definition of K_∞, and by 5.6.5.2 there exists a unique intermediate field K_n of K_∞/K such that $\mathsf{k}_{K_n} = \mathsf{k}_n$. It has the following properties:

$[K_n : K] = n$, K_n/K is Galois, and by 5.6.5.5 (with $L = T = K_n$) the assignment $\sigma \mapsto \overline{\sigma}$ defines an isomorphism $\mathrm{Gal}(K_n/K) \to \mathrm{Gal}(\mathsf{k}_n/\mathsf{k})$.

Let $\varphi_{K_n/K} \in \mathrm{Gal}(K_n/K)$ be the unique automorphism satisfying $\overline{\varphi_{K_n/K}} = \phi_n$. Then $\mathrm{Gal}(K_n/K) = \langle \varphi_{K_n/K}\rangle$, and $\varphi_{K_n/K}(x) \equiv x^q \mod \mathfrak{p}_{K_n}$ for all $x \in \mathcal{O}_{K_n}$.

By 5.4.3.2(b) the residue class epimorphism $\mathcal{O}_{K_n} \to \mathsf{k}_n$ induces an isomorphism $\mu_{q^n-1}(K_n) \overset{\sim}{\to} \mu_{q^n-1}(\mathsf{k}_n) = \mathsf{k}_n^\times$. If $\mu_{q^n-1} = \langle \zeta \rangle$ for some primitive (q^n-1)-th root of unity ζ, then $\mathsf{k}_n^\times = \langle \overline{\zeta} \rangle$, hence $\mathsf{k}_n = \mathsf{k}_K(\overline{\zeta})$, and thus $K_n = K(\zeta)$ by 5.6.3.1. If $\xi \in \mu_{q^n-1}(K_n)$ is any (q^n-1)-th root of unity, then $\varphi_{K_n/K}(\xi) \in \mu_{q^n-1}(K_n)$ satisfies $\varphi_{K_n/K}(\xi) \equiv \xi^q \mod \mathfrak{p}_{K_n}$, and therefore $\varphi_{K_n/K}(\xi) = \xi^q$.

If $m, n \in \mathbb{N}$, then 5.6.5.2 implies that $K_m \subset K_n$ if and only if $\mathsf{k}_m \subset \mathsf{k}_n$, and by 1.7.1.3 this holds if and only if $m \mid n$. If $x \in K_\infty$ and $[K(x){:}K] = n$, then $K(x) = K_n$, and therefore K_∞ is the union of $(K_n)_{n\in\mathbb{N}}$. If $\sigma \in \mathrm{Gal}(K_\infty/K)$, $n \in \mathbb{N}$ and $x \in \mathcal{O}_{K_n}$, then

$$(\overline{\sigma} \upharpoonright \mathsf{k}_n)(\overline{x}) = \overline{\sigma}(\overline{x}) = \overline{\sigma(x)} = \overline{(\sigma \upharpoonright K_n)(x)} = \overline{\sigma \upharpoonright K_n}(\overline{x}),$$

and consequently $\overline{\sigma} \upharpoonright \mathsf{k}_n = \overline{\sigma \upharpoonright K_n}$.

If $(K, |\cdot|)$ is discrete, then $\varphi \upharpoonright K_n$ is the Frobenius automorphism of $\mathfrak{p}_{K_n}$ over K by its very definiton (see 2.14.5).

2. If $m, n \in \mathbb{N}$ and $K_m \subset K_n$, then we assert that $\varphi_{K_n/K} \upharpoonright K_m = \varphi_{K_m/K}$. Indeed, $\overline{\varphi_{K_n/K} \upharpoonright K_m} = \overline{\varphi_{K_n/K}} \upharpoonright \mathsf{k}_m = \phi_{\mathsf{k}} \upharpoonright \mathsf{k}_m = \overline{\varphi_{K_m/K}}$, and therefore we obtain $\varphi_{K_n/K} \upharpoonright K_m = \varphi_{K_m/K}$ by 1.(a). As K_∞ is the union of $\{K_n \mid n \in \mathbb{N}\}$, there exists a unique automorphism $\varphi_K \in \mathrm{Gal}(K_\infty/K)$ satisfying $\varphi_K \upharpoonright K_n = \varphi_{K_n/K}$ for all $n \in \mathbb{N}$.

Next we prove the equivalence of the defining properties for φ_K using 1.(a).

$$\begin{aligned}\varphi\restriction K_n = \varphi_{K_n/K} \text{ for all } n\in\mathbb{N} \\ \iff \overline{\varphi}\restriction \mathsf{k}_n = \overline{\varphi\restriction K_n} = \overline{\varphi_{K_n/K}} = \phi_{\mathsf{k}}\restriction \mathsf{k}_n \text{ for all } n\in\mathbb{N} \\ \iff \overline{\varphi} = \phi_{\mathsf{k}}.\end{aligned}$$

If $x \in \mathcal{O}_{K_\infty}$, then $x\in\mathcal{O}_{K_n}$ for some $n\in\mathbb{N}$, and consequently

$$\varphi_K(x) - x^q = \varphi_{K_n/K}(x) - x^q \in \mathfrak{p}_{K_n} \subset \mathfrak{p}_{K_\infty} \quad \text{implies } \varphi_K(x) \equiv x^q \mod \mathfrak{p}_{K_\infty}.$$

We postpone the proof of $\varphi_{K_n} = \varphi_K^n$ until we have proved 3.

3. Let $\zeta\in\mu'$ and $\operatorname{ord}(\zeta) = m$. Then $(m,q) = 1$, and there exists some $f\in\mathbb{N}$ such that $q^f\equiv 1 \mod m$. Then ζ is a (q^f-1)-th root of unity, hence $\zeta\in K_f\subset K_\infty$ by 1.(b), and therefore $K(\mu')\subset K_\infty$. Conversely, if $a\in K_\infty$, then there is some $n\in\mathbb{N}$ such that $a\in K_n = K(\zeta)\subset K(\mu')$ (where ζ is a primitive (q^n-1)-th root of unity).

If $\xi\in\mu'$ and $n\in\mathbb{N}$ is such that $\xi\in K_n$, then $\varphi_K(\xi) = \varphi_{K_n/K}(\xi) = \xi^q$. By 5.4.3.3 we obtain $|\mu'\cap K_n| = |\mathsf{k}_n^\times| = q^n-1$.

Now we can prove that $\varphi_{K_n} = \varphi_K^n$ for $n\in\mathbb{N}$. If $\xi\in\mu'$, then $|\mathsf{k}_n| = q^n$ implies $\varphi_{K_n}(\xi) = \xi^{q^n} = \varphi_K^n(\xi)$, and as $K_\infty = K(\mu')$, we get $\varphi_{K_n} = \varphi_K^n$.

4. Let $L = K(\zeta_m)$. Then $L\subset K_\infty$, and consequently $L\subset K_n = K^{(q^n-1)}$ for every $n\in\mathbb{N}$ such that $m\,|\,q^n-1$ by 3. Therefore $[L:K] = f$ if $f\in\mathbb{N}$ is the smallest integer satisfying $q^f\equiv 1 \mod m$. By 5.4.3.3 it follows that $\overline{\zeta}_m = \zeta+\mathfrak{p}_L\in\mathsf{k}_L$ is a primitive m-th root of unity in k_L, and $[\mathsf{k}_K(\overline{\zeta}_m):\mathsf{k}_K] = f$ by 1.7.5.4. Hence $\mathsf{k}_L = \mathsf{k}_K(\overline{\zeta}_m)$ and therefore $\mathcal{O}_L = \mathcal{O}_K[\zeta_m]$ by 5.6.3.1.

Finally, 1.(b) and 2. yield $\operatorname{Gal}(K(\zeta_m)/K) = \langle\varphi_{K(\zeta_m)/K}\rangle = \langle\varphi_K\restriction K(\zeta_m)\rangle$. □

The Galois-theoretic significance of φ_K will be disclosed by the infinite Galois theory to be developed in the forthcoming volume on class field theory.

Corollary 5.6.8. *Let K be a complete non-Archimedian valued field and L/K a finite unramified field extension.*

1. *Let L/K be Galois, $G = \operatorname{Gal}(L/K)$, and for $\sigma\in G$ we denote by $\overline{\sigma}\in\operatorname{Gal}(\mathsf{k}_L/\mathsf{k}_K)$ be the residue class automorphism induced by σ. Then the assignment $\sigma\mapsto\overline{\sigma}$ defines an isomorphism $G\overset{\sim}{\to}\operatorname{Gal}(\mathsf{k}_L/\mathsf{k}_K)$, and if $c\in\mathcal{O}_L$, then*

$$\mathsf{Tr}_{L/K}(c)+\mathfrak{p}_K = \mathsf{Tr}_{\mathsf{k}_L/\mathsf{k}_K}(c+\mathfrak{p}_L) \text{ and } \mathsf{N}_{L/K}(c)+\mathfrak{p}_K = \mathsf{N}_{\mathsf{k}_L/\mathsf{k}_K}(c+\mathfrak{p}_L).$$

2. *If K is a local field, $|\mathsf{k}_K| = q<\infty$ and $[L:K] = n$, then L/K is (up to isomorphisms) the unique unramified extensions of degree n, L/K is cyclic, and $\operatorname{Gal}(L/K) = \langle\varphi_{L/K}\rangle$, where $\varphi_{L/K}$ is the Frobenius automorphism of L/K satisfying $\varphi_{L/K}(x)\equiv x^q \mod \mathfrak{p}_L$ for all $x\in\mathcal{O}_L$* (see 5.6.7.1(a)).

Proof. 1. By 5.6.5, $L = T$ is the inertia field of L/K, and the assignment $\sigma \mapsto \overline{\sigma}$ defines an isomorphism $G \overset{\sim}{\to} \mathrm{Gal}(\mathsf{k}_L/\mathsf{k}_K)$. If $c \in \mathcal{O}_L$, then

$$\mathsf{N}_{L/K}(c)+\mathfrak{p}_K = \mathsf{N}_{L/K}(c)+\mathfrak{p}_L = \prod_{\sigma\in G} \sigma(c)+\mathfrak{p}_L = \prod_{\sigma\in G} \overline{\sigma}(c+\mathfrak{p}_L) = \mathsf{N}_{\mathsf{k}_L/\mathsf{k}_K}(c+\mathfrak{p}_L),$$

and alike for the trace.

2. If $L \subset \overline{K}$ for some algebraic closure $\overline{K}$ of K, then the assertions follow by 5.6.7.1 with $L = K_n$. □

Theorem 5.6.9. *Let K be a complete discrete valued field and L/K be a finite Galois unramified field extension.*

1. $\mathsf{N}_{L/K}U_L^{(n)} = U_K^{(n)}$ *for all* $n \geq 1$.
2. *The residue class map* $a \mapsto a + \mathfrak{p}_K$ *induces an isomorphism*

$$U_K/\mathsf{N}_{L/K}U_L \overset{\sim}{\to} \mathsf{k}_K^\times/\mathsf{N}_{\mathsf{k}_L/\mathsf{k}_K}\mathsf{k}_L^\times.$$

3. *Let K be a local field, $[L\colon K] = f$ and $\pi \in \mathcal{O}_K$ a prime element. Then $U_K = \mathsf{N}_{L/K}U_L$, $K^\times/\mathsf{N}_{L/K}L^\times = \langle \pi \mathsf{N}_{L/K}L^\times\rangle$, and the map*

$$\phi\colon K^\times/\mathsf{N}_{L/K}L^\times \overset{\sim}{\to} \mathbb{Z}/f\mathbb{Z}, \ \ \text{defined by} \ \ \phi(a\mathsf{N}_{L/K}L^\times) = v_K(a)+f\mathbb{Z},$$

is an isomorphism.

Proof. Let $G = \mathrm{Gal}(L/K)$ and $\pi \in \mathcal{O}_K$ a prime element. As L/K is unramified, we obtain $v_L = v_K$, $\mathfrak{p}_L = \pi\mathcal{O}_L$, and $\mathsf{k}_L/\mathsf{k}_K$ is separable.

1. Let $n \geq 1$ and $u \in U_L^{(n)}$. Then $u = 1 + \pi^n y$ for some $y \in \mathcal{O}_L$,

$$\mathsf{N}_{L/K}(u) = \prod_{\sigma\in G}(1 + \pi^n\sigma(y)) \equiv 1 + \pi^n\mathsf{Tr}_{L/K}(y) \mod \mathfrak{p}_L^{2n},$$

and since $\mathfrak{p}_L^{2n} \cap K = \mathfrak{p}_K^{2n}$, it follows that $\mathsf{N}_{L/K}(u) \equiv 1+\pi^n\mathsf{Tr}_{L/K}(y) \mod \mathfrak{p}_K^{2n}$, hence $\mathsf{N}_{L/K}(u) \in U_K^{(n)}$, and therefore $\mathsf{N}_{L/K}U_L^{(n)} \subset U_K^{(n)}$. Next we prove:

A. $U_K^{(i)} \subset \mathsf{N}_{L/K}(U_L^{(i)})U_K^{(i+1)}$ for all $i \geq 1$.

Proof of **A.** Let $i \geq 1$ and $u = 1 + \pi^i a \in U_K^{(i)}$, where $a \in \mathcal{O}_K$. Since $\mathsf{Tr}_{\mathsf{k}_L/\mathsf{k}_K}(\mathsf{k}_L) = \mathsf{k}_K$, there exists some $c \in \mathcal{O}_L$ such that

$$a + \mathfrak{p}_K = \mathsf{Tr}_{\mathsf{k}_L/\mathsf{k}_K}(c + \mathfrak{p}_L) = \mathsf{Tr}_{L/K}(c) + \mathfrak{p}_K \ \text{(see 5.6.8.1)}.$$

Hence $1 + \pi^i c \in U_L^{(i)}$, $\mathsf{Tr}_{L/K}(c) \equiv a \mod \mathfrak{p}_K$,

$$\mathsf{N}_{L/K}(1 + \pi^i c) \equiv 1 + \pi^i\mathsf{Tr}_{L/K}(c) \equiv 1 + \pi^i a \mod \mathfrak{p}_K^{i+1},$$

and $u = 1 + \pi^i a = \mathsf{N}_{L/K}(1 + \pi^i c)v \in \mathsf{N}_{L/K}(U_L^{(i)})U_K^{(i+1)}$ for some $v \in U_K^{(i+1)}$. □[**A.**]

Now we can prove that $U_K^{(n)} \subset \mathsf{N}_{L/K}U_L^{(n)}$. Let $a \in U_K^{(n)}$. We construct recursively two sequences $(u_\nu)_{\nu\ge 0}$ and $(a_\nu)_{\nu\ge 0}$ such that

$$u_\nu \in U_L^{(n+\nu)}, \quad a_\nu \in U_K^{(n+\nu)} \quad \text{and} \quad a = \mathsf{N}_{L/K}(u_0u_1\cdot\ldots u_{\nu-1})a_\nu \quad \text{for all} \quad \nu \ge 0.$$

Let $a_0 = a$, $u_0 = 1$, and for $\nu \ge 0$ assume that $a = \mathsf{N}_{L/K}(u_0u_1\cdot\ldots\cdot u_{\nu-1})a_\nu$, where $a_\nu \in U_K^{(n+\nu)}$ and $u_i \in U_L^{(n+i)}$ for all $i \in [0, \nu-1]$. By **A** there exist $u_\nu \in U_L^{(n+\nu)}$ and $a_{\nu+1} \in U_K^{(n+\nu+1)}$ such that $a_\nu = \mathsf{N}_{L/K}(u_\nu)a_{\nu+1}$, and thus $a = \mathsf{N}_{L/K}(u_0\cdot\ldots\cdot u_\nu)a_{\nu+1}$.

Since $u_0\cdot\ldots\cdot u_\nu \in U_L^{(n)}$,

$$u_0\cdot\ldots\cdot u_\nu - u_0\cdot\ldots\cdot u_{\nu-1} = u_0\cdot\ldots\cdot u_{\nu-1}(u_\nu - 1) \in \mathfrak{p}_L^{n+\nu} \text{ for all } \nu \ge 1$$

and $U_L^{(n)}$ is closed in L, the sequence $(u_0\cdot\ldots\cdot u_\nu)_{\nu\ge 0}$ converges, and

$$u = \lim_{\nu\to\infty} u_0\cdot\ldots\cdot u_\nu \in U_L^{(n)}.$$

Since $(a_\nu)_{\nu\ge 0} \to 1$ and the map $\mathsf{N}_{L/K}\colon L \to K$ is continuous (see 5.5.3.1) it follows that $a = \mathsf{N}_{L/K}(u)$.

2. The residue class map induces an epimorphism $\psi\colon U_K \to \mathsf{k}_K^\times/\mathsf{N}_{\mathsf{k}_L/\mathsf{k}_K}\mathsf{k}_L^\times$, and it suffices to prove that $\operatorname{Ker}(\psi) = \mathsf{N}_{L/K}(U_L)$. If $c \in U_L$, then 5.6.8.1 implies $\mathsf{N}_{L/K}(c) + \mathfrak{p}_K = \mathsf{N}_{\mathsf{k}_L/\mathsf{k}_K}(c + \mathfrak{p}_L) \in \mathsf{N}_{\mathsf{k}_L/\mathsf{k}_K}\mathsf{k}_L^\times$ and consequently $\mathsf{N}_{L/K}U_L \subset \operatorname{Ker}(\psi)$. Conversely, if $u \in \operatorname{Ker}(\psi)$, and $c \in U_L$ is such that

$$u + \mathfrak{p}_K = \mathsf{N}_{\mathsf{k}_L/\mathsf{k}_K}(c + \mathfrak{p}_L) = \mathsf{N}_{L/K}(c) + \mathfrak{p}_K,$$

then $u = \mathsf{N}_{L/K}(c)u_1$ for some $u_1 \in U_K^{(1)}$, and as $U_K^{(1)} \subset \mathsf{N}_{L/K}U_L^{(1)} \subset \mathsf{N}_{L/K}U_L$ by 1., it follows that $u \in \mathsf{N}_{L/K}U_L$.

3. If k_K is finite, then $\mathsf{N}_{\mathsf{k}_L/\mathsf{k}_K}\mathsf{k}_L^\times = \mathsf{k}_K^\times$ by 1.7.1.2, and thus $\mathsf{N}_{L/K}U_L = U_K$ by 2. The map $\phi_0\colon K^\times \to \mathbb{Z}/f\mathbb{Z}$, defined by $\phi_0(x) = v_K(x) + f\mathbb{Z}$, is an epimorphism, and

$$\operatorname{Ker}(\phi_0) = \langle\pi^f\rangle\cdot U_K = \langle\mathsf{N}_{L/K}(\pi)\rangle\cdot\mathsf{N}_{L/K}U_L = \mathsf{N}_{L/K}(\langle\pi\rangle\cdot U_L) = \mathsf{N}_{L/K}L^\times.$$

Hence ϕ_0 induces an isomorphism as asserted. □

5.7 Ramified field extensions

We continue to use the notation from 5.3.1 *and* 5.5.1.

Let K be a discrete valued field. A monic polynomial

$$f = X^n + a_{n-1}X^{n-1} + \ldots + a_1X + a_0 \in \mathcal{O}_K[X]$$

is called an **Eisenstein polynomial** of K if $v_K(a_i) \ge 1$ for all $i \in [1, n-1]$ and $v_K(a_0) = 1$ (then f is a $\mathfrak{p}_K$-Eisenstein polynomial according to 1.1.11).

Theorem 5.7.1. *Let K be a complete discrete valued field and L/K a finite field extension.*

1. *Let $g \in \mathcal{O}_K[X]$ be an Eisenstein polynomial of K, and suppose that $L = K(\Pi)$, where $g(\Pi) = 0$. Then g is irreducible in $K[X]$, $[L:K] = \partial(g)$, L/K is fully ramified and $v_L(\Pi) = 1$.*

2. *Let L/K be fully ramified, let $\Pi \in \mathcal{O}_L$ be a prime element and $g \in \mathcal{O}_K[X]$ the minimal polynomial of Π over K. Then g is an Eisenstein polynomial of K, $L = K(\Pi)$ and $\mathcal{O}_L = \mathcal{O}_K[\Pi]$.*

Proof. By 2.12.8, applied with $\mathfrak{o} = \mathcal{O}_K$ and $\mathfrak{O} = \mathcal{O}_L$. □

We proceed with the investigation of finite separable extensions of fields of formal Laurent series over perfect fields. In this case, the splitting up into an unramified and a fully ramified extension as in 5.6.6 has an extremely simple form.

Theorem 5.7.2. *Let $K = F((X))$ be the field of formal Laurent series in X over a perfect field F, $\mathsf{o} = \mathsf{o}_X \colon K \to \mathbb{Z} \cup \{\infty\}$ the order valuation of K with respect to X and $x \in K$ such that $\mathsf{o}(x) = 1$ (then $K = F((x))$ and $\mathsf{o} = \mathsf{o}_x$,* see 5.3.7).

1. *Let $t \in K$, $\mathsf{o}(t) = m \ge 1$ and $F((t)) \subset K$ the field of formal Laurent series in t* (see 5.3.7.3(b)). *Then $K/F((t))$ is fully ramified, and $[K:F((t))] = m$.*

2. *Let K'/K be a finite separable field extension, o' the discrete valuation of K' and $x' \in K'$ such that $\mathsf{o}'(x') = 1$. Then there exists an intermediate field F' of K'/F such that*

$$K = F((x)) \subset F'((x)) \subset F'((x')) = K',$$

$F'((x))/F((x))$ is unramified of degree $f = [F':F]$, $K'/F'((x))$ is fully ramified of degree $m = \mathsf{o}'(x)$, and $x = x'^m w$ for some power series $w \in F'[\![x']\!]^\times$. In particular, $F'((x))$ is the inertia field of K'/K.

If F is algebraically closed, then (by a suitable choice of x') we may achieve $w \equiv 1 \bmod x'F[\![x']\!]$, and if additionally $\mathrm{char}(F) \nmid m$, then we may even achieve $w = 1$.

Proof. 1. By 5.3.7.2(b) we get $\mathsf{o} \restriction F((t)) = m\mathsf{o}_t$, hence $e(K/F((t))) = m$, and since F is the residue class field of both K and $F((t))$, we get $f(K/F((t))) = 1$. Therefore 5.5.6.1 implies $[K:F((t))] = e(K/F((t))) = m$.

2. By 5.5.3.2 K' is a complete discrete valued field. Let o' be its discrete valuation and $x' \in K'$ such that $\mathsf{o}'(x') = 1$. The residue class extension $\mathsf{k}_{K'}/F$ of K'/K is finite separable, and $\mathrm{char}(K') = \mathrm{char}(\mathsf{k}_{K'})$. Therefore K' possesses a coefficient field F' by 5.4.10, and we may assume that $K' = F'(\!(x')\!)$ and $\mathsf{k}_{K'} = F'$. Since x is a common prime element of the valuation domains $F[\![x]\!] = \mathcal{O}_{F(\!(x)\!)}$ and $F'[\![x]\!] = \mathcal{O}_{F'(\!(x)\!)}$ and the residue class extension is separable, it follows that $F'(\!(x)\!)/F(\!(x)\!)$ is unramified. Since $\mathsf{k}_{K'} = F' = \mathsf{k}_{F'(\!(x)\!)}$, we obtain $f(K'/F'(\!(x)\!)) = 1$, and as $e(K'/F'(\!(x)\!)) = \mathsf{o}'(x) = m$, it follows that $m = [K' : F'(\!(x)\!)]$. Hence $K'/F'(\!(x)\!)]$ is fully ramified of degree m, and $x = x'^m w$ for some $w \in \mathcal{O}_{K'}^\times = F'[\![x']\!]^\times$.

Assume now that F is algebraically closed. Then $F = F'$, and we set $w = w_0 w_1$, where $w_0 \in F^\times$, $w_1 \in F[\![x']\!]^\times$ and $w_1 \equiv 1 \mod x'F[\![x']\!]$. If $y \in F$ is such that $y^m = w_0$ and $x_1 = x'y$, then it follows that $\mathsf{o}'(x_1) = \mathsf{o}'(x') = 1$, $K' = F(\!(x_1)\!)$ and $x = x'^m w = (x_1 y^{-1})^m y^m w_1 = x_1^m w_1$. We may replace x' with x_1 and consequently w with w_1.

Assume additionally that $\mathrm{char}(F) \nmid m$. Then $X^m - w_0 \in F[X]$ is the residue class polynomial of $X^m - w \in F[\![x']\!][X]$, and $X^m - w_0$ has in F simple zeros. By 5.4.3.1 there exists some $y \in F[\![x']\!]$ such that $y^m = w$. If $x_1 = x'y$, then $\mathsf{o}'(x_1) = \mathsf{o}'(x') = 1$, $K' = F(\!(x_1)\!)$ and $x = x_1^m$. We may replace x' with x_1 and consequently w with 1. □

The following shifting result for ramifications yields a new proof and a far-reaching generalization of 2.13.5.3.

Theorem 5.7.3 (Shifting theorem for the ramification). *Let $K \subset \overline{K}$ be fields, let L and K' be over K finite intermediate fields of $\overline{K}/K$ and $L' = LK'$.*

1. *Let $|\cdot|\colon K \to \mathbb{R}_{\ge 0}$ be a discrete absolute value of K such that $(K, |\cdot|)$ is complete and k_K is perfect. Then $e(L'/K') \le e(L/K)$.*

2. *Let $\mathfrak{o}$ be a Dedekind domain such that $K = \mathsf{q}(\mathfrak{o})$, $\mathfrak{o}' = \mathrm{cl}_{K'}(\mathfrak{o})$, $\mathfrak{O} = \mathrm{cl}_L(\mathfrak{o})$, $\mathfrak{O}' = \mathrm{cl}_{L'}(\mathfrak{o})$ and $\mathfrak{p} \in \mathcal{P}(\mathfrak{o})$ be such that $\mathsf{k}_\mathfrak{p}$ is perfect. Let $\mathfrak{p}' \in \mathcal{P}(\mathfrak{o}')$ such that $\mathfrak{p}' \cap K = \mathfrak{p}$, $\mathfrak{P}' \in \mathcal{P}(\mathfrak{O}')$ such that $\mathfrak{P}' \cap K' = \mathfrak{p}'$ and $\mathfrak{P}' \cap L = \mathfrak{P}$.*

   Then $\mathfrak{P} \cap K = \mathfrak{p}$ and $e(\mathfrak{P}'/\mathfrak{p}') \le e(\mathfrak{P}/\mathfrak{p})$. In particular, if $\mathfrak{p}$ is unramified in L, then $\mathfrak{p}'$ is unramified in L'.

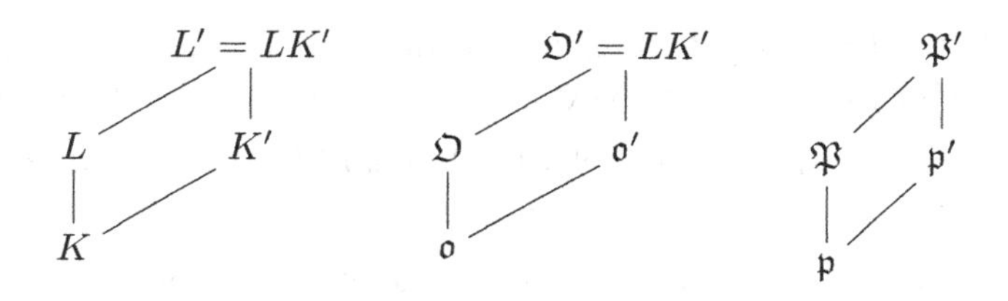

Proof. 1. Let T be the inertia field of L/K and T' the inertia field of L'/K' (see 5.6.5). By 5.6.4.1 the field extension TK'/K' is unramified, hence $TK' \subset T'$ and $e(L'/K') = [L':T'] \le [L':K'T] \le [L:T] = e(L/K)$ by 5.6.6.

2. By definition, $\mathfrak{P} \cap K = \mathfrak{P}' \cap L \cap K = \mathfrak{P}' \cap K' \cap K = \mathfrak{p}' \cap K = \mathfrak{p}$. We choose completions such that $L_{\mathfrak{P}} \subset L'_{\mathfrak{P}'}$, $K'_{\mathfrak{p}'} \subset L'_{\mathfrak{P}'}$ and $K_{\mathfrak{p}} \subset L_{\mathfrak{P}} \cap K'_{\mathfrak{p}'}$. Then we obtain $K \subset L' = LK' \subset L_{\mathfrak{P}}K'_{\mathfrak{p}'} \subset L'_{\mathfrak{P}'}$, hence $L_{\mathfrak{P}}K'_{\mathfrak{p}'}$ is dense in $L'_{\mathfrak{P}'}$, and as $[L_{\mathfrak{P}}K'_{\mathfrak{p}'} : K'_{\mathfrak{p}'}] < \infty$, it follows that $L_{\mathfrak{P}}K'_{\mathfrak{p}'}$ is complete and thus closed in $L'_{\mathfrak{P}'}$. Therefore $L'_{\mathfrak{P}'} = L_{\mathfrak{P}}K'_{\mathfrak{p}'}$, and by 5.5.9.2 and 1. it follows that $e(\mathfrak{P}'/\mathfrak{p}') = e(L'_{\mathfrak{P}'}/K'_{\mathfrak{p}'}) \le e(L_{\mathfrak{P}}/K_{\mathfrak{p}}) = e(\mathfrak{P}/\mathfrak{p})$.

If $\mathfrak{p}$ is unramified in L, then $e(\widetilde{\mathfrak{P}}/\mathfrak{p}) = 1$ for all $\widetilde{\mathfrak{P}} \in \mathcal{P}(\mathfrak{O})$ satisfying $\widetilde{\mathfrak{P}} \cap K = \mathfrak{p}$. Then $e(\widetilde{\mathfrak{P}}'/\mathfrak{p}') = 1$ for all $\widetilde{\mathfrak{P}}' \in \mathcal{P}(\mathfrak{O}')$ satisfying $\widetilde{\mathfrak{P}}' \cap K' = \mathfrak{p}'$, and thus $\mathfrak{p}'$ is unramified in L'. □

Definitions and Remarks 5.7.4. Let $L = (L, |\cdot|)$ be a non-Archimedian valued field and K a subfield of L such that L/K is algebraic. Then $|\cdot|\restriction K$ is an absolute value of K by 5.1.7.2(d), and $e = e(L/K) = (|L^\times| : |K^\times|)$.

1. Let $[L:K] = n < \infty$ (then $e \le n$ by 5.5.2.1). The field extension L/K is called **tamely ramified** if $\mathrm{char}(\mathsf{k}_K) \nmid e$ and $\mathsf{k}_L/\mathsf{k}_K$ is separable. If M is an intermediate field of L/K, then L/K is tamely ramified if and only if both L/M and M/K are tamely ramified.

2. Let L/K be arbitrary. Then L/K is called **tamely ramified** if E/K is tamely ramified for all over K finite intermediate fields of L/K. Otherwise L/K is called **wildly ramified**.

If L/K is unramified, then L/K is tamely ramified. If $\mathrm{char}(\mathsf{k}_K) = 0$, then every algebraic field extension L/K is tamely ramified.

3. Let $\widehat{K}$ be a completion of K and $\widehat{L}$ a completion of L such that $\widehat{K} \subset \widehat{L}$. Then $e = e(\widehat{L}/\widehat{K})$, $\mathsf{k}_K = \mathsf{k}_{\widehat{K}}$ and $\mathsf{k}_L = \mathsf{k}_{\widehat{L}}$. Hence L/K is tamely ramified if and only if $\widehat{L}/\widehat{K}$ is tamely ramified.

4. Let $(L, |\cdot|)$ be discrete. Then $(K, |\cdot|)$ is also discrete, $e = e(\mathfrak{p}_L/\mathfrak{p}_K)$, $\mathsf{k}_K = \mathsf{k}_{\mathfrak{p}_K}$ and $\mathsf{k}_L = \mathsf{k}_{\mathfrak{p}_L}$. Hence L/K is tamely ramified if and only if $\mathfrak{p}_L/\mathfrak{p}_K$ is tamely ramified (see 2.12.1).

Theorem 5.7.5 (Characterizing tamely ramified field extensions). *Let K be a complete discrete valued field and L/K a finite field extension.*

1. Let L/K be fully and tamely ramified, and $e = [L : K]$. Then $\mathrm{char}(\mathsf{k}_K) \nmid e$, and $L = K(\sqrt[e]{\pi})$ for some prime element $\pi \in \mathcal{O}_K$.

2. Let $b \in K^\times$, $e \in \mathbb{N}$, $\mathrm{char}(\mathsf{k}_K) \nmid e$ and $L = K(\beta)$, where $\beta^e = b$. Then L/K is tamely ramified.

Proof. Let $\pi_K \in \mathcal{O}_K$ and $\pi_L \in \mathcal{O}_L$ be prime elements, and for a polynomial $P \in \mathcal{O}_L[X]$ let $\overline{P} \in \mathsf{k}_L[X]$ be the residue class polynomial.

1. As $e = e(L/K)$, we get $f = 1$, $\operatorname{char}(\mathsf{k}_K) \nmid e$ and $\pi_L^e = \pi_K u$ for some $u \in U_L$. Since $\mathsf{k}_K = \mathsf{k}_L$, there exists some $u_0 \in U_K$ such that $u \equiv u_0 \mod \mathfrak{p}_L$. It follows that $u_0^{-1}u \in U_L^{(1)}$ and therefore $u = u_0 c^e$ for some $c \in U_L^{(1)}$ by 5.4.3.2(a). Hence $L = K(\sqrt[e]{\pi_K u_0 c^e}\,) = K(\sqrt[e]{\pi}\,)$, where $\pi = \pi_K u_0 \in \mathcal{O}_K$ is a prime element.

2. Let $\pi \in \mathcal{O}_K$ be a prime element and $b = \pi^a u$ for some $a \in \mathbb{Z}$ and $u \in U_K$. We set $g = X^e - u \in \mathcal{O}_K[X]$ and $K_1 = K(x)$, where $g(x) = 0$. Since $\overline{g} \in \mathsf{k}_K[X]$ is separable, the field extension K_1/K is unramified by 5.6.3.2. Let $K' = K_1^{(e)} = K_1(\zeta)$, where ζ is a primitive e-th root of unity. By 5.6.7.4 the field extension K'/K_1 is unramified. Hence K'/K is unramified, π is a prime element of $\mathcal{O}_{K'}$, and $K'(\sqrt[e]{\pi}\,)/K'$ is fully ramified of degree $e = [K'(\sqrt[e]{\pi}\,){:}K']$ by 5.7.1.1, since $X^e - \pi \in K'[X]$ is an Eisenstein polynomial. As $\operatorname{char}(\mathsf{k}_K) \nmid e$, the field extension $K'(\sqrt[e]{\pi}\,)/K'$ and thus also $K'(\sqrt[e]{\pi}\,)/K$ is tamely ramified. Since $\beta \in K(\zeta, x, \sqrt[e]{\pi}\,) = K'(\sqrt[e]{\pi}\,)$, the field extension $K(\beta)/K$ is tamely ramified, too. □

Definition and Remarks 5.7.6. Let K be a complete discrete valued field. For an algebraic field extension L/K let $\mathcal{E}(L/K)$ be the set of all over K finite intermediate fields of L/K. Then the group

$$\mathcal{N}(L/K) = \bigcap_{E \in \mathcal{E}(L/K} \mathsf{N}_{E/K}E^\times \subset K^\times$$

is called the **group of universal norms** of L/K. If $\mathcal{E}' \subset \mathcal{E}(L/K)$ is a directed subset such that L is the compositum of $\mathcal{E}'$, then

$$\mathcal{N}(L/K) = \bigcap_{E \in \mathcal{E}'} \mathsf{N}_{E/K}E^\times .$$

If $[L{:}K] < \infty$, then $\mathcal{N}(L/K) = \mathsf{N}_{L/K}L^\times$.

Theorem 5.7.7. *Let K be a non-Archimedian local field and L/K an algebraic field extension. Then L/K is fully ramified if and only if $\mathcal{N}(L/K)$ contains a prime element of $\mathcal{O}_K$.*

Proof. Let $\mathcal{E}(L/K)$ be the set of all over K finite intermediate fields of L/K.

If L/K is not fully ramified, then E/K is not fully ramified for some intermediate field $E \in \mathcal{E}(L/K)$. It follows that $f(E/K) > 1$ by 5.6.2.2, and consequently $v_K(\mathsf{N}_{L/K}(x)) = f(E/K)v_L(x) \ne 1$ for all $x \in E^\times$. Hence $\mathsf{N}_{E/K}E^\times$ and all the more $\mathsf{N}_{L/K}L^\times$ contains no prime element of $\mathcal{O}_K$.

Let now L/K be fully ramified, and for $E \in \mathcal{E}(L/K)$ let $P(E)$ be the set of all prime elements of $\mathcal{O}_E$. If $E \in \mathcal{E}(L/K)$ and $a \in E^\times$, then it follows that $v_K(\mathsf{N}_{E/K}(a)) = v_E(a)$, and therefore $\mathsf{N}_{E/K}E^\times \cap P(K) = \mathsf{N}_{E/K}(P(E))$. Consequently, we must prove that

$$\mathcal{N}(L/K) \cap P(K) = \bigcap_{E \in \mathcal{E}(L/K)} \mathsf{N}_{E/K}(P(E)) \ne \emptyset.$$

If $E \in \mathcal{E}(L/K)$ and $\pi_E \in \mathcal{O}_E$ is a prime element then $P(E) = \pi_E U_E$ is compact since E is local by 5.5.3.2, and as $\mathsf{N}_{E/K}$ is continuous, the set $\mathsf{N}_{E/K}(P(E))$ is compact and thus a closed subset of the compact set $P(K)$. Hence it suffices to prove that the set $\{\mathsf{N}_{E/K}P(E) \mid E \in \mathcal{E}(L/K)\}$ has the finite intersection property.

If $E, E' \in \mathcal{E}(L/K)$ and $E \subset E'$, then E'/E is fully ramified, hence we get $\mathsf{N}_{E'/K}(P(E')) = \mathsf{N}_{E/K}(\mathsf{N}_{E'/E}(P(E'))) \subset \mathsf{N}_{E/K}(P(E))$. If $m \in \mathbb{N}$, $E_1, \ldots, E_m \in \mathcal{E}(L/K)$, and $E = E_1 \cdot \ldots \cdot E_m$, then $E \in \mathcal{E}(L/K)$ and $\emptyset \neq \mathsf{N}_{E/K}(P(E)) \subset \mathsf{N}_{E_1/K}(P(E_1)) \cap \ldots \cap \mathsf{N}_{E_m/K}(P(E_m))$. This is the required finite intersection property. □

5.8 Non-Archimedian valued fields 2

We continue to use the notation of the preceding sections.

In this section we expand the structure theory of non-Archimedian valued and in particular of $\mathfrak{p}$-adic fields.

Theorem 5.8.1. *Let $K = (K, |\cdot|)$ be a complete non-Archimedian valued field and $\overline{K}$ an algebraic closure of K.*

1. (Krasner's lemma) *Let $\alpha \in \overline{K}$ be separable over K, and let $\alpha = \alpha_1, \ldots, \alpha_n$ be its conjugates over K. If $\beta \in \overline{K}$ is such that $|\beta - \alpha| < |\alpha_i - \alpha|$ for all $i \in [2, n]$, then $\alpha \in K(\beta)$.*

2. *Let $f \in K[X]$ be monic and separable, $\partial(f) = n$, and let $\alpha_1, \ldots, \alpha_n \in \overline{K}$ be the zeros of f. Then there exists some $\delta \in \mathbb{R}_{>0}$ such that every polynomial $g \in K[X]$ satisfying $\partial(g) = n$ and $|f - g| < \delta$ has the following properties:*

   (a) *g is separable and possesses zeros $\beta_1, \ldots, \beta_n \in \overline{K}$ such that $K(\beta_i) = K(\alpha_i)$ for all $i \in [1, n]$, and $|\beta_i - \alpha_i| < |\alpha_k - \alpha_i|$ for all i, $k \in [1, n]$ with $i \neq k$.*

   (b) *g and f have in $K[X]$ the same factorization type, i.e., for every $m \in \mathbb{N}$ both f and g have the same number of irreducible factors of degree m.*
   In particular, g is irreducible if and only if f is irreducible.

Proof. 1. Assume to the contrary that $\alpha \notin K(\beta)$. Then there exists some $i \in [2, n]$ such that α and α_i are conjugate over $K(\beta)$. Hence $\beta - \alpha$ and $\beta - \alpha_i$ are also conjugate over $K(\beta)$. It follows that $|\beta - \alpha| = |\beta - \alpha_i|$, and

$$|\alpha_i - \alpha| = |(\beta - \alpha) - (\beta - \alpha_i)| \leq |\beta - \alpha| < |\alpha_i - \alpha|, \quad \text{a contradiction.}$$

2. As f is monic, it follows that $\alpha_i \in \mathcal{O}_{K(\alpha_i)}$, hence $|\alpha_i| \le 1$, and therefore $|f'(\alpha_i)| \le |f| \le 1$ for all $i \in [1,n]$. Let $\delta \in \mathbb{R}_{>0}$ be such that $\delta < |f'(\alpha_i)|^2$ for all $i \in [1,n]$, and let $g \in K[X]$ be such that $|f-g| < \delta$.

Since $|f-g| < \delta < |f'(\alpha_i)|^2 \le |f|$, we obtain $|g| = |f-(f-g)| = |f| \le 1$ and thus $g \in \mathcal{O}_K[X]$. Since

$$|f'(\alpha_i) - g'(\alpha_i)| \le |f'-g'| \le |f-g| < \delta < |f'(\alpha_i)|,$$

it follows that $|g'(\alpha_i)| = |f'(\alpha_i) - (f'(\alpha_i) - g'(\alpha_i)| = |f'(\alpha_i)|$ and

$$|g(\alpha_i)| = |g(\alpha_i) - f(\alpha_i)| \le |g-f| < \delta < |f'(\alpha_i)|^2 = |g'(\alpha_i)|^2.$$

By 5.4.2 there exists some $\beta_i \in \mathcal{O}_{K(\alpha_i)}$ such that

$$g(\beta_i) = 0 \quad \text{and} \quad |\beta_i - \alpha_i| \le \frac{|g(\alpha_i)|}{|g'(\alpha_i)|} < \frac{\delta}{|f'(\alpha_i)|} < |f'(\alpha_i)| = \prod_{\substack{j=1\\ j\ne i}}^{n} |\alpha_j - \alpha_i|.$$

If $k,\ i \in [1,n]$ and $k \ne i$, then

$$|\alpha_j - \alpha_i| \le \max\{|\alpha_j|, |\alpha_i\} \le 1 \text{ for all } j \in [1,n] \setminus \{i\},$$

and consequently

$$|\beta_i - \alpha_i| < |\alpha_k - \alpha_i| \prod_{\substack{j=1\\ j\ne i,k}}^{n} |\alpha_j - \alpha_i| \le |\alpha_k - \alpha_i|.$$

In particular, if $i,\ j \in [1,n]$ and $i \ne j$, then $|\beta_j - \alpha_j| < |\alpha_i - \alpha_j|$, and

$$|\beta_j - \alpha_i| = (|(\beta_j - \alpha_j) - (\alpha_i - \alpha_j)| = |\alpha_i - \alpha_j| > |\beta_i - \alpha_i|, \text{ hence } \beta_i \ne \beta_j,$$

and g is separable. By 1. we obtain $\alpha_i \in K(\beta_i)$ and thus $K(\alpha_i) = K(\beta_i)$ since $\beta_i \in K(\alpha_i)$ for all $i \in [1,n]$. Hence (a) follows.

For $m \in \mathbb{N}$ let $A_m(f)$ resp. $A_m(g)$ be the number of irreducible factors of f resp. g of degree m in $K[X]$. Then

$$\begin{aligned} A_m(f) &= \frac{|\{i \in [1,n] \mid [K(\alpha_i){:}K] = m\}|}{m} \\ &= \frac{|\{i \in [1,n] \mid [K(\beta_i){:}K] = m\}|}{m} = A_m(g). \end{aligned}$$

Hence (b) follows. □

Theorem 5.8.2 (Completion and algebraic closure). *Let $K = (K, |\cdot|)$ be an algebraically closed non-Archimedian valued field and $\widehat{K}$ a completion of K. Then $\widehat{K}$ is algebraically closed, too.*

Proof. It suffices to prove:

A. Every monic irreducible separable polynomial $f \in \widehat{K}[X]$ has a zero in $\widehat{K}$.

B. $\widehat{K}$ is perfect.

Proof of **A.** Let $f \in \widehat{K}[X]$ be monic, irreducible and separable, $\widehat{K}^*$ an algebraic closure of $\widehat{K}$ and $\alpha \in \widehat{K}^*$ a zero of f. By 5.8.1.2 there exists some $\delta \in \mathbb{R}_{>0}$ with the following property: Every polynomial $g \in \widehat{K}[X]$ such that $|g - f| < \delta$ has a zero $\beta \in \widehat{K}^*$ satisfying $\widehat{K}(\alpha) = \widehat{K}(\beta)$. As K is dense in $\widehat{K}$, there exists some polynomial $g \in K[X]$ such that $|g - f| < \delta$. Then g has a zero $\beta \in \widehat{K}^*$ such that $\widehat{K}(\beta) = \widehat{K}(\alpha)$, and since K is algebraically closed, it follows that $\beta \in K$ and thus $\alpha \in \widehat{K}$. $\square$[**A.**]

Proof of **B.** We may assume that $\text{char}(K) = p > 0$, and by 1.4.3.1(b) we must prove that $\widehat{K} = \widehat{K}^p$. Thus let $a \in \widehat{K}^\times$, and choose $b,\, t \in K^\times$ such that $|b^p a| \le 1$ and $|t| < 1$. For $n \ge 1$ let $f_n = X^p + t^n X - b^p a \in \mathcal{O}_{\widehat{K}}[X]$. Then $f_n' = t^n \in K^\times$, hence f_n is separable. By **A** and since $\mathcal{O}_{\widehat{K}}$ is integrally closed, there exists some $\alpha_n \in \mathcal{O}_{\widehat{K}}$ such that $f_n(\alpha_n) = 0$. If $n \ge 1$, then we obtain $|\alpha_n^p - b^p a| = |t|^n |\alpha_n| \le |t|^n$, hence $(\alpha_n^p)_{n\ge 1} \to b^p a$. Since

$$|\alpha_{n+1} - \alpha_n| = |\alpha_{n+1}^p - \alpha_n^p|^{1/p} \le \max\{|\alpha_{n+1}^p - b^p a|, |\alpha_n^p - b^p a|\}^{1/p} \le |t|^{n/p},$$

it follows that $(\alpha_n)_{n\ge 1}$ is a Cauchy sequence in $\widehat{K}$, and if $(\alpha_n)_{n\ge 1} \to \alpha \in \widehat{K}$, then $(\alpha_n^p)_{n\ge 1} \to \alpha^p$, hence $\alpha^p = b^p a$, and $a = (b^{-1}\alpha)^p \in \widehat{K}^p$. $\square$[**B.**] $\square$

Theorem 5.8.3 (Non-Archimedian local fields)**.** *Let K be a field.*

1. *Let* $\text{char}(K) = 0$.

(a) K is a non-Archimedian local field if and only if it is (up to isomorphisms) a finite extension of $\mathbb{Q}_p$ for some prime p.
In this case we always assume that $\mathbb{Q}_p \subset K$.

(b) If K is a finite extension of $\mathbb{Q}_p$ for some prime p, then there exists an algebraic number field L and a prime ideal $\mathfrak{p} \in \mathcal{P}_L$ such that $\mathfrak{p} \,|\, p$ and $K \cong L_\mathfrak{p}$.
In this case, K is called a **$\mathfrak{p}$-adic number field**.

2. *Let* $\text{char}(K) = p > 0$. *Then K is a local field if and only if there exists a finite field F such that $K \cong F((X))$ (and then $F \cong \mathsf{k}_K$).*
In this case we always assume that $K = F((X))$, $v_K = \mathsf{o}_X$ and $\mathsf{k}_K = F$ (see 5.4.8).

Proof. 1. (a) $\mathbb{Q}_p$ is a local field, and by 5.5.3.2 every finite extension of a local field is again a local field.

Now let $K = (K, |\cdot|)$ be a local field, $\mathbb{Q} \subset K$, $\operatorname{char}(\mathsf{k}_K) = p > 0$ and $|\mathsf{k}_K| = p^f$ for some $f \in \mathbb{N}$. Then $|p| < 1$, $|\cdot| \restriction \mathbb{Q}$ is equivalent to $|\cdot|_p$, the (topological) closure of $\mathbb{Q}$ in K is isomorphic to $\mathbb{Q}_p$, and we may assume that $\mathbb{Q}_p \subset K$. Then $|\mathsf{k}_K| = p^f$ implies $f = f(K/\mathbb{Q}_p) < \infty$, and $e = v_K(p) = e(K/\mathbb{Q}_p)$. By 5.5.6.1 it follows that $[K:\mathbb{Q}_p] = ef < \infty$.

(b) By (a) we may assume that $K = \mathbb{Q}_p(\alpha) = (K, |\cdot|_p)$, where α is algebraic over $\mathbb{Q}_p$, and we denote by $g \in \mathbb{Q}_p[X]$ the minimal polynomial of α over $\mathbb{Q}_p$. By 5.8.1.2 there exist a monic polynomial $h \in \mathbb{Q}[X]$ which is irreducible in $\mathbb{Q}_p[X]$ and a zero $\beta \in K$ of h such that $K = \mathbb{Q}_p(\beta)$. Then $L = \mathbb{Q}(\beta)$ is an algebraic number field, L is dense in K, and $|\cdot|_p \restriction K_0$ is a $\mathfrak{p}$-adic absolute value for some $\mathfrak{p} \in \mathcal{P}_L$ by 5.5.10.2. Hence $K = L_\mathfrak{p}$.

2. If F is a finite field, then $F((X))$ is a local field by 5.3.7. Conversely, if K is a local field, then 5.4.10 implies that $K = F((X))$ for some finite field F. □

Next we investigate explicitly the structure of $\mathfrak{p}$-adic number fields (i. e., of finite extensions of $\mathbb{Q}_p$). Throughout, we denote by $\overline{\mathbb{Q}}_p$ an algebraic closure of $\mathbb{Q}_p$ and assume that all finite extensions of $\mathbb{Q}_p$ are inside $\overline{\mathbb{Q}}_p$.

Theorem 5.8.4. *Let p be a prime, $K/\mathbb{Q}_p$ a finite extension, $f = f(K/\mathbb{Q}_p)$ and $e = v_K(p) = e(K/\mathbb{Q}_p)$.*

1. Let $K/\mathbb{Q}_p$ be unramified (hence $e = 1$, $f = [K:\mathbb{Q}_p]$, $|\mathsf{k}_K| = p^f$, $v_K \restriction \mathbb{Q}_p = v_p$ and $\mathfrak{p}_K = p\mathcal{O}_K$).

(a) Let $r \in \mathbb{N}$, $p^r > 2$ and $K^{(p^r)} = K(\zeta)$, where $\zeta \in \overline{\mathbb{Q}}_p$ is a primitive p^r-th root of unity. Then $[K^{(p^r)}:K] = p^{r-1}(p-1)$, $K^{(p^r)}/K$ is fully ramified, $|\mu(K^{(p^r)})| = p^r(p^f-1)$,

$$v_{K^{(p^r)}}(1-\zeta) = 1 \text{ and } \mathsf{N}_{K^{(p^r)}/K}(1-\zeta) = p.$$

(b) Let $p > 2$, $\zeta \in \overline{\mathbb{Q}}_p$ be a primitive p-th root of unity, and let $K^{(p)} = K(\zeta)$. Then $-p = [(1-\zeta)u]^{p-1}$, where $u \in U_{K^{(p)}}$, and $K^{(p)} = K(\sqrt[p-1]{-p})$.

(c) If $\mu(K)$ denotes the group of roots of unity in K, then

$$|\mu(K)| = \begin{cases} p^f - 1 & \text{if } p \neq 2, \\ 2(2^f-1) & \text{if } p = 2, \end{cases} \quad \text{and} \quad |\mu(\mathbb{Q}_p)| = \max\{2, p-1\}.$$

2. Let T be the inertia field of $K/\mathbb{Q}_p$ (see 5.6.5). *Then $T/\mathbb{Q}_p$ is cyclic and unramified, $[T:\mathbb{Q}_p] = f$, $\mathsf{k}_T = \mathsf{k}_K = \mathbb{F}_{p^f}$, K/T is fully ramified, and $[K:T] = e$.*

3. $\mu_{p^\infty}(K) \subset U_K^{(1)}$. If $m \in \mathbb{N}$ and $m(p-1) > e$, then

$$\mu_{p^\infty}(K) \cap U_K^{(m)} = \mathbf{1}.$$

4. Let $\pi \in \mathcal{O}_K$ be a prime element and $p = \pi^e \varepsilon$, where $\varepsilon \in U_K$. Then K contains a primitive p-th root of unity if and only if $p-1 \mid e$ and $-\varepsilon \in U_K^{p-1}$.

Proof. For every extension field L of K we denote by $\mu'(L)$ the group of roots of unity of order coprime to p in L (then $\mu(L) = \mu_{p^\infty}(L) \cdot \mu'(L)$).

1. (a) Let $L = K^{(p^r)} = K(\zeta)$. Then ζ is a zero of the polynomial

$$g = \frac{X^{p^r} - 1}{X^{p^{r-1}} - 1} = \sum_{j=0}^{p-1} X^{p^{r-1}j} \in \mathbb{Z}[X],$$

$\zeta - 1$ is a zero of $g_1 = g(X+1) \in \mathbb{Z}[X]$, and $[(X+1)^{p^{r-1}} - 1]g_1 = (X+1)^{p^r} - 1$ implies $g_1 X^{p^{r-1}} \equiv X^{p^r} \bmod p\mathbb{Z}[X]$. Hence $g_1 \equiv X^{p^{r-1}(p-1)} \bmod p\mathcal{O}_K[X]$, and g_1 is an Eisenstein polynomial of K, since $g_1(0) = g(1) = p$. Hence g_1 is irreducible, $[L:K] = \partial(g_1) = \partial(g) = p^{r-1}(p-1)$, L/K is fully ramified, and $v_L(\zeta - 1) = 1$. By 1.6.2.1 we obtain $\mathsf{N}_{L/K}(1-\zeta) = (-1)^{[L:K]}p = p$.

As $|\mathsf{k}_L| = |\mathsf{k}_K| = p^f$, it follows by 5.4.3.3 that $|\mu'(L)| = |\mu'(K)| = p^f - 1$. If $m \in \mathbb{N}$ and $\xi \in L$ is a primitive p^m-th root of unity, then it follows that $[K(\xi):K] = p^{m-1}(p-1)$, hence $m \le r$, and

$$|\mu(K^{(p^r)})| = |\mu_{p^\infty}(K^{(p^r)})|\,|\mu'(K^{(p^r)})| = p^r(p^f - 1).$$

(b) Let $L = K^{(p)} = K(\zeta)$. Then

$$p = \mathsf{N}_{L/K}(1-\zeta) = \prod_{j=1}^{p-1}(1-\zeta^j) = (1-\zeta)^{p-1}\eta,$$

where

$$\eta = \prod_{j=2}^{p-1} \frac{1-\zeta^j}{1-\zeta} = \prod_{j=2}^{p-1}(1+\zeta+\ldots+\zeta^{j-1}) \equiv (p-1)! \equiv -1 \mod \mathfrak{p}_L,$$

since $\zeta \equiv 1 \mod \mathfrak{p}_L$, and $(p-1)! \equiv -1 \mod p$ by Wilson's theorem. By 5.4.3.2(a) there exists some $u \in U_L$ such that $-\eta = u^{p-1} \in U_L^{p-1}$, hence $-p = (1-\zeta)^{p-1}u^{p-1}$ and therefore $K(\sqrt[p-1]{-p}) \subset K^{(p)}$ (note that K contains a primitive $(p-1)$-th root of unity). Since $\sqrt[p-1]{-p}$ is a zero of the Eisenstein polynomial $X^{p-1} + p \in K[X]$, 5.7.1.1 implies $[K(\sqrt[p-1]{-p}) : K] = p - 1 = [K^{(p)}:K]$, and therefore $K^{(p)} = K(\sqrt[p-1]{-p})$.

(c) We have seen in (a) that $|\mu'(K)| = p^f - 1$. If $m \in \mathbb{N}$ and $p^m > 2$, then $[K^{(p^m)}:K] = p^{m-1}(p-1) > 1$ and thus $\mu_{p^\infty}(K) \subset \mu_2(K) = \{\pm 1\}$. Hence

$$|\mu(K)| = |\mu_{p^\infty(K)}|\,|\mu'(K)| = \begin{cases} p^f - 1 & \text{if } p \ne 2, \\ 2(2^f - 1) & \text{if } p = 2. \end{cases}$$

2. By definition, T is the largest over $\mathbb{Q}_p$ unramified intermediate field of $K/\mathbb{Q}_p$ and $|\mathsf{k}_T| = |\mathsf{k}_K| = p^f$. By 5.6.6, $[T:\mathbb{Q}_p] = f$, $[K:T] = e$, and K/T is fully ramified. By 5.6.8.2 the field extension $T/\mathbb{Q}_p$ is cyclic, and by 5.6.7.1(b) we obtain $T = \mathbb{Q}_p(\zeta)$, where ζ is a primitive (p^f-1)-th root of unity.

3. Let $\zeta \in \mu_{p^\infty}(K) \setminus \{1\}$ and $K_0 = T(\zeta)$. By 1. we obtain

$$v_K(\zeta - 1) = e(K/K_0)v_{K_0}(\zeta - 1) = e(K/K_0) \geq 1,$$

hence $\zeta \in U_K^{(1)}$, and $\zeta \notin U_K^{(m)}$ if $m > e(K/K_0)$. Since K_0/T is fully ramified and $[K_0:T] = p-1$ by 1., it follows that $e = e(K/K_0)(p-1)$, and therefore $\zeta \notin U_K^{(m)}$ if $m(p-1) > e$. In all we get $\mu_{p^\infty}(K) \subset U_K^{(1)}$ and $\mu_{p^\infty}(K) \cap U_K^{(m)} = \mathbf{1}$ if $m(p-1) > e$.

4. For $p = 2$ there is nothing to do. Thus let $p > 2$, $\zeta \in \overline{K}$ a primitive p-th root of unity and T the inertia field of $K/\mathbb{Q}_p$.

If $\zeta \in K$, then $\mathbb{Q}_p(\zeta) \subset K$ implies $e(\mathbb{Q}_p(\zeta)/\mathbb{Q}_p) = p-1 \,|\, e$, and by 1.(b) (applied for K/T) there exists some $u \in U_K$ such that $-p = [(1-\zeta)u]^{p-1}$. Consequently

$$-\varepsilon = \frac{-p}{\pi^e} = \left[\frac{(1-\zeta)u}{\pi^{e/(p-1)}}\right]^{p-1} \in U_K^{p-1}.$$

Conversely, suppose that $p-1 \,|\, e$ and $-\varepsilon = u^{p-1}$ for some $u \in U_K$. Then $-p = \pi^e u^{p-1} = (\pi^{e/(p-1)}u)^{p-1} \in K^{p-1}$, and $\mathbb{Q}_p(\zeta) = \mathbb{Q}_p(\sqrt[p-1]{-p}) \subset K$. $\square$

For the convenience of the reader we summarize the properties of cyclotomic fields over $\mathbb{Q}_p$.

Corollary 5.8.5. *Let p be a prime. For $m \in \mathbb{N}$, let $\zeta_m \in \overline{\mathbb{Q}}_p$ be a primitive m-th root of unity, $\mathbb{Q}_p^{(m)} = \mathbb{Q}_p(\zeta_m)$ and $\mu_m = \langle \zeta_m \rangle \subset \mathbb{Q}_p^{(m)}$.*

1. *If $p \nmid m$, then $[\mathbb{Q}_p^{(m)} : \mathbb{Q}_p] = \min\{f \in \mathbb{N} \mid p^f \equiv 1 \bmod m\}$, $\mathbb{Q}_p^{(m)}/\mathbb{Q}_p$ is unramified, $\mathcal{O}_{\mathbb{Q}_p^{(m)}} = \mathbb{Z}_p[\zeta_m]$ and $\mathrm{Gal}(\mathbb{Q}_p^{(m)}/\mathbb{Q}_p) = \langle \varphi \rangle$, where $\varphi(\xi) \equiv \xi^p \bmod p\mathbb{Z}_p[\zeta_m]$ for all $\xi \in \mu_m$.*

2. *If $r \in \mathbb{N}$, then $[\mathbb{Q}_p^{(p^r)} : \mathbb{Q}_p] = p^{r-1}(p-1)$, $\mathbb{Q}_p^{(p^r)}/\mathbb{Q}_p$ is fully ramified and $\mathcal{O}_{\mathbb{Q}_p^{(p^r)}} = \mathbb{Z}_p[\zeta_{p^r}]$.*

Proof. By 5.6.7 (3. and 4.), and 5.8.4.1. $\square$

Let K be a non-Archimedean local field and $p = \mathrm{char}(\mathsf{k}_K)$. By 5.4.9 there is a topological isomorphism $K^\times \overset{\sim}{\to} \mathbb{Z} \times \mathsf{k}_K^\times \times U_K^{(1)}$ (where $\mathbb{Z} \times \mathsf{k}_K$ has the discrete topology), and it remains to disclose the structure of $U_K^{(1)}$. For this we make $U_K^{(1)}$ into a $\mathbb{Z}_p$-module and determine its structure using the basics of the theory of $\mathfrak{p}$-adic logarithms.

Definition and Theorem 5.8.6. *Let $K = (K, |\cdot|)$ be a complete discrete valued field such that $\operatorname{char}(\mathsf{k}_K) = p > 0$, $u \in U_K^{(1)}$, $\alpha \in \mathbb{Z}_p$ and $(a_n)_{n\ge 0}$ a sequence in $\mathbb{Z}$ such that $(a_n)_{n\ge 0} \to \alpha$. Then the sequence $(u^{a_n})_{n\ge 0}$ converges in $U_K^{(1)}$, and its limit depends only on u and α.* We define

$$u^\alpha = |\cdot|_p\text{-}\lim_{n\to\infty} u^{a_n} \in U_K^{(1)}.$$

The map $\mathbb{Z}_p \times U_K^{(1)} \to U_K^{(1)}$, $(\alpha, u) \mapsto u^\alpha$, is a continuous $\mathbb{Z}_p$-module structure of $U_K^{(1)}$ satisfying $v_K(u^\alpha - 1) \ge v_p(\alpha) + v_K(u-1)$ for all $\alpha \in \mathbb{Z}_p$ and $u \in U_K^{(1)}$. In particular, if $n \in \mathbb{N}$, then $U_K^{(n)}$ is a $\mathbb{Z}_p$-submodule of $U_K^{(1)}$.

Proof. We show first:

A. If $u \in U_K^{(1)}$ and $a \in \mathbb{Z}$, then $v_K(u^a - 1) \ge v_p(a) + v_K(u-1)$.

Proof of **A.** Let $u \in U_K^{(1)}$ and $a \in \mathbb{Z}$. We may assume that $u \ne 1$ and $a \ne 0$. Let $m = v_K(u-1)$, hence $m \ge 1$ and $u \in U_K^{(m)}$, and let $n = v_p(a)$. We must prove that $u^a \in U_K^{(m+n)}$, and we use induction on n. For $n = 0$ there is nothing to do.

$n \ge 1$, $n-1 \to n$: Let $a = pc$, where $c \in \mathbb{Z}$ and $v_p(c) = n-1$. The induction hypothesis yields $u^a = (u^c)^p \in (U_K^{(n-1+m)})^p$, and we consider the isomorphism $\psi_{n-1+m} \colon U_K^{(n-1+m)}/U_K^{(n+m)} \overset{\sim}{\to} \mathsf{k}_K$ (see 2.10.5.4). By its very definition we obtain $\psi_{n-1+m}(u^a U_K^{(n+m)}) = p\psi_n(u^c U_K^{(n+m)}) = 0$, and thus $u^a \in U_K^{(n+m)}$. □[**A.**]

For $n \ge 1$, **A** implies

$$v_K(u^{a_{n+1}} - u^{a_n}) = v_K(u^{a_n}) + v_K(u^{a_{n+1}-a_n} - 1) \ge v_p(a_{n+1} - a_n).$$

Therefore $(u^{a_n})_{n\ge 0}$ is a Cauchy sequence in $U_K^{(1)}$, hence it has a limit in $U_K^{(1)}$, and we show that this limit does not depend on the sequence $(a_n)_{n\ge 0}$. Thus let $(b_n)_{n\ge 0}$ be another sequence in $\mathbb{Z}$ such that $(b_n)_{n\ge 0} \to \alpha$ in $\mathbb{Z}_p$. Then $(b_n - a_n)_{n\ge 0} \to 0$ in $\mathbb{Z}_p$, and since

$$v_K(u^{b_n} - u^{a_n}) = v_K(u^{a_n}) + v_K(u^{b_n - a_n} - 1) \ge v_p(b_n - a_n) \text{ for all } n \ge 0,$$

it follows that $(u^{b_n} - u^{a_n})_{n\ge 0} \to 0$. Hence the sequences $(u^{a_n})_{n\ge 0}$ and $(u^{b_n})_{n\ge 0}$ have the same limit

$$u^\alpha = |\cdot|\text{-}\lim_{n\to\infty} u^{a_n} = |\cdot|\text{-}\lim_{n\to\infty} u^{b_n}.$$

Next we prove:

B. If $u \in U_K^{(1)}$ and $\alpha \in \mathbb{Z}_p$, then $v_K(u^\alpha - 1) \ge v_p(\alpha) + v_K(u-1)$.

Proof of **B.** Let $u \in U_K^{(1)}$, $\alpha \in \mathbb{Z}_p$, and let $(a_n)_{n\ge 0}$ be a sequence in $\mathbb{Z}$ satisfying $(a_n)_{n\ge 0} \to \alpha$ in $\mathbb{Z}_p$. We may assume that $u \ne 1$, $\alpha \ne 0$ and $u^\alpha \ne 1$. Then $(u^{a_n} - 1)_{n\ge 0} \to u^\alpha - 1$, and for $n \gg 1$ we obtain

$$v_K(u^\alpha - 1) = v_K(u^{a_n} - 1) \ge v_p(a_n) + v_K(u-1) = v_p(\alpha) + v_K(u-1). \qquad \square[\mathbf{B}.]$$

Next we prove that the map $\mathbb{Z}_p \times U_K^{(1)} \to U_K^{(1)}$ is continuous. Let $(\alpha_n)_{n\ge 0}$ be a sequence in $\mathbb{Z}_p$ and $(u_n)_{n\ge 0}$ a sequence in $U_K^{(1)}$ such that $(\alpha_n)_{n\ge 0} \to \alpha \in \mathbb{Z}_p$ and $(u_n)_{n\ge 0} \to u \in U_K^{(1)}$. Then $u_n^{\alpha_n} - u^\alpha = u^{\alpha_n}[(u^{-1}u_n)^{\alpha_n} - 1] + u^\alpha(u^{\alpha_n - \alpha} - 1)$, and consequently

$$\begin{aligned} v_K(u_n^{\alpha_n} - u^\alpha) &\ge \min\{v_K((u^{-1}u_n)^{\alpha_n} - 1), v_K(u^{\alpha_n - \alpha} - 1)\} \\ &\ge \min\{v_p(\alpha_n) + v_K(u^{-1}u_n - 1),\ v_p(\alpha_n - \alpha) + v_K(u - 1)\} \\ &= \min\{v_p(\alpha_n) + v_K(u_n - u),\ v_p(\alpha_n - \alpha) + v_K(u - 1)\} \to \infty, \end{aligned}$$

hence $(u_n^{\alpha_n})_{n\ge 0} \to u^\alpha$. Using continuity, it is now easy to check the module axioms: If α, $\beta \in \mathbb{Z}_p$ and u, $v \in U_K^{(1)}$, then $u^{\alpha+\beta} = u^\alpha u^\beta$, $(uv)^\alpha = u^\alpha v^\alpha$ and $(u^\alpha)^\beta = u^{\alpha\beta}$.

If $n \ge 1$, $u \in U_K^{(n)}$ and $\alpha \in \mathbb{Z}_p$, then $v_K(u^\alpha - 1) \ge v_K(u-1) \ge n$, hence $u^\alpha \in U_K^{(n)}$, and thus $U_K^{(n)}$ is a $\mathbb{Z}_p$-submodule of $U_K^{(1)}$. $\square$

If $\operatorname{char}(K) = 0$, there is a close connection between the additive and the multiplicative structure of K with is (similar to the Archimedian case) provided by the p-adic exponential and the p-adic logarithm.

Theorem 5.8.7 (Convergence of power series). *Let $(K, |\cdot|)$ be a complete valued field such that $|\cdot|$ satisfies the triangle inequality, and let $(a_n)_{n\ge 0}$ be a sequence in K.*

1. *If $\sum_{n=0}^\infty |a_n| < \infty$, then the series $\sum_{n\ge 0} a_n$ converges.*
2. *Suppose that*

$$r = \frac{1}{\limsup_{n\to\infty} \sqrt[n]{|a_n|}} \in [0, \infty].$$

Then the power series

$$f(x) = \sum_{n=0}^\infty a_n x^n \quad \textit{converges in} \quad B_r(0) = \{x \in K \mid |x| < r\}$$

and defines a continuous function $f\colon B_r(0) \to K$.
For $x \in K \setminus \overline{B_r(0)}$, the power series diverges.

Proof. Literally as in complex analysis. $\square$

Definition and Theorem 5.8.8 (p-adic exponential and logarithm). *Let p be a prime, $K/\mathbb{Q}_p$ a finite field extension, $e = v_K(p) = e(K/\mathbb{Q}_p)$,*

$$\rho = \left\lfloor \frac{e}{p-1} \right\rfloor + 1.$$

and consider ther formal power series

$$L(X) = \sum_{n=1}^{\infty} \frac{(-1)^{n-1}}{n} X^n \in \mathbb{Q}[\![X]\!] \quad \textit{and} \quad E(X) = \sum_{n=0}^{\infty} \frac{1}{n!} X^n \in \mathbb{Q}[\![X]\!]$$

as defined in 5.3.8.

1. *For $x \in \mathfrak{p}_K$, the series $L(x)$ converges, and we define*

$$\log\colon U_K^{(1)} \to K \quad \textit{by} \quad \log(1+x) = L(x) \quad \textit{for all} \quad x \in \mathfrak{p}_K.$$

We call $\log = \log_{\mathfrak{p}_K}$ the **$\mathfrak{p}_K$-adic logarithm** and $\log_p = \log_{p\mathbb{Z}}\colon U_p^{(1)} \to \mathbb{Q}_p$ the **p-adic logarithm.**

(a) $\log$ *is continuous, and $v_K(\log(1+x)) = v_K(x)$ for all $x \in \mathfrak{p}_K^{\rho}$.*

(b) If $u, v \in U_K^{(1)}$ and $\alpha \in \mathbb{Z}_p$, then

$$\log(uv) = \log u + \log v \quad \textit{and} \quad \log(u^\alpha) = \alpha \log u.$$

2. *For $x \in \mathfrak{p}_K^{\rho}$, the series $E(x)$ converges, and we define*

$$\exp\colon \mathfrak{p}_K^{\rho} \to K \quad \textit{by} \quad \exp(x) = E(x).$$

We call $\exp = \exp_{\mathfrak{p}_K}$ the **$\mathfrak{p}_K$-adic exponential** and $\exp_p = \exp_{p\mathbb{Z}}$ the **p-adic exponential.**

(a) $\exp$ *is continuous, and $v_K(\exp(x) - 1) = v_K(x)$ for all $x \in \mathfrak{p}_K^{\rho}$. In particular, if $m \in \mathbb{N}$ and $m \ge \rho$, then* $\exp(\mathfrak{p}_K^m) \subset U_K^{(m)}$.

(b) If x, $y \in \mathfrak{p}_K^{\rho}$ and $\alpha \in \mathbb{Z}_p$, then

$$\exp(x+y) = \exp(x)\exp(y) \quad \textit{and} \quad \exp(\alpha x) = \exp(x)^\alpha.$$

3. *If $m \ge \rho$, then* $\exp\colon \mathfrak{p}_K^m \to U_K^{(m)}$ *and* $\log\colon U_K^{(m)} \to \mathfrak{p}_K^m$ *are mutually inverse topological $\mathbb{Z}_p$-isomorphisms.*

4. *If $u \in U_K^{(\rho)}$ and $\alpha \in \mathbb{Z}_p$, then $u^\alpha = \exp(\alpha \log u)$.*

Proof. For the proof of 1. and 2. we show first:

A. If $n \in \mathbb{N}$, then

$$v_p(n!) \le \frac{n-1}{p-1} \qquad \text{and} \qquad v_p(n) \le \min\Big\{\frac{\log n}{\log p}, \frac{n-1}{p-1}\Big\}$$

Proof of **A.** Let $n \in \mathbb{N}$ and $r \geq 0$ such that $p^r \leq n < p^{r+1}$. Then

$$v_p(n) \leq r \leq \frac{\log n}{\log p},$$

and as $v_p(n) \leq v_p(n!)$, it suffices to estimate $v_p(n!)$. Since $p^r \leq n < p^{r+1}$, we get

$$n = \sum_{i=0}^{r} a_i p^i, \quad \text{where} \quad a_0, \ldots, a_r \in [0, p-1] \ \text{ and } \ a_r \neq 0.$$

If $k \in [1, r]$, then

$$\Big\lfloor \frac{n}{p^k} \Big\rfloor = \Big\lfloor \frac{1}{p^k} \sum_{i=0}^{k-1} a_i p^i \Big\rfloor + \sum_{i=k}^{r} a_i p^{i-k} = \sum_{i=k}^{r} a_i p^{i-k},$$

and consequently

$$\begin{aligned} v_p(n!) &= \sum_{k=1}^{r} \Big\lfloor \frac{n}{p^k} \Big\rfloor = \sum_{k=1}^{r} \sum_{i=k}^{r} a_i p^{i-k} = \sum_{i=1}^{r} \sum_{k=1}^{i} a_i p^{i-k} = \frac{1}{p-1} \sum_{i=1}^{r} a_i (p^i - 1) \\ &= \frac{1}{p-1} \Big[\sum_{i=0}^{r} a_i p^i - \sum_{i=0}^{r} a_i \Big] = \frac{1}{p-1} \Big[n - \sum_{i=0}^{r} a_i \Big] \leq \frac{n-1}{p-1}. \qquad \square[\mathbf{A}.] \end{aligned}$$

1. If $x \in \mathfrak{p}_K$ and $n \in \mathbb{N}$, then

$$\begin{aligned} v_K\Big(\frac{(-1)^{n-1} x^n}{n} \Big) &= n v_K(x) - v_K(n) = n v_K(x) - e v_p(n) \\ &\geq n \Big[1 - \frac{e \log n}{n \log p} \Big] \to \infty \ \text{ for } \ n \to \infty. \end{aligned}$$

Hence $L(x)$ converges, and the map $\log \colon U_K^{(1)} \to K$ is continuous. If $x \in \mathfrak{p}_K^{\rho}$ and $n \geq 2$, the **A** implies

$$\begin{aligned} v_K\Big(\frac{(-1)^{n-1} x^n}{n} \Big) - v_K(x) &= (n-1) v_K(x) - e v_p(n) \geq (n-1)\rho - e\,\frac{n-1}{p-1} \\ &= (n-1) \Big[\rho - \frac{e}{p-1} \Big] > 0 \end{aligned}$$

and therefore

$$\begin{aligned} v_K(\log(1+x)) = v_K(L(x)) &= v_K\Big(x + \sum_{n=2}^{\infty} \frac{(-1)^{n-1} x^n}{n} x^n \Big) \\ &= \lim_{N \to \infty} v_K\Big(x + \sum_{n=2}^{N} \frac{(-1)^{n-1} x^n}{n} x^n \Big) = v_K(x). \end{aligned}$$

If $u = 1 + x \in U_K^{(1)}$ and $v = 1 + y \in U_K^1$, then

$$\log(uv) = L(x + y + xy) = L(x) + L(y) = \log u + \log v.$$

Hence $\log(u^a) = a\log u$ for all $a \in \mathbb{Z}$, and thus $\log(u^\alpha) = \alpha\log u$ for all $\alpha \in \mathbb{Z}_p$, since log is continuous.

2. If $x \in \mathfrak{p}_K^\rho$ and $n \geq 2$, then

$$v_K\Big(\frac{x^n}{n!}\Big) - v_K(x) = (n-1)v_K(x) - ev_p(n!) \geq (n-1)\Big[\rho - \frac{e}{p-1}\Big] > 0,$$

and in particular

$$\lim_{n\to\infty} v_K\Big(\frac{x^n}{n!}\Big) = \infty.$$

Hence $E(x)$ converges, and the map $\exp\colon \mathfrak{p}_K^\rho \to K$ is continuous. If $x \in \mathfrak{p}_K^\rho$, then

$$v_K(\exp(x)-1) = v_K(E(x)-1) = v_K\Big(x + \sum_{n=2}^\infty \frac{x^n}{n!}\Big)$$

$$= \lim_{N\to\infty} v_K\Big(x + \sum_{n=2}^N \frac{x^n}{n!}\Big) = v_K(x).$$

If $x,\ y \in \mathfrak{p}_K^\rho$, then $\exp(x+y) = \exp(x)\exp(y)$ by 1. Hence $\exp(ax) = \exp(x)^a$ for all $a \in \mathbb{Z}$, and as exp is continuous, it follows that $\exp(\alpha x) = \exp(x)^\alpha$ for all $\alpha \in \mathbb{Z}_p$.

3. If $x \in \mathfrak{p}_K^\rho$, then $v_K(\exp(x)-1) = v_K(\log(x+1)) = v_K(x)$, and consequently $\exp(\mathfrak{p}_K^m) \subset U_K^{(m)}$ and $\log(U_K^{(m)}) \subset \mathfrak{p}_K^m$ for all $m \geq \rho$. By 5.8.8.3 we obtain

$$\log(\exp(x)) = L(E(x)-1) = x$$

and

$$\exp(\log(x)) = E(L(x-1)) = 1 + (x-1) = x.$$

Consequently, for every $m \geq \rho$ the maps $\exp\colon \mathfrak{p}_K^m \to U_K^{(m)}$ and $\log\colon U_K^{(m)} \to \mathfrak{p}_K^m$ are mutually inverse topological $\mathbb{Z}_p$-isomorphisms.

4. If $u \in U_K^{(\rho)}$ and $\alpha \in \mathbb{Z}_p$, then $u^\alpha \in U_K^{(\rho)}$ by 5.8.6, and by 1. and 3. it follows that $\exp(\alpha\log u) = \exp\circ\log(u^\alpha) = u^\alpha$. □

From 5.8.8 we almost immediately obtain the following structural result.

Theorem 5.8.9 (Local Unit Theorem). *Let p be a prime, $K/\mathbb{Q}_p$ a finite field extension, $d = [K\colon\mathbb{Q}_p]$, $e = v_K(p) = e(K/\mathbb{Q}_p)$ and $n \in \mathbb{N}$ such that*

$$n \geq \Big\lfloor \frac{e}{p-1} \Big\rfloor + 1.$$

1. $U_K^{(n)}$ is a free $\mathbb{Z}_p$-module of rank d, $U_K^{(n)}/U_K^{(n+e)}$ is a d-dimensional vector space over $\mathbb{F}_p$, and a d-tuple $(u_1,\ldots,u_d)$ in $U_K^{(n)}$ is a $\mathbb{Z}_p$-basis of $U_K^{(n)}$ if and only if $(u_1U_K^{(n+e)},\ldots,u_dU_K^{(n+e)})$ is an $\mathbb{F}_p$-basis of $U_K^{(n)}/U_K^{(n+e)}$.

2. $U_K^{(1)} = \mu_{p^\infty}(K)\cdot V$ for some (multiplicative) free $\mathbb{Z}_p$-module V of rank d.

Proof. 1. By 5.8.8.3, $\exp\colon \mathfrak{p}_K^n \overset{\sim}{\to} U_K^{(n)}$ is a topological $\mathbb{Z}_p$-isomorphism, and by 5.5.6.1, $\mathcal{O}_K$ is a free $\mathbb{Z}_p$-module of rank d. Since $\mathfrak{p}_K^n = \pi^n\mathcal{O}_K$ for some prime element $\pi \in \mathcal{O}_K$, it follows that $\mathfrak{p}_K^n$ and thus $U_K^{(n)}$ are free $\mathbb{Z}_p$-modules of rank d, too. Since log induces an $\mathbb{F}_p$-isomorphism $U_K^{(n)}/U_K^{(n+e)} \overset{\sim}{\to} \mathfrak{p}_K^n/\mathfrak{p}_K^{n+e}$ and since $p\mathfrak{p}_K^n = \mathfrak{p}_k^{n+e}$ the assertion concerning bases follows from 2.10.11.2.

2. From the exact sequence $\mathbf{1} \to U_K^{(n)} \hookrightarrow U_K^{(1)} \to U_K^{(1)}/U_K^{(n)} \to \mathbf{1}$ and since

$$(U_K^{(1)}\!:\!U_K^{(n)}) = \prod_{j=1}^{n-1}(U_K^{(j)}\!:\!U_K^{(j+1)}) = q^{n-1} < \infty \quad \text{(by 2.10.5.4)}$$

it follows from 1. that $U_K^{(1)}$ is a finitely generated $\mathbb{Z}_p$-module of rank d. Therefore 2.3.5.2 implies that $U_K^{(1)} = T{\cdot}V$, where $V \cong \mathbb{Z}_p^d$ and T is a direct product of finite residue class rings of $\mathbb{Z}_p$. Hence T is a finite p-group, and by 5.8.4.3 we obtain $T = \mu_{p^\infty}(K) \cap U_K^{(1)} = \mu_{p^\infty}(K)$. □

In the following Theorem 5.8.10 we summarize our knowledge concerning the multiplicative structure of a non-Archimedian local field.

Theorem 5.8.10. *Let K be a non-Archimedian local field such that $\mathrm{char}(\mathsf{k}_K) = p$, $|\mathsf{k}_K| = q$, and let $\pi \in \mathcal{O}_K$ be a prime element.*

1. *$U_K = \mu_{q-1}(K) \cdot U_K^{(1)}$ and $K^\times = \langle\pi\rangle \cdot \mu_{q-1}(K) \cdot U_K^{(1)}$ are topological inner direct products if we endow the groups $\langle\pi\rangle \cong \mathbb{Z}$ and $\mu_{q-1}(K) \cong \mathbb{Z}/(q-1)\mathbb{Z}$ with the discrete topology.*

2. *$\mu_{q-1}(K)$ is the group of all roots of unity in K of some order coprime to p. In particular, if $n \in \mathbb{N}$ and $p \nmid n$, then $|\mu_n(K)| = n$ if and only if $q \equiv 1 \bmod n$.*

3. *Let $m \in \mathbb{N}$. Then*

$$U_K/U_K^n \cong \mathbb{Z}/(q-1,n)\mathbb{Z} \times U_K^{(1)}/(U_K^{(1)})^n,$$

and $K^\times/K^{\times n} \cong \mathbb{Z}/n\mathbb{Z} \times U_K/U_K^n$.

(a) If $p \nmid n$, then $U_K^{(1)} = (U_K^{(1)})^n$.

(b) If $\mathrm{char}(K) = p \,|\, n$, then $(U_K^{(1)}\!:\!(U_K^{(1)})^n) = \infty$.

(c) Let $p \,|\, n$, $\mathrm{char}(K) = 0$, $[K\!:\!\mathbb{Q}_p] = d$ and $|\mu_{p^\infty}(K)| = p^t$. Then

$$U_K^{(1)}/(U_K^{(1)})^n \cong \mathbb{Z}/(p^t,n)\mathbb{Z} \times (\mathbb{Z}/p^{v_p(n)}\mathbb{Z})^d.$$

In any case, $(K^\times\!:\!K^{\times n}) = n(U_K\!:\!U_K^n)$, and

$$(U_K\!:\!U_K^n) = \begin{cases} (q-1,n) & \text{if } p \nmid n, \\ \infty & \text{if } \mathrm{char}(K) = p \,|\, n, \\ (p^t(q-1),n)p^{dv_p(n)} & \text{if } p \,|\, n,\ \mathrm{char}(K) = 0 \text{ and } (K\!:\!\mathbb{Q}_p) = d. \end{cases}$$

4. If $n \in \mathbb{N}$ and $\operatorname{char}(K) \nmid n$, then $(U_K^{(1)})^n$ in $U_K^{(1)}$ and $K^{\times n}$ are open subgroups of $K^\times$ of finite index, and $\{(U_K^{(1)})^n \mid n \in \mathbb{N}\}$ is a fundamental system of neighborhoods of 1 in $K^\times$.

Proof. 1. holds by 5.4.9 and 2. by 5.4.3.3.

3. By 1. we obtain $U_K/U_K^n \cong \mu_{q-1}(K)/\mu_{q-1}(K)^n \times U_K^{(1)}/(U_K^{(1)})^n$, and since $\mu_{q-1}(K)$ is cyclic of order $q-1$, the group $\mu_{q-1}(K)/\mu_{q-1}(K)^n$ is isomorphic to $\mathbb{Z}/(q-1,n)\mathbb{Z}$. Again by 1., we obtain $K^\times/K^{\times n} \cong \langle\pi\rangle/\langle\pi^n\rangle \times U_K/U_K^n$, and $\langle\pi\rangle/\langle\pi^n\rangle \cong \mathbb{Z}/n\mathbb{Z}$.

(a) By 5.4.3.2(a).

(b) Assume that $\operatorname{char}(K) = p \,|\, n$ and $K = \mathbb{F}_q((X))$ for some p-power q. It suffices to prove that $(U_K^{(1)} : (U_K^{(1)})^p) = \infty$. If $\mu,\ \nu \in \mathbb{N},\ \ \mu \neq \nu$ and $p \nmid \mu\nu$, then

$$h_{\mu,\nu} = \frac{1 - X^\mu}{1 - X^\nu} = (1 - X^\mu)\sum_{i=0}^{\infty} X^{\nu i} \in U_K^{(1)} \setminus (U_K^{(1)})^p,$$

since $h'_{\mu,\nu} \neq 0$, but $h' = 0$ for all $h \in (U_K^{(1)})^p$. Hence the cosets $(1-X^\nu)(U_K^{(1)})^p$ with $\nu \in \mathbb{N}$ and $p \nmid \nu$ are distinct, and thus $(U_K^{(1)} : (U_K^{(1)})^p) = \infty$.

(c) By 5.8.9.2 we obtain $U_K^{(1)}/(U_K^{(1)})^n \cong \mu_{p^\infty}(K)/\mu_{p^\infty}(K)^n \times V/V^n$, where $V \cong \mathbb{Z}_p^d$. Since $\mu_{p^\infty}(K)$ is cyclic of order p^t, it follows that $\mu_{p^\infty}(K)/\mu_{p^\infty}(K)^n$ is isomorphic to $\mathbb{Z}/(p^t,n)\mathbb{Z}$. Moreover, $V/V^n \cong \mathbb{Z}_p^d/n\mathbb{Z}_p^d \cong (\mathbb{Z}/p^{v_p(n)}\mathbb{Z})^d$ by 5.3.6.

4. Note that $(U_K^{(1)} : (U_K^{(1)})^n) < \infty$ by 3. Since $U_K^{(1)}$ is compact, it follows that $(U_K^{(1)})^n$ is compact, hence closed and therefore open in $U_K^{(1)}$. Since $U_K^{(1)}$ is open in $K^\times$, it follows that $(U_K^{(1)})^n$ is open in $K^\times$, and consequently $K^{\times n}$ is open in $K^\times$ (proofs of these simple facts concerning topological groups are left as exercises; the announced forthcoming volume on class field theory will contain a detailed chapter on topological groups).

By 5.3.2.3(a) $\{U_K^{(m)} \mid m \in \mathbb{N}\}$ is a fundamental system of neighborhoods of 1 in $K^\times$. If $m \in \mathbb{N}$, then $(U_K^{(1)} : U_K^{(m)}) < \infty$ and thus there exists some $n \in \mathbb{N}$ such that $(U_K^{(1)})^n \subset U_K^{(m)}$. Hence $\{(U_K^{(1)})^n \mid n \in \mathbb{N}\}$ is also a fundamental system of neighborhoods of 1 in $K^\times$. □

As an important consequence of the general theory, we highlight the case $\mathbb{Q}_p$.

Theorem 5.8.11 (Structure of $\mathbb{Q}_p^\times$). *Let p be a prime, $U_p = \mathbb{Z}_p^\times$, and for $n \in \mathbb{N}$ let $U_p^{(n)} = 1 + p^n\mathbb{Z}_p$.*

1. If $p \neq 2$ and ζ_{p-1} is a primitive $(p-1)$-th root of unity, then $\mu(\mathbb{Q}_p) = \langle\zeta_{p-1}\rangle,\ \ U_p = \langle\zeta_{p-1}\rangle \cdot U_p^{(1)}$, and if $n \geq 1$, then the group

$U_p^{(n)} = (1+p)^{p^{n-1}\mathbb{Z}_p} = U_p^{(p-1)p^{n-1}}$ is a free (multiplicative) $\mathbb{Z}_p$-module of rank 1 with basis element $(1+p)^{p^{n-1}}$.

2. $\mu(\mathbb{Q}_2) = \{\pm 1\}$, $U_2 = U_2^{(1)} = 1+2\mathbb{Z}_2 = \{\pm 1\}{\cdot}5^{\mathbb{Z}_2}$, and if $n \geq 2$, then $U_2^{(n)} = 1+2^n\mathbb{Z}_2 = 5^{2^{n-2}\mathbb{Z}_2}$ is a free (multiplicative) $\mathbb{Z}_2$-module of rank 1 with basis element $5^{2^{n-2}}$.

Proof. By 5.8.9 and 5.8.10. Observe that $p^n\mathbb{Z}_p$ is the only subgroup of index p^n of $\mathbb{Z}_p$ and that

$$(U_p^{(n)} : U_p^{(n+i)}) = \prod_{\nu=1}^{i} (U_p^{(n+\nu-1)} : U_p^{(n+\nu)}) = p^i$$

for all n, $i \in \mathbb{N}$ and all primes p by 2.10.5.4. □

5.9 Different and discriminant

We develop the theory of differents and discriminants for extensions of Dedekind domains. Although we should rather have done this in Sections 2.12 and 2.13, we decided to postpone this theory until we have valuation-theoretic tools at our disposal. As already announced, we did this not least because only the use of methods from valuation theory makes the notions coherent and the proofs transparent. Consequently, the present section could also be used to demonstrate the power of valuation-theoretic methods in the purely algebraic theory of Dedekind domains.

We freely use the structural results concerning extensions of Dedekind domains and their arithmetic as derived in Chapter 2. Only at the end of this section we shall discuss the special cases of complete discrete valued fields and algebraic number fields.

Theorem and Definition 5.9.1. *Let $\mathfrak{o}$ be a Dedekind domain, $K = \mathsf{q}(\mathfrak{o})$, L/K a finite separable field extension, $[L:K] = n$ and $\mathfrak{O} = \mathrm{cl}_L(\mathfrak{o})$.*

An **$\mathfrak{o}$-lattice** in L is a finitely generated $\mathfrak{o}$-submodule $\mathfrak{A}$ of L such that $K\mathfrak{A} = L$ [equivalently, $\mathfrak{A}$ contains a K-basis of L]. Every fractional ideal $\mathfrak{A} \in \mathcal{J}(\mathfrak{O})$ is an $\mathfrak{o}$-lattice in L. If $\mathfrak{A}$ is an $\mathfrak{o}$-lattice in L and $\mathfrak{p} \in \mathcal{P}(\mathfrak{o})$, then $\mathfrak{A}_\mathfrak{p}$ is a free $\mathfrak{o}_\mathfrak{p}$-module of rank n by 2.3.5.2.

Let $\mathfrak{A}$ and $\mathfrak{B}$ be $\mathfrak{o}$-lattices in L.

1. There exists some $q \in \mathfrak{o}^\bullet$ such that $q\mathfrak{B} \subset \mathfrak{A}$, and $\mathfrak{A}_\mathfrak{p} = \mathfrak{B}_\mathfrak{p}$ for almost all $\mathfrak{p} \in \mathcal{P}(\mathfrak{o})$.

2. There exists a unique fractional ideal $[\mathfrak{A} : \mathfrak{B}]_\mathfrak{o} \in \mathcal{J}(\mathfrak{o})$ such that for all $\mathfrak{p} \in \mathcal{P}(\mathfrak{o})$ the following assertion holds:

If $\boldsymbol{u}_\mathfrak{p} = (u_{\mathfrak{p},1}, \ldots, u_{\mathfrak{p},n})$ *is an* $\mathfrak{o}_\mathfrak{p}$*-basis of* $\mathfrak{A}_\mathfrak{p}$*,* $\boldsymbol{v}_\mathfrak{p} = (v_{\mathfrak{p},1}, \ldots, v_{\mathfrak{p},n})$ *is an* $\mathfrak{o}_\mathfrak{p}$*-basis of* $\mathfrak{B}_\mathfrak{p}$ *and* $T_\mathfrak{p} \in \mathsf{GL}_n(K)$ *is such that* $\boldsymbol{v}_\mathfrak{p} = \boldsymbol{u}_\mathfrak{p} T_\mathfrak{p}$*, then* $[\mathfrak{A} : \mathfrak{B}]_\mathfrak{o} \mathfrak{o}_\mathfrak{p} = \det(T_\mathfrak{p})\mathfrak{o}_\mathfrak{p}$.

The fractional ideal $[\mathfrak{A} : \mathfrak{B}]_\mathfrak{o} \in \mathcal{J}(\mathfrak{o})$ is called the **$\mathfrak{o}$-module index** of $\mathfrak{B}$ in $\mathfrak{A}$.

3. (a) Let $\mathfrak{A}$ *and* $\mathfrak{B}$ *be* $\mathfrak{o}$*-free,* $\boldsymbol{u} = (u_1, \ldots, u_n)$ *an* $\mathfrak{o}$*-basis of* $\mathfrak{A}$*,* $\boldsymbol{v} = (v_1, \ldots, v_n)$ *an* $\mathfrak{o}$*-basis of* $\mathfrak{B}$*, and* $T \in \mathsf{GL}_n(K)$ *such that* $\boldsymbol{v} = \boldsymbol{u}T$*. Then it* $[\mathfrak{A} : \mathfrak{B}]_\mathfrak{o} = \det(T)\mathfrak{o}$.

(b) If $\mathfrak{p} \in \mathcal{P}(\mathfrak{o})$*, then* $[\mathfrak{A} : \mathfrak{B}]_\mathfrak{o} \mathfrak{o}_\mathfrak{p} = [\mathfrak{A}_\mathfrak{p} : \mathfrak{B}_\mathfrak{p}]_{\mathfrak{o}_\mathfrak{p}}$

(c) If $\mathfrak{o} = \mathbb{Z}$ *and* $\mathfrak{B} \subset \mathfrak{A}$*, then* $[\mathfrak{A} : \mathfrak{B}]_\mathbb{Z} = (\mathfrak{A} : \mathfrak{B})\mathbb{Z}$.

(d) $[\mathfrak{A} : \mathfrak{A}]_\mathfrak{o} = \mathfrak{o}$*,* $[\mathfrak{B} : \mathfrak{A}]_\mathfrak{o} = [\mathfrak{A} : \mathfrak{B}]_\mathfrak{o}^{-1}$*, and if* $\mathfrak{C}$ *is another* $\mathfrak{o}$*-lattice in* L*, then* $[\mathfrak{A} : \mathfrak{C}]_\mathfrak{o} = [\mathfrak{A} : \mathfrak{B}]_\mathfrak{o} [\mathfrak{B} : \mathfrak{C}]_\mathfrak{o}$.

4. Let $\mathfrak{A}$ *and* $\mathfrak{B}$ *be fractional ideals of* $\mathfrak{O}$*. Then*

$$[\mathfrak{A} : \mathfrak{B}]_\mathfrak{o} = \mathcal{N}_{\mathfrak{O}/\mathfrak{o}}(\mathfrak{A}^{-1}\mathfrak{B}),$$
$$[\mathfrak{A} : \mathfrak{B}]_\mathfrak{o} = [\mathfrak{B}^{-1} : \mathfrak{A}^{-1}]_\mathfrak{o} \text{ and } [\mathfrak{O} : \mathfrak{B}]_\mathfrak{o} = \mathcal{N}_{\mathfrak{O}/\mathfrak{o}}(\mathfrak{B}).$$

Proof. 1. Let $(u_1, \ldots, u_n) \in \mathfrak{A}^n$ be a K-basis of L and $\mathfrak{B} = {}_\mathfrak{o}(v_1, \ldots, v_m)$. For $j \in [1, m]$ let $c_{j,1}, \ldots, c_{j,n} \in K$ be such that $v_j = c_{j,1}u_1 + \ldots + c_{j,n}u_n$. If $q \in \mathfrak{o}^\bullet$ is such that $qc_{j,\nu} \in \mathfrak{o}$ for all $j \in [1, m]$ and $\nu \in [1, n]$, then $qv_j \in \mathfrak{A}$ for all $j \in [1, m]$ and thus $q\mathfrak{B} \subset \mathfrak{A}$.

Now let $q, t \in \mathfrak{o}^\bullet$ be such that $q\mathfrak{B} \subset \mathfrak{A}$ and $t\mathfrak{A} \subset \mathfrak{B}$. If $\mathfrak{p} \in \mathcal{P}(\mathfrak{o})$ and $v_\mathfrak{p}(qt) = 0$, then $\mathfrak{B}_\mathfrak{p} = q\mathfrak{B}_\mathfrak{p} = (q\mathfrak{B})_\mathfrak{p} \subset \mathfrak{A}_\mathfrak{p}$ and $\mathfrak{A}_\mathfrak{p} = t\mathfrak{A}_\mathfrak{p} = (t\mathfrak{A})_\mathfrak{p} \subset \mathfrak{B}_\mathfrak{p}$, hence $\mathfrak{A}_\mathfrak{p} = \mathfrak{B}_\mathfrak{p}$. Consequently, $\mathfrak{A}_\mathfrak{p} = \mathfrak{B}_\mathfrak{p}$ for almost all $\mathfrak{p} \in \mathcal{P}(\mathfrak{o})$.

2. For $\mathfrak{p} \in \mathcal{P}(\mathfrak{o})$ let $\boldsymbol{u}_\mathfrak{p} = (u_{\mathfrak{p},1}, \ldots, u_{\mathfrak{p},n})$ be an $\mathfrak{o}_\mathfrak{p}$-basis of $\mathfrak{A}_\mathfrak{p}$, let $\boldsymbol{v}_\mathfrak{p} = (v_{\mathfrak{p},1}, \ldots, v_{\mathfrak{p},n})$ be an $\mathfrak{o}_\mathfrak{p}$-basis of $\mathfrak{B}_\mathfrak{p}$, and let $T_\mathfrak{p} \in \mathsf{GL}_n(K)$ be such that $\boldsymbol{v}_\mathfrak{p} = \boldsymbol{u}_\mathfrak{p} T_\mathfrak{p}$. We assert that the principal ideal $\det(T_\mathfrak{p})\mathfrak{o}_\mathfrak{p}$ does not depend on $\boldsymbol{u}_\mathfrak{p}$ and $\boldsymbol{v}_\mathfrak{p}$. Indeed, if $\boldsymbol{u}'_\mathfrak{p} = (u'_{\mathfrak{p},1}, \ldots, u'_{\mathfrak{p},n})$ resp. $\boldsymbol{v}'_\mathfrak{p} = (v'_{\mathfrak{p},1}, \ldots, v'_{\mathfrak{p},n})$ are other bases of $\mathfrak{A}_\mathfrak{p}$ resp. $\mathfrak{B}_\mathfrak{p}$, then there exist matrices $M, N \in \mathsf{GL}_n(\mathfrak{o}_\mathfrak{p})$ such that $\boldsymbol{u}'_\mathfrak{p} = \boldsymbol{u}_\mathfrak{p} M$ and $\boldsymbol{v}'_\mathfrak{p} = \boldsymbol{v}_\mathfrak{p} N$. Hence $\boldsymbol{v}'_\mathfrak{p} = \boldsymbol{u}'_\mathfrak{p} M^{-1} T_\mathfrak{p} N$, and $\det(M^{-1}T_\mathfrak{p}N)\mathfrak{o}_\mathfrak{p} = \det(T_\mathfrak{p})\mathfrak{o}_\mathfrak{p}$, since $\det(M)$, $\det(N) \in \mathfrak{o}_\mathfrak{p}^\times$. In particular, if $\mathfrak{A}_\mathfrak{p} = \mathfrak{B}_\mathfrak{p}$, then $\det(T_\mathfrak{p})\mathfrak{o}_\mathfrak{p} = \mathfrak{o}_\mathfrak{p}$.

Since $\mathfrak{A}_\mathfrak{p} = \mathfrak{B}_\mathfrak{p}$ and thus $\det(T_\mathfrak{p})\mathfrak{o}_\mathfrak{p} = \mathfrak{o}_\mathfrak{p}$ for almost all $\mathfrak{p} \in \mathcal{P}(\mathfrak{o})$, there exists a unique fractional ideal $[\mathfrak{A} : \mathfrak{B}]_\mathfrak{o} \in \mathcal{J}(\mathfrak{o})$ such that $[\mathfrak{A} : \mathfrak{B}]_\mathfrak{o}\mathfrak{o}_\mathfrak{p} = \det(T_\mathfrak{p})\mathfrak{o}_\mathfrak{p}$ for all $\mathfrak{p} \in \mathcal{P}(\mathfrak{o})$, namely

$$[\mathfrak{A} : \mathfrak{B}]_\mathfrak{o} = \prod_{\mathfrak{p} \in \mathcal{P}(\mathfrak{o})} \mathfrak{p}^{v_\mathfrak{p}(\det(T_\mathfrak{p}))}.$$

3. (a) and (b) are obvious by the definitions.

(c) Let $\mathfrak{o} = \mathbb{Z}$ and $\mathfrak{B} \subset \mathfrak{A}$. Then $\mathfrak{A}$ and $\mathfrak{B}$ are free. If $\boldsymbol{u} = (u_1, \ldots, u_n)$ is a basis of $\mathfrak{A}$, $\boldsymbol{v} = (v_1, \ldots, v_n)$ is a basis of $\mathfrak{B}$ and $T \in \mathsf{GL}_n(\mathbb{Q})$ is such that $\boldsymbol{v} = \boldsymbol{u}T$, then $T \in \mathsf{M}_n(\mathbb{Z})$ and $(\mathfrak{A} : \mathfrak{B}) = |\det(T)|$ by 2.3.6.3. Hence $[\mathfrak{A} : \mathfrak{B}]_\mathbb{Z} = \det(T)\mathbb{Z} = (\mathfrak{A} : \mathfrak{B})\mathbb{Z}$.

(d) The assertions are obvious if $\mathfrak{A}$, $\mathfrak{B}$ and $\mathfrak{C}$ are $\mathfrak{o}$-free. Consequently, for all $\mathfrak{p} \in \mathcal{P}(\mathfrak{o})$ we obtain $[\mathfrak{A}:\mathfrak{A}]_\mathfrak{o}\mathfrak{o}_\mathfrak{p} = [\mathfrak{A}_\mathfrak{p}:\mathfrak{A}_\mathfrak{p}]_{\mathfrak{o}_\mathfrak{p}} = \mathfrak{o}_\mathfrak{p}$,

$$[\mathfrak{B}:\mathfrak{A}]_\mathfrak{o}\mathfrak{o}_\mathfrak{p} = [\mathfrak{B}_\mathfrak{p}:\mathfrak{A}_\mathfrak{p}]_{\mathfrak{o}_\mathfrak{p}} = ([\mathfrak{A}_\mathfrak{p}:\mathfrak{B}_\mathfrak{p}]_{\mathfrak{o}_\mathfrak{p}})^{-1} = ([\mathfrak{A}:\mathfrak{B}]_\mathfrak{o}\mathfrak{o}_\mathfrak{p})^{-1} = [\mathfrak{A}:\mathfrak{B}]_\mathfrak{o}^{-1}\mathfrak{o}_\mathfrak{p},$$

and $[\mathfrak{A}:\mathfrak{C}]_\mathfrak{o}\mathfrak{o}_\mathfrak{p} = [\mathfrak{A}_\mathfrak{p}:\mathfrak{C}_\mathfrak{p}]_{\mathfrak{o}_\mathfrak{p}} = [\mathfrak{A}_\mathfrak{p}:\mathfrak{B}_\mathfrak{p}]_{\mathfrak{o}_\mathfrak{p}}[\mathfrak{B}_\mathfrak{p}:\mathfrak{C}_\mathfrak{p}]_{\mathfrak{o}_\mathfrak{p}} = ([\mathfrak{A}:\mathfrak{B}]_\mathfrak{o}[\mathfrak{B}:\mathfrak{C}]_\mathfrak{o})\mathfrak{o}_\mathfrak{p}$. Therefore the assertions follow by 2.7.5.

4. Let $\mathfrak{p} \in \mathcal{P}(\mathfrak{o})$. Then $\mathfrak{O}_\mathfrak{p}$ is a semilocal principal ideal domain and a free $\mathfrak{o}_\mathfrak{p}$-module. Thus let $\mathfrak{A}_\mathfrak{p} = a\mathfrak{O}_\mathfrak{p}$, $\mathfrak{B}_\mathfrak{p} = b\mathfrak{O}_\mathfrak{p}$ for some $a,\, b \in \mathfrak{O}_\mathfrak{p}^\bullet$, and $(\mathfrak{A}\mathfrak{B}^{-1})_\mathfrak{p} = \mathfrak{A}_\mathfrak{p}^{-1}\mathfrak{B}_\mathfrak{p} = a^{-1}b\mathfrak{O}_\mathfrak{p}$. Let $\boldsymbol{u} = (u_1, \ldots, u_n)$ be an $\mathfrak{o}_\mathfrak{p}$-basis of $\mathfrak{O}_\mathfrak{p}$. Then $a\boldsymbol{u}$ resp. $b\boldsymbol{u}$ is an $\mathfrak{o}_\mathfrak{p}$-basis of $\mathfrak{A}$ resp. $\mathfrak{B}$. Let $A,\, B \in \mathsf{GL}_n(K)$ such that $a\boldsymbol{u} = \boldsymbol{u}A$ and $b\boldsymbol{u} = \boldsymbol{u}B$. Then $b\boldsymbol{u} = a\boldsymbol{u}A^{-1}B$, and therefore

$$[\mathfrak{A}:\mathfrak{B}]_\mathfrak{o}\mathfrak{o}_\mathfrak{p} = [\mathfrak{A}_\mathfrak{p}:\mathfrak{B}_\mathfrak{p}]_{\mathfrak{o}_\mathfrak{p}} = \det(A^{-1}B)\mathfrak{o}_\mathfrak{p}.$$

Since $a^{-1}b\boldsymbol{u} = \boldsymbol{u}A^{-1}B$, we get $\mathsf{N}_{L/K}(a^{-1}b) = \det(A^{-1}B)$, and therefore, using 2.14.2.4 and 2.14.4.3,

$$\begin{aligned}[\mathfrak{A}:\mathfrak{B}]_\mathfrak{o}\mathfrak{o}_\mathfrak{p} &= \det(A^{-1}B)\mathfrak{o}_\mathfrak{p} = \mathsf{N}_{L/K}(a^{-1}b)\mathfrak{o}_\mathfrak{p} = \mathcal{N}_{\mathfrak{O}_\mathfrak{p}/\mathfrak{o}_\mathfrak{p}}(a^{-1}b\mathfrak{O}_\mathfrak{p})\\ &= \mathcal{N}_{\mathfrak{O}_\mathfrak{p}/\mathfrak{o}_\mathfrak{p}}((\mathfrak{A}^{-1}\mathfrak{B})_\mathfrak{p}) = \mathcal{N}_{\mathfrak{O}/\mathfrak{o}}(\mathfrak{A}^{-1}\mathfrak{B})\mathfrak{o}_\mathfrak{p}.\end{aligned}$$

Since this holds for all $\mathfrak{p} \in \mathcal{P}(\mathfrak{o})$, we obtain $[\mathfrak{A}:\mathfrak{B}]_\mathfrak{o} = \mathcal{N}_{\mathfrak{O}/\mathfrak{o}}(\mathfrak{A}^{-1}\mathfrak{B})$ by 2.7.5. Now it follows that

$$[\mathfrak{B}^{-1}:\mathfrak{A}^{-1}]_\mathfrak{o} = \mathcal{N}_{\mathfrak{O}/\mathfrak{o}}(\mathfrak{B}\mathfrak{A}^{-1}) = \mathcal{N}_{\mathfrak{O}/\mathfrak{o}}(\mathfrak{A}^{-1}\mathfrak{B})^{-1} = [\mathfrak{A}:\mathfrak{B}]_\mathfrak{o}^{-1}$$

and $[\mathfrak{O}:\mathfrak{B}]_\mathfrak{o} = \mathcal{N}_{\mathfrak{O}/\mathfrak{o}}(\mathfrak{B})$. □

Theorem and Definition 5.9.2. *Let $\mathfrak{o}$ be a Dedekind domain, $K = \mathsf{q}(\mathfrak{o})$, L/K a finite separable field extension, $[L:K] = n$ and $\mathfrak{O} = \mathrm{cl}_L(\mathfrak{o})$.*
For an $\mathfrak{o}$-lattice $\mathfrak{A}$ in L we define the **Dedekind complementary module** $\mathfrak{C}_{\mathfrak{A}/\mathfrak{o}}$ of $\mathfrak{A}/\mathfrak{o}$ with respect to the trace form $(x, y) \mapsto \mathsf{Tr}_{L/K}(xy)$ by

$$\mathfrak{C}_{\mathfrak{A}/\mathfrak{o}} = \{x \in L \mid \mathsf{Tr}_{L/K}(x\mathfrak{A}) \subset \mathfrak{o}\}$$

and the **discriminant** $\mathfrak{d}_{\mathfrak{A}/\mathfrak{o}}$ of $\mathfrak{A}$ over $\mathfrak{o}$ by

$$\mathfrak{d}_{\mathfrak{A}/\mathfrak{o}} = [\mathfrak{C}_{\mathfrak{A}/\mathfrak{o}}:\mathfrak{A}]_\mathfrak{o} \in \mathcal{J}(\mathfrak{o}).$$

(we shall see in an moment that $\mathfrak{C}_{\mathfrak{A}/\mathfrak{o}}$ is also an $\mathfrak{o}$-lattice in L).

1. Let $\mathfrak{A}$ be an $\mathfrak{o}$-lattice in L and $q \in L^\times$. Then $\mathfrak{C}_{q\mathfrak{A}/\mathfrak{o}} = q^{-1}\mathfrak{C}_{\mathfrak{A}/\mathfrak{o}}$, and $\mathfrak{C}_{\mathfrak{A}/\mathfrak{o}}$ is an $\mathfrak{o}$-lattice. If $\mathfrak{A}$ is $\mathfrak{o}$-free with basis $(u_1, \ldots, u_n)$ and $(u_1^*, \ldots, u_n^*)$ is the dual basis with respect to $\mathsf{Tr}_{L/K}$ (see 1.6.10), then $\mathfrak{C}_{\mathfrak{A}/\mathfrak{o}}$ is the free $\mathfrak{o}$-module with basis $(u_1^*, \ldots, u_n^*)$.

2. Let $\mathfrak{A}$ be an $\mathfrak{o}$-lattice in L. Then $(\mathfrak{C}_{\mathfrak{A}/\mathfrak{o}})_\mathfrak{p} = \mathfrak{C}_{\mathfrak{A}_\mathfrak{p}/\mathfrak{o}_\mathfrak{p}}$ and $(\mathfrak{d}_{\mathfrak{A}/\mathfrak{o}})_\mathfrak{p} = \mathfrak{d}_{\mathfrak{A}_\mathfrak{p}/\mathfrak{o}_\mathfrak{p}}$ for all $\mathfrak{p} \in \mathcal{P}(\mathfrak{o})$.

3. Let $\mathfrak{U}$ be the set of all K-bases of L lying in $\mathfrak{A}$. Then

$$\mathfrak{d}_{\mathfrak{A}/\mathfrak{o}} = \sum_{(u_1,\ldots,u_n)\in\mathfrak{U}} d_{L/K}(u_1,\ldots,u_n)\mathfrak{o},$$

and if $(u_1,\ldots,u_n)$ is an $\mathfrak{o}$-basis of $\mathfrak{A}$, then $\mathfrak{d}_{\mathfrak{A}/\mathfrak{o}} = d_{L/K}(u_1,\ldots,u_n)\mathfrak{o}$.

4. Let $\mathfrak{A}$ and $\mathfrak{B}$ be $\mathfrak{o}$-lattices in L. Then $\mathfrak{d}_{\mathfrak{A}/\mathfrak{o}} = [\mathfrak{A}\!:\!\mathfrak{B}]_{\mathfrak{o}}^2\,\mathfrak{d}_{\mathfrak{B}/\mathfrak{o}}$, and the ideal class $[\mathfrak{d}_{\mathfrak{A}/\mathfrak{o}}] \in \mathcal{C}(\mathfrak{o})$ is a square in $\mathcal{C}(\mathfrak{o})$.

Proof. 1. If $x \in L^\times$, then

$$x \in \mathfrak{C}_{q\mathfrak{A}/\mathfrak{o}} \iff \mathsf{Tr}_{L/K}(xq\mathfrak{A}) \subset \mathfrak{o} \iff xq \in \mathfrak{C}_{\mathfrak{A}/\mathfrak{o}} \iff x \in q^{-1}\mathfrak{C}_{\mathfrak{A}/\mathfrak{o}}.$$

Hence $\mathfrak{C}_{q\mathfrak{A}/\mathfrak{o}} = q^{-1}\mathfrak{C}_{\mathfrak{A}/\mathfrak{o}}$.

Assume now that $\mathfrak{A}$ is $\mathfrak{o}$-free with basis $(u_1,\ldots,u_n)$, let $(u_1^*,\ldots,u_n^*)$ be the dual basis and $\mathfrak{A}^* = {}_{\mathfrak{o}}(u_1^*,\ldots,u_n^*)$. We shall prove that $\mathfrak{C}_{\mathfrak{A}/\mathfrak{o}} = \mathfrak{A}^*$.

$\mathfrak{C}_{\mathfrak{A}/\mathfrak{o}} \subset \mathfrak{A}^*$: Let $x = c_1u_1^* + \ldots + c_nu_n^* \in \mathfrak{C}_{\mathfrak{A}/\mathfrak{o}} \subset L$, where $c_1,\ldots,c_n \in K$. Then

$$\mathsf{Tr}_{L/K}(xu_i) = \sum_{\nu=1}^n c_\nu \mathsf{Tr}_{L/K}(u_\nu^*u_i) = c_i \in \mathfrak{o} \quad \text{for all} \quad i \in [1,n],$$

and therefore $x \in \mathfrak{A}^*$.

$\mathfrak{A}^* \subset \mathfrak{C}_{\mathfrak{A}/\mathfrak{o}}$: Let $x = c_1u_1^* + \ldots + c_nu_n^* \in \mathfrak{A}^*$, where $c_1,\ldots,c_n \in \mathfrak{o}$. If $a \in \mathfrak{A}$, say $a = a_1u_1 + \ldots + a_nu_n$ for some $a_1,\ldots,a_n \in \mathfrak{o}$, then

$$\mathsf{Tr}_{L/K}(xa) = \sum_{\nu=1}^n\sum_{i=1}^n c_\nu a_i \mathsf{Tr}_{L/K}(u_\nu^*u_i) = \sum_{i=1}^n c_ia_i \in \mathfrak{o},$$

and therefore $x \in \mathfrak{C}_{\mathfrak{A}/\mathfrak{o}}$.

Let again $\mathfrak{A}$ be any $\mathfrak{o}$-lattice in L, $(u_1,\ldots,u_n) \in \mathfrak{A}^n$ a K-basis of L, and let $(u_1^*,\ldots,u_n^*)$ be the dual basis. Let $\mathfrak{B} = {}_{\mathfrak{o}}(u_1,\ldots,u_n)$, and let $q \in \mathfrak{o}^\bullet$ be such that $q\mathfrak{A} \subset \mathfrak{B} \subset \mathfrak{A}$. Then

$$\mathfrak{C}_{\mathfrak{A}/\mathfrak{o}} \subset \mathfrak{C}_{\mathfrak{B}/\mathfrak{o}} = {}_{\mathfrak{o}}(u_1^*,\ldots,u_n^*) \subset \mathfrak{C}_{q\mathfrak{A}/\mathfrak{o}} = q^{-1}\mathfrak{C}_{\mathfrak{A}/\mathfrak{o}},$$

and $q\mathfrak{C}_{\mathfrak{B}/\mathfrak{o}} = {}_{\mathfrak{o}}(qu_1^*,\ldots,qu_n^*) \subset \mathfrak{C}_{\mathfrak{A}/\mathfrak{o}}$.

Hence $\mathfrak{C}_{\mathfrak{A}/\mathfrak{o}}$ is a finitely generated $\mathfrak{o}$-module and contains the K-basis $(qu_1^*,\ldots,qu_n^*)$ of L. Therefore it is an $\mathfrak{o}$-lattice.

2. Let $\mathfrak{p} \in \mathcal{P}(\mathfrak{o})$ and $\mathfrak{A} = {}_{\mathfrak{o}}(u_1,\ldots,u_m) \subset L$.

$(\mathfrak{C}_{\mathfrak{A}/\mathfrak{o}})_{\mathfrak{p}} \subset \mathfrak{C}_{\mathfrak{A}_{\mathfrak{p}}/\mathfrak{o}_{\mathfrak{p}}}$: Suppose that $s^{-1}x \in (\mathfrak{C}_{\mathfrak{A}/\mathfrak{o}})_{\mathfrak{p}}$, where $x \in \mathfrak{C}_{\mathfrak{A}/\mathfrak{o}}$ and $s \in \mathfrak{o}\setminus\mathfrak{p}$. If $t^{-1}c \in \mathfrak{A}_{\mathfrak{p}}$ for some $c \in \mathfrak{A}$ and $t \in \mathfrak{o}\setminus\mathfrak{p}$, then it follows that $\mathsf{Tr}_{L/K}((s^{-1}x)(t^{-1}c)) = (st)^{-1}\mathsf{Tr}(xc) \in \mathfrak{o}_{\mathfrak{p}}$, and therefore $s^{-1}x \in \mathfrak{C}_{\mathfrak{A}_{\mathfrak{p}}/\mathfrak{o}_{\mathfrak{p}}}$.

$\mathfrak{C}_{\mathfrak{A}_{\mathfrak{p}}/\mathfrak{o}_{\mathfrak{p}}} \subset (\mathfrak{C}_{\mathfrak{A}/\mathfrak{o}})_{\mathfrak{p}}$: Let $x \in \mathfrak{C}_{\mathfrak{A}_{\mathfrak{p}}/\mathfrak{o}_{\mathfrak{p}}}$. As $\mathsf{Tr}_{L/K}(xu_j) \in \mathsf{Tr}_{L/K}(x\mathfrak{A}_{\mathfrak{p}}) \subset \mathfrak{o}_{\mathfrak{p}}$ for all $j \in [1,m]$, there exists some $s \in \mathfrak{o}\setminus\mathfrak{p}$ such that

$$\mathsf{Tr}_{L/K}(sxu_j) = s\mathsf{Tr}_{L/K}(xu_j) \in \mathfrak{o} \text{ for all } j \in [1,m].$$

It follows that $\mathsf{Tr}_{L/K}(sx\mathfrak{A}) \subset \mathfrak{o}$, hence $sx \in \mathfrak{C}_{\mathfrak{A}/\mathfrak{o}}$ and $x \in (\mathfrak{C}_{\mathfrak{A}/\mathfrak{o}})_{\mathfrak{p}}$.

$(\mathfrak{d}_{\mathfrak{A}/\mathfrak{o}})_{\mathfrak{p}} = [\mathfrak{C}_{\mathfrak{A}/\mathfrak{o}} : \mathfrak{A}]_{\mathfrak{o}}\, \mathfrak{o}_{\mathfrak{p}} = [(\mathfrak{C}_{\mathfrak{A}/\mathfrak{o}})_{\mathfrak{p}} : \mathfrak{A}_{\mathfrak{p}}]_{\mathfrak{o}_{\mathfrak{p}}} = [\mathfrak{C}_{\mathfrak{A}_{\mathfrak{p}}/\mathfrak{o}_{\mathfrak{p}}} : \mathfrak{A}_{\mathfrak{p}}]_{\mathfrak{o}_{\mathfrak{p}}} = \mathfrak{d}_{\mathfrak{A}_{\mathfrak{p}}/\mathfrak{o}_{\mathfrak{p}}}$.

3. Assume first that $(u_1, \ldots, u_n)$ is an $\mathfrak{o}$-basis of $\mathfrak{A}$, $(u_1^*, \ldots, u_n^*)$ is the dual basis and $T \in \mathsf{GL}_n(K)$ is such that $(u_1, \ldots, u_n) = (u_1^*, \ldots, u_n^*)T$. Then $(u_1^*, \ldots, u_n^*)$ is an $\mathfrak{o}$-basis of $\mathfrak{C}_{\mathfrak{A}/\mathfrak{o}}$, and $\mathfrak{d}_{\mathfrak{A}/\mathfrak{o}} = [\mathfrak{C}_{\mathfrak{A}/\mathfrak{o}} : \mathfrak{A}]_{\mathfrak{o}} = \det(T)\mathfrak{o}$. On the other hand, using 1.6.9.3 and 1.6.10.1, we obtain

$$d_{L/K}(u_1, \ldots, u_n) = d_{L/K}(u_1^*, \ldots, u_n^*) \det(T)^2 = d_{L/K}(u_1, \ldots, u_n)^{-1} \det(T)^2$$

hence $\det(T)^2 = d_{L/K}(u_1, \ldots, u_n)^2$ and therefore $\det(T)\mathfrak{o} = d_{L/K}(u_1, \ldots, u_n)\mathfrak{o}$.

Now let $(u_1, \ldots, u_n) \in \mathfrak{U}$ be arbitrary, $\mathfrak{p} \in \mathcal{P}(\mathfrak{o})$ and $(b_1, \ldots, b_n) \in \mathfrak{A}^n$ an $\mathfrak{o}_{\mathfrak{p}}$-basis of $\mathfrak{A}_{\mathfrak{p}}$. Let $T \in \mathsf{M}_n(\mathfrak{o}_{\mathfrak{p}})$ be such that $(u_1, \ldots, u_n) = (b_1, \ldots, b_n)T$, and therefore $d_{L/K}(u_1, \ldots, u_n) = d_{L/K}(b_1, \ldots, b_n) \det(T)^2 \in \mathfrak{d}_{\mathfrak{A}/\mathfrak{o}}\mathfrak{o}_{\mathfrak{p}}$. Consequently,

$$\mathfrak{d}_{\mathfrak{A}/\mathfrak{o}}\mathfrak{o}_{\mathfrak{p}} = d_{L/K}(b_1, \ldots, b_n)\mathfrak{o}_{\mathfrak{p}} \subset \sum_{(u_1,\ldots,u_n)\in\mathfrak{U}} d_{L/K}(u_1, \ldots, u_n)\mathfrak{o}_{\mathfrak{p}} \subset \mathfrak{d}_{\mathfrak{A}/\mathfrak{o}}\mathfrak{o}_{\mathfrak{p}},$$

hence equality holds, and since this is true for all $\mathfrak{p} \in \mathcal{P}(\mathfrak{o})$, it follows by 2.7.5 that

$$\mathfrak{d}_{\mathfrak{A}/\mathfrak{o}} = \sum_{(u_1,\ldots,u_n)\in\mathfrak{U}} d_{L/K}(u_1, \ldots, u_n)\mathfrak{o}.$$

4. Let first $\mathfrak{A}$ and $\mathfrak{B}$ be $\mathfrak{o}$-free. Let $(u_1, \ldots, u_n)$ be an $\mathfrak{o}$-basis of $\mathfrak{A}$, $(v_1, \ldots, v_n)$ an $\mathfrak{o}$-basis of $\mathfrak{B}$ and $(v_1, \ldots, v_n) = (u_1, \ldots, u_n)T$ for some matrix $T \in \mathsf{GL}_n(K)$. Then it follows that

$$\mathfrak{d}_{\mathfrak{A}/\mathfrak{o}}[\mathfrak{A} : \mathfrak{B}]_{\mathfrak{o}}^2 = d_{L/K}(u_1, \ldots, u_n) \det(T)^2\mathfrak{o} = d_{L/K}(v_1, \ldots, v_n)\mathfrak{o} = \mathfrak{d}_{\mathfrak{B}/\mathfrak{o}}.$$

If $\mathfrak{A}$ and $\mathfrak{B}$ are arbitrary and $\mathfrak{p} \in \mathcal{P}(\mathfrak{o})$, then $\mathfrak{A}_{\mathfrak{p}}$ and $\mathfrak{B}_{\mathfrak{p}}$ are $\mathfrak{o}_{\mathfrak{p}}$-free, and therefore, as above, $(\mathfrak{d}_{\mathfrak{B}/\mathfrak{o}})_{\mathfrak{p}} = \mathfrak{d}_{\mathfrak{B}_{\mathfrak{p}}/\mathfrak{o}_{\mathfrak{p}}} = [\mathfrak{B}_{\mathfrak{p}} : \mathfrak{A}_{\mathfrak{p}}]_{\mathfrak{o}_{\mathfrak{p}}}^2\, \mathfrak{d}_{\mathfrak{A}_{\mathfrak{p}}/\mathfrak{o}_{\mathfrak{p}}} = [\mathfrak{B} : \mathfrak{A}]_{\mathfrak{o}}^2\, \mathfrak{d}_{\mathfrak{A}/\mathfrak{o}}\mathfrak{o}_{\mathfrak{p}}$. Since this holds for all $\mathfrak{p} \in \mathcal{P}(\mathfrak{o})$, the assertion follows by 2.7.5.

If $\mathfrak{A}$ is arbitrary and $\mathfrak{B}$ is $\mathfrak{o}$-free with basis $(v_1, \ldots, v_n)$, then

$$\mathfrak{d}_{\mathfrak{A}/\mathfrak{o}} = [\mathfrak{A} : \mathfrak{B}]_{\mathfrak{o}}^2 \mathfrak{d}_{\mathfrak{B}/\mathfrak{o}} = [\mathfrak{A} : \mathfrak{B}]_{\mathfrak{o}}^2\, d_{L/K}(v_1, \ldots, v_n),$$

and consequently $[\mathfrak{d}_{\mathfrak{A}/\mathfrak{o}}] = \big[[\mathfrak{A} : \mathfrak{B}]_{\mathfrak{o}}^2\big] = \big[[\mathfrak{A} : \mathfrak{B}]_{\mathfrak{o}}\big]^2 \in \mathcal{C}(\mathfrak{o})^2$. □

Theorem and Definition 5.9.3. *Let* $\mathfrak{o}$ *be a Dedekind domain,* $K = \mathsf{q}(\mathfrak{o})$, L/K *a finite separable field extension,* $[L:K] = n$ *and* $\mathfrak{O} = \mathrm{cl}_L(\mathfrak{o})$.

1. *Let* $\mathfrak{A} \in \mathcal{J}(\mathfrak{O})$ *be a fractional ideal of* $\mathfrak{O}$. *Then* $\mathfrak{C}_{\mathfrak{A}/\mathfrak{o}}$ *is also a fractional ideal of* $\mathfrak{O}$, *and* $\mathfrak{C}_{\mathfrak{A}/\mathfrak{o}} \supset \mathfrak{A}^{-1}$.

The fractional ideal $\mathfrak{D}_{\mathfrak{A}/\mathfrak{o}} = \mathfrak{C}_{\mathfrak{A}/\mathfrak{o}}^{-1} \in \mathcal{J}(\mathfrak{O})$ is called the **different** of $\mathfrak{A}/\mathfrak{o}$.

(a) *$\mathfrak{C}_{\mathfrak{A}/\mathfrak{o}} = \mathfrak{A}^{-1}\mathfrak{C}_{\mathfrak{O}/\mathfrak{o}}$, $\mathfrak{D}_{\mathfrak{A}/\mathfrak{o}} = \mathfrak{A}\mathfrak{D}_{\mathfrak{O}/\mathfrak{o}} \subset \mathfrak{A}$ and $(\mathfrak{D}_{\mathfrak{A}/\mathfrak{o}})_{\mathfrak{p}} = \mathfrak{D}_{\mathfrak{A}_{\mathfrak{p}}/\mathfrak{o}_{\mathfrak{p}}}$ for all $\mathfrak{p} \in \mathcal{P}(\mathfrak{o})$.*

(b) *$\mathfrak{d}_{\mathfrak{O}/\mathfrak{o}} = \mathcal{N}_{\mathfrak{O}/\mathfrak{o}}(\mathfrak{D}_{\mathfrak{O}/\mathfrak{o}})$, and $\mathfrak{d}_{\mathfrak{A}/\mathfrak{o}} = \mathcal{N}_{\mathfrak{O}/\mathfrak{o}}(\mathfrak{A})^2\, \mathfrak{d}_{\mathfrak{O}/\mathfrak{o}}$.*

In particular, $\mathfrak{D}_{\mathfrak{O}/\mathfrak{o}} \in \mathcal{J}'(\mathfrak{O})$ and $\mathfrak{d}_{\mathfrak{O}/\mathfrak{o}} \in \mathcal{J}'(\mathfrak{o})$ are (integral) ideals.

2. (Different and discriminant tower theorem) *Let K' be an intermediate field of L/K and $\mathfrak{o}' = \mathrm{cl}_{K'}(\mathfrak{o})$. Then*

$$\mathfrak{D}_{\mathfrak{O}/\mathfrak{o}} = \mathfrak{D}_{\mathfrak{o}'/\mathfrak{o}}\mathfrak{D}_{\mathfrak{O}/\mathfrak{o}'} \quad \textit{and} \quad \mathfrak{d}_{\mathfrak{O}/\mathfrak{o}} = \mathfrak{d}_{\mathfrak{o}'/\mathfrak{o}}^{[L:K']}\mathcal{N}_{\mathfrak{o}'/\mathfrak{o}}(\mathfrak{d}_{\mathfrak{O}/\mathfrak{o}'}).$$

Proof. 1. Since $\mathsf{Tr}_{L/K}(\mathfrak{O}\mathfrak{C}_{\mathfrak{A}/\mathfrak{o}}\mathfrak{A}) = \mathsf{Tr}_{L/K}(\mathfrak{C}_{\mathfrak{A}/\mathfrak{o}}\mathfrak{A}) \subset \mathfrak{o}$, it follows that $\mathfrak{O}\mathfrak{C}_{\mathfrak{A}/\mathfrak{o}} \subset \mathfrak{C}_{\mathfrak{A}/\mathfrak{o}}$. As $\mathfrak{C}_{\mathfrak{A}/\mathfrak{o}}$ is a finitely generated $\mathfrak{o}$-module, it is a finitely generated $\mathfrak{O}$-module and thus a fractional ideal of $\mathfrak{O}$.

Finally, $\mathsf{Tr}_{L/K}(\mathfrak{A}^{-1}\mathfrak{A}) = \mathsf{Tr}_{L/K}(\mathfrak{O}) \subset \mathfrak{o}$ implies $\mathfrak{A}^{-1} \subset \mathfrak{C}_{\mathfrak{A}/\mathfrak{o}}$.

(a) If $\mathfrak{p} \in \mathcal{P}(\mathfrak{o})$, then $(\mathfrak{D}_{\mathfrak{A}/\mathfrak{o}})_{\mathfrak{p}} = (\mathfrak{C}_{\mathfrak{A}/\mathfrak{o}}^{-1})_{\mathfrak{p}} = (\mathfrak{C}_{\mathfrak{A}/\mathfrak{o}})_{\mathfrak{p}}^{-1} = \mathfrak{C}_{\mathfrak{A}_{\mathfrak{p}}/\mathfrak{o}_{\mathfrak{p}}}^{-1} = \mathfrak{D}_{\mathfrak{A}_{\mathfrak{p}}/\mathfrak{o}_{\mathfrak{p}}}$. As $\mathfrak{O}_{\mathfrak{p}}$ is a semilocal principal ideal domain, we obtain $\mathfrak{A}_{\mathfrak{p}} = a\mathfrak{O}_{\mathfrak{p}}$ for some $a \in L^\times$, and $(\mathfrak{C}_{\mathfrak{A}/\mathfrak{o}})_{\mathfrak{p}} = \mathfrak{C}_{\mathfrak{A}_{\mathfrak{p}}/\mathfrak{o}_{\mathfrak{p}}} = \mathfrak{C}_{a\mathfrak{O}_{\mathfrak{p}}/\mathfrak{o}_{\mathfrak{p}}} = a^{-1}\mathfrak{C}_{\mathfrak{O}_{\mathfrak{p}}/\mathfrak{o}_{\mathfrak{p}}} = (\mathfrak{A}^{-1}\mathfrak{C}_{\mathfrak{O}/\mathfrak{o}})_{\mathfrak{p}}$. Hence $\mathfrak{C}_{\mathfrak{A}/\mathfrak{o}} = \mathfrak{A}^{-1}\mathfrak{C}_{\mathfrak{O}/\mathfrak{o}}$ by 2.7.5, and $\mathfrak{D}_{\mathfrak{A}/\mathfrak{o}} = \mathfrak{C}_{\mathfrak{A}/\mathfrak{o}}^{-1} = \mathfrak{A}\mathfrak{C}_{\mathfrak{O}/\mathfrak{o}}^{-1} = \mathfrak{A}\mathfrak{D}_{\mathfrak{O}/\mathfrak{o}}$. Finally, $\mathfrak{A}^{-1} \subset \mathfrak{C}_{\mathfrak{A}/\mathfrak{o}}$ implies $\mathfrak{D}_{\mathfrak{A}/\mathfrak{o}} \subset \mathfrak{A}$.

(b) By 5.9.1.4 we obtain $\mathcal{N}_{\mathfrak{O}/\mathfrak{o}}(\mathfrak{D}_{\mathfrak{O}/\mathfrak{o}}) = [\mathfrak{O}:\mathfrak{D}_{\mathfrak{O}/\mathfrak{o}}]_{\mathfrak{o}} = [\mathfrak{C}_{\mathfrak{O}/\mathfrak{o}}:\mathfrak{O}]_{\mathfrak{o}} = \mathfrak{d}_{\mathfrak{O}/\mathfrak{o}}$, and $\mathfrak{d}_{\mathfrak{A}/\mathfrak{o}} = [\mathfrak{C}_{\mathfrak{A}/\mathfrak{o}}:\mathfrak{A}]_{\mathfrak{o}} = \mathcal{N}_{\mathfrak{O}/\mathfrak{o}}(\mathfrak{C}_{\mathfrak{A}/\mathfrak{o}}^{-1}\mathfrak{A}) = \mathcal{N}_{\mathfrak{O}/\mathfrak{o}}(\mathfrak{A}^2\mathfrak{D}_{\mathfrak{O}/\mathfrak{o}}) = \mathcal{N}_{\mathfrak{O}/\mathfrak{o}}(\mathfrak{A})^2\,\mathfrak{d}_{\mathfrak{O}/\mathfrak{o}}$.

2. It suffices to prove that $\mathfrak{C}_{\mathfrak{O}/\mathfrak{o}} = \mathfrak{C}_{\mathfrak{o}'/\mathfrak{o}}\mathfrak{C}_{\mathfrak{O}/\mathfrak{o}'}$. Indeed, once this is done, we obtain $\mathfrak{D}_{\mathfrak{O}/\mathfrak{o}} = \mathfrak{C}_{\mathfrak{O}/\mathfrak{o}}^{-1} = \mathfrak{C}_{\mathfrak{o}'/\mathfrak{o}}^{-1}\mathfrak{C}_{\mathfrak{O}/\mathfrak{o}'}^{-1} = \mathfrak{D}_{\mathfrak{o}'/\mathfrak{o}}\mathfrak{D}_{\mathfrak{O}/\mathfrak{o}'}$ and, using 2.14.2.1,

$$\begin{aligned}\mathfrak{d}_{\mathfrak{O}/\mathfrak{o}} = \mathcal{N}_{\mathfrak{O}/\mathfrak{o}}(\mathfrak{D}_{\mathfrak{o}'/\mathfrak{o}}\mathfrak{D}_{\mathfrak{O}/\mathfrak{o}'}) &= \mathcal{N}_{\mathfrak{o}'/\mathfrak{o}}\circ\mathcal{N}_{\mathfrak{O}/\mathfrak{o}'}(\mathfrak{D}_{\mathfrak{o}'/\mathfrak{o}}\mathfrak{D}_{\mathfrak{O}/\mathfrak{o}'}) \\ &= \mathcal{N}_{\mathfrak{o}'/\mathfrak{o}}(\mathfrak{D}_{\mathfrak{o}'/\mathfrak{o}}^{[L:K']}\mathcal{N}_{\mathfrak{O}/\mathfrak{o}'}(\mathfrak{D}_{\mathfrak{O}/\mathfrak{o}'})) \\ &= \mathfrak{d}_{\mathfrak{o}'/\mathfrak{o}}^{[L:K']}\mathcal{N}_{\mathfrak{o}'/\mathfrak{o}}(\mathfrak{d}_{\mathfrak{O}/\mathfrak{o}'}).\end{aligned}$$

- $\mathfrak{C}_{\mathfrak{O}/\mathfrak{o}} \subset \mathfrak{C}_{\mathfrak{o}'/\mathfrak{o}}\mathfrak{C}_{\mathfrak{O}/\mathfrak{o}'}$: Let $x \in \mathfrak{C}_{\mathfrak{O}/\mathfrak{o}}$. Then

$$\mathfrak{o} \supset \mathsf{Tr}_{L/K}(x\mathfrak{O}) = \mathsf{Tr}_{K'/K}\circ\mathsf{Tr}_{L/K'}(x\mathfrak{O}\mathfrak{o}') = \mathsf{Tr}_{K'/K}(\mathsf{Tr}_{L/K'}(x\mathfrak{O})\mathfrak{o}'),$$

and therefore $\mathsf{Tr}_{L/K'}(x\mathfrak{O}) \subset \mathfrak{C}_{\mathfrak{o}'/\mathfrak{o}}$. It follows that

$$\mathfrak{o}' = \mathfrak{C}_{\mathfrak{o}'/\mathfrak{o}}^{-1}\mathfrak{C}_{\mathfrak{o}'/\mathfrak{o}} \supset \mathfrak{C}_{\mathfrak{o}'/\mathfrak{o}}^{-1}\mathsf{Tr}_{L/K'}(x\mathfrak{O}) = \mathsf{Tr}_{L/K'}(x\mathfrak{C}_{\mathfrak{o}'/\mathfrak{o}}^{-1}\mathfrak{O}),$$

hence $x\mathfrak{C}_{\mathfrak{o}'/\mathfrak{o}}^{-1}\mathfrak{O} = x(\mathfrak{C}_{\mathfrak{o}'/\mathfrak{o}}\mathfrak{O})^{-1} \subset \mathfrak{C}_{\mathfrak{O}/\mathfrak{o}'}$ and $x \in (\mathfrak{C}_{\mathfrak{o}'/\mathfrak{o}}\mathfrak{O})\mathfrak{C}_{\mathfrak{O}/\mathfrak{o}'} = \mathfrak{C}_{\mathfrak{o}'/\mathfrak{o}}\mathfrak{C}_{\mathfrak{O}/\mathfrak{o}'}$.

- $\mathfrak{C}_{\mathfrak{o}'/\mathfrak{o}}\mathfrak{C}_{\mathfrak{O}/\mathfrak{o}'} \subset \mathfrak{C}_{\mathfrak{O}/\mathfrak{o}}$: Let $x \in \mathfrak{C}_{\mathfrak{o}'/\mathfrak{o}}$ and $y \in \mathfrak{C}_{\mathfrak{O}/\mathfrak{o}'}$. Then

$$\mathsf{Tr}_{L/K}(xy\mathfrak{O}) = \mathsf{Tr}_{K'/K}(x\mathsf{Tr}_{L/K'}(y\mathfrak{O})) \subset \mathsf{Tr}_{K'/K}(x\mathfrak{o}') \subset \mathfrak{o},$$

and consequently $xy \in \mathfrak{C}_{\mathfrak{O}/\mathfrak{o}}$. □

Theorem 5.9.4. *Let $\mathfrak{o}$ be a Dedekind domain, $K = \mathsf{q}(\mathfrak{o})$, L/K a finite separable field extension, $\mathfrak{O} = \mathrm{cl}_L(\mathfrak{o})$, $L = K(a)$, and let $g \in K[X]$ be the minimal polynomial of a over K.*

1. *Suppose that $a \in \mathfrak{O}$, and let $\mathfrak{f} = (\mathfrak{o}[a]:\mathfrak{O}) = \{x \in L \mid x\mathfrak{O} \subset \mathfrak{o}[a]\}$ be the conductor of $\mathfrak{o}[a]$ (see 2.11.1). Then*

$$\mathfrak{C}_{\mathfrak{o}[a]/\mathfrak{o}} = \frac{1}{g'(a)}\,\mathfrak{o}[a] \quad \textit{and} \quad \mathfrak{D}_{\mathfrak{O}/\mathfrak{o}}\mathfrak{f} = g'(a)\mathfrak{O}.$$

In particular, $\mathfrak{O} = \mathfrak{o}[a]$ if and only if $\mathfrak{D}_{\mathfrak{O}/\mathfrak{o}} = g'(a)\mathfrak{O}$.

2. *Let $\mathfrak{p} \in \mathcal{P}(\mathfrak{O})$ and $g \in \mathfrak{o}_\mathfrak{p}[X]$.*

(a) *If $\mathfrak{P} \in \mathcal{P}(\mathfrak{O})$, $\mathfrak{P} \cap \mathfrak{o} = \mathfrak{p}$ and $v_\mathfrak{P}(g'(a)) = 0$, then it follows that $v_\mathfrak{P}(\mathfrak{D}_{\mathfrak{O}/\mathfrak{o}}) = 0$.*

(b) *If $v_\mathfrak{P}(g'(a)) = 0$ for all $\mathfrak{P} \in \mathcal{P}(\mathfrak{O})$ lying above $\mathfrak{p}$, then it follows that $\mathfrak{O}_\mathfrak{p} = \mathfrak{o}_\mathfrak{p}[a]$.*

Proof. 1. Let $g = X^n + a_{n-1}X^{n-1} + \ldots + a_0 \in \mathfrak{o}[X]$. By 1.6.10.2 the dual basis of $(1, a, \ldots, a^{n-1})$ is given by

$$\Big(\frac{\beta_0}{g'(a)}, \ldots, \frac{\beta_{n-1}}{g'(a)}\Big), \text{ where } \beta_i = a^{n-i-1} + \sum_{\nu=1}^{n-i-1} a_{i+\nu}a^{\nu-1} \text{ for all } i \in [0, n-1],$$

and since $(1, a, \ldots, a^{n-1})$ is an $\mathfrak{o}$-basis of $\mathfrak{o}[a]$, 5.9.2.1 implies

$$\mathfrak{C}_{\mathfrak{o}[a]/\mathfrak{o}} = \sum_{i=0}^{n-1} \mathfrak{o}\,\frac{\beta_i}{g'(a)} = \frac{1}{g'(a)} \sum_{i=0}^{n-1} \mathfrak{o}\beta_i = \frac{1}{g'(a)} \sum_{i=0}^{n-1} \mathfrak{o}a^i = \frac{1}{g'(a)}\,\mathfrak{o}[a].$$

For $x \in L$ we obtain

$$\begin{aligned} x \in \mathfrak{f} \iff x\mathfrak{O} \subset \mathfrak{o}[a] &\iff \frac{1}{g'(a)}x\mathfrak{O} \subset \frac{1}{g'(a)}\mathfrak{o}[a] = \mathfrak{C}_{\mathfrak{o}[a]/\mathfrak{o}} \\ &\iff \mathsf{Tr}_{L/K}\Big(\frac{1}{g'(a)}x\mathfrak{O}\Big) = \mathsf{Tr}_{L/K}\Big(\frac{1}{g'(a)}x\mathfrak{O}\mathfrak{o}[a]\Big) \subset \mathfrak{o} \\ &\iff \frac{1}{g'(a)}x \in \mathfrak{C}_{\mathfrak{O}/\mathfrak{o}} \iff x \in g'(a)\mathfrak{D}_{\mathfrak{O}/\mathfrak{o}}^{-1}. \end{aligned}$$

Therewith we obtain $\mathfrak{D}_{\mathfrak{O}/\mathfrak{o}}\mathfrak{f} = g'(a)\mathfrak{O}$, and consequently

$$\mathfrak{D}_{\mathfrak{O}/\mathfrak{o}} = g'(a)\mathfrak{O} \iff \mathfrak{f} = \mathfrak{O} \iff \mathfrak{O} = \mathfrak{o}[a].$$

2. Since $g \in \mathfrak{o}_\mathfrak{p}[X]$, we get $a \in \mathfrak{O}_\mathfrak{p}$ and $(\mathfrak{o}_\mathfrak{p}[a] : \mathfrak{O}_\mathfrak{p}) = (\mathfrak{o}[a] : \mathfrak{O})_\mathfrak{p} = \mathfrak{f}_\mathfrak{p}$. Hence 1. (applied with $\mathfrak{o}_\mathfrak{p}$) shows that $(\mathfrak{D}_{\mathfrak{O}/\mathfrak{o}}\mathfrak{f})_\mathfrak{p} = \mathfrak{D}_{\mathfrak{O}_\mathfrak{p}/\mathfrak{o}_\mathfrak{p}}\mathfrak{f}_\mathfrak{p} = g'(a)\mathfrak{O}_\mathfrak{p}$, and if $\mathfrak{P} \in \mathcal{P}(\mathfrak{O})$ lies above $\mathfrak{p}$, then $g'(a)\mathfrak{O}_\mathfrak{P} = (\mathfrak{D}_{\mathfrak{O}/\mathfrak{o}})_\mathfrak{P}\mathfrak{f}_\mathfrak{P}$, and therefore we obtain $v_\mathfrak{P}(g'(a)) = v_\mathfrak{P}(\mathfrak{D}_{\mathfrak{O}/\mathfrak{o}}) + v_\mathfrak{P}(\mathfrak{f})$. In particular, if $v_\mathfrak{P}(g'(a)) = 0$ then it apparently follows that $v_\mathfrak{P}(\mathfrak{D}_{\mathfrak{O}/\mathfrak{o}}) = v_\mathfrak{P}(\mathfrak{f}) = 0$. Now (a) is obvious and (b) follows by 2.7.9, since $\mathfrak{f}_\mathfrak{p} = \mathfrak{O}_\mathfrak{p}$ implies $\mathfrak{O}_\mathfrak{p} = \mathfrak{o}_\mathfrak{p}[a]$. $\square$

Our next aim is to study the behavior of differents and discriminants under completion. As already announced, this will enable us to use the powerful methods of valuation theory for their investigation. We tacitly use the notation and results from 5.3.4 and 5.5.9.

Theorem 5.9.5. *Let $\mathfrak{o}$ be a Dedekind domain, $K = \mathsf{q}(\mathfrak{o})$, L/K a finite separable field extension, $\mathfrak{O} = \mathrm{cl}_L(\mathfrak{o})$ and $\mathfrak{p} \in \mathcal{P}(\mathfrak{o})$. Then*

1. $\mathfrak{D}_{\mathfrak{O}/\mathfrak{o}}\widehat{\mathfrak{O}}_{\mathfrak{P}} = \mathfrak{D}_{\widehat{\mathfrak{O}}_{\mathfrak{P}}/\widehat{\mathfrak{o}}_{\mathfrak{p}}}$ and $v_{\mathfrak{P}}(\mathfrak{D}_{\mathfrak{O}/\mathfrak{o}}) = v_{\mathfrak{P}}(\mathfrak{D}_{\widehat{\mathfrak{O}}_{\mathfrak{P}}/\widehat{\mathfrak{o}}_{\mathfrak{p}}})$ for all $\mathfrak{P} \in \mathcal{P}(\mathfrak{O})$ above $\mathfrak{p}$.

2. We have

$$\mathfrak{d}_{\mathfrak{O}/\mathfrak{o}}\widehat{\mathfrak{o}}_{\mathfrak{p}} = \prod_{\substack{\mathfrak{P}\in\mathcal{P}(\mathfrak{O})\\ \mathfrak{P}\supset\mathfrak{p}}} \mathfrak{d}_{\widehat{\mathfrak{O}}_{\mathfrak{P}}/\widehat{\mathfrak{o}}_{\mathfrak{p}}} \quad \text{and} \quad v_{\mathfrak{p}}(\mathfrak{d}_{\mathfrak{O}/\mathfrak{o}}) = \sum_{\substack{\mathfrak{P}\in\mathcal{P}(\mathfrak{O})\\ \mathfrak{P}\supset\mathfrak{p}}} v_{\mathfrak{p}}(\mathfrak{d}_{\widehat{\mathfrak{O}}_{\mathfrak{P}}/\widehat{\mathfrak{o}}_{\mathfrak{p}}}).$$

Proof. In both cases it suffices to prove the first assertion.

We start with the formula for the different. Let $\mathfrak{P} = \mathfrak{P}_1, \ldots, \mathfrak{P}_r \in \mathcal{P}(\mathfrak{O})$ be the prime ideals of $\mathfrak{O}$ lying above $\mathfrak{p}$. Let $|\cdot|_{\mathfrak{p}}$ be a $\mathfrak{p}$-adic absolute value of K, and for $i \in [1,r]$ let $|\cdot|_{\mathfrak{P}_i}$ be a $\mathfrak{P}_i$-adic absolute value of L extending $|\cdot|_{\mathfrak{p}}$. By 5.5.9.1, $|\cdot|_{\mathfrak{P}_1}, \ldots, |\cdot|_{\mathfrak{P}_r}$ are the distinct extensions of $|\cdot|_{\mathfrak{p}}$ to L, and by 5.5.7.4(b) the image of the diagonal embedding $\delta\colon L \to \widehat{L}_{\mathfrak{P}_1} \times \ldots \times \widehat{L}_{\mathfrak{P}_r}$ is dense with respect to the product topology. We shall repeatedly use the weak approximation theorem 5.1.6.2(a).

We show first that $\mathfrak{C}_{\mathfrak{O}_{\mathfrak{p}}/\mathfrak{o}_{\mathfrak{p}}}$ is dense in $\mathfrak{C}_{\widehat{\mathfrak{O}}_{\mathfrak{P}}/\widehat{\mathfrak{o}}_{\mathfrak{p}}}$; we proceed in two steps.

a. $\mathfrak{C}_{\mathfrak{O}_{\mathfrak{p}}/\mathfrak{o}_{\mathfrak{p}}} \subset \mathfrak{C}_{\widehat{\mathfrak{O}}_{\mathfrak{P}}/\widehat{\mathfrak{o}}_{\mathfrak{p}}}$:

Let $x \in \mathfrak{C}_{\mathfrak{O}_{\mathfrak{p}}/\mathfrak{o}_{\mathfrak{p}}}$. We must prove that $\mathsf{Tr}_{L_{\mathfrak{P}}/K_{\mathfrak{p}}}(x\widehat{\mathfrak{O}}_{\mathfrak{P}}) \subset \widehat{\mathfrak{o}}_{\mathfrak{p}}$. Thus let $y \in \widehat{\mathfrak{O}}_{\mathfrak{P}}$, and let $(y_n)_{n\ge 0}$ be a sequence in L such that

$$(|y_n - y|_{\mathfrak{P}})_{n\ge 0} \to 0 \quad \text{and} \quad (|y_n|_{\mathfrak{P}_i})_{n\ge 0} \to 0 \ \text{ for all } \ i \in [2,r].$$

As $\widehat{\mathfrak{O}}_{\mathfrak{P}_i}$ is open in $L_{\mathfrak{P}_i}$, we may assume that $y_n \in \widehat{\mathfrak{O}}_{\mathfrak{P}_i} \cap L = \mathfrak{O}_{\mathfrak{P}_i}$ for all $n \ge 0$ and all $i \in [1,r]$. By 2.7.9.1 it follows that

$$y_n \in \bigcap_{i=1}^{r} \mathfrak{O}_{\mathfrak{P}_i} = \mathfrak{O}_{\mathfrak{p}} \quad \text{and therefore} \quad \mathsf{Tr}_{L/K}(xy_n) \in \mathfrak{o}_{\mathfrak{p}} \subset \widehat{\mathfrak{o}}_{\mathfrak{p}} \ \text{ for all } \ n \ge 0.$$

Since

$$|\cdot|_{\mathfrak{P}}\text{-}\lim_{n\to\infty} xy_n = xy, \qquad |\cdot|_{\mathfrak{P}_i}\text{-}\lim_{n\to\infty} xy_n = 0 \ \text{ for all } \ i \in [2,r]$$

and $\mathsf{Tr}_{L_{\mathfrak{P}_i}/K_{\mathfrak{p}}}$ is continuous (see 5.5.3.1), we obtain

$$|\cdot|_{\mathfrak{p}}\text{-}\lim_{n\to\infty} \mathsf{Tr}_{L_{\mathfrak{P}_i}/K_{\mathfrak{p}}}(xy_n) = \begin{cases} \mathsf{Tr}_{L_{\mathfrak{P}}/K_{\mathfrak{p}}}(xy) & \text{if } i = 1, \\ 0 & \text{if } i \in [2,r]. \end{cases}$$

Since

$$\mathsf{Tr}_{L/K}(xy_n) = \mathsf{Tr}_{L_\mathfrak{P}/K_\mathfrak{p}}(xy_n) + \sum_{i=2}^{r} \mathsf{Tr}_{L_{\mathfrak{P}_i}/K_\mathfrak{p}}(xy_n) \qquad \text{(see 5.5.7.4(a))},$$

it follows that

$$\begin{aligned}\mathsf{Tr}_{L_\mathfrak{P}/K_\mathfrak{p}}(xy) &= |\cdot|_\mathfrak{p}\text{-}\lim_{n\to\infty} \mathsf{Tr}_{L_\mathfrak{P}/K_\mathfrak{p}}(xy_n) \\ &= |\cdot|_\mathfrak{p}\text{-}\lim_{n\to\infty} \Big[\mathsf{Tr}_{L_\mathfrak{P}/K_\mathfrak{p}}(xy_n) + \sum_{i=2}^{r} \mathsf{Tr}_{L_{\mathfrak{P}_i}/K_\mathfrak{p}}(xy_n)\Big] \\ &= |\cdot|_\mathfrak{p}\text{-}\lim_{n\to\infty} \mathsf{Tr}_{L/K}(xy_n) \in \widehat{\mathfrak{o}}_\mathfrak{p}. \end{aligned} \qquad \square[\mathbf{a}.]$$

b. $\mathfrak{C}_{\widehat{\mathfrak{O}}_\mathfrak{P}/\widehat{\mathfrak{o}}_\mathfrak{p}} \subset \overline{\mathfrak{C}_{\mathfrak{O}_\mathfrak{p}/\mathfrak{o}_\mathfrak{p}}} \subset L_\mathfrak{P}$:
Let $x \in \mathfrak{C}_{\widehat{\mathfrak{O}}_\mathfrak{P}/\widehat{\mathfrak{o}}_\mathfrak{p}} \subset L_\mathfrak{P}$, and let $(x_n)_{n\ge0}$ be a sequence in L such that

$$(|x_n - x|_\mathfrak{P})_{n\ge0} \to 0 \qquad \text{and} \qquad (|x_n|_{\mathfrak{P}_i})_{n\ge0} \to 0 \ \text{ for all } \ i \in [2,r].$$

Let $(u_1, \ldots, u_m)$ be an $\mathfrak{o}_\mathfrak{p}$-basis of $\mathfrak{O}_\mathfrak{p}$. For all $j \in [1,m]$ we obtain

$$(\mathsf{Tr}_{L_{\mathfrak{P}_i}/K_\mathfrak{p}}(x_nu_j))_{n\ge0} \to \begin{cases} \mathsf{Tr}_{L_\mathfrak{P}/K_\mathfrak{p}}(xu_j) \in \widehat{\mathfrak{o}}_\mathfrak{p} & \text{if } i = 1, \\ 0 & \text{if } i \in [2,r], \end{cases}$$

and consequently

$$\begin{aligned}(\mathsf{Tr}_{L/K}(x_nu_j))_{n\ge0} &= \Big(\mathsf{Tr}_{L_\mathfrak{P}/K_\mathfrak{p}}(x_nu_j) + \sum_{i=2}^{r} \mathsf{Tr}_{L_{\mathfrak{P}_i}/K_\mathfrak{p}}(x_nu_j)\Big)_{n\ge0} \\ &\to \mathsf{Tr}_{L_\mathfrak{P}/K_\mathfrak{p}}(xu_j) \in \widehat{\mathfrak{o}}_\mathfrak{p}.\end{aligned}$$

As $\widehat{\mathfrak{o}}_\mathfrak{p}$ is open in $K_\mathfrak{p}$, we may assume that $\mathsf{Tr}_{L/K}(x_nu_j) \in \widehat{\mathfrak{o}}_\mathfrak{p} \cap K = \mathfrak{o}_\mathfrak{p}$ for all $n \ge 0$ and all $j \in [1,r]$. It follows that $\mathsf{Tr}_{L/K}(x_n\mathfrak{O}_\mathfrak{p}) \subset \mathfrak{o}_\mathfrak{p}$, hence $x_n \in \mathfrak{C}_{\mathfrak{O}_\mathfrak{p}/\mathfrak{o}_\mathfrak{p}}$ for all $n \ge 0$, and therefore $x \in \overline{\mathfrak{C}_{\mathfrak{O}_\mathfrak{p}/\mathfrak{o}_\mathfrak{p}}}$. $\square$[**b.**]

By **a** and **b** we obtain $\mathfrak{C}_{\mathfrak{O}_\mathfrak{P}/\mathfrak{o}_\mathfrak{p}}\widehat{\mathfrak{o}}_\mathfrak{p} \subset \mathfrak{C}_{\widehat{\mathfrak{O}}_\mathfrak{P}/\widehat{\mathfrak{o}}_\mathfrak{p}}\widehat{\mathfrak{o}}_\mathfrak{p} = \mathfrak{C}_{\widehat{\mathfrak{O}}_\mathfrak{P}/\widehat{\mathfrak{o}}_\mathfrak{p}} \subset \overline{\mathfrak{C}_{\mathfrak{O}_\mathfrak{P}/\mathfrak{o}_\mathfrak{p}}} \subset L_\mathfrak{P}$. But $\mathfrak{C}_{\mathfrak{O}_\mathfrak{P}/\mathfrak{o}_\mathfrak{p}}\widehat{\mathfrak{o}}_\mathfrak{p}$ is closed in $L_\mathfrak{P}$ by 5.2.9.2(b), hence

$$\mathfrak{C}_{\widehat{\mathfrak{O}}_\mathfrak{P}/\widehat{\mathfrak{o}}_\mathfrak{p}} \subset \overline{\mathfrak{C}_{\mathcal{O}_\mathfrak{P}/\mathfrak{o}_\mathfrak{p}}} \subset \mathfrak{C}_{\mathfrak{O}_\mathfrak{P}/\mathfrak{o}_\mathfrak{p}}\widehat{\mathfrak{o}}_\mathfrak{p} \subset \mathfrak{C}_{\widehat{\mathfrak{O}}_\mathfrak{P}/\widehat{\mathfrak{o}}_\mathfrak{p}},$$

and thus equality holds. Added together, it follows that

$$\mathfrak{C}_{\mathfrak{O}/\mathfrak{o}}\widehat{\mathfrak{O}}_\mathfrak{P} = (\mathfrak{C}_{\mathfrak{O}/\mathfrak{o}})_\mathfrak{p}\widehat{\mathfrak{O}}_\mathfrak{P} = \mathfrak{C}_{\mathfrak{O}_\mathfrak{p}/\mathfrak{o}_\mathfrak{p}}\widehat{\mathfrak{o}}_\mathfrak{p}\widehat{\mathfrak{O}}_\mathfrak{P} = \mathfrak{C}_{\widehat{\mathfrak{O}}_\mathfrak{P}/\widehat{\mathfrak{o}}_\mathfrak{p}}\widehat{\mathfrak{O}}_\mathfrak{P} = \mathfrak{C}_{\widehat{\mathfrak{O}}_\mathfrak{P}/\widehat{\mathfrak{o}}_\mathfrak{p}},$$

and

$$\mathfrak{D}_{\mathfrak{O}/\mathfrak{o}}\widehat{\mathfrak{O}}_\mathfrak{P} = \mathfrak{C}_{\mathfrak{O}/\mathfrak{o}}^{-1}\widehat{\mathfrak{O}}_\mathfrak{P} = (\mathfrak{C}_{\mathfrak{O}/\mathfrak{o}}\widehat{\mathfrak{O}}_\mathfrak{P})^{-1} = \mathfrak{C}_{\widehat{\mathfrak{O}}_\mathfrak{P}/\widehat{\mathfrak{o}}_\mathfrak{p}}^{-1} = \mathfrak{D}_{\widehat{\mathfrak{O}}_\mathfrak{P}/\widehat{\mathfrak{o}}_\mathfrak{p}}.$$

Now we prove the formula for the discriminant. Since $\mathfrak{O}_\mathfrak{p}$ is a principal ideal domain, $\mathfrak{D}_{\mathfrak{O}_\mathfrak{p}/\mathfrak{o}_\mathfrak{p}} = D\mathfrak{O}_\mathfrak{p}$ for some $D \in \mathfrak{O}_\mathfrak{p}$, and therefore

$$D\widehat{\mathfrak{O}}_\mathfrak{P} = \mathfrak{D}_{\mathfrak{O}_\mathfrak{p}/\mathfrak{o}_\mathfrak{p}}\widehat{\mathfrak{O}}_\mathfrak{P} = (\mathfrak{D}_{\mathfrak{O}/\mathfrak{o}})_\mathfrak{p}\widehat{\mathfrak{O}}_\mathfrak{P} = \mathfrak{D}_{\mathfrak{O}/\mathfrak{o}}\widehat{\mathfrak{O}}_\mathfrak{P} = \mathfrak{D}_{\widehat{\mathfrak{O}}_\mathfrak{P}/\widehat{\mathfrak{o}}_\mathfrak{p}}.$$

Using 5.5.7.4(a) and 2.14.4.3, we obtain

$$\begin{aligned}
\mathfrak{d}_{\mathfrak{O}/\mathfrak{o}}\widehat{\mathfrak{o}}_{\mathfrak{p}} &= \mathcal{N}_{\mathfrak{O}/\mathfrak{o}}(\mathfrak{D}_{\mathfrak{O}/\mathfrak{o}})\widehat{\mathfrak{o}}_{\mathfrak{p}} = \mathcal{N}_{\mathfrak{O}/\mathfrak{o}}(\mathfrak{D}_{\mathfrak{O}/\mathfrak{o}})\mathfrak{o}_{\mathfrak{p}}\widehat{\mathfrak{o}}_{\mathfrak{p}} = \mathcal{N}_{\mathfrak{O}_{\mathfrak{p}}/\mathfrak{o}_{\mathfrak{p}}}(\mathfrak{D}_{\mathfrak{O}_{\mathfrak{p}}/\mathfrak{o}_{\mathfrak{p}}})\widehat{\mathfrak{o}}_{\mathfrak{p}} \\
&= \mathsf{N}_{L/K}(D)\widehat{\mathfrak{o}}_{\mathfrak{p}} = \prod_{i=1}^{r} \mathsf{N}_{L_{\mathfrak{P}_i}/K_{\mathfrak{p}}}(D)\mathfrak{o}_{\mathfrak{p}} = \prod_{i=1}^{r} \mathcal{N}_{\widehat{\mathfrak{O}}_{\mathfrak{P}_i}/\widehat{\mathfrak{o}}_{\mathfrak{p}}}(D\widehat{\mathfrak{O}}_{\mathfrak{P}_i}) \\
&= \prod_{i=1}^{r} \mathcal{N}_{\widehat{\mathfrak{O}}_{\mathfrak{P}_i}/\widehat{\mathfrak{o}}_{\mathfrak{p}}}(\mathfrak{D}_{\widehat{\mathfrak{O}}_{\mathfrak{P}_i}/\widehat{\mathfrak{o}}_{\mathfrak{p}}}) = \prod_{i=1}^{r} \mathfrak{d}_{\widehat{\mathfrak{O}}_{\mathfrak{P}_i}/\widehat{\mathfrak{o}}_{\mathfrak{p}}}. \qquad \square
\end{aligned}$$

Definition and Theorem 5.9.6. (Dedekind's different theorem, local case) *Let L/K be a finite separable extension of complete discrete valued fields, $n = [L\colon K]$, $e = e(L/K)$ and $f = f(L/K)$.*
We tacitly use the notation and results from 5.3.1, 5.3.4, 5.5.1, and 5.6.1. In addition, we define

$$\mathfrak{D}_{L/K} = \mathfrak{D}_{\mathcal{O}_L/\mathcal{O}_K}, \quad \mathfrak{d}_{L/K} = \mathfrak{d}_{\mathcal{O}_L/\mathcal{O}_K} \text{ and } \mathcal{N}_{L/K} = \mathcal{N}_{\mathcal{O}_L/\mathcal{O}_K}.$$

We call $\mathfrak{D}_{L/K}$ the **different**, $\mathfrak{d}_{L/K}$ the **discriminant** and $\mathcal{N}_{L/K}$ the **ideal norm** of the field extension L/K.

1. *If $c \in \mathcal{O}_L$, then*

$$\begin{aligned}
\mathsf{Tr}_{L/K}(c) + \mathfrak{p}_K &= e\,\mathsf{Tr}_{\mathsf{k}_L/\mathsf{k}_K}(c + \mathfrak{p}_L) \quad \text{and} \\
\mathsf{N}_{L/K}(c) + \mathfrak{p}_K &= \mathsf{N}_{\mathsf{k}_L/\mathsf{k}_K}(c + \mathfrak{p}_L)^e.
\end{aligned}$$

2. $v_L(\mathfrak{D}_{L/K}) \ge e - 1$ *and* $v_K(\mathfrak{d}_{L/K}) \ge (e-1)f$.

3. *The following assertions are equivalent:*

(a) *L/K is tamely ramified.*

(b) $\mathsf{Tr}_{L/K}(\mathcal{O}_L) = \mathcal{O}_K$.

(c) $v_L(\mathfrak{D}_{L/K}) < e$.

(d) $v_L(\mathfrak{D}_{L/K}) = e - 1$.

4. *The following assertions are equivalent:*

(a) *L/K is unramified.*

(b) $v_L(\mathfrak{D}_{L/K}) = 0$.

(c) $v_K(\mathfrak{d}_{L/K}) = 0$.

5. *Let L/K be wildly ramified,* $\mathrm{char}(K) = 0$, *and suppose that $\mathsf{k}_L/\mathsf{k}_K$ is separable. Then*

$$v_L(\mathfrak{D}_{L/K}) \le e - 1 + e v_K(e).$$

Proof. 1. Let $c \in \mathcal{O}_L$ and $\boldsymbol{\omega} = (\omega_1, \ldots, \omega_f) \in \mathsf{k}_L^f$ be a k_K-basis of k_L. For $i \in [0, e]$ set $\mathcal{A}_i = \mathfrak{p}_L^{e-i}/\mathfrak{p}_K\mathcal{O}_L$. Then

$$\mathbf{0} = \mathfrak{p}_L^e/\mathfrak{p}_K\mathcal{O}_L = \mathcal{A}_0 \subsetneq \mathcal{A}_1 \subsetneq \ldots \subsetneq \mathcal{A}_{e-1} = \mathfrak{p}_L/\mathfrak{p}_K\mathcal{O}_L \subsetneq \mathcal{A}_e = \mathcal{O}_L/\mathfrak{p}_K\mathcal{O}_L$$

is a sequence of $\mathcal{O}_L$-modules and vector spaces over k_K. For $i \in [1, e]$ let

$$\psi_i \colon \mathcal{A}_i/\mathcal{A}_{i-1} = (\mathfrak{p}_L^{e-i}/\mathfrak{p}_K\mathcal{O}_L)/(\mathfrak{p}_L^{e-i+1}/\mathfrak{p}_K\mathcal{O}_L) = \mathfrak{p}_L^{e-i}/\mathfrak{p}_L^{e-i+1} \overset{\sim}{\to} \mathsf{k}_L$$

be the natural epimorphism, given by $\psi_i(\pi_L^{e-i}a + \mathfrak{p}_L^{e-i+1}) = a + \mathfrak{p}_L$ for all $a \in \mathcal{O}_L$, where $\pi_L \in \mathcal{O}_L$ is a prime element (see 2.10.5.1). Then

$$\dim_{\mathsf{k}_K} \mathcal{O}_L/\mathfrak{p}_K\mathcal{O}_L = \sum_{i=1}^{e} \dim_{\mathsf{k}_K}(\mathcal{A}_i/\mathcal{A}_{i-1}) = ef = n.$$

Let $\psi_i^f : \mathcal{A}_i^f \to \mathsf{k}_K^f$ be defined component-wise. If $\boldsymbol{w}^{(i)} = (w_1^{(i)}, \ldots, w_f^{(i)}) \in \mathcal{A}_i^f$ is such that $\psi_i^f \boldsymbol{w}^{(i)}) = \boldsymbol{\omega}$, then $(\boldsymbol{w}^{(1)}, \ldots, \boldsymbol{w}^{(i)})$ is a k_K-basis of $\mathcal{A}_i$, $\psi_i^f(\boldsymbol{w}^{(j)}) = \mathbf{0}$ for all $j \in [1, i-1]$, and

$$(c\boldsymbol{w}^{(1)}, \ldots, c\boldsymbol{w}^{(i)}) = (\boldsymbol{w}^{(1)}, \ldots, \boldsymbol{w}^{(i)})C_i, \quad \text{where } C_i = \begin{pmatrix} C_{1,1} & C_{1,2} & \ldots & C_{1,i} \\ 0 & C_{2,2} & \ldots & C_{2,i} \\ . & . & \ddots & . \\ 0 & 0 & \ldots & C_{i,i} \end{pmatrix}$$

is a matrix in $\mathsf{M}_{if}(\mathsf{k}_K)$, built with boxes $C_{i,\nu} \in \mathsf{M}_f(\mathsf{k}_K)$ for $1 \le j \le \nu \le i$. Explicitly,

$$c\boldsymbol{w}^{(i)} = \sum_{\nu=1}^{i} \boldsymbol{w}^{(\nu)}C_{\nu,i}, \qquad \psi_i^f(c\boldsymbol{w}^{(i)}) = c\boldsymbol{\omega} = \psi_i^f(\boldsymbol{w}^{(i)}C_{i,i}) = \boldsymbol{\omega}C_{i,i},$$

which implies

$$\mathsf{Tr}_{\mathsf{k}_L/\mathsf{k}_K}(c+\mathfrak{p}_L) = \mathrm{trace}(C_{i,i}) \text{ and } \mathsf{N}_{\mathsf{k}_L/\mathsf{k}_K}(c+\mathfrak{p}_L) = \det(C_{i,i}) \text{ for all } i \in [1, e].$$

Now $\dim_{\mathsf{k}_K}(\mathcal{O}_L/\mathfrak{p}_K\mathcal{O}_L) = n$, and $\boldsymbol{w} = (\boldsymbol{w}^{(1)}, \ldots, \boldsymbol{w}^{(e)}) \in (\mathcal{O}_L/\mathfrak{p}_K\mathcal{O}_L)^n$ is a k_K-basis of $\mathcal{A}_e = \mathcal{O}_L/\mathfrak{p}_K\mathcal{O}_L$. If $\boldsymbol{u} = (u_1, \ldots, u_n) \in \mathcal{O}_L^n$ is a representative for $\boldsymbol{w}$, then $\boldsymbol{u}$ is an $\mathcal{O}_K$-basis of $\mathcal{O}_L$ by 2.10.11.2, and therefore $\boldsymbol{u}$ is a K-basis of L.

Let $A \in \mathsf{M}_n(\mathcal{O}_L)$ be such that $c\boldsymbol{u} = \boldsymbol{u}A$. Then we get $\mathsf{Tr}_{L/K}(c) = \mathrm{trace}(A)$, $\mathsf{N}_{L/K}(c) = \det(A)$ and $c\boldsymbol{w} = \boldsymbol{w}\overline{A}$, where $\overline{A} = A + \mathfrak{p}_K\mathsf{M}_n(\mathcal{O}_L) \in \mathsf{M}_n(\mathsf{k}_K)$. On the other hand, $c\boldsymbol{w} = \boldsymbol{w}C_e$ implies $\overline{A} = C_e$. Hence we finally obtain

$$\mathsf{Tr}_{L/K}(c) + \mathfrak{p}_K = \mathrm{trace}(\overline{A}) = \mathrm{trace}(C_e) = \sum_{i=1}^{e} \mathrm{trace}(C_{i,i}) = e\mathsf{Tr}_{\mathsf{k}_L/\mathsf{k}_K}(c + \mathfrak{p}_L)$$

and

$$\mathsf{N}_{L/K}(c) + \mathfrak{p}_K = \det(\overline{A}) = \det(C_e) = \prod_{i=1}^{e} \det(C_{i,i}) = \mathsf{N}_{\mathsf{k}_L/\mathsf{k}_K}(c + \mathfrak{p}_L)^e.$$

2. Since $\mathfrak{p}_L/\mathfrak{p}_K\mathcal{O}_L \subset \mathcal{O}_L/\mathfrak{p}_K\mathcal{O}_L$ and $\dim_{\mathsf{k}_K}(\mathfrak{p}_L/\mathfrak{p}_K\mathcal{O}_L) = (e-1)f$, there exists a k_K-basis $(\omega_1, \ldots, \omega_n)$ of $\mathcal{O}_L/\mathfrak{p}_K\mathcal{O}_L$ such that $\omega_i \in \mathfrak{p}_L/\mathfrak{p}_K\mathcal{O}_L$ for all $i \in [1, (e-1)f]$. For $i \in [1, n]$ let $u_i \in \mathcal{O}_L$ such that $u_i + \mathfrak{p}_K\mathcal{O}_L = \omega_i$. Then $(u_1, \ldots, u_n)$ is an $\mathcal{O}_K$-basis of $\mathcal{O}_L$ by 2.10.11.2, and

$$\mathfrak{d}_{L/K} = d_{L/K}(u_1, \ldots, u_n)\mathcal{O}_K = \det(\mathsf{Tr}_{L/K}(u_iu_j))_{i,j\in[1,n]}\mathcal{O}_K$$

by 5.9.2.3. By 1., the first $(e-1)f$ rows of the matrix $(\mathsf{Tr}_{L/K}(u_iu_j))_{i,j\in[1,n]}$ lie in $\mathsf{Tr}_{L/K}(\mathfrak{p}_L) \subset \mathfrak{p}_K$. Hence $\mathfrak{d}_{L/K} \subset \mathfrak{p}_K^{(e-1)f}$, which implies

$$(e-1)f \le v_K(\mathfrak{d}_{L/K}) = v_K(\mathcal{N}_{L/K}(\mathfrak{D}_{L/K}) = f v_L(\mathfrak{D}_{L/K}),$$

and $v_L(\mathfrak{D}_{L/K}) \ge e - 1$.

3. (a) $\Leftrightarrow$ (b) Since $\mathsf{Tr}_{L/K}(\mathcal{O}_L) \in \mathcal{J}'(\mathcal{O}_K)$, it follows that $\mathsf{Tr}_{L/K}(\mathcal{O}_L) = \mathcal{O}_K$ if and only if there exists some $a \in \mathcal{O}_L$ such that $\mathsf{Tr}_{L/K}(a) \notin \mathfrak{p}$, and by 1. this holds if and only if there exists some $\alpha \in \mathsf{k}_L$ such that $e\,\mathsf{Tr}_{\mathsf{k}_L/\mathsf{k}_K}(\alpha) \ne 0$.

If L/K is wildly ramified, then either $\mathsf{k}_L/\mathsf{k}_K$ is inseparable or $\mathrm{char}(\mathsf{k}_K) \mid e$, and in either case we obtain $e\,\mathsf{Tr}_{\mathsf{k}_L/\mathsf{k}_K}(\alpha) = 0$ for all $\alpha \in \mathsf{k}_L$.

If L/K is tamely ramified, then $\mathrm{char}(\mathsf{k}_K) \nmid e$ and $\mathsf{k}_L/\mathsf{k}_K$ is separable. Hence 1.6.2.2(a) implies the existence of some $\alpha \in \mathsf{k}_L$ such that $e\,\mathsf{Tr}_{\mathsf{k}_L/\mathsf{k}_K}(\alpha) \ne 0$.

(b) $\Leftrightarrow$ (c) By definition

$$\begin{aligned} v_L(\mathfrak{D}_{L/K}) \ge e &\Longleftrightarrow \mathfrak{D}_{L/K} \subset \pi_K\mathcal{O}_L \\ &\Longleftrightarrow \mathfrak{C}_{L/K} \supset \pi_K^{-1}\mathcal{O}_L \Longleftrightarrow \mathsf{Tr}_{L/K}(\pi_K^{-1}\mathcal{O}_L) = \pi_K^{-1}\mathsf{Tr}_{L/K}(\mathcal{O}_L) \subset \mathcal{O}_K \\ &\Longleftrightarrow \mathsf{Tr}_{L/K}(\mathcal{O}_L) \subset \pi_K\mathcal{O}_K = \mathfrak{p}_K \Longleftrightarrow \mathsf{Tr}_{L/K}(\mathcal{O}_L) \ne \mathcal{O}_K. \end{aligned}$$

(c) $\Leftrightarrow$ (d) Obvious by 2.

4. (a) $\Rightarrow$ (b) If L/K is unramified, then it is tamely ramified and $e = 1$, which implies $v_L(\mathfrak{D}_{L/K}) = 0$ by 3.

(b) $\Rightarrow$ (a) If $v_L(\mathfrak{D}_{L/K}) = 0$, then $e = 1$ by 2., and L/K is tamely ramified by 3. Hence $\mathsf{k}_L/\mathsf{k}_K$ is separable, and L/K is unramified.

(b) $\Leftrightarrow$ (c) Obvious, since $v_K(\mathfrak{d}_{L/K}) = f v_L(\mathfrak{D}_{L/K})$.

5. Let T be the inertia field of L/K. Then T/K is unramified, hence $v_K = v_T$ and $\mathfrak{D}_{T/K} = \mathcal{O}_T$ by 2. L/T is fully ramified, $[L:T] = e$, and $\mathfrak{D}_{L/K} = \mathfrak{D}_{T/K}\mathfrak{D}_{L/T} = \mathfrak{D}_{L/T}$ by 5.9.3.2. By 5.7.1.2 we obtain $\mathcal{O}_L = \mathcal{O}_T[\pi]$ for some prime element $\pi \in \mathcal{O}_L$, and the minimal polynomial g of π over T is an Eisenstein polynomial, say $g = X^e + a_{e-1}X^{e-1} + \ldots + a_1X + a_0 \in \mathcal{O}_T[X]$,

where $v_K(a_i) \ge 1$ for all $i \in [1, e-1]$ and $v_K(a_0) = 1$. Since $\mathfrak{D}_{L/K} = g'(\pi)\mathcal{O}_L$ by 5.9.3.1, we obtain $v_L(\mathfrak{D}_{L/K}) = v_L(g'(\pi))$. Now

$$g'(\pi) = \sum_{i=1}^{e-1} i a_i \pi^{i-1} + e\pi^{e-1},$$

and $\operatorname{char}(K) = 0$ implies that $v_L(e\pi^{e-1}) = ev_K(e) + e - 1 \equiv -1 \bmod e$. If $i \in [1, e-1]$, then $v_L(ia_i\pi^{i-1}) \equiv i - 1 \bmod e$. Hence $v_L(e\pi^{e-1})$ and the finite values among $v_L(ia_i\pi^{i-1})$ for $i \in [1, e-1]$ are all distinct, and therefore

$$\begin{aligned} v_L(g'(\pi)) &= \min\{v_L(e\pi^{e-1}), v_L(ia_i\pi^{i-1}) \mid i \in [1, e-1]\} \\ &\le v_L(e\pi^{e-1}) = e - 1 + ev_K(e). \end{aligned}$$

□

Theorem 5.9.7 (Dedekind's different and discriminant theorem, global case). *Let $\mathfrak{o}$ be a Dedekind domain, $K = \mathsf{q}(\mathfrak{o})$, L/K a finite separable field extension and $\mathfrak{O} = \operatorname{cl}_L(\mathfrak{o})$.*

1. *Let $\mathfrak{P} \in \mathcal{P}(\mathfrak{O})$ and $\mathfrak{P} \cap \mathfrak{o} = \mathfrak{p}$. Then $v_{\mathfrak{P}}(\mathfrak{D}_{\mathfrak{O}/\mathfrak{o}}) \ge e(\mathfrak{P}/\mathfrak{p}) - 1$, and equality holds if and only if $\mathfrak{P}/\mathfrak{p}$ is tamely ramified. Moreover, $\mathfrak{P}/\mathfrak{p}$ is unramified if and only if $v_{\mathfrak{P}}(\mathfrak{D}_{\mathfrak{O}/\mathfrak{o}}) = 0$.*
2. *Let $\mathfrak{p} \in \mathcal{P}(\mathfrak{o})$. Then*

$$v_{\mathfrak{p}}(\mathfrak{d}_{\mathfrak{O}/\mathfrak{o}}) \ge [L\colon K] - \sum_{\substack{\mathfrak{P}\in\mathcal{P}(\mathfrak{O}) \\ \mathfrak{P}\supset\mathfrak{p}}} f(\mathfrak{P}/\mathfrak{p}),$$

and equality holds if and only if $\mathfrak{p}$ is tamely ramified in L. Moreover, $\mathfrak{p}$ is unramified in L if and only if $v_{\mathfrak{p}}(\mathfrak{d}_{L/K}) = 0$.

3. *The ideal class $[\mathfrak{d}_{\mathfrak{O}/\mathfrak{o}}] \in \mathcal{C}(\mathfrak{o})$ is a square in $\mathcal{C}(\mathfrak{o})$.*

Proof. 1. Recall from 5.6.1 that $e(\mathfrak{P}/\mathfrak{p}) = e(K_{\mathfrak{P}}/K_{\mathfrak{p}})$. Hence $\mathfrak{P}/\mathfrak{p}$ is unramified [tamely ramified] if and only if $L_{\mathfrak{P}}/K_{\mathfrak{p}}$ is unramified [tamely ramified]. By 5.9.5 we obtain $\mathfrak{D}_{\mathfrak{O}/\mathfrak{o}}\widehat{\mathfrak{O}}_{\mathfrak{P}} = \mathfrak{D}_{\widehat{\mathfrak{O}}_{\mathfrak{P}}/\widehat{\mathfrak{o}}_{\mathfrak{p}}} = \mathfrak{D}_{L_{\mathfrak{P}}/K_{\mathfrak{p}}}$ and consequently $v_{\mathfrak{P}}(\mathfrak{D}_{\mathfrak{O}/\mathfrak{o}}) = v_{\mathfrak{P}}(\mathfrak{D}_{L_{\mathfrak{P}}/K_{\mathfrak{p}}})$. Now the assertions follow by 5.9.6.

2. Using 5.5.9.2 and 2.14.4.1, we obtain

$$\begin{aligned} v_{\mathfrak{p}}(\mathfrak{d}_{\mathfrak{O}/\mathfrak{o}}) = v_{\mathfrak{p}}(\mathcal{N}_{\mathfrak{O}/\mathfrak{o}}(\mathfrak{D}_{\mathfrak{O}/\mathfrak{o}})) &= \sum_{\substack{\mathfrak{P}\in\mathcal{P}(\mathfrak{O}) \\ \mathfrak{P}\supset\mathfrak{p}}} f(\mathfrak{P}/\mathfrak{p}) v_{\mathfrak{P}}(\mathfrak{D}_{\mathfrak{O}/\mathfrak{o}}) \\ &\ge \sum_{\substack{\mathfrak{P}\in\mathcal{P}(\mathfrak{O}) \\ \mathfrak{P}\supset\mathfrak{p}}} f(\mathfrak{P}/\mathfrak{p})[e(\mathfrak{P}/\mathfrak{p}) - 1] = [L\colon K] - \sum_{\substack{\mathfrak{P}\in\mathcal{P}(\mathfrak{O}) \\ \mathfrak{P}\supset\mathfrak{p}}} f(\mathfrak{P}/\mathfrak{p}), \end{aligned}$$

and equality holds if and only if $v_{\mathfrak{P}}(\mathfrak{D}_{\mathfrak{O}/\mathfrak{o}}) = e(\mathfrak{P}/\mathfrak{p}) - 1$ and thus $\mathfrak{P}/\mathfrak{p}$ is tamely ramified for all $\mathfrak{P} \in \mathcal{P}(\mathfrak{O})$ lying above $\mathfrak{p}$. Hence equality holds if and only if $\mathfrak{p}$ is tamely ramified in L. Moreover, $v_{\mathfrak{p}}(\mathfrak{d}_{\mathfrak{O}/\mathfrak{o}}) = 0$ if and only if $e(\mathfrak{P}/\mathfrak{p}) = 1$ for all $\mathfrak{P} \in \mathcal{P}(\mathfrak{O})$ lying above $\mathfrak{p}$, that is, if and only if $\mathfrak{p}$ is unramified in L.

3. By 5.9.2.4. □

Definition and Theorem 5.9.8. For a finite extension L/K of algebraic number fields we tacitly use the notation of Section 3. We define

$$\mathfrak{D}_{L/K} = \mathfrak{D}_{\mathfrak{o}_L/\mathfrak{o}_K}, \quad \mathfrak{d}_{L/K} = \mathsf{d}_{\mathfrak{o}_L/\mathfrak{o}_K} \text{ and } \mathcal{N}_{L/K} = \mathcal{N}_{\mathfrak{o}_L/\mathfrak{o}_K} : \mathcal{I}'_L \to \mathcal{I}'_K.$$

We call $\mathfrak{D}_{L/K}$ the **different**, $\mathfrak{d}_{L/K}$ the **discriminant** and $\mathcal{N}_{L/K}$ the **ideal norm** of the field extension L/K.

1. Let L/K be a finite extension of algebraic number fields, and let K' be an intermediate field of L/K. Then

$$\mathfrak{D}_{L/K} = \mathfrak{D}_{K'/K}\mathfrak{D}_{L/K'} \quad \text{and} \quad \mathfrak{d}_{L/K} = \mathfrak{d}_{K'/K}^{[L:K']}\mathcal{N}_{K'/K}(\mathfrak{d}_{L/K'}).$$

2. Let K be an algebraic number field. We call $\mathfrak{D}_K = \mathfrak{D}_{K/\mathbb{Q}}$ the **different** of K. *Then $\mathfrak{N}(\mathfrak{D}_K) = |\mathsf{d}_K|$ and $\mathfrak{d}_{K/\mathbb{Q}} = \mathsf{d}_K\mathbb{Z}$. In particular, if $K \neq \mathbb{Q}$, then*

- *a prime p is ramified in K if and only if $p \,|\, \mathsf{d}_K$.*
- *$|\mathsf{d}_K| > 1$ and $K/\mathbb{Q}$ is ramified.*
- *If L/K is a finite field extension, then $\mathsf{d}_K^{[L:K]} \,|\, \mathsf{d}_L$.*

3. Let L/K be a finite extension of algebraic number fields and $\mathfrak{p} \in \mathcal{P}_K$. Then

$$\mathfrak{D}_{L/K}\widehat{\mathfrak{o}}_{\mathfrak{P}} = \mathfrak{D}_{L_{\mathfrak{P}}/K_{\mathfrak{p}}} \quad \text{for all } \mathfrak{P} \in \mathcal{P}_L \text{ such that } \mathfrak{P} \supset \mathfrak{p},$$

and

$$\mathfrak{d}_{L/K}\widehat{\mathfrak{o}}_{\mathfrak{p}} = \prod_{\substack{\mathfrak{P} \in \mathcal{P}_L \\ \mathfrak{P} \supset \mathfrak{p}}} \mathfrak{d}_{L_{\mathfrak{P}}/K_{\mathfrak{p}}}.$$

Proof. 1. By 5.9.3.2.

2. Let $(u_1, \ldots, u_n)$ be an integral basis of K. Then 5.9.2.3 implies

$$\mathfrak{d}_{K/\mathbb{Q}} = \mathfrak{d}_{\mathfrak{o}_K/\mathbb{Z}} = d_{K/\mathbb{Q}}(u_1, \ldots, u_n)\mathbb{Z} = \mathsf{d}_K\mathbb{Z},$$

and since $\mathfrak{d}_{K/\mathbb{Q}} = \mathcal{N}_{K/\mathbb{Q}}(\mathfrak{D}_K)$, it follows that $\mathfrak{N}(\mathfrak{D}_K) = |\mathsf{d}_K|$. If $K \neq \mathbb{Q}$, then $|\mathsf{d}_K| > 1$ by 3.5.6, and by 5.9.7.2 a prime p is ramified in K if and only if $p \,|\, \mathsf{d}_K$.

If L/K is a finite field extension, then

$$\mathsf{d}_L\mathbb{Z} = \mathfrak{d}_{L/\mathbb{Q}} = \mathfrak{d}_{K/\mathbb{Q}}^{[L:K]}\mathcal{N}_{K/\mathbb{Q}}(\mathfrak{d}_{L/K}) \subset \mathsf{d}_K^{[L:K]}, \text{ and therefore } \mathsf{d}_K^{[L:K]} \,|\, \mathsf{d}_L.$$

3. By 5.9.5. □

We end this chapter with an investigation of the arithmetic of radical extensions $K(\sqrt[n]{a}\,)$ of a field K for some $n \geq 2$. By Capelli's Theorem 1.7.9 the polynomial $X^n - a$ is irreducible in $K[X]$ if and only if $a \notin K^p$ for all primes p dividing n, and additionally $a \notin -4K^4$ if $4 \mid n$. If K contains a primitive 4-th root of unity, then $-4K^4 \subset K^2$, and the latter condition is redundant.

Theorem 5.9.9 (Arithmetic of radical extensions). *Let $\mathfrak{o}$ be a Dedekind domain, $K = \mathsf{q}(\mathfrak{o})$, $a \in K$ and $n \in \mathbb{N}$ such that the polynomial $X^n - a \in K[X]$ is irreducible. Suppose that $L = K(\alpha)$, where $\alpha^n = a$, let $\mathfrak{O} = \mathrm{cl}_L(\mathfrak{o})$, $\mathfrak{p} \in \mathcal{P}(\mathfrak{o})$ and* $\mathrm{char}(\mathsf{k}_\mathfrak{p}) \nmid n$.

1. *Let $r = (n, v_\mathfrak{p}(a))$, $n = re$, $K' = K(\alpha^e)$ and $\mathfrak{o}' = \mathrm{cl}_{K'}(\mathfrak{o})$. Then $[L : K'] = e$, $\mathfrak{p}$ is unramified in K', and every prime ideal $\mathfrak{p}' \in \mathcal{P}(\mathfrak{o}')$ lying above $\mathfrak{p}$ is fully ramified in L. In particular, $\mathfrak{p}$ is unramified in L if and only if $n \mid v_\mathfrak{p}(a)$.*

2. *Let $a \in \mathfrak{o}$, $v_\mathfrak{p}(a) = 0$ and $\zeta \in K$ be a primitive n-th root of unity. Let $t \in \mathbb{N}$ be the greatest divisor of n such that $\overline{a} = a + \mathfrak{p} \in \mathsf{k}_\mathfrak{p}^{\times t}$, and set $n = tf$. Then $\mathfrak{p}\mathfrak{O} = \mathfrak{P}_1 \cdot \ldots \cdot \mathfrak{P}_t$, where $\mathfrak{P}_1, \ldots, \mathfrak{P}_t \in \mathcal{P}(\mathfrak{O})$ are distinct, and $f(\mathfrak{P}_i/\mathfrak{p}) = f$ for all $i \in [1, t]$.*

Proof. 1. Let $v_\mathfrak{p}(a) = rd$, $\mathfrak{P} \in \mathcal{P}(\mathfrak{O})$ and $\mathfrak{p}' = \mathfrak{P} \cap \mathfrak{o}'$. Since $(\alpha^e)^r = a$, we obtain $[K' : K] = r$ and $[L : K'] = e$. If $b \in \mathfrak{p} \setminus \mathfrak{p}^2$ and $\alpha_1 = b^{-d}\alpha^e$, then $K' = K(\alpha_1)$, $\alpha_1^r = b^{-rd}a$ and $v_\mathfrak{p}(b^{-rd}a) = -rd + \mathsf{v}_\mathfrak{p}(a) = 0$. The polynomial $g_1 = X^r - b^{-dr}a \in \mathfrak{o}_\mathfrak{p}[X]$ is the minimal polynomial of α_1 over K, $g_1'(\alpha_1) = r\alpha_1^{r-1}$, and for all $\mathfrak{p}' \in \mathcal{P}(\mathfrak{o}')$ lying above $\mathfrak{p}$ we obtain $v_{\mathfrak{p}'}(g'(\alpha)) = 0$. By 5.9.4.2(a) this implies $v_{\mathfrak{p}'}(\mathfrak{D}_{\mathfrak{o}'/\mathfrak{o}}) = 0$, and thus $\mathfrak{p}'/\mathfrak{p}$ is unramified by 5.9.7.1. Hence $\mathfrak{p}$ is unramified in K'.

Now it follows immediately that $L = K'(\alpha)$, $\alpha^e \in K'$, $[L\!:\!K'] = e$ and $rv_{\mathfrak{p}'}(\alpha^e) = v_{\mathfrak{p}'}(\alpha^n) = v_{\mathfrak{p}'}(a) = v_\mathfrak{p}(a)$, hence $\mathsf{v}_{\mathfrak{p}'}(\alpha^e) = d$.

Since $ev_\mathfrak{P}(\alpha) = v_\mathfrak{P}(\alpha^e) = e(\mathfrak{P}/\mathfrak{p}')v_{\mathfrak{p}'}(\alpha^e) = e(\mathfrak{P}/\mathfrak{p}')d$ and $(d, e) = 1$, we get $e \mid e(\mathfrak{P}/\mathfrak{p}')$ and therefore $e = e(\mathfrak{P}/\mathfrak{p}')$.

2. Let $\alpha \in \mathsf{k}_\mathfrak{p}$ be such that $\overline{a} = \alpha^t$. Since $\mathrm{char}(\mathsf{k}_K) \nmid n$, $\overline{\zeta} = \zeta + \mathfrak{p} \in \mathsf{k}_\mathfrak{p}$ is also a primitive n-th root of unity, and

$$X^n - \overline{a} = \prod_{j=0}^{t-1}(X^f - \overline{\zeta}^{fj}\alpha)$$

By the maximal choice of t it follows that $\overline{\zeta}^{fj}\alpha \notin \mathsf{k}_\mathfrak{p}^p$ for all primes p dividing f, and thus the polynomials $X^f - \overline{\zeta}^{fj}\alpha \in \mathsf{k}_\mathfrak{p}[X]$ for $j \in [0, t-1]$ are irreducible.

If $g = X^n - a$, then $g'(\alpha) = n\alpha^{n-1}$, and $v_\mathfrak{p}(na) = 0$ implies $v_\mathfrak{P}(g'(\alpha)) = 0$ for all $\mathfrak{P} \in \mathcal{P}(\mathfrak{O})$ lying above $\mathfrak{p}$. By 5.9.4.2(b) we obtain $\mathfrak{O}_\mathfrak{p} = \mathfrak{o}_\mathfrak{p}[\alpha]$, and therefore we can apply 2.12.6 to complete the proof.

$\square$

5.10 Higher ramification groups

Throughout this section let L/K be a finite Galois extension of complete discrete valued fields such that $\mathsf{k}_L/\mathsf{k}_K$ is separable, $G = \mathrm{Gal}(L/K)$ and $1 = \mathrm{id}_L \in G$.

We tacitly use the notation and results of Sections 5.3, 5.5, and 5.6.

For a real number $s \geq -1$ we define the s-th **ramification group** of L/K by

$$G_s = \mathsf{G}_s(L/K) = \{\sigma \in G \mid v_L(\sigma a - a) \geq s + 1 \ \text{ for all } \ a \in \mathcal{O}_L\}.$$

By definition, $G_s = G_{\lceil s \rceil}$ for all $s \geq -1$, and $G_{-1} = G$.

In 2.13.2 we introduced the Hilbert subgroup series containing the ramification groups with integral index in the general context of Dedekind domains. Now we continue these investigations and deepen them using valuation-theoretic methods. To start with we recall the previous results in the context of discrete valued fields.

In the setting of 2.13.2, $G = G_{\mathfrak{p}_L}$ is the decompostion group, $G_0 = I_{\mathfrak{p}_L}$ is the inertia group, and $G_i = G_{\mathfrak{p}_L, i}$ for $i \geq 1$ is the i-th ramification group of $\mathfrak{p}_L$ over K (considered for the extension $\mathfrak{o}_L \supset \mathfrak{o}_K$ of local Dedekind domains).

G_i is a normal subgroup of G for all $i \geq 0$, and $G_i = \mathbf{1}$ for all $i \gg 1$. The field $T = L^{G_0}$ is the inertia field and for $i \in \mathbb{N}$ the field $V_i = L^{G_i}$ is the i-th ramification field. There exist (natural) isomorphisms

$$G/G_0 \xrightarrow{\sim} \mathrm{Gal}(T/K) \xrightarrow{\sim} \mathrm{Gal}(\mathsf{k}_L/\mathsf{k}_K),, \quad \text{defined by } \ \sigma G_0 \mapsto \sigma \restriction T \mapsto \overline{\sigma},$$

where $\overline{\sigma} \in \mathrm{Gal}(\mathsf{k}_L/\mathsf{k}_K)$ is the residue class homomorphism induced by σ (see 5.5.3.4). Furthermore,

$$(G{:}G_0) = f(L/K) = f(\mathfrak{p}_L/\mathfrak{p}_K) \quad \text{and} \quad |G_0| = e(L/K) = e(\mathfrak{p}_L/\mathfrak{p}_K).$$

To proceed, we define $i_{L/K}\colon G \to \mathbb{N}_0 \cup \{\infty\}$ by

$$i_{L/K}(\sigma) = \inf\{v_L(\sigma a - a) \mid a \in \mathcal{O}_L\} \in \mathbb{N}_0 \cup \{\infty\}.$$

Then $i_{L/K}(1) = \infty$, $i_{L/K}(\sigma) \in \mathbb{N}_0$ for all $\sigma \in G \setminus \{1\}$, and

$$G_s = \{\sigma \in G \mid i_{L/K}(\sigma) \geq s + 1\} \quad \text{for all } \ s \geq -1.$$

If $\pi \in \mathcal{O}_L$ is a prime element, the following assertions hold by 2.13.8:

- $i_{L/K}(\sigma) = v_L(\sigma\pi - \pi)$ for all $\sigma \in G_0$.

- If $i \in \mathbb{N}_0$, then
$$G_i = \Big\{\sigma \in G_0 \;\Big|\; \frac{\sigma\pi}{\pi} \in U_L^{(i)}\Big\},$$
and $\Phi_i \colon G_i/G_{i+1} \to U_L^{(i)}/U_L^{(i+1)}$, defined by
$$\Phi_i(\sigma G_{i+1}) = \frac{\sigma\pi}{\pi}U_L^{(i+1)} \quad \text{for all} \quad \sigma \in G_i,$$
is a monomorphism which does not depend on π.

- G_0/G_1 is cyclic, and G_i/G_{i+1} is abelian for all $i \in \mathbb{N}$.

- If $\operatorname{char}(\mathsf{k}_K) = p > 0$, then $p \nmid (G_0 : G_1)$, G_1 is a p-group, and G_i/G_{i+1} is an elementary abelian p-group for all $i \in \mathbb{N}$.

- If k_K is finite, then G/G_0 is cyclic, and G is solvable.

- If $\operatorname{char}(\mathsf{k}_K) = 0$, then G_0 is cyclic, and $G_i = \mathbf{1}$ for all $i \geq 1$.

In 5.9.6 we got an idea of how the size of the different measures the degree of ramification of a field extension. The following result makes this precise.

Theorem 5.10.1. *For $i \in \mathbb{N}_0$ let $G_i = \mathsf{G}_i(L/K)$ and $g_i = |G_i|$. Then*
$$v_L(\mathfrak{D}_{L/K}) = \sum_{\sigma \in G \setminus \{1\}} i_{L/K}(\sigma) = \sum_{i \geq 0}(g_i - 1).$$

Proof. Let T be the inertia field of L/K, $\pi \in \mathcal{O}_L$ be a prime element and $g \in \mathcal{O}_T[X]$ the minimal polynomial of π over T. Then $\mathcal{O}_L = \mathcal{O}_T[\pi]$ by 5.7.1.2, and since $G_0 = \operatorname{Gal}(L/T)$, we obtain (using 5.9.3.2, 5.9.6.4 and 5.9.4.1)
$$\mathfrak{D}_{L/K} = \mathfrak{D}_{T/K}\mathfrak{D}_{L/T} = \mathfrak{D}_{L/T} = g'(\pi)\mathcal{O}_L = \prod_{\sigma \in G_0 \setminus \{1\}} (\sigma\pi - \pi)\mathcal{O}_L,$$
and
$$v_L(\mathfrak{D}_{L/K}) = \sum_{\sigma \in G_0 \setminus \{1\}} v_L(\sigma\pi - \pi) = \sum_{\sigma \in G_0 \setminus \{1\}} i_{L/K}(\sigma) = \sum_{\sigma \in G \setminus \{1\}} i_{L/K}(\sigma),$$
since $i_{L/K} \restriction G \setminus G_0 = 0$. Let $N \in \mathbb{N}$ be such that $g_i = 1$ for all $i \geq N$. Then
$$\begin{aligned} v_L(\mathfrak{D}_{L/K}) &= \sum_{\sigma \in G_0 \setminus \{1\}} i_{L/K}(\sigma) = \sum_{i=0}^{N-1} \sum_{\sigma \in G_i \setminus G_{i+1}} i_{L/K}(\sigma) \\ &= \sum_{i=0}^{N-1} (g_i - g_{i+1})(i+1) \\ &= \sum_{i=0}^{N-1} g_i(i+1) - \sum_{i=1}^{N} g_i i = \sum_{i=0}^{N-1} g_i - N \\ &= \sum_{i=0}^{N-1} (g_i - 1) = \sum_{i \geq 0} (g_i - 1). \end{aligned}$$
$\square$

We proceed with the formal properties of the function $i_{L/K}$. Recall from 5.5.6.4 that there exists some $x \in \mathcal{O}_L$ such that $\mathcal{O}_L = \mathcal{O}_K[x]$.

Theorem 5.10.2.

1. If $x \in \mathcal{O}_L$ is such that $\mathcal{O}_L = \mathcal{O}_K[x]$, then $i_{L/K}(\sigma) = v_L(\sigma x - x)$ for all $\sigma \in G$.

2. If $\sigma, \tau \in G$, then

$$i_{L/K}(\tau\sigma\tau^{-1}) = i_{L/K}(\sigma), \; i_{L/K}(\sigma\tau) \geq \min\{i_{L/K}(\sigma), i_{L/K}(\tau)\},$$

and if $i_{L/K}(\sigma) \neq i_{L/K}(\tau)$, then $i_{L/K}(\sigma\tau) = \min\{i_{L/K}(\sigma), i_{L/K}(\tau)\}$.

3. Let H be a subgroup of G and $M = L^H$.

(a) $i_{L/M} = i_{L/K} \upharpoonright H$, and $\mathsf{G}_i(L/M) = \mathsf{G}_i(L/K) \cap H$ for all $i \geq -1$.

(b) Let H be a normal subgroup of G. Then

$$i_{M/K}(\sigma \upharpoonright M) = \frac{1}{e(L/M)} \sum_{\tau \in H} i_{L/K}(\sigma\tau) \quad \text{for all} \quad \sigma \in G.$$

Proof. 1. Let $[L:K] = n$, $\mathcal{O}_L = \mathcal{O}_K[x]$ and $\sigma \in G$. We must prove that $v_L(\sigma a - a) \geq v_L(\sigma x - x)$ for all $a \in \mathcal{O}_L$. Let $a = c_0 + c_1 x + \ldots + c_{n-1}x^{n-1} \in \mathcal{O}_L$, where $c_0, \ldots, c_{n-1} \in \mathcal{O}_K$. Then

$$\sigma a - a = \sum_{i=1}^{n-1} c_i(\sigma x^i - x^i) \in (\sigma x - x)\mathcal{O}_L,$$

and therefore $v_L(\sigma a - a) \geq v_L(\sigma x - x)$.

2. Let $\sigma, \tau \in G$ and $x \in \mathcal{O}_L$ be such that $\mathcal{O}_L = \mathcal{O}_K[x]$. Then $\mathcal{O}_L = \mathcal{O}_K[\tau^{-1}x]$, $\tau\sigma\tau^{-1}x - x = \tau(\sigma\tau^{-1}x - \tau^{-1}x)$, and

$$i_{L/K}(\tau\sigma\tau^{-1}) = v_L(\tau\sigma\tau^{-1}x - x) = v_L(\sigma\tau^{-1}x - \tau^{-1}x) = i_{L/K}(\sigma).$$

Since $\sigma\tau x - x = \sigma(\tau x - x) + (\sigma x - x)$, we obtain

$$\begin{aligned} i_{L/K}(\sigma\tau) &= v_L(\sigma\tau x - x) \geq \min\{v_L(\tau x - x), \, v_L(\sigma x - x)\} \\ &= \min\{i_{L/K}(\sigma), i_{L/K}(\tau)\}, \end{aligned}$$

and equality holds if $i_{L/K}(\sigma) \neq i_{L/K}/(\tau)$.

3. (a) Obvious by the very definition.

(b) Let $\sigma \in G$, $x \in \mathcal{O}_L$ and $y \in \mathcal{O}_M$ be such that $\mathcal{O}_L = \mathcal{O}_K[x]$ and $\mathcal{O}_M = \mathcal{O}_K[y]$. Then $e(L/M)i_{M/K}(\sigma \upharpoonright M) = e(L/M)v_M(\sigma y - y) = v_L(\sigma y - y)$, and

$$f = \prod_{\tau \in H} (X - \tau x) = \sum_{j=0}^{m} c_j X^j \in \mathcal{O}_M[X]$$

is the minimal polynomial of x over M. For $j \in [0, m]$ we set

$$c_j = \sum_{\mu=0}^{d} a_{j,\mu} y^\mu, \quad \text{where } d = [M:K] \text{ and } a_{j,\mu} \in K,$$

and then

$$\sigma c_j - c_j = \sum_{\mu=0}^{d} a_{j,\mu}((\sigma y)^\mu - y^\mu) \in (\sigma y - y)\mathcal{O}_M.$$

Now

$$\sigma f = \prod_{\tau \in H} (X - \sigma\tau x) \quad \text{implies}$$

$$v_L((\sigma f)(x)) = \sum_{\tau \in H} v_L(x - \sigma\tau x) = \sum_{\tau \in H} i_{L/K}(\sigma\tau),$$

and therefore we must prove that $v_L((\sigma f)(x)) = v_L(\sigma y - y)$. Since

$$(\sigma f)(x) = (\sigma f - f)(x) = \sum_{j=0}^{m} (\sigma c_j - c_j)x^j \in (\sigma y - y)\mathcal{O}_L,$$

it follows that $v_L((\sigma f)(x)) \geq v_L(\sigma y - y)$. As to the reverse inequality, let $g \in \mathcal{O}_K[X]$ be such that $y = g(x)$. Then $y - g \in \mathcal{O}_M[X]$, $(y - g)(x) = 0$, and therefore $y - g = fh$ for some $h \in \mathcal{O}_M[X]$. Hence $\sigma y - g = \sigma(y - g) = (\sigma f)(\sigma h)$, and consequently $\sigma y - y = (\sigma y - g)(x) = (\sigma f)(x)(\sigma h)(x)$ which implies $v_L(\sigma y - y) \geq v_L((\sigma f)(x))$. □

If M is an intermediate field of L/K and M/K is Galois, then it is by no means obvious how to determine the sequence of ramification groups of M/K. To manage this we introduce the Herbrand function and the upper numbering of the ramification groups.

Definition and Theorem 5.10.3. *For $s \geq -1$, let $G_s = \mathsf{G}_s(L/K)$. We define the* **Herbrand function** $\eta_{L/K} \colon [-1, \infty) \to \mathbb{R}$ *by*

$$\eta_{L/K}(s) = s \quad \text{for} \quad s \in [-1, 0] \quad \text{and}$$

$$\eta_{L/K}(s) = \int_0^s \frac{dx}{(G_0 : G_x)} \quad \text{for} \quad s \geq 0.$$

1. $\eta_{L/K} \colon [-1, \infty) \to \mathbb{R}_{\geq 0}$ is continuous and strictly increasing, $\eta_{L/K}(0) = 0$ and $\eta_{L/K}([-1, \infty)) = [-1, \infty)$. If $m \in \mathbb{Z}$ and $m \geq -1$, then $\eta_{L/K} \restriction [m, m+1]$ is linear, and

$$\eta'_{L/K}(s) = \frac{1}{(G_0 : G_s)} = \frac{|G_s|}{e(L/K)} \quad \textit{for all} \quad s \in (-1, \infty) \setminus \mathbb{Z}.$$

2. Let $g_i = |G_i|$ for all $i \geq 0$. If $s \in \mathbb{R}_{\geq 0}$ and $m = \lfloor s \rfloor$, then

$$\eta_{L/K}(s) = \frac{1}{g_0}\Big[\sum_{j=0}^{m} g_j + (s-m)g_{m+1}\Big] - 1 = \frac{1}{g_0}\sum_{\sigma \in G} \min\{i_{L/K}(\sigma), s+1\} - 1$$

3. If $s \in [-1, \infty)$ and $\eta_{L/K}(s) \in \mathbb{Z}$, then $s \in \mathbb{Z}$.

Proof. 1. Elementary analysis.

2. Let $s \in \mathbb{R}_{\geq 0}$ and $m = \lfloor s \rfloor$.

$$\eta_{L/K}(s) = \int_0^s \frac{dx}{(G_0 : G_x)} = \sum_{j=0}^{m-1} \int_j^{j+1} \frac{dx}{(G_0 : G_x)} + \int_m^s \frac{dx}{(G_0 : G_x)}$$

$$= \sum_{j=0}^{m-1} \frac{g_{j+1}}{g_0} + \frac{g_{m+1}}{g_0}(s-m) = \frac{1}{g_0}\Big[\sum_{i=0}^{m} g_i + (s-m)g_{m+1}\Big] - 1.$$

If $\sigma \in G$, then $i_{L/K}(\sigma) > s+1$ if and only if $i_{L/K}(\sigma) \geq m+2$, and this holds if and only if $\sigma \in G_{m+1}$. Consequently,

$$\sum_{\sigma \in G} \min\{i_{L/K}(\sigma), s+1\} = \sum_{\sigma \in G_{m+1}} (s+1) + \sum_{i=-1}^{m} \sum_{\sigma \in G_i \setminus G_{i+1}} i_{L/K}(\sigma)$$

$$= (s+1)g_{m+1} + \sum_{i=-1}^{m} (i+1)(g_i - g_{i+1})$$

$$= (s+1)g_{m+1} + \sum_{i=0}^{m} (i+1)g_i - \sum_{i=1}^{m+1} i g_i$$

$$= (s+1)g_{m+1} + g_0 - (m+1)g_{m+1} + \sum_{i=1}^{m} g_i = \sum_{i=0}^{m} g_i + (s-m)g_{m+1}$$

$$= g_0[\eta_{L/K}(s) + 1]$$

and therefore

$$\eta_{L/K}(s) = \frac{1}{g_0}\sum_{\sigma \in G} \min\{i_{L/K}(\sigma), s+1\} - 1.$$

3. If $s \in [-1, \infty)$, $\eta_{L/K}(s) \in \mathbb{Z}$ and $m \in \mathbb{Z}$ such that $-1 \leq m \leq s < m+1$, then 2. implies

$$s = m + \frac{g_0 \eta_{L/K}(s)}{g_{m+1}} - \sum_{j=1}^{m} \frac{g_j}{g_{m+1}} \in \mathbb{Z}. \qquad \square$$

Theorem 5.10.4 (Theorem of Herbrand). *Let H be a normal subgroup of G, $M = L^H$, $s \in [-1, \infty)$ and $t = \eta_{L/M}(s)$. Then*

$$\mathsf{G}_t(M/K) = \{\sigma \restriction M \mid \sigma \in \mathsf{G}_s(L/K)\}.$$

If we identify $\mathrm{Gal}(M/K) = G/H$, *then* $\mathsf{G}_t(M/K) = \mathsf{G}_s(L/K)H/H$ *or, in simplified terms,* $(G/H)_t = G_sH/H$.

Proof. We tacitly use 5.10.2. For $\rho \in \mathrm{Gal}(M/K)$ we choose $\rho^* \in G$ such that $\rho^* \restriction M = \rho$ and $i_{L/K}(\rho^*) = \max\{i_{L/K}(\sigma) \mid \sigma \in G,\ \sigma \restriction M = \rho\}$. Then it follows that $i_{L/K}(\rho^*) \geq i_{L/K}(\rho^*\tau)$ for all $\tau \in H$, and we shall prove:

a. $i_{L/K}(\rho^*\tau) = \min\{i_{L/M}(\tau), i_{L/K}(\rho^*)\}$ for all $\tau \in H$.

b. $i_{M/K}(\rho) = \eta_{L/M}(i_{L/K}(\rho^*) - 1) + 1$.

Proof of **a.** Let $\tau \in H$.
If $i_{L/K}(\tau) < i_{L/K}(\rho^*)$, then $i_{L/K}(\rho^*\tau) = i_{L/K}(\tau) = i_{L/M}(\tau)$.
If $i_{L/M}(\tau) = i_{L/K}(\tau) \geq i_{L/K}(\rho^*)$, then the maximal choice of ρ^* implies that $i_{L/K}(\rho^*) \geq i_{L/K}(\rho^*\tau) \geq \min\{i_{L/K}(\rho^*), i_{L/K}(\tau)\} = i_{L/K}(\rho^*)$. □[**a.**]

Proof of **b.** By 5.10.3.2 (applied for L/K with $s = i_{L/K}(\rho^*)-1$) it follows that

$$\begin{aligned} i_{M/K}(\rho) = i_{M/K}(\rho^* \restriction M) &= \frac{1}{e(L/M)} \sum_{\tau \in H} i_{L/K}(\rho^*\tau) \\ &= \frac{1}{e(L/M)} \sum_{\tau \in H} \min\{i_{L/M}(\tau), i_{L/K}(\rho^*)\} \\ &= \eta_{L/M}(i_{L/K}(\rho^*) - 1) + 1. \end{aligned}$$ □[**b.**]

By means of **a** and **b** we obtain: If $\rho \in \mathrm{Gal}(M/K)$, then

$$\begin{aligned} \rho \in \mathsf{G}_t(M/K) &\iff i_{M/K}(\rho) - 1 \geq t \iff \eta_{L/M}(i_{L/K}(\rho^*) - 1) \geq t = \eta_{L/M}(s) \\ &\iff i_{L/K}(\rho^*) - 1 \geq s \iff \rho^* \in \mathsf{G}_s(L/K). \end{aligned}$$

However, $\rho^* \in \mathsf{G}_s(L/K)$ holds if and only if there exists some $\sigma \in \mathsf{G}_s(L/K)$ such that $\sigma \restriction M = \rho$. Consequently, $\mathsf{G}_t(M/K) = \mathsf{G}_s(L/K)H/H$ (if we identify $\sigma H = \sigma \restriction M$). □

Definition and Theorem 5.10.5 (Upper numbering).
Let $\psi_{L/K} = \eta_{L/K}^{-1} \colon [-1,\infty) \to \mathbb{R}$ be the inverse function of $\eta_{L/K}$ and define the upper numbering of the ramification groups of L/K by

$$G^t = \mathsf{G}^t(L/K) = \mathsf{G}_{\psi_{L/K}(t)}(L/K) = G_{\psi_{L/K}(t)} \quad \textit{for all}\ \ t \in [-1,\infty).$$

1. $\psi_{L/K} \colon [-1,\infty) \to \mathbb{R}$ *is continuous, strictly increasing and piecewise linear,* $\psi_{L/K}(-1) = -1$, $\psi_{L/K}(0) = 0$, $G^{-1} = G_{-1} = G$ *and* $G^0 = G_0$.

2. If $s \in [-1,\infty)$ *and* $t = \eta_{L/K}(s)$, *then the following assertions are equivalent:*

(a) $s \in \mathbb{Z}$ *and* $G_s \neq G_{s+1}$.

(b) There exists some $\delta_1 \in \mathbb{R}_{>0}$ *such that* $G^t \neq G^{t+\delta} = G^{t+\delta_1}$ *for all* $\delta \in (0,\delta_1]$.

If these conditions are fulfilled, then t is called a **jump in the upper numbering** of the ramification groups of L/K.

3. Let H be a normal subgroup of G, $M = L^H$, and identify $G/H = \mathrm{Gal}(M/K)$.

(a) $\eta_{L/K} = \eta_{M/K} \circ \eta_{L/M}$ and $\psi_{L/K} = \psi_{L/M} \circ \psi_{M/K}$.

(b) If $t \in [-1, \infty)$, then $\mathsf{G}^t(M/K) = \mathsf{G}^t(L/K)H/H$, or, in simplified terms, $(G/H)^t = G^t H/H$.

Proof. 1. Obvious by 5.10.3.1.

2. By definition, (a) is equivalent to

(a)′ There exists some $\varepsilon_1 \in \mathbb{R}_{>0}$ such that $G_s \ne G_{s+\varepsilon} = G_{s+\varepsilon_1}$ for all $\varepsilon \in (0, \varepsilon_1]$.

As $\eta_{L/K}$ is strictly increasing and piecewise linear, (a)′ and (b) are equivalent.

3. (a) If $m \in \mathbb{Z}_{\ge -1}$, then $\eta_{L/M} \restriction (m, m{+}1)$ and $\eta_{L/K} \restriction (m, m{+}1)$ are linear, and $\eta_{L/M}((m, m+1)) \cap \mathbb{Z} = \emptyset$ by 5.10.3.3. Hence $\eta_{M/K} \circ \eta_{L/M} \restriction (m, m+1)$ is linear for all $m \in \mathbb{Z}_{\ge -1}$, and $\eta_{M/K} \circ \eta_{L/(M)}(-1) = -1 = \eta_{L/K}(-1)$. If $s \in (-1, \infty) \setminus \mathbb{Z}$, then $t = \eta_{L/M}(s) \in (-1, \infty) \setminus \mathbb{Z}$, and

$$(\eta_{M/K} \circ \eta_{L/M})'(s) = \eta'_{M/K}(t)\eta'_{L/M}(s) = \frac{|(G/H)_t|}{e(M/K)} \frac{|H_s|}{e(L/M)}$$
$$= \frac{|G_s|}{e(L/K} = \eta'_{L/K}(s),$$

since $|(G/H)_t|\,|H_s| = |G_s H/H|\,|G_s \cap H| = |G_s/G_s \cap H|\,|G_s \cap H| = |G_s|$ by 5.10.4. Hence $\eta_{M/K} \circ \eta_{L/(M} = \eta_{L/K}$, and $\psi_{L/K} = \psi_{L/(M} \circ \psi_{M/K}$.

(b) If $t \in [-1, \infty)$, then 5.10.4 implies

$$G^t H/H = G_{\psi_{L/K}(t)} H/H = (G/H)_{\eta_{L/M} \circ \psi_{L/K}(t)}$$
$$= (G/H)_{\eta_{L/M} \circ \psi_{L/M} \circ \psi_{M/K}(t)} = (G/H)_{\psi_{M/K}(t)} = (G/H)^t. \qquad \square$$

We conclude this chapter with the Theorem of Hasse-Arf concerning the jumps in the upper numbering of the ramification groups of an abelian field extension (see 5.10.7). Here we follow [13, Ch. III.4], using the subsequent Theorem of Sen 5.10.6.

Theorem 5.10.6 (Theorem of Sen). *Assume that* $\mathrm{char}(\mathsf{k}_K) = p > 0$. *Let L/K be cyclic and fully ramified, $[L{:}K] = p^m$ for some $m \ge 2$ and $\sigma \in G$.*

1. Let k_1, $k_2 \in \mathbb{Z}$.

(a) *If* $v_p(k_1) = v_p(k_2)$, *then* $i_{L/K}(\sigma^{k_1}) = i_{L/K}(\sigma^{k_2})$.

(b) *If* $v_p(k_1) < v_p(k_2)$, *then* $i_{L/K}(\sigma^{k_1}) < i_{L/K}(\sigma^{k_2})$.

2. *For every* $k \in \mathbb{Z}$ *there exists some* $a_k \in L$ *such that*

$$v_L(a_k) = k \quad \textit{and} \quad v_L(\sigma a_k - a_k) = i_{L/K}(\sigma^k) + k - 1.$$

3. *If* $n \in [1, m-1]$ *and* $\sigma^{p^n} \neq 1$, *then*

$$i_{L/K}(\sigma^{p^{n-1}}) \equiv i_{L/K}(\sigma^{p^n}) \mod p^n.$$

Proof. Let $\pi \in \mathcal{O}_L$ be a prime element. As G is a p-group, we obtain $G = G_1$, hence $i_{L/K}(\sigma) = v_L(\sigma\pi - \pi) \geq 2$ for all $\sigma \in G$,

$$v_L\Big(\frac{\sigma\pi}{\pi} - 1\Big) = i_{L/K}(\sigma) - 1 \geq 1, \quad \text{and therefore} \quad \frac{\sigma\pi}{\pi} \in U_L^{(1)}.$$

Let $\mathcal{R} \subset \mathcal{O}_K$ be a set of representatives for $\mathsf{k}_L = \mathsf{k}_K$ such that $0 \in \mathcal{R}$. Then every $a \in L^\times$ has a unique representation $a = \pi^{v_L(a)}\rho u$, where $\rho \in \mathcal{R}^\bullet$ and $u \in U_L^{(1)}$. It follows that

$$\frac{\sigma a}{a} \in U_L^{(1)} \quad \text{and} \quad v_L(\sigma a - a) = v_L\Big(\frac{\sigma a}{a} - 1\Big) + v_L(a) > v_L(a)$$

for all $a \in L^\times$ and $\sigma \in G$.

If $a \in \mathcal{O}_L$, then by 5.3.5 there exists a sequence $(a_n)_{n \geq 0}$ in $\mathcal{R}$ such that

$$a = \sum_{n=0}^{\infty} a_n\pi^n \quad \text{and} \quad \sigma a - a = \sum_{n=1}^{\infty} a_n(\sigma\pi^n - \pi^n) = (\sigma\pi - \pi)\sum_{n=1}^{\infty} a_n \sum_{\nu=0}^{n-1} \sigma\pi^\nu,$$

and consequently $v_L(\sigma a - a) \geq v_L(\sigma\pi - \pi)$.

1. (a) It suffices to prove: If $k = ap^j \in \mathbb{Z}$, where $j \in \mathbb{N}_0$, $a \in \mathbb{Z}$ and $p \nmid a$, then $i_{L/K}(\sigma^k) = i_{L/K}(\sigma^{p^j})$.

Thus suppose that $k = ap^j$, where $a \in \mathbb{Z}$, $p \nmid a$, $j \in \mathbb{N}_0$, and let $a_0 \in \mathbb{N}$ be such that $a_0 \equiv a \mod p^m$. If $\nu \in [0, a_0 - 1]$, then $\sigma^{\nu p^j}\pi \equiv \pi \mod \mathfrak{p}_L^2$, hence

$$\pi_1 = \sum_{\nu=0}^{a_0-1} \sigma^{\nu p^j}\pi \equiv a_0\pi \mod \mathfrak{p}_L^2 \quad \text{and consequently} \quad v_L(\pi_1) = 1.$$

Now

$$\sigma^k - 1 = \sigma^{a_0p^j} - 1 = (\sigma^{p^j} - 1)\sum_{\nu=0}^{a_0-1} \sigma^{\nu p^j}$$

implies

$$i_{L/K}(\sigma^k) = v_L((\sigma^k - 1)\pi) = v_L((\sigma^{p^j} - 1)\pi_1) = i_{L/K}(\sigma^{p^j}).$$

(b) It suffices to prove that $i_{L/K}(\tau) < i_{L/K}(\tau^p)$ for all $\tau \in G$. Thus suppose that $\tau \in G$. Then

$$\tau^p\pi - \pi = (\tau^p - 1)\pi = (\tau - 1)\sum_{j=0}^{p-1} \tau^j\pi.$$

If $j \in [0, p-1]$, then $\tau^j\pi = \pi(1 + u_j)$ for some $u_j \in \mathfrak{p}_L$, hence

$$(\tau - 1)(\tau^j\pi) = (\tau\pi - \pi) + \tau u_j(\tau\pi - \pi) + \pi(\tau u_j - u_j),$$

and therefore

$$\frac{\tau^p\pi - \pi}{\tau\pi - \pi} = p + \sum_{j=0}^{p-1} \tau u_j + \pi \sum_{j=0}^{p-1} \frac{\tau u_j - u_j}{\tau\pi - \pi}.$$

Since $v_L(p) \geq 1$, $v_L(\tau u_j) = v_L(u_j) \geq 1$ and $v_L(\tau u_j - u_j) \geq v_L(\tau\pi - \pi)$ for all $j \in [0, p-1]$, it follows that

$$i_{L/K}(\tau^p) - i_{L/K}(\tau) = v_L(\tau^p\pi - \pi) - v_L(\tau\pi - \pi) = v_L\Big(\frac{\tau^p\pi - \pi}{\tau\pi - \pi}\Big) \geq 1.$$

2. Obviously $a_0 = 1$ has the desired property. Thus let $k \in \mathbb{N}$, set

$$a_k = \prod_{i=0}^{k-1} \sigma^i\pi \quad \text{and} \quad a_{-k} = a_k^{-1}.$$

Then it follows that $v_L(a_k) = k$, $v_L(a_{-k}) = -v_L(a_k) = -k$,

$$\begin{aligned} v_L(\sigma a_k - a_k) &= v_L\Big(\prod_{i=1}^{k} \sigma^i\pi - \prod_{i=0}^{k-1} \sigma^i\pi\Big) = v_L(\sigma^k\pi - \pi) + v_L\Big(\prod_{i=1}^{k-1} \sigma^i\pi\Big) \\ &= i_{L/K}(\sigma^k) + k - 1 \end{aligned}$$

and

$$\begin{aligned} v_L(\sigma a_{-k} - a_{-k}) &= v_L(\sigma a_k^{-1} - a_k^{-1}) = v_L(\sigma a_k - a_k) - 2v_L(a_k) \\ &= i_{L/K}(\sigma^k) + k - 1 - 2k = i_{L/K}(\sigma^{-k}) - k - 1. \end{aligned}$$

3. By induction on n using 1. and 2. Let $n \in [1, m-1]$, suppose that

$$i_{L/K}(\tau^{p^{j-1}}) \equiv i_{L/K}(\tau^{p^j}) \mod p^j$$

for all $j \in [1, n-1]$ and $\tau \in G$ satisfying $\tau^{p^j} \neq 1$, and assume to the contrary that

$$\sigma^{p^n} \neq 1 \quad \text{and} \quad i_{L/K}(\sigma^{p^{n-1}}) \not\equiv i_{L/K}(\sigma^{p^n}) \mod p^n.$$

If $d = i_{L/K}(\sigma^{p^{n-1}}) - i_{L/K}(\sigma^{p^n})$, then $\mathsf{v}_p(d) < n$; we assert that $\mathsf{v}_p(d) = n-1$. This is obvious if $n = 1$. If $n > 1$, then $(\sigma^p)^{p^{n-1}} \neq 1$, and by the induction hypothesis $i_{L/K}((\sigma^p)^{p^{n-2}}) \equiv i_{L/K}((\sigma^p)^{p^{n-1}}) \mod p^{n-1}$. Hence it follows that $i_{L/K}(\sigma^{p^{n-1}}) \equiv i_{L/K}(\sigma^{p^n}) \mod p^{n-1}$, and consequently $v_p(d) = n - 1$.

By 2., there exists some $a \in L$ such that $v_L(a) = d$ and

$$v_L(\sigma^p a - a) = i_{L/K}(\sigma^{pd}) + d - 1 = i_{L/K}(\sigma^{p^n}) + d - 1 = i_{L/K}(\sigma^{p^{n-1}}) - 1.$$

For $\nu \in [0, p-1]$ we have $\sigma^\nu a = a(1 + u_\nu)$ for some $u_\nu \in \mathfrak{p}_L$, and we set

$$b = \sum_{\nu=0}^{p-1} \sigma^\nu a = a\Big(p + \sum_{\nu=0}^{p-1} u_\nu\Big).$$

Then $v_L(b) > v_L(a) = d$ and $v_L(\sigma b - b) = v_L(\sigma^p a - a) = i_{L/K}(\sigma^{p^{n-1}}) - 1$.

For every $k \ge v_L(b)$, again 2. implies that there exists some $b'_k \in L$ such that $v_L(b'_k) = k$ and $v_L(\sigma b'_k - b'_k) = i_{L/K}(\sigma^k) + k - 1$. By 5.3.5 we obtain a representation

$$b = \sum_{k=v_L(b)}^{\infty} r_k b'_k \quad \text{with coefficients } r_k \in \mathcal{R},$$

and for $k \ge v_L(b)$ we set $b_k = r_k b'_k$. Then it follows that either $r_k = b_k = 0$, or $v_L(b_k) = k$ and $v_L(\sigma b_k - b_k) = v_L(\sigma b'_k - b'_k) = i_{L/K}(\sigma^k) + k - 1$. Now we consider the partition

$$\sigma b - b = \sum_{\substack{k \ge v_L(b) \\ \mathsf{v}_p(k) \le n-1}} (\sigma b_k - b_k) + \sum_{\substack{k \ge v_L(b) \\ \mathsf{v}_p(k) \ge n}} (\sigma b_k - b_k),$$

and we shall prove:

B. The numbers $v_L(\sigma b_k - b_k)$ (for $k \ge v_L(b)$ such that $b_k \ne 0$ and $\mathsf{v}_p(k) \le n-1$) and $v_L(\sigma b - b)$ are all distinct.

Proof of **B.** Note that $i_{L/K}(\sigma^{p^{\nu-1}}) \equiv i_{L/K}(\sigma^{p^\nu}) \bmod p^\nu$ holds for all $\nu \in [1, n-1]$ by the induction hypothesis.

First let $k_1, k_2 \ge v_L(b)$ be such that $b_{k_\nu} \ne 0$ and $l_\nu = \mathsf{v}_p(k_\nu) \le n-1$ for $\nu \in \{1, 2\}$, and suppose that $v_L(\sigma b_{k_1} - b_{k_1}) = v_L(\sigma b_{k_2} - b_{k_2})$, hence $i_{L/K}(\sigma^{k_1}) + k_1 = i_{L/K}(\sigma^{k_2}) + k_2$. If $l_1 = l_2$, then $i_{L/K}(\sigma^{k_1}) = i_{L/K}(\sigma^{k_2})$ and $k_1 = k_2$. If $l_1 \ne l_2$, say $l_1 < l_2$, then $i_{L/K}(\sigma^{p^{\nu-1}}) \equiv i_{L/K}(\sigma^{p^\nu}) \bmod p^\nu$ for all $\nu \in [l_1 + 1, l_2]$, and consequently

$$\begin{aligned} l_1 &= \mathsf{v}_p(k_1 - k_2) = \mathsf{v}_p(i_{L/K}(\sigma^{k_1}) - i_{L/K}(\sigma^{k_2})) \\ &= \mathsf{v}_p(i_{L/K}(\sigma^{p^{l_1}}) - i_{L/K}(\sigma^{p^{l_2}})) = \mathsf{v}_p\Big(\sum_{\nu=l_1+1}^{l_2} [i_{L/K}(\sigma^{p^{\nu-1}}) - i_{L/K}(\sigma^{p^\nu})]\Big) \\ &\ge \min\{\mathsf{v}_p(i_{L/K}(\sigma^{p^{\nu-1}}) - i_{L/K}(\sigma^{p^\nu})) \mid \nu \in [l_1+1, l_2]\} \ \ge\ l_1 + 1, \end{aligned}$$

a contradiction.

Now let $k \geq v_L(b)$, $b_k \neq 0$, $l = \mathsf{v}_p(k) \leq n-1$ and $v_L(\sigma b_k - b_k) = v_L(\sigma b - b)$. Then we obtain

$$\begin{aligned} 0 = v_L(\sigma b_k - b_k) - v_L(\sigma b - b) &= [i_{L/K}(\sigma^k) + k - 1] - [i_{L/K}(\sigma^{p^{n-1}}) - 1] \\ &= k + i_{L/K}(\sigma^{p^l}) - i_{L/K}(\sigma^{p^{n-1}}) = k + \sum_{\nu=l+1}^{n-1} [i_{L/K}(\sigma^{p^{\nu-1}}) - i_{L/K}(\sigma^{p^\nu})]. \end{aligned}$$

However, if $\nu \in [l+1, n-1]$, then

$$\mathsf{v}_p(i_{L/K}(\sigma^{p^{\nu-1}}) - i_{L/K}(\sigma^{p^\nu})) \geq \nu \geq l+1 > l = \mathsf{v}_p(k),$$

a contradiction. □[**B.**]

From **B** we deduce

$$\begin{aligned} v_L\Big(\sum_{\substack{k \geq v_L(b) \\ \mathsf{v}_p(k) \geq n}} (\tau b_k - b_k) \Big) &= v_L\Big((\tau b - b) - \sum_{\substack{k \geq v_L(b) \\ \mathsf{v}_p(k) \leq n-1}} (\tau b_k - b_k) \Big) \\ &= \min\{ v_L(\tau b - b),\ v_L(\tau b_k - b_k) \mid k \geq v_L(b), \mathsf{v}_p(k) \leq n-1 \} \leq v_L(\tau b - b). \end{aligned}$$

On the contrary, for all $k \geq v_L(b)$ such that $\mathsf{v}_p(k) = l \geq n$ we obtain

$$\begin{aligned} v_L(\tau b_k - b_k) &= i_{L/K}(\sigma^k) + k - 1 \\ &= i_{L/K}(\sigma^{p^l}) + k - 1 \geq i_{L/K}(\sigma^{p^n}) + v_L(b) - 1 \\ &> i_{L/K}(\sigma^{p^n}) + d - 1 = v_L(\sigma - b), \end{aligned}$$

a contradiction. □

Theorem 5.10.7 (Theorem of Hasse - Arf). *Let* $\mathrm{char}(\mathsf{k}_K) = p > 0$, *and let* L/K *be abelian. Then the jumps in the upper numbering of the ramification groups of* L/K *are integers.*

Proof. Let T be the inertia field of L/K. Then $\mathsf{G}_s(L/T) = \mathsf{G}_s(L/K)$ for all $s \geq 0$, and $\eta_{T/K}(s) = s$ for all $s \in [-1, \infty)$. Hence $\eta_{L/K} = \eta_{L/T}$ and $\mathsf{G}^t(L/T) = \mathsf{G}^t(L/K)$ for all $t \geq 0$. Consequently, we may assume that L/K is fully ramified and thus $T = L$. Let $G = \mathrm{Gal}(L/K)$, $G_s = \mathsf{G}_s(L/K)$ and $G^s = \mathsf{G}^s(L/K)$ for all $s \geq -1$.

Special case: L/K is cyclic and $[L{:}K] = p^m$ for some $m \in \mathbb{N}$.

Let $G = \langle \tau \rangle = G_1$ and $n \in \mathbb{N}$ such that $G_n \neq G_{n+1}$. We must prove that $\eta_{L/K}(n) \in \mathbb{Z}$. Let $r \geq 0$ and $0 \leq n_0 < n_1 < \ldots < n_r = n < n_{r+1}$ be such that, for all $i \in [0, r]$,

$$G_{n_i} \neq G_{n_i+1} = G_{n_{i+1}} \quad \text{and} \quad G_{n_i} = \langle \tau^{p^{m_i}} \rangle \ \text{for some } m_i \in [0, m-1].$$

It follows that

$$\eta_{L/K}(n) = \sum_{j=0}^{n} \frac{|G_j|}{|G|} = n_0 + 1 + \sum_{i=1}^{r} (n_i - n_{i-1}) \frac{|G_{n_i}|}{|G|} = n_0 + 1 + \sum_{i=1}^{r} \frac{n_i - n_{i-1}}{p^{m_i}}.$$

For $i \in [0, r]$ we obtain $(G_{n_i} : G_{n_{i+1}}) = (G_{n_i} : G_{n_i+1}) = p$, hence $m_{i+1} = m_i + 1$, and $\tau^{p^{m_i}} \in G_{n_i} \setminus G_{n_i+1}$ implies $i_{L/K}(\tau^{p^{m_i}}) = n_i + 1$. By 5.10.6.3 it follows that

$$n_i - n_{i-1} = i(\tau^{p^{m_i}}) - i(\tau^{p^{m_i - 1}}) \equiv 0 \mod p^{m_i} \quad \text{for all} \quad i \in [1, r],$$

and thus $\eta_{L/K}(n) \in \mathbb{Z}$.

General case: As L/K is fully ramified, it follows that $G_s = G^s = G$ for all $s \in [-1, 0]$ and $G_s = G_1 = G^{s|G_1|/|G_0|}$ for all $s \in (0, 1)$. Hence 0 is a jump. Thus now let $t > 0$ be a jump. By 5.10.5.2 there exist some $s \in \mathbb{N}$ and $\delta_1 \in \mathbb{R}_{>0}$ such that $G_s = G^t \neq G^{t+\delta} = G^{t+\delta_1}$ for all $\delta \in (0, \delta_1]$. Hence there exists some character $\chi \in \mathsf{X}(G)$ such that $\chi \upharpoonright G^t \neq 1$ and $\chi \upharpoonright G^{t+\delta_1} = 1$. We set $H = \mathrm{Ker}(\chi)$ and $M = L^H$. Since $G_s = G^t$ is a p-group, we may assume that χ is of p-power order. Then M/K is cyclic of prime power degree, $\mathrm{Gal}(M/K) = G/H$, and (by 5.10.5.4) $(G/H)^t = G^t H/H \neq \mathbf{1}$ and $(G/H)^{t+\delta} = G^{t+\delta} H/H = \mathbf{1}$ for all $\delta \in \mathbb{R}_{>0}$. Hence t is a jump in the upper numbering of the ramification groups of M/K and thus $t \in \mathbb{N}$ by the special case. □

In general, the theorem of Hasse-Arf 5.10.7 fails if L/K is not abelian. For an example (e. g., if $\mathrm{Gal}(L/K)$ is the quaternion group of order 8), and for more details on higher ramification groups we refer to [59, Ch. IV].

5.11 Exercises for Chapter 5

1. Let p be a prime. Let $a \in \mathbb{Q}$, $a = m/n$ such that $m, n \in \mathbb{Z}$, $n \neq 0$ and $mn = p^s c$, where $s \in \mathbb{N}_0$, $c \in \mathbb{Z}$ and $p \nmid c$. Then $a \in \mathbb{Q}_p^{\times 2}$ if and only if

$$2 \,|\, s \text{ and } \Big(\frac{c}{p}\Big) = 1 \quad \text{(where } \Big(\frac{c}{p}\Big) \text{ denotes the Kronecker symbol).}$$

2. Theorem 5.8.1.2 continues to hold if f is not necessarily monic.

3. Let $K_1 = \mathbb{Q}(\sqrt[3]{2}\,) \subset \mathbb{C}$ and $K_2 = \mathbb{Q}(\sqrt{1 + \sqrt{2}}\,) \subset \mathbb{C}$. Describe explicitly the real and complex embeddings and the corresponding normalized Archimedian absolute values of K_1 and of K_2 (compare 5.2.5 and 5.2.6).

4. Let L_1/K and L_2/K be finite extensions of complete discrete valued fields (inside a fixed algebraic closure), and assume that $\mathsf{k}_{L_1}/\mathsf{k}_K$ and $\mathsf{k}_{L_2}/\mathsf{k}_K$ are separable. Then $\mathsf{k}_{L_1 \cap L_2} = \mathsf{k}_{L_1} \cap \mathsf{k}_{L_2}$.

5. Let L/K be a finite separable extension of complete discrete valued fields. Then $\mathsf{N}_{L/K}(U_L^{(1)}) \subset U_K^{(1)}$, and if L/K is tamely ramified, then equality holds.

6. Let $(K, |\cdot|^*)$ be a complete valued field. Let $|\cdot|\colon K \to \mathbb{R}_{\geq 0}$ be an absolute value of K which is not equivalent to $|\cdot|^*$, and let $\widehat{K}$ be a completion of $(K, |\cdot|)$. Then $\widehat{K}$ is algebraically closed. Hint: It suffices to prove that every separable polynomial in $K[X]$ splits.

7. The map

$$\rho\colon \mathbb{Z}[\![X]\!] \to \mathbb{Z}_p, \quad \text{defined by } \rho\Big(\sum_{n=0}^{\infty} a_n X^n\Big) = \sum_{n=0}^{\infty} a_n p^n$$

induces a ring isomorphism $\mathbb{Z}[\![X]\!]/(X-p) \xrightarrow{\sim} \mathbb{Z}_p$.

8. Let $x \in \mathbb{Z}_p^\bullet$. The p-adic digit sequence of x is ultimately periodic if and only if $x \in \mathbb{Z}_{(p)}$. Determine the p-adic digit sequence of $\frac{2}{3}$ and $-\frac{2}{3}$ for $p \in \{2, 5, 7\}$.

9. If $N \in \mathbb{N}$, $N \geq 2$ and p is any prime, then sequence $(N^{-n})_{n\geq 0}$ is not convergent in $\mathbb{Q}_p$.

10. Let K be a complete discrete valued field, $\overline{K}/K$ an algebraic field extension and L an intermediate field of $\overline{K}/K$.

a. Let L/K be tamely ramified and K' an over K finite intermediate field of $\overline{K}/K$. Then LK'/K' is tamely ramified.

b. Let $(L_i)_{i\in I}$ be a family of over K tamely ramified intermediate fields of $\overline{K}/K$ and L its compositum. Then L is tamely ramified over K. In particular, there exists a largest over K tamely ramified intermediate field of $\overline{K}/K$.

c. Let L/K be finite Galois and $\mathsf{k}_L/\mathsf{k}_K$ separable. Then the first ramification field of L/K is the largest over K tamely ramified intermediate field of L/K.

11. Let p be a prime and K a quadratic extension of $\mathbb{Q}_p$ (i. e., $[K:\mathbb{Q}_p] = 2$).

a. If $p \neq 2$ and $\zeta \in \mathbb{Q}_p$ is a primitive $(p-1)$-th root of unity, then $K = \mathbb{Q}_p(\sqrt{d})$ for some $d \in \{\zeta, p, p\zeta\}$. At that $\mathbb{Q}_p(\sqrt{\zeta})$ is the only unramified quadratic extension of $\mathbb{Q}_p$, and in any case $\mathcal{O}_K = \mathbb{Z}_p[\sqrt{d}]$.

b. If $p = 2$, then $K = \mathbb{Q}_2(\sqrt{d})$ for some $d \in \{-1, \pm 2, \pm 5, \pm 10\}$. At that $\mathbb{Q}_2(\sqrt{5})$ is the only unramified quadratic extension of $\mathbb{Q}_2$. It follows that $\mathcal{O}_{\mathbb{Q}_2(\sqrt{5})} = \mathbb{Z}_2[(1+\sqrt{5})/2]$, and $\mathcal{O}_K = \mathbb{Z}_2[\sqrt{d}]$ in all other cases.

c. Calculate $\mathfrak{d}_{K/\mathbb{Q}_p}$ in all cases.

12. Principal units in quadratic extensions of $\mathbb{Q}_p$ (see [19]). Let p be a prime and $K = \mathbb{Q}_p(\sqrt{d})$, where $d \in \mathbb{Q}_p^\times \setminus \mathbb{Q}_p^{\times 2}$. Then $U_K^{(1)} = \mu_{p^\infty}(K)\cdot V$, where V is a free (multiplicative) $\mathbb{Z}_p$-module of rank 2. Explicitly, we have $U_K^{(1)} = {}_{\mathbb{Z}_p}(\zeta, w_1, w_2)$, where $\mu_{p^\infty}(K) = \langle \zeta \rangle$ and (w_1, w_2) is a $\mathbb{Z}_p$-basis of V. Possible values are given in the following two tables.

$p=2$	ζ	w_1	w_2
$d \equiv -1 \bmod 8$	i	$-1+2\sqrt{d}$	5
$d \equiv \pm 5 \bmod 8$	-1	$\sqrt{d}$	$-1+2\sqrt{d}$
$d \equiv \pm 2 \bmod 4$	-1	$1+\sqrt{d}$	5

(i denotes a primitive 4-th root of unity in K).

$p \neq 2$	ζ	w_1	w_2
$p \nmid d$	1	$1+p$	$1+p\sqrt{d}$
$p=3$ and $d \equiv -3 \bmod 9$	ρ	4	$1+3\sqrt{d}$
$p \neq 3$ or $d \equiv 3 \bmod 9$	1	$1+\sqrt{d}$	$1+p$

(ρ denotes a primitive 3-rd root of unity in K).

13. Let $K = \mathbb{Q}_3(x,y)$, where $x^3 = 3$ and $y^2 = 2 - x$. Then $e(K/\mathbb{Q}_3) = 3$ and $f(K/\mathbb{Q}_3) = 2$. Determine the inertia field of $K/\mathbb{Q}_3$ and calculate $\mathfrak{d}_{K/\mathbb{Q}_3}$.

14. Let K be a $\mathfrak{p}$-adic number field and $n \in \mathbb{N}$. Then (inside a fixed algebraic closure) there exist only finitely many field extensions L/K of degree n. Hint: It suffices to prove that there exist only finitely many fully ramified field extensions of degree at most n, and for this use Krasner's lemma and a compactness argument.

15. An example of E. Artin. Let $f = X^5 - X + 1 \in \mathbb{Q}[X]$ and $K = \mathbb{Q}(\alpha)$, where $f(\alpha) = 0$.

a. $\mathsf{D}(f) = 19 \cdot 151$, and $\mathrm{Gal}_{\mathbb{Q}}(f) = \mathfrak{S}_5$. Hint: Factorize f modulo 2 and modulo 5, and apply 2.5.5. Alternatively or additionally use the hints in [44, Aufgabe 15.25].

b. $\mathcal{O}_K = \mathbb{Z}[\alpha]$ and $h_K = 1$. Hint: Use 3.6.1.

c. $K_0 = \mathbb{Q}(\sqrt{19 \cdot 151}\,) \subset K$, $\mathrm{Gal}(K/K_0) \cong \mathfrak{A}_5$, and all $\mathfrak{p} \in \mathcal{P}_{K_0}$ are unramified in K. Hint: Consider d_K and $\mathfrak{d}_{K/K_0}$.

16. let L/K be a finite Galois extension of complete discrete valued fields such that $\mathrm{char}(\mathsf{k}_K) = p > 0$ and $\mathsf{k}_L/\mathsf{k}_K$ is separable. Let $G = \mathrm{Gal}(L/K)$, $i, j \in \mathbb{N}$, $\sigma \in G_i \setminus G_{i+1}$ and $\tau \in G_j \setminus G_{j+1}$. Then $\sigma\tau\sigma^{-1}\tau^{-1} \in G_{i+j+1}$ and $i \equiv j \bmod p$.
Hint:

$$\frac{\sigma\pi}{\pi} = 1+\alpha\pi^i \text{ and } \frac{\tau\pi}{\pi} = 1+\beta\pi^j \implies \frac{\sigma\tau\sigma^{-1}\tau^{-1}\pi}{\pi} \equiv 1+(j-i)\alpha\beta\pi^{i+j} \mod \mathfrak{p}_L^{i+j+1}.$$

17. ([13, III. 2.3]). Let p be a prime, $K/\mathbb{Q}_p$ a finite field extension and L/K a cyclic field extension of degree p. Let $e = v_K(p) = e(K/\mathbb{Q}_p)$, set $G = \mathrm{Gal}(L/K)$, and let $s \in \mathbb{N}$ be such that $G = G_s$ and $G_{s+1} = \mathbf{1}$. Then it follows that $s = pe/(p-1)$, and K contains a primitive p-th root of unity.

18. a. If p is a prime and $x \in p\mathbb{Z}_p$, then

$$\log_p(1+x) = \lim_{n\to\infty} \frac{(1+x)^{p^n} - 1}{p^n} \in \mathbb{Z}_p.$$

b. If $x \in \mathbb{Z}_2$ and $v_2(x) = 1$, then $v_2(\log_2(1+x)) \geq 2$ and

$$\exp_2 \circ \log_2(1+x) = -(1+x).$$

Hint: $x = -1 - 4y$ and $\log(-1) = 0$. Compare 5.8.8.3.

19. This exercise deals again with Witt vectors (see Ch.1, Ex. **8** and [57, Ch. V]).

Let F be a perfect field of characteristic $p > 0$ and $\mathsf{W}(F)$ the Witt ring over F as defined in Ch.1, Ex. **8**, and let $\sigma\colon F \to \mathsf{W}(F)$ and V, $\mathsf{F}\colon \mathsf{W}(F) \to \mathsf{W}(F)$ be defined as there.

a. Every $x \in \mathsf{W}(F)^\bullet$ has a unique representation $x = p^r u$, where $r \in \mathbb{N}_0$ and $u \in \mathsf{W}(F)$ such that $u_0 \neq 0$. Hint: If $r \in \mathbb{N}_0$ and $x \in \mathsf{W}(F)$, then $F^{p^r} = F$ and

$$(p^r x)_n = \begin{cases} 0 & \text{if } n \le r, \\ x_{n-r}^{p^r} & \text{if } n > r. \end{cases}$$

b. For $x \in \mathsf{W}(F)$ we define $w_0(x) = \inf\{n \in \mathbb{N}_0 \cup \{\infty\} \mid x_n \neq 0\}$. Then $w_0(x \cdot y) = w_0(x) + w_0(y)$ and $w_0(x+y) \ge \min\{w_0(x), w_0(y)\}$ for all $x, y \in \mathsf{W}(F)$. In particular, $\mathsf{W}(F)$ is a domain of characteristic 0, we assume that $\mathbb{Z} \subset \mathsf{W}(F)$, and we set $\mathbb{W}(F) = \mathsf{q}(\mathsf{W}(F))$. If $x \in \mathsf{W}(F)$ and $q \in \mathbb{Z}$, then $qx = x \cdot q$.

c. If $x \in \mathsf{W}(F)$, then

$$x = \sum_{i=0}^{\infty} p^i \sigma\big(x_i^{p^{-i}}\big) = \sum_{i=0}^{\infty} \sigma\big(x_i^{p^{-i}}\big) \cdot p^i;$$

if $x_0 \neq 0$ and $1 - x\sigma(x_0^{-1}) = \mathsf{V}y$ for some $y \in \mathsf{W}(F)$, then

$$x\sigma(x_0^{-1}) \sum_{i=0}^{\infty} (\mathsf{V}y)^i = 1, \quad \text{hence } x \in \mathsf{W}(F)^\times$$

(observe that

$$\sum_{i=0}^{\infty} (\mathsf{V}x)^i \in \mathsf{W}(F), \quad \text{indeed, } ((Vx)^i)_n = 0 \text{ if } i > n);$$

d. $\mathsf{W}(F)$ is a discrete valuation domain. If $v\colon \mathbb{W}(F) \to \mathbb{Z} \cup \{\infty\}$ is the associated discrete valuation, then $v \restriction \mathsf{W}(F) = w_0$. $(\mathbb{W}(F), v)$ is a complete discrete valued field with valuation domain $\mathsf{W}(F)$, valuation ideal $p\mathsf{W}(F)$, residue class field $\mathsf{W}(F)/p\mathsf{W}(F) \cong F$ (identify!), and $\sigma(F)$ is a multiplicative residue system.
Hint: Prove first that $(\mathbb{W}(F), v)$ is a discrete valued field and only afterwards that it coincides with its completion using **c**.

e. Let (K, v) be a complete discrete valued field of characteristic 0 such that p is a prime element and F is the residue class field of K. Then there is a unique isomorphism $\Phi\colon \mathbb{W}(F) \to K$ such that $\Phi \circ \sigma(F)$ is the multiplicative residue system of (K, v).

20. Let S be a finite set of primes and $n \in \mathbb{N}$. Then there are only finitely many algebraic number fields K such that $[K:\mathbb{Q}] = n$ and at most the primes in S are ramified in K. Hint: $v_p(\mathsf{d}_K) \le n - 1 + n\log n/\log p$ for every prime p.

6

Algebraic Function Fields

There is a deep analogy between algebraic number fields and algebraic function fields, in particular over finite fields of constants. Therefore several authors, mainly in the German school, presented a unified treatment of the subjects based on valuation theory as initiated in the famous lecture notes "Algebraic Numbers and Algebraic Functions" by E. Artin [1] (see [27], [43], [37]). In all these volumes, the theory around the theorem of Riemann-Roch remains exclusively reserved for algebraic function fields and it seems that this is the crucial point of distinction. Only recently, in the course of modern algebraic geometry (Arakelov theory) it was possible to rephrase the theory around the Riemann-Roch theorem in a form also applicable to algebraic number fields. An account of this theory culminating in the definition of the genus of an algebraic number field can be found in J. Neukirch's book [52].

In my feeling, however, a Riemann-Roch theory for number fields without its embedding into the setting of higher algebraic geometry is difficult to understand and thus I renounced it. Also beyond Riemann-Roch theory there are plenty of results which are distinctive for algebraic function fields and justify a separate treatment. This is the motivation for the present chapter. My main references are [60], [11], [38], [14] and [48]. For a wealth of additional results which are special for function fields we refer to [58].

The ideal-theoretic and valuation-theoretic tools presented in Chapters 2 and 5 are indispensable for the actual exposition of the theory.

We start with some algebraic preliminaries and sketch the connection with the theory of algebraic curves. A first highlight is the theorem of Riemann-Roch in Section 6.4. The preceding sections are essentially devoted to the introduction of the necessary tools for this famous theorem (divisors and differentials and their properties). The following two sections deal with the arithmetic of function field extensions. In particular, we apply the general theory of differents and discriminants from Section 5.10 to function fields and prove the Hurwitz genus formula. Sections 6.7 and 6.8 contain a detailed analysis of the connection between Weil differentials and classical differentials and a proof of the residue theorem. Finally in Section 6.9 we consider function fields with finite fields of constants and their zeta functions. A proof of the Hasse-Weil theorem ("Riemann conjecture") following [14] forms the concluding highlight of the present volume.

6.1 Field theoretic properties

We start with some elementary facts supplementing Chapter 1.

Lemma 6.1.1. *Let $K \subset K' \subset L$ be fields, and let $x \in L$ be transcendental over K'. Then $[K'(x):K(x)] = [K':K]$.*

Proof. If $(u_i)_{i\ge 0}$ is a family in K' which is linearly independent over K, then $(u_i)_{i\in I}$ is linearly independent over $K[x]$ and thus over $K(x)$. Indeed, let $(f_i)_{i\ge 0}$ be a family in $K[x]$ such that $f_i = 0$ for almost all $i \ge 0$, say

$$f_i = \sum_{\nu\ge 0} a_{i,\nu}x^\nu. \text{ Then } \sum_{i\in I} f_i u_i = 0 \text{ implies } \sum_{i\in I} a_{i,\nu}u_i = 0 \text{ for all } \nu \ge 0,$$

hence $a_{i,\nu} = 0$ for all $i \in I$ and $\nu \ge 0$, and therefore $f_i = 0$ for all $i \in I$. In particular, $[K'(x) : K(x)] \ge [K' : K]$, and we may presuppose $[K' : K] < \infty$. We proceed by induction on $[K':K]$ and assume that the assertion holds for all field extensions of smaller degree. Let $\alpha \in K' \setminus K$ and $[K(\alpha) : K] = n$. Then $(1, \alpha, \ldots, \alpha^{n-1})$ is linearly independent over $K(x)$, and it follows that $[K(x)(\alpha):K(x)] \ge n$. Since obviously $[K(\alpha)(x):K(x)] = [K(x)(\alpha):K(x)] \le n$, equality holds.

Now $[K'(x) : K(\alpha)(x)] = [K' : K(\alpha)]$ by the induction hypothesis, and consequently

$$\begin{aligned}[K'(x):K(x)] &= [K'(x):K(\alpha)(x)]\,[K(\alpha)(x):K(x)] = [K':K(\alpha)]\,[K(\alpha):K] \\ &= [K':K]. \end{aligned}$$

□

Definition 6.1.2. A field extension K/F is called an (algebraic) **function field** (in one variable) if there exists some $x \in K$ such that x is transcendental over F and $[K:F(x)] < \infty$. Then we also call K a function field over F. By definition, K/F is a function field if and only if K/F is finitely generated and $\mathsf{tr}(K/F) = 1$ (see Section 1.8).

If K/F is a function field, then the relative algebraic closure $\widetilde{F}$ of F in K is called its **field of constants**.

An element $x \in K$ is called **separating** (for K/F) if $K/F(x)$ is finite and separable. Then it follows by 1.5.1 that $K = F(x, y)$, where $y \in K$ is separable over $F(x)$.

If $K = F(x)$ and x is transcendental over F, then the function field $F(x)/F$ is called the **rational function field** in x over F.

Theorem 6.1.3. *Let K/F be a function field, $\widetilde{F}$ its field of constants and K'/K a finite field extension.*

1. *$K/\widetilde{F}$ is a function field and $\widetilde{F}$ is its field of constants.*
2. *K'/F is a function field.*
3. *If $t \in K$ is transcendental over F, then $[K\!:\!F(t)] < \infty$.*

Proof. Let $x \in K$ be transcendental over F such that $[K\!:\!F(x)] < \infty$.

1. Since $\widetilde{F}/F$ is algebraic, it follows that x is transcendental over $\widetilde{F}$, and $[K:\widetilde{F}] \le [K:F] < \infty$. By 1.2.6.2, $\overline{F}$ is relatively algebraically closed in K, and therefore $\widetilde{F}$ is the field of constants of $K/\widetilde{F}$.

2. $[K'\!:\!F(x)] = [K'\!:\!K][K\!:\!F(x)] < \infty$.

3. Let $t \in K$ be transcendental over F. Then t is algebraic over $F(x)$, and thus $f(t) = 0$ for some polynomial $f \in F(x)[T]^\bullet$. After multiplication with a common denominator we may assume that $f \in F[x][T]^\bullet$, and then there exists a polynomial $\Phi \in F[X,T]^\bullet$ such that $\Phi(x,T) = f$ and thus $\Phi(x,t) = 0$, say

$$\Phi = \Phi(X,T) = \sum_{\nu=0}^{n} a_\nu(T)X^\nu,$$

where $n \in \mathbb{N}_0$, $a_\nu(T) \in F[T]$ for all $\nu \in [0,n]$ and $a_n(T) \ne 0$. As t is transcendental over F, we obtain $a_n(t) \ne 0$, hence $\Phi(X,t) \in F(t)[X]^\bullet$, and since $\Phi(x,t) = 0$ it follows that x is algebraic over $F(t)$. Consequently,

$$[K\!:\!F(t)] = [K\!:\!F(t,x)]\,[F(t,x)\!:\!F(t)] \le [K\!:\!F(x)]\,[F(t)(x)\!:\!F(t)] < \infty. \quad \square$$

Theorem 6.1.4. *Let $K = F(x)$ be a rational function field over F, and*

$$h = \frac{f(x)}{g(x)} \in K \setminus F,$$

where $f,\ g \in F[X]^\bullet$ are coprime polynomials such that $\max\{\partial(f), \partial(g)\} \ge 1$.

Then $[K : F(h)] = \max\{\partial(f), \partial(g)\}$, and F is the field of constants of K/F. In particular, $[K\!:\!F(x^n)] = n$, and $x \notin K^n$ for all $n \in \mathbb{N}_{\ge 2}$.

Proof. We consider the polynomial $P = f - gh \in F(h)[X] \setminus F(h)$. Since $P(x) = 0$, it follows that x and thus K is algebraic over $F(h)$, and therefore h is transcendental over F. Since f and g are coprime, it follows that the polynomial $f - gZ \in F[Z,X]$ is irreducible. Hence P is irreducible in $F[h][X]$ and thus in $F(h)[X]$. If $c \in F(h)$ is the leading coefficient of P in X, then $c^{-1}P$ is the minimal polynomial of x over $F(h)$, and

$$[K\!:\!F(h)] = \partial(c^{-1}P) = \partial(P) = \max\{\partial(f), \partial(g)\}.$$

In particular, $[K : F(x^n)] = n$ for all $n \in \mathbb{N}$. Assume that $n \geq 2$ and $x = z^n$ for some $z \in K$. Then z is transcendental over F, $K(z) = K(x)$, and therefore $[K(z) : K(x)] = [K(z) : K(z^n)] = n$, a contradiction.

Since every $h \in K \setminus F$ is transcendental over F, it follows that F is relatively algebraically closed in K and thus it is the field of constants of K/F. □

The following Theorem 6.1.5 provides the link between algebraic function fields and algebraic curves. We sketch this in the simplest case of an affine plane algebraic curve. Let $\overline{F}$ be an algebraic closure of a field F. A subset C of $\overline{F}^2$ is called an over F defined (**plane affine algebraic**) **curve** if there exists a polynomial $f \in F[X, Y] \setminus F$ such that

$$C = \mathsf{V}(f) = \{(\alpha, \beta) \in \overline{F}^2 \mid f(\alpha, \beta) = 0\}.$$

If x, y: $C \to \overline{F}$ are the coordinate functions, defined by $x(\alpha, \beta) = \alpha$ and $y(\alpha, \beta) = \beta$, then x and y are not both algebraic over F, and we call

$$F[C] = F[x, y]$$

the **coordinate ring** of C. The curve C is called **irreducible** if $F[C]$ is a domain [equivalently, $C = \mathsf{V}(f)$ for an irreducible polynomial $f \in F[X, Y]$ or, geometically, C is not the union of two curves different from C]. If C is irreducible, then the field $F(C) = \mathsf{q}(F[C]) = F(x, y)$ is called the **function field** of C. If $C = \mathsf{V}(f)$ for some irreducible polynomial $f \in F[X, Y]$, then $F[C] \cong F[X, Y]/(f)$.

An irreducible polynomial $f \in F[X, Y]$ is called **absolutely irreducible** if f is irreducible in $\overline{F}[X, Y]$ [equivalently, f is irreducible in $L[X, Y]$ for every algebraic extension field L of K].

Theorem 6.1.5.

1. *Let C be an over F defined irreducible algebraic curve. Then $F(C)/F$ is a function field.*

2. *Let K/F be a function field, and let x, $y \in K$ be such that $K = F(x, y)$. Then there exists an up to factors from $F^\times$ uniquely determined irreducible polynomial $f \in F[X, Y]$ such that $f(x, y) = 0$.*
If f is absolutely irreducible, then F is the field of constants of K/F.
If $C = \mathsf{V}(f)$ and x', y': $C \to \overline{F}$ are the coordinate functions of C, then there exists a unique F-isomorphism Φ: $K \cong F(C)$ such that $\Phi(x) = x'$ and $\Phi(y) = y'$.

Proof. 1. Let $C = \mathsf{V}(f)$ for some irreducible polynomial $f \in F[X, Y]$, and let x, y: $C \to \overline{F}$ be the coordinate functions of C. Then $F(C) = F(x, y)$, $f(x, y) = 0$, and we may assume that x is transcendental over F. Then y is algebraic over $F(x)$, hence $[K : F(x)] < \infty$, and K/F is a function field.

2. We may assume that x is transcendental over F. Then $[K\!:\!F(x)] < \infty$ by 6.1.3.3, hence y is algebraic over $F(x)$, we set $J = \{g \in F[X,Y] \mid g(x,y) = 0\}$ of $F[X,Y]$, and we prove first:

> **A.** There exists an irreducible polynomial $f \in F[X,Y]$ such that $J = (f)$.

Proof of **A.** Let $g \in F(x)[Y]^\bullet$ be such that $g(y) = 0$. After multiplication with a common denominator we may assume that $g \in F[x,Y]$, and then there exists a polynomial $q \in F[X,Y]^\bullet$ such that $q(x,Y) = g$ and thus $q(x,y) = 0$. Since $K[X,Y]$ if factorial, there exists an irreducible polynomial $f \in K[X,Y]$ such that $f \,|\, q$ and $f(x,y) = 0$. Then $f \in J$, hence $(f) \subset J$ and we assert equality. We assume to the contrary that there exists some $h \in J \setminus (f)$.

If $\mathfrak{o} = F[X]$, then h and f are polynomials in $\mathfrak{o}[Y]$ without a common factor in $\mathfrak{o}[Y] \setminus \mathfrak{o} = F[X,Y] \setminus F[X]$. By 2.2.7.2 there exist $p,\, p_1 \in \mathfrak{o}[Y] = F[X,Y]$ such that $pf + p_1 h = r \in \mathfrak{o}^\bullet = F[X]^\bullet$, and since

$$r(x) = p(x,y)f(x,y) + p_1(x,y)h(x,y) = 0,$$

this contradicts the transcendence of x over F. □[**A.**]

If $J = (f)$ for some irreducible polynomial $f \in F[X,Y]$, then $f(x,y) = 0$, and if $f_1 \in F[X,Y]$ is another irreducible polynomial such that $f_1(x,y) = 0$, then $f_1 \in J$, hence $f \,|\, f_1$ and therefore $f_1 = cf$ for some $c \in F^\times$. Thus existence and uniqueness of f up to factors from $F^\times$ are proved.

We assume now that f is absolutely irreducible and $\widetilde{F}$ is the field of constants of K/F. Then f is irreducible in $\widetilde{F}[X,Y]$, x is transcendental over $\widetilde{F}$ and $K = \widetilde{F}(x,y)$. Let $c \in F[x]$ be the leading coefficient of the polynomial $f(x,Y) \in F[x][Y]$. Since f is irreducible in $\widetilde{F}[X,Y]$, it follows that $f(x,Y)$ is irreducible in $\widetilde{F}[x,Y] = \widetilde{F}[x][Y]$, and by 2.2.5.3 $f(x,Y)$ is irreducible in $\widetilde{F}(x)[Y]$. Hence $c^{-1}f(x,Y)$ is the minimal polynomial of y over $F(x)$ and also over $\widetilde{F}(x)$. Hence $[K : F(x)] = [K : \widetilde{F}(x)]$ and by 6.1.1 it follows that $1 = [\widetilde{F}(x)\!:\!F(x)] = [\widetilde{F}\!:\!F]$.

The assertions concerning $C = \mathsf{V}(f)$ follow from the very definitions. □

Below in 6.1.7 we shall prove that every function field K/F over a perfect field F is of the form $K = F(x,y)$ for some $x,\, y \in K$.

Theorem 6.1.6. *Let K/F be a function field, and assume that F is perfect and* $\text{char}(F) = p > 0$.

1. *If $n \in \mathbb{N}$, then K^{p^n}/F is a function field, and $[K\!:\!K^{p^n}] = p^n$.*

2. *Let $x \in K$ be transcendental over F and K_1 the separable closure of $F(x)$ in K. Then $K_1 = K^{p^n}$ for some $n \in \mathbb{N}_0$, and x is separating for K/F if and only if $x \notin K^p$.*

Proof. Let $x \in K$ be transcendental over F. Then $[K:F(x)] < \infty$ by 6.1.3.3.

1. Let $n \in \mathbb{N}$. Then x^{p^n} is transcendental over F, the map $\varphi\colon K \to K^{p^n}$, defined by $\varphi(z) = z^{p^n}$ for all $z \in K$, is an isomorphism, and as F is perfect, we obtain $\varphi(F) = F$ and $\varphi(F(x)) = F(x)^{p^n} = F(x^{p^n})$. Hence K^{p^n}/F is a function field, and $[K^{p^n} : F(x^{p^n})] = [K : F(x)]$. Since $[F(x) : F(x^{p^n})] = p^n$ by 6.1.4, we obtain

$$\begin{aligned}[K:F(x^{p^n})] &= [K:F(x)]\,[F(x):F(x^{p^n})] = [K:F(x)]\,p^n \\ &= [K:K^{p^n}]\,[K^{p^n}:F(x^{p^n})] = [K:K^{p^n}]\,[K:F(x)],\end{aligned}$$

and consequently $[K:K^{p^n}] = p^n$.

2. As K/K_1 is purely inseparable and $[K : K_1] \le [K : F(x)] < \infty$, it follows that $[K : K_1] = p^n$ for some $n \in \mathbb{N}_0$ and $K^{p^n} \subset K_1$ by 1.4.7.2. Since $[K:K^{p^n}] = p^n$ by 1., we obtain $K_1 = K^{p^n}$.

If $K/F(x)$ is inseparable, then $K_1 \ne K$, $n \ge 1$ and $x \in K_1 \subset K^p$. Conversely, if $x \in K^p$, then $F(x) \subset K^p \subsetneq K$, and thus $K/F(x)$ is inseparable. □

Corollary 6.1.7. *Let K/F be a function field, and let F be perfect. Then K contains a separating element for K/F and, in particular, $K = F(x, y)$ for some x, $y \in K$ such that y is separable over $F(x)$.*

Proof. If $\mathrm{char}(F) = 0$, there is nothing to do. Thus let $\mathrm{char}(F) = p > 0$, let $x \in K$ be transcendental over F and K_1 the separable closure of $F(x)$ in K. By 6.1.6.2 we obtain $[K:K_1] = p^n$ for $n \in \mathbb{N}_0$, hence $x = t^{p^n}$ for some $t \in K$, and it suffices to prove that $t \notin K^p$, for then t is separating for K/F.

Assume to the contrary that $t = u^p$ for some $u \in K$. Then we obtain $t^{p^{n-1}} = u^{p^n} \in K_1$ and $(t^{p^{n-1}})^p = x$. As $x \notin F(x^p) = F(x)^p$ (by 6.1.4), it follows that $t^{p^{n-1}}$ is inseparable over $F(x)$, a contradiction.

If x is separating for K/F, then we have already seen that $K = F(x, y)$ for some $y \in K$ by which is separable over $F(x)$. □

Definition and Theorem 6.1.8. *Let K/F and K'/F' be function fields such that $K \subset K'$, $F \subset F'$, and K'/K is algebraic.* Then we call K'/F' an **algebraic function field extension** of K/F. For short, we will also say that K'/K is an algebraic extension of function fields.

1. F'/F is algebraic, and $[K':K] < \infty$ if and only if $[F':F] < \infty$

2. KF'/F' is a function field, and if $[F':F] < \infty$, then K'/F and KF'/F are function fields, too.

3. If K'/K is separable and x is separating for K/F, then x is separating for K'/F', too.

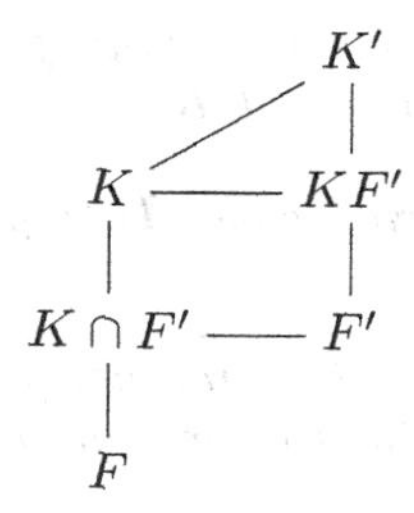

The field extension K'/K is called a **constant field extension** if $K' = KF'$, and it is called **geometric** if $F' = F$.

Proof. We tacitly use 6.1.3.3 and 6.1.1. Let $x \in K$ be transcendental over F.

1. By 1.8.4 we obtain

$$\begin{aligned}\mathsf{tr}(K'/F) &= \mathsf{tr}(K'/K) + \mathsf{tr}(K/F) = 0 + 1 = 1 \\ &= \mathsf{tr}(K'/F') + \mathsf{tr}(F'/F) = 1 + \mathsf{tr}(F'/F),\end{aligned}$$

hence $\mathsf{tr}(F'/F) = 0$ and F'/F is algebraic. Therefore x is transcendental over F', and $[K':K]\,[K:F(x)] = [K':F'(x)]\,[F'(x):F(x)] = [K':F'(x)]\,[F':F]$. If follows that $[F':F] < \infty$ if and only if $[K':K] < \infty$.

2. KF'/F' is a function field, since $[KF' : F'(x)] \le [K : F(x)] < \infty$. If $[F':F] < \infty$, then

$$\begin{aligned}[KF':F(x)] \le [K':F(x)] &= [K':F'(x)]\,[F'(x):F(x)] \\ &= [K':F'(x)]\,[F':F] < \infty,\end{aligned}$$

and therefore KF'/F and K'/F are function fields.

3. If $K/F(x)$ is separable, then the extensions $K'/F(x)$ and $K'/F'(x)$ are separable by 1.4.4.3(c). $\square$

Algebraic function field extensions are ubiquitous in the theory of algebraic function fields. We give two examples.

a. If K/F is a function field, $\widetilde{F}$ is its field of constants and $x \in K \setminus \widetilde{F}$, then $F(x)$ is a rational function field, and $K/\widetilde{F}$ is an algebraic function field extension of $F(x)/F$.

b. Let $F(x)$ be a rational function field, $K \supset F(x)$ an extension field and F' an intermediate field of K/F such that F'/F is algebraic. Then $F'(x)/F(x)$ is algebraic by 1.2.3.2(a), and thus $F'(x)/F'$ is a constant field extension of $F(x)/F$ (a constant field extension of a rational function field is a rational function field).

We close this section by summarizing the algebraic properties of constant field extensions. The following Theorem 6.1.9 will frequently be used in Sections 6.5 and 6.6.

Theorem 6.1.9. *Let K'/F' be an algebraic function field extension of K/F, and let F be the field of constants of K.*

1. Let $g \in F[X]$ be irreducible. Then g is irreducible in $K[X]$, too. In particular:

(a) If $\alpha \in F'$ and g is the minimal polynomial of α over F, then g is also the minimal polynomial of α over K,

$$[F(\alpha):F] = [K(\alpha):K],$$

and α is separable over F if and only if α is separable over K.

(b) If K'/K is (finite) separable, then F'/F is (finite) separable, too.

2. Let F'/F be separable.

(a) Every family in K which is linearly independent over F remains linearly independent over F'.

(b) If $x \in K \setminus F$, then $[K:F(x)] = [KF':F'(x)] \le [K':F'(x)]$.

(c) F' is the field of constants of KF'/F'.

3. Assume that either $F' = F(\alpha)$ for some $\alpha \in F'$, or that F'/F is separable. Then $[F':F] = [KF':K] \le [K':K]$.

Proof. Recall that F'/F is algebraic by 6.1.8.1.

1. We may assume that g is monic, and we suppose to the contrary that $g = g_1 g_2$, where $g_1, g_2 \in K[X] \setminus K$ are monic. Let $\overline{K} \supset K$ be a splitting field of g, and let $g_1 = (X-\alpha_1)\cdot\ldots\cdot(X-\alpha_r)$ and $g_2 = (X-\alpha_{r+1})\cdot\ldots\cdot(X-\alpha_n)$, where $\alpha_1,\ldots,\alpha_n \in \overline{K}$ and $r \in [1, n-1]$. Then $\alpha_1,\ldots,\alpha_n$ and accordingly the coefficients of g_1 and g_2 are algebraic over F. As $g_1, g_2 \in K[X]$ and F is relatively algebraic closed in K, it follows that $g_1, g_2 \in F[X]$, a contradiction.

(a) Obvious.

(b) By 6.1.8.1 K'/K is finite if and only if F'/F is finite. Thus suppose that F'/F is inseparable. If $\alpha \in F'$ is inseparable over F, then its minimal polynomial is inseparable, and thus α is also inseparable over K by (a).

2. (a) Observe that a family is linearly independent over F resp. F' if and only if every finite subfamily is linearly independent over F resp. F'. Hence it suffices to prove: If $n \in \mathbb{N}$ and an n-tuple $(u_1,\ldots,u_n) \in K^n$ is linearly independent over F, then it is linearly independent over F'.

Thus let $(u_1,\ldots,u_n) \in K^n$ be linearly independent over F and $\alpha_1,\ldots,\alpha_n \in F'$ such that $\alpha_1 u_1 + \ldots + \alpha_n u_n = 0$. As F'/F is separable, 1.5.1 implies that there exists some $\alpha \in F'$ such that $F(\alpha_1,\ldots,\alpha_n) = F(\alpha)$, and we suppose that for all $i \in [1,n]$

$$\alpha_i = \sum_{\nu=0}^{d-1} a_{i,\nu}\alpha^\nu, \quad \text{where} \quad d = [F(\alpha):F] \text{ and } a_{i,\nu} \in F.$$

By 1. we obtain $[F(\alpha):F] = [K(\alpha):K]$, and therefore $(1, \alpha, \ldots, \alpha^{d-1})$ is linearly independent over K. Since

$$0 = \sum_{i=1}^{n} \alpha_i u_i = \sum_{i=1}^{n} \Big(\sum_{\nu=0}^{d-1} a_{i,\nu} \alpha^\nu \Big) u_i = \sum_{\nu=0}^{d-1} \Big(\sum_{i=1}^{n} a_{i,\nu} u_i \Big) \alpha^\nu,$$

it follows that

$$\sum_{i=1}^{n} a_{i,\nu} u_i = 0 \quad \text{for all} \quad \nu \in [0, d-1],$$

hence $a_{i,\nu} = 0$ for all $i \in [1,n]$ and $\nu \in [0, d-1]$, and therefore $\alpha_i = 0$ for all $i \in [1,n]$.

(b) Obviously, $[KF' : F'(x)] \le [K : F(x)]$ and $[KF' : F'(x)] \le [K' : F'(x)]$. Let $(u_1, \ldots, u_n)$ be an $F(x)$-basis of K. It suffices to prove that $u_1, \ldots, u_n$ are linearly independent over $F'[x]$ (then they are linearly independent over $F'(x)$, and thus $[KF' : F'(x)] \ge n = [K : F(x)]$). Suppose that $f_1, \ldots, f_n \in F'[x]$ are such that $f_1 u_1 + \ldots + f_n u_n = 0$. If

$$f_i = \sum_{\nu=0}^{N} a_{i,\nu} x^\nu \quad \text{for all} \quad i \in [1,n], \quad \text{where} \quad N \in \mathbb{N} \quad \text{and} \quad a_{i,\nu} \in F',$$

then

$$\sum_{i=1}^{n} \sum_{\nu=0}^{N} a_{i,\nu} x^\nu u_i = 0.$$

Since $(u_1, \ldots, u_n)$ is linearly independent over $F[x]$, the family $(x^\nu u_i)_{\nu \in [0,N],\, i \in [1,n]}$ is linearly independent over F and thus by 1. also over F'. It follows that $a_{i,\nu} = 0$ for all $i \in [1,n]$ and $\nu \in [0,N]$, and consequently $f_i = 0$ for all $i \in [1,n]$.

(c) Let $\alpha \in KF'$ be algebraic over F' (and thus over F). Then there exists a finite subset S of F' such that $\alpha \in K(S)$, and since $F(S)/F$ is finite separable, there exists some $\beta \in F(S)$ such that $F(S) = F(\beta)$. Then $K(S) = K(\beta)$ and thus $\alpha \in K(\beta)$. Since $F(\alpha, \beta)/F$ is finite separable, there exists some $\gamma \in F(\alpha, \beta)$ such that $F(\alpha, \beta) = F(\gamma)$, which implies that $K(\gamma) = K(\alpha, \beta) = K(\beta)$. Using 1. we obtain

$$[F(\gamma):F] = [K(\gamma):K] = [K(\beta):K] = [F(\beta):F],$$

and since $F(\beta) \subset F(\gamma)$, it follows that $F(\gamma) = F(\beta) \subset F'$ and therefore $\alpha \in F'$.

3. If $F' = F(\alpha)$, the assertion follows from 1. Thus assume that F'/F is separable. It suffices to prove that $[KF' : K] = [F' : F]$ for every finite field extension F'/F. However, if $[F' : F] < \infty$, then $F' = F(\alpha)$ for some $\alpha \in F'$. □

The following result will be used in 6.9.8 for the proof of the Riemann conjecture.

Theorem 6.1.10. *Let K/F be a function field, F its field of constants, F'/F a finite separable field extension and $K' = KF'$. If $\tau \in \mathrm{Gal}(K/F)$ is an automorphism of finite order, then τ has a unique extension to an automorphism $\tau' \in \mathrm{Gal}(K'/F')$.*

Proof. Let E be the fixed field of τ. Then K/E is cyclic and $\mathrm{Gal}(K/E) = \langle\tau\rangle$ by 1.5.3. Let $x \in E \setminus F$. By 6.1.9.2(b) we obtain $[K:F(x)] = [K':F'(x)]$ and $[E:F(x)] = [EF':F'(x)]$, and consequently $[K:E] = [K':EF']$.

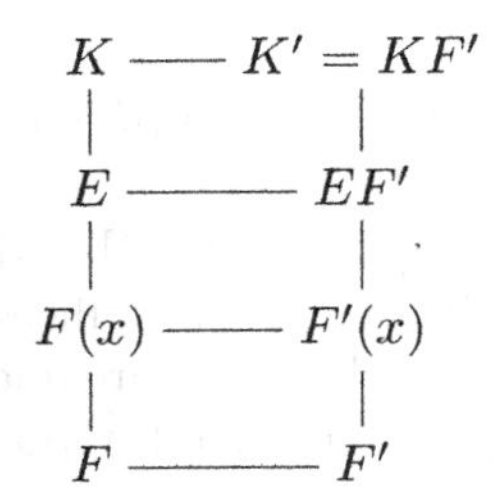

Since $K' = KEF'$, 1.5.6.2 shows that K'/EF' is cyclic, and the map

$$\mathrm{Gal}(K'/EF') \to \mathrm{Gal}(K/K \cap EF'), \quad \text{defined by} \quad \nu \mapsto \nu \restriction K,$$

is an isomorphism. Hence $[K:K \cap EF'] = [K':EF'] = [K:E]$, and therefore $E = K \cap EF'$. Hence $\nu\colon \mathrm{Gal}(K'/EF') \stackrel{\sim}{\to} \mathrm{Gal}(K/E)$ is an isomorphism, and $\tau' = \nu^{-1}(\tau)$ is the unique extension of τ. □

6.2 Divisors

Theorem 6.2.1. *Let K/F be a function field and $\widetilde{F}$ its field of constants.*

1. *Let $|\cdot|\colon K \to \mathbb{R}_{\ge 0}$ be an absolute value such that $|x| = 1$ for all $x \in F^\times$. Then $|\cdot|$ is discrete, and $|x| = 1$ for all $x \in \widetilde{F}^\times$.*

2. *Let $\mathcal{O}$ be a discrete valuation domain such that $\mathsf{q}(\mathcal{O}) = K$ and $F \subset \mathcal{O}$. Then $\widetilde{F} \subset \mathcal{O}$.*

Proof. 1. By 5.1.7.1, $|\cdot|$ is non-Archimedean. If $x \in \widetilde{F}$, then there exist $a_0, \ldots, a_{n-1} \in F$ such that $x^n + a_{n-1}x^{n-1} + \ldots + a_0 = 0$. As $|a_i| \le 1$ for all $i \in [0, n-1]$, 5.1.7.2(c) implies $|x| \le 1$. If $x \in \widetilde{F}^\times$, then $x^{-1} \in \widetilde{F}$, hence $|x^{-1}| = |x|^{-1} \le 1$ and therefore $|x| = 1$.

Let $\mathfrak{p} = \{x \in K \mid |x| < 1\}$ its valuation ideal of $(K, \cdot|)$ and $z \in \mathfrak{p}^\bullet$. Then z is transcendental over F, and $\mathfrak{p} \cap F[z] = pF[z]$ for some prime element $p \in F[z]$, and thus $|\cdot| \restriction F(z)$ is a $pF[z]$-adic absolute value by 5.1.8.2. In particular, $|\cdot| \restriction F(z)$ is discrete, and as $[K:F(z)] < \infty$, 5.5.9.1 implies that $|\cdot|$ itself is discrete.

2. Let $|\cdot|$ be an $\mathcal{O}$ defining absolute value. Since $F \subset \mathcal{O}$, it follows that $|x| \le 1$ for all $x \in F$ and thus $|x| = 1$ for all $x \in F^\times$. By 1. we obtain $|x| = 1$ for all $x \in \widetilde{F}^\times$, and therefore $\widetilde{F} \subset \mathcal{O}$. □

Definition 6.2.2. Let K/F be a function field and $\widetilde{F}$ its field of constants. A subset P of K is called a **prime divisor** (of K/F) if P is the maximal ideal of a discrete valuation domain $\mathcal{O}$ such that $\mathsf{q}(\mathcal{O}) = K$ and $F \subset \mathcal{O}$ [equivalently, there exists an absolute value $|\cdot|$ of K such that $|\cdot| \upharpoonright F^\times = 1$ and $P = \{x \in K \mid |x| < 1\}$]. We denote by $\mathbb{P}_K = \mathbb{P}_{K/F}$ the set of all prime divisors of K/F. We postpone the proof of the existence of prime divisors until 6.2.7 where we shall even prove that $|\mathbb{P}_K| = \infty$.

In the course of this chapter we consider the notion of a prime divisor as a central one, and as to the other notions of valuation theory we shall tacitly use 2.10.4.3 again and again.

Let $P \in \mathbb{P}_K$ be the maximal ideal of the discrete valuation domain $\mathcal{O}$. As $\mathcal{O}$ is uniquely determined by P, we set $\mathcal{O} = \mathcal{O}_P$, and we let $v_P \colon K \to \mathbb{Z} \cup \{\infty\}$ be the discrete valuation defining $\mathcal{O}_P$. Conversely,

- if $\mathcal{O}$ is a discrete valuation domain such that $\mathsf{q}(\mathcal{O}) = K$ and $F \subset \mathcal{O}$, then there exists a unique prime divisor $P \in \mathbb{P}_K$ such that $\mathcal{O} = \mathcal{O}_P$;

- if $v \colon K \to \mathbb{Z} \cup \{\infty\}$ is a discrete valuation such that $F \subset \mathcal{O}_v$, then there exists a unique prime divisor $P \in \mathbb{P}_K$ such that $v = v_P$.

Let $P \in \mathbb{P}_K$. A prime element $\pi \in \mathcal{O}_P$ is called a **local uniforming parameter** of P, and the field $\mathsf{k}_P = \mathcal{O}_P/P$ is called the **residue class field** of P. Since $\widetilde{F} \subset \mathcal{O}_P$ (see 6.2.1.2) and $P \cap \widetilde{F} = \mathbf{0}$, the inclusion $\widetilde{F} \hookrightarrow \mathcal{O}_P$ induces a monomorphism $\widetilde{F} \to \mathsf{k}_P$, $a \mapsto a + P$. We identify. Then $\widetilde{F} \subset \mathsf{k}_P$, and we call

$$\deg(P) = [\mathsf{k}_P : \widetilde{F}] \in \mathbb{N} \cup \{\infty\}$$

the **degree** of P. In particular, if $\dim_F(\mathsf{k}_P) = 1$, then $F \subset \widetilde{F} \subset \mathsf{k}_P$ implies $\widetilde{F} = F$ and $\deg(P) = 1$. In 6.2.7.2 we shall prove that $\deg(P) < \infty$ for all $P \in \mathbb{P}_K$. For $d \in \mathbb{Z}$, we set $\mathbb{P}_K^d = \{P \in \mathbb{P}_K \mid \deg(P) = d\}$.

We view the elements of K as functions on $\mathbb{P}_K$ with values in the respective residue class fields. Explicitly, for $z \in K$ and $P \in \mathbb{P}_K$ we define the **value of z at P** by

$$z(P) = \begin{cases} z + P \in \mathsf{k}_P & \text{if } z \in \mathcal{O}_P, \\ \infty & \text{if } z \notin \mathcal{O}_P. \end{cases}$$

The map $\varphi_P \colon K \to \mathsf{k}_P \cup \{\infty\}$, defined by $\varphi_P(z) = z(P)$ for all $z \in K$, is called the **place** associated with P (for a foundation of the theory of function fields by means of places we refer to [11] or [14, Ch. 3]).

Owing to the embedding $\widetilde{F} \subset \mathcal{O}_P$ we obtain $z(P) = z$ for all $z \in \widetilde{F}$ and $P \in \mathbb{P}_K$ (hence the elements of $\widetilde{F}$ are constant functions in the usual sense). If $z(P) = 0$ [equivalently, $z \in P$ or $v_P(z) > 0$], then we call P a **zero** of

z and $v_P(z)$ its **order**. If $z(P) = \infty$ [equivalently, $z \notin \mathcal{O}_P$ or $v_P(z) < 0$], then we call P a **pole** of z and $-v_P(z)$ its **order**. If $z \in \mathcal{O}_P$, then z is called **regular** or **holomorphic** in P. We denote by $\mathcal{N}(z) = \mathcal{N}^K(z)$ the set of all zeros and by $\mathcal{P}(z) = \mathcal{P}^K(z)$ the set of all poles of z.

By definition, $\mathcal{N}(z^{-1}) = \mathcal{P}(z)$ for all $z \in K^\times$, and $\mathcal{N}(z) = \mathcal{P}(z) = \emptyset$ for all $z \in \widetilde{F}^\times$. In 6.2.7.1 we shall prove that conversely $\mathcal{N}(z) \ne \emptyset$ and $\mathcal{P}(z) \ne \emptyset$ for all $z \in K \setminus \widetilde{F}$ (a non-constant function has at least one pole and at least one zero).

Definition and Remarks 6.2.3. If K/F is any function field, then every $x \in K$ which is transcendental over F generates a rational function field $F(x)$ inside K. In 2.10.9 we discussed the rational function field $F(X)$, built with an indeterminate X, which we now view as the prototype of a rational function field. Recall that $\mathcal{P}(F[X]) = \{pF[X] \mid p \in \mathsf{P}(F, X)\}$ is the set of all prime ideals of $F[X]$ (where $\mathsf{P}(F, X)$ denotes the set of all monic irreducible polynomials in $F[X]$).

Now let $K = F(x)$ be any rational function field. Then there is an isomorphism $\iota_x \colon F(X) \overset{\sim}{\to} F(x)$, given by $\iota_x(h) = h(x)$ for all $h \in K(X)$. Clearly, $\iota_x(F[X]) = F[x]$, and thus $\{p(x) \mid p \in \mathsf{P}(F, X)\}$ is a set of representatives for the prime elements of $F[x]$.

By 6.1.4, F is the field of constants of K. For a prime element $p(x) \in F[x]$ we set $(p(x)) = p(x)F[X]$, and we denote by $v_{p(x)} \colon F(x) \to \mathbb{Z} \cup \{\infty\}$ the $p(x)$-adic valuation of $F(x)$ (see 2.2.2). Then $\mathcal{O}_{v_{p(x)}} = F[x]_{(p(x))}$ is a discrete valuation domain with quotient field $\mathsf{q}(\mathcal{O}_{v_{p(x)}}) = K$ and maximal ideal $\mathfrak{p}_{v_{p(x)}} = p(x)F[x]_{(p(x))}$ (by 2.10.7 and the remarks preceding 2.10.9). We call $P_{p(x)} = \mathfrak{p}_{v_{p(x)}} \in \mathbb{P}_K$ the **prime divisor associated with** $p(x)$. Then $\mathcal{O}_{P_{p(x)}} = \mathcal{O}_{v_{p(x)}}$, $P_{p(x)} = p(x)\mathcal{O}_{P_{p(x)}}$, $v_{P_{p(x)}} = v_{p(x)}$, and

$$\mathsf{k}_{P_{p(x)}} = \mathcal{O}_{P_{p(x)}}/P_{p(x)} = F[x]/(p(x)) = F(\xi),$$

where $\xi = x + (p(x)) = x(P_{p(x)}) \in \mathsf{k}_{P_{p(x)}}$, $p(\xi) = 0$, and

$$\deg(P_{p(x)}) = \dim_F(\mathsf{k}_{P_{p(x)}}) = \partial(p).$$

If $z \in K^\times$ and $z = p(x)^n g(x)^{-1} h(x)$, where $n \in \mathbb{Z}$, $g, h \in F[X]$, $(g, h) = 1$ and $p \nmid gh$, then $v_{p(x)}(z) = n$,

$$z(P_{p(x)}) = \begin{cases} 0 & \text{if } n > 0, \\ \infty & \text{if } n < 0, \end{cases}$$

and if $n = 0$, then

$$z(P_{p(x)}) = \frac{h(P_{p(x)})}{g(P_{p(x)})} = \frac{h(x) + (p(x))}{g(x) + (p(x))} = \frac{h(\xi)}{g(\xi)} \in \mathsf{k}_{P_{p(x)}}.$$

If $c \in F$, then $x - c \in F[x]$ is a prime element, $\deg(P_{x-c}) = \partial(X - c) = 1$, $\mathsf{k}_{P_{x-c}} = F$, and if $g \in F[x]$, then $g(P_{x-c}) = g + (x - c)F[x]_{(x-c)} = g(c)$. From $\mathsf{k}_{P_{x-c}} = F$ it follows again that F is the field of constants of K.

Let $v_\infty \colon K \to \mathbb{Z} \cup \{\infty\}$ be the degree valuation of K with respect to x, which is given by $v_\infty(g(x)^{-1}h(x)) = \partial(g) - \partial(h)$ for all $g \in F[X]$ and $h \in F[X]^\bullet$ (see 2.10.9). Then $\mathcal{O}_{v_\infty} = F[x^{-1}]_{(x^{-1})}$, $\mathfrak{p}_{v_\infty} = x^{-1}\mathcal{O}_{v_\infty}$, and we call $P_\infty = \mathfrak{p}_{v_\infty} \in \mathbb{P}_K$ the **infinite prime divisor** with respect to x. Then $\mathcal{O}_{P_\infty} = \mathcal{O}_{v_\infty}$, $v_{P_\infty} = v_\infty$, $\mathsf{k}_{P_\infty} = F$ and $\deg(P_\infty) = 1$. If $z \in K^\times$, we set

$$z = \frac{b_m x^m + b_{m-1}x^{m-1} + \ldots + b_0}{a_n x^n + a_{n-1}x^{n-1} + \ldots + a_0} = x^{m-n}\frac{b_m + b_{m-1}x^{-1} + \ldots + b_0 x^{-m}}{a_n + a_{n-1}x^{-1} + \ldots + a_0 x^{-n}},$$

where $m, n \in \mathbb{N}_0$, $a_0, \ldots, a_n, b_0, \ldots, b_m \in K$ and $a_n b_m \neq 0$. Then it follows that $v_\infty(z) = n - m$, and

$$z(P_\infty) = \begin{cases} a_n^{-1}b_m & \text{if } m = n, \\ 0 & \text{if } m < n, \\ \infty & \text{if } m > n. \end{cases}$$

We summarize the results of 6.2.3 (making allowance for 2.10.9) in the following theorem.

Theorem 6.2.4. *Let $K = F(x)$ be a rational function field.*

1. F is the field of constants of K/F, $\{p(x) \mid p \in \mathsf{P}(F, X)\}$ is a set of representatives for the prime elements of $F[X]$, and

$$\mathbb{P}_K = \{P_{p(x)} \mid p \in \mathsf{P}(F, X)\} \cup \{P_\infty\}$$

is the set of prime divisors of K.

2. If $p \in \mathsf{P}(F, X)$, then $\deg(P_{p(x)}) = \partial(p)$, and $\deg(P_\infty) = 1$. If $g \in F[X]$, then $\mathcal{N}(g(x)) = \{P_{p(x)} \mid p \in \mathsf{P}(F, X),\ p \,|\, g\}$, and $\mathcal{P}(g(x)) = \{P_\infty\}$.

3. Let $P \in \mathbb{P}_K$.

(a) If $p \in \mathsf{P}(K, X)$, then $P = P_{p(x)}$ if and only if $p(x)(P) = 0$.

(b) $P = P_\infty$ if and only if $x^{-1}(P) = 0$.

Having fixed the terminology for rational function fields, we proceed now with the theory of algebraic function field extension.

Definition and Theorem 6.2.5 (Extension of prime divisors). *Let K'/F' be an algebraic function field extension of K/F.*

1. Let $P' \in \mathbb{P}_{K'}$ and $P = P' \cap K$. Then $K \not\subset \mathcal{O}_{P'}$, $P \in \mathbb{P}_K$, $\mathcal{O}_P = \mathcal{O}_{P'} \cap K$, and $v_{P'} \restriction K = e v_P$, where $e = e(P'/P) \in \mathbb{N}$.
We call P' an **extension** of P and say that P' **lies above** P. The embedding $\mathcal{O}_P \hookrightarrow \mathcal{O}_{P'}$ induces a monomorphism

$$\mathsf{k}_P \to \mathsf{k}_{P'}, \quad x + P \mapsto x + P'.$$

We identify:

$$\mathsf{k}_P \subset \mathsf{k}_{P'}.$$

We call $f(P'/P) = [\mathsf{k}_{P'} : \mathsf{k}_P] \in \mathbb{N} \cup \{\infty\}$ the **residue class degree** and $e(P'/P)$ the **ramification index** of P'/P. Our definitions are consistent with those in 2.12.1, applied for the extension $\mathcal{O}_P \subset \mathcal{O}_{P'}$ of (local) Dedekind domains.
If $x \in K$, then $x(P) = x(P') \in \mathsf{k}_P \cup \{\infty\} \subset \mathsf{k}_{P'} \cup \{\infty\}$.

2. For $P \in \mathbb{P}_K$ and $P' \in \mathbb{P}_{K'}$ the following assertions are equivalent:

(a) $P \subset P'$.

(b) $\mathcal{O}_P \subset \mathcal{O}_{P'}$.

(c) $P = P' \cap K$.

(d) $\mathcal{O}_P = \mathcal{O}_{P'} \cap K$.

3. Suppose that $[K' : K] = n < \infty$, let $P \in \mathbb{P}_K$ and $\mathcal{O}'_P = \mathrm{cl}_{K'}(\mathcal{O}_P)$ the integral closure of $\mathcal{O}_P$ in K'.

(a) $\mathcal{O}'_P$ is a semilocal principal ideal domain, and $\mathsf{q}(\mathcal{O}'_P) = K'$; $\mathcal{O}'_P \cap K = \mathcal{O}_P$, and if $P' \in \mathbb{P}_{K'}$ lies above P, then $\mathcal{O}'_P \subset \mathcal{O}_{P'}$.

(b) The maps

$$\theta \colon \{P' \in \mathbb{P}_{K'} \mid P' \cap K = P\} \to \mathcal{P}(\mathcal{O}'_P)$$

defined by $\theta(P') = P' \cap \mathcal{O}'_P$ for all $P' \in \mathbb{P}_{K'}$ lying above P, and

$$\theta_1 \colon \mathcal{P}(\mathcal{O}'_P) \to \{P' \in \mathbb{P}_{K'} \mid P' \cap K = P\},$$

defined by $\theta_1(\mathfrak{P}) = \mathfrak{P}(\mathcal{O}'_P)_{\mathfrak{P}}$ for all $\mathfrak{P} \in \mathcal{P}(\mathcal{O}'_P)$, are mutually inverse bijections. In particular,

$$\mathcal{O}_{P'} = (\mathcal{O}'_P)_{P' \cap \mathcal{O}'_P}, \quad \text{and} \quad \mathcal{O}'_P = \bigcap_{\substack{P' \in \mathbb{P}_{K'} \\ P' \cap K = P}} \mathcal{O}_{P'}.$$

If $P' \in \mathbb{P}_{K'}$ lies above P, then $v_{P'} = v_{P \cap \mathcal{O}'_P}$, $\mathsf{k}_{P'} = \mathsf{k}_{P \cap \mathcal{O}'_P}$, $e(P'/P) = e(P' \cap \mathcal{O}'_P/P) \le n$ and $f(P'/P) = f(P' \cap \mathcal{O}'_P/P) \le n$.

(c) $1 \le |\{P' \in \mathbb{P}_{K'} \mid P' \cap K = P\}| \le n$.

Proof. 1. We assume to the contrary that $K \subset \mathcal{O}_{P'}$. Let $x \in K' \setminus \mathcal{O}_{P'}$ and apply 5.1.7.2(c) with an absolute value $|\cdot|$ belonging to $v_{P'}$. As x is algebraic over K, we obtain $x^n + a_{n-1}x^{n-1} + \ldots + a_1 x + a_0 = 0$ for some $n \in \mathbb{N}$ and $a_0, \ldots, a_{n-1} \in K \subset \mathcal{O}_{P'}$. Hence $|a_i| \le 1$ for all $i \in [0, n-1]$, consequently $|x| \le 1$ and thus $x \in \mathcal{O}_{P'}$, a contradiction.

$v_{P'} \restriction K^\times \colon K^\times \to \mathbb{Z}$ is a homomorphism, and as $K \not\subset \mathcal{O}_{P'}$, we obtain $v_{P'} \restriction K^\times \neq 0$. Hence $v_{P'}(K^\times) = e\mathbb{Z}$ for some $e \in \mathbb{N}$,

$$v_0 = e^{-1} v_{P'} \restriction K \colon K \to \mathbb{Z} \cup \{\infty\}$$

is a discrete valuation,

$$P = P' \cap K = \{x \in K \mid v_{P'}(x) > 0\} = \{x \in K \mid v_0(x) > 0\} = \mathfrak{p}_{v_0} \in \mathbb{P}_K,$$

hence $v_P = v_0$, and

$$\mathcal{O}_P = \mathcal{O}_{v_0} = \{x \in K \mid v_0(x) \geq 0\} = \{x \in K \mid v_{P'}(x) \geq 0\} = \mathcal{O}_{P'} \cap K.$$

Consequently, $\mathcal{O}'_P = \mathrm{cl}_{K'}(\mathcal{O}_P) \subset \mathrm{cl}_{K'}(\mathcal{O}_{P'}) = \mathcal{O}_{P'}$.

If $x \in \mathcal{O}_P$, then due to the identification $x + P = x + P'$ we obtain $x(P) \in \mathsf{k}_P \subset \mathsf{k}_{P'}$ and therefore $x(P) = x(P')$. If $x \in K \setminus \mathcal{O}_P$, then $x \notin \mathcal{O}_{P'}$ and therefore $x(P) = x(P') = \infty$.

2. By 1. it follows that $P_1 = P' \cap K \in \mathbb{P}_K$ and $\mathcal{O}_{P_1} = \mathcal{O}_{P'} \cap K$.

(a) implies $P \subset P' \cap K = P_1$, hence $P = P_1$ by 2.10.4.3 and consequently (b), (c) and (d).

(b) implies $\mathcal{O}_P \subset \mathcal{O}_{P'} \cap K = \mathcal{O}_{P_1}$, hence $\mathcal{O}_P = \mathcal{O}_{P_1}$, and therefore (a), (c) and (d) hold.

The implications (c) $\Rightarrow$ (a) and (d) $\Rightarrow$ (b) are obvious.

3. (a) By 2.12.3.1 we get $\mathsf{q}(\mathcal{O}'_P) = K'$ and $\mathcal{O}'_P \cap K = \mathcal{O}_P$. Therefore it follows that $\mathcal{O}'_P = \mathrm{cl}_{K'}(\mathcal{O}_P) \subset \mathrm{cl}_{K'}(\mathcal{O}_{P'}) = \mathcal{O}_{P'}$, and $\mathcal{O}'_P$ is a semilocal principal ideal domain by 2.12.4.

(b), (c) If $P' \in \mathbb{P}_{K'}$ lies above P, then $\mathcal{O}'_P \subset \mathcal{O}_{P'}$ by (a), hence $v_{P'}(x) \geq 0$ for all $x \in \mathcal{O}'_P$, and by 2.10.7.1 there exists some $\mathfrak{P} \in \mathcal{P}(\mathcal{O}'_P)$ such that $\mathcal{O}_{P'} = (\mathcal{O}'_P)_\mathfrak{P}$, $P' = \mathfrak{P}(\mathcal{O}'_P)_\mathfrak{P}$, and $v_P = v_\mathfrak{P}$. Conversely, if $\mathfrak{P} \in \mathcal{P}(\mathcal{O}'_P)$, then $P' = \mathfrak{P}(\mathcal{O}'_P)_\mathfrak{P} \in \mathbb{P}_{K'}$ and $P' \cap \mathcal{O}'_P = \mathfrak{P}$. Hence θ and θ_1 are mutually inverse bijections.

The equalities $e(P'/P) = e(P' \cap \mathcal{O}'_P/P)$ and $f(P'/P) = f(P' \cap \mathcal{O}'_P/P)$ hold by 2.12.2.5, $\mathsf{k}_{P'} = \mathsf{k}_{\theta(P')}$ holds by 2.7.3.3 (since $P' = \theta(P')(\mathcal{O}'_P)_{\theta(P')}$), and 2.7.5 implies

$$\mathcal{O}'_P = \bigcap_{\substack{P' \in \mathbb{P}_{K'} \\ P' \cap K = P}} \mathcal{O}_{P'},$$

By 2.12.4 we get

$$\sum_{\substack{P' \in \mathbb{P}_{K'} \\ P' \cap K = P}} e(P'/P) f(P'/P) = \sum_{\mathfrak{P} \in \mathcal{P}(\mathcal{O}'_P)} e(\mathfrak{P}/P) f(\mathfrak{P}/P) \leq n.$$

Hence it follows that $1 \leq |\{P' \in \mathbb{P}_{K'} \mid P' \supset P\}| \leq n$. Moreover, if $P' \in \mathbb{P}_{K'}$ lies above P, then $e(P/P') \leq n$ and $f(P'/P) \leq n$. $\square$

Definition 6.2.6. Let K'/F' be an algebraic function field extension of K/F. Let $P' \in \mathbb{P}_{K'}$ and $P = P' \cap K$. Then P'/P or P' over P is called

- **unramified** if $e(P'/P) = 1$ and $\mathsf{k}_{P'}/\mathsf{k}_P$ is separable, otherwise **ramified**;

- **tamely ramified** if $\mathrm{char}(F) \nmid e(P'/P)$ and $\mathsf{k}_{P'}/\mathsf{k}_P$ is separable;

A prime divisor $P \in \mathbb{P}_K$ is called in K'

- **unramified** if P'/P is unramified for all $P' \in \mathbb{P}_{K'}$ lying above P, otherwise **ramified**;

- **tamely ramified** if P'/P is tamely ramified for all $P' \in \mathbb{P}_{K'}$ lying above P;

- **fully ramified** if $e(P_1/P) = [K_1 : K]$ for every over K finite intermediate field K_1 of K'/K.

If $[K' : K] < \infty$, then the above definitions are consistent with those of 2.12.5, applied for the extension $\mathcal{O}_P \subset \mathcal{O}'_P = \mathrm{cl}_{K'}(\mathcal{O}_P)$ resp. $\mathcal{O}_P \subset \mathcal{O}_{P'}$. If $K = K'$ then (as in 2.12.5) all these notions collapse.

If K_1 is a intermediate field of K'/K and $P_1 \in \mathbb{P}_{K_1}$ such that $P \subset P_1 \subset P'$, then $e(P'/P) = e(P'/P_1)e(P_1/P)$ and $f(P'/P) = f(P'/P_1)f(P_1/P)$.

Theorem 6.2.7. *Let K/F be a function field and $\widetilde{F}$ its field of constants.*

1. Let $P \in \mathbb{P}_K$, $x \in K \setminus \widetilde{F}$, and let $P_x \in \mathbb{P}_{F(x)}$ be the prime divisor associated with x (see 6.2.3). *Then $P \in \mathcal{N}(x)$ if and only if $P \cap K(x) = P_x$.*

2. If $x \in K \setminus \widetilde{F}$, then

$$1 \le |\mathcal{N}(x)| \le [K : F(x)], \ 1 \le |\mathcal{P}(x)| \le [K : F(x)],$$

and $v_P(x) = 0$ for almost all $P \in \mathbb{P}_K$.

3. $|\mathbb{P}_K| = \infty$, $[\widetilde{F} : F] < \infty$, $\deg(P) < \infty$ and $\dim_F \mathsf{k}_P < \infty$ for all $P \in \mathbb{P}_K$.

4. Let K'/F' be an algebraic function field extension of K/F, $\widetilde{F}$ resp. $\widetilde{F}'$ the field of constants of K/F resp. K'/F', $P' \in \mathbb{P}_{K'}$ and $P = P' \cap K$. Then $\widetilde{F}' \cap K = \widetilde{F}$, and

$$\deg(P')[\widetilde{F}' : \widetilde{F}] = \deg(P)f(P'/P).$$

In particular, $f(P'/P) < \infty$ if and only if $[\widetilde{F}' : \widetilde{F}] < \infty$.

Proof. 1. If $P_* = P \cap K(x)$, then $x(P) = x(P_*)$ by 6.2.5.1, and by 6.2.4.3 we obtain $x(P) = 0$ if and only if $P_* = P_x$.

2. Let $x \in K \setminus \overline{F}$. Then $\mathcal{N}(x) = \{P \in \mathbb{P}_K \mid P \cap K = P_x\}$ by 1. and therefore $1 \le |\mathcal{N}(x)| \le [K : F(x)]$ by 6.2.5.3(c). As a consequence to $\mathcal{P}(x) = \mathcal{N}(x^{-1})$ and $F(x) = F(x^{-1})$, it follows that also $1 \le |\mathcal{P}(x)| \le [K : F(x)]$. If $v_P(x) \ne 0$, then $P \in \mathcal{N}(x) \cup \mathcal{P}(x)$, and thus $v_P(x) = 0$ for almost all $P \in \mathbb{P}_K$.

3. By 2. we know that $\mathbb{P}_K \ne \emptyset$, and we assume that $|\mathbb{P}_K| < \infty$. By the weak approximation theorem 5.1.6.2(c) there exists some $x \in K$ such that $v_P(x) = -1$ for all $P \in \mathbb{P}_K$, hence $x \notin \widetilde{F}$ and $\mathcal{N}(x) = \emptyset$, a contradiction.

If $P \in \mathbb{P}_K$, then $F \subset \widetilde{F} \subset \mathsf{k}_P$, and therefore it suffices to prove that $[\mathsf{k}_P : F] < \infty$. If $x \in P^\bullet$, then $x \notin \widetilde{F}$ and $x(P) = 0$. Hence $P \cap F(x) = P_x$, $F = \mathsf{k}_{P_x} \subset \mathsf{k}_P$ and $[\mathsf{k}_P : F] = f(P/P_x) < \infty$ by 6.2.5.3(b).

4. By 6.1.8.1, F'/F is algebraic. Hence also $\widetilde{F}'/F$ and $\widetilde{F}' \cap K/F$ are algebraic, and therefore $\widetilde{F}' \cap K \subset \widetilde{F}$. Since $\widetilde{F}/F$ is algebraic, it follows that $\widetilde{F}$ is algebraic over F', hence $\widetilde{F} \subset \widetilde{F}' \cap K$ and thus $\widetilde{F} = \widetilde{F}' \cap K$.

$$
\begin{array}{ccc}
K & \text{———} & K' \\
| & & | \\
\widetilde{F} = \widetilde{F}' \cap K & \text{—} & \widetilde{F}' \\
| & & | \\
F & \text{———} & F'
\end{array}
$$

Now we obtain

$$
f(P'/P)\deg(P) = [\mathsf{k}_{P'} : \mathsf{k}_P][\mathsf{k}_P : \widetilde{F}] = [\mathsf{k}_{P'} : \widetilde{F}'][\widetilde{F}' : \widetilde{F}] = \deg(P')[\widetilde{F}' : \widetilde{F}],
$$

and since both $\deg(P)$ and $\deg(P')$ are finite, it follows that $f(P'/P) < \infty$ if and only if $[\widetilde{F}' : \widetilde{F}] < \infty$. □

Theorem 6.2.8. *Let K/F be a function field and $P \in \mathbb{P}_K$. Then the discrete valued field (K, v_P) has a completion $K_P = (K_P, v_P)$ which is a complete discrete valued field. If F is perfect and t is a prime element of $\mathcal{O}_P$, then $K_P = \mathsf{k}_P((t))$ and $v_P = \mathsf{o}_t$.*

Proof. By 5.3.3 the completion $K_P = (K_P, v_P)$ is a complete discrete valued field, and $\mathsf{k}_{K_P} = \mathsf{k}_P$.

Let F be perfect. Then $\mathsf{k}_P = \mathsf{k}_{K_P}$ is perfect by 1.4.6.2(a) on account of $[\mathsf{k}_P : F] < \infty$ (see 6.2.7.3). Therefore $K_P = \mathsf{k}_P((t))$ (the field of formal Laurent series in t over k_P) and $v_P = \mathsf{o}_t$ by 5.4.10. □

If K/F is a function field and $\widetilde{F}$ is its field of constants, then $K/\widetilde{F}$ is again a function field and $\widetilde{F}$ is its field of constants by 6.1.3.1. Since all concepts studied for K/F in fact refer to the function field $K/\widetilde{F}$, there is no restriction and it is convenient to consider function fields K/F for which F is the field of constants, and we shall do this from now on.

Definition 6.2.9. Let K/F be a function field and F its field of constants. We denote by $\mathbb{D}_K = \mathbb{D}_{K/F}$ the free abelian group with basis $\mathbb{P}_K$. Its elements are called **divisors** of K/F. By definition, every divisor $D \in \mathbb{D}_K$ has a unique representation

$$D = \sum_{P \in \mathbb{P}_K} n_P P, \quad \text{where } n_P \in \mathbb{Z}, \quad n_P = 0 \ \text{ for almost all } \ P \in \mathbb{P}_K,$$

for $P \in \mathbb{P}_K$ we call $v_P(D) = n_P$ the **P-adic value** and we call

$$\operatorname{supp}(D) = \{P \in \mathbb{P}_K \mid v_P(D) \neq 0\}$$

the **support** of D. The zero $0 \in \mathbb{D}_K$ is called the **zero divisor**. If $D \in \mathbb{D}_K$, then $D = 0$ if and only if $\operatorname{supp}(D) = \emptyset$. For divisors $D_1, D_2 \in \mathbb{D}_K$ we define $D_1 \le D_2$ if $v_P(D_1) \le v_P(D_2)$ for all $P \in \mathbb{P}_K$, and we set $D_1 < D_2$ if $D_1 \le D_2$ and $D_1 \neq D_2$. Apparently, $D_1 \le D_2$ holds if and only if $D_2 - D_1 \ge 0$, and if $D_1 \le D_2$, then $D_1 + D \le D_2 + D$ for all $D \in \mathbb{D}_K$. A divisor $D \in \mathbb{D}_K$ is called **effective** if $D \ge 0$. The set $\mathbb{D}'_K$ of all effective divisors is a free monoid with basis $\mathbb{P}_K$.

For a divisor $D \in \mathbb{D}_K$ we define its **positive part** D_+ and its **negative part** D_- by

$$D_+ = \sum_{\substack{P \in \mathbb{P}_K \\ v_P(D) > 0}} v_P(D) P \quad \text{and} \quad D_- = \sum_{\substack{P \in \mathbb{P}_K \\ v_P(D) < 0}} -v_P(D) P.$$

By definition, D_+ and D_- are effective divisors such that $D = D_+ - D_-$ and $-D_- \le D \le D_+$.

Then degree map $\deg \colon \mathbb{P}_K \to \mathbb{N}$ has a unique extension to a homomorphism $\deg \colon \mathbb{D}_K \to \mathbb{Z}$. Explicitly, if $D \in \mathbb{D}_K$, then

$$\deg(D) = \sum_{P \in \mathbb{P}_K} v_P(D) \deg(P) \in \mathbb{Z},$$

and we call $\deg(D)$ the **degree** of D. For $n \in \mathbb{Z}$ we denote by $\mathbb{D}_K^n$ the set of all divisors of degree n. In particular, $\mathbb{D}_K^0 = \operatorname{Ker}(\deg)$, and if $n \in \mathbb{Z}$ and $D \in \mathbb{D}_K^n$, then $\mathbb{D}_K^n = D + \mathbb{D}_K^0$ is a coset of $\mathbb{D}_K^0$. We define

$$\partial_K = \gcd(\{\deg(P) \mid P \in \mathbb{P}_K\}).$$

Then $\partial_K \mathbb{Z} = \operatorname{Im}(\deg \colon \mathbb{D}_K \to \mathbb{Z})$, $\mathbb{D}_K^n = \emptyset$ if $n \notin \partial_K \mathbb{Z}$, and

$$\mathbb{D}_K = \bigcup_{n \in \partial_K \mathbb{Z}} \mathbb{D}_K^n.$$

For $x \in K^\times$ we define the **divisor of zeros** $(x)_0 = (x)_0^K$, the **divisor of poles** $(x)_\infty = (x)_\infty^K$ and the **principal divisor** $(x) = (x)^K$ of x by

$$(x)_0 = \sum_{P \in \mathcal{N}(x)} v_P(x) P, \quad (x)_\infty = \sum_{P \in \mathcal{P}(x)} -v_P(x) P \quad \text{and} \quad (x) = \sum_{P \in \mathbb{P}_K} v_P(x) P.$$

Then $(x) = (x)_0 - (x)_\infty$, $(x)_0 = (x)_+$, $(x)_\infty = (x)_- = (x^{-1})_0$, and $v_P((x)) = v_P(x)$ for all $P \in \mathbb{P}_K$. Moreover, $(x) = 0$ if and only if $x \in F^\times$.

The map $K^\times \to \mathbb{D}_K$, defined by $x \mapsto (x)$, is a group homomorphism with kernel $F^\times$ and image $(K^\times) = \{(x) \mid x \in K^\times\} \subset \mathbb{D}_K$. The group $(K^\times)$ is called the **group of principal divisors**, and the factor group $\mathcal{C}_K = \mathbb{D}_K/(K^\times)$ is called the **divisor class group** of K. For $D \in \mathbb{D}_K$ we denote by

$$[D] = D + (K^\times) \in \mathcal{C}_K$$

the **divisor class** of D. Divisors $D_1, D_2 \in \mathbb{D}_K$ are called **linearly equivalent** and we write $D_1 \sim D_2$ if $[D_1] = [D_2]$ [equivalently, $D_2 = D_1 + (x)$ for some $x \in K^\times$]. Consequently, a divisor $D \in \mathbb{D}_K$ is a principal divisor if and only if $D \sim 0$, and there exists a (natural) exact sequence

$$\mathbf{1} \to F^\times \to K^\times \overset{(\cdot)}{\to} \mathbb{D}_K \to \mathcal{C}_K \to \mathbf{0}.$$

Definition and Theorem 6.2.10. *Let K/F be a function field and F its field of constants.* For a divisor $D \in \mathbb{D}_K$ we define the **space of multiples** $\mathcal{L}(D)$ of $-D$ by

$$\begin{aligned}\mathcal{L}(D) &= \{x \in K^\times \mid (x) \ge -D\} \cup \{0\}\\ &= \{x \in K \mid v_P(x) \ge -v_P(D) \ \text{ for all } \ P \in \mathbb{P}_K\}.\end{aligned}$$

$\mathcal{L}(D)$ is an F-subspace of K, and $\dim(D) = \dim_F \mathcal{L}(D)$ is called the **dimension** of D. Only in 6.3.3.2 we shall prove that $\dim(D) < \infty$ for all $D \in \mathbb{D}_K$.

1. *$\mathcal{L}(0) = F$ and $\dim(0) = 1$.*

2. *Let $D, D' \in \mathbb{D}_K$.*

(a) *If $D \le D'$, then $\mathcal{L}(D) \subset \mathcal{L}(D')$ and $\dim(D) \le \dim(D')$.*

(b) *If $D = D' + (z)$ for some $z \in K^\times$, then the map $x \mapsto xz$ defines an F-isomorphism $\mathcal{L}(D) \overset{\sim}{\to} \mathcal{L}(D')$. In particular, if $D \sim D'$, then $\dim(D) = \dim(D')$, and if $D \sim 0$, then $\dim(D) = 1$.*

If $\mathbf{c} = [D] \in \mathcal{C}_K$, then we call $\dim(\mathbf{c}) = \dim(D)$ the **dimension** of the divisor class $\mathbf{c}$.

3. *Let $D \in \mathbb{D}_K$.*

(a) *If $D < 0$, then $\mathcal{L}(D) = \mathbf{0}$ and $\dim(D) = 0$.*

(b) *$\dim(D) > 0$ if and only if $D \sim D'$ for some $D' \in \mathbb{D}'_K$.*

Proof. If $D \in \mathbb{D}_K$, $x, y \in \mathcal{L}(D)$ and $c \in F$, then

$$v_P(x+y) \ge \min\{v_P(x), v_P(y)\} \ge -v_P(D) \ \text{ for all } \ P \in P_K,$$

hence $x+y \in \mathcal{L}(D)$, and

$$v_P(cx) = v_P(c) + v_P(x) = v_P(x) \geq -v_P(D) \quad \text{for all} \quad P \in P_K,$$

hence $cx \in \mathcal{L}(D)$. Therefore $\mathcal{L}(D)$ is an F-subspace of K.

1. $\mathcal{L}(0) = \{x \in K^\times \mid (x) \geq 0\} \cup \{0\} = \{x \in K^\times \mid \mathcal{P}(x) = \emptyset\} \cup \{0\} = F$.

2. (a) Obvious.

(b) Let $z \in K^\times$ and $D = D' + (z)$. Obviously, $x \mapsto xz$ defines and F-isomorphism $K \to K$. If $x \in \mathcal{L}(D)^\bullet$, then $(xz) = (x)+(z) \geq -D+(z) = -D'$, hence $x \in \mathcal{L}(D')$. Similarly, if $y \in \mathcal{L}(D')$, then $z^{-1}y \in \mathcal{L}(D)$, and $z(z^{-1}y) = y$. Therefore $x \mapsto xz$ defines an F-isomorphism $\mathcal{L}(D) \overset{\sim}{\to} \mathcal{L}(D')$.

3. (a) Let $D < 0$. Then $\mathcal{L}(D) \subset \mathcal{L}(0) = F$, but if $c \in F^\times$, then $(c) = 0$ and thus $c \notin \mathcal{L}(D)$. Hence $\mathcal{L}(D) = \mathbf{0}$.

(b) If $\dim(D) > 0$ and $x \in \mathcal{L}(D)^\bullet$, then $(x) \geq -D$, $D' = D + (x) \geq 0$ and $D' \sim D$.

Conversely, if $D' \in \mathbb{D}'_K$ and $D' \sim D$, then $D' = D + (x)$ for some $x \in K^\times$, and as $(x) = -D + D' \geq -D$, it follows that $x \in \mathcal{L}(D)^\bullet$. □

If K/F is a function field, then we call $\mathrm{Gal}(K/F)$ the **automorphism group** of K/F. In the following Theorem 6.2.11 we show that the automorphism group acts on the divisor group in a natural way.

Theorem 6.2.11. *Let K/F be a function field, $\tau \in \mathrm{Gal}(K/F)$, and $\widetilde{F}$ the field of constants of K/F.*

1. $\tau\widetilde{F} = \widetilde{F}$.

2. If $P \in \mathbb{P}_K$ and t is a prime element of $\mathcal{O}_P$, then $\tau P \in \mathbb{P}_K$, $\mathcal{O}_{\tau P} = \tau\mathcal{O}_P$, $\tau(t)$ is a prime element of $\mathcal{O}_{\tau P}$, $v_{\tau P} = v_P \circ \tau^{-1}$, and τ induces an F-isomorphism $\tau_P\colon \mathsf{k}_P \to \mathsf{k}_{\tau P}$ such that

$$\tau_P(x(P)) = \tau(x)(\tau P) \text{ for all } x \in \mathcal{O}_P.$$

3. For $D \in \mathbb{D}_K$ we set

$$\tau D = \sum_{P \in \mathbb{P}_K} v_P(D)\tau P \in \mathbb{D}_K.$$

Then we obtain $\deg(\tau D) = \deg(D)$, $v_P(\tau D) = v_{\tau^{-1}P}(D)$ for all $P \in \mathbb{P}_K$, $\tau\mathcal{L}(D) = \mathcal{L}(\tau D)$ and $\dim(\tau D) = \dim(D)$.

4. If $x \in K^\times$, then $(\tau(x))_0 = \tau((x)_0)$, $(\tau(x))_\infty = \tau((x)_\infty)$ and $\tau((x)) = (\tau(x))$.

Proof. Straightforward. □

6.3 Repartitions and definition of the genus

Throughout this section let K/F be a function field and F its field of constants.

Definition 6.3.1. A **repartition** or an **incomplete adele** of K/F is a family $(\alpha_P)_{P\in\mathbb{P}_K}$ in K such that $\alpha_P \in \mathcal{O}_P$ for almost all $P \in \mathbb{P}_K$. We denote by $\mathbb{A}_K = \mathbb{A}_{K/F}$ be the set of all repartitions of K/F. Equipped with component-wise addition and multiplication, $\mathbb{A}_K$ is a commutative ring.

For $\alpha = (\alpha_P)_{P\in\mathbb{P}_K} \in \mathbb{A}_K$ and $Q \in \mathbb{P}_K$, we define $v_Q(\alpha) = v_Q(\alpha_Q)$. Apparenetly, $v_Q(\alpha\beta) = v_Q(\alpha) + v_Q(\beta)$ and $v_Q(\alpha+\beta) \ge \min\{v_Q(\alpha), v_Q(\beta)\}$ for all α, $\beta \in \mathbb{A}_K$. If $a \in K^\times$, then $v_P(a) = 0$ for almost all $P \in \mathbb{P}_K$, thus the constant vector $(a)_{P\in\mathbb{P}_K}$ is a repartition satisfying $v_Q(a) = v_Q((a))$ for all $Q \in \mathbb{P}_K$, and the assignment $K \to \mathbb{A}_K$, defined by $a \mapsto (a)_{P\in\mathbb{P}_K}$, is a ring monomorphism. We identify: $a = (a)_{P\in\mathbb{P}_K}$. Then K is a subring of $\mathbb{A}_K$.

For $D \in \mathbb{D}_K$ we set

$$\mathbb{A}_K(D) = \{\alpha \in \mathbb{A}_K \mid v_Q(\alpha) \ge -v_Q(D) \text{ for all } Q \in \mathbb{P}_K\}.$$

$\mathbb{A}_K(D)$ is an F-subspace of $\mathbb{A}_K$, $\mathbb{A}_K(D) \cap K = \mathcal{L}(D)$, and if D, $D' \in \mathbb{D}_K$ and $D \le D'$, then $\mathbb{A}_K(D) \subset \mathbb{A}_K(D')$. By definition, it follows that

$$\mathbb{A}_K = \bigcup_{D\in\mathbb{D}_K} \mathbb{A}_K(D).$$

Theorem 6.3.2. *Let D, $D' \in \mathbb{D}_K$ be divisors such that $D \le D'$. Then there is an exact sequence of F-homomorphisms*

$$\mathbf{0} \to \mathcal{L}(D) \xrightarrow{j} \mathcal{L}(D') \xrightarrow{\tau} \mathbb{A}_K(D')/\mathbb{A}_K(D) \xrightarrow{\sigma} (\mathbb{A}_K(D')+K)/(\mathbb{A}_K(D)+K) \to \mathbf{0},$$

which are given by $j = (\mathcal{L}(D) \hookrightarrow \mathcal{L}(D'))$, $\tau(x) = x + \mathbb{A}_K(D)$ for all $x \in \mathcal{L}(D')$ and $\sigma(\alpha + \mathbb{A}_K(D)) = \alpha + (\mathbb{A}_K(D) + K)$ for all $\alpha \in \mathbb{A}_K(D')$.

Proof. Obviously, j, σ and τ are F-linear, j is injective, σ is surjective, and

$$\begin{aligned}\operatorname{Ker}(\tau) &= \mathcal{L}(D') \cap \mathbb{A}_K(D) = \mathcal{L}(D') \cap K \cap \mathbb{A}_K(D)\\ &= \mathcal{L}(D') \cap \mathcal{L}(D) = \mathcal{L}(D) = \operatorname{Im}(j).\end{aligned}$$

If $x \in \mathcal{L}(D')$, then $\sigma\circ\tau(x) = \sigma(x + \mathbb{A}_K(D)) = x + (\mathbb{A}_K(D) + K) = 0$, hence $\operatorname{Im}(\tau) \subset \operatorname{Ker}(\sigma)$, and it remains to prove that $\operatorname{Ker}(\sigma) \subset \operatorname{Im}(\tau)$.

Let $\alpha \in \mathbb{A}_K(D')$ be such that $\alpha + \mathbb{A}_K(D) \in \operatorname{Ker}(\sigma)$. Then $\alpha \in \mathbb{A}_K(D) + K$, say $\alpha = \beta + x$, where $\beta \in \mathbb{A}_K(D)$ and $x \in K$. Hence $x = \alpha - \beta \in K \cap \mathbb{A}_K(D') = \mathcal{L}(D')$ and $\alpha + \mathbb{A}_K(D) = x + \mathbb{A}_K(D) = \tau(x) \in \operatorname{Im}(\tau)$. □

Theorem 6.3.3. *Let $D,\ D' \in \mathbb{D}_K$ such that $D \le D'$.*

1. $\dim_F \mathcal{L}(D')/\mathcal{L}(D) \le \dim_F \mathbb{A}_K(D')/\mathbb{A}_K(D) = \deg(D'-D)$.
2. $\dim(D) \le \deg(D_+) + 1 < \infty$.
3. $$\dim_F(\mathbb{A}_K(D') + K)/(\mathbb{A}_K(D)+K) = [\deg(D') - \dim(D')] - [\deg(D) - \dim(D)].$$

In particular, $\deg(D') - \dim(D') \ge \deg(D) - \dim(D)$.

Proof. 1. By 6.3.2 we obtain $\dim_F \mathcal{L}(D')/\mathcal{L}(D) \le \dim_F \mathbb{A}_K(D')/\mathbb{A}_K(D)$. We set $D' = D + Q_1 + \ldots + Q_r$, where $r \in \mathbb{N}_0$ and $Q_1, \ldots, Q_r \in \mathbb{P}_K$ and prove by induction on r that

$$\dim_F \mathbb{A}_K(D')/\mathbb{A}_K(D) = \deg(D'-D) = \deg(Q_1) + \ldots + \deg(Q_r).$$

For $r = 0$ there is nothing to do.

$r \ge 1,\ r-1 \to r$: We set $D_0 = D + P_1 + \ldots + P_{r-1},\ \ P = P_r$, and we must prove that $\dim_F(\mathbb{A}_K(D_0+P)/\mathbb{A}_K(D_0)) = \deg(P)$.

Let $t \in K$ be such that $v_Q(t) = v_Q(D_0+Q) = v_Q(D_0)+1$. If now $\alpha \in \mathbb{A}_K(D_0+Q)$, then $v_Q(\alpha) \ge -v_Q(D_0+Q) = -v_Q(D_0) - 1$ and therefore $v_Q(t\alpha_Q) = v_Q(t) + v_Q(\alpha) \ge 0$. Hence $t\alpha_Q \in \mathcal{O}_Q$, and we define

$$\psi \colon \mathbb{A}_K(D_0+Q) \to \mathsf{k}_Q = \mathcal{O}_Q/Q \quad \text{by} \quad \psi(\alpha) = (t\alpha_Q)(Q).$$

Then ψ is F-linear, and it suffices to prove that

$$\psi \text{ is surjective and } \operatorname{Ker}(\psi) = \mathbb{A}_K(D_0).$$

Indeed, then ψ induces an F-isomorphism $\mathbb{A}_K(D_0+Q)/\mathbb{A}_K(D_0) \xrightarrow{\sim} \mathsf{k}_Q$, which shows that $\dim_F(\mathbb{A}_K(D_0+Q)/\mathbb{A}_K(D_0)) = \dim_F \mathsf{k}_Q = \deg(Q)$.

ψ is surjective: Let $\xi = z(Q) \in \mathsf{k}_Q$, where $z \in \mathcal{O}_Q$, and set $\alpha_Q = t^{-1}z \in K$. Then $v_Q(\alpha_Q) = -v_Q(t) + v_Q(z) \ge -v_Q(D_0+Q)$. For $P \in \mathbb{P}_K \setminus \{Q\}$ we choose $\alpha_P \in K$ such that $v_P(\alpha_P) = -v_P(D_0) = -v_P(D_0+Q)$. Then we get $\alpha = (\alpha_P)_{P \in \mathbb{P}_K} \in \mathbb{A}_K(D_0+Q)$ and $\psi(\alpha) = \xi$.

$\operatorname{Ker}(\psi) = \mathbb{A}_K(D_0)$: Let $\alpha \in \mathbb{A}_K(D_0+Q)$. Then $v_P(\alpha) \ge -v_P(D_0+Q) = -v_P(D_0)$ for all $P \in \mathbb{P}_K \setminus \{Q\}$ and $v_Q(\alpha) \ge -v_Q(D_0+Q) = -v_Q(D_0)-1$. By definition we have $\alpha \in \operatorname{Ker}(\psi)$ if and only if $v_Q(t\alpha_Q) \ge 1$, and this holds if and only if $v_Q(\alpha_Q) \ge -v_Q(D_0)$, that is, $\alpha \in \mathbb{A}_K(D_0)$.

2. Since $D_+ \ge 0$, 6.3.2 implies $\dim_F \mathcal{L}(D_+)/\mathcal{L}(0) \le \deg(D_+)$, and consequently $\dim(D_+) = \dim_F \mathcal{L}(D_+) \le \deg(D_+) + \dim_F \mathcal{L}(0) = \deg(D_+) + 1$. Finally, $D \le D_+$ implies $\dim(D) \le \dim(D_+) \le \deg(D_+) + 1$.

3. By 6.3.2 and 1. we obtain

$$\begin{aligned} &\dim_F(\mathbb{A}_K(D') + K)/(\mathbb{A}_K(D) + K) \\ &\qquad = \dim_F \mathbb{A}_K(D')/\mathbb{A}_K(D) - \dim_F \mathcal{L}(D') + \dim_F \mathcal{L}(D) \\ &\qquad = \deg(D') - \deg(D) - \dim(D') + \dim(D) \\ &\qquad = [\deg(D') - \dim(D')] - [\deg(D) - \dim(D)], \end{aligned}$$

and consequently $\deg(D') - \dim(D') \ge \deg(D) - \dim(D)$. □

Theorem and Definition 6.3.4.

1. If $x \in K \setminus F$, then $\deg(x)_0 = \deg(x)_\infty = [K : F(x)]$, and there exists some $C \in \mathbb{D}'_K$ such that $\dim(l(x)_\infty + C) \ge (l+1)\deg(x)_\infty$ for all $l \in \mathbb{N}_0$.

2. If $x \in K^\times$, then $\deg((x)) = 0$, and consequently $(K^\times) \subset \mathbb{D}^0_K$.

3. If $D, D' \in \mathbb{D}_K$ and $D \sim D'$, then $\deg(D) = \deg(D')$. Consequently, deg induces an (equally denoted) epimorphism

$$\deg\colon \mathcal{C}_K \to \partial_K\mathbb{Z} \quad \textit{satisfying} \quad \deg([D]) = \deg(D) \ \textit{for all} \ D \in \mathbb{D}_K.$$

If $\mathbf{c} \in \mathcal{C}_K$, then we call $\deg(\mathbf{c})$ the **degree** of $\mathbf{c}$, and for $n \in \mathbb{Z}$ we denote by $\mathcal{C}^n_K = \{[D] \mid D \in \mathbb{D}^n_K\}$ the set of all divisor classes of degree n.

Then $\mathcal{C}^0_K = \mathrm{Ker}(\deg\colon \mathcal{C}_K \to \mathbb{Z}) = \mathbb{D}^0_K/(K^\times))$, and deg induces the split exact sequence

$$\mathbf{0} \to \mathcal{C}^0_K \to \mathcal{C}_K \to \partial_K\mathbb{Z} \to \mathbf{0}.$$

In particular, $\mathcal{C}_K \cong \mathcal{C}^0_K \oplus \mathbb{Z}$. If $n \in \mathbb{Z}$ and $\mathbf{c} \in \mathcal{C}^n_K$, then $\mathcal{C}^n_K = \mathbf{c} + \mathcal{C}^0_K$, and

$$|\mathcal{C}^n_K| = \begin{cases} |\mathcal{C}^0_K| & \textit{if } \partial_K \mid n, \\ 0 & \textit{otherwise} \end{cases}$$

We call $\mathcal{C}^0_K$ the **0-divisor class group** and $h_K = |\mathcal{C}^0_K| \in \mathbb{N} \cup \{\infty\}$ the **class number** of K.

Proof. 1. Let $x \in K \setminus F$ and $n = [K : F(x)]$. If $P \in \mathbb{P}_K$, then $P \in \mathcal{N}(x)$ if and only if $P \cap F(x) = P_x$ by 6.2.7.1. If $P \in \mathcal{N}(x)$, then

$$v_P(x) = e(P/P_x)v_{P_x}(x) = e(P/P_x) \quad \text{and}$$
$$\deg(P) = [\mathsf{k}_P : F] = [\mathsf{k}_P : \mathsf{k}_{P_x}] = f(P/P_x)$$

Therefore we obtain, using 6.2.5.3,

$$\deg(x)_0 = \sum_{P \in \mathcal{N}(x)} v_P(x)\deg(P) = \sum_{\substack{P \in \mathbb{P}_K \\ P \cap F(x) = P_x}} e(P/P_x)f(P/P_x) \le n.$$

To prove the reverse inequality, let $(u_1, \ldots, u_n)$ be an $F(x)$-basis of K, $l \in \mathbb{N}_0$ and $C \in \mathbb{D}'_K$ such that $C \ge -(u_i)$ for all $i \in [1,n]$. If $i \in [1,n]$ and $j \in [0,l]$, then $x^j u_i \in K$ and $(x^j u_i) = j(x) + (u_i) \ge -j(x)_\infty - C \ge -l(x)_\infty - C$, and therefore $x^j u_i \in \mathcal{L}(l(x)_\infty + C)$. Since $\{x^j u_i \mid j \in [0,l],\ i \in [1,n]\}$ is linearly independent over F, it follows that

$$\dim(l(x)_\infty + C) \ge (l+1)n. \qquad (*)$$

This is the asserted inequality as soon as we have proved that $n = \deg(x)_\infty$. For all $l \geq 0$ we obtain, using 6.3.3.2,

$$(l+1)n \leq \dim(l(x)_\infty + C) \leq \deg(l(x)_\infty + C)_+ + 1 = l\deg(x)_\infty + \deg(C) + 1,$$

and therefore $l(n - \deg(x)_\infty) \leq \deg(C) + 1 - n$. For sufficiently large l this implies $\deg(x)_\infty \geq n$.

Up to now we have proved that $\deg(x)_0 \leq n$ and $\deg(x)_\infty \geq n$. If we replace x with x^{-1}, then we obtain $\deg(x)_\infty = \deg(x^{-1})_0 \leq n$ and therefore $\deg(x)_0 = \deg(x^{-1})_\infty \geq n$, since $n = [K : F(x)] = [K : F(x^{-1}]$. Thus it eventually follows that $\deg(x)_0 = \deg(x)_\infty = n$.

2. If $x \in K \setminus F$, then $\deg(x) = \deg(x)_0 - \deg(x)_\infty = 0$ by 1. If $x \in F^\times$, then $\deg((x)) = \deg(0) = 0$.

3. If D, $D' \in \mathbb{D}_K$ and $D \sim D'$, then $D' = D + (x)$ for some $x \in K^\times$ and thus $\deg(D') = \deg(D) + \deg((x)) = \deg(D)$. As $(K^\times) \subset \mathbb{D}_K^0 = \mathrm{Ker}(\deg)$ and $\mathrm{Im}(\deg) = \partial_K\mathbb{Z}$, the homomorphism $\deg\colon \mathbb{D}_K \to \mathbb{Z}$ induces an epimorphism $\deg\colon \mathcal{C}_K = \mathbb{D}_K/(K^\times) \to \partial_K\mathbb{Z}$ with kernel $\mathcal{C}_K^0$ as asserted. Since $\partial_K\mathbb{Z} \cong \mathbb{Z}$, the epimorphism $\deg\colon \mathcal{C}_K \to \partial_K\mathbb{Z}$ splits, and therefore $\mathcal{C}_K \cong \mathcal{C}_K^0 \oplus \mathbb{Z}$ by 1.1.1. $\square$

The following simple Corollary 6.3.5 will be used again and again without further reference.

Corollary 6.3.5. *Let $D \in \mathbb{D}_K$.*

1. *If $\deg(D) \leq 0$, then $D \sim 0$ if and only if $\dim(D) \geq 1$.*
2. *If $\deg(D) < 0$, then $\dim(D) = 0$.*
3. *If $\deg(D) > 0$, then $\dim(D) \leq \deg(D) + 1$.*

Proof. We apply 6.2.10.

1. Let $\deg(D) \leq 0$. If $D \sim 0$, then $\dim(D) = 1$. Thus assume that $\dim(D) \geq 1$. Then there exists some $D' \in \mathbb{D}_K'$ such that $D' \sim D$. Since $\deg(D') = \deg(D) \leq 0$, it follows that $D' = 0$ and $D \sim 0$.

2. If $\dim(D) \geq 1$, then there exists some $D' \in \mathbb{D}_K'$ such that $D' \sim D$, and consequently $\deg(D) = \deg(D') \geq 0$.

3. If $\dim(D) = 0$, there is nothing to do. Thus suppose that $\dim(D) \geq 1$, and let $D' \in \mathbb{D}_K'$ be such that $D' \sim D$. Then

$$\dim(D) = \dim(D') \leq \deg(D') + 1 = \deg(D) + 1 \text{ by } 6.3.3.2. \qquad \square$$

Remark 6.3.6 (Existence problem for functions with prescribed poles and zeros). We consider a typical instance.

Let r, $s \in \mathbb{N}_0$, $P_1, \ldots, P_r$, $Q_1, \ldots, Q_s \in \mathbb{P}_K$ and $d_1, \ldots, d_r$, $e_1, \ldots, e_s \in \mathbb{N}$. We ask whether there exists a function $x \in K$ satisfying $v_{P_i}(x) \geq d_i$ for all

$i \in [1, r]$, $\mathcal{P}(x) \subset \{Q_1, \ldots, Q_r\}$, and $v_{Q_i}(x) \geq -e_i$ for all $i \in [1, s]$. We consider the divisor

$$D = \sum_{i=1}^{r} d_i P_i - \sum_{i=1}^{s} e_i Q_i.$$

Then we ask for a function $x \in K^\times$ such that $(x) \geq D$, that is, whether $\mathcal{L}(-D) \neq \mathbf{0}$. Now it is easy to calculate $\deg(D)$, but we want to know $\dim(D)$. The theorem of Riemann-Roch accomplishes the passage from deg to dim. We start with preliminary versions of this famous theorem in 6.3.7 and 6.3.8 and present the full version in 6.4.5 after we have introduced differentials.

Definition and Theorem 6.3.7 (Theorem of Riemann-Roch, weak version). *There exists a unique integer $g \geq 0$ such that*

$$\dim(D) = \deg(D) - g + 1 + \dim_F \mathbb{A}_K/(\mathbb{A}_K(D) + K) \quad \textit{for all} \quad D \in \mathbb{D}_K.$$

If $\deg(D) \gg 1$, *then* $\mathbb{A}_K = \mathbb{A}_K(D) + K$ *and* $\dim(D) = \deg(D) - g + 1$.
The integer $g = g_K = g_{K/F}$ is called the **genus** of K.

Proof. In 4 steps. The main tool is 6.3.3.3 which we use without further reference.

I. The set $\{\deg(D) - \dim(D) \mid D \in \mathbb{D}_K\}$ is bounded from above.

Proof of **I.** Let $x \in K \setminus F$ and $B = (x)_\infty$. By 6.3.4.1 there exists some $C \in \mathbb{D}'_K$ such that $\dim(lB + C) \geq (l+1)\deg(B)$ for all $l \in \mathbb{N}_0$, and we shall prove that $\deg(D) - \dim(D) \leq \deg(C)$ for all $D \in \mathbb{D}_K$.

Let $D \in \mathbb{D}_K$, $l \geq 0$ and $C_1 \in \mathbb{D}'_K$ such that $C_1 \geq D$. Then

$$\begin{aligned}\deg(lB) - \dim(lB) &\leq \deg(lB + C) - \dim(lB + C) \\ &\leq l\deg(B) + \deg(C) - (l+1)\deg(B) \\ &= \deg(C) - \deg(B) < \deg(C),\end{aligned}$$

hence

$$\begin{aligned}\deg(lB) - \deg(C) &< \dim(lB) \\ &\leq \deg(lB - C_1 + C_1) - \deg(lB - C_1) + \dim(lB - C_1) \\ &\leq \deg(C_1) + \dim(lB - C_1)\end{aligned}$$

and therefore

$$\dim(lB - C_1) \geq -\deg(C) - \deg(C_1) + \deg(lB) = -\deg(C + C_1) + l\deg(B) \geq 1,$$

provided that $l \gg 1$.

Let $l \geq 1$ be such that $\dim(lB - C_1) \geq 1$, and let $z \in \mathcal{L}(lB - C_1)^\bullet$. As $D \leq C_1$ and $C_1 - (z) \leq lB$, we obtain

$$\begin{aligned}\deg(D) - \dim(D) &\leq \deg(C_1) - \dim(C_1) = \deg(C_1 - (z)) - \dim(C_1 - (z)) \\ &\leq \deg(lB) - \dim(lB) < \deg(C). \qquad \square[\mathbf{I.}]\end{aligned}$$

Now we define

$$g = \max\{\deg(D) - \dim(D) \mid D \in \mathbb{D}_K\} + 1.$$

II. Let $D_0, D \in \mathbb{D}_K$ be such that $g = \deg(D_0) - \dim(D_0) + 1$ and $\deg(D) \geq \deg(D_0) + g$. Then $\dim(D) = \deg(D) - g + 1$.

Proof of **II.** By definition, $\deg(D - D_0) - \dim(D - D_0) \leq g - 1$ and therefore $\dim(D - D_0) \geq \deg(D) - \deg(D_0) - g + 1 \geq 1$. If $z \in \mathcal{L}(D - D_0)^\bullet$, then $(z) \geq -D + D_0$, hence $D + (z) \geq D_0$ and

$$\begin{aligned} g - 1 &= \deg(D_0) - \dim(D_0) \leq \deg(D + (z)) - \dim(D + (z)) \\ &= \deg(D) - \dim(D) \leq g - 1. \end{aligned}$$

Hence equality holds, and $\dim(D) = \deg(D) - g + 1$. □[**II.**]

III. Let $D \in \mathbb{D}_K$ be such that $\dim(D) = \deg(D) - g + 1$. Then $\mathbb{A}_K = \mathbb{A}_K(D) + K$.

Proof of **III.** Let $\alpha \in \mathbb{A}_K$, $D_1 \in \mathbb{D}_K$ such that $D_1 \geq D$ and $\alpha \in \mathbb{A}_K(D_1)$. Then

$$\begin{aligned} \dim(D_1) &= \dim(D) + \dim[D + (D_1 - D)] - \dim(D) \\ &\leq \dim(D) + \deg[D + (D_1 - D)] - \deg(D) = \dim(D) + \deg(D_1 - D) \\ &= \deg(D) - g + 1 + \deg(D_1) - \deg(D) = \deg(D_1) - g + 1 \leq \dim(D_1), \end{aligned}$$

hence equality holds, and therefore $\dim(D_1) = \deg(D_1) - g + 1$. Since

$$\begin{aligned} &\dim_F(\mathbb{A}_K(D_1) + K)/(\mathbb{A}_K(D) + K) \\ &\quad = [\deg(D_1) - \dim(D_1)] - [\deg(D) - \dim(D)] = (g - 1) - (g - 1) = 0, \end{aligned}$$

it follows that $\alpha \in \mathbb{A}_K(D_1) \subset \mathbb{A}_K(D_1) + K = \mathbb{A}_K(D) + K$, and therefore we obtain $\mathbb{A}_K = \mathbb{A}_K(D) + K$. □[**III.**]

IV. End of proof. Let $D \in \mathbb{D}_K$. It remains to prove that

$$\dim(D) - \deg(D) + g - 1 = \dim_F \mathbb{A}_K/(\mathbb{A}_K(D) + K).$$

By **II** there exists some $D_1 \in \mathbb{D}_K$ such that

$$D_1 \geq D \text{ and } \dim(D_1) = \deg(D_1) - g + 1,$$

III implies $\mathbb{A}_K = \mathbb{A}_K(D_1) + K$, and eventually

$$\begin{aligned} \dim_F \mathbb{A}_K/(\mathbb{A}_K(D) + K) &= \dim_F(\mathbb{A}_K(D_1) + K)/(\mathbb{A}_K(D) + K) \\ &= [\deg(D_1) - \dim(D_1)] - [\deg(D) - \dim(D)] \\ &= g - 1 - \deg(D) + \dim(D). \end{aligned}$$

The uniqueness of g is obvious. □

Corollary 6.3.8 (Riemann's inequality). *If $D \in \mathbb{D}_K$, then*

$$\dim(D) \geq \deg(D) - g_K + 1, \quad \textit{and equality holds if } \deg(D) \gg 1.$$

Proof. Obvious by the Riemann-Roch theorem 6.3.7. □

6.4 Weil differentials and the theorem of Riemann-Roch

Throughout this section, let K/F be a function field, F its field of constants and $g = g_{K/F}$ its genus.

In this section we investigate the correction term $\mathbb{A}_K/(\mathbb{A}_K(D)+K)$ in the weak theorem of Riemann-Roch 6.3.7 in order to obtain the full version of this fundamental theorem. For this purpose we introduce the notion of Weil differentials, an algebraic substitute for the notion of differentials in geometry which allows us to work over an arbitrary (not necessarily perfect) field of constants. The connection with the classical theory of differentials in the style of analysis and geometry will be provided in the Sections 6.7 and 6.8. For a proof of the Riemann-Roch theorem using the classical notion of differentials we refer to [37, Ch. 5].

Definition 6.4.1. A **Weil differential** of K/F is an F-linear map

$$\sigma\colon \mathbb{A}_K \to F \text{ such that } \sigma \restriction \mathbb{A}_K(D) + K = 0 \text{ for some } D \in \mathbb{D}_K.$$

We denote by $\Sigma_K = \Sigma_{K/F}$ the set of all Weil differentials of K/F, and for $D \in \mathbb{D}_K$ we define

$$\Sigma_K(D) = \{\sigma \in \Sigma_K \mid \sigma \restriction \mathbb{A}_K(D) + K = 0\}.$$

Apparently, $\Sigma_K(D)$ is an F-subspace of $\mathrm{Hom}_F(\mathbb{A}_K, F)$. If D, $D' \in \mathbb{D}_K$ and $D \leq D'$, then $\mathbb{A}_K(D) + K \subset \mathbb{A}_K(D') + K$ and thus $\Sigma_K(D') \subset \Sigma_K(D)$. For any D_1, $D_2 \in \mathbb{D}_K$ there exists some $D \in \mathbb{D}_K$ such that $D \leq D_1$ and $D \leq D_2$, which implies $\Sigma_K(D_1) \cup \Sigma_K(D_2) \subset \Sigma_K(D)$. Consequently,

$$\Sigma_K = \bigcup_{D \in \mathbb{D}_K} \Sigma_K(D)$$

is the union of a directed set of F-subspaces of $\mathrm{Hom}_F(\mathbb{A}_K, F)$, and thus Σ_K itself is an F-subspace of $\mathrm{Hom}_F(\mathbb{A}_K, F)$.

For $\sigma \in \Sigma_K$ and $x \in K$ we define $x\sigma\colon \mathbb{A}_K \to F$ by $(x\sigma)(\alpha) = \sigma(x\alpha)$. The following Theorem 6.4.2 shows among others that the operation $(x, \sigma) \mapsto x\sigma$ defines on Σ_K the structure of a one-dimensional vector space over K.

Theorem 6.4.2.

1. If $D \in \mathbb{D}_K$, then

$$\dim_F \Sigma_K(D) = \dim_F \mathbb{A}_K/(\mathbb{A}_K(D)+K)) = \dim(D)-\deg(D)+g-1.$$

2. If $x \in K$ and $\sigma \in \Sigma_K$, then $x\sigma \in \Sigma_K$.
More precisely: If $D \in \mathbb{D}_K$ and $\sigma \in \Sigma_K(D)$, then

(a) $x\sigma \in \Sigma_K(D-B)$ for all $B \in \mathbb{D}_K$ and $x \in \mathcal{L}(B)$.

(b) $x\sigma \in \Sigma_K(D+(x))$ for all $x \in K^\times$.

3. The operation $K \times \Sigma_K \to \Sigma_K$, $(x,\sigma) \mapsto x\sigma$ makes Σ_K into a one-dimensional vector space over K.

Proof. 1. By definition, there is a (natural) F-isomorphism

$$\Sigma_K(D) \overset{\sim}{\to} \mathrm{Hom}_F(\mathbb{A}_K/(\mathbb{A}_K(D)+K), F),$$

and $\dim_F \Sigma_K(D) = \dim_F(\mathbb{A}_K/(\mathbb{A}_K(D)+K)) = \dim(D) - \deg(D) + g - 1$ by 6.3.7.

2. If $x \in K$ and $\sigma \in \Sigma_K$, then it is easily checked that $x\sigma \colon \mathbb{A}_K \to F$ is F-linear.

Next we prove (a). Let $B,\, D \in \mathbb{D}_K$, $\sigma \in \Sigma_K(D)$ and $x \in \mathcal{L}(B)$. We must show that $x\sigma \upharpoonright \mathbb{A}_K(D-B)+K = 0$.

If $\alpha \in \mathbb{A}_K(D-B)$ and $z \in K$, then $(x\sigma)(\alpha+z) = \sigma(x\alpha+xz) = \sigma(x\alpha)$ and $v_P(x\alpha) = v_P(x)+v_P(\alpha) \ge -B-(D-B) = -D$ for all $P \in \mathbb{P}_K$, hence $x\alpha \in \Sigma_K(D)$ and therefore $(x\sigma)(\alpha+z) = \sigma(x\alpha) = 0$.

Now (b) follows with $B = -(x)$, since $x \in \mathcal{L}(-(x))$.

It remains to prove that $x\sigma \in \Sigma_K$ for all $x \in K$ and $\sigma \in \Sigma_K$. If $x \in K^\times$ and $\sigma \in \Sigma_K$, then there exist $B,\, D \in \mathbb{D}_K$ such that $x \in \mathcal{L}(B)$, $\sigma \in \Sigma_K(D)$, and then (a) implies $x\sigma \in \Sigma_K(D-B) \subset \Sigma_K$.

3. If $\sigma,\, \sigma' \in \Sigma_K$ and $x,\, x' \in K$, then $1\sigma = \sigma$, $x(\sigma+\sigma') = x\sigma + x\sigma'$, $(x+x')\sigma = x\sigma + x'\sigma$ and $(xx')\sigma = x(x'\sigma)$. Hence Σ_K is a vector space over K.

Now let $\sigma_1, \sigma_2 \in \Sigma_K^\bullet$. We must prove that there exists some $x \in K$ such that $\sigma_2 = x\sigma_1$. For $i \in \{1,2\}$ let $D_i \in \mathbb{D}_K$ be such that $\sigma_i \in \Sigma_K(D_i)$. By 6.3.7 there exists some $B \in \mathbb{D}'_K$ such that $\deg(B) > 3g-3-\deg(D_1+D_2)$ and

$$\dim(D_i+B) = \deg(D_i+B) - g + 1 \quad \text{for} \quad i \in \{1,2\}.$$

Then 2.(a) implies $x\sigma_i \in \Sigma_K(-B)$ for all $x \in \mathcal{L}(D_i+B)$, and we define

$$\varphi_i \colon \mathcal{L}(D_i+B) \to \Sigma_K(-B) \;\text{ by }\; \varphi_i(x) = x\sigma_i \;\text{ for all }\; x \in \mathcal{L}(D_i+B).$$

φ_i is an F-monomorphism, and we set $U_i = \mathrm{Im}(\varphi_i) \subset \Sigma_K(-B)$. By 1. we obtain

$$\begin{aligned}\dim_F(U_1+U_2) \le \dim_F(\Sigma_K(-B)) &= \dim(-B) - \deg(-B) + g - 1 \\ &= \deg(B) + g - 1,\end{aligned}$$

and therefore

$$\begin{aligned}\dim_F(U_1 \cap U_2) &= \dim_F(U_1) + \dim_F(U_2) - \dim_F(U_1 + U_2) \\ &\geq \dim(D_1 + B) + \dim(D_2 + B) - (\deg(B) + g - 1) \\ &\geq [\deg(D_1 + B) - g + 1] + [\deg(D_2 + B) - g + 1] - [\deg(B) + g - 1] \\ &= \deg(D_1 + D_2) + \deg(B) - 3g + 3 > 0.\end{aligned}$$

Hence $U_1 \cap U_2 \neq \mathbf{0}$, and there exist $x_1 \in \mathcal{L}(D_1 + B)^\bullet$ and $x_2 \in \mathcal{L}(D_2 + B)^\bullet$ such that $\varphi_1(x_1) = \varphi_2(x_2)$. Hence it follows that $x_1\sigma_1 = x_2\sigma_2$, and therefore $\sigma_2 = (x_2^{-1}x_1)\sigma_1 \in K\sigma_1$. □

Definition and Theorem 6.4.3.

1. Let $\sigma \in \Sigma_K^\bullet$. There exists a (unique) greatest divisor $W \in \mathbb{D}_K$ such that $\sigma \in \Sigma_K(W)$. Explicitly, W has the following property:

If $D \in \mathbb{D}_K$, then $\sigma \in \Sigma_K(D)$ if and only if $D \leq W$.

We set $W = (\sigma) = \operatorname{div}(\sigma)$ and call (σ) the **divisor of** σ. For $P \in \mathbb{P}_K$ we define $v_P(\sigma) = v_P((\sigma))$. We call P a **zero** [a **pole**] of σ if $v_P(\sigma) > 0$ [$v_P(\sigma) < 0$], and we call σ **regular** in P if $v_P(\sigma) \geq 0$. A Weil differential $\sigma \in \Sigma_K$ is called **regular** or **holomorphic** if $v_P(\sigma) \geq 0$ for all $P \in \mathbb{P}_K$. A divisor $W \in \mathbb{D}_K$ is called **canonical** if $W = (\sigma)$ for some $\sigma \in \Sigma_K^\bullet$. A divisor class $\boldsymbol{w} \in \mathcal{C}_K$ is called **canonical** if it contains a canonical divisor.

2. If $\sigma \in \Sigma_K^\bullet$ and $x \in K^\times$, then $(x\sigma) = (x) + (\sigma)$.

3. Let $W \in \mathbb{D}_K$ be canonical and $D \in \mathbb{D}_K$. Then D is canonical if and only if $D \sim W$. In particular, there exists precisely one canonical divisor class $\boldsymbol{w}$, and $\boldsymbol{w}$ contains all canonical divisors.

4. Let $D \in \mathbb{D}_K$, $\sigma \in \Sigma_K^\bullet$ and $W = (\sigma)$. Then the assignment $x \mapsto x\sigma$ defines an F-isomorphism $\mu_\sigma \colon \mathcal{L}(W - D) \to \Sigma_K(D)$. In particular, $\dim(W - D) = \dim_F \Sigma_K(D)$.

Proof. 1. If $D \in \mathbb{D}_K$ and $\deg(D) \gg 1$, then $\mathbb{A}_K = \mathbb{A}_K(D) + K$ and thus $\Sigma_K(D) = \mathbf{0}$ by 6.3.7. Hence there exists some divisor $W \in \mathbb{D}_K$ of maximal degree such that $\sigma \in \Sigma_K(W)$, and it suffices to prove:

If $D \in \mathbb{D}_K$, then $\sigma \in \Sigma_K(D)$ if and only if $D \leq W$.

If $D \in \mathbb{D}_K$ and $D \leq W$, then $\Sigma_K(W) \subset \Sigma_K(D)$ and thus $\sigma \in \Sigma_K(D)$. As to the converse, we assume to the contrary that there exists some divisor $D \in \mathbb{D}_K$ such that $\sigma \in \Sigma_K(D)$ and yet $D \not\leq W$. Then there exists some $Q \in \mathbb{P}_K$ such that $v_Q(D) > v_Q(W)$, and we assert that $\sigma \restriction \mathbb{A}_K(Q + W) = 0$ (which contradicts the maximal choice of W).

Let $\alpha \in \mathbb{A}_K(Q+W)$, and define α', $\alpha'' \in \mathbb{A}_K$ by

$$\alpha'_P = \begin{cases} \alpha_P & \text{if } P \ne Q, \\ 0 & \text{if } P = Q, \end{cases} \quad \text{and} \quad \alpha''_P = \begin{cases} 0 & \text{if } P \ne Q, \\ \alpha_Q & \text{if } P = Q, \end{cases} \quad \text{hence } \alpha = \alpha' + \alpha''.$$

Then $v_Q(\alpha') = \infty \ge -v_Q(W)$ and $v_P(\alpha') = v_P(\alpha) \ge -v_P(Q+W) = -v_P(W)$ for all $P \in \mathbb{P}_K \setminus \{Q\}$. Hence $\alpha' \in \mathbb{A}_K(W)$ and $\sigma(\alpha') = 0$. As well we obtain $v_Q(\alpha'') = v_Q(\alpha) \ge -v_Q(Q+W) = -1 - v_Q(W) \ge -v_Q(D)$, and alike $v_P(\alpha'') = \infty \ge -v_P(D)$ for all $P \in \mathbb{P}_K \setminus \{Q\}$. Hence $\alpha'' \in \mathbb{A}_K(D)$, $\sigma(\alpha'') = 0$, and thus finally $\sigma(\alpha) = \sigma(\alpha') + \sigma(\alpha'') = 0$.

2. Let $\sigma \in \Sigma_K^\bullet$ and $x \in K^\times$. As $\sigma \in \Sigma_K((\sigma))$, we obtain $x\sigma \in \Sigma_K((x)+(\sigma))$ by 6.4.2.2(b), and consequently $(x\sigma) \ge (x) + (\sigma)$. On the other hand, it follows that $(\sigma) = (x^{-1}(x\sigma)) \ge (x^{-1}) + (x\sigma) = -(x) + (x\sigma)$ and therefore $(x\sigma) \le (x) + (\sigma)$. Hence equality holds.

3. Let $W = (\sigma)$ for some $\sigma \in \Sigma_K^\bullet$. If D is canonical, say $D = (\sigma_1)$ for some $\sigma_1 \in \Sigma_K^\bullet$, then $\sigma_1 = x\sigma$ for some $x \in K^\times$ by 6.4.2.3, and therefore $D = (x\sigma) = (x) + W \sim W$. Conversely, if $D \sim W$, say $D = (x) + W$ for some $x \in K^\times$, then $D = (x) + (\sigma) = (x\sigma)$ is canonical.

4. If $x \in \mathcal{L}(W-D)$ and $\sigma \in \Sigma_K(W)$, then $x\sigma \in \Sigma_K(D)$ by 6.4.2.2(a). Hence there is an F-monomorphism $\mu_\sigma \colon \mathcal{L}(W-D) \to \Sigma_K(D)$ satisfying $\mu_\sigma(x) = x\sigma$ for all $x \in \mathcal{L}(W-D)$, and we must prove that it is surjective. If $\sigma_1 \in \Sigma_K(D)^\bullet$, then $\dim_K \Sigma_K = 1$ implies $\sigma_1 = x\sigma$ for some $x \in K^\times$, and it follows that $(x) = (\sigma_1) - (\sigma) \ge D - W$. Hence $x \in \mathcal{L}(W-D)$ and $\sigma_1 = \mu_\sigma(x)$. □

Definition and Theorem 6.4.4 (Local components of Weil differentials). For $x \in K$ and $P \in \mathbb{P}_K$ we define the **local embedding** $\iota_P \colon K \to \mathbb{A}_K$ by

$$\iota_P(x)_Q = \begin{cases} x & \text{if } Q = P, \\ 0 & \text{if } Q \ne P. \end{cases}$$

If $D \in \mathbb{D}_K$, then $\iota_P(x) \in \mathbb{A}_K(D)$ if and only if $v_P(x) \ge -v_P(D)$.

For a Weil differential $\sigma \in \Sigma_K$ and $P \in \mathbb{P}_K$ the map $\sigma_P = \sigma \circ \iota_P \colon K \to F$ is called the ***P*-component** of σ.

1. *If $P \in \mathbb{P}_K$, then $\iota_P \colon K \to \mathbb{A}_K$ and $\sigma_P \colon K \to F$ are F-linear maps satisfying*

$$z\iota_P(x) = \iota_P(zx), \quad (z\sigma)_P(x) = \sigma_P(zx) \quad \text{and} \quad (\sigma+\sigma')_P = \sigma_P + \sigma'_P$$

for all σ, $\sigma' \in \Sigma_K$ and x, $z \in K$.

2. *Let $\sigma \in \Sigma_K$ and $\alpha = (\alpha_P)_{P \in \mathbb{P}_K} \in \mathbb{A}_K$. Then $\sigma_P(\alpha_P) = 0$ for almost all $P \in \mathbb{P}_K$, and*

$$\sigma(\alpha) = \sum_{P \in \mathbb{P}_K} \sigma_P(\alpha_P).$$

In particular, if $a \in K$, then

$$\sum_{P \in \mathbb{P}_K} \sigma_P(a) = 0 \qquad \text{(abstract residue theorem).}$$

3. Let $\sigma \in \Sigma_K^\bullet$ and $P \in \mathbb{P}_K$. Then $\sigma_P \neq 0$, and

$$v_P((\sigma)) = \max\{m \in \mathbb{Z} \mid \sigma_P(x) = 0 \text{ for all } x \in K \text{ such that } v_P(x) \geq -m\}.$$

4. Let $\sigma, \sigma' \in \Sigma_K$ and $P \in \mathbb{P}_K$ such that $\sigma_P = \sigma'_P$. Then $\sigma = \sigma'$.

Proof. 1. Obvious.

2. For $\sigma = 0$ there is nothing to do. Thus let $\sigma \neq 0$, $W = (\sigma)$, and let S be a finite subset of $\mathbb{P}_K$ such that $v_P(W) = 0$ and $v_P(\alpha_P) \geq 0$ for all $P \in \mathbb{P}_K \setminus S$. Let

$$\beta = (\beta_P)_{P \in \mathbb{P}_K} \in \mathbb{A}_K \text{ be defined by } \beta_P = \begin{cases} \alpha_P & \text{if } P \notin S, \\ 0 & \text{if } P \in S. \end{cases}$$

If $P \in \mathbb{P}_K \setminus S$, then $v_P(\beta) = v_P(\alpha_P) \geq 0 = -v_P(W)$, hence $\iota_P(\alpha_P) \in \mathbb{A}_K(W)$ and $\sigma_P(\alpha_P) = 0$. Moreover, as $v_P(\beta) = \infty \geq -v_P(W)$ for all $P \in S$, we obtain $\beta \in \mathbb{A}_K(W)$ and thus $\sigma(\beta) = 0$. Therefore

$$\alpha = \beta + \sum_{P \in S} \iota_P(\alpha_P) \text{ implies } \sigma(\alpha) = \sigma(\beta) + \sum_{P \in S} \sigma_P(\alpha_P) = \sum_{P \in \mathbb{P}_K} \sigma_P(\alpha_P).$$

3. Let $W = (\sigma)$. If $x \in K$ and $v_P(x) \geq -v_P(W)$, then $\iota_P(x) \in \mathbb{A}_K(W)$ and $\sigma_P(x) = \sigma(\iota_P(x)) = 0$. Thus we must prove that there exists some $x \in K$ such that $v_P(x) = -v_P(W) - 1$ and $\sigma_P(x) \neq 0$.

By definition, $\sigma \restriction \mathbb{A}_K(W+P) \neq 0$. Thus let $\alpha = (\alpha_Q)_{Q \in \mathbb{P}_K} \in \mathbb{A}_K(W+P)$ be such that $\sigma(\alpha) \neq 0$. Then $\alpha \notin \mathbb{A}_K(W)$, and therefore $v_P(\alpha_P) = -v_P(W) - 1$. Since $\alpha = (\alpha - \iota_P(\alpha_P)) + \iota_P(\alpha_P)$ and $\alpha - \iota_P(\alpha_P) \in \mathbb{A}_K(W)$, it follows that

$$0 \neq \sigma(\alpha) = \sigma(\iota_P(\alpha_P)) = \sigma_P(\alpha_P).$$

Hence $x = \alpha_P$ has the desired property.

4. Since $(\sigma - \sigma')_P = \sigma_P - \sigma'_P = 0$, 3. implies $\sigma = \sigma'$. □

Theorem 6.4.5.

1. (Theorem of Riemann-Roch) *Let $W \in \mathbb{D}_K$ be a canonical divisor and $D \in \mathbb{D}_K$. Then*

$$\dim(D) = \deg(D) - g + 1 + \dim(W - D).$$

2. For a divisor $D \in \mathbb{D}_K$ the following assertions are equivalent:

(a) D is canonical.

(b) $\dim(D) = g$ and $\deg(D) = 2g - 2$.

(c) $\dim(D) \geq g$, and $\deg(D) = 2g - 2$.

(d) $\deg(D) \geq 2g - 2$ and $\dim(D) \geq \deg(D) - g + 2$.

3. If $\deg(D) \ge 2g-1$, *then* $\dim(D) = \deg(D) - g + 1$.

4. If $g = 0$, *then* $h_K = |\mathcal{C}_K^0| = 1$.

Proof. 1. By 6.4.3.4 and 6.4.2.1 we obtain

$$\dim(W - D) = \dim_F \Sigma_K(D) = \dim(D) - \deg(D) + g - 1.$$

The reader might wonder at the short proof of this celebrated theorem. But one should notice that almost all that we did after defining the degree and the dimension of a divisor was done to prepare the Rieman-Roch theorem.

2. (a) $\Rightarrow$ (b) By 1., applied with $(0, D)$ instead of (D, W), it follows that

$$1 = \dim(0) = \deg(0) - g + 1 + \dim(D) = -g + 1 + \dim(D),$$

hence $\dim(D) = g$. Again 1., this time applied with $D = W$, shows that

$$g = \dim(D) = \deg(D) - g + 1 + \dim(0) = \deg(D) - g + 2,$$

and consequently $\deg(D) = 2g - 2$.

(b) $\Rightarrow$ (c) $\Rightarrow$ (d) Obvious.

(d) $\Rightarrow$ (a) Let $W \in \mathbb{D}_K$ be canonical. As we have already shown in (a) $\Rightarrow$ (b), $\deg(W) = 2g-2 \le \deg(D)$, hence $\deg(W-D) \le 0$. On the other hand, 1. implies $\dim(W - D) = g - 1 + \dim(D) - \deg(D) \ge 1$. Hence $W - D \sim 0$, and $D \sim W$ implies that D is canonical by 6.4.3.3.

3. Let $W \in \mathbb{D}_K$ be canonical. Then $\deg(W - D) = 2g - 2 - \deg(D) < 0$, hence $\dim(W - D) = 0$, and therefore $\dim(D) = \deg(D) - g + 1$ by 1.

4. Let $g = 0$ and $D \in \mathbb{D}_K^0$, then $\dim(D) = 1$ by 3., and therefore $D \sim 0$. Hence $\mathbb{D}_K^0 = (K^\times)$ and thus $h_K = 1$. □

Corollary 6.4.6. *Let* $P \in \mathbb{P}_K$ *and* $n \in \mathbb{N}$ *such that* $(n-1)\deg(P) \ge 2g - 1$. *Then there exists some* $x \in K^\times$ *such that* $(x)_\infty = nP$. *In particular, for every* $P \in \mathbb{P}_K$ *there exist* $x, z \in K^\times$ *such that* $\mathcal{P}(x) = \mathcal{N}(z) = \{P\}$.

Proof. Since $\deg((n-1)P) = (n-1)\deg(P) \ge 2g - 1$, the Riemann-Roch theorem 6.4.5.3 implies

$$\dim(nP) = n\deg(P) - g + 1 > (n-1)\deg(P) - g + 1 = \dim((n-1)P),$$

hence $\mathcal{L}((n-1)P) \subsetneq \mathcal{L}(nP)$. If $x \in \mathcal{L}(nP) \setminus \mathcal{L}((n-1)P)$, then $v_P(x) = -n$ and $v_Q(x) \ge 0$ for all $Q \in \mathbb{P}_K \setminus \{P\}$. Therefore it follows that $(x)_\infty = nP$, and $\mathcal{P}(x) = \mathcal{N}(x^{-1}) = \{P\}$. □

Theorem 6.4.7 (Rational function fields).

1. K is a rational function field if and only if $\mathbb{P}_K^1 \ne \emptyset$ *and* $g = 0$.

2. *Let $K = F(x)$ be a rational function field and $P_\infty \in \mathbb{P}_K$ the infinite prime divisor with respect to x* (see 6.2.3 and 6.2.4).

(a) *$-2P_\infty$ is a canonical divisor of K.*

(b) *Let $z \in K^\times$. Then z has a unique factorization*

$$z = c\prod_{i=1}^{r} p_i(x)^{e_i},$$

where $c \in F^\times$, $r \in \mathbb{N}_0$, $e_1, \ldots, e_r \in \mathbb{Z}^\bullet$, $p_1, \ldots, p_r \in \mathsf{P}(F, X)$ are distinct monic irreducible polynomials. If z is as above, then

$$(z) = \sum_{i=1}^{r} e_i(P_{p_i(x)} - \partial(p_i)P_\infty).$$

In particular:

- *If $z = g(x)$ for some $g \in F[X]^\bullet$, then $(z)_\infty = \partial(g)P_\infty$.*
- *$(x) = P_x - P_\infty$.*

Proof. We prove first 2.(b). Let $K = F(x)$.

Recall that $K = \mathsf{q}(F[x])$, $F[x]$ is factorial, and $\{p(x) \mid p \in \mathsf{P}(F, X)\}$ is a set of representatives for the prime elements of $F[x]$. Hence every $z \in K^\times$ has a unique factorization as asserted. If $p \in \mathsf{P}(F, X)$ and $P \in \mathsf{P}_{F(x)}$, then

$$v_P(p(x)) = \begin{cases} 1 & \text{if } P = P_{p(x)}, \\ -\partial(p) & \text{if } P = P_\infty, \\ 0 & \text{otherwise.} \end{cases}$$

From this it follows that

$$\Big(c\prod_{i=1}^{r} p_i(x)^{e_i}\Big) = \sum_{i=1}^{r} e_i(P_{p_i(x)} - \partial(p_i)P_\infty).$$

The special cases are now obvious.

Next we prove 1., and therefore we tacitly use the Riemann-Roch theorem 6.4.5.

Let first $K = F(x)$ be a rational function field and P_∞ the infinite prime divisor with respect to x. Then $\deg(P_\infty) = 1$, and we must prove that $g = 0$. Let $r \in \mathbb{N}$ and $r \geq 2g - 1$. Then $\dim(rP_\infty) = \deg(rP_\infty) - g + 1 = r - g + 1$. If $n \in [0, r]$, then $(x^n) = n(x) \geq -nP_\infty \geq -rP_\infty$, hence $x^n \in \mathcal{L}(rP_\infty)$, and as $(1, x, \ldots, x^r)$ is linearly independent over F, we get

$$r + 1 \leq \dim(rP_\infty) = r - g + 1, \text{ hence } g \leq 0 \text{ and thus } g = 0.$$

Assume now that $g = 0$ and let $P \in \mathbb{P}_K^1$. Since $\deg(P) = 1 \geq 2g - 1 = -1$, we get $\dim(P) = \deg(P) - g + 1 = 2$ and thus $\mathcal{L}(P) \supsetneq F$. If $x \in \mathcal{L}(P) \setminus F$, then $(x)_\infty = P$, and 6.3.4.1 implies $[K : F(x)] = \deg(P) = 1$. Hence $K = F(x)$.

It remains to prove 2.(a) Since $\deg(-2P_\infty) = -2 = 2g - 2$, it follows that $\dim(-2P_\infty) = 0 = g$, and therefore $-2P_\infty$ is canonical. □

If $P_1, \ldots, P_r \in \mathbb{P}_K$ are distinct, $x_1, \ldots, x_r \in K$ and $c_1, \ldots, c_r \in \mathbb{Z}$, then by the weak approximation theorem 5.1.6 there exists some $x \in K$ such that $v_{P_i}(x - x_i) = c_i$ for all $i \in [1, r]$. The subsequent strong approximation theorem asserts that we can even find such an $x \in K$ which in addition is regular for all prime divisors $P \in \mathbb{P}_K \setminus \{P_1, \ldots, P_r\}$ with at most one prescribed exception.

Theorem 6.4.8 (Strong approximation theorem). *Let $r \in \mathbb{N}_0$, and suppose that $P_0, P_1, \ldots, P_r \in \mathbb{P}_K$ are distinct. Let $x_1, \ldots, x_r \in K$ and $n_1, \ldots, n_r \in \mathbb{Z}$. Then there exists some $x \in K$ such that $v_{P_i}(x - x_i) = n_i$ for all $i \in [1, r]$ and $v_P(x) \ge 0$ for all $P \in \mathbb{P}_K \setminus \{P_0, P_1, \ldots, P_r\}$.*

Proof. Let $\alpha \in \mathbb{A}_K$ be defined by

$$\alpha_P = \begin{cases} x_i & \text{if } P = P_i \text{ for some } i \in [1, r], \\ 0 & \text{if } P \in \mathbb{P}_K \setminus \{P_1, \ldots, P_r\}, \end{cases}$$

and for $m \in \mathbb{N}$ let

$$D_m = mP_0 - \sum_{i=1}^{r} (n_i + 1) P_i \in \mathbb{D}_K.$$

Let m be so large that $\mathbb{A}_K = \mathbb{A}_K(D_m) + K$, and let $z \in K$ be such that $z - \alpha \in \mathbb{A}_K(D_m)$. Then $v_{P_i}(z - x_i) \ge n_i + 1$ for all $i \in [1, r]$ and $v_P(z) \ge 0$ for all $P \in \mathbb{P}_K \setminus \{P_0, P_1, \ldots, P_r\}$.

Let $\beta = (\beta_P)_{P \in \mathbb{P}_K} \in \mathbb{A}_K$ be such that $v_{P_i}(\beta_{P_i}) = n_i$ for all $i \in [1, r]$, $\beta_P = 0$ for all $P \in \mathbb{P}_K \setminus \{P_0, P_1, \ldots, P_r\}$, and let $y \in K$ be such that $y - \beta \in \mathbb{A}_K(D_m)$. Then $v_{P_i}(y - \beta_{P_i}) \ge n_i + 1$ for all $i \in [1, r]$ and $v_P(y) \ge 0$ for all $P \in \mathbb{P}_K \setminus \{P_0, P_1, \ldots, P_r\}$. If $x = y + z$, then

$$v_{P_i}(x - x_i) = v_{P_i}((y - \beta_{P_i}) + (z - x_i) + \beta_{P_i}) = n_i \text{ for all } i \in [1, r],$$

and $v_P(x) \ge 0$ for all $P \in \mathbb{P}_K \setminus \{P_0, P_1, \ldots, P_r\}$. □

6.5 Algebraic function field extensions 1

Throughout this section, let K'/F' be an algebraic function field extension of K/F, let F be the field of constants of K/F and F' the field of constants of K'/F'.

Theorem 6.5.1. *$K \cap F' = F$, and f or every $P \in \mathbb{P}_K$ there are only finitely many $P' \in \mathbb{P}_{K'}$ lying above P.*

Proof. By 6.2.7.4 we get $K \cap F' = F$. Thus let $P \in \mathbb{P}_K$. By 6.4.6 there exists some $x \in K$ such that $\mathcal{N}^K(x) = \{P\}$. If $P' \in \mathbb{P}_{K'}$, then $x(P') = x(P' \cap K)$, and thus $x(P') = 0$ if and only if $P' \cap K = P$. But the set

$$\{P' \in \mathbb{P}_{K'} \mid P' \cap K = P\} = \mathcal{N}^{K'}(x)$$

is finite due to 6.2.7.2. □

Definition 6.5.2. The **divisor embedding** for K'/K is the unique group homomorphism $j_{K'/K} \colon \mathbb{D}_K \to \mathbb{D}_{K'}$ satisfying

$$j_{K'/K}(P) = \sum_{\substack{P' \in \mathbb{P}_{K'} \\ P' \cap K = P}} e(P'/P)P' \quad \text{for all} \quad P \in \mathbb{P}_K.$$

By definition, if $D \in \mathbb{D}_K$, then

$$j_{K'/K}(D) = \sum_{P \in \mathbb{P}_K} v_P(D) \sum_{\substack{P' \in \mathbb{P}_{K'} \\ P' \cap K = P}} e(P'/P)P = \sum_{P' \in \mathbb{P}_{K'}} e(P'/P' \cap K) v_{P' \cap K}(D)P'$$

Consequently, if $P' \in \mathbb{P}_{K'}$ and $P = P' \cap K$, then

$$v_{P'}(j_{K'/K}(D)) = e(P'/P)v_P(D).$$

Apparently, $j_{K'/K} \colon \mathbb{D}_K \to \mathbb{D}_{K'}$ is injective. If $D_1, D_2 \in \mathbb{D}_K$, then $D_1 \le D_2$ if and only if $j_{K'/K}(D_1) \le j_{K'/K}(D_2)$, and if K^* is an intermediate field of K'/K, then $j_{K'/K} = j_{K'/K^*} \circ j_{K^*/K}$.

Theorem 6.5.3. *For $P \in \mathbb{P}_K$, let $\mathcal{O}'_P = \mathrm{cl}_{K'}(\mathcal{O}_P)$ (see 6.2.5.3).*

1. *If $x \in K^\times$, then $j_{K'/K}((x)^K) = (x)^{K'}$, $j_{K'/K}((x)_0^K) = (x)_0^{K'}$, and $j_{K'/K}((x)_\infty^K) = (x)_\infty^{K'}$.*

2. *Let $[K' \colon K] = n < \infty$.*

(a) If $P \in \mathbb{P}_K$, then $\mathcal{O}'_P$ is a finitely generated $\mathcal{O}_P$-module, and

$$\sum_{\substack{P' \in \mathbb{P}_{K'} \\ P' \cap K = P}} e(P'/P) f(P'/P) = n.$$

(b) If $D \in \mathbb{D}_K$, then

$$\deg(j_{K'/K}(D)) = \frac{n \deg(D)}{[F' \colon F]}.$$

3. Let $\tau \in \mathrm{Gal}(K'/K)$, $P' \in \mathbb{P}_{K'}$, t' a prime element of $\mathcal{O}_{P'}$ and $P = P' \cap K$.

(a) $\tau P' \in \mathbb{P}_{K'}$, $\tau P' \cap K = P$, $\mathcal{O}_{\tau P'} = \tau\mathcal{O}_{P'}$, $\tau(t')$ is a prime element of $\mathcal{O}_{\tau P'}$,

$$v_{\tau P'} = v_{P'} \circ \tau^{-1} \colon K' \to \mathbb{Z} \cup \{\infty\},\ \tau\mathcal{O}'_P = \mathcal{O}'_P,$$

$$f(\tau P'/P) = f(P'/P) \text{ and } e(\tau P'/P) = e(P'/P).$$

(b) τ has a unique extension to a topological K_P-isomorphism

$$\tau^*_{P'} \colon K'_{P'} \to K'_{\tau P'}$$

and induces a k_P-isomorphism $\tau_{P'} \colon \mathsf{k}_{P'} \to \mathsf{k}_{\tau P'}$ such that $\tau_{P'}(x(P')) = \tau(x)(\tau P')$ for all $x \in \mathcal{O}_{P'}$.
If F is perfect, then $K'_{P'} = \mathsf{k}_{P'}((t'))$, $K'_{\tau P'} = \mathsf{k}_{\tau P'}((\tau(t')))$ and

$$\tau^*_{P'}\Big(\sum_{n\in\mathbb{Z}} a_n t'^n\Big) = \sum_{n\in\mathbb{Z}} \tau_{P'}(a_n)\tau(t')^n$$

for every sequence $(a_n)_{n\in\mathbb{Z}}$ in $\mathsf{k}_{P'}$ such that $a_{-n} = 0$ for all $n \gg 1$.

Proof. 1. If $x \in K^\times$, then

$$\begin{aligned}(x)_0^{K'} &= \sum_{\substack{P'\in\mathbb{P}_{K'}\\ v_{P'}(x)>0}} v_{P'}(x)P' = \sum_{\substack{P\in\mathbb{P}_K\\ v_P(x)>0}} \sum_{\substack{P'\in\mathbb{P}_{K'}\\ P'\cap K=P}} e(P'/P)v_P(x)P' \\ &= \sum_{\substack{P\in\mathbb{P}_K\\ v_P(x)>0}} v_P(x) \sum_{\substack{P'\in\mathbb{P}_{K'}\\ P'\cap K=P}} e(P'/P)P' = \sum_{\substack{P\in\mathbb{P}_K\\ v_P(x)>0}} v_P(x) j_{K'/K}(P) \\ &= j_{K'/K}\Big(\sum_{\substack{P\in\mathbb{P}_K\\ v_P(x)>0}} v_P(x)P\Big) = j_{K'/K}((x)_0^K).\end{aligned}$$

It follows that $j_{K'/K}((x)_\infty^K) = j_{K'/K}((x^{-1})_0^K) = (x^{-1})_0^{K'} = (x)_\infty^{K'}$ and

$$j_{K'/K}((x)^K) = j_{K'/K}((x)_0^K) - j_{K'/K}((x)_\infty^K) = (x)_0^{K'} - (x)_\infty^{K'} = (x)^{K'}.$$

2. (a) Let $P \in \mathbb{P}_K$ and $x \in K^\times$ such that $\mathcal{N}^K(x) = \{P\}$. If $P' \in \mathbb{P}_{K'}$, then $x(P') = x(P' \cap K)$, hence $x(P') = 0$ if and only if $P' \cap K = P$. It follows that $\mathcal{N}^{K'}(x) = \{P' \in \mathbb{P}_{K'} \mid P' \cap K = P\}$ and, using 6.1.1 and 6.3.4.1,

$$\begin{aligned}[K' \colon F(x)] &= [K' \colon F'(x)]\,[F'(x) \colon F(x)] = \deg((x)_0^{K'})\,[F' \colon F] \\ &= \sum_{\substack{P'\in\mathbb{P}_{K'}\\ P'\cap K=P}} v_{P'}(x)\deg(P')\,[F' \colon F] = \sum_{\substack{P'\in\mathbb{P}_{K'}\\ P'\cap K=P}} e(P'/P)v_P(x)\deg(P)f(P'/P) \\ &= v_P(x)\deg(P) \sum_{\substack{P'\in\mathbb{P}_{K'}\\ P'\cap K=P}} e(P'/P)f(P'/P).\end{aligned}$$

On the other hand, $(x)_0^K = v_P(x)P$ implies

$$[K':F(x)] = [K':K]\,[K:F(x)] = n\deg((x)_0^K) = nv_P(x)\deg(P),$$

and comparison implies

$$\sum_{\substack{P'\in\mathbb{P}_{K'}\\ P'\cap K=P}} e(P'/P)f(P'/P) = n.$$

By 6.2.5.3(b) and 2.12.4, $\mathcal{P}(\mathcal{O}'_P) = \{P'\cap\mathcal{O}'_P \mid P'\in\mathbb{P}_{K'},\ P'\cap K = P\}$,

$$\begin{aligned}\sum_{\substack{P'\in\mathbb{P}_{K'}\\ P'\cap K=P}} e(P'/P)f(P'/P) &= \sum_{\substack{P'\in\mathbb{P}_{K'}\\ P'\cap K=P}} e(P'\cap\mathcal{O}'_P/P)f(P'\cap\mathcal{O}'_P/P)\\ &= \sum_{\mathfrak{P}\in\mathcal{P}(\mathcal{O}'_P)} e(\mathfrak{P}/P)f(\mathfrak{P}/P)\ \le\ n,\end{aligned}$$

and since equality holds, $\mathcal{O}'_P$ is a finitely generated $\mathcal{O}_P$-module.

(b) Since $D\mapsto \deg(j_{K'/K}(D))$ and $D\mapsto n[F':F]^{-1}\deg(D)$ are homomorphisms, it suffices to prove the assertion for $D = P\in\mathbb{P}_K$. By 6.2.7.4 we get

$$\begin{aligned}\deg(j_{K'/K}(P)) &= \sum_{\substack{P'\in\mathbb{P}_{K'}\\ P'\cap K=P}} e(P'/P)\deg(P')\\ &= \sum_{\substack{P'\in\mathbb{P}_{K'}\\ P'\cap K=P}} e(P'/P)\frac{f(P'/P)\deg(P)}{[F':F]} = \frac{n\deg(P)}{[F':F]}.\end{aligned}$$

3. Straightforward (for (b) use 6.2.8). □

In the following Theorem 6.5.4 we refine and amend 6.5.3.3 in the case if a finite Galois extension along the lines of Section 2.13 and Theorem 5.5.8.

Definition and Theorem 6.5.4. *Suppose that K'/K is a finite Galois extension and $G = \mathrm{Gal}(K'/K)$. Let $P\in\mathbb{P}_K$, $\mathcal{O}'_P = \mathrm{cl}_{K'}(\mathcal{O}_P)$, $P'\in\mathbb{P}_{K'}$ such that $P'\cap K = P$, $e = e(P'/P)$, and $f = f(P'/P)$. Let $G_{P'} = \{\tau\in G \mid \tau P' = P'\}$ be the isotropy group of P', and assume that $\mathsf{k}_{P'}/\mathsf{k}_P$ is separable.*
The group $G_{P'}$ is called the **decomposition group** of P' over K.

1. G operates transitively on the set $\mathcal{P}_P$ of all prime divisors of P in K'. If $(G:G_{P'}) = r$ and $G/G_{P'} = \{\tau_1G_{P'},\ldots,\tau_rG_{P'}\}$, then $\mathcal{P}_P = \{\tau_1P',\ldots,\tau_rP'\}$, $j_{K'/K}(P) = e(\tau_1P'+\ldots+\tau_rP')$, $|G_{P'}| = ef$, and $[K':K] = ref$. If $\sigma\in G$, then

$$G_{\sigma P'} = \sigma G_{P'}\sigma^{-1},\ e(\sigma P'/P) = e \text{ and } f(\sigma P'/P) = f.$$

Moreover, $G_{P'} = G_{P'\cap\mathcal{O}'_P}$ is the decomposition group of the prime ideal $P'\cap\mathcal{O}'_P\in\mathcal{P}(\mathcal{O}'_P)$ over K, see 6.2.5.3 and 2.5.3).

2. The residue class extension $\mathsf{k}_{P'}/\mathsf{k}_P$ *is Galois, and the map* $\rho_{P'}\colon G_{P'} \to \mathrm{Gal}(\mathsf{k}_{P'}/\mathsf{k}_P)$, *defined by* $\rho_{P'}(\tau) = \tau_{P'}$ *for all* $\tau \in G_{P'}$, *is a group epimorphism.*
The automorphism $\tau_{P'} \in \mathrm{Gal}(\mathsf{k}_{P'}/\mathsf{k}_P)$ is called the **residue class automorphism** induced by τ. The group $I_{P'} = \mathrm{Ker}(\rho_{P'})$ is called the **inertia group** of P' over K.
$|I_{P'}| = e$, $(G_{P'} : I_{P'}) = f$, $I_{\sigma P'} = \sigma^{-1} I_{P'} \sigma$ *for all* $\sigma \in G$, *and* $I_{P'} = I_{P' \cap \mathcal{O}'_P}$ *is the inertia group of the prime ideal* $P' \cap \mathcal{O}'_P$ *of* $\mathcal{O}'_P$ *over* K, see 6.2.5.3 and 2.13.2,

3. Every $\sigma \in G_{P'}$ *has a unique extension to an automorphism* $\widehat{\sigma} \in \mathrm{Gal}(K'_{P'}/K_P)$. *The extension* $K'_{P'}/K_P$ *is Galois, and the assignment* $\sigma \mapsto \widehat{\sigma}$ *defines an isomorphism* $G_{P'} \overset{\sim}{\to} \mathrm{Gal}(K'_{P'}/K_P)$.

4. Let k_P *be finite, let* P *be unramified in* K' *and* $\sigma \in G_P$. *Then the following assertions are equivalent:*

(a) *σ is a Frobenius automorphism of* $P' \cap \mathcal{O}'_P$ *over* K (as defined in 2.14.5 for the Dedekind domains $\mathcal{O}_P \subset \mathcal{O}'_P$).

(b) *σ is a Frobenius automorphism of* $K'_{P'}/K_P$ (as defined in 5.6.7.1).

If these conditions are fulfilled, we call σ a **Frobenius automorphism** of P' over K.

Proof. We tacitly use 6.5.3.3, 6.2.5.3 and the results from Section 2.12.2.2.

1. and 2. The map $\theta\colon \mathcal{P}_P \to \mathcal{P}(\mathcal{O}'_P)$, defined by $\theta(P_1) = P_1 \cap \mathcal{O}'_P$, is bijective, and $\theta^{-1}(\mathfrak{P}) = \mathfrak{P}(\mathcal{O}'_P)_{\mathfrak{P}}$ for all $\mathfrak{P} \in \mathcal{P}(\mathcal{O}'_P)$. If $P_1 \in \mathcal{P}_P$, then $e(\theta(P_1)/P) = e(P_1/P)$, $f(\theta(P_1)/P) = f(P_1/P)$,

$$x(P_1) = x + \theta(P_1) \in \mathsf{k}_{\theta(P_1)} = \mathsf{k}_{P_1} \text{ for all } x \in \mathcal{O}_{P_1} = (\mathcal{O}'_P)_{\theta(P_1)},$$

and $\tau\theta(P_1) = \theta(\tau(P_1)$ for all $\tau \in G$. Consequently, $G_{P_1} = G_{\theta(P_1)}$, the homomorphism $\rho_{P'}\colon G_{P'} \to \mathsf{k}_{P'}/\mathsf{k}_P$ coincides with the homomorphism $\rho_{\theta(P')}\colon G_{\theta(P')} \to \mathsf{k}_{\theta(P')/\mathsf{k}_P}$. Consequently $\tau_{P'} = \tau_{\theta(P')}$ and $I_{P'} = I_{\theta(P')}$. With these observations 1. and 2. follow from 2.5.4 and 2.13.2.

3. By 5.5.8.2, applied for an absolute value belonging to v_P (see 5.1.3.4).

4. If k_P is finite, then F is finite and thus perfect. Observing 1., the assertion follows from 2.14.6, 5.5.9.5 and 5.6.7.1(a). □

We proceed with the comparatively simple case of purely inseparable extensions.

Theorem 6.5.5. *Let* F *be perfect,* $\mathrm{char}(F) = p > 0$, K'/K *purely inseparable and* $[K' : K] = p^n$ *for some* $n \in \mathbb{N}_0$.

1. $F = F'$.

2. Let $P \in \mathbb{P}_K$ *and* $P' = \{z \in K' \mid z^{p^n} \in P\}$. *Then* P' *is the only prime divisor of* K' *lying above* P, $\mathcal{O}_{P'} = \{z \in K' \mid z^{p^n} \in \mathcal{O}_P\}$, $e(P'/P) = p^n$, $f(P'/P) = 1$ *and* $\deg(P') = \deg(P)$.

Proof. 1. If $z \in F'$, then $z^{p^n} \in K \cap F' = F$ by 1.4.7.2 and 6.5.1. As F is perfect, we obtain $z \in F$.

2. By 2.12.7.2 it follows that $\mathcal{O}'_P$ is local with maximal ideal P', and $f(P'/P) = 1$ (indeed, k_P is perfect by 1.4.6.2, since $[\mathsf{k}_P : F] = \deg(P) < \infty$). Hence 6.2.5.3 implies that P' is the only prime divisor of K' lying above P, $\mathcal{O}_{P'} = \mathcal{O}'_P$, and thus $e(P'/P) = p^n$ by 6.5.3.2(a). □

Next we define the divisor norm for an algebraic function field extension. It is the (almost) complete analogon to the ideal norm for extensions of Dedekind domains (see 2.14.2 and 2.14.4). For separable extensions the following Theorem 6.5.6 can easily be deduced from 2.14.2 and 2.14.4. Here we give independent proofs with only weak assumptions concerning separability.

Definition and Theorem 6.5.6. *Let* $n = [K' : K] < \infty$, *and suppose that either* K'/K *is separable or* F *is perfect.*
The **divisor norm** $\mathcal{N}_{K'/K} \colon \mathbb{D}_{K'} \to \mathbb{D}_K$ is the unique group homomorphism satisfying

$$\mathcal{N}_{K'/K}(P') = f(P'/P)P \quad \text{for all} \quad P' \in \mathbb{P}_{K'} \quad \text{and} \quad P = P' \cap K.$$

If K^* is an intermediate field of K'/K, then $\mathcal{N}_{K'/K} = \mathcal{N}_{K^*/K} \circ \mathcal{N}_{K'/K^*}$. By definition, if K'/K is separable, then $\mathcal{N}_{K'/K}(P') = \mathcal{N}_{\mathcal{O}'_P/\mathcal{O}_P}(P' \cap \mathcal{O}'_P)$ (see 2.14.2).

1. If $D' \in \mathbb{D}_{K'}$ *and* $P \in \mathbb{P}_K$, *then*

$$v_P(\mathcal{N}_{K'/K}(D')) = \sum_{\substack{P' \in \mathbb{P}_{K'} \\ P' \cap K = P}} f(P'/P) v_{P'}(D'),$$

and $\deg(\mathcal{N}_{K'/K}(D')) = \deg(D')$.

2. If $D \in \mathbb{D}_K$, *then* $\mathcal{N}_{K'/K} \circ j_{K'/K}(D) = nD$.

3. Let K'/K *be Galois,* $G = \mathrm{Gal}(K'/K)$ *and* $D' \in \mathbb{D}_{K'}$. *Then*

$$j_{K'/K} \circ \mathcal{N}_{K'/K}(D') = \sum_{\sigma \in G} \sigma D'.$$

4. Let $x \in K'^{\times}$. *Then* $(\mathsf{N}_{K'/K}(x))^K = \mathcal{N}_{K'/K}((x)^{K'})$, *and*

$$v_P(\mathsf{N}_{K'/K}(x)) = \sum_{\substack{P' \in \mathbb{P}_{K'} \\ P' \cap K = P}} f(P'/P) v_{P'}(x) \quad \textit{for all} \quad P \in \mathbb{P}_K.$$

Proof. 1. If $D' \in \mathbb{D}_{K'}$, then

$$\mathcal{N}_{K'/K}(D') = \sum_{P \in \mathbb{P}_K} \sum_{\substack{P' \in \mathbb{P}_{K'} \\ P' \cap K = P}} v_{P'}(D') \mathcal{N}_{K'/K}(P') = \sum_{P \in \mathbb{P}_K} \sum_{\substack{P' \in \mathbb{P}_{K'} \\ P' \cap K = P}} f(P'/P) v_{P'}(D') P,$$

and therefore, for all $P \in \mathbb{P}_K$,

$$v_P(\mathcal{N}_{K'/K}(D')) = \sum_{\substack{P' \in \mathbb{P}_{K'} \\ P' \cap K = P}} f(P'/P) v_{P'}(D').$$

If $P' \in \mathbb{P}_{K'}$ and $P = P' \cap K$, then

$$\deg(\mathcal{N}_{K'/K}(P')) = f(P'/P)\deg(P) = [\mathsf{k}_{P'} : \mathsf{k}_P][\mathsf{k}_P : F] = [\mathsf{k}_{P'} : F] = \deg(P').$$

Since $\deg \circ \mathcal{N}_{K'/K} \colon \mathbb{D}_{K'} \to \mathbb{Z}$ is a homomorphism, the equality follows for all $D' \in \mathbb{D}_{K'}$.

2. As $\mathcal{N}_{K'/K} \circ j_{K'/K} \colon \mathbb{D}_K \to \mathbb{D}_K$ is a homomorphism, it suffices to consider the case $D = P \in \mathbb{P}_K$. By 6.5.3.2(a) we obtain

$$\mathcal{N}_{K'/K} \circ j_{K'/K}(P) = \mathcal{N}_{K'/K}\Big(\sum_{\substack{P' \in \mathbb{P}_{K'} \\ P' \cap K = P}} e(P'/P) P' \Big) = \sum_{\substack{P' \in \mathbb{P}_{K'} \\ P' \cap K = P}} e(P'/P) f(P'/P) P = nP.$$

3. Again it suffices to consider the case $D' = P' \in \mathbb{P}_{K'}$. We use 6.5.4.1. Let $P = P' \cap K$, $G_{P'}$ be the decomposition group of P' over K, $(G : G_{P'}) = r$ and $G/G_{P'} = \{\sigma_1 G_{P'}, \ldots, \sigma_r G_{P'}\}$. Then $|G_{P'}| = e(P'/P) f(P'/P)$,

$$j_{K'/K} \circ \mathcal{N}_{K'/K}(P') = j_{K'/K}(f(P'/P)P) = f(P'/P) e(P'/P) \sum_{i=1}^{r} \sigma_i P' = |G_{P'}| \sum_{i=1}^{r} \sigma_i P'$$

and

$$\sum_{\sigma \in G} \sigma P' = \sum_{i=1}^{r} \sum_{\tau \in G_{P'}} \sigma_i \tau P' = |G_{P'}| \sum_{i=1}^{r} \sigma_i P'.$$

4. Let $x \in K'^{\times}$.

CASE 1: K'/K is Galois and $G = \mathrm{Gal}(K'/K)$. Then, using 6.2.11.4 and 3.,

$$\jmath_{K'/K}\big((\mathsf{N}_{K'/K}(x))^K\big) = (\mathsf{N}_{K'/K}(x))^{K'} = \Big(\prod_{\sigma\in G}\sigma x\Big)^{K'}$$
$$= \sum_{\sigma\in G}(\sigma x)^{K'} = \sum_{\sigma\in G}\sigma((x)^{K'}) = \jmath_{K'/K}\circ\mathcal{N}_{K'/K}((x)^{K'}),$$

and as $\jmath_{K'/K}$ is injective, we obtain $(\mathsf{N}_{K'/K}(x))^K = \mathcal{N}_{K'/K}((x)^{K'})$.

CASE 2: K'/K is separable. Let K^* be a Galois closure of K'/K and $d = [K^*:K']$. Then $(\mathsf{N}_{K^*/K}(x))^K = \mathcal{N}_{K^*/K}((x)^{K^*})$ by CASE 1,

$$(\mathsf{N}_{K^*/K}(x))^K = (\mathsf{N}_{K'/K}\circ\mathsf{N}_{K^*/K'}(x))^K = (\mathsf{N}_{K'/K}(x)^d)^K = d(\mathsf{N}_{K'/K}(x))^K$$

and (using 2.)

$$\mathcal{N}_{K^*/K}((x)^{K^*}) = \mathcal{N}_{K'/K}\circ\mathcal{N}_{K^*/K'}(\jmath_{K^*/K'}((x)^{K'})) = \mathcal{N}_{K'/K}(d(x)^{K'})$$
$$= d\mathcal{N}_{K'/K}((x)^{K'}).$$

Hence it follows that $(\mathsf{N}_{K'/K}(x))^K = \mathcal{N}_{K'/K}((x)^{K'})$.

CASE 3: K'/K is inseparable. Let $\mathrm{char}(K) = p$, K^* be the separable closure of K in K' and $[K':K^*] = p^n$ for some $n \in \mathbb{N}$. If $P' \in \mathbb{P}_{K'}$ and $P^* = P'\cap K^*$, then 6.5.5.2 implies that P' is the only prime divisor of K' above P^*, $f(P'/P^*) = 1$ and $e(P'/P^*) = p^n$. Hence $p^n v_{P'}(x) = v_{P'}(x^{p^n}) = p^n v_{P^*}(x^{p^n})$, and consequently $v_{P'}(x) = v_{P^*}(x^{p^n}) = v_{P^*}(\mathsf{N}_{K'/K^*}(x))$. Therefore we obtain

$$\mathcal{N}_{K'/K^*}\big((x)^{K'}\big) = \mathcal{N}_{K'/K^*}\Big(\sum_{P'\in\mathbb{P}_{K'}} v_{P'}(x)P'\Big) = \sum_{P'\in\mathbb{P}_{K'}} v_{P'}(x)(P'\cap K^*)$$
$$= \sum_{P^*\in\mathbb{P}_{K^*}} v_{P^*}(\mathsf{N}_{K'/K^*}(x))P^* = \big(\mathsf{N}_{K'/K^*}(x)\big)^{K^*},$$

and together with CASE 2 it follows that

$$\mathcal{N}_{K'/K}\big((x)^{K'}\big) = \mathcal{N}_{K^*/K'}\circ\mathcal{N}_{K'/K^*}\big((x)^{K'}\big) = \mathcal{N}_{K^*/K}\big((\mathsf{N}_{K'/K^*}(x))^{K^*}\big)$$
$$= \big(\mathsf{N}_{K^*/K}\circ\mathsf{N}_{K'/K^*}\big)^K = \big(\mathsf{N}_{K'/K}(x)\big)^K.$$

In any case, together with 1. we obtain

$$v_P(\mathsf{N}_{K'/K}(x)) = v_P\big((\mathsf{N}_{K'/K}(x))^{K'}\big) = v_P\big(\mathcal{N}_{K'/K}((x)^{K'})\big)$$
$$= \sum_{\substack{P'\in\mathbb{P}_{K'}\\ P'\cap K=P}} f(P'/P)v_{P'}\big((x)^{K'}\big) = \sum_{\substack{P'\in\mathbb{P}_{K'}\\ P'\cap K=P}} f(P'/P)v_{P'}(x). \qquad \square$$

Our next aim is to adjust the general theory of differents and discriminants from Section 5.9 for a finite separable extension K'/K of function fields. We will start locally with a prime divisor $P \in \mathbb{P}_K$, the extension of Dedekind domains $\mathcal{O}'_P/\mathcal{O}_P$ and the associated extensions $K'_{P'}/K_P$ of their completions. Here we will apply the theory of Section 5.9 and then we will perform the globalization with genuine methods from the theory of function fields.

Definitions and Remarks 6.5.7. Let K'/K be finite separable and $P \in \mathbb{P}_K$. Recall from 6.2.5.3 that $\mathcal{O}'_P = \mathrm{cl}_{K'}(\mathcal{O}_P)$ is a semilocal principal ideal domain,

$$\mathcal{O}'_P = \bigcap_{\substack{P' \in \mathbb{P}_{K'} \\ P' \supset P}} \mathcal{O}_{P'} \quad \text{and} \quad \mathcal{P}(\mathcal{O}'_P) = \{P' \cap \mathcal{O}'_P \mid P' \in \mathbb{P}_{K'},\ P' \cap K = P\}.$$

If $P' \in \mathbb{P}_{K'}$ and $P' \cap K = P$, then $\mathcal{O}_{P'} = (\mathcal{O}'_P)_{P' \cap \mathcal{O}'_P}$, $v_{P'} = v_{P' \cap \mathcal{O}'_P}$, $e(P'/P) = e(P' \cap \mathcal{O}'_P/P)$ and $f(P'/P) = f(P' \cap \mathcal{O}'_P/P)$.

We recall from 5.9.2 and 5.9.3.1 the definition of the Dedekind complementary module $\mathfrak{C}_{\mathcal{O}'_P/\mathcal{O}_P}$ and the different $\mathfrak{D}_{\mathcal{O}'_P/\mathcal{O}_P} = \mathfrak{C}^{-1}_{\mathcal{O}'_P/\mathcal{O}_P}$ of $\mathcal{O}'_P/\mathcal{O}_P$.

If $\mathfrak{D}_{\mathcal{O}'_P/\mathcal{O}_P} = t_P\mathcal{O}'_P$ for some $t_P \in \mathcal{O}'_P$, then $\mathfrak{C}_{\mathcal{O}'_P/\mathcal{O}_P} = t_P^{-1}\mathcal{O}'_P$, and for $P' \in \mathbb{P}_{K'}$ such that $P = P' \cap K$ we call

$$d(P'/P) = v_{P'}(t_P) = v_{P' \cap \mathcal{O}'_P}(t_P) = v_{P' \cap \mathcal{O}'_P}(\mathfrak{D}_{\mathcal{O}'_P/\mathcal{O}_P}) \in \mathbb{N}_0$$

the **different exponent** of P'/P. It follows that

$$\begin{aligned}\mathfrak{D}_{\mathcal{O}'_P/\mathcal{O}_P} &= \prod_{\substack{P' \in \mathbb{P}_{K'} \\ P' \cap K = P}} (P' \cap \mathcal{O}_{P'})^{d(P'/P)} \\ &= \{x \in K' \mid v_{P'}(x) \ge d(P'/P) \text{ for all } P' \in \mathbb{P}_{K'} \text{ above } P\}\end{aligned}$$

and

$$\begin{aligned}\mathfrak{C}_{\mathcal{O}'_P/\mathcal{O}_P} &= \prod_{\substack{P' \in \mathbb{P}_{K'} \\ P' \cap K = P}} (P' \cap \mathcal{O}_{P'})^{-d(P'/P)} \\ &= \{x \in K' \mid v_{P'}(x) \ge -d(P'/P) \text{ for all } P' \in \mathbb{P}_{K'} \text{ above } P\}.\end{aligned}$$

Theorem and Definition 6.5.8. *Let K'/K be finite separable; for $P \in \mathbb{P}_K$ let $\mathcal{O}'_P = \mathrm{cl}_{K'}(\mathcal{O}_P)$.*

1. *Let $(u_1, \ldots, u_n)$ be a K-basis of K' and $(u_1^*, \ldots, u_n^*)$ its dual. Then for almost all $P \in \mathbb{P}_K$ and all $P' \in \mathbb{P}_{K'}$ lying above P we have*

$$\mathcal{O}'_P = \sum_{i=1}^n \mathcal{O}_P u_i = \sum_{i=1}^n \mathcal{O}_P u_i^*, \quad \textit{and} \quad d(P'/P) = 0.$$

The divisor

$$\mathrm{D}_{K'/K} = \sum_{\substack{P' \in \mathbb{P}_{K'} \\ P' \cap K = P}} d(P'/P)P' \in \mathbb{D}_{K'}$$

is called the **different** and the divisor

$$\mathrm{d}_{K'/K} = \mathcal{N}_{K'/K}(\mathrm{D}_{K'/K}) \in \mathbb{D}_K$$

is called the **discriminant** of K'/K.

2. If $P \in \mathbb{P}_K$ and $P' \in \mathbb{P}_{K'}$ lies above P, then

$$v_{P'}(\mathrm{D}_{K'/K}) = v_{P' \cap \mathcal{O}'_P}(\mathfrak{D}_{\mathcal{O}'_P/\mathcal{O}_P}) = d(P'/P)$$

and

$$v_P(\mathrm{d}_{K'/K}) = v_P(\mathfrak{d}_{\mathcal{O}'_P/\mathcal{O}_P}) = \sum_{\substack{P' \in \mathbb{P}_{K'} \\ P' \cap K = P}} f(P'/P)d(P'/P).$$

In particular, $v_P(\mathrm{d}_{K'/K}) = 0$ if and only if $d(P'/P) = 0$ for all $P' \in \mathbb{P}_{K'}$ lying above P.

3. If $P \in \mathbb{P}_K$ and $(u_1, \ldots, u_n)$ is an $\mathcal{O}_P$-basis of $\mathcal{O}'_P$, then

$$v_P(\mathrm{d}_{K'/K}) = v_P(d_{K'/K}(u_1, \ldots, u_n)).$$

Proof. 1. Let $\mathcal{S}$ be the (finite) subset of $\mathbb{P}_K$ consisting of all poles of the coefficients of the minimal polynomials of $u_1, \ldots, u_n, u_1^*, \ldots, u_n^*$.

Let $P \in \mathbb{P}_K \setminus \mathcal{S}$, $M_P = \mathcal{O}_P u_1 + \ldots + \mathcal{O}_P u_n$ and $M_P^* = \mathcal{O}_P u_1^* + \ldots + \mathcal{O}_P u_n^*$. Then $M_P \subset \mathcal{O}'_P$, $M_P^* \subset M'_P$, and by 5.9.2.1 we obtain

$$\mathfrak{C}_{M_P/\mathcal{O}_P} = M_P^* \subset \mathcal{O}'_P \subset \mathfrak{C}_{\mathcal{O}'_P/\mathcal{O}_P} \subset \mathfrak{C}_{M_P/\mathcal{O}_P}.$$

Hence $\mathfrak{C}_{\mathcal{O}'_P/\mathcal{O}_P} = \mathcal{O}'_P$, $\mathfrak{D}_{\mathcal{O}'_P/\mathcal{O}_P} = \mathcal{O}'_P$ and $d(P'/P) = 0$ for all $P' \in \mathbb{P}_{K'}$ above P.

2. Let $P \in \mathbb{P}_K$ and $t_P \in K'$ such that $\mathfrak{D}_{\mathcal{O}'_P/\mathcal{O}_P} = t_P\mathcal{O}'_P$. If $P' \in \mathbb{P}_{K'}$ lies above P, then $v_{P'}(\mathrm{D}_{K'/K}) = d(P'/P) = v_{P'}(t_P) = v_{P' \cap \mathcal{O}'_P}(\mathfrak{D}_{\mathcal{O}'_P/\mathcal{O}_P})$ by definition. Using 6.5.6 and 2.14.4.1 we obtain

$$v_P(\mathfrak{d}_{\mathcal{O}'_P/\mathcal{O}_P}) = v_P(\mathcal{N}_{\mathcal{O}'_P/\mathcal{O}_P}(t\mathcal{O}'_P)) = v_P(\mathsf{N}_{K'/K}(t)) = \sum_{\substack{P' \in \mathbb{P}_{K'} \\ P' \cap K = P}} f(P'/P)v_{P'}(t)$$

$$= \sum_{\substack{P' \in \mathbb{P}_{K'} \\ P' \cap K = P}} f(P'/P)d(P'/P) = v_P\big(\mathcal{N}_{K'/K}(\mathrm{D}_{K'/K})\big) = v_P(\mathrm{d}_{K'/K}).$$

3. By 5.9.2.3 it follows that $\mathfrak{d}_{\mathcal{O}'_P/\mathcal{O}_P} = d_{K'/K}(u_1, \ldots, u_n)\mathcal{O}_P$, and by 2. we obtain $v_P(\mathrm{d}_{K'/K}) = v_P(\mathfrak{d}_{\mathcal{O}'_P/\mathcal{O}_P}) = v_P(d_{K'/K}(u_1, \ldots, u_n)$. □

Theorem 6.5.9. *Le K'/K be finite separable.*

1. (Different and discriminant tower theorem) *Let K^* be an intermediate field of K'/K. Then*

$$\mathrm{D}_{K'/K} = j_{K'/K^*}(\mathrm{D}_{K^*/K}) + \mathrm{D}_{K'/K^*}$$

and

$$\mathrm{d}_{K'/K} = [K' \colon K^*]\mathrm{d}_{K^*/K} + \mathcal{N}_{K^*/K}(\mathrm{d}_{K'/K^*}).$$

2. (Dedekind's different theorem) *If $P \in \mathbb{P}_K$, $P' \in \mathbb{P}_{K'}$ and $P' \cap K = P$, then $d(P'/P) \ge e(P'/P) - 1$, and equality holds if and only if P'/P is tamely ramified. Moreover, P'/P is unramified if and only if $d(P'/P) = 0$.*

3. *Let $P \in \mathbb{P}_K$, $K' = K(a)$, and let $g \in \mathcal{O}_P[X]$ be the minimal polynomial of a over K. If $P' \in \mathbb{P}_{K'}$ lies above P, then we have $d(P'/P) \le v_{P'}(g'(a))$, and equality holds for all $P' \in \mathbb{P}_{K'}$ above P if and only if $\mathcal{O}'_P = \mathcal{O}_P[a]$.*

4. *A prime divisor $P \in \mathbb{P}_K$ is unramified in K' if and only if $v_P(\mathrm{d}_{K'/K}) = 0$.*

Proof. By means of 6.5.8 the assertions follow from the corresponding ones of Section 5.9. In particular, 1. holds by 5.9.3.2, 2. by 5.9.7.1, 3. by 5.9.4, and 4. follows from 2. using 6.5.8.2. □

Theorem 6.5.10. *Let K'/K be finite separable, and for $P \in \mathbb{P}_K$ and $P' \in \mathbb{P}_{K'}$ above P let $K'_{P'}/K_P$ be the associated extension of the associated completions. Let $\mathfrak{D}_{K'_{P'}/K_P}$ be the different and $\mathfrak{d}_{K'_{P'}/K_P}$ the discriminant of $K'_{P'}/K_P$* (see 5.9.6). *Then*

$$v_{P'}(\mathrm{D}_{K'/K}) = v_{P'}(\mathfrak{D}_{K'_{P'}/K_P}) \quad \textit{and} \quad v_P(\mathrm{d}_{K'/K}) = \sum_{\substack{P' \in \mathbb{P}_{K'} \\ P' \cap K = P}} v_P(\mathfrak{d}_{K'_{P'}/K_P}).$$

Proof. Let $\widehat{\mathcal{O}}_P$ be the valuation domain of K_P, $\widehat{\mathcal{O}}_{P'}$ the valuation domain of $K'_{P'}$ and $\mathcal{O}'_P = \mathrm{cl}_{K'}\mathcal{O}_P$. Then $\widehat{\mathcal{O}}_{P'} \cap K = \mathcal{O}_{P'} = (\mathcal{O}'_P)_{P' \cap \mathcal{O}'_P}$ by 6.2.5.3 and therefore $\widehat{\mathcal{O}}_{P'} = \overline{(\mathcal{O}'_P)_{P' \cap \mathcal{O}'_P}} \subset K'_{P'}$. Now we apply 6.5.8.2, 5.9.5 and 5.9.6. It follows that

$$v_{P'}(\mathrm{D}_{K'/K}) = v_{P'}(\mathfrak{D}_{\mathcal{O}'_P/\mathcal{O}_P}) = v_{P'}(\mathfrak{D}_{\widehat{\mathcal{O}}_{P'}/\widehat{\mathcal{O}}_P}) = v_{P'}(\mathfrak{D}_{K'_{P'}/K_P})$$

and

$$v_P(\mathrm{d}_{K'/K}) = v_P(\mathfrak{d}_{\mathcal{O}'_P/\mathcal{O}_P}) = \sum_{\substack{P' \in \mathbb{P}_{K'} \\ P' \cap K = P}} v_P(\mathfrak{d}_{\widehat{\mathcal{O}}_{P'}/\widehat{\mathcal{O}}_P}) = \sum_{\substack{P' \in \mathbb{P}_{K'} \\ P' \cap K = P}} v_P(\mathfrak{d}_{K'_{P'}/K_P}).$$

□

6.6 Algebraic function field extensions 2

We start with the investigation of Weil differentials in finite separable function field extensions. We need two preparations: The concept of relative repartitions (see 6.6.1) and a tool from linear algebra (see 6.6.2).

Remark and Definition 6.6.1. Let K'/F' be an algebraic function field extension of K/F such that $[K':K] < \infty$. We denote by $\mathbb{A}_{K'/K}$ the set of all repartitions $\alpha = (\alpha_{P'})_{P' \in \mathbb{P}_{K'}} \in \mathbb{A}_{K'}$ such that $\alpha_{P'} = \alpha_{Q'}$ for all P', $Q' \in \mathbb{P}_{K'}$ satisfying $P' \cap K = Q' \cap K$. The elements of $\mathbb{A}_{K'/K}$ are called **relative repartitions** of K'/K. Note that $K' \subset \mathbb{A}_{K'/K}$.
For $C' \in \mathbb{D}_{K'}$ we define $\mathbb{A}_{K'/K}(C') = \mathbb{A}_{K'/K} \cap \mathbb{A}_{K'}(C')$, and we assert that

$$\mathbb{A}_{K'} = \mathbb{A}_{K'/K} + \mathbb{A}_{K'}(C') \qquad (*)$$

Proof of $(*)$: Let $\alpha = (\alpha_{P'})_{P' \in \mathbb{P}_{K'}} \in \mathbb{A}_{K'}$. For $P \in \mathbb{P}_K$, by the weak approximation theorem 5.1.6.2(c) there exists some $x_P \in K'$ such that $v_{P'}(\alpha_{P'} - x_P) \geq -v_{P'}(C')$ for all $P' \in \mathbb{P}_{K'}$ lying above P, and we define $\beta = (\beta_{P'})_{P' \in \mathbb{P}_{K'}} \in \mathbb{A}_{K'/K}$ by $\beta_{P'} = x_{P' \cap K}$ for all $P' \in \mathbb{P}_{K'}$. It follows that $\alpha - \beta \in \mathbb{A}_{K'}(C')$ and $\alpha = \beta + (\alpha - \beta)$. $\square[(*)]$

For $\alpha = (\alpha_{P'})_{P' \in \mathbb{P}_{K'}} \in \mathbb{A}_{K'/K}$ we set $\alpha_P = \alpha_{P'}$ for all $P \in \mathbb{P}_K$ and $P' \in \mathbb{P}_{K'}$ above P. For almost all $P \in \mathbb{P}_K$ we have

$$\alpha_P \in \bigcap_{\substack{P' \in \mathbb{P}_{K'} \\ P' \cap K = P}} \mathcal{O}_{P'} = \mathcal{O}_{P'} = \mathrm{cl}_{K'}(\mathcal{O}_P)$$

and $\mathsf{Tr}_{K'/K}(\alpha_P) \in \mathcal{O}_P$ by 2.5.1.3. Hence $(\mathsf{Tr}_{K'/K}(\alpha_P))_{P \in \mathbb{P}_K} \in \mathbb{A}_K$, and we define

$$\mathsf{Tr}_{K'/K} \colon \mathbb{A}_{K'/K} \to \mathbb{A}_K \quad \text{by} \quad \mathsf{Tr}_{K'/K}(\alpha)_P = \mathsf{Tr}_{K'/K}(\alpha_{P'})$$

for all $\alpha = (\alpha_{P'})_{P' \in \mathbb{P}_{K'}} \in \mathbb{A}_{K'/K}$, $P \in \mathbb{P}_K$ and $P' \in \mathbb{P}_{K'}$ above P.

Lemma 6.6.2 (Linear algebra). *Let L/M be a finite separable field extension, V a vector space over L and $\mu \in \mathrm{Hom}_M(V, M)$. Then there exists a unique homomorphism $\mu' \in \mathrm{Hom}_L(V, L)$ such that $\mu = \mathsf{Tr}_{L/M} \circ \mu'$.*

Proof. Let $[L:M] = n$, and observe that $\mathrm{Hom}_M(L, M)$ is a vector space over L by means of $(z\varphi)(a) = \varphi(za)$ for all $z \in L$, $\varphi \in \mathrm{Hom}_M(L, M)$ and $a \in L$. This L-structure on $\mathrm{Hom}_M(L, M)$ induces the usual M-structure on the M-dual space $\mathrm{Hom}_M(L, M)$. Hence $n = \dim_M \mathrm{Hom}_M(L, M) = n \dim_L \mathrm{Hom}_M(L, M)$ and consequently $\dim_L \mathrm{Hom}_M(L, M) = 1$. We proceed in 3 steps.

I. The map $\theta\colon L \to \mathrm{Hom}_M(L, M)$, defined by $\theta(z)(x) = \mathsf{Tr}_{L/M}(zx)$ for all $z,\ x \in L$, is L-linear and non-zero, hence an L-isomorphism, and it induces an L-isomorphism $\theta^*\colon \mathrm{Hom}_L(V, L) \overset{\sim}{\to} \mathrm{Hom}_L(V, \mathrm{Hom}_M(L, M))$, which is given by $\theta^*(\varphi) = \theta{\circ}\varphi$.

II. The maps

$$\Psi\colon \mathrm{Hom}_L(V, \mathrm{Hom}_M(L, M)) \to \mathrm{Hom}_M(V, M),$$

defined by $\Psi(\psi)(v) = \psi(v)(1)$, and

$$\Phi\colon \mathrm{Hom}_M(V, M) \to \mathrm{Hom}_L(V, \mathrm{Hom}_M(L, M)),$$

defined by $\Phi(\varphi)(v)(z) = \varphi(zv)$, are mutually invese L-isomorphisms, and $\Psi{\circ}\theta^*\colon \mathrm{Hom}_L(V, L) \to \mathrm{Hom}_M(V, M)$ is an L-isomorphism.

III. Let $\mu \in \mathrm{Hom}_M(V, M)$. Then there exists a unique $\mu' \in \mathrm{Hom}_L(V, L)$ such that $\mu = \Psi{\circ}\theta^*(\mu')$. For $v \in V$ we obtain

$$\mu(v) = \theta^*(\mu')(v)(1) = \theta{\circ}\mu'(v)(1) = \mathsf{Tr}_{L/M}(\mu'(v))$$

and therefore $\mu = \mathsf{Tr}_{L/M}{\circ}\mu'$. □

Definition and Theorem 6.6.3 (Embedding of Weil differentials). *Let K'/K be an finite separable extension of algebraic function fields, F the field of constants of K and F' the field of constants of K' (then F'/F is finite separable by* 6.1.9.1(b)). *For every Weil differential $\sigma \in \Sigma_K$ there exists a unique Weil differential $\sigma' = \iota_{K'/K}(\sigma) \in \Sigma_{K'}$ such that*

$$\mathsf{Tr}_{F'/F}(\sigma'(\alpha)) = \sigma(\mathsf{Tr}_{K'/K}(\alpha)) \ \ \textit{for all} \ \ \alpha \in \mathbb{A}_{K'/K}.$$

The map $\iota_{K'/K}\colon \Sigma_K \to \Sigma_{K'}$ is a K-monomorphism.

It is called the **embedding of Weil differentials** *and has the following properties:*

(a) If $\sigma \in \Sigma_K$, $P \in \mathbb{P}_K$ and $y \in K'$, then

$$\sigma_P(\mathsf{Tr}_{K'/K}(y)) = \mathsf{Tr}_{F'/F}\Big(\sum_{\substack{P' \in \mathbb{P}_{K'} \\ P' \cap K = P}} \iota_{K'/K}(\sigma)_{P'}(y)\Big).$$

(b) If $\sigma \in \Sigma_K^\bullet$ and $W = (\sigma)$, then $(\iota_{K'/K}(\sigma)) = j_{K'/K}(W) + \mathrm{D}_{K'/K}$.

If K^* is an intermediate field of K'/K, then $\iota_{K'/K} = \iota_{K'/K^*}{\circ}\iota_{K^*/K}$.

Proof. For $P \in \mathbb{P}_K$ we fix a prime element $t_P \in \mathcal{O}_P$. Set $\mathcal{O}'_P = \mathrm{cl}_{K'}(\mathcal{O}_P)$, and let $\sigma \in \Sigma_K$. We proceed in 5 steps.

I. *Uniqueness.* Let $\sigma',\ \sigma'' \in \Sigma_{K'}$, $\mathsf{Tr}_{F'/F}(\sigma'(\alpha)) = \mathsf{Tr}_{F'/F}(\sigma''(\alpha))$ for all $\alpha \in \mathbb{A}_{K'/K}$, and set $\sigma^* = \sigma' - \sigma'' \in \Sigma_{K'}$. Then $\mathsf{Tr}_{F'/F}(\sigma^*(\alpha)) = 0$ for all $\alpha \in \mathbb{A}_{K'/K}$, and we assume that $\sigma^* \neq 0$. Then $\sigma^*(\mathbb{A}_{K'}) = F'$, and as

F'/F is separable, there exists some $\beta \in \mathbb{A}_{K'}$ such that $\mathsf{Tr}_{F'/F}(\sigma^*(\beta)) \neq 0$. Let $C' \in \mathbb{D}_{K'}$ be such that $\sigma^* \restriction \mathbb{A}_{K'}(C') = 0$. Then $\beta = \alpha + \beta'$ for some $\alpha \in \mathbb{A}_{K'/K}$ and $\beta' \in \mathbb{A}_{K'}(C')$ by 6.6.1. Hence it follows that $\sigma^*(\beta) = \sigma^*(\alpha)$, and consequently $\mathsf{Tr}_{F'/F}(\sigma^*(\beta)) = \mathsf{Tr}_{F'/F}(\sigma^*(\alpha)) = 0$, a contradiction. Hence $\sigma' = \sigma''$. $\square$[**I.**]

II. *Proof of* (a). Let $\sigma' \in \Sigma_{K'}$ and $\mathsf{Tr}_{F'/F}(\sigma'(\alpha)) = \sigma(\mathsf{Tr}_{K'/K}(\alpha))$ for all $\alpha \in \mathbb{A}_{K'/K}$. We assert that

$$\sigma_P(\mathsf{Tr}_{K'/K}(y)) = \mathsf{Tr}_{F'/F}\Big(\sum_{\substack{P' \in \mathbb{P}_{K'} \\ P' \cap K = P}} \sigma'_{P'}(y)\Big) \quad \text{for all } \ P \in \mathbb{P}_K \ \text{ and } \ y \in K'.$$

Let $P \in \mathbb{P}_K$ and $y \in K'$. Then

$$\sum_{\substack{P' \in \mathbb{P}_{K'} \\ P' \cap K = P}} \iota_{P'}(y) \in \mathbb{A}_{K'/K}, \ \ \iota_P(\mathsf{Tr}_{K'/K}(y)) = \mathsf{Tr}_{K'/K}\Big(\sum_{\substack{P' \in \mathbb{P}_{K'} \\ P' \cap K = P}} \iota_{P'}(y)\Big),$$

and therefore

$$\begin{aligned}\sigma_P(\mathsf{Tr}_{K'/K}(y)) &= \sigma \circ \iota_P(\mathsf{Tr}_{K'/K}(y)) = \sigma \circ \mathsf{Tr}_{K'/K}\Big(\sum_{\substack{P' \in \mathbb{P}_{K'} \\ P' \cap K = P}} \iota_{P'}(y)\Big) \\ &= \mathsf{Tr}_{F'/F} \circ \sigma'\Big(\sum_{\substack{P' \in \mathbb{P}_{K'} \\ P' \cap K = P}} \iota_{P'}(y)\Big) = \mathsf{Tr}_{F'/F}\Big(\sum_{\substack{P' \in \mathbb{P}_{K'} \\ P' \cap K = P}} \sigma' \circ \iota_{P'}(y)\Big) \\ &= \mathsf{Tr}_{F'/F}\Big(\sum_{\substack{P' \in \mathbb{P}_{K'} \\ P' \cap K = P}} \sigma'_{P'}(y)\Big).\end{aligned}$$

$\square$[**II.**]

III. (Main step) Let $\sigma \neq 0$, $W = (\sigma)$, $W' = j_{K'/K}(W) + \mathrm{D}_{K'/K}$, and define

$$\sigma_1 = \sigma \circ \mathsf{Tr}_{K'/K} \colon \mathbb{A}_{K'/K} \to F.$$

a. σ_1 is F-linear, $\sigma_1 \restriction K' = 0$, and if $B' \in \mathbb{D}_{K'}$, then

$$\sigma_1 \restriction \mathbb{A}_{K'/K}(B') = 0 \quad \text{if and only if} \quad B' \le W'.$$

b. If $\alpha \in \mathbb{A}_{K'}$ and $\alpha = \beta + \gamma$, where $\beta \in \mathbb{A}_{K'/K}$ and $\gamma \in \mathbb{A}_{K'}(W')$, then $\sigma_1(\beta)$ only depends on α, and we set $\sigma_2(\alpha) = \sigma_1(\beta)$. The map $\sigma_2 \colon \mathbb{A}_{K'} \to F$ defined in this way is F-linear, $\sigma_2 \restriction K' = 0$, and if $B' \in \mathbb{D}_{K'}$, then $\sigma_2 \restriction \mathbb{A}_{K'/K}(B') = 0$ if and only if $B' \le W'$. In particular, $\sigma_2 \neq 0$.

Proof of **III. a.** Apparently σ_1 is F-linear. If $y \in K'$, then $\mathsf{Tr}_{K'/K}(y) \in K$ and therefore $\sigma_1(y) = \sigma(\mathsf{Tr}_{K'/K}(y)) = 0$. Now let $B' \in \mathbb{D}_{K'}$. We shall prove:

$$\textbf{1)}\ B' \not\leq W' \implies \sigma_1 \restriction \mathbb{A}_{K'/K}(B') \neq 0;$$
$$\textbf{2)}\ B' \leq W' \implies \sigma_1 \restriction \mathbb{A}_{K'/K}(B') = 0.$$

1) Let $B' \not\leq W'$, let $P^* \in \mathbb{P}_{K'}$ such that $v_{P^*}(B') > v_{P^*}(W')$, and set $P = P^* \cap K$. Then

$$v_{P^*}(B') > v_{P^*}(j_{K'/K}(W) + \mathrm{D}_{K'/K}) = v_{P^*}(j_{K'/K}(W)) + d(P^*/P),$$

and therefore $v_{P^*}(j_{K'/K}(W) - B') < -d(P^*/P)$.

By the weak approximation theorem 5.1.6.2(c) there exists some $u \in K'$ such that $v_{P'}(u) = v_{P'}(j_{K'/K}(W) - B')$ for all $P' \in \mathbb{P}_{K'}$ lying above P. In particular, $v_{P^*}(u) < -d(P^*/P)$ and thus $u \notin \mathfrak{C}_{\mathcal{O}'_P/\mathcal{O}_P}$ by 6.5.7. Hence $\mathsf{Tr}_{K'/K}(u\mathcal{O}'_P) \not\subset \mathcal{O}_P$, and since $\mathsf{Tr}_{K'/K}(u\mathcal{O}'_P) \in \mathfrak{I}(\mathcal{O}_P)$, we obtain $t_P^{-1}\mathcal{O}_P \subset \mathsf{Tr}_{K'/K}(u\mathcal{O}'_P)$.

By 6.4.4.3 there exists some $x \in K$ such that $v_P(x) = -v_P(W) - 1$ and $\sigma_P(x) \neq 0$. Let $y \in K$ be such that $v_P(y) = v_P(W)$. It follows that $v_P(xy) = -1$ and consequently $xy \in t^{-1}\mathcal{O}_P \subset \mathsf{Tr}_{K'/K}(u\mathcal{O}'_P)$. Let $z \in u\mathcal{O}'_P$ be such that $\mathsf{Tr}_{K'/K}(z) = xy$, and define $\beta = (\beta_{P'})_{P' \in \mathbb{P}_{K'}} \in \mathbb{A}_{K'/K}$ by $\beta_{P'} = y^{-1}z$ if $P' \cap K = P$ and $\beta_{P'} = 0$ if $P' \cap K \neq P$. Then we get $\mathsf{Tr}_{K'/K}(\beta) = \iota_P(\mathsf{Tr}_{K'/K}(y^{-1}z)) = \iota_P(y^{-1}\mathsf{Tr}_{K'/K}(z)) = \iota_P(x)$. If $P' \in \mathbb{P}_{K'}$ lies above P, then $v_{P'}(y) = e(P'/P)v_P(y) = e(P'/P)v_P(W) = v_{P'}(j_{K'/K}(W))$, and therefore

$$\begin{aligned} v_{P'}(\beta) &= -v_{P'}(y) + v_{P'}(z) \geq -v_{P'}(j_{K'/K}(W)) + v_{P'}(u) \\ &= -v_{P'}(j_{K'/K}(W)) + v_{P'}(j_{K'/K}(W) - B') = -v_{P'}(B'). \end{aligned}$$

Hence $\beta \in \mathbb{A}_{K'/K}(B')$, and $\sigma_1(\beta) = \sigma(\mathsf{Tr}_{K'/K}(\beta)) = \sigma(\iota_P(x)) = \sigma_P(x) \neq 0$.

2) Let $B' \leq W'$. Then $\mathbb{A}_{K'/K}(B') \subset \mathbb{A}_{K'/K}(W')$, and therefore it suffices to prove that $\sigma_1 \restriction \mathbb{A}_{K'/K}(W') = 0$. Let $\alpha = (\alpha_{P'})_{P' \in \mathbb{P}_{K'}} \in \mathbb{A}_{K'/K}(W')$. We shall prove that $v_P(\mathsf{Tr}_{K'/K}(\alpha)) \geq -v_P(W)$ for all $P \in \mathbb{P}_K$. Then it follows that $\mathsf{Tr}_{K'/K}(\alpha) \in \mathbb{A}_K(W)$ and therefore $\sigma_1(\alpha) = \sigma \circ \mathsf{Tr}_{K'/K}(\alpha) = 0$.

Thus let $P \in \mathbb{P}_K$, $x \in K$ such that $v_P(x) = v_P(W)$. For all $P' \in \mathbb{P}_{K'}$ above P we set $\alpha_P = \alpha_{P'}$ and obtain

$$\begin{aligned} v_{P'}(x\alpha_P) &= v_{P'}(x\alpha_{P'}) \geq v_{P'}(x) - v_{P'}(W') \\ &= e(P'/P)v_P(x) - v_{P'}(j_{K'/K}(W)) - d(P'/P) = -d(P'/P). \end{aligned}$$

By 6.5.7 it follows that $x\alpha_P \in \mathfrak{C}_{\mathcal{O}'_P/\mathcal{O}_P}$, hence

$$0 \leq v_P(\mathsf{Tr}_{K'/K}(x\alpha_P)) = v_P(x\mathsf{Tr}_{K'/K}(\alpha)) = v_P(x) + v_P(\mathsf{Tr}_{K'/K}(\alpha)),$$

and consequently $v_P(\mathsf{Tr}_{K'/K}(\alpha)) \geq -v_P(x) = -v_P(W)$. □[**III.a.**]

Proof of **III. b.** Let $\alpha \in \mathbb{A}_{K'}$ and $\alpha = \beta + \gamma = \beta' + \gamma'$, where $\beta, \beta' \in \mathbb{A}_{K'/K}$ and $\gamma, \gamma' \in \mathbb{A}_{K'}(W')$. Then $\sigma_1(\gamma) = \sigma_1(\gamma') = 0$ by **a**, hence $\sigma_1(\beta) = \sigma_1(\beta')$

depends only on α, $\sigma_2\colon \mathbb{A}_{K'} \to F$ is well-defined, and it is easily checked that it is F-linear.

Since $K' \subset \mathbb{A}_{K'/K}$, it follows that $\sigma_2 \restriction K' = \sigma_1 \restriction K' = 0$. If $B' \in \mathbb{D}_{K'}$, then $\sigma_1 \restriction \mathbb{A}_{K'/K}(B') = 0$ if and only if $B' \le W'$ by **a**, and the same is true for σ_2. □[**III.b.**]

IV. *End of the existence proof.* Let $\sigma_2\colon \mathbb{A}_{K'} \to F$ be as in **III**. By 6.6.2 there exists a unique F'-linear map $\sigma'\colon \mathbb{A}_{K'} \to F'$ such that $\mathsf{Tr}_{F'/F}\circ\sigma' = \sigma_2$. If $\alpha \in \mathbb{A}_{K'/K}$, then

$$\sigma\circ\mathsf{Tr}_{K'/K}(\alpha) = \sigma_1(\alpha) = \sigma_2(\alpha) = \mathsf{Tr}_{F'/F}(\sigma'(\alpha)).$$

Assume that $\sigma \ne 0$. Then $\sigma_2 \ne 0$, hence $\sigma' \ne 0$, and in order to prove $(\sigma') = W'$, we must show: $\sigma' \restriction \mathbb{A}_K(W') + K' = 0$, and if $B' \in \mathbb{A}_{K'}$ and $B' \not\le W'$, then $\sigma' \restriction \mathbb{A}_K(B') \ne 0$.

If $\sigma' \restriction \mathbb{A}_K(W') + K' \ne 0$, then $\sigma'(\mathbb{A}_K(W') + K') = F'$ since σ' is F'-linear, and consequently $\sigma_2(\mathbb{A}_K(W') + K') \ne 0$, a contradiction. If $B' \in \mathbb{D}_K$ and $B' \not\le W'$, then $\sigma_2 \restriction \mathbb{A}_{K'}(B') \ne 0$ by **III.b** and thus also $\sigma' \restriction \mathbb{A}_{K'}(B') \ne 0$.

V. *Proof that $\iota_{K'/K}$ is a K-monomorphism.* Let σ, $\sigma_1 \in \Sigma_K$ and $c \in K$. For $\alpha \in \mathbb{A}_{K'/K}$ it is easily checked that

$$\mathsf{Tr}_{F'/F}(\iota_{K'/K}(\sigma+\sigma_1)(\alpha)) = \mathsf{Tr}_{F'/F}\big((\iota_{K'/K}(\sigma) + \iota_{K'/K}(\sigma_1))(\alpha)\big)$$

and

$$\mathsf{Tr}_{F'/F}(\iota_{K'/K}(c\sigma)(\alpha)) = \mathsf{Tr}_{F'/F}((c\iota_{K'/K}(\sigma))(\alpha)).$$

From **I** it follows that $\iota_{K'/K}(\sigma+\sigma_1) = \iota_{K'/K}(\sigma)+\iota_{K'/K}(\sigma_1)$ and $\iota_{K'/K}(c\sigma) = c\iota_{K'/K}(\sigma)$. Hence $\iota_{K'/K}$ is K-linear. If $\sigma \ne 0$, then we have seen in **IV** that $\iota_{K'/K}(\sigma) \ne 0$. Hence $\iota_{K'/K}$ is a K-monomorphism.

Now let K^* be an intermediate field of K'/K and $F^* = F' \cap K^*$ its field of constants. Then $\mathbb{A}_{K'/K} \subset \mathbb{A}_{K'/K^*}$, and for $\alpha \in \mathbb{A}_{K'/K}$ and $\sigma \in \Sigma_{K'}$ it is easily checked that $\mathsf{Tr}_{F'/F}(\iota_{K'/K^*}\circ\iota_{K^*/K}(\sigma)(\alpha)) = \sigma(\mathsf{Tr}_{K'/K}(\alpha))$. From **I** it follows that $\iota_{K'/K^*}\circ\iota_{K^*/K} = \iota_{K'/K}$. □

Theorem 6.6.4 (Hurwitz' genus formula). *Let K'/K be an finite separable extension of algebraic function fields, F the field of constants of K and F' the field of constants of K'. Then*

$$\begin{aligned} 2g_{K'} - 2 &= \frac{[K':K]}{[F':F]}(2g_K - 2) + \deg(\mathrm{D}_{K'/K}) \\ &\ge \frac{[K':K]}{[F':F]}(2g_K - 2) + \sum_{P\in\mathbb{P}_K} \sum_{\substack{P'\in\mathbb{P}_{K'} \\ P'\cap K=P}} [e(P'/P) - 1]\deg(P'), \end{aligned}$$

and $g_{K'} \ge g_K$.

Proof. Let $\sigma \in \Sigma_K^\bullet$, $W = (\sigma)$ and $W' = (\iota_{K'/K}(\sigma)) = j_{K'/K}(W) + \mathrm{D}_{K'/K}$. Using the Riemann-Roch theorem 6.4.5.2, 6.6.3(b) and 6.5.3.2(b), we obtain

$$\begin{aligned} 2g_{K'} - 2 &= \deg(W') = \deg(j_{K'/K}(W)) + \deg(\mathrm{D}_{K'/K}) \\ &= \frac{[K':K]}{[F':F]} \deg(W) + \deg(\mathrm{D}_{K'/K}) = \frac{[K':K]}{[F':F]} (2g_K - 2) + \deg(\mathrm{D}_{K'/K}). \end{aligned}$$

The inequality follows from 6.5.9.2. By 6.1.9.3 we obtain $[K':K] \geq [F':F]$ and therefore $g_{K'} \geq g_K$ (recall that F'/F is finite separable by 6.1.9.1(b)). □

The following Corollary 6.6.5 is the counterpart to Hermite's theorem for number fields (see 3.5.6 and 5.9.8.2).

Corollary 6.6.5. *Let K/F be a function field, F its field of constants, and let $x \in K$ be separating for K/F. Then*

$$g_K = 1 - [K:F(x)] + \frac{1}{2} \deg(\mathrm{D}_{K/F(x)}),$$

and if $K \neq F(x)$, then $\deg(\mathrm{D}_{K/F(x)}) \geq 2$ and $K/F(x)$ is ramified.

Proof. Since $g_{F(x)} = 0$, 6.6.4 implies $2g_K - 2 = [K:F(x)](-2) + \deg(\mathrm{D}_{K/F(x)})$, and therefore

$$g_K = 1 - [K:F(x)] + \frac{1}{2} \deg(\mathrm{D}_{K/F(x)}).$$

If $[K:F(x)] \geq 2$, then $g_K \geq 0$ implies $\deg(\mathrm{D}_{K/F(x)}) \geq 2$, and thus $K/F(x)$ is ramified by 6.5.9.4. □

Theorem 6.6.6 (Theorem of Lüroth). *Let F be a field, $F(x)$ a rational function field over F and $F \subsetneq K \subset F(x)$ an intermediate field. Then $K = F(y)$ for some $y \in K$.*

Proof. CASE 1: $F(x)/K$ is separable. Then $g_K \leq g_{F(x)}$ by 6.6.4, and therefore $g_{F(x)} = 0$ implies $g_K = 0$.

Let $P_x \in \mathbb{P}_{F(x)}$ be the prime divisor associated with x (see 6.2.3) and $P = P_x \cap K$. Then $1 = \deg(P_x) = f(P_x/P) \deg(P)$, hence $\deg(P) = 1$, and thus K/F is a rational function field by 6.4.7.1.

CASE 2: $\mathrm{char}(F) = p > 0$ and $F(x)/K$ is inseparable. Let K' be the separable closure of K in $F(x)$. Then $F(x)/K'$ is purely inseparable, and by CASE 1 it suffices to prove that K'/F is a rational function field. Suppose that $[F(x) : K'] = p^r$, where $r \in \mathbb{N}$. Then 1.4.7.2 implies $x^{p^r} \in K'$ and therefore $F(x^{p^r}) \subset K' \subset F(x)$. Since $[F(x) : F(x^{p^r})] = \deg((x^{p^r})_\infty^{F(x)}) = p^r$ by 6.3.4.1, it follows that $K' = F(x^{p^r})$ is a rational function field. □

Our next goal is to study the arithmetic of (not necessarily finite) constant field extensions (see 6.1.8).

Theorem 6.6.7. *Let K/F be a function field and F its field of constants. Let F'/F be a (not necessarily finite) separable field extension and $K' = KF'$ the constant field extension of K with F' (then K'/F' is a function field by* 6.1.8.2, *K'/K is separable by* 1.4.5.1 *and F' is the field of constants of K' by* 6.1.9.2(c)*).*

1. Let $F' = F(\alpha)$. If $P \in \mathbb{P}_K$, then $\mathcal{O}'_P = \mathrm{cl}_{K'}(\mathcal{O}_P) = \mathcal{O}_P[\alpha]$, and if $P' \in \mathbb{P}_{K'}$ lies above P, then $d(P'/P) = 0$. In particular, $\mathsf{D}_{K'/K} = 0$.

2. Let $P' \in \mathbb{P}_{K'}$ and $P = P' \cap K$. Then we have $e(P'/P) = 1$ and $\mathsf{k}_{P'} = \mathsf{k}_P F'$. In particular, if $F' \supset \mathsf{k}_P$, then $\deg(P') = 1$.

3. $g_{K'} = g_K$, and if $D \in \mathbb{D}_K$, then

$$\deg(j_{K'/K}(D)) = \deg(D) \quad \text{and} \quad \dim(j_{K'/K}(D)) = \dim(D).$$

More precisely, every F-basis of $\mathcal{L}(D)$ is an F'-basis of $\mathcal{L}(j_{K'/K}(D))$.

4. If W is a canonical divisor of K, then $j_{K'/K}(W)$ is a canonical divisor of K'.

Proof. 1. Let $g \in F[X]$ be the minimal polynomial of α over F. Then g is irreducible in $K[X]$ by 6.1.9.1, and $K' = K(\alpha)$. Now $F \subset \mathcal{O}_P$ implies $F' \subset \mathcal{O}'_P$, hence $g'(\alpha) \in F'^{\times} \subset \mathcal{O}'^{\times}_P$ and $v_{P'}(g'(\alpha)) = 0$ for all $P' \in \mathbb{P}_{K'}$ above P. By 6.5.9.3 it follows that $d(P'/P) = 0$ for all $P' \in \mathbb{P}_{K'}$ above P, hence $\mathsf{D}_{K'/K} = 0$ and $\mathcal{O}'_P = \mathcal{O}_P[\alpha]$.

2. Let $t \in \mathcal{O}_{P'}$ be a prime element and F_1 an over F finite intermediate field of F'/F such that $t \in K_1 = KF_1$. Then $F_1 = F(\alpha)$ for some $\alpha \in F_1$ by 1.5.1. If $P_1 = P' \cap K_1$, then $e(P'/P_1) = 1$, $d(P_1/P) = 0$ by 1. and $e(P_1/P) = 1$ by 6.5.9.2. Consequently, $e(P'/P) = e(P'/P_1)e(P_1/P) = 1$.

Obviously, $\mathsf{k}_P F' \subset \mathsf{k}_{P'}$. To prove the reverse inclusion, let $\xi = z(P') \in \mathsf{k}_{P'}$, where $z \in \mathcal{O}_{P'}$, let F_1 be an over F finite intermediate field of F'/F such that $z \in K_1 = KF_1$, and set $P_1 = P' \cap K_1$. By the weak approximation theorem 5.1.6.2(c) there exists some $u \in K_1$ such that $v_{P_1}(z - u) > 0$ and $v_{P^*}(u) \geq 0$ for all $P^* \in \mathbb{P}_{K_1} \setminus \{P_1\}$ lying above P. Since $z \in \mathcal{O}_{P_1}$, it follows that $u = z - (z - u) \in \mathcal{O}_{P_1}$, hence $u \in \mathcal{O}_{P^*}$ for all $P^* \in \mathbb{P}_{K_1}$ lying above P, and thus $u \in \mathrm{cl}_{K_1}\mathcal{O}_P$. If $F_1 = F(\alpha_1)$, then $\mathrm{cl}_{K_1}\mathcal{O}_P = \mathcal{O}_P[\alpha_1]$ by 1., and therefore $\xi = z(P') = z(P_1) = u(P_1) \in \mathsf{k}_P[\alpha_1] \subset \mathsf{k}_P F_1 \subset \mathsf{k}_P F'$.

3. We prove first that $\deg(j_{K'/K}(D)) = \deg(D)$ for all $D \in \mathbb{P}_K$, and for this we may assume that $D = P \in \mathbb{P}_K$.

Thus let $P \in \mathbb{P}_K$ and $x \in K^\times$ such that $\mathcal{N}^K(x) = \{P\}$. Then $(x)_0^K = rP$ for some $r \in \mathbb{N}$, and $(x)_0^{K'} = j_{K'/K}((x)_0^K) = rj_{K'/K}(P)$. By 6.1.9.2(b) and 6.3.4.1, we obtain

$$\begin{aligned} r\deg(P) &= \deg((x)_0^K) = [K:F(x)] = [K':F'(x)] \\ &= \deg((x)_0^{K'}) = r\deg(j_{K'/K}(P)), \end{aligned}$$

and consequently $\deg(j_{K'/K}(P)) = \deg(P)$.

We finish the proof with the following three assertions.

A. Let $D \in \mathbb{D}_K$ and $(u_1, \ldots, u_r)$ an F-Basis of $\mathcal{L}(D)$. Then $u_i \in \mathcal{L}(j_{K'/K}(D))$ for all $i \in [1,r]$, and $(u_1, \ldots, u_r)$ is linearly independent over F'. In particular, $\dim(D) \le \dim(j_{K'/K}(D))$.

B. $g_{K'} = g_K$.

C. $\dim(D) = \dim(j_{K'/K}(D))$.

Proof of **A.** Let $i \in [1,r]$ and $u_i \in \mathcal{L}(D)$. Then $(u_i)^K \ge -D$, and

$$(u_i)^{K'} = j_{K'/K}((u_i)^K) \ge j_{K'/K}(-D) = -j_{K'/K}(D),$$

hence $u_i \in \mathcal{L}(j_{K'/K}(D))$. By 6.1.9.2(a) it follows that $(u_1, \ldots, u_r)$ is linearly independent over F'. □[**A.**]

Proof of **B.** If $D \in \mathbb{D}_K$ and

$$\deg(D) = \deg(j_{K'/K}(D)) \ge \max\{2g_K - 1, 2g_{K'} - 1\},$$

then the Riemann-Roch theorem 6.4.5.3. and **A** yield

$$\deg(D) + 1 - g_K = \dim(D) \le \dim(j_{K'/K}(D)) = \deg(D) + 1 - g_{K'}$$

and therefore $g_{K'} \le g_K$.

To prove that $g_{K'} \ge g_K$, let $(u_1, \ldots, u_r)$ be an F'-basis of $\mathcal{L}(j_{K'/K}(D))$, and let F_1 be an over F finite intermediate field of F'/F such that $\{u_1, \ldots, u_r\} \subset K_1 = KF_1$. If $i \in [1,r]$, then

$$(u_i)^{K'} = j_{K'/K_1}((u_i)^{K_1}) \ge -j_{K'/K}(D) = j_{K'/K_1}(-j_{K_1/K}(D)),$$

hence $(u_i)^{K_1} \ge -j_{K_1/K}(D)$ by the monotonicity of j_{K'/K_1} and thus $u_i \in \mathcal{L}(j_{K_1/K}(D))$. As $(u_1, \ldots, u_r)$ is linearly independent over F_1, we get $\dim(j_{K_1/K}(D)) \ge \dim(j_{K'/K}(D))$.

Since $\mathrm{D}_{K_1/K} = 0$ by 1. and $[K_1 : K] = [F_1 : F]$ by 6.1.9.3, the Hurwitz genus formula 6.6.4 implies

$$2g_{K_1} - 2 = \frac{[K_1:K]}{[F_1:F]}(2g_K - 2) + \deg(\mathrm{D}_{K_1/K}) = 2g_K - 2,$$

and thus $g_{K_1} = g_K$.

Since $\deg(j_{K_1/K}(D)) = \deg(D) \ge 2g_K - 1 = 2g_{K_1} - 1$, it follows that

$$\begin{aligned} \deg(D) + 1 - g_K = \deg(D) + 1 - g_{K_1} &= \dim(j_{K_1/K}(D)) \ge \dim(j_{K'/K}(D)) \\ &= \deg(D) + 1 - g_{K'} \end{aligned}$$

and therefore $g_{K'} \ge g_K$ □[**B.**].

Proof of **C.** Assume first that $D \in \mathbb{D}_K$ and $\deg(D) \geq 2g_K - 1$. Since $g_K = g_{K'}$ and $\deg(D) = \deg(j_{K'/K}(D))$, it follows that

$$\dim(j_{K'/K}(D)) = \deg(j_{K'/K}(D)) + 1 - g_{K'} = \deg(D) + 1 - g_K = \dim(D).$$

Let now $D \in \mathbb{D}_K$ be arbitrary and $(u_1, \ldots, u_r)$ an F-basis of $\mathcal{L}(D)$. Then **A** implies that $(u_1, \ldots, u_r) \in \mathcal{L}(j_{K'/K}(D))^r$ is linearly independent over F', and we must prove that every $z \in \mathcal{L}(j_{K'/K}(D))$ is an F'-linear combination of $u_1, \ldots, u_r$. Let $P_1, P_2 \in \mathbb{P}_K$ and $n_1, n_2 \in \mathbb{N}_0$ be such that $P_1 \neq P_2$ and $\deg(D + n_iP_i) \geq 2g_K - 1$ for $i \in \{1, 2\}$. If $x \in K^\times$, then $(x)^K \geq -D$ if and only if $(x)^K \geq -(D + n_1P_1)$ and $(x)^K \geq -(D + n_2P_2)$, which implies $\mathcal{L}(D) = \mathcal{L}(D + n_1P_1) \cap \mathcal{L}(D + n_2D_2)$. Let $y_1, \ldots, y_m \in \mathcal{L}(D + n_1P_1)$ and $z_1, \ldots, z_n \in \mathcal{L}(D + n_2D_2)$ be such that $(u_1, \ldots, u_r, y_1, \ldots, y_m)$ is an F-basis of $\mathcal{L}(D + n_1P_1)$ and $(u_1, \ldots, u_r, z_1, \ldots, z_n)$ is an F-Basis of $\mathcal{L}(D + n_2D_2)$. We assert that

$$(u_1, \ldots, u_r, y_1, \ldots, y_m, z_1, \ldots, z_n) \text{ is linearly independent over } F. \qquad (*)$$

For the proof of $(*)$ assume that there is a relation

$$\sum_{i=1}^r a_iu_i + \sum_{j=1}^m b_jy_j + \sum_{k=1}^n c_kz_k = 0 \quad \text{with coefficients } a_i,\ b_j,\ c_k \in F. \qquad (1)$$

Then it follows that

$$\sum_{i=1}^r a_iu_i + \sum_{j=1}^m b_jy_j = -\sum_{k=1}^n c_kz_k \in \mathcal{L}(D + n_1P_1) \cap \mathcal{L}(D + n_2P_2) = \mathcal{L}(D),$$

and consequently

$$\sum_{i=1}^r a_iu_i + \sum_{j=1}^m b_jy_j = \sum_{i=1}^r a_i'u_i \quad \text{with coefficients } a_i' \in F.$$

Since $(u_1, \ldots, u_r, y_1, \ldots, y_m)$ is linearly independent over F, we get $b_j = 0$ for all $j \in [1, m]$. From (1) and the linear independence of $(u_1, \ldots, u_r, z_1, \ldots, z_n)$ over F it follows that $a_i = 0$ for all $i \in [1, r]$ and $c_k = 0$ for all $k \in [1, n]$. $\square[(*)]$

Let now $z \in \mathcal{L}(j_{K'/K}(D))$ and $i \in \{1, 2\}$. Then $z \in \mathcal{L}(j_{K'/K}(D + n_iP_i))$, and since $\deg(D + n_iP_i) \geq 2g_K - 1$, we obtain

$$\begin{aligned} \dim_F \mathcal{L}(D + n_iP_i) &= \dim(D + n_iP_i) = \dim(j_{K'/K}(D + n_iP_i)) \\ &= \dim_{F'} \mathcal{L}(j_{K'/K}(D + n_iP_i). \end{aligned}$$

As $(u_1, \ldots, u_r, y_1, \ldots, y_m)$ is an F-basis of $\mathcal{L}(D + n_1P_1)$ and is linearly independent over F', it is an F'-basis of $\mathcal{L}(j_{K'/K}(D + n_1P_1))$. Alike, $(u_1, \ldots, u_r, z_1, \ldots, z_n)$ is an F'-basis of $\mathcal{L}(j_{K'/K}(D+n_2P_2))$, and consequently

$$z = \sum_{i=1}^r a_i'u_i + \sum_{j=1}^m b_j'y_j = \sum_{i=1}^r a_i''u_i + \sum_{k=1}^n c_k'z_k$$

with coefficients $a_i',\ a_i'',\ b_j',\ c_k' \in F'$.

As $(u_1,\ldots,u_r,\, y_1,\ldots,y_m,\, z_1,\ldots,z_n)$ is linearly independent over F, it is also linearly independent over F', hence $b'_j = c'_k = 0$ for all $j \in [1,m]$ and $k \in [1,n]$, and z is an F'-linear combination of $(u_1,\ldots,u_r)$. $\square$[**C.**]

4. Let W be a canonical divisor of K. Since

$$\deg(j_{K'/K}(W)) = \deg(W) = 2g_K - 2, \quad \dim(j_{K'/K}(W)) = \dim(W) = g_K,$$

and $g_K = g_{K'}$, it follows that $j_{K'/K}(W)$ is a canonical divisor of K' by the Riemann-Roch theorem 6.4.5.2. $\square$

We close this section with some finiteness results for function fields with a finite field of constants.

Theorem 6.6.8. *Let q be a prime power, $K/\mathbb{F}_q$ a function field, $\mathbb{F}_q$ its field of constants and $g = g_K$.*

1. For $n \in \mathbb{Z}$ let $A_n = |\mathbb{D}_K^n \cap \mathbb{D}'_K|$ the number of effective divisors of degree n.

(a) If $n < 0$ or $n \notin \partial_K\mathbb{Z}$, then $A_n = 0$; $A_0 = 1$, $A_1 = |\mathbb{P}_K^1|$, and $A_n < \infty$ for all $n \in \mathbb{Z}$. In particular, for every $c \in \mathbb{R}_{\geq 1}$ the set $\{P \in \mathbb{P}_K \mid \deg(P) \leq c\}$ is finite.

(b) $h_K = |\mathcal{C}_K^0| < \infty$,

$$|\mathbf{c} \cap \mathbb{D}'_K| = \frac{q^{\dim(\mathbf{c})} - 1}{q-1} \quad \text{for all } \mathbf{c} \in \mathcal{C}_K,$$

and if $n \in \partial_K\mathbb{N}$ and $n \geq 2g-1$, then

$$A_n = \frac{h_K(q^{n+1-g} - 1)}{q-1}.$$

In particular, $A_n \ll q^n$ for all $n \geq 0$.

2. Let $\overline{K}$ be an algebraic closure of K, $r \in \mathbb{N}$, $\mathbb{F}_q \subset \mathbb{F}_{q^r} \subset \overline{K}$, and $K_r = K\mathbb{F}_{q^r}$ the constant extension of degree r.

(a) $[K_r\!:\!K] = r$, $\mathbb{F}_{q^r}$ is the field of constants of $K_r/\mathbb{F}_{q^r}$, $g_{K_r} = g$, and

$$|\mathbb{P}_{K_r}^1| = \sum_{1 \leq d \,|\, r} d\,|\mathbb{P}_K^d|.$$

(b) Let $m \in \mathbb{N}$, $P \in \mathbb{P}_K^m$ and $d = (m,r)$. Then it follows that $j_{K_r/K}(P) = P_1 + \ldots + P_d$, where $P_1,\ldots,P_d \in \mathbb{P}_{K_r}$ are distinct, and

$$\deg(P_i) = \frac{m}{d} \quad \text{for all } i \in [1,d].$$

Proof. 1.(a) By definition, $A_n = 0$ if either $n < 0$ or $\partial_K \nmid n$, $A_0 = 1$ and $A_1 = |\mathbb{P}_K^1|$.

We prove first that $|\mathbb{P}_K^n| < \infty$ for all $n \in \mathbb{N}$. Thus let $n \in \mathbb{N}$, $x \in K \setminus \mathbb{F}_q$, $m = [K:\mathbb{F}_q(x)]$, $d \in \mathbb{N}$ and $B_d(q) = |\{p \in \mathsf{P}(\mathbb{F}_q, X) \mid \partial(p) = d\}|$ the (finite) number of monic irreducible polynomials $p \in \mathbb{F}_q[X]$ of degree d (see 4.1.3). By 6.2.3 we get $|\mathbb{P}^d_{\mathbb{F}_q(x)}| = B_d(q) < \infty$ if $d \geq 2$, and $|\mathbb{P}^1_{\mathbb{F}_q(x)}| = B_1(q) + 1 = q + 1$. If $P \in \mathbb{P}_K$ and $P_0 = P \cap \mathbb{F}_q(x)$, then $\deg(P) = \deg(P_0)f(P/P_0) \leq \deg(P_0)m$ by 6.2.7.4. On the other hand, for every $P_0 \in \mathbb{P}_{\mathbb{F}_q(x)}$ at most m prime divisors of K lie above P_0, and therefore

$$|\mathbb{P}_K^n| \leq |\{P \in \mathbb{P}_K \mid \deg(P) \leq n\}| \leq n\, |\{P_0 \in \mathbb{P}_{\mathbb{F}_q(x)} \mid \deg(P_0) \leq n/m\}| < \infty.$$

If $n \in \mathbb{N}$ and $D \in \mathbb{D}_K^n \cap \mathbb{D}'_K$, then

$$D = \sum_{P \in \mathbb{P}_K} n_P P, \quad \text{where} \quad n_P \in \mathbb{N}_0 \quad \text{and} \quad \sum_{P \in \mathbb{P}_K} n_P \leq \sum_{P \in \mathbb{P}_K} n_P \deg(P) = n.$$

This implies the (very rough) estimate

$$A_n = |\mathbb{D}_K^n \cap \mathbb{D}'_K| \leq n \sum_{d=1}^{n} |\mathbb{P}_K^d| < \infty.$$

(b) Let $n \in \partial_K\mathbb{N}$, $n \geq g$, $\boldsymbol{c} \in \mathcal{C}_K^n$ and $C \in \boldsymbol{c}$. Then Riemann's inequality 6.3.8 implies $\dim(C) \geq \deg(C) + 1 - g = n + 1 - g \geq 1$, and by 6.2.10.3(b) there exists some $A \in \mathbb{D}'_K$ such that $A \sim C$ and thus $A \in \boldsymbol{c}$. Hence it follows that $h_K = |\mathcal{C}_K^0| = |\mathcal{C}_K^n| \leq A_n < \infty$

Now let $\boldsymbol{c} \in \mathcal{C}_K$ and $C \in \boldsymbol{c}$, then $\boldsymbol{c} \cap \mathbb{D}'_K = \{C + (x) \mid x \in \mathcal{L}(C)^\bullet\}$, and

$$|\mathcal{L}(C)^\bullet| = q^{\dim(C)} - 1 = q^{\dim(\boldsymbol{c})} - 1.$$

If $x, x' \in \mathcal{L}(C)^\bullet$, then $(x) = (x')$ if and only if $x' \in x\mathbb{F}_q^\times$. Consequently, we obtain

$$|\boldsymbol{c} \cap \mathbb{D}'_K| = \frac{q^{\dim(\boldsymbol{c})-1}}{q-1}.$$

If $\deg(\boldsymbol{c}) \geq 2g - 1$, then $\dim(\boldsymbol{c}) = n + 1 - g$ by the Riemann-Roch theorem 6.4.5.3. If $n \in \partial_K\mathbb{N}$ and $n \geq 2g - 1$, then

$$A_n = \sum_{\boldsymbol{c} \in \mathcal{C}_K^n} |\boldsymbol{c} \cap \mathbb{D}'_K| = \frac{h_K(q^{n+1-g} - 1)}{q-1}. \qquad \square$$

2. From 6.1.9 and 6.6.7 it follows that $[K_r : K] = r$, $\mathbb{F}_{q^r}$ is the field of constants of $K_r/\mathbb{F}_{q^r}$ and $g_{K_r} = g$. Next we prove (b).

Let $P \in \mathbb{P}_K^m$ and $P' \in \mathbb{P}_{K_r}$ above P. Then $e(P'/P) = 1$ and $\mathsf{k}_{P'} = \mathsf{k}_P\mathbb{F}_{q^r}$ by 6.6.7.2, and using 1.7.1.3 we get

$$f(P'/P) = [\mathsf{k}_{P'} : \mathsf{k}_P] = \frac{[\mathsf{k}_P\mathbb{F}_{q^r} : \mathbb{F}_q]}{[\mathsf{k}_P : \mathbb{F}_q]} = \frac{\operatorname{lcm}(m, r)}{m} = \frac{r}{d}.$$

By 6.5.3.2(a) we obtain

$$r = [K_r : K] = \sum_{\substack{P' \in \mathbb{P}_{K'} \\ P' \cap K = P}} e(P'/P) f(P'/P) = \frac{r}{d} |\{P' \in \mathbb{P}_{K_r} \mid P' \cap K = P\}|,$$

hence $j_{K'/K}(P) = P_1 + \ldots + P_d$, where $P_1, \ldots, P_d \in \mathbb{P}_{K_r}$ are distinct, and

$$\deg(P_i) = \frac{\deg(P) f(P_i/P)}{r} = \frac{m}{d} \quad \text{for all } \ i \in [1, d] \quad \text{by 6.2.7.4.}$$

Finally, we prove the formula for $\mathbb{P}^1_{K_r}$. If $P' \in \mathbb{P}^1_{K_r}$ and $d = \deg(P' \cap K)$, then

$$1 = \deg(P') = \frac{d}{(d, r)}, \quad \text{and consequently } \ (d, r) = d \,|\, r.$$

Conversely, if $1 \le d \,|\, r$ and $P \in \mathbb{P}^d_K$, then there exist precisely d prime divisors $P' \in \mathbb{P}_{K_r}$ lying above P, and they all have degree 1. Hence

$$|\mathbb{P}^1_{K_r}| = \sum_{1 \le d \,|\, r} d \, |\mathbb{P}^d_K|.$$

6.7 Derivations and differentials

In the following two sections we disclose the connection between Weil differentials and the classical notion of differentials in analysis and algebraic geometry. We start with the basic notions of derivations and differentials in commutative algebra.

Definition 6.7.1. Let $F \subset K$ be commutative rings and M a K-module. A map $D\colon K \to M$ is called a **derivation** (of K into M) if

$$D(a + b) = D(a) + D(b) \quad \text{and} \quad D(ab) = aD(b) + bD(a) \quad \text{for all } \ a, b \in K.$$

We call a derivation $D\colon K \to M$ an **F-derivation** if it is F-linear. We denote by $\mathrm{Der}_F(K, M)$ be the set of all F-derivations of K into M, and we set $\mathrm{Der}_F(K) = \mathrm{Der}_F(K, K)$.

Let K^* be a commutative overring of K and M^* a K^*-module. For F-derivations D, $D' \in \mathrm{Der}_F(K, M^*)$ and $c \in K^*$, we define $D + D'\colon K \to M^*$ and $cD\colon K \to M^*$ by

$$(D + D')(x) = D(x) + D'(x) \quad \text{and} \quad (cD)(x) = cD(x) \quad \text{for all } \ x \in K.$$

Then $D + D'$ and cD are F-derivations, and with this pointwise addition and scalar multiplication $\mathrm{Der}_F(K, M^*)$ is a K^*-module. If M is a K-submodule of M^*, then $\mathrm{Der}_F(K, M) = \{D \in \mathrm{Der}_F(K, M^*) \mid D(K) \subset M\}$ is a K-submodule of $\mathrm{Der}_F(K, M^*)$. In particular, $\mathrm{Der}_F(K)$ is a K-submodule of $\mathrm{Der}_F(K, K^*)$.

For a derivation $D\colon K \to K^*$ and a polynomial

$$f = \sum_{i=0}^{n} a_i X^i \in K[X] \;\text{ we define }\; f^D = \sum_{i=0}^{n} D(a_i) X^i \in K^*[X].$$

Theorem 6.7.2. *Let K be a commutative ring.*

1. Let $D\colon K \to M$ be a derivation of K into a K-module M and $a \in K$.

(a) $D(a^n) = na^{n-1}D(a)$ for all $n \in \mathbb{N}_0$. In particular, $D(1) = 0$.

(b) If $a \in K^\times$, then $D(a^n) = na^{n-1}D(a)$ for all $n \in \mathbb{Z}$, and

$$D\Big(\frac{b}{a}\Big) = \frac{aD(b) - bD(a)}{a^2} \quad \text{for all } b \in K.$$

(c) Let F be a subring of K. Then D is an F-derivation if and only if $D \restriction F = 0$. In particular, if F is the prime ring of K, then D is an F-derivation.

2. Let K^* be a commutative overring of K and $D\colon K \to K^*$ a derivation.

(a) If $f \in K[X]$ and $a \in K^*$, then $D(f(a)) = f^D(a) + f'(a)D(a)$.

(b) Let $D^*\colon K[X] \to K^*[X]$ be defined by $D^*(f) = f^D$ for all $f \in K[X]$. Then D^* is a derivation and $D^* \restriction K = D$.

3. Let K be a domain, $K' = \mathsf{q}(K)$, M a vector space over K' and $D\colon K \to M$ a derivation. Then D has a unique extension to a derivation $D'\colon K' \to M$.

Proof. 1. (a) Clearly, $D(1) = D(1 \cdot 1) = D(1) + D(1)$ implies $D(1) = 0$. Now we prove $D(a^n) = na^{n-1}D(a)$ by induction on n. For $n \in \{0, 1\}$ there is nothing to do.

$n \ge 1,\ n \to n+1$: Using the induction hypothesis, we obtain

$$D(a^{n+1}) = a^n D(a) + aD(a^n) = a^n D(a) + a[na^{n-1}D(a)] = (n+1)a^n D(a).$$

(b) If $n \in \mathbb{N}$, then $0 = D(1) = D(a^{-n}a^n) = a^{-n}D(a^n) + a^n D(a^{-n})$, and therefore

$$D(a^{-n}) = -a^{-2n}D(a^n) = -a^{-2n}[na^{n-1}D(a)] = -na^{-n-1}D(a).$$

If $b \in R$, then

$$D(a^{-1}b) = a^{-1}D(b) + bD(a^{-1}) = a^{-1}D(b) - ba^{-2}D(a) = a^{-2}[aD(b) - bD(a)].$$

(c) If D is an F-derivation and $c \in F$, then $D(c) = cD(1) = 0$. Conversely, if $D \restriction F = 0$, $c \in F$ and $x \in D$, then $D(cx) = cD(x) + xD(c) = cD(x)$, and thus D is F-linear.

2. (a) If $f \in K[X]$ and $a \in K^*$, then

$$\begin{aligned} D(f(a)) = D\Big(\sum_{i=0}^{n} a_i a^i\Big) &= \sum_{i=0}^{n}[a^i D(a_i) + a_i D(a^i)] \\ &= \sum_{i=0}^{n}[a^i D(a_i) + i a_i a^{i-1} D(a)] \\ &= f^D(a) + f'(a)D(a). \end{aligned}$$

(b) Obviously, $D^* \restriction K = D$, and $(f+g)^D = f^D + g^D$ for all $f, g \in K[X]$. Now let

$$f = \sum_{n \ge 0} a_n X^n \in K[X] \quad \text{and} \quad g = \sum_{n \ge 0} b_n X^n \in K[X],$$

where $a_n, b_n \in K$ for all $n \ge 0$ and $a_n = b_n = 0$ for all $n \gg 1$. Then

$$\begin{aligned} (fg)^D &= \sum_{n\ge0}\Big(\sum_{\substack{i,j\ge0\\ i+j=n}} D(a_i b_j)\Big)X^n = \sum_{n\ge0}\Big(\sum_{\substack{i,j\ge0\\ i+j=n}} [a_i D(b_j) + b_j D(a_i)]\Big)X^n \\ &= \Big(\sum_{i\ge0} a_i X^i\Big)\Big(\sum_{j\ge0} D(b_j)X^j\Big) + \Big(\sum_{j\ge0} b_j X^j\Big)\Big(\sum_{i\ge0} D(a_i)X^i\Big) \\ &= fg^D + f^D g. \end{aligned}$$

3. Uniqueness is obvious by 1.(b). As to existence, we define $D' \colon K' \to M$ by

$$D'(x) = \frac{aD(b) - bD(a)}{a^2} \quad \text{if } x = \frac{b}{a}, \quad \text{where } a \in K^\times \text{ and } b \in K.$$

It is easily checked that D' is a well defined derivation, and $D' \restriction K = D$. □

The following surprising result turns out to be crucial for the class field theory in characteristic p.

Theorem 6.7.3. *Let K be a field, $D\colon K \to K$ a derivation and $n \in \mathbb{N}$ such that $D^n = D \circ \ldots \circ D = 0$. For $a \in K$ let $D + a\colon K \to K$ be defined by*

$$(D+a)(x) = D(x) + ax.$$

1. *For $a \in K$ the following assertions are equivalent:*

(a) There exists some $b \in K^\times$ such that $a = b^{-1}D(b)$.

(b) $(D+a)^n(1) = 0$.

2. Let K' be a subfield of K, $D(K') \subset K'$, $a \in K'$ and $b \in K^\times$ such that $a = b^{-1}D(b)$. Then there exists some $b' \in K'^\times$ such that $a = b'^{-1}D(b')$.

Proof. 1. (a) $\Rightarrow$ (b) It suffices to prove that $(D+a)^i(x) = b^{-1}D^i(bx)$ for all $i \in \mathbb{N}_0$ and $x \in K$ (the assertion follows for $i = n$ and $x = 1$). For $i = 0$ there is nothing to do, since $D^0 = (D+a)^0 = \mathrm{id}_K$.

$i \geq 0$, $i \to i+1$: Suppose that $(D+a)^i(x) = b^{-1}D^i(bx)$. Then

$$\begin{aligned}(D+a)^{i+1}(x) &= (D+a)(b^{-1}D^i(bx)) = D(b^{-1}D^i(bx)) + ab^{-1}D^i(bx) \\ &= -b^{-2}D(b)D^i(bx) + b^{-1}D^{i+1}(bx) + b^{-1}D(b)b^{-1}D^i(bx) \\ &= b^{-1}D^{i+1}D(bx).\end{aligned}$$

(b) $\Rightarrow$ (a) If $a = 0$, we set $b = 1$.

If $a \neq 0$, then $(D+a)(1) = D(1) + a = a \neq 0$, and there exists some $i \in [1, n-1]$ such that $(D+a)^i(1) \neq 0$ and $(D+a)^{i+1}(1) = 0$. If $b = (D+a)^i(1)^{-1}$, then

$$0 = (D+a)^{i+1}(1) = (D+a)\circ(D+a)^i(1) = (D+a)(b^{-1}) = -b^{-2}D(b) + ab^{-1}$$

and therefore $a = b^{-1}D(b)$.

2. By 1., since 1.(b) does not depend on K. □

Theorem 6.7.4. *Let $K \subset K^*$ be fields, $D\colon K \to K^*$ a derivation, $x \in K^*$ and $f \in K[X]$ such that $\{g \in K[X] \mid g(x) = 0\} = fK[X]$.*

1. For $u \in K^*$, the following assertions are equivalent:

(a) There exists a (unique) derivation $D'\colon K(x) \to K^*$ such that $D' \restriction K = D$ and $D'(x) = u$.

(b) $f^D(x) + f'(x)u = 0$.

2. If x is transcendental over K and $u \in K^*$, then there exists a unique derivation $D'\colon K(x) \to K^*$ such that $D' \restriction K = D$ and $D'(x) = u$.

3. Let x be separable over K and $g \in K[X]$ the minimal polynomial of x over K. Then there exists a unique derivation $D'\colon K(x) \to K^*$ such that $D' \restriction K = D$; it satisfies

$$D'(x) = -\frac{g^D(x)}{g'(x)}.$$

Proof. 1. (a) $\Rightarrow$ (b) Let $D'\colon K(x) \to K^*$ be a derivation such that $D' \restriction K = D$ and $D'(x) = u$. As $f(x) = 0$, we obtain

$$0 = D'f(x) = f^D(x) + f'(x)D'(x) = f^D(x) + f'(x)u \quad \text{by 6.7.2.2(a).}$$

(b) $\Rightarrow$ (a) By 6.7.2.3 it suffices to prove that there exists a unique derivation $D_1\colon K[x] \to K^*$ such that $D_1 \restriction K = D$ and $D_1(x) = u$.

Uniqueness: Let $D_1\colon K[x] \to K^*$ be a derivation such that $D_1 \restriction K = D$ and $D_1(x) = u$. If $z \in K[x]$, then $z = g(x)$ for some $g \in K[X]$, and again 6.7.2.2(a) implies $D_1(z) = g^{D_1}(x) + g'(x)u = g^D(x) + g'(x)u$.

Existence: Let $z = g(x) = g_1(x) \in K[x]$, where $g, g_1 \in K[X]$. Then $g_1 - g = fq$ for some $q \in K[X]$,

$$\begin{aligned} g_1^D(x) + g_1'(x)u &= (g+fq)^D(x) + (g+fq)'(x)u \\ &= g^D(x) + f(x)q^D(x) + f^D(x)q(x) + g'(x)u + f'(x)q(x)u + f(x)q'(x)u \\ &= g^D(x) + g'(x)u + (f^D(x) + f'(x)u)q(x) = g^D(x) + g'(x)u, \end{aligned}$$

and we define $D_1(z) = g^D(x) + g'(x)u$ (which now only depends on z). In particular, it follows that $D_1(x) = u$ (with $g = X$), $D_1(z) = D(z)$ for all $z \in K$ (with $g = z$), and it is now easily checked that D_1 is a derivation.

2. By 1. with $f = 0$.

3. By 1. with $f = g$ and $u = -g'(x)^{-1}g^D(x)$ (note that $g'(x) \neq 0$ by 1.4.2.3). □

Theorem 6.7.5. *Let K/F be a function field and F its field of constants. Let $x \in K$ be separating for K/F and K^* an extension field of K.*

1. *For every $u \in K^*$ there exists a unique F-derivation*

$$D \in \mathrm{Der}_F(K, K^*) \text{ such that } D(x) = u,$$

and if $u \in K$, then $D \in \mathrm{Der}_F(K)$. In particular, there exists a unique F-derivation $D_x \in \mathrm{Der}_F(K)$ such that $D_x(x) = 1$.

2. *Let $D_x \in \mathrm{Der}_F(K)$ be the unique derivation satisfying $D_x(x) = 1$.*

   (a) *If $D \in \mathrm{Der}_F(K, K^*)$, then $D = D(x)D_x$. In particular, it follows that $\mathrm{Der}_F(K, K^*) = K^*D_x$ and $\dim_K \mathrm{Der}_F(K) = 1$.*

   (b) *Let F be perfect and $z \in K$. Then the following assertions are equivalent:*

      i. *$D(z) \neq 0$ for every F-derivation $D \in \mathrm{Der}_F(K, K^*)^\bullet$.*

      ii. *$D_x(z) \neq 0$.*

      iii. *z is separating for K/F.*

Proof. 1. By 6.1.7 $K = F(x)(x_1)$, where x_1 is separable over $F(x)$. By 6.7.4.2 there exists a unique F-derivation $D_1\colon F(x) \to K^*$ such that $D_1(x) = u$, and by 6.7.4.3 D_1 has a unique extension to an F-derivation $D\colon K \to K^*$. Conversely, if $D'\colon K \to K^*$ is any F-derivation such that $D'(x) = u$, then $D' \restriction F(x) = D_1$ and thus $D = D'$. Clearly, $u \in K$ implies $D(K) \subset K$.

2. (a) If $D \in \mathrm{Der}_F(K, K^*)$, then $D(x) = (D(x)D_x)(x)$ and therefore we get $D = D(x)D_x$ by the uniqueness in 1. In particular, $\mathrm{Der}_F(K, K^*) = K^*D_x$.

(b) (i) $\Rightarrow$ (ii) Obvious.

(ii) $\Rightarrow$ (iii) Assume that z is not separating for K/F. Then either $z \in F$ (and then $D_x(z) = 0$) or $\mathrm{char}(K) = p > 0$ and $z \in K^p$ by 6.1.6.2. In the latter case, $z = u^p$ for some $u \in K$, and $D_x(z) = pu^{p-1}D_x(u) = 0$.

(iii) $\Rightarrow$ (i) If $D \in \mathrm{Der}_F(K, K^*)^\bullet$, then $D = D(z)D_z$ by (a), and consequently $D(z) \neq 0$. □

Definition and Theorem 6.7.6. *Let K/F be a function field, F its field of constants and $x \in K$ a separating element for K/F.* The dual space

$$\Omega_K = \Omega_{K/F} = \mathrm{Der}_F(K)^* = \mathrm{Hom}_K(\mathrm{Der}_F(K), K)$$

is called the **global differential module** of K. Its elements are called **global differentials**. The map

$$d = d_K\colon K \to \Omega_K, \quad \text{defined by } \ dz(D) = D(z) \ \text{ for all } \ D \in \mathrm{Der}_F(K),$$

is called the **universal global derivation** of K. By definition, Ω_K is a one-dimensional vector space over K. Hence $\Omega_K = K dz$ for all $z \in K$ such that $dz \neq 0$.

1. *d is an F-derivation of K.*

2. *$\Omega_K = K dx$, and if $z \in K$, then $dz = D_x(z)dx$.* In the style of differential calculus we write this relation in the form

$$D_x(z) = \frac{dz}{dx}.$$

3. *Let F be perfect and $z \in K$. Then $dz \neq 0$ if and only if z is separating for K/F.*

4. *If $u, v, f, g \in K$, then*

$$f du = g dv \ \text{ if and only if } \ fD_x(u) = gD_x(v).$$

5. (Universal property of the global derivation)
For every F-derivation $D\colon K \to M$ into a vector space M over K there exists a unique K-linear map $\vartheta\colon \Omega_K \to M$ such that $D = \vartheta \circ d$.

Proof. 1. If $u, v \in K$ and $c \in F$, then it is easily checked that $d(u+v) = du+dv$, $d(cu) = cdu$ and $d(uv) = udv+vdu$.

2. Since $dx(D_x) = D_x(x) = 1$, it follows that $dx \neq 0$ and thus $\Omega_K = Kdx$. If $z \in K$ and $D \in \mathrm{Der}_F(K)$, then

$$[D_x(z)dx](D) = D_x(z)D(x) = D(z) = dz(D) \text{ by } 6.7.5.2(\mathrm{a}),$$

and consequently $D_x(z)dx = dz$.

3. By 2. $dz \neq 0$ if and only if $D_x(z) \neq 0$, and by 6.7.5.2(b) this holds if and only if z is separating.

4. By 2., we obtain:

$$fdu = gdv \iff fD_x(u)dx = gD_x(v)dx \iff fD_x(u) = gD_x(v).$$

5. As $\Omega_K = Kdx$, a K-linear map $\vartheta\colon \Omega_K \to M$ is uniquely determined by $\vartheta(dx)$. If $D \in \mathrm{Der}_F(K,M)$ and $\vartheta \in \mathrm{Hom}_K(\Omega_K, M)$, then $\vartheta\circ d = D$ if and only if $\vartheta(dx) = D(x)$. This implies the existence and uniqueness of ϑ. □

While the extension of a Weil differential to a finite separable extension is a rather delicious problem (see 6.6.3), the corresponding procedure for differentials is almost obvious.

Remark and Definition 6.7.7. Let K'/F' be an algebraic function field extension of K/F such that K'/K is separable. Let F resp. F' be the field of constants of K/F resp. K'/F', and let $x \in K$ be separating for K/F. Then x is separating for K'/F' by 6.1.8.3, and therefore $\Omega_K = Kd_Kx$ and $\Omega_{K'} = K'd_{K'}x$ by 6.7.6.2. The unique K-monomorphism

$$\eta_{K'/K}\colon \Omega_K \to \Omega_{K'} \text{ satisfying } \eta_{K'/K}(fd_Kx) = fd_{K'}x \text{ for all } f \in K$$

is called the **embedding of differentials**. It apparently has the following properties:

- If $(f,u) \in K \times K$, then

$$\eta_{K'/K}(fd_Ku) = \eta_{K'/K}(fD_x(u)d_Kx) = fD_x(u)d_{K'}x = fd_{K'}u.$$

- $d_{K'}\restriction K = \eta_{K'/K}\circ d_K\colon K \to \Omega_{K'}$.

- If K^* is an intermediate field of K'/K, then $\eta_{K'/K} = \eta_{K'/K^*}\circ\eta_{K^*/K}$.

We apply the abstract (and up to now rather shallow) theory to the field of formal Laurent series (see 2.10.10 and 5.3.7).

Definition and Theorem 6.7.8. *Let $K = F((X))$ be the field of formal Laurent series in X over a field F, $\mathsf{o}_X = \mathsf{o}\colon K \to \mathbb{Z} \cup \{\infty\}$ the order valuation with respect to X and $x \in K$ such that $\mathsf{o}(x) = 1$ (then $K = F((x))$ and $\mathsf{o} = \mathsf{o}_x$).*
For $u \in K$, say

$$u = \sum_{n\in\mathbb{Z}} a_n x^n, \quad \text{where } (a_n)_{n\in\mathbb{Z}} \in F^{\mathbb{Z}} \text{ and } a_{-n} = 0 \text{ for all } n \gg 1, \qquad (*)$$

we define

$$\delta_x(u) = \sum_{n\in\mathbb{Z}} n a_n x^{n-1} \in K \qquad \text{and} \qquad \mathrm{Res}_x(u) = a_{-1} \in F.$$

$\delta_x(u)$ is called that **derivative** of u with respect to x, and $\mathrm{Res}_x(u)$ is called the **residue** of u with respect to x.

1. $\delta_x\colon K \to K$ is a continuous F-derivation, $\mathrm{Res}_x\colon K \to F$ is F-linear and continuous (with respect to the discrete topology on F), and the following assertions hold for all $u \in K$:

(a) $\mathrm{Res}_x(\delta_x(u)) = 0$.

(b) If $\mathsf{o}(u) \ge 0$, then $\mathrm{Res}_x(u) = 0$.

(c) If $\mathsf{o}(u) = 1$, then $\mathsf{o}(\delta_x(u)) = 0$.

2. Let $z \in K^\times$ and $\mathsf{o}(z) \ge 1$. Then $\delta_x(u) = \delta_x(z)\delta_z(u)$ for all $u \in F((z))$, and thus $\delta_x \restriction F((z)) = \delta_x(z)\delta_z\colon F((z)) \to F((z))$. In particular, if $\mathsf{o}(z) = 1$, then $K = F((z))$, $\delta_x = \delta_x(z)\delta_z\colon K \to K$, and $\delta_z(x) = \delta_x(z)^{-1}$.

3. If $u, z \in K$ and $\mathsf{o}(z) = 1$, then $\mathrm{Res}_z(u) = \mathrm{Res}_x(u\delta_x(z))$.

4. Let $u \in K$.

(a) $\delta_x(u) = 0$ if and only if either $u \in F$, or $\mathrm{char}(F) = p > 0$ and $u \in F((x^p))$.

(b) Suppose that $\mathrm{char}(F) = p > 0$. Then $\delta_x^p = \delta_x \circ \ldots \circ \delta_x = 0$. If moreover F is perfect, then $K^p = F((x^p))$, and $\delta_x(u) = 0$ if and only if $u \in K^p$.

Proof. 1. By definition, δ_x and Res_x are F-linear, and $\mathsf{o}(\delta_x(u)) \ge \mathsf{o}(u) - 1$ for all $u \in K$. If $(u^{(k)})_{k\ge 0}$ is a sequence in K such that $(u^{(k)})_{k\ge 0} \to u \in K$, then $\mathsf{o}(\delta_x(u^{(k)}) - \delta_x(u)) = \mathsf{o}(\delta_x(u^{(k)} - u)) \ge \mathsf{o}(u^{(k)} - u) - 1$ for all $k \gg 1$, and thus $(\delta_x(u^{(k)}))_{k\ge 0} \to \delta_x(u)$. Hence δ_x is continuous. If

$$u^{(k)} = \sum_{n\in\mathbb{Z}} a_n^{(k)} x^n \quad \text{and} \quad u = \sum_{n\in\mathbb{Z}} a_n x^n,$$

then $\mathsf{o}(u^{(k)}-u) \ge -1$, hence $\mathrm{Res}_x(u^{(k)}) = a_{-1}^{(k)} = a_{-1} = \mathrm{Res}_x(u)$ for all $k \gg 1$, whence $(\mathrm{Res}_x(u^{(k)}))_{k\ge 0} \to \mathrm{Res}_x(u)$. Hence $\mathrm{Res}_x\colon K \to F$ is continuous.

Next we prove that δ_x is a derivation. Let $y,\, z \in K$, say

$$y = \sum_{n\in\mathbb{Z}} b_n x^n \quad \text{and} \quad z = \sum_{n\in\mathbb{Z}} c_n x^n,$$

where $(b_n)_{n\in\mathbb{Z}},\, (c_n)_{n\in\mathbb{Z}} \in F^{\mathbb{Z}}$ and $b_{-n} = c_{-n} = 0$ for all $n \gg 1$. For $m \in \mathbb{Z}$ we obtain

$$\begin{aligned}\delta_x(x^m z) &= \sum_{n\in\mathbb{Z}} (n+m)c_n x^{n+m-1} = x^m \sum_{n\in\mathbb{Z}} nc_n x^{n-1} + mx^{m-1} \sum_{n\in\mathbb{Z}} c_n x^n \\ &= x^m \delta_x(z) + \delta_x(x^m) z,\end{aligned}$$

and as δ_x is F-linear and continuous, it follows that

$$\delta_x(yz) = \sum_{n\in\mathbb{Z}} \delta_x(b_n x^n z) = \sum_{n\in\mathbb{Z}} [b_n x^n \delta_x(z) + nb_n x^{n-1} z] = y\delta_x(z) + \delta_x(y) z.$$

(a), (b) and (c) are obvious.

2. Let $u \in K\,(\!(z)\!)$, say

$$u = \sum_{n\in\mathbb{Z}} b_n z^n, \quad \text{where } (b_n)_{n\in\mathbb{Z}} \in F^{\mathbb{Z}} \text{ such that } b_{-n} = 0 \text{ for all } n \gg 1.$$

Since δ_x is continuous and F-linear, it follows that

$$\delta_x(u) = \sum_{n\in\mathbb{Z}} nb_n z^{n-1} \delta_x(z) = \delta_z(u)\delta_x(z).$$

If $\mathsf{o}(z) = 1$, then $K = K\,(\!(z)\!)$, $\delta_x = \delta_x(z)\delta_z\colon K \to K$, and we may interchange x and z. Then $\delta_z = \delta_z(x)\delta_x$, and $1 = \delta_z(z) = \delta_x(z)\delta_z(x)$.

3. Suppose that

$$u = \sum_{d=1}^{N} a_{-d} z^{-d} + u_0 \ \text{ along with } \ u_0 = \sum_{n\ge 0} a_n z^n, \quad \text{and} \quad z = x \sum_{n=0}^{\infty} c_n x^n,$$

where $N \in \mathbb{N}_0$, and $(a_n)_{n\ge -N}$ and $(c_n)_{n\in\mathbb{Z}}$ are sequences in F such that $c_0 \ne 0$.

We show first that it suffices to prove the formula if $c_0 = 1$. Indeed, suppose that

$$z_1 = c_0^{-1} z = x\Big(1 + \sum_{n=1}^{\infty} c_0^{-1} c_n z^n\Big) \quad \text{and} \quad \mathrm{Res}_{z_1}(u) = \mathrm{Res}_x(u\delta_x(z_1)).$$

Then

$$u = \sum_{n\ge -N} a_n z^n = \sum_{n\ge -N} a_n c_0^n z_1^n \quad \text{and therefore } \mathrm{Res}_{z_1}(u) = c_0^{-1} a_{-1}.$$

Hence it follows that

$$\begin{aligned}\mathrm{Res}_z(u) &= a_{-1} = c_0\mathrm{Res}_{z_1}(u) = c_0\mathrm{Res}_x(u\delta_x(z_1)) \\ &= \mathrm{Res}_x(u\delta_x(cz_1)) = \mathrm{Res}_x(u\delta_x(z)).\end{aligned}$$

Assume now that $c_0 = 1$. By 1.(c) we obtain $\delta_x(z) \in F[\![x]\!]^\times$, and therefore $\mathsf{o}(u_0\delta_x(z)) = \mathsf{o}(u_0) \geq 0$. Hence $\mathrm{Res}_x(u_0\delta_x(z)) = 0$ and

$$\mathrm{Res}_x(u\delta_x(z)) = \sum_{d=1}^{N} a_{-d}\mathrm{Res}_x(z^{-d}\delta_x(z)).$$

Since $\mathrm{Res}_z(u) = a_{-1}$, it suffices to prove that for all $d \in \mathbb{N}$

$$\mathrm{Res}_x(z^{-d}\delta_x(z)) = \begin{cases} 1 & \text{if } d = 1, \\ 0 & \text{if } d \geq 2. \end{cases}$$

For $d \in \mathbb{N}$ we obtain

$$\begin{aligned} z^{-d}\delta_x(z) &= x^{-d}\Big(1 + \sum_{n=1}^{\infty} c_n x^n\Big)^{-d}\Big(1 + \sum_{n=1}^{\infty}(n+1)c_n x^n\Big) \\ &= x^{-d}\Big[1 + \sum_{j=1}^{\infty}\binom{-d}{j}\Big(\sum_{n=1}^{\infty} c_n x^n\Big)^j\Big]\Big(1 + \sum_{n=1}^{\infty}(n+1)c_n x^n\Big) \\ &= x^{-d}\Big[1 + \sum_{n=1}^{\infty} G_{n,d}(c_1, \dots, c_n)x^n\Big], \end{aligned}$$

where $G_{n,d} \in \mathbb{Z}[X_1, \dots, X_n]$ are polynomials which come into being if we expand the product. Apparently, these polynomials do not depend on the field F, and we obtain

$$\mathrm{Res}_x(z^{-1}\delta_x(z)) = 1 \text{ and } \mathrm{Res}_x(z^{-d}\delta_x(z)) = G_{d-1,d}(c_1, \dots, c_{d-1}) \text{ for all } d \geq 2.$$

If $\mathrm{char}(K) = 0$ and $d \geq 2$, then

$$z^{-d}\delta_x(z) = \frac{1}{-d+1}\delta_x(z^{-d+1}) \text{ and therefore } \mathrm{Res}_x(z^{-d}\delta_x(z)) = 0 \text{ by 1(a)}.$$

Since this holds for all $z \in x(1 + xK[\![x]\!])$ built with sequences $(c_n)_{n\geq 1}$ in F, it follows that $G_{d-1,d}(c_1, \dots, c_{d-1}) = 0$ for all $(c_1, \dots, c_{d-1}) \in K^{d-1}$. In particular, $G_{d-1,d} = 0$ for all $d \geq 2$, and thus the assertions also hold if $\mathrm{char}(K) > 0$.

4. (a) This is proved as in the polynomial case (see 1.4.2.1).

(b) As δ_x is continuous and F-linear, it suffices to prove that $\delta^p(x^n) = 0$ for all $n \in \mathbb{Z}$. If $n \in \mathbb{Z}$, then $n(n-1) \cdot \ldots \cdot (n-p+1) \equiv 0 \mod p$, and thus we obtain $\delta^p(x^n) = n(n-1) \cdot \ldots \cdot (n-p+1)x^{n-p} = 0$.

Assume now that F is perfect. It suffices to prove that $K^p = F((x^p))$. Let $u \in K$ be given by $(*)$. Then

$$u^p = \lim_{N\to\infty}\Big(\sum_{\substack{n\in\mathbb{Z}\\ n\le N}} a_n x^n\Big)^p = \lim_{N\to\infty}\sum_{\substack{n\in\mathbb{Z}\\ n\le N}} a_n^p x^{np} = \sum_{n\in\mathbb{Z}} a_n^p x^{np},$$

and consequently $K^p = F^p((x^p)) = F((x^p))$. □

The following Theorem 6.7.9 provides the connection between abstract derivations and concrete derivatives in the context of algebraic function fields.

Theorem 6.7.9. *Let K/F be a function field, F its perfect field of constants and $P \in \mathbb{P}_K$. For a separating element t for K/F let $D_t\colon K \to K$ be the unique F-derivation satisfying $D_t(t) = 1$* (see 6.7.5.1).

1. *Let $x \in \mathcal{O}_P$ be a prime element. Then x is separating for K/F, $K_P = \mathsf{k}_P((x))$ is the completion of (K, v_P),*

$$v_P = \mathsf{o}_x\colon K_P \to \mathbb{Z}\cup\{\infty\}, \quad \text{and} \quad \delta_x \restriction K = D_x.$$

2. *Let $t \in K$ be separating for K/F. Then there exists a unique k_P-derivation $\delta_{t,P}\colon K_P \to K_P$ such that $\delta_{t,P}\restriction K = D_t$. If $u \in K_P$ is such that $v_P(u) = 1$, then $\delta_u(t) \ne 0$, and*

$$\delta_{t,P} = \delta_u(t)^{-1}\delta_u\colon K_P \to K_P.$$

Proof. 1. As $v_P(x) = 1$, it follows by 6.1.6.2 that x is separating. Recall that $K_P = \mathsf{k}_P((x))$ and $v_P = \mathsf{o}_x\colon K_P \to \mathbb{Z}\cup\{\infty\}$ by 6.2.8. By definition, $\delta_x \restriction K\colon K \to K_P$ is an F-derivation and $\delta_x(F(x)) \subset F(x)$. Hence $\delta_x \restriction F(x)\colon F(x) \to F(x) \hookrightarrow K$ is an F-derivation, and as $K/F(x)$ is separable, 6.7.4.3 implies $\delta_x(K) \subset K$. Eventually $\delta_x(x) = 1$ entails $\delta_x\restriction K = D_x$.

2. Let $t_P \in K$ and $v_P(t_P) = 1$. Then $K_P = \mathsf{k}_P((t_P))$, $\delta_{t_P}\restriction K = D_{t_P}$, and $\delta_{t_P}(t) \ne 0$ by 6.7.5.2(b). Therefore $\delta_{t,P} = \delta_{t_P}(t)^{-1}\delta_{t_P}\colon K_P \to K_P$ is a k_P-derivation, and $\delta_{t,P}\restriction K = D_{t_P}(t)^{-1}D_{t_P} = D_t\colon K \to K$ by 6.7.5.2(a). The uniqueness of $\delta_{t,P}$ follows from continuity since K is dense in K_P.

If $u \in K_P$ and $v_P(u) = 1$, then $\delta_{t_P} = \delta_{t_P}(u)\delta_u$ by 6.7.8.2, hence $\delta_{t_P}(t) \ne 0$ implies $\delta_u(t) \ne 0$, and consequently

$$\delta_{t,P} = \delta_{t_P}(t)^{-1}\delta_{t_P} = [\delta_{t_P}(u)\delta_u(t)]^{-1}\delta_{t_P}(u)\delta_u = \delta_u(t)^{-1}\delta_u\colon K_P \to K_P. \ \square$$

We proceed with the local theory by the definition of differentials and their residues. The globalization of these notions in the context of algebraic function fields will be performed in 6.8.1.

Definition and Theorem 6.7.10. *Let $K = F((X))$ be the field of formal Laurent series in X over a field F, $\mathsf{o}_X = \mathsf{o}\colon K \to \mathbb{Z} \cup \{\infty\}$ the order valuation and $x \in K$ such that $\mathsf{o}(x) = 1$ (then $K = F((x))$ and $\mathsf{o} = \mathsf{o}_x$).*

1. For (f, u), $(g, v) \in K \times K$ we define

$$(f, u) \sim (g, v) \quad \text{if} \quad f\delta_x(u) = g\delta_x(v).$$

Then $\sim$ is an equivalence relation on $K \times K$ which does not depend on x.
The set of equivalence classes $\Omega_K = K \times K/\sim$ is called the **local differential module**, and its elements are called **local differentials** of K. For a pair $(f, u) \in K \times K$ we let $fd_K u = fdu \in \Omega_K$ be its equivalence class.

(a) If f, g, $u \in K$, then $fdu = gdx$ if and only if $g = f\delta_x(u)$. In particular, for every $\omega \in \Omega_K$ there exists a unique $g \in K$ such that $\omega = g\,dx$.
We call $\mathsf{o}(\omega) = \mathsf{o}(g)$ the **order** of ω. *The definition of $\mathsf{o}(\omega)$ does not depend on x.*

(b) Let $\omega = gdx$, $\omega_1 = g_1dx \in \Omega_K$, where g, $g_1 \in K$ and $c \in K$. Then we define $\omega + \omega_1 = (g + g_1)dx$ and $c\omega = (cg)dx$.
With these operations, $\Omega_K = Kdx$ is a one-dimensional vector space over K which does not depend on x, and the map $d = d_K\colon K \to \Omega_K$, $u \mapsto du$, is an F-derivation, called the **universal (local) derivation** of K.

(c) If $\omega \in \Omega_K$ and $u \in K$, then

$$\mathsf{o}(u\omega) = \mathsf{o}(u) + \mathsf{o}(\omega), \; du = \delta_x(u)dx,$$

and $du \neq 0$ if and only if $\delta_x(u) \neq 0$.

2. Let $f \in K$ and $\omega = f\,dx \in \Omega_K$. Then we call

$$\mathrm{Res}_K(\omega) = \mathrm{Res}_K(fdx) = \mathrm{Res}_x(f) \in F$$

the **residue** of ω. *It only depends on ω (and not on f or x). The map $\mathrm{Res}_K\colon \Omega_K \to F$ is F-linear.* It is called the **residue map** of K.

3. Let F be perfect, $z \in K^\times$ and $\mathsf{o}(z) \geq 1$. Then $[K : F((z))] < \infty$, and

$$K/F((z)) \text{ is separable} \iff \delta_x(z) \neq 0 \iff dz \neq 0.$$

Proof. 1. Obviously, $\sim$ is an equivalence relation on $K \times K$. If $(f,u) \sim (g,v)$ and $z \in K$ satisfies $\mathsf{o}(z) = 1$, then $f\delta_z(u) = f\delta_x(u)\delta_z(x) = g\delta_x(v)\delta_z(x) = g\delta_z(v)$ by 6.7.8.2. Hence $\sim$ does not depend on x.

(a) If $f,\, g,\, u \in K$, then $fdu = gdx$ if and only if $f\delta_x(u) = g\delta_x(x) = g$.

Let $z \in K$ be such that $\mathsf{o}(z) = 1$ and $\omega = gdx \in \Omega_K$. Then $\omega = g\delta_z(x)dz$, and $\mathsf{o}(g\delta_x(z)) = \mathsf{o}(g) + \mathsf{o}(\delta_z(x)) = \mathsf{o}(g\delta_z(x))$ by 6.7.8.1(c). Hence the definition of $\mathsf{o}(\omega)$ does not depend on x.

(b) By definition, $\Omega_K = Kdx$ is a vector space over K, and it apparently does not depend on x. As $\delta_x(x) = 1 \neq 0$, it follows that $dx \neq 0$, and therefore $\dim_K \Omega_K = 1$. Using 6.7.8.1, it is easily checked that $d\colon K \to \Omega_K$ is an F-derivation.

(c) Let $u \in K$ and $\omega = gdx \in \Omega_K$, where $g \in K$. Then $u\omega = ugdx$ and therefore $\mathsf{o}(u\omega) = \mathsf{o}(ug) = \mathsf{o}(u) + \mathsf{o}(g) = \mathsf{o}(u) + \mathsf{o}(\omega)$. By definition, $du = \delta_x(u)dx$ and therefore $du \neq 0$ if and only if $\delta_x(u) \neq 0$.

2. Let $z \in K$ be such that $\mathsf{o}(z) = 1$ and $\omega = fdx = gdz$. Then $f = g\delta_x(z)$ and $\operatorname{Res}_z(g) = \operatorname{Res}_x(g\delta_x(z)) = \operatorname{Res}_x(f)$ by 6.7.8.3. Obviously, the map $\operatorname{Res}_K\colon \Omega_K \to F$ is F-linear.

3. Recall that $[K : F((z))] = \mathsf{o}(z) < \infty$ by 5.7.2.1 and $dz \neq 0$ if and only if $\delta_x(z) \neq 0$ by 1., which holds if $\operatorname{char}(F) = 0$.

Thus let $\operatorname{char}(F) = p > 0$. Then $\delta_x(z) = 0$ if and only if $z \in K^p = F((x^p))$ by 6.7.8.4(b). If $z \in K^p$, then $F((z)) \subset K^p$ and $K/F((z))$ is inseparable. Assume conversely that $K/F((z))$ is inseparable and let K_1 be the separable closure of $F((z))$ in K. Then K/K_1 is purely inseparable, and if $[K : K_1] = p^m$ for some $m \in \mathbb{N}$, then $K^{p^m} = F((x^{p^m})) \subset K_1$ by 1.4.7.2. On the other hand, $[K : F((x^{p^m}))] = p^m$ by 5.7.2.1, hence $K_1 = K^{p^m}$ and thus $z \in K^{p^m} \subset K^p$. □

Next we investigate the behavior of the local differential module in finite separable field extensions. The results of 5.7.2 are essential for the following theorem.

Definition and Theorem 6.7.11. *Let $K = F((X))$ be the field of formal Laurent series in X over a perfect field F, $\mathsf{o}_X = \mathsf{o}\colon K \to \mathbb{Z} \cup \{\infty\}$ the order valuation and $x \in K$ such that $\mathsf{o}(x) = 1$ (then $K = F((x))$ and $\mathsf{o} = \mathsf{o}_x$).*

Let K'/K be a finite separable extension, $[K' : K] = m$, o' the discrete valuation of K' and $x' \in K'$ such that $\mathsf{o}'(x') = 1$. Then $K' = F'((x'))$, where $F' = \mathsf{k}_{K'}$ is an over F finite intermediate field of K'/F (according to 5.7.2). *Let $d_K\colon K \to \Omega_K$ and $d_{K'}\colon K' \to \Omega_{K'}$ be the universal local derivations.*

1. *There exists a unique K-monomorphism $\eta_{K'/K}\colon \Omega_K \to \Omega_{K'}$ such that*

$$\eta_{K'/K}(fd_Ku) = fd_{K'}u \quad \textit{for all} \quad (f,u) \in K \times K.$$

We call $\eta_{K'/K}\colon \Omega_K \to \Omega_{K'}$ the **embedding of local differentials**.

In particular:

- $d_{K'} \upharpoonright K = \eta_{K'/K} \circ d_K$.
- *If* K^* *is an intermediate field of* K'/K, *then*

$$\eta_{K'/K} = \eta_{K'/K^*} \circ \eta_{K^*/K}.$$

- *If* $z \in K$, $\mathsf{o}(z) \ge 1$ *and* $K/F\,((z))$ *is separable, then*

$$\Omega_K = K d_K z \quad \textit{and} \quad \Omega_{K'} = K' d_{K'} z.$$

We identify: $\Omega_K \subset \Omega_{K'}$, and we set $du = d_K u = d_{K'} u$ for all $u \in K$.

2. Let $\omega = u dx = u \delta_{x'}(x) dx' \in \Omega_{K'}$. Then we call

$$\mathsf{Tr}_{K'/K}(\omega) = \mathsf{Tr}_{K'/K}(u) dx \in \Omega_K \quad \text{the } \textbf{trace} \text{ of } \omega.$$

For every $\omega \in \Omega_K$, $\mathsf{Tr}_{K'/K}(\omega)$ *only depends on* ω *(and not on* u *or* x*), and*

$$\operatorname{Res}_K(\mathsf{Tr}_{K'/K}(\omega)) = \mathsf{Tr}_{F'/F}(\operatorname{Res}_{K'}(\omega))$$

or, equivalently, $\operatorname{Res}_x(\mathsf{Tr}_{K'/K}(u)) = \mathsf{Tr}_{F'/F}(\operatorname{Res}_{x'}(u\delta_{x'}(x)))$.

Proof. 1. Since $K = F\,((x)) \subset F'\,((x)) \subset F'\,((x')) = K'$ and K'/K is separable, it follows that $K'/F'\,((x'))$ is separable, and therefore $\delta_{x'}(x) \ne 0$ by 6.7.10.3. Consequently, if (f,u), $(g,v) \in K \times K$, then

$$\begin{aligned} f d_{K'} u = g d_{K'} v \iff f\delta_{x'}(u) = g\delta_{x'}(v) &\iff f\delta_x(u)\delta_{x'}(x) = g\delta_x(v)\delta_{x'}(x) \\ \iff f\delta_x(u) = g\delta_x(v) &\iff f d_K u = g d_K v. \end{aligned}$$

This shows the existence and uniqueness of a K-monomorphism $\eta_{K'/K}$ as asserted. If $z \in K$, $\mathsf{o}(z) \ge 1$ and $K/F\,((z))$ is separable, then $dz \ne 0$ by 6.7.10.3, hence $\Omega_K = K dz$ and $\Omega_{K'} = K' dz$. The remaining assertions of 1. are obvious.

2. We prove first that the definition of $\mathsf{Tr}_{K'/K}(\omega)$ does not depend on u or x. Let $z \in K$ and $v \in K'$ such that $\mathsf{o}(z) = 1$ and $u dx = v dz$. By definition and 6.7.8.2 we obtain $u\delta_{x'}(x) = v\delta_{x'}(z) = v\delta_x(z)\delta_{x'}(x)$ and therefore $u = v\delta_x(z)$. Since $dz = \delta_x(z) dx$, it follows that

$$\mathsf{Tr}_{K'/K}(v) dz = \mathsf{Tr}_{K'/K}(v)\delta_x(z) dx = \mathsf{Tr}_{K'/K}(v\delta_x(z)) dx = \mathsf{Tr}_{K'/K}(u) dx.$$

We proceed with the proof of the trace formula in a special case.

I. Special case: $F = F'$ is algebraically closed and $[K'\!:\!K] = m$. By 5.7.2.2 the extension K'/K is fully ramified, and we may assume that $x = x'^m w$, where $w \in F\,[\![x']\!]^\times$ and $w \equiv 1 \bmod x'F\,[\![x']\!]$. Then

$$x = x'^m\Big(1 + \sum_{i=1}^{\infty} a_i x'^i\Big) \quad \text{for some sequence } (a_i)_{i \ge 1} \text{ in } F.$$

Recall from 2.10.10.2 and 5.5.3.3 that $F[\![x']\!] = \mathcal{O}_{K'} = \mathrm{cl}_{K'}\mathcal{O}_K = \mathrm{cl}_{K'}F[\![x]\!]$. Hence x' is integral over $F[\![x]\!]$, and there exists an integral equation

$$x'^m + \sum_{j=0}^{m-1} b_j(x)x'^j = 0, \text{ where } b_j(x) = \sum_{\nu=0}^{\infty} b_{j,\nu}x^\nu \in F[\![x]\!] \text{ for all } j \in [0, m-1].$$

The following fact is the core of the proof:

A. There exist polynomials $B_{j,\nu} \in \mathbb{Z}[X_1, X_2, \ldots]$ which do not depend on the field F such that $b_{j,\nu} = B_{j,\nu}(a_1, a_2, \ldots)$ for all $j \in [0, m-1]$ and $\nu \geq 0$.

Proof of **A.** If $j \in [0, m-1]$, then multiplying out yields

$$b_j(x) = \sum_{\nu=0}^{\infty} b_{j,\nu}x'^{m\nu}\Big(1 + \sum_{i=1}^{\infty} a_i x'^i\Big)^\nu = \sum_{\nu=0}^{\infty} b_{j,\nu}x'^{m\nu} \sum_{i=0}^{\infty} g_{\nu,i}x'^i,$$

where (for all ν, $i \geq 0$) $g_{\nu,i} = G_{\nu,i}(a_1, \ldots, a_i)$ with suitable polynomials $G_{\nu,i} \in \mathbb{Z}[X_1, \ldots, X_i]$ which do not depend on F, and $g_{\nu,0} = G_{\nu,0} = 1$. Now the integral equation takes the form

$$0 = x'^m + \sum_{j=0}^{m-1}\sum_{\nu=0}^{\infty} b_{j,\nu}x'^{m\nu+j} \sum_{i=0}^{\infty} g_{\nu,i}x'^i.$$

For $n \in \mathbb{N}_0$ we set $n = m\nu + j$ (where $\nu \in \mathbb{N}_0$ and $j \in [0, m-1]$), $b_n = b_{j,\nu}$ and $g_{n,i} = g_{\nu,i}$. Then it follows that

$$0 = x'^m + \sum_{n=0}^{\infty} b_n x'^n \sum_{i=0}^{\infty} g_{n,i}x'^i = x'^m + \sum_{k=0}^{\infty}\Big(\sum_{n=0}^{k} b_n g_{n,k-n}\Big)x'^k,$$

and comparing coefficients yields, for all $k \geq 0$,

$$b_0g_{0,k} + b_1g_{1,k-1} + \ldots + b_{k-1}g_{k-1,1} + b_k = \begin{cases} -1 & \text{if } k = m, \\ 0 & \text{if } k \neq m. \end{cases}$$

We solve this system of equations recursively. Since $g_{\nu,i} = G_{\nu,i}(a_1, \ldots, a_i)$ for all i, $\nu \geq 0$, we obtain $b_{j,\nu} = b_{m\nu+j} = B_{j,\nu}(a_1, a_2, \ldots)$ (for all $\nu \in \mathbb{N}_0$ and $j \in [0, m-1]$), where $B_{j,\nu} \in \mathbb{Z}[X_1, X_2, \ldots]$ are polynomials which do not depend on F. $\square$[**A.**]

Now let $u \in K'$, say

$$u = \sum_{j=0}^{m-1} c_j(x)x'^j, \quad \text{where } c_j(x) = \sum_{\mu=M}^{\infty} c_{j,\mu}x^\mu \in F(\!(x)\!)$$

for some $M \in \mathbb{Z}$ and all $j \in [0, m-1]$. Then

$$u\begin{pmatrix}1 & x' & \ldots & x'^{m-1}\end{pmatrix} = \begin{pmatrix}1 & x' & \ldots & x'^{m-1}\end{pmatrix} D$$

for some matrix $D = (d_{k,l}(x))_{k,l\in[0,m-1]} \in \mathsf{M}_m(K)$. Calculating this matrix explicitly implies that

$$d_{k,l}(x) = P_{k,l}(b_0(x), \ldots, b_{m-1}(x), c_0(x), \ldots, c_{m-1}(x))$$

with polynomials $P_{k,l} \in \mathbb{Z}[Y_0, \ldots, Y_{m-1}, Z_0, \ldots, Z_{m-1}]$ which do not depend on F and are linear in $Z_0, \ldots, Z_{m-1}$. Hence $\mathsf{o}(d_{k,l}) \geq M$ for all $k,\ l \in [0, m-1]$, and we set

$$d_{k,l}(x) = \sum_{\mu=M}^{\infty} d_{k,l,\mu}x^{\mu} \in F\,(\!(x)\!).$$

Now we apply **A** and get $d_{k,l,\mu} = P_{k,l,\mu}(\boldsymbol{a}, \boldsymbol{c}) \in F$ for all $k,\ l \in [0, m-1]$ and $\mu \geq M$, where $P_{k,l,\mu} \in \mathbb{Z}[\boldsymbol{X}, \boldsymbol{Z}]$ are polynomials in the families of indeterminates $\boldsymbol{X} = (X_i)_{i\geq 1}$ and $\boldsymbol{Z} = (Z_{j,\mu})_{j\in[0,m-1],\,\mu\geq M}$ which do not depend on F, $\boldsymbol{a} = (a_i)_{i\geq 1}$ and $\boldsymbol{c} = (c_{j,\mu})_{j\in[0,m-1],\,\mu\geq M}$. Now it follows that

$$\mathsf{Tr}_{K'/K}(u) = \sum_{j=0}^{m-1} d_{j,j}(x) \quad \text{and}$$

$$\mathrm{Res}_x(\mathsf{Tr}_{K'/K}(u)) = \sum_{j=0}^{m-1} d_{j,j,-1} = \sum_{j=0}^{m-1} P_{j,j,-1}(\boldsymbol{a}, \boldsymbol{c}).$$

On the other hand, again multiplying out implies

$$u = \sum_{j=0}^{m-1} c_j(x)x'^j = \sum_{j=0}^{m-1}\sum_{\mu\geq M} c_{j,\mu}x'^{m\mu}\Big(1 + \sum_{i=1}^{\infty} a_i x'^i\Big)^{\mu} = \sum_{\mu\geq M} Q_{\mu}(\boldsymbol{a}, \boldsymbol{c})x'^{\mu},$$

where $Q_\mu \in \mathbb{Z}[\boldsymbol{X}, \boldsymbol{Z}]$ (for $\mu \geq M$) are polynomials which do not depend on F, and we obtain $\mathrm{Res}_{x'}(u) = Q_{-1}(\boldsymbol{a}, \boldsymbol{c})$.

Now the assertion of the Theorem is equivalent to the system of equations

$$Q_{-1}(\boldsymbol{a}, \boldsymbol{c}) = \sum_{j=0}^{m-1} P_{j,j,-1}(\boldsymbol{a}, \boldsymbol{c})$$

for all families $\boldsymbol{a} = (a_i)_{i\geq 1}$ and $\boldsymbol{c} = (c_{l,\mu})_{l\in[0,m-1],\,\mu\geq M}$ in F. Hence it suffices to prove the assertion in the case $\mathrm{char}(F) = 0$. Indeed, then it follows that

$$Q_{-1} = \sum_{j=0}^{m-1} P_{j,j,-1} \in \mathbb{Z}[\boldsymbol{X}, \boldsymbol{Z}],$$

and consequently the assertion holds for all algebraically closed fields.

Now let F be an algebraically closed field such that $\mathrm{char}(F) = 0$. By 5.7.2.2 we may assume that $x = x'^m$. Then $\delta_{x'}(x) = mx'^{m-1}$, and we must prove that

$$\mathrm{Res}_x(\mathsf{Tr}_{K'/K}(u)) = \mathrm{Res}_{x'}(mux'^{m-1}).$$

Since

$$u = \sum_{j=0}^{m-1} c_j(x)x'^j = \sum_{j=0}^{m-1}\sum_{\mu=M}^{\infty} c_{j,\mu}x^\mu x'^j = \sum_{j=0}^{m-1}\sum_{\mu=M}^{\infty} c_{j,\mu}x'^{m\mu+j}$$

and since the maps $\mathsf{Tr}_{K'/K}$, $\mathrm{Res}_{x'}$ and Res_x are continuous and F-linear, it follows that

$$\mathrm{Res}_x(\mathsf{Tr}_{K'/K}(u)) = \sum_{j=0}^{m-1}\sum_{\mu=M}^{\infty} c_{j,\mu}\mathrm{Res}_x(x^\mu \mathsf{Tr}_{K'/K}(x'^j))$$

and

$$\mathrm{Res}_{x'}(mux'^{m-1}) = \sum_{j=0}^{m-1}\sum_{\mu=M}^{\infty} c_{j,\mu}m\mathrm{Res}_{x'}(x'^{m(\mu+1)+j-1}).$$

If $j \in [1, m-1]$, then $X^{m/(m,j)} - x^{j/(m,j)} \in K[X]$ is the minimal polynomial of x'^j over K, and 1.6.2.1 implies

$$\mathsf{Tr}_{K'/K}(x'^j) = \begin{cases} 0 & \text{if } j \neq 0, \\ m & \text{if } j = 0. \end{cases}$$

For all $j \in [0, m-1]$ and $\mu \geq M$ it follows that

$$\mathrm{Res}_x(x^\mu \mathsf{Tr}_{K'/K}(x'^j)) = \begin{cases} m & \text{if } (j,\mu) = (0,-1), \\ 0 & \text{otherwise,} \end{cases} \quad \text{and}$$

$$\mathrm{Res}_x(\mathsf{Tr}_{K'/K}(u)) = c_{0,-1}m.$$

Since $m(\mu+1) + j - 1 = -1$ if and only if $(j,\mu) = (0,-1)$, we also obtain

$$\mathrm{Res}_{x'}(mux'^{m-1}) = c_{0,-1}m.$$

II. Let F be arbitrary. Let $\widetilde{K}$ be an algebraically closed extension field of K' and $\overline{F}$ the algebraic closure of F in $\widetilde{K}$.

IIa. Let K'/K be fully ramified. Then $F = F'$ according to 5.7.2.2, we set $\overline{K} = \overline{F}((x))$ and $\overline{K'} = \overline{F}((x')) = \overline{K}K'$. As $x = x'^m w$ and $w \in F[\![x']\!]^\times \subset \overline{F}[\![x']\!]^\times$, it follows that $[\overline{K'} : \overline{K}] = m = [K' : K]$ by 5.7.2.2. Hence if $g \in K[X]$ is the minimal polynomial of x' over K, then g is also the minimal polynomial of x' over $\overline{K}$. Consequently, if $x'_1, \ldots, x'_m \in \widetilde{K}$ are the conjugates of x' over K, then they are also the conjugates of x' over $\overline{K}$, and we obtain bijective maps

$$\begin{cases} \mathrm{Hom}^K(K', \widetilde{K}) & \to \{x'_1, \ldots, x'_m\} \\ \sigma & \mapsto \sigma(x') \end{cases}, \quad \begin{cases} \mathrm{Hom}^{\overline{K}}(\overline{K'}, \widetilde{K}) & \to \{x'_1, \ldots, x'_m\} \\ \sigma & \mapsto \sigma(x'), \end{cases}$$

and

$$\begin{cases} \mathrm{Hom}^{\overline{K}}(\overline{K'}, \widetilde{K}) & \to \mathrm{Hom}^K(K', \widetilde{K}) \\ \sigma & \mapsto \sigma \restriction K' \end{cases}.$$

Consequently $\mathsf{Tr}_{\overline{K'}/\overline{K}}(z) = \mathsf{Tr}_{K'/K}(z)$ for all $z \in K'$. By means of **I** it follows that, for all $u \in K'$, $\mathrm{Res}_x(\mathsf{Tr}_{K'/K}(u)) = \mathrm{Res}_x(\mathsf{Tr}_{\overline{K'}/\overline{K}}(u)) = \mathrm{Res}_{x'}(u\delta_{x'}(x))$.

Here it does not matter whether $\mathrm{Res}_{x'}(u\delta_{x'}(x))$ is calculated in K' or in $\overline{K'}$.

IIb. Let K'/K be unramified. In this case we may assume that $x = x'$, and $K' = F'((x)) = KF'$ by 5.7.2.2. Then $[F' : F] = [K' : K]$, and for $\sigma \in \mathrm{Hom}^F(F', \widetilde{K})$ we define

$$\overline{\sigma}\Big(\sum_{n\in\mathbb{Z}} a_n x^n\Big) = \sum_{n\in\mathbb{Z}} \sigma(a_n)x^n \quad \text{for all } z = \sum_{n\in\mathbb{Z}} a_n x^n \in K' = F'((x)).$$

Then $\sigma \mapsto \overline{\sigma}$ induces an injective map $\mathrm{Hom}^F(F', \widetilde{K}) \to \mathrm{Hom}^K(K', \widetilde{K})$ which is even bijective since $[F' : F] = [K' : K]$. Hence it follows that

$$\mathsf{Tr}_{K'/K}\Big(\sum_{n\in\mathbb{Z}} a_n x^n\Big) = \sum_{n\in\mathbb{Z}} \mathsf{Tr}_{F'/F}(a_n)x^n \quad \text{for all } u = \sum_{n\in\mathbb{Z}} a_n x^n \in K' = F'((x))$$

and therefore $\mathrm{Res}_x(\mathsf{Tr}_{K'/K}(u)) = \mathsf{Tr}_{F'/F}(a_{-1}) = \mathsf{Tr}_{F'/F}(\mathrm{Res}_x(u))$.

IIc. Let $K = F((x)) \subset F'((x)) \subset F'((x'))$ according to 5.7.2.2. By **IIa.** and **IIb.** we obtain, for all $u \in K'$,

$$\begin{aligned}\mathrm{Res}_x(\mathsf{Tr}_{K'/K}(u)) &= \mathrm{Res}_x(\mathsf{Tr}_{F'((x))/K} \circ \mathsf{Tr}_{K'/F'((x))}(u)) \\ &= \mathsf{Tr}_{F'/F}(\mathrm{Res}_x(\mathsf{Tr}_{K'/F((x))}(u)) = \mathsf{Tr}_{F'/F}(\mathrm{Res}_{x'}(u\delta_{x'}(x))).\end{aligned}$$

□

6.8 Differentials and Weil differentials

Throughout this section, let K/F be a function field and F its perfect field of constants.

We start with the connection between local and global differentials and a global definition of residues.

Definition and Remarks 6.8.1. Let $P \in \mathbb{P}_K$, $x \in \mathcal{O}_P$ be a prime element and $K_P = (K_P, v_P)$ the completion of the discrete valued field (K, v_P). Recall that x is separating (by 6.7.9.1), $K_P = \mathsf{k}_P((x))$ and $v_P = \mathsf{o}_x \colon K_P \to \mathbb{Z} \cup \{\infty\}$ (by 6.2.8), $\mathcal{O}_{K_P} = \mathsf{k}_P[\![x]\!]$ (by 2.10.10.2) and $\delta_x \restriction K = D_x \in \mathrm{Der}_F(K)$ (by 6.7.9.1)

Let $\Omega_K = K\,dx$ be the global differential module and $d = d_K \colon K \to \Omega_K$ the universal global derivation of K/F (see 6.7.6). Let $\Omega_P = \Omega_{K_P} = K_P d_P x$

be the local differential module and $d_P = d_{K_P} \colon K_P \to \Omega_P$ the universal local derivation of K_P (see 6.7.10.1).

1. Since $d_P \restriction K \colon K \to \Omega_P$ is an F-derivation, the universal property of (Ω_K, d) (see 6.7.6.5) implies that there exists a unique K-linear map

$$\iota_P \colon \Omega_K \to \Omega_K \quad \text{satisfying} \quad \iota_P \circ d = d_P \restriction K.$$

In particular, $\iota_P(dx) = d_P x$, and thus ι_P is injective. We call ι_P the **local embedding of differentials**. Explicitly, we have $\iota_P(fdu) = fd_P u$ for all $(f, u) \in K \times K$.

2. Let $\omega = fdu = f\delta_x(u)dx \in \Omega_K$, where $(f, u) \in K \times K$. According to 6.7.10 we call

$$v_P(\omega) = \mathsf{o}(\iota_P(\omega)) = \mathsf{o}_x(f\delta_x(u))$$

the **P-adic value** and

$$\mathrm{Res}_P(\omega) = \mathrm{Res}_{K_P}(\iota_P(\omega)) = \mathrm{Res}_x(f\delta_x(u))$$

the **residue of ω at P.**

If $u \in K^\times$, then

$$v_P(u\omega) = \mathsf{o}_x(u\iota_P(\omega)) = \mathsf{o}_x(u) + \mathsf{o}_x(\iota_P(\omega)) = v_P(u) + v_P(\omega)$$

by 6.7.10.1. Apparently, $v_P(\omega + \omega') \geq \min\{v_P(\omega), v_P(\omega')\}$ for all ω, $\omega' \in \Omega_K$, and if $v_P(\omega) \geq 0$, then $\mathrm{Res}_P(\omega) = 0$.

Definition and Theorem 6.8.2. *If $\omega \in \Omega_K^\bullet$, then $v_P(\omega) = 0$ for almost all $P \in \mathbb{P}_K$.* We call

$$(\omega) = \sum_{P \in \mathbb{P}_K} v_P(\omega)P \in \mathbb{D}_K$$

the **divisor** of ω, and we call ω **holomorphic** or **regular** if $(\omega) \geq 0$.
If $g \in K^\times$, then $(g\omega) = (g) + (\omega)$, and if $z \in K$ is a separating element for K/F, then

$$(dz) = \mathrm{D}_{K/F(z)} - 2(z)_\infty^K.$$

Proof. Let $z \in K$ be separating, $P \in \mathbb{P}_K$, $P_0 = P \cap F(z)$ and $e = e(P/P_0)$. We first prove that

$$v_P(dz) = \begin{cases} v_P(\mathrm{D}_{K/F(z)}) & \text{if } P_0 \neq P_\infty, \\ v_P(\mathrm{D}_{K/F(z)}) - 2e & \text{if } P_0 = P_\infty. \end{cases}$$

Let $x \in \mathcal{O}_P$ and $u \in \mathcal{O}_{P_0}$ be prime elements. Then it follows that $K_P = \mathsf{k}_P((x))$, $F(z)_{P_0} = \mathsf{k}_{P_0}((u))$, and 5.7.2.2 shows that

$$F(z)_{P_0} = \mathsf{k}_{P_0}((u)) \subset \mathsf{k}_P((u)) \subset \mathsf{k}_P((x)) = K_P,$$

$\mathsf{k}_P\,(\!(u)\!)/\mathsf{k}_{P_0}\,(\!(u)\!)$ is unramified of degree $f(P/P_0) = [\mathsf{k}_P : \mathsf{k}_{P_0}]$, and the extension $\mathsf{k}_P\,(\!(x)\!)/\mathsf{k}_P\,(\!(u)\!)$ is fully ramified of degree $e = \mathsf{o}_x(u)$.

Let $g \in \mathsf{k}_P\,[\![u]\!][X]$ be the minimal polynomial of x over $\mathsf{k}_P\,(\!(u)\!)$. By 5.7.1.2 it is an Eisenstein polynomial of $\mathsf{k}_P\,(\!(u)\!)$ and $\mathsf{k}_P\,[\![x]\!] = \mathsf{k}_P\,[\![u]\!][x]$. Now it follows that $\mathfrak{D}_{\mathsf{k}_P\,[\![x]\!]/\mathsf{k}_P\,[\![u]\!]} = g'(x)\mathsf{k}_P\,[\![x]\!]$ by 5.9.4.1, $\mathfrak{D}_{\mathsf{k}_P\,[\![u]\!]/\mathsf{k}_{P_0}\,[\![u]\!]} = \mathsf{k}_P\,[\![u]\!]$ by 5.9.6.4 and $\mathfrak{D}_{\mathsf{k}_P\,[\![x]\!]/\mathsf{k}_{P_0}\,[\![u]\!]} = g'(x)\mathsf{k}_P\,[\![x]\!]$ by 5.9.3.2. If $\mathcal{O}'_{P_0} = \mathrm{cl}_K(\mathcal{O}_{P_0})$, then 5.9.5 and 6.5.9.1 yield

$$v_P(\mathfrak{D}_{\mathsf{k}_P\,[\![x]\!]/\mathsf{k}_{P_0}\,[\![u]\!]}) = v_P(\mathfrak{D}_{\mathcal{O}'_{P_0}/\mathcal{O}_{P_0}}) = d(P/P_0) = v_P(\mathrm{D}_{K/F(z)}),$$

and thus finally $v_P(\mathrm{D}_{K/F(z)}) = v_P(g'(x))$.

Let $g = X^e + \alpha_{e-1}X^{e-1} + \ldots + \alpha_1 X + \alpha_0$, where $\alpha_i \in u\mathsf{k}_P\,[\![u]\!]$ for all $i \in [0, e-1]$ and $\mathsf{o}_u(\alpha_0) = 1$. As $g(x) = 0$, 6.7.2.2(a) yields

$$0 = \delta_x g(x) = g^{\delta_x}(x) + g'(x) = \delta_x(\alpha_{e-1})x^{e-1} + \ldots + \delta_x(\alpha_0) + g'(x).$$

If $j \in [0, e-1]$ then $\delta_x(\alpha_j) = \delta_u(\alpha_j)D_z(u)D_x(z)$, and it follows that

$$0 = \sum_{j=0}^{e-1} \delta_x(\alpha_j)x^j + g'(x) = \Big(\sum_{j=0}^{e-1} \delta_u(\alpha_j)x^j\Big) D_z(u)D_x(z) \ + g'(x).$$

As $\mathsf{o}_x(\delta_u(\alpha_0)) = e\mathsf{o}_u(\delta_u(\alpha_0)) = 0$ and $\mathsf{o}_x(\delta_u(\alpha_i x^i)) = e\mathsf{o}_u(\delta_u(\alpha_i)) + i \geq 1$ for all $i \in [1, e-1]$, we get

$$\mathsf{o}_x\Big(\sum_{j=0}^{e-1} \delta_u(\alpha_j)x^j\Big) = 0,$$

and consequently $v_P(g'(x)) = v_P(D_z(u)) + v_P(D_x(z))$.

Now we make an appropriate choice of u. If $P \cap F(z) = P_{q(z)}$, where $q \in F[X]$ is monic and irreducible, we set $u = q(z)$. Then $D_z(u) = q'(z)$ and $v_P(D_z(u)) = 0$, since q is separable. If $P \cap F(z) = P_\infty$, we set $u = z^{-1}$. Then $D_z(u) = -u^2$ and $v_P(D_z(u)) = 2e$, hence

$$\begin{aligned} v_P(dz) &= v_P(D_x(z)) = v_P(g'(x)) - v_P(D_z(u)) \\ &= \begin{cases} v_P(\mathrm{D}_{K/F(z)}) & \text{if } P_0 \neq P_\infty, \\ v_P(\mathrm{D}_{K/F(z)}) - 2e & \text{if } P_0 = P_\infty, \end{cases} \end{aligned}$$

and consequently $v_P(dz) = 0$ for almost all $P \in \mathbb{P}_K$. Since

$$v_P((z)_\infty^K) = v_P(j_{K/F(z)}(z)_\infty^{F(z)})) = \begin{cases} 0 & \text{if } P_0 \neq P_\infty, \\ -e & \text{if } P_0 = P_\infty. \end{cases}$$

we eventually obtain $(dz) = \mathrm{D}_{K/F(z)} - 2(z)_\infty^K$. If $\omega \in \Omega_K^\bullet$, then $\omega = f dz$ for some $f \in K^\times$, and $v_P(\omega) = v_P(f) + v_P(dz) = 0$ for almost all $P \in \mathbb{P}_K$. If $g \in K^\times$, then $v_P(g\omega) = v_P(g) + v_P(\omega)$ for all $P \in \mathbb{P}_K$ by 6.8.1.2 and therefore $(g\omega) = (g) + (\omega)$. □

Our next aim is the proof of the residue theorem 6.8.4. We place in front a preliminary result which allows the reduction to an algebraically closed field of constants.

Theorem 6.8.3. *Let $\widetilde{K}$ be an algebraically closed extension field of K, $\overline{F}$ an algebraic closure of F in $\widetilde{K}$, $\overline{K} = K\overline{F}$, $\eta_{\overline{K}/K}\colon \Omega_K \to \Omega_{\overline{K}}$ the embedding of differentials, $\omega \in \Omega_K$, $\overline{\omega} = \eta_{\overline{K}/K}(\omega)$ and $P \in \mathbb{P}_K$. Then*

$$\mathsf{Tr}_{\mathsf{k}_P/F}(\mathrm{Res}_P\,\omega) = \sum_{\substack{\overline{P}\in\mathbb{P}_{\overline{K}}\\ \overline{P}\cap K=P}} \mathrm{Res}_{\overline{P}}\,(\overline{\omega}).$$

Proof. Let $\deg(P) = d$ and $F \subset \mathsf{k}_P \subset \overline{F}$. Let k_P^* be the Galois closure of k_P/F in $\overline{F}$, $K' = K\mathsf{k}_P$ and $K^* = K\mathsf{k}_P^*$. Hence $K \subset K' \subset K^* \subset \overline{K} = K\overline{F}$, and K^*/K is a finite Galois extension by 1.5.6. We set $G = \mathrm{Gal}(K^*/K)$, $G' = \mathrm{Gal}(K^*/K') \subset G$, and we define $\rho\colon G \to \mathrm{Gal}(\mathsf{k}_P^*/F)$ by $\rho(\sigma) = \sigma \restriction \mathsf{k}_P^*$ for all $\sigma \in G$. Since $\mathsf{k}_P^* \cap K = F$, it follows that ρ is an isomorphism, and $\rho(G') = \mathrm{Gal}(\mathsf{k}_P^*/\mathsf{k}_P)$. Observe that k_P is the field of constants of K' and k_P^* is the field of constants of K^* by 6.1.9.3.

Let $P' \in \mathbb{P}_{K'}$, $P^* \in \mathbb{P}_{K^*}$ and $\overline{P} \in \mathbb{P}_{\overline{K}}$ be such that $P \subset P' \subset P^* \subset \overline{P}$. Then clearly $\mathsf{k}_{\overline{P}} = \overline{F}$, and we may assume that $\mathsf{k}_P \subset \mathsf{k}_{P^*} \subset \overline{F}$. By 6.6.7.2 we obtain $e(\overline{P}/P) = 1$, $\mathsf{k}_{P'} = \mathsf{k}_P$, $\mathsf{k}_{P^*} = \mathsf{k}_P^*$, and consequently likewise $\deg(P') = \deg(P^*) = \deg(\overline{P}) = 1$.

The following diagram shows the involved fields and prime divisors.

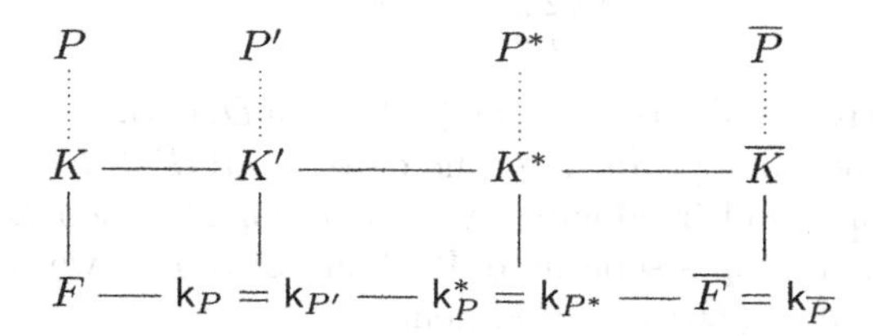

Since $\deg(j_{K'/K}(P)) = \deg(j_{K^*/K}(P)) = \deg(j_{\overline{K}/K}(P)) = d$ by 6.6.7.3, it follows that $j_{K'/K}(P) = P_1' + \ldots P_d'$, $j_{K^*/K}(P) = P_1^* + \ldots + P_d^*$ and $j_{\overline{K}/K}(P) = \overline{P}_1 + \ldots + \overline{P}_d$, where $P_1' = P'$, $P_1^* = P^*$, $\overline{P}_1 = \overline{P}$, $P_i' \in \mathbb{P}_{K'}$, $P_i^* \in \mathbb{P}_{K^*}$, $\overline{P}_i \in \mathbb{P}_{\overline{K}}$. $P_i' \subset P_i^* \subset \overline{P}_i$ and $\mathsf{k}_{P_i'} \subset \mathsf{k}_{P_i^*} \subset \mathsf{k}_{\overline{P}_i} = \overline{F}$ for all $i \in [1,d]$.

If $\sigma \in G'$, then $\sigma(P') = P'$, hence $\sigma(P^*) = P^*$, and thus G' is contained in the decomposition group G_{P^*} of P^* over K. Since

$$(G\colon G_{P^*}) = d = \deg(P) = [\mathsf{k}_P\colon F] = [K'\colon K] = (G\colon G'),$$

it follows that $G_{P^*} = G'$. Suppose that $G/G' = \{\sigma_1 G', \ldots, \sigma_d G'\}$. Then we obtain $\mathrm{Hom}^K(K', K^*) = \{\sigma_1 \restriction K', \ldots, \sigma_d \restriction K'\}$, and by means of the isomorphism ρ it follows that $\mathrm{Hom}^F(\mathsf{k}_P, \mathsf{k}_{P^*}) = \{\sigma_1 \restriction \mathsf{k}_P, \ldots, \sigma_d \restriction \mathsf{k}_P\}$. After renumbering we may assume that $P_i^* = \sigma_i(P^*)$ for all $i \in [1,d]$.

Let $t \in \mathcal{O}_P$ be a prime element and $u \in K$ such that $\omega = ud_K t$. For all $i \in [1, d]$ we have $v_{\overline{P}_i}(t) = v_{P_i^*}(t) = 1$, hence $\overline{\omega} = \eta_{\overline{K}/K}(\omega) = ud_{\overline{K}}t$, $\omega^* = \eta_{K^*/K}(\omega) = ud_{K^*}t$ and $\operatorname{Res}_{\overline{P}_i}(\overline{\omega}) = \operatorname{Res}_{P_i^*}(\omega^*) = \operatorname{Res}_t(u)$. It follows that

$$\sum_{\substack{\overline{P} \in \mathbb{P}_{\overline{K}} \\ \overline{P} \cap K = P}} \operatorname{Res}_{\overline{P}}(\overline{\omega}) = \sum_{i=1}^{d} \operatorname{Res}_{\overline{P}_i}(\overline{\omega}) = \sum_{i=1}^{d} \operatorname{Res}_{P_i^*}(\omega^*).$$

On the other hand, since $\operatorname{Hom}^F(\mathsf{k}_P, \mathsf{k}_{P^*}) = \{\sigma_1 \restriction \mathsf{k}_P, \ldots, \sigma_d \restriction \mathsf{k}_P\}$, we obtain

$$\mathsf{Tr}_{\mathsf{k}_P/F}(\operatorname{Res}_P(\omega)) = \sum_{i=1}^{d} \sigma_i(\operatorname{Res}_P(\omega)),$$

and therefore we must prove that $\operatorname{Res}_{P_i^*}(\omega^*) = \sigma_i(\operatorname{Res}_P(\omega))$ for all $i \in [1, d]$.

For $i \in [1, d]$ let $\sigma_i^* \colon K_{P^*} \xrightarrow{\sim} K_{K_i^*}$ be the topological K_P-isomorphism satisfying $\sigma_i^* \restriction K = \sigma_i$, and let $\sigma_{iP^*} \colon \mathsf{k}_{P^*} \xrightarrow{\sim} \mathsf{k}_{P_i^*}$ be the residue class isomorphism induced by σ_i satisfying $\sigma_{iP^*}(x(P^*)) = \sigma_i(x)(\sigma_i P^*)$ for all $x \in \mathcal{O}_{P^*}$. Then $\sigma_{iP^*} = \sigma_i^* \restriction \mathsf{k}_P$ by 6.5.3.3(b), and since $\mathsf{k}_P \subset K^*$, it follows that $\sigma_i^* \restriction \mathsf{k}_P = \sigma_i \restriction \mathsf{k}_P$. Now let $\omega = ud_K t$ again and

$$u = \sum_{n \in \mathbb{Z}} a_n t^n \in K \subset K_P, \quad \text{where } a_n \in \mathsf{k}_P \text{ and } a_{-n} = 0 \text{ for all } n \gg 1.$$

Then $\operatorname{Res}_P(\omega) = \operatorname{Res}_t(u) = a_{-1}$. On the other hand, inside $K_{P_i^*}^*$ we calculate

$$u = \sigma_i^*(u) = \sum_{n \in \mathbb{Z}} \sigma_i(a_n) t^n \in K_{P_i^*}^*, \quad \text{and}$$

$$\operatorname{Res}_{P_i^*}(\omega^*) = \operatorname{Res}_{P_i^*}(ud_{K^*}t) = \sigma_i(a_{-1}).$$

Hence it follows that

$$\mathsf{Tr}_{\mathsf{k}_P/F}(\operatorname{Res}_P(\omega)) = \sum_{i=1}^{d} \sigma_i(a_{-1}) = \sum_{i=1}^{r} \operatorname{Res}_{P_i^*}(\omega^*). \qquad \square$$

Theorem 6.8.4 (Residue theorem). *If $\omega \in \Omega_K$, then*

$$\sum_{P \in \mathbb{P}_K} \mathsf{Tr}_{\mathsf{k}_P/F}(\operatorname{Res}_P(\omega)) = 0.$$

Proof. We proceed in 3 steps.

I. Let F be algebraically closed, $K = F(x)$ a rational function field over F and $dx = d_K x \in \Omega_K$. Then $\mathbb{P}_K = \{P_{x-a} \mid a \in K\} \cup \{P_\infty\}$ by 6.2.3, and we suppose that $\omega = udx \in \Omega_K$ for some rational function $u = u(x) \in K$. By 2.3.2.2 we obtain the partial fraction expansion

$$u(x) = \sum_{c \in F} \sum_{j \geq 1} \frac{b_{c,j}}{(x-c)^j} + \sum_{i=0}^{\infty} b_i x^i,$$

where $b_{c,j},\, b_i \in F$ and $b_{c,j} = b_i = 0$ for almost all $i \ge 0,\ j \ge 1$ and $c \in F$. Hence it suffices to prove that for all $c \in F,\ j \ge 1$ and $i \ge 0$ we have

$$\sum_{P\in\mathbb{P}_K} \operatorname{Res}_P\Big(\frac{1}{(x-c)^j}\,dx\Big) = \sum_{P\in\mathbb{P}_K} \operatorname{Res}_P(x^i\,dx) = 0. \qquad (\dagger)$$

First let $c \in F$ and $j \ge 1$. If $a \in F$, then

$$\operatorname{Res}_{P_{x-a}}\Big(\frac{1}{(x-c)^j}\,dx\Big) = \operatorname{Res}_{x-a}\Big(\frac{1}{(x-c)^j}\,\delta_{x-a}(x)\Big) = \operatorname{Res}_{x-a}\Big(\frac{1}{(x-c)^j}\Big)$$
$$= \begin{cases} 1 & \text{if } a = c \text{ and } j = 1, \\ 0 & \text{otherwise,} \end{cases} \quad \text{since} \quad v_{P_{x-a}}\Big(\frac{1}{(x-c)^j}\Big) = 0 \quad \text{if } a \ne c.$$

If $t = x^{-1}$, then

$$\operatorname{Res}_{P_\infty}\Big(\frac{1}{(x-c)^j}\,dx\Big) = \operatorname{Res}_t\Big(\frac{1}{(t^{-1}-c)^j}\,\delta_t(x)\Big) = \operatorname{Res}_t\Big(\frac{t^j}{(1-tc)^j}\,\frac{-1}{t^2}\Big)$$
$$= -\operatorname{Res}_t\Big(\frac{t^{j-2}}{(1-ct)^j}\Big) = \begin{cases} -1 & \text{if } j = 1, \\ 0 & \text{otherwise,} \end{cases} \quad \text{since} \quad (1-ct)^j \in F\,[\![t]\!]^\times.$$

Now let $i \ge 0$ and $a \in F$. Then

$$\begin{aligned} \operatorname{Res}_{P_{x-a}}(x^i\,d_Kx) &= \operatorname{Res}_{x-a}(x^i\,\delta_{x-a}(x)) \\ &= \operatorname{Res}_{x-a}(x^i) = 0, \quad \text{since } v_{P_{x-a}}(x^i) \ge 0. \end{aligned}$$

If $t = x^{-1}$, then $\operatorname{Res}_{P_\infty}(x^i\,dx) = \operatorname{Res}_t(t^{-i}\delta_t(x)dt) = \operatorname{Res}_t(-t^{-i-2}) = 0$.
All in all, $(\dagger)$ follows.

II. Let F be algebraically closed and $x \in K$ a separating element for K/F. We prove first:

A. If $P_0 \in \mathbb{P}_{F(x)}$, then

$$\sum_{\substack{P\in\mathbb{P}_K \\ P\cap F(x)=P_0}} \operatorname{Res}_P(\omega) = \operatorname{Res}_{P_0}(\omega_0), \quad \text{where } \omega_0 = \mathsf{Tr}_{K/F(x)}(y)dx \in \Omega_K.$$

Proof of **A.** Let $P_0 \in \mathbb{P}_{F(x)},\ \{P_1,\ldots,P_r\} = \{P \in \mathbb{P}_K \mid P \cap F(x) = P_0\}$ and $i \in [1,r]$. Let $t \in \mathcal{O}_{P_0}$ and $t_i \in \mathcal{O}_{P_i}$ be prime elements. Since F is algebraically closed, we get

$$\mathsf{k}_{P_0} = \mathsf{k}_{P_i} = F \quad \text{and} \quad K_{P_i} = F\,(\!(t_i)\!) \supset F\,(\!(t)\!) = F(x)_{P_0}.$$

Suppose that $\omega = yd_Kx$ for some $y \in K$. Then 6.7.11.2 implies

$$\begin{aligned} \operatorname{Res}_{P_i}(\omega) &= \operatorname{Res}_{t_i}(y\delta_{t_i}(x)) = \operatorname{Res}_{t_i}\big(y\delta_{t_i}(t)\delta_t(x)\big) = \operatorname{Res}_t\big(\mathsf{Tr}_{K_{P_i}/F(x)_{P_0}}(y\delta_t(x))\big) \\ &= \operatorname{Res}_t\big(\mathsf{Tr}_{K_{P_i}/F(x)_{P_0}}(y)\delta_t(x)\big), \quad \text{since } \delta_t(x) \in F(x) \text{ by 6.7.9.1,} \end{aligned}$$

and, using 5.5.9.2,

$$\sum_{\substack{P\in\mathbb{P}_K\\ P\cap F(x)=P_0}} \operatorname{Res}_P(\omega) = \sum_{i=1}^{r} \operatorname{Res}_{P_i}(\omega) = \operatorname{Res}_t\Big(\sum_{i=1}^{r} \mathsf{Tr}_{K_{P_i}/F(x)_{P_0}}(y)\delta_t(x)\Big)$$

$$= \operatorname{Res}_t\big(\mathsf{Tr}_{K/F(x)}(y)\delta_t(x)\big) = \operatorname{Res}_{P_0}(\omega_0). \qquad \square[\mathbf{A}.]$$

Using **A** we obtain

$$\sum_{P\in\mathbb{P}_K} \operatorname{Res}_P(\omega) = \sum_{P_0\in\mathbb{P}_{F(x)}} \sum_{\substack{P\in\mathbb{P}_K\\ P\cap F(x)=P_0}} \operatorname{Res}_P(\omega) = \sum_{P_0\in\mathbb{P}_{F(x)}} \operatorname{Res}_{P_0}(\omega_0) = 0 \quad \text{by } \mathbf{I}.$$

III. *Proof in the general case.* Let $\overline{F}$ be an algebraic closure of F, $\overline{K} = K\overline{F}$ and $\overline{\omega} = \eta_{\overline{K}/K}(\omega)$. By **II** and 6.8.3 it follows that

$$\sum_{P\in\mathbb{P}_K} \mathsf{Tr}_{\mathsf{k}_P/F}(\operatorname{Res}_P(\omega)) = \sum_{P\in\mathbb{P}_K} \sum_{\substack{\overline{P}\in\mathbb{P}_{\overline{K}}\\ \overline{P}\cap K=P}} \operatorname{Res}_{\overline{P}}(\overline{\omega}) = \sum_{\overline{P}\in\mathbb{P}_{\overline{K}}} \operatorname{Res}_{\overline{P}}(\overline{\omega}) = 0. \quad \square$$

Now we are in a position to highlight the connection betweem differentials and Weil differentials.

Theorem 6.8.5 (Comparison theorem for differentials). *For $\omega \in \Omega_K$ we define*

$$\sigma_\omega\colon \mathbb{A}_K \to F \quad \textit{by} \quad \sigma_\omega(\alpha) = \sum_{P\in\mathbb{P}_K} \mathsf{Tr}_{\mathsf{k}_P/F}(\operatorname{Res}_P(\alpha_P\omega)) \quad \textit{for all} \quad \alpha \in \mathbb{A}_K.$$

Then $\sigma_\omega \in \Sigma_K$, $(\sigma_\omega) = (\omega) \in \mathbb{D}_K$ and $(\sigma_\omega)_P(y) = \mathsf{Tr}_{\mathsf{k}_P/F}(\operatorname{Res}_P(y\omega))$ for all $P \in \mathbb{P}_K$ and $y \in K$. The map

$$\lambda_K\colon \Omega_K \to \Sigma_K, \quad \textit{defined by} \quad \lambda_K(\omega) = \sigma_\omega,$$

is a K-isomorphism.

Proof. If $\alpha = (\alpha_P)_{P\in\mathbb{P}_K} \in \mathbb{A}_K$, then $v_P(\alpha_P\omega) = v_P(\alpha_P) + v_P(\omega) \geq 0$ and thus $\operatorname{Res}_P(\alpha_P\omega) = 0$ for almost all $P \in \mathbb{P}_K$. Hence the sum defining $\sigma_\omega(\alpha)$ is finite. By definition, $\sigma_\omega\colon \mathbb{A}_K \to F$ is F-linear, and if $x \in K$, then

$$\sigma_\omega(x) = \sum_{P\in\mathbb{P}_K} \mathsf{Tr}_{\mathsf{k}_P/F}(\operatorname{Res}_P(x\omega)) = 0$$

by the residue theorem 6.8.4. The main task is to prove the following assertion.

> **A.** Let $\omega \in \Omega_K^\bullet$, $W = (\omega)$ and $B \in \mathbb{D}_K$. Then $\sigma_\omega \restriction \mathbb{A}_K(B) = 0$ if and only if $B \leq W$.

By **A** we obtain $\sigma_\omega \in \Sigma_K$, $(\sigma_\omega) = W$ and

$$(\sigma_\omega)_P(y) = \sigma_\omega \circ \iota_P(y) = \mathsf{Tr}_{\mathsf{k}_P/F}(\operatorname{Res}_P(y\omega))$$

for all $y \in K$ (see 6.4.4).

Proof of **A.** Assume first that $B \le W$. Then $\mathbb{A}_K(B) \subset \mathbb{A}_K(W)$, and it suffices to prove that $\sigma_\omega \restriction \mathbb{A}_K(W) = 0$. If $\alpha \in \mathbb{A}_K(W)$, then $(\alpha) \ge -(\omega)$, hence it follows that $v_P(\alpha_P\omega) = v_P(\alpha) + v_P(\omega) \ge 0$ and thus $\mathrm{Res}_P(\alpha_P\omega) = 0$ for all $P \in \mathbb{P}_K$. Consequently $\sigma_\omega(\alpha) = 0$.

Now let that $B \not\le W$, and let $Q \in \mathbb{P}_K$ be such that $m = v_Q(W) < v_Q(B)$. Let $t \in \mathcal{O}_Q$ be a prime element and $\omega = t^m d_K t \in \Omega_K$. As k_Q/F is separable, there exists some $\alpha_0 \in \mathsf{k}_Q$ such that $\mathsf{Tr}_{\mathsf{k}_Q/F}(\alpha_0) \ne 0$, and we let $a_0 \in \mathcal{O}_Q$ be such that $a_0(Q) = \alpha_0$. Then

$$a_0 = \alpha_0 + \sum_{i=1}^{\infty} \alpha_i t^i, \quad \text{where } \alpha_i \in \mathsf{k}_Q,$$

and therefore $\mathrm{Res}_Q(a_0 t^{-1} d_K t) = \mathrm{Res}_t(\alpha_0 t^{-1}) = \alpha_0$.

Let $\alpha \in \mathbb{A}_K$ be defined by $\alpha_Q = a_0 t^{-m-1}$ and $\alpha_P = 0$ for all $P \in \mathbb{P}_K \setminus \{Q\}$. Then $v_Q(\alpha) = -m-1 = -v_Q(W) - 1 \ge -v_Q(B)$, hence $\alpha \in \mathbb{A}_K(B)$ and

$$\begin{aligned}\sigma_\omega(\alpha) &= \mathsf{Tr}_{\mathsf{k}_Q/F}\big(\mathrm{Res}_Q(\alpha_Q\omega)\big) = \mathsf{Tr}_{\mathsf{k}_Q/F}(\mathrm{Res}_Q(a_0 t^{-1} d_K t)) \\ &= \mathsf{Tr}_{\mathsf{k}_Q/F}(\alpha_0) \ne 0. \qquad \square[\mathbf{A.}]\end{aligned}$$

It is now easily checked that $\lambda_K \colon \Omega_K \to \Sigma_K$ is K-linear, and as $\lambda_K \ne 0$ and $\dim_K(\Omega_K) = \dim_K(\Sigma_K) = 1$, it follows that λ_K is an isomorphism. □

Theorem 6.8.6. *Let K'/F' be an algebraic function field extension of K/F such that K'/K is finite separable and F' is the field of constants of K'/F'. Then there is a commutative diagram*

$$\begin{array}{ccc}\Omega_K & \xrightarrow{\eta_{K'/K}} & \Omega_{K'} \\ \lambda_K\big\downarrow & & \big\downarrow\lambda_{K'} \\ \Sigma_K & \xrightarrow{\iota_{K'/K}} & \Sigma_{K'}\end{array} .$$

Proof. By 6.6.3 we must prove that

$$\mathsf{Tr}_{F'/F}(\lambda_{K'} \circ \eta_{K'/K}(\omega)(\alpha)) = \lambda_K(\omega)(\mathsf{Tr}_{K'/K}(\alpha)) \quad \text{for all } \alpha \in \mathbb{A}_{K'/K}.$$

First let K'/K be Galois and $G = \mathrm{Gal}(K'/K)$. Let $\alpha = (\alpha_{P'})_{P' \in \mathbb{P}_{K'}} \in \mathbb{A}_{K'/K}$, let $x \in K$ be separating for K/F and $\omega = u d_K x \in \Omega_K$ for some $u \in K$. Then x is separating for K'/F', $\eta_{K'/K}(\omega) = u d_{K'} x$, and we must prove that

$$\begin{aligned}&\mathsf{Tr}_{F'/F}\Big(\sum_{P' \in \mathbb{P}_{K'}} \mathsf{Tr}_{\mathsf{k}_{P'}/F'}\big(\mathrm{Res}_{P'}(u\alpha_{P'} d_{K'} x)\big)\Big) \\ &\qquad = \sum_{P \in \mathbb{P}_K} \mathsf{Tr}_{\mathsf{k}_P/F}\big(\mathrm{Res}_P(\mathsf{Tr}_{K'/K}(\alpha)_P u d_K x)\big).\end{aligned}$$

Since

$$\mathsf{Tr}_{F'/F}\Big(\sum_{P'\in\mathbb{P}_{K'}}\mathsf{Tr}_{\mathsf{k}_{P'}/F'}\big(\mathrm{Res}_{P'}(u\alpha_{P'}d_{K'}x)\big)\Big) = \sum_{P\in\mathbb{P}_K}\sum_{\substack{P'\in\mathbb{P}_{K'}\\ P'\cap K=P}}\mathsf{Tr}_{\mathsf{k}_{P'}/F}\big(\mathrm{Res}_{P'}(u\alpha_{P'}d_{K'}x)\big),$$

it suffices to prove that, for all $p\in\mathbb{P}_K$,

$$\sum_{\substack{P'\in\mathbb{P}_{K'}\\ P'\cap K=P}}\mathsf{Tr}_{\mathsf{k}_{P'}/F}\big(\mathrm{Res}_{P'}(u\alpha_{P'}d_{K'}x)\big) = \mathsf{Tr}_{\mathsf{k}_P/F}\big(\mathrm{Res}_P(\mathsf{Tr}_{K'/K}(\alpha)_P u d_K x\big).$$

Let $P\in\mathbb{P}_K$, fix some $P_0'\in\mathbb{P}_{K'}$ above P, let G' be the decompostiton group of P_0' over K, $(G:G')=r$, $G/G'=\{\sigma_1G',\ldots,\sigma_rG'\}$, and for $i\in[1,r]$ let $P_i'=\sigma_iP_0'$. Then $P_1',\ldots,P_r'$ are precisely the prime divisors of K' lying above P,

$$\sum_{\substack{P'\in\mathbb{P}_{K'}\\ P'\cap K=P}}\mathsf{Tr}_{\mathsf{k}_{P'}/F}\big(\mathrm{Res}_{P'}(u\alpha_{P'}d_{K'}x)\big) = \mathsf{Tr}_{\mathsf{k}_P/F}\Big(\sum_{i=1}^r\mathsf{Tr}_{\mathsf{k}_{P_i'}/\mathsf{k}_P}\big(\mathrm{Res}_{P_i'}(u\alpha d_{K'}x)\big)\Big),$$

and it suffices to prove that

$$\sum_{i=1}^r\mathsf{Tr}_{\mathsf{k}_{P_i'}/\mathsf{k}_P}\big(\mathrm{Res}_{P_i'}(u\alpha d_{K'}x)\big)\big) = \mathrm{Res}_P(\mathsf{Tr}_{K'/K}(\alpha)_P u d_K x).$$

Let $t\in\mathcal{O}_P$ and $t_i\in\mathcal{O}_{P_i'}$ for $i\in[1,r]$ be prime elements. By 5.5.7.4(a) we obtain

$$\begin{aligned}\mathrm{Res}_P(\mathsf{Tr}_{K'/K}(\alpha)_P u d_K x) &= \mathrm{Res}_t(\mathsf{Tr}_{K'/K}(\alpha)u\delta_t(x)) = \mathrm{Res}_t(\mathsf{Tr}_{K'(K}(\alpha u\delta_t(x))\\ &= \mathrm{Res}_t\Big(\sum_{i=1}^r\mathsf{Tr}_{K'_{P_i'}/K_P}(\alpha u\delta_t(x)\Big) = \sum_{i=1}^r\mathrm{Res}_t\big(\mathsf{Tr}_{K'_{P_i'}/K_P}(\alpha u\delta_t(x)\big),\end{aligned}$$

and thus we must prove that

$$\mathsf{Tr}_{\mathsf{k}_{P_i'}/\mathsf{k}_P}(\mathrm{Res}_{P_i'}(u\alpha d_{K'}x)) = \mathrm{Res}_t\big(\mathsf{Tr}_{K'_{P_i'}/K_P}(\alpha u\delta_t(x))\big)\quad\text{for all } i\in[1,r].$$

By means of 6.7.8.2 and 6.7.11 we obtain, for all $i\in[1,r]$,

$$\begin{aligned}\mathsf{Tr}_{\mathsf{k}_{P_i'}/\mathsf{k}_P}(\mathrm{Res}_{P_i'}(u\alpha d_{K'}x)) &= \mathsf{Tr}_{\mathsf{k}_{P_i'}/\mathsf{k}_P}(\mathrm{Res}_{t_i'}(u\alpha\delta_{t_i'}(x)))\\ &= \mathsf{Tr}_{\mathsf{k}_{P_i'}/\mathsf{k}_P}(\mathrm{Res}_{t_i'}(u\alpha\delta_t(x)\delta_{t_i'}(t))) = \mathrm{Res}_t(\mathsf{Tr}_{K'_{P_i'}/K_P}(u\alpha\delta_t(x))).\end{aligned}$$

Finally let K'/K be an arbitrary finite extension and K'' a Galois closure of K'/K. Then $\iota_{K''/K} = \iota_{K''/K'} \circ \iota_{K'/K}$ by 6.6.3, and $\eta_{K''/K} = \eta_{K''/K'} \circ \eta_{K'/K}$ by 6.7.7. Since K''/K and K''/K' are Galois, we obtain

$$\begin{aligned}\lambda_{K''} \circ \eta_{K''/K} = \lambda_{K''} \circ \eta_{K''/K'} \circ \eta_{K'/K} &= \iota_{K''/K'} \circ \lambda_{K'} \circ \eta_{K'/K} \\ = \iota_{K''/K} \circ \lambda_K &= \iota_{K''/K'} \circ \iota_{K'/K} \circ \lambda_K,\end{aligned}$$

and as $\iota_{K''/K'}$ is injective, it follows that $\lambda_{K'} \circ \eta_{K'/K} = \iota_{K'/K} \circ \lambda_K$. □

6.9 Zeta functions

Throughout this section, let q be a prime power, $K/\mathbb{F}_q$ a function field, $\mathbb{F}_q$ its field of constants, $g = g_K$ and $h = h_K = |\mathcal{C}_K^0|$.

We introduce ray class characters, zeta functions and L functions formally completely analogous to the corresponding notions for number fields (see 3.3.4 and 4.2.2).

Definitions and Remarks 6.9.1.

1. For an effective divisor $M \in \mathbb{D}_K'$ we set

$$\mathbb{D}_K^M = \{D \in \mathbb{D}_K \mid \operatorname{supp}(D) \cap \operatorname{supp}(M) = \emptyset\},$$

$$K^M = \{a \in K^\times \mid v_P(a-1) \ge v_P(M) \text{ for all } P \in \operatorname{supp}(M)\},$$

and we call $\mathcal{C}_K^M = \mathbb{D}_K^M/(K^M)$ the **ray class group modulo** M.

For $D \in \mathbb{D}_K^M$, we denote by $[D]^M = D + (K^M) \in \mathcal{C}_K^M$ the ray class of D modulo M. In particular, for $M = 0$ we obtain $\mathbb{D}_K^0 = \mathbb{D}_{K,\emptyset} = \mathbb{D}_K$, $K^0 = K^\times$ and $\mathcal{C}_K^0 = \mathcal{C}_K$.

2. Let $M \in \mathbb{D}_K'$. A character $\chi \in \mathsf{X}(\mathcal{C}_K^M)$ is called a **ray class character modulo** M. Note that contrary to the number field case a ray class character need not be of finite order. Every ray class character $\chi \in \mathsf{X}(\mathcal{C}_K^M)$ induces an (equally denoted) homomorphism $\chi\colon \mathbb{D}_K^M \to \mathbb{T}$ satisfying $\chi \restriction (K^M) = 1$, where $\chi(D) = \chi([D]^M)$ for all $D \in \mathbb{D}_K^M$, and additionally we set $\chi(D) = 0$ for all $D \in \mathbb{D}_K \setminus \mathbb{D}_K^M$. Conversely every homomorphism $\chi\colon \mathbb{D}_K^M \to \mathbb{T}$ with $\chi \restriction (K^M) = 1$ is induced by a ray class character modulo M and is itself called a **ray class character modulo** M.

3. For a divisor $D \in \mathbb{D}_K$ we call $\mathfrak{N}(D) = q^{\deg(D)}$ the **absolute norm** of D. Then (as for algebraic number fields) $\mathfrak{N}\colon \mathbb{D}_K \to \mathbb{Q}^\times$ is the unique group homomorphism satisfying $\mathfrak{N}(P) = q^{\deg(P)} = |\mathsf{k}_P|$ for all $P \in \mathbb{P}_K$.

4. Let $M \in \mathbb{D}_K'$ and χ be a ray class character modulo M. We define (in complete analogy with the number field case) for $s \in \mathcal{H}_1$ the L **function**

$L(s, \chi)$ associated with χ and the **zeta function** $\zeta_K(s)$ of K by

$$L(s,\chi) = \sum_{D\in\mathbb{D}'_K} \frac{\chi(D)}{\mathfrak{N}(D)^s} = \sum_{D\in\mathbb{D}'_K} \chi(D)q^{-s\deg(D)}$$

and

$$\zeta_K(s) = L(s,1) = \sum_{D\in\mathbb{D}'_K} \frac{1}{\mathfrak{N}(D)^s} = \sum_{D\in\mathbb{D}'_K} q^{-s\deg(D)}.$$

For $n \in \mathbb{N}_0$ we set $A_n = |\mathbb{D}_K^n \cap \mathbb{D}'_K|$. Recall from 6.6.8.1(b) that $A_n \ll q^n$ for all $n \ge 0$. Thus, if $s \in \mathcal{H}_1$ and $\Re(s) \ge \sigma > 1$, then

$$\sum_{D\in\mathbb{D}'_K} |\chi(D)q^{-s\deg(D)}| \le \sum_{D\in\mathbb{D}'_K} q^{-\sigma\deg(D)} = \sum_{n=0}^{\infty} A_n q^{-n\sigma} < \infty.$$

In particular, the series defining $L(s,\chi)$ converges absolutely and locally uniformly in $\mathcal{H}_1$, and therefore $L(\,\cdot\,,\chi)\colon \mathcal{H}_1 \to \mathbb{C}$ is a holomorphic function.

The defining series of $L(s,\chi)$ is a power series in $t = q^{-s}$ which converges for $|t| < q^{-1}$, and it is advantageous to deal with it as a power series in t. We shall do this in the sequel, and we stress the strong correspondence between zeta and L functions by choosing a common denotation. Thus we define, for $t \in \mathbb{C}$ such that $|t| < q^{-1}$,

$$\mathcal{Z}(t,\chi) = L(s,\chi) = \sum_{D\in\mathbb{D}'_K} \chi(D)t^{\deg(D)} \quad \text{and}$$

$$\mathcal{Z}_K(t) = \zeta_K(s) = \sum_{D\in\mathbb{D}'_K} t^{\deg(D)} = \sum_{n=0}^{\infty} A_n t^n.$$

The function $\mathcal{Z}_K(t)$ is called the **(congruence) zeta function** of K.

Theorem 6.9.2. *Let $M \in \mathbb{D}'_K$, χ a ray class character modulo M, $t \in \mathbb{C}$ and $|t| < q^{-1}$. Then*

$$\mathcal{Z}(t,\chi) = \sum_{D\in\mathbb{D}'_K} \chi(D)t^{\deg(D)} = \prod_{P\in\mathbb{P}_K} \frac{1}{1-\chi(P)t^{\deg(P)}} \ne 0.$$

Proof. By 4.1.2.2. Observe that the function $f\colon \mathbb{D}'_K = \mathcal{F}(\mathbb{P}_K) \to \mathbb{C}$, defined by $f(D) = \chi(D)t^{\deg(D)}$, is completely multiplicative, and

$$\sum_{D\in\mathbb{D}'_K} |\chi(D)t^{\deg(D)}| \le \sum_{n=0}^{\infty} A_n|t|^n < \infty \quad \text{if} \quad |t| < q^{-1}. \qquad \square$$

For the remainder of this section we concentrate on the case $M = 0$. The case $M \ne 0$ of proper ray class characters will be treated in the forthcoming volume on class field theory. Ray class characters modulo 0 are called **divisor class characters**.

Theorem 6.9.3 (Structure and functional equation of $\mathcal{Z}(t,\chi)$ for $M=0$).

1. Let $r \in \mathbb{N}$, $K_r = K\mathbb{F}_{q^r}$, μ_r the group of r-th roots of unity in $\mathbb{C}$, $t \in \mathbb{C}$ and $|t| < q^{-1}$. Then

$$\mathcal{Z}_{K_r}(t^r) = \prod_{\zeta\in\mu_r} \mathcal{Z}_K(\zeta t).$$

2. (F.K. Schmidt) $\partial_K = \gcd(\deg(\mathbb{D}_K)) = 1$. *In particular,* $\deg\colon \mathbb{D}_K \to \mathbb{Z}$ *is surjective.*

3. (Analytic class number formula)

$$\lim_{t\to 1}(1-t)\mathcal{Z}_K(t) = \frac{h}{1-q}.$$

4. Let $\chi \in \mathsf{X}(\mathcal{C}_K)$ be a divisor class character, $\mathbf{e} \in \mathcal{C}_K^1$ a divisor class of degree 1, and let $\chi(\mathbf{e}) = \eta \in \mathbb{T}$.

- *If $\chi \restriction \mathcal{C}_K^0 \neq 1$ then $\mathcal{Z}(\,\cdot\,,\chi)$ is a polynomial of degree $2g-2$.*
- *If $\chi \restriction \mathcal{C}_K^0 = 1$, then $\chi(D) = \eta^{\deg(D)}$ for all $D \in \mathbb{D}_K$. In this case, the function $\mathcal{Z}(\,\cdot\,,\chi)$ has an extension to a rational function on $\mathbb{C} \setminus \{\eta^{-1}, (q\eta)^{-1}\}$ with simple poles in η^{-1} and $(q\eta)^{-1}$, and $\mathcal{Z}(t,\chi) = \mathcal{Z}_K(\eta t)$ for all $t \in \mathbb{C}\setminus\{\eta^{-1},(q\eta)^{-1}\}$.*

(a) If $g=0$, then $h=1$ and

$$\mathcal{Z}(t,\chi) = \frac{1}{(1-\eta t)(1-q\eta t)}.$$

(b) If $g \geq 1$, then $\mathcal{Z}(t,\chi) = F(t,\chi) + G(t,\chi)$, where

$$F(t,\chi) = \frac{1}{q-1}\sum_{n=0}^{2g-2}\Big(\sum_{c\in\mathcal{C}_K^n} \chi(\mathbf{c})q^{\dim(\mathbf{c})}\Big)t^n$$

and

$$G(t,\chi) = \frac{h\varepsilon(\chi)}{q-1}\Big(\frac{q^g(\eta t)^{2g-1}}{1-q\eta t} - \frac{1}{1-\eta t}\Big)$$

$$\text{with } \varepsilon(\chi) = \begin{cases} 1 & \text{if } \chi\restriction\mathcal{C}_K^0 = 1, \\ 0 & \text{if } \chi\restriction\mathcal{C}_K^0 \neq 1. \end{cases}$$

(c) In any case, $\mathcal{Z}(\,\cdot\,,\chi)$ satisfies the functional equation

$$\mathcal{Z}(t,\chi) = \chi(\boldsymbol{w})\big(t\sqrt{q}\big)^{2g-2}\mathcal{Z}\Big(\frac{1}{qt},\overline{\chi}\Big) \quad \text{for all } t\in\mathbb{C},$$

where $\boldsymbol{w} \in \mathcal{C}_K$ denotes the canonical class.

Proof. 1. We need the following identity for roots of unity.

W. If $m \in \mathbb{N}$ and $d = (r, m)$, then

$$\left(1 - X^{mr/d}\right)^d = \prod_{\zeta \in \mu_r} \left(1 - (\zeta X)^m\right) \in \mathbb{C}[X].$$

Proof of **W.** If $\zeta \in \mu_r$, then $(\zeta^m)^{r/d} = (\zeta^r)^{m/d} = 1$, and $\zeta^m = 1$ if and only if $\zeta^d = 1$. We consider the homomorphism

$$\theta \colon \mu_r \to \mu_{r/d}, \qquad \text{defined by } \theta(\zeta) = \zeta^m \text{ for all } \zeta \in \mu_r.$$

Since $\mathrm{Ker}(\theta) = \mu_d$, it follows that $|\mathrm{Im}(\theta)| = r/d$. Hence θ is surjective, and we obtain

$$\prod_{\zeta \in \mu_r} (1 - \zeta^m X^m) = \prod_{\zeta \in \mu_{r/d}} (1 - \zeta X^m)^d = (1 - X^{mr/d})^d. \qquad \square[\mathbf{W}.]$$

Let $P \in \mathbb{P}_K$, $\deg(P) = m$ and $d = \mathrm{ggT}(m, r)$. By 6.6.8.2(b) there exist exactly d prime divisors $P' \in \mathbb{P}_{K_r}$ above P, and they all have degree m/d. Hence **W** implies

$$\prod_{\substack{P' \in \mathbb{P}_{K_r} \\ P' \cap K = P}} \left(1 - t^{r \deg(P')}\right) = \left(1 - t^{rm/d}\right)^d = \prod_{\zeta \in \mu_r} \left(1 - (\zeta t)^m\right),$$

and consequently

$$\begin{aligned} \mathcal{Z}_{K_r}(t^r) &= \prod_{P \in \mathbb{P}_K} \prod_{\substack{P' \in \mathbb{P}_{K_r} \\ P' \cap K = P}} \frac{1}{1 - t^{r \deg(P')}} \\ &= \prod_{P \in \mathbb{P}_K} \prod_{\zeta \in \mu_r} \frac{1}{1 - (\zeta t)^{\deg(P)}} = \prod_{\zeta \in \mu_r} \mathcal{Z}_K(\zeta t). \end{aligned}$$

2., 3. and 4. We prove first preliminary formulas for $\mathcal{Z}(t, \chi)$ with an arbitrary value for $\partial = \partial_K$, and only then we use these formulas for a proof of $\partial = 1$.

Let $D_1 \in \mathbb{D}_K^\partial$ and set $\chi(D_1) = \eta$. In every divisor class $\boldsymbol{c} \in \mathcal{C}_K^0$ we fix a divisor $D_{\boldsymbol{c}} \in \boldsymbol{c}$. Then $\{D_{\boldsymbol{c}} + nD_1 \mid n \in \mathbb{Z},\ \boldsymbol{c} \in \mathcal{C}_K^0\} \subset \mathbb{D}_K$ is a set of representatives for $\mathcal{C}_K$. If $D_0 \in \mathbb{D}_K$, then

$$|\{D \in \mathbb{D}_K' \mid D \sim D_0\}| = \frac{q^{\dim(D_0)} - 1}{q - 1} \quad \text{by 6.6.8.2(b).}$$

With these preliminaries it follows that

$$\mathcal{Z}(t, \chi) = \sum_{D \in \mathbb{D}_K'} \chi(D) t^{\deg(D)} = \sum_{n \in \mathbb{Z}} \sum_{\boldsymbol{c} \in \mathcal{C}_K^0} \chi(D_{\boldsymbol{c}} + nD_1) t^{n\partial} \, \frac{q^{\dim(D_{\boldsymbol{c}} + nD_1)} - 1}{q - 1}.$$

If $\boldsymbol{c} \in \mathcal{C}_K^0$ and $n \in \mathbb{N}$, then $\deg(D_{\boldsymbol{c}} + nD_1) = n\partial$ and $\chi(D_{\boldsymbol{c}} + nD_1) = \chi(\boldsymbol{c})\eta^n$. In particular, if $\chi \restriction \mathcal{C}_K^0 = 1$, then $\chi(D) = \eta^{\deg(D)/\partial}$ for all $D \in \mathbb{D}_K$, and

$\eta = \chi(\boldsymbol{e})$ does not depend on $\boldsymbol{e}$. If $n < 0$, then $\dim(D_{\boldsymbol{c}} + nD_1) = 0$, and if $n\partial \ge 2g-1$, then $\dim(D_{\boldsymbol{c}}+nD_1) = n\partial - g+1$ by the Riemann-Roch theorem 6.4.5.3.

If $g = 0$, then $h = 1$ by 6.4.5.4, and therefore

$$\mathcal{Z}(t,\chi) = \sum_{n=0}^{\infty} \eta^n t^{n\partial}\, \frac{q^{n\partial+1}-1}{q-1} = \frac{1}{q-1}\Big(\frac{q}{1-\eta(qt)^{\partial}} - \frac{1}{1-\eta t^{\partial}}\Big).$$

Thus suppose that $g \ge 1$. We set

$$k = \Big\lceil \frac{2g-1}{\partial} \Big\rceil \quad \text{and observe that} \quad \sum_{\boldsymbol{c}\in\mathcal{C}_K^0} \chi(\boldsymbol{c}) = \varepsilon(\chi)h.$$

Then we obtain

$$\begin{aligned}\mathcal{Z}(t,\chi) &= \sum_{n=0}^{\infty} \sum_{\boldsymbol{c}\in\mathcal{C}_K^0} \chi(\boldsymbol{c})\eta^n t^{n\partial}\, \frac{q^{\dim(D_{\boldsymbol{c}}+nD_1)}-1}{q-1} \\ &= \frac{1}{q-1}\sum_{n=0}^{k-1} \sum_{\boldsymbol{c}\in\mathcal{C}_K^0} \chi(\boldsymbol{c})\eta^n q^{\dim(D_{\boldsymbol{c}}+nD_1)} t^{n\partial} \\ &\quad + \frac{h\varepsilon(\chi)}{q-1}\Big[\sum_{n=k}^{\infty} \eta^n t^{n\partial} q^{n\partial-g+1} - \sum_{n=0}^{\infty} \eta^n t^{n\partial}\Big] \\ &= F(t,\chi) + \frac{h\varepsilon(\chi)}{q-1}\Big[\frac{q^{-g+1+\partial k}(\eta t^{\partial})^k}{1-\eta(qt)^{\partial}} - \frac{1}{1-\eta t^{\partial}}\Big],\end{aligned}$$

where $F(X,\chi) \in \mathbb{Q}[X]$ is a polynomial of degree at most $2g-2$. In particular, we see that $\mathcal{Z}(t,\chi)$ has an extension to a rational function in $\mathbb{C}$ with at most two poles, and that it is even a polynomial of degree at most $2g-2$ if $\chi \restriction \mathcal{C}_K^0 \ne 1$.

For $\chi = 1$ we obtain (whether or not $g = 0$)

$$\lim_{t\to 1}(1-t)\mathcal{Z}_K(t) = \frac{-h}{q-1}\lim_{t\to 1}\frac{1-t}{1-t^{\partial}} = \frac{h}{\partial(1-q)} \ne 0,$$

and we apply this result to $K_\partial = K\mathbb{F}_{q^\partial}$ instead of K. For a root of unity $\zeta \in \mu_\partial$ we have $\mathcal{Z}_K(\zeta t) = \mathcal{Z}_K(t)$, and therefore

$$\mathcal{Z}_{K_\partial}(t^{\partial}) = \prod_{\zeta\in\mu_\partial} \mathcal{Z}_K(\zeta t) = \mathcal{Z}_K(t)^{\partial}.$$

Since

$$\begin{aligned}\lim_{t\to 1}(1-t)^{\partial}\mathcal{Z}_K(t)^{\partial} = \Big(\frac{h}{\partial(1-q)}\Big)^{\partial} \ne 0 \quad &\text{and} \quad \lim_{t\to 1}(1-t^{\partial})\mathcal{Z}_{K_\partial}(t^{\partial}) \\ &= \lim_{t\to 1}(1-t)\mathcal{Z}_{K_\partial}(t) \ne 0,\end{aligned}$$

it follows that

$$0 \ne \lim_{t\to 1}\frac{(1-t)^{\partial}\mathcal{Z}_K(t)^{\partial}}{(1-t^{\partial})\mathcal{Z}_{K_\partial}(t^{\partial})} = \lim_{t\to 1}\frac{(1-t)^{\partial}}{1-t^{\partial}} = \lim_{t\to 1}\frac{-\partial(1-t)^{\partial-1}}{-\partial t^{\partial-1}} = \begin{cases} 0 & \text{if } \partial > 1, \\ 1 & \text{if } \partial = 1,\end{cases}$$

and consequently $\partial = 1$.

We insert this result into the formulas above. If $g = 0$, then

$$\mathcal{Z}(t, \chi) = \frac{1}{q-1}\Big(\frac{q}{1-q\eta t} - \frac{1}{1-\eta t}\Big) = \frac{1}{(1-\eta t)(1-q\eta t)}.$$

If $g \ge 1$, then $\mathcal{Z}(t,\chi) = F(t,\chi) + G(t,\chi)$, where

$$F(t,\chi) = \frac{1}{q-1}\sum_{n=0}^{2g-2}\sum_{\boldsymbol{c}\in\mathcal{C}_K^0} \chi(\boldsymbol{c})\eta^n q^{\dim(D_{\boldsymbol{c}}+nD_1)} t^n$$
$$= \frac{1}{q-1}\sum_{n=0}^{2g-2}\Big(\sum_{\boldsymbol{c}\in\mathcal{C}_K^n} \chi(\boldsymbol{c}) q^{\dim(\boldsymbol{c})}\Big) t^n$$

and

$$G(t,\chi) = \frac{h\varepsilon(\chi)}{q-1}\Big(\frac{q^g(\eta t)^{2g-1}}{1-q\eta t} - \frac{1}{1-\eta t}\Big).$$

In any case (whether or not $g = 0$),

$$\lim_{t\to 1}(1-t)\mathcal{Z}_K(t) = \frac{h}{q-1}.$$

If $\chi \restriction \mathcal{C}_K^0 = 1$, then the above formulas show that $\mathcal{Z}(t,\chi) = \mathcal{Z}_K(\eta t)$ holds for all $t \in \mathbb{C} \setminus \{\eta^{-1}, (q\eta)^{-1}\}$.

It remains to prove the functional equation.

If $g = 0$, then $\deg(\boldsymbol{w}) = -2$, $\chi(\boldsymbol{w}) = \eta^{-2}$, and

$$\chi(\boldsymbol{w})\big(t\sqrt{q}\big)^{-2}\mathcal{Z}\Big(\frac{1}{qt}, \overline{\chi}\Big) = \frac{1}{q(\eta t)^2\big(1-\frac{1}{q\eta t}\big)\big(1-\frac{1}{\eta t}\big)}$$
$$= \frac{1}{(1-\eta q t)(1-\eta t)} = \mathcal{Z}(t,\chi).$$

Thus assume that $g \ge 1$. Then the assignment $\boldsymbol{c} \mapsto \boldsymbol{w}-\boldsymbol{c}$ defines a permutation of the set $\{\boldsymbol{c} \in \mathcal{C}_K \mid \deg(\boldsymbol{c}) \in [0, 2g-2]\}$. Using the Riemann-Roch theorem 6.4.5.1, we obtain

$$F(t,\chi) = \frac{1}{q-1}\sum_{n=0}^{2g-2}\Big(\sum_{\boldsymbol{c}\in\mathcal{C}_M^n}\chi(\boldsymbol{c})q^{\dim(\boldsymbol{c})}\Big)t^n = \frac{1}{q-1}\sum_{\substack{\boldsymbol{c}\in\mathcal{C}_K\\ \deg(\boldsymbol{c})\in[0,2g-2]}} \chi(\boldsymbol{c})q^{\dim(\boldsymbol{c})}t^{\deg(\boldsymbol{c})}$$
$$= \frac{1}{q-1}\sum_{\substack{\boldsymbol{c}\in\mathcal{C}_K\\ \deg(\boldsymbol{c})\in[0,2g-2]}} \chi(\boldsymbol{w}-\boldsymbol{c})q^{\dim(\boldsymbol{w}-\boldsymbol{c})}\, t^{deg(\boldsymbol{w}-\boldsymbol{c})}$$
$$= \frac{\chi(\boldsymbol{w})}{q-1}\sum_{\substack{\boldsymbol{c}\in\mathcal{C}_K\\ \deg(\boldsymbol{c})\in[0,2g-2]}} \overline{\chi}(\boldsymbol{c})q^{g-1-\deg(\boldsymbol{c})+\dim(\boldsymbol{c})}\, t^{2g-2-\deg(\boldsymbol{c})}$$
$$= \frac{\chi(\boldsymbol{w})q^{g-1}t^{2g-2}}{q-1}\sum_{\substack{\boldsymbol{c}\in\mathcal{C}_K\\ \deg(\boldsymbol{c})\in[0,2g-2]}} \overline{\chi}(\boldsymbol{c})q^{\dim(\boldsymbol{c})}\Big(\frac{1}{qt}\Big)^{\deg(\boldsymbol{c})}$$
$$= \chi(\boldsymbol{w})(t\sqrt{q}\,)^{2g-2}F\Big(\frac{1}{qt}, \overline{\chi}\Big).$$

On the other hand,

$$\begin{aligned}\chi(\boldsymbol{w})(t\sqrt{q})^{2g-2}G\Big(\frac{1}{qt},\overline{\chi}\Big) &= \frac{(\eta t)^{2g-2}q^{g-1}h\varepsilon(\chi)}{q-1}\Big(\frac{q^g(q\eta t)^{1-2g}}{1-\frac{1}{\eta t}} - \frac{1}{1-\frac{1}{q\eta t}}\Big)\\ &= \frac{h\varepsilon(\chi)}{q-1}\Big(\frac{(\eta t)^{-1}}{1-\frac{1}{\eta t}} - \frac{(\eta t)^{2g-2}q^{g-1}}{1-\frac{1}{q\eta t}}\Big)\\ &= \frac{h\varepsilon(\chi)}{q-1}\Big(\frac{q^g(\eta t)^{2g-1}}{1-q\eta t} - \frac{1}{1-\eta t}\Big) = G(t,\chi).\end{aligned}$$

Combining these formulas implies the assertion. If $\chi\restriction\mathcal{C}_K^0 \neq 1$, then the functional equation shows that $\mathcal{Z}(t,\chi)$ is a polynomial of exact degree $2g-2$. □

Theorem and Definition 6.9.4. *For $r \in \mathbb{N}$ let $K_r = K\mathbb{F}_{q^r}$ be the constant field extension of degree r.*
The **$\mathcal{L}$-polynomial** of K is defined by

$$\mathcal{L}_K(t) = (1-t)(1-qt)\mathcal{Z}_K(t) \quad \textit{for} \ \ t \in \mathbb{C}.$$

1. $\mathcal{L}_K(t) \in \mathbb{Z}[t]$, $\mathcal{L}_K(1) = h$, and

$$\mathcal{L}_K(t) = q^g t^{2g}\mathcal{L}_K\Big(\frac{1}{qt}\Big) \quad \textit{for all} \ \ t \in \mathbb{C}^\times.$$

2. $\mathcal{L}_K(t) = a_0 + a_1 t + \ldots + a_{2g}t^{2g}$, where $a_0 = 1$, $a_1 = |\mathbb{P}_K^1| - (q+1)$, $a_{2g} = q^g$, and $a_{2g-i} = q^{g-i}a_i$ for all $i \in [0,g]$. In particular $\mathcal{L}_K(t) \in \mathbb{Z}[t]$ is a polynomial of degree $2g$.

3. There exist algebraic integers $\alpha_1,\ldots,\alpha_{2g}$ such that $\alpha_i\alpha_{g+i} = q$ for all $i \in [1,g]$,

$$q^g = \prod_{i=1}^{2g}\alpha_i,$$

and if $r \in \mathbb{N}$, then

$$\mathcal{L}_{K_r}(t) = \prod_{i=1}^{2g}(1-\alpha_i^r t) \ \ \textit{and} \ \ q^r + 1 - |\mathbb{P}_{K_r}^1| = \sum_{i=1}^{2g}\alpha_i^r.$$

In particular, $\{\alpha_1^{-1},\ldots,\alpha_{2g}^{-1}\} = \{t \in \mathbb{C} \mid \mathcal{Z}_K(t) = 0\} = \{t \in \mathbb{C} \mid \mathcal{L}_K(t) = 0\}$. $\alpha_1,\ldots,\alpha_{2g}$ are called the **distinguished numbers** of K.
If $\alpha_1,\ldots,\alpha_{2g}$ are the distinguished numbers of K, then $\alpha_1^r,\ldots,\alpha_{2g}^r$ are the distinguished numbers of K_r.

Proof. If $g = 0$, then

$$\mathcal{Z}_K(t) = \frac{1}{(1-t)(1-qt)}, \quad \mathcal{L}_K(t) = 1 \quad \text{and} \quad h = 1 \quad \text{(see also 6.4.5.4.)}$$

Thus let $g \geq 1$.

1. By definition,

$$\mathcal{L}_K(t) = (1-t)(1-qt)F(t) + \frac{h}{q-1}[(1-t)q^g t^{2g-1} - (1-qt)],$$

where

$$F(t) = \frac{1}{q-1} \sum_{n=0}^{2g-2} \Big(\sum_{c \in \mathcal{C}_K^n} q^{\dim(\mathbf{c})} \Big) t^n.$$

Therefore $\mathcal{L}_K(1) = h$, $\mathcal{L}_K(t) \in \mathbb{Q}[t]$, and even $\mathcal{L}_K(t) \in \mathbb{Z}[t]$, since

$$\mathcal{L}_K(t) = (1-t)(1-qt) \sum_{n=0}^{\infty} A_n t^n \in \mathbb{Z}\,[\![t]\!],$$

where $A_n = |\mathbb{D}_K^n \cap \mathbb{D}'_K|$ for all $n \geq 0$.

The functional equation for $\mathcal{Z}_K(t)$ implies

$$\begin{aligned} q^g t^{2g} \mathcal{L}_K\Big(\frac{1}{qt}\Big) &= q^g t^{2g} \Big(1 - \frac{1}{qt}\Big)\Big(1 - \frac{1}{t}\Big) \mathcal{Z}_K\Big(\frac{1}{qt}\Big) \\ &= q^g t^{2g} \Big(1 - \frac{1}{qt}\Big)\Big(1 - \frac{1}{t}\Big) q^{1-g} t^{2-2g} \mathcal{Z}_K(t) \\ &= (1-t)(1-qt)\mathcal{Z}_K(t) = \mathcal{L}_K(t). \end{aligned}$$

2. By 6.9.3.4(b), $\mathcal{L}_K(t)$ is a polynomial of degree at most $2g$, and as

$$\mathcal{L}_K(t) = \sum_{i=0}^{2g} a_i t^i = (1-t)(1-qt) \sum_{n=0}^{\infty} A_n t^n = A_0 + [\,A_1 - (q+1)A_0\,]t + \sum_{i=2}^{2g} a_i t^i,$$

we get $a_0 = A_0 = 1$ and $a_1 = A_1 - (q+1)A_0 = |\mathbb{P}_K^1| - (q+1)$. By the functional equation given in 1. it follows that

$$\mathcal{L}_K(t) = q^g t^{2g} \sum_{i=0}^{2g} a_i \Big(\frac{1}{qt}\Big)^i = \sum_{i=0}^{2g} a_i q^{g-i} t^{2g-i} = \sum_{i=0}^{2g} a_{2g-i} t^{2g-i}$$

and therefore $a_{2g-i} = a_i q^{g-i}$ for all $i \in [0, g]$. In particular, $a_{2g} = a_0 q^g = q^g$, and thus $\mathcal{L}_K(t)$ is a polynomial of degree $2g$.

3. We consider the reciprocal $\mathcal{L}_K^*$, given by

$$\mathcal{L}_K^*(t) = t^{2g} \mathcal{L}_K\Big(\frac{1}{t}\Big) = \sum_{i=0}^{2g} a_i t^{2g-i} = t^{2g} + a_1 t^{2g-1} + \ldots + q^g = \prod_{i=1}^{2g} (t - \alpha_i),$$

where $\alpha_1, \ldots, \alpha_{2g} \in \mathbb{C}$ are algebraic integers such that

$$-a_1 = q + 1 - |\mathbb{P}_K^1| = \sum_{i=1}^{2g} \alpha_i \quad \text{and} \quad q^g = \prod_{i=1}^{2g} \alpha_i.$$

The functional equation implies

$$\mathcal{L}_K(t) = t^{2g} \mathcal{L}_K^*\Big(\frac{1}{t}\Big) = \prod_{i=1}^{2g} (1 - \alpha_i t) = q^g \prod_{i=1}^{2g} \Big(t - \frac{1}{\alpha_i}\Big)$$
$$= q^g t^{2g} \mathcal{L}_K\Big(\frac{1}{qt}\Big) = q^g \prod_{i=1}^{2g} \Big(t - \frac{\alpha_i}{q}\Big).$$

Hence there exists a permutation $\sigma \in \mathfrak{S}_{2g}$ such that $\alpha_i \alpha_{\sigma(i)} = q$ for all $i \in [1, 2g]$. After a suitable renumbering we obtain

$$\mathcal{L}_K(t) = \prod_{i=1}^{k} (1 - \alpha_i t)\Big(1 - \frac{q}{\alpha_i} t\Big)(1 - \sqrt{q}\, t)^l (1 + \sqrt{q}\, t)^m,$$

where $k,\ l,\ m \in \mathbb{N}_0$ and $2k + l + m = 2g$. Since $a_{2g} = q^g = q^k (-1)^l \sqrt{q}^{\,l+m}$ is the leading coefficient of $\mathcal{L}_K(t)$, it follows that $l = 2l'$ and $m = 2m'$, where $l',\ m' \in \mathbb{N}_0$ and $g = k + l' + m'$. We set $\alpha_i = \sqrt{q}$ for $i \in [k+1, k+l']$, $\alpha_i = -\sqrt{q}$ for $i \in [k + l' + 1, k + l' + m']$ and $\alpha_{i+g} = q/\alpha_i$ for all $i \in [1, g]$.

Let $r \in \mathbb{N}$, $\mu_r \subset \mathbb{C}^\times$ the group of r-th roots of unity, $t \in \mathbb{C}$ and $|t| < q^{-1}$. Then 6.9.3.1 implies

$$\mathcal{L}_{K_r}(t^r) = (1 - t^r)(1 - q^r t^r) \mathcal{Z}_{K_r}(t^r) = \prod_{\zeta \in \mu_r} (1 - \zeta t)(1 - \zeta q t) \mathcal{Z}_K(\zeta t)$$
$$= \prod_{\zeta \in \mu_r} \mathcal{L}_K(\zeta t) = \prod_{i=1}^{2g} \prod_{\zeta \in \mu_r} (1 - \alpha_i \zeta t) = \prod_{i=1}^{2g} (1 - \alpha_i^r t^r),$$

hence

$$\mathcal{L}_{K_r}(t) = \prod_{i=1}^{2g} (1 - \alpha_i^r t), \quad \text{and therefore} \quad |\mathbb{P}_{K_r}^1| = q^r + 1 - \sum_{i=1}^{2g} \alpha_i^r.$$

In particular, $\{\alpha_1^{-1}, \ldots, \alpha_{2g}^{-1}\} = \{t \in \mathbb{C} \mid \mathcal{L}_K(t) = 0\} = \{t \in \mathbb{C} \mid \mathcal{Z}_K(t) = 0\}$, and if $\alpha_1, \ldots, \alpha_{2g}$ are the distinguished numbers of K, then $\alpha_1^r, \ldots, \alpha_{2g}^r$ are the distinguished numbers of K_r. □

Theorem 6.9.5. *Let $\alpha_1, \ldots, \alpha_{2g}$ be the distinguished numbers of K, and for $r \in \mathbb{N}$ let $K_r = K\mathbb{F}_{q^r}$. Then the following assertions are equivalent:*

(a) $|\alpha_i| = \sqrt{q}$ *for all* $i \in [1, 2g]$.

(b) $\big|q^r + 1 - |\mathbb{P}_{K_r}^1|\big| \le 2g q^{r/2}$ *for all* $r \in \mathbb{N}$.

(c) There exists some $c \in \mathbb{R}_{>0}$ *such that* $\big|q^r + 1 - |\mathbb{P}_{K_r}^1|\big| \le c q^{r/2}$ *for all* $r \in \mathbb{N}$.

(d) *There exist some $c \in \mathbb{R}_{>0}$ and $N \in \mathbb{N}$ such that*

$$\big|q^r + 1 - |\mathbb{P}^1_{K_r}|\big| \le cq^{r/2} \text{ for all sufficiently large } r \in N\mathbb{N}.$$

Proof. (a) $\Rightarrow$ (b) By 6.9.4.3 we obtain

$$\big|q^r + 1 - |\mathbb{P}^1_{K_r}|\big| = \Big|\sum_{i=1}^{2g} \alpha_i^r\Big| \le \sum_{i=1}^{2g} |\alpha_i|^r = 2gq^{r/2}.$$

(b) $\Rightarrow$ (c) $\Rightarrow$ (d) Obvious.

(c) $\Rightarrow$ (a) Let $c > 0$ be such that $\big|q^r + 1 - |\mathbb{P}^1_{K_r}|\big| \le cq^{r/2}$ for all $r \in \mathbb{N}$. If $t \in \mathbb{C}$ and $|t| \le \min\{|\alpha_i|^{-1} \mid i \in [1, 2g]\}$, then

$$\sum_{i=1}^{2g} \frac{\alpha_i t}{1 - \alpha_i t} = \sum_{r=1}^{\infty}\Big(\sum_{i=1}^{2g} \alpha_i^r\Big)t^r = \sum_{r=1}^{\infty}(q^r + 1 - |\mathbb{P}^1_{K_r}|)t^r$$

by 6.9.4.3. If R is the radius of convergence of the latter power series, then

$$\frac{1}{R} = \limsup_{r\to\infty} \sqrt[r]{\big|q^r + 1 - |P^1_{K_r}|\big|} \le \limsup_{r\to\infty} \sqrt[r]{cq^{r/2}} = \sqrt{q}, \quad \text{hence} \quad R \ge \frac{1}{\sqrt{q}}.$$

On the other hand, $\alpha_1^{-1}, \ldots, \alpha_{2g}^{-1}$ are the only poles of the rational function

$$g(t) = \sum_{i=1}^{2g} \frac{\alpha_i t}{1 - \alpha_i t},$$

which implies $R = \min\{|\alpha_i|^{-1} \mid i \in [1, 2g]\} \ge q^{-1/2}$. It follows that $|\alpha_i| \le \sqrt{q}$ for all $i \in [1, 2g]$, and as $\alpha_1 \cdot \ldots \cdot \alpha_{2g} = q^g$, we finally obtain $|\alpha_i| = \sqrt{q}$ for all $i \in [1, 2g]$.

(d) $\Rightarrow$ (a) Let $c \in \mathbb{R}_{>0}$ and $N \in \mathbb{N}$ such that $\big|q^r + 1 - |\mathbb{P}^1_{K_r}|\big| \le cq^{r/2}$ for almost all $r \in N\mathbb{N}$ and $\overline{c} \in \mathbb{R}_{>0}$ such that $\big|q^{Nd} + 1 - |\mathbb{P}^1_{K_{Nd}}|\big| \le \overline{c}q^{Nd/2}$ for all $d \in \mathbb{N}$. We set $\overline{q} = q^N$ and $\overline{K} = K_N$. Then it follows that

$$\big|\overline{q}^d + 1 - |\mathbb{P}^1_{\overline{K}_d}|\big| \le \overline{c}\,\overline{q}^{d/2} \quad \text{for all} \quad d \in \mathbb{N}.$$

Now we apply the implication (c) $\Rightarrow$ (a) with $\overline{q}$ and $\overline{K} = K_N$ instead of q and K and obtain $|\alpha_i^N| \le \sqrt{q^N}$ for all $i \in [1, 2g]$, since $\alpha_1^N, \ldots, \alpha_{2g}^N$ are the distinguished numbers of K_N. Eventually, $|\alpha_i| \le \sqrt{q}$ for all $i \in [1, 2g]$ follows. □

The validity of the equivalent assertions of 6.9.4.5 are the content of the theorem of Hasse-Weil (see 6.9.8). This is an analog to the Riemann conjecture which states that the classical zeta and L functions have their zeros on the line $\Re(s) = 1/2$. If we write the zeta function of K in terms of s, say

$$\zeta_K(s) = \mathcal{Z}_K(q^{-s}) = \sum_{D \in \mathbb{D}'_K} \frac{1}{q^{s\deg(D)}}$$

and if $s \in \mathbb{C}$ is a zero of $\zeta_K(s)$, then $|q^{-s}| = q^{-1/2}$ and thus $\Re(s) = 1/2$.

We proceed with some preparations for the proof of the theorem of Hasse-Weil. Let $\overline{\mathbb{F}}_q$ be an algebraic closure of $\mathbb{F}_q$, and assume tacitly that all finite extensions of $\mathbb{F}_q$ are subfields of $\overline{\mathbb{F}}_q$. Let $\phi_q \colon \overline{\mathbb{F}}_q \to \overline{\mathbb{F}}_q$ be the Frobenius automorphism over $\mathbb{F}_q$, defined by $\phi_q(x) = x^q$ for all $x \in \overline{\mathbb{F}}_q$. Then $\text{Gal}(E/\mathbb{F}_q) = \langle \phi_q \restriction E \rangle$ for every finite field extension $E/\mathbb{F}$ (see 1.7.1).

If $P \in \mathbb{P}_K$ and $\tau \in \text{Gal}(K/\mathbb{F}_q)$ such that $\tau P = P$, then 6.2.11 implies $v_P = v_P \circ \tau$, and τ induces an automorphism $\tau_P \in \text{Gal}(\mathsf{k}_P/\mathbb{F}_q$ such that $\tau_P(x(P)) = \tau(x)(P)$ for all $x \in \mathcal{O}_P$. By the very definiton of ϕ_q it follows that $\tau_P = \phi_q \restriction \mathsf{k}_P$ if and only if $\tau(x) \equiv x^q \mod P$ for all $x \in \mathcal{O}_P$. We set

$$\mathbb{P}_K^{(\tau)} = \{P \in \mathbb{P}_K \mid \tau P = P \text{ and } \tau_P = \phi_q \restriction \mathsf{k}_P\} \quad \text{and} \quad N_K^{(\tau)} = \sum_{P \in \mathbb{P}_K^{(\tau)}} \deg(P).$$

If $\tau = \text{id}_K$, then $\tau_P = \text{id}_{\mathsf{k}_P}$, $\mathbb{P}_K^{(\text{id}_K)} = \mathbb{P}_K^1$ and $N_K^{(\text{id}_K)} = |\mathbb{P}_K^1|$.

The following two auxiliary theorems 6.9.6 and 6.9.7 are the centerpiece in the proof of the Hasse-Weil theorem 6.9.8.

Theorem 6.9.6. *Let* $a \in \mathbb{N}$, $q = a^2 > (g+1)^4$, $\mathbb{P}_K^1 \neq \emptyset$ *and* $\tau \in \text{Gal}(K/\mathbb{F}_q)$. *Then*

$$N_K^{(\tau)} < q + 1 + (2g+1)\sqrt{q}.$$

Proof. Let $m = a-1$, $n = a + 2g$, $r = m + an$ and $Q \in \mathbb{P}_K^1$. By 6.3.3.1 we obtain $\dim(iQ) - \dim((i-1)Q) = \dim_{\mathbb{F}_q} \mathcal{L}(iQ)/\mathcal{L}((i-1)Q) \le \deg(Q) = 1$ for all $i \ge 0$. For $k \in \mathbb{N}_0$ we set

$$I_k = \{i \in [0,k] \mid \dim_{\mathbb{F}_q} \mathcal{L}(iQ)/\mathcal{L}((i-1)Q) = 1\},$$

and for every $i \in I_k$ we fix some $u_i \in \mathcal{L}(iQ) \setminus \mathcal{L}((i-1)Q)$. Then $v_Q(u_i) = -i$, $(u_i)_\infty = iQ$, and $\{u_i \mid i \in I_k\}$ is an $\mathbb{F}$-basis of $\mathcal{L}(kQ)$. In particular, the set $\{u_i \mid i \in I_m\}$ is an $\mathbb{F}_q$-basis of $\mathcal{L}(mQ)$.

As a is a power of $p = \text{char}(\mathbb{F}_q)$, the assignment $x \mapsto x^a$ defines a monomorphism of K into itself. Hence the set $\mathcal{L}(nQ)^a = \{y^a \mid y \in \mathcal{L}(nQ)\}$ is a vector space over $\mathbb{F}_q$ with basis $\{u_j^a \mid j \in I_n\}$, $\dim_{\mathbb{F}_q} \mathcal{L}(nQ) = \dim_{\mathbb{F}_q} \mathcal{L}(nQ)^a$, and

$$\mathcal{L} = \Big\{ \sum_{i \in I_m} u_i y_i^a \;\Big|\; (y_i)_{i \in I_m} \in \mathcal{L}(nQ)^{I_m} \Big\}$$

is a vector space over $\mathbb{F}_q$.

If $i \in I_m$ and $y_i \in \mathcal{L}(nQ)$, then $v_Q(u_i y_i^a) \ge -i - an \ge -m - an = -r$ and $v_P(u_i y_i^a) \ge 0 \ge -r$ for all $P \in \mathbb{P}_K \setminus \{Q\}$. Hence $u_i y_i^a \in \mathcal{L}(rQ)$, and therefore $\mathcal{L} \subset \mathcal{L}(rQ)$. We prove next:

A. $L = \{u_i u_j^a \mid i \in I_m,\ j \in I_n\}$ is an $\mathbb{F}_q$-basis of $\mathcal{L}$, and

$$\dim_{\mathbb{F}_q} \mathcal{L} > q + g + 1.$$

Proof of **A.** As $\mathcal{L} = {}_{\mathbb{F}_q}(L)$, it suffices to prove that L is linearly independent over $\mathbb{F}_q$, and for this it suffices to prove that $\{u_i \mid i \in I_m\}$ is linearly independent over K^a. Assume to the contrary that there is a relation

$$\sum_{i\in I_m} u_i y_i^a = 0, \quad \text{where } \ y_i \in K, \ \text{ not all equal } \ 0.$$

Then there exist indices $i,\ j \in I_m$ such that $i \ne j$ and $v_Q(u_i y_i^a) = v_Q(u_j y_j^a)$. It follows that $-i + a v_Q(y_i) = -j + a v_Q(y_j)$, hence $i \equiv j \mod a$, a contradiction, since $I_m \in [1, a-1]$. It remains to estimate $\dim_{\mathbb{F}_q} \mathcal{L}$. By Riemann's inequality 6.3.8 we obtain

$$\begin{aligned} \dim_{\mathbb{F}_q} \mathcal{L} &= |I_m|\,|I_n| = \dim(mQ)\dim(nQ) \ge (m-g+1)(n-g+1) \\ &= (a-g)(a+g+1) = a^2 - g^2 + a - g \\ &= q + \sqrt{q} - g(g+1) > q + g + 1, \end{aligned}$$

since $\sqrt{q} - g(g+1) > (g+1)^2 - g(g+1) = g+1$. □[**A.**]

Now we consider the vector space

$$\mathcal{L}' = \Big\{ \sum_{i\in I_m} \tau^{-1} u_i^a\, y_i \ \Big|\ y_i \in \mathcal{L}(nQ) \Big\},$$

over $\mathbb{F}_q$, and we prove:

B. $\mathcal{L}' \subset \mathcal{L}(ma\,\tau^{-1}Q + nQ)$, and $\dim_{\mathbb{F}_q} \mathcal{L}' \le q + g + 1$.

Proof of **B.** For the proof of $\mathcal{L}' \subset \mathcal{L}(ma\,\tau^{-1}Q + nQ)$ we must show that

$$v_P(\tau^{-1}(u_i)^a\, y_i) \ge -v_P(ma\,\tau^{-1}Q + nQ) \quad \text{for all } \ i \in I_m \ \text{ and } \ P \in \mathbb{P}_K. \quad (*)$$

Let $i \in I_m$ and $P \in \mathbb{P}_K$. Then

$$\begin{aligned} v_P(\tau^{-1}(u_i)^a\, y_i) &= a v_{\tau P}(u_i) + v_P(y_i), \\ -v_P(am\tau^{-1}Q + nQ) &= -am v_{\tau P}(Q) - n v_P(Q), \end{aligned}$$

and $(*)$ follows, since

$$v_{\tau P}(u_i) \ge \begin{cases} -i \ge -m = -m v_{\tau P}(Q) & \text{if } \ \tau P = Q, \\ \quad 0 = -m v_{\tau P}(Q) & \text{if } \ \tau P \ne Q \end{cases}$$

and

$$v_P(y_i) \ge \begin{cases} -n = -n v_P(Q) & \text{if } \ P = Q, \\ \ \ 0 = -n v_P(Q) & \text{if } \ P \ne Q. \end{cases}$$

Since $\deg(am\,\tau^{-1}Q + nQ) = am + n = a(a-1) + a + 2g = q + 2g > 2g - 1$, the Riemann-Roch theorem 6.4.5.3 implies

$$\dim(am\,\tau^{-1}Q + nQ) = \deg(am\,\tau^{-1}Q + nQ) - g + 1 = q + g + 1,$$

and consequently $\dim_{\mathbb{F}_q} \mathcal{L}' \le \dim(am\,\tau^{-1}Q + nQ) = q + g + 1$. □[**B.**]

We define $\psi\colon \mathcal{L} \to \mathcal{L}'$ by

$$\psi\Big(\sum_{i\in I_m} u_i y_i^a\Big) = \sum_{i\in I_m} \tau^{-1}(u_i)^a\, y_i \quad \text{for all families } (y_i)_{i\in I_m} \in \mathcal{L}(nQ)^{I_m}.$$

ψ is $\mathbb{F}_q$-linear, and as $\dim_{\mathbb{F}_q}\mathcal{L} > q+g+1 \ge \dim_{\mathbb{F}_q}\mathcal{L}'$, it follows that ψ is not injective. If $0 \ne u \in \mathrm{Ker}(\psi) \subset \mathcal{L}(rQ)$, then there exists a family $(y_i)_{i\in I_m} \in \mathcal{L}(nQ)^{I_m} \setminus \{\mathbf{0}\}$ such that

$$\sum_{i\in I_m} \tau^{-1}(u_i)^a\, y_i = 0 \quad\text{and}\quad u = \sum_{i\in I_m} u_i y_i^a \in \mathcal{L}(rQ)^\bullet.$$

Let $P \in \mathbb{P}_K^{(\tau)} \setminus \{Q\}$. If $i \in I_m$, then $\{u_i, y_i\} \subset \mathcal{O}_P$, and observing $q = a^2$, we obtain

$$u(P) = \sum_{i\in I_m} u_i(P) y_i(P)^a = \sum_{i\in I_m} (\tau^{-1}u_i^q)(P) y_i(P)^a$$
$$= \Big(\sum_{i\in I_m} \tau^{-1}(u_i)^a\, y_i\Big)^a (P) = 0,$$

hence $P \in \mathcal{N}^K(u)$. It follows that

$$N_K^{(\tau)} - 1 \le \sum_{P\in\mathbb{P}_K^{(\tau)}\setminus\{Q\}} \deg(P) \le \deg((u)_0) = \deg((u)_\infty)$$
$$\le r = a-1+a(a+2g) = a^2-1+a(1+2g) = q-1+(1+2g)\sqrt{q},$$

and therefore $N_K^{(\tau)} < q+1+(2g+1)\sqrt{q}$. □

Theorem 6.9.7. *Let $x \in K \setminus \mathbb{F}_q$ and K'/K be a finite field extension such that $\mathbb{F}_q$ is the field of constants of K' and $K'/\mathbb{F}_q(x)$ is Galois. Assume that $\mathbb{P}_{K'}^1 \ne \emptyset$, set $g' = g_{K'}$, and let $a \in \mathbb{N}$ be such that $q = a^2 > (g'+1)^4$.*
Let $\tau \in \mathrm{Gal}(K/\mathbb{F}_q(x))$, $m = [K'\colon K]$ and $n = [K'\colon\mathbb{F}_q(x)]$. Then

$$N_K^{(\tau)} \ge q+1-\frac{n-m}{m}(2g'+1)\sqrt{q}.$$

Proof. Let $G = \mathrm{Gal}(K'/\mathbb{F}_q(x))$, $H = \mathrm{Gal}(K'/K)$, and let $\tau' \in G$ be such that $\tau' \restriction K = \tau$. We prove first:

A. Let $P \in \mathbb{P}_K$, $P' \in \mathbb{P}_{K'}$ above P, and let $H_{P'}$ be the decomposition group of P' over K.

(a) If $\nu \in H$ and $P' \in \mathbb{P}_{K'}^{(\tau'\nu)}$, then $P \in \mathbb{P}_K^{(\tau)}$.

(b) If $P \in \mathbb{P}_K^{(\tau)}$, then there exists some $\nu \in H$ such that $P' \in \mathbb{P}_{K'}^{(\tau'\nu)}$, and

$$|\{\mu \in H \mid P' \in \mathbb{P}_{K'}^{(\tau'\mu)}\}| = e(P'/P).$$

Proof of **A.** (a) Suppose that $\nu \in H$ and $P' \in \mathbb{P}_{K'}^{(\tau'\nu)}$. Then it follows that

$$\tau P = \tau'\nu(P' \cap K) = \tau'\nu P' \cap K = P' \cap K = P$$

and

$$\tau(x) - x^q = \tau'\nu(x) - x^q \in P' \cap K = P \quad \text{for all} \quad x \in \mathcal{O}_P,$$

hence $P \in \mathbb{P}_K^{(\tau)}$.

Since $\tau'\nu(x) \equiv x^q \bmod P'$ for all $x \in \mathcal{O}_{P'}$, we get $\tau'\nu P' = P'$ and $\tau'\nu \in H_{P'}$. Since $\mathcal{O}_P = \mathcal{O}_{P'} \cap K$, it follows that $\tau'\nu(x) = \tau(x) \equiv x^q \bmod P$ for all $x \in \mathcal{O}_P$, and consequently $P \in \mathbb{P}_K^{(\tau)}$.

(b) Let $P \in \mathbb{P}_K^{(\tau)}$. We tacitly use 6.2.11 and 6.5.4. Since

$$\tau'^{-1}(P') \cap K = \tau'^{-1}(P' \cap K) = \tau'^{-1}P = \tau^{-1}P = P,$$

there exists some $\theta \in H$ such that $\theta(P') = \tau'^{-1}(P')$, hence $\tau'\theta(P') = P'$. We consider the residue class automorphism $(\tau'\theta)_{P'} \in \mathrm{Gal}(\mathsf{k}_{P'}/\mathbb{F}_q)$. Since $(\tau'\theta)_{P'} \restriction \mathsf{k}_P = (\tau'\theta \restriction K)_P = \tau_P = \phi_q \restriction \mathsf{k}_P = (\phi_q \restriction \mathsf{k}_{P'}) \restriction \mathsf{k}_P$, there exists some $\overline{\nu} \in \mathrm{Gal}(\mathsf{k}_{P'}/\mathsf{k}_P)$ such that $(\tau'\theta)_{P'}\overline{\nu} = \phi_q \restriction \mathsf{k}_{P'}$. The residue class map $H_{P'} \to \mathrm{Gal}(\mathsf{k}_{P'}/\mathsf{k}_P)$, $\sigma \mapsto \sigma_{P'}$, is surjective, and therefore there exists some $\nu' \in H_{P'}$ such that $\nu'_{P'} = \overline{\nu}$. It follows that $\nu = \theta\nu' \in H$,

$$\tau'\nu(P') = \tau'\theta\nu'(P') = \tau'\theta(P') = P' \quad \text{and} \quad (\tau'\nu)_{P'} = (\tau'\theta)_{P'}\overline{\nu} = \phi_q \restriction \mathsf{k}_{P'},$$

which implies $P' \in \mathbb{P}_{K'}^{(\tau'\nu)}$.

Let $I_{P'}$ be the inertia group of P' over K. Then $|I_{P'}| = e(P'/P)$, and therefore it suffices to prove that

$$\{\mu \in H \mid P' \in \mathbb{P}_{K'}^{(\tau_\mu)}\} = \nu I_{P'}.$$

Thus let $\mu \in H$. Then

$$\begin{aligned} P' \in \mathbb{P}_{K'}^{(\tau'\mu)} &\iff \tau'\mu(P') = P' = \tau'\nu(P') \text{ and } (\tau'\mu)_{P'} = \phi_q \restriction \mathsf{k}_{P'} = (\tau'\nu)_{P'} \\ &\iff \nu^{-1}\mu(P') = (\tau'\nu)^{-1}(\tau'\mu)(P') = P' \\ &\qquad\qquad \text{and } (\nu^{-1}\mu)_{P'} = (\tau'\nu)^{-1})_{P'} \circ (\tau'\mu)_{P'} = \mathrm{id}_{\mathsf{k}_{P'}} \\ &\iff \nu \in I_{P'} \end{aligned}$$

$\square$[**A.**]

By means of **A** we achieve

$$\begin{aligned} \sum_{\nu \in H} N_{K'}^{(\tau'\nu)} &= \sum_{\nu \in H} \sum_{P' \in \mathbb{P}_{K'}^{(\tau'\nu)}} \deg(P') = \sum_{P \in \mathbb{P}_K^{(\tau)}} \sum_{\substack{P' \in \mathbb{P}_{K'} \\ P' \cap K = P}} e(P'/P) \deg(P') \\ &= \sum_{P \in \mathbb{P}_K^{(\tau)}} \sum_{\substack{P' \in \mathbb{P}_{K'} \\ P' \cap K = P}} e(P'/P) f(P'/P) \deg(P) = m N_K^{(\tau)}. \end{aligned}$$

The same reasoning, applied for $(\mathbb{F}_q(x), K', \mathrm{id}_{\mathbb{F}_q(x)}, \mathrm{id}_{K'})$ instead of (K, K', τ, τ'), implies

$$\sum_{\theta \in G} N_{K'}^{(\theta)} = n N_{\mathbb{F}_q(x)}^{(\mathrm{id}_{\mathbb{F}_q(x)})} = n|\mathbb{P}^1_{\mathbb{F}_q(x)}| = n(q+1).$$

Now we apply 6.9.6 and obtain

$$\begin{aligned} n(q+1) &= \sum_{\theta \in G} N_{K'}^{(\theta)} = \sum_{\nu \in H} N_{K'}^{(\tau'\nu)} + \sum_{\theta \in G \setminus \tau' H} N_{K'}^{(\theta)} \\ &\le \sum_{\nu \in H} N_{K'}^{(\tau'\nu)} + \sum_{\theta \in G \setminus \tau' H} [q+1+(2g'+1)\sqrt{q}\,] \\ &= \sum_{\nu \in H} N_{K'}^{(\tau'\nu)} + (n-m)[q+1+(2g'+1)\sqrt{q}\,], \end{aligned}$$

hence

$$\begin{aligned} \sum_{\nu \in H} N_{K'}^{(\tau'\nu)} &\ge n(q+1) - (n-m)[q+1+2(g'+1)\sqrt{q}\,] \\ &= m(q+1) - (n-m)(2g'+1)\sqrt{q}, \end{aligned}$$

and consequently

$$N_K^{(\tau)} = \frac{1}{m} \sum_{\nu \in H} N_{K'}^{(\tau'\nu)} \ge q+1 - \frac{n-m}{m}(2g'+1)\sqrt{q}. \qquad \square$$

Theorem 6.9.8 (Theorem of Hasse-Weil, "Riemann's conjecture"). *Let $\tau \in \mathrm{Gal}(K/\mathbb{F}_q)$ be of finite order. For $r \in \mathbb{N}$ let $K_r = K\mathbb{F}_{q^r}$, and denote the unique extension of τ to K_r again by τ* (see 6.1.10). *Then there exists some real constant $c \in \mathbb{R}_{>0}$ such that*

$$|q^r + 1 - N_{K_r}^{(\tau)}| \le c q^{r/2} \quad \textit{for all sufficiently large even } r \in \mathbb{N}.$$

In particular,

$$\big|q^r + 1 - |\mathbb{P}^1_{K_r}|\big| \le 2g q^{r/2} \quad \textit{for all } r \in \mathbb{N}.$$

Proof. Let E be the fixed field of τ and $x \in E$ a separating element for $E/\mathbb{F}_q$ (see 6.1.7). Then $K/\mathbb{F}_q(x)$ is finite separable, and we denote by K' a Galois closure of $K/\mathbb{F}_q(x)$. Let F' be the field of constants of $K'/\mathbb{F}_q$, $g' = g_{K'}$, $g = g_K$, $m = [K' : KF']$ and $n = [K' : F'(x)]$. Let F''/F' be a finite field extension such

$$|F''| = q^{r''} = a^2 > [g'+1]^4, \quad \text{where } a,\, r'' \in \mathbb{N} \text{ and } \mathbb{P}^1_{K'F''} \ne \emptyset \text{ (see 6.6.7.2)}.$$

Now let $r \ge r''$ be even, and consider the constant field extensions $K'_r = K'\mathbb{F}_{q^r}$ and $K_r = K\mathbb{F}_{q^r}$. Then it follows that $g = g_{K_r}$, $g' = g_{K'_r}$ (see 6.6.7.4),

$q^r = a'^2 > [g'+1]^4 = [g(K_r')+1]^4$ for some $a' \in \mathbb{N}$, and $\mathbb{P}^1_{K_r'} \neq \emptyset$. As $K'/\mathbb{F}'(x)$ is a finite Galois extension, the same is true for $K_r'/\mathbb{F}_{q^r}(x)$ by 1.5.6. Now it follows that

$$[K_r' : \mathbb{F}_{q^r}(x)] = [K' : F'(x)] = n, \ [K_r : \mathbb{F}_{q^r}(x)] = [K : \mathbb{F}_q(x)] = [KF' : F'(x)],$$

and consequently $[K_r' : K_r] = [K' : KF'] = m$.

$$\begin{array}{ccccccc}
 & & K' & \text{---} & K'F'' & \text{---} & K_r' = K'\mathbb{F}_{q^r} \\
 & \diagup & | & & | & & | \\
K & \text{---} & KF' & \text{---} & KF'' & \text{---} & K_r = K\mathbb{F}_{q^r} \\
| & & | & & | & & | \\
F(x) & \text{---} & F'(x) & \text{---} & F''(x)] & \text{---} & \mathbb{F}_{q^r}(x) \\
| & & | & & | & & | \\
\mathbb{F}_q & \text{---} & F' & \text{---} & F'' & \text{---} & \mathbb{F}_{q^r}
\end{array}$$

From 6.9.6 and 6.9.7, applied for $\mathbb{F}_{q^r} \subset K_r \subset K_r'$ we get

$$q^r + 1 - \frac{n-m}{m}(2g'+1)q^{r/2} \leq N_{K_r}^{(\tau)} < q^r + 1 + (2g+1)q^{r/2}.$$

Overall it follows that, for all sufficiently large even $r \in \mathbb{N}$ we have

$$|q^r + 1 - N_{K_r}^{(\tau)}| \leq cq^{r/2}, \quad \text{where} \ \ c = \max\Big\{2g+1, \frac{n-m}{m}(2g'+1)\Big\}.$$

Applied with $\tau = \mathrm{id}_K$, it follows that there exists some $c \in \mathbb{R}_{>0}$ such that

$$\big|q^r + 1 - |\mathbb{P}^1_{K_r}|\big| \leq cq^{r/2} \ \text{ for all sufficiently large even } \ r \in \mathbb{N},$$

and from 6.9.5 we obtain the Hasse-Weil theorem

$$\big|q^r + 1 - |\mathbb{P}^1_{K_r}|\big| \leq 2gq^{r/2} \quad \text{for all} \ \ r \in \mathbb{N}. \qquad \square$$

Theorem 6.9.9 (Prime divisor theorem).

$$|\mathbb{P}^r_K| = \frac{q^r}{r} + O\Big(\frac{q^{r/2}}{r}\Big) \quad \textit{for all} \ \ r \in \mathbb{N}.$$

Proof. Let $r \in \mathbb{N}$. By 6.9.8 we obtain $|\mathbb{P}^1_{K_r}| = q^r + O(q^{r/2})$, and 6.6.8.1(a) implies

$$|\mathbb{P}^1_{K_r}| = \sum_{1 \leq d \,|\, r} d\, |\mathbb{P}^d_K|.$$

Now we apply the additive version of 1.7.2.2 and (once more) 6.9.8 to obtain

$$r|\mathbb{P}^r_K| = \sum_{1 \leq d \,|\, r} \mu\Big(\frac{r}{d}\Big)|\mathbb{P}^1_{K_d}| \leq \sum_{d=1}^{r} |\mathbb{P}^1_{K_d}| \leq \sum_{d=1}^{r} [\,q^d + O(q^{d/2}] = O(q^r).$$

For d instead of r this implies

$$\sum_{\substack{1\le d\,|\,r\\ d<r}} d\,|\mathbb{P}_K^d| \le \sum_{d=1}^{r/2} O(q^d) = O(q^{r/2}),$$

hence

$$r|\mathbb{P}_K^r| = |\mathbb{P}_{K_r}^1| - \sum_{\substack{1\le d\,|\,r\\ d<r}} d\,|\mathbb{P}_K^d| = q^r + O(q^{r/2}) \quad \text{and} \quad |\mathbb{P}_K^r| = \frac{q^r}{r} + O\Big(\frac{q^{r/2}}{r}\Big). \ \square$$

6.10 Exercises for Chapter 6

1. Let K/F be a function field, $K = F(x,y)$ and $f \in F[X,Y]$ an irreducible polynomial such that $f(x,y) = 0$. If F is the field of constants of K/F and F is perfect, then f is absolutely irreducible.

2. Let K/F be a function field and F its field of constants. The following assertions are equivalent:

(a) $g_K = 0$.

(b) There exists some $D \in \mathbb{D}_K$ satisfying $\deg(D) = 2$ and $\dim(D) = 3$.

(c) There exists some $D \in \mathbb{D}_K$ satisfying $\deg(D) \ge 1$ and

$$\dim(D) > \deg(D).$$

(d) There exists some $D \in \mathbb{D}_K$ satisfying $\deg(D) \ge 1$ and

$$\dim(D) = \deg(D) + 1.$$

If $\mathrm{char}(F) \ne 2$, then these conditions are also equivalent to

(e) $K = F(x,y)$ for some $x,\, y \in K$ such that $y^2 = ax^2 + b$ for some $a,\, b \in F^\times$.

3. Let K'/K be a finite separable extension of function fields with perfect fields of constants, and let $t \in K$ be a separating element. For a differential $\omega = u d_{K'}t \in \Omega_{K'}$ (with $u \in K'$) we define its trace by

$$\mathsf{Tr}_{K'/K}(\omega) = \mathsf{Tr}_{K'/K}(u) d_K t.$$

This definition does not depend on t. If $P \in \mathbb{P}_K$ and $P' \in \mathbb{P}_{K'}$ lies above P, then $\mathrm{Res}_P(\mathsf{Tr}_{K'/K}(\omega)) = \mathsf{Tr}_{\mathsf{k}_{P'}/\mathsf{k}_P}(\mathrm{Res}_{P'}(\omega))$.

4. Let $K = \mathbb{R}(x,y)$, where x is transcendental over $\mathbb{R}$ and $x^2+y^2+1=0$.

a. $K = \mathsf{q}(\mathbb{R}[X,Y]/(X^2+Y^2+1))$ is a function field, $\mathbb{R}$ is its field of constants, and $[K:\mathbb{R}(x)] = 2$.

b. $g_K = 0$ and $\deg(P) = 2$ for all $P \in \mathbb{P}_K$. In particular, K is not a rational function field.

5. Let K/F be a function field, g its genus and $P \in \mathbb{P}_K$. An integer $n \in \mathbb{N}$ is called a **pole number** of P if $nP = (x)_\infty$ for some $x \in K^\times$. Otherwise n is called a **gap number** of P.

a. The set of pole numbers of P is a finitely generated (additive) subsemigroup of $\mathbb{N}$, and every $n \ge 2g$ is a pole number.

b. (Weierstrass gap theorem) There are precisely g gap numbers of P.

6. Let q be a prime power and $K/\mathbb{F}_q$ a function field. Recall from Exercise **5** the Weierstrass gap theorem: For every $P \in \mathbb{P}_K$ there are precisely g integers $k \in \mathbb{N}$ such that there is no $x \in K^\times$ satisfying $(x)_\infty = kP$.

a. Let $P \in \mathbb{P}_K^1$ and $x \in K^\times$ such that $(x)_\infty = kP$ for some $k \in \mathbb{N}$. Then $kq \ge |\mathbb{P}_K^1| - 1$.

b. $|\mathbb{P}_K^1| \le g_K q + q + 1$. Compare with the Hasse-Weil bound.

7. Let K/F be a function field, F its field of constants and $W \in \mathbb{D}_K$ a canonical divisor. A divisor $D \in \mathbb{D}_K$ is called **special** if $\dim(W-D) \ge 1$. Otherwise D is called **non-special**.

a. W is special, and every $D \in \mathbb{D}_K$ with $\deg(D) \ge 2g-1$ is non-special.

b. If $C,\ D \in \mathbb{D}_K,\ C \ge D$ and D is non-special, then C is non-special.

c. If $D \in \mathbb{D}_K,\ \dim(D) \ge 1$ and $\deg(D) \le g-1$, then D is special.

d. Let $D \in \mathbb{D}_K',\ \dim(D) = 1$ and $\deg(D) \le g-1$ (so D is special by c). If $P_1,\ldots,P_g \in \mathbb{P}_K^1$ are distinct, then there exists some $i \in [1,g]$ such that $D + P_i$ is non-special.

e. Let $C,\ D \in \mathbb{D}_K$. If D is special, then $\dim(D-C) \le \dim(W-C)$ and $\dim(D) \le g$. Hint: There is a monomorphism $\mathcal{L}(D-C) \to \Sigma_K(C)$.

8. Let $K = \mathbb{F}_3(x,y)$, where x is transcendental over $\mathbb{F}_3$ and

$$y^2 + x^4 - x^2 + 1 = 0.$$

Then $K/\mathbb{F}_3$ is a function field, and if F is its field of constants, then $[F:\mathbb{F}_3] = 2$ and $K = F(x)$.

9. (Elliptic function fields). A function field K/F is called **elliptic** if $\mathbb{P}_k^1 \ne \emptyset$ and $g_K = 1$.

a. Let K/F be an elliptic function field.

(i) For every $A \in \mathbb{D}_K^1$ there exists a unique $P \in \mathbb{P}_K^1$ such that $A \sim P$.

(ii) Let $O \in \mathbb{P}_K^1$. Then the map

$$\Phi\colon \mathbb{P}_K^1 \to \mathcal{C}_K^0, \quad \text{defined by} \quad \Phi(P) = [P - O],$$

is bijective. If $\oplus\colon \mathbb{P}_K^1 \times \mathbb{P}_K^1 \to \mathbb{P}_K^1$ is the law of composition which makes Φ into an isomorphism, then $P \oplus Q = R$ if and only if $P + Q \sim R + O$ for all P, Q, $R \in \mathbb{P}_K^1$.

(iii) Let $\mathrm{char}(F) \notin \{2,3\}$. Then there exist a, $b \in F$ such that

$$K = F(x,y), \quad \text{where} \quad y^2 = x^3 + ax + b.$$

Hint: If $P \in \mathbb{P}_K^1$, then $F = \mathcal{L}(P) \subsetneq \mathcal{L}(2P) \subsetneq \mathcal{L}(3P)$, and there exist x_1, $y_1 \in K$ such that $(x_1)_\infty = 2P$, $(y_1)_\infty = 3P$, $K = F(x_1, y_1)$ and $(y_1^2, x_1y_1, y_1, x_1^3, x_1^2, x_1, 1)$ is linearly dependent. Then a suitable substitution $(x_1, y_1) \mapsto (x, y)$ yields the assertion.

b. Let $\mathrm{char}(F) \ne 2$ and $f = Y^2 - g \in F[X,Y]$, where $g \in F[X]$ is monic and $\partial(g) = 3$ (then g is called a **Weierstrass polynomial**). Let $K = F(x,y)$, where $y^2 = g(x)$, and let $\mathsf{D}(g)$ be the discriminant of g.

(i) f is absolutely irreducible (hence F is the field of constants of K/F).

(ii) If $\mathsf{D}(f) = 0$, then K/F is a rational function field, and if $\mathsf{D}(f) \ne 0$, then K/F is an elliptic function field.

10. Let K/F be a function field, F its field of constants and $g_K \ge 2$. Then the following assertions are equivalent:

(a) K contains a rational function field $F(x)$ such that $[K\colon F(x)] = 2$.

(b) There exists some $D \in \mathbb{D}_K$ such that $\deg(D) = 2$ and $\dim(D) \ge 2$.

A function field satisfying these conditions is called **hyperelliptic**. Every function field K/F with $g_K = 2$ is hyperelliptic.

11. Let $F = F_0(T)$ be the rational function field in T over a field F_0 such that $\mathrm{char}(F_0) = p > 2$. Let $K = F(x,y)$, where x is transcendental over F and $y^2 = x^p - T$. Let $F' = F(T^{1/p})$ and $K' = KF'$ the constant field extension of K with F'. Then $g_K = (p-1)/2$ and K' is a rational function field. Compare with Theorem 6.6.7.

12. Let K/F be a function field, F its field of constants, $x \in K$ a separating element, $[K\colon F(x)] = n > 1$ and $\mathcal{P}(K,x)$ the set of all $P \in \mathbb{P}_{F(x)}$ which are ramified in K. Then $\mathcal{P}(K,x) \ne \emptyset$.

Let now $\mathrm{char}(F) = 0$ or $\mathrm{char}(F) > n$, and set

$$d(K,x) = \sum_{P \in \mathcal{P}(K,x)} \deg(P).$$

a. $d(K,x) \ge 2$, and if $g_K > 0$, then $d(K,x) \ge 3$.

b. Suppose that F is algebraically closed and $d(K, x) = 2$. Then there exists some $y \in K$ such that

$$K = F(x, y) \quad \text{and} \quad y^n = \frac{ax + b}{cx + d} \quad \text{with} \quad \begin{pmatrix} a & b \\ c & d \end{pmatrix} \in \mathsf{GL}_2(F).$$

13. For every finite field F there exists a function field K/F such that F is its field of constants and $\mathbb{P}_K^1 = \emptyset$.

14. (An explicit prime divisor theorem) Let q be a prime power, $K/\mathbb{F}_q$ an algebraic function field, $\mathbb{F}_q$ its field of constants, $g = g_K$, $\{\alpha_1^{-1}, \ldots, \alpha_{2g}^{-1}\}$ the set of zeros of $\mathcal{L}_K$, and $r \in \mathbb{N}$. Then

$$|\mathbb{P}_K^r| = \frac{1}{r} \sum_{1 \le d \,|\, r} \mu\Big(\frac{r}{d}\Big)(r^d - S_r), \quad \text{where } S_r = \sum_{i=1}^{2g} \alpha_i^r,$$

and

$$\Big| |\mathbb{P}_K^r| - \frac{q^r}{r} \Big| \le \Big(\frac{q}{q-1} + 2g\frac{\sqrt{q}}{\sqrt{q}-1}\Big)\frac{q^{1/2} - 1}{r} < (2 + 7g)\,\frac{q^{r/2}}{r}.$$

In particular, if either $g = 0$ or $2g + 1 \le q^{(r-1)/2}(\sqrt{q} - 1)$, then $|\mathbb{P}_K^r| \ge 1$.

15. Let q be a prime power, $K/\mathbb{F}_q$ a function field, $g = g_K$ and $N = |\mathbb{P}_K^1|$.

a. There exist algebraic integers $\alpha_1, \ldots, \alpha_g$ such that $|\alpha_i| = \sqrt{q}$ for all $i \in [1, g]$,

$$\mathcal{L}_K(t) = \prod_{i=1}^{g} (1 - 2\Re(\alpha_i)t + qt^2) \quad \text{and} \quad |\mathbb{P}_{K_r}^1| = q + 1 - 2\sum_{i=1}^{g} \Re(\alpha_i^r).$$

b. (Serre bound) $\big|q + 1 - |\mathbb{P}_K^1|\big| \le g\lceil 2\sqrt{q}\rceil$, and equality holds if and only if

$$\mathcal{L}_K(t) = \prod_{i=1}^{g} \big(1 - \lceil 2\sqrt{q}\rceil t + qt^2\big).$$

Hint: For $i \in [1, g]$ let $\gamma_i^{\pm} = \pm(2\Re(\alpha_i)) + \lceil 2\sqrt{q}\rceil + 1$, show that

$$\prod_{i=1}^{g} \gamma_i^{\pm} \in \mathbb{N},$$

and apply the inequality between the arithmetic and the geometric mean of $\gamma_1^{\pm}, \ldots, \gamma_g^{\pm}$.

c. Let $q = s^2$ with $\in \mathbb{N}$ and $|P_K^1| = q + 1 - 2gs$. Then $\alpha_i = -\sqrt{q}$ for all $i \in [1, g]$ and $g \le s(s - 1)/2$. Hint: Use $|\mathbb{P}_K^1| \le |\mathbb{P}_{K_2}|$.

16. Let K/F be a function field, F its field of constants, x, $y \in K$ and $c \in F$. Then $dy = cdx$ if and only if $y - cx$ is not separating.

17. Let K/F be a function field, F its field of constants, $x \in K$ a separating element and $P \in \mathbb{P}_K^1$. Then

$$\mathrm{Res}_P\left(\frac{dx}{x}\right) = v_P(x).$$

18. Let $F(x)$ be the rational function field over a field F such that $\mathrm{char}(F) \neq 2$. Let $h \in K[X]$ be squarefree, $\partial(h) = d \geq 1$, and suppose that $K = F(x, y)$, where $y^2 = h(x)$.

Let $P_\infty \in \mathbb{P}_{F(x)}$ be the infinite prime divisor with respect to x, and let $P'_\infty \in \mathbb{P}_K$ be such that $P'_\infty \cap F(x) = P_\infty$. Then

$$e(P'_\infty/P_\infty) = 2, \quad \mathrm{D}_{K/F(x)} = (y)_0^K + P'_\infty \quad \text{and } g_K = \frac{d-1}{2} \quad \text{if } d \equiv 1 \bmod 2,$$

and

$$e(P'_\infty/P_\infty) = 1, \quad \mathrm{D}_{K/F(x)} = (y)_0^K \quad \text{and } g_K = \frac{d-2}{2} \quad \text{if } d \equiv 0 \bmod 2.$$

19. Let $n \in \mathbb{N}$, $n > 1$, let $F(x)$ be the rational function field over a field F such that $\mathrm{char}(F) \nmid n$, and $K = F(x, y)$, where $x^n + y^n = 1$. Then

$$g_K = \frac{(n-1)(n-2)}{2}.$$

Hint: Consider the constant field extension $K(\zeta) = KF(\zeta)$, where ζ is a primitive n-th root of unity, and apply 5.9.9, 6.6.4 and 6.5.9.

20. Let K/F be a function field, F its field of constants and x a separating element for K/F. Let $P \in \mathbb{P}_{K(x)}^1$ be the only prime divisor which is ramified in K. If P is tamely ramified, then $K = F(x)$.

Bibliography

[1] E. Artin, *Algebraic Numbers and Algebraic Functions*, Gordon and Breach, 1967.

[2] S.I. Borewicz and I.R. Šafarevič, *Zahlentheorie*, Birkhäuser, 1966.

[3] N. Bourbaki, *Algebra II*, ch. 4-7, Masson, 1990.

[4] J. W. S. Cassels, *Global Fields*, Algebraic Number Theory (J. W. S. Cassels and A. Fröhlich, eds.), Academic Press, 1967.

[5] J. W. S. Cassels, *Local Fields*, London Math. Soc., 1986.

[6] N. Childress, *Class Field Theory*, Springer, 2007.

[7] H. Cohen, *Number Theory. Volume I: Tools and Diophantine Equations, Volume II: Analytic and Modern Tools*, Springer, 2007.

[8] P. M. Cohn, *Algebraic Numbers and Algebraic Functions*, Chapman and Hall/CRC, 1991.

[9] D. A. Cox, *Primes of the form* $x^2 + ny^2$, John Wiley, 1989.

[10] H. Delange, *Généralisation du théorème de Ikehara*, Ann. Sci. Éc. Norm. Supér. **71** (1954), 213 – 242.

[11] M. Deuring, *Lectures on the Theory of Algebraic Functions of One Variable*, Springer, 1973.

[12] J. Dieudonné, *Treatise on Analysis, Vol. I, II*, Academic Press, 1976.

[13] I. B. Fesenko and S. V. Vostokov, *Local Fields and Their Extensions*, AMS, 2002.

[14] M. Fried and M. Jarden, *Field Arithmetic*, Springer, 2008.

[15] A. Fröhlich and M. J. Taylor, *Algebraic number theory*, Cambridge University Press, 1991.

[16] P. Furtwängler, *Über die Führer von Zahlringen*, S.-Ber. Akad. Wien **128** (1919), 239 – 245.

[17] A. Geroldinger and F. Halter-Koch, *Non-Unique Factorizations. Algebraic, Combinatorial and Analytic Theory*, Pure and Applied Mathematics, vol. 278, Chapman & Hall/CRC, 2006.

[18] R. Gilmer, *Multiplicative Ideal Theory*, Marcel Dekker, 1972.

[19] F. Halter-Koch, *Einseinheitengruppen und prime Restklassengruppen in quadratischen Zahlkörpern*, J. Number Theory **4** (1972), 70 – 77.

[20] F. Halter-Koch, *Die Klassengruppe einer kommutativen Ordnung*, Math. Nachr. **168** (1994), 97 – 108.

[21] F. Halter-Koch, *Ideal Systems. An Introduction to Multiplicative Ideal Theory*, Marcel Dekker, 1998.

[22] F. Halter-Koch, *Quadratic Irrationals*, Pure and Applied Mathematics, vol. 306, Chapman & Hall/CRC, 2013.

[23] H. Hasse, *Arithmetische Theorie der kubischen Zahlkörper auf klassenkörpertheoretischer Grundlage*, Math. Z. **31** (1930), 550 – 564.

[24] H. Hasse, *Über die Klassenzahl abelscher Zahlkörper*, Springer, 1963.

[25] H. Hasse, *Bericht über neuere Untersuchungen und Probleme aus der Theorie der algebraischen Zahlkörper, Teil I, Ia und II*, Physica-Verlag, 1965.

[26] H. Hasse, *Vorlesungen über Klassenkörpertheorie*, Physica-Verlag, 1967.

[27] H. Hasse, *Number Theory*, Springer, 1980.

[28] L. K. Hua, *Introduction to Number Theory*, Springer, 1956.

[29] T. W. Hungerford, *Algebra*, Springer, 1974.

[30] A . Ivić, *The Riemann Zeta Function*, Wiley (Reprint Dover 2003), 1985.

[31] S. Iyanaga, *The Theory of Numbers*, North-Holland, 1975.

[32] N. Jacobson, *Basic Algebra, I, II*, Freeman, 1980.

[33] G. J. Janusz, *Algebraic Number Fields*, Academic Press, 1996.

[34] N. Klingen, *Arithmetic Similarities*, Oxford Science Publ., 1998.

[35] K. Knopp, *Theorie und Anwendung der unendlichen Reihen*, Springer, 1947.

[36] H. Koch, *Galoissche Theorie der p-Erweiterungen*, VEB Deutscher Verlag der Wissenschaften, 1970.

[37] H. Koch, *Zahlentheorie*, Vieweg, 1997.

[38] S. Lang, *Introduction to Algebraic and Abelian Functions*, Springer, 1982.

[39] S. Lang, *Algebra*, Addison-Wesley, 1993.

[40] S. Lang, *Algebraic Number Theory*, Springer, 1994.

[41] M.D. Larsen and P.J. McCarthy, *Multiplicative Theory of Ideals*, Academic Press, 1971.

[42] F. Lorenz, *Einführung in die Algebra, II*, B. I., 1990.

[43] F. Lorenz, *Algebraische Zahlentheorie*, B.I., 1993.

[44] F. Lorenz and F. Lemmermeyer, *Algebra I*, Elsevier, 2007.

[45] D. A. Marcus, *Number Fields*, Springer, 1977.

[46] H. Matsumura, *Commutative ring theory*, Cambridge University Press, 1997.

[47] R. Mollin, *Quadratics*, Chapman & Hall/CRC, 1996.

[48] C. Moreno, *Algebraic Curves over Finite Fields*, Cambridge University Press, 1991.

[49] W. Narkiewicz, *Number Theory*, World Scientific, 1983.

[50] W. Narkiewicz, *Elementary and Analytic Theory of Algebraic Numbers*, Springer, 2004.

[51] W. Narkiewicz, *The Story of Algebraic Numbers in the First Half of the 20th Century*, Springer, 2018.

[52] J. Neukirch, *Algebraic Number Theory*, Springer, 1999.

[53] C. Prabpayak and G. Lettl, *Conductor ideals of orders in algebraic number fields*, Arch. Math. **103** (2014), 133 – 138.

[54] R. Remmert, *Theory of Complex Functions*, Springer, 1991.

[55] R. Remmert, *Classical Topics in Complex Function Theory*, Springer, 1998.

[56] P. Ribenboim, *Classical Theory of Algebraic Numbers*, Springer, 1972.

[57] P. Ribenboim, *L'arithmétique des corps*, Hermann, 1972.

[58] M. Rosen, *Number Theory in Function Fields*, Springer, 2002.

[59] J.-P. Serre, *Local Fields*, Springer, 1995.

[60] H. Stichtenoth, *Algebraic Function Fields and Codes*, Springer, 2009.

[61] L. C. Washington, *Introduction to Cyclotomic Fields*, Springer, 1982.

[62] A. Weil, *Basic Number Theory*, Springer, 1967.

[63] E. Weiss, *Algebraic Number Theory*, McGraw-Hill, 1963.

[64] O. Zariski and P. Samuel, *Commutative Algebra, Vol. I, II*, Springer, 1960.

Index

K-homomorphism, 19
P-adic value of a divisor, 482
σ-positive, 130, 256
Čebotarev's density theorem, 335
$\mathfrak{m}$-integral, 128
$\mathfrak{o}$-lattice, 435
$\mathfrak{o}$-module index, 436
$\mathfrak{p}$-adic absolute value, 347
$\mathfrak{p}$-adic logarithm and exponential, 430
$\mathfrak{p}$-adic number field, 424
$\mathfrak{p}$-adic valuation, 121, 124
π-adic digit expansion, 139
π-adic valuation, 79
p-adic absolute value, 347
p-adic integer, p-adic unit, p-adic valuation, 372
p-adic number field, 372, 425, 427
p-integral, 116, 141

squarefree kernel, 81

abelian number field, 224
abscissa of absolute convergence, 297
absolute Galois group, 34
absolute norm, 175
absolute value, 344
Absolute values of a number field, 406
absolutely irreducible polynomial, 468
algebra, 7
algebraic closure, 21
algebraic element, 16
algebraic field extension, 16
algebraic integer, 100
algebraic number, 35
algebraic number field, 190
algebraically closed, 21
algebraically independent, 68
ambiguous ideals of a quadratic order, 280
analytic class number formula, 307
annihilator, 92
Archimedian absolute value, 344
Arithmetic of radical extensions, 449
Artin reciprocity law, 267
Artin symbol, Artin map, Artin isomorphism, 181, 206, 212, 267
Artin-Schreier extension, 67
Artinian ring, 185
associated elements, 76
atom, 77
atomic monoid, 77

basis of a free monoid, 83
basis of a module, 2, 91
Basis theorem for complete modules, 195
Bezout domain, 188
Blichfeld constant, 244

canonical divisor (class), 493
character group (dual group), 216
characteristic of a ring, 15
Chinese remainder theorem, 4
class field, 267
class group of a number field, 191
class group of a quadratic order, 277
class number of a function field, 487
class number of a number field, 191
class number of a quadratic order, 277

CM-field, 289
coefficient field, 388
Comparison theorem for differentials, 543
complete module, 192
complete valued field, 356
completely multiplicative function, 294
completion, 356
completion and algebraic closure, 423
Completion theorem, 356
compositum of fields, 16
conductor of a Dirichlet character, 221
conductor of a quadratic discriminant, 232
conductor of an order, 145, 160
congruence zeta function, 547
Congruences with signatures, 130
conjugate algebraic elements, 16
conjugate quadratic irrationals, 271
constant field extension, 471, 515
content ideal of a polynomial, 45
content of a polynomial, 85
coprime elements, 8, 77
coprime ideals, 4, 127
cycle of a permutation, cycle type, 40
cyclic field extension, 27
cyclotomic field, cyclotomic polynomial, 61
cyclotomic number field, 202, 211

decomposed prime ideal, 155
decomposed prime number, 205
decomposition group (field), 105, 501
Dedekind arithmetic of ideals, 121, 122
Dedekind complementary module, 437
Dedekind domain, 120
Dedekind zeta function, 303
Dedekind's independence theorem, 36
defining valuation, 136
degee of a prime divisor, 475
degree of a divisor, 482
degree of a divisor class, 487
degree of a polynomial, 10
degree of a prime ideal, 105, 205
degree of an algebraic element, 16
degree of an algebraic number field, 190
degree valuation, 141
derivation, 520
derivative, 527
different and discriminant
of a function field extension, 506–508
different and discriminant of a Dedekind domain extension, 439–448
different and discriminant of a local field extension, 444
different and discriminant of a number field extension, 448
different exponent, 506
differential module, 525, 531
dimension of a divisor (class), 483
Dirichlet character, 219, 303
Dirichlet density, 315
Dirichlet L function, 303
Dirichlet series and integrals, 297
Dirichlet's theorem for ideal classes, 332
Dirichlet's theorem on arithmetic progressions, 329
Dirichlet's unit theorem, 248
discrete absolute value, 344
discrete valuation, 84, 347
discrete valuation domain, 133
Discrete valuations of $\mathbb{Q}$ and $F(X)$, 141
discrete valued field, 347
discriminant of an $\mathfrak{o}$-lattice, 437
discriminant of a complete module, 192
discriminant of a polynomial, 45
discriminant of a quadratic irrational, 271

discriminant of a quadratic order, 272
discriminant of an algebraic number field, 192
divide, 8, 76
division algorithm, 11
division with remainder, 11
divisor class character, 547
divisor class group, 483
divisor embedding, 499
divisor norm, 503
divisor of a differential, 538
divisor of a function field, 482
divisor of a Weil differential, 493
divisor of zeros (poles), 482
domain, 1
domain of algebaic integers, 100
domain with finite residue class rings, 175
double coset space, 171
dual basis, 51
dual homomorphism, 218

effective divisor, 482
Eisenstein's criterion, 13
Eisenstein polynomial, 13, 418
Eisenstein's factorization law, 158
elliptic function field, 563
embedding of differentials, 526, 532, 538
embedding of Weil differentials, 510
embeddings of a number field, 190
equivalent complete modules, 192
essentially unique factorization, 77
Euclidean algorithm, 11
Euclidean domain, 186
Euler phi function, 57, 175
Euler products, 294
even (odd) Dirichlet character, 224
Extension of $\mathfrak{p}$-adic absolute values, 403
Extension of absolute values, complete case, 393, 396
Extension of absolute values, Galois case, 401
Extension of absolute values, general case, 398
Extensions of Dedekind domains, 152

Factorial Dedekind domains, 125
factorial domain, 9, 84
factorial monoid, 77
factorization, 77
factorization law, 206
Factorization law of class field theory, 269
factorization of a prime number, 205
Factorization of prime ideals, 154
factorization properties, 77
factorization type of a prime ideal, 183
field embedding, 241
field extension, 15
field homomorphism, 2
field of algebraic numbers, 35
field of constants, 466
finite abelian groups, 97
finite field extension, 16
finitely generated abelian groups, 96
finitely generated field extension, 15
finitely generated module, 2
Finiteness of the class number, 247
fixed field, 19
formal exponential and logarithm, 375
formal Laurent and power series, 142, 143, 373
fractional (principal) ideal, 109
free abelian group, 2, 96
free module, 2
free monoid, 83
Frobenius automorphism, 53, 179, 206, 208, 212, 413, 502
Frobenius symbol, 181
fully decomposed prime ideal, 155
fully decomposed prime number, 205
fully ramified field extension, 407
fully ramified prime divisor, 480

fully ramified prime ideal, 156
fully ramified prime number, 205
function field, 466
function field extension, 470
function field of an algebraic curve, 468
functional equation of the congruence zeta function, 548
fundamental discriminant, 232
fundamental parallelotope, 238
fundamental units, 251

Galois closure, 34
Galois field extension, 27
Galois group, 19
Galois group of a polynomial, 41
Gauss sum, 235
Gauss's lemma, 14, 86, 87
gcd, 8, 77
genus group of a quadratic order, 280
genus of a function field, 489
geometric constant field extension, 471
global differential module, 525

Hensel's Lemma, 386
Herbrand function, 453
Hilbert class field, 269
Hilbert subgroup series, 166
holomorphic (regular) differential, 538
homomorphism of valued fields, 344
Hurwitz zeta function, 341
Hurwitz' genus formula, 513

ideal, 2
Ideal asymptotic, 261
ideal class group (Picard group), 112, 130, 191
ideal norm, 176, 191
independent units, 251
induced Dirichlet character, 221
inert prime ideal, 156, 205
inert prime number, 205
inertia group (field), 166, 410, 502
infinite prime divisor, 477
inseparable degree of a field extension, 34
inseparable element, 27
inseparable field extension, 27
integral basis, 190
integral element, integral closure, integral equation, 98
integral ideal, 109
integrality in localizations, 119
Integrality in quotient domains, 118
integrally closed domain, 98
invertible ideal, 109
irreducible polynomial, 11

Krasner's lemma, 422
Kronecker symbol, 230
Kronecker's existence theorem, 20
Krull dimension, 102
Krull's existence theorem, 6
Kummer extension, 64
Kummer's factorization law, 156

L-polynomial of a function field, 552
lattice, 238
Laurent series field extension, 418
lcm, 8, 77
leading coefficient, 10, 185
Legendre symbol, 207
length of a factorization, 184
level of a principal unit, 138
linear equivalence of divisors, 483
linearly independent, 2, 91
local differential module, 531
local field, 377, 424
local ring, 6
Local unit theorem, 432
Local-global principle for invertibility, 117
localization, 116
lying above, 8, 151, 205, 477

Möbius mu-function, 55
Main theorem of algebra, 35

Main theorem of class field theory, 267
Main theorem of finite fields, 53
Main theorem of finite Galois theory, 37
Main theorem of genus theory, 282
maximal ideal, 6
maximal order of a number field, 190
maximal unramified field extension, 410
minimal polynomial, 16
Minkowski constant, 244
Minkowski's convex body theorem, 240
modular group, 271
modules over principal ideal domains, 92, 94
modulus of definition of a character, 221
monic polynomial, 10
monoid, 1, 76
monoid homomorphism, 76
multiple zero, 27
multiplicative function, 55, 293
multiplicative residue system, 388
multiplicatively closed set, 6, 76

narrow class group, 257
Noetherian ring, 3
non-Archimedean absolute value, 344
norm of an element, 41
normal closure, 26
normal field extension, 24
normalized absolute value, 346
Normindex inequality, 330

order in a Dedekind domain, 145, 160
order in an algebraic number field, 192
order of a differential, 531
order of a formal Laurent series, 143
order of a number modulo m, 57
order valuation, 143
Overrings of Dedekind domains, 135

partial fractions, 88
partial fractions of rational functions, 89
partial zeta function, 303
Pell equation, 273
perfect field, 27
permutations, 40
positive (negative) part of a divisor, 482
power residues vs. local powers, 387
Prüfer domain, 188
Prime avoidance theorem, 7
prime discriminant, 288
prime divisor of a function field, 475
prime divisor of an ideal, 151
Prime divisor theorem, 561
prime element, 8, 77
prime field, prime ring, 15
prime ideal, 6
Prime ideal theorem, 328
Prime ideal theorem for orders, 332
Prime ideals in a quadratic order, 274
Prime ideals in quotient domains, 116
Prime number theorem, 328
primitive Dirichlet character, 221
Primitive element theorem, 36
primitive ideal of a quadratic order, 274
primitive polynomial, 85
primitive root, 57
primitive root of unity, 59
principal divisor, 482
principal ideal, 3, 76
principal ideal domain, 3
principal units, 138
product formula, 347, 406
projective module, 184
proper ideal, 4
properties of the Artin symbol, 182
pure cubic number field, 199, 209
purely inseparable element, 27
purely inseparable field extension, 27

quadratic character, 227, 233
quadratic discriminant, 232, 271
quadratic irrational, 271
quadratic number field, 198, 208
quadratic order, 270
Quadratic reciprocity law, 228, 237
quadratic residue, 207
quotient group, 76
quotient homomorphism, 76
Quotients of Dedekind domains, 126

ramification group (field), 166, 450
ramification index, 151, 205, 391, 478
ramified field extension, 407, 417
ramified prime divisor, 480
ramified prime ideal, 151, 155, 205
ramified prime number, 205
rank of a module, 92
rational function field, 141, 466, 496
ray class character, 303
ray class character of a function field, 546
ray class field, 267
ray class group, 131, 256
ray class group of a function field, 546
real embedding, 130
Reciprocity law for the Kronecker symbol, 230
reduced monoid, 76
reducible polynomial, 11
regular ideal of a quadratic order, 274
regular Weil differential, 493
regulator, 250, 251
repartition (incomplete adele), 485
representatives for prime elements, 79
residue, 527, 531
residue class automorphism, 105, 394, 502
residue class degree, 105, 391, 478
residue class extension, 105, 391
residue class field, 6, 136, 365, 475
residue class isomorphism, 105
Residue Theorem, 541
resultant of polynomials, 45
Riemann zeta function, 295
Riemann's inequality, 491
ring, 1
ring homomorphism, 2
ring of multipliers of a complete module, 192
root of unity, 59

saturated submonoid, 114
semigroup, 1
semilocal ring, 6
separable closure, 34
separable degree of a field extension, 30
separable polynomial (field extension), 27
separating element, 466
Shifting theorem of finite Galois theory, 39
signature of a number field, 190
solvable field extension, 27
space of multipliers of a divisor, 483
split homomorphism, 2
splitting field, 23
Splitting theorem, 267
squarefree, 81
Strong approximation theorem, 498
Structure of discrete valuation domains, 137
submonoid, 1, 76
subring, 1
Summation of L series, 313
support of a divisor, 482
support of an ideal, 121

tamely ramified field extension, 420
tamely ramified prime divisor, 480
tamely ramified prime ideal, 151, 155
Tauber theorem of Ikehara - Delange, 321
The class group determines the arithmetic, 333
Theorem of Artin, 37

Theorem of Bauer, 319, 338
Theorem of Capelli - Redei, 66
Theorem of Hasse - Arf, 460
Theorem of Hasse - Weil, 560
Theorem of Hensel - Newton, 379
Theorem of Hensel - Ore, 382
Theorem of Herbrand, 454
Theorem of Kronecker - Weber, 269
Theorem of Krull-Cohen-Seidenberg, 101
Theorem of Lüroth, 514
Theorem of Minkowski - Hermite, 246
Theorem of Riemann - Roch, 489, 495
Theorem of Sen, 456
torsion module, torsion-free, 92
totally positive algebraic number, 190, 256
totally real (complex) number field, 190
trace of a local differential, 533
trace of an element, 41
transcendence basis (degree), 68
transcendental element, 16
transcendental field extension, 16
type of a quadratic irrational, 271

undecomposed prime ideal, 155
undecomposed prime number, 205
unit group of a monoid, 76
unit group of a number field, 190
unit group of a quadratic order, 272
unit group of a ring, 1
unit group of a valued field, 136, 365
unit ideal, 109
unit rank, 250
universal global derivation, 525
universal local derivation, 531
universal norms, 421
unramified field extension, 407
unramified prime divisor, 480
unramified prime ideal, 151, 155, 205
unramified prime number, 205
upper numbering of ramification groups , 455

valuation domain, 133, 136, 365
valuation ideal, 136, 365
valuation topology, 354
value preserving homomorphism, 344
valued field, 344

Weak approximation theorem, 348
Weierstrass polynomial, 564
Weil differential, 491
Weil differential, P-component, 494
wildly ramified field extension, 420
wildly ramified prime ideal, 151, 155
Witt vectors, 70, 71, 464

zero divisor, 482
zeta and L function of a function field, 547
Zorn's Lemma, xiv

List of Symbols

$(\mathfrak{P}, L/K)$, $\left(\frac{L/K}{\mathfrak{p}}\right)$, $[\mathfrak{p}, L/K]$, $\left[\frac{L/K}{\mathfrak{P}}\right]$, 181
$(\mathfrak{b}:\mathfrak{a})$, X^{-1}, 109
(σ), σ_P, 493, 494
$(a_1, \ldots, a_n) = 1$, 8, 77
(p, K), 206
(x), $(x)_0$, $(x)_\infty$, 482
$A \oplus B$, $A \cdot B$, 3
$A^{(I)}$, 2
$A^\bullet$, $A \subset B$, $A \subsetneq B$, $\restriction$, xiii
$D^\times$, D°, D_{red}, $\mathsf{q}(D)$, $\mathsf{q}(\varphi)$, 76
G_K, 34
$H^\perp$, 216
$I(\xi)$, 273
$K(\mathfrak{m})$, $\mathfrak{o}(\mathfrak{m})$, $K^{\circ\mathfrak{m}}$, $\pi_\mathfrak{m}$, 128, 129
$K[\mathcal{G}], K[\mathfrak{m}]$, 267
K^+, $K^\mathfrak{m}$, $K^{\mathfrak{m},\sigma}$, U_K^+, $U_K^\mathfrak{m}$, $U_K^{\mathfrak{m},\sigma}$, $U_K^{\circ\mathfrak{m}}$, 256
$K^{(n)}$, 61
K^σ, $\mathcal{H}^\sigma(\mathfrak{o})$, $\mathcal{C}^\sigma(\mathfrak{o})$, $[\mathfrak{a}]^\sigma$, 130
$K^{\mathfrak{m},\sigma}$, $\mathcal{H}^{\mathfrak{m},\sigma}(\mathfrak{o})$, $\mathcal{C}^{\mathfrak{m},\sigma}(\mathfrak{o})$, $[\mathfrak{a}]^{\mathfrak{m},\sigma}$, 131
$K_\mathfrak{p}$, $\widehat{\mathfrak{o}}_\mathfrak{p}$, $\widehat{\mathfrak{p}}$, 369
K_{sep}, 34
$L(s, \chi)$, $\zeta_K(s)$, 303
M_{tor}, $\text{Ann}_\mathfrak{o}(M)$, $\text{rk}(M)$, 92
O, o, $\ll$, xiv
$R[X_1, \ldots, X_n]$, $R[X]$, $R[X, Y]$, $R[\boldsymbol{X}]$, etc., 10
$R^\bullet$, $R^\times$, $z(R)$, $\mathsf{q}(\mathfrak{o})$, 1
$T^{-1}X$, $X_\mathfrak{p}$, 113, 116
$U^{(n)}$, 138
$[L:K]_\mathsf{s}$, 30
$[L:K]_\mathsf{i}$, 34
$[\mathfrak{A}:\mathfrak{B}]_\mathfrak{o}$, 436
$[a, b]$, xiii
$\mathbb{A}_K$, $\mathbb{A}_K(D)$, 485
$\Delta(\mathfrak{a})$, $\mathcal{R}(\mathfrak{a})$, 192
$\mathbb{F}_p$, xiii
Ω_K, 525
$\mathbb{P}$, $\mathbb{N}$, $\mathbb{N}_0$, $\mathbb{Z}$, $\mathbb{Q}$, $\mathbb{R}$, $\mathbb{C}$, $\mathbb{R}_{\geq 0}$, $\mathbb{Q}_{>0}, \ldots$, xiii
$\mathbb{P}_K$, 475
$\mathbb{D}_K$, $\mathbb{D}'_K$, $\mathbb{D}^0_K$, $\mathbb{D}^n_K$, 482
$\mathbb{Q}^+$, 191
$\mathbb{Q}(m)$, $\mathbb{Q}^+(m)$, $\mathbb{Z}(m)$, $\mathbb{Q}^{\circ m}$, $\mathbb{Q}^m$, $\mathbb{Q}^m_\pm$, 192
$\mathbb{Q}_p$, $\mathbb{Z}_p$, U_p, 372
$\Re(z)$, $\Im(z)$, $\sqrt{z}$, $\text{sgn}(x)$, xiii
Σ_K, $\Sigma_K(D)$, 491
$\mathbb{Z}_{(p)}$, 116
$\left(\frac{a}{p}\right)$, $\left(\frac{m}{n}\right)$, 207, 230
$\mathbf{0}$, $\mathbf{1}$, xiv
$\boldsymbol{\sigma}(K)$, $\mathcal{H}_K^{\mathfrak{m},\sigma}$, $\mathcal{C}_K^{\mathfrak{m},\sigma}$, $[\mathfrak{a}]^{\mathfrak{m},\sigma}$, 256, 257
χ_K, χ_d, 232
$\deg(P)$, $\deg(D)$, 475, 482
$\delta_x(u)$, $\text{Res}_x(u)$, 527
$\dim(R)$, 102
$\dim_K L = [L:K]$, 16
$\eta_{K'/K}$, 526
$\text{Gal}(L/K)$, 19
$\text{Gal}_K(g)$, 41
$\text{Hom}(M, M')$, $\text{Hom}_R(M, M')$, $\text{End}(M)$, $\text{End}_R(M)$, 2
$\text{Hom}^R(A, B)$, 7
∞, 79
$\lambda(X)$, 240
$\lfloor x \rfloor$, $\lceil x \rceil$, xiii
$\mathbb{T}$, 216
$\mathsf{M}_{m,n}(X)$, $\mathsf{M}_n(X)$, A^t, 2
$\mathsf{N}_{A/K}$, $\mathsf{Tr}_{A/K}$, 41
$\mathsf{P}(K, X)$, $\mathsf{D}(K, X)$, 84
$\mathsf{R}(f, g)$, $\mathsf{D}(g)$, 45
$\mathsf{X}(G)$, $\mathsf{X}(m)$, 216, 219
d_K, 192

e, i, xiii
$\mathsf{o}(f)$, 142
$\mathsf{tr}(L/K)$, 69
$\mathcal{C}(\mathfrak{o})$, $[\mathfrak{a}]$, 112
$\mathcal{C}_K$, $\mathcal{C}_K^n$, $(K^\times)$, 483, 487
$\mathcal{F}(P)$, 83
$\mathcal{L}(D)$, $\dim(D)$, 483
$\mathcal{N}^K(z)$, $\mathcal{P}^K(z)$, 476
$\mathcal{N}_{L/K}$, $\mathfrak{D}_{L/(K}$, $\mathfrak{d}_{L/K}$, 444, 448
$\mathcal{N}^{\mathfrak{m}}_{L/K}$, 267
$\mathcal{N}_{\mathfrak{O}/\mathfrak{o}}$, $\mathcal{N}_{L/K}$, 176, 191
$\mathcal{O}(D)$, $f \sim g$, 299
$\mathcal{O}_K$, $\mathfrak{p}_K$, k_K, U_K, $U_K^{(n)}$, 366
$\mathcal{O}_v$, U_v, $\mathfrak{p}_v$, k_v, 136
$\mathcal{Z}_K(t)$, 547
$\mathfrak{C}_{\mathfrak{A}/\mathfrak{o}}$, 437
$\mathfrak{D}_{\mathfrak{A}/\mathfrak{o}}$, 439
$\mathfrak{N}(\mathfrak{a})$, 175
$\mathfrak{P} \mid \mathfrak{p}$, $e(\mathfrak{P}/\mathfrak{p})$, $f(\mathfrak{P}/\mathfrak{p})$, 105, 151
$\mathfrak{S}(A)$, 40
$\mathfrak{a} \mid \mathfrak{b}$, 121
$\mathfrak{a} \sim \mathfrak{b}$, 192
$\mathfrak{c}(f)$, 45
$\mathfrak{d}_{\mathfrak{A}/\mathfrak{o}}$, 437
$\mathfrak{o}_K$, $\mathcal{P}_K$, $\mathcal{J}_K$, $\mathcal{J}'_K$, $\mathcal{H}_K$, $\mathcal{H}'_K$,
$\mathcal{C}_K$, h_K, U_K, 190, 191
$\mathcal{H}'(D)$, 76
$\mathcal{H}_K^+$, $\mathcal{C}_K^+$, h_K^+, 257
$\mathcal{H}_d^{(+)}$, $\mathcal{C}_d^{(+)}$, $h_d^{(+)}$, $[\mathfrak{a}]^{(+)}$, 277
$\mathcal{C}_d^{(+)}[2]$, 280
$\mathcal{J}(\mathfrak{o})$, $\mathcal{J}'(\mathfrak{o})$, $\mathcal{H}(\mathfrak{o})$, $\mathcal{H}'(\mathfrak{o})$, 109, 110
$\mathcal{J}^{\mathfrak{m}}(\mathfrak{o})$, $\mathcal{J}'^{\mathfrak{m}}(\mathfrak{o})$, $\mathcal{P}^{\mathfrak{m}}(\mathfrak{o})$, 127
$\mathcal{J}_K^{\mathfrak{m}}$, $\mathcal{H}_K^{\mathfrak{m}}$, $\mathcal{H}_K^{\circ\mathfrak{m}}$, $\mathcal{C}_K^{\mathfrak{m}}$, $\mathcal{C}_K^{\mathfrak{m},\sigma}$, $[\mathfrak{a}]^{\mathfrak{m},\sigma}$,
256, 257
$\mathcal{P}(R)$, $\max(R)$, $\mathsf{k}_{\mathfrak{p}}$, 6
$\mu(K)$, $\mu_n(K)$, $\mu_{p^\infty}(K)$, 59
$\underline{\mathrm{ord}}_m(a)$, 57
$\overline{\mathbb{Q}}$, $\overline{\mathbb{Z}}$, 35, 190
$\partial(f)$, 10
∂_K, 482
σ_d, ω_d, $\mathcal{O}_d$, U_d, U_d^+, ε_d, 272, 273
$\mathrm{supp}(\mathfrak{a})$, $\mathrm{supp}(D)$, 121, 482
$\tau(\chi, a)$, $\tau(\chi)$, 235
$\varphi(m)$, $\varphi(\mathfrak{a})$, 57, 175
$\wp$, 67
${}_R(X)$, ${}_R(x_1, \dots, x_n)$, $\langle X \rangle$,
$\langle x_1, \dots, x_n \rangle$, 2
$a \mid b$, $a \simeq b$, 8, 76
$d(P'/P)$, 506
$d_{L/K}(u_1, \dots, u_n)$, 50
$e(L/K)$, $f(L/K)$, 391
$e(P'/P)$, $f(P'/P)$, 478
g_K, 489
$j_{\mathfrak{O}/\mathfrak{o}}$, $j^*_{\mathfrak{O}/\mathfrak{o}}$, 113
v_π, $v_{\mathfrak{p}}$, 79, 121, 124
w_K, R_K, 251
$z(P)$, 475
$\mathrm{D}_{K'/K}$, $\mathrm{d}_{K'/K}$, 507
$\mathrm{Split}(L/K)$, $\mathrm{Ram}(L/K)$, 266
$\mathrm{cl}_S(R)$, 98
$\mathrm{vol}(\Gamma)$, 238

gcd, lcm, 8, 77

9781138583610